Springer Collected Works in Mathematics

More information about this series at http://www.springer.com/series/11104

E. E. Kummer

Ernst Eduard Kummer

Collected Papers I

Contributions to Number Theory

Editor
André Weil

Reprint of the 1975 Edition

Author
Ernst Eduard Kummer (1810 – 1893)
Universität Berlin
Berlin
Germany

Editor
André Weil (1906 – 1998)
Institute for Advanced Study
Princeton, NJ
USA

ISSN 2194-9875
Springer Collected Works in Mathematics
ISBN 978-3-662-48832-4 (Softcover)
978-3-540-06835-8 (Hardcover)

Library of Congress Control Number: 2012954381

Springer Heidelberg New York Dordrecht London

Printed on acid-free paper

Springer-Verlag GmbH Berlin Heidelberg is part of Springer Science+Business Media
(www.springer.com)

ERNST EDUARD KUMMER
COLLECTED PAPERS

VOLUME I

CONTRIBUTIONS TO NUMBER THEORY

Edited
by André Weil

SPRINGER-VERLAG
BERLIN HEIDELBERG NEW YORK 1975

ISBN 978-3-540-06835-8 Springer-Verlag Berlin Heidelberg New York

Printing: Julius Beltz, Hemsbach/Bergstr. Binding: Konrad Triltsch, Würzburg.

Library of Congress Cataloging in Publication Data. Kummer, Ernst Eduard, 1810–1893. Number-theoretic papers. (His Collected works; v. 1) Papers in German, French, and Latin; introd. in English. 1. Numbers, Theory of - Addresses, essays, lectures. I. Title. QA3. K98. 1975. vol. 1 [QA241] 510'. 8s [512'. 72'08]. 74-23839.

Table of Contents

Introduction

by André Weil

Note. In Kummer's days, French was gradually replacing Latin as the international scientific language, and Kummer used both in his early publications, concurrently with German; eventually he wrote for *Liouville's Journal* an important expository paper in French [39], which, it must be confessed, had no perceptible effect on French mathematics. For much the same reasons, this Introduction has been written in English at the request of the publisher, in the hope that it may help to make Kummer's work more attractive to the international mathematical public of to-day.

The letters *L* and *F* refer respectively to E. Lampe's biographical notice (*Jahresbericht der deutschen Mathematiker-Vereinigung* 3 (1892–93), 13–21 = this vol., pp. 15–23) and to the *Festschrift* published in 1910 by K. Hensel (this vol., pp. 33–133). Numbers in square brackets refer to the bibliography appended to L (this vol., pp. 23–30); a reference such as [40; 120] designates paper [40], page 120 *of the original pagination* (not the pagination of this volume); similarly, [F; 46] refers to page 46 of the original pagination of the *Festschrift.*

The great number-theorists of the last century are a small and select group of men. The names of Gauss, Jacobi, Dirichlet, Kummer, Hermite, Eisenstein, Kronecker, Dedekind, Minkowski, Hilbert spring to mind at once. To these one may add a few more, such as the universal Cauchy, H. Smith, H. Weber, Frobenius, Hurwitz. Some of them were solely or chiefly number-theorists, while the work of others extends over a wide range of mathematical subjects. Most of them were no sooner dead than the publication of their collected papers was undertaken and in due course brought to completion. To this there are two notable exceptions: Kummer and Eisenstein. Did one die too young and the other live too long? Were there other reasons for this neglect, more personal and idiosyncratic perhaps than scientific? Hilbert dominated German mathematics for many years after Kummer's death. More than half of his famous *Zahlbericht* (viz., parts IV and V) is little more than an account of Kummer's number-theoretical work, with inessential improvements; but his lack of sympathy for his predecessor's mathematical style, and more specifically for his brilliant use of p-adic analysis, shows clearly through many of the somewhat grudging references to Kummer in that volume. Minkowski might at least have promoted the publication of Eisenstein, to which he refers in his correspondence; but he, too, died rather young.

It must now be a matter of high gratification to mathematicians, and particularly to all number-theorists, that the Springer-Verlag plans to fill this gap on our shelves and is printing the complete works of Kummer. Wisely they have decided to begin with Kummer's number-theory, which occupies the present volume.

It is not necessary here to add anything to Lampe's brief but excellent biographical note [L], nor to the extensive biography [F; 1–37] composed in 1910 by Kummer's pupil Hensel for the centenary of his teacher's birth. The latter is fittingly supplemented by some delightful letters from Kummer to his mother [F; 41–46], and, more valuable still, by his letters to Kronecker [F; 46–102], where we can follow, sometimes almost day by day, his progress in number-theory.

As will be apparent from the above sources and from Lampe's bibliography, Kummer's mathematical career is made up of three distinct segments, with little overlap. Until the age of thirty-two, he concerned himself almost exclusively, and with notable success, with problems in what was then known as "function-theory". His papers deal with specific series and definite integrals, differential equations and series expansions; the most famous one (and deservedly so) treats the theory of the hypergeometric series as thoroughly as could be done before Riemann, i.e. without analytic continuation in the complex plane; actually, as Kronecker pointed out [L; 16], he left little for Riemann to do in this topic in the way of concrete results. All these papers show, not only an extraordinary skill in the handling of formulas, but a no less impressive sureness of instinct in following the right track in a maze of analytic transformations. It is possible that even now some new insights could be gained by a reconsideration of some of this work, e.g. the relations he finds between the hypergeometric series, elliptic integrals and the gamma-function. However that may be, it may safely be assumed that the experience he had thus gained stood him in good stead for his work on number-theory, to which we now turn.

During his first period, he had first of all been a brilliant student in Halle, then a teacher in his native town of Sorau, then (with an interruption of one year for his military service as a musketeer) a teacher in the Gymnasium in Liegnitz, where eventually Kronecker became his pupil. A short paper [8] on Fermat's equation, written in 1835, is the only indication, during those years, of a budding interest in number-theory. He took up this subject in earnest in 1842, when his appointment as a professor in Breslau appeared certain [F; 46]; to it he was to dedicate almost wholly the next seventeen years of his life, and in the work of those seventeen years lies his main claim to fame, even though his third and last period, from 1859 onwards, also produced some memorable results; most notable perhaps was the discovery of the family of surfaces (quartics with sixteen double points in P^3) which are still known as "Kummer's surfaces".

In 1842, his first footsteps in the arithmetical field, while hardly timid, are still rather uncertain [F; 46–52]. In the theory of λ-th roots of unity (where λ is any odd prime), Gauss had introduced the cyclotomic "periods" of order e:

$$\eta_i = \sum_j \alpha^{\gamma^{i+ej}}$$

where α is a primitive root of $\alpha^\lambda = 1$, e is any divisor of $\lambda - 1$, γ is a "primitive root modulo λ" (i.e. a generator of the multiplicative group $\mathbf{F}_\lambda^\times$ of the prime field $\mathbf{F}_\lambda = \mathbf{Z}/\lambda\mathbf{Z}$) and where one takes $0 \leqq i < e$, $0 \leqq j < (\lambda - 1)/e$ (the notations are not Gauss's, but they are those invariably followed by Kummer from 1844 onwards). In 1805, Lagrange, using an old idea of his, improved upon this by introducing the "Lagrange resolvents"

$$g = \sum_i \eta_i \zeta^i ,$$

where ζ is any e-th root of unity. Actually they occur also in Gauss's early work, but Gauss steadfastly refused to acknowledge their superiority, probably because they involve the "extraneous" irrationality ζ. In modern parlance, they have become known as the "Gaussian sums" belonging to the prime field $\mathbf{F}_\lambda$, and it will be convenient for us to use that name for them from now on.

In the case $e=2$, $\zeta = -1$, these "Gaussians sums" involve no extraneous irrationality, and they occur very prominently in Gauss under the guise

$$\sum_{n=0}^{\lambda-1} \alpha^{n^2} = 1 + 2\eta_0 = \eta_0 - \eta_1 = g .$$

In this case, Gauss, after showing that $g^2 = \pm\lambda$, had solved the far more difficult problem of determining the sign of g and had shown the importance of this result for the theory of quadratic residues. Gauss had also obtained results for the "periods" of order 3; these are more conveniently expressed in terms of the "Gaussian sum" g and imply that, for $\lambda \equiv 1 \pmod 3$, $g^3 = -\lambda\varrho$, where ϱ is a prime factor of λ in the field $\mathbf{Q}(\zeta)$ of a cubic root ζ of 1, further determined by $\varrho \equiv 1 \pmod 3$; in particular, we have $|g|^2 = \lambda$. It thus seemed natural to take such periods (or sums) as a starting point for the investigation of cubic residues, and no less natural to expect that the key to the problem would be found in the determination of the argument of the Gaussian sums. This, in effect, is what Kummer attempted in 1842 ([16], [17]). Needless to say, he did not succeed; not only is the determination of the argument of the Gaussian sums (even in the case of order 3) a still unsolved question, but few mathematicians would expect that any answer can be given to it, other than statistical. But Kummer was not wasting his time; apart from providing him with enough material to fulfil the rather modest requirements of a "Habilitationsschrift" for Breslau University [17], this episode equipped him with valuable experience for the major tasks he was soon to undertake.

Another episode, not so well documented, must have been more painful. Without proper equipment, he attacked Fermat's last theorem, stating the impossibility of $x^n + y^n = z^n$ in (non-zero) integers. We know, from Gauss's private papers (cf. *Werke*, vol. II, p. 387–391) and his correspondence, that he had taken some interest in this problem; he saw that the key to it must lie in the theory of cyclotomic fields, worked out a proof for $n=3$ which included the impossibility in the field of cubic roots of unity, and sought to extend this to $n=5$. Legendre and Dirichlet dealt with some further cases. Some time in the early forties (cf. [F; 22]), Kummer submitted what purported to be a full solution to Dirichlet, who at once objected that it assumed the unique decomposition of integers into irreducible factors in the cyclotomic field, an unproved assumption about which he had serious doubts.[1]

This might have proved a crushing blow to a less determined young man. Indeed we get some inkling of his feelings in a passage of his very next paper ([20a; 202]: "*Maxime dolendum videtur* ..."); "it is greatly to be lamented, he writes, that this virtue of the real numbers" (i.e. of the rational integers), "to be decomposable into prime factors, always the same ones for a given number, does not also belong to the complex numbers" (i.e. to the integers in cyclotomic fields); "were this the case, the whole theory, which is still laboring under such difficulties, could easily be brought to its conclusion. For this reason the complex numbers we have been considering seem imperfect, and one may well ask whether one ought not to look for another kind which would preserve the analogy with the real numbers with respect to such

[1] Eisenstein seems to allude ironically to this in his letter of 1844 to M. A. Stern (Abh. z. Gesch. d. math. Wiss. VI (1895) p. 173).

a fundamental property. Nevertheless, the complex numbers generated by the roots of unity and the real integers are not arbitrarily made up ...".

But, instead of being crushed, Kummer saw the need for properly laying the foundations of the theory of cyclotomic fields, and this is the object of the paper we have just quoted ([20a]; cf. also [F; 53–63]) and of the next ones.[2]

With impressive insight, he begins by seeing that there is at least one case where one can expect an integer in the cyclotomic field $\mathbf{Q}(\alpha)$ to behave like a "true" prime; it is the case when its norm is a rational prime p. Here, as always, α is a primitive λ-th root of unity, λ being an odd rational prime; by an integer ("a complex number" in Kummer's language) we understand an element of the ring $\mathbf{Z}[\alpha]$; we note once for all that neither Dirichlet nor Kummer ever had the general concept of an algebraic integer (to be discovered later by Kronecker and Dedekind), so that it was a lucky accident for Kummer that $\mathbf{Z}[\alpha]$ happens to be the ring of all integers of $\mathbf{Q}(\alpha)$. Now he takes up the systematic investigation of the norm of integers in $\mathbf{Q}(\alpha)$; writing such an integer as $f(\alpha)$, where f is a polynomial over $\mathbf{Z}$, he shows that $Nf(\alpha)$ is a multiple of λ if $f(1) \equiv 0$ (mod. λ), and otherwise is $\equiv 1$ mod. λ. He makes a preliminary investigation of the units, showing that every unit is of the form $\alpha^k\varepsilon$ where ε is in the real subfield of $\mathbf{Q}(\alpha)$, but referring to Dirichlet's and Kronecker's forthcoming work for deeper results (the reference is here to Dirichlet's famous work on units, and to Kronecker's thesis, none of which had yet appeared in print). He then proceeds to study the case $p = Nf(\alpha)$, where p is a rational prime other than λ, which of course must then be $\equiv 1$ mod. λ. If we put $\pi = f(\alpha)$, then, in our language, (π) is a prime ideal of degree 1. This is indeed what Kummer proves in the following form ("theorema insigne", [20a; 197]): every integer in $\mathbf{Z}[\alpha]$ is congruent to some rational integer modulo π. Also, for this case, he determines, in terms of π and its conjugates, the prime factor decomposition of the Gaussian sums for $\mathbf{F}_p$.

"Unfortunately" (as Kummer says), not every prime $p \equiv 1$ mod. λ is a norm in $\mathbf{Z}[\alpha]$, and he is able (using Gauss's theory of quadratic forms) to give examples of this fact, for $\lambda = 23$. What happens next, i.e. when p is not a norm? and what happens when p is not $\equiv 1$ mod. λ? The latter questions is taken up in [23] (cf. [F; 57–63]). A first answer is supplied by the concept, due chiefly to Euler, of the "prime divisors of a form"; a prime p is called a divisor of a "form" F (i.e. a polynomial, not necessarily homogeneous, in one or more variables) if $F(x) \equiv 0$ mod. p has non-trivial solutions; for instance p is a divisor of $x^n - ay^n$ if and only if a is an n-th power residue modulo p. It is now clear (and it was already known to Euler) that p is a divisor of $x^{\lambda-1} + x^{\lambda-2} + \ldots + x + 1$ if and only if $p = \lambda$ or λ divides $p-1$. Similarly, if the η_i, for $0 \leqq i < e$, are the "periods" of order e, Kummer finds that, except possibly for finitely many primes, the prime divisors of the "form" $\Pi_i(y - \eta_i)$ are precisely those

[2] For the chronology of Kummer's discoveries, it is important to correct a misprint in *Crelle's Journal* and an error in Lampe's bibliography [L; 23–24]. Kummer's first publication on the "ideal complex numbers" was the note in the *Monatsberichte* for 1846, pp. 87–96, reproducing a communication made to the Berlin Academy on 26 March 1846. This was reprinted in *Crelle's Journal*, vol. 35 (1846), p. 319–326, with a reference to the *Monatsberichte* of "March 1845" (obviously a misprint). Failing to notice this, Lampe listed it twice (erroneously for 1845, as [21], then correctly for 1846, as [22]); moreover, he further confused this with a later communication [33] to the Academy, and listed the same paper again for 1850 as [32].

which are e-th power residues modulo λ. Results of the same kind are also obtained for the "very remarkable" form:

$$F(z) = N(\sum_i z_i \eta_i),$$

where the norm is taken from $\mathbf{Q}(\eta_0)$ to $\mathbf{Q}$. The proof rests essentially on the construction of a homomorphism of the ring $\mathbf{Z}[\eta_0, \ldots, \eta_{e-1}]$ onto the prime field $\mathbf{F}_p$ when such a homomorphism exists, i.e. when p decomposes in $\mathbf{Q}(\eta_0)$ into prime ideals of degree 1.

And now comes the decisive and triumphant letter of 18 October 1845 ([F; 64–67]; cf. [21]) where Kummer, "chiefly in order to achieve clarity in his own thought", describes to Kronecker the giant step he has just taken. No longer need one deplore the lamentable behavior of complex numbers [21; 323]; mathematicians should imitate the boldness of the chemists, who have no misgivings about introducing elements such as fluor, even though they know they cannot isolate them ([F; 68]; cf. [27; 360]). The irreducible factors which make up the "complex numbers" need not be "existing" ones ("wirklich"); we may not be able to isolate them as numbers, but they are there all the same. They are the "ideal prime factors".

To appreciate the boldness of this step, one must realize that such men as Gauss, Dirichlet and Eisenstein had shied away from it. We know now that Gauss's theory of the composition of quadratic forms is essentially a theory of the ideals in quadratic fields; as Dirichlet told Kummer [F; 68], Gauss had had some inkling of this, but had been unable to put this idea into shape. Both Dirichlet and Eisenstein had obtained many fundamental results which properly concern the group of ideal-classes in cyclotomic fields; but they had, somewhat unhappily, come to the conclusion that this must be couched in the language of "decomposable forms", i.e. homogeneous forms of degree n in n variables, with rational coefficients, which decompose into products of n linear forms over some cyclotomic field. As we know from Kummer himself [34; 93], Dirichlet, using this rather unsuitable language, had even completed what amounted to the determination of the class-number for cyclotomic fields. Now Kummer had changed all that.

As happens so frequently in mathematics, this was achieved by inverting the procedure which Kummer had followed at first. He had found, firstly, that, if $p \equiv 1$ mod. λ, a prime factor π of p in $\mathbf{Z}[\alpha]$, if one exists, determines a homomorphism φ of $\mathbf{Z}[\alpha]$ onto $\mathbf{F}_p$, and that $\varphi(\xi)$ is 0 if and only if ξ is a multiple of π in $\mathbf{Z}[\alpha]$. He had then found that, again for $p \equiv 1$ mod. λ, such homomorphisms do exist, even if π does not. But who cares about π? *The homomorphism is the prime.* Call it $\mathfrak{p}$; then we know what it means to say that ξ is a multiple of $\mathfrak{p}$; it means that $\varphi(\xi)$ is 0, and this is all we want to know.

Here we have been going too fast; it is not quite all we want to know. Firstly we want to know also the prime factors of primes p which are not $\equiv 1$ mod. λ; this is where we need the periods. Secondly we want to know, not only whether ξ is a multiple of $\mathfrak{p}$, but also, for every n, whether it is a multiple of $\mathfrak{p}^n$. In other words, Kummer constructs explicitly all the valuations of the field $\mathbf{Q}(\alpha)$; of course he cannot do this without at the same time obtaining the laws of decomposition of the rational primes in that field, and it is lucky for him that he is dealing with a field where those laws can be described explicitly. The results, as stated in his letter of 18 October 1845, are almost complete; he only makes an exception for those rational primes p, necessarily in finite number, which belong to an exponent $f=(\lambda-1)/e$ modulo λ (this means that p^f, but no lower power of p, is $\equiv 1$ mod. λ), but are not prime to the norm of $\eta_0-\eta_1$. This last case, for a long time, gave him trouble; in fact, it is only quite late, in 1856 [46] that he gives a full and correct treatment, virtually acknowledging the gap in his earlier proof as given in [27] and in [39]. Let those of us who have never been guilty of a more serious error cast the first stone.

Now, for perhaps a year, Kummer is "idle" (cf. [F; 75]); he amuses himself with a diophantine problem [30], with the gamma function [25], with some geometry [26], with a review of Jacobi's Works [29]. But this is not the shy young man of 1842, respectfully seeking Dirichlet's advice through Kronecker. He has become Dirichlet's "sworn brother" [F; 69]; Dirichlet and Jacobi treat him as their equal. He is now conscious of his strength. In 1847 he takes up again his "old equation" $x^\lambda+y^\lambda=z^\lambda$; in a few weeks [F; 75–80] he constructs a proof, resting upon two assumptions concerning the class-number and the units, assumptions which he knows to be true for some low values of λ (at any rate $\lambda=3, 5, 7$), and which he expects to be true for "most" primes λ, if not for all. Actually these are the so-called "regular" primes, and even now we are not sure that there are infinitely many of them; but, after such a step forward, Kummer was entitled to some optimism. Now he sees the need for more precise results on the class-number. He knew, of course, that in substance Dirichlet had already solved the problem in a different language, and for a while he kept waiting for Dirichlet to publish his solution. Obviously Dirichlet, perceiving the superiority of the language of ideals, preferred to step aside and told Kummer to go ahead. The results appear only in 1850 [34], but they must have been in Kummer's possession in 1847 or earlier, since otherwise he could not have perceived the connection between the class-number and the Bernoulli numbers [F; 80], suspected by him since 1846 [F; 69].

But this is only the beginning; from 1847 onwards, the flow of Kummer's ideas, as he communicates them to Kronecker, becomes so torrential that the biographer has trouble keeping pace with it. Methods of p-adic analysis, Bernoulli numbers, Gaussian sums, repeated attacks on the higher laws of reciprocity (his "arch-enemy", he says in 1848; cf. [F; 82]), the so-called "Theorem 90" of Hilbert's *Zahlbericht*, all this hits us in quick succession and leaves us breathless. Instead, we will try to analyze his major themes as they appear in his publications of the fifties ([33] to [60]).

What strikes us first (and what, incidentally, appears to have been most distasteful to Hilbert) is the gradual appearance of powerful methods of p-adic analysis. Of course he never introduced the concept of p-adic fields; the credit for this goes to his pupil, or rather his pupil's pupil Hensel.[3] Perhaps this concept could only occur to someone like Hensel, who had also been Weierstrass's pupil and who was familiar, not only with Cantor's definition of the real numbers, but with the ideas of Dedekind and

[3] The genealogy of p-adics is clearly marked and is worth noting here. Kronecker was Kummer's pupil in the Gymnasium at Liegnitz and remained his lifelong admirer and friend (cf. his touching letter of 1881 to Kummer, [F; 102–103]); his thesis of 1845, on the units of cyclotomic fields, seems clearly inspired by Kummer. Hensel was Kronecker's student, and through him came close to Kummer (cf. [F; 1–2]). Hasse, who perhaps did more than anyone else to establish p-adics firmly as an important branch of mathematics, was Hensel's student and wrote his thesis under Hensel in Marburg.

Weber on the analogies between number-fields and function-fields. But Kummer did not really miss the concept. He gradually came to realize that there are formal power-series (such as the logarithmic and exponential series) whose partial sums can be used to define a number modulo p^n for arbitrarily high values of n (cf. [40; 130], where Kummer indicates that also Eisenstein had independently made some use of p-adic series). This, coupled with the fact that very soon he felt quite free, in his p-adic calculations, to write fractions whose denominators are prime to p, gave him a fully adequate substitute for what we would rather couch in the language of p-adic fields. Even better, he does not even miss the fact that $\mathbf{Z}_p$ contains the roots of unity of order $p-1$; we would prove this by observing that, for every integer a prime to p, the sequence a^{p^n} converges p-adically towards such a root. Instead of that, Kummer simply writes a^{p^n} when he is calculating modulo p^{n+1}. For instance, take the Gaussian sum

$$g = \sum_{i=1}^{\lambda-1} \beta^i \alpha^{\gamma^i},$$

where, as always, λ is an odd prime, α is a primitive root of $\alpha^\lambda = 1$, γ is a primitive root modulo λ in $\mathbf{Z}$, and β is a root of $\beta^{\lambda-1} = 1$; we know that this is an integer in the local field $\mathbf{Q}_\lambda(\alpha)$, but Kummer, in order to be able to work modulo λ^{n+1} in $\mathbf{Z}[\alpha]$, simply substitutes for g the sum

$$g_k' = \sum_{i=1}^{\lambda-1} \gamma^{-ik\lambda^n} \alpha^{\gamma^i} \quad (1 \leqq k < \lambda - 1),$$

this being, as he says, "fully analogous" to g [40; 132]. The use of the λ-adic field would bring more elegance but would not affect the substance of Kummer's treatment.

Kummer's initial motivation for "p-adics" seems to have come from Fermat's theorem (see [24], [35]; cf. also [F; 75–81]). For his proof of that theorem, there were two prerequisites: (A) with λ, α as before, the class-number h for the field $K = \mathbf{Q}(\alpha)$ should be prime to λ; (B) every unit in K which is $\equiv c$ mod. λ, with $c \in \mathbf{Z}$, should be a λ-th power in K. As to (A), it can be subdivided into two parts, since h is in a natural way the product of two factors h_1, h_2, and we can write (A_1), (A_2) for the conditions $h_1 \not\equiv 0$ (mod. λ), $h_2 \not\equiv 0$ (mod. λ), respectively. Here the "first factor" h_1 is an explicit product of elementary factors, each of which takes its values in the field generated over $\mathbf{Q}$ by the roots of unity of order $\lambda - 1$; as we have seen above, they can be embedded in the λ-adic field $\mathbf{Q}_\lambda$. On the other hand, the "second factor" h_2 is no other than the class-number of the real subfield K_0 of K, and Kummer showed that it can also be expressed as the index, within the group of all the units in K, of the group of the "cyclotomic units", i.e. the group generated by the units of the form $\pm \alpha^i$, $(1-\alpha^i)/(1-\alpha)$. As to (A_1), Kummer found that it is equivalent to his famous condition (known as the "regularity condition" for λ): (R) λ must not divide any of the Bernoulli numbers B_n for $1 \leqq n \leqq (\lambda-3)/2$. The calculation may conveniently be carried out in $\mathbf{Q}_\lambda$, but it can also be presented in a purely elementary manner, because of the fact that the field of $(\lambda-1)$-th roots of unity is unramified at λ, and it is so presented by Kummer. On the other hand, when it comes to (A_2) and (B), one has in effect to work in $\mathbf{Q}(\alpha)$ at the "place" $(1-\alpha)$ (since $(1-\alpha)$ is the unique prime dividing λ in that field): in other words, one has to work in $\mathbf{Q}_\lambda(\alpha)$, which is ramified (more precisely, it is "tamely ramified") over Q_λ. Actually, if we express in modern language what Kummer does at this point, it is neither more nor less than to show that (R) is equivalent to the following: (R′) the λ-adic regulator of the group of cyclotomic units is $\not\equiv 0 \pmod{\lambda}$ ([34]; cf. [40]). At this point, λ-adic methods become essential; if one compares, say, [40] with [34], one can see how the use of such methods, experimental and fumbling at first, is gradually forged by Kummer into a powerful instrument for arithmetical research. It is impressive, in fact, to see how in [40] Kummer develops methods for obtaining results modulo λ^{n+1} for arbitrarily high values of n, even though, as he remarks [40; 134] the applications he has in mind hardly ever go beyond $n=0$ and 1. Even the case $n=1$ seems to occur only once [48; 46] in connection with a successful effort to extend the proof of Fermat's theorem at least to some irregular primes, after Kummer had given up the hope of obtaining a general proof and thereby winning 3000 francs.[4] What interests us here is less the additional benefit in extending the validity of Fermat's theorem (even Kummer seems to have attached little value to this) than the concrete example provided by Kummer for the main theorems of classfield theory; we are dealing here with a case where the class-number is a multiple of λ but not of λ^2, so that $Q(\alpha)$ has exactly one unramified cyclic extension of degree λ; in effect, Kummer provides it for us, in the form $Q(\alpha, e^{1/\lambda})$, where e is a suitably defined cyclotomic unit.

Except for this solitary foray into uncharted territory, Kummer always restricted his field of investigation to the "regular case" where λ does not divide the class-number h of $Q(\alpha)$. Very soon Kronecker, and later Hilbert, were to become fascinated by the unramified abelian extensions of number-fields and their magical virtue of converting ideals into numbers; not so Kummer, who, while always dealing (explicitly or implicitly) with cyclic extensions of $Q(\alpha)$ of degree λ, seems to have regarded the "irregular case" as a more or less pathological nuisance. This led him first to conjecture [33] and then ([50], [52], [60]) to prove the reciprocity law in a peculiar form which must be explained now.

With λ and α as before, take in $Q(\alpha)$ a prime ideal $\mathfrak{p}$ prime to λ and an integer μ prime to $\mathfrak{p}$. Put $q = N\mathfrak{p}$; this is $\equiv 1 \pmod{\lambda}$, and we may put $q-1=\lambda n$. Then one writes $(\frac{\mu}{\mathfrak{p}})$ or $(\mu/\mathfrak{p})$ for the power α^m of α which is $\equiv \mu^n (\text{mod. } \mathfrak{p})$. For any ideal $\mathfrak{a}$ prime to μ, one defines then the symbol $(\mu/\mathfrak{a})$ by prescribing that it is to be multiplicative in $\mathfrak{a}$, i.e. that $(\mu/\mathfrak{a}\mathfrak{b}) = (\mu/\mathfrak{a})\cdot(\mu/\mathfrak{b})$. In introducing this symbol, Kummer and Eisenstein were of course only following in the footsteps of Gauss and Jacobi.

What is peculiar to Kummer is that now, concentrating his attention on the regular case, he introduces symbols $(\mathfrak{m}/\mathfrak{p})$, $(\mathfrak{m}/\mathfrak{a})$, where $\mathfrak{m}$ is an ideal; he does this as follows (cf. e.g. [33]). Since h is the class-number, $\mathfrak{m}^h$ is a principal ideal (μ), where μ is determined only up to a unit. Let us determine μ further by some suitable congruence condition, so that then it is uniquely determined up to the λ-th power of a unit; then $(\mu/\mathfrak{a})$ is well-determined, and, since h is prime to λ, there is a well-determined power α^m of α such that $(\alpha^m)^h = (\mu/\mathfrak{a})$. This, by definition, is $(\mathfrak{m}/\mathfrak{a})$. As to the condition to be imposed on μ, when μ is prime to λ, it is given by $\mu \equiv b \pmod{(1-\alpha)^2}$, $\mu\bar{\mu} \equiv c$ (mod. λ), where b, c are rational integers; then μ is called *primary*. Of course it has

[4] Goldfrancs, of course; cf. [F; 84] and [F; 88]. He did get them, nevertheless; cf. [L; 17, footnote].

to be shown that this does determine μ, and that it determines it uniquely up to the λ-th power of a unit; this is so in the regular case (but not otherwise), as Kummer shows by the use of λ-adic methods. Thus his symbol is well-defined, and his procedure is fully justified in his eyes by the simple form taken by the reciprocity law: $(\mathfrak{p}/\mathfrak{p}') = (\mathfrak{p}'/\mathfrak{p})$ for any two distinct primes $\mathfrak{p}$, $\mathfrak{p}'$, other than $(1-\alpha)$.

From the vantage point of our modern knowledge, we know that this is misleading. Eisenstein, who was not particularly interested in the "regular case", had in some sense a better appreciation of the problem. Let us for a moment drop the assumption about "regularity"; then $(\mathfrak{m}/\mathfrak{a})$ is meaningless, but $(\mu/\mathfrak{a})$ is well-defined for any integer μ prime to $\mathfrak{a}$; let us write (μ/ν), instead of $(\mu/(\nu))$, for two mutually prime integers μ, ν, both prime to λ. Then, if we also use Hilbert's normresidue symbol $\{\mu, \nu/\mathfrak{p}\}$, we have

$$(\mu/\nu)\cdot(\nu/\mu)^{-1} = \Pi\{\mu,\nu/\mathfrak{p}\}\,,$$

where the product in the right-hand side is extended to all the primes dividing $\mu\nu$, or, what amounts to the same, to all the "places" of $\mathbf{Q}(\alpha)$ except $(1-\alpha)$. The reciprocity law gives now, as we know:

$$(\mu/\nu)\cdot(\nu/\mu)^{-1} = \{\mu,\nu/(1-\alpha)\}^{-1}\,. \tag{1}$$

In particular, the left-hand side is λ-adically continuous in μ and ν, and it is in this form that Eisenstein conjectures the reciprocity law; moreover, he gives the procedure by which, if this is assumed, the right-hand side can be completely calculated.

On the other hand, in the regular case, this right-hand side turns out to be 1 provided μ and ν are both primary; in view of this, and of the definition of Kummer's symbol, the above formula gives at once the reciprocity law in Kummer's form.

This is the peak he sighted for the first time in January 1848 (cf. [33; 155]), and whose summit he reached ten years later ([50]; cf. [52] and [60]). Perhaps he thought of this as the most lasting achievement of his career. In our eyes, it is more like the story of the first ascent of the Matterhorn to one who has learnt to climb Mount Everest. It is of greater interest to us to look at the rich territories which he had to explore in the course of this ten-year campaign.

At the outset, the most promising tool seemed to be the Gaussian sums; this was of course suggested by the previous special cases treated by Gauss, Cauchy, Jacobi, Eisenstein. Actually, using no other tool, Eisenstein succeeded in proving, in 1850, the special case which bears his name; with the same notations as before, this says (without any assumption about the "regularity" of λ) that $(\mu/m) = (m/\mu)$ provided m is a rational integer and μ is congruent to a rational integer modulo $(1-\alpha)^2$. Kummer greatly perfected the use of this instrument with the help of his methods of λ-adic analysis, and also extended the concept of the Gaussian sums (or rather of the "Jacobi sums", but for such purposes this amounts to the same thing) from the prime field $\mathbf{F}_p$ to any finite field (cf. [33], [40] and [F; 89]). This gave him a fairly complete and explicit set of formulas for the so-called "complementary laws" ("Ergänzungssätze"; cf. [40] and [53]) in the regular case, culminating in an explicit formula [40; 144] for the "correction factor" given by (1) when μ, ν are not primary. More precisely, let us write (μ, ν) for the left-hand side of (1); Kummer had shown that μ, ν can be written as $\mu=\mu_1\eta$, $\nu=\nu_1\zeta$, where η, ζ are units and μ_1, ν_1 are primary. Then the "correction factor" is

$$\alpha^{\Sigma}=(\mu,\nu)\cdot(\mu_1,\nu_1)^{-1}\,,$$

and it is for the exponent Σ that Kummer gives an explicit bilinear formula in terms of the λ-adic logarithms of μ, ν. In view of what has been said above, α^{Σ} is of course nothing else than Hilbert's norm-residue symbol. Incidentally and by the way, Kummer had plucked a "nice little fruit" ("ein ganz nettes Früchtchen", [F; 88]; cf. [F; 86–87]), no more and no less than his famous congruences for Bernoulli numbers [38], which contain, as we know now, a large chunk of the modern theory of p-adic zeta-functions and L-functions.

Eventually, Kummer becomes convinced that Eisenstein's law is the utmost that Gaussian sums can supply in the way of a general reciprocity law. He is still far from the main summit; but this just spurs him on. A new instrument is forged [42], generalizing in some sense the Gaussian sums or rather, as Kummer says, the Lagrangian resolvents. This is no other than what is now generally known as "Hilbert's Theorem 90", the grandfather of cohomology theory. In [42], its basic features are described, and it is applied to still another special case of the reciprocity law.

Is the goal at last within reach? Alas, no. "I have something sad to tell you", he writes to Kronecker in 1853 ([F; 93]; cf. [42; 212] and [42; 232]). Even in the particular case to which he has applied the new instrument, its use is subject to some awkward exceptions. Still another base-camp has to be elaborately organized before the final assault.

Of course what was missing is clear to us. Even for the law of quadratic reciprocity in $\mathbf{Z}$, one can give elementary proofs, but it cannot be understood properly within the narrow confines of the rational numberfield; one has to consider also its quadratic extensions, or equivalently, as Gauss did, the theory of binary quadratic forms. It had taken Gauss and then Jacobi to see that the reciprocity laws for λ-th powers cannot even be formulated within $\mathbf{Q}$, and that one has to reach out at least to $\mathbf{Q}(\alpha)$ with $\alpha^{\lambda}=1$. But to prove them is a different matter, except in the lowest cases where λ is 4 or 3. Here the analogue of the quadratic extensions of $\mathbf{Q}$ is given by the cyclic extensions of $\mathbf{Q}(\alpha)$ of degree λ, and this is the ridge which Kummer had to follow.

It is fascinating to watch him doing it with what any beginner would now regard as most inadequate equipment. He lacks the general concept of an algebraic integer, and is therefore unable to develop a proper ideal theory for the fields he must now study. For the cyclotomic fields generated by roots of unity of any order, he eventually developed a complete ideal-theory [44], similar to the theory for roots of prime order λ. He also knew that Kronecker was at work on the general case of an arbitrary number-field [52; 57], but he does not feel that he really needs this. To our surprise (and to Hilbert's; *Zahlbericht*, p. 494) he manages to turn his lack of knowledge to his own advantage; this happens as follows.

Put $K=\mathbf{Q}(\alpha)$, with λ and α as before; call K' the "Kummer extension" of K generated by a root w of $w^{\lambda}=D$, where D is an integer in K; K' is ramified over K at most at the "bad places" entering into λ and D. In K', Kummer introduces two rings of integers; one is $R_w=\mathbf{Z}[\alpha,w]$, and the other is a certain subring R_z of R_w

containing 1 and λR_w (cf. [52; 32–44]); in modern language, the "conductors" of these rings are confined to the bad places. To construct the valuations of K' outside the bad places is well within Kummer's powers; in fact, outside such places, a prime ideal $\mathfrak{p}$ in K splits into λ factors or remains prime in K' according as D is a λ-th power residue modulo λ or not, so that the connection with Kummer's main problem becomes immediately clear. This supplies Kummer with an ideal theory for R_w and R_z which turns out to be perfectly adequate for his purposes. As to the "bad places", he just remarks in passing that there cannot be any proper ideal theory there; apparently he has discovered experimentally some of the phenomena connected with the primary ideals at the places in the conductor. In his language, the ideals, in R_w and R_z, which belong, or rather which ought to belong to the bad places are simply "not defined". Miraculously, this provides him with a simpler theory of genera, particularly in R_z, than could be obtained in the ring of all integers of K'. In retrospect, this also provides some justification for Gauss's insistence on writing $aX^2 + 2bXY + cY^2$, rather than $aX^2 + bXY + cY^2$ as Lagrange and Legendre did and as we do again, for binary quadratic forms over $\mathbf{Z}$. Having reached this point, Kummer boldly sets out to extend Gauss's theory of genera of quadratic forms to the ideals in R_z, and this time full success crowns his efforts. Gauss's "second proof" for the quadratic reciprocity law, the one which is based on the theory of the genera, can indeed be generalized; Kummer's flag can be planted on the summit [52; 157].

But Gauss had given proof after proof of the quadratic reciprocity law, and Kummer cannot rest until he has done the same. He constructs two more proofs, one of which generalizes Legendre's, while the other is altogether new ([60]; cf. [52; 158]). We need not follow him there, noting only that the new proofs dispense with R_z and use only the somewhat simpler ring R_w. But two more remarks may be made. Firstly, in the course of his investigation [52; 82], he restricts the choice of the integer D by prescribing that it should be $\equiv 1$ mod. $(1-\alpha)$ but not mod. $(1-\alpha)^2$. That being so, the completion of the field $K' = K(D^{1/\lambda})$ at the place λ is not tamely ramified any more over $\mathbf{Q}_\lambda$, and this requires a further refinement of Kummer's λ-adic methods. Secondly, he is naturally led to investigate the local condition, at the place $(1-\alpha)$, for a given integer μ of K to be the norm of an integer of K' over K; in modern terms, this is the same as to calculate the norm-residue symbol $\{\mu, D/(1-\alpha)\}$; this is in effect what he does, and what leads him once more, to his own surprise, to the bilinear formula he had found earlier for the "correction factor" in the reciprocity law. In view of this, nothing is missing for the formulation of Hilbert's reciprocity law, except the language and a clear perception of the relationship between "local" and "global" facts.

Here, except for some shorter notes (interesting but of comparatively minor import), Kummer takes leave of number-theory; and here we leave him. As the reader must realize, we have necessarily had to confine ourselves to the most superficial description of the contents of this volume. Even after a hundred years, an attentive study can richly repay his efforts.

Kummer's Language and Notations

Kummer's writings seem exceptionally free from mathematical errors (even if one counts as an error a faulty argument for a correct conclusion); when errors have been noticed, they are indicated at the end of the volume (pp. 955–57). On the other hand, there are not a few misprints, virtually all of which are quite harmless and should be immediately obvious to any moderately attentive reader. No systematic effort has been made to search for them; some (and, one hopes, the worst ones) are indicated on pp. 955–57, but the reader should use common sense whenever he has doubts about a particular formula.

As to Kummer's language and notations for number-theory, they are essentially lucid and free from ambiguity once one has become used to his style, but they are perhaps such as to discourage some modern readers. The following remarks may prove helpful.

Kummer is invariably dealing at every stage with an explicitly given ring of algebraic integers. If for instance he is concerned with the cyclotomic field generated by a primitive λ-th root α of unity (λ always denotes an odd prime), then he will speak of "numbers in α", by which he means elements of the ring $\mathbf{Z}[\alpha]$, and he writes $F(\alpha)$ for such an element, where F is always thought of as a polynomial with rational integral coefficients. He can then not only write $F(\alpha^2), \ldots, F(\alpha^{\lambda-1})$ for the conjugates of $F(\alpha)$, but he can also meaningfully speak of $F(1)$; of course he knows that, for a given element of $\mathbf{Z}[\alpha]$, the polynomial F is only determined modulo $1+X+\ldots+X^{\lambda-1}$, so that e.g. $F(1)$ is defined only as an integer modulo λ. It is all-important for him, actually, that he can formally substitute e^v for α, expand $F(e^v)$ into a formal power-series at $v=0$, take the n-th logarithmic differential, etc.; naturally, he must sometimes make sure that the main conclusions are independent of the choice of F modulo $1+X+\ldots+X^{\lambda-1}$; this is done with proper care when necessary.

Of course $\mathbf{Z}[\alpha]$ happens to be *the* ring of integers in $\mathbf{Q}(\alpha)$, but Kummer is unaware of this fact, since he lacks the general concept of an algebraic integer. On the other hand, he is well acquainted with the Galois theory of the cyclotomic field $\mathbf{Q}(\alpha)$; certainly he never read Galois, but all the relevant facts were known to him from Gauss's *Disquisitiones.* As Gauss had shown, for each divisor e of $\lambda-1$ (where λ is again an odd prime), $\mathbf{Q}(\alpha)$ has a subfield of degree e over $\mathbf{Q}$, generated by any one of the Gaussian "periods" $\eta_i = \sum_j \alpha^{\gamma^{i+ej}}$, where γ is a primitive root modulo λ, $0 \leqq i < e$, and $0 \leqq j < (\lambda-1)/e$ (Kummer omits the subscript 0, writing η for η_0). Clearly every element of $\mathbf{Z}[\alpha]$, contained in $\mathbf{Q}(\eta)$, can be written as a polynomial in the η_i and more precisely as a linear combination of the η_i, with coefficients in $\mathbf{Z}$, but such elements, as Kummer also knows, are not always in $\mathbf{Z}[\eta]$; logically he ought to write them as $f(\eta, \eta_1, \ldots \eta_{e-1})$, but, without warning, he always writes them as $f(\eta)$, so that he can write simply $f(\eta_1), \ldots, f(\eta_{e-1})$ for the conjugates of $f(\eta)$; he knows that the automorphisms of $\mathbf{Q}(\alpha)$ over $\mathbf{Q}$ permute the η_i cyclically. This "abuse of notation" (and of an already ambiguous notation) involves the reader, and in fact Kummer himself, in difficulties when Kummer seeks to define the prime ideals of degree 1 in the ring in question by means of homomorphisms of that ring onto the prime field modulo p, where p is a suitable prime; such a homomorphism can only be given by mapping each η_i onto a suitable integer u_i modulo p, but Kummer writes it as $f(\eta) \to f(u)$, and gets into trouble when p is not relatively prime to all the $\eta - \eta_i$.

Eventually Kummer also considers various extensions of the field $\mathbf{Q}(\alpha)$, particularly those generated by a p-th root x of unity (p being a prime other than λ) and the "Kummer fields" generated over $\mathbf{Q}(\alpha)$ by a root w of an equation $w^\lambda = D(\alpha)$, with $D(\alpha)$ in $\mathbf{Z}[\alpha]$; he needs the former for the constructions of "Gaussian sums" (called by Kummer "Lagrange resolvents", which historically is alone correct), while the latter play the main role in his treatment of the reciprocity laws. He also extends part of his theory of $\mathbf{Q}(\alpha)$ to fields generated by n-th roots of unity, for any integer n. In each case he concentrates his attention on rings generated over $\mathbf{Z}$ by certain specified elements, and writes the elements of such rings accordingly, i.e. as polynomials in the generators. For instance he will write $F(\alpha, x)$ for an element of $\mathbf{Z}[\alpha, x]$, where $\alpha^\lambda = 1$, $x^p = 1$; here it is understood that F is a polynomial with coefficients in $\mathbf{Z}$, so that he can meaningfully write $F(1, x)$ or even $F(e^v, x)$ as explained above. In the case of the "Kummer fields" $\mathbf{Q}(\alpha, w)$, he simplifies his notation further by regarding $\mathbf{Z}[\alpha]$ as the groundring for $\mathbf{Z}[\alpha, w]$; consequently he will write $F(w)$ for an element of the latter ring, with the understanding that F is now a polynomial with coefficients in $\mathbf{Z}[\alpha]$; the conjugates of $F(w)$ over $\mathbf{Q}(\alpha)$ are then $F(w\alpha), \ldots, F(w\alpha^{\lambda-1})$. For the Kummer field, Kummer also needs a subring of $\mathbf{Z}[\alpha, w]$ generated over $\mathbf{Z}[\alpha]$ by λ elements $z_0, z_1, \ldots, z_{\lambda-1}$, which are permuted cyclically by the Galois group of $\mathbf{Q}(\alpha, w)$ over $\mathbf{Q}(\alpha)$, and any $\lambda - 1$ among which, together with 1, make up a basis for $\mathbf{Z}[\alpha, z_0, \ldots. z_{\lambda-1}]$ over $\mathbf{Z}[\alpha]$. The elements of this latter ring are written as $F(z)$, the understanding being now that F is either a polynomial in $z_0, \ldots, z_{\lambda-1}$ or simply a linear form in $1, z_0, \ldots, z_{\lambda-1}$, with coefficients in $\mathbf{Z}[\alpha]$; they are called "numbers in z" or "complex numbers in z", while the elements of $\mathbf{Z}[\alpha, w]$ and those of $\mathbf{Z}[\alpha]$ are called numbers, or complex numbers, "in w" and "in α" respectively. Somewhat misleadingly again, Kummer often writes z instead of z_0, and $F(z_1), \ldots, F(z_{\lambda-1})$ for the conjugates of $F(z)$ over $\mathbf{Q}(\alpha)$. Similar notations, such as $F(\alpha)$, $F(w)$, $F(z)$, are also used for the elements of the corresponding fields, which are then called "fractional complex numbers" ("gebrochene complexe Zahlen") in α, in w, in z; the word "wirklich" (in French, "existant") is used to indicate elements of rings, as distinct from (non-principal) ideals in such rings. Rational integers are variously called "ordinary" ("gewöhnliche ganze Zahlen"), "real" ("reelle g. Z.") or "non-complex" ("nichtcomplexe g. Z.").

Sometimes Kummer, after introducing e.g. an element $F(\alpha)$ of $\mathbf{Z}[\alpha]$, uses the notation $F(w)$, not (as one might have thought) for the result of the substitution of w for α in $F(\alpha)$, but for an element of $\mathbf{Z}[\alpha, w]$, possibly quite unrelated with $F(\alpha)$; actually the context never leaves any doubt about the intended meaning.

Kummer uses the sign N for the norm, whether absolute, i.e. over $\mathbf{Q}$, or relative, e.g. in $\mathbf{Q}(\alpha, w)$ over $\mathbf{Q}(\alpha)$; in this matter, too, the context is always decisive; when necessary, Kummer adds explanatory words, e.g. "norm in η" for the norm taken in $\mathbf{Q}(\eta)$ over $\mathbf{Q}$, where η is the Gaussian period as written above; "norm in w" or "norm in z" for the norm in $\mathbf{Q}(\alpha, w)$ over $\mathbf{Q}(\alpha)$; in the latter case, he will then sometimes write NN for the norm in $\mathbf{Q}(\alpha, w)$ over $\mathbf{Q}$, the second N denoting the norm in $\mathbf{Q}(\alpha, w)$ over $\mathbf{Q}(\alpha)$ and the first one the norm in $\mathbf{Q}(\alpha)$ over $\mathbf{Q}$.

The reader will have to get accustomed to Kummer's usage in dealing with ideals. The major difficulty is here that Kummer never got rid of his habit of regarding, say, the non-zero integers in $\mathbf{Z}[\alpha]$ as making up a subset of the set of non-zero ideals in the same ring. In fact, he compares his use of the word "ideal", as applied to the "complex numbers in α", to the normal usage for the word "imaginary" in classical algebra. There, he says, the phrase "imaginary number" may be used in reference to any number $a + b\sqrt{-1}$ with a, b real, or it may be used to specify that b is not 0 and that the number in question is not real. Thus, Kummer has no language to express the distinction between an algebraic integer, say $F(\alpha)$, and the principal ideal it defines. Of course he knows that, in some circumstances (when a certain set of conditions determines only a "principal ideal") a number $F(\alpha)$ may be determined only up to a unit; then he writes it as $e(\alpha)F(\alpha)$, where $e(\alpha)$ is an arbitrary unit in $\mathbf{Z}[\alpha]$. Similarly, he will use the notation $f(\alpha)$ for a non-zero ideal in $\mathbf{Z}[\alpha]$, calling this "an ideal complex number", and again regarding it as determined up to a unit. This notation has the advantage that he can then write $f(\alpha^2), \ldots, f(\alpha^{\lambda-1})$ for the conjugates of $f(\alpha)$, i.e. for its transforms under the automorphisms of $\mathbf{Q}(\alpha)$ over $\mathbf{Q}$. But he will also sometimes write $e(\alpha)f(\alpha)$, where $e(\alpha)$ is a unit, and the significance of this notation has to be disengaged from the context; in many cases it is as follows. Call h the class-number for $\mathbf{Q}(\alpha)$; then, if $\mathfrak{a}$, in modern notation, is an ideal in $\mathbf{Z}[\alpha]$, $\mathfrak{a}^h$ is a principal ideal which can be written as $(\varphi(\alpha))$, where $\varphi(\alpha)$ is in $\mathbf{Z}[\alpha]$, and is well determined up to a unit $e(\alpha)$; symbolically, therefore, one may put $f(\alpha) = \varphi(\alpha)^{1/h}$ and use $f(\alpha)$ as a symbol for the ideal $\mathfrak{a}$, and it is meaningful to determine the unit in $\varphi(\alpha)$ by prescribing for $\varphi(\alpha)$ some suitable congruence condition, e.g. modulo λ. Kummer's main purpose is usually to obtain reciprocity laws for the λ-th power residue symbol in $\mathbf{Z}[\alpha]$, and his basic assumption is invariably that the class-number h is prime to λ (i.e. that λ is "a regular prime"); raising multiplicative formulas to the h-th power is then harmless, as Kummer well knows, and this remark is all that is needed in order to justify his procedures. We again note that the use of one and the same notation for numbers and for ideals makes it all the more necessary to have a word to indicate whether a given symbol $f(\alpha)$ denotes "an ideal complex number", an element of $\mathbf{Z}[\alpha]$, or an element of $\mathbf{Q}(\alpha)$, and this is done by the words "ideal", "wirklich", "gebrochen", as noted above.

One more complication occurs in the case of the rings $\mathbf{Z}[\alpha, w]$, $\mathbf{Z}[\alpha, z]$ in the Kummer extension $\mathbf{Q}(\alpha, w) = \mathbf{Q}(\alpha, D(\alpha)^{1/\lambda})$ of $\mathbf{Q}(\alpha)$; in modern language, these rings have conductors, containing the prime ideals entering into $\lambda D(\alpha)$; Kummer wisely makes no attempt to look at the behavior of those rings at such places, and regards as "defined" in those rings only those ideals which, in modern language, would be prime to the conductor, i.e. to $\lambda D(\alpha)$; for those ideals, the unique decomposition into powers of prime ideals is valid, and the concept of equivalence can be defined in the usual way (but the class-number depends of course upon the choice of the ring).

Finally, it should be noted that the word "to contain" ("enthalten") is to be understood in the sense of "to be a multiple of", as is natural from the point of view of the prime ideal decomposition; thus, a "number" ("wirkliche Zahl") $F(\alpha)$ is said to contain the ideal factor $f(\alpha)$ if and only if $F(\alpha) \equiv 0 \bmod. f(\alpha)$. In the theory of the ring $\mathbf{Z}[\alpha, z]$ and of its "ambiguous" ideals (i.e., those which are equivalent to their conjugates), the word "to contain" is give a much more special sense [52; 77] and is then used as a convenient technical term for that theory; but this should cause no difficulty to the reader.

Note to the Reader

Kummer's papers have been arranged in this volume in the chronological order, not of publication but of composition (which is always easy to establish, by means of Kummer's own dating and of his correspondence).

An asterisk in the margin (*) refers to the Notes at the end of the volume.

Nachruf für Ernst Eduard Kummer

Von E. Lampe

Jahresbericht der Deutschen Mathematiker-Vereinigung 3, 13-28 (1892-1893)

Zu Interlaken wohnte im Sommer 1875 mit dem Schreiber dieser Zeilen in derselben Pension der greise Mathematiker Scherk nebst seinen zwei Töchtern. Von Halle, wo derselbe als Professor der Mathematik gewirkt hatte, war er nach Kiel berufen worden; bei der Wiederherstellung der dänischen Herrschaft in den deutschen Herzogtümern musste er aber seine dortige Professur aufgeben und fand als Gymnasiallehrer in Bremen eine Zufluchtsstätte. Trotz seines hohen Alters war er noch immer lebhaften Geistes, fand viel Gefallen am Schachspiel und freute sich, einem Schüler Kummer's zu begegnen, dessen Lehrer in Halle gewesen zu sein eine seiner schönsten Erinnerungen war. In dem gemeinsamen Gefühle inniger Verehrung des damals im 66. Lebensjahre stehenden Berliner Gelehrten fanden sich sein alter Lehrer mit seinem Schüler und Nachfolger an der Kriegsakademie zusammen, und ein herzlicher Gruss des halb erblindeten Scherk wurde an Kummer gern übermittelt.

Wie es damals dem jüngeren Mathematiker wunderbar vorkam, dass der Lehrer Kummer's noch schaffte und wirkte, so dürfte die jetzige heranwachsende mathematische Jugend sich erst besinnen müssen, um sich zu erinnern, dass die ehrwürdige Gestalt Kummer's bis zum 14. Mai des Jahres 1893 unter den Lebenden gewandelt habe. Seit neun Jahren aller Thätigkeit entsagend, führte er im Schosse seiner Familie ein ver-

*) Dieser Nachruf ist zuerst in der Naturwissenschaftlichen Rundschau vom 15. Juli 1893 erschienen (ohne das jetzt zugefügte Verzeichnis der Schriften Kummer's und ohne die Fussnote auf S. 17).

borgenes Leben, in welches nur die nächsten Freunde zuweilen einen Einblick thun durften. Keine Zeile von seiner Hand, kein in der Oeffentlichkeit gesprochenes Wort gab Zeugnis von dem Dasein des früher so rastlos thätigen, gewaltigen Geistes. Und dies geschah mit vollem Vorbedacht. Als in den sechziger Jahren bei der Veröffentlichung des Nachlasses von Gauss in den gesammelten Werken desselben ein jüngerer Mathematiker bemerkte, nun seien ja auch die nachlässigen Werke von Gauss zugänglich geworden, sprach sich Kummer, der dem Witzworte schmunzelnd sein Ohr geliehen hatte, dahin aus, er werde dafür sorgen, dass von ihm nichts Nachlässiges vorgefunden werde, und er werde aufhören, etwas zu veröffentlichen, sobald er die Abnahme seiner geistigen Kräfte spüre. Diesem Vorsatze ist er treu nachgekommen; vielleicht zu früh ist er von der Bühne des Schaffens abgetreten. Aber in diesem Zuge tritt uns der schlichte und klare Sinn des Verewigten anschaulich vor die Augen. Wie ein antiker Charakter wachte er über sein Thun und hat daher, so wie es sein Wunsch war, der Mitwelt den ungetrübten Eindruck einer hochstehenden, wissenschaftlichen Persönlichkeit und eines unantastbaren, sittlichen Charakters hinterlassen.

Als Sohn eines Arztes wurde Ernst Eduard Kummer zu Sorau in der Niederlausitz innerhalb der jetzigen Provinz Brandenburg, aber hart an der schlesischen Grenze, am 29. Januar 1810 geboren. Der Typhus, welchen die aus Russland zurückkehrenden Reste der grossen Napoleonischen Armee mit sich brachten, raffte den Vater im Jahre 1813 hinweg, und die Mutter hatte ihn und seinen älteren Bruder bei sehr knapp zugemessenen Mitteln zu erziehen. Wem fiele da nicht die Schilderung des Lebens jener bewegten Zeit aus Gustav Freytag's letztem Bande der Ahnen „aus einer kleinen Stadt“ ein? Dieselben Bedrängnisse, welche der Dichter nach den elterlichen Erzählungen aus seiner Vaterstadt schildert, haben auch die Jugend Kummer's eingeengt; derselbe auf ein reiches inneres Leben und auf das geliebte Vaterland gerichtete Sinn ist ihm angeboren und anerzogen. Dieselbe Liebe zur engeren Heimat, als welche er Schlesien betrachtete, hat er bis an sein Ende bewahrt.

Das Gymnasium seiner Vaterstadt darf sich rühmen, ihn bis zur Reife für die Universität gefördert zu haben. Die Mutter wusste es zu ermöglichen, dass er 1828 die Universität Halle bezog, um dort Theologie zu studiren. Gewissensbedenken und philosophische Studien, zu denen er durch sein Fach geführt wurde, bewirkten allmählich, dass er sich der Mathematik ergab. „Der allgemeine Grund dafür, das mathematisches und philosophisches Talent sich oft vereint finden, liegt darin, dass es nur die eine Befähigung und Neigung für das rein abstracte Denken ist, welcher die beiden verschiedenen Wege der mathematischen sowie der philosophischen Speculation gleichmässig offen stehen; ob ein mit diesem Talente begabter wissenschaftlicher Forscher sich mehr der einen oder der anderen dieser verwandten Wissenschaften zuwendet, scheint mehr nur von äusseren Bedingungen abhängig zu sein.“ So spricht sich Kummer in seiner Festrede vom 26. Januar 1865 hierüber aus; er selbst wählte unter Scherk's Leitung die Mathematik, weil in ihr „allein Irrtümer und falsche Ansichten nicht vorkommen können“, also in dem Streben nach der Erkenntnis reiner Wahrheit. Der Student, welcher aus Rücksicht auf die Knappheit seiner Mittel den Weg zwischen Sorau und Halle mit dem Ränzel auf dem Rücken zu Fuss zurücklegte, schritt in der mathematischen Erkenntnis rasch fort, löste im dritten Studienjahre eine mathematische Preisfrage und wurde auf Grund seiner Arbeit am 10. September 1831 in Halle zum Doctor promovirt. Als Gymnasiallehrer war er zuerst in seiner Vaterstadt, dann von 1832 bis 1842 in Liegnitz thätig. Die akademische Laufbahn wurde ihm durch seine Berufung in die Professur für Mathematik an der Universität Breslau eröffnet (1842), und 1855 wurde er der Nachfolger Dirichlet's in Berlin, sowohl an der Universität und in der Akademie, als auch an der damaligen allgemeinen Kriegsschule, jetzigen Kriegsakademie, nachdem er schon seit seinem 29. Lebensjahre der Akademie als correspondirendes Mitglied angehört hatte.

Ueber die Ziele seiner Forschungen berichten wir am besten mit den Worten seiner Antrittsrede in der Leibnizsitzung vom 3. Juli 1856. „Der deutsche Geist, getrieben von dem ihm eigenen Drange nach Erkenntnis hat mit verjüngter Kraft den ewigen Formen und Gesetzen des Mathematischen sich zugewendet und in denselben ein reiches Feld seiner Thätigkeit gefunden. Es ist darum jetzt in der Mathematik, in ähnlicher Weise wie in den ihr verwandten Wissenschaften, die wissenschaftliche Forschung die vorherrschende Richtung, die Forschung, welche weniger im Wissen als im Erkennen ihre Befriedigung findet und darum in die Tiefe der Wissenschaft zu dringen sucht, wo sie die Lösung vorhandener Rätsel findet und wo neue Rätsel ihr entgegentreten . . . . Wenn ich meinen wissenschaftlichen Standpunkt noch näher angeben soll, so kann ich ihn füglich als einen theoretischen bezeichnen, und zwar nicht allein darum, weil die Erkenntnis allein das Endziel meiner Studien ist, sondern namentlich auch darum, weil ich vorzüglich nur diejenige Erkenntnis in der Mathematik erstrebt habe, welche sie innerhalb der ihr eigentümlichen Sphäre ohne Rücksicht auf ihre Anwendungen gewährt."

Das grosse Vorbild, dem er nachstrebte und das er seinen Schülern zu empfehlen nicht abliess, war der unvergleichliche Gauss. Im Mittelpunkte der ersten Periode seiner wissenschaftlichen Schöpfungen steht die Abhandlung über die hypergeometrische Reihe, „eine würdige Ergänzung jener fundamentalen, nur in ihrem ersten Teile erschienenen Gauss'schen Arbeit, gegründet auf tiefstes, in einem Liegnitzer Programm zuerst dargelegtes Erkennen der für die Vergleichung von Transcendenten massgebenden Principien und durchgeführt in solcher Vollständigkeit, dass bei viel später mit ganz neuen Mitteln von Riemann aufgenommenen Untersuchungen sich nur eine kleine Nachlese an Resultaten ergeben hat" (Kronecker). Als nicht mehr ganz junger Einjährig-Freiwilliger sandte Kummer die ersten Ergebnisse seiner Forschungen über die hypergeometrische Reihe an Jacobi in einem Soldatenbriefe, und dieser zeigte die Sendung in Königsberg mit den Worten: „Sieh da, jetzt machen schon preussische Musketiere mit ihren mathematischen Arbeiten den Professoren Concurrenz!" Gern erinnerte sich Kummer in seinen späteren Jahren dieser Anknüpfung seines Briefwechsels mit Jacobi und Dirichlet, und Encke ergötzt sich daran, diesen Umstand in der Erwiderung auf Kummer's akademische Antrittsrede zu erwähnen.

Wenn nun während der elfjährigen Amtsperiode am Liegnitzer Gymnasium der wachsende Ruhm Kummer's sich auf die Arbeiten gründete, welche überwiegend der Functionentheorie angehörten, so dass er auf Grund derselben zum correspondirenden Mitgliede der Akademie erwählt und in die Professur nach Breslau berufen wurde, so fallen doch auch in jene Zeit schon einige Abhandlungen, welche die neue Richtung der die zweite Periode seiner Thätigkeit umfassenden, tiefsinnigen und fruchtbaren Forschungen andeuteten, und zu denen ausser den Disquisitiones arithmeticae besonders die Abhandlungen über die biquadratischen Reste von Gauss die ersten Gesichtspunkte geliefert haben. Auf diesem zahlentheoretischen Gebiete offenbarte sich jetzt vor allem sein eindringender Scharfsinn und die schöpferische Kraft seines Geistes. Die von ihm ersonnenen, idealen Factoren wurden nicht bloss in seiner Hand ein Instrument, das zur Aufhellung alter Probleme, zur Entdeckung neuer Gesetze führte, sondern sie bildeten auch später den Ausgangspunkt weiterer Begriffsbildungen durch Kronecker, Dedekind und Weber. Bald floss aus diesen Untersuchungen der Beweis des Fermat'schen Satzes, grosses und gerechtes Aufsehen erregend, weil ungeachtet so vieler Bestrebungen bedeutendster Forscher der Beweis bis dahin nur in einigen wenigen Fällen gelungen war, und nach weiteren umfassenden Forschungen gelang auch der theoretische Beweis der vorher durch Induction gefundenen höheren Reciprocitätsgesetze. Mit den Arbeiten aus diesem Gebiete gewann Kummer im Jahre 1857 den grossen mathematischen Preis in Paris, ohne sich um ihn beworben zu haben*).

Nach zwanzigjähriger angestrengtester Arbeit auf den abstractesten Gebieten der Zahlentheorie entnahm Kummer den Disquisitiones generales circa superficies curvas von Gauss die Grundgedanken für eine neue Reihe von Forschungen in der Geometrie. Diese geometrischen Abhandlungen, welche an Hamilton's Untersuchungen über Strahlensysteme anknüpfen, füllen hauptsächlich die letzte Periode der Schöpfungen Kummer's aus, und weil sie für viele neuere Arbeiten zur Anknüpfung gedient haben, auch die von ihm im Jahre 1864 entdeckte Fläche vierter Ordnung mit sechzehn Knotenpunkten seinen Namen erhalten hat, so ist durch diese geometrischen Forschungen der Name Kummer's in den weitesten Kreisen der Mathematiker populär geworden. Die Universalität des Kummer'schen Genius bekundet sich aber weiter in dieser letzten Periode seines Schaffens darin, dass er neben der Erweiterung und dem Ausbau der früheren Ideenkreise auch auf concrete Probleme der Physik eingeht. Die Abhandlung über atmosphärische Strahlenbrechung, welche mit seinen Untersuchungen über die Strahlensysteme zusammenhängt, ist ein merkwürdiges Zeugnis für seine schöpferische Phantasie auf einem ganz neuen Gebiete; und die grosse Arbeit über die Wirkung des Luftwiderstandes auf Körper von verschiedener Gestalt, insbesondere auf die Geschosse, hervorgerufen durch die wiederholte Behandlung des ballistischen Problems an der Kriegsakademie, zeigte der staunenden Welt, dass der Mathematiker aus den abstractesten Sphären seiner Ueberlegungen

*) Der lakonische Bericht in C. R. XLIV, 158 (1857) lautet: Rapport sur le concours pour le grand prix de sciences mathématiques. Déjà remis au concours pour 1853 et prorogé jusqu' en 1856. La commission, n'ayant trouvé parmi les pièces adressées au concours, aucun travail qui lui ait paru digne du prix, a proposé à l'Académie de l'accorder à M. Kummer, pour ses belles recherches sur les nombres complexes composés de racines de l'unité et de nombres entiers. L'Académie a adopté cette proposition. — Darauf folgt S. 573 desselben Bandes die Notiz: M. Kummer remercie l'Académie qui lui a décerné un des grands prix de sciences mathématiques de 1856, pour ses Recherches sur les nombres complexes composés de racines de l'unité et de nombres entiers.

in das concrete Reich des experimentirenden Physikers herabgestiegen war. „Wenn", wie er sich zu einem jüngeren Freunde äusserte, „ich alter Mathematiker zum Experiment greife, so ist dies ein Beweis, dass der Frage auf mathematischem Wege nicht beizukommen ist." So ist er in dieser Arbeit wie in allen seinen Schriften von unbestechlicher Aufrichtigkeit gegen sich selbst, von durchsichtiger Klarheit in Wort und Gedanken, ein bewundernswertes Muster eines deutschen Gelehrten. Weder verbirgt er die Wege, auf denen er gewandelt ist, noch schielt er nach dem Beifalle besonderer Freunde und Anhänger oder der grossen Menge, noch ist sein hoher Sinn durch Nebenrücksichten von den gesteckten Zielen abgelenkt. Nur auf die Erkenntnis und Erforschung der Wahrheit geht sein Trachten.

Ein Mann von solcher Geistesrichtung, von solchem Charakter musste ein vortrefflicher Lehrer sein. In anschaulichster Weise und mit freundlichem Wesen hielt er in Berlin lange Jahre seine Privatvorlesungen an der Universität über analytische Geometrie, über krumme Oberflächen, über Zahlentheorie und über analytische Mechanik. Sein Ziel war nicht etwa, die Fortgeschrittenen in die Kreise seiner schöpferischen Gedanken einzuweihen, sondern vielmehr die Anfänger mit sicherer Hand in die Mathematik einzuführen. Das danken ihm die Tausende, welche in der Universität und in der Kriegsakademie seinen Vorträgen mit Verständnis und Erfolg gelauscht haben. Dagegen widmete er vor der Gründung des mathematischen Seminars die öffentlichen Vorlesungen dem Vortrage über die Ergebnisse seiner eigenen Arbeiten und regte dadurch in erfolgreichster Weise zu weiteren Forschungen in denselben Gebieten an. Später verlegte er, unter Aufgebung dieser öffentlichen Vorlesungen, die Anregung zu Forschungen seiner Schüler in jenes auf seinen Antrag gegründete Seminar. Den streng sachlichen und unübertrefflich klaren, daher leicht fasslichen Vortrag würzte er zuweilen durch kleine philosophische oder humoristische Bemerkungen.

„Anziehend ist ein Problem nur, so lange es ungelöst ist, und Freude empfindet der Forscher allein bei der ersten Entdeckung des Weges zur Lösung. Eine ähnliche Freude empfindet der Entdecker höchstens, wenn er als Lehrer seine Schüler auf die erkannte Wahrheit leitet."

„Wer diesen Factor unbeachtet lässt, ist einem Menschen zu vergleichen, der eine Pflaume isst und den Kern verschluckt, während er das Fleisch ausspuckt."

„Die Franzosen konnten den Namen Potential nicht erfinden; denn er hätte sie an potence — den Galgen — erinnert."

Solche und ähnliche Aeusserungen werden in grosser Menge bei seinen Schülern fortleben. Seine herzliche Freude an dem Gelingen der Arbeiten seiner Zuhörer, seine aufrichtige Teilnahme an ihren Freuden und Leiden hat er oft bekundet. Als ihm ein junger Doctor kurz vor Weihnachten das eben gebundene Exemplar der Dissertation überbrachte und auf die Frage, wo derselbe das Weihnachtsfest verleben würde, erwiderte, in der Heimat bei den Eltern, die von der vollzogenen Promotion noch nichts wüssten und die das Diplom nebst der Dissertation unverhofft auf dem Weihnachtstische finden sollten, nahm er erfreut die Hand des jungen Mannes zwischen seine beiden Hände und sagte gerührt: „Das thun Sie; dies ist das schönste Geschenk, das ein Sohn seinen Eltern machen kann."

Die streng sachliche Art seines Denkens und Handelns machten ihn vorzugsweise für die Verwaltung von Aemtern in der Akademie und an der Universität geeignet. Die Acten beider Institute bewahren zahlreiche Schriftstücke von seiner schönen und klaren Handschrift, dem Abbilde seines Wesens. Als immerwährender Secretar der Akademie, als Decan und als Rector der Universität erfreute er sich des unbedingten Vertrauens seiner Collegen, die, wie Emil du Bois-Reymond bei dem Festessen zum fünfzigjährigen Doctorjubiläum Kummer's ausführte, immer das sichere Gefühl hatten, wenn Kummer die Geschäfte führe, so wache Achill für die Völkerscharen. Und er selbst fühlte sich auch in dieser Thätigkeit wohl und behaglich, betrachtete die Geschäftsführung als eine Erholung, deren der Gelehrte bedürfe, um neue Kräfte für wissenschaftliche Arbeiten zu sammeln. Dabei war er dafür bekannt, dass er sein unbestechliches Urteil ohne Scheu mit Festigkeit aussprach und vertrat. Weil aber alle davon überzeugt waren, dass er einzig und allein von sachlichen Gesichtspunkten geleitet wurde, so gestand man ihm einen grossen Einfluss zu und liess sich gern von ihm leiten.

Wir erwähnen nur im Vorübergehen die vielen Ehrenbezeugungen, welche dem verehrten Manne im Laufe seines langen Lebens ungesucht zugefallen sind.

Mit zunehmendem Alter zog er sich allmählich aus den verschiedenen Aemtern zurück, die ihm anvertraut waren, am frühesten von seiner Lehrthätigkeit an der Kriegsakademie 1874. Der General von Ollech, welcher damals der Director dieses Instituts war, wollte es versuchen, dem scheidenden verdienten Lehrer nach neunzehnjähriger, erspriesslicher Amtszeit an der Anstalt eine Pension zu erwirken, obschon die Stelle keine pensionsfähige war. Kummer lehnte es ab, zu diesem Zwecke

2*

irgend welche Schritte zu thun. Er habe diese Stellung stets als eine solche angesehen, die in Folge von unberechenbaren Zufällen plötzlich aufgegeben werden könne, und habe deshalb das Honorar nie in seinen Etat mit aufgenommen. Die Zinsen der so entstandenen Ersparnisse seien nun gerade so hoch gewachsen wie die bisherigen Einnahmen an der Kriegsakademie. Diese mit innerem Behagen abgegebene Erklärung bildete noch längere Zeit den Gegenstand der Bewunderung an der Kriegsakademie. Der Zug zeigt uns aber auch den grossen Gelehrten als sorgsamen Haushalter, während ein anderer Vorfall seine liebenswürdige Freigebigkeit und seine Fürsorge für seine Schüler beleuchten möge. Ein junger Mathematiker, der eben das Doctorexamen bestanden hatte, erkrankte an den Pocken und reiste in seine Heimat in der Provinz Posen an der russischen Grenze. Da keine Nachricht von ihm ankam und es bekannt war, dass derselbe in bedrängten Vermögensverhältnissen lebte, so entstand die Besorgnis, ob er auch die nötige Pflege fände. Sobald Kummer von diesen Befürchtungen hörte, versah er einen Freund des Abgereisten mit den nötigen Geldmitteln und entsandte ihn nach Posen mit dem Auftrage, für den Erkrankten in jeder Beziehung zu sorgen.

Den Huldigungen zum fünfzigjährigen Doctorjubiläum im Jahre 1881 wollte er sich entziehen, indem er eine Reise nach der sächsischen Schweiz unternahm und, wie er später erzählte, den Tag mit seiner Gattin im Kuhstall feierte. Seine vielen Verehrer bereiteten ihm im Herbste des Jahres dennoch ein Fest, das er dann freundlich annahm und in bester Laune erwiderte. Die Stiftung, die bei dieser Gelegenheit gemacht und deren Verwendung ihm überlassen wurde, wandte er der Universität Halle zu, wo er durch Scherk für die Mathematik gewonnen war. Drei Jahre später stellte er als Vierundsiebziger die Vorlesungen an der Universität ein und lebte in stiller Zurückgezogenheit nur noch für seine Familie. Im Sommer zog er wieder und wieder in die geliebten Berge Schlesiens, wohin seine Jugendjahre wiesen, und bewahrheitete damit den Ausspruch, den er einmal gethan hatte, er reise nur dorthin, wo er schon gewesen sei. Das Vergnügen an der einfachen, ihm lieb gewordenen Umgebung leuchtet auch aus der Aeusserung hervor, die er zu seiner Tochter machte, als er sie in Zürich besuchte und von dem Hügel hinter ihrem Hause auf den Züricher See hinabschaute: „Beinahe so schön wie der Blick von meinem Balkon auf den Garten in der Schöneberger Strasse.“ Die Freude an seiner zahlreichen Familie verschönte seine letzten Lebensjahre. Neun Kinder überleben ihn; die Geburt eines Urenkels, welche unlängst erfolgte, erfreute ihn herzlich, und den ältesten Sohn, der kurz vor seinem Hinscheiden zum Geheimrat ernannt war, begrüsste er neckend als seinen Collegen. Die hohe, schlanke Gestalt war allmählich gebeugt, das klare, durchdringende und doch so freundlich blickende Auge trübe geworden; langsam schwanden die Kräfte. Endlich raffte ihn die Influenza am 14. Mai hinweg. An einem schönen sonnigen Maitage, als der Flieder duftete und die Singvögel jubelten, wurden seine irdischen Reste auf dem Jacobikirchhofe beim Rollkruge in die Gruft gesenkt, um welche sich alle versammelt hatten, die ihn lieb und wert hielten, und über welche sich die umflorte Fahne des mathematischen Vereins der Universität Berlin senkte, den er von seiner Gründung an begünstigt hatte.

Unsterblich wird sein Bild fortleben unter seinen Freunden, seinen Schülern, den Mathematikern aller Zeiten als eine Idealgestalt eines deutschen Forschers und Gelehrten.

Schriften von **E. E. Kummer.**

1832.

1. De cosinuum et sinuum potestatibus secundum cosinus et sinus arcuum multiplicium evolvendis. Dissertatio quam Universitatis Halensis Amplissimus Philosophorum Ordo praemio regio ornavit. Halae in libraria Orphanotrophei 1832. 31 S. 4°.

1834.

2. De generali quadam aequatione differentiali tertii ordinis. Oster-Progr. Liegnitz 1834. 10 S. 4°.
   Abgedruckt in J. für Math. C, 1—9 (1886).
3. Sur l'intégration générale de l'équation de Riccati par des intégrales définies. J. für Math. XII, 144—147.

1835.

4. Ueber die Convergenz und Divergenz der unendlichen Reihen. J. für Math. XIII. 171—184, datirt Liegnitz, d. 29. Jan. 1833.
5. Ueber unendlich verschiedene Entwickelungen der Potenzen der Cosinus und Sinus. J. für Math. XIV, 110—122, datirt Sorau, August 1832.

1836.

6. Ueber die hypergeometrische Reihe
$$1+\frac{\alpha.\beta}{1.\gamma}x+\frac{\alpha(\alpha+1)\beta(\beta+1)}{1.2.\gamma(\gamma+1)}x^2+\frac{\alpha(\alpha+1)(\alpha+2)\beta(\beta+1)(\beta+2)}{1.2.3.\gamma(\gamma+1)(\gamma+2)}x^3+\ldots$$
J. für Math. XV, 39—83, 127—172.
   Ins Englische übersetzt durch H. Nagaoka: On the hypergeometric series etc. Tokio Math. Ges. IV, 273—325, 353—405 (1891).

1837.

7. Eine neue Methode, die numerischen Summen langsam convergirender Reihen zu berechnen. J. für Math. XVI, 206—214, datirt Liegnitz, den 10. Novbr. 1834.
8. De aequatione $x^{2\lambda}+y^{2\lambda}=z^{2\lambda}$ per numeros integros resolvenda. J. für Math. XVII, 203—209, datirt Lignicii Oct. 1835.
9. De integralibus definitis et seriebus infinitis. J. für Math. XVII, 210—227, datirt Lignicii m. Majo 1836.
10. De integralibus quibusdam definitis et seriebus infinitis. J. für Math. XVII, 228—242, datirt Lignicii, mense Aprili a. 1837.

1838.

11. Recension von: J. M. C. Bartels, Vorlesungen über mathematische Analysis. Jahrb. für wiss. Kritik 1838², 271, 273, 281, 289.

1839.

12. Recension von: Burhenne, Die Mathematik als System betrachtet. Jahrb. für wiss. Kritik 1839¹, 101, 105.
13. Note sur l'intégration de l'équation $\frac{d^n y}{dx^n}=x^m.y$ par des intégrales définies. J. für Math. XIX, 286—288.
13a. Sur l'intégration de l'équation $\frac{d^n y}{dx^n}=x^m.y$. Journ. de Math. IV, 390—391 (1839).

1840.

14. Sur quelques transformations générales des intégrales définies. J. für Math. XX, 1—10.
15. Ueber die Transcendenten, welche aus wiederholten Integrationen rationaler Formeln entstehen. J. für Math. XXI, 74—90, 193—225, 328—371.
15a. Ueber die Transcendenten, welche aus wiederholten Integrationen rationaler Formeln entstehen. Progr. Liegnitz 19 S. 4⁰.

1842.

16. Eine Aufgabe, betreffend die Theorie der kubischen Reste. J. für Math. XXIII, 285—286.
17. De residuis cubicis disquisitiones nonnullae analyticae quas auctoritate Amplissimi Philosophorum Ordinis in Academia Vratislaviensi pro loco professoris publici ordinarii rite obtinendo die XXVI. M. Octobris A. MDCCCXLII publice defendet Ernestus Eduardus Kummer Phil. Dr. Prof. Publ. Ord. Des. Vratislaviae 1842.

    Abgedruckt in J. für Math. XXXII, 341—359 (1846).
18. Recension von: Ohm, Der Geist der mathematischen Analysis. Jahrb. für wiss. Kritik. Berlin, Bd. II, 209—216.

1843.

19. Bemerkungen über die kubische Gleichung, durch welche die Haupt-Axen der Flächen zweiten Grades bestimmt werden. J. für Math. XXVI, 268—272, Giorn. Arcad. XCVIII, 71—82 (1844).

1844.

20. De numeris complexis, qui radicibus unitatis et numeris integris realibus constant. Gratulationsschrift der Univ. Breslau zur Jubelfeier der Univ. Königsberg. (Academiae Albertinae Regiomontanae secularia tertia celebranti gratulatur Academia Vratislaviensis.) 28 S. 4°.

20a. Sur les nombres complexes qui sont formés avec les nombres entiers réels et les racines de l'unité. Journ. de Math. XII, 185—212 (1847). Abdruck des lateinischen Originaltextes.

1845.

* 21. Zur Theorie der complexen Zahlen. Berlin. Monatsber. 1845, 87—96, abgedruckt in J. für Math. XXXV, 319—326 (1847).

1846.

* 22. Vervollständigung der Theorie der complexen Zahlen. Berlin. Monatsber. 1846, 87—96.

23. Ueber die Divisoren gewisser Formen der Zahlen, welche aus der Theorie der Kreisteilung entstehen. J. für Math. XXX, 107—116.

23a. Sur les diviseurs de certaines formes de nombres qui résultent de la théorie de la division du cercle. Traduction de M. Houël. Journ. de Math. (2) V, 369—386 (1860).

1847.

24. Beweis des Fermat'schen Satzes der Unmöglichkeit von $x^\lambda + y^\lambda = z^\lambda$ für eine unendliche Anzahl Primzahlen λ. Berlin. Monatsber. 1847, 132—139, 140—141, 305—319.

25. Beitrag zur Theorie der Function $\Gamma(x) = \int_0^\infty e^{-v} v^{x-1}\, dv$. J. für Math. XXXV, 1—4.

26. Ueber Systeme von Curven, welche einander überall rechtwinklig durchschneiden. J. für Math. XXXV, 5—12, datirt Breslau im September 1847.

26a. Ins Französische übersetzt: Sur les systèmes de courbes algébriques planes qui se coupent orthogonalement et sur leur confocalité. Nouv. Ann. XI, 426—434 (1852). [D'après M. le professeur E. E. Kummer (Journal de M. Crelle) t. XXXV, p. 5; 1847.]

27. Ueber die Zerlegung der aus Wurzeln der Einheit gebildeten complexen Zahlen in ihre Primfactoren. J. für Math. XXXV, 327—367, datirt Breslau im September 1846.

28. Extrait d'une lettre de M. Kummer à M. Liouville. (Breslau, le 28 avril 1847.) C. R. XXIV, 899—900 u. Journ. de Math. XII, 136.

29. Anzeige des ersten Bandes der Mathematischen Werke von C. G. J. Jacobi. Neue Jenaische Allgemeine Literaturzeitung VI, No. 201, 202, 203. 12 S. 4°.

1848.

30. Ueber die Vierecke, deren Seiten und Diagonalen rational sind. J. für Math. XXXVII, 1—20, datirt Breslau im December 1846.

31. Ueber die akademische Freiheit. Eine Rede, gehalten bei der Uebernahme des Rectorats der Universität Breslau, am 15. October 1848. 10 S. 8°.

1850.

* 32. Auszug aus den neuesten zahlentheoretischen Untersuchungen. Berlin, Monatsber. 1850, 87—96; vorgelegt 26. März 1850.

33. Allgemeine Reciprocitätsgesetze für beliebig hohe Potenzreste. Berlin. Monatsber. 1850, 154—165; vorgelegt 27. Mai 1850.

34. Bestimmung der Anzahl nicht aequivalenter Klassen für die aus λten Wurzeln der Einheit gebildeten complexen Zahlen und die idealen Factoren derselben. J. für Math. XL, 93–116, datirt Breslau, den 16. Juni 1849.

35. Zwei besondere Untersuchungen über die Klassen-Anzahl und über die Einheiten der aus λten Wurzeln der Einheit gebildeten complexen Zahlen. J. für Math. XL, 117—129, datirt Breslau, den 18. Juni 1849.

36. Allgemeiner Beweis des Fermat'schen Satzes, dass die Gleichung $x^\lambda + y^\lambda = z^\lambda$ durch ganze Zahlen unlösbar ist, für alle diejenigen Potenz-Exponenten λ, welche ungerade Primzahlen sind und in den Zählern der ersten $\frac{1}{2}(\lambda - 3)$ Bernoulli'schen Zahlen als Factoren nicht vorkommen. J. für Math. XL, 130—138, datirt Breslau, den 19. Juni 1849.

37. Théorème de Fermat et manuscrit arabe. Nouv. Ann. IX, 386—392.

Französischer Bericht über die Kummer'schen Arbeiten aus Bd. XL des Journ. für Math. sowie über den Woepcke schen Aufsatz ebenda.

1851.

38. Ueber eine allgemeine Eigenschaft der rationalen Entwickelungscoefficienten einer bestimmten Gattung analytischer Functionen. J. für Math. XLI, 368—372, datirt Breslau, den 30. Juli 1850.

39. Mémoire sur la théorie des nombres complexes composés de racines de l'unité et de nombres entiers. Journ. de Math. XVI, 377—498.

1852.

40. Ueber die Ergänzungssätze zu den allgemeinen Reciprocitätsgesetzen. J. für Math. XLIV, 93—146; datirt Breslau, den 30. Novbr. 1851.

1853.

* 41. Note sur une expression analogue à la résolvante de Lagrange pour l'équation $z^p - 1$. Atti dell'Acc. Pont. de' Nuovi Lincei VI, 237—241.

1855.

42. Ueber eine besondere Art, aus complexen Einheiten gebildeter Ausdrücke. J. für Math. L, 212—232, datirt Breslau, den 31. Aug. 1854.

1856.

43. Antrittsrede als ordentliches Mitglied der Akademie der Wissenschaften. Berlin. Monatsber. 1856, 377—379, 3. Juli 1856.

44. Theorie der idealen Primfactoren der complexen Zahlen, welche aus den Wurzeln der Gleichung $\omega^n = 1$ gebildet sind, wenn n eine zusammengesetzte Zahl ist. Berlin. Abh. 1856, 1—47, gelesen 18. Decbr. 1856.

Blosser Titel, ohne Auszug in Berlin. Monatsber. 1856, 648.

1857.

45. Einige Sätze über die aus den Wurzeln der Gleichung $\alpha^\lambda = 1$ gebildeten complexen Zahlen, für den Fall, dass die Klassenanzahl durch λ theilbar ist, nebst Anwendung derselben auf einen weiteren Beweis des letzten

Fermat'schen Lehrsatzes. Berlin. Monatsber. 1857, 275—282; gelesen 4. Mai 1857.

Auszug aus der unter Nr. 48 angeführten Abhandlung.

46. Ueber die den Gaussischen Perioden der Kreisteilung entsprechenden Congruenzwurzeln. J. für Math. LIII, 142—148, datirt Berlin, den 5. Juni 1856.

47. Anzeige einer Schrift des Herrn Reuschle in Stuttgart. J. für Math. LIII, 379.

48. Einige Sätze über die aus den Wurzeln der Gleichung $\alpha^\lambda = 1$ gebildeten complexen Zahlen, für den Fall, dass die Klassenanzahl durch λ teilbar ist, nebst Anwendung derselben auf einen weiteren Beweis des letzten Fermat'schen Lehrsatzes. Berl. Abh. 1857, 41—74.

49. Nouvelles recherches de M. Kummer. (Communication de M. Hermite.) C. R. XLV, 1010—1011.

Auch unter dem Titel: Sur la théorie des nombres complexes avec application à la démonstration du théorème de Fermat.

Bericht über Kummer'sche Arbeiten, die der Akademie übersandt sind, aus dem Jahre 1857.

1858.

50. Ueber die allgemeinen Reciprocitätsgesetze der Potenzreste. Berliner Monatsber. 1858, 158—171, gelesen 18. Feb. 1858.

1859.

51. Vorlegung von: Reuschle. Tafel der aus fünften Einheitswurzeln züsammengesetzten primären complexen Primfactoren aller reellen Primzahlen von der Form $5\mu+1$ in der ersten Viertelmyriade. Berliner Monatsber. 1859, 488—499, 694—697; 1860, 190—199, 714—734; Sitzungen vom 30. Juni 1859, 14. Novbr. 1859; 26. April 1860, 29. Novbr. 1860.

52. Ueber die allgemeinen Reciprocitätsgesetze unter den Resten und Nichtresten der Potenzen, deren Grad eine Primzahl ist. Berlin. Abh. 1859, 19—158.

Blosser Titel, ohne Auszug in Monatsber. 1859, 367.

Ueber die allgemeine Theorie der geradlinigen Strahlensysteme. Blosser Titel, Berl. Monatsber. 1859, 637.

53. Ueber die Ergänzungssätze zu den allgemeinen Reciprocitätsgesetzen. J. für Math. LVI, 270—279, datirt Berlin, im Decbr. 1858.

1860.

54. Ueber atmosphärische Strahlenbrechung. Berl. Monatsber. 1860, 405—420; gelesen 12. Juli 1860; abgedruckt in J. für Math. LXI, 263—275 (1863).

55. Bemerkungen über die aus 29ten Einheitswurzeln gebildeten complexen Zahlen. Berlin. Monatsber. 1860, 734—735; gelesen 29. Novbr. 1860.

56. Gedächtnisrede auf Gustav Peter Lejeune Dirichlet. Berlin. Abhandl. 1860, 1—36.

Blosser Titel, ohne Auszug in Berlin. Monatsber. 1860, 405; 5. Juli 1860.

56a. Discorso commemorativo su Giustavo Pietro Lejeune Dirichlet. Annali di Mat. III, 221—231, 283—297 (1860).

57. Allgemeine Theorie der geradlinigen Strahlensysteme. J. für Math. LVII, 189—230, datirt October 1859.

57a. Théorie générale des systèmes des rayons rectilignes. Traduit par M. E. Dewulf. Nouv. Ann. XIX, 362—371; XX, 72—76, 255—260, 359 bis 365; (2) I, 31—40, 82—89.

58. Ueber drei aus Fäden verfertigte Modelle der allgemeinen, unendlich dünnen, geradlinigen Strahlenbündel. Berlin. Monatsber. 1860, 469—474; gelesen 30. Juli 1860.

1861.

59. Ueber die Klassenanzahl der aus nten Einheitswurzeln gebildeten complexen Zahlen. Berlin. Monatsber. 1861, 1051—1053; gelesen 9. Decbr. 1861.

60. Zwei neue Beweise der allgemeinen Reciprocitätsgesetze unter den Resten und Nichtresten der Potenzen, deren Grad eine Primzahl ist. Berlin. Abh. 1861, 81—122; J. für Math. C, 10—50 (1886).
Blosser Titel, ohne Auszug in Berlin. Monatsber. 1861, 656; 11. Juli 1861.

1862.

61. Ueber ein von Hrn. stud. phil. Schwarz angefertigtes, in Gips gegossenes Modell der Krümmungsmittelpunktsfläche des dreiaxigen Ellipsoids. Berlin. Monatsber. 1862, 426—428; gelesen 30. Juni 1862.

1863.

62. Ueber die Klassenanzahl der aus zusammengesetzten Einheitswurzeln gebildeten idealen complexen Zahlen. Berlin. Monatsber. 1863, 21—28; gelesen 8. Jan. 1863.

63. Ueber die Flächen vierten Grades, auf welchen Scharen von Kegelschnitten liegen. Berlin. Monatsber. 1863, 324—336; gelesen 16. Juli 1863.
Abgedruckt im J. für Math. LXIV, 66—76 (1865).

64. Gipsmodell der Steiner'schen Fläche. Berlin. Monatsber. 1863, 539; gelesen 26. Novbr. 1863.

1864.

65. Ueber die Flächen vierten Grades mit sechzehn singulären Punkten. Berlin. Monatsber. 1864, 246—260; gelesen 18. April 1864.

66. Ueber die Strahlensysteme, deren Brennflächen Flächen vierten Grades mit sechzehn singulären Punkten sind. Berlin. Monatsber. 1864, 495—499; gelesen 18. Juli 1864.

1865.

67. Rede zur Gedächtnisfeier Königs Friedrichs II. Berlin. Monatsber. 1865, 62—75; gelesen 26. Jan. 1865.

68. Ueber die algebraischen Strahlensysteme, insbesondere über die der ersten und der zweiten Ordnung. Berlin. Monatsber. 1865, 288—293; gelesen 22. Juni 1865.

69. Erwiderung auf die Antrittsrede A. W. Hofmann's. Berlin. Monatsber. 1865, 324—326; gelesen 6. Juli 1865.

1866.

70. Rede zur Feier des Geburtstages Sr. Majestät des Königs. Berlin. Monatsber. 1866, 171—184; gelesen 22. März 1866.

71. Ueber zwei merkwürdige Flächen vierten Grades und Gipsmodelle derselben. Berlin. Monatsber. 1866, 216—220; gelesen 23. April 1866.

72. Ueber die algebraischen Strahlensysteme, insbesondere über die der ersten und zweiten Ordnung. Berlin. Abh. 1866, 1—120.

Ueber die algebraischen Strahlensysteme, insbesondere über die der ersten Ordnung und der ersten Klasse. Monatsberichte 1864, 282; der dritten Ordnung 1870, 584; bloss Titel ohne Auszug.

1867.

73. Rede zur Feier des Leibnizischen Jahrestages. Berlin. Monatsber. 1867, 387 bis 395; gelesen 4. Juli 1867.

74. Antwort auf die Antrittsreden der Herren Auwers und Roth. Berlin. Monatsber. 1867, 410—414; gelesen 4. Juli 1867.

Ueber die Krümmungsmittelpunktsfläche des Ellipsoids und Paraboloids. Blosser Titel ohne Auszug in Berlin. Monatsber. 1867, 772.

Einige Anwendungen der Zahlentheorie auf die Geometrie. Blosser Titel ohne Auszug in Berlin. Monatsber. 1868, 183.

1869.

75. Rede zur Feier des Jahrestages Friedrichs II. Berlin. Monatsber. 1869, 67—78; gelesen 28. Januar 1869.

76. Festrede zum Andenken Friedrich Wilhelms des Dritten gehalten am 3. August 1869 in der Aula der Friedrich-Wilhelms-Universität. Berlin. 12 S. 4°.

1870.

77. Rede zur Nachfeier des Geburtsfestes Seiner Majestät des Königs. (Bloss kurzer Auszug.) Berlin. Monatsber. 1870, 183; gelesen 24. März 1870.

77a. Separat: Berlin 1870, 8 S. 8°.

78. Ueber die einfachste Darstellung der aus Einheitswurzeln gebildeten complexen Zahlen, welche durch Multiplication mit Einheiten bewirkt werden kann. Berlin. Monatsber. 1870, 409—420; gelesen 16. Juni 1870.

79. Ueber die aus 31sten Wurzeln der Einheit gebildeten complexen Zahlen. Berlin. Monatsber. 1870, 755—766; gelesen 10. Oct. 1870.

80. Ueber eine Eigenschaft der Einheiten der aus den Wurzeln der Gleichung $\alpha^{\lambda} = 1$ gebildeten complexen Zahlen, und über den zweiten Factor der Klassenzahl. Berlin. Monatsber. 1870, 855—880; gelesen 1. Dec. 1870.

1871.

81. Rede zur Feier des Leibnizischen Jahrestages. Berlin. Monatsber., 1871, 351—355; gelesen 6. Juli 1871.

1872.

82. Ueber ein von Hrn. Prof. Schwarz angefertigtes Gipsmodell einer Minimalfläche. Berlin. Monatsber. 1872, 122—123; vorgelegt am 19. Februar 1872.

83. Ueber einige besondere Arten von Flächen vierten Grades. Berl. Monatsber. 1872, 474—483; gelesen 20. Juni 1872.

1873.

84. Rede zur Gedächtnisfeier König Friedrichs des Zweiten. Berlin. Monatsber. 1873, 71—87; gelesen 23. Jan. 1873.

84a. Rede zur Feier des Geburtstages Friedrichs des Grossen, gehalten in der öffentlichen Sitzung der Königlichen Akademie der Wissenschaften am 23. Januar 1873. Berlin 1873. 20 S. 4°.

1874.

85. Ueber diejenigen Primzahlen λ, für welche die Klassenzahl der aus λten Einheitswurzeln gebildeten complexen Zahlen durch λ teilbar ist. Berlin. Monatsber. 1874, 239—248; gelesen 19. März 1874.

1875.

86. Rede zur Feier des Leibnizischen Jahrestages. Berlin. Monatsber. 1875, 425—433; gelesen 1. Juli 1875.

87. Ueber die Wirkung des Luftwiderstandes auf Körper von verschiedener Gestalt, insbesondere auch auf die Geschosse. Berlin. Abh. 1875, 1—57.

87a. Abgedruckt im Archiv für die Artillerie und Ingenieur-Offiziere des deutschen Reichsheeres. Jahrgang 40, Bd. 79, 192—248. Mit Vorbemerkung von General-Lieut. v. Neumann, S. 189—192.

Blosse Titel, ohne Auszug in Berlin. Monatsber. 1874, 703 und 1875, 286; gelesen 27. Mai 1875; 1876, 825.

1876.

88. Neue Versuche zur Bestimmung des Angriffspunktes der Resultante des Luftwiderstandes gegen rechteckige schiefe Ebenen.

Zusatz zu der Abhandlung: Ueber die Wirkung des Luftwiderstandes etc., Jahrgang 1875 der Abhandlungen. Berlin. Abh. 1876, 1—9.

1877.

89. Festrede zur Feier des Geburtstages Sr. Majestät des Kaisers und Königs am 22. März 1877. Berlin. Monatsber.

Sep. Berlin 1877. 12 S. 8°.

1878.

90. Ueber diejenigen Flächen, welche mit ihren reciprok polaren Flächen von derselben Ordnung sind und die gleichen Singularitäten besitzen. Berlin. Monatsber. 1878, 25—36; gelesen 17. Jan. 1878.

91. Neuer elementarer Beweis des Satzes, dass die Anzahl aller Primzahlen eine unendliche ist. Berlin. Monatsber. 1878, 777—778; gelesen 25. November 1878.

1880.

92. Ueber die kubischen und biquadratischen Gleichungen, für welche die zu ihrer Auflösung nötigen Quadrat- und Kubikwurzelausziehungen alle rational auszuführen sind. Berlin. Monatsber. 1880, 930—936; gelesen 15. Novbr. 1880.

Festschrift zur Feier des 100. Geburtstages Eduard Kummers

mit Briefen an seine Mutter und an Leopold Kronecker
Leipzig und Berlin: B. G. Teubner 1910

VORREDE.

Die Berliner Mathematische Gesellschaft hat zur Feier des 100sten Geburtstages EDUARD KUMMERS (29. Januar 1910) am Sonnabend den 8. Januar 1910 im großen Hörsaal des Physikalischen Instituts der hiesigen Universität eine Festsitzung veranstaltet. Herr HENSEL übernahm die Festrede. Diese, als erste Gedächtnisrede großen Stiles, die zum Andenken des großen Berliner Mathematikers gehalten worden ist, gelangt hier zum Abdruck.

Hinzugekommen sind Briefe KUMMERS an seine Mutter und an LEOPOLD KRONECKER, die hier zum erstenmal mit gütiger Genehmigung ihrer Besitzer der Öffentlichkeit übergeben werden. Die Berliner Mathematische Gesellschaft sagt Frau Geheimrat KUMMER und Herrn Geh. Justizrat Dr. ERNST KRONECKER verbindlichsten Dank für die liebenswürdige Überlassung dieses Briefwechsels, der als kostbare Fundgrube für die Charakteristik des Menschen wie des Forschers bezeichnet werden muß.

Beigegeben ist ein Bildnis KUMMERS aus dem Jahre 1875, das aus einem Kabinettporträt des Photographen ERNST MILSTER-Berlin reproduziert worden ist.

Es ist uns eine angenehme Pflicht, der Verlagsbuchhandlung für ihr Entgegenkommen auf unsere und des Autors mannigfachen Wünsche den besten Dank auszusprechen.

Der Vorstand
der Berliner Mathematischen Gesellschaft:

E. JAHNKE.

R. GÜNTSCHE. C. FÄRBER.

INHALT.

Verehrte Damen und Herren!

Der heutige Vortrag über ERNST EDUARD KUMMER und sein Lebenswerk verdankt der Tatsache seine Entstehung, daß wir in den nächsten Tagen, am 29. Januar 1910, den hundertjährigen Geburtstag dieses großen Mannes feiern. Manchem unter uns wird es vielleicht wunderbar erscheinen, daß schon eine so lange Zeit seit seiner Geburt verstrichen ist, während uns seine Persönlichkeit und sein Werk noch so nahe stehen.

Mir selbst erscheint dies besonders in bezug auf seine Persönlichkeit merkwürdig: habe ich ihn doch noch gut und nahe gekannt von der Zeit an, als ich, ein achtzehnjähriger Student an der Berliner Universität, sein Schüler sein durfte, bis zu seinem Tode. Allerdings war KUMMER damals ein siebzigjähriger Ehrfurcht gebietender Greis, und bei dem wärmsten Interesse für uns junge Leute war es nicht seine Art, uns einen Einblick in sein innerstes Leben und in seine eigene reiche Gedankenwelt zu gewähren, aber es war mir vergönnt, seinem besten und liebsten Freunde, seinem begeistertsten Schüler LEOPOLD KRONECKER nahe zu stehen, der schon ganz jung auf dem Liegnitzer Gymnasium zu seinen Füßen gesessen hatte und durch ihn für die Mathematik gewonnen worden war. Seit dieser Schulzeit erwuchs zwischen beiden Männern ein Freundschaftsverhältnis, wie es sich nur selten gleich innig, rein und ungetrübt durch so lange Zeit erhält. Von dieser Freundschaft legt eine Reihe von Briefen das schönste Zeugnis ab, welche KUMMER in einer zwölfjährigen Trennungszeit an KRONECKER gerichtet hat, und in denen er ihm in der bedeutsamsten Epoche seiner wissenschaftlichen Entwicklung fast von Monat zu Monat die Fortschritte der Erkenntnis berichtet, zu denen ihn sein Nachdenken geführt hat. In diesen Briefen, welche mir durch die Güte von LEOPOLD KRONECKERS ältestem Sohne, dem Geh. Justizrat ERNST KRONECKER zugänglich wurden, konnte ich deutlich die innere Entwicklung der größten Gedanken KUMMERS verfolgen von den ersten arithmetischen Versuchen an durch Zweifel und Bedenken hindurch bis zu dem wunderbaren Briefe vom 18. Oktober 1845, in welchem er mit gerechtem Stolze dem Freunde seine ganze gewaltige Theorie der idealen Zahlen in einer Reihe von lapidaren Sätzen vortrug, wie sie seitdem die Grundlage für die Entwicklung der modernen Zahlenlehre geworden ist.

In der Zeit, als ich KRONECKER näher trat, waren beide Männer schon lange in Berlin vereinigt, aber obwohl die Jahre den Altersunterschied mehr hätten verwischen können, war die Pietät und Bewunderung des jüngeren Mannes für den älteren nur vertieft und verschönt worden. Es war KRONECKER ein Bedürfnis, von der wissenschaftlichen Entwicklung, von dem festen edlen Charakter seines Freundes zu sprechen, und so durfte ich durch seine Vermittlung auch dem bewunderten Lehrer KUMMER so nahe treten, wie dies wohl wenigen anderen meines Alters vergönnt gewesen ist.

Da mich auch meine eigene Neigung früh auf das Arbeitsgebiet KUMMERS hinführte, so nahm ich mit dankbarer Freude die Aufforderung der Berliner Mathematischen Gesellschaft an, in dieser Festsitzung ein Bild des Mannes zu entwerfen, dem wir und unsere Wissenschaft so vieles verdanken. Ich tat dies um so lieber, als ich auch als Leiter des CRELLEschen Journales eine besondere Dankesschuld an KUMMER abzutragen habe. Hat er doch vom zwölften bis zum hundertsten Bande dreiunddreißig größere und kleinere Abhandlungen in dieser Zeitschrift veröffentlicht, welche zusammengenommen fast drei ihrer Bände füllen würden und die einen ihrer schönsten Ruhmestitel bilden; und hat er doch vom Jahre 1856 bis 1880 den verdienstvollen Leiter des Journales, C. W. BORCHARDT, in seinem verantwortungsreichen Amte unterstützt.

Bei der Aufgabe, ein möglichst lebensvolles Bild von der Persönlichkeit KUMMERS zu gewinnen und Ihnen wiederzugeben, wurde ich in wirksamster Weise besonders durch seine verehrte Gattin, Frau Geheimrat KUMMER, und seine Töchter, Frau Geheimrat SCHWARZ und Frl. EMMA KUMMER, unterstützt. Durch ihre Güte konnte ich die Briefe KUMMERS an seine Mutter lesen, welche in rührender Weise sein innerstes Wesen enthüllen.

Von diesen ihm nächststehenden und teuersten Angehörigen erhielt ich ein getreues Bild des edlen Menschen; über KUMMER als Lehrer und als Gelehrten konnte ich neben meinen eigenen Erinnerungen diejenigen meiner verehrten Kollegen FROBENIUS, GUNDELFINGER und HETTNER und den schönen Nekrolog auf KUMMER von Herrn LAMPE benutzen. So habe ich selbst durch ein Zusammentreffen günstiger

Umstände ein lebensvolles und wahres Bild dieses großen Mannes gewinnen können; möchte es mir gelingen, Ihnen seine Persönlichkeit durch die folgenden Worte nahe zu bringen und die Erinnerung an sein Werk aufs neue in Ihnen aufleben zu lassen.

Ernst Eduard Kummer wurde am 29. Januar 1810 in Sorau nahe der schlesischen Grenze als zweiter und jüngster Sohn des Stadt- und Landphysikus Carl Gotthelf Kummer geboren. Sein Vater wurde durch den Typhus, den die aus Rußland zurückkehrenden Reste der Napoleonischen Armee mit sich brachten, noch in jüngeren Jahren hinweggerafft, und so blieb die junge Frau Friederike Sophie Kummer geb. Rothe mit ihren beiden Söhnen in äußerst bedrängten Verhältnissen zurück. Glücklicherweise hatte ihr der Himmel auch die Fähigkeiten verliehen, das zu ertragen, was er ihr auferlegt hatte. Sie stammte aus einer alten Predigerfamilie und besaß eine für die damalige Zeit gute Bildung; als das große Unglück über sie hereinbrach, nahm sie, um ihre kleinen Kinder durchzubringen, jede Arbeit an; lange Zeit nähte sie Soldatenhemden, für deren jedes sie drei Groschen erhielt und von denen sie drei am Tage fertig stellte. Ihre große, mit ein wenig schlesischem Aberglauben gewürzte Frömmigkeit, ihre innere Heiterkeit und ihr guter Humor führten sie über die Zeiten größter Not glücklich hinweg, und besonders diese Eigenschaften halfen ihr dazu, daß sie ihren beiden Söhnen in der ersten Kinderzeit und auch später stets innerlich nahe stand und besonders auf Ernst Eduard einen großen Einfluß ausüben konnte. Die Pietät, welche Kummer immer gegen ältere Leute gehabt hat, äußert sich am schönsten in den Briefen, die er als junger Student aus Halle an die Mutter schreibt. In einem langen Leben hat die Mutter im Hause ihres geliebten Sohnes in Breslau und in Berlin an seinem Glücke teilnehmen und sich an seinen Kindern als gute und richtige Großmutter erfreuen dürfen.

Trotz ihrer bedrängten Lage setzte Frau Kummer es durch, daß ihre beiden Söhne nach vorhergehender privater Ausbildung das Gymnasium in Sorau besuchen konnten. Ernst, der damals erst neun Jahre alt war und etwa die Kenntnisse eines Quartaners hatte, wurde, da sich die untersten Klassen dieser Anstalt damals keines besonders guten Rufes erfreuten, auf den Wunsch seiner Mutter sogleich in die Sekunda aufgenommen, ein Erfolg, der heute wohl selbst einer sehr energischen Mutter nicht mehr beschieden sein dürfte. Nach drei Jahren in Sekunda hatte er sich bis zum Ersten dieser Klasse emporgearbeitet, da er aber noch zu jung und zu klein war, wurde er auch

1*

jetzt nicht nach Prima versetzt, sondern mußte noch zwei, also im ganzen fünf Jahre Sekundaner und nachher noch vier Jahre in Prima bleiben. In seinem Abgangszeugnis wird schon seine große natürliche Begabung und sein launiger Witz gerühmt, auch wird bereits hervorgehoben, daß er sich in den mathematischen Wissenschaften zu seinem Vorteil ausgezeichnet habe, daß er sich im Deutschen ohne Schwulst zum Poetischen zu erheben und daß er im Lateinischen ziemlich geläufig, wenn auch nicht ciceronianisch zu disputieren vermöge. Dieser letzte Vorwurf darf dem achtzehnjährigen Jüngling billig verziehen werden, denn Cicero selbst wird in diesem Alter wohl auch noch nicht ganz ciceronianisch disputiert haben.

Im Jahre 1828 ging er nach Halle, um dort ebenso wie sein Bruder Karl, auch mit auf den Wunsch seiner Mutter, Theologie zu studieren. Sehr bald regten sich aber in ihm Bedenken dagegen, ob die schroffe rationalistische Art, wie diese Wissenschaft damals in Halle studiert werden mußte, für ihn das Richtige sei. Die Theologie und Philosophie führen zwar nach seiner Ansicht unmittelbar zum höchsten Ziele des Menschen, der Erkenntnis des Wahren, Guten und Schönen, aber die erstere, wie sie zur Zeit betrieben werde, lege dem Geiste des Menschen Fesseln an, welche er nicht ertragen könne, und welche seiner unwürdig seien. „Die Philosophie", so fährt er in einem Briefe an die Mutter fort, „führt den Menschen ebenso sicher zu seinem Ziele, und in ihr werden seinem Geiste keine Fesseln angelegt. Um aber Philosophie treiben zu können, ist das Studium der Mathematik als der Gesetze des menschlichen Geistes und der ganzen Natur die trefflichste Vorbereitungswissenschaft. Diesen beiden Wissenschaften will ich auch daher meine Kräfte weihen, und das Studium der Mathematik wird mir auch mein Brot verschaffen, und mehr verlange ich nicht, denn es kommt mir nicht darauf an, in der Welt zu glänzen oder bequem und ruhig zu leben, sondern einmal nicht umsonst gelebt zu haben ist mein Zweck und meinen Geist veredelt und gebildet zu haben."

Die Philosophie ist für Kummer immer eine Lieblingswissenschaft geblieben, trotzdem er sich in seiner Bescheidenheit hier als einen Dilettanten bezeichnete; allerdings hat ihn auch die von ihm damals als Vorbereitungswissenschaft bezeichnete Mathematik etwas länger in Anspruch genommen und mehr ausgefüllt, als er es mit seinen neunzehn Jahren voraussah. Man kann aber, wie es Kummer ein Vierteljahrhundert später öffentlich aussprach, den Grund dafür, daß mathematisches und philosophisches Talent sich oft vereint finden, darin sehen, daß es nur die eine Befähigung für das rein abstrakte Denken ist, welcher diese beiden Betätigungen gleichmäßig offen stehen; ob ein mit diesem Talent begabter wissenschaftlicher Forscher sich mehr der einen oder der anderen dieser verwandten Wissenschaften zuwendet, scheint mehr nur von äußeren Bedingungen abhängig zu sein.

Von diesen mehr äußeren Bedingungen war im Falle des jungen KUMMER wohl die wichtigste die, daß er in dem anregenden Mathematiker SCHERK einen Lehrer fand, der die schwierige Kunst verstanden hat, seinen Schüler gleich in eigene mathematische Untersuchungen hineinzuführen; ist doch der Übergang vom aufnehmenden zum schaffenden Denker gerade in unserer Wissenschaft der schwerste. Ist dieser Schritt von einem bedeutenden Menschen aber einmal getan, so ist kaum mehr zu befürchten, daß er den Wunsch hegen oder die Möglichkeit erhalten wird, sich einer andern Wissenschaft zuzuwenden.

SCHERK selber war sich dieses seines großen Verdienstes um die Wissenschaft wohl bewußt und sprach es später mit Stolz aus, er habe nicht vergebens gelebt, da er einen KUMMER zum Schüler gehabt habe. Wohl im Hinblick auf ihn stellte er 1831 im dritten Jahre von KUMMERS Studium eine Preisarbeit „de cosinuum et sinuum potestatibus secundum cosinus et sinus arcuum multiplicium evolvendis“, welche von KUMMER in ausgezeichneter Weise gelöst wurde und die Grundlage aller weiteren Arbeiten seiner ersten funktionentheoretischen Periode bildete. Nach der Urteilsverkündigung und der Zuerkennung des Preises von 50 Reichstalern an den Studiosus ERNST EDUARD KUMMER umarmte und küßte SCHERK in der Freude seines Herzens seinen Schüler in der Aula der Universität. Nach diesem ersten wissenschaftlichen Erfolge war ihm natürlich der Weg zum Staats- und Doktorexamen sehr geebnet, und beide Prüfungen verliefen nach seiner eigenen humoristischen Schilderung an seine Mutter höchst angenehm und fast behaglich; ich muß aber gestehen, daß die von KUMMER berichteten Anforderungen im Staatsexamen entschieden höher waren als die jetzigen.

So wurde er schon am 10. September 1831, also mit einundzwanzig Jahren, Doktor, trat dann sofort beim Gymnasium seiner Vaterstadt zur Ableistung des Probejahres ein, siedelte aber schon im folgenden Jahre als sechster ordentlicher Lehrer an das Gymnasium in Liegnitz über, wo er bis zum Jahre 1842 sehr zu seiner eigenen Befriedigung verblieb. Eine schwere Zeit müssen diese elf Jahre aber für den jungen Lehrer gewesen sein, denn neben seinen 22—24 Unterrichtsstunden, die in den späteren Jahren den gesamten Rechen- und

Mathematikunterricht des Gymnasiums und den Physikunterricht auf den höheren Klassen umfaßten, schrieb er in rascher Folge jene Arbeiten auf dem Gebiete der Funktionentheorie, auf Grund deren er schon mit neunundzwanzig Jahren zum korrespondierenden Mitgliede der Berliner Akademie ernannt und 1842 nach dem plötzlichen Tode des Professors SCHOLZ als Professor an die Breslauer Universität berufen wurde. KUMMER selbst sagt mit Recht in einem späteren Briefe: „Ich habe zu der Zeit, wo ich in Liegnitz mit 24 Lehrstunden belastet war, unter denen damals noch fast gar keine mathematischen waren, in den ersten Jahren meines dortigen Aufenthaltes wenigstens ebensoviel geleistet als später unter den äußerlich günstigsten Verhältnissen." Immer gedenkt KUMMER mit besonderer Dankbarkeit dieser Liegnitzer Zeit, die ihm auch das ersehnte häusliche Glück in der Ehe mit seiner ersten Frau brachte. Hier wurde ihm auch sein erster Sohn geboren. Nur acht Jahre blieb ihm dieses Glück beschieden; am 30. Juli 1848 wurde ihm seine geliebte Frau nach einer zuerst leichten Krankheit, die schnell in ein schweres Nervenfieber überging, durch den Tod entrissen.

Aber es war für ihn natürlich in wissenschaftlicher Beziehung eine Lebensfrage, daß er die anstrengende und für seine eigenen Arbeiten nicht nutzbare Lehrertätigkeit mit der akademischen Lehrtätigkeit vertauschen durfte, und er schreibt mit innerer Befriedigung an KRONECKER, wie sehr ihm dieser Tausch behagt. „Ich habe hier weder Vorgesetzte, noch Untergebene, ich kann tun und lassen was ich will, denn ich habe niemand etwas zu befehlen."

Es ist so, als ob schon die Aussicht auf die größere geistige Freiheit KUMMER die Anregung dazu gab, sich zum ersten Male auf das Gebiet zu wagen, welches ihm die größte Förderung verdanken sollte, ja dessen Entwicklung ohne ihn gar nicht gedacht werden kann, ich meine das Gebiet der höheren Zahlentheorie. Er schreibt am 16. Januar 1842 an KRONECKER:

„Seit ich bei meiner letzten Anwesenheit in Berlin merkte, es könne mit Breslau Ernst werden, so setzte ich mich zu Hause hin und arbeitete sehr fleißig, um so etwas wie eine Dissertation zur Habilitierung zu arbeiten, und ich fing bei etwas mir ganz Neuem, nämlich bei den cubischen Resten der Primzahlen $6n+1$ an."

So neu ist ihm dieses Gebiet, daß er KRONECKER bittet, ihm eventuell doch die Literatur, besonders die JACOBI betreffende, darüber zusammenzusuchen; so sind auch einige Resultate, die sich KUMMER selbst erarbeitet hat, gar nicht neu, aber schon hier regt sich seine

Phantasie und treibt ihn von der Betrachtung dieser speziellen Zahlkörper zu allgemeineren Bereichen. Immer aber geht der Ausdruck des Bedauerns hindurch, daß er sich erst das Handwerkszeug selbst schmieden muß, während andere es schon fertig erhalten können. So schreibt er an Kronecker:

„Wenn Dirichlet Ihnen im nächsten Sommer komplexe Zahlentheorie liest, so nehmen Sie dies dankbar an; ich selbst würde gern mit zuhören, wenn ich könnte, denn Sie können glauben, es ist etwas mühsam, sich dergleichen alles allein aufzusuchen, wenn man wie ich nur etwa die Hauptideeen von anderen erfahren hat."

Erst später will ich genauer auf die schönste und größte geistige Tat Kummers, seine Schöpfung der idealen Zahlen, eingehen, welche in dieser Breslauer Zeit vollständig ausgeführt wurde. Hier setzt eben die Reihe seiner Briefe an Kronecker ein, welche aber außerdem noch viel des Interessanten gibt. Zuerst handelt es sich um die Antrittsvorlesung und die Disputation über drei Thesen, welcher sich der ordentliche Professor damals noch unterziehen mußte, und hier ist die weitere große Schwierigkeit die, daß außer Kummer selbst niemand in Breslau ist, der Mathematik, geschweige denn Zahlentheorie versteht.

„So stelle ich", schreibt er an Kronecker, „allgemeine Thesen auf, über welche sich reden läßt:

1) In der Mathematik ist vieles, was ebensowohl bejaht, als auch verneint werden muß; eine höchst philosophische These", fügt er hinzu, „welche indirekt behauptet, daß die Mathematik über den endlichen Verstand zur Vernunft vordringt.

2) Theorie und Praxis in der Mechanik widersprechen sich nie (und zwar weil die eine von ganz anderen Dingen handelt, als die andere)."

Die Freude an philosophischer Betrachtung der Mathematik, die sich hier wieder zeigt, tritt noch stärker in dem schon damals von Kummer ernsthaft erwogenen Plan einer Enzyklopädie der mathematischen Wissenschaften hervor, welche den Zweck haben würde, „die ganze Summe mathematischer Erkenntniß der Gegenwart als ein Ganzes erscheinen zu lassen und womöglich überall den Gedanken bloßzulegen, welcher oft ziemlich versteckt liegt. Es müßte eine Enzyklopädie im Hegelschen Sinne werden, zwar nicht philosophisch, aber eine Grundlage der Philosophie der Mathematik. Sie müßte auch im Grunde echt philosophisch sein, aber niemals die philosophische Seite nach außen kehren". Es scheint mir höchst interessant, daß der in unserer Zeit mit so glänzendem Erfolge ausgeführte Plan einer allgemeinen Enzyklopädie schon damals Kummer so stark beschäftigt hat, und daß er sich allein die Kraft zutraute, dieses große Werk auszuführen. Aber es will mir scheinen, als ob Kummer selbst eingesehen hat, daß dieser Plan jedenfalls für Deutschland und ganz besonders für ihn selbst ein verfrühter wäre. In der zwei Jahre später in der Jenaischen allgemeinen Litteraturzeitung vom 24. August 1847 veröffentlichten Anzeige des ersten Bandes von Jacobis mathematischen Werken sieht er es direkt als einen Vorzug der deutschen Mathematik an, daß hier vorläufig noch nicht, wie in Frankreich, eine so große Anzahl zusammenfassender Lehrbücher, sondern im wesentlichen nur Abhandlungen produziert werden. Er meint, daß wir noch in der ersten frischeren Periode der mathematischen Entwicklung stehen, und fährt dann fort:

„An das erste Stadium, in welchem der Geist, gleichsam eroberungssüchtig, vorzüglich nur auf Erweiterung der Grenzen der Wissenschaft gerichtet ist, schließt sich das zweite an, in welchem die schaffende Tätigkeit zwar keineswegs ausgeschlossen ist, aber die formgebende die Herrschaft über dieselbe erlangt. . . . Auch wir werden in der Folge unsere Tätigkeit mehr auf die Form wenden und werden, in das zweite Stadium der Entwicklung dieser Periode unserer Wissenschaft eintretend, umfangreichere Meisterwerke produzieren, wie sie dem umfassenden und zu strenger Systematik geneigten Geiste der deutschen Nation entsprechen. Von da an aber, in dem dritten Stadium, dem des Verfalls, werden wir anfangs noch gelehrte Sammelwerke schaffen und in dem weiteren Verlaufe desselben uns vielleicht damit begnügen, nur das in besseren Zeiten Erarbeitete verständlicher oder gar flacher zurechtzulegen. . . . So vegetiert dann die Wissenschaft fort, bis mit einem neu erwachenden Geist wieder eine neue Periode beginnt."

Darum, fügt Kummer hinzu, wollen wir dahin arbeiten, daß uns erst noch der gegenwärtige in uns lebendige Geist die reichsten Schätze entfalten möge. Dieser Auffassung ist Kummer auch durch sein ganzes Leben treu geblieben. Auch später hat er der Versuchung, ein zusammenfassendes Werk über die Gleichungstheorie zu veröffentlichen, mannhaft widerstanden, und er hat, wie ein Konversationslexikon gewissermaßen bedauernd schrieb, „nur" Abhandlungen veröffentlicht. Allerdings schließt Kummer bei seiner Warnung ausdrücklich solche Werke und Lehrbücher aus, welche, wie die Fundamenta nova von Jacobi, im wesentlichen größere Abhandlungen sind, da sie nur die eigenen Arbeiten des Verfassers enthalten. In dieser Begrenzung ist aber die eindringliche Mahnung KUMMERS gerade in der heutigen Zeit höchst beachtenswert.

Die soeben erwähnte Anzeige der JACOBIschen Abhandlungen führt mich zu dem Verhältnis KUMMERS zu seinen wissenschaftlichen Zeitgenossen, in welche diese und seine zahlreichen Briefe uns einen fesselnden Einblick gewähren. Unter den damals lebenden Heroen unserer Wissenschaft stand ihm GAUSS unbedingt am höchsten; sein Beispiel war es, welches er seinen Schülern immer und immer wieder vor Augen stellte, und es ist kein Zufall, daß im Anfange der drei großen Perioden seines wissenschaftlichen Schaffens jedesmal eine Arbeit von GAUSS steht, an welche KUMMER anknüpft. Im Mittelpunkte der ersten elfjährigen der Funktionentheorie gewidmeten Periode seiner wissenschaftlichen Arbeit steht seine Abhandlung über die hypergeometrische Reihe, von der KRONECKER in der Gratulationsadresse der Akademie mit Recht sagen konnte, sie ist „eine würdige Ergänzung jener fundamentalen nur in ihrem ersten Teile erschienenen GAUSS'schen Arbeit, gegründet auf tiefstes in einem Liegnitzer Programm zuerst dargelegtes Erkennen der für die Vergleichung von Transzendenten maßgebenden Prinzipien und durchgeführt mit solcher Vollständigkeit, daß bei viel später mit ganz neuen Mitteln von RIEMANN aufgenommenen Untersuchungen sich nur eine ganz kleine Nachlese von Resultaten ergeben hat.“

Die Untersuchungen seiner zweiten fast zwanzigjährigen arithmetischen Periode knüpfen an die GAUSS'schen Abhandlungen über die biquadratischen Reste an, in welchen zuerst die Zahlen von der Form $a + bi$ mit rationalen Koeffizienten arithmetisch untersucht werden, wo also der erste algebraische Körper, und zwar in der erklärten Absicht betrachtet wird, die Methoden, welche in der Theorie der rationalen Zahlen zum quadratischen Reziprozitätsgesetze geführt hatten, nun zum Beweise des biquadratischen Reziprozitätsgesetzes zu benutzen.

Endlich nehmen die Arbeiten der dritten wesentlich geometrischen Periode von KUMMERS Schaffenszeit ihren Ausgang von Grundgedanken, welche sich in GAUSS' berühmter Abhandlung „Disquisitiones generales circa superficies curvas“ finden. Im Anschluß an Untersuchungen von HAMILTON begründet KUMMER hier eine Theorie der allgemeinen Strahlensysteme, von der die Lehre von den Flächen und ihren Normalensystemen ein spezieller Fall ist, so daß er von diesem höheren Standpunkte aus die GAUSSischen Theorieen wesentlich erweitert und ihnen einen neuen Reiz verliehen hat. An sie schließen sich dann

Kummers Arbeiten zur Behandlung der speziellen algebraischen Strahlensysteme an, die seinen Namen gerade in den weitesten Kreisen so sehr bekannt gemacht haben.

Nach Gauss sind es Jacobi und Dirichlet, welche nach Kummers Ausspruch allein der deutschen Mathematik um diese Zeit das Übergewicht über die französische verleihen, „wo uns drei Sterne erster Größe glänzen, jenen nur einer, nämlich Cauchy".

Schon sein bereits erwähntes Liegnitzer Osterprogramm über die hypergeometrische Reihe schickte Kummer als Einjährig-Freiwilliger in einem Soldatenbriefe an Jacobi nach Königsberg: „Dies kann mir nun freilich sonst nichts nutzen", schreibt er mit der ihm eigenen Heiterkeit an seine Mutter, „aber es ist dafür um so besser, denn man muß ja nicht alles des Nutzens wegen tun. Ich überschicke es ihm bloß darum, weil er gerade die größten Entdeckungen in dem Fache gemacht hat, worüber mein Programm handelt, und er wird sich darüber wundern und freuen zugleich, daß ein Musketier dieselben Gegenstände behandelt als er." Dieser Brief, der die für ihn so wichtigen Beziehungen mit Jacobi und Dirichlet anknüpfte, hatte auch die von Kummer gewünschte Wirkung, denn Jacobi soll den Soldatenbrief in Königsberg mit den Worten gezeigt haben: „Wenn schon Musketiere jetzt so ausgezeichnete mathematische Arbeiten machen, dann möchte ich einmal die der Herren Unteroffiziere sehen."

Besonders haben ihn Jacobis Untersuchungen über die Kreisteilungsgleichungen im Anfang seiner zweiten Schaffensperiode stark beeinflußt; eine ganze Reihe von Resultaten hatte er selbst gefunden, um nachher zu erfahren, daß sie schon Jacobi erkannt und in seinen Vorlesungen vorgetragen hatte.

Sehr eng wurde Kummers Verhältnis zu Dirichlet, der schon früh die Bedeutung Kummers klar erkannt hatte; auf ihn besonders ist wohl die so frühe Ernennung zum Korrespondenten der Akademie zurückzuführen, er war es wohl auch, der Kummer zuerst in Breslau empfahl und ihn später in Berlin als seinen Nachfolger vorschlug. Die größte Verehrung und Bewunderung für diesen Mann und diesen Gelehrten erfüllte Kummer stets; fast jeder Brief an Kronecker enthält seinen Namen; das schönste Denkmal hat er Dirichlet in der Gedächtnisrede gesetzt, welche er am 5. Juli 1860 seinem großen Lehrer an der Berliner Akademie gehalten hat. Denn Kummers Lehrer ist Dirichlet, obwohl er nur fünf Jahre älter war und obwohl Kummer nie eine Vorlesung bei ihm gehört hat, im höchsten Sinne gewesen, dafür legen die Werke Kummers das schönste Zeugnis ab. Ist doch in seiner Theorie der Kreiskörper die Frage der Unabhängigkeit der Kreisteilungseinheiten und die der Klassenzahlbestimmung vollkommen auf der Grundlage gegeben, welche DIRICHLET in seinen berühmten Beweisen über die arithmetische Reihe und in der Klassenzahlbestimmung für quadratische Formen von gegebener Determinante verwendet.

Von gleichaltrigen und jüngeren Mathematikern stand ihm durch sein ganzes Leben hindurch KRONECKER am nächsten; in seiner Breslauer Zeit schloß er sich besonders an JOACHIMSTHAL an; weniger persönlich nahe trat er EISENSTEIN, aber aus seinen Briefen ergibt sich deutlich, wie hoch er ihn als Gelehrten einschätzte. Es war KUMMER, welcher auf eine Anregung von JACOBI die Ernennung EISENSTEINS zum Ehrendoktor an der Breslauer Universität beantragte und unter den größten Schwierigkeiten, die ihm durch Rückständigkeiten seiner Kollegen gemacht wurden, auch wirklich durchsetzte. Besonders bewunderte KUMMER, wie er in der Einleitung der großen Akademieabhandlung vom Jahre 1859 über die allgemeinen Reziprozitätsgesetze selbst hervorhebt, den bekannten EISENSTEINschen Beweis dieses Gesetzes durch Vermittlung der Kreisfunktionen, der so leicht auf die kubischen und biquadratischen Reziprozitätsgesetze ausgedehnt werden kann, wenn er auch die Versuche EISENSTEINS, auf demselben Wege zu den allgemeinen Gesetzen zu kommen, als mißlungen bezeichnen mußte.

Endlich möchte ich noch mit ein paar Worten auf KUMMERS Verhältnis zu WEIERSTRASS hinweisen. Als KUMMER 1855 als Nachfolger DIRICHLETS nach Berlin berufen wurde, hatte die Breslauer Universität für ihn einen Nachfolger zu bestimmen. Hier hatte KUMMER den Antrag gestellt, drei bewährte akademische Gelehrte vorzuschlagen und WEIERSTRASS und ROSENHAIN zwar rühmend zu nennen, aber nicht vorzuschlagen. In bezug auf WEIERSTRASS begründete KUMMER in einem Briefe an KRONECKER diesen Antrag damit, daß WEIERSTRASS bisher eine einzige Abhandlung „Zur Theorie der ABELschen Funktionen“ habe erscheinen lassen. Diese wissenschaftliche Leistung sei allerdings derart, daß er auf Grund derselben sogleich mit Ehren Mitglied der Berliner oder einer anderen Akademie werden könne; aber zur Ausfüllung der Professur für Mathematik an einer Universität wie die Breslauer, in welcher er nicht nach seinem Belieben lesen könne, sondern, da er im wesentlichen allein sei, für alles stehen müsse, was zur Ausbildung junger Mathematiker gehört, böte sie noch nicht die nötigen Garantieen.

Aber schon zu dieser Zeit stand es bei Kummer fest, daß Weierstrass nach Berlin kommen müsse, und so sehen wir ihn denn nur wenige Wochen später nach den Mitteilungen eines Briefes an Kronecker vom 25. August 1855 in Berlin im Ministerium eifrig dafür wirken, daß Weierstrass mit dem bei Dirichlet ersparten Gehalte gleich in Berlin eine seiner Bedeutung angemessene Stellung erhielte. Es waren also nur die Garantieen dafür, daß Weierstrass geeignete Übung dafür besäße, alle Studierende einer kleinen Universität auch durch die elementaren Vorlesungen im vollen Umfange in die Mathematik einzuführen, welche Kummer vermißte, aber er selbst wollte diesen großen Mann neben sich an der ersten preußischen Universität haben, wie es denn auch später wirklich geschah. Weierstrass hat ja dann in Berlin Kummer sehr nahe gestanden, wie Kronecker in einem Briefe bezeugt: „Die etwa 20 Jahre, von 1856 bis nahe 1876, in denen wir drei, Kummer, Weierstrass und ich, des engsten und lebhaftesten wissenschaftlichen Verkehrs uns erfreuten, haben uns reiche Früchte und den Segen wahrer geistiger Erbauung gebracht."

Nach dem schweren Verluste, der Kummer in Breslau betroffen hatte, führte ein günstiges Geschick den Bund fürs Leben mit seiner zweiten Gattin Bertha aus der Künstler- und Gelehrtenfamilie Cauer herbei, mit welcher ihn und seine erste Frau schon längere Zeit eine nahe und innige Freundschaft verbunden hatte. Es war ihr beschieden, in 45jähriger glücklichster Ehe mit Kummer vereint alle die reichen Jahre in Breslau und Berlin zu verleben, und mit ihm seine zahlreichen Kinder (neun Kinder überlebten ihn) zu wertvollen Menschen zu erziehen. Auch heute dürfen wir mit Dank und Freude diese verehrte Frau in unserem Kreise begrüßen.

Wir müssen jetzt Breslau verlassen und Kummer nach Berlin begleiten, wo er nach vielem Suchen in der Köthener Straße 14 eine Wohnung fand, welche er dann nach einigem Wechsel später mit der ihm so lieb gewordenen Wohnung Schöneberger Straße 10 vertauschte, in der er bis zu seinem Tode geblieben ist; die Aussicht von seinem Balkon in den Garten dieser letzten Wohnung stellte er, allerdings scherzhaft, beinahe höher als den Blick vom Hause seines Schwiegersohnes über den im vollen Sonnenglanze vor ihm liegenden Züricher See.

Kummer wurde im vollen Umfange Dirichlets Nachfolger in Berlin. Auch ihm wurde die Professur an der damaligen allgemeinen Kriegsschule, der jetzigen Kriegsakademie übertragen. Während

Dirichlet aber allmählich gerade unter dieser Verpflichtung etwas litt, hat sie Kummer in den neunzehn Jahren bis 1874, wo er sich bei zunehmendem Alter von ihr zurückzog, gern erfüllt, und in so ausgezeichneter Weise, daß die Oberleitung dieses Institutes ihm, wie Herr Lampe in seinem Nekrologe erzählt, eine Pension aussetzen wollte, obschon diese Stelle nicht pensionsberechtigt war. Es ist charakteristisch für Kummers Denkart, daß er bat, von einem solchen Schritte Abstand zu nehmen; er habe diese Stelle stets als eine solche angesehen, die infolge von unberechenbaren Zufällen plötzlich aufgegeben werden könnte, und habe deshalb das Honorar nie in seinen Etat mit aufgenommen. Die Zinsen der so entstandenen Ersparnisse seien nun gerade so hoch gewachsen, wie die bisherigen Einnahmen an der Kriegsakademie. Diese mit innerem Behagen abgegebene Erklärung wurde noch längere Zeit bewundernd an der Kriegsakademie besprochen.

Noch wichtiger war es, daß Kummer Dirichlets Nachfolger in der Berliner Akademie der Wissenschaften wurde; am 3. Juli 1856 hielt er am Leibniztage mit Borchardt zusammen seine Antrittsrede. Nur wenige Sätze sind es, in denen er hier sein Glaubensbekenntnis entwickelt, aber sie sind höchst charakteristisch für den Mann, der sie ausspricht:

„Die mathematischen Wissenschaften haben in unserem Vaterlande seit mehreren Dezennien einen neuen Aufschwung genommen. Der deutsche Geist, getrieben von dem ihm eigenen Drange nach Erkenntnis, hat mit verjüngter Kraft den ewigen Formen und Gesetzen des Mathematischen sich zugewendet und in demselben ein reiches Feld seiner Tätigkeit gefunden. Es ist darum jetzt in der Mathematik die wissenschaftliche *Forschung* die vorherrschende Richtung, die Forschung, welche weniger im *Wissen* als *Erkennen* ihre Befriedigung findet und darum in die Tiefe der Wissenschaft zu dringen sucht, wo sie die Lösung vorhandener Rätsel findet, und wo neue Rätsel ihr entgegentreten. Zu dieser Richtung habe auch ich aus innerer Neigung mich hingezogen gefühlt; wenn ich aber meinen wissenschaftlichen Standpunkt noch weiter angeben soll, so kann ich ihn füglich als einen *theoretischen* bezeichnen, und zwar nicht allein darum, weil die Erkenntnis allein das Endziel meiner Studien ist, sondern namentlich auch darum, weil ich vorzüglich nur diejenige Erkenntnis in der Mathematik erstrebt habe, welche sie innerhalb der ihr eigentümlichen Sphäre ohne Rücksicht auf ihre Anwendungen gewährt. Ihre höchste Blüte kann sie nach meinem Dafürhalten nur in dem ihr eigenen Elemente des abstrakten reinen Quantums entfalten, wo sie unabhängig von der äußeren Wirklichkeit nur sich selbst zum Zwecke hat."

Sowohl die Bedeutung von Kummers wissenschaftlichen Arbeiten, welche er nun in großer Zahl in den Sitzungsberichten und Abhandlungen der Akademie veröffentlichte, als auch die streng sachliche Art seines Denkens, die Rechtlichkeit und Selbständigkeit seines Handelns machten ihn sehr bald zu einem der ausschlaggebenden Mitglieder dieser Körperschaft. Dies trat auch äußerlich darin hervor, daß er schon im Jahre 1865 zum ständigen Sekretar der physikalisch-mathematischen Klasse ernannt wurde. Ihm war es keine Last, neben seinen wissenschaftlichen Arbeiten auch die Geschäfte der Akademie zu führen, ja er betrachtete diese Tätigkeit als eine Art von Erholung, deren er bedürfe, um neue Kräfte zu sammeln. Durch die Festigkeit seines Charakters, durch die Unbestechlichkeit seines Urteiles, über dessen lautere Motive nie ein Zweifel bestehen konnte, eignete er sich ganz besonders für dieses schwierige Amt. Auch den Reden, welche er in den öffentlichen Sitzungen zur Gedächtnisfeier Friedrich des Großen, zum Leibniztage und zum Geburtstage des Königs zu halten hatte, widmete er sich mit besonderem Interesse. Sie gaben ihm Gelegenheit, sich über die zahlreichen nicht rein mathematischen, besonders philosophischen und historischen Fragen auszusprechen, welche ihn auch in seinem späteren Leben im reichsten Maße fesselten. Seine kurzen klaren Gedenkreden auf Friedrich den Großen und sein Verhältnis zu den mathematischen Wissenschaften, ferner über dessen Stellung zur Leibniz-Wolffschen Philosophie und den Enzyklopädisten zeigen, wie eingehend er sich auch jetzt noch mit der ihm so teuren Philosophie beschäftigte, seine Rede über Friedrich den Großen und die militärischen Bildungsanstalten läßt erkennen, wie tief er sich in die Geschichte und die Anschauungen dieses von ihm am höchsten gestellten Fürsten hineingelebt hatte.

Die wichtigste äußere Aufgabe, welche an Kummer als Nachfolger Dirichlets herantrat, war die Vertretung der Mathematik an der größten Universität Preußens, nachdem er vorher in dem kleineren Breslau als im wesentlichen einziger Vertreter seines Faches Gelegenheit gehabt hatte, sich als vielseitiger Lehrer der Jugend zu bewähren. Seiner ganzen Geistesrichtung, seiner Freude am Unterricht und an dem Verkehr mit der Jugend nach mußte Kummer ein vortrefflicher akademischer Lehrer sein. Aber es ist recht interessant zu sehen, wie er diese Pflicht auffaßte, welche Ziele er beim akademischen Unterricht hauptsächlich verfolgte. Während Weierstrass, Kronecker, Helm-

HOLTZ Wert darauf legten, die jugendlichen Studenten auf schwierigem Wege zu den höchsten erreichbaren Höhen der Erkenntnis zu führen, zu Fragen, mit deren Lösung sie selber zum Teil erst beschäftigt waren, hielt es KUMMER mehr für seine Pflicht als akademischer Lehrer, seine Schüler mit sicherer Hand in den schon begründeten Besitz der mathematischen Wahrheiten einzuführen; hierin aber war er Meister, und die vielen vielen Hunderte von Mathematikern, denen er für ihr Leben das Beste und Sicherste gegeben hat, die feste Grundlage, auf der sich ihr geistiger Besitz aufbaute, danken ihm die weise Beschränkung, welche sich dieser große phantasievolle Denker auferlegte, um ganze Generationen von wertvollen Gelehrten und Lehrern heranzuziehen. So ist es zwar gekommen, daß vielleicht z. B. WEIERSTRASS durch seine Lehrtätigkeit mehr individuelle Schüler herangebildet hat, denn er sagte in seinen Vorträgen manches, was außer ihm kein anderer zu sagen vermochte, aber KUMMER glaubte, daß diese Einwirkung durch seine Arbeiten geschehen würde, die er stets so ausführlich und pädagogisch schrieb, daß sie jeder genügend vorgebildete Leser verhältnismäßig leicht und mit Genuß lesen konnte. Und als Schüler seiner Schriften muß sich doch wohl heutzutage jeder Mathematiker, insbesondere jeder Arithmetiker, ansehen; der beste Beweis dafür liegt vielleicht in der statistischen Tatsache, daß der HILBERTsche Zahlbericht, die beste zusammenfassende Darstellung der höheren Arithmetik, die wir heute haben und wohl auch auf lange Zeit haben werden, in dem sehr knappen Verzeichnis der benutzten Abhandlungen 26 Arbeiten von KUMMER aufführen muß, mehr als bei irgend einem anderen Gelehrten.

KUMMER las in der späteren Berliner Zeit einen regelmäßigen zweijährigen Kursus von vier vierstündigen Vorlesungen: Analytische Geometrie, Mechanik, Flächentheorie, Zahlentheorie. Er war immer ein sehr beliebter Lehrer; in der früheren Zeit hatte er gewöhnlich 40—50 Zuhörer, aber in der Zeit des großen Andranges habe ich bei ihm eine Vorlesung über Zahlentheorie gehört, welche im Barackenauditorium vor mindestens 250 Zuhörern gehalten wurde.

Was KUMMER in seinen Vorlesungen in hohem Grade auszeichnete, war seine Klarheit, seine Ruhe beim Vortrage, die durchsichtige Disposition und auch die angeregte Art seines Sprechens; man hatte immer den Eindruck, als freue er sich, seinen Zuhörern diese Dinge vortragen zu können, und dies war auch wirklich der Fall; hat er es doch selbst ausgesprochen, daß nach der Erkenntnis einer neuen Wahrheit sein größter Genuß darin bestanden hätte, eine solche empfänglichen Menschen darzulegen. Der Minister von Gossler erzählte, Kummer habe ihm im hohen Alter gesagt, daß er sich noch immer sorgfältig auf jede einzelne seiner Vorlesungsstunden vorbereite.

Er hatte in hohem Maße das Talent, Figuren in der Luft mit der Hand vor seinen Zuhörern gleichsam entstehen zu lassen. Modelle benutzte er nie, sprach sich sogar über deren Gebrauch mißbilligend aus. Er liebte es, kleine witzige Geschichten zu erzählen, jedoch kamen diese keineswegs oft vor. Eine solche Geschichte kündigte sich schon vorher durch ein Lächeln bei ihm an, und schon dadurch wurden seine Zuhörer unwillkürlich heiter gestimmt. In solchen witzigen Bemerkungen geißelte er gern die voreiligen Schlüsse der Physiker, welche mitunter auf verhältnismäßig nur wenige Experimente gestützt überraschende Naturgesetze aufstellen und sie damit für streng bewiesen halten, oder er spottete über die Mathematiker, die gelegentlich durch Einführung neuer Bezeichnungen den Mangel neuer Gedanken zu verbergen suchen. Ein großer Reiz bei seinen Vorlesungen lag auch in der Persönlichkeit des Vortragenden; immer hatte man das Gefühl, nicht nur einem großen Gelehrten, sondern auch einem Charakter gegenüberzustehen.

Während Kummer in seinen großen Vorlesungen prinzipiell vom Eingehen auf eigene Untersuchungen absah, hielt er es in dem mathematischen Seminar, welches er wohl als einer der ersten in Deutschland eingerichtet hat, im Gegenteil für geboten, vor den Mitgliedern, welche er zu eigenen Untersuchungen anregen wollte, Fragen zu behandeln, in welchen er bereits schöpferisch tätig gewesen war, und ihnen den Weg zu weiterem Vordringen zu weisen. Meistens begannen jene Seminarübungen damit, daß Kummer eine Anzahl Aufgaben vorlegte, welche er eingehend besprach, und die fast immer mit seinen eigenen, besonders seinen geometrischen Arbeiten in engem Zusammenhange standen; er hatte aber auch den Humor, uns unter diesen auch die vollständige Lösung des großen Fermatschen Problemes für dieses Semester zu empfehlen; ihm selbst sei dieses zwar nicht gelungen, aber warum sollten wir nicht glücklicher sein? Dann wandte er sich dem eigentlichen Thema seiner Besprechungen zu.

Hier gab er z. B. eine Darstellung seiner Abhandlung vom 16. Juli 1863 „Über die Flächen vierten Grades, auf welchen Scharen von Kegelschnitten liegen". Diese führte ihn zu einer Besprechung der sog. Steinerschen Fläche und dadurch wurde ihm Gelegenheit gegeben, höchst merkwürdige Einzelheiten über die Persönlichkeit ihres Entdeckers Steiner und seine Arbeiten anzugeben, dann aber auch eine

Anzahl interessanter verwandter Fragen vorzuführen. Ein anderes Mal knüpfte er die Besprechungen an seine Arbeit im CRELLEschen Journale Bd. 35 „Über Systeme von Curven, welche einander überall rechtwinklig durchschneiden“ an.

In den späteren Stunden wurden dann die Lösungen der Aufgaben von den Mitgliedern vorgetragen. Nur wenn sich niemand fand, der, wie er sagte, die Kosten der Unterhaltung trug, hielt er selber einen Vortrag, in dem er meisterhaft anschaulich und schnell in ein Gebiet der höheren Mathematik einführte und darauf hinwies, in welcher Richtung wohl diese Probleme ausgebaut werden könnten.

Sehr interessierten ihn in seiner späteren Berliner Zeit im Jahre 1875 auch seine Untersuchungen über den Luftwiderstand der Geschosse, und er hat damals mitunter auch über die mathematischen Grundlagen dieser Theorie, soweit diese sich aufstellen ließen, gesprochen. Da er aber erkannte, daß dieser Frage, wie er sich selber ausdrückte, auf rein mathematischem Wege noch nicht beizukommen sei, so mußte er hier schon einmal zum Experimente greifen, um die Hauptfrage zu entscheiden, in welcher Weise der Angriffspunkt der Resultante des Luftdruckes von dem Winkel abhängt, den die Achse des Geschosses mit der Richtung der fortschreitenden Bewegung bildet. Einmal lud er auch die Seminarmitglieder zu sich ein, um ihnen den zu diesem Zwecke von ihm konstruierten Rotationsapparat zu zeigen.

Die Stellung eines Themas für eine Doktorarbeit lehnte er prinzipiell ab; das wäre gerade so, meinte er, wie wenn mich ein junger Mann fragte, ob ich ihm nicht ein hübsches junges Mädchen empfehlen wollte, das er heiraten sollte. Auch die Bearbeitung der von Akademieen gestellten Preisarbeiten verwarf er, wenn man sich nicht zufällig vorher so wie so mit dem Gegenstande beschäftigt hatte; die Untersuchungen müßten aus den eigenen Studien herauswachsen.

Aus den vorhin gemachten Bemerkungen über die große Wirkung, welche KUMMERS Arbeiten auf die ganze Entwicklung seiner Wissenschaft ausgeübt haben, erkennen Sie, verehrte Anwesende, daß meine Darstellung des Lebens und des Lebenswerkes von KUMMER unvollständig sein würde, wenn ich nicht seiner Werke gedenken wollte, durch die er ein Lehrer und ein Pfadfinder für seine ganze Wissenschaft bei der Mit- und Nachwelt geworden ist. Die mir zu Gebote stehende Zeit verbietet mir leider, auf seine ganze Lebensarbeit einzugehen; ich muß und will mich beschränken auf den Teil derselben, durch den er am stärksten auf die Wissenschaft gewirkt hat, aus dem man am besten die Art seiner schöpferischen Phantasie beurteilen kann, ich meine seine Entdeckungen in der höheren Zahlenlehre und seine Schöpfung der idealen Zahlen.

Ich erwähnte schon, daß der Beginn seiner zwanzigjährigen Entdeckerarbeit im Gebiete der Zahlenlehre, welche ihn zuletzt an die Spitze nicht nur der deutschen, sondern aller Arithmetiker seiner Zeit stellte, fast genau mit seiner Berufung an die Breslauer Universität im Jahre 1842 zusammenfällt, und daß gerade hier der Briefwechsel mit Kronecker einsetzt, welcher das Werden und Wachsen seiner Ideen so deutlich erkennen läßt.

Da ist es nun höchst interessant zu sehen, wie schwer es Kummer zuerst gemacht wurde, in das Gebiet hineinzukommen, welches später das Feld seiner größten wissenschaftlichen Triumphe werden sollte.

Ich möchte zum besseren Verständnis des Weiteren auf die wichtige Frage der elementaren Zahlenlehre hinweisen, von der Kummer später ausging. Die Zahlentheorie oder Arithmetik beschäftigt sich bekanntlich mit den gewöhnlichen ganzen Zahlen $1, 2, 3, \ldots$ und sucht, soweit sie systematisch begründet ist, diejenigen ihrer Eigenschaften zu ergründen, welche auf ihrer multiplikativen Zerlegung beruhen. Die Eigenschaften der Zahlen, die auf ihrer Zerlegung durch Addition beruhen, müssen leider fast ganz ausgeschlossen werden, da ihrer direkten wissenschaftlichen Behandlung vorläufig noch unüberwindliche Schwierigkeiten entgegenstehen. Die Grundlage dieser ganzen multiplikativen Zahlenlehre beruht nun auf dem Satze, daß jede Zahl stets und nur auf eine Weise in ein Produkt von gleichen oder verschiedenen Primfaktoren zerlegt werden kann, daß sie sich also alle auf eine einzige Art multiplikativ aus den Primzahlen $2, 3, 5, 7, 11, \ldots$ zusammensetzen lassen.

Fast alle lange ungelösten großen Probleme der Zahlenlehre sind solche, welche auf additive Eigenschaften der Zahlen gegründet sind, und fast immer gelang ihre Lösung dadurch, daß es möglich war, diese Frage in eine Frage der Multiplikation umzuformen. Besonders war dies der Fall bei der Aufgabe, welche ich jetzt mit Bewußtsein in den Vordergrund stelle, alle rechtwinkligen Dreiecke mit ganzzahligen Seiten, oder was dasselbe ist, alle ganzzahligen Lösungen der einfachsten Fermatschen Gleichung

$$x^2 = y^2 + z^2 \tag{1}$$

zu finden; an sie knüpft nämlich die ganze Schöpfung Kummers an, durch sie wurde sie hervorgerufen. Schreibt man diese Gleichung in der Form

$$(1a) \qquad z^2 = x^2 - y^2 = (x + y)(x - y) = b_0 b_1,$$

so ist das additive Problem in das multiplikative verwandelt, alle ganzen Zahlen x und y zu finden, für welche das Produkt $b_0 b_1$ eine Quadratzahl ist. In dieser Form kann die Aufgabe leicht vollständig gelöst werden; schon die Pythagoräer konnten unendlich viele, die indischen Mathematiker alle ganzzahligen Lösungen dieser Gleichung aufstellen. So ist z. B.

$$5^2 = 3^2 + 4^2, \quad 13^2 = 5^2 + 12^2, \quad 25^2 = 7^2 + 24^2, \cdots.$$

Es wird Ihnen nun bekannt sein, daß der große französische Mathematiker, der Parlamentsrat in Toulouse PIERRE FERMAT, in einer Randbemerkung zu den sechs arithmetischen Büchern des alexandrinischen Mathematikers DIOPHANT die Behauptung ausgesprochen hat, daß die obige einfachste unter den Gleichungen

$$(2) \qquad x^\lambda = y^\lambda + z^\lambda$$

für $\lambda = 2, 3, 4, \ldots$ die einzige ist, welche überhaupt ganzzahlige Lösungen besitzt, daß es also z. B. keinen ganzzahligen Kubus gibt, der die Summen zweier Kuben, keine vierte Potenz, die die Summe zweier vierten Potenzen wäre usw. Er fügte hinzu, der Rand des Buches sei zu klein, um den von ihm gefundenen wunderbaren Beweis dieser überraschenden Tatsache aufzunehmen.

Zunächst erscheint es als ein Unglück, daß FERMAT hier nicht mehr freies Papier zu seiner Verfügung gehabt hat; bei näherer Betrachtung möchte man diesen Umstand eher als einen für die Entwicklung der Wissenschaft glücklichen bezeichnen, denn ohne ihn würden wir zwar vielleicht den vollständigen Beweis dieses an sich gar nicht so wichtigen FERMATschen Satzes, aber wohl sicher nicht die KUMMERsche Idealtheorie besitzen, welche allein aus den Bemühungen, diesen Satz zu beweisen, erwachsen ist.

Man erkennt ohne Schwierigkeit, daß es genügt, die Unmöglichkeit der Lösbarkeit der sog. FERMATschen Gleichungen

$$x^\lambda = y^\lambda + z^\lambda$$

in ganzen Zahlen nur für diejenigen unter ihnen zu beweisen, in denen der Exponent λ entweder vier oder eine ungerade Primzahl 3, 5, 7, 11, ... ist. In dieser Form ist das FERMATsche Problem ein Prüfstein für unsere Zahlentheorie geworden, und als solcher hat es den größten Einfluß auf die Entwicklung unserer Wissenschaft gehabt.

2*

Man könnte nun auch z. B. im Falle $\lambda = 4$ die Umwandlung aus dem additiven in ein multiplikatives Problem versuchen, jene Gleichung (2) also in der Form

$$z^4 = x^4 - y^4 \tag{3}$$

schreiben; denn auch hier läßt sich die rechte Seite in vier Faktoren zerlegen und in der Form

$$z^4 = (x-y)(x+y)(x-iy)(x+iy) = b_0 b_1 b_2 b_3 \tag{3a}$$

schreiben, wo $i = \sqrt{-1}$ ist. Man kann die Frage auch in dieser Form behandeln, muß dann aber wohl beachten, daß die Faktoren rechts nicht mehr ganze reelle Zahlen, daß vielmehr wenigstens die beiden letzten b_2, b_3 ganze komplexe Zahlen, nämlich Zahlen von der Form $a + bi$ sind, wo a und b reelle ganze Zahlen bedeuten. Will man auf sie also die Ergebnisse unserer Arithmetik anwenden, so muß man diese komplexen Zahlen ganz ebenso behandeln, wie die Zahlen $\pm 1, \pm 2, \pm 3, \cdots$. Gauss hat diesen großen Schritt in der bereits vorher erwähnten Abhandlung getan; er hat gezeigt, daß in diesem größeren Gebiete aller Zahlen $a + bi$ genau dieselben Gesetze bestehen, wie in dem kleineren der reellen ganzen Zahlen. Insbesondere wies er nach, daß nun jede reelle und jede komplexe Zahl in diesem Bereiche auf eine einzige Weise als Produkt nicht weiter zerlegbarer Primzahlen darstellbar ist. In diesem Zahlenbereiche sind nämlich alle reellen Primzahlen von der Form $4n - 1$, also die Zahlen

$$3,\ 7,\ 11,\ 19,\ 23,\ 31,\ \ldots$$

unzerlegbar, dagegen zerfällt jede reelle Primzahl von der Form $4n + 1$, also jede der Zahlen

$$5,\ 13,\ 17,\ 29,\ 37,\ \ldots$$

in ein Produkt von zwei voneinander verschiedenen unzerlegbaren komplexen Zahlen; so ist z. B.:

$$5 = (2+i)(2-i),\quad 13 = (3+2i)(3-2i),\quad 17 = (4+i)(4-i),\ \ldots;$$

die gerade Primzahl 2 endlich ist, abgesehen von einer Einheit, gleich dem Quadrate des Primfaktors $1 - i$. Dies sind nun aber auch alle Primzahlen im Bereiche dieser komplexen Zahlen, und hier, wie in der Theorie der reellen Zahlen besteht der Fundamentalsatz:

Jede komplexe Zahl kann stets und nur auf eine einzige Weise in ein Produkt von komplexen Primzahlen zerlegt werden.

So besteht z. B. folgende eindeutig bestimmte Zerlegung:

$$24 - 3i = 3(2+i)(3-2i).$$

Deshalb allein gelten auch in diesem erweiterten Gebiete alle Sätze unserer gewöhnlichen Zahlenlehre und deshalb kann man auch ohne wesentliche Schwierigkeit die Unmöglichkeit der FERMATschen Gleichung für $\lambda = 4$ allein mit Hilfe der umgeformten Gleichung (3a) beweisen.

Es lag nun nahe, auch bei der FERMATschen Gleichung $x^\lambda = y^\lambda + z^\lambda$, wenn λ eine beliebige ungerade Primzahl ist, etwa für $\lambda = 5$, dieselbe Umwandlung aus dem additiven in ein multiplikatives Problem zu versuchen. Schreibt man diese Gleichung in der Form

$$z^\lambda = x^\lambda - y^\lambda,$$

so kann man nämlich die rechte Seite wieder in ein Produkt von λ Faktoren zerlegen, wenn man sich nicht scheut, zu den rationalen Zahlen ebenso wie vorher die Zahl $i = \sqrt{-1}$ jetzt eine andere sog. algebraische Zahl hinzuzunehmen. Ist nämlich α eine sog. primitive Wurzel der Gleichung

$$\alpha^\lambda = 1,$$

so läßt sich die rechte Seite jetzt als das Produkt von λ Faktoren

$$\text{(4)} \quad z^\lambda = (x - y)(x - \alpha y)(x - \alpha^2 y) \cdots (x - \alpha^{\lambda-1} y) = b_0 b_1 \cdots b_{\lambda-1}$$

darstellen, und man könnte nun versuchen, dieselben Methoden anzuwenden, wie sie vorher für den Fall $\lambda = 4$ benutzt wurden. Hier ist aber zu beachten, daß wir in diesen Faktoren b_k nicht gewöhnliche ganze Zahlen, sondern Zahlen zu untersuchen haben, welche aus α und ganzen Zahlen rational zusammengesetzt sind; diese lassen sich analog wie vorher die Zahlen $a_0 + a_1 i$ in der allgemeinen Form

$$b = a_0 + a_1 \alpha + a_2 \alpha^2 + \cdots + a_{\lambda-1} \alpha^{\lambda-1}$$

mit ganzzahligen Koeffizienten darstellen.

Will man also auf dieses allgemeine FERMATsche Problem die Methoden und Ergebnisse der Zahlenlehre anwenden, so muß man zunächst fragen, ob auch für diese Zahlen „in α“, wie ich sie nennen will, die Grundsätze der Arithmetik gelten, zunächst also, ob sich jede solche Zahl stets als Produkt von nicht weiter zerlegbaren Zahlen in α darstellen läßt.

KUMMER wandte sich diesen Fragen mit der ganzen Begeisterung des geborenen Arithmetikers zu, trotzdem er zuerst gar keine Vorkenntnisse in diesem Gebiete hatte. Bald war er zu dem wichtigen Resultate gelangt, daß sich in der Tat die ganzen Zahlen in α stets in unzerlegbare Faktoren zerfällen lassen, und auf Grund dieser Tatsache hielt er die Zeit für gekommen, um mit einem Schlage dieses vielumworbene heißumstrittene FERMATsche Problem ganz allgemein zu lösen, während dies vorher unter großen Schwierigkeiten nur für ganz wenige Exponenten $\lambda = 4, 3, 5$ und einige andere und zwar durch Methoden geschehen war, die scheinbar allein gerade auf diese Exponenten zugeschnitten waren.

Es ist wohl nur wenigen bekannt, aber durch ganz einwandfreie Zeugnisse, unter anderen durch das des Herren GUNDELFINGER belegt, der die Mitteilung dem Mathematiker GRASSMANN verdankte, daß KUMMER in dieser Zeit des Vorwärtsstrebens wirklich einen vollständigen Beweis des großen FERMATschen Satzes gefunden zu haben glaubte und ihn im Manuskript DIRICHLET vorgelegt hat, welcher schon als Verfasser des unübertrefflich schönen Beweises für den Satz im Falle $\lambda = 5$ der beste Kritiker für diese Frage sein mußte. Nach einigen Tagen gab ihn DIRICHLET mit dem Urteile zurück, der Beweis sei ganz ausgezeichnet, und sicher richtig, wenn es feststände, daß die Zahlen in α nicht bloß stets, was KUMMER ja bewiesen hatte, in unzerlegbare Faktoren zerfielen, sondern daß dies auch nur auf eine Weise möglich wäre. Wäre diese zweite Annahme aber nicht richtig, so wären die meisten Sätze der Arithmetik für die Zahlen in α unbewiesen und der ganze KUMMERsche Beweis sei völlig hinfällig; leider schienen ihm die Zahlen in α jene Fundamentaleigenschaft wirklich nicht allgemein zu haben.

Die Art nun, wie KUMMER diesen niederschmetternden Einwand, von dessen Berechtigung er sich leider bald überzeugen mußte, aufnahm, ist höchst charakteristisch für den Geist und für die Energie dieses wahrhaft großen Gelehrten. Er sieht vor allem ein, und spricht dies auch in der Gratulationsschrift für die Königsberger Universität aus, daß es nicht angehe, die arithmetische Untersuchung der Zahlen in α aufzugeben, weil sie dieser Grundforderung unserer wissenschaftlichen Arithmetik nicht genügen, denn sie sind von der Natur gegeben, und ihre Erkenntnis ist unentbehrlich für das Eindringen in die Natur der Kreisteilungsgleichungen, überhaupt für die gesamte Algebra und besonders für den FERMATschen Satz und die höheren Reziprozitätsgesetze, welche schon damals KUMMER als sein höchstes wissenschaftliches Ziel vorschwebten. Aber ein weniger tiefgehender Denker würde vielleicht für seine Person sich von dieser Aufgabe zurückgezogen haben; dieser Gedanke kommt KUMMER nicht. Andererseits würde vielleicht mancher andere versucht haben, demselben Probleme von einer anderen Seite beizukommen, aber auch das widerstrebt KUMMER;

er versucht es vielmehr, den Bereich dieser Zahlen in α so zu modifizieren, daß das Fundamentalgesetz der wissenschaftlichen Zahlenlehre, die eindeutige Zerlegbarkeit der Zahlen in Primfaktoren, doch erhalten bleibe.

Allmählich schält sich aus den jahrelangen Bemühungen KUMMERS um diese ungeheuerliche Eigenschaft der Zahlen in α, daß sie auf verschiedene Arten in Primzahlen zerlegt werden können, die Überzeugung heraus, der Grund liege einfach darin, daß der Bereich der Zahlen in α zu klein sei, um ihre wirklichen Primzahlen, ihre einfachsten Elemente zu enthalten. Uns, die wir auf den Schultern KUMMERS stehen, ist es verhältnismäßig leicht, schon aus dem Bereiche unserer gewöhnlichen ganzen Zahlen 1, 2, 3, . . . durch Weglassung gewisser Zahlen einen engeren Bereich herzustellen, welcher genau dieselben merkwürdigen Eigenschaften besitzt, wie der Bereich der Zahlen in α, und bei dem auch weiter die Richtigkeit aller KUMMERschen Sätze über jene Zahlen ganz einfach erkannt werden kann. Deshalb will ich Ihnen einen solchen Bereich bilden.

Denken Sie sich jede von den Zahlen 1, 2, 3, . . . in ihre gleichen oder verschiedenen Primfaktoren zerlegt, so kann man sie in zwei Klassen K_0 und K_1 scheiden, wenn man festsetzt, K_0 soll aus allen Zahlen bestehen, welche eine gerade Anzahl von gleichen oder verschiedenen Primfaktoren enthalten, K_1 aus denjenigen, welche eine ungerade Anzahl von Primfaktoren besitzen. Also K_0 und K_1 enthalten die Zahlen:

$K_0 = ($1,	4,	6,	9,	10,	14,	15,	16,	21,	22,	24, . . .)
	2·2	2·3	3·3	2·5	2·7	3·5	2·2·2·2	3·7	2·11	2·2·2·3

$K_1 = ($2,	3,	5,	7,	8,	11,	12,	13,	17,	18,	19,	20, . . .).
				2·2·2		2·2·3			2·3·3		2·2·5

Die Klasse K_1 enthält unter anderen alle Primzahlen im gewöhnlichen Sinne. Nun wollen wir nur die Zahlen der ersten Klasse K_0 betrachten, also annehmen, daß uns die von K_1 gar nicht zugänglich wären, aber wir wollen trotzdem manchmal von ihnen sprechen. Dann sind unzerlegbare Zahlen in K_0 selbst alle und nur die, welche nur aus zwei gleichen oder verschiedenen Primfaktoren bestehen, also z. B. die Zahlen 4, 6, 9, 10, 14, 15, 21, . . .; denn zerfiele eine solche Zahl in zwei Faktoren in K_0, so müßte jeder einzelne von diesen Faktoren mindestens zwei, jene Zahl selbst also mindestens vier Primzahlen enthalten, während diese selbst nur aus zwei Primteilern bestehen. Dagegen sind zerlegbar in K_0 alle und nur diejenigen, welche, wie

z. B. 16, 24, 36, 40, ..., vier, sechs oder mehr Primfaktoren enthalten. Es ist nun klar, daß hier jede zusammengesetzte Zahl stets aber im allgemeinen auf mehrere Arten in unzerlegbare Faktoren in K_0 dekomponiert werden kann. So bestehen z. B. für die Zahl $210 = 2 \cdot 3 \cdot 5 \cdot 7$ von K_0 die folgenden drei Zerlegungen in unzerlegbare Faktoren:

$$210 = (2 \cdot 3) \cdot (5 \cdot 7) = (2 \cdot 5) \cdot (3 \cdot 7) = (2 \cdot 7) \cdot (3 \cdot 5)$$
$$= 6 \cdot 35 = 10 \cdot 21 = 14 \cdot 15.$$

Es tritt also hier genau dasselbe Unglück ein, wie im Gebiete der Zahlen in α, und daher gelten die Gesetze der Arithmetik im Bereiche K_0 ebensowenig, wie im Bereiche der Zahlen in α.

Kummer stellte sich jetzt das Problem so: Wie muß man den Bereich der Zahlen in α erweitern, damit in dem größeren Bereiche die Zerlegung der Zahlen in Primfaktoren eine eindeutige ist? Sie sehen, in unserem Gebiete K_0 würde die Aufgabe vollständig gelöst sein, wenn wir noch die Zahlen von K_1 hinzunähmen; tut man das nämlich, so sind ja z. B. alle jene drei Zerlegungen von 210 identisch, denn sie sind alle gleich

$$210 = 2 \cdot 3 \cdot 5 \cdot 7.$$

In unserem einfachen Falle würde also die Lösung des Kummerschen Problems folgendermaßen lauten:

Nimmt man zu den Zahlen 1, 4, 6, 9, 10, ... des Bereiches K_0 alle und nur die Zahlen 2, 3, 5, 7, 8, ... des Bereiches K_1 hinzu, so gelten in dem erweiterten Zahlenreiche alle Gesetze der elementaren Arithmetik, speziell der Satz von der eindeutigen Zerlegbarkeit der Zahlen in Primfaktoren.

Dieselbe Frage hat nun Kummer für den Bereich der Zahlen in α vollständig gelöst, und er nannte die neuen Divisoren, welche hier den Zahlen in K_1 entsprechen würden, ideale Divisoren, weil sie eben im Gebiete der Zahlen in α nicht existieren; speziell nannte er die Primdivisoren, welche also unseren Primzahlen 2, 3, 5, ... innerhalb K_1 entsprechen, die idealen Primfaktoren; die Zahlen oder Divisoren in α nannte er wirkliche Zahlen. Auch in unserem Beispiele will ich in der Folge die Zahlen von K_0 die wirklichen, die von K_1 die idealen Zahlen nennen, und speziell sollen die in K_1 befindlichen gewöhnlichen Primzahlen 2, 3, 5, 7, 11, ... die idealen Primfaktoren oder Primdivisoren heißen.

Kummer zeigte also, daß man den Bereich der wirklichen Zahlen oder der Divisoren in α durch Hinzunahme der idealen Divisoren so erweitern kann, daß im erweiterten Gebiete aller wirklichen und idealen

Zahlen jede Zahl und jeder Divisor auf eine und jetzt nur auf eine Weise als Produkt gleicher oder verschiedener Primfaktoren dargestellt werden kann, und daß also in diesem erweiterten Gebiete nun im wesentlichen alle multiplikativen Gesetze der elementaren Arithmetik gelten.

Die Untersuchung der Zahlen in α lehrt dann wirklich, daß der Grund, warum in diesem ursprünglichen Bereiche jede Zahl im allgemeinen auf mehrere Arten in unzerlegbare Faktoren zerfällt, wörtlich derselbe ist, wie der soeben für die Zahlen innerhalb K_0 gefundene.

Will man mit den idealen Primzahlen genau ebenso rechnen, wie mit den gewöhnlichen rationalen Primzahlen, so muß man zuerst ein Mittel angeben können, um zu entscheiden, ob und wann eine beliebige wirkliche Zahl durch einen bestimmten idealen Primfaktor teilbar ist. Die geniale Art, auf welche KUMMER dieses Grundproblem im Bereiche der Zahlen in α gelöst und damit die Existenz der idealen Zahlen erst in Evidenz gesetzt hat, läßt sich in unserem einfachen Beispiele leicht anschaulich machen. Wir stellen uns die Aufgabe, allein im Bereiche der „wirklichen" Zahlen von K_0 zu entscheiden, ob irgend eine derselben a einen idealen Primteiler, etwa die Primzahl 2, enthält oder nicht, ob also der Quotient $\frac{a}{2}$ ganz oder gebrochen ist. Nun ist aber $\frac{a}{2}$ dann und nur dann ganz, wenn auch

$$\frac{5a}{2} = \frac{(5 \cdot 3) \cdot a}{(2 \cdot 3)}$$

ganz ist, und da in dem rechts stehenden Bruche im Zähler und Nenner allein „wirkliche" Zahlen des Bereiches K_0 stehen, so kann diese Frage allein in diesem Bereiche durch Probieren entschieden werden. Dasselbe Resultat läßt sich offenbar auch so aussprechen: Anstatt a durch die ganze ideale Zahl 2 zu dividieren, teile ich sie durch die gebrochene Zahl

$$\frac{2}{5} = \frac{(2 \cdot 3)}{(5 \cdot 3)}$$

welche, wie die rechte Seite dieser Gleichung zeigt, als wirkliche gebrochene Zahl darstellbar ist und welche in ihrer gehobenen Form allein den idealen Primteiler 2 als Zähler besitzt. Dann ist die Zahl a stets und nur dann durch 2 teilbar, wenn sie durch den wirklichen Bruch $\frac{2}{5} = \frac{2 \cdot 3}{5 \cdot 3}$ divisibel ist. Allgemeiner ist a stets und nur dann z. B. durch 2^7 teilbar, wenn dasselbe für den Bruch

$$\left(\frac{2}{5}\right)^7 = \frac{(2 \cdot 3)^7}{(5 \cdot 3)^7}$$

gilt, wenn also der Quotient

$$\frac{(5\cdot 3)^7\cdot a}{(2\cdot 3)^7}$$

eine ganze Zahl von K_0 ist.

Genau ebenso kann man für den Bereich der wirklichen Zahlen in α, anstatt eine wirkliche Zahl a in α durch eine ideale Primzahl $\mathfrak{p}$ zu dividieren, als Teiler irgendeinen gebrochenen Divisor $\frac{\mathfrak{p}}{\mathfrak{r}}$ wählen, dessen Zähler $\mathfrak{p}$ ist, während sein Nenner nur $\mathfrak{p}$ nicht enthalten darf. Immer kann man dann diesen Nenner außerdem noch so wählen, daß der Bruch $\frac{\mathfrak{p}}{\mathfrak{r}}$ durch Erweiterung mit einem geeigneten Divisor $\mathfrak{s}$ in einen wirklichen Bruch $\frac{p}{r}$ in α übergeht, so daß also

$$\frac{\mathfrak{p}}{\mathfrak{r}}=\frac{\mathfrak{p}\mathfrak{s}}{\mathfrak{r}\mathfrak{s}}=\frac{p}{r}$$

ist. Daher kann man genau wie in unserem Beispiel den Satz aussprechen:

Eine wirkliche Zahl a in α ist dann und nur dann durch den idealen Primdivisor $\mathfrak{p}$ teilbar, wenn sie den wirklichen gebrochenen Divisor $\frac{p}{r}$ enthält, oder, was dasselbe ist, wenn ra durch p teilbar ist. Allgemeiner ist a durch $\mathfrak{p}^\varrho$ teilbar, wenn $r^\varrho a$ die wirkliche Zahl p^ϱ enthält.

Damit wäre nun aber nicht viel gewonnen, wenn man nicht diese idealen Divisoren allein durch die wirklichen Zahlen in α auch darstellen könnte; aber diese Darstellung gelingt Kummer ebenfalls auf einem bewundernswert einfachen Wege, und auch dieser läßt sich in unserem Beispiele in sehr anschaulicher Weise angeben.

In diesem Beispiele besteht nämlich der Bereich K_0 aller wirklichen Zahlen aus den Zahlen

$$1,\quad 2\cdot 2,\quad 2\cdot 3,\quad 3\cdot 3,\ldots$$

mit einer geraden Anzahl von Primfaktoren, der Bereich K_1 aller idealen Zahlen aus denjenigen:

$$2,\ 3,\ 5,\ 7,\ 8,\ 11,\ldots,$$

welche eine ungerade Anzahl von Primfaktoren enthalten, und die Aufgabe ist hier die, alle Zahlen des Bereiches K_1 durch Zahlen von K_0 darzustellen. Hier liegt die Lösung auf der Hand: Ist z. B. $35=2\cdot 3\cdot 5$ eine solche ideale Zahl von K_1, so ist ihr Quadrat $35^2=2\cdot 2\cdot 3\cdot 3\cdot 5\cdot 5$ eine Zahl von K_0, d. h. eine wirkliche Zahl, da sie eine gerade Anzahl von Primfaktoren enthält.

Das Quadrat a^2 einer jeden idealen Zahl ist also eine wirkliche Zahl A, jede ideale Zahl ist gleich der Quadratwurzel aus einer wirklichen Zahl.

Ist schon a selbst eine wirkliche Zahl, so ist natürlich ihr Quadrat $a^2 = A$ auch wirklich. Das Quadrat jeder idealen oder wirklichen Zahl ist demnach stets eine wirkliche Zahl. Um also alle idealen und wirklichen Zahlen zu erhalten, braucht man zu den wirklichen Zahlen nur noch gewisse Quadratwurzeln aus denselben hinzuzunehmen. Eine sehr leichte Überlegung zeigt auch, was hier nur erwähnt sei, aus welchen Zahlen von K_0 die Quadratwurzel ausgezogen werden muß, um alle idealen Zahlen von K_1 zu erhalten.

Damit ist nun die Frage, wie man entscheiden kann, ob irgendeine wirkliche Zahl b von K_0 durch eine ideale oder eine wirkliche Zahl a teilbar ist oder nicht, in anderer Weise als vorher vollständig gelöst. Ist nämlich a auch wirklich, so kann diese Frage durch einfaches Probieren entschieden werden. Ist das aber nicht der Fall, so ist ja $\frac{b}{a}$ nur dann ganz, wenn $\left(\frac{b}{a}\right)^2 = \frac{b^2}{a^2}$ ganz ist, und da a^2 dann wirklich ist, so kann auch in diesem Falle die Frage durch einfaches Probieren gelöst werden.

Wir können dieses Ergebnis auch so aussprechen: Wir haben alle wirklichen und idealen Zahlen in die beiden Klassen K_0 und K_1 gesondert, von denen die erste, die sog. Hauptklasse, alle wirklichen, die zweite alle idealen Zahlen enthält. Multipliziert man jede Zahl von K_0 mit jeder Zahl von K_1, so gehören alle diese Produkte einer und derselben Klasse, nämlich der Klasse K_1 an, da jedesmal der erste Faktor eine gerade, der zweite eine ungerade Anzahl von Primfaktoren enthält. Ebenso gehört jedes Produkt aus je zwei Faktoren von K_1 zu K_0, und jedes Produkt von je zwei Faktoren von K_0 ebenfalls zu K_0. Diese Klasseneinteilung ist also derart, daß alle Produkte, welche ich erhalte, wenn ich jedes Element einer Klasse mit jedem Elemente derselben oder der anderen Klasse multipliziere, stets in einer und derselben Klasse vorkommen.

Kummer hat nun gezeigt, daß wörtlich dieselben Verhältnisse für den Bereich der wirklichen und idealen Zahlen in α bestehen; nur sind hier die Beziehungen viel komplizierter, und der Beweis, daß hier dasselbe einfache Einteilungsgesetz für diese Zahlen herrscht wie in unserem Beispiel, war wohl eine der größten mathematischen Leistungen aller Zeiten.

Nimmt man zum Bereich der Zahlen in α oder zu den wirklichen Divisoren noch die Gesamtheit aller idealen Divisoren hinzu, so erhält man nämlich einen sehr viel größeren Divisorenbereich, welchen man genau wie in unserem einfachen Beispiele ebenfalls in Klassen einteilen kann. Nur ist die Anzahl dieser Klassen nicht wie dort gleich zwei, sondern im allgemeinen größer, aber, was besonders wichtig ist, immer eine endliche Zahl. Ich will mit KUMMER die Anzahl dieser Klassen durch H und diese Klassen selbst durch

$$K_0, K_1, K_2, \ldots K_{H-1}$$

bezeichnen. Die Klasse K_0, die sog. Hauptklasse, enthält wie in unserem Beispiele alle wirklichen Divisoren, d. h. alle diejenigen, welche den Zahlen in α selbst entsprechen, die übrigen Divisoren verteilen sich auf die $H-1$ übrigen Klassen.

Auch hier ist jedes Produkt $\beta\gamma$ irgend zweier Zahlen der Hauptklasse K_0 wieder eine Zahl in K_0, da dieses ja wieder eine Zahl in α ist. Sind aber allgemein H_g und H_k zwei beliebige Divisorenklassen, und bedeuten $\mathfrak{D}_g$ und $\mathfrak{D}_k$ alle Divisoren derselben, so ist auch hier die Klasseneinteilung so gemacht, daß alle Produkte $\mathfrak{D}_l = \mathfrak{D}_g \mathfrak{D}_k$ wiederum einer und derselben Klasse H_l angehören, welche durch die Klassen H_g und H_k bestimmt ist, genau wie in unserem einfachen Beispiel.

Ebenso wie oben das Quadrat jeder Zahl aus K_0 oder aus K_1 der Hauptklasse K_0 angehörte, kann hier allein aus der soeben angegebenen Fundamentaleigenschaft unserer Klasseneinteilung gefolgert werden, daß jeder ideale Divisor $\mathfrak{D}$ zur H^{ten} Potenz erhoben sicher einen Divisor D der Hauptklasse K_0, also eine wirkliche Zahl in α ergibt. Jeder ideale Divisor $\mathfrak{D}$ selbst ist also gleich $\sqrt[H]{D}$, d. h. gleich der H^{ten} Wurzel aus einer wirklichen Zahl. Man erhält also alle wirklichen oder idealen Zahlen in α, wenn man zu den Zahlen in α noch H^{te} Wurzeln aus gewissen Zahlen hinzunimmt.

Dieses Resultat gibt uns ein neues Mittel, um zu erkennen, ob eine beliebige Zahl β durch einen Divisor $\mathfrak{D}$ teilbar ist oder nicht, denn der Quotient $\frac{\beta}{\mathfrak{D}}$ ist dann und nur dann ganz, wenn das gleiche für seine H^{te} Potenz $\frac{\beta^H}{\mathfrak{D}^H} = \frac{\beta^H}{D}$ gilt, und diese Frage kann durch einfache Division entschieden werden.

Wie wir gesehen haben, muß die H^{te} Potenz jeder idealen Zahl wirklich sein. Dies braucht aber auch hier nicht die niedrigste Potenz zu sein, denn wenn z. B. $\mathfrak{D}$ selbst wirklich ist, ist es eben schon die erste. Ist aber $\mathfrak{D}^h$ die kleinste Potenz von $\mathfrak{D}$, die wirklich ist, so erkennt man leicht, daß h ein Teiler von H sein muß. Ebenso leicht kann man schließen: Ist $\mathfrak{D}^r$ wirklich und hat r mit H keinen gemeinsamen Teiler, so muß $\mathfrak{D}$ selbst wirklich sein. So erkennt man in unserem Beispiele, daß, falls für eine gewöhnliche rationale Zahl $\mathfrak{D}$ $\mathfrak{D}^r$ zur Hauptklasse K_0 gehört, also eine gerade Anzahl von Primfaktoren besitzt und r ungerade ist, $\mathfrak{D}$ selber der Hauptklasse angehören muß, da, falls $\mathfrak{D}$ eine ungerade Anzahl von Primfaktoren hätte, dasselbe für jede Potenz $\mathfrak{D}^r$ mit ungeradem Exponenten gelten würde. Auf dieser wichtigen Tatsache beruht fast allein die Möglichkeit, den FERMATschen Satz zu beweisen.

Die Klassenanzahl H der Divisoren in α ist die wichtigste Zahl der ganzen Theorie. Ist sie gleich 1, so gibt es eben gar keine idealen Zahlen, denn dann existiert nur die eine Klasse K_0 der wirklichen Zahlen. Dann enthält diese auch alle Primfaktoren; dann und nur dann sind alle Zahlen in α auf eine einzige Weise in wirkliche Primfaktoren zerlegbar. Nur in diesem Falle gelten alle Gesetze der elementaren Arithmetik schon für die wirklichen Zahlen in α, nur dann ist also die Theorie der idealen Zahlen überflüssig.

Die wirkliche Berechnung dieser Klassenanzahl für bestimmte Gleichungen $\alpha^\lambda = 1$ ist eine der schwierigsten Aufgaben der ganzen Zahlentheorie. Ihre Lösung ist KUMMER, wie bereits erwähnt, durch die Methoden seines großen Lehrers DIRICHLET gelungen. Ich muß es mir versagen, an dieser Stelle auf die wunderbar feinen Methoden DIRICHLETS einzugehen, durch welche er der Zahlentheorie ganz neue Wege eröffnete, indem er die gewaltigen Hilfsmittel der Analysis in ihren Dienst zwang. Es sind das diejenigen Methoden, welche der uns leider zu früh entrissene HERMANN MINKOWSKI durch seine Geometrie der Zahlen zu so hoher Einfachheit und vollendeter Schönheit ausgebildet hat.

Ich möchte hier nur erwähnen, daß nach der KUMMERschen Darstellung die Klassenzahl H aus zwei Faktoren besteht, von denen der erste verhältnismäßig sehr leicht, der letzte sehr schwer zu berechnen ist. Es ist von großer Wichtigkeit, daß für die meisten Anwendungen nur die Kenntnis des ersten Faktors der Klassenzahl nötig ist.

Nur ein großer Geist und eine reiche Phantasie konnte in einem Gebiete, in welchem alles regellos jeder Ordnung zu spotten schien, das tiefverborgene aber so einfache Gesetz ahnen; es erforderte eine unbeugsame Energie, auf diesem mühevollen Wege nicht mutlos stehen zu bleiben; haben doch die arithmetischen Arbeiten KUMMERS zwanzig seiner besten Jahre gefordert. Aber der Erfolg war den hohen Einsatz wert, denn die hier dargelegten Prinzipien KUMMERS haben seitdem die ganze Arithmetik und Algebra beherrscht; es hat sich nämlich gezeigt, daß sie nicht bloß für die hier betrachteten Bereiche der Zahlen in α gelten, sondern für alle algebraischen Bereiche. KUMMER selbst hat zwar seine Theorie nur so weit ausgedehnt, als es die von ihm behandelten Probleme nötig machten; er war sich wohl bewußt, daß sie allgemeingültig waren, er sprach es aber aus, daß es ihm genüge, diese Schönheiten im einzelnen geschaut und völlig erkannt zu haben.

Bei allen seinen zwanzigjährigen Arbeiten über die Arithmetik waren es immer zwei bestimmte Probleme, deren vollständige Lösung ihm als höchstes Ziel vor Augen stand; um ihretwillen hat er sein gewaltiges Lehrgebäude aufgerichtet. Er hat es selbst ausgesprochen, daß er seine Theorie der idealen Zahlen trotz ihrer Erfolge vielleicht doch verworfen haben würde, wenn er ein anderes Mittel gewußt hätte, den FERMATschen Satz und das allgemeine Reziprozitätsgesetz zu beweisen. Wir, denen die KUMMERsche Theorie als der festeste Grundstein der Arithmetik erscheint, werden wohl anderer Ansicht sein; wir werden es umgekehrt dem FERMATschen Probleme danken, daß durch dieses der Geist eines KUMMER so gewaltig angeregt worden ist, daß wir die Idealtheorie erhalten haben.

Lassen Sie mich nun einen Blick auf die Art werfen, wie KUMMER mit Hilfe der Idealtheorie der allgemeine Beweis des FERMATschen Satzes gelang. KUMMER führt noch allgemeiner den Beweis, daß es unmöglich ist, die Gleichung

$$x^\lambda = y^\lambda + z^\lambda$$

nicht bloß im Bereiche der ganzen Zahlen, sondern sogar in dem größeren Gebiete der Zahlen in α zu lösen. Schreibt man diese Gleichung wieder in der multiplikativen Form

$$z^\lambda = x^\lambda - y^\lambda = (x - y)(x - \alpha y) \cdots (x - \alpha^{\lambda-1} y) = b_0 b_1 \cdots b_{\lambda-1},$$

so haben wir die λ^{te} Potenz z^λ als Produkt von λ Zahlen b_r in α dargestellt. Von diesen λ Zahlen wollen wir jetzt annehmen, daß nicht zwei von ihnen eine und dieselbe Primzahl gemeinsam haben, daß je zwei also aus lauter verschiedenen idealen oder wirklichen Primfaktoren bestehen. Man kann zeigen, daß diese Annahme entweder erfüllt ist, oder daß, falls dies nicht der Fall sein sollte, die ursprüngliche Gleichung anderweitig so umgeformt werden kann, daß eine entsprechende Voraussetzung erfüllt ist.

Ist aber das Produkt $b_0 b_1 \cdots b_{\lambda-1}$, dessen Faktoren sämtlich verschiedene Primteiler enthalten, eine λ^{te} Potenz, so muß jeder Faktor für sich eine λ^{te} Potenz sein.*) Zu jedem Faktor b_r gehört also ein Divisor $\mathfrak{D}_r^\lambda$, welcher die λ^{te} Potenz eines anderen ist. Von diesem Divisor $\mathfrak{D}_r$ kann man nun folgendes aussagen: Seine λ^{te} Potenz ist die wirkliche Zahl b_r in α. Ist also λ kein Teiler der Klassenzahl H, so folgt nach dem vorher erwähnten Hauptsatz, daß auch $\mathfrak{D}_r$ selbst wirklich sein muß; allerdings ist diese Folgerung auch nur in diesem Falle gestattet. Ist also λ kein Teiler der Klassenzahl, so ist b_r gleich der λ^{ten} Potenz einer wirklichen Zahl d_r, eventuell multipliziert mit einer andern wirklichen Zahl, welche keinen einzigen Teiler mehr hat, einer sog. Einheit. Es ist also für alle λ Faktoren

$$b_r = x - \alpha^r y = e_r d_r^\lambda \qquad (r = 0, 1, \ldots \lambda - 1),$$

wo e_r eine Einheit bedeutet; und aus diesen λ Gleichungen für die beiden Unbekannten x und y folgt durch einfache Rechnung, daß sie unmöglich bestehen können, weil durch sie x und y überbestimmt sind. Damit ist der Fermatsche Satz bewiesen.

Nur dann ist also das Fermatsche Problem durch Kummers Betrachtungen gelöst, wenn die Klassenzahl H für die λ^{ten} Einheitswurzeln nicht durch λ teilbar ist. Daraus folgt, daß man für jede Primzahl λ die zugehörige Klassenzahl auf ihre Teilbarkeit oder Nichtteilbarkeit durch λ untersuchen muß. Hier gelingt es nun Kummer, vor allen Dingen zu beweisen, daß H dann und nur dann durch λ teilbar ist, wenn ihr leicht zu bestimmender erster Faktor bereits λ enthält. Nur dieser muß also für alle Primzahlen $\lambda = 3, 5, 7, 11, \ldots$ berechnet werden, und diese von Kummer für die im ersten Hundert auftretenden Primzahlen durchgeführte Rechnung ergab, daß hier nur für die drei Primzahlen

$$\lambda = 37, \quad \lambda = 59, \quad \lambda = 67$$

die Klassenzahl durch λ teilbar ist. So ist es also z. B. hiernach nicht

*) Der Satz, daß ein Produkt mehrerer Faktoren, welche keinen unzerlegbaren Bestandteil gemeinsam haben, nur dann eine λ^{te} Potenz sein kann, wenn dasselbe für jeden dieser Faktoren der Fall ist, gilt nur dann, wenn die Zahlen des Bereiches nur auf eine Weise in unzerlegbare Faktoren dekomponiert werden können. So gilt in unserem einfachen Beispiel die Gleichung:

$$(2 \cdot 3)(3 \cdot 5)(5 \cdot 7)(7 \cdot 2) = (2 \cdot 3 \cdot 5 \cdot 7)^2,$$

d. h. das Produkt der vier links stehenden Faktoren von K_0 ist einer Quadratzahl gleich, während die vier Faktoren $(2 \cdot 3), \ldots$ innerhalb K_0 sogar unzerlegbar, also sicher keine Quadratzahlen sind.

sicher, ob die Fermatsche Gleichung

$$x^{37} = y^{37} + z^{37}$$

niemals eine Lösung haben kann. Für alle übrigen 22 Primzahlen aus dem ersten Hundert ist dieser wichtige Satz jedoch mit einem Schlage bewiesen, aber auch für sehr viele andere Exponenten λ über Hundert hinaus.

Allerdings zeigte es sich, daß diese Ausnahmezahlen, für welche H durch λ teilbar ist, um so häufiger auftraten, je weiter man in der Reihe der Primzahlen ging; so finden sich unter den ersten 13 Primzahlen des zweiten Hundert schon 5 Ausnahmezahlen, und nach einer Bemerkung Kummers kann es als wahrscheinlich angesehen werden, daß zuletzt ungefähr ebensoviele Primzahlen existieren, welche regulär, als solche, welche irregulär sind. Dann konnte Kummer sich also rühmen, den Fermatschen Satz in der Hälfte aller Fälle bewiesen zu haben.

Es ist nun noch Kummer gelungen, durch eine bewunderungswürdige Verfeinerung seiner Methode auch für eine große Zahl von den bisher auszunehmenden Primzahlen, insbesondere für $\lambda = 37, 59, 67$ die Unmöglichkeit der Fermatschen Gleichung zu beweisen. Aber immer noch harren sehr viele, wahrscheinlich sogar unendlich viele einzelne Fermatsche Probleme ihrer endgültigen Lösung, und man muß es sagen, daß es den heißen Bemühungen der besten Arithmetiker seit Kummer bisher nicht gelungen ist, die Unlösbarkeit auch nur einer der von Kummer ausgeschlossenen Fermatschen Gleichungen vollständig zu erweisen. Allerdings wurden in den letzten Jahrzehnten einerseits die großen Kummerschen Gedanken über die idealen Zahlen so ausgebildet und verfeinert, andererseits die speziellen Kriterien Kummers für die Ausnahmezahlen λ zum Teil so wesentlich vereinfacht, daß die Hoffnung berechtigt erscheint, daß dieses große Werk Kummers auch einmal ganz zu Ende geführt werden wird.

Nur kurz kann ich hier darauf eingehen, wie die Theorie der idealen Zahlen es Kummer nun auch ermöglichte, das zweite Ziel seiner Sehnsucht, die Aufstellung und den Beweis der in der Arithmetik vorkommenden allgemeinen Reziprozitätsgesetze in demselben Umfange zu erreichen, wie ihm die Lösung des Fermatschen Problems gelungen war.

Das quadratische Reziprozitätsgesetz im Bereiche der natürlichen ganzen Zahlen, welches ich hier als bekannt voraussetze, stellt sich in seiner vollen Einfachheit

$$\left(\frac{p}{q}\right) = \left(\frac{q}{p}\right)$$

erst dann dar, wenn die beiden untersuchten Zahlen p und q einmal als Primzahlen und zweitens als primär vorausgesetzt werden, wenn sie nämlich mit der Einheit $+1$ oder mit der Einheit -1 multipliziert, d. h. positiv oder negativ angenommen werden, je nachdem sie absolut genommen von der Form $4n+1$ oder $4n-1$ sind.

Gauss war es, welcher erkannt hatte, daß die entsprechende Frage für die biquadratischen Reste und Nichtreste erst dann einfach gelöst werden kann, wenn man anstatt des Bereiches der gewöhnlichen ganzen Zahlen den ihn umfassenden der komplexen Zahlen $a+bi$ arithmetisch untersucht; nur mußten hier wieder statt der reellen Primzahlen p und q zwei der bereits früher a. S. 20 erwähnten komplexen Primzahlen π und $\varkappa$ dieses Bereiches zugrunde gelegt werden. Ferner aber treten in dieser Theorie statt der zwei Einheiten ± 1 deren vier, nämlich $+1, -1, +i, -i$ auf, und unter den vier hier möglichen Werten, welche aus jeder der beiden Primzahlen durch Multiplikation mit einer dieser Einheiten hervorgehen, mußte wieder die sog. primäre Form für diese Primzahl so ausgewählt werden, daß der Ausdruck für das biquadratische Gesetz möglichst einfach wurde; denn auch hier haben jene Primzahlen in Beziehung darauf, ob sie Reste oder Nichtreste füreinander sind, ganz andere Charaktere, je nachdem man die Einheit wählt, mit der sie behaftet sind.

Auf dieser Grundlage gelang es Gauss und Dirichlet, zu zeigen, daß für geeignet gewählte primäre Primzahlen genau dasselbe biquadratische Reziprozitätsgesetz

$$\left(\frac{\pi}{\varkappa}\right) = \left(\frac{\varkappa}{\pi}\right)$$

gilt; und ganz dasselbe ist, wie Jacobi und später Eisenstein nachwiesen, für das kubische Reziprozitätsgesetz der Fall, sobald man hier alle ganzen Zahlen des Bereiches der dritten Einheitswurzeln betrachtet, welche sich ebenfalls eindeutig in wirkliche Primzahlen zerlegen lassen, und wenn man dann jede dieser Primzahlen in einer geeigneten primären Form annimmt, indem man die zu ihr gehörige Einheit unter den sechs hier vorhandenen passend auswählt.

Die Gründe, warum die aus der Anregung von Gauss hervorgegangenen Beweise des kubischen und des biquadratischen Reziprozitätsgesetzes trotz der Anstrengungen von Männern wie Jacobi und Eisenstein nicht auf die höheren Gesetze ausgedehnt werden konnten, liegen darin, daß für die allgemeinen λ^{ten} Potenzreste, wo λ eine beliebige Primzahl ist, die Untersuchung erst einfach wird, wenn man wieder den vorher schon betrachteten Bereich der Zahlen in α oder der λ^{ten} Wurzeln der Einheit zugrunde legt. In diesem ist aber erstens die Zerlegung in Primfaktoren, wie bereits oben erwähnt wurde, im allgemeinen keine eindeutige, und zweitens ist von $\lambda = 5$ an die Anzahl der Einheiten in α unendlich groß; und so fehlten hier zunächst die beiden Vorbedingungen für die Aufstellung eines einfachen Reziprozitätsgesetzes.

KUMMER hatte nun die erste Schwierigkeit durch die Einführung der idealen Primfaktoren und durch den Nachweis vollkommen beseitigt, daß sich dieselben als H^{te} Wurzeln aus wirklichen Zahlen darstellen lassen, wenn H die Klassenzahl bedeutet. Die zweite Schwierigkeit, welche in der geeigneten Wahl der den Primzahlen zuzuordnenden Einheiten bestand, konnte er überwinden, nachdem er mit DIRICHLETs Prinzipien gezeigt hatte, daß man alle Einheiten in α durch ein System von Fundamentaleinheiten einfach darstellen und sie so völlig beherrschen kann. Vollständig durchführen konnte er aber die Untersuchung auch hier nur unter der Voraussetzung, daß im Bereiche der gerade betrachteten Zahlen in α niemals die λ^{te} Potenz einer idealen Zahl eine wirkliche Zahl sein kann, oder was nach dem auf S. 28 erwähnten Satze ganz dasselbe ist, wenn λ keine der vorhin erwähnten Ausnahmezahlen ist, für welche die Klassenzahl H durch λ teilbar ist.

So hatte sich KUMMER selbst die Mittel geliefert, um für alle regulären Primzahlen λ die Reziprozitätsgesetze für die λ^{ten} Potenzreste wenigstens aufzustellen. Die wesentliche Schwierigkeit lag hier in der Bestimmung derjenigen primären Form für die Primfaktoren, für welche das allgemeinste Reziprozitätsgesetz womöglich ebenso einfach würde, wie in den schon bekannten Fällen $\lambda = 2, 3, 4$. Mit einer geradezu staunenswerten Divinationsgabe hat KUMMER diese Frage gelöst, und schon im Jahre 1847 stellte er dieses schönste und tiefste Gesetz der Zahlenlehre in seiner großartigen Einfachheit auf, ohne es allerdings damals schon beweisen zu können; aber er hatte seine Richtigkeit an einer Unzahl von Beispielen geprüft und zunächst diese Resultate im Jahre 1850 veröffentlicht. In dieser Fassung spricht sich das allgemeinste Theorem für alle regulären Primzahlen λ wörtlich ebenso aus, wie das gewöhnliche quadratische Fundamentaltheorem:

Sind nämlich π und $\varkappa$ zwei ideale oder wirkliche aber primäre Primzahlen in α, so besteht für das dem LEGENDREschen analog gebildete Zeichen $\left(\frac{\pi}{\varkappa}\right)$ die Reziprozitätsgleichung

$$\left(\frac{\pi}{\varkappa}\right)=\left(\frac{\varkappa}{\pi}\right).$$

Von der Aufstellung dieses wundervollen Gesetzes bis zu seinem vollständigen Beweise lag noch ein weiter steiler Weg, die angestrengte Arbeit von neun Jahren. KUMMER mußte sehr bald erkennen, daß für diesen Beweis das Gebiet der wirklichen und idealen Zahlen in α zu klein war, obwohl das Gesetz selbst in diesem Bereich in vollster Einfachheit ausgesprochen werden konnte, daß es vielmehr nötig war, über diesem Zahlgebiete ein höheres dieses umfassendes Reich algebraischer Zahlen in derselben Weise aufzubauen, wie früher den Bereich der Zahlen in α über demjenigen der natürlichen Zahlen.

Dieses höhere Gebiet umfaßt außer den Zahlen in α noch die λ^{te} Wurzel aus einer bestimmten Zahl in α, und in ihm gelten alle Ergebnisse der KUMMERschen Idealtheorie in geeigneter Modifikation. Es gelang nun KUMMER nach mehrjähriger intensiver geistiger Arbeit in diesem noch nie durchforschten Gebiete, den schönsten und tiefsten zweiten GAUSSschen Beweis des quadratischen Reziprozitätsgesetzes, welcher auf der Theorie der quadratischen Formen und der Einteilung ihrer Klassen in Genera beruht, so zu verallgemeinern, daß er auch in diesem höheren Zahlenreiche anwendbar bleibt und hier den vollständigen Beweis des allgemeinsten Reziprozitätsgesetzes liefert.

Ebenso wie GAUSS bei seiner immer erneuten eindringenden Beschäftigung mit dem quadratischen Reziprozitätssatze nach dem ersten Beweise, in dem er in steilem Anstieg das Problem bezwungen hatte, noch andere und andere Beweise desselben Satzes erschließen konnte, fügte auch KUMMER diesem ersten schwierigsten Beweise seines allgemeinsten Satzes noch zwei andere in mancher Hinsicht einfachere hinzu. Der erste von ihnen beruht wesentlich auf der Theorie der zu dem erweiterten Zahlbereich gehörigen Einheiten und kann als eine Verallgemeinerung des zuerst unvollständig von LEGENDRE für das quadratische Gesetz gegebenen Beweises angesehen werden; der zweite hatte noch kein Vorbild unter den damals bekannten Beweisen für den Fall $\lambda = 2$, er liefert vielmehr auf diesen Fall angewendet einen neuen höchst einfachen Beweis dieses Gesetzes, und er selbst ist wohl der gedanklich einfachste unter den drei Beweisen KUMMERS. Während uns aber GAUSS im ganzen acht Beweise seines theorema fundamentale gegeben hat und die Anzahl der mehr oder weniger verschiedenen Beweise dieses einfacheren Problemes auf

3*

etwa 50 gestiegen ist, sind diese drei großen Beweise Kummers für das allgemeine Reziprozitätsgesetz lange Zeit hindurch die einzigen geblieben, welche wir besitzen.

Erst in neuester Zeit sind auch diese Kummerschen Untersuchungen, wesentlich durch die Arbeiten Hilberts, in hohem Maße vereinfacht und gefördert worden, und es besteht die Hoffnung, daß auch hier die von Kummer uns hinterlassenen ungelösten Probleme einmal ihrer vollständigen Lösung entgegengeführt werden können.

Es ist ein Zeichen für seine hohe Auffassung des Gelehrten- und Lehrerberufes, daß Kummer drei Jahre nach der Feier seines fünfzigjährigen Doktorjubiläums, also im Jahre 1884, als Vierundsiebzigjähriger seine Vorlesungen an der Universität einstellte, nachdem er schon im Jahre 1878 das Sekretariat der Akademie niedergelegt hatte. Er wollte nur so lange als Lehrer wirken, als er dies im höchsten Sinne tun konnte, er wollte nach beispielloser Arbeit und beispiellosen Erfolgen die letzten ihm noch beschiedenen Jahre seiner geliebten Frau, seinen Kindern leben, denen durch sein ganzes Leben seine größte Liebe und Sorgfalt gewidmet gewesen war. Jetzt wollte er sich noch einmal in die Schätze der deutschen und englischen Literatur, besonders in Goethe und Shakespeare, die ihn durch sein ganzes Leben begleitet hatten, versenken, jetzt noch, so oft es ihm vergönnt wäre, die geliebten Berge Schlesiens aufsuchen, zu denen ihn immer und immer wieder seine Sehnsucht hinzog. Und ein gütiges Geschick hat es gewollt, daß auch diese seine letzten Wünsche schön und voll in Erfüllung gingen. Ebenso wie es Kummer beschieden war, uns in seiner Wissenschaft alles geben zu können, was er ersehnte, so daß sich nach seinem Tode nichts Angefangenes oder Unfertiges vorfand, konnte er seinen Lebensabend, sein wohlerworbenes ruhiges abgeklärtes Glück noch neun Jahre in voller geistiger Frische genießen.

Langsam nur, ganz langsam beugte sich die hohe schlanke Gestalt des ehrwürdigen Greises, wurde sein klares durchdringendes und doch so freundlich blickendes Auge trübe, langsam schwanden die Kräfte. Endlich raffte ihn die Influenza inmitten seiner Lieben am 14. Mai 1893 hinweg. An einem herrlichen sonnigen Tage im Mai wurden seine sterblichen Reste auf dem Jacobikirchhofe in die Gruft gesenkt, um welche sich alle versammelt hatten, die ihn lieb und wert hielten.

Aber wenn er auch den Seinigen, wenn er auch allen denen unter uns, die ihn kannten und verehrten, auf immer entrissen ist, so wird doch das Größte, was er erstrebt und erreicht hat, niemals verloren gehen. In fast allen Natur- und Geisteswissenschaften ist alles in ewigem Wechsel; das, was in ihnen die Größten geschaffen, wodurch sie die Mitwelt unwiderstehlich mit sich fortgerissen haben, nach kurzer Zeit ist es zu seinem besten Teile veraltet, fast vergessen; es kann nicht weiter wirken, weil der ganze Bau der Wissenschaft, den eine Generation aufgeführt hat, von der nächsten niedergerissen werden muß, um von ihr neu und schöner wieder aufgeführt zu werden.

Das aber ist das Große in unserer Wissenschaft, daß alles, was in ihr einmal wahr und erhaben gewesen ist, immer wahr und erhaben bleiben wird, mag auch der Leib dessen, der es uns erkennen lehrte, lange in Staub zerfallen sein. Immer höher, schöner und einfacher steigt das Gebäude der Mathematik im Laufe der Jahrhunderte empor, und die großen Männer, denen es vergönnt war, dieses Gebäude in seiner Schönheit aufzuführen, werden nie und nimmer vergessen werden.

Wir werden wohl dahin kommen mit dem Rüstzeug, welches uns KUMMER gegeben hat, manches zu erreichen, was er vielleicht ersehnt hat, manches anders und vielleicht einfacher zu sehen, als er es getan oder auch gewollt hat, aber für alle Zeiten werden die großen Ideen, die er in seiner vorwärtsstürmenden Jugend konzipierte, die er in seinem langen Gelehrtenleben mit wunderbarer geistiger Kraft durchdachte und ausreifen ließ, zu den Grundpfeilern gehören, auf denen das hehre Gebäude unserer Wissenschaft ruht und ohne die es nicht gedacht werden kann.

Berlin, den 7. Januar 1910.

Aus dem reichen Schatze der in der vorstehenden Rede erwähnten Briefe KUMMERS an seine Mutter und an LEOPOLD KRONECKER veröffentliche ich mit gütiger Genehmigung der Besitzer, Frau Geheimrat KUMMER und Geh. Justizrat ERNST KRONECKER, diejenigen, welche mir für die Charakteristik des Mannes und des Forschers besonders wertvoll erscheinen.

Ich habe auch solche Briefe ausgewählt, welche erkennen lassen, in wie anstrengender Gedankenarbeit KUMMER zu den großen Resultaten aufsteigen mußte, welche uns aus seinen klassischen Schriften wohlbekannt sind, wie sogar manchmal seine Hoffnung, eine geahnte Gesetzmäßigkeit bestätigt zu finden, enttäuscht wurde, so aber daß sich an ihrer Stelle eine schönere und tiefere enthüllte. Mir scheint, daß gerade durch solche Einblicke in die Werkstatt seines Denkens die Ergebnisse seiner Arbeit einen neuen Reiz, einen höheren Wert erhalten.

Von einem Kommentar der Briefe glaubte ich absehen zu sollen, da durch ihn der Charakter dieser ganzen Festschrift wesentlich geändert worden wäre, ohne daß für denjenigen, der die Idealtheorie KUMMERS nicht kennt, durch solche Erläuterungen viel zum Verständnis beigetragen werden könnte. Dem Leser der Gedächtnisrede wird, so hoffe ich, vieles von dem mathematischen Inhalt der Briefe verständlich sein, und für den genauen Kenner der KUMMERschen Arbeiten wird ein besonderer Reiz dieser Briefe gerade in der Vergleichung der hier zuerst auftretenden manchmal noch unfertigen Gedanken mit der klassischen Form liegen, welche sie nachher in den Abhandlungen gewonnen haben.

Als schönsten Abschluß dieser Briefe habe ich das Schreiben veröffentlicht, mit welchem KRONECKER die Überreichung seiner Festschrift zu KUMMERS fünfzigjährigem Doktorjubiläum begleitete — ein seltenes Denkmal für einen seltenen Freundschaftsbund.

Marburg, den 17. Juli 1910.

K. Hensel.

Kummers Briefe an seine Mutter.

Halle, d. 8. Juli 1828.

Meine theuerste Mutter!

Sie haben uns durch die Ueberschickung des Geldes sehr erfreut, was mir aber das liebste war, das war, daß Sie meinen Entschluß nicht für Schwäche und Wankelmüthigkeit gehalten haben. Wahrlich es ist wohl nichts weniger als dieß, denn wäre ich wankelmüthig und schwach, so würde ich wohl nicht eine sichere Aussicht auf ein ruhiges sorgenloses bequemes Leben mit der unsicheren Aussicht auf eine Versorgung als Lehrer der Mathematik vertauschen, auch würde ich nicht das bei weitem leichtere Studium, was für ein Theologen-Examen gehört mit dem schwereren mathematischen vertauschen. — Glauben sie nicht daß ich von ängstigenden Zweifeln umstrickt sei, nein, es ist nie klarer vor meine Seele getreten daß der Mensch unter jeder Bedingung recht handeln soll ohne auf irgend einen Lohn zu sehen, aber ich halte nicht das äußere Glück für das höchste Gut des Menschen, sondern die Seelenruhe, welche aus dem Bewußtsein hervorgeht recht gehandelt zu haben. Solange ich dieß Bewußtsein habe werde ich mich nie von einem niedrigen Unmuthe hinreißen lassen an einem Gott und einer Unsterblichkeit zu verzweifeln, wenn ich auch auf dem Wege der Vernunft erkannt habe daß der Geist unsterblich ist, und daß ein Gott ist, welcher diesen Geist ins Dasein gerufen hat nicht um ihn zu vernichten, sondern um ihn zu seiner höchsten Vollkommenheit sich erheben zu lassen, in welcher seine Seeligkeit bestehen wird. Jetzt kann ich mit gutem Gewissen nicht fortfahren Theologie zu studiren, darum habe ich es aufgegeben, und habe mir die Mathematik erwählt, weil es die Wissenschaft ist, in welcher der tiefer forschende von andern nicht mißverstanden, oder für gottlos und schlecht gehalten wird sondern in welcher was einer wahres findet von allen anerkannt werden muß und anerkannt wird. Ich glaube sie werden meinen Entschluß billigen, denn wahrlich er stammt aus reinen Beweggründen. . . . Nun leben Sie wohl meine theuerste Mutter; sehnlich hofft auf eine baldige Antwort Ihr Sie innig liebender Sohn

EDUARD KUMMER.

Halle, d. 5. Aug. 1831.

Meine innigst geliebte Mutter!

Vorgestern war ja nun jener wichtige Tag meines Ruhmes. Früh beendigte ich zuerst meine zweite Examenarbeit, ganz und gar, auch schon eingeschrieben, sodaß ich nun die Freude habe schon 2 Arbeiten fertig vor mir zu sehen. Um eilf Uhr sodann ging ich mit KARLEN auf das Waisenhaus, wo in einem sehr großen Saale eine lateinische Rede von einem der Professoren gehalten ward. Es war alles höchst feierlich, als wir ein Weilchen da waren, so kamen zuerst die 3 Pedelle jeder einen feuerrothen langen Mantel um und einen silbernen Scepter in der Hand, hinter diesen kam der Hr. Prorector HEFFTER (der Bruder des Sorauer HEFFTER) mit breitem Hut und Staatsdegen, und sodann die übrigen Professoren. Zuerst wurde mit Trompeten und Pauken ein Lied an den König gesungen, sodann bestieg der Redner den Katheder. Dieser brachte nun in seiner lateinischen Rede alle seit Jahrtausenden abgedroschenen Floskeln alle wieder einmal zum Vorschein, sodaß die lange Weile sehr arg war, besonders da er so undeutlich sprach, daß man auch von diesen nur wenig verstehen konnte. Als er endlich mit seiner Rede nach einer Stunde fertig war so kam er daran die eingegangenen Preisfragen zu beurtheilen, und die noch versiegelten Namen der Autoren zu entsiegeln und vorzulesen. Zuerst fielen die Theologen schmählich weg, denn 4 Theologische Arbeiten welche eingegangen waren fielen alle durch, die medicinische hatte keiner gemacht, die Juristische wurde unter zwei getheilt, welche ebenfalls nicht sonderlich gelobt wurden, Philologische waren zwei gegeben von denen eine nur bearbeitet worden war welche nur aus Gnaden den Preis bekam. Endlich unter allen zuletzt kam die Mathematische und der Redner drückte sich so aus, daß durch irgend ein Ohngefähr grade diejenige Arbeit zuletzt dran komme, welche von allen zuerst verdient hätte erwähnt zu werden. Er las nun das Urtheil über dieselbe vor, ich paßte zwar nicht wenig auf konnte aber von dem was er vorlas fast gar nichts hören, weil ich zu entfernt saß. Nachdem er das Urtheil vorgelesen hatte erbrach er meine Kapsel und las: ERNESTUS EDUARDUS

Kummer, Soranus. Nach diesem glänzenden Schlusse wurde noch ein Lied gesungen und als das ganze aus war so kam zuerst der Professor Scherk auf mich zu, umarmte und küßte mich und gratulirte mir zuerst, dann gratulirten mir auch noch einige andere Professoren und ich ging nach Hause ohne jedoch noch das Urtheil über meine Arbeit zu kennen, denn Karl, Jacobi, und mehrere andere hatten ebensowenig verstanden als ich. Heute aber war ich beim Hr. Prof. Scherk und dieser sagte mir dann das Urtheil hätte ohngefähr so gelautet: Die Arbeit wäre in 2 Theile getheilt gewesen, im ersten Theile hätte der Verfasser das wiedergegeben was die Mathematiker schon früher über dieses Problem geschrieben hätten, und hätte dieß berichtigt, und vervollständigt. Im zweiten Theile sodann hätte der Verfasser seine eigenen Untersuchungen mit so großem Scharfsinn entwickelt, daß ihm schon um dieses einen Theiles willen der Preis hätte zuerkannt werden müssen. Dies ist aber das Urtheil nur ohngefähr, denn es war weit länger. In acht oder vierzehn Tagen erscheint nun ein Programm über das ganze Fest, worin die Sieger und die Urtheile abgedruckt sind, sobald dieß erschienen ist, so kaufe ich mir es natürlich und dann will ich ihnen auch eine genaue wörtliche Uebersetzung meines Urtheils zukommen lassen, denn dieß ist in lateinischer Sprache geschrieben. Ich werde mir nun nächstens meine 50 Thaler auszahlen lassen, und ruhig bis zu meiner Doctor-Promotion aufheben. Sobald ich nun mein Oberlehrerexamen fertig habe, so werde ich ihnen auf der Stelle meine Zeugnisse schicken, daß sie diese dem Hr. Rector geben und der kann dann unterdessen immer an das Ministerium schreiben während ich hier mein zweites Examen mache, denn es wäre sehr gut, wenn ich schon zu Michaelis, oder wenigstens bald nachher Stunden an der Schule geben könnte, daß ich wenn es nöthig ist nachher noch vor dem 23. Jahre mein Militärjahr abdienen kann. Ich mache nun noch frisch über meiner philologischen Arbeit, die philosophische ist wieder 11 Bogen stark. Meinen Preis können Sie dem Hr. Rector mittheilen welcher sich gewiß sehr darüber freuen wird. . . . Nun leben Sie recht wohl meine liebe Mutter, freuen Sie sich über meinen Preis, und noch mehr auf meine Ankunft. Denn es wird ja nun nicht mehr lange dauern, so wird in Ihren Armen liegen

Ihr Sie innig liebender Sohn

Ed. Kummer.

Für das schöne Tuch welches Sie uns geschickt haben und für das andere danke ich von Herzen.

[Halle, den 30. August 1831.]

Meine herzlich geliebte Mutter!

Es freut mich daß ich Ihnen nun auch mein glücklich überstandenes Examen melden kann, welches am Sonnabende war. Schon vor einer Woche hatte mir SCHERK gesagt daß auf den Sonnabend mein Examen sein sollte, aber ich wartete immer auf eine Citation bis Freitag früh, dann warf ich mich in Frack und ging selbst zum Vorsteher der Prüfungscommission, welcher vergessen hatte mich citiren zu lassen. Er sagte mir nun wer mich alles examiniren würde, und so hatte ich grade nur noch Zeit genug an diesem Morgen zu allen herumzulaufen. Hier kam mir das erstemal meine Preisschrift zustatten, denn da die Examinatoren alle bei der Vertheilung derselben zugegen gewesen waren und ihr großes Lob (welches immer noch nicht im Druck erschienen ist) gehört hatten, so schnitten sie mir erst darüber Elogen und meinten sie würden es nun in dem übrigen nicht so streng nehmen. Sonnabend früh um zehn Uhr wurde ich zuerst vom Professor GRUBER im Deutschen examinirt. Dieser lobte zuerst wieder meine philosophische Examenarbeit, daß ich eine große Kenntniß der Sachen und ein selbstständiges Urtheil vereinigt hätte und examinirte dann ein bischen deutsche Literatur. Sodann wurde ich von 11—12 in der Theologie examinirt, wo ich so glänzend bestand, daß ich ohne Spaß höchstwahrscheinlich den Religionsunterricht durch alle Klassen erhalten werde. Ich mußte aus dem neuen Testamente übersetzen und weil dort von der Rechtfertigung durch Christum die Rede war, so mußte ich diese entwickeln. Ferner wurde ich in der Moral und Kirchengeschichte examinirt, wo ich zufällig alles wußte was gefragt wurde, zuletzt entwickelte ich noch da ich den zweiten Psalm aus dem Hebräischen übersetzen mußte (den ich früher einmal hebräisch auswendig gekonnt hatte) eine große Kenntniß des Hebräischen und der Examinator war so zufrieden daß er mir zum Schlusse die Hand drückte. Nachmittag hatte ich nun noch 3 Stunden in der Philologie Geschichte und Mathematik. Ich kam um 2 Uhr zuerst zu dem Philologen Prof. BERNHARDI (denn alle 3, welche mich Nachmittag examinirten hatten sichs bequem gemacht und ließen mich auf ihre Stuben kommen). Dieser kam zuerst mit seinen echtphilologischen Fragen hervor, ich gestand ihm aber gleich immer wo ich nichts wußte, und so legte er mir denn ein griechisches Buch vor was ich übersetzen sollte. Dieß war nun grade meine stärkste Seite, und ich übersetzte es ihm deutsch und lateinisch ganz gut. Er meinte wenn ich ein Wort nicht wisse so sollte ich ihn nur fragen, ich wußte sie aber alle außer einem, von dem er selbst sagte daß es ein sehr seltenes Wort sei. Kurz wir unterhielten uns recht gut, weil ihm wie es mir schien meine Offenherzigkeit gefiel und er sah daß ich mein Wissen grade so gab wie es war. Er meinte er werde mir den Unterricht in den unteren Klassen geben, was ich auch bloß gewünscht hatte. In der Geschichte wurde ich sowohl in der ältsten als neusten und mittlern examinirt, in den beiden ersten ging es recht gut, aber in der mittlern wußte ich leider nicht viel. Beim Professor SCHERK sodann, zu dem ich ging, setzten wir uns etwas aufs Sopha und unterhielten uns, und er meinte er brauche mich eigentlich gar nicht zu examiniren, wenn er nicht darüber Bericht erstatten müßte. Er hatte eben einen Brief von einem seiner ehemaligen Schüler erhalten, welcher ihn bei einer Aufgabe um Rath frug, er gab sie mir, und ich löste denn die Schwierigkeit sogleich; sodann frug er mich noch in der Mechanik und Physik und ich blieb ihm keine einzige Antwort schuldig. Ich dachte Ende gut alles gut! . . . Nun leben Sie indessen recht wohl und freuen Sie Sich immer darauf Ihre beiden Söhne recht bald wiederzusehen.

EDUARD KUMMER.

Breslau, d. 30. Jan. 1834.

Meine herzlich geliebte Mutter!

Sie haben mir zu meinem Geburtstage durch das niedliche Feuerzeugtäschchen eine sehr große Freude gemacht und ich sage Ihnen meinen herzlichsten Dank dafür. Um uns zu Weihnachten und zu unseren Geburtstagen beschenken zu können dürfen Sie sich nicht mehr wünschen in der Lotterie zu gewinnen, denn über eine Arbeit von Ihrer Hand freuen wir uns doch mehr, als über pomphafte Geschenke. Beiläufig gesagt: ich habe nicht in die Lotterie gesetzt, weil ich denke der Sperling in der Hand ist besser als die Taube auf dem Dache oder weil die 9 Thaler die ich mir dadurch gewiß erspare besser sind als alle Gewinne zusammen wenn ich keinen derselben bekomme. Ich soll nun einmal ein armer Schlucker bleiben, denn wenn ich es auch einmal dahin bringe Professor zu werden so habe ich dann auch wahrscheinlich keinen großen Gehalt. Mein Osterprogramm ist nun fertig es ist zwar klein und füllt nur anderthalb Bogen aber es ist inhaltsschwer. Ich habe auch schon einen Brief an einen großen Mathematiker in Königsberg fertig, welchem ich es überschicken werde. Dieß kann mir nun freilich sonst nichts nutzen, aber es ist dafür um so besser, denn man muß ja nicht alles des Nutzens wegen thun. Ich überschicke es ihm bloß darum weil er gerade die größten Entdeckungen in dem Fache gemacht hat worüber mein Programm handelt und er wird sich darüber wundern und freuen zugleich, daß ein Musketier dieselben Gegenstände behandelt als er. ...

Ihr Sie innig liebender Sohn

EDUARD.

Kummers Briefe an Leopold Kronecker.

Mein herzlich geliebter Freund und Schüler!

Wenn Sie Ihren alten Lehrer D. K. wieder sehen wollen, woran ich nicht zweifle, so kommen Sie Sonntag früh zur Ankunft der Personenpost (wie ich glaube um 5 Uhr) in das Berliner Postgebäude, dort werden Sie ihn vom Wagen absteigen sehen, und zwar ist das mein voller Ernst. Alles übrige mündlich.

Ihr Sie herzlich liebender

E. KUMMER.

Liegnitz, d. 16. Januar 1842.

Herzlich geliebter Freund!

Ich komme mit der Beantwortung Ihres mir sehr lieben Briefes über SCHELLINGS Vorlesungen und über andere wichtige Gegenstände etwas spät, ebenso auch mit der Meldung meines neusten Avancements zum ordentlichen Professor an der Universität Breslau, ich verschwende aber die Zeit nicht erst mit Entschuldigungen. — Seit ich bei meiner letzten Anwesenheit in Berlin merkte, es könne mit Breslau Ernst werden, so setzte ich mich zu Hause hin und arbeitete sehr fleißig um so etwas wie eine Dissertation zur Habilitirung zu arbeiten, und ich fing bei etwas mir ganz neuem an, nämlich bei den Cubischen Resten der Primzahlen $6n+1$. Ich habe seitdem einige bemerkenswerthe Dinge darin gefunden, bin aber mit den früheren oder vielmehr bisherigen Leistungen Anderer in diesem Fache ganz unbekannt, und um damit nicht unbekannt zu bleiben, so richte ich an Sie die Bitte einmal DIRICHLET hierüber zu befragen, sich recht genau darüber zu instruiren, und mir es nachher brieflich mitzutheilen. Wenn ich mich selbst an DIRICHLET wendete so würde ich diesen zu einem ihm unangenehmen Geschäfte zum Briefschreiben nothzüchtigen und das will ich ihm nicht anthun; Sie besorgen es mir ja gern und gut. Wenn Sie sich deshalb an DIRICHLET wenden, so grüßen Sie ihn bestens von mir, versichern ihn meiner Hochachtung u. s. w. und sagen Sie ihm ich werde nächstens an ihn schreiben, nächstens heißt so viel als etwa vor Ostern. Ich könnte Ihnen nun etwas von meiner Arbeit mittheilen und will es auch thun immer im Vergleich mit quadratischen Resten. So wie dort Zahlen a und b (Reste und Nichtreste) zu unterscheiden sind, so gibt es hier drei Zahlenreihen α, β und γ, so wie dort $\left(\frac{a}{p}\right) = +1$, $\left(\frac{b}{p}\right) = -1$, wenn p die Primzahl ist, so ist hier nach einer ähnlichen Bezeichnung, die ich anwende $\left(\frac{\alpha}{p}\right) = +1$, $\left(\frac{\beta}{p}\right) = \frac{-1+\sqrt{-3}}{2}$, $\left(\frac{\gamma}{p}\right) = \frac{-1-\sqrt{-3}}{2}$, so wie dort die Reihen $\sum \cos \frac{2a\pi}{p}$, $\sum \cos \frac{2b\pi}{p}$, eine wichtige Rolle spielen, ebenso hier die Reihen $\sum \cos \frac{2\alpha\pi}{p}$, $\sum \cos \frac{2\beta\pi}{p}$, $\sum \cos \frac{2\gamma\pi}{p}$, von denen GAUSS im letzten Abschnitte der disq. arith. gezeigt hat, daß sie die drei reellen Wurzeln einer cubischen Gleichung sind. Aus jeder Reihe $A + A_1 \cos v + A_2 \cos 2v + A_3 \cos 3v + \cdots$ kann man die Summe der Glieder ΣA_a deren Indices quadratische Reste sind und ΣA_b. apart finden, ebenso hier die drei Summen ΣA_α, ΣA_β und ΣA_γ. So wie $4 \cdot \frac{1-x^p}{1-x}$ in die Form $X^2 \pm p Y^2$ gesetzt werden kann, ebenso kann $27 \cdot \frac{1-x^p}{1-x}$ in eine ternäre cubische Form gesetzt werden von der Art wie DIRICHLET dieselben nimmt, $\Sigma\alpha$, $\Sigma\beta$ und $\Sigma\gamma$, auch $\Sigma\alpha^n$, $\Sigma\beta^n$, $\Sigma\gamma^n$ fehlen hier ebenfalls nicht und geben sehr interessante Resultate. Kurz es ist in der Theorie der quadratischen Reste nichts was nicht sein Analogon in der Theorie der cubischen Reste hätte. In specie hätte ich auch gern Auskunft darüber, ob der Satz bekannt ist: wenn $4p = t^2 + 27u^2$, daß t und u beides cubische Reste der Primzahl p sind. Eine zweite Schwierigkeit, wenn Sie mir die Quellen der Geschichte der cubischen Reste werden mitgetheilt haben, wird die sein, wie ich mir dieselben werde verschaffen können, aber ich rechne hierin auch schon im Voraus auf Ihren Beistand. Die Einführung der Zeichen und Begriffe $\left(\frac{n}{p}\right) = 1$, wenn $n^{\frac{p-1}{3}} \equiv 1$ und $\left(\frac{n}{p}\right) = \frac{-1 \pm \sqrt{-3}}{2}$ wenn $n^{\frac{p-1}{3}}$ nicht $\equiv 1$ nehme ich, bis ich darüber werde enttäuscht worden sein, für eine meiner Erfindungen und zwar für eine recht fruchtbare.

Somit beschließe ich den mathematischen Theil meiner Epistel und komme nun zu dem philosophischen, d. i. zu SCHELLING. Nach Allem, was Sie mir schreiben liegt der wesentlichste Unterschied SCHELLINGscher und HEGELscher Philosophie darin, daß SCHELLING das Sein und Denken als ursprünglich verschieden setzt und sie sich erst einander nähern läßt oder vielmehr nur das Denken dem Sein zu nähern sucht, während sie bei HEGEL ursprünglich eine ungetrennte Einheit ausmachen. Mir scheint die Trennung des Denkens und Seins der früheren niederen Bildungsstufe der Philosophie anzugehören und so auch SCHELLING insofern er an dieser Trennung fest hält. Sonst aber ist SCHELLING gewiß die interessanteste und größte philosophische Person gegenwärtiger Zeit und in seinem Colleg zu sitzen und zuzuhören einer der größten philosophischen Genüsse die man in gegenwärtiger Zeit haben kann. Denn wenn auch ein anderer Philosoph besser vorträgt oder auch ebenso wahre und wahrere Dinge sagen kann als er, so ist er doch der einzige welthistorische Philosoph der noch lebt und insofern ist ein Wort von ihm gesprochen gewichtiger als von irgend einem Andern. Wenn ich ehe ich SCHELLING gehört hatte der Meinung war es reiche hin für die Ehre Preußens und für die Belebung philosophischen Interesses, wenn er auf dem Oranienburger Kirchhofe begraben würde, so bin ich jetzt anderer Meinung, denn ein Colleg wie das seinige setzt mehr als irgend eines eine Masse geistiger Kräfte in Thätigkeit und Bewegung und ist schon insofern von unschätzbarem Gewinne. . . .

Leben Sie recht wohl und behalten sie immer lieb Ihren Sie von ganzem Herzen liebenden

E. KUMMER.

Liegnitz, d. 9. Febr. 1842.

Herzlich geliebter Freund!

Meinen besten Dank für ihre gütige prompte Besorgung, die ich Ihnen in meinem vorigen Briefe aufgetragen hatte. JACOBIs Aufsatz commentatio de resid. cub. im 2. Bande von Crelles Journal habe ich seitdem gelesen, bei weitem begieriger aber wäre ich auf JACOBIS Abhandlung in der Berliner Academie wo er die complexen Zahlen anwendet, wenn Sie mir diese doch irgendwie verschaffen könnten. Auf LEBESQUE bin ich nicht so piquirt, denn meine Weise ist den Franzosen noch fremd, da sie mehr der Weise von DIRICHLET wie derselbe die den 2. Grad betreffenden Reste und Geschichten behandelt sich anschließt. Was ich Ihnen in meinem vorigen Briefe geschrieben habe, konnte ich selbst gar nicht alles für neu halten, da vieles sich jedem, der die cubischen Reste behandelt, nothwendig darbieten muß; es war vielmehr nur meine Absicht Ihnen eine Uebersicht über die zu behandelnden Gegenstände zu geben, und nur einiges namentlich bezeichnete hielt ich für neu. Heute nun will ich fortfahren Ihnen was ich seitdem gearbeitet habe mitzutheilen und da Sie so wie auch DIRICHLET auf $\Sigma\alpha$, $\Sigma\beta$, $\Sigma\gamma$ begierig scheinen so will ich zuerst mittheilen was ich über diese weiß: Dieß gründet sich auf die Lösung folgender Aufgabe: Wenn $f(v) = A_1 \cos v + A_2 \cos 2v + A_3 \cos 3v + \cdots$ etc. gegeben ist, die Summe folgender Reihe zu finden:

$$\left(\frac{1}{p}\right)_3 A_1 + \left(\frac{2}{p}\right)_3 A_2 + \left(\frac{3}{p}\right)_3 A_3 + \cdots \quad \text{wo} \quad \left(\frac{\varkappa}{p}\right)_3 = 1,\ \frac{-1+\sqrt{-3}}{2},\ \frac{-1-\sqrt{-3}}{2}$$

jenachdem $\varkappa$ der Reihe der α, β, oder γ resp. angehört. Die Lösung ist sehr leicht es ist nämlich

$$3\sum_{1}^{p-1}{}_{\varkappa} \left(\frac{\varkappa}{p}\right)_3^2 f\left(\frac{2\varkappa\pi}{p}\right) = (z_1 + h^2 z_2 + h z_3)\cdot \sum_{1}^{\infty}{}_{m} \left(\frac{m}{p}\right)_3 A_m$$

wo

$$h = \frac{-1+\sqrt{-3}}{2}, \quad \text{oder } h^3 = 1$$

und

$$z_1 = \sum_{0}^{p-1}{}_{\varkappa} \cos\frac{2\alpha\varkappa^3\pi}{p}, \quad z_2 = \sum_{0}^{p-1}{}_{\varkappa} \cos\frac{2\beta\varkappa^3\pi}{p}, \quad z_3 = \sum_{0}^{p-1}{}_{\varkappa} \cos\frac{2\gamma\varkappa^3\pi}{p}.$$

Dieser Satz ist die Quelle folgender Resultate, für welche ich noch einige Bezeichnungen mittheilen muß: Wenn man unter $\Sigma\alpha$, $\Sigma\beta$, $\Sigma\gamma$ die Summen aller α, β, γ zwischen 0 und p versteht so sind diese Summen ohne Interesse, alle einander gleich, nämlich $= \frac{p(p-1)}{6}$, es sind also darunter nur die Summen zwischen 0 und $\frac{p}{2}$ zu verstehen; die Summen der Quadrate aber $\Sigma\alpha^2$, $\Sigma\beta^2$, $\Sigma\gamma^2$ sind verschieden, und können in den Grenzen 0 und p genommen werden. Unter dieser Voraussetzung lassen sich $\Sigma\alpha^2$, $\Sigma\beta^2$, $\Sigma\gamma^2$ durch $\Sigma\alpha$, $\Sigma\beta$, $\Sigma\gamma$ ausdrücken, und zwar sind die Ausdrücke ganz verschieden jenachdem die Zahl 2 der Reihe der α, β oder γ angehört. Es verhält sich ähnlich wie bei den quadratischen Resten der Zahlen $4n+1$. Die Summation der Reihe $\sum_0 \left(\frac{2m+1}{p}\right)_3 \frac{1}{(2m+1)^2}$ nach dem obigen Satze lehrt nun wie die dreimal drei Größen z_1, z_2, z_3; $\Sigma\alpha$, $\Sigma\beta$, $\Sigma\gamma$, und $\sum \frac{1}{\alpha_1^2}$, $\sum \frac{1}{\beta_1^2}$, $\sum \frac{1}{\gamma_1^2}$ in welchen letzteren α_1 alle ungeraden cubischen Reste von 1 bis ins unendliche β_1 alle etc. etc. bezeichnet, so zusammen hängen, daß aus z_1, z_2, z_3 und $\Sigma\alpha$, $\Sigma\beta$, $\Sigma\gamma$, die übrigen drei nämlich $\sum \frac{1}{\alpha_1^2}$, $\sum \frac{1}{\beta_1^2}$, $\sum \frac{1}{\gamma_1^2}$ bestimmt werden können und ebenso $\Sigma\alpha$, $\Sigma\beta$, $\Sigma\gamma$ aus den übrigen sechs, und auch z_1, z_2, z_3 aus den übrigen sechs. Dieß ist ein Beispiel von der Anwendung der obigen Methode und zwar eins der interessantesten. Mit Hilfe der hierdurch erlangten Formeln kann man auch gewissermaßen die Aufgabe lösen: welche von den drei Wurzeln der cubischen Gleichung dem z_1, welche dem z_2, und welche dem z_3 zugehört oder gleich ist, ich sage gewissermaßen, weil man dazu die berechneten Werthe der $\Sigma\alpha$, $\Sigma\beta$ und $\Sigma\gamma$ braucht, welches nicht gut ist, da man lieber aus der Zahl p oder aus $4p = t^2 + 27u^2$ schon entscheiden möchte welche Wurzel jedem der z_1, z_2 und z_3 angehört. Ich habe die Werthe der $\Sigma\alpha$, $\Sigma\beta$ und $\Sigma\gamma$ für die Primzahlen von der Form $6n+1$ unter 200 alle berechnet, und die Werthe der z_1, z_2, z_3 für alle diese bestimmt, um durch Induction zu finden welches Gesetz obwaltet, habe aber aus den Factoren von $p-1$ oder aus den Zahlen t und u keines finden können. Die drei Wurzeln der cubischen Gleichung liegen eine in den Grenzen $-2\sqrt{p}$ und $-\sqrt{p}$, die andere in den Grenzen $-\sqrt{p}$ und $+\sqrt{p}$, die dritte in den Grenzen $+\sqrt{p}$ und $+2\sqrt{p}$, und es liegt $z_1 = \sum_0^{p-1}{}_{x} \cos\frac{2x^3\pi}{p}$ in den Grenzen $-2\sqrt{p}$ bis $-\sqrt{p}$ für $p = 97$, 139, 151, 199, aber z_1 in den Grenzen $-\sqrt{p}$ bis $+\sqrt{p}$ für $p = 13$, 19, 37, 61, 109, 157, 193 und z_1 in den Grenzen $+\sqrt{p}$ bis $+2\sqrt{p}$ für $p = 31$, 43, 67, 73, 79, 103, 127, 163, 181. Ich schreibe Ihnen dieß, damit Sie sich an einer Induction versuchen können, wenn Sie wollen. Durch die lineäre Form des p wie bei $\sum \cos\frac{2x^2\pi}{p}$, oder bei den quadratischen Resten, ist hier nichts gethan, auch gelten die Resultate hier nicht so wie dort auch für zusammengesetzte Zahlen, sondern stellen sich da ganz anders; wenn nämlich P eine zusammengesetzte Zahl bedeutet, so ist $\sum_0^{P-1} \cos \frac{2x^3\pi}{P}$ wenn $P = 2^r \cdot 3^s \cdot a^\alpha \cdot b^\beta \cdot c^\gamma \cdots$ in folgenden Fällen gleich Null: 1) sobald einer der Primfactoren a, b, c die Form $6n-1$ hat, 2) wenn der Exponent der 2 nämlich r von der Form $3n+1$ ist, und 3) wenn s von der Form $3n+1$ ist. Wenn keiner der Exponenten r, s, α, β, $\gamma \ldots$ die Form $3n+1$ hat, so ist $\sum_0^{P-1} \cos \frac{2x^3\pi}{P} =$ einer ganzen Zahl, welche nur die Factoren 2, 3, a, b, c, ... enthält, wenn aber irgend welche der Factoren a, b, c einen Exponenten von der Form $3n+1$ haben, z. B. wenn $\alpha = 3n+1$ und $\gamma = 3m+1$, so ist

$$\sum_0^{P-1} \cos \frac{2x^3\pi}{P} = M \sum_0^{a-1} \cos \frac{2x^3\pi}{a} \cdot \sum_0^{c-1} \cos \frac{2x^3\pi}{c},$$

wo M eine ganze Zahl ist. Von den drei unendlichen Reihen

$$\sum \frac{1}{\alpha_1^2}, \quad \sum \frac{1}{\beta_1^2}, \quad \sum \frac{1}{\gamma_1^2}$$

ist zu bemerken, daß sie ebenfalls die drei Wurzeln einer cubischen Gleichung mit ganzzahligen Coefficienten sind und daß sie folgenden endlichen Reihen gleich sind

$$c \cdot \sum_0^{p-1} \frac{1}{x\left(\cos \frac{2\alpha x^3\pi}{p}\right)^2}, \quad c \cdot \sum_0^{p-1} \frac{1}{x\left(\cos \frac{2\beta x^3\pi}{p}\right)^2}, \quad c \cdot \sum_0^{p-1} \frac{1}{x\left(\cos \frac{2\gamma x^3\pi}{p}\right)^2}.$$

Einem Reciprocitätsgesetze für cubische Reste habe ich ebenfalls gleich anfangs nachgestrebt, als ich die cubischen Reste zum Gegenstande meiner Untersuchungen machte, jetzt da ich weiß daß Jacobi dieses in der einfachsten Gestalt für complexe Zahlen aufgestellt hat, habe ich vorläufig mit complexen Zahlen von der Form $\frac{a+b\sqrt{-3}}{2}$ mich befaßt, denn diese Art gehört den cubischen Resten an oder die Art $a+bh+ch^2$, wo $h^3=1$ ist. Dieß entfernte mich aber von meinen Reihen $z_1 = \sum \cos \frac{2x^3\pi}{p}$ etc. die ich durch viele darauf verwandte Mühe lieb gewonnen habe. Erst in der ganz neuen Zeit habe ich die complexen Zahlen $\frac{a+b\sqrt{-3}}{2}$ wieder mit mehr Liebe angesehen, seit sie mir ein Aequivalent für meine verloren gegangenen z_1, z_2, z_3 zu bieten versprechen, nämlich in den Reihen die aus elliptischen Functionen auf ähnliche Weise gebildet sind, wie jene aus Kreisfunctionen. Die Elliptischen Functionen haben nämlich die reale und imaginäre Periode und so wie $\cos\frac{2\varkappa\pi}{p}$ dasselbe bleibt wenn man für $\varkappa$ eine andere Zahl $\varkappa'$ nimmt, so daß $\varkappa \equiv \varkappa' \mod p$, so ist dieß ähnlich für complexe Zahlen bei $\cos\operatorname{am}\frac{Km+K'ni}{p}$. So wie aber für complexe Zahlen von der Form $a+b\sqrt{-1}$ die Lemniscate oder das Integral $\int\frac{dx}{\sqrt{1-x^4}}$ aus der Theorie der Elliptischen Functionen zugeordnet ist, so habe ich gefunden daß für die complexen Zahlen von der Form $\frac{a+b\sqrt{-3}}{2}$ das Integral $\int\frac{dx}{\sqrt{1-x^6}}$ dieselbe Rolle spielt, dieß ist nämlich dasjenige elliptische Integral der ersten Art dessen Modul $=\sqrt{\frac{2-\sqrt{3}}{4}}=\sin 15^0$. Es sollte mich sehr wundern, wenn dieß JACOBI entgangen sein sollte, überhaupt wenn so einer einen Gegenstand schon behandelt hat, und unser einer macht sich daran, so bleibt einem wenig mehr übrig als eine Ausführung derjenigen Ideen, welche von ihm schon ausgesprochen, oder wenigstens schon gehegt worden sind. Ich studiere jetzt, in Rücksicht auf diese complexen Zahlen, von neuem elliptische Functionen, namentlich die erste Abhandlung von ABEL in Crelles Journal und ich glaube es wird sich etwas machen lassen, gemacht aber ist bis jetzt noch nichts, nämlich von mir in diesem Fache. Ich habe Sie nun wieder einmal ganz von dem Stande meiner Angelegenheiten au fait gesetzt, und wenn DIRICHLET dafür sich auch interessiert, so theilen sie ihm davon mit was Sie wollen, ich wollte nur es wäre etwas besseres, was seiner Aufmerksamkeit mehr werth wäre. Wenn Sie einmal zufällig in seinem Ciguarren-Kasten seinen rothen Adler sehen (denn dort soll sein gewöhnlicher Aufenthaltsort sein, wie mir ARNOLD geschrieben hat) so gratuliren Sie ihm in meinem Namen zu dieser Ritterschaft. Wenn DIRICHLET Ihnen im nächsten Sommer complexe Zahlen-Theorie liest, so nehmen Sie dies dankbar an, ich selbst würde gern mit zuhören, wenn ich könnte, denn Sie können glauben es ist etwas mühsam sich dergleichen alles allein zu machen, wenn man wie ich nur etwa die Haupt-Ideen von anderen erfahren hat. . . .

Leben Sie recht wohl und schreiben Sie mir recht bald wieder. Daß Sie DIRICHLET und meine Vettern in Berlin von mir grüßen, versteht sich von selbst. Ihr Sie herzlich liebender

E. KUMMER.

Breslau, den 10. Apr. 1844.

Herzlich geliebter Freund!

Weil ich Sie jetzt nicht hier habe um über unsere complexen Zahlen mündlich conferiren zu können, so theile ich Ihnen einiges schriftlich mit, welches wie ich glaube für Sie um so mehr Interesse haben wird weil es speciell das betrifft worüber Sie hier zuletzt arbeiteten. Ich gehe aber erst von einer anderen Sache aus, nämlich von dem Satze:

Wenn $f(\alpha)$ die Norm p hat (p Primzahl $\lambda n+1$), so ist jede complexe Zahl einer reellen congruent für den Modul $f(\alpha)$.

Hierbei ist nur zu zeigen, daß $\alpha \equiv \xi \bmod f(\alpha)$, wo ξ reell. Dieß scheint sich von selbst zu verstehen, weil $\xi - \alpha$ wenn

$$1+\xi+\xi^2+\cdots+\xi^{\lambda-1} \equiv 0 \mod p$$

stets einen Factor mit p gemein hat, wie in dem Beweise, daß jede Primzahl p sich in $\lambda-1$ complexe Factoren zerlegen läßt gezeigt wird. Es versteht sich aber bei näherer Betrachtung nicht von selbst, sondern bedarf folgenden Beweises. Ich fange den Beweis mit einem Hülfssatze an, der wenn Sie ihn nicht schon selbst allgemein bewiesen haben Ihnen erwünscht sein wird, nämlich wenn

$$f\alpha^2 \cdot f\alpha^3 \cdot f\alpha^4 \cdots f\alpha^{\lambda-1} = A + A_1\alpha + A_2\alpha^2 + \cdots + A_{\lambda-1}\alpha^{\lambda-1}$$

so ist allgemein für jeden Werth des n

$$(A_n - A_{n+1})^2 \equiv (A_{n-1}-A_n)(A_{n+1}-A_{n+2}) \mod p.$$

Sei

$$A + A_1\alpha + A_2\alpha^2 + \cdots + A_{\lambda-1}\alpha^{\lambda-1} = f\alpha^2 f\alpha^3 f\alpha^4 \cdots f\alpha^{\lambda-1} = \Psi(\alpha)$$

so folgt nach sehr bekannten Methoden

$$\lambda A_n - (A + A_1 + A_2 + \cdots + A_{\lambda-1})$$
$$= \alpha^n \Psi\alpha + \alpha^{2n}\Psi\alpha^2 + \alpha^{3n}\Psi\alpha^3 + \cdots + \alpha^{(\lambda-1)n}\Psi\alpha^{\lambda-1}$$

also wenn $n+1$ statt n gesetzt und subtrahirt wird

$$\lambda(A_n - A_{n+1}) = \alpha^n(1-\alpha)\Psi\alpha$$
$$+ \alpha^{2n}(1-\alpha^2)\Psi\alpha^2 + \alpha^{3n}(1-\alpha^3)\Psi\alpha^3 + \cdots + \alpha^{(\lambda-1)n}(1-\alpha^{\lambda-1})\Psi\alpha^{\lambda-1}.$$

Setzt man hierin wieder erstens $n+1$ statt n, zweitens $n-1$ statt n und bildet den Ausdruck

$$\lambda^2(A_n - A_{n+1})^2 - \lambda^2(A_{n-1}-A_n)(A_{n+1}-A_{n+2}),$$

so übersieht man in der That sehr leicht, daß diejenigen Glieder welche das Quadrat eines Ψ enthalten, nämlich $(\Psi\alpha)^2$, $(\Psi\alpha^2)^2$, $(\Psi\alpha^3)^2$ etc. alle verschwinden, und daß nur diejenigen bleiben, welche Produkte zweier verschiedenen Ψ enthalten. Jedes Produkt zweier verschiedener Ψ enthält aber die Factoren $f\alpha f\alpha^2 f\alpha^3 \cdots f\alpha^{\lambda-1}$ also den Factor p. Dieser Ausdruck ist daher durch p theilbar, also die obige Congruenz erwiesen.

Von hier manövrire ich weiter zu meinem Ziele hin. Ich setze:

$$\frac{A_n - A_{n+1}}{A_{n+1} - A_{n+2}} \equiv \xi \mod p,$$

so ist nach der obigen Congruenz ξ von n unabhängig, man hat also die Congruenzen

$$A\xi - A_{\lambda-1} \equiv A_1\xi - A \equiv A_2\xi - A_1 \equiv A_3\xi - A_2 \equiv \cdots$$
$$\equiv A_{\lambda-1}\xi - A_{\lambda-2}, \mod p,$$

welche ich um des Folgenden willen auch so darstelle

$$\text{(A)}\quad \begin{array}{ll} A_{\lambda-1} - A \equiv (A - A_1)\xi, & A_{\lambda-2} - A_{\lambda-1} \equiv (A - A_1)\xi^2, \\ A_{\lambda-3} - A_{\lambda-2} \equiv (A - A_1)\xi^3, & \ldots\ A_1 - A_2 \equiv (A - A_1)\xi^{\lambda-1}; \end{array}$$

Ich schreibe diese Congruenzen auch als Gleichungen

$$\begin{aligned} A\xi - A_{\lambda-1} &= mp + \mu \\ A_1\xi - A &= m_1p + \mu \\ A_2\xi - A_1 &= m_2p + \mu \\ &\vdots \\ A_{\lambda-1}\xi - A_{\lambda-2} &= m_{\lambda-1}p + \mu. \end{aligned}$$

Diese Gleichungen multiplicire ich resp. mit $1, \alpha, \alpha^2, \ldots \alpha^{\lambda-1}$ und addire, wodurch ich erhalte:

$$(\xi - \alpha)(A + A_1\alpha + A_2\alpha^2 + \cdots + A_{\lambda-1}\alpha^{\lambda-1})$$
$$= p(m + m_1\alpha + m_2\alpha^2 + \cdots + m_{\lambda-1}\alpha^{\lambda-1})$$

und wenn für p sein Werth

$$p = f(\alpha) \cdot (A + A_1\alpha + A_2\alpha^2 + \cdots + A_{\lambda-1}\alpha^{\lambda-1})$$

gesetzt wird und aufgehoben wird so hat man endlich

$$\xi - \alpha = f(\alpha) \cdot (m + m_1\alpha + m_2\alpha^2 + \cdots + m_{\lambda-1}\alpha^{\lambda-1})$$

oder

$$\xi - \alpha \equiv 0 \mod f(\alpha).$$

Daß die Zahl ξ eine Wurzel der Congruenz $1 + \xi + \xi^2 + \cdots + \xi^{\lambda-1} \equiv 0 \mod p$ ist versteht sich von selbst.

Ich rücke jetzt Ihrem Satze daß zwei verschiedene Zerlegungen des p sich nur durch Einheiten unterscheiden, wieder einen Schritt näher, aber noch zuvor mache ich eine andere wichtige Angelegenheit ab.

Sei

$$f(\alpha) = a + a_1\alpha + a_2\alpha^2 + \cdots + a_{\lambda-1}\alpha^{\lambda-1};$$
$$f\alpha^2 f\alpha^3 f\alpha^4 \cdots f\alpha^{\lambda-1} = A + A_1\alpha + A_2\alpha^2 + \cdots + A_{\lambda-1}\alpha^{\lambda-1}$$

so ist

$$(a + a_1\alpha + a_2\alpha^2 + \cdots + a_{\lambda-1}\alpha^{\lambda-1})(A + A_1\alpha + A_2\alpha^2 + \cdots + A_{\lambda-1}\alpha^{\lambda-1}) = p,$$

entwickelt man dieses Product und zieht das Glied, welches α enthält, von dem ab, welches kein α enthält, so erhält man offenbar:

$$(A - A_1)a + (A_{\lambda-1} - A)a_2 + (A_{\lambda-2} - A_{\lambda-1})a_2 + \cdots + (A_1 - A_2)a_{\lambda-1} = p;$$

macht man nun von den bei (A) angegebenen Congruenzen Gebrauch so erhält man

$$(A - A_1)(a + a_1\xi + a_2\xi^2 + \cdots + a_{\lambda-1}\xi^{\lambda-1}) \equiv 0 \mod p,$$

also auch

$$a + a_1\xi + a_2\xi^2 + \cdots + a_{\lambda-1}\xi^{\lambda-1} \equiv 0 \mod p$$

oder wenn $f(\alpha)$ ein Divisor von $\xi - \alpha$, so ist $f(\xi)$ durch p theilbar.

Jetzt endlich betrachte ich zwei verschiedene Zerlegungen der Primzahl p nämlich

$$p = f\alpha f\alpha^2 f\alpha^3 \cdots f\alpha^{\lambda-1} \quad \text{und} \quad p = \varphi(\alpha)\varphi(\alpha^2)\varphi(\alpha^3)\cdots\varphi(\alpha^{\lambda-1}).$$

Nun ist oben erwiesen daß jeder Factor f ein Divisor von $\xi - \alpha$ oder $\xi^2 - \alpha$ oder $\xi^3 - \alpha \ldots \xi^{\lambda-1} - \alpha$ sein muß, wo $1 + \xi + \xi^2 + \cdots + \xi^{\lambda-1} \equiv 0$, mod p, ebenso muß jeder Factor φ ein Divisor einer dieser Größen sein, und zwar so, daß für einen bestimmten Werth des ξ, die Zahl $\xi - \alpha$ sowohl einen Divisor unter den f als auch einen Divisor unter den φ haben muß. Als diese Divisoren welche beide in $\xi - \alpha$ theilbar sind nehme ich $f(\alpha)$ und $\varphi(\alpha)$, (welches im Grunde dasselbe ist als wenn ich $f(\alpha^\mu)$ und $\varphi(\alpha^\nu)$ genommen hätte). Ich bilde nun das Product

$$\varphi(\alpha) \cdot f\alpha^2 f\alpha^3 f\alpha^4 \ldots f\alpha^{\lambda-1}$$
$$= (b + b_1\alpha + b_2\alpha^2 + \cdots + b_{\lambda-1}\alpha^{\lambda-1})(A + A_1\alpha + A_2\alpha^2 + \cdots + A_{\lambda-1}\alpha^{\lambda-1})$$

dessen Entwicklung sei

$$C + C_1\alpha + C_2\alpha^2 + \cdots + C_{\lambda-1}\alpha^{\lambda-1}.$$

Durch Ausführung der Multiplication erhält man hiernach folgende Werthe der $C, C_1, C_2, \ldots C_{\lambda-1}$:

$$\begin{aligned}
Ab + A_{\lambda-1}b_1 + A_{\lambda-2}b_2 + \cdots + A_1 b_{\lambda-1} &= C\\
A_1 b + \quad Ab_1 + A_{\lambda-1}b_2 + \cdots + A_2 b_{\lambda-1} &= C_1\\
A_2 b + \quad A_1 b_1 + \quad Ab_2 + \cdots + A_3 b_{\lambda-1} &= C_2\\
\vdots\quad&\\
A_{\lambda-1} b + A_{\lambda-2}b_1 + A_{\lambda-3}b_2 + \cdots + \quad Ab_{\lambda-1} &= C_{\lambda-1}
\end{aligned}$$

und hieraus durch Subtraction:

$$\begin{aligned}
(A-A_1)b + (A_{\lambda-1}-A)b_1 + (A_{\lambda-2}-A_{\lambda-1})b_2 + \cdots + (A_1-A_2)b_{\lambda-1} &= C-C_1\\
(A_1-A_2)b + (A-A_1)b_1 + (A_{\lambda-1}-A)b_2 + \cdots + (A_2-A_3)b_{\lambda-1} &= C_1-C_2\\
(A_2-A_3)b + (A_1-A_2)b_1 + (A-A_1)b_2 + \cdots + (A_3-A_4)b_{\lambda-1} &= C_2-C_3\\
\vdots\quad&\\
(A_{\lambda-2}-A_{\lambda-1})b + (A_{\lambda-3}-A_{\lambda-2})b_1 + (A_{\lambda-4}-A_{\lambda-3})b_2 + \cdots + (A_{\lambda-1}-A)b_{\lambda-1} &= C_{\lambda-2}-C_{\lambda-1}.
\end{aligned}$$

Durch Anwendung der Congruenzen (A) erhält man hieraus

$$\begin{aligned}
(A-A_1)(b+b_1\xi+b_2\xi^2+\cdots+b_{\lambda-1}\xi^{\lambda-1}) &\equiv C - C_1\\
\xi^{\lambda-1}(A-A_1)(b+b_1\xi+b_2\xi^2+\cdots+b_{\lambda-1}\xi^{\lambda-1}) &\equiv C_1 - C_2\\
\xi^{\lambda-2}(A-A_1)(b+b_1\xi+b_2\xi^2+\cdots+b_{\lambda-1}\xi^{\lambda-1}) &\equiv C_2 - C_3\\
\vdots\quad&\\
\xi^{2}(A-A_1)(b+b_1\xi+b_2\xi^2+\cdots+b_{\lambda-1}\xi^{\lambda-1}) &\equiv C_{\lambda-2} - C_{\lambda-1}.
\end{aligned}$$

Nun ist aber nach der Voraussetzung $\varphi(\alpha)$ ein Divisor von $\xi-\alpha$, also $\varphi(\xi)\equiv 0$, d. i.

$$b + b_1\xi + b_2\xi^2 + \cdots + b_{\lambda-1}\xi^{\lambda-1} \equiv 0 \mod p,$$

also

$$C \equiv C_1 \equiv C_2 \equiv C_3 \equiv \cdots \equiv C_{\lambda-1} \mod p$$

Das Product $\varphi(\alpha)\cdot f\alpha^2 f\alpha^3 f\alpha^4 \ldots f\alpha^{\lambda-1}$ ist also ein vielfaches von p. Darum darf man setzen

$$\varphi(\alpha) f\alpha^2 f\alpha^3 f\alpha^4 \cdots f\alpha^{\lambda-1} = p\cdot\Psi(\alpha)$$

und wenn mit $f(\alpha)$ multiplicirt und p weggehoben wird hat man

$$\varphi(\alpha) = f(\alpha)\cdot\Psi(\alpha).$$

Bildet man die Norm und hebt alsdann auf beiden Seiten p, so hat man

$$1 = \Psi\alpha\,\Psi\alpha^2\,\Psi\alpha^3\cdots\,\Psi\alpha^{\lambda-1}$$

also $\Psi(\alpha)$ ist eine Einheit und es unterscheiden sich in der That die Factoren der einen Zerlegung des p von denen einer anderen Zerlegung nur dadurch, daß Einheiten als Factoren hinzutreten können.

Wenn ich nun weder bescheiden noch unbescheiden, sondern so wie es mir vorkommt selbst mein Urtheil über diese Methoden sagen soll, so finde ich sie trotz dem daß es nur Beweise von Dingen sind, die sich gewissermaßen von selbst verstehen, recht erbaulich und elegant und wenn sich noch manches andere ebenso leicht gestaltet, so gedenke ich diese Dinge im nächsten Semester meinen Zuhörern der Kreistheilung mitzutheilen.

Ich erwarte zur Belohnung für diesen langen Brief, daß Sie mir auch bald einmal von Liegnitz aus schreiben und wenn Sie durch Umstände wie sich allenfalls erwarten läßt gehindert sind hübsche Dinge über unsere complexen Zahlen zu suchen und zu finden so schreiben Sie mir wenigstens den unmathematischen Theil eines Briefes als da ist: Nachrichten über die socialen Zustände von Liegnitz, über bestimmte Personen, über Gymnasium und Academie, über KÖHLER und MÜLLER, über MATTHÄUS abgeschriebenes Programm, über Ihr und der Ihrigen Befinden, wobei ich zugleich bemerke, daß mein ganzer Hausstand in sehr erwünschtem Wohlsein ist. Einen Rückfall meines Schnupfens habe ich mir neulich mit einem guten warmen Punsche sehr gründlich curirt, so daß ich von Stund an wieder ganz wohl war, und ich rathe Ihnen in ähnlichen Fällen dasselbe Mittel anzuwenden.... Nun leben Sie recht wohl, grüßen Sie Ihre Eltern und alle anderen Bekannten und Freunde. Ihr Sie herzlich liebender

E. KUMMER.

Breslau, d. 2. Oct. 44.

Geliebter Freund!

Da Sie wohl auf Ihren Reisen die Arbeiten über complexe Zahlen und resp. Einheiten etwas werden vernachläßigt haben und mir darum vielleicht hierüber noch nichts Neues zu schreiben haben, so fange ich selbst die Fortsetzung des neuen Briefwechsels an.

Seit unserer Trennung habe ich einiges erarbeitet, und zwar zunächst ein kleines Aufsätzchen für CRELLE: „Ueber die Divisoren gewisser Formen der Zahlen, welche aus der Theorie der Kreistheilung entstehen" . . . Ist nämlich $p-1=e\cdot f$, und man theilt die Wurzeln der Gleichung $x^p=1$ in e Perioden von je f Gliedern, welche ich mit $\eta, \eta_1, \eta_2, \ldots \eta_{e-1}$ bezeichne, so ist bekanntlich

$$(y-\eta)(y-\eta_1)(y-\eta_2)\cdots(y-\eta_{e-1})=y^e+c_1y^{e-1}+c_2y^{e-2}+\cdots+c_e=\Phi(y)$$

eine ganze rationale Function mit ganzzahligen Coefficienten. Die

Divisoren dieser Form $\Phi(y)$ untersuche ich zunächst und finde folgende Sätze: I. Jede Primzahl q, welche ein e^{ter} Potenzrest von p ist, ist Divisor von $\Phi(y)$ oder genauer die Congruenz $\Phi(y) \equiv 0$ Mod q hat wenn q ein e^{ter} Potenzrest von p und Primzahl ist immer e reale Wurzeln. II. p selbst ist auch Divisor von $\Phi(y)$. III. Wenn der Grad e der Form $\Phi(y)$ eine Primzahl ist, so hat $\Phi(y)$ keine anderen Divisoren als solche welche e^{te} Potenzreste von p sind und den Divisor p selbst. IV. Wenn e eine zusammengesetzte Zahl ist und die von Eins verschiedenen Divisoren $d, d', d'' \ldots$ hat, so kann $\Phi(y)$ außer den Primfactoren welche e^{te} Potenzreste des p sind und p selbst auch eine endliche Anzahl anderer Primfactoren enthalten, welche nur d^{te} oder d'^{te} etc. Potenzreste von p sind. Dieselben Sätze gelten auch von der Form

$$F(\eta)\, F(\eta_1)\, F(\eta_2) \cdots F(\eta_{e-1}) = \Psi$$

in welcher $F(\eta) = z\eta + z_1\eta_1 + z_2\eta_2 + \cdots + z_{e-1}\eta_{e-1}$, auch kommt nebenbei noch manches in dem Aufsätzchen vor.

Ferner habe ich in diesen Tagen einen wahrhaftigen strengen Beweis dafür gesucht und gefunden, daß jede Primzahl $p = 5m + 1$ in die Form $p = f(\alpha) f(\alpha^2) f(\alpha^3) f(\alpha^4)$ gesetzt werden kann. Nach § 9 meiner Dissertation hängt alles nur von dem Beweise des Satzes ab, daß wenn c, c_1, c_2, c_3, c_4 beliebige Größen sind die complexe Zahl $c + c_1\alpha + c_2\alpha^2 + c_3\alpha^3 + c_4\alpha^4$, dadurch daß man diese Größen um ganze Zahlen vermehrt oder vermindert immer dahin gebracht werden kann daß ihre Norm kleiner als Eins wird. Die hierzu hinreichenden ganzen Zahlen werden zunächst so gewählt, daß $c - \varkappa, c_1 - \varkappa_1, c_2 - \varkappa_2, c_3 - \varkappa_3, c_4 - \varkappa_4$ alle in einem Intervalle liegen welches kleiner als Eins ist oder daß der Unterschied der größten und kleinsten kleiner als Eins wird. Dieß kann auf 5 wesentlich verschiedene Arten geschehen, wie leicht einzusehen ist; von diesen 5 Arten wähle ich diejenige für welche das ganze Intervall das möglichst kleinste wird. Oder wenn man die um ganze Zahlen veränderten Coefficienten der Größe nach ordnet, und den Unterschied des größten vom nächst kleineren δ_1 nennt, den Unterschied dieses vom nächsten δ_2, etc., so ist das ganze Intervall $\delta_1 + \delta_2 + \delta_3 + \delta_4$, welches < 1 ist, (die δ sind alle positiv). Macht man den größten Coefficienten durch Subtraction von 1 zum kleinsten so bleibt das Intervall in welchem alle liegen kleiner als 1, etc. Setzt man nun $1 = \delta + \delta_1 + \delta_2 + \delta_3 + \delta_4$, so kann man immer die Sache so einrichten, daß δ größer wird als alle δ_1, δ_2, δ_3, δ_4. Alles dieß war mir und auch Ihnen schon bekannt, nun kommt aber erst die Anwendung.

Es ist
$$f(\alpha)f(\alpha^4) = -P(\alpha+\alpha^4) - Q(\alpha^2+\alpha^3),$$
wenn
$$f(\alpha) = a + a_1\alpha + a_2\alpha^2 + a_3\alpha^3 + a_4\alpha^4$$
wo
$$P = a^2 + a_1^2 + a_2^2 + a_3^2 + a_4^2 - aa_1 - a_1a_2 - a_2a_3 - a_3a_4 - a_4a$$
$$Q = a^2 + a_1^2 + a_2^2 + a_3^2 + a_4^2 - aa_2 - a_1a_3 - a_2a_4 - a_3a_5 - a_4a_1$$
oder
$$2P = (a-a_1)^2 + (a_1-a_2)^2 + (a_2-a_3)^2 + (a_3-a_4)^2 + (a_4-a)^2$$
$$2Q = (a-a_2)^2 + (a_1-a_3)^2 + (a_2-a_4)^2 + (a_3-a)^2 + (a_4-a_1)^2$$
ferner wird $Nf(\alpha) = -P^2 - Q^2 + 3PQ = PQ - (P-Q)^2$. Hieraus folgt zunächst, daß wenn P und Q beide kleiner als 1 sind, auch $Nf(\alpha)$ kleiner als 1 sein muß. Löst man ferner die Gleichung $-P^2 - Q^2 + 3PQ - Nf(\alpha) = 0$ auf, so wird $P = \frac{3Q \pm \sqrt{5Q^2 - 4Nf(\alpha)}}{2}$ woraus folgt, daß $Nf(\alpha) < \frac{5}{4}Q^2$ ist, also wenn $Q^2 < \frac{4}{5}$, so ist $Nf(\alpha) < 1$, ebenso wenn $P^2 < \frac{4}{5}$ ist $Nf(\alpha) < 1$.

Nun kann man ohne daß $Nf(\alpha)$ sich ändert den ersten Coefficienten zum größten machen und den zweiten zum nächstgrößten (nämlich durch Multiplikation mit einer passenden Potenz von α und durch Substitution von $\alpha^\varkappa$ statt α), es sei also a der größte a_1 der nächstgrößte Coefficient. Wegen der Größe der übrigen sind nun folgende 6 Fälle zu unterscheiden:

1. $a > a_1 > a_2 > a_3 > a_4$; $a - a_1 = \delta_1$, $a_1 - a_2 = \delta_2$, $a_2 - a_3 = \delta_3$, $a_3 - a_4 = \delta_4$;
$2P = \delta_1^2 + \delta_2^2 + \delta_3^2 + \delta_4^2 + (\delta_1+\delta_2+\delta_3+\delta_4)^2$;
$2Q = (\delta_1+\delta_2)^2 + (\delta_2+\delta_3)^2 + (\delta_3+\delta_4)^2 + (\delta_1+\delta_2+\delta_3)^2 + (\delta_2+\delta_3+\delta_4)^2$.

2. $a > a_1 > a_2 > a_4 > a_3$, $a - a_1 = \delta_1$, $a_1 - a_2 = \delta_2$, $a_2 - a_4 = \delta_3$, $a_4 - a_3 = \delta_4$;
$2P = \delta_1^2 + \delta_2^2 + (\delta_3+\delta_4)^2 + \delta_4^2 + (\delta_1+\delta_2+\delta_3)^2$;
$2Q = (\delta_1+\delta_2)^2 + (\delta_2+\delta_3+\delta_4)^2 + \delta_3^2 + (\delta_1+\delta_2+\delta_3+\delta_4)^2 + (\delta_2+\delta_3)^2$.

3. $a > a_1 > a_3 > a_2 > a_4$; $a - a_1 = \delta_1$, $a_1 - a_3 = \delta_2$, $a_3 - a_2 = \delta_3$, $a_2 - a_4 = \delta_4$;
$2P = \delta_1^2 + (\delta_2+\delta_3)^2 + \delta_3^2 + (\delta_3+\delta_4)^2 + (\delta_1+\delta_2+\delta_3+\delta_4)^2$;
$2Q = (\delta_1+\delta_2+\delta_3)^2 + \delta_2^2 + \delta_4^2 + (\delta_1+\delta_2)^2 + (\delta_2+\delta_3+\delta_4)^2$.

4. $a > a_1 > a_3 > a_4 > a_2$, $a - a_1 = \delta_1$, $a_1 - a_3 = \delta_2$, $a_3 - a_4 = \delta_3$, $a_4 - a_2 = \delta_4$;
$2P = \delta_1^2 + (\delta_2+\delta_3+\delta_4)^2 + (\delta_3+\delta_4)^2 + \delta_3^2 + (\delta_1+\delta_2+\delta_3)^2$;
$2Q = (\delta_1+\delta_2+\delta_3+\delta_4)^2 + \delta_2^2 + \delta_4^2 + (\delta_1+\delta_2)^2 + (\delta_2+\delta_3)^2$.

5. $a > a_1 > a_4 > a_2 > a_3$, $a - a_1 = \delta_1$, $a_1 - a_4 = \delta_2$, $a_4 - a_2 = \delta_3$, $a_2 - a_3 = \delta_4$;
$2P = \delta_1^2 + (\delta_2+\delta_3)^2 + \delta_4^2 + (\delta_3+\delta_4)^2 + (\delta_1+\delta_2)^2$;
$2Q = (\delta_1+\delta_2+\delta_3)^2 + (\delta_2+\delta_3+\delta_4)^2 + \delta_3^2 + (\delta_1+\delta_2+\delta_3+\delta_4)^2 + \delta_2^2$.

6. $a > a_1 > a_4 > a_3 > a_2$, $a - a_1 = \delta_1$, $a_1 - a_4 = \delta_2$, $a_4 - a_3 = \delta_3$, $a_3 - a_2 = \delta_4$;
$2P = \delta_1^2 + (\delta_2 + \delta_3 + \delta_4)^2 + \delta_4^2 + \delta_3^2 + (\delta_1 + \delta_2)^2$;
$2Q = (\delta_1 + \delta_2 + \delta_3 + \delta_4)^2 + (\delta_2 + \delta_3)^2 + (\delta_3 + \delta_4)^2 + (\delta_1 + \delta_2 + \delta_3)^2 + \delta_2^2$.

Für die Fälle 1, 2, 5, 6 beweist man sehr leicht daß $P < (\delta_1 + \delta_2 + \delta_3 + \delta_4)^2$, weil $\delta_1, \delta_2, \delta_3, \delta_4$ in allen Fällen positiv sind. Es ist aber $\delta_1 + \delta_2 + \delta_3 + \delta_4$ immer kleiner als $\frac{4}{5}$, weil $1 = \delta + \delta_1 + \delta_2 + \delta_3 + \delta_4$ und δ größer als $\delta_1, \delta_2, \delta_3, \delta_4$, also $P < (\frac{4}{5})^2$, also auch $P < \frac{4}{5}$, also $Nf(\alpha) < 1$.

Für die Fälle 3 und 4 reicht dieses nicht aus. Hier wird aber ebenso leicht bewiesen daß P und Q beide < 1 sind. Es ist nämlich für den Fall 3

$$P = \delta_1^2 + \delta_2^2 + 2\delta_3^2 + \delta_4^2 + 2\delta_2\delta_3 + 2\delta_3\delta_4 + \delta_1\delta_2 + \delta_1\delta_3 + \delta_1\delta_4 + \delta_2\delta_4,$$
$$\text{also } P < (\delta_1 + \delta_2 + 2\delta_3 + \delta_4)^2$$
$$Q = \delta_1^2 + 2\delta_2^2 + \delta_3^2 + \delta_4^2 + 2\delta_1\delta_2 + 2\delta_2\delta_3 + \delta_1\delta_3 + \delta_2\delta_4 + \delta_3\delta_4,$$
$$\text{also } Q < (\delta_1 + 2\delta_2 + \delta_3 + \delta_4)^2,$$

für den Fall 4

$$P = \delta_1^2 + \delta_2^2 + 2\delta_3^2 + \delta_4^2 + 2\delta_2\delta_3 + 2\delta_3\delta_4 + \delta_2\delta_4 + \delta_1\delta_2 + \delta_1\delta_3,$$
$$\text{also } P < (\delta_1 + \delta_2 + 2\delta_3 + \delta_4)^2,$$
$$Q = \delta_1^2 + 2\delta_2^2 + \delta_3^2 + \delta_4^2 + 2\delta_1\delta_2 + \delta_1\delta_3 + \delta_1\delta_4 + 2\delta_2\delta_3 + \delta_2\delta_4 + \delta_3\delta_4,$$
$$\text{also } Q < (\delta_1 + 2\delta_2 + \delta_3 + \delta_4)^2,$$

es ist aber

$$\delta_1 + 2\delta_2 + \delta_3 + \delta_4 < \delta + \delta_1 + \delta_2 + \delta_3 + \delta_4 \quad \text{also } < 1,$$

ebenso

$$\delta_1 + \delta_2 + 2\delta_3 + \delta_4 < (\delta + \delta_1 + \delta_2 + \delta_3 + \delta_4) \quad \text{also } < 1,$$

da also in beiden Fällen P und Q kleiner als 1 sind so ist auch hier $Nf(\alpha) < 1$, w. z. b. w.

Ich habe schon etwas daran gearbeitet einen ähnlichen Beweis für $\lambda = 7$ zu machen, natürlich müßten dabei die 120 verschiedenen Fälle welche den obigen 6 analog sind, unter gemeinsame Gesichtspunkte gefaßt werden. Es wird mir zum Beweise glaube ich folgender einfache Satz behilflich sein. Es ist immer

$$c^2 + c_1^2 + c_2^2 + \cdots + c_n^2 > \frac{(c + c_1 + c_2 + \cdots + c_n)^2}{n+1},$$

was auch die $n + 1$ Größen c, c_1, etc. sein mögen.

Elegant ist der gefundene Beweis der Zerlegbarkeit der Primzahlen $p = 5m + 1$ in vier complexe Factoren nicht, indessen es ist doch ein strenger Beweis, und wenn es mir gelingen sollte auf diesem Wege weiterzugehen, so glaube ich wird schon der Beweis für den

Fall $\lambda = 7$ auf den für $\lambda = 5$ günstig zurückwirken, denn beim weitergehen wird man genöthigt das mehr Zufällige außer Acht zu lassen und das Wesentliche festzuhalten.

Ich erwarte nun nächstens einen recht ausführlichen Brief von Ihnen nicht allein über Ihre Arbeiten in unserem Gebiete, sondern auch über die Berliner Mathematiker überhaupt und über deren Bestrebungen u. s. w. auch über Ihren Besuch bei Scherk in Kiel, und wie Sie mein kleines Pathchen daselbst gefunden haben. Jacobi hat mir noch nicht geschrieben, hat mir aber durch Rosenhain versprechen lassen es bald nach seiner Ankunft in Berlin zu thun. Ich beneide Sie eigentlich darum, daß Sie jetzt in Berlin an der Quelle mathematischen Wissens sitzen, obgleich es mir andererseits auch gar nicht unlieb ist, daß ich fern von unmittelbarer Einwirkung der großen Mathematiker meine mathematische Selbstständigkeit besser habe bewahren können als mancher andere meines Gleichen. Nun leben Sie recht wohl mein herzlich geliebter Freund und schreiben Sie recht bald

Ihrem

Kummer.

Breslau den 16. Octbr. 1844.

Herzlich geliebter Freund!

Ich bin zu Ende der Ferien noch ziemlich fleißig gewesen und theile Ihnen als meinem Mitarbeiter im Reiche der complexen Zahlen einige Resultate meines Fleißes mit, welche Sie in dem beiliegenden Nachtrage zu meinem Programme finden. Nachdem Sie denselben durchgelesen haben bitte ich Sie ihn mit den gehörigen Empfehlungen von mir, an Jacobi zu übergeben, welchem dieser Nachtrag ebenso wie das Programm selbst gewidmet ist. Sie werden aus demselben ersehen, daß der Beweis der Zerfällbarkeit in 4 compl. Factoren der Primzahlen $5m + 1$ in der That noch schöner Vereinfachungen fähig war, und daß der Fall $\lambda = 7$ sich ganz nach denselben Principien abmachen läßt. Es ist in diesen beiden Fällen möglich gemacht worden die verschiedene Stellung der Coefficienten $a, a_1, a_2, \ldots a_{\lambda-1}$ in der complexen Zahl $a + a_1\alpha + a_2\alpha^2 + \cdots + a_{\lambda-1}\alpha^{\lambda-1}$ ganz außer Acht zu lassen, weil der Beweis sich auf eine Größe stützt welche eine symmetrische Function aller dieser Coefficienten, und auch der Differenzen je zweier derselben ist. Weiter auszudehnen geht diese Art des Beweises nicht, ich zweifle aber nicht daran, daß auch für $\lambda = 11, 13, 17, 19$ sich ähnliche einfache Beweise werden finden lassen.

Ferner habe ich alle Hauptresultate meines Programmes jetzt auch für diejenigen complexen Zahlen bewiesen, welche nicht aus den Wurzeln α der Gleichung $\alpha^\lambda = 1$, sondern aus den Perioden dieser Wurzeln gebildet sind. Es gestalten sich auch diese allgemeineren Resultate ebenso einfach und zum Theil noch eleganter als die für den speciellen Fall, wo die Perioden eingliedrig also die Wurzeln selbst sind.

Es ist nun aber nachdem ich Ihnen so fleißig Nachrichten gegeben habe wohl auch an der Zeit, daß Sie mir einmal schreiben. Thun Sie dieß ja recht bald, sonst drohe ich Ihnen mit meiner ganzen Ungnade. Grüßen Sie Joachimsthal, Arnold und alle welche mir lieb und werth sind, von Ihrem Sie herzlich liebenden

E. Kummer.

Breslau den 26. Decbr. 1844.

Herzlich geliebter Freund!

Tausend Dank für Ihre beiden Briefe. Leider werde ich Ihnen dießmal wenig oder gar nichts mathematisches schreiben können, denn ich bin seit längerer Zeit sehr faul gewesen, und habe mich damit begnügt meine beiden Collegien ordentlich nicht nur zu lesen, sondern auch auszuarbeiten. Außerdem habe ich in der Philomathie, die Sie ja wohl dem Namen nach kennen (einem Vereine von Professoren die alle 14 Tage einmal zusammen kommen von einem einen Vortrag oder Abhandlung anhören und dann essen) einen Vortrag gehalten und zwar über das quantitativ Unendliche. Derselbe bestand aus zwei Theilen einem allgemeinen welchen ich aufgeschrieben hatte und ablas und einem speciellen welchen ich frei vortrug. Dieser Vortrag hat mich einige Zeit ganz beschäftigt und sehr interessirt, er hatte überall philosophische Form und es ist mir wirklich gelungen bei denen meiner Zuhörer welche überhaupt zu denken sich nicht scheuen nicht nur Interesse sondern auch Achtung vor dem mathematischen zu erwecken.

Ehe ich nun speciell auf den Inhalt Ihrer Mittheilungen eingehe will ich Ihnen noch einmal den guten Rath wiederholen, mit ihren mathematischen Untersuchungen Epoche zu machen, nämlich recht zu verstehen ἐπ-οχή von ἐπέχω anhalten abschließen. Halten Sie einen Augenblick an bei dem was Sie haben, formen Sie es zu einem Ganzen, wenn auch nicht alles, doch das Hauptsächlichste, alsdann lassen Sie sich so rasch als möglich zum Doctor machen. Ich rathe Ihnen zu dieser Eile hauptsächlich darum weil ich sehe daß Sie sich in Berlin nicht nur übel befinden sondern auch langweilen, trotz aller Arbeit. Es fehlt Ihnen dort ein Haupterforderniß des guten Gelingens: das Arbeiten mit Muße und mit Freude an allem was man zu Tage fördert, diese kann und muß man rein genießen und hat dabei durchaus nicht nöthig eitel zu sein, oder seine eigenen Productionen zu überschätzen. Vor ein paar Tagen erhielt ich als Ergänzung zu einem Theile Ihrer Mittheilungen das CRELLEsche Journal mit der Abhandlung von EISENSTEIN über die cubischen Formen. Es ist eine vortreffliche Abhandlung, welche uns manches aufklärt was wir bisher nur unbestimmt gleichsam vermuthet hatten, nur eins habe ich als eine Hauptsache bei oberflächlicher Lectüre noch vermißt, nämlich daß jede Primzahl welche cubischer Rest von p ist sich entweder durch die Hauptform oder durch eine der nichtäquivalenten associirten Formen darstellen läßt und nur durch eine.

Was übrigens die Anzahl der Zahlen $\Psi(\alpha)$ betrifft welche die Kreistheilung gewährt, so ist dieselbe durch folgende höchst einfache Regel gegeben: „Die Anzahl der wesentlich verschiedenen Zahlen $\Psi(\alpha)$ ist immer gleich derjenigen ganzen Zahl welche nur um einen echten Bruch größer ist als $\frac{\lambda}{6}$"; also für $\lambda = 5$ ist die Anzahl 1, für $\lambda = 7$ Anzahl 2, für $\lambda = 11$ Anzahl 2, $\lambda = 13$ giebt diese Anzahl 3, $\lambda = 17$ ebenfalls 3, $\lambda = 19$ Anzahl 4 etc. Schon bei $\lambda = 11$ giebt die Kreistheilung nicht mehr alle möglicherweise durch die $f(\alpha)$ zu bildenden $\Psi(\alpha)$ welche der Gleichung $\Psi(\alpha)\,\Psi(\alpha^{-1}) = p$ genügen.

Von besonderer Wichtigkeit scheint mir der Gesichtspunkt den Sie mir mitgetheilt haben: wenn p nicht als $Nf(\alpha)$ darstellbar ist, daß alsdann immer einige Zahlen (endlich viele und zwar wenige) existiren von der Art daß mp als $Nf(\alpha)$ darstellbar sei für alle möglichen Werthe des $p = 2\varkappa\lambda + 1$. Es ist auch leicht anderweitig zu zeigen daß eine endliche Anzahl dieser m ausreichend ist, wie viele aber nöthig und ausreichend sind mag wohl denselben Rang der Schwierigkeit haben als die Untersuchung der Anzahl quadratischer Formen für eine Determinante.

EISENSTEINs Art und Weise zu den associirten Formen zu gelangen gefällt mir nicht recht, ich glaube man müsse dazu durch die Lösung folgender Aufgabe kommen nämlich: zu untersuchen welche Formen des e^{ten} Grades mit e Unbestimmten haben die Eigenschaft daß sie nur solche Divisoren haben, welche e^{te} Potenzreste von einer gegebenen Primzahl p sind (wo e auch vorläufig Primzahl ist) und $p \equiv 1 \bmod e$.

Grüßen Sie JOACHIMSTHAL, RÜHLE und EISENSTEIN recht herzlich von mir, ich werde ihnen nichts schuldig bleiben sondern nächstens bei passender Gelegenheit als z. B. zu meinem Geburtstage auch auf deren und Ihr Wohl trinken. Ich fasse nun ihren Geburtstag und das Neujahr zusammen und schenke Ihnen nichts weniger als meine herzliche Liebe wie immer

Ihr

KUMMER.

Breslau den 18. October 1845.

Herzlich geliebter Freund!

Nicht allein in der Absicht, daß Sie meine Arbeiten kennen lernen um sie wo es sein kann bei Ihren eigenen zu benutzen, sondern besonders auch um meiner selbst Willen, um für mich etwas Ordnung und Klarheit hineinzubringen schreibe ich Ihnen schon wieder davon, und zwar heute nur über allgemeine Theorie der complexen Zahlen, ohne Anwendungen. Ich werde Ihnen vorzüglich nur eine Sammlung von Sätzen geben können aber in einer wohlgeordneten Folge, so daß Sie sich allenfalls Schritt für Schritt die Beweise selbst machen können, auch werde ich dieselben wohl hier und da mit Glossen versehen. Die Bezeichnungen, welche ich habe, werden Ihnen wohl aus dem vorigen Briefe und sonst schon bekannt sein, nämlich $\lambda - 1 = e \cdot f$ und q zum Exponenten f gehörig modulo λ und Primzahl aber kein Factor von $N(\eta - \eta_r)$. Ich unterscheide auch hier gar nicht mehr Primzahlen p von den q, sondern nehme die zum Exponenten 1 gehörigen Primzahlen als eben in demselben Rechte befindlich als die andern, es sind für diese nur die Perioden eingliedrig und $f = 1$.

Nun beginnt die Reihe von Sätzen:

1. Lehrsatz: Jede complexe Zahl $f(\alpha)$ läßt sich und zwar nur auf eine einzige Weise in die Form setzen

$$f(\alpha) = \Phi(\eta) + \alpha\Phi_1(\eta) + \alpha^2\Phi_2(\eta) + \cdots + \alpha^{f-1}\Phi_{f-1}(\eta).$$

2. Erklärung: Wenn für $\eta = u$ (u Wurzel der bekannten Congruenz) zugleich folgende Congruenzen Statt haben:

$$\Phi(u) \equiv 0, \ \Phi_1(u) \equiv 0, \ \Phi_2(u) \equiv 0, \ldots \Phi_{f-1}(u) \equiv 0 \bmod q$$

so soll dieses kurz ausgedrückt werden durch: $f(\alpha) \equiv 0 \bmod q$, für $u = \eta$.

3. Lehrsatz: Wenn $f(\alpha) \equiv 0 \bmod q$ für $u = \eta$, so ist $Nf(\alpha) \equiv 0 \bmod q^f$ und umgekehrt; wenn $Nf(\alpha) \equiv 0 \bmod q^f$, so ist $f(\alpha) \equiv 0 \bmod q$ für $\eta = u$ (d. h. nicht für eine bestimmte, sondern für irgend eine Congruenzwurzel u). Dieß ist der Satz welchen ich Ihnen schon bei Ihrer Anwesenheit in Breslau mitgetheilt habe. . . .

Bemerkung: Die jetzt folgenden Sätze über Theilbarkeit einer complexen Zahl durch eine ganze Zahl beziehen sich alle nur auf den Fall, wo die complexe Zahl nicht in ihrem entwickelten Zustande fertig ist, sondern als Product complexer Factoren gegeben.

4. Lehrsatz: Wenn eine complexe Zahl aus Factoren besteht, und es wird ein Factor $\equiv 0 \bmod q$ für $u = \eta_r$, so behält auch die Zahl nach Ausführung der Multiplication diese Eigenschaft; und umgekehrt: Wenn eine entwickelte complexe Zahl $\equiv 0 \bmod q$ für $u = \eta_r$ ist, und dieselbe läßt sich in Factoren zerlegen, so muß auch einer der Factoren $\equiv 0 \bmod q$ für $u = \eta_r$ sein.

Ein wichtiger Satz, welcher sich fast von selbst zu verstehen scheint, aber doch eines Beweises bedarf so gut als die übrigen; daß ich diesen Satz als sich von selbst verstehend angenommen, und auch auf den Fall stillschweigend übertragen hatte, wo zwei Factoren für $u = \eta$ congruent Null werden mod q, wo also das Product $\equiv 0 \bmod q^2$ wird, hat mich lange an einer richtigen Erkenntnis des Gegenstandes gehindert, in diesem Falle ist nämlich für $u = \eta$ die Form des unentwickelten Productes durch q^2 theilbar in der entwickelten Form aber ist es nur durch q theilbar.

Lehrsatz: Wenn eine (in Form eines Productes erscheinende) complexe Zahl für alle verschiedenen Congruenzwurzeln $u = \eta$, $u_1 = \eta$, $u_2 = \eta, \ldots u_{e-1} = \eta$ einzeln congruent Null wird mod q, so ist sie selbst durch q theilbar, wenn aber diese complexe Zahl für irgend eine Congruenzwurzel $u_r = \eta$ nicht $\equiv 0$ wird mod q, so ist diese Zahl auch nicht durch q theilbar.

Es bezeichne jetzt $f(\eta)$ eine complexe Zahl nur aus Perioden gebildet, von der Art daß $f(\eta)f(\eta_1)f(\eta_2)\ldots f(\eta_{e-1})$ durch q, aber nicht durch q^2, theilbar ist, es sei ferner $f(\eta_1)f(\eta_2)\ldots f(\eta_{e-1}) = F(\eta)$ und u sei diejenige Congruenzwurzel welche der Bedingung $f(\eta) \equiv 0 \bmod q$

für $\eta = u$ genügt (eine dieser Wurzeln muß bekanntlich dieser Bedingung genügen).

Erklärung des Ausdruckes idealer Primfactor einer complexen Zahl:

Wenn eine complexe Zahl, nachdem sie mit $F(\eta_r)$ multiplicirt ist, durch q theilbar ist, so wollen wir dieß so ausdrücken: sie enthält den zu $\eta_r = u$ gehörigen idealen Primfactor des q.

Wenn ferner eine complexe Zahl, nachdem sie durch eine beliebig hohe Potenz von $F(\eta_r)$ als $\{F(\eta_r)\}^m$ multiplicirt ist, den realen Factor q^μ, aber nicht $q^{\mu+1}$ enthält, wo $\mu < m$ ist so wollen wir dieß so ausdrücken: sie enthält den zu $\eta_r = u$ gehörigen idealen Primfactor des q μ mal.

Es wird Sie diese Definition beim ersten Anblick ziemlich kalt lassen, und Sie können vielleicht denken es wäre einfacher zu definiren daß wenn $\Phi(\alpha)$ für $\eta = u$ den realen Factor q μ mal enthält sie als den entsprechenden idealen Factor ebenso oft enthaltend anzusehen sei. Dieß stimmt für den Fall $\mu = 1$ vollkommen mit der gegebenen Erklärung; für $\mu > 1$ aber ist es ganz nichtsnutzig, weil dann eine und dieselbe complexe Zahl jenachdem sie als Product oder entwickelt erscheint den idealen Factor das einemal öfter als das anderemal enthalten würde. Bei der von mir gewählten Definition sind die idealen Factoren einer Zahl vollkommen fest, welches in den folgenden Sätzen festgestellt wird.

Lehrsatz: Wenn man durch Multiplication mit $(F(\eta)_r)^m$ gefunden hat, daß eine complexe Zahl den zu $u = \eta_r$ gehörigen idealen Primfaktor des q μ mal enthält, so bleibt dasselbe Resultat auch feststehen, wenn man m beliebig anders jedoch immer $m \geqq \mu$ wählt.

Lehrsatz: Wenn man zur Aufsuchung der idealen Primfactoren statt der complexen Zahl $f(\eta)$ aus welcher $F(\eta)$ gebildet ist, irgend eine andere mit denselben oben angegebenen Eigenschaften wählt, so erhält man immer dieselben idealen Primfactoren.

Lehrsatz: Das entwickelte Product mehrer complexen Zahlen hat genau dieselben idealen Primfactoren und auch jeden genau eben so oft, als die Factoren des Productes zusammengenommen.

Lehrsatz: Wenn $\Phi(\alpha)$ m ideale Primfactoren des q (verschiedene oder gleiche) enthält, so enthält $N\Phi(\alpha)$ den Factor

q^{mf}, und umgekehrt wenn $N\Phi(\alpha)$ den Factor q^{mf} enthält, so enthält $\Phi(\alpha)$ m ideale Primfactoren des q.

Lehrsatz: Jede bestimmte complexe Zahl enthält eine endliche Anzahl idealer Primfactoren, welche nur auf eine Weise in ihr vorhanden sind (die complexen Einheiten machen hiervon eine Ausnahme).

Lehrsatz: Wenn zwei complexe Zahlen $\Phi(\alpha)$ und $\Psi(\alpha)$ dieselben idealen Primfactoren haben, so sind sie einander gleich bis auf eine complexe Einheit welche als Factor der einen Zahl hinzutreten kann.

Lehrsatz: Eine complexe Zahl $\Phi(\alpha)$ ist durch eine andere $\Psi(\alpha)$ theilbar, wenn jene alle idealen Primfactoren von dieser, und zwar alle wenigstens ebenso oft enthält, und es ist diese Bedingung nicht allein hinreichend zur Theilbarkeit, sondern auch nothwendig. Der Quotient enthält den Ueberschuß der idealen Factoren des Dividendus über die des Divisors.

Dieser letzte Satz rechtfertigt die Benennung idealer Primfactor vollkommen, denn man kann mit ihnen nun wie mit numerischen ganzzahligen Primfactoren rechnen. Die idealen Primfactoren sind für die ganzen complexen Zahlen ohngefähr ebenso nothwendig wie die imaginären Wurzeln für die Theorie der Gleichungen oder der Zerlegung einer ganzen rationalen Function von x in ihre lineären Factoren. Sie sind, will man sie begreifen im philosophischen Sinne, ganz abstruse Dinge, sonst aber sind sie in mathematischer Hinsicht ganz einfache Ausdrücke bestimmter Eigenschaften gegebener complexer Zahlen. Ich bemerke noch folgendes: Wenn eine complexe Zahl nur einen einzigen zu $u = \eta_r$ gehörigen idealen Primfactor des q enthält, so ist sie statt eines idealen ein realer Primfactor des q. Es vereinigen sich auch oft mehrere in einer complexen Zahl steckende ideale Factoren mannigfaltig zu einem einzigen realen Factor. Die Gesetze dieser Vereinigung aber, denen noch nachzuspüren ist, sind nicht so simpel als die Vereinigung der beiden imaginären Factoren $a + bi$ und $a - bi$ zu dem realen $a^2 + b^2$. Die ganze Theorie der complexen Zahlen, welche aus λ^{ten} Wurzeln der Einheit oder aus Perioden derselben gebildet sind, scheint mir nun in diesem letzten Satze in so weit vollendet und abgemacht zu sein, als man die Frage nach der Realität oder bloßen Idealität der Primfactoren unberücksichtigt läßt, und noch einige Nebensachen und Ausführungen abgerechnet. . . .

Uebermorgen, als den Mondtag gehen meine Vorlesungen an und zwar gleich drei auf einmal zu denen ich mich noch präpariren soll, was bei dem Anfange allemal Schwierigkeiten hat, wenn man sich nicht gleich in medias res hineinstürzen will oder kann.

Leben Sie recht wohl

Ihr

E. KUMMER.

Breslau d. 14. Juni 1846.

Herzlich geliebter Freund!

... Ueber die idealen Factoren habe ich mehreres mit DIRICHLET und einiges mit JACOBI verhandelt. Ich gebrauchte dabei immer die bildliche Ausdrucksweise, aus der Chemie entnommen: Die Primfactoren sind die Elemente, die idealen Primfactoren sind diejenigen Elemente welche nicht für sich darstellbar nur in Verbindung mit anderen vorkommen, äquivalente complexe ideale Zahlen sind an sich dasselbe als äquivalente Gewichtsmengen der chemischen Stoffe. Die Aufsuchung der idealen Primfactoren ist die chemische Analyse, die in meinem Aufsatze mit ψ oder Ψ bezeichneten complexen Zahlen sind die Reagentien und die ganze Zahl q, welche als realer Factor heraustritt, ist der Niederschlag welcher nach Anwendung des richtigen Reagens sich zeigt. Kurz die ganze Begriffssphäre der Chemie stimmt auf eine eclatante Weise mit derjenigen zusammen in welcher sich die Lehre von den complexen Zahlen bewegt. DIRICHLET hat mich sehr ermahnt die Theorie bald fertig auszuarbeiten und CRELLE zum Drucke zu übergeben. Auch hat er mir erzählt und gezeigt, nämlich aus mündlichen und schriftlichen Aeußerungen von GAUSS, daß GAUSS schon bei Anfertigung des Abschnittes de compositione formarum aus den Disqu. arith. etwas ähnliches wie ideale Factoren zu seinem Privatgebrauche gehabt hat, daß er dieselben aber nicht auf sicheren Grund zurückgeführt hat, er sagt nämlich in einer Note seiner Abhandlung über die Zerfällung der ganzen rat. Functionen in lineäre Factoren ohngefähr so: „Wenn ich hätte auf dieselbe Weise verfahren wollen wie die früheren Mathematiker mit dem imaginären, so würde eine andere meiner Untersuchungen die sehr schwierig ist sich auf sehr leichte Weise haben machen lassen.“ Daß hier die compositio formarum gemeint ist, hat DIRICHLET später mündlich von GAUSS erfahren. Ich habe ferner DIRICHLET meine Vermuthung mitgetheilt daß zwischen den Formenanzahlen wie sie DIRICHLET gefunden hat, und zwischen den Summen der Potenzreste gewisse einfache Beziehungen Statt haben möchten, und ich habe ihm versprochen ihm meine Arbeit über diejenigen Exponenten, zu welchen die idealen Primfactoren erhoben zu wirklichen complexen Zahlen werden, mitzutheilen. Noch ist zu bemerken daß ich mit DIRICHLET als meinem Vetter Brüderschaft gemacht habe. BORCHARD habe ich besucht und kennen gelernt. EISENSTEIN hat keinen recht guten Eindruck dießmal auf mich gemacht, er spielt zu sehr den Sonderling, er sucht noch seine vierte Dimension und ich habe als sein väterlicher Freund ihn überhaupt vor der Art des Götzendienstes gewarnt, nach welcher man das, was rein analytischer Natur ist, sich räumlich zu versinnbildlichen trachtet, dasselbe kann dadurch leichter faßlich und anschaulich werden, aber es wird dadurch aus seiner höheren Stellung herabgezogen. JOACHIMSTHAL ist inzwischen ein sehr berühmter Mathematiker geworden, die comptes rendus sind seiner Ehre voll und LIOUVILLES Journal hat einen Aufsatz wo sein Name im Titel florirt. . . .

Von meinen eigenen Arbeiten erwähne ich noch eine recht nette Sache. Ich habe nämlich ganz allgemein die Aufgabe gelöst Vierecke zu finden, deren Seiten und Diagonalen rational sind. Dieselbe wird dadurch interessant daß CHASLES (fälschlich) meint die Inder hätten sich diese Aufgabe gestellt und in ihrer Weise gelöst, ferner dadurch daß die Pointe des ganzen darauf hinauskommt eine Wurzel $\sqrt{a + bx + cx^2 + dx^3 + ex^4}$ rational zu machen, welches EULER schon in der Algebra behandelt und wovon JACOBI später gezeigt hat daß die Theorie der elliptischen Functionen diese Aufgabe löst. Jede einzelne Lösung dieser Aufgabe giebt eine allgemeine, drei willkürliche rationale Zahlen enthaltende Lösung der obigen Aufgabe über das Viereck. Die Vierecke, welche man so erhält, sind Vierecke ohne Eigenschaften oder besser ohne andere gewöhnliche Eigenschaften, neben anderen, welche z. B. einem Kreise eingeschrieben sind und dergl. Nun aber schreibe ich nichts weiter als daß ich Sie nochmals bitte recht bald hierher zu kommen, damit wir alles was wir einander mitzutheilen haben mündlich besprechen können.

Ihr Sie herzlich liebender Freund

E. KUMMER.

Breslau d. 13. Aug. 1846.

Herzlich geliebter Freund!

... Ihre Untersuchungen über die allgemeinsten zerlegbaren Formen, sind mir von sehr großem Interesse, weil dadurch die Willkürlichkeit in den zu untersuchenden Formen aufgehoben wird, ebenso die Untersuchungen über die ABELschen Gleichungen.

Unsere mathematische Literatur besteht wie Sie wissen aus Abhandlungen, kleineren und größeren. Darum wollte ich Ihnen den wohlmeinenden lehrerlichen Rath geben, Ihre mathematischen Studien bald anfangs so zu betreiben, daß Abhandlungen entstehen, d. h. daß Sie gewisse Stoffe zu einer gewissen Abrundung verarbeiten, sodaß sie wenn auch nach vielen Seiten hin das Bedürfniß nach weiterem Fortschritte in sich tragend, doch als etwas abgeschlossenes Ganzes gelten können. Was ich hier sage ist von einer allgemeinen Bedeutung, es betrifft nämlich überhaupt alles was Entwickelung ist, also auch die Weltgeschichte das Leben der Staaten der Individuen etc. Ueberall hier ist dies unbefriedigende jedes besonderen Zustandes oder des bisher erreichten vorhanden und zwar nothwendig weil dieses den Fortschritt nothwendig bedingt, aber es ist auch ebenso das Bedürfniß da die errungenen Zustände zu fixiren in der relativen Vollendung, welche sie erreicht haben. Es ist überall nicht ein bloßer Fortschritt sondern auch ein Bestehen, überall mouvement und Stabilität. Huldigen Sie darum auch in richtigem Maaße der letzteren, so werden schöne Abhandlungen entstehen, und damit sie wie DIRICHLET sagt lesbar werden, so müssen Sie etwas episches Moment hineinbringen, nur nicht zu viel, sonst werden sie langweilig wie so manche epische Versuche in CRELLES Journal.

Ich arbeite jetzt mein rationales Viereck aus. Außerdem habe ich noch etwas anderes gefunden. Sie wissen vielleicht, daß ich vor zwei Jahren einem meiner Zuhörer Namens WITTIBER die Aufgabe zu einer zu fertigenden Dissertation gab „Systeme von Curven zu suchen, welche sich selbst überall rechtwinklig schneiden“, so wie die Kegelschnitte welche dieselben beiden Brennpunkte haben. Dieser hat nun endlich nach zweijähriger Arbeit ein solches System von Curven vierten Grades gefunden. Dasselbe interessirte mich sehr und ich griff selbst noch einmal die Aufgabe an und fand eine allgemeine Formel für dergleichen Systeme welche noch eine ganz willkürliche Function in sich enthält. Die in dieser allgemeinen Formel enthaltenen Systeme haben unter andern stets die Eigenschaft eine Anzahl Brennpunkte (nach PLÜCKERS Definition) zu enthalten, welche für alle Curven eines Systems dieselben sind, wie bei den confocalen Kegelschnitten. . . .

Ihr Sie herzlich liebender

E. KUMMER.

Geliebter Freund!

Ihre freundliche Einladung nehme ich an, wie sich von selbst versteht, ich komme Sonnabend mit dem ersten Zuge etwa um 10 Uhr in Haynau an, wo es mich sehr freuen wird Sie mit „drachenbespannter Kalesche“ zu finden. Ich bringe mir erstens gute Wasserstiefeln mit und zweitens drei Abhandlungen die ich seit Ihrem letzten Hiersein ins Reine geschrieben habe: 1. über Gamma, 2. Systeme sich rechtwinklig schneidender Curven, 3. die Zerlegung der complexen. Die Zeit wird uns nicht lang werden, selbst wenn das Wetter zur Jagd nicht immer günstig sein sollte. Die Fortsetzung mündlich auf den Sonnabend d. 26.

Ihr

E. KUMMER.

Breslau d. 23. Septbr. 1846.

Herzlich geliebter Freund!

Ihre mir versprochene Gelegenheit um die Bücher, welche ich richtig besorgt habe, Ihnen zuzuschicken währt mir etwas zu lange, darum entschließe ich mich jetzt Ihnen dieselben per Post zu schicken und Ihnen zugleich zu melden, daß mir die Strapazen der Jagd sehr gut bekommen sind. Einer der beiden Hasen, die Sie mir mitgegeben haben und welcher beiläufig gesagt sehr gut geschmeckt hat wie auch CAUER bezeugen kann, mußte offenbar auch von mir geschossen oder wenigstens angeschossen sein, denn seine Wunden hatten unterwegs wo er zu meinen Füßen lag von neuem so stark geblutet, daß es sogar durch die Tasche hindurchgedrungen war, welches bekanntlich ein sicheres Zeichen von der Gegenwart des Mörders ist. Da ich nun einmal bei der Jagd stehe, so füge ich hier sogleich die Bitte bei beiliegendes Jagdmesser dem Herrn Förster HÖNEL in meinem Namen zu übergeben und ihm für die Mühe die er sich unverdrossen um meinetwillen gegeben hat nochmals zu danken.

Die Abhandlung von VANDERMONDE habe ich, wenn auch noch nicht mit der gehörigen Andacht, durchgelesen und sie auch, was die Darstellung betrifft, gedankenreich und vortrefflich gefunden, weniger wird Ihnen der Artikel BERNOULLIsche Zahlen in KLÜGELS math. Wörterbuch genügen, ich lege darum außerdem SCHERKs Abhandlung bei welche, so viel ich weiß, mehr enthält.

Die Factorenzerfällung des (α, x) habe ich vollständig ergründet zunächst das Beispiel wo $\beta^{21} = 1$, $\alpha^3 = 1$, $x^7 = 1$, oder $\alpha = \beta^7$, $x = \beta^3$, hier ist

$$(\alpha, x) = x + x^6 + \alpha(x^3 + x^4) + \alpha^2(x^2 + x^5) = (\alpha^2 - x - x^2)^2(\alpha - x - x^2)^4 E(\beta)$$

wo $E(\beta)$ eine complexe Einheit ist. Die sechs zusammengehörigen Zahlen

$$\alpha - x - x^2,\quad \alpha - x^2 - x^4,\quad \alpha - x^3 - x^6,\quad \alpha - x^4 - x,\quad \alpha - x^5 - x^3,\quad \alpha - x^6 - x^5,$$

sind nämlich als gleich zu erachten, weil sie sich nur durch complexe Einheiten unterscheiden, oder jede derselben ist durch jede andere theilbar, und der Quotient eine Einheit, ebenso wie $1 - x$, $1 - x^2$, $1 - x^3 \ldots$ Doch nun zum allgemeinen: sei $\beta^{p\lambda} = 1$, $\alpha^\lambda = 1$, $x^p = 1$, $p = \mu\lambda + 1$, p und λ Primzahlen $\alpha = \beta^p$, $x = \beta^\lambda$, so läßt sich p wirklich oder ideal stets folgendermaßen in ein Product von $(p-1)\cdot(\lambda-1)$ Factoren zerlegen:

$$p = \{f(\alpha x) f(\alpha x^2) f(\alpha x^3) \cdots f(\alpha x^{p-1})\} \cdot \{f(\alpha^2 x) f(\alpha^2 x^2) f(\alpha^2 x^3) \cdots f(\alpha^2 x^{p-1})\} \ldots$$
$$\{f(\alpha^{\lambda-1} x) f(\alpha^{\lambda-1} x^2) f(\alpha^{\lambda-1} x^3) \cdots f(\alpha^{\lambda-1} x^{p-1})\}.$$

Hierbei unterscheiden sich die je $p - 1$ in Klammern eingeschlossenen Factoren nur durch complexe Einheiten, oder es ist allgemein $f(\alpha^\varkappa x^m) = f(\alpha^\varkappa x^n) E(\beta)$. Dieselben lassen sich also durch eine $(p-1)^{\text{te}}$ Potenz ersetzen, sodaß man auch hat

$$p = f(\alpha x)^{p-1} f(\alpha^2 x)^{p-1} f(\alpha^3 x)^{p-1} \cdots f(\alpha^{\lambda-1} x)^{p-1} \cdot E(\beta)\,^{*)}$$

oder

$$p = \{f(\alpha x) f(\alpha^2 x) f(\alpha^3 x) \cdots f(\alpha^{\lambda-1} x)\}^{p-1} E(\beta).$$

Mit Hilfe dieser wirklichen oder idealen Factoren des p wird nun (α, x) folgendermaßen zerlegt

$$(\alpha, x) = f(\alpha x)^{\mu m_1} \cdot f(\alpha^2 x)^{\mu m_2} \cdot f(\alpha^3 x)^{\mu m_3} \cdots f(\alpha^{\lambda-1} x)^{\mu m_{\lambda-1}} \cdot E(\beta),$$

wo $\mu = \dfrac{p-1}{\lambda}$ und $m_1, m_2, m_3 \ldots m_{\lambda-1}$ dieselben Zahlen sind als im

---

*) NB. Die mit $E(\beta)$ bezeichneten complexen Einheiten sind nicht überall dieselben, sondern nur überhaupt Einheiten, quod vix monendum erat.

Königsberger Programm, nämlich $\varkappa \cdot m_\varkappa \equiv 1$, mod λ und alle in den Grenzen 0 und λ excl.

Was Ihre Bemerkung betrifft durch welche Sie die Widersinnigkeit einer solchen Zerlegung darthun wollten, so wollen wir jetzt (α, x) als Function von β allein ansehen, so daß $(\alpha, x) = F(\beta)$ ist, so ist allemal $F(\beta^{p(\lambda-2)+1}) = (\alpha^{-1}, x)$; der reciproke Ausdruck (α^{-1}, x) läßt sich also wirklich immer darstellen, indem man nur statt β eine passende Potenz von β setzt. Uebrigens enthält wirklich wie Sie sehen (α, x) gewisse Factoren mit (α^{-1}, x) gemein, welches aber nichts schadet, weil die Factoren nur gruppenweise verschieden sind.

Mir scheint vorläufig, daß dieser Zerlegung des (α, x) nicht dieselbe Bedeutung zukommt, als der Zerlegung des $(\alpha, x)^\lambda$ in die von α allein abhängigen Primfactoren, es ist aber ganz angenehm sie einmal ergründet zu haben. . . .

Leben Sie wohl besuchen Sie mich möglichst bald einmal und lassen Sie wenn dies nicht geschehen kann wenigstens schriftlich bald etwas von sich hören.

Ihr Sie herzlich liebender Freund

E. Kummer

Breslau, d. 17. Octbr. 1846.

Breslau, d. 7. Decbr. 1846.

Herzlich geliebter Freund!

Ich gratulire zum Geburtstage! und wenn Sie meine Gratulation auch erst morgen bekommen, so habe ich doch zur rechten Zeit gratulirt. ... Ihre mathematischen Mittheilungen haben mich sehr erfreut. Treiben Sie nur das Geschäft der Redaction Ihrer Arbeiten mit rechter Liebe und Sorgfalt, und erinnern Sie sich der epischen Breite, welche wenn sie mit Maaß angewendet wird sehr vorzüglich ist, namentlich das Aussprechen des Gedankengehalts der Arbeiten sodaß der Leser alsbald weiß worauf es überall ankommt. Den 19. Band von Crelles Journal hatte ich an Morbach verliehen, habe mir ihn aber seit einiger Zeit wiedergeholt, um, wenn ein Bote von Ihnen kommt, denselben bereit zu haben. Die Bücher aus der hiesigen Bibliothek schicken Sie mir wohl der Verabredung gemäß entweder noch vor Weihnachten oder bringen Sie selbst zu Weihnachten mit. Vor einigen Tagen habe ich aus unserer Bibliothek endlich den Band der Petersb. Academie erhalten in welchem EULERs Abhandlung über die Rationalität von $\sqrt{A+Bx+Cx^2+Dx^3+Ex^4}$ steht. EULERs Methode ist zweckmäßiger als die, welche ich selbst gefunden habe, auch viel einfacher als die Anwendung der Elliptischen Functionen, mit welcher sie nur in gewissem Sinne zusammenfällt, welches ich nicht sogleich näher angeben kann. Ich arbeite immerfort noch über dem rationalen Vierecke, und werde noch lange daran arbeiten können, wenn ich auch interessante Besonderheiten mit einschließe.

Meine zahlentheoretischen Nachdenkungen führen mich soeben dahin über die Anzahl Ihrer Jahre, nämlich 23 Betrachtungen anzustellen. Sie sind jetzt in der Zahl der Jahre, für welche notorisch die Zerlegbarkeit in 22 complexe Factoren nicht mehr Statt hat. Die Zerlegbarkeit in zwei reciproke complexe Factoren findet aber immer Statt, wenn Sie daher in diesem Jahre sich in zwei reciproke complexe Factoren zerlegen wollen, so steht von zahlentheoretischem Standpunkte nichts entgegen. ... Nun leben Sie wohl und behalten Sie immer lieb Ihren Sie herzlich liebenden Freund

E. KUMMER.

Breslau, d. 20. Febr. 1847.

Herzlich geliebter Freund!

Als Antwort auf Ihr letztes liebes Schreiben kommt dieß Briefchen zwar etwas spät, ich will Ihnen aber dennoch namentlich über Ihren vorläufig auf unbestimmte Zeit veränderten Lebensplan etwas schreiben, und zwar zunächst nur dieß, daß ich Sie darum nur um so höher schätze, daß Sie um Ihrer Familie willen dieses Opfer bringen. Es ist auch gar nicht nöthig, daß Ihre Banquier-Geschäfte auf Ihre Wissenschaft auf die Dauer einen nachtheiligen Einfluß üben, nehmen Sie als Beispiel mich selbst, ich habe zu der Zeit wo ich in Liegnitz mit 24 Stunden unter denen fast gar keine mathematischen waren, in den ersten Jahren meines dasigen Aufenthaltes wenigstens ebenso viel geleistet als später unter den äußerlich günstigsten Verhältnissen. Ferner LAVOISIER, der berühmte Chemiker war in Paris vor der Revolution Generalpächter und nebenbei eine der ersten Zierden der Academie. Wenn Sie für einige Zeit sich jetzt ganz den Geschäften widmen um sich ordentlich einzuarbeiten, so werden Sie um so eher wieder sich der Wissenschaft widmen können. Die Landwirtschaft erscheint allerdings günstiger für wissenschaftliche Studien, aber Geldgeschäfte schließen dieselben auch nicht aus. Ich selbst habe in diesem Jahre vielleicht noch weniger mathematisch gearbeitet als Sie. Ich habe nämlich noch keinen neuen Stoff gefunden, und um doch etwas zu thun, werde ich eine Recension über die von JACOBI herausgegebenen dem Könige dedicirten Abhandlungen schreiben. ... — Ich hoffte immer Sie in kurzem einmal hier zu sehen, denn ich denke, die Geschäfte müssen Sie bald einmal herführen, sollten diese es nicht thun, so thun Sie es doch selbst recht bald. Die besten Grüße von den Meinen an Sie und die Ihrigen Ihr Sie herzlich liebender

E. KUMMER.

Breslau, d. 2. April 1847.

Herzlich geliebter Freund!

Versetzen Sie sich wieder einmal in die complexen Zahlen und Einheiten, wo λ, α, u. s. w. lauter bekannte Zeichen sind. Nehmen Sie auch vorläufig einmal folgenden Satz als bewiesen an:

I. „Wenn eine Einheit $E(\alpha)$ die Form hat $E(\alpha) = c + \lambda f(\alpha)$, ($c$ reale g. Z.), so ist $E(\alpha)$ eine λ^{te} Potenz einer andern Einheit.“

Der umgekehrte Satz versteht sich ganz von selbst, aber auch dieser ist für $\lambda = 5$ und $\lambda = 7$ sehr leicht zu beweisen, und so überall wo man die Fundamental-Einheiten kennt. Einen allgemeinen Beweis habe ich bisher noch nicht.

II. Es sei ferner λ eine solche Primzahl, für welche die Anzahl aller nicht äquivalenten Formen (oder nach meiner Auffassung der nicht äquivalenten idealen complexen Zahlen) nicht durch λ selbst theilbar ist. Dieß gilt offenbar wieder für $\lambda = 5$, $\lambda = 7$, und für unendlich viele Primzahlen λ, ich weiß nicht ob für alle. Es wird unter dieser Voraussetzung, wenn $(f(\alpha))^\lambda$ eine wirkliche complexe Zahl ist, allemal auch $f(\alpha)$ selbst eine wirkliche complexe Zahl sein; denn die Potenz, welche die ideale Zahl zur wirklichen macht, hat stets mit der Anzahl der nicht äquivalenten Formen einen gemeinschaftlichen Factor.

Für alle diejenigen Primzahlen λ, welche diesen beiden Bedingungen I und II genügen kann ich nun die Unmöglichkeit der Gleichung $x^\lambda - y^\lambda = z^\lambda$ vollständig beweisen wie folgt.

Zunächst beweise ich, daß wenn $x^\lambda - y^\lambda = z^\lambda$ Statt haben soll, eine der drei Zahlen durch λ theilbar sein muß. Sei x nicht durch λ theilbar, so giebt $x^\lambda = z^\lambda + y^\lambda$ folgende Gleichungen:

$$z + y = a^\lambda \quad \text{und} \quad z + \alpha^r y = E(\alpha) f(\alpha)^\lambda$$

ich verwandle α in α^{-1}, wodurch

$$z + \alpha^{-r} y = E(\alpha^{-1}) f(\alpha^{-1})^\lambda$$

es ist aber

$$E(\alpha^{-1}) = \pm \alpha^\varkappa E(\alpha), \quad \text{also} \quad z + \alpha^{-r} y = \pm \alpha^\varkappa E(\alpha) f(\alpha^{-1})^\lambda$$

also wenn $E(\alpha)$ eliminirt wird

$$\pm \alpha^\varkappa (z + \alpha^r y) f(\alpha^{-1})^\lambda = (z + \alpha^{-r} y) f(\alpha)^\lambda.$$

Hieraus eine Congruenz mod λ gemacht, giebt $f(\alpha)^\lambda \equiv c$ mod λ, ebenso $f(\alpha^{-1}) \equiv c$ mod λ also, weil c nicht durch λ theilbar ist,

$$\pm \alpha^\varkappa (z + \alpha^r y) \equiv z + \alpha^{-r} y \mod \lambda$$

oder

$$0 \equiv z + \alpha^{-r} y \mp \alpha^\varkappa z \mp \alpha^{\varkappa + r} y \mod \lambda.$$

Diese Congruenz kann nicht bestehen, ohne daß eine der Zahlen z oder y durch λ theilbar ist. q. e. d. Es sei also z die durch λ theilbare Zahl.

Anstatt der Gleichung $x^\lambda - y^\lambda = z^\lambda$, wo z durch λ theilbar ist, behandle ich die allgemeinere Gleichung für complexe Zahlen:

1) $$u^\lambda - v^\lambda = E(\alpha)(1-\alpha)^{m\lambda} w^\lambda \quad (E(\alpha) \text{ Einheit})$$

wo u, v, w complexe Zahlen sind, w den Factor $1 - \alpha$ nicht weiter enthaltend, und ich setze von u und v nur das voraus, daß sie in folgende Form gebracht werden können:

2) $$u = c + (1-\alpha)^{m\lambda - \lambda + 1} \cdot \Phi(\alpha); \quad v = c + (1-\alpha)^{m\lambda - \lambda + 1} \cdot \Psi(\alpha). \quad (c \text{ real}).$$

Ich zerlege nun $u^\lambda - v^\lambda$ in seine λ complexen Factoren, $u - v$, $u - \alpha v$, $u - \alpha^2 v$, etc. Diese Factoren haben unter sich keinen gemeinschaftlichen Factor außer $1 - \alpha$, diesen aber haben sie alle und zwar jedes nur einmal, eins aber hat ihn alle übrigen male, und dieß ist nach der Voraussetzung (siehe 2)) $u - v$ es ist also

3) $$u - v = e(\alpha)(1-\alpha)^{m\lambda - \lambda + 1} \cdot w_1^\lambda$$

4) $$u - \alpha^r v = e_r(\alpha)(1 - \alpha^r) t_r^\lambda$$

$e(\alpha)$ und $e_r(\alpha)$ sind Einheiten, w_1 und t_r complexe Zahlen.

(Der Satz: „wenn eine Potenz einer complexen Zahl in Factoren zerlegt wird, welche relative Primzahlen sind, so müssen diese Factoren selbst ebensolche Potenzen sein, multiplicirt mit Einheiten", folgt klar aus meinen früheren Untersuchungen, noch ist zu bemerken, daß weil w_1^λ und t_r^λ wirkliche complexe Zahlen sind, auch w_1 und t_r selbst solche sein müssen, nach der obigen Voraussetzung.)

Ich substituire in 4) die Werthe des u und v aus 2), so wird, wenn durch $1-\alpha^r$ dividirt ist:

$$5)\qquad c+(1-\alpha)^{m\lambda-\lambda}\cdot\left(\frac{1-\alpha}{1-\alpha^r}\right)(\Phi(\alpha)-\alpha^r\Psi(\alpha))=e_r(\alpha)t_r^\lambda$$

Hieraus mache ich eine Congruenz modulo λ und bemerke, daß $(1-\alpha)^{m\lambda-\lambda}$ durch λ theilbar ist, wenn $m>1$, welches hier vorausgesetzt wird, so ist

$$c\equiv e_r(\alpha)t_r^\lambda \mod \lambda;$$

es ist aber die λ^{te} Potenz der complexen Zahl allemal einer realen Zahl congruent, also $t_r^\lambda\equiv b \mod \lambda$ folglich

$$c\equiv e_r(\alpha)b \mod \lambda,$$

und weil $e_r(\alpha)$ einer realen Zahl congruent ist modulo λ, so ist, nach dem oben angenommenen Satze, $e_r(\alpha)$ gleich einer λ^{ten} Potenz einer andern Einheit, also $e_r(\alpha)t_r^\lambda$ gleich einer λ^{ten} Potenz, gleich u_1^λ.

Dieß in der Gleichung (4) substituirt, giebt

$$6)\qquad u-\alpha^r v=(1-\alpha^r)u_1^\lambda$$

ebenso hat man für irgend einen anderen Werth des r, welchen ich s nenne

$$7)\qquad u-\alpha^s v=(1-\alpha^s)v_1^\lambda$$

und wenn noch die Gleichung 3) hinzugenommen wird:

$$3)\qquad u-v=e(\alpha)(1-\alpha)^{m\lambda-\lambda+1}\cdot w_1^\lambda$$

und aus diesen u und v eliminirt werden, so erhält man

$$u_1^\lambda-v_1^\lambda=\frac{(\alpha^r-\alpha^s)e(\alpha)(1-\alpha)^{m\lambda-\lambda+1}\cdot w_1^\lambda}{(1-\alpha^r)(1-\alpha^s)}$$

und wenn

$$\frac{(\alpha^r-\alpha^s)(1-\alpha)}{(1-\alpha^r)(1-\alpha^s)}e(\alpha)=E_1(\alpha)$$

gesetzt wird

$$8)\qquad u_1^\lambda-v_1^\lambda=E_1(\alpha)(1-\alpha)^{(m-1)\lambda}\cdot w_1^\lambda.$$

Diese Gleichung 8) ist nun dieselbe als 1) nur $m-1$ statt m gesetzt. Um aber zu zeigen, daß dieselbe Verwandlung sich wieder mit dieser Gleichung vornehmen läßt, müssen wir auch noch beweisen, daß die

Bedingungen (2), welchen u und v unterworfen sind, analog auch für u_1 und v_1 gelten, oder dass

$$u_1 = c_1 + (1-\alpha)^{(m-2)\lambda+1}\Phi_1(\alpha), \quad v_1 = c_1 + (1-\alpha)^{(m-2)\lambda+1}\Psi_1(\alpha).$$

Zu diesem Zwecke wende ich die Gleichung (5) an, in welcher, wie oben gezeigt worden, $e_r(\alpha)t_r^\lambda = u_1^\lambda$ zu setzen ist, dieselbe giebt so:

9) $$c \equiv u_1^\lambda \mod (1-\alpha)^{(m-1)\lambda}.$$

Ich setze ferner

$$u_1 = a + (1-\alpha)\Theta,$$

(wo a real, Θ complex, in diese Form kann jede complexe Zahl gesetzt werden), so ist:

10) $$u_1^\lambda = a^\lambda + \lambda(1-\alpha)a^{\lambda-1}\Theta + \lambda\left(\frac{\lambda-1}{2}\right)(1-\alpha)^2 a^{\lambda-2}\Theta^2 + \cdots + (1-\alpha)^\lambda\Theta^\lambda$$

und weil

$$\lambda = (1-\alpha)^{\lambda-1}e(\alpha),$$

so folgt

11) $$u_1^\lambda \equiv a^\lambda \mod (1-\alpha)^\lambda.$$

Aus (9) und (11) folgt nun

$$c \equiv a^\lambda \mod (1-\alpha)^\lambda$$

und hieraus

$$c \equiv a^\lambda \mod \lambda^2.$$

(Ich glaube diese Folgerung wird Ihnen klar sein.)

Wenn aber $c \equiv a^\lambda \mod \lambda^2$ so hat man auch $c \equiv c_1^\lambda \mod \lambda^n$, wo n jede beliebig große Zahl bedeutet also auch

$$c \equiv c_1^\lambda \mod (1-\alpha)^{(m-1)\lambda}.$$

Diese Congruenz mit 9 verbunden giebt

$$u_1^\lambda \equiv c_1^\lambda, \mod (1-\alpha)^{(m-1)\lambda}$$

hieraus folgt nun, daß von den λ Factoren des $u_1^\lambda - c_1^\lambda$ jeder einen Factor $1-\alpha$ enthalten muß, nur einer dieser Factoren, für welchen man offenbar den Factor $u_1 - c_1$ nehmen kann, muß den Ueberschuß enthalten, also den Factor $1-\alpha$ $((m-1)\lambda - \lambda + 1)$ mal, sodaß

$$u_1 = c_1 + (1-\alpha)^{(m-2)\lambda+1}\Phi_1(\alpha).$$

Was nun für u_1 bewiesen ist gilt offenbar ebenso für v_1, also hat man auch

$$v_1 = c_1 + (1-\alpha)^{(m-2)\lambda+1}\Psi_1(\alpha).$$

Aus der Gleichung (1) mit ihren Bedingungen für u und v folgt also eine ebensolche Gleichung (8) in welcher m um eine Einheit kleiner geworden ist, mit den entsprechenden Bedingungen für u_1 und v_1. Aus dieser folgt nun ebenso eine neue Gleichung mit ihren Bedingungen und so fort, bis man endlich zu dem Werthe $m = 1$ gelangt, oder zu der Gleichung

$$u^\lambda - v^\lambda = E_{m-1}(\alpha)(1-\alpha)^\lambda w^\lambda$$

wo w keinen Factor $1-\alpha$ enthält. Diese Gleichung aber ist unmöglich weil $u^\lambda - v^\lambda$, wenn es überhaupt durch $1-\alpha$ theilbar ist, diesen Factor wenigstens $\lambda + 1$ mal enthalten muß.

Dieß beweise ich endlich so: Wenn von den λ Factoren des $u^\lambda - v^\lambda$ nämlich $u - v$, $u - \alpha v$, $u - \alpha^2 v$, etc. einer durch $1-\alpha$ theilbar ist, so sind es alle, man hat also

$$u - v = (1-\alpha)\Theta, \quad u - \alpha v = (1-\alpha)\Theta_1, \quad u - \alpha^2 v = (1-\alpha)\Theta_2 \quad \text{etc.}$$

aus $u = v + (1-\alpha)\Theta$ folgt aber

$$u^\lambda = v^\lambda + \lambda(1-\alpha)v^{\lambda-1}\Theta + \lambda\left(\frac{\lambda-1}{2}\right)(1-\alpha)^2 v^{\lambda-2}\Theta^2 + \cdots$$
$$+ \lambda(1-\alpha)^{\lambda-1} v\Theta^{\lambda-1} + (1-\alpha)^\lambda\Theta^\lambda,$$

hieraus mache ich eine Congruenz modulo $(1-\alpha)^{\lambda+1}$, so ist

$$u^\lambda \equiv v^\lambda + \lambda(1-\alpha)v^{\lambda-1}\Theta + (1-\alpha)^\lambda\Theta^\lambda;$$

$$u^\lambda - v^\lambda \equiv (1-\alpha)^\lambda\left\{\frac{(1-\alpha)(1-\alpha^2)(1-\alpha^3)\ldots(1-\alpha^{\lambda-1})}{(1-\alpha)(1-\alpha)(1-\alpha)\ldots(1-\alpha)}v^{\lambda-1}\Theta + \Theta^\lambda\right\} \bmod (1-\alpha)^{\lambda+1},$$

der in Klammern eingeschlossene Ausdruck wird nun noch einmal modulo $1-\alpha$ untersucht, wo sich findet, daß er durch dasselbe theilbar ist. Es ist nämlich

$$\frac{(1-\alpha)(1-\alpha^2)(1-\alpha^3)\ldots(1-\alpha^{\lambda-1})}{(1-\alpha)(1-\alpha)(1-\alpha)\ldots(1-\alpha)} \equiv 1\cdot 2\cdot 3\cdots(\lambda-1) \bmod (1-\alpha)$$

also nach dem Wilsonschen Satze congruent -1; ferner ist $\Theta \equiv b \bmod (1-\alpha)$, $v \equiv c \bmod (1-\alpha)$, also der in Klammern eingeschlossene Ausdruck, wie leicht zu sehen, durch $1-\alpha$ theilbar, also $u^\lambda - v^\lambda$ durch $(1-\alpha)^{\lambda+1}$ theilbar.

Die Gleichung

$$u^\lambda - v^\lambda = E(\alpha)(1-\alpha)^\lambda w^\lambda,$$

in welcher w nicht durch $1-\alpha$ theilbar ist, kann also nicht Statt haben, also auch nicht die Gleichung

$$u^\lambda - v^\lambda = E(\alpha)(1-\alpha)^{m\lambda} w^\lambda$$

wenn

$$u = c + (1-\alpha)^{m\lambda-\lambda+1}\,.\,\Phi \quad \text{und} \quad v = c_{\prime} + (1-\alpha)^{m\lambda-\lambda+1}\,.\,\Psi$$

also auch nicht die Gleichung, welche ein spezieller Fall von dieser ist

$$x^\lambda - y^\lambda = z^\lambda,$$

wo z durch λ theilbar ist. q. e. d.

Wenn Sie mir einen Gefallen thun wollen, so lesen Sie diesen Beweis nur mit dem größten Mißtrauen durch, und mit der Absicht, irgendwo einen Fehler, etwas unhaltbares, zu finden. Ich selbst halte ihn für so sicher bewiesen als irgend einen Satz der Zahlentheorie....

Der obige Beweis von $x^\lambda + y^\lambda = z^\lambda$ ist erst drei Tage alt, denn erst nach Beendigung der Recension fiel es mir ein wieder einmal diese alte Gleichung vorzunehmen, und ich kam diesmal bald auf den richtigen Weg. Ich wiederhole meine Ermahnungen zur Pünktlichkeit mit der Bemerkung, daß mir eigentlich weit mehr daran liegt von Ihnen Nachrichten zu erhalten, als gerade die Recension zu haben, welche wohl noch ein paar Tage warten kann ehe sie nach Jena geschickt wird.

Empfehlen Sie mich Ihren Eltern, grüßen Sie AULICH

Ihr

E. KUMMER.

Breslau d. 17. Maj 1847.

Herzlich geliebter Freund!

... Nun noch einiges über Voraussetzung II, betreffend die Formenzahl, und Voraussetzung I betreffend die Einheiten welche λ^{te} Potenzen sind. Ich glaube seit Ihrer Abreise von hier darin wieder eine wichtige Einsicht erlangt zu haben welche in dem noch nicht vollständig von mir bewiesenen Satze liegt:

„Für jede Primzahl λ, welche als Factor einer der ersten $\frac{\lambda-3}{2}$ BERNOUILLIschen Zahlen vorkommt, ist die Formenanzahl durch λ theilbar.“

Vollständig bewiesen aber habe ich folgenden Satz:

„Wenn $f(\alpha)$ eine ideale complexe Zahl ist und $f(\alpha)$ nicht äquivalent $f(\alpha^{-1})$, sondern $f(\alpha)\varphi(\alpha)$ äquivalent $f(\alpha^{-1})$ wo $\varphi(\alpha)$ ideal ist, so kann die λ^{te} Potenz von $\varphi(\alpha)$ d. i. $\varphi(\alpha)^\lambda$ nur dann zu einer wirklichen complexen Zahl werden, wenn λ eine solche Primzahl ist, welche als Factor einer der ersten $\frac{\lambda-3}{2}$ BERNOUILLIschen Zahlen vorkommt.“

Sollte der erste dieser beiden Sätze sich als wahr bestätigen, wie er im speciellen Falle welchen der zweite Satz enthält wirklich wahr ist, so wäre die Identität der Voraussetzung I mit II bewiesen und zugleich wüßte man, für welche Primzahlen mein Beweis giltig ist.

Ihr

E. KUMMER.

Breslau d. 25. Febr. 1848.

Herzlich geliebter Freund!

Wenn Sie mich jetzt bald einmal besuchen könnten, so würde ich Sie außer unserer sonstigen freundschaftlichen Unterhaltung auch noch mit schönen und interessanten mathematischen Sachen bewirthen können. Ich habe nämlich den Werth von

$$\left(\frac{e(\alpha^{\gamma^i})}{f(\alpha)}\right),\ \text{wo}\ e(\alpha)=\sqrt{\frac{(1-\alpha^{\gamma})(1-\alpha^{-\gamma})}{(1-\alpha)(1-\alpha^{-1})}}\ \text{und}\ Nf(\alpha)=p\ \text{ist},$$

allgemein gefunden und bewiesen, ja nicht nur bewiesen, sondern durch eine genetische Methode gefunden. Dieser Werth ist folgender:

Setzt man

$$\left(\frac{e(\alpha^{\gamma^i})}{f(\alpha)}\right)\equiv e(\alpha^{\gamma^i})^{\frac{p-1}{\lambda}}\equiv\alpha^{\varkappa},$$

so ist (mod. λ)

$$\varkappa\equiv\frac{\gamma^{2i}(\gamma^2-1)B_1m_{\lambda-3}}{2\cdot s^2\cdot h}-\frac{\gamma^{4i}(\gamma^4-1)B_2m_{\lambda-5}}{4\cdot s^2\cdot h}+\cdots+\frac{(-1)^{\frac{\lambda-3}{2}}\gamma^{(\lambda-3)i}(\gamma^{\lambda-3}-1)B_{\frac{\lambda-3}{2}}m_2}{(\lambda-3)s^2\cdot h},$$

wo $B_1, B_2, \ldots B_{\frac{\lambda-3}{2}}$ die BERNOUILLIschen Zahlen sind, h der Exponent der Potenz, zu welcher $f(\alpha)$ erhoben werden muß, um wirklich zu werden

$$f(\alpha)^h=c+c_1\alpha+c_2\alpha^2+\cdots+c_{\lambda-1}\alpha^{\lambda-1}$$

$$s=c+c_1+c_2+\cdots+c_{\lambda-1}$$

$$m_{2\varkappa}=K_1+2^{2\varkappa}K_2+3^{2\varkappa}K_3+\cdots+(\lambda-1)^{2\varkappa}K_{\lambda-1}$$

$$K_r=rcc_r+(r+1)c_1c_{r+1}+\cdots+(r+\lambda-1)c_{\lambda-1}c_{r+\lambda-1}$$

oder

$$m_{2\varkappa}=\sum_{0}^{\lambda-1}{}_r\sum_{0}^{\lambda-1}{}_t\, r^{2\varkappa}(r+t)c_rc_{r+t}$$

γ primit. Wurzel von λ.

$\varkappa$ ist in Beziehung auf die Coefficienten von $f(\alpha)^h$ vom zweiten Grade. Auch läßt sich $\varkappa$ durch die Coefficienten von $\psi_1(\alpha)\,\psi_2(\alpha)$ etc. wo $\psi_r(\alpha) = \frac{(\alpha, x)(\alpha^r, x)}{(\alpha^{r+1}, x)}$ ausdrücken und zwar als lineäre Function derselben. Ich ziehe aber den Ausdruck durch die Coefficienten von $f(\alpha)^h$ vor.

Da ich nun diesen nicht unwichtigen Punkt erobert und der Herrschaft der Wissenschaft unterthänig gemacht habe, so können Sie sich denken, daß ich jetzt versuche von hier aus weiter gegen meinen Hauptfeind, das simple Reciprocitätsgesetz, zu operiren. Es fehlt mir auch nicht an Muth dazu, da ich durch die bisherigen Erfolge kühner gemacht worden bin, und da ich mir bewußt bin bis jetzt noch täglich an der gründlicheren Kenntnis meines Gegenstandes zu gewinnen.

... Leben Sie wohl, empfehlen Sie mich Ihrer Fräulein Braut und den Ihrigen allen und kommen Sie ja recht bald zu Ihrem Freunde

E. Kummer.

Breslau d. 5. Maj 1848.

... Können Sie sich wohl vorstellen, daß ich seit acht Tagen mich zweimal als Volksredner versucht habe? Bei einer Versammlung unseres Wahlbezirks trat ich zuerst auf, und sprach über die Eigenschaften eines guten Wahlmannes, welches sehr großen Anklang fand. Ich wurde darauf einstimmig zum Vorsitzenden für die nächste Versammlung gewählt. Auch hatte ich bei dieser ersten Versammlung mein Terrain recognoscirt und gefunden, daß der demokratische Klubb ganz dominierte und zwar nur durch unbedeutende Personen, welche sich zu Wahlmännern aufwerfen wollten. Diese schmeichelten den Arbeitern um zu reüssiren, verdächtigten die Beamten und gebrauchten alle die gewöhnlichen Kunstgriffe. Ich faßte darum den Entschluß wenigstens einen dieser Leute durch meine Person zu verdrängen und hielt in der zweiten Versammlung eine zweite Rede vorzüglich an die Arbeiter gerichtet. Obgleich ich nun gerade die entgegengesetzten Mittel anwendete, als jene Demokraten, nämlich den Arbeitern zu zeigen, was sie seit dem 18. März wirklich erreicht hätten, und ihnen Vertrauen zu der gegenwärtigen Regierung einzuflößen, so reüssirte ich doch vollständig. Die Demokraten hielten zwar noch eine Versammlung am Sonntage, wo sie mich zu verdrängen suchten, es gelang ihnen aber nicht, wie Sie aus der Liste der Wahlmänner ersehen haben. Neben mir sind außer zwei hiesigen Bürgern allerdings nur Mitglieder des demokratischen Klubbs für Berlin und Frankfurt gewählt worden; überhaupt haben die Demokraten hier durchgängig gesiegt. Ich selbst bin auch gar nicht gegen die Demokraten überhaupt eingenommen, wenn sie es nur gegenwärtig mit der Befestigung einer durchaus freisinnigen constitutionellen Monarchie redlich meinen, und nicht gegen das Königthum zu Felde ziehen, auch nicht streben es heimlich zu untergraben, so sind mir die Demokraten im Grunde lieber als die philisterhaften Bürger, welche an den Wahlen für Frankfurt fast gar nicht mehr Theil nahmen, weil sie für diese wenig oder gar kein Interesse hatten. Die Anforderungen, die ich an einen Deputirten nach Berlin stelle sind 1. wahre Vaterlandsliebe, 2. Einsicht und Verstand, 3. Charakterfestigkeit. Speciellere Anforderungen stelle ich nicht, weil wir die Candidaten nehmen müssen wie sie eben zuletzt bei den engeren und engsten Wahlen übrig bleiben. Wohl uns, wenn wir zuletzt aus zwei guten den besten wählen können, es kann aber auch kommen, daß wir zuletzt aus zwei Uebeln noch das geringere zu wählen haben. Für einen Deputirten nach Frankfurt würden die Anforderungen dieselben sein, nur daß seine Vaterlandsliebe mehr in dem einigen Deutschland als in Preußen ihre Hauptwurzel haben müsse, und daß auch seine Einsicht sich mehr auf das allgemeinere erstrecken möchte. — Ich bin auf meine Würde als Wahlmann sehr stolz wie Sie daraus ersehen können, daß ich mich in Fürstenstein, wo wir am Mittwoch waren, als E. Kummer, Wahlmann eingeschrieben habe, meine Frau als Wahlweib, den Vetter als Urwähler und Louise Cauer als Wahlverwandtschaft. Ich freue mich aber wirklich, daß es mir gelungen ist, besonders darum weil mich nur wahrer Patriotismus dazu vermocht hat meine Schüchternheit zu überwinden, und als Redner vor einer solchen gemischten Versammlung aufzutreten. Sobald ich meinen Pflichten als Bürger werde genügt haben, nämlich unmittelbar nach den Wahlen für die Frankfurter Versammlung, werde ich sogleich wieder meine mathematischen Arbeiten mit voller Kraft vornehmen, denn dann habe ich für Politik für den Augenblick nichts weiter zu thun. Wenn Sie mich in nächster Woche besuchen, worauf ich mich sehr freue so erzähle ich Ihnen das nähere über die hiesigen Wahlen und will Ihnen auch das Concept meiner ersten Rede mittheilen, die zweite war fast ganz frei gesprochen, und nur im allgemeinen prämeditirt. Leben Sie wohl, . . . und empfangen Sie . . . die herzlichsten Grüße der meinigen

Ihr Sie herzlich liebender Freund

E. Kummer.

6*

Breslau d. 17. Sept. 1849.

... Die von der Pariser Akademie gestellte Preisfrage lautet: Trouver pour un exposant entier quelconque n les solutions en nombres entiers et inégaux de l'équation $x^n + y^n = z^n$, ou prouver qu'elle n'en a pas. Sie verlangt also etwas, was ich nicht kann und was wahrscheinlich auch keiner außer mir leisten wird, da ich selbst doch bisher in dieser Sache am weitesten vorgeschritten bin. Ich werde aber doch noch einige Versuche machen, wozu eine genauere Ergründung der complexen Zahlen für den Fall gehört, daß mein λ im Zähler einer der ersten BERNOUILLIschen Zahlen als Faktor vorkommt. ...

Breslau, d. 28. Decbr. 1849.

Herzlich geliebter Freund!

1. Wenn λ keine Ausnahmszahl wie 37, so haben die beiden Voraussetzungen I und II immer Statt. 2. Wenn λ eine solche Ausnahmszahl wie 37 ist, so ist die Klassenanzahl, und zwar der erste Faktor derselben allemal durch λ theilbar, die Voraussetzung I kann Statt haben oder auch nicht, dies habe ich noch nicht ergründet, für $\lambda = 37$ habe ich berechnet daß die Voraussetzung I nicht Statt findet, weil es hier wirklich eine Einheit giebt, welche congruent der ganzen Zahl c ist, Mod λ, ohne eine λ^{te} Potenz einer anderen Einheit zu sein. 3. Für die Ausnahmszahlen ist der Ausdruck $\frac{D}{\Delta}$ nicht nothwendig durch λ theilbar, ich gehe ja sogar darauf aus zu beweisen, daß dieser zweite Faktor der Formenanzahl niemals also auch für keine der bekannten Ausnahmszahlen durch λ theilbar ist.

Wenn Du mir schreibst „vermuthest Du nicht ferner, daß grade für alle Zahlen λ deine Voraussetzung I stattfände und deshalb auch überall $\frac{D}{\Delta}$ nicht durch λ theilbar sei?" so liegt hierin ein großer Irrthum deinerseits, denn das steht felsenfest: Wenn für eine Ausnahmezahl λ die Voraussetzung I Statt hat, so ist allemal auch $\frac{D}{\Delta}$ durch λ theilbar. Wenn die n^{te} BERNOUILLIsche Zahl wo $n < \frac{\lambda-1}{2}$ durch λ theilbar ist, γ eine primitive Wurzel von λ, so ist allemal

$$e(\alpha)\, e(\alpha^{\gamma})^{\gamma^{-2n}}\, e(\alpha^{\gamma^2})^{\gamma^{-4n}} \ldots e(\alpha^{\gamma^{\mu-1}})^{\gamma^{-2(\mu-1)n}} \equiv c, \quad \text{Mod } \lambda, \quad \left(\mu = \frac{\lambda-1}{2}\right)$$

wäre aber

$$e(\alpha)e(\alpha^{\gamma})^{\gamma^{-2n}}\ldots e\left(\alpha^{\gamma^{\mu-1}}\right)^{\gamma^{-2(\mu-1)n}} = E(\alpha)^{\lambda}$$

so wäre $\frac{D}{\Delta}$ durch λ theilbar.

Deine auf dieser falschen Ansicht beruhenden Folgerungen fallen somit von selbst weg. Ich gedenke vielmehr den Beweis des FERMATschen Satzes auf folgendes zu gründen:

1. Auf den noch zu beweisenden Satz, daß es für die Ausnahmszahlen λ stets Einheiten giebt, welche ganzen Zahlen congruent sind für den Modul λ, ohne darum λ^{te} Potenzen anderer Einheiten zu sein, oder was dasselbe ist, daß hier niemals $\frac{D}{\Delta}$ durch λ theilbar wird.
2. Daß es aber für diese Ausnahmszahlen λ niemals eine Einheit giebt, welche congruent einer ganzen Zahl ist, für den Modul λ^2, ohne eine λ^{te} Potenz einer anderen Einheit zu sein.

Für den Fall, daß es einige Zahlen n, n', n'', giebt welche $< \frac{\lambda-1}{2}$ sind und für welche alle $B_n \equiv 0$ $B_{n'} \equiv 0$, $B_{n''} \equiv 0$ ist mod λ, kann ich leicht beweisen, wenn gesetzt wird

$$m(\beta) = 1 + \gamma^{-2n}\beta + \gamma^{-4n}\beta^2 + \cdots + \gamma^{-2(\mu-1)n}\beta^{\mu-1}$$

und analog

$$m'(\beta) = 1 + \gamma^{-2n'}\beta + \gamma^{-4n'}\beta^2 + \cdots + \gamma^{-2(\mu-1)n'}\beta^{\mu-1},$$

und ebenso

$$m''(\beta) = 1 + \gamma^{-2n''}\beta + \cdots$$

daß immer

$$e(\alpha)^{r\cdot m(\beta)}\cdot e(\alpha)^{r'\cdot m'(\beta)}\cdot e(\alpha)^{r''\cdot m''(\beta)}\ldots \equiv c \text{ Mod } \lambda \text{ ist,}$$

wo r, r', r'' beliebige ganze Zahlen sind. Die Einheit

$$e(\alpha)^{r\cdot m(\beta)}\cdot e(\alpha)^{r'\cdot m'(\beta)}\cdot e(\alpha)^{r''\cdot m''(\beta)}\ldots$$

kann aber niemals eine λ^{te} Potenz einer anderen Einheit werden, wenn nicht $e(\alpha)^{r\cdot m(\beta)}$, $e(\alpha)^{r'\cdot m'(\beta)}$, $e(\alpha)^{r''\cdot m''(\beta)}$ einzeln für sich gleich λ^{ten} Potenzen von Einheiten sind. (Sehr wichtiger Satz, welcher eine große Schwierigkeit hebt!)

Dieß wird, glaube ich, Dir über Deine an mich gerichteten Fragen die nöthige Aufklärung geben. Außer diesem hätte ich Dir noch sehr viel zu sagen, aber ich weiß nicht wo ich anfangen soll. Die Untersuchung der Logarithmen complexer Zahlen, oder derjenigen complexen Zahlen, welchen die Logarithmen anderer congruent sind für den Modul λ, oder λ^2, oder λ^3 etc. habe ich abgeschlossen und auf die einfachsten Ausdrücke gebracht. Hierüber theile ich Dir folgenden Satz mit

Sei

$$F(\alpha) = A + A_1(\alpha + \alpha^{-1}) + A_2(\alpha^2 + \alpha^{-2}) + \cdots + A_\mu(\alpha^\mu + \alpha^{-\mu})$$

sei ferner

$$\Psi_r(\alpha) = \alpha + \alpha^{-1} + \gamma^{-2r\lambda}(\alpha^\gamma + \alpha^{-\gamma}) + \gamma^{-4r\lambda}(\alpha^{\gamma^2} + \alpha^{-\gamma^2}) + \cdots$$
$$+ \gamma^{-2(\mu-1)r\lambda}\left(\alpha^{\gamma^{\mu-1}} + \alpha^{-\gamma^{\mu-1}}\right)$$

(sodaß eigentlich $\Psi_r(\alpha) = (\beta, \alpha)$ ist wenn für β gesetzt wird $\gamma^{-2r\lambda}$).

Sei ferner

$$\Phi(v) = A + 2A_1 \cos v + 2A_2 \cos 2v + \cdots + 2A_\mu \cos \mu v$$

wo v eine Variable ist, so ist

$$\sum_0^{\mu-1}{}_h \gamma^{-2r\lambda h}\, l\left(\frac{F(\alpha^{\gamma^h})}{F(1)}\right) \equiv (-1)^r\, \Psi_r(\alpha)\, \frac{d^{2n\lambda}\, l\left(\frac{\Phi(v)}{\Phi(0)}\right)}{dv^{2n\lambda}} \qquad \text{Mod } \lambda^2$$

wenn noch der Differenziation $v = 0$ gesetzt wird.

Diese Congruenz repräsentirt für $r = 1, 2, 3, \ldots \mu - 1$, $\mu - 1$ Congruenzen, aus denen man unmittelbar auch $l\left(\frac{F(\alpha)}{F(1)}\right)$ erhält, wenn man dieses apart zu haben wünscht.

Für den Fall, daß $F(\alpha) = e(\alpha)$ die Kreistheilungs-Einheit ist, wird $\frac{\Phi(v)}{\Phi(0)} = \frac{\sin \gamma \frac{v}{2}}{\gamma \sin \frac{v}{2}}$ und weil die Entwicklung von $l\left(\frac{\sin \gamma \frac{v}{2}}{\gamma \sin \frac{v}{2}}\right)$ nach Potenzen von v bekannt ist, so erhält man unmittelbar

$$\sum_0^{\mu-1}{}_h \gamma^{-2r\lambda h}\, l\left(\frac{e(\alpha^{\gamma^h})}{e(1)}\right) \equiv (-1)^{n-1}\, \Psi_r(\alpha)\, \frac{B_{r\lambda}(\gamma^{2r\lambda} - 1)}{2r\lambda} \;\ldots\; \text{Mod } \lambda^2.$$

Hierbei ist zu bemerken, daß $B_{r\lambda}$ (wo $B_{r\lambda}$ die $r\lambda^{\text{te}}$ BERNOUILLIsche Zahl) immer durch λ theilbar ist, (außer für $\lambda = 3$), und daß

$$\frac{B_{r\lambda}}{r\lambda} \equiv \frac{(-1)^\mu \cdot r B_{r+\mu}}{r+\mu} - \frac{(2r-1)B_r}{r} \quad \text{Mod } \lambda^2 \qquad (r < \mu).$$

Wenn Du mir doch von dieser Congruenz einen einfachen Beweis liefern könntest, und auch einen von der Congruenz

$$\frac{B_r}{r} \equiv \frac{(-1)^\mu B_{r+\mu}}{r+\mu}, \quad \text{Mod } \lambda.$$

Du bist mir auch noch Deinen Ausdruck der BERNOUILLIschen Zahlen schuldig!

Schreibe mir nur ja recht bald wieder über dieses und über alle Deine Observationen in dieser ganzen Sache, denn in Ermangelung mündlicher Unterhaltung mit Dir schreibe ich Dir sehr gern was mich beschäftigt. . . . Bald nach Empfang Deines Briefes, welchen ich erwarte, werde ich Dir einen weiteren Bericht über meine Studien geben; bis dahin lebe wohl. Dein treuer Freund

E. Kummer.

Breslau, den 14. Jan. 1850.

Geliebter Freund!

Indem ich für Deine Mittheilung in dem heutigen Schreiben über die Sätze von den Bernouillischen Zahlen Dir herzlich danke, bemerke ich, daß ich eine ähnliche Art des Beweises wie Du sie mir vorschlägst schon angewendet habe. Ich habe aber die Sache alsdann weit tiefer ergründet und erkannt, daß diese Eigenschaft der Bernouillischen Zahlen eine ganz allgemeine Eigenschaft der Entwicklungs-Coefficienten der Functionen ist. Hier hast Du alles in folgendem Satze.

Lehrsatz: Wenn $\Phi(v)$ eine Function von v ist, welche in eine nach aufsteigenden ganzen Potenzen von $\left(2\sin\frac{v}{2}\right)^2$ geordnete Reihe so entwickelt werden kann, daß die Coefficienten dieser Entwicklung rationale Zahlen sind, in deren Nennern die Primzahl λ nicht als Faktor vorkommt, wenn ferner außerdem

$$\Phi(v) = A + \frac{A_1 v^2}{1\cdot 2} + \frac{A_2 v^4}{1\cdot 2\cdot 3\cdot 4} + \text{ etc.}$$

so ist

$$A_m - (-1)^\mu A_{m+\mu} \equiv 0, \quad \text{Mod } \lambda \qquad \left(\mu = \frac{\lambda-1}{2}\right)$$

$$A_m - 2\cdot(-1)^\mu A_{m+\mu} + A_{m+2\mu} \equiv 0, \quad \text{Mod } \lambda^2 \qquad \text{wenn } m > 1$$

$$A_m - 3\cdot(-1)^\mu A_{m+\mu} + 3A_{m+2\mu} - (-1)^\mu A_{m+3\mu} \equiv 0, \quad \text{Mod } \lambda^3 \quad m > 1$$

allgemein

$$A_m - \frac{r}{1}(-1)^\mu A_{m+\mu} + \frac{r(r-1)}{1\cdot 2}A_{m+2\mu} - (-1)^\mu \frac{r(r-1)(r-2)}{1\cdot 2\cdot 3}A_{m+3\mu} - \cdots$$

$$\pm A_{m+r\mu} \equiv 0, \quad \text{Mod } \lambda^r \qquad \text{wenn } 2m > r$$

Dieses hat auf die Sekanten-Coefficienten unmittelbare Anwendung.

Auf die BERNOUILLIschen Zahlen wende ich es an indem ich

$$\Phi(v) = \frac{d^2 l\left(\dfrac{\sin\gamma\dfrac{v}{2}}{\gamma\sin\dfrac{v}{2}}\right)}{dv^2}$$

nehme (γ primitive Wurzel von λ). Außerdem kann man noch manches damit machen.

Die Congruenz der BERNOUILLIschen Zahlen

$$\frac{B_{n\lambda}}{n\lambda} \equiv (-1)^{\mu}\frac{2nB_{n+\mu}}{n+\mu} - \frac{(2n-1)B_n}{n}, \quad \text{Mod } \lambda^2$$

entspringt aus einer Wiederholung der Congruenz

$$\frac{B_n}{n} - 2(-1)^{\mu}\frac{B_{n+\mu}}{n+\mu} + \frac{B_{n+2\mu}}{n+2\mu} \equiv 0, \quad \text{Mod } \lambda^2,$$

welche durch den allgemeinen Satz geliefert wird.

Ich halte den allgemeinen Satz welchen ich Dir mitgetheilt habe für ein ganz nettes Früchtchen meiner Untersuchungen. Auf die Frucht von 3000 Franks aber werde ich, so weit ich es jetzt übersehe, wohl verzichten müssen, denn ich werde zwar wohl damit zu Stande kommen den FERMATschen Satz weiter zu beweisen auch für die Primzahlen λ, für welche $B_n \equiv 0$ mod λ ist und $n < \frac{\lambda-1}{2}$, aber der Beweis wird sich wieder nicht auf alle diese Primzahlen λ erstrecken, es werden hier wieder diejenigen die Ausnahme machen, für welche außerdem $B_{n\lambda} \equiv 0$ mod λ^2 ist und $B_n \equiv 0$, $n < \frac{\lambda-1}{2}$. Wenn ich dieß erst noch werde ergründet haben, ob es so ist oder nicht, dann werde ich die habsüchtigen Pläne gänzlich fallen lassen, und wieder nur für die Wissenschaft arbeiten, namentlich für die Reciprocitätsgesetze, für welche ich schon neue Ideen gefaßt habe.

Lebe wohl empfiehl mich Deiner lieben Frau und Deinen Eltern freue Dich immer über Deinen kleinen Jungen so recht von Herzen, Du weißt wie hoch auch ich solche Liebe schätze. Die Meinigen sind alle wohl und grüßen herzlich. Dein treuer Freund

E. KUMMER.

Breslau, den 4. Decbr. 1851.

Geliebter Freund!

Ich gratulire Dir zu Deinem Geburtstage und wünsche Dir von Herzen alles Gute sowohl für Dich als für Deine ganze Familie. Namentlich aber wünsche ich Dir außer den allgemeinen Wünschen als Gesundheit, Wohlergehen und dergleichen insbesondere auch Zeit, Zeit für wissenschaftliche Arbeiten und Zeit daß Du einmal Deinen Freund KUMMER ordentlich besuchen könnest, denn ich habe Dich wer weiß wie lange nicht mehr gesehen und in Ruhe gesprochen schon fast seit einem Jahre nicht mehr. Seit ich Dich das letztemal gesehen habe, habe ich wieder eine größere Abhandlung ablaufen lassen, und ich hätte sehr gewünscht, sie Dir vorher einmal zeigen zu können. Eine genaue Entwicklung ihres Inhalts und der Fortschritte in der Erkenntniß der Reciprocitätsgesetze, die sie enthält, kann ich Dir nur mündlich geben. Hier theile ich Dir die Ueberschriften der fünf Paragraphen mit, die sie enthält. Der erste Paragraph enthält die Bestimmung der Symbole $\left(\frac{\lambda}{f(\alpha)}\right)$, $\left(\frac{1-\alpha^{\varkappa}}{f(\alpha)}\right)$, $\left(\frac{\varepsilon(\alpha)}{f(\alpha)}\right)$ für den Fall daß $f(\alpha)$ eine ideale complexe Primzahl von der Art ist, daß $Nf(\alpha) = p$ eine Primzahl der Form $n\lambda + 1$ ist. Es ist dieser Paragraph die Ausführung dessen, was ich in meiner letzten Mittheilung an die Academie in Berlin im Abrisse gegeben habe. Der zweite Paragraph führt den Titel: Eine Erweiterung der Theorie der Kreistheilung. Die durch die Kreistheilung erzeugten complexen Zahlen enthalten bekanntlich alle nur Primfactoren der realen Primzahl p von der Form $n\lambda + 1$. Hier aber erzeuge ich auf ganz analoge Weise complexe Zahlen welche nur complexe Primfactoren des q enthalten, wo q eine zu irgend einem Exponenten t gehörende Primzahl ist, $q^t \equiv 1$, Mod λ. Die dem $\psi_r(\alpha)$ der Kreistheilung entsprechenden Zahlen dieser neuen Methode haben die analogen Eigenschaften wie diese und das analoge Bildungsgesetz. Der dritte Paragraph giebt, auf die im vorigen § entwickelte Methode sich stützend, die Bestimmungen der Symbole $\left(\frac{\lambda}{f(\alpha)}\right)$, $\left(\frac{1-\alpha^{\varkappa}}{f(\alpha)}\right)$, $\left(\frac{E(\alpha)}{f\alpha}\right)$ für den allgemeinen Fall daß $f(\alpha)$ ein beliebiger idealer Primfactor ist, doch allgemein ein Primfactor einer Primzahl q, welche zum Exponenten t gehört für den Modul λ. Dieser Paragraph ist, so wie der vorhergehende ganz neuen Ursprungs. Der vierte Paragraph enthält meine, Dir bekannte Methode die Logarithmen der complexen Zahlen für den Modul λ oder λ^n zu entwickeln und anzuwenden.

Der fünfte Paragraph enthält eine Anwendung der gefundenen Resultate auf das allgemeine Reciprocitätsgesetz, bestehend in der vollständigen Lösung der bestimmten Aufgabe. Wenn das Reciprocitäts-Verhältniß zweier complexen Primzahlen in der primären Form derselben (nach meiner Definition des Primären) als gegeben angesehen wird, daraus das Reciprocitäts-Verhältniß derselben beiden Zahlen zu finden, wenn dieselben nicht primär sondern beliebig mit Einheiten belastet sind. Das Resultat dieser Aufgabe ist sehr merkwürdig, nämlich wie folgt: Wenn $f(\alpha)$ und $\varphi(\alpha)$ zwei beliebige complexe Zahlen sind, $f_1(\alpha)$, $\varphi_1(\alpha)$ dieselben in der primären Form, so ist

$$\frac{\left(\frac{f(\alpha)}{\varphi(\alpha)}\right)}{\left(\frac{\varphi(\alpha)}{f(\alpha)}\right)} = \frac{\left(\frac{f_1(\alpha)}{\varphi_1(\alpha)}\right)}{\left(\frac{\varphi_1(\alpha)}{f_1(\alpha)}\right)} \cdot \alpha^{\Sigma}$$

wo

$$\sum = \frac{d_0^2 l f(e^v)}{dv^2} \cdot \frac{d_0^{\lambda-2} l \varphi(e^v)}{dv^{\lambda-2}} - \frac{d_0^3 l f(e^v)}{dv^3} \cdot \frac{d_0^{\lambda-3} l \varphi(e^v)}{dv^{\lambda-3}} + \frac{d_0^4 l f(e^v)}{dv^4} \cdot \frac{d_0^{\lambda-4} l \varphi(e^v)}{dv^{\lambda-4}}$$

$$- \cdots - \frac{d_0^{\lambda-2} l f(e^v)}{dv^{\lambda-2}} \cdot \frac{d_0^2 l \varphi(e^v)}{dv^2};$$

die Differentialquotienten für den Werth $v = 0$ genommen.

NB. Wenn $f(\alpha)$ oder $\varphi(\alpha)$ ideal sind, so ist dafür $\sqrt[H]{(f(\alpha)^H)}$ zu nehmen wo $f(\alpha)^H$ wirklich ist. Ebenso in den Differentialquotienten $\frac{d_0^{\varkappa} l f(e^v)}{dv^{\varkappa}}$ etc.

Wenn das von mir gefundene Reciprocitäts-Gesetz richtig ist, nach welchem $\left(\frac{f_1(\alpha)}{\varphi_1(\alpha)}\right) = \left(\frac{\varphi_1(\alpha)}{f_1(\alpha)}\right)$, so ist allgemein für beliebige complexe Primzahlen, die nicht primär sind

$$\left(\frac{f(\alpha)}{\varphi(\alpha)}\right) = \left(\frac{\varphi(\alpha)}{f(\alpha)}\right) \cdot \alpha^{\Sigma}$$

Nach meiner Vermuthung stimmt dieses mit dem EISENSTEINschen Resultate im 39. Bande von CRELLES Journal überein, wo es jedoch weder in seiner einfachsten noch besten Gestalt so verworren dargestellt ist, daß es schwer fällt die Übereinstimmung zu ergründen. EISENSTEIN hätte, wenn dieß wahr ist, durch seine Methode also wirklich mein einfaches Reciprocitätsgesetz finden können, wenn er es geschickter angefangen hätte und wenn er etwas mehr von dem Einflusse der Einheiten auf die complexen Zahlen gewußt hätte. Seine Bedingungen aber, die er dafür angiebt, daß die einfache Reciprocität zwischen zwei Zahlen Statt habe, sind zwar nicht falsch aber unerfüllbar.

Ein Concept der ganzen Abhandlung wird dir vorgelegt werden sobald Du einmal zu mir kommst. Jetzt nur noch die Nachricht, daß ich gestern einen großen schwarz gesiegelten Brief erhalten habe, aus welchem ein Schreiben mit einem breiten schwarzen Trauerrande sich entwickelte, nebst einem Diplome. Die Trauer bezog sich auf den seeligen Ernst August von Hannover, das Diplom, mit Gauss Unterschrift, ernennt mich zum correspondirenden Mitgliede der Göttinger Akademie. Lebe wohl empfiehl mich Deiner lieben Frau und besuche bald einmal Deinen Freund

E. Kummer.

Breslau, den 2. Jan. 1852.

Geliebtester Freund!

Ich bin Dir sehr dankbar für Deine interessanten mathematischen Mittheilungen, durch welche, wenn ich sie ein halbes Jahr eher erhalten hätte, meine neue Redaction der Theorie der complexen Zahlen, die in Liouville erscheinen soll, nicht unbedeutend hätte abgekürzt und verbessert werden können. Besonders begierig aber bin ich auf den versprochenen Beweis, daß in $x^\lambda + y^\lambda = z^\lambda$ eine der Zahlen x, y, z durch λ theilbar sein muß. Ein allgemeiner Beweis dieses Satzes existiert noch nicht. Meine Voraussetzung daß λ nicht eine Ausnahmszahl sei, daß B_n nicht $\equiv 0$ Mod λ sei für einen der Werthe $n = 1, 2, 3, \ldots \frac{\lambda - 3}{2}$, ist von Cauchy schon dahin zurückgeführt, daß die Summe $1^{\lambda-4} + 2^{\lambda-4} + \cdots + \left(\frac{\lambda - 1}{2}\right)^{\lambda-4}$ nicht durch λ theilbar sei, welches darauf hinausläuft, daß nur $B_{\frac{\lambda-3}{2}}$ nicht $\equiv 0$ Mod λ sein muß, ohne diese Voraussetzung aber geht es nicht. Ich habe dieselbe später noch weiter eingeschränkt, nämlich dahin, wenn nur nicht beide Bernouillischen Zahlen $B_{\frac{\lambda-3}{2}}$ und $B_{\frac{\lambda-5}{2}}$ zugleich durch λ theilbar sind so muß eine der Zahlen x, y, z durch λ theilbar sein, ich kann auch noch mehr Bedingungen hinzufügen, aber ein voraussetzungsloser Beweis ist auch nach meiner Methode nicht hervorgegangen, denn ich kann nicht beweisen, daß $B_{\frac{\lambda-3}{2}}$ niemals durch λ theilbar ist, auch nicht daß keine Zahl λ existirt, für welche $B_{\frac{\lambda-3}{2}}$ und $B_{\frac{\lambda-5}{2}}$ zugleich durch λ theilbar sind.

. . .

Dein Dich herzlich liebender

E. Kummer.

Breslau d. 12. März 1853.

Geliebtester Freund!

Ich habe mir schon die ganze Woche selbst Vorwürfe gemacht, daß ich Deinen ersten Brief noch nicht beantwortet habe und bin nun doppelt beschämt, daß Du mir noch einen zweiten nachschickst . . . Ueber den Fortschritt Deiner Arbeit freue ich mich so sehr, als ob es meine eigene wäre. Daß die idealen Primzahlen für eine algebraische Untersuchung nicht wesentlich, sondern allenfalls Erleichterungs- und Hilfsmittel sein könnten, war mir eigentlich von vornherein klar. Ich kann es daher nur billigen, daß Du dieselben hier als unwesentlich ausscheidest, namentlich da ohne dieselben der Zweck einfacher erreicht wird. Erst dann, wenn man die gefundenen algebraischen Resultate auf Zahlentheorie wird anwenden wollen, werden dieselben zu ihrem vollen Rechte gelangen, weil dieses ihr eigentlicher Beruf ist, den sie aber auch in so ausgezeichneter Weise erfüllen, daß ich nicht zweifle sie werden eine bleibende Errungenschaft der mathematischen Wissenschaften für alle folgenden Zeiten sein. Daß ich die von Dir gefundenen Resultate für richtig halte, versteht sich von selbst, ich habe nicht den geringsten Grund daran zu zweifeln, wenn auch die rein objective Unzweifelbarkeit erst dann vorhanden ist, wenn man die vollständig ausgearbeiteten Beweise derselben vor sich hat, und wenn man dieselben Schritt für Schritt in seinem eigenen Geiste reproducirt hat. Bis dahin kann man immer noch meinen, daß einem etwas entgangen sein könnte, oder daß irgend eine nothwendige Grundlage der ganzen Schlußfolge fehlen könnte. Die einfachste Form der Wurzeln der Abelschen Gleichungen zu finden, deren Coefficienten bestimmte Irrationalitäten sind, scheint mir vollständig ausgeführt noch eine sehr große Arbeit auszumachen, namentlich wenn Du nicht etwa auch hier eine Methode findest, durch welche das Ideale vermieden wird, denn von den idealen Primfactoren der complexen Zahlen, welche aus den Wurzeln einer beliebigen irreduktiblen Gleichung des n^{ten} Grades gebildet sind, habe ich selbst keine recht klare Vorstellung. Es fällt mir hierbei ein, daß ich Dir noch eine Antwort auf eine früher mir gestellte Frage schuldig bin, wie ein idealer Primfactor als Wurzel einer wirklichen complexen Zahl definirt werden könne. Diese Definition ist höchst einfach die folgende: Wenn man eine wirkliche complexe Zahl $\Phi(\eta)$ hat, wo η Periode von t Gliedern ist, deren Norm, in Beziehung auf die Perioden genommen, gleich der h^{ten} Potenz der Primzahl q ist, welche zum Exponenten t gehört, und wenn das Produkt

$$\Phi(\eta_1)\,\Phi(\eta_2)\ldots\Phi\left(\eta_{\frac{p-1}{t}-1}\right)$$

(d. h. das Produkt aller Faktoren der Norm mit Ausschluß eines einzigen), nicht durch q theilbar ist, so ist $\sqrt[h]{\Phi(\eta)}$ ein idealer Primfaktor des q. . . .

Nun lebe wohl, komme bald zu mir, begrenze Dir Deine mathematische Arbeit in der Art, daß sie ein vollständiges Ganze bildet, und daß sie für eine baldige Vollendung nicht zu viel Arbeit erfordert und arbeite dann innerhalb der Dir selbst gesteckten Grenzen bis Du alles vollständig erkannt hast und was Du erkannt hast das arbeite so bald als möglich aus.

Dein

E. Kummer.

Breslau den 24. April 1853.

Geliebtester Freund!

. . . Ueber meine eigenen mathematischen Arbeiten habe ich Dir etwas trauriges zu melden, nämlich eine ganze ziemlich lange Zeit verlorener Mühe. Ich ging an die Ausarbeitung des Beweises des Reciprocitätsgesetzes für conjugirte complexe Primzahlen und war schon mit der Reinschrift für Crelles Journal beschäftigt, als ich ein ganz kleines Versehen bemerkte, als ich demselben weiter nachforschte, entdeckte ich denn, daß dieser Beweis wirklich nicht für alle Primzahlen giltig ist, sondern nur für diejenigen, für welche Ind $E_n(\alpha)$ nicht congruent Null ist, obgleich meine Zahlenbeispiele, die ich für diesen Zweck noch besonders zahlreich gemacht habe, immer die Richtigkeit des Reciprocitätsgesetzes zeigen. Hiermit hängt auch zusammen, daß eines meiner Hauptresultate, auf welches ich seit einem Vierteljahre gebaut hatte, daß der zweite Faktor der Klassenzahl $\frac{D}{\Delta}$ niemals durch λ theilbar ist, falsch ist oder wenigstens unbewiesen. Das ganz kleine Versehen, welches dieses alles hervorgebracht hatte, besteht nämlich lediglich darin, daß ich Einheiten $E(\alpha, \eta)$ deren Normen in Beziehung auf die Perioden genommen wirklich nicht gleich Eins sind, sondern Einheiten $E(\alpha)$, dem zu bildenden Ausdrucke $P\varepsilon(\alpha,\eta)$ zu Grunde gelegt hatte in der Meinung daß $N_\eta\,\varepsilon(\alpha,\eta) = 1$ sei, nämlich in den Fällen, wo die Kreistheilungs-Einheiten in diesem Ausdrucke nicht den erwünschten Erfolg geben und wo immer fundamentalere Einheiten existiren. Daß diese dann wirklich existiren ist ganz richtig, aber leider eben so richtig, daß sie zwar an sich recht schöne Einheiten sind aber die hier nothwendige Eigenschaft nicht besitzen, daß ihre Normen in Beziehung auf die Perioden allein genommen schon gleich Eins sein müssen. Einen Fortschritt der Erkenntniß habe ich auch durch diese Enttäuschung gemacht, welche mir noch lange verborgen geblieben wäre, wenn ich nicht daran gegangen wäre meine gefundenen Resultate vollständig auszuarbeiten. Ich habe ferner hieraus ersehen, daß meine Ausdrücke $P\varepsilon(\alpha, \eta)$ so schön sie auch sind, doch für einen hinreichenden Beweis des Reciprocitätsgesetzes nicht ausreichen, daß ich also wenn ich diese Aufgabe, die mich seit langer Zeit beschäftigt, nicht ganz will fallen lassen, andere fruchtbare neue Gedanken fassen muß, welches eine Sache ist, die ich nicht in meiner Gewalt habe. Ich werde also vorläufig hauptsächlich meinen Fleiß nur auf die Weiterführung der Theorie der complexen Zahlen wenden, und dann sehen, ob etwas daraus entsteht, was auch über jene Aufgabe Licht verbreitet. . . .

Dein

E. Kummer.

Berlin den 29. Nov. 1856.

Geliebtester Freund!

. . . Mit Deinen mathematischen Arbeiten hast Du eine getäuschte Erwartung gehabt aber eine Erkenntniß gewonnen, also doch einen Fortschritt gehabt. Uebrigens betrifft diese getäuschte Erwartung ja nur einen untergeordneten Punkt, denn so viel ich verstanden habe wird dadurch Deine Aussicht in den Abelschen Transcendenten die wahren auflösenden Gleichungen für alle Grade, namentlich für den 5ten Grad in den Modulargleichungen der elliptischen Functionen zu finden, nicht alterirt. Ehe ich Dir etwas Näheres über den Zusammenhang der Einheiten $e(z)$ mit den ambigen Einheiten $e(w)$ schreibe muß ich Dir mittheilen, daß ich ein Avancement gehabt habe, indem ich in der Göttinger Akademie vom correspondierenden zum auswärtigen Mitgliede befördert worden bin. Du weißt in welchem Sinne ich so etwas auffasse, es macht mir manchmal etwas bange, weil ich meine Verdienste zu meinen Ehren auch ohne weitere Beförderungen schon im richtigen Verhältniß weiß, so besorge ich es könnte dieses richtige

Verhältniß durch weitere Beförderung gestört werden, oder es möchten mir daraus Verpflichtungen erwachsen, denen zu genügen ich nicht im Stande bin.

Nun über die Einheiten. Jede Einheit $e(z)$ hat wie jede wirkliche Zahl $f(z)$ die Eigenschaft, daß $Ne(z) \equiv c \bmod \lambda$ wo c eine nichtcomplexe weder z noch α enthaltende ganze Zahl ist; weil nun $Ne(z)$ auch $= E(\alpha)$ ist also $E(\alpha) \equiv c \bmod \lambda$, so muß $E(\alpha)$ eine λ^{te} Potenz einer Einheit sein, also $Ne(z) = \varepsilon(\alpha)^\lambda$ darum ist $\frac{e(z)}{\varepsilon(\alpha)}$ eine Einheit $e(z)$ deren Norm gleich Eins ist. Nun ist, wenn

$$Pe(z) = 1 + \alpha^r e(z) + \alpha^{2r} e(z) e(z_1) + \cdots + \alpha^{(\lambda-1)r} e(z) e(z_1) \cdots e(z_{\lambda-2})$$

$$e(z) \cdot Pe(z_1) = Pe(z), \quad \text{also } e(z) = \frac{Pe(z)}{Pe(z_1)}.$$

Ich stelle $Pe(z)$ als durch w ausgedrückt dar, so treten aus demselben gewisse Factoren $1 - \alpha$ heraus. Sei

$$Pe(z) = \varrho^\varkappa F(w)$$

so ist

$$Pe(z_1) = \varrho^\varkappa F(w\alpha)$$

also

$$e(z) = \frac{F(w)}{F(w\alpha)}.$$

Aus dieser Gleichung geht hervor daß $F(w)$, nachdem die mit $F(w\alpha)$ gemeinschaftl. Factoren, welche nur α enthalten, hinweggehoben sind, nur noch von der Art sein kann, daß $NF(w)$ nichts als $1 - \alpha$ und die Factoren von w^λ enthält. Ich zeige ferner, daß wenn w^λ nur Potenz einer Primzahl $f(\alpha)$, ist $NF(w)$ den Factor $f(\alpha)$ nicht enthält, wenn man in $Pe(z)$ die simple Einheit α^r richtig bestimmt, und zweitens daß $NF(w)$ nicht durch $\varrho = 1 - \alpha$ theilbar sein kann, sodaß $F(w)$ nichts als eine Einheit $E(w)$ ist, und zwar eine ambige, weil

$$e(z) = \frac{E(w)}{E(w\alpha)}.$$

Ich bin damit beschäftigt in ähnlicher Weise von jeder Einheit $e(w)$ deren Norm gleich Eins ist (oder was dasselbe ist $Ne(w) = E(\alpha)^\lambda$) zu einer Einheit $\varepsilon(w)$ zurückzusteigen

$$e(w) = \frac{\varepsilon(w)}{\varepsilon(w\alpha)}$$

welche Zurücksteigung zu fundamentaleren Einheiten so weit fortzusetzen sein wird, bis man auf eine Einheit $\varepsilon(w)$ kommt, deren Norm eine Einheit $\varepsilon(\alpha)$ ist welche nicht eine λ^{te} Potenz ist. Es fehlt mir dazu einzig und allein nur der vollkommene Beweis des Satzes daß wenn $Ne(w) = 1$ ist, $e(w) = 1 + (1 - \alpha)\,\Phi(w)$ sein muß, ein Satz den ich für durchaus richtig halte, da ich leicht beweisen kann, daß wenn $Ne(w) = 1$ ist, und man setzt $e(w) = 1 + \Psi(w)$, die Norm $N\Psi(w) \equiv 0$ mod λ sein muß.

Lebe wohl geliebter Freund empfiehl mich Deiner Frau und Mutter und nimm die herzlichsten Grüße von meiner Frau.

Dein

E. Kummer.

Berlin d. 10. Decbr. 1856.

Geliebtester Freund!

... In meinen mathematischen Studien habe ich wieder manches ergründet und dabei unter andern auch gefunden, daß die Art und Weise wie ich aus der wirklichen Darstellung der unabhängigen ambiguae das Reciprocitätsgesetz für 5te Potenzreste bewiesen habe, eben nur für die 5ten Potenzreste anwendbar ist, für die höheren aber nicht. Der Grund hiervon liegt darin, daß wenn für das allgemeine $\lambda > 5$ und wenn $w^\lambda = D(\alpha)$ nur eine einzige Primzahl enthält oder eine Potenz einer einzigen, von den $\frac{\lambda - 1}{2}$ unabhängigen ambiguis nur die zwei auf der Hand liegenden, nämlich w und $1 - w$ nicht dem Genus principale angehören, alle übrigen aber nothwendig dem Genus principale angehören, da sie sogar die Form haben

$$f(w) = a + \varrho\,\Phi(w), \text{ also } Nf(w) \equiv a^\lambda \mod \lambda$$

für dieselben ist, so daß die Charaktere alle $\equiv 0$ sind. Uebrigens muß ich bemerken, daß mir zum vollen Beweise dieses nicht grade angenehmen Factums noch das Niederschreiben oder Ausarbeiten des im Kopfe fertigen fehlt und daß doch noch vielleicht ein Loch für andere ambiguas gelassen sein könnte, welches mir entgangen wäre und welches sich noch beim Ausarbeiten ergeben könnte. Ich habe nämlich jenes System von $(\lambda - 1) + (\lambda - 2) + (\lambda - 3) + \cdots + 2 + 1$ Congruenzen jetzt gebändigt, wenn auch nur für den Fall, auf den hier zunächst alles ankommt, nämlich wo alle diese Congruenzen nur in Beziehung auf den einfachen Modul ϱ betrachtet werden. Es ist mir so mit meiner Untersuchung ähnlich ergangen wie Dir mit der

Deinen, aber ich lasse darum auch den Muth eben so wenig sinken als Du. Ich habe noch die feste Ueberzeugung daß ich in meiner jetzigen Arbeit die Ingredienzien zur Ergründung der Reciprocitätsgesetze besitze, und daß es wenn nicht auf die erste so doch auf eine andere Weise gelingen wird auch für $\lambda = 7, 11, 13$, etc. aus der Betrachtung der complexen Zahlen $F(z)$ mit Zuziehung der complexen Zahlen $f(w)$ das Reciprocitätsgesetz zu beweisen. Es fehlt nur noch an der vollen Erkenntniß der ambiguae und der mit denselben verwandten Einheiten, darum steuere ich jetzt auf diese los. . . .

Dein

E. Kummer.

Berlin den 6. Juli 1860.

Geliebter Freund!

. . . Ich habe . . . das vollständig ausgearbeitet und ins Reine geschrieben, was ich den nächsten Donnerstag in der Akademie vorzutragen habe: über atmosphärische Strahlenbrechung, dasselbe ist sehr kurz behandelt, so daß es nur für die Monatsberichte eingerichtet ist, wo es etwa einen Bogen füllen wird. Leider kann ich es Dir vor dem Drucke nicht zur Durchsicht vorlegen. In drei Wochen, in unserer nächsten Klassensitzung, will ich noch etwas vortragen, nämlich über die drei Arten von Strahlenbündeln (deren Modelle ich vorzeigen werde), in wiefern dieselben in der Natur wirklich vorkommen. Ich weiß nicht ob ich Dir den Hauptsatz, der dieses vollständig und höchst anschaulich löst, schon mitgetheilt habe. Derselbe lautet etwa so: Wenn einem Krystalle die Wellenfläche $W = 0$ zukommt (welche die Fresnelsche oder das Rotationsellipsoid, oder jede andere sein kann), so denke man sich diese Wellenfläche in beliebiger Größe um den Krystall herum passend beschrieben (so daß die Axen die richtige Lage haben). An einem beliebigen Punkte der Wellenfläche denke man sich den unendlich kleinen Dupinschen Kegelschnitt, die Indikatrix und für diesen irgend zwei conjugirte Durchmesser, so sind die beiden Ebenen, welche durch den Mittelpunkt der Wellenfläche und durch diese beiden conjugirten Durchmesser hindurchgelegt werden, die Fokalebenen eines Strahlenbündels, welches in der betreffenden Richtung (d. h. in der Richtung vom Mittelpunkte des Dupinschen Kegelschnitts nach dem Mittelpunkte der Wellenfläche) im Krystalle wirklich existiert und durch ein passend angebrachtes auf den Krystall auffallendes normales Strahlenbündel im Krystalle hervorgebracht werden kann. In der betreffenden Richtung existiert auch kein anderes Strahlenbündel im Krystalle als ein solches dessen Fokalebenen aus der Wellenfläche diese conjugirten Richtungen ausschneiden. Ist die Indikatrix eine Hyperbel, so finden auch imaginäre conjugirte Richtungen Statt, also auch Strahlenbündel mit imaginären Fokalebenen.

... Mit herzlichen Grüßen von mir und den Meinen an Dich und Deine Frau

Dein

E. Kummer.

Berlin, d. 2. Decbr. 1860 (Abends).

Geliebter Freund!

Das Produkt

$$P_\varkappa = \sin\frac{2\varkappa\pi}{29}\cdot\sin\frac{2\cdot 5\varkappa\pi}{29}\cdot\sin\frac{2\cdot 11\varkappa\pi}{29}\cdot\sin\frac{2\cdot 12\varkappa\pi}{29}\cdot\sin\frac{2\cdot 15\varkappa\pi}{29}\cdot\sin\frac{2\cdot 16\varkappa\pi}{29}\cdot\sin\frac{2\cdot 23\varkappa\pi}{29}\cdot\sin\frac{2\cdot 27\varkappa\pi}{29}$$

hat für alle Werthe des $\varkappa = 1, 2, 3, \ldots 14$, und darum auch für die übrigen $\varkappa = 15, 16, \ldots 28$ stets einen positiven Werth; es wird dieß am leichtesten gezeigt, wenn man die Zahlen 1, 5, 11, 12, 15, 16, 23, 27 nach einander mit $1, 3, 3^2, 3^3 \ldots$ multiplicirt statt mit 1, 2, 3, 4 ... und die Vielfachen von 29 wegläßt. Man erhält so:

$$\begin{array}{rrrrrrrrl}
1, & 5, & 11, & 12, & \underline{15}, & \underline{16}, & \underline{23}, & \underline{27}, & -\ 4 \text{ neg.} \\
3, & \underline{15}, & 4, & 7, & \underline{16}, & \underline{19}, & 11, & \underline{23}, & -\ 4 \text{ „} \\
9, & \underline{16}, & 12, & \underline{21}, & \underline{19}, & \underline{28}, & 4, & 11, & -\ 4 \text{ „} \\
\underline{27}, & \underline{19}, & 7, & 5, & \underline{28}, & \underline{26}, & 12, & 4, & -\ 4 \text{ „} \\
\underline{23}, & \underline{28}, & \underline{21}, & \underline{15}, & \underline{26}, & \underline{20}, & 7, & 12, & -\ 6 \text{ „} \\
11, & \underline{26}, & 5, & \underline{16}, & \underline{20}, & 2, & \underline{21}, & 7, & -\ 4 \text{ „} \\
4, & \underline{20}, & \underline{15}, & \underline{19}, & 2, & 6, & 5, & \underline{21}, & -\ 4 \text{ „} \\
12, & 2, & \underline{16}, & \underline{28}, & 6, & \underline{18}, & \underline{15}, & 5, & -\ 4 \text{ „} \\
7, & 6, & \underline{19}, & \underline{26}, & \underline{18}, & \underline{25}, & \underline{16}, & \underline{15}, & -\ 6 \text{ „} \\
\underline{21}, & \underline{18}, & \underline{28}, & \underline{20}, & \underline{25}, & \underline{17}, & \underline{19}, & \underline{16}, & -\ 8 \text{ „} \\
5, & \underline{25}, & \underline{26}, & 2, & \underline{17}, & \underline{22}, & \underline{28}, & \underline{19}, & -\ 6 \text{ „} \\
\underline{15}, & \underline{17}, & \underline{20}, & 6, & \underline{22}, & 8, & \underline{26}, & \underline{28}, & -\ 6 \text{ „} \\
\underline{16}, & \underline{22}, & 2, & \underline{18}, & 8, & \underline{24}, & \underline{20}, & \underline{26}, & -\ 6 \text{ „} \\
\underline{19}, & 8, & 6, & \underline{25}, & \underline{24}, & 14, & 2, & \underline{20}, & -\ 4 \text{ „}
\end{array}$$

Die Zahlen größer als $\frac{29}{2}$ geben negative Sinus, die übrigen positive, die Anzahl der negativen, welche unterstrichen sind, ist stets eine grade.

Du wirst vielleicht längst selbst gefunden haben, daß die Bedingung, nach welcher die Kreistheilungseinheiten die ein Quadrat einer Einheit geben sollen stets positiv für alle conjugirten sein müssen nichts weiter ergiebt, als das uns beiden bekannte, daß, wenn

$$\varepsilon(\alpha)^2 = e(\alpha)^{m(\beta)} = e(\alpha)^m e(\alpha^\gamma)^{m_1} \dots e(\alpha^{\gamma^{\mu-1}})^{m_{\mu-1}}$$

gesetzt wird, wo $e(\alpha)$ die Kreistheilungseinheit ist, und β irgend eine Wurzel der Gleichung $\beta^\mu = +1$ $\left(\mu = \frac{\lambda-1}{2}\right)$, $m(\beta)$ einen idealen oder wirklichen Primfaktor von 2 enthalten muß. Aus der Forderung, daß

$$e(\alpha^{\gamma^h})^m e(\alpha^{\gamma^{h+1}})^{m_1} e(\alpha^{\gamma^{h+2}})^{m_2} \dots e(\alpha^{\gamma^{h+\mu-1}})^{m_{\mu-1}}$$

für alle Werthe des h positiv sein soll habe ich direct gefolgert, daß

$$m + m_1\beta + m_2\beta^2 + \cdots + m_{\mu-1}\beta^{\mu-1}$$

einen complexen Primfactor von 2 enthalten muß, aber umgekehrt findet dieser Satz nicht Statt, sondern nur der folgende:

Wenn $b_0 + b_1\beta + b_2\beta^2 + \cdots + b_{\mu-1}\beta^{\mu-1}$, einen complexen Primfactor von 2 enthält, (wo $b_0, b_1, \dots b_{\mu-1}$ dieselben Zahlen sind, wie in meinem Memoir, nämlich $\lambda b_\varkappa = \gamma\gamma_{\varkappa-1} - \gamma_\varkappa$ und die primitive Wurzel γ ungerade angenommen ist) so kann man stets die Exponenten $m, m_1, \dots m_{\mu-1}$ so bestimmen, daß

$$e(\alpha)^m e(\alpha^\gamma)^{m_1} \dots e(\alpha^{\gamma^{\mu-1}})^{m_{\mu-1}}$$

mit allen seinen conjugirten stets positiv ist. Es folgt hieraus auch, daß dieß niemals der Fall ist, wenn nicht der erste Faktor der Klassenanzahl durch 2 theilbar ist. Darum habe ich als schlagendes Beispiel nur $\lambda = 29$ nehmen müssen.

Ich habe übrigens dieß alles in den zwei Stunden ergründet, die seit der Absendung meines ersten Stadtpostbriefes an Dich verflossen sind, woraus Du abnehmen kannst, daß es leicht zu zeigen war.

Lebe wohl, ich erwarte bald Nachricht von Dir.

Dein

E. Kummer.

Berlin, d. 14. Jan. 1861, 12 Uhr früh.

Geliebter Freund!

Mit herzlichem Danke für Deine Mittheilung, die ich soeben erhalte, melde ich Dir, daß ich gestern Abend das was Du andeutest vollständig ausgeführt habe. Der Ausdruck der Klassenanzahl für compl. Zahlen aus n^{ten} Einheitswurzeln ist wenn $n = p^\pi p_1^{\pi_1} p_2^{\pi_2} \ldots$ (den Fall wo n gerade ist habe ich vorläufig bei Seite liegen lassen) ist folgender

$$H = \frac{(p-1)\,(p_1-1)\,\ldots}{(p^\tau-1)^\theta (p_1^{\tau_1}-1)^{\theta_1}\ldots} \, \frac{P}{(2n)^{\frac{1}{2}\varphi(n)-1}} \cdot \frac{D}{\Delta}$$

$$\text{wenn } p \text{ zum Exp. } \tau \text{ gehört mod } \varphi\left(\frac{n}{p^\pi}\right) \quad \tau\theta = \varphi\left(\frac{n}{p^\pi}\right)$$

$$\text{,,} \quad p_1 \quad \text{,,} \quad \text{,,} \quad \tau_1 \quad \text{,,} \quad \text{mod } \varphi\left(\frac{n}{p_1^{\pi_1}}\right) \quad \tau_1\theta_1 = \varphi\left(\frac{n}{p_1^{\pi_1}}\right)$$

etc. etc.

$$P = \Pi \Sigma_m w^{c \operatorname{ind} m} w_1^{c_1 \operatorname{ind}_1 m} w_2^{c_2 \operatorname{ind}_2 m} \ldots m$$

wo das Produktzeichen Π auf alle Werthverbindungen der $c, c_1, c_2 \ldots$ geht für

$$c = 0, 1, 2, \ldots, \varphi(p^\pi) - 1, \quad c_1 = 0, 1, 2, \ldots \varphi(p_1^{\pi_1}) - 1, \text{ etc.}$$

für welche $c + c_1 + c_2 \ldots$ ungrade ist, w prim. Wurzel von $w^{\varphi(p^\pi)} = 1$ etc. m alle Zahlen kleiner als n u. rel. Prz. zu n.

Es ist ganz angenehm so alles unter einem Hute zu haben, nur der erste Faktor, der nicht das n sondern das p, p_1, p_2 einzeln enthält, ist störend. Zum Rechnen aber ist diese Form nicht wohl zu gebrauchen, weil die im P enthaltenen überflüssigen Faktoren zu störend sind. Dein E. Kummer.

Berlin, den 25. Juli 1862.

Geliebter Freund!

... Ich habe Dir nun einige wissenschaftliche Mittheilungen zu machen, aber nicht über meine eigenen Arbeiten, die keine nennenswerthen Ergebnisse gebracht haben, sondern mehr litterarische. Salmon in Dublin hat eine Geometry of three dimensions herausgegeben, welche ich mir angeschafft habe und in welcher ich mit Fleiß und mit Lust studire. Ich finde darin vielfach Aufschlüsse grade über die Fragen die für mich ein besonderes Interesse haben, unter andern auch über die Theorie der algebraischen Kurven doppelter Krümmung und die damit zusammenhängenden algebraischen Fragen. SALMON stützt diese Theorie auf einen sehr einfachen Satz, der aber bei ihm nicht hinlänglich bewiesen wird, so daß ich nicht weiß, ob er auch allgemein wahr sein mag, nämlich auf den Satz: Eine Raum-Curve des r^{ten} Grades hat mit einer Fläche des p^{ten} Grades pr Punkte gemein (hat sie mehr mit derselben gemein, so liegt sie ganz auf der Fläche, wenn nämlich Curve und Fläche irreductibel sind). Erschöpfend ist SALMONs Art der Behandlung der algebraischen Raumcurven nicht, aber es ist vieles von dem, was wir damals erörtert haben, sehr nett und einfach entwickelt. Schaffe Dir doch auch dieses Buch gleich an und außerdem vielleicht auch noch die anderen desselben Verfasser treatise on conic sections, und treatise on higher plane curves, die ich besitze, und Lessons on higher Algebra, die ich mir bestellt habe, auf welche er in der Geometrie vielfach verweist und welche nicht bloß die formalen Geschichten von CAYLEY und SYLVESTER, sondern die realeren Fragen behandelt, wie ich aus den Zitaten schließe. Ein anderes Werk desselben Verf. Sermons preached in the chapel of Trinity College wird für uns weniger Interesse haben. . . .

Der Dr. QUINCKE hat eine experimentelle Untersuchuug der dünnen optischen Strahlenbündel ausgeführt, in welcher überall nachgewiesen wird, daß meine theoretisch ermittelten Resultate über die Winkel der beiden Fokalebenen in einfach brechenden Medien im einaxigen Kalkspath und im zweiaxigen Aragonit richtig sind; seine beobachteten Winkel weichen von meinen berechneten höchstens um einen halben Grad ab. Er hat mit ausgezeichneter Genauigkeit experimentirt. Seine sehr gut geschriebene Abhandlung hierüber habe ich am vorigen Donnerstag in der Akademie vorgetragen und sie erscheint in den Monatsberichten für den Juli. . . .

Herzliche Grüße an Dich und Deine Frau von den Meinen Dein

E. KUMMER.

Berlin den 10. Juni 1865.

Geliebter Freund!

Die sechs Strahlensysteme zweiter Ordnung und zweiter Klasse, welche auf einer Fläche vierten Grades mit 16 singulären Punkten liegen, sind in der That so beschaffen, daß ein jedes derselben 16 singuläre Punkte hat, von welchen Strahlbüschel ausgehen, welche in den 16 singulären Tangentialebenen liegen. Ein einziges bestimmtes dieser sechs Strahlensysteme hat in jedem der 16 singulären Punkte nur ein Strahlbüschel, welches in einer der 6 durch diesen Punkt gehenden Ebenen liegt, die übrigen fünf Strahlensysteme vertheilen ihre von diesem singulären Punkte ausgehenden Strahlbüschel in die übrigen fünf durch diesen singulären Punkt gehenden sing. Tangentialebenen. Nähert man sich von einem beliebigen Punkte der Fläche ausgehend irgend einem Punkte der Curve in welcher eine singuläre Tangentialebene die Fläche 4ten Grades berührt, so nähern sich die sechs von diesem Punkte ausgehenden, den verschiedenen 6 Systemen angehörenden Strahlen den Richtungen welche nach den 6 in dieser singulären Tangentialebene liegenden singulären Punkten führen, und sobald man in diese Berührungscurve hineintritt, gehen die 6 Strahlen nach diesen sechs singulären Punkten. Diese 6 singulären Punkte sind daher die Mittelpunkte von 6 Strahlbüscheln welche in dieser singulären Tangentialebene liegen und welche den 6 verschiedenen Systemen angehören. Dieß ist die vollständige Lösung der Schwierigkeit, welche wir in unseren Ueberlegungen dieser Sache fanden.

Daß ein jedes System 2ter Ordnung und 2ter Klasse singuläre Punkte mit ebenen Strahlbüscheln haben muß, habe ich a priori noch nicht beweisen können.

Dein treuer Freund

E. Kummer.

Leopold Kronecker an E. Kummer.

Nach Mitternacht des 9. September 1881,
noch in Berlin, Bellevuestr. 13.

Geliebtester Freund!

In der ersten Stunde des Tages Deines Doctor-Jubiläums schreibe ich diese Zeilen, die ich im Laufe des Tages selber in Deine Wohnung tragen will. In der ersten Stunde dieses Tages will ich Dir meinen Glückwunsch aufschreiben, da es mir versagt ist, Dir ihn persönlich zu überbringen. Wie herzlich mein Glückwunsch ist, weißt Du, aber weit mehr als Wunsch — denn Du brauchst ihn kaum — bringe ich Dir Dank! Früh hast Du mir Liebe, dann Freundschaft geschenkt, Du hast mir mein mathematisches, ja überhaupt den wesentlichsten Theil meines geistigen Lebens gegeben, und so habe ich Dich immer ganz wie einen Vater geliebt und verehrt, und es hat mich gedrängt, meinen Empfindungen der Pietät zu Deinem Gedenktage Ausdruck zu geben. Darum habe ich es unternommen, aus allem meinem mathematischen Denken, das Du mich einst gelehrt, die Summe zu ziehen und eine Festschrift zu Deinem Jubiläum zu veröffentlichen, in welcher die Liebe zu Dir in der Liebe zur Wissenschaft erscheint, der Du mich geweiht hast. Ich habe mit aller Freudigkeit und mit aller Anstrengung und auch mit Erfolg gearbeitet — denn auf den Werken der Liebe und Pietät ruht der Segen! Theuerster Freund, weniger als das Beste, was ich zu leisten vermag, wollte ich Dir nicht widmen; darum habe ich seit Monaten gedanklich geschafft und seit Wochen geschrieben, bis ich nun wirklich zum Tage fertig geworden bin und schon den ersten Druckbogen bei mir habe. . . . Mit der Abhandlung, die ich Dir jetzt zu Deinem Gedenktage widme, und die den Titel hat „Grundzüge eine arithmetischen Theorie der algebraischen Größen“, erscheint zusammen ein neuer und vervollständigter Abdruck meiner Dir gewidmeten Dissertation. Das früheste und dieses späte Zeichen meiner Liebe und Dankbarkeit sollen verbunden Zeugniß ablegen von der Innigkeit und Stetigkeit derselben, sie sollen die zwei um 36 Jahre auseinanderliegenden Punkte eines seltenen Freundschaftslebens bezeichnen, das einen wesentlichen Theil meines Glückes bildet.

Dein ältester treuer Freund

LEOPOLD KRONECKER.

De aequatione $x^{2\lambda} + y^{2\lambda} = z^{2\lambda}$ per numeros integros resolvenda

Journal für die reine und angewandte Mathematik 17, 203–209 (1837)

Quod clarissimus *Fermat* contendit: aequationem $x^{n+2}+y^{n+2}=z^{n+2}$ per numeros integros resolvi non posse, haud dubie ad elegantissima theoremata referendum est, quae de numerorum proprietatibus hactenus proposita sunt, cujus autem demonstratio gravissimis difficultatibus videtur laborare. Quamquam enim incrementa permagna nostris temporibus theoria numerorum accepit, tamen geometrae clarissimi, qui huic theoremati operam tribuerunt, paucos solummodo casus simpliciores demonstrationibus munire potuerunt. *Cl. Euler, Legendre* et *Lejeune Dirichlet* pro potestatibus tertiis, quartis, quintis et decimis quartis theorematis hujus demonstrationes invenerunt, quae in eo conveniunt, ut ex aequatione proposita alia ejusdem formae aequatio eliciatur, cujus numeri variabiles minores sint quam aequationis datae variabiles; artificia autem per quae ad hanc aequationem similem pervenerunt, pro potestatibus diversis maxime diversa sunt, neque ad alios casus applicationes patiuntur. Itaque res non multum profecit. In re tam difficili, nisi omni proventu carere volumus, a facilioribus incipiendum esse nobis necessarium videtur, itaque aequationem Fermatianam pro potestatum indicibus paribus, nobis tractandam proponimus. Hanc etiam disquisitionem faciliorem ad finem perducere nondum nobis contigit, attamen summas aliquas, quae hanc rem quodammodo promovere videntur cum geometris communicabimus.

Disquisitio nostra praesertim huic theoremati innititur:

Theorema 1. „Si n est numerus primus, atque a et b inter se primi, quantitates $a\pm b$ et $\frac{a^n\pm b^n}{a\pm b}$ non habet factorem communem, nisi numerum n, si vero $a^n\pm b^n$ habet factorem n, eundem etiam $a\pm b$ habere debet, et numerus factorum n in $a^n\pm b^n$ numerum factorum n in $a\pm b$ unitate superat."

Hujus theorematis veritas facile probatur ex aequatione identica

$$1.\quad \frac{a^n\pm b^n}{a\pm b}=(a\pm b)^{n-1}\mp n(a\pm b)^{n-3}ab+\frac{n(n-3)}{1.2.}(a\pm b)^{n-5}a^2b^2\mp\dots$$
$$\dots(\mp 1)^h\frac{n(n-h-1)(n-h-2)\dots(n-2h+1)}{1.2.3\dots h}(a\pm b)^{n-2h-1}a^hb^h+\dots(\mp 1)^{\frac{n-1}{2}}n(ab)^{\frac{n-1}{2}}.$$

Si enim $\frac{a^n\pm b^n}{a\pm b}$ et $a\pm b$ factorem communem habent, etiam $n(ab)^{\frac{n-1}{2}}$ (terminus solus ad dextram aequationis (1.), qui factorem $a\pm b$ non continet) per eundem factorem divisibilis esse debet, et quia ab et $a\pm b$ inter se primi sunt, maximus factor communis quem quantitates $\frac{a^n\pm b^n}{a\pm b}$ et $a\pm b$ habere possunt, erit numerus n. Ad alteram theorematis partem demonstrandam observo coefficientes omnes

$$\frac{n}{1},\quad \frac{n(n-3)}{1.2.},\ \dots\ \frac{n(n-h-1)(n-h-2)\dots(n-2h+1)}{1.2.3\dots h},\ \dots$$

quia integri sunt, et numerus primus n e numeratore per denominatoris factores minores tolli nequit, per n esse divisibiles. Inde sequitur $a^n\pm b^n$ factorem n continere non posse, nisi simul $a\pm b$ per n est divisibilis, positisque $a^n\pm b^n=C.n^{\varkappa}$ et $a\pm b=c.n^{\lambda}$ ex aequatione (1.) sequitur $\lambda=\varkappa-1$, id quod demonstrandum erat.

Quibus praeparatis ad aequationem propositam vertamur:

$$2.\quad x^{2\lambda}+y^{2\lambda}=z^{2\lambda}.$$

Salva quaestionis generalitate numeros x, y, z inter se primos accipimus, et λ numerum primum, si enim duo numerorum x, y, z factorem communem haberent, per eundem etiam tertius numerus divisibilis esset, atque hic factor omnium communis tolleretur, porro si λ esset numerus compositus e factoribus primis $\lambda=\alpha.\beta.\gamma\dots$, aequatione $x^{2\lambda}+y^{2\lambda}=z^{2\lambda}$ satisfieri non posset, nisi aequationes $x'^{2\alpha}+y'^{2\alpha}=z'^{2\alpha}$, $x''^{2\beta}+y''^{2\beta}=z''^{2\beta}$ etc. omnes simul per numeros integros solvi possent. Praeterea patet numerorum x, y, z unum parem ceteros impares esse et quia summa duorum quadratorum inter se primorum per altiorem potestatem ipsius 2 non est divisibilis, sequitur hunc numerum parem non esse z, sed alterum numerorum x et y. Hunc numerum parem nos ubique accipiemus esse y.

Jam theorema supra demonstratum ad aequationem propositam applicemus. Cui si forma datur:

$$3.\quad (z^2-y^2)\left(\frac{z^{2\lambda}-y^{2\lambda}}{z^2-y^2}\right)=x^{2\lambda}.$$

patet primo, si x factorem λ non continet, quia z^2-y^2 et $\frac{z^{2\lambda}-y^{2\lambda}}{z^2-y^2}$ inter se primi sunt, esse

$$4. \quad z^2-y^2=a^{2\lambda}$$

et quia $z+y$ et $z-y$ factorem communem non habent

$$5. \quad z+y=\upsilon^{2\lambda}, \quad z-y=\omega^{2\lambda}.$$

Si vero x per λ divisibilis est, maximaque potestas ipsius λ quae in x continetur est λ^{μ}, $x^{2\lambda}$, ideoque $z^{2\lambda}-y^{2\lambda}$ habent factorem $\lambda^{2\lambda\mu}$, itaque per theorema (1.) z^2-y^2 continebit factorem $\lambda^{2\lambda\mu-1}$, denique quia solo factore communi λ excepto z^2-y^2 et $\frac{z^{2\lambda}-y^{2\lambda}}{z^2-y^2}$ inter se primi sunt, esse debet

$$6. \quad z^2-y^2=\lambda^{2\lambda\mu-1}a^{2\lambda}$$

unde

$$7. \quad z\pm y=\lambda^{2\lambda\mu-1}\upsilon^{2\lambda}, \quad z\mp y=\omega^{2\lambda}.$$

Simili modo ex aequatione $z^{2\lambda}-x^{2\lambda}=y^{2\lambda}$, si y factorem λ non continet, sequitur

$$8. \quad z^2-x^2=b^{2\lambda}.$$

Per hypothesin est y numerus par, z et x impares, itaque si maxima potestas ipsius 2, quae in y continetur est 2^{ν}, $z^{2\lambda}$ habet factorem $2^{2\lambda\nu}$, eundem factorem habet z^2-x^2, inde quia maximus divisor communis numerorum $z+x$ et $z-x$ est 2, sequitur

$$9. \quad z\pm x=2.p^{2\lambda}, \quad z\mp x=2^{2\lambda\nu-1}.q^{2\lambda}.$$

Si vero y per λ divisibilis est, et maxima potestas ipsius λ quae in y continetur est λ^{μ}, per theorema (1.) erit

$$10. \quad z^2-x^2=\lambda^{2\lambda\mu-1}b^{2\lambda}.$$

Praeterea si accipimus maximam potestatem numeri 2 quae in y continetur esse 2^{ν}, quia etiam b eundem factorem 2^{ν} habere debet, erit

$$11. \quad \text{sive} \quad z\pm x=2.p^{2\lambda} \quad \text{et} \quad z\mp x=2^{2\lambda\nu-1}.\lambda^{2\lambda\mu-1}.q,$$

$$12. \quad \text{sive} \quad z\pm x=2.\lambda^{2\lambda\mu-1}p^{2\lambda} \quad \text{et} \quad z\pm x=2^{2\lambda\nu-1}.q^{2\lambda}.$$

Inde quatuor casus speciales erunt discernendi, primus quo neuter numerorum x et y per λ est divisibilis, secundus quo numerus impar x factorem λ habet, tertius et quartus casus, quibus numerus par y per λ divisibilis est. Pro singulis iis casibus est:

$$13. \left\{\begin{array}{llll}
\text{I. } z+y=\upsilon^{2\lambda}, & z-y=\omega^{2\lambda}, & z\pm x=2p^{2\lambda}, & z\mp x=2^{2\lambda\nu-1}q^{2\lambda}, \\
\text{II. } z\pm y=\lambda^{2\lambda\mu-1}.\upsilon^{2\lambda}, & z\mp y=\omega^{2\lambda}, & z\pm x=2p^{2\lambda}, & z\mp x=2^{2\lambda\nu-1}q^{2\lambda}, \\
\text{III. } z+y=\upsilon^{2\lambda}, & z-y=\omega^{2\lambda}, & z\pm x=2p^{2\lambda}, & z\mp x=2^{2\lambda\nu-1}.\lambda^{2\lambda\mu-1}.q^{2\lambda}, \\
\text{IV. } z+y=\upsilon^{2\lambda}, & z+y=\omega^{2\lambda}, & z\pm x=2.\lambda^{2\lambda\mu-1}.p^{2\lambda}, & z\mp x=2^{2\lambda\nu-1}.q^{2\lambda},
\end{array}\right.$$

ex quibus deducuntur formae numerorum x, y, z:

$$
14.\quad \begin{cases}
\text{I. } z = \dfrac{\upsilon^{2\lambda}+\omega^{2\lambda}}{2}, & y = \dfrac{\upsilon^{2\lambda}-\omega^{2\lambda}}{2}, \\
\quad z = p^{2\lambda}+2^{2\lambda n-2}.q^{2\lambda}, & \pm x = p^{2\lambda}-2^{2\lambda n-2}.q^{2\lambda}; \\
\text{II. } z = \dfrac{\lambda^{2\lambda\mu-1}.\upsilon^{2\lambda}+\omega^{2\lambda}}{2}, & \pm y = \dfrac{\lambda^{2\lambda\mu-1}\upsilon^{2\lambda}-\omega^{2\lambda}}{2}, \\
\quad z = p^{2\lambda}+2^{2\lambda\nu-2}q^{2\lambda}, & \pm x = p^{2\lambda}-2^{2\lambda\nu-2}q^{2\lambda}; \\
\text{III. } z = \dfrac{\upsilon^{2\lambda}+\omega^{2\lambda}}{2}, & y = \dfrac{\upsilon^{2\lambda}-\omega^{2\lambda}}{2}, \\
\quad z = p^{2\lambda}+2^{2\lambda n-2}.\lambda^{2\lambda\mu-1}.q^{2\lambda}, & \pm x = p^{2\lambda}-2^{2\lambda\nu-2}.\lambda^{2\nu\mu-1}.q^{2\lambda}; \\
\text{IV. } z = \dfrac{\upsilon^{2\lambda}+\omega^{2\lambda}}{2}, & y = \dfrac{\upsilon^{2\lambda}-\omega^{2\lambda}}{2}, \\
\quad z = \lambda^{2\lambda\mu-1}.p^{2\lambda}+2^{2\lambda\nu-2}.q^{2\lambda}, & \pm x = \lambda^{2\lambda\mu-1}.p^{2\lambda}-2^{2\lambda\nu-2}.q^{2\lambda};
\end{cases}
$$

formisque binis ipsius z aequalibus positis est

$$
15.\quad \begin{cases}
\text{I. } \upsilon^{2\lambda}+\omega^{2\lambda} = 2p^{2\lambda}+2^{2\lambda\nu-1}.q^{2\lambda}, \\
\text{II. } \lambda^{2\lambda\mu-1}.\upsilon^{2\lambda}+\omega^{2\lambda} = 2p^{2\lambda}+2^{2\lambda\nu-1}.q^{2\lambda}, \\
\text{III. } \upsilon^{2\lambda}+\omega^{2\lambda} = 2p^{2\lambda}+2^{2\lambda n-1}.\lambda^{2\lambda\mu-1}.q^{2\lambda}, \\
\text{IV. } \upsilon^{2\lambda}+\omega^{2\lambda} = 2.\lambda^{2\lambda\mu-1}.p^{2\lambda}+2^{2\lambda\nu-1}.q^{2\lambda}.
\end{cases}
$$

In omnibus iis aequationibus numeri υ, ω, p et q impares et inter se primi sunt, numeri υ et ω factores ipsius x, et p et q factores numeri y.

Ex aequatione proposita $x^{2\lambda}+y^{2\lambda}=z^{2\lambda}$, sequitur etiam $(z^{\lambda}+y^{\lambda})(z^{\lambda}-y^{\lambda})=x^{2\lambda}$, et quia factores $z^{\lambda}+y^{\lambda}$ et $z^{\lambda}-y^{\lambda}$ inter se primi sunt:

$$16.\quad z^{\lambda}+y^{\lambda} = A^{2\lambda}, \qquad z^{\lambda}-y^{\lambda} = B^{2\lambda},$$

simili modo est $(z^{\lambda}\pm x^{\lambda})(z^{\lambda}\mp x^{\lambda})=y^{2\lambda}$, unde quia maximus factor communis numerorum $z^{\lambda}\pm x^{\lambda}$ et $z^{\lambda}\mp x^{\lambda}$ est 2, et per hypothesin maxima potestas ipsius 2, quam y continet est 2^{ν}, habetur

$$17.\quad z^{\lambda}\pm x^{\lambda} = 2.C^{2\lambda}, \qquad z^{\lambda}\mp x^{\lambda} = 2^{2\lambda\nu-1}.D^{2\lambda}.$$

Signa ambigua $\pm$ et $\mp$ ita accipienda sunt, ut cum signis aequationum (13.), (14.) et (15.) conveniant, ubi enim in illis aequationibus signa superiora vel inferiora valent, eadem etiam in his valebunt.

Quum probabile sit omnes quatuor casus quos supra separavimus non pro omnibus numeris λ locum habituros esse, dijudicandum videtur: quinam casus ad certos numeros λ possint pertinere. Primum accipiamus numerum primum λ talem esse ut etiam $2\lambda+1$ sit numerus primus, id quod ex. gr. evenit pro numeris $\lambda=3$, 5, 11, 23, 29, 41, 53, ... et pro alliis innumeris. Quo posito inquiramus an aequationum (13.) utraeque partes secundum modulum $2\lambda+1$ congruae esse possint. Quia per cognitum theorema omnis potestas $2\lambda^{\text{ta}}$ unitati congrua est modulo $2\lambda+1$ (numero primo) nisi per $2\lambda+1$ est divisibilis, facile cognosci potest aequationum (15.) casus I. et III. consistere non posse, nisi q per $2\lambda+1$ divisibilis sit, sed casus II. et IV. nullomodo locum habere (casu $\lambda=3$ excepto). Praeterea demonstrari potest etiam primum casum rejiciendum esse, est enim identice

$$18.\quad \frac{z^{2\lambda}-x^{2\lambda}}{z^2-x^2}=(z^2-x^2)^{\lambda-1}+\lambda(z^2-x^2)^{\lambda-3}z^2x^2+\dots+\lambda(zx)^{\lambda-1},$$

porro est $z^{2\lambda}-x^{2\lambda}=y^{2\lambda}$, et casu primo, de quo agitur, $z^2-x^2=2^{2\lambda\nu}.p^{2\lambda}_{\mathbf{2}}.q^{2\lambda}$, ergo $\frac{z^{2\lambda}-x^{2\lambda}}{z^2-x^2}$ est potestas $2\lambda^{\text{ta}}$, quae sit $y'^{2\lambda}$. Cum supra inventum sit casu primo numerum q per $2\lambda+1$ divisibilem esse debere, etiam z^2-x^2 hunc factorem contineat necesse est; itaque ex aequatione (18.), terminis per z^2-x^2 sive per $2\lambda+1$ divisibilibus omissis, habemus congruentiam:

$$19.\quad y'^{2\lambda}\equiv\lambda(zx)^{\lambda-1}\qquad(\text{mod. }2\lambda+1).$$

Denique ex aequationibus $z=p^{2\lambda}+2^{2\lambda\nu-2}q^{2\lambda}$ et $\pm x=p^{2\lambda}-2^{2\lambda\nu-2}q^{2\lambda}$ sequitur $\pm x\equiv1$ et $z\equiv1$ modulo $2\lambda+1$, unde $(zx)^{\lambda-1}\equiv1$ (mod. $2\lambda+1$); itaque congruentia 19. mutatur in

$$20.\quad y'^{2\lambda}\equiv\lambda\qquad(\text{mod. }2\lambda+1),$$

quae congruentia nullomodo locum habere potest. Solus igitur remanet casus tertius, atque habemus

Theorema 2. „Si praeter λ etiam $2\lambda+1$ est numerus primus, aequatio $x^{2\lambda}+y^{2\lambda}=z^{2\lambda}$ per numeros integros solvi nequit, nisi y, qui est numerus par, simul per λ et per $2\lambda+1$ divisibilis est, et numerorum x, y, z formae sunt: $z=\frac{v^{2\lambda}+\omega^{2\lambda}}{2}$, $y=\frac{v^{2\lambda}-\omega^{2\lambda}}{2}$, $z=p^{2\lambda}+2^{2\lambda\nu-2}.\lambda^{2\lambda\mu-1}.q^{2\lambda}$, $\pm x=p^{2\lambda}-2^{2\lambda\nu-2}\lambda^{2\lambda\mu-1}.q^{2\lambda}$."

Consideramus etiam residua, quae aequationes (15.) dant modulo 8. Quum numerorum imparium p, q, v, ω quadrata sive potestates pares unitati congrua sint modulo 8, ex aequationibus illis habemus congruentias: pro casu secundo: $1+\lambda\equiv2$ (mod. 8.), et pro quarto casu: $1+1\equiv2\lambda$ (mod. 8.), unde elucet casum secundum non posse locum habere nisi λ habeat formam $8n+1$, neque casum quartum nisi sit $\lambda=4n+1$. Generalius autem demonstrari potest.

Theorema 3. „Casus primus, secundus et quartus non possunt locum habere nisi λ habet formam $8n+1$, pro ceteris formis numeri λ, $8n+3$, $8n+5$ et $8n+7$ solus casus tertius locum habere potest."

Hoc theorema demonstratur ex aequatione identica

$$21.\quad \frac{z^\lambda \mp x^\lambda}{z\mp x} = (z\mp x)^{\lambda-1} \pm \lambda(z\mp x)^{\lambda-3}xz + \ldots\ldots + \lambda(\pm xz)^{\frac{\lambda-1}{2}}$$

est enim pro casibus I., II. et IV. $z^\lambda \mp x^\lambda = 2^{2\lambda\nu-2}D^{2\lambda}$ et $z\mp x = 2^{2\lambda\nu-1}q^{2\lambda}$, ergo

$$22.\quad \frac{z^\lambda\mp x^\lambda}{z\mp x} = E^{2\lambda}$$

inde, terminis per 8 divisibilibus omissis, aequatio (21.) mutatur in congruentiam

$$23.\quad E^{2\lambda} \equiv \lambda(\pm xz)^{\frac{\lambda-1}{2}} \quad (\text{mod. } 8)$$

porro e formis numerorum $\mp x$ et z, ad (14.) notatis sequitur esse ubique $\pm xz \equiv 1$ (mod. 8), itaque congruentia (23.) mutatur in

$$24.\quad 1 \equiv \lambda \quad (\text{mod. } 8)$$

quae congruentia continet theorema pronunciatum.

Revertimur ad aequationes (17.) quae in hanc formam redigi possunt:

$$25.\quad z^\lambda - C^{2\lambda} = 2^{2\lambda\nu-2}.D^{2\lambda},\quad C^{2\lambda}\mp x^\lambda = 2^{2\lambda\nu-2}D^{2\lambda}.$$

Casibus I., II., et IV. $z\mp x$ factorem λ non continet, pro iis igitur casibus neque $z^\lambda \mp x$, neque D factorem λ potest continere, itaque per theorema primum ex aequationibus (25.) sequuntur:

$$26.\quad z - C^2 = 2^{2\lambda\nu-2}r^{2\lambda},\quad C^2\mp x = 2^{2\lambda\nu-2}.s^{2\lambda},$$

ex iisque additis:

$$27.\quad z\mp x = 2^{2\lambda\nu-2}(r^{2\lambda}+s^{2\lambda}),$$

et quia pro casibus I. II. et IV. est $z\mp x = 2^{2\lambda\nu-2}q^{2\lambda}$, habemus

$$28.\quad r^{2\lambda}+s^{2\lambda} = 2.q^{2\lambda}.$$

Pro casu tertio $z\mp x$ continet factorem $\lambda^{2\lambda\mu-1}$, ergo $z^\lambda\mp x^\lambda$ factorem habebit $\lambda^{2\lambda\mu}$, et D factorem λ^μ, inde per theorema primum ex aequationibus (25.) pro hoc tertio casu deducuntur

$$29.\quad z - C^2 = 2^{2\lambda\nu-2}.\lambda^{2\lambda\mu-1}.r^{2\lambda},\quad C^{2\lambda}\mp x = 2^{2\lambda\nu-2}.\lambda^{2\lambda\mu-1}.s^{2\lambda},$$

quibus additis:

$$30.\quad z\mp x = 2^{2\lambda\nu-2}.\lambda^{2\lambda\mu-2}.(r^{2\lambda}+s^{2\lambda})$$

et quia pro casu tertio invenimus $z\mp x = 2^{2\lambda\nu-1}.\lambda^{2\lambda\mu-1}.q^{2\lambda}$ est

$$31.\quad r^{2\lambda}+s^{2\lambda} = 2.q^{2k}.$$

Numeri r, s, et q aequationum (28.) et (31.) factores sunt numeri D, ideoque etiam numeri D, ideoque etiam numeri y, eaeque aequationes continent theorema insigne:

Theorema 4. „Si aequatio $x^{2\lambda}+y^{2\lambda}=z^{2\lambda}$ per numeros integros solvi potest, semper inveniri possunt numeri tres, r, s et q, numeri y, ejusmodi ut satisfaciant aequationi $r^{2\lambda}+s^{2\lambda}=2.q^{2\lambda}$."

De numeris r, s, et q pauca adjicienda esse videntur. Per numeros x, y et z determinantur hoc modo:

casu I., II. et IV.

$$32.\quad z-\left(\frac{z^\lambda \pm x^\lambda}{2}\right)^{\frac{1}{\lambda}} = 2^{2\lambda\nu-2}.r^{2\lambda},$$

$$33.\quad \left(\frac{z^\lambda \pm x^\lambda}{2}\right)^{\frac{1}{\lambda}} \mp x = 2^{2\lambda\nu-2}.s^{2\lambda},$$

$$34.\quad z\mp x = 2^{2\lambda\nu-1}.q^{2\lambda},$$

casu III.

$$35.\quad z-\left(\frac{z^\lambda \pm x^\lambda}{2}\right)^{\frac{1}{\lambda}} = 2^{2\lambda\nu-2}.\lambda^{2\lambda\mu-1}.r^{2\nu},$$

$$36.\quad \left(\frac{z^\lambda \pm x^\lambda}{2}\right)^{\frac{1}{\lambda}} \pm x = 2^{2\lambda\nu-2}.\lambda^{2\lambda\mu-1}.s^{2\lambda},$$

$$37.\quad z\mp x = 2^{2\lambda\nu-1}.\lambda^{2\lambda\mu-1}.q^{2\lambda}.$$

Fieri potest ut numeri r, s, et q, quos hae aequationes praebent, factores communes habeant, qui vero, cum omnium trium communes esse debeant, ex aequatione $r^{2\lambda}+s^{2\lambda}=2.q^{2\lambda}$ tolli poterunt. Praeterae contendo, iis factoribus communibus sublatis, numerorum r et s factores omnes formam $2\lambda n+1$ habere. Notum est enim formae $z^\lambda \mp x^\lambda$ factores omnes, qui non sunt factores ipsius $z\mp x$, hanc formam habere, unde sequitur omnes etiam factores ipsius D, qui non sint factores ipsius $z\mp x$, sive ipsius q, eandem formam habere. Quum vero numeri r et s factores sint numeri D, e quibus per hyp. factores cum q communes sublati sunt, sequitur omnes eorum factores formam $2\lambda n+1$ habere. Si certum aliquem factorem primum numeri r accipimus esse $2\lambda m+1$, ex aequatione $r^{2\lambda}+s^{2\lambda}=2q^{2\lambda}$ habemus congruentiam

$$38.\quad s^{2\lambda} \equiv 2.q^{2\lambda} \quad (\text{mod. } 2\lambda m+1)$$

unde

$$39.\quad s^{2\lambda m} \equiv 2^m.q^{2\lambda m} \quad (\text{mod. } 2\lambda m+1)$$

et quia per hyp. numerus $2\lambda m+1$ est primus, esse debet

$$s^{2\lambda m}\equiv 1 \quad \text{et} \quad q^{2\lambda m}\equiv 1 \quad \text{modulo } 2\lambda m+1,$$

itaque

$$40.\quad 2^m \equiv 1 \quad (\text{mod. } 2\lambda m+1)$$

huic igitur congruentiae omnes factores primi numeri r satisfacere debent, et apertum est hoc idem de factoribus primis numeri s valere.

Lignicii Oct. 1835.

Eine Aufgabe, betreffend die Theorie der cubischen Reste

Journal für die reine und angewandte Mathematik 23, 285–286 (1842)

Wenn p eine Primzahl von der Form $3n+1$ ist, und g eine primitive Wurzel derselben, so kann die Reihe $1, g, g^2, g^3, \ldots. g^{n-2}$ in drei verschiedene Reihen geordnet werden, nämlich $1, g^3, g^6, \ldots. g^{p-4}$, ferner $g, g^4, g^7, \ldots. g^{p-3}$ und $g^2, g^5, g^8, \ldots. g^{p-2}$. Die Reste der ersten Reihe für den Modul p sind die cubischen Reste, von denen wir einen beliebigen mit α bezeichnen; die Reste der zweiten und dritten Reihe, welche wir resp. mit β und γ bezeichnen, sind die cubischen Nichtreste. Wenn nun α, β und γ die angegebene Bedeutung haben, so sind, wie *Gauſs* gezeigt hat, folgende drei Reihen:

$$z_1 = \sum_{0}^{p-1}{}_k \cos\frac{2\alpha k^3\pi}{p}, \quad z_2 = \sum_{0}^{p-1}{}_k \cos\frac{2\beta k^3\pi}{p}, \quad z_3 = \sum_{0}^{p-1}{}_k \cos\frac{2\gamma k^3\pi}{p}$$

die drei Wurzeln von folgender cubischen Gleichung:

* $$z^3 = 2pz + pt,$$

wo t durch die in ganzen Zahlen aufzulösende Gleichung $4p = t^2 + 27u^2$ und durch die Bedingung, daſs es positiv oder negativ zu nehmen sei, je nachdem es die Form $3h+1$ oder $3h-1$ hat, vollständig bestimmt ist. Die drei Reihen z_1, z_2, z_3 spielen in der Theorie der cubischen Reste eine sehr wichtige Rolle, sind aber durch diese cubische Gleichung noch nicht genau bestimmt, da es ganz unentschieden bleibt, welche der drei Wurzeln dieser Gleichung einer jeden dieser Reihen gleich ist; mit dieser Unbestimmtheit sind daher auch alle Resultate behaftet, welche man mit Hülfe dieser Reihen über cubische Reste gewinnt. Ich habe nun die Aufgabe, diese Unbestimmtheit aufzuheben, allerdings in einem gewissen Sinne gelöset, aber die Lösung genügt nicht recht, weil sie die Kenntniſs der Summe aller cubischen Reste, welche kleiner sind als $\frac{1}{2}p$, und eben so die Kenntniſs der Summen der beiden verschiedenen Arten von Nichtresten voraussetzt. Aus diesen berechneten Summen habe ich denn auch die Werthe der drei Reihen z_1, z_2, z_3 für alle Primzahlen von der Form $3n+1$, unter 400, vollständig bestimmt und will die Resultate hier mittheilen, damit vielleicht ein Anderer durch Induction das allgemeine Gesetz finden könne, welches mir noch verborgen geblieben ist.

Zunächst bemerke ich, daſs, da t in den Grenzen $-2\sqrt{p}$ und $+2\sqrt{p}$ liegt, die drei Wurzeln der cubischen Gleichung stets in folgenden drei Intervallen enthalten sein müssen: die eine in den Grenzen $-2\sqrt{p}$ und $-\sqrt{p}$, eine der andern in den Grenzen $-\sqrt{p}$ und $+\sqrt{p}$ und die dritte in den Grenzen $+\sqrt{p}$ und $+2\sqrt{p}$. Ferner bemerke ich, daſs, wenn eine der drei Reihen vollständig bestimmt ist, die Bestimmung der beiden anderen keine Schwierigkeiten weiter hat, da diese sich rational durch jene ausdrücken lassen; daher bestimme ich hier nur die Reihe $z_1 = \sum_0^{p-1}{}_k \cos\frac{2\alpha k^3\pi}{p}$, in welcher auch $\alpha = 1$ genommen werden kann. Diese Reihe aber liegt nach meiner Berechnung

1) in den Grenzen $-2\sqrt{p}$ und $-\sqrt{p}$ für die Primzahlen
97, 139, 151, 199, 211, 331;

2) in den Grenzen $-\sqrt{p}$ und $+\sqrt{p}$ für die Primzahlen
13, 19, 37, 61, 109, 157, 193, 241, 283, 367, 373, 379, 397;

3) in den Grenzen $+\sqrt{p}$ und $+2\sqrt{p}$ für die Primzahlen
7, 31, 43, 67, 73, 79, 103, 127, 163, 181, 223, 229, 271, 277, 307, 313, 337, 349.

Es käme nun darauf an, zu suchen, welche Eigenthümlichkeiten jede dieser drei Reihen habe, die keine Primzahl der beiden anderen Reihen theilte. Die lineäre Form der Primzahlen scheint hierbei keine Bedeutung zu haben, wohl aber die quadratische Form $4p = t^2 + 27u^2$; vielleicht auch die Form $p = r^2 + 3.s^2$. Da ich aber auch aus diesen kein Gesetz entdecken konnte, so nahm ich meine Zuflucht zu den Zahlen, welchen $\beta^{\frac{p-1}{3}}$ und $\gamma^{\frac{p-1}{3}}$ congruent sind, aber mit eben so wenig Erfolg; auch ob gewisse Zahlen, namentlich 2 und 3 cubische Reste sind, oder nicht, entschied hierbei nichts. Jedenfalls scheint das Gesetz etwas tief zu liegen, genauer Nachforschungen aber wohl werth zu sein.

De residuis cubicis disquisitiones nonnullae analyticae

Journal für die reine und angewandte Mathematik 32, 341–359 (1846)

Quum clarissimus *Gauss* disquisitiones suas arithmeticas edidisset, abhinc annos XLI, perpauci erant geometrae, qui profundissimas ejus cogitationes mathematicas mente capere possent, admirationes autem omnium effecerunt theoremata nova de divisionibus circuli geometrice perficiendis, exempli gratia in septendecim partes, quae omnibus patefecerat. Iis praesertim inventis clarissimis adducti geometrae in doctrinam numerorum, imprimis autem in opus illud Gaussianum, operam maximam impenderunt, sed paucis tantum successit inceptum, rei ipsius enim difficultas eo augetur, quod in hoc libro omnia ita inter se cohaerent, ut these aliqua omissa, ea quae sequuntur vix perspici possint. Sero tandem, quum in Germania litterae mathematicae magis effloruissent, nonnulli exstiterunt qui *Gaussii* disquisitiones non solum penitus intelligerent, sed ipsi inventis novis doctrinam numerorum ulterius promoverent. Inter quos prae ceteris duo viri summi commemorandi sunt *Jacobi* et *Lejeune Dirichlet,* quorum inventa et methodi cum iis quae nos hac dissertatione tractabimus arcte connexa sunt. Cl. *Jacobi* pro residuis cubicis primus invenit reciprocitatis legem simplicissimam, qua totius huius doctrinae summa continetur, et Cl. *Lejeune Dirichlet* analysin quantitatum continuarum ad doctrinam numerorum tam prospero successu applicavit, ut omnium admirationem sibi pararet et nomen suum celeberrimum redderet. Disquisitiones, quas hactenus in publicum edidit, praecipue in residuis et formis quadraticis versantur, ideoque a nostris, quae de residuis cubicis instituemus, alienae videntur esse; nihilominus tamen confitemur in omnibus fere nos illius vestigia secutos esse. Magna enim residuorum cubicorum analogia est cum residuis quadraticis, praecipue numerorum formae $4n+1$, multaque theoremata et methodi, quae pro hisce locum habent, mutatis mutandis ad illa transferri possunt. Nuper etiam audivimus a geometra quodam Gallico *Lebesque* commentationem de residuis cubicis conscriptam esse, anno 1837, quam vero comparare non potuimus, praeterea de hac re in *Crellii* diario adnotationes nonnullae exstant minoris momenti, quarum auctores sunt viri clarissimi *Clausen* et *Stern.*

§. 1.

Ante omnia notiones et propositiones nonnullae elementares de residuis cubicis nobis repetendae sunt, quibus disquisitiones nostrae nituntur. Quilibet numerus α, qui cubo alicui congruus est modulo p, numero primo formae $6n+1$, hujus numeri primi residuum cubicum appellatur, qui vero cubo congrui esse non possunt, appellantur nonresidua; sive si congruentia $x^3 \equiv k$, modulo p, radicem aliquam habet, est k residuum, si vero haec congruentia radicem nullam habet, est k nonresiduum numeri primi p. Quia numeri inter se congrui semper simul sunt residua, vel nonresidua, hic soli numeri modulo p minores considerandi sunt. Potestas $x^{\frac{1}{3}(p-1)}$ pro valoribus variis numeri x non nisi tria residua incongrua habet, quorum unum est unitas, alterum si ponitur $=f$, est tertium $\equiv f^2$ et $1+f+f^2 \equiv 0$. Eos numeros x, pro quibus est $x^{\frac{1}{3}(p-1)} \equiv 1$, per α, eos qui dant $x^{\frac{1}{3}(p-1)} \equiv f$, per β, et eos qui dant $x^{\frac{1}{3}(p-1)} \equiv f^2$ per γ designamus, quibus literis ubi opus erit indices addemus. Omnes numeri modulo p minores aequaliter inter has tres classes distributae inveniuntur, ita ut numerus ipsorum α sit $\frac{1}{3}(p-1)$, idemque numerus ipsorum β et γ. Si g est radix primitiva pro modulo p, residua potestatum g^0, g^3, g^6, $\ldots\, g^{p-4}$ cum numeris α, α_1, α_2, ... consentiunt, residua potestatum g^1, g^4, g^7, ..., g^{p-3} cum numeris β, β_1, β_2, ... et residua potestatum g^2, g^5, g^8, ... g^{p-2} cum numeris γ, γ_1, γ_2, ... In serie numerorum α, α_1, α_2, ... reperiuntur etiam $p-\alpha$, $p-\alpha_1$, $p-\alpha_2$, ..., qua de causa eos tantum numeros α, α_1, α_2, ... cognovisse sufficit, qui minores sunt quam $\frac{1}{2}p$, cum quibus altera semissis simul data est, idem cadit in numeros β, β_1, β_2, ... nec non in numeros γ, γ_1, γ_2, ... Seriei α, α_1, α_2, ... termini unitate aucti: $\alpha+1$, α_1+1, α_2+1, ... excepto ultimo, qui est p, cujus loco ponimus 1, omnes in aliqua serierum α, β, et γ reperiuntur, numerus eorum qui continentur in serie α sit a, eorum qui in serie β continentur $=b$, ceterorum qui in serie γ continentur $=c$. Numeri a, b, c hoc modo definiti satisfaciunt aequationibus duabus

$$a+b+c=\tfrac{1}{3}(p-1), \quad \text{et} \quad (6a-3b-3c-2)^2+27(c-b)^2=4p;$$

itaque posito $6a-3b-3c-2=t$, $c-b=u$, est $4p=t^2+27u^2$, et quum numerus $4p$ unico tantum modo in hanc formam redigi possit, t et u, neglectis signis $\pm$, omnino determinati sunt, signum numeri t eo determinatur, quod $t=6a-3b-3c-2$ esse debet numerus formae $3m+1$, sed signum numeri u hoc modo definiri nequit, quia nondum distinximus quaenam nonresiduorum cubicorum literis β et quae literis γ designandae sint; nam serie numerorum β permutata cum serie numerorum γ, signum numeri u simul mutatur. Hinc autem nos serierum $\beta, \beta_1, \beta_2, \ldots$ et $\gamma, \gamma_1, \gamma_2, \ldots$ distinctionem accuratam petimus, quas ita semper eligendas esse constituimus, ut $u = c - b$ sit numerus positivus. Summas tres, quae in theoria residuorum cubicorum maximi momenti sunt:

$$\Sigma \cos \frac{2a\pi}{p} = y_1, \quad \Sigma \cos \frac{2\beta\pi}{p} = y_2, \quad \Sigma \cos \frac{2\gamma\pi}{p} = y_3,$$

cl. *Gauss* invenit radices esse aequationis cubicae

$$y^3 + y^2 - \tfrac{1}{3}(p-1)y - \tfrac{1}{27}(pt + 3p - 1) = 0;$$

nos autem in iis quae sequuntur illarum summarum loco accipiemus has:

$$\Sigma \cos \frac{2\alpha x^3 \pi}{p} = z_1, \quad \Sigma \cos \frac{2\beta x^3 \pi}{p} = z_2, \quad \Sigma \cos \frac{2\gamma x^3 \pi}{p} = z_3,$$

in quibus signa summatoria ad valores $x = 0, 1, 2, \ldots p-1$ referenda sunt. Hae summae z_1, z_2 et z_3 sunt radices aequationis simplicioris

$$1. \quad z^3 = 3pz + pt,$$

et cum summis y_1, y_2, y_3 ita cohaerent, ut sit $z_1 = 1 + 3y_1$, $z_2 = 1 + 3y_2$, $z_3 = 1 + 3y_3$. Inde efficiuntur aequationes

$$2. \quad \begin{cases} z_1 + z_2 + z_3 = 0, \quad z_1 z_2 + z_2 z_3 + z_3 z_1 = -3p, \quad z_1 z_2 z_3 = pt, \\ z_1^2 + z_2^2 + z_3^2 = 6p, \quad z_1^3 + z_2^3 + z_3^3 = 3pt. \end{cases}$$

Porro habentur sex aequationes, quarum ope quaelibet radicum z_1, z_2, z_3 rationaliter per aliam determinatur:

$$3. \quad \begin{cases} 3uz_2 = -z_1^2 + \tfrac{1}{2}(t-3u)z_1 + 2p, & 3uz_3 = z_1^2 - \tfrac{1}{2}(t+3u)z_1 - 2p, \\ 3uz_3 = -z_2^2 + \tfrac{1}{2}(t-3u)z_2 + 2p, & 3uz_1 = z_2^2 - \tfrac{1}{2}(t+3u)z_2 - 2p, \\ 3uz_1 = -z_3^2 + \tfrac{1}{2}(t-3u)z_3 + 2p, & 3uz_2 = z_3^2 - \tfrac{1}{2}(t+3u)z_3 - 2p; \end{cases}$$

quae binae per subtractionem conjunctae dant:

$$4. \quad \begin{cases} 3u(z_2 - z_3) = -2z_1^2 + tz_1 + 4p, \\ 3u(z_3 - z_1) = -2z_2^2 + tz_2 + 4p, \\ 3u(z_1 - z_2) = -2z_3^2 + tz_3 + 4p, \end{cases}$$

iisque multiplicatis, per faciles reductiones fit

$$5. \quad (z_1 - z_2)(z_2 - z_3)(z_3 - z_1) = 27pu.$$

Aequatio cubica (1.) rite soluta dat radices tres:

$$z = -\sqrt{p} \cdot \sin\tfrac{1}{3}\nu - \sqrt{(3p)}\cos\tfrac{1}{3}\nu, \quad z = 2\sqrt{p} \cdot \sin\tfrac{1}{3}\nu, \quad z = -\sqrt{p} \cdot \sin\tfrac{1}{3}\nu + \sqrt{(3p)}\cos\tfrac{1}{3}\nu,$$

in quibus ν est arcus minimus cujus sinus $= \frac{-t}{2\sqrt{p}}$, qui intra limites $-\frac{1}{2}\pi$ et $+\frac{1}{2}\pi$ versatur. Inde concluditur harum radicum unam intra limites $-2\sqrt{p}$ et $-\sqrt{p}$, alteram intra limites $-\sqrt{p}$ et $+\sqrt{p}$, tertiam intra limites $+\sqrt{p}$ et $+2\sqrt{p}$ sitam esse, quae autem radicum definitarum z_1, z_2, z_3, ad hos singulos limites pertineat, quaestio est difficillima, de qua infra copiosius agemus.

45 *

§. 2.

In theoria residuorum quadraticorum summae serierum duarum $\Sigma\cos\frac{2a\varkappa^2\pi}{p}$ et $\Sigma\cos\frac{2b\varkappa^2\pi}{p}$, quae seriebus z_1, z_2 et z_3 respondent, facile adeo extenduntur, ut p sit numerus quicunque compositus; eandem summarum amplificationem in seriebus z_1, z_2 et z_3 perficiemus, unde magna earum differentia ab iis, quae ad residua quadratica pertinent, elucebit. Quum enim theoremata, quae de illis pro modulis primis inventa sunt, aucta etiam elegantia pro modulis compositis pronunciari possint, et res tota vertatur in residuis, quae numerus compositus p habeat, modulo 4: hic non ipsius numeri compositi, sed factorum ejus primorum formae lineares, et numerus factorum inter se aequalium summas determinant. Antequam vero de modulis quomodocunque compositis agemus, casum peculiarem absolvemus, quo modulus est potestas numeri primi, atque in eo seorsim considerabimus factores primos formarum $6r+1$ et $6r-1$, et praeterea numeros primos 2 et 3.

Si p est numerus primus formae $6r+1$, et n numerus integer unitate major, residua numerorum 1^3, 2^3, 3^3, 4^3, ... $(p^n-1)^3$, modulo p^n, separatis multiplis numeri p, omnia continentur forma $\alpha+\mu p$, ubi α quodvis residuum cubicum numeri p, et μ quemvis numerorum $0, 1, 2, 3, \ldots (p^{n-1}-1)$ significat. Nam congruentiae $x^3\equiv A$, modulo p^n, satisfieri nequit, nisi sit $x^3\equiv A$, modulo p, sive, quod idem est, $A=\alpha+\mu p$, numerus autem residuorum cubicorum α est $\frac{1}{3}(p-1)$, et numerus omnium valorum numeri μ, qui formam $\alpha+\mu p$ modulo p^n minorem reddunt, est p^{n-1}, unde colligitur numerum omnium residuorum cubicorum, pro modulo p^n, majorem quam $\frac{1}{3}p^{n-1}(p-1)$ esse non posse. Praeterea notum est congruentiam $x^3\equiv A$, modulo p^n, plures quam tres radices incongruas non habere, qua re, quum terni numeri 1^3, 2^3, 3^3, ... $(p^n-1)^3$ idem residuum dare possint, neque plures, numerus omnium residuorum non minor esse potest tertia parte numerorum illorum, quae est $\frac{1}{3}p^{n-1}(p-1)$. Ex iis concluditur, omnes numeros, quos forma $\alpha+\mu p$ contineat, et praeter hos nullos, residua esse numerorum 1^3, 2^3, 3^3, ... $(p^n-1)^3$, modulo p^n, eaque residua singula ter inveniri. Multipla numeri p, quae series 0^3, 1^3, 2^3, 3^3, ... $(p^n-1)^3$ continet, sunt 0^3, $1^3\cdot p^3$, $3^3\cdot p^3$, ... $(p^{n-1}-1)^3p^3$, eorumque residua, modulo p^n, si n non majus quam 3, omnia sunt aequalia nihilo; si vero $n>3$, factore communi p^3 omisso, quaerenda sunt residua numerorum 0, 1^3, 2^3, 3^3, ... $(p^{n-1}-1)^3$, modulo p^{n-3}, quae esse residua numerorum 0, 1^3, 2^3, 3^3, ... $(p^{n-3}-1)^3$ toties iterata quot p^2 continet unitates, sponte apparet. Quibus positis, si $n>3$ et m numerus arbitrarius ad p^n primus, summa quaesita tali modo exhiberi potest:

$$\Sigma \cos \frac{2m\varkappa^3\pi}{p^n} = p^2 \Sigma \cos \frac{2m\lambda^3\pi}{p^{n-3}} + 3\Sigma\Sigma \cos \frac{2m(\alpha+\mu p)\pi}{p^n};$$

in qua formula signa summatoria hanc vim habent, ut literae $\varkappa$ tribuendi sint valores $0, 1, 2, \ldots (p^n-1)$, literae λ valores $0, 1, 2, \ldots (p^{n-3}-1)$, literae μ valores $0, 1, 2, 3, \ldots (p^{n-1}-1)$ et literae α valores omnium residuorum cubicorum numeri p. Consummatio secundum literam μ, quae per notas formulas trigonometricas facillime perficitur, summam dat zero, qua re haec aequationis pars tota evanescit, et formula efficitur simplicissima

$$6. \quad \Sigma \cos \frac{2m\varkappa^3\pi}{p^n} = p^2 \Sigma \cos \frac{2m\lambda^3\pi}{p^{n-3}}.$$

Inde pro quolibet numero n hujus seriei summa reducta est ad casus quibus $n=1$, 2 et 3. Pro $n=2$ et $n=3$ ex iis, quae modo exposuimus, facile concluditur

$$\Sigma \cos \frac{2m\varkappa^3\pi}{p^2} = p, \qquad \Sigma \cos \frac{2m\varkappa^3\pi}{p^3} = p^2,$$

itaque formula (6.) adhibita est generaliter

$$7. \quad \begin{cases} \Sigma \cos \dfrac{2m\varkappa^3\pi}{p^n} = p^{\frac{1}{3}(2n-2)} \Sigma \cos \dfrac{2m\varkappa^3\pi}{p}, & \text{si } n \equiv 1 \text{ modulo } 3, \\ \Sigma \cos \dfrac{2m\varkappa^3\pi}{p^n} = p^{\frac{1}{3}(2n-1)}, & \text{si } n \equiv 2, \text{ modulo } 3, \\ \Sigma \cos \dfrac{2m\varkappa^3\pi}{p^n} = p^{\frac{2}{3}n}, & \text{si } n \equiv 0, \text{ modulo } 3. \end{cases}$$

Deinde si q est numerus primus formae $6r-1$, residua cuborum 1^3, 2^3, 3^3, $\ldots (q^n-1)^3$, omissis multiplis numeri q, eadem sunt ac numeri omnes minores quam q^n et ad q primi, cujus propositionis demonstrationem eo petimus, quod illorum cuborum duo x^3 et y^3 eadem residua dare non possunt. Posito enim $x^3 \equiv y^3$, esset $(x-y)(x^2+xy+y^2) \equiv 0$, et quia forma x^2+xy+y^2 factorem $q=6r-1$ habere non potest, esse deberet $x-y \equiv 0, \ldots,$ modulo q^n, quod, quia x et y minores quam q^n supponuntur, nullo modo fieri potest. Facile inde concluditur, haec residua omnia forma $\nu+\mu p$ contineri, in qua $\nu=1, 2, 3, \ldots q^{n-1}-1$. De residuis cuborum qui per q divisibiles sunt idem valet, quod supra de modulo p^n invenimus, scilicet, si n non >3, haec residua, quorum numerus est q^{n-1} omnia nihilo aequalia sunt, si vero $n>3$, factore communi q^3 omisso, residuis numerorum 0^3, 1^3, 2^3, 3^3, $\ldots$ $(q^{n-3}-1)^3$, totidem sumtis quot q^2 continet unitates. Itaque est simili modo ac supra:

$$\Sigma \cos\frac{2m\varkappa^3\pi}{q^n} = q^2 \Sigma \cos\frac{2m\lambda^3\pi}{q^{n-1}} + \Sigma\Sigma\cos\frac{2m(\nu+\mu q)\pi}{q^n},$$

ubi $\varkappa = 0, 1, 2, 3, \ldots. q^n - 1$, $\lambda = 0, 1, 2, 3, \ldots. q^{n-3} - 1$, $\mu = 0, 1, 2, \ldots. q^{n-1} - 1$, $\nu = 1, 2, 3, \ldots. q - 1$. Consummatione secundum omnes valores literae μ perfecta, haec summa duplex evanescit, et prodit formula simplex

$$\Sigma \cos\frac{2m\varkappa^3\pi}{q^n} = q^2 \Sigma \cos\frac{2m\lambda^3\pi}{q^{n-3}}.$$

Denique, quum sit

$$\Sigma\cos\frac{2m\varkappa^3\pi}{q} = 0, \quad \Sigma\cos\frac{2m\varkappa^3\pi}{q^2} = q, \quad \Sigma\cos\frac{2m\varkappa^3\pi}{q^3} = q^2,$$

est generaliter:

$$8. \quad \begin{cases} \Sigma\cos\dfrac{2m\varkappa^3\pi}{q^n} = 0, & \text{si } n \equiv 1, \text{ modulo } 3, \\ \Sigma\cos\dfrac{2m\varkappa^3\pi}{q^n} = q^{\frac{1}{3}(2n-1)}, & \text{si } n \equiv 2, \text{ modulo } 3, \\ \Sigma\cos\dfrac{2m\varkappa^3\pi}{q^n} = q^{\frac{2}{3}n}, & \text{si } n \equiv 0, \text{ modulo } 3. \end{cases}$$

Eaedem summae valent etiam pro $q = 2$, neque alia huius rei est demonstratio, nisi quod forma $x^2 + xy + y^2$ hic tanquam summa trium numerorum imparium factorem $q = 2$ habere non potest.

Denique pro modulo 3^n residua cuborum per 3 non divisibilium 1^3, 2^3, 4^3, 5^3, ... $(3^n - 1)^3$ formam habent $9\mu \pm 1$; nam si horum cuborum aliquis accipitur x^3, x habet formam $3r \pm 1$, et congruentia $x^3 \equiv A$, modulo 3^n, dat $27r^3 + 27r^2 + 9r \pm 1 \equiv A$, ergo A divisum per 9, residuum dat ± 1, uti contendimus. Porro congruentia $x^3 \equiv y^3$ habet tres radices incongruas, terna igitur residua numerorum 1^3, 2^3, 4^3, ... $(3^n - 1)^3$ inter se aequalia sunt, et numerus inaequalium est $2 \cdot 3^{n-2}$, qui, quum sit numerus omnium numerorum formae $9\mu \pm 1$, modulo 3^n minorum, concluditur hos omnes in residuis illis reperiri. Cuborum per 3 divisibilium 0, $1^3 \cdot 3^3$, $2^3 \cdot 3^3$, $3^3 \cdot 3^3$, ... $(3^{n-1} - 1)^3 \cdot 3^3$ residua, factore communi 3^3 omisso, non alia sunt ac residua cuborum 0, 1^3, 2^3, 3^3, ... $(3^{n-3} - 1)^3$, quae novies iterantur, unde sequitur

$$\Sigma\cos\frac{2m\varkappa^3\pi}{3^n} = 9\Sigma\cos\frac{2m\lambda^3\pi}{3^{n-3}} + 3\Sigma\cos\frac{2m(9\mu \pm 1)\pi}{3^n},$$

ubi $\varkappa = 0, 1, 2, \ldots. (3^n - 1)$, $\lambda = 0, 1, 2, \ldots. 3^{n-3} - 1$, $\mu = 0, 1, 2, \ldots. 3^{n-2} - 1$, et m est numerus arbitrarius factorem 3 non complectens. Etiam hoc casu terminus tertius, consummatione secundum valores literae μ peracta, evanescit, unde fit:

$$\Sigma\cos\frac{2m\varkappa^3\pi}{3^n} = 9\Sigma\cos\frac{2m\lambda^3\pi}{3^{n-3}},$$

et quia summa quaesita pro $n=1$, $n=2$ et $n=3$ facile invenitur

$$\Sigma\cos\frac{2m\varkappa^3\pi}{3}=0,\quad \Sigma\cos\frac{2m\varkappa^3\pi}{9}=3\left(1+2\cos\frac{2m\pi}{9}\right),\quad \Sigma\cos\frac{2m\varkappa^3\pi}{27}=9,$$

habemus generaliter:

$$9.\quad \begin{cases} \Sigma\cos\dfrac{2m\varkappa^3\pi}{3^n}=0, & \text{si } n\equiv 1, \text{ modulo } 3,\\ \Sigma\cos\dfrac{2m\varkappa^3\pi}{3^n}=3^{\frac{1}{3}(2n-1)}\left(1+2\cos\dfrac{2m\pi}{9}\right), & \text{si } n\equiv 2, \text{ modulo } 3,\\ \Sigma\cos\dfrac{2m\varkappa^3\pi}{3^n}=3^{\frac{2}{3}n}, & \text{si } n\equiv 0, \text{ modulo } 3. \end{cases}$$

Casibus iis specialibus absolutis, seriei propositae summa pro modulis quomodocunque compositis facile invenitur, theorematis ope quod in hac aequatione continetur:

$$10.\quad \Sigma\cos\frac{2mQ^2\varkappa^3\pi}{P}\cdot\Sigma\cos\frac{2mP^2\lambda^3\pi}{Q}=\Sigma\cos\frac{2m\mu^3\pi}{PQ},$$

in qua summae extendendae sunt ad numeros $\varkappa=0,1,2,3,\ldots P-1$; $\lambda=0,1,2,3,\ldots Q-1$; $\mu=0,1,2,3,\ldots PQ-1$, et literae P, Q, m numeros quoscunque inter se primos designant. Primum observamus seriem $\Sigma\cos\frac{2m\varkappa^3\pi}{P}$ eandem esse ac $\Sigma e^{\frac{2m\varkappa^3\pi\sqrt{-1}}{P}}$, cujus pars imaginaria per se evanescit, qua de causa in seriebus illis cosinuum loco quantitatibus exponentialibus imaginariis uti licet. Inde per multiplicationem simplicem habemus

$$\Sigma e^{\frac{2mQ^2\varkappa^3\pi\sqrt{-1}}{P}}\cdot\Sigma e^{\frac{2mP^2\lambda^3\pi\sqrt{-1}}{Q}}=\Sigma\Sigma e^{\frac{2m(Q^3\varkappa^3+P^3\lambda^3)\pi\sqrt{-1}}{PQ}},$$

adjecto factore $e^{6m\varkappa\lambda(Q\varkappa+P\lambda)\pi\sqrt{-1}}$, qui est $=1$, haec summa duplex induit formam

$$\Sigma\Sigma e^{\frac{2m(Q\varkappa+P\lambda)^3\pi\sqrt{-1}}{PQ}}.$$

Formula $Q\varkappa+P\lambda$ valores congruos, modulo PQ, non habet. Si enim esset $Q\varkappa+P\lambda\equiv Q\varkappa^1+P\lambda^1$, modulo PQ, inde sequeretur $Q(\varkappa-\varkappa^1)+P(\lambda-\lambda^1)\equiv 0$, et quum P et Q per hypothesin sint inter se primi, haec congruentia subsistere nequit, nisi $\varkappa-\varkappa^1\equiv 0$, modulo P, et $\lambda-\lambda^1\equiv 0$, modulo Q, quod, quia $\varkappa$ et $\varkappa^1$ sunt minores quam P et inaequales, λ et λ^1 minores quam Q et inaequales, absurdum est. Praeterea patet, numerum omnium valorum formae $Q\varkappa+P\lambda$ esse PQ, unde sequitur ut, multiplis moduli PQ omissis, hi valores consentiant cum numeris $0, 1, 2, 3, \ldots PQ-1$. Inde ista summa duplex in simplicem mutatur, partibusque imaginariis, quarum altera alteram tollit, rejectis, provenit aequatio supra allata. Jam si $R=2^\alpha\cdot 3^\beta\cdot p^\gamma\cdot p'^{\gamma'}\ldots q^\delta\cdot q'^{\delta'}\ldots$ est numerus quomodocunque compositus, in quo $p, \overset{1}{p}, \dots$ designant numeros primos formae $6m+1$ et $q, \overset{1}{q}, \dots$ numeros primos formae $6r-1$, summa $\Sigma\cos\frac{2mx^3\pi}{R}$, per hoc theorema in productum tot summarum diffinditur, quot R factores habet inter se primos, scilicet

$$11. \quad \Sigma\cos\frac{2mx^3\pi}{R} = \Sigma\cos\frac{2max^3\pi}{2^\alpha}\cdot\Sigma\cos\frac{2mbx^3\pi}{3^\beta}\cdot\Sigma\cos\frac{2mcx^3\pi}{p^\gamma}\cdot\Sigma\cos\frac{2m\overset{1}{c}x^3\pi}{\overset{1}{p}{}^{\overset{1}{\gamma}}}\dots$$
$$\dots\Sigma\cos\frac{2mdx^3\pi}{q^\delta}\cdot\Sigma\cos\frac{2m\overset{1}{d}x^3\pi}{\overset{1}{q}{}^{\overset{1}{\delta}}}\dots;$$

in qua formula brevitatis causa scriptum est $\left(\frac{R}{2^\alpha}\right)^2 = a$, $\left(\frac{R}{3^\beta}\right)^2 = b$, $\left(\frac{R}{p^\gamma}\right)^2 = c$, etc. easque summas singulas, quarum moduli sunt numerorum primorum potestates, supra invenimus. Indoles igitur summae $\Sigma\cos\frac{2mx^3\pi}{R}$ maxime a residuis pendet quae factorum primorum numeri R exponentes, $\alpha, \beta, \gamma, \overset{1}{\gamma}, \dots \delta, \overset{1}{\delta}, \dots$ dant modulo 3. Nam si quis numerorum $\alpha, \beta, \delta, \overset{1}{\delta}, \dots$ habet formam $3r+1$, semper est $\Sigma\cos\frac{2mx^3\pi}{R} = 0$; si vero exponentium $\alpha, \beta, \gamma, \overset{1}{\gamma}, \dots \delta, \overset{1}{\delta}, \dots$ nullus habet formam $3r+1$, haec summa aequatur numero integro C, numeri R divisori, cui, si $\beta = 3r+2$, praeterae adjicendus est factor $1+2\cos\frac{2m\pi}{9}$; deniqne si qui numerorum $\gamma, \overset{1}{\gamma}, \dots$ habent formam $3r+1$, exempli causa ipsi duo numeri γ et $\overset{1}{\gamma}$, haec summa numero illi integro C aequatur, multiplicato per summas $\Sigma\cos\frac{2mx^3\pi}{\overset{1}{p}}\cdot\Sigma\cos\frac{2mx^3\pi}{p}$. Posito $R = 42375500 = 2^2\cdot3^3\cdot5^3\cdot7^4\cdot13$, est:

$$\Sigma\cos\frac{2mx^3\pi}{R} = 22050\,\Sigma\cos\frac{6mx^3\pi}{7}\cdot\Sigma\cos\frac{4mx^3\pi}{13};$$

posito $R = 1000 = 2^3\cdot5^3$ est

$$\Sigma\cos\frac{2mx^3\pi}{1000} = 100.$$

§. 3.

Notissimum est signum $\left(\frac{m}{p}\right)$ a Cl. ***Legendre*** residuis quadraticis adhibitum, et propter eximiam elegantiam, quam formulae describendae ideo accipiunt, postea ab omnibus fere geometris receptum, quod statuitur $=+1$ vel $=-1$, prout m est residuum vel nonresiduum quadraticum numeri primi p. Idem signum etiam residuis cubicis maximam utilitatem affert; pro iis vero $\left(\frac{m}{p}\right)$ non radici quadraticae, sed radici cubicae unitatis aequale statuendum est, ita ut sit $\left(\frac{m}{p}\right) = 1$, si m est residuum cubicum, sive si m est aliquis numerorum $\alpha, \alpha_1, \alpha_2, \ldots$; porro sit $\left(\frac{m}{p}\right) = \frac{-1+\sqrt{-3}}{2}$, si m est aliquis numerorum $\beta, \beta_1, \beta_2, \ldots$; denique sit $\left(\frac{m}{p}\right) = \frac{-1-\sqrt{-3}}{2}$, si m est inter numeros $\gamma, \gamma_1, \gamma_2, \ldots$. Haec significationis amplificatio Clo. *Jacobi* debetur, qui in literis de circuli sectionibus earumque applicationibus ad doctrinam numerorum, regiae litterarum academiae Berolinensi mense octobri anni 1837 traditis, hujus signi ope residuorum cubicorum legem reciprocitatis forma simplicissima exhibuit. Pro hoc signo valent aequationes fundamentales:

$$12. \quad \left(\frac{m}{p}\right)\cdot\left(\frac{n}{p}\right) = \left(\frac{mn}{p}\right), \quad \left(\frac{p-m}{p}\right) = \left(\frac{m}{p}\right), \quad \left(\frac{m+\varkappa p}{p}\right) = \left(\frac{m}{p}\right);$$

casu quo $m = \varkappa p$ est multiplum numeri p, signo $\left(\frac{m}{p}\right)$ valorem zero tribuimus. Jam nobis problema, cui disquisitiones ulteriores maxime innituntur, solvendum proponimus:

Data summa seriei

$$\varphi(v) = A_1 \cos v + A_2 \cos 2v + A_3 \cos 3v + A_4 \cos 4v + \ldots$$

invenire hujus seriei summam:

$$\left(\frac{1}{p}\right)A_1 + \left(\frac{2}{p}\right)A_2 + \left(\frac{3}{p}\right)A_3 + \left(\frac{4}{p}\right)A_4 + \ldots$$

Quem in finem consideremus summam $\Sigma\left(\frac{m\varkappa}{p}\right)^2 \cos\frac{2m\varkappa\pi}{p}$ pro $\varkappa = 0, 1, 2, \ldots (p-1)$, quae, casibus, quibus $m\varkappa$ ad seriem numerorum α vel β vel γ pertinet, separatis, in has tres dilabitur:

$$\Sigma\left(\frac{m\varkappa}{p}\right)^2 \cos\frac{2m\varkappa\pi}{p} = \Sigma\cos\frac{2\alpha\pi}{p} + \frac{-1-\sqrt{-3}}{2}\Sigma\cos\frac{2\beta\pi}{p} + \frac{-1+\sqrt{-3}}{2}\Sigma\cos\frac{2\gamma\pi}{p}.$$

Dehinc radicem cubicam unitatis $\frac{-1+\sqrt{-3}}{2}$ nos ubique litera h designabimus, unde sequitur ut alteri radici imaginariae $\frac{-1-\sqrt{-3}}{2}$ signum h^2 tribuendum sit. Summas singulas, quas haec formula continet, supra literis y_1, y_2, y_3 designavimus, earumque loco has novas $z_1 = 1+3y_1$, $z_2 = 1+3y_2$, $z_3 = 1+3y_3$ substituimus, quibus signis adhibitis haec formula simpliciorem formam accipit:

$$13. \quad 3\Sigma\left(\frac{m\varkappa}{p}\right)^2 \cos\frac{2m\varkappa\pi}{p} = z_1 + h^2 z_2 + h z_3,$$

quia $\left(\frac{m\varkappa}{p}\right)^2 = \left(\frac{m}{p}\right)^2 \cdot \left(\frac{\varkappa}{p}\right)^2$ et $\left(\frac{m}{p}\right)^3 = 1$; multiplicando per $\left(\frac{m}{p}\right)$ habemus

$$14.\quad 3\Sigma\left(\frac{\varkappa}{p}\right)^2 \cos\frac{2m\varkappa\pi}{p} = \left(\frac{m}{p}\right)(z_1 + h^2 z_2 + h z_3);\ \varkappa = 0, 1, 2, \ldots (p-1).$$

Hujus summae termini bini, ab initio et a fine aeque distantes, aequales sunt, quibus binis in unum conjunctis, est

$$15.\quad 6\Sigma\left(\frac{\varkappa}{p}\right)^2 \cos\frac{2m\varkappa\pi}{p} = \left(\frac{m}{p}\right)(z_1 + h^2 z_2 + h z_3);\ \varkappa = 0, 1, 2, \ldots \tfrac{1}{2}(p-1).$$

Jam si in aequatione

$$\varphi(v) = A_1 \cos v + A_2 \cos 2v + A_3 \cos 3v + A_4 \cos 4v,$$

quae intra limites $v = 0$ et $v = 2\pi$ valere debet, $v = \frac{2\varkappa\pi}{p}$ ponitur, per $3\cdot\left(\frac{\varkappa}{p}\right)^2$ multiplicatur et summae formantur pro $\varkappa = 1, 2, 3, \ldots p-1$, per formulam (14.) fit

$$16.\quad 3\Sigma\left(\frac{\varkappa}{p}\right)^2 \varphi\left(\frac{2\varkappa\pi}{p}\right) = (z_1 + h^2 z_2 + h z_3)\left(\left(\frac{1}{p}\right)A_1 + \left(\frac{2}{p}\right)A_2 + \left(\frac{3}{p}\right)A_3 + \ldots\right)$$

pro $\varkappa = 1, 2, 3, 4, \ldots (p-1)$. Simili modo, si aequatio

$$\varphi(v) = A \cos v + A_2 \cos 2v + A_3 \cos 3v + \ldots$$

intra limites $v = 0$ et $v = \pi$ valet, ex formula (15.) deducitur

$$17.\quad 6\Sigma\left(\frac{\varkappa}{p}\right)^2 \varphi\left(\frac{2\varkappa\pi}{p}\right) = (z_1 + h^2 z_2 + h z_3)\left(\left(\frac{1}{p}\right)A_1 + \left(\frac{2}{p}\right)A_2 + \left(\frac{3}{p}\right)A_3 + \ldots\right)$$

pro $\varkappa = 1, 2, 3, \ldots \frac{1}{2}(p-1)$, eaeque formulae (16. et 17.) problematis propositi solutionem continent.

§. 4.

Formularum generalium usum exemplis nonnullis illustraturi primum faciamus

$$\varphi(v) = \cos v + \frac{\cos 3v}{3^2} + \frac{\cos 5v}{5^2} + \ldots,$$

cujus seriei summa intra limites $v = 0$ et $v = \pi$ est $\varphi(v) = -\frac{1}{4}\pi v + \frac{1}{8}\pi^2$, quibus in formula (17.) substitutis, est

$$6\Sigma\left(\frac{\varkappa}{p}\right)^2\left(\frac{\pi^2}{8} - \frac{\pi^2\varkappa}{2p}\right) = (z_1 + h^2 z_2 + h z_3)\left(\left(\frac{1}{p}\right) + \left(\frac{3}{p}\right)\frac{1}{3^2} + \left(\frac{5}{p}\right)\frac{1}{5^2} + \ldots\right)$$

pro $\varkappa = 0, 1, 2, 3, \ldots \frac{1}{2}(p-1)$. Separatis casibus, quibus $\left(\frac{\varkappa}{p}\right) = 1$, $\left(\frac{\varkappa}{p}\right) = h$ et $\left(\frac{\varkappa}{p}\right) = h^2$, fit:

$$18.\quad -\frac{3\pi^2}{p}(\Sigma\alpha + h^2\Sigma\beta + h\Sigma\gamma) = (z_1 + h^2 z_2 + h z_3)\left(\left(\frac{1}{p}\right) + \left(\frac{3}{p}\right)\frac{1}{3^2} + \left(\frac{5}{p}\right)\frac{1}{5^2} + \ldots\right)$$

signaque summarum ad omnes valores numerorum α, β, γ, minores quam $\frac{1}{2}p$ extendenda sunt. Simili modo etiam haec series infinita in tres series singulas dilabitur. Posito enim

$$\Sigma\frac{1}{\alpha^2}=A,\quad \Sigma\frac{1}{\beta^2}=B,\quad \Sigma\frac{1}{\gamma^2}=C,$$

ubi summarum signa ad omnes valores impares numerorum α, β, γ in infinitum sunt extendenda, fit

$$\left(\frac{1}{p}\right)+\left(\frac{3}{p}\right)\frac{1}{3^2}+\left(\frac{5}{p}\right)\frac{1}{5^2}+\dots = A+hB+h^2C,$$

ideoque est

$$19.\quad \frac{-3\pi^2}{p}(\Sigma\alpha+h^2\Sigma\beta+h\Sigma\gamma) = (z+h^2z_2+hz^3)(A+hB+h^2C).$$

Formulae quae sequuntur multo simpliciores fiunt, si loco ipsarum summarum $\Sigma\alpha$, $\Sigma\beta$, $\Sigma\gamma$ earum differentiis a valore medio arithmetico utimur, quas literis m_1, m_2 et m_3 designabimus. Ponimus igitur

$$\tfrac{1}{3}(\Sigma\beta+\Sigma\gamma-2\Sigma\alpha)=m_1,\ \tfrac{1}{3}(\Sigma\alpha+\Sigma\gamma-2\Sigma\beta)=m_2,\ \tfrac{1}{3}(\Sigma\alpha+\Sigma\beta-2\Sigma\gamma)=m_3,$$

sive, quia est $\Sigma\alpha+\Sigma\beta+\Sigma\gamma = 1+2+3+\dots+\frac{1}{2}(p-1) = \frac{1}{8}(p^2-1)$:

$$\tfrac{1}{24}(p^2-1)-\Sigma\alpha=m_1,\ \tfrac{1}{24}(p^2-1)-\Sigma\beta=m_2,\ \tfrac{1}{24}(p^2-1)-\Sigma\gamma=m_3.$$

Numeri m_1, m_2, m_3, quos integros esse patet, ita comparati sunt, ut eorum summa nihilo aequalis sit, sive $m_1+m_2+m_3=0$, iisque summarum $\Sigma\alpha$, $\Sigma\beta$, $\Sigma\gamma$ loco substitutis, formula (19.) hanc formam accipit:

$$20.\quad \frac{3\pi^2}{p}(m_1+h^2m_2+hm_3) = (z_1+h^2z_2+hz_3)(A+hB+h^2C).$$

Quia h est forma imaginaria, haec aequatio duas reales complectitur, quibus, si adduntur aequationes $m_1+m_2+m_3=0$, $A+B+C=\frac{\pi^2(p^2-1)}{8p^2}$ et $z_1+z_2+z_3=0$, hae aequationes reales triplici modo exhiberi possunt, prout m_1, m_2, m_3, vel A, B, C, vel z_1, z_2, z_3 pro incognitis habentur. Inde prodeunt aequationes elegantes:

$$21.\quad \begin{cases} \dfrac{3\pi^2}{p}m_1 = Az_1+Bz_2+Cz_3, \\ \dfrac{3\pi^2}{p}m_2 = Az_2+Bz_3+Cz_1, \\ \dfrac{3\pi^2}{p}m_3 = Az_3+Bz_1+Cz_2, \end{cases}$$

46*

$$22.\quad \begin{cases} A = \dfrac{\pi^2(p^2-1)}{24p^2} + \dfrac{\pi^2}{3p^2}(m_1 z_1 + m_2 z_2 + m_3 z_3), \\ B = \dfrac{\pi^2(p^2-1)}{24p^2} + \dfrac{\pi^2}{3p^2}(m_1 z_2 + m_2 z_3 + m_3 z_1), \\ C = \dfrac{\pi^2(p^2-1)}{24p^2} + \dfrac{\pi^2}{3p^2}(m_1 z_3 + m_2 z_1 + m_3 z_2), \end{cases}$$

$$23.\quad \begin{cases} z_1 = \dfrac{2p^2(Am_1+Bm_3+Cm_2)}{\pi^2(m_1^2+m_2^2+m_3^2)} = \dfrac{-2p((A-B)m_3+(A-C)m_2)}{\pi(m_1^2+m_2^2+m_3^2)}, \\ z_2 = \dfrac{2p^2(Am_2+Bm_1+Cm_3)}{\pi^2(m_1^2+m_2^2+m_3^2)} = \dfrac{-2p((A-B)m_1+(A-C)m_3)}{\pi(m_1^2+m_2^2+m_3^2)}, \\ z_3 = \dfrac{2p^2(Am_3+Bm_2+Cm_1)}{\pi^2(m_1^2+m_2^2+m_3^2)} = \dfrac{-2p((A-B)m_2+(A-C)m_1)}{\pi(m_1^2+m_2^2+m_3^2)}. \end{cases}$$

His etiam differentias binarum quantitatum z_1, z_2 et z_3 adjungimus, quae consilio nostro inservient:

$$24.\quad \begin{cases} z_1 - z_2 = \dfrac{2p^2((A-2B+C)m_1-(A-2C+B)m_2)}{\pi^2(m_1^2+m_2^2+m_3^2)}, \\ z_2 - z_3 = \dfrac{2p^2((A-2B+C)m_2-(A-2C+B)m_3)}{\pi^2(m_1^2+m_2^2+m_3^2)}, \\ z_3 - z_1 = \dfrac{2p^2((A-2B+C)m_3-(A-2C+B)m_1)}{\pi^2(m_1^2+m_2^2+m_3^2)}. \end{cases}$$

Jam si praeter numeros m_1, m_2, m_3, etiam series infinitae A, B, C cognitae essent, per aequationes (23.) valores quantitatum z_1, z_2, z_3 accurate determinari possent; in quaestione vero praesenti sufficit cognovisse $A-B$, $A-C$, $A-2B+C$, $A-2C+B$ omnes esse positivos, quod inde sequitur ut sit $A = 1+\frac{1}{\alpha_1^2}+\frac{1}{\alpha_2^2}+\dots$, ergo $A>1$ et $A+B+C<\frac{1}{8}\pi^2<\frac{5}{4}$, itaque $B+C<\frac{1}{4}$, nec non $B<\frac{1}{4}$, et $C<\frac{1}{4}$. Numerorum m_1, m_2, m_3, quorum summa est nihilo aequalis, unus est absolute maximus, reliqui duo minores eadem signa habent, opposita signo maximi. Ponamus primo, m_1 esse numerum maximum, eumque positivum, unde m_2 et m_3 minores et negativi, hinc per aequationes (23.) et (24.) est z_1 positivum, z_1-z_2 positivum et z_1-z_3 positivum, itaque z_1 est aequationis illius cubicae radix maxima positiva, quae intra limites $+2\sqrt{p}$ et $+\sqrt{p}$ sita est. Simili modo, si m_1 est numerus absolute maximus, sed negativus, m_2 et m_3 erunt positivi, et per aequationes (23.) et (24.) z_1, z_1-z_2, z_1-z_3 omnes erunt negativi; unde concluditur z_1 esse radicem minimam aequationis cubicae, quae intra limites $-\sqrt{p}$ et $-2\sqrt{p}$ sita est. Pro numero absolute maximo m_2 habemus eodem modo, si m_2 est positivus, z_2 intra limites $+2\sqrt{p}$ et $+\sqrt{p}$, si vero m_2 est negativus, z_2 intra limites

$-\sqrt{p}$ et $-2\sqrt{p}$; denique pro numero absolute maximo m_3 est z_3 intra limites $+2\sqrt{p}$ et $\sqrt{p}$, si m_3 est positivus, et z_3 intra limites $-\sqrt{p}$ et $-2\sqrt{p}$, si m_3 est negativus. Semper igitur una radicum z_1, z_2, z_3 omnino determinata est per numeros m_1, m_2, m_3 et quia per unam radicem reliquae rationaliter expressae sunt, omnes jam determinatae sunt. Ceterum notandum est, ad reliquarum duarum radicum limites definiendos non opus esse aequationibus (3.), sed hanc quaestionem facillime per aequationem (5.) absolvi. In universum enim sex modi dantur quibus singulae tres radices z_1, z_2, z_3 ad tria intervalla $-2\sqrt{p}$ et $-\sqrt{p}$, $-\sqrt{p}$ et $+\sqrt{p}$, $+\sqrt{p}$ et $+2\sqrt{p}$ pertinere possunt. Quum vero per aequationem (5.) $(z_1-z_2)(z_2-z_3)(z_3-z_1) = 27pu$ sit, et u numerus positivus, haec intervalla, quae singulis radicibus respondent, ita eligi debent, ut $(z_1-z_2)(z_2-z_3)(z_3-z_1)$ sit quantitas positiva, unde fit ut tres casus rejiciendi sint, et soli tres sequentes locum habere possint:
z_1 intra lim. $-2\sqrt{p}$ et $-\sqrt{p}$, z_2 intra lim. $-\sqrt{p}$ et $+\sqrt{p}$, z_3 intra lim. $+\sqrt{p}$ et $+2\sqrt{p}$,
z_1 intra lim. $-\sqrt{p}$ et $+\sqrt{p}$, z_2 intra lim. $+\sqrt{p}$ et $+2\sqrt{p}$, z_3 intra lim. $-2\sqrt{p}$ et $-\sqrt{p}$,
z_1 intra lim. $+\sqrt{p}$ et $+2\sqrt{p}$, z_2 intra lim. $-2\sqrt{p}$ et $-\sqrt{p}$, z_3 intra lim. $-\sqrt{p}$ et $+\sqrt{p}$.

Hoc modo aequationis illius radices singulae accurate determinatae sunt; sed haec problematis solutio non ea est, quae peritis ab omni parte satisfacere possit; nam hoc desideratur, ut ex ispsius numeri primi p indole dijudicari possit, ad quae intervalla radices z_1, z_2, z_3 referendae sint, neque vero ex indole numerorum m_1, m_2, m_3, qui pro dato numero p non sine labore computantur. Hunc calculum numerorum m_1, m_2, m_3, pro omnibus numeris primis formae $6n+1$ usque ad 499, ope tabularum canonis arithmetici a Clo. *Jacobi* editi perfecimus et per methodum modo traditam ex iis invenimus esse I°, z_1 intra limites $-2\sqrt{p}$ et $-\sqrt{p}$ pro numeris primis $p=97$, 139, 151, 199, 211, 331, 433. II°, z_1 intra limites $-\sqrt{p}$ et $+\sqrt{p}$ pro numeris primis $p=13$, 19, 37, 61, 109, 157, 193, 241, 283, 367, 373, 379, 397, 487. III°, z_1 intra limites $+\sqrt{p}$ et $+2\sqrt{p}$ pro numeris primis $p=7$, 31, 43, 67, 73, 79, 103, 127, 163, 181, 223, 229, 271, 277, 307, 313, 337, 349, 409, 421, 439, 457, 463, 499. Omnes numeri primi, formae $6n+1$, talimodo in tres classes dividuntur, atque ex 45 numeris primis infra 500 ad classem primam pertinent 7, ad classem secundam 14, ed ad classem tertiam 24, quorum numerorum ratio proxime exprimitur per 1:2:3; nec improbabile est, eandem rationem etiam pro majore numero semper servatum iri. Quum doctrinae numerorum historia multis exemplis doceat, veritates arithmeticas per inductionem inventas et tum demum demonstratas esse, nos talem inductionem pro his tribus classibus numerorum primorum dignoscendis multis modis tentavimus, sed hactenus frustra criterium quaesivimus, quo alicujus classis numeri a reliquis numeris primis formae $6n+1$ discerni possint.

§. 5.

Ex aequationibus supra allatis (20—23) etiam numerorum m_1, m_2, m_3, et serierum infinitarum A, B, C proprietates nonnullas deducemus. Trium aequationum (21.) quadrata addita, adhibitis aequationibus $z_1^2+z_2^2+z_3^2=6p$ et $z_1z_2+z_2z_3+z_3z_1=-3p$, dant

$$25.\quad \frac{3\pi^4}{2p^3}(m_1^2+m_2^2+m_3^2) = A^2+B^2+C^2-AB-BC-CA.$$

Quantitas $A^2+B^2+C^2-AB-BC-CA$, quam breviter litera δ designabimus, limitibus arctis circumscripta est, quos facillime per repraesentationem hujus quantitatis δ in forma producti infiniti invenimus. Quem in finem litera r designamus omnes numeros primos impares, qui sunt residua cubica numeri p, et litera n omnes numeros primos impares, qui sunt nonresidua cubica ejusdem numeri p. Inde, per principia, quae jam summus **Eulerus** in introductione in analysin infinitorum exposuit, concluditur

$$\Pi\frac{1}{1-\frac{1}{r^2}}\,\Pi\frac{1}{1-\left(\frac{n}{p}\right)\frac{1}{n^2}} = 1+\left(\frac{3}{p}\right)\frac{1}{3^2}+\left(\frac{5}{p}\right)\frac{1}{5^2}+\left(\frac{7}{p}\right)\frac{1}{7^2}+\cdots,$$

$$\Pi\frac{1}{1-\frac{1}{r^2}}\,\Pi\frac{1}{1-\left(\frac{n}{p}\right)^2\frac{1}{n^2}} = 1+\left(\frac{3}{p}\right)^2\frac{1}{3^2}+\left(\frac{5}{p}\right)^2\frac{1}{5^2}+\left(\frac{7}{p}\right)^2\frac{1}{7^2}+\cdots.$$

Harum serierum altera est $A+hB+h^2C$, altera $A+h^2B+hC$, earumque productum est δ; itaque, duabus iis formulis inter se multiplicatis, habemus

$$\Pi\frac{1}{1-\frac{2}{r^2}+\frac{1}{r^4}}\,\Pi\frac{1}{1-\frac{1}{n^2}+\frac{1}{n^4}} = \delta.$$

Praeterea est

$$\Pi\frac{1}{1-\frac{1}{r^2}}\,\Pi\frac{1}{1-\frac{1}{n^2}} = 1+\frac{1}{3^2}+\frac{1}{5^2}+\frac{1}{7^2}+\cdots = \frac{\pi^2}{8}\left(1-\frac{1}{p^2}\right),$$

$$\Pi\frac{1}{1-\frac{1}{r^6}}\,\Pi\frac{1}{1-\frac{1}{n^6}} = 1+\frac{1}{3^6}+\frac{1}{5^6}+\frac{1}{7^6}+\cdots = \frac{\pi^6}{960}\left(1-\frac{1}{p^6}\right),$$

earumque aequationum altera per alteram divisa:

$$\Pi\frac{1}{1+\frac{1}{r^2}+\frac{1}{r^4}}\,\Pi\frac{1}{1+\frac{1}{n^2}+\frac{1}{n^4}} = \frac{\pi^4}{120}\left(1+\frac{1}{p^2}+\frac{1}{p^4}\right).$$

Per hanc aequationem nonresidua ex repraesentatione quantitatis δ tolli possunt, unde provenit

$$26.\quad \delta = \frac{\pi^4}{120}\left(1+\frac{1}{p^2}+\frac{1}{p^4}\right)\Pi\left(\frac{1+\frac{1}{r^2}+\frac{1}{r^4}}{1-\frac{2}{r^2}+\frac{1}{r^4}}\right);$$

simili modo δ per nonresidua n exprimitur

$$27.\quad \delta = \frac{\pi^4}{64}\left(1-\frac{1}{p^2}\right)^2\Pi\left(\frac{1-\frac{2}{n^2}+\frac{1}{n^4}}{1+\frac{1}{n^2}+\frac{1}{n^4}}\right).$$

Horum productorum infinitorum alterum est majus unitate, alterum minus unitate; qua de causa est $\delta > \frac{\pi^4}{120}\left(1+\frac{1}{p^2}+\frac{1}{p^4}\right) > \frac{\pi^4}{120}$, et $\delta < \frac{\pi^4}{64}\left(1-\frac{1}{p^2}\right)^2 < \frac{\pi^4}{64}$; limites igitur, intra quos δ continetur, sunt $\frac{\pi^4}{120}$ et $\frac{\pi^4}{64}$, et limites summae quadratorum $m_1^2+m_2^2+m_3^2$ sunt

$$28.\quad m_1^2+m_2^2+m_3^2 > \frac{p^3}{180},\quad m_1^2+m_2^2+m_3^2 < \frac{p^3}{96}.$$

Hinc facile deducuntur limites, quos maximus numerorum m_1, m_2, m_3, quem m_1 esse accipimus, excedere nequit. Est enim $m_1+m_2+m_3=0$, itaque $m_1^2+m_2^2+m_3^2 = 2(m_1^2+m_1m_2+m_2^2)$; porro m_2 habet signum contrarium signo numeri m_1, qua re m_1m_2 est negativum; nec non est $m_1m_2+m_2^2 = -m_2m_3$ quantitas negativa, unde sequitur ut sit $m_1^2+m_2^2+m_3^2 < 2m_1^2$; valorem autem minimum formula $m_1^2+m_1m_2+m_2^2$ obtinet pro $m_2 = -\frac{1}{2}m_1$, unde habemus $m_1^2+m_2^2+m_3^2 > \frac{3}{2}m_1^2$. Quibus cum formulis (28.) comparatis, est

$$2m_1^2 > \frac{p^3}{180} \quad\text{et}\quad \tfrac{3}{2}m_1^2 < \frac{p^3}{96},$$

$$\pm m_1 > \frac{p\sqrt{p}}{6\sqrt{10}} \quad\text{et}\quad \pm m_1 < \frac{p\sqrt{p}}{12}.$$

§. 6.

Series A, B, C hoc proprium habent, ut aequationis cubicae radices sint, cujus coëfficientes, si radices A, B et C per $\frac{p^2}{\pi^2}$ multiplicatae accipiuntur, omnes sunt numeri integri. Formam autem simplicissimam haec aequatio habet, si quantitates

$$\frac{p^2}{\pi^2}A - \frac{p^2-1}{24} = \xi_1,\quad \frac{p^2}{\pi^2}B - \frac{p^2-1}{24} = \xi_2,\quad \frac{p^2}{\pi^2}C - \frac{p^2-1}{24} = \xi_3$$

pro radicibus accipiuntur, quarum summa est nihilo aequalis. Aequationem quaesitam ex formula (20.) deducemus, lemmatis hujus facile demonstrandi auxilio:

Si a, b, c quantitates quascunque designant, quarum summa est nihilo aequalis, et h est radix imaginaria aequationis $h^3 = 1$, est

$$(a+h^2b+hc)^3 = \tfrac{1}{2}\cdot 27abc + \tfrac{1}{2}\cdot 3\sqrt{-3}(a-b)(b-c)(c-a),$$
$$(a+h^2b+hc)(a+hb+h^2c) = -3(ab+bc+ca).$$

Formula (20.), si loco quantitatum A, B et C novae ξ_1, ξ_2 et ξ_3 introducuntur, hanc formam accipit:

$$3p(m_1+h^2m_2+hm_3) = (z_1+h^2z_2+hz_3)(\xi_1+h\xi_2+h^2\xi_3),$$

et utraque parte per $(z_1+hz_2+h^2z_3)$ multiplicata, fit

$$29.\quad (z_1+hz_2+h^2z_3)(m_1+h^2m_2+hm_3) = 3(\xi_1+h\xi_2+h^2\xi_3).$$

Mutato h in h^2, unde h^2 mutatur in $h^4=h$, haec aequatio cum priori multiplicata dat

$$30.\quad p(m_1m_2+m_2m_3+m_3m_1) = \xi_1\xi_2+\xi_2\xi_3+\xi_3\xi_1.$$

Porro per lemma supra propositum est

$$(m_1+h^2m_2+hm_3)^3 = \tfrac{1}{2}\cdot 27m_1m_2m_3 + \tfrac{1}{2}\cdot 3\sqrt{-3}(m_1-m_2)(m_2-m_3)(m_3-m_1),$$
$$(\xi_1+h\xi_2+h^2\xi_3)^3 = \tfrac{1}{2}\cdot 27\xi_1\xi_2\xi_3 - \tfrac{1}{2}\cdot 3\sqrt{-3}(\xi_1-\xi_2)(\xi_2-\xi_3)(\xi_3-\xi_1),$$
$$(z_1+hz_2+h^2z_3)^3 = \tfrac{1}{2}\cdot 27p(t-3u\sqrt{-3}),$$

unde aequationis (29.) utraque parte ad tertiam potestatem elevata, efficitur

$$\tfrac{1}{2}p(t-3u\sqrt{-3})(\tfrac{1}{2}\cdot 27m_1m_2m_3+\tfrac{1}{2}\cdot 3\sqrt{-3}(m_1-m_2)(m_2-m_3)(m_3-m_1))$$
$$= \tfrac{1}{2}.27\xi_1\xi_2\xi_3 - \tfrac{1}{2}\cdot 3\sqrt{-3}(\xi_1-\xi_2)(\xi_2-\xi_3)(\xi_3-\xi_1);$$

quae aequatio, partibus realibus et imaginariis separatis, in has duas dilabitur:

$$\xi_1\xi_2\xi_3 = \tfrac{1}{2}p(tm_1m_2m_3+u(m_1-m_2)(m_2-m_3)(m_3-m_1)),$$
$$(\xi_1-\xi_2)(\xi_2-\xi_3)(\xi_3-\xi_1) = \tfrac{1}{2}p(27um_1m_2m_3-t(m_1-m_2)(m_2-m_3)(m_3-m_1)).$$

Quantitatibus $\xi_1+\xi_2+\xi_3$, $\xi_1\xi_2+\xi_2\xi_3+\xi_3\xi_1$, $\xi_1\xi_2\xi_3$ inventis, statim formatur aequatio cubica

$$31.\quad x^3+p(m_1m_2+m_2m_3+m_3m_1)x-\tfrac{1}{2}p(tm_1m_2m_3+u(m_1-m_2)(m_2-m_3)(m_3-m_1)) = 0,$$

cujus radices sunt $x=\frac{p^2}{\pi^2}A-\frac{p^2-1}{24}$, $x=\frac{p^2}{\pi^2}B-\frac{p^2-1}{24}$, $x=\frac{p^2}{\pi^2}C-\frac{p^2-1}{24}$.

Per formulam generalem (17.) serierum infinitarum A, B et C summae notatu dignae iuveniuntur, statuendo

$$\varphi(v) = \cos 8v+2\cos 16v+3\cos 24v+\ldots+\tfrac{1}{2}(p-1)\cdot\cos 4(p-1)v,$$

cujus seriei summa est

$$\varphi(v) = \frac{p+1}{4}\cdot\frac{\sin 4pv}{\sin 4v}-\frac{1}{2}\left(\frac{\sin 2(p+1)v}{\sin 4v}\right)^2,$$

quae, posito $v=\frac{2\varkappa\pi}{p}$, dat

$$\varphi\left(\frac{2\varkappa\pi}{p}\right) = -\tfrac{1}{8}\sec^2\frac{4\varkappa\pi}{p},$$

quibus in formula (17.) substitutis, habemus

$$-\tfrac{3}{4}\Sigma\left(\frac{\varkappa}{\pi}\right)^2\sec^2\frac{4\varkappa\pi}{p}$$
$$= (z_1+h^2z_2+hz_3)\left(\left(\frac{1}{p}\right)+\left(\frac{2}{p}\right)\cdot 2+\left(\frac{3}{p}\right)\cdot 3+\ldots.+\left(\frac{\frac{1}{2}(p-1)}{p}\right)\tfrac{1}{2}(p-1)\right);$$

sejunctis casibus, quibus $\left(\frac{\varkappa}{p}\right)=1\cdot\left(\frac{\varkappa}{p}\right)=h$ et $\left(\frac{\varkappa}{p}\right)=h^2$, haec formula hoc modo repraesentari potest:

$$-\tfrac{3}{4}\left(\Sigma\sec^2\frac{4\varkappa\pi}{p}+h^2\Sigma\sec^2\frac{4\varkappa\pi}{p}+h\Sigma\sec^2\frac{4\varkappa\pi}{p}\right)$$
$$= (z_1+h^2z_2+hz_3)(\Sigma\alpha+h\Sigma\beta+h^2\Sigma\gamma).$$

Substituto valore formulae $\Sigma\alpha+h\Sigma\beta+h^2\Sigma\gamma$, quem aequatio (19.) praebet, in qua h mutandum est cum h^2, haec formula transit in hanc:

$$\Sigma\sec^2\frac{4\alpha\pi}{p}+h^2\Sigma\sec^2\frac{4\beta\pi}{p}+h\Sigma\sec^2\frac{4\gamma\pi}{p} = \frac{4p^2}{\pi^2}(A+h^2B+hC),$$

quae, conjuncta cum aequatione facile demonstranda

$$\Sigma\sec^2\frac{4\alpha\pi}{p}+\Sigma\sec^2\frac{4\beta\pi}{p}+\Sigma\sec^2\frac{4\gamma\pi}{p} = \frac{4p^2}{\pi^2}(A+B+C),$$

in has aequationes simplices dilabitur:

$$32.\quad A=\frac{\pi^2}{4p^2}\Sigma\sec^2\frac{4\alpha\pi}{p},\quad B=\frac{\pi^2}{4p^2}\Sigma\sec^2\frac{4\beta\pi}{p},\quad C=\frac{\pi^2}{4p^2}\Sigma\sec^2\frac{4\gamma\pi}{p}.$$

§. 7.

Summae numerorum α, β, γ, qui minores sunt quam $\frac{1}{2}p$, scilicet $\Sigma\alpha$, $\Sigma\beta$, $\Sigma\gamma$, si numerus p magnus est, non sine multo labore computantur; singulae enim $\frac{1}{2}(p-1)$ terminis constant. Hic autem per applicationem singularem methodi generalis in §. 3. expositae, formulas simplices inveniemus, quibus hae summae ad similes summas reducuntur, qui eos tantum numeros α, β et γ continent, qui sunt minores quam $\frac{1}{4}p$, qua re hic labor dimidio minuitur. Ad hunc finem adhibemus seriem infinitam

$$\varphi(v) = \cos v+\frac{\cos 2v}{2^2}+\frac{\cos 3v}{3^2}+\frac{\cos 5v}{5^2}+\frac{\cos 6v}{6^2}+\ldots.,$$

in qua omnes termini desunt, quorum denominatores sunt multipla numeri 4.

Cujus seriei summa est:

$$\varphi(v) = \frac{-3\pi v}{8} + \frac{5\pi^2}{32}, \quad \text{si } v \text{ est intra limites } 0 \text{ et } \tfrac{1}{2}\pi,$$

$$\varphi(v) = \frac{-\pi v}{8} + \frac{\pi^2}{32}, \quad \text{si } v \text{ est intra limites } \tfrac{1}{2}\pi \text{ et } \pi;$$

est igitur

$$\varphi\left(\frac{2\varkappa\pi}{p}\right) = \frac{-3\pi^2\varkappa}{4} + \frac{5\pi^2}{32}, \quad \text{si } \varkappa \text{ est intra limites } 0 \text{ et } \tfrac{1}{4}p,$$

$$\varphi\left(\frac{2\varkappa\pi}{p}\right) = \frac{-\pi^2\varkappa}{4} + \frac{\pi^2}{32}, \quad \text{si } \varkappa \text{ est intra limites } \tfrac{1}{4}p \text{ et } \tfrac{1}{2}p,$$

quibus in formula generali (17.) substitutis, habemus

$$33. \quad 6\pi^2\Sigma\left(\frac{\varkappa}{p}\right)^2\left(\frac{-3\varkappa}{4p} + \frac{5}{32}\right) + 6\pi^2\Sigma\left(\frac{\varkappa}{p}\right)^2\left(\frac{-\varkappa}{4p} + \frac{1}{32}\right) +$$
$$= (z_1 + h^2 z_2 + h z_3)\left(\left(\frac{1}{p}\right) + \left(\frac{2}{p}\right)\frac{1}{2^2} + \left(\frac{3}{p}\right)\frac{1}{3^2} + \left(\frac{5}{p}\right)\frac{1}{5^2} + \cdots\right),$$

in qua formula signum summae alterum ad omnes numeros integros $\varkappa$ extendendum est, qui intra limites 0 et $\frac{1}{4}p$, alterum autem ad eos qui intra limites $\frac{1}{4}p$ et $\frac{1}{2}p$ jacent. Altera pars hujus aequationis facile in hanc formam redigitur:

$$(z_1 + h^2 z_2 + h z_3)\left(1 + \left(\frac{2}{p}\right)\frac{1}{4}\right)\left(\left(\frac{1}{p}\right) + \left(\frac{3}{p}\right)\frac{1}{3^2} + \left(\frac{5}{p}\right)\frac{1}{5^2} + \left(\frac{7}{p}\right)\frac{1}{7^2} + \cdots\right),$$

quae per formulam (18.) transformatur in hanc:

$$-\frac{3\pi^2}{p}\left(1 + \left(\frac{2}{p}\right)\frac{1}{4}\right)(\Sigma\alpha + h^2\Sigma\beta + h\Sigma\gamma).$$

Etiam in altera parte formulae (33.) sejungendi sunt valores literae $\varkappa$, pro quibus $\left(\frac{\varkappa}{p}\right) = 1$, $\left(\frac{\varkappa}{p}\right) = h$ et $\left(\frac{\varkappa}{p}\right) = h^2$. Quem in finem per characterem $\Sigma_1\alpha$ designamus summam eorum numerorum α, qui sunt minores quam $\frac{1}{4}p$, et per $\Sigma_2\alpha$ summam eorum, qui intra limites $\frac{1}{4}p$ et $\frac{1}{2}p$ reperiuntur, unde etiam signorum $\Sigma_1\beta$, $\Sigma_2\beta$, $\Sigma_1\gamma$ et $\Sigma_2\gamma$ vis perspicua est. Praeterea sit λ numerus terminorum summae $\Sigma_1\alpha$, λ_1 numerus terminorum summae $\Sigma_2\alpha$, et eodem modo sint μ et μ_1, ν et ν_1 numeri terminorum quos summae $\Sigma_1\beta$ et $\Sigma_2\beta$, $\Sigma_1\gamma$ et $\Sigma_2\gamma$ habent. Quibus positis formula (33.) hanc formam accipit:

$$\left(-\tfrac{3}{2}\Sigma_1\alpha + \frac{5\lambda p}{16} - \tfrac{1}{2}\Sigma_2\alpha + \frac{\lambda_1 p}{16}\right) + h^2\left(-\tfrac{3}{2}\Sigma_1\beta + \frac{5\mu p}{16} - \tfrac{1}{2}\Sigma_2\beta + \frac{\mu_1 p}{16}\right)$$
$$+ h\left(-\tfrac{3}{2}\Sigma_1\gamma + \frac{5\nu p}{16} - \tfrac{1}{2}\Sigma_2\gamma + \frac{\nu_1 p}{16}\right) = -\left(1 + \left(\frac{2}{p}\right)\frac{1}{4}\right)(\Sigma\alpha + h^2\Sigma\beta + h\Sigma\gamma).$$

Haec formula multo simplicior redditur per aequationes

$$\Sigma_1\alpha + \Sigma_2\alpha = \Sigma\alpha, \quad \Sigma_1\beta + \Sigma_2\beta = \Sigma\beta, \quad \Sigma_1\gamma + \Sigma_2\gamma = \Sigma\gamma,$$
$$\lambda + \lambda_1 = \tfrac{1}{6}(p-1), \quad \mu + \mu_1 = \tfrac{1}{6}(p-1), \quad \nu + \nu_1 = \tfrac{1}{6}(p-1),$$

quorum auxilio fit:

$$(4\Sigma_1\alpha-p\lambda)+h^2(4\Sigma_1\beta-p\mu)+h(4\Sigma_1\gamma-p\nu)=\left(2+\left(\frac{2}{p}\right)\right)(\Sigma\alpha+h^2\Sigma\beta+h\Sigma\gamma).$$

Haec aequatio, unitatis radicem cubicam imaginariam h involvens, duas aequationes reales continet, quae cum aequatione $\Sigma\alpha+\Sigma\beta+\Sigma\gamma=\frac{1}{8}(p^2-1)$ conjunctae ad determinandas singulas quantitates $\Sigma\alpha$, $\Sigma\beta$ et $\Sigma\gamma$ sufficiunt. In hac re tres casus distinguendi sunt: primus quo $\left(\frac{2}{p}\right)=1$, secundus quo $\left(\frac{2}{p}\right)=h$ et tertius quo $\left(\frac{2}{p}\right)=h^2$, pro quibus singulis calculi indicati sine ulla difficultate perficiuntur, quibus igitur perscribendis supersedebimus. Summae autem hae sunt:

I. Si $\left(\frac{2}{p}\right)=1$, sive si 2 est residuum cubicum:

$$\begin{aligned} 3\Sigma\alpha &= 4\Sigma_1\alpha-p\lambda+\tfrac{1}{6}p(p-1),\\ 3\Sigma\beta &= 4\Sigma_1\beta-p\mu+\tfrac{1}{6}p(p-1),\\ 3\Sigma\gamma &= 4\Sigma_1\gamma-p\nu+\tfrac{1}{6}p(p-1). \end{aligned}$$

II. Si $\left(\frac{2}{p}\right)=h$, sive si 2 in serie numerorm β reperitur:

$$\begin{aligned} 3\Sigma\alpha &= 4\Sigma_1\alpha-4\Sigma_1\beta-p\lambda+p\mu+\tfrac{1}{8}(p^2-1),\\ 3\Sigma\beta &= 4\Sigma_1\beta-4\Sigma_1\gamma-p\mu+p\nu+\tfrac{1}{8}(p^2-1),\\ 3\Sigma\gamma &= 4\Sigma_1\gamma-4\Sigma_1\alpha-p\nu+p\lambda+\tfrac{1}{8}(p^2-1). \end{aligned}$$

III. Si $\left(\frac{2}{p}\right)=h^2$, sive si 2 in serie numerorum γ reperitur:

$$\begin{aligned} 3\Sigma\alpha &= 4\Sigma_1\alpha-4\Sigma_1\gamma-p\lambda+p\nu+\tfrac{1}{8}(p^2-1),\\ 3\Sigma\beta &= 4\Sigma_1\beta-4\Sigma_1\alpha-p\mu+p\lambda+\tfrac{1}{8}(p^2-1),\\ 3\Sigma\gamma &= 4\Sigma_1\gamma-4\Sigma_1\beta-p\nu+p\mu+\tfrac{1}{8}(p^2-1). \end{aligned}$$

Eaedem formulae etiam per methodos directas, in doctrina numerorum usitatas, demonstrari possunt, sed hic earum deductionem methodo analytica tradere placuit, qua primum a nobis inventae sunt. Hae autem disquisitiones, quas de residuis cubicis instituimus, tanquam specimina sufficiant.

De numeris complexis, qui radicibus unitatis et numeris integris realibus constant

Journal de mathématiques pures et appliquées XII, 185–212 (1847)

Numeri complexi, quos summus Gaussius primus in doctrinam numerorum introduxit, et quorum auxilio residuorum biquadraticorum theoriam absolvit, formam habent $a + b\sqrt{-1}$. Præter hos autem numeros complexos alii etiam innumeri fingi possunt, qui ad alia doctrinæ numerorum capita eodem modo pertineant, quo hoc genus simplicissimum numerorum complexorum præcipue ad residua quadratica et biquadratica referendum est. Inter hos præcipue notatu digni videntur numeri complexi altioribus unitatis radicibus, per numeros integros reales multiplicatis, compositi, qui doctrinæ de sectione circuli et de residuis potestatum altiorum inserviunt, et cum iis disciplinis tam arcte conjuncti sunt, ut ab ipsis quasi generentur. Quæ de iis numeris hactenus in publicum edita sunt summo geometræ Cl. Jacobi debentur, qui primus demonstravit quemlibet numerum primum formæ $m\lambda + 1$ in duos factores complexos ejus generis discerpi posse. Quod idem numerus primus p pluribus modis diversis in factores duos diffinditur, et quod producta certa ex iis factoribus formata per alios factores divisibiles fiunt, neque tamen hi ipsi factores cum illis compensari possunt, res maximi momenti, indicat hos factores non esse primos sed compositos. Ulteriorem factorum dissolutionem in factores primos Cl. Jacobi pro iis numeris perfecit, qui radices unitatis quintas, octavas et duodecimas continent, ejusque rei notitiam cum regia Academia litterarum Berolinensi communicavit. In hoc quæstionum genere etiam ea versantur, quæ vobis almæ Universitatis Albertinæ viris doctis illustrissimis tanquam magnæ meæ erga vos observantiæ et reverentiæ documentum, hac occasione solemni data, tradere audeo.

---

[*] C'est le Mémoire que nous avons annoncé à la page 136 : il a été imprimé pour la première fois, comme nous l'avons dit, en 1844, à Breslau, sous ce titre : *De numeris complexis, qui radicibus unitatis et numeris integris realibus constant*, et adressé par l'Université de Breslau à l'Université de Kœnigsberg, à l'occasion du troisième jubilé séculaire de cette dernière Université. Nous donnons ici le texte latin. Nous n'avons pas eu le temps d'en faire la traduction. (J. Liouville.)

§ I.

Si α est radix quædam primitiva æquationis $\alpha^\lambda = 1$, quemlibet functionem rationalem integram radicis α, cujus omnes coefficientes numeri integri sint, numerum integrum complexum voco. Talis functio, æquationis $\alpha^\lambda = 1$ ope, statim ad gradum $\lambda - 1$ reducitur, itaque numerorum complexorum quos tractaturi sumus forma generalis est hæc

$$f(\alpha) = a + a_1\alpha + a_2\alpha^2 + \ldots + a_{\lambda-1}\alpha^{\lambda-1};$$

numerum λ hic ubique numerum primum accipimus, qui est casus maximi momenti, et quasi fons a quo tota hæc doctrina derivatur. Inde rejecta radice $\alpha = 1$, quæ hac conditione est sola radix non primitiva, habemus æquationem

$$1 + \alpha + \alpha^2 + \ldots + \alpha^{\lambda-1} = 0,$$

cujus ope ex repræsentatione numeri complexi $f(\alpha)$ unus terminus removeri potest, ex. gr. ultimus, quo facto numerus complexus respectu radicis α ad gradum $\lambda - 2$ deprimitur. Hanc autem reductionem, quæ summarum symetriam turbaret, in universum non adoptabimus, sed retentis omnibus terminis et coefficientibus a, a_1, a_2, etc., æquatione

$$1 + \alpha + \alpha^2 + \ldots + \alpha^{\lambda-1} = 0$$

ita utemur, ut summa omnium coefficientium $a + a_1 + a_2 + \ldots + a_{\lambda-1}$ quodam modo in potestate nostra sit. Nam, *si coefficientes numeri complexi $f(\alpha)$ omnes eodem numero augentur vel minuuntur, hic ipse numerus $f(\alpha)$ non mutatur,* quia nihil accedit nisi multiplum formæ $1 + \alpha + \alpha^2 + \ldots + \alpha^{\lambda-1}$, quæ nihilo æqualis est. Vice versa, *duo numeri complexi æquales esse nequeunt nisi, coefficientibus omnibus eodem numero auctis vel minutis, alter prorsus idem fit atque alter.* Positis enim

$$f(\alpha) = a + a_1\alpha + a_2\alpha^2 + \ldots + a_{\lambda-1}\alpha^{\lambda-1},$$
$$\varphi(\alpha) = b + b_1\alpha + b_2\alpha^2 + \ldots + b_{\lambda-1}\alpha^{\lambda-1},$$

si est $f(\alpha) = \varphi(\alpha)$, habemus

$$0 = a - b + (a_1 - b_1)\alpha + (a_2 - b_2)\alpha^2 + \ldots + (a_{\lambda-1} - b_{\lambda-1})\alpha^{\lambda-1}.$$

Hæc autem æquatio gradus $\lambda - 1$ idem valere debet atque æquatio

$$1 + \alpha + \alpha^2 + \ldots + \alpha^{\lambda-1} = 0,$$

quia, si res aliter se haberet, ex utraque æquatione conjuncta alia æquatio gradus minoris prodiret, cujus coefficientes rationales essent, quod fieri non posse ex Gaussii disquisitionibus arithmeticis notum est. Itaque ex æqualitate numerorum $f(\alpha)$ et $\varphi(\alpha)$, sequitur

$$a - b = a_1 - b_1 = a_2 - b_2 = \ldots = a_{\lambda-1} - b_{\lambda-1}.$$

In numero complexo $f(\alpha)$ est α radix quædam æquationis

$$1 + \alpha + \alpha^2 + \ldots + \alpha^{\lambda-1} = 0,$$

cujus ceteræ radices unius α potestates sunt: omnibus iis radicibus loco α in $f(\alpha)$ substitutis habemus $\lambda - 1$ numeros complexos $f(\alpha), f(\alpha^2), f(\alpha^3), \ldots, f(\alpha^{\lambda-1})$, quos *numeros conjunctos* appellabimus. Inter hos bini ad radices reciprocas pertinent, scilicet $f(\alpha^{\mu})$ et $f(\alpha^{-\mu})$, quos *numeros reciprocos* vocare convenit. Productum omnium numerorum conjunctorum tanquam functio invariabilis integra omnium radicum æquationis

$$1 + \alpha + \alpha^2 + \ldots + \alpha^{\lambda-1} = 0,$$

semper est numerus realis integer, qui numeri complexi *norma* appellatur. Ex ipsa normæ definitione statim perspici possunt propositiones simplices: *numeros conjunctos eamdem normam habere*, et *normam producti æqualem esse producto ex normis singulorum factorum*. Numeri complexi $f(\alpha)$ normam, præeunte Cl. Lejeune-Dirichlet, præposita littera N designamus, ita ut sit

$$\mathrm{N}f(\alpha) = f(\alpha) f(\alpha^2) f(\alpha^3) \ldots f(\alpha^{\lambda-1});$$

unde hæ propositiones tali modo exhiberi possunt

$$\mathrm{N}f(\alpha^r) = \mathrm{N}f(\alpha) \quad \text{et} \quad \mathrm{N}[f(\alpha)\varphi(\alpha)] = \mathrm{N}f(\alpha) . \mathrm{N}\varphi(\alpha).$$

§ II.

Si omnes coefficientes numeri complexi $f(\alpha)$ pro indeterminatis habentur, et productum factorum omnium qui normam constituunt evolvitur, forma homogenea gradus $\lambda - 1$ et λ indeterminatorum prodit, quorum vero unus ex arbitrio eligi potest, ita ut $\lambda - 1$ numeri indeterminati remaneant. Omnes igitur disquisitiones de numeris complexis pro disquisitionibus de talibus formis gradus $\lambda - 1$ totidemque indeterminatorum haberi possunt. De formatione hujus formæ notare convenit eam pro æquatione haberi posse, quæ ex æquationibus duabus

$$0 = a + a_1 \alpha + a_2 \alpha^2 + \ldots + a_{\lambda-1} \alpha^{\lambda-1},$$

$$0 = 1 + \alpha + \alpha^2 + \ldots + \alpha^{\lambda-1},$$

quantitate α eliminata, efficitur, quam eamdem esse atque æquationem

$$f(\alpha) f(\alpha^2) \ldots f(\alpha^{\lambda-1}) = 0,$$

ex notissimis regulis algebraicis constat. Alio modo, norma tanquam denominator communis invenitur, quem systematis æquationum linearium incognitæ evolutæ habent. Quales denominatores Cl. Jacobi determinantium nomine ornavit et pluribus locis in-

24..

geniosissime tractavit. Accipimus systema æquationum linearium hocce :

$$
(\mathrm{A})\quad
\begin{cases}
a\,b + a_{\lambda-1} b_1 + a_{\lambda-2} b_2 + \ldots + a_1 b_{\lambda-1} = c + m,\\
a_1 b + \;a\, b_1 \;+ a_{\lambda-1} b_2 + \ldots + a_2 b_{\lambda-1} = c_1 + m,\\
a_2 b + \;a_1 b_1 \;+ a\, b_2 \quad + \ldots + a_3 b_{\lambda-1} = c_2 + m,\\
\ldots\ldots\ldots\ldots\ldots\ldots\ldots\ldots\ldots\\
a_{\lambda-1} b + a_{\lambda-2} b_1 + a_{\lambda-3} b_2 + \ldots + a\, b_{\lambda-1} = c_{\lambda-1} + m,
\end{cases}
$$

cujus incognitæ sint b, b_1, $b_2,\ldots$, $b_{\lambda-1}$, easque æquationes secundum ordinem multiplicamus per 1, α, $\alpha^2,\ldots$, $\alpha^{\lambda-1}$, quo facto summa omnium facile in hanc formam redigitur

$$
\left(a + a_1\alpha + a_2\alpha^2 + \ldots + a_{\lambda-1}\alpha^{\lambda-1}\right)\left(b + b_1\alpha + b_2\alpha^2 + \ldots + b_{\lambda-1}\alpha^{\lambda-1}\right)
$$
$$
= c + c_1\alpha + c_2\alpha^2 + \ldots + c_{\lambda-1}\alpha^{\lambda-1};
$$

inde positis

$$
f(\alpha) = a + a_1\alpha + a_2\alpha^2 + \ldots + a_{\lambda-1}\alpha^{\lambda-1},
$$
$$
\varphi(\alpha) = b + b_1\alpha + b_2\alpha^2 + \ldots + b_{\lambda-1}\alpha^{\lambda-1},
$$
$$
\psi(\alpha) = c + c_1\alpha + c_2\alpha^2 + \ldots + c_{\lambda-1}\alpha^{\lambda-1},
$$

est

$$
f(\alpha).\varphi(\alpha) = \psi(\alpha),
$$

et utraque parte per $f(\alpha^2)f(\alpha^3)\ldots f\left(\alpha^{\lambda-1}\right)$ multiplicata, fit

$$
\varphi(\alpha).\mathrm{N}f(\alpha) = \psi(\alpha)f(\alpha^2)f(\alpha^3)\ldots f\left(\alpha^{\lambda-1}\right),
$$

sive

$$
\varphi(\alpha) = \frac{\psi(\alpha)f(\alpha^2)f(\alpha^3)\ldots f\left(\alpha^{\lambda-1}\right)}{\mathrm{N}f(\alpha)},
$$

ex quo apparet quantitatum b, b_1, $b_2,\ldots$, $b_{\lambda-1}$, quæ formulam $\varphi(\alpha)$ constituunt, denominatorem generalem esse normam $\mathrm{N}f(\alpha)$, uti contendimus.

Quum norma sit forma aliqua gradus $\lambda - 1$ et $\lambda - 1$ indeterminatorum, statim quæstio oboritur de numeris qui hac forma repræsentari possint et qui non possint. Hanc autem quæstionem gravissimam, quæ sine dubio inter mysteria doctrinæ numerorum maxime recondita referenda est, hactenus non potuimus absolvere; sôla hæc propositio elementaris ad hanc quæstionem spectans : *normam semper habere formam* $m\lambda + 1$, *vel* $m\lambda$, jam hoc loco, quasi in limine disquisitionum nostrarum, facile demonstrari potest. Quem ad finem adhibemus tres numeros complexos $f(\alpha)$, $\varphi(\alpha)$ et $\psi(\alpha)$, eosdem quibus modo usi sumus, quorum vero coefficientes omnes integri sint. Si est

$$
f(\alpha).\varphi(\alpha) = \psi(\alpha),
$$

inter coefficientes horum numerorum complexorum æquationes lineares (A) locum habere debent, iisque additis fit

$$\begin{aligned}(a + a_1 + a_2 + \ldots + a_{\lambda-1})\,(b + b_1 + b_2 + \ldots + b_{\lambda-1})\\ = c + c_1 + c_2 + \ldots + c_{\lambda-1} + \lambda m,\end{aligned}$$

quæ æquatio in congruentiam pro modulo λ mutata docet : *summam coefficientium producti duorum numerorum complexorum producto e summis coefficientium utriusque factoris congruam esse, modulo* λ. Quæ propositio facillime ad productum quotcunque factorum extenditur. Norma semper est productum $\lambda - 1$ factorum complexorum, qui easdem coefficientium summas habent, pro ea igitur productum e summis coefficientium omnium factorum conflatum in potestatem exponentis $\lambda - 1$ abit, unde habemus

$$(a + a_1 + a_2 + \ldots + a_{\lambda-1})^{\lambda-1} \equiv \mathrm{N}f(\alpha) \quad (\text{mod. } \lambda),$$

itaque, per theorema Fermatianum,

$$\mathrm{N}\,f(\alpha) \equiv 1 \quad (\text{mod. } \lambda),$$

nisi forte sit

$$a + a_1 + a_2 + \ldots + a_{\lambda-1} \equiv 0 \quad (\text{mod. } \lambda),$$

qua conditione fit

$$\mathrm{N}f(\alpha) \equiv 0 \quad (\text{mod. } \lambda).$$

§ III.

Normæ usus insignis est in divisione numerorum complexorum, quippe cujus auxilio divisor complexus semper in divisorem realem mutari potest. Posito enim

$$\mathrm{F}(\alpha) = f(\alpha^2)\, f(\alpha^3) \ldots f(\alpha^{\lambda-1})$$

fit

$$\frac{\varphi(\alpha)}{f(\alpha)} = \frac{\varphi(\alpha)\,\mathrm{F}(\alpha)}{\mathrm{N}\,f(\alpha)}.$$

Si $\varphi(\alpha)$ per $f(\alpha)$ divisibilis est, i. e. si hic quotiens numero integro complexo æqualis est, $\varphi(\alpha)\,\mathrm{F}(\alpha)$ per numerum integrum realem $\mathrm{N}\,f(\alpha)$ dividi possit necesse est, numerus autem complexus per numerum realem divisibilis non est, nisi coefficientes ejus singuli, per hunc numerum divisi, eadem residua habeant. Inde nacti sumus hoc criterium generale, quo dijudicari possit, utrum numerus complexus per alium numerum complexum divisibilis sit, an non : *Numerus complexus* $\varphi(\alpha)$ *per alium numerum* $f(\alpha)$ *divisibilis est, si in producto evoluto* $\varphi(\alpha)\, f(\alpha^2)\, f(\alpha^3) \ldots f(\alpha^{\lambda-1})$ *omnes coefficientes pro modulo* $\mathrm{N}\,f(\alpha)$ *congrui sunt, sin vero hi coefficientes non omnes congrui sunt,* $\varphi(\alpha)$ *certo non divisibilis est per* $f(\alpha)$.

Alio etiam modo numerorum complexorum divisio perfici potest, ita quidem, ut ad divisionem fonctionum rationalium integrarum revocetur. Altioribus enim potestatibus radicis α admissis, numerus $\varphi(\alpha)$ infinitis modis diversis representari potest, qui omnes hac forma generali continentur

$$\varphi(\alpha) - (1 + \alpha + \alpha^2 + \ldots + \alpha^{\lambda-1})\,\psi(\alpha),$$

in qua $\psi(\alpha)$ functionem integram quamcunque designat, quæ pro fine singulorum problematum apte eligi poterit. Jam dico, *si* $\varphi(\alpha)$ *per* $f(\alpha)$ *divisibilis sit, huic numero* $\varphi(\alpha)$ *semper formam ejusmodi dari posse, ut pro quolibet valore quantitatis* α, *quæ tanquam variabilis spectanda est, divisio succedat et residuum relinquatur nullum.* Per hypothesin quotiens $\dfrac{\varphi(\alpha)}{f(\alpha)}$ æqualis est numero alicui complexo $F(\alpha)$, itaque

$$\frac{\varphi(\alpha)}{f(\alpha)} = F(\alpha), \quad \text{sive } \varphi(\alpha) = f(\alpha)\,F(\alpha).$$

Jam signo α in x mutato, videmus functionem rationalem integram variabilis x, $\varphi(x) - f(x)\,F(x)$ evanescere, si x cuilibet radici æquationis

$$1 + x + x^2 + \ldots + x^{\lambda-1} = 0$$

æqualis fit; hæc igitur functio integra factorem $1 + x + x^2 + \ldots + x^{\lambda-1}$ habeat necesse est: quare in hanc formam redigi potest

$$\varphi(x) - f(x)\,F(x) = (1 + x + x^2 + \ldots + x^{\lambda-1})\,\psi(x),$$

ex qua theorema enuntiatum sponte manat.

§ IV.

Inter numeros complexos præcipue notatu digni sunt ii, quorum norma est unitas, quos omnes unitatum complexarum nomine designamus. Hæ unitates in doctrina numerorum complexorum easdem fere partes suscipiunt, quas æquationis Pellianæ

$$x^2 - Dy^2 = \pm 1$$

solutiones in doctrina formarum secundi gradus determinantis positivi agunt. Numerus harum unitatum, excepto solo casu $\lambda = 3$, semper infinitus est, et simili modo ut in solvenda æquatione Pelliana semper unitates quædam fundamentales dantur, ex quibus ad potestates evectis et inter se multiplicatis infinitus numerus aliarum unitatum deducitur. Simplicissimæ unitates sunt $\pm 1, \pm\alpha, \pm\alpha^2, \ldots, \pm\alpha^{\lambda-1}$, quæ sponte se offerunt, præter has autem aliæ facile inveniuntur, ratione habita numeri complexi $1 + \alpha + \alpha^2 + \ldots + \alpha^{r-1}$, in quo r est numerus quilibet integer minor quam λ; hic numerus fractionis forma exhibetur hoc modo

$$1 + \alpha + \alpha^2 + \ldots + \alpha^{r-1} = \frac{1 - \alpha^r}{1 - \alpha},$$

ejusque norma est

$$\frac{(1-\alpha^r)(1-\alpha^{2r})(1-\alpha^{3r})\ldots(1-\alpha^{(\lambda-1)r})}{(1-\alpha)(1-\alpha^2)(1-\alpha^3)\ldots(1-\alpha^{\lambda-1})},$$

in qua, loco exponentium r, $2r$, $3r$, ..., $(\lambda - 1)r$ residuis minimis modulo λ substitutis, singuli factores numeratoris iidem fiunt atque denominatoris, quæ igitur unitati æqualis est. Quum jam $1 + \alpha + \alpha^2 + \ldots + \alpha^{r-1}$ sit unitas, et norma producti æqualis producto e normis factorum composito, sequitur ut forma generalis

$$\pm \alpha^k (1 + \alpha + \alpha^2 + .. + \alpha^{r-1})^l (1 + \alpha + \alpha^2 + .. + \alpha^{s-1})^m (1 + \alpha + \alpha^2 + .. + \alpha^{t-1})^n ..,$$

pro omnibus numeris k, l, m, n,..., r, s, t,..., unitates complexas contineat. Numerus factorum diversorum, qui ad potestates evehendi et multiplicandi sunt, ut diversæ unitates obtineantur, itemque numeri r, s, t,..., pro singulis numeris λ accurate definiri possunt, quam vero disquisitionem hoc loco prætereuntes, unitatum proprietatem generalem demonstrabimus, qua in posterum utemur, scilicet: *unitates complexas non tam singularum radicum* 1, α, α^2,..., $\alpha^{\lambda-1}$, *quam periodorum ex binis earum*, $\alpha + \alpha^{\lambda-1}$, $\alpha^2 + \alpha^{\lambda-2}$, *etc., functiones lineares esse, si ad factorem accedentem* $\pm \alpha^k$ *non respiciatur,* sive *omnes unitates hanc formam habere*

$$\pm \alpha^k \left\{ c + c_1 (\alpha + \alpha^{\lambda-1}) + c_2 (\alpha^2 + \alpha^{\lambda-2}) + \ldots + c_{\frac{\lambda-1}{2}} \left(\alpha^{\frac{\lambda-1}{2}} + \alpha^{\frac{\lambda+1}{2}} \right) \right\}.$$

E producto

$$1 = \varphi(\alpha)\varphi(\alpha^2)\varphi(\alpha^3)\ldots\varphi(\alpha^{\lambda-1})$$

factorum semissem ex arbitrio eligo, ita tamen ut reciproci in eadem semissi non insint, sed cujuslibet factoris reciprocus in altera semissi reperiatur. Horum factorum productum sit $\psi(\alpha)$, unde alterius semissis productum est $\psi(\alpha^{-1})$, et $\psi(\alpha)\psi(\alpha^{-1}) = 1$. Ponatur

$$\psi(\alpha) = a + a_1\alpha + a_2\alpha^2 + \ldots + a_{\lambda-1}\alpha^{\lambda-1},$$

unde

$$\psi(\alpha^{-1}) = a + a_1\alpha^{\lambda-1} + a_2\alpha^{\lambda-2} + \ldots + a_{\lambda-1}\alpha,$$

atque ex multiplicatione oriatur

$$\psi(\alpha)\psi(\alpha^{-1}) = \mathrm{A} + \mathrm{A}_1\alpha + \mathrm{A}_2\alpha^2 + \ldots + \mathrm{A}_{\lambda-1}\alpha^{\lambda-1},$$

erit

$$\mathrm{A} = a^2 + a_1^2 + a_2^2 + \ldots + a_{\lambda-1}^2,$$

$$\mathrm{A}_1 = aa_1 + a_1a_2 + a_2a_3 + \ldots + a_{\lambda-1}a,$$

$$\mathrm{A}_2 = aa_2 + a_1a_3 + a_2a_4 + \ldots + a_{\lambda-1}a_1,$$

etc., etc.

et summa coefficientium fit

$$\mathrm{A} + \mathrm{A}_1 + \mathrm{A}_2 + \ldots + \mathrm{A}_{\lambda-1} = (a + a_1 + a_2 + \ldots + a_{\lambda-1})^2;$$

præterea quum sit

$$A + A_1 \alpha + A_2 \alpha^2 + \ldots + A_{\lambda-1} \alpha^{\lambda-1} = 1,$$

erit

$$A_1 = A_2 = A_3 = \ldots = A_{\lambda-1} = m$$

et

$$A = m + 1,$$

itaque

$$m\lambda + 1 = (a + a_1 + a_2 + \ldots + a_{\lambda-1})^2$$

et

$$\pm 1 \equiv a + a_1 + a_2 + \ldots + a_{\lambda-1} \quad (\text{mod. } \lambda);$$

horum coefficientium summa, quæ congrua est ± 1 modulo λ, etiam æqualis ± 1 accipi potest, quo facto fit $m = 0$, $A = 1$, et quum A sit summa quadratorum positivorum integrorum

$$A = a^2 + a_1^2 + a_2^2 + \ldots + a_{\lambda-1}^2,$$

unitati æqualis fieri non potest nisi unus numerorum $a, a_1, a_2, \ldots$, unitati æqualis, ceteri nihilo æquales fiunt, habemus igitur

$$\psi(\alpha) = \pm \alpha^k.$$

Quum $\psi(\alpha)$ alteram factorum semissem normæ 1 contineat, hac sola conditione eligendam, ut factores reciproci non insint, semper plura producta $\psi(\alpha)$ erunt, quæ unitati simplici $\pm \alpha^k$ æqualia sint. Quorum productorum duo eligo, $\psi(\alpha)$ et $\psi_1(\alpha)$, quæ excepto uno factores ceteros omnes eosdem habeant, atque hic solus factor, quo $\psi(\alpha)$ differt a $\psi_1(\alpha)$, sit $\varphi(\alpha^\mu)$, unde factorquem $\psi_1(\alpha)$ continet, neque tamen $\psi(\alpha)$, erit $\varphi(\alpha^{-\mu})$. Inde, ex conjunctis æquationibus

$$\psi(\alpha) = \pm \alpha^k$$

et

$$\psi_1(\alpha) = \pm \alpha^h,$$

fit

$$\psi_1(\alpha) = \pm \alpha^{h-k} \psi(\alpha),$$

et factoribus communibus sublatis et posito

$$h - k = m,$$

est

$$\varphi(\alpha^{-\mu}) = \pm \alpha^m \varphi(\alpha^\mu);$$

quælibet igitur unitas complexa hoc proprium habet, ut a reciproca sua non differat nisi adjecto factore qui ipse est unitas simplex. Facillime inde theorematis supra propositi veritas perspicitur.

Pro $\lambda = 3$, unitates omnes habere debent formam hanc :

$$\varphi(\alpha) = \pm \alpha^k [c + c_1(\alpha + \alpha^2)],$$

et quia $\alpha + \alpha^2 = -1$,

$$\varphi(\alpha) = \pm \alpha^k (c - c_1),$$

ex quo sequitur ut sit

$$c - c_1 = 1;$$

pro hoc igitur casu præter unitates simplices ± 1, $\pm \alpha$, $\pm \alpha^2$, aliæ non dantur.

Pro $\lambda = 5$, unitatum forma generalis hæc est

$$\varphi(\alpha) = \pm \alpha^k [c + c_1(\alpha + \alpha^4) + c_2(\alpha^2 + \alpha^3)],$$

quæ adhibita æquatione

$$1 + \alpha + \alpha^2 + \alpha^3 + \alpha^4 = 0,$$

et posito

$$c - c_2 = t, \quad c_1 - c_2 = u,$$

in hanc simpliciorem mutatur

$$\varphi(\alpha) = \pm \alpha^k [t + u(\alpha + \alpha^4)],$$

ex qua fit

$$\varphi(\alpha)\varphi(\alpha^2) = \alpha^{3k}(t^2 - tu - u^2),$$

itaque

$$t^2 - tu - u^2 = \pm 1.$$

Cognitæ hujus æquationis solutiones omnes continentur formis

$$t = \frac{\left(\frac{-1+\sqrt{5}}{2}\right)^{n-1} - \left(\frac{-1-\sqrt{5}}{2}\right)^{n-1}}{\sqrt{5}}, \quad u = \frac{\left(\frac{-1+\sqrt{5}}{2}\right)^{n} - \left(\frac{-1-\sqrt{5}}{2}\right)^{n}}{\sqrt{5}},$$

quæ, quia

$$\alpha + \alpha^4 = \frac{-1+\sqrt{5}}{2}, \quad \alpha^2 + \alpha^3 = \frac{-1-\sqrt{5}}{2},$$

etiam hoc modo repræsentari possunt

$$t = \frac{(\alpha + \alpha^4)^{n-1} - (\alpha^2 + \alpha^3)^{n-1}}{\alpha + \alpha^4 - \alpha^2 - \alpha^3}, \qquad u = \frac{(\alpha + \alpha^4)^{n} - (\alpha^2 + \alpha^3)^{n}}{\alpha + \alpha^4 - \alpha^2 - \alpha^3};$$

per faciles transformationes ex iis fit

$$t + u(\alpha + \alpha^4) = (\alpha + \alpha^4)^n, \quad \varphi(\alpha) = \pm \alpha^k (\alpha + \alpha^4)^n,$$

itaque pro $\lambda = 5$, unitates complexæ omnes unius unitatis fundamentalis potestates existunt, quæ præterea unitatibus simplicibus formæ $\pm \alpha^k$ multiplicatæ esse possunt, præter has autem aliæ non dantur. Pro majoribus numeris λ unitatum omnium indagatio multo difficilior est, et principia peculiaria poscit, quæ nos hactenus nondum satis perscrutati sumus. Eo magis hac quæstione nobis supersedendum videtur, quum compertum habeamus Cl. Lejeune-Dirichlet recentissimo tempore in Italia, ubi etiam nunc versatur, de iis unitatibus complexis theoremata fundamentalia elaborasse, quæ ut propediem in publicum edat cum desiderio exspectamus. Hic etiam adnotabo geometram juvenilem Leopoldum Kronecker, qui nunc Vratislaviæ litteris mathematicis studet, pro absolvendis iis unitatibus quæ ad numerum $\lambda = 7$ pertinent methodum subtilem invenisse, quæ ni fallor felici successu ad altiorum ordinum unitates pertractandas applicari poterit.

§ V.

Progredimur ad perscrutandos eos numeros complexos, quorum norma sit numerus primus p, ita ut habeatur

$$p = f(\alpha) f(\alpha^2) f(\alpha^3) \ldots f(\alpha^{\lambda-1}).$$

Numerus primus, qui tali modo in $\lambda - 1$ factores dissolvi potest, secundum theorema paragrapho tertia propositum, formam linearem $m\lambda + 1$ habere debet, nisi est ipse numerus $\lambda = p$, qui tali modo in $\lambda - 1$ factores dissolvitur

$$\lambda = (1-\alpha)(1-\alpha^2)(1-\alpha^3)\ldots(1-\alpha^{\lambda-1}).$$

Factores $f(\alpha)$, $f(\alpha^2)$, etc., *sunt numeri complexi primi qui, si unitates complexæ excluduntur, ulterius in factores dissolvi non possunt.* Nam si ponimus $f(\alpha)$ e factoribus $\varphi(\alpha).\psi(\alpha)$ constare, habemus

$$f(\alpha) = \varphi(\alpha).\psi(\alpha),$$

et

$$\mathrm{N}f(\alpha) = \mathrm{N}\varphi(\alpha).\mathrm{N}\psi(\alpha),$$

et quia $\mathrm{N}f(\alpha) = p$ est numerus primus, alter factorum realium integrorum $\mathrm{N}\varphi(\alpha)$ et $\mathrm{N}\psi(\alpha)$ unitati æqualis esse debet, itaque alter factorum $\varphi(\alpha)$ et $\psi(\alpha)$ erit unitas complexa, et $f(\alpha)$ numerus primus complexus. Porro demonstramus *hos factores* $f(\alpha)$, $f(\alpha^2)$, etc., *omnes inter se diversos esse.* Ex æqualitate duorum factorum

$$f(\alpha^\mu) = f(\alpha^\nu),$$

alia radice idonea loco α substituta, sequeretur

$$f(\alpha) = f(\alpha^n),$$

et repetita substitutione α^n loco α, esset

$$f(\alpha) = f(\alpha^n) = f(\alpha^{n^2}) = f(\alpha^{n^3}) = \ldots.$$

Jam si infima potestas numeri n, quæ unitati congrua sit modulo λ, est n^h, in producto

$\lambda - 1$ factorum numeri p erunt h factores æquales; alia deinde radice substituta, quæ inter radices α, α^n, α^{n^2}, etc., non invenitur, statim alios h factores æquales obtinemus, et ita porro, usque dum omnes factores exhausti sunt. Igitur ex æqualitate duorum factorum primum sequeretur ut p sit potestas h^{ta} producti eorum factorum complexorum qui inter se diversi sint. Radices α, α^n, α^{n^2}, etc., periodum h terminorum efficiunt, atque si functio rationalis integra $f(\alpha)$, loco α omnibus periodi radicibus substitutis, immutata manet, hanc ipsam omnium similium periodorum, itaque etiam unius earum, functionem rationalem integram esse constat. Ceteri factores diversi ab hoc non differre possunt, nisi quod ceterarum periodorum functiones sunt, et omnium factorum diversorum productum, tanquam functio invariabilis omnium periodorum similium, esse debet numerus integer realis. Hinc tandem sequeretur ut p esset potestas numeri realis et exponentis h, quod quum absurdum sit concludimus omnes illos factores primos complexos numeri p inter se diversos esse.

Si tali factore primo complexo numeri p tanquam modulo sive divisore utimur, simul productum ceterorum factorum

$$F(\alpha) = f(\alpha^2) f(\alpha^3) \ldots f(\alpha^{\lambda-1}),$$

cujus auxilio divisio per numerum complexum $f(\alpha)$ ad divisionem per numerum realem $\mathrm{N} f(\alpha)$ reducitur, magni momenti est, quam ob rem hujus producti proprietates principales præterire non possumus. Sit

$$p = f(\alpha) F(\alpha),$$

et producto $F(\alpha)$ per multiplicationem evoluto habeatur

$$F(\alpha) = A + A_1 \alpha + A_2 \alpha^2 + \ldots + A_{\lambda-1} \alpha^{\lambda-1},$$

unde etiam

$$F(\alpha^2) = A + A_1 \alpha^2 + A_2 \alpha^4 + \ldots + A_{\lambda-1} \alpha^{2\lambda-2},$$
$$F(\alpha^3) = A + A_1 \alpha^3 + A_2 \alpha^6 + \ldots + A_{\lambda-1} \alpha^{3\lambda-3},$$
$$\ldots\ldots\ldots\ldots\ldots\ldots\ldots\ldots$$
$$F(\alpha^{\lambda-1}) = A + A_1 \alpha^{\lambda-1} + A_2 \alpha^{2\lambda-2} + \ldots + A_{\lambda-1} \alpha^{(\lambda-1)(\lambda-1)};$$

iis æquationibus secundum ordinem per α^{-n}, α^{-2n}, $\alpha^{-3n}, \ldots, \alpha^{-(\lambda-1)n}$ multiplicatis et additione conjunctis, habemus

$$\alpha^{-n} F(\alpha) + \alpha^{-2n} F(\alpha^2) + \ldots + \alpha^{-(\lambda-1)n} F(\alpha^{\lambda-1}) = \lambda A_n - (A + A_1 + A_2 + \ldots + A_{\lambda-1}),$$

inde mutando n in $n-1$, et subtrahendo,

$$\alpha^{-n}(1-\alpha) F(\alpha) + \alpha^{-2n}(1-\alpha^2) F(\alpha^2) + \ldots + \alpha^{-(\lambda-1)n} F(\alpha^{\lambda-1}) = \lambda (A_n - A_{n-1});$$

ex hac forma, denuo mutato n in $n+1$ et $n+2$, prodeunt similes formæ pro

$\lambda(A_{n+1}-A_n)$ et $\lambda(A_{n+2}-A_{n+1})$, quibus inter se multiplicatis formetur evolutio formæ

$$\lambda^2(A_{n+1}-A_n)^2-\lambda^2(A_{n+2}-A_{n+1})(A_n-A_{n-1}).$$

Facile perspicitur in hac evolutione quadrata numerorum complexorum $F(\alpha)$, $F(\alpha^2)$, etc., non reperiri, sed sola producta binorum diversorum, quæ factores $f(\alpha)f(\alpha^2)f(\alpha^3)\ldots f(\alpha^{\lambda-1})$ omnes simul, ideoque factorem realem p continent. Quum igitur p sit factor communis omnium terminorum hujus formulæ evolutæ, factore λ^2 omisso, habemus congruentiam

$$(A_{n+1}-A_n)^2\equiv(A_{n+2}-A_{n+1})(A_n-A_{n-1})\pmod{p},$$

quæ etiam hoc modo repræsentari potest

$$\frac{A_{n+1}-A_n}{A_{n+2}-A_{n+1}}\equiv\frac{A_n-A_{n-1}}{A_{n+1}-A_n}\pmod{p};$$

posito igitur

$$\frac{A_{n+1}-A_n}{A_{n+2}-A_{n+1}}\equiv\xi\pmod{p},$$

videmus numerum ξ pro omnibus diversis numeris n eumdem manere. Hanc congruentiam si hoc modo scribimus

$$A_{n+1}\xi-A_n\equiv A_{n+2}\xi-A_{n+1}\pmod{p},$$

videmus etiam

$$A_{n+1}\xi-A_n\equiv\eta\pmod{p}$$

pro omnibus diversis numeris n non variari, itaque habemus congruentias

$$(A)\qquad\left\{\begin{array}{l}A\xi-A_{\lambda-1}\equiv\eta,\\ A_1\xi-A\equiv\eta,\\ A_2\xi-A_1\equiv\eta\\ \ldots\ldots\ldots\ldots\\ A_{\lambda-1}\xi-A_{\lambda-2}\equiv\eta.\end{array}\right.\pmod{p},$$

Aliud etiam systema congruentiarum, quibus coefficientes producti $F(\alpha)$ satisfacere debent, facile e congruentia

$$A_{n+1}\xi-A_n\equiv A_{n+2}\xi-A_{n+1}\pmod{p}$$

deducitur:

$$(B)\qquad\left\{\begin{array}{l}A_{\lambda-1}-A\equiv(A-A_1)\xi,\\ A_{\lambda-2}-A_{\lambda-1}\equiv(A-A_1)\xi^2,\\ A_{\lambda-3}-A_{\lambda-2}\equiv(A-A_1)\xi^3,\\ \ldots\ldots\ldots\ldots\ldots\\ A_1-A_2\equiv(A-A_1)\xi^{\lambda-1},\end{array}\right.$$

quibus additis, omisso factore $A - A_1$, qui nihilo congruus esse non potest, sequitur ut ξ sit una $\lambda - 1$ radicum realium congruentiæ

$$1 + \xi + \xi^2 + \xi^3 + \ldots + \xi^{\lambda - 1} \equiv 0 \pmod{p}.$$

§ VI.

Singulis congruentiis (A) paragraphi antecedentis secundum ordinem quo scriptæ sunt factoribus $1, \alpha, \alpha^2, \ldots, \alpha^{\lambda-1}$ multiplicatis, earum summa facile in hanc formam redigitur :

$$(\xi - \alpha)(A + A_1\alpha + A_2\alpha^2 + \ldots + A_{\lambda-1}\alpha^{\lambda-1}) \equiv 0 \pmod{p},$$

sive

$$(\xi - \alpha) F(\alpha) \equiv 0 \pmod{p},$$

et factore communi $F(\alpha)$ e modulo $p = f(\alpha) F(\alpha)$ et ex ipsa congruentia sublato habemus

$$\xi - \alpha \equiv 0 \quad [\text{mod.}\ f(\alpha)]$$

unde elucet etiam pro quolibet numero complexo $\varphi(\alpha)$ semper esse

$$\varphi(\alpha) \equiv \varphi(\xi) \quad [\text{mod.}\ f(\alpha)];$$

itaque nacti sumus theorema insigne : *Pro modulo complexo $f(\alpha)$, cujus norma est numerus primus, omnes numeri complexi realibus numeris congrui sunt.* Hoc theorema omnibus congruentiis numerorum complexorum, quarum modulus normam habet numerum primum, viam patefacit, quam vero patefactam hic statim relinquimus.

E congruentia

$$\alpha \equiv \xi \quad [\text{mod.}\ f(\alpha)]$$

sequitur

$$f(\alpha) \equiv f(\xi) \quad [\text{mod.}\ f(\alpha)],$$

itaque

$$f(\xi) \equiv 0 \quad [\text{mod.}\ f(\alpha)];$$

numerus autem integer realis $f(\xi)$, qui factorem $f(\alpha)$ continet, omnes etiam factores huic conjunctos, ideoque factorem p continere debet, unde concludimus eam esse indolem factorum complexorum numeri p, ut certa radice congruentiæ

$$1 + \xi + \xi^2 + \ldots + \xi^{\lambda-1} \equiv 0 \pmod{p},$$

loco α substituta, horum factorum unus per p divisibilis fiat. Præterea adnotamus pro quolibet numero ξ unum, non plures, factorum $f(\xi), f(\xi^2), f(\xi^3)$, etc., per p divisibilem esse; si enim simul esset

$$f(\xi) \equiv 0 \quad \text{et} \quad f(\xi^r) \equiv 0 \pmod{p},$$

per congruentiam

$$\xi \equiv \alpha \quad [\text{mod.}\ f(\alpha)]$$

etiam esse deberet

$$f(\alpha^r) \equiv 0 \quad [\text{mod.}\ f(\alpha)],$$

duo igitur factores $f(\alpha)$ et $f(\alpha^r)$ æquales esse deberent, quod fieri non posse supra demonstravimus. Quum radices congruentiæ

$$1+\xi+\xi^2+\ldots+\xi^{\lambda-1}\equiv 0 \pmod{p}$$

omnes tanquam potestates unius radicis repræsentari possint, alia quacunque radice loco ξ substituta, alium etiam factorem producti $f(\xi).f(\xi^2).f(\xi^3)\ldots f(\xi^{\lambda-1})$ per p divisibilem fieri patet, *singuli igitur* $\lambda-1$ *factores* $f(\alpha), f(\alpha^2)\ldots f(\alpha^{\lambda-1})$, *cum singulis* $\lambda-1$ *radicibus congruentiæ*

$$1+\xi+\xi^2+\ .\ +\xi^{\lambda-1}\equiv 0,$$

ita conjuncti sunt, ut pro certa radice ξ *certus etiam illorum factorum, si* ξ *loco* α *substituitur, per p divisibilis fiat, idemque sit divisor numeri* $\xi-\alpha$.

Jam eo pervenimus ut quæstionem gravissimam absolvere possimus : utrum plures numeri complexi eamdem normam, quæ numerus primus est, habere possint an non, sive an idem numerus primus p, pluribus modis diversis in $\lambda-1$ factores primos complexos dissolvi possit. Fingamus duos numeros complexos $f(\alpha)$ et $\varphi(\alpha)$ eamdem normam p, numerum primum, habere, ita ut sit

$$p=f(\alpha)f(\alpha^2)f(\alpha^3)\ldots f(\alpha^{\lambda-1}),$$
$$p=\psi(\alpha)\psi(\alpha^2)\psi(\alpha^3)\ldots\psi(\alpha^{\lambda-1}).$$

Quum singuli factores horum productorum ad singulas radices congruentiæ

$$1+\xi+\xi^2+\ldots+\xi^{\lambda-1}\equiv 0 \pmod{p}$$

pertineant, alterius producti factores ita dispositos accipiamus, ut $\psi(\alpha)$ et $f(\alpha)$ ad eamdem radicem ξ pertineant, quo facto uterque etiam erit divisor numeri $\xi-\alpha$, atque

$$\psi(\xi)\equiv 0,\quad f(\xi)\equiv 0 \pmod{p}.$$

Inde positis

$$\psi(\alpha)=c+c_1\alpha+c_2\alpha^2+\ldots+c_{\lambda-1}\alpha^{\lambda-1},$$
$$F(\alpha)=f(\alpha^2)f(\alpha^3)\ldots f(\alpha^{\lambda-1})=A+A_1\alpha+A_2\alpha^2+\ldots+A_{\lambda-1}\alpha^{\lambda-1},$$

et

$$\psi(\alpha).F(\alpha)=C+C_1\alpha+C_2\alpha^2+\ldots+C_{\lambda-1}\alpha^{\lambda-1},$$

hoc producto per multiplicationem evoluto habemus :

$$C=Ac+A_{\lambda-1}c_1+A_{\lambda-2}c_2+\ldots+A_1c_{\lambda-1},$$
$$C_1=A_1c+Ac_1+A_{\lambda-1}c_2+\ldots+A_2c_{\lambda-1},$$
$$\ldots\ldots\ldots\ldots\ldots\ldots\ldots\ldots\ldots$$
$$C_{\lambda-1}=A_{\lambda-1}c+A_{\lambda-2}c_1+A_{\lambda-3}c_2+\ldots+Ac_{\lambda-1},$$

et si harum æquationum binæ contiguæ subtrahuntur :

$$C - C_1 = (A - A_1)c + (A_{\lambda-1} - A)c_1 + (A_{\lambda-2} - A_{\lambda-1})c_2 + \ldots + (A_1 - A_2)c_{\lambda-1},$$

$$C_1 - C_2 = (A_1 - A_2)c + (A - A_1)c_1 + (A_{\lambda-1} - A)c_2 + \ldots + (A_2 - A_3)c_{\lambda-1},$$

. . . .

$$C_{\lambda-2} - C_{\lambda-1} = (A_{\lambda-2} - A_{\lambda-1})c + (A_{\lambda-3} - A_{\lambda-2})c_1 + (A_{\lambda-1} - A_{\lambda-3})c_2 + \ldots + (A_{\lambda-1} - A)c_{\lambda-1}$$

quæ per congruentias (B), § V, mutantur in

$$C - C_1 \equiv (A - A_1)(c + c_1\xi + c_2\xi^2 + \ldots + c_{\lambda-1}\xi^{\lambda-1}),$$

$$C_1 - C_2 \equiv (A_1 - A_2)(c + c_1\xi + c_2\xi^2 + \ldots + c_{\lambda-1}\xi^{\lambda-1}),$$

. . . .

$$C_{\lambda-2} - C_{\lambda-1} \equiv (A_{\lambda-2} - A_{\lambda-1})(c + c_1\xi + c_2\xi^2 + \ldots + c_{\lambda-1}\xi^{\lambda-1}),$$

et quum sit

$$\psi(\xi) = c + c_1\xi + c_2\xi^2 + \ldots + c_{\lambda-1}\xi^{\lambda-1} \equiv 0 \pmod{p},$$

habemus

$$C \equiv C_1 \equiv C_2 \equiv \ldots \equiv C_{\lambda-1} \pmod{p},$$

unde sequitur ut productum $\psi(\alpha)F(\alpha)$ per p divisibile sit ; itaque ponere licet

$$\psi(\alpha).F(\alpha) = p\,\varphi(\alpha),$$

et quum sit $p = f(\alpha)F(\alpha)$, factore communi $F(\alpha)$ sublato habemus

$$\psi(\alpha) = f(\alpha)\varphi(\alpha),$$

unde etiam

$$N\psi(\alpha) = Nf(\alpha)\,N\varphi(\alpha),$$

et quia $Nf(\alpha) = N\psi(\alpha) = p$, hoc factore omisso est

$$N\varphi(\alpha) = 1;$$

ergo $\varphi(\alpha)$ est unitas complexa, atque habemus theorema : *Omnes numeri complexi ejusdem normæ, quæ numerus primus est, non inter se differunt nisi unitatibus complexis, quibus multiplicatæ esse possunt.* Ad diversitatem numerorum conjunctorum, quæ obvia est, hic non respicimus. Idem theorema etiam hoc modo enuntiari potest : *Si numerus primus realis aliquo modo in $\lambda - 1$ factores complexos dissolvi potest, idem semper infinitis aliis modis tanquam productum $\lambda - 1$ factorum complexorum repræsentari potest, qui vero e factoribus unius repræsentationis unitatum complexarum ope, omnes eruuntur.*

§ VII.

Facile inveniuntur numeri complexi, quorum normæ factorem p, numerum primum formæ $m\lambda + 1$, implicant. Nam *si* ξ *est radix aliqua congruentiæ*

$$1 + \xi + \xi^2 + \ldots + \xi^{\lambda-1} \equiv 0 \pmod{p},$$

et si coefficientes numeri complexi

$$\psi(\alpha) = a + a_1\alpha + a_2\alpha^2 + \ldots + a_{\lambda-1}\alpha^{\lambda-1}$$

congruentiæ

$$a + a_1\xi + a_2\xi^2 + \ldots + a_{\lambda-1}\xi^{\lambda-1} \equiv 0$$

satisfaciunt, semper est

$$\mathrm{N}\psi(\alpha) \equiv 0 \pmod{p}.$$

Etenim, si ponimus

$$a_1(\xi - \alpha) + a_2(\xi^2 - \alpha^2) + a_3(\xi^3 - \alpha^3) + \ldots + a_{\lambda-1}(\xi^{\lambda-1} - \alpha^{\lambda-1}) = p\mathrm{P} - \psi(\alpha),$$

et si alteri parti hujus æquationis forma datur $(\xi - \alpha)\,\varphi(\alpha)$, est

$$\psi(\alpha) = p\mathrm{P} - (\xi - \alpha)\,\varphi(\alpha),$$

cujus numeri complexi normam factorem p habere patet siquidem, quod posuimus, norma numeri $\xi - \alpha$, i. e. $1 + \xi + \xi^2 + \ldots \xi^{\lambda-1}$, per p divisibilis est.

Hujus theorematis ope *omnes* numeros complexos inveniri, quorum normæ factorem p habeant, inde apparet, quod theorema inversum etiam valet, scilicet : *Si norma numeri complexi*

$$\psi(\alpha) = a + a_1\alpha + a_2\alpha^2 + \ldots + a_{\lambda-1}\alpha^{\lambda-1},$$

factorem p implicat, semper datur radix aliqua ξ *congruentiæ*

$$1 + \xi + \xi^2 + \ldots + \xi^{\lambda-1} \equiv 0 \pmod{p},$$

quæ numerum

$$\psi(\xi) = a + a_1\xi + a_2\xi^2 + \ldots + a_{\lambda-1}\xi_{\lambda-1}$$

per p divisibilem reddat. Consideremus hanc functionem rationalem integram variabilis x :

$$\psi(x)\,\psi(x^2)\,\psi(x^3)\ldots\psi(x^{\lambda-1}) - \mathrm{N}\psi(\alpha),$$

quæ quum manifesto evanescat pro

$$x = \alpha,\ \alpha^2,\ \alpha^3,\ \ldots,\ \alpha^{\lambda-1},$$

factorem $1 + x + x^2 + \ldots + x^{\lambda-1}$ habere debet, eamque ob causam, si x radici alicui congruentiæ

$$1 + x + x^2 + \ldots + x^{\lambda-1} \equiv 0 \pmod{p}$$

æqualis ponitur, per p divisibilis est; inde si hujus congruentiæ radicem aliquam, uti supra fecimus, littera ξ denotamus, semper unus factorum producti $\psi(\xi)\,\psi(\xi^2)\,\psi(\xi^3)\ldots\psi(\xi^{\lambda-1})$ per p divisibilis esse debet. Præterea, quia omnes radices illius congruentiæ unius radicis ξ potestates sunt, sponte apparet pro alia radice ξ alium etiam factorem hujus producti per p divisibilem fieri, itaque pro quolibet factore certa quædam radix ξ exstare debet, qua substituta nihilo congruus reddatur.

§ VIII.

Postquam demonstravimus quo modo infinitus numerus complexorum numerorum omnium inveniatur, quorum normæ factorem p habeant, quærendum nobis videtur an semper inter hos numeros complexos existent quarum norma sit ipse numerus p, sive, quod idem est, an quilibet numerus primus p formæ $m\lambda + 1$ in $\lambda - 1$ factores primos conjunctos diffindi possit. Ad hunc finem ponamus esse

$$p = f(\alpha)\,f(\alpha^2)\,f(\alpha^3)\ldots f(\alpha^{\lambda-1}),$$

et videamus quæ inde pro numero p sequantur. Hujus producti factores secundum radices unitatis α, α^2, α^3,..., $\alpha^{\lambda-1}$, quas continent in periodos dividamus. Productum factorum eorum qui ad eamdem periodum pertinent, ex notis Cl. Gaussii theorematis, erit functio rationalis integra linearis omnium periodorum similium. Jam si $\lambda - 1$ factoribus e et f constat, et e periodi f terminorum accipiuntur, quas Gaussii signis designamus $(f, 1)$, (f, g), (f, g^2),..., (f, g^{e-1}), productum eorum factorum, qui eamdem periodum constituunt hanc formam habet

$$b + b_1(f, 1) + b_2(f, g) + \ldots + b_e(f, g^{e-1}),$$

et simili modo producta factorum qui cæteras periodos conficiunt

$$\begin{gathered} b + b_1(f, g\ ) + b_2(f, g^2) + \ldots + b_e(f, 1), \\ b + b_1(f, g^2) + b_2(f, g^3) + \ldots + b_e(f, g), \\ \text{etc,} \qquad\qquad \text{etc.}; \end{gathered}$$

quarum omnium productum quum sit functio invariabilis (symetrica) omnium periodorum, ab iis periodis liberum est, et formam certam gradus e et $e + 1$ indeterminatorum b, b_1, b_2,..., b_e efficit, qui facile ad e indeterminatos numeros reducuntur. Igitur si e, e', e'',... sunt divisores numeri $\lambda - 1$, numerus p repræsentari debet per formam certam gradus e et e indeterminatorum, idemque per formam gradus e' et e' indeterminatorum, etc. Hoc autem pro certis numeris p et λ fieri non posse facillime intelligitur ex forma secundi gradus et duorum indeterminatorum, quam p habere debet, si in $\lambda - 1$ factores primos complexos diffindi potest; duæ enim periodi, quarum altera ad residua quadratica, altera ad non residua pertinet, valores habent $\dfrac{-1 + \sqrt{\pm\lambda}}{2}$ et $\dfrac{-1 - \sqrt{\pm\lambda}}{2}$, inde productum ex altera semissi factorum complexorum numeri p

compositum hanc formam habebit $\frac{a \pm b\sqrt{\pm\lambda}}{2}$, et ipse numerus p habebit formam $p = \frac{a^2 \mp \lambda b^2}{4}$, seu $4p = a^2 \mp \lambda b^2$, quam numeri formæ $m\lambda + 1$ non semper patiuntur, siquidem præter formam principalem aliæ quoque formæ non æquivalentes secundi gradus, determinantis $\mp\lambda$, locum habent. Simili modo etiam altiorum graduum formæ impedimento esse possunt, quominus numerus p in $\lambda - 1$ factores primos complexos dissolvi queat.

Quia numeri primi reales formæ $m\lambda + 1$ non semper tanquam producta $\lambda - 1$ factorum complexorum repræsentari possunt, multis etiam numerorum integrorum realium proprietatibus simplicibus numeri complexi carent. Pro iis generaliter non valet propositio fundamentalis ut quilibet numerus sit productum factorum complexorum simplicium, qui neglectis unitatibus complexis semper iidem sint, re enim vera nonnunquam idem numerus compositus pluribus modis diversis in factores simplices complexos diffindi potest. Eadem res etiam hoc modo enuntiari potest : Si numerus complexus per alium numerum complexum ita dividi potest, ut quotiens sit integer complexus, factores simplices divisoris non ubique cum factoribus simplicibus dividendi compensari possunt. Quod ut demonstremus denuo numerum ξ cognitæ illius congruentiæ radicem in auxilium vocamus, quæ hoc modo in factores dissolvitur :

$$(\xi - \alpha)(\xi - \alpha^2)\ldots(\xi - \alpha^{\lambda-1}) \equiv 0 \pmod{p}.$$

Jam si factores divisoris p cum factoribus dividendi tollerentur, p cum aliquo factorum $\xi - \alpha$, $\xi - \alpha^2$, etc., factorem communem haberet, ideoque etiam cum singulo quoque. Sit $f(\alpha)$ hic factor communis numerorum p et $\xi - \alpha$, $f(\alpha^2)$ erit factor communis numerorum p et $\xi - \alpha^2$, et ita porro; omnes igitur numeri complexi conjuncti $f(\alpha), f(\alpha^2)\ldots f(\alpha^{\lambda-1})$, factores essent numeri p, omnesque inter se diversi, quia $\xi - \alpha$, $\xi - \alpha^2$, etc., non possunt factores communes habere nisi eos quorum norma sit 1 vel λ, scilicet $\alpha - \alpha^2$, $\alpha - \alpha^3$, etc. Inde sequeretur ut quilibet numerus primus $p = m\lambda + 1$ esset productum $\lambda - 1$ factorum complexorum conjunctorum, quod in universum non pro omnibus valoribus numerorum p et λ valere supra demonstravimus. Maxime dolendum videtur, quod hæc numerorum realium virtus, ut in factores primos dissolvi possint, qui pro eodem numero semper iidem sint, non eadem est numerorum complexorum, quæ si esset, tota hæc doctrina, quæ magnis adhuc difficultatibus laborat, facile absolvi et ad finem perduci posset. Eam ipsam ob causam numeri complexi, quos hic tractamus, imperfecti esse videntur, et dubium inde oriri posset, utrum hi numeri complexis ceteris qui fingi possint præferendi, an alii quærendi essent, qui in hac re fundamentali analogiam cum numeris integris realibus servarent. Attamen hi numeri complexi, qui unitatis radicibus et numeris integris realibus componuntur, non ex arbitrio facti sunt, sed ex ipsa doctrina numerorum procreati, atque ipsorum ea ratio est, ut in doctrina sectionis circuli et residuorum potestatum altiorum ulterius promovenda iis carere nullo modo possimus.

§ IX.

Quum numerorum primorum formæ $m\lambda+1$ alii in $\lambda-1$ factores complexos discerpi possint, alii non possint, e re est ut inveniatur qui et quales sint ii numeri λ et p, pro quibus talis repræsentatio locum habet, atque ut pro iis qui minorem tantum numerum factorum primorum habent, horum factorum numerus et forma propria indagetur, quod vero problema ulterioribus virorum doctorum perscrutationibus relinquendum est. Ipse ut hanc rem accuratius cognoscerem, et ut exemplis docerer, omnium numerorum primorum infra mille, qui formas habent

$$5m+1,\quad 7m+1,\quad 11m+1,\quad 13m+1,\quad 17m+1,\quad 19m+1 \text{ et } 23m+1$$

factores primos computavi, et methodos quibus usus sum, et ipsos factores primos computatos hoc loco in publicum edam.

Altera methodus factore communi duorum numerorum complexorum inveniendo nititur, quod problema simili modo solvendum est atque pro numeris integris realibus. Si $\varphi(\alpha)$ et $f(\alpha)$ sunt numeri complexi quorum factor communis maximus investigandus est, et $\mathrm{N}\varphi(\alpha) > \mathrm{N}f(\alpha)$, a numero fracto $\frac{\varphi(\alpha)}{f(\alpha)}$ numerum certum integrum subtraho $\psi(\alpha)$, quem siquidem fieri potest ita eligo, ut norma residui sit minor unitate, sive

$$\mathrm{N}\left(\frac{\varphi(\alpha)}{f(\alpha)}-\psi(\alpha)\right)<1.$$

Ut hoc efficiatur, evolvo productum

$$\mathrm{F}(\alpha)=f(\alpha^2)f(\alpha^3)\ldots f(\alpha^{\lambda-1}),$$

ejusque auxilio etiam productum

$$\varphi(\alpha)\,\mathrm{F}(\alpha)=\mathrm{C}+\mathrm{C}_1\alpha+\mathrm{C}_2\alpha^2+\ldots+\mathrm{C}_{\lambda-1}\alpha^{\lambda-1};$$

porro accipio

$$\psi(\alpha)=c+c_1\alpha+c_2\alpha^2+\ldots+c_{\lambda-1}\alpha^{\lambda-1},$$

et simpliciter scribo n, loco $\mathrm{N}f(\alpha)$. Inde erit

$$\frac{\varphi(\alpha)}{f(\alpha)}-\psi(\alpha)=\frac{\varphi(\alpha)\,\mathrm{F}(\alpha)}{n}-\psi(\alpha),$$

et posito

$$\frac{\varphi(\alpha)\,\mathrm{F}(\alpha)}{n}-\psi(\alpha)=k+k_1\alpha+k_2\alpha^2+\ldots+k_{\lambda-1}\alpha^{\lambda-1},$$

erit

$$k=\frac{\mathrm{C}}{n}-c,\quad k_1=\frac{\mathrm{C}_1}{n}-c_1,\quad k_2=\frac{\mathrm{C}_2}{n}-c_2,\quad \text{etc.}$$

Jam numeros c, c_1, c_2, etc., ita eligo ut integri maximi sint, qui singulis fractionibus

26..

$\frac{C}{n}$, $\frac{C_1}{n}$, $\frac{C_2}{n}$, etc., contineantur, quo facto numeri k, k_1, k_2, etc., omnes erunt positivi et minores unitate. Certo talis numeri complexi $k + k_1\alpha + k_2\alpha^2 + \ldots + k_{\lambda-1}\alpha^{\lambda-1}$, norma parva erit; sed semper eam unitate minorem fore nondum liquet, neque etiam semper res ita se habet. Hoc autem loco nobis notandum est numeros k, k_1, k_2, etc., iis conditionibus quas posuimus nondum plane determinatos esse, omnibus enim coefficientibus C, C_1, C_2, etc., eodem numero auctis, $\varphi(\alpha).F(\alpha)$ non mutatur, sed numeri integri maximi, qui fractionibus $\frac{C}{n}$, $\frac{C_1}{n}$, $\frac{C_2}{n}$, etc., insunt, revera mutari possunt. Nam si numeros C, C_1, C_2, etc., omnes æqualiter crescentes accipimus, numeri k, k_1, k_2, etc., omnes eodem modo crescent, usque dum maximus eorum unitatem superaverit, quo facto unitate minuendus est, et subito omnium minimus fit. Eadem mutatione repetita patet in universum obtineri λ numeros complexos $k+k_1\alpha+k_2\alpha^2+\ldots+k_{\lambda-1}\alpha^{\lambda-1}$, quorum normæ diversæ sint. Jam si vel omnium harum normarum nulla unitate minor existet, methodus nostra nos deficit, hunc autem defectum non methodo vicio dandum, sed rei ipsius natura necessarium esse, infra demonstrabitur. Si vero, quod fere semper evenire solet, talis norma unitate minor fit, habemus

$$N\left[\frac{\varphi(\alpha)}{f(\alpha)} - \psi(\alpha)\right] < 1, \quad \text{ideoque} \quad N[\varphi(\alpha) - f(\alpha)\psi(\alpha)] < Nf(\alpha).$$

Posito $\varphi(\alpha) - f(\alpha)\psi(\alpha) = R(\alpha)$, patet factorem maximum communem numerorum $\varphi(\alpha)$ et $f(\alpha)$ eumdem esse numerorum $f(\alpha)$ et $R(\alpha)$; unde indagatio factoris communis eo reducta est, ut aliorum duorum numerorum, quorum normæ minores sunt, factor communis quærendus sit. Itaque, hac methodo repetita, nisi forte casus ille adversus evenit, quem supra commemoravimus, tandem ad duos numeros pervenimus, quorum alter factor alterius, ideoque hic ipse factor communis est quem quærimus; cujus norma si unitas est, numeri illi sunt inter se primi.

Facile hæc methodus ad factores primos numeri $p = m\lambda + 1$ indagandos applicari potest, siquidem p revera est productum $\lambda - 1$ factorum conjunctorum. Quem ad finem quæratur radix aliqua ξ congruentiæ

$$1 + \xi + \xi^2 + \ldots + \xi^{\lambda-1} \equiv 0 \pmod{p},$$

quæ si p in $\lambda - 1$ factores conjunctos diffindi potest, semper talis est, ut $\xi - \alpha$ et p factorem complexum communem habeant. Hic igitur, secundum methodum traditam inventus, erit factor simplex complexus numeri p. Supra vidimus pro certis numeris p et λ talem factorem communem numerorum $\xi - \alpha$ et p non adesse, quamvis norma numeri $\xi - \alpha$ per p divisibilis sit; porro si per methodum traditam numeri complexi normarum minorum quæruntur, patet eorum omnium normas per p divisibiles esse, quam ob rem nullo modo ad normam unitatem pervenire possumus; semper igitur factor communis ab unitate diversus inveniretur, etiam ubi talem factorem non adesse demonstravimus, nisi in omnibus iis calculis eveniret ut norma istius numeri complexi fracti $k + k_1\alpha + k_2\alpha^2 + \ldots + k_{\lambda-1}\alpha^{\lambda-1}$ minor unitate fieri non posset, qua conditione methodus nostra ad finem propositum perducere non potest.

Altera methodus minus quidem directa tamen multo faciliori negotio factores primos complexos numeri $p = m\lambda + 1$ præbet. In hac omnes congruentiæ

$$1 + \xi + \xi^2 + \ldots + \xi^{\lambda-1} \equiv 0 \quad (\text{mod. } p),$$

radices ξ, ξ^2, ξ^3, etc., in usum vocamus, quibus e canone arithmetico a Cl. Jacobi edito depromptis, sive alio modo inventis, solutiones congruentiæ

$$a + a_1\xi + a_2\xi^2 + \ldots + a_{\lambda-1}\xi^{\lambda-1} \equiv 0 \quad (\text{mod. } p)$$

quærimus, ex quovis systemate numerorum a, a_1, $a_2, \ldots, a_{\lambda-1}$, qui huic congruentiæ satisfaciunt numerum complexum $f(\alpha) = a + a_1\alpha + a_2\alpha^2 + \ldots + a_{\lambda-1}\alpha^{\lambda-1}$ componimus, et ex omnibus iis numeris, quarum normæ per theorema supra demonstratum, § VII, factorem p habent, eam eligimus quæ simplicissima sit, et normam quam minimam habere videatur, quæ jam ipsa computanda est, ut appareat utrum revera ipsi numero p an multiplo ejus æqualis sit. Si eveniret hanc normam non esse ipsum p, sed ejus multiplum e numeris complexis quarum normæ per p divisibiles sunt alius quærendus et examinandus esset, et ita porro. Si vero horum numerorum complexorum nullus ipsam normam p habet, hic numerus primus p iis adnumerandus est, qui $\lambda - 1$ factoribus complexis conjunctis non sunt compositi.

§ X.

Antequam ipsos numerorum primorum realium factores primos complexos quos invenimus litteris consignamus pauca de iis præmittere convenit. Pro $\lambda = 5, 7, 11, 13, 17$ et 19, omnes numeros primos formæ $m\lambda + 1$ in primo mille contentos in $\lambda - 1$ factores dissolvimus: primus numerus λ, pro quo hoc genus factorum primorum non semper datur, est numerus $\lambda = 23$; inter octo enim numeros primos formæ $23m + 1$, qui minores sunt quam mille, tres sunt qui viginti duobus factoribus primis constant, reliquos autem quinque, qui quum formam quadraticam $4p = a^2 + 23b^2$ non patiantur, in viginti duo factores conjunctos dissolvi non possunt, in undecim factores primos complexos discerpere nobis contigit. Tales numeri eumdem characterem habere videntur ac numeri primi qui non sunt formæ $m\lambda + 1$, quos de hac nostra commentatione exclusimus, hi omnes habent minorem numerum factorum complexorum primorum, qui non tam radicum singularum æquationis $\alpha^\lambda = 1$ quam periodorum functiones lineares sunt. Simili enim modo horum quinque numerorum primorum factores primi complexi, quos invenimus, periodos binarum radicum continent. Quibus præmissis ipsos factores inventos tradimus.

(1) Si $\lambda = 5$, et α radix æquationis $\alpha^5 = 1$.

$11 = N(2+\alpha)$

$31 = N(2-\alpha)$

$41 = N(3+2\alpha+\alpha^2)$

$61 = N(3+\alpha)$

$71 = N(3-\alpha+\alpha^2)$

$101 = N(3+\alpha-\alpha^2)$

$131 = N(3+\alpha-\alpha^4)$

$151 = N(3+2\alpha-\alpha^4)$

$181 = N(4+3\alpha)$

$191 = N(4+\alpha+2\alpha^2)$

$211 = N(3-2\alpha)$

$241 = N(4-\alpha+\alpha^2)$

$251 = N(5+2\alpha+\alpha^4)$

$271 = N(3-3\alpha+\alpha^2)$

$281 = N(4+\alpha-\alpha^2)$

$311 = N(3+2\alpha+2\alpha^2+\alpha^3)$

$331 = N(4-2\alpha+\alpha^2)$

$401 = N(4+3\alpha-\alpha^4)$

$421 = N(5+2\alpha+2\alpha^2)$

$431 = N(4-2\alpha-\alpha^4)$

$461 = N(4-\alpha-\alpha^2)$

$491 = N(5+3\alpha+\alpha^3)$

$521 = N(5+\alpha)$

$541 = N(3-3\alpha-\alpha^2)$

$571 = N(6+5\alpha+3\alpha^2)$

$601 = N(5+2\alpha-\alpha^2)$

$631 = N(4-2\alpha-\alpha^3)$

$641 = N(5+3\alpha+4\alpha^2)$

$661 = N(5+\alpha-\alpha^2+3\alpha^3)$

$691 = N(3-3\alpha-2\alpha^2)$

$701 = N(4-\alpha-2\alpha^2+\alpha^3)$

$751 = N(6+4\alpha+3\alpha^2)$

$761 = N(5-2\alpha+\alpha^2)$

$811 = N(3-3\alpha-2\alpha^2+\alpha^3)$

$821 = N(4-\alpha-2\alpha^2+2\alpha^3)$

$881 = N(6+2\alpha+\alpha^2)$

$911 = N(5+\alpha^2-2\alpha^4)$

$941 = N(4+3\alpha-3\alpha^2-\alpha^3)$

$971 = N(5-2\alpha-\alpha^4)$

$991 = N(6+\alpha+\alpha^3)$

(2) Si $\lambda = 7$, et α est radix æquationis $\alpha^7 = 1$.

$29 = N(1+\alpha-\alpha^2)$

$43 = N(2+\alpha)$

$71 = N(2+\alpha+\alpha^3)$

$113 = N(2-\alpha+\alpha^5)$

$127 = N(2-\alpha)$

$197 = N(3+\alpha+\alpha^5+\alpha^6)$

$211 = N(3+\alpha+2\alpha^2)$

$239 = N(3+2\alpha+2\alpha^2+\alpha^3)$

$281 = N(2-\alpha-2\alpha^3)$

$337 = N(2+\alpha-\alpha^2-\alpha^4)$

$379 = N(3+2\alpha+\alpha^2)$

$421 = N(3+\alpha+\alpha^2)$

$449 = N(2+\alpha-\alpha^3-\alpha^6)$

$463 = N(3+2\alpha)$

$491 = N(3+\alpha+\alpha^3-\alpha^5)$

$547 = N(3+\alpha)$

$617 = N(2+\alpha+\alpha^2-\alpha^5)$

$631 = N(2+2\alpha-\alpha^2+\alpha^3+\alpha^6)$

$659 = N(2 + 2\alpha - \alpha^2 + \alpha^5)$

$673 = N(4 + 3\alpha + 2\alpha^2 + \alpha^4 + 2\alpha^6)$

$701 = N(3 + \alpha + \alpha^4 - \alpha^5 + \alpha^6)$

$743 = N(3 + 2\alpha - \alpha^3 - \alpha^4)$

$757 = N(3 + 2\alpha + \alpha^3)$

$827 = N(2 + 2\alpha - \alpha^4 - \alpha^6)$

$883 = N(2 - \alpha^2 - 2\alpha^3 - \alpha^5)$

$911 = N(3 + 2\alpha - \alpha^3 + \alpha^4)$

$953 = N(3 + \alpha - \alpha^2 - \alpha^3)$

$967 = N(2 + 2\alpha - \alpha^3 + 2\alpha^5)$

(3) Si $\lambda = 11$, et α est radix æquationis $\alpha^{11} = 1$.

$23 = N(1 + \alpha + \alpha^9)$

$67 = N(1 + \alpha + \alpha^2 + \alpha^4 + \alpha^5)$

$89 = N(1 + \alpha + \alpha^4 + \alpha^6)$

$199 = N(1 + \alpha - \alpha^2)$

$331 = N(1 - \alpha + \alpha^3 + \alpha^5)$

$353 = N(1 + \alpha + \alpha^3 + \alpha^4 - \alpha^7)$

$397 = N(1 + \alpha + \alpha^6 - \alpha^7)$

$419 = N(1 + \alpha - \alpha^2 + \alpha^3)$

$463 = N(1 - \alpha - \alpha^2 + \alpha^5 + \alpha^6)$

$617 = N(2 + \alpha + \alpha^3 + \alpha^{10})$

$661 = N(1 + \alpha - \alpha^2 + \alpha^4 - \alpha^8)$

$683 = N(2 + \alpha)$

$727 = N(1 + \alpha + \alpha^3 - \alpha^8 - \alpha^9)$

$859 = N(1 + \alpha + \alpha^2 + \alpha^3 + \alpha^7 - \alpha^8)$

$881 = N(1 + \alpha + \alpha^2 + \alpha^3 - \alpha^4 - \alpha^7 - \alpha^9)$

$947 = N(2 + \alpha^3 - \alpha^4 - \alpha^6)$

$991 = N(2 + \alpha + \alpha^3)$

(4) Si $\lambda = 13$, et α est radix æquationis $\alpha^{13} = 1$.

$53 = N(1 + \alpha + \alpha^3)$

$79 = N(1 - \alpha + \alpha^{10})$

$131 = N(1 - \alpha + \alpha^{11})$

$157 = N(1 + \alpha + \alpha^2 + \alpha^5)$

$313 = N(1 - \alpha + \alpha^3 + \alpha^6)$

$443 = N(1 + \alpha - \alpha^3 + \alpha^8)$

$521 = N(1 + \alpha - \alpha^{12})$

$547 = N(1 - \alpha - \alpha^2 + \alpha^3 + \alpha^6)$

$599 = N(1 + \alpha - \alpha^7 + \alpha^8 + \alpha^{11})$

$677 = N(1 - \alpha - \alpha^4 + \alpha^8 + \alpha^9)$

$959 = N(1 + \alpha - \alpha^2 - \alpha^5 + \alpha^7)$

$911 = N(1 + \alpha^3 + \alpha^5 - \alpha^7 - \alpha^{11})$

$937 = N(1 + \alpha^3 - \alpha^7 + \alpha^8 - \alpha^{10})$

(5) Si $\lambda = 17$, et α est radix æquationis $\alpha^{17} = 1$.

$103 = N(1 + \alpha^2 + \alpha^9)$

$137 = N(1 + \alpha - \alpha^3)$

$239 = N(1 + \alpha + \alpha^3)$

$307 = N(1 - \alpha + \alpha^7)$

$409 = N(1 - \alpha^3 + \alpha^8)$

$443 = N(1 + \alpha + \alpha^2 + \alpha^3 - \alpha^{15})$

$613 = N(1 + \alpha^2 - \alpha^3)$

$647 = N(1 + \alpha + \alpha^{13} + \alpha^{15})$

$919 = N(1 + \alpha + \alpha^4 + \alpha^5 + \alpha^9)$

$953 = N(1 + \alpha + \alpha^9 - \alpha^{13})$

(6) Si $\lambda = 19$, et α est radix æquationis $\alpha^{19} = 1$.

$191 = N(1 + \alpha + \alpha^{16})$ $\qquad$ $457 = N(1 + \alpha + \alpha^{3})$

$229 = N(1 - \alpha - \alpha^{5})$ $\qquad$ $571 = N(1 + \alpha + \alpha^{2} + \alpha^{3} - \alpha^{5})$

$419 = N(1 + \alpha - \alpha^{8})$ $\qquad$ $647 = N(1 - \alpha^{2} + \alpha^{9})$

$$761 = N(1 - \alpha^{2} + \alpha^{12})$$

(7) Si $\lambda = 23$, et α est radix æquationis $\alpha^{23} = 1$.

$599 = N(1 + \alpha^{15} - \alpha^{16})$ $\qquad$ $691 = N(1 + \alpha + \alpha^{5})$

$$829 = N(1 + \alpha^{11} + \alpha^{20})$$

Reliqui numeri primi formæ $23m + 1$ infra mille undecim factoribus primis constant, habet

47 factorem $\alpha^{10} + \alpha^{13} + \alpha^{8} + \alpha^{15} + \alpha^{7} + \alpha^{16}$

139 » $\alpha^{10} + \alpha^{13} + \alpha^{8} + \alpha^{15} + \alpha^{4} + \alpha^{19}$

277 » $2 + \alpha + \alpha^{2\square}$ [illegible] $+ \alpha^{7} + \alpha^{16}$

461 » $\alpha + \alpha^{22} + \alpha^{10} + \alpha^{13} + \alpha^{8} + \alpha^{15} + \alpha^{9} + \alpha^{14}$

967 » $2 + \alpha^{11} + \alpha^{12} + \alpha^{4} + \alpha^{19}$.

§ XI.

Quæ de numeris complexis et de eorum factoribus primis commentati sumus ad doctrinam de sectione circuli felicissimo successu applicari possunt. In hac enim doctrina tales numeri complexi eorumque producta maximi momenti sunt, quorum vera indoles in luce clarissima ponitur si in factores primos diffinduntur.

Sit p numerus primus realis formæ $m\lambda + 1$, α radix imaginaria æquationis $\alpha^{\lambda} = 1$, g radix primitiva numeri primi p, et

$$(\alpha, x) = x + \alpha x^{g} + \alpha^{2} x^{g^{2}} + \ldots + \alpha^{p-2} x^{g^{p-2}}$$

Totius fere doctrinæ de circuli sectione caput est formæ hujus (α, x) potestas exponentis λ, quæ a radice x non pendet, sed radicis α functio rationalis integra est, ideoque numerus complexus ejus generis quod supra tractavimus. Ipsa hæc formula (α, x), quam Cl. Lagrange primus adhibuit, proprietatibus insignibus gaudet, quarum maximas Cl. Jacobi primus invenit

$$(\alpha, x)(\alpha^{-1}, x) = p,$$

$$\frac{(\alpha^{m}, x)(\alpha^{n}, x)}{(\alpha^{m+n}, x)} = \psi(\alpha) = A + A_{1}\alpha + A_{2}\alpha^{2} + \ldots + A_{\lambda-1}\alpha^{\lambda-1},$$

numerus complexus $\psi(\alpha)$ ita semper comparatus est, ut sit

$$\psi(\alpha)\,\psi(\alpha^{-1}) = p.$$

Inde positis

$$(\alpha,x)\,(\alpha,x) = \psi_1(\alpha)\,(\alpha^2,x),$$
$$(\alpha,x)\,(\alpha^2,x) = \psi_2(\alpha)\,(\alpha^3,x),$$
$$(\alpha,x)\,(\alpha^3,x) = \psi_3(\alpha)\,(\alpha^4,x),$$
$$\cdots\cdots\cdots\cdots\cdots\cdots$$
$$(\alpha,x)\,(\alpha^{\lambda-2},x) = \psi_{\lambda-2}(\alpha)\,(\alpha^{\lambda-1},x),$$

iisque æquationibus inter se multiplicatis, adhibita formula

$$(\alpha,x)\,(\alpha^{-1},x) = p,$$

fit

$$(\alpha,x)^\lambda = p\,.\,\psi_1(\alpha)\,\psi_2(\alpha)\ldots\psi_{\lambda-1}(\alpha).$$

Numeri integri complexi, quibus hoc productum constat, $\psi_1(\alpha)$, $\psi_2(\alpha)$, etc., a se invicem ita pendent, ut quamvis non singuli, tamen productum omnium per unum eorum exprimi possit. Tali autem reductione non indigemus, si pro numeris complexis p, $\psi_1(\alpha)$, $\psi_2(\alpha)$, etc., qui compositi sunt, eorum factores primos adhibemus, quo facto repræsentatio formæ $(\alpha,x)^\lambda$ solos factores primos conjunctos numeri p continebit. Disquisitionem nostram ad tales numeros primos p restringentes, qui in $\lambda - 1$ factores complexos conjunctos diffindi possunt, habemus

$$p = f(\alpha)\,f(\alpha^2)\,f(\alpha^3)\ldots f(\alpha^{\lambda-1}):$$

etiam numeri complexi $\psi_1(\alpha)$, $\psi_2(\alpha)$, etc., alios factores primos habere non possunt nisi eos qui in p reperiuntur; est enim, pro quolibet numero r, $\psi_r(\alpha)\,\psi_r(\alpha^{-1}) = p$, et supra, § VI, demonstravimus numerum p, neglectis unitatibus complexis, quibus factores affecti esse possunt, pluribus modis diversis in $\lambda - 1$ factores primos dissolvi non posse. Quilibet igitur numerorum $\psi_r(\alpha)$ est productum alterius semissis factorum $f(\alpha)$, $f(\alpha^2)$, etc., et unitas complexa, quæ factor accedere potest, si per $\varphi(\alpha)$ designatur, conditioni $\varphi(\alpha)\,\varphi(\alpha^{-1}) = 1$ satisfacere, ideoque secundum ea quæ § IV invenimus, simplex unitas $\pm\,\alpha^x$ esse debet. Inde loco factorum compositorum factoribus simplicibus substitutis, hanc formam habemus:

$$(\alpha,x)^\lambda = \pm\,\alpha^x f^{m_1}(\alpha)\,f^{m_2}(\alpha^2)\,f^{m_3}(\alpha^3)\ldots f^{m_{\lambda-1}}(\alpha^{\lambda-1}),$$

in qua m_1, m_2, m_3, etc., sunt exponentes integri positivi, quos jam determinaturi sumus. Primum adhibita formula simplici

$$(\alpha,x)\,(\alpha^{-1},x) = p = f(\alpha)\,f(\alpha^2)\,f(\alpha^3)\ldots f(\alpha^{\lambda-1}),$$

sponte elucet esse

$$m_1 + m_{\lambda-1} = \lambda,\quad m_2 + m_{\lambda-2} = \lambda,\quad \text{etc.},$$

i. e. summam binorum exponentium, ab initio et a fine æque distantium, constantem esse et numero λ æqualem; unde sequitur ut omnes hi exponentes numero λ minores sint. Deinde per formulam generaliorem

$$\frac{(\alpha, x)(\alpha^r, x)}{(\alpha^{r+1}, x)} = \psi_r(\alpha),$$

fit

$$\frac{f^{m_1}(\alpha)f^{m_2}(\alpha^2)\ldots f^{m_{\lambda-1}}(\alpha^{\lambda-1}).f^{m_1}(\alpha^r)f^{m_2}(\alpha^{2r})\ldots f^{m_{\lambda-1}}(\alpha^{r\lambda-r})}{f^{m_1}(\alpha^{r+1})f^{m_2}(\alpha^{2r+2})\ldots f^{m_{\lambda-1}}(\alpha^{(\lambda-1)(r+1)})} = [\psi_r(\alpha)]^\lambda,$$

et quia quotiens talium numerorum complexorum, quorum norma est numerus primus, non potest integer esse, nisi singuli factores denominatoris cum factoribus numeratoris compensantur, et quia hic quotiens potestati $\lambda^{\text{tæ}}$ numeri complexi æqualis est, singulis tribus potestatibus numeri primi $f(\alpha^k)$ in unam conjunctis, facile colligitur pro quolibet numero k esse debere

$$m_k + m_\mu - m_\nu \equiv 0, \quad \text{si } \mu \equiv \frac{k}{r}, \quad \nu \equiv \frac{k}{r+1} \quad (\text{mod. } \lambda);$$

inde, posito

$$km_k \equiv n_k, \quad \text{sive } m_k \equiv \frac{n_k}{k} \quad (\text{mod. } \lambda),$$

fit

$$m_\mu \equiv \frac{n_\mu}{\mu} \equiv \frac{rn_\mu}{k} \quad \text{et } m_\nu \equiv \frac{n_\nu}{\nu} \equiv \frac{(r+1)n_\nu}{k} \quad (\text{mod. } \lambda.),$$

iisque substitutis, congruentia

$$m_k + m_\mu - m_\nu \equiv 0$$

mutatur in

$$\frac{n_k}{k} + \frac{r.n_k}{k} - \frac{(r+1)n_\nu}{k} \equiv 0 \quad (\text{mod. } \lambda),$$

unde posito $k = 1$, habemus

$$n_1 + rn_\mu - (r+1)n_\nu \equiv 0, \quad \text{si } \mu \equiv \frac{1}{r} \quad \text{et } \nu \equiv \frac{1}{r+1} \quad (\text{mod. } \lambda).$$

Jam si primum facimus $r = 1$, fit n_1 æquale numero n cujus index congruus est $\frac{1}{2}$; deinde, posito $r=2$, idem æqualis invenitur numero n cujus index congruus est $\frac{1}{3}$: tum, posito $r = 3$, etiam n cujus index $\frac{1}{4}$ illis æqualis invenitur, et ita porro; numeri autem fracti $\frac{1}{1}$, $\frac{1}{2}$, $\frac{1}{3}$, etc., omnibus numeris integris 1, 2, 3, etc., etsi alio ordine, congrui sunt, modulo λ; itaque numeri n pro omnibus indicibus diversis iidem sunt et indices omitti possunt, quo facto habemus

$$m_k \equiv \frac{n}{k} \quad (\text{mod. } \lambda),$$

quæ determinatio revera congruentiæ

$$m + m_\mu - m_\nu \equiv 0,$$

pro quolibet valore numerorum k et r satisfacit. Inde habemus theorema insigne : *Si $f(\alpha)$ est factor primus complexus numeri p, cujus norma est ipse numerus p, est*

$$\text{(C)} \qquad (\alpha, x)^\lambda = \pm \alpha^x f^{m_1}(\alpha) f^{m_2}(\alpha^2) f^{m_3}(\alpha^3) \ldots f^{m_{\lambda-1}}(\alpha^{\lambda-1}),$$

et exponentes m_1, m_2, m_3, etc., ita determinantur ut sint numeri minimi positivi, qui per exponentes potestatum radicis α, ad quos pertinent, multiplicati omnes eidem numero congrui fiant pro modulo λ. Numerus primus complexus $f(\alpha)$, alia radice imaginaria æquationis $\alpha^\lambda = 1$ accepta, etiam signo $f(\alpha^r)$ designari potest, in quo r est numerus integer arbitrarius, quem si ita eligimus ut sit $nr \equiv 1$ (mod. λ) exponentes m_1, m_2, m_3, etc., non jam per congruentiam $m_k \equiv \frac{n}{k}$, sed per hanc simpliciorem $m_k \equiv \frac{1}{k}$ (mod. λ) determinatur; præterea, quum omnes debeant esse minores quam λ, nihil amplius indeterminati relictum est nisi radix α, quæ ex omnibus æquationis $\alpha^\lambda = 1$ radicibus imaginariis eligenda sit, et unitas simplex $\pm \alpha^x$ qua hoc productum multiplicatum sit. Inde formæ (α, x) repræsentatio, quæ sectionis circuli caput est, pro omnibus iis numeris p, qui in $\lambda - 1$ factores complexos conjunctos diffindi possunt, ad simplicitatem maximam perducta est. Omnes enim difficultates et calculi longiores eo revocati sunt, ut numeri $p = m\lambda + 1$ factores primi complexi inveniantur, quod methodorum supra traditarum ope facile perficitur. Quum numerus $f(\alpha)$, cujus norma est p, siquidem revera talis numerus datur, non plane determinatus sit, sed semper infinite multis modis diversis exhiberi possit, dubium inde oriri posset, quisnam omnium horum numerorum accipiendus sit, nisi productum illud hac virtute gauderet, ut pro omnibus iis diversis numeris semper idem sit. Facile hoc theorematum paragraphi quartæ auxilio demonstratur; nam si $f(\alpha)$ est aliquis numerorum quorum norma est p, ceteros omnes demonstravimus hac forma contineri $f(\alpha)\,\varphi(\alpha)$, in qua $\varphi(\alpha)$ est unitas complexa : inde, si $f(\alpha)\,\varphi(\alpha)$ loco $f(\alpha)$ in formula (C) substituitur, accedit factor

$$\varphi^{m_1}(\alpha)\,\varphi^{m_2}(\alpha^2)\,\varphi^{m_3}(\alpha^3) \ldots \varphi^{m_{\lambda-1}}(\alpha^{\lambda-1}) = \Phi(\alpha).$$

Hac unitate $\Phi(\alpha)$ cum reciproca $\Phi(\alpha^{-1})$ multiplicata, quia

$$m_1 + m_{\lambda-1} = \lambda, \quad m_2 + m_{\lambda-2} = \lambda, \quad \text{etc.},$$

fit

$$\Phi(\alpha)\,\Phi(\alpha^{-1}) = [\varphi(\alpha)\,\varphi(\alpha^2)\,\varphi(\alpha^3) \ldots \varphi(\alpha^{\lambda-1})]^\lambda = 1;$$

unitas autem complexa, quæ per reciprocam suam multiplicata unitatem realem efficit, simplici unitati $\pm \alpha^s$ æqualis est; itaque quum sit $\Phi(\alpha) = \pm \alpha^s$ videmus solam mutationem levem quam, substitutis aliis numeris complexis $f(\alpha)$ æquationi $\mathrm{N}f(\alpha) = p$ sa-

27..

tisfacientibus, formula (C) pati possit, eam esse, ut loco factoris α^k alia quæcunque radix æquationis $\alpha^\lambda = 1$ substituenda sit.

Ut methodi in hac paragrapho explicatæ indolem veram melius cognoscamus, ad eam respiciamus qua Cl. Gauss usus est in sectione septima disquisitionum arithmeticarum, § CCCLVIII, ubi casum $\lambda = 3$ ingeniosissime pertractavit. Hic totam rem eo reduxit ut numerus $4p$ in formam $t^2 + 27u^2$ redigatur, et quum eo pervenisset, problema ab omni parte absolutum esse censuit. Simili modo secundum methodum nostram numerus p in formam certam gradus $\lambda - 1$, totidemque indeterminatorum redigi debet, quo facto difficultates omnes sublatæ sunt. Quum enim norma sit forma certa gradus $\lambda - 1$ et $\lambda - 1$ indeterminatorum, talem factorem primum complexum numeri p invenire idem est, atque hunc numerum in formam dictam redigere. Si omnes numeri primi p formæ $m\lambda + 1$ in hanc formam redigi possent, sive, quod idem est, si in $\lambda - 1$ factores complexos conjunctos discerpi possent, totius doctrinæ de circuli sectione pars major confecta esset; quum vero non omnes numeri p repræsentationem per formam illam patiantur, restat ut etiam pro iis forma propria expressionis $(\alpha, x)^\lambda$ inveniatur. Hæc autem res maximis difficultatibus obnoxia est, quæ ut superentur magno etiam virorum doctorum labore opus erit.

Über die Divisoren gewisser Formen der Zahlen, welche aus der Theorie der Kreistheilung entstehen

Journal für die reine und angewandte Mathematik 30, 107–116 (1846)

Wenn man die imaginären Wurzeln der Gleichung $x^p = 1$, in welcher p eine Primzahl ist, in Perioden theilt, und zwar in e Perioden von je f Gliedern, wenn $p-1 = e \cdot f$ ist, so sind diese bekanntlich die Wurzeln einer algebraischen Gleichung $e$ten Grades, in welcher der Coëfficient des höchsten Gliedes gleich Eins ist und alle übrigen ganze Zahlen sind. Ist x irgend eine imaginäre Wurzel der Gleichung $x^p = 1$ und g eine primitive Wurzel für die Primzahl p, so haben diese Perioden, welche kurz durch η, η_1, η_2, ... η_{e-1} bezeichnet werden sollen, folgende Werthe:

$$\eta = x + x^{g^e} + x^{g^{2e}} + \ldots + x^{g^{(f-1)e}},$$
$$\eta_1 = x^g + x^{g^{e+1}} + x^{g^{2e+1}} + \ldots + x^{g^{(f-1)e+1}},$$
$$\ldots\ldots\ldots\ldots\ldots\ldots$$
$$\eta_{e-1} = x^{g^{e-1}} + x^{g^{2e-1}} + x^{g^{3e-1}} + \ldots + x^{g^{fe-1}}.$$

Setzt man nun

$$\varphi(y) = (y-\eta)(y-\eta_1)(y-\eta_2)\ldots(y-\eta_{e-1}),$$

so hat man auch

$$\varphi(y) = y^e + a_1 y^{e-1} + a_2 y^{e-2} + \ldots + a_{e-1} y + a_e,$$

wo a_1, a_2, a_3, ... a_e ganze Zahlen sind, welche in jedem speciellen Fall bestimmt werden können. Wir werden nun diese ganze rationale Function $\varphi(y)$ vom $e$ten Grade als Form betrachten, unter welcher gewisse Zahlen dargestellt werden können, und namentlich die Divisoren untersuchen, welche diese Form haben kann. Zu diesem Zwecke werden wir Congruenzen anwenden, in welchen nicht nur reale ganze Zahlen, sondern auch die irrationalen, oft imaginären Perioden η, η_1, η_2, ... η_e vorkommen, und darum müssen wir zunächst den Sinn festhalten, welchen solche Congruenzen haben sollen. Bekanntlich läſst sich jede ganze rationale Function dieser Perioden, welche ganze Zahlen zu Coëfficienten hat, auf die lineäre Form $c\eta + c_1\eta_1 + c_2\eta_2 + \ldots + c_{e-1}\eta_{e-1}$ bringen, und zwar nur auf eine einzige Weise. Wir wollen also

14 *

den Begriff der Congruenz für den gegenwärtigen Zweck dahin ausdehnen, dafs zwei ganze rationale Functionen der Perioden für den Modul q, welcher eine ganze Zahl sein soll, congruent heifsen sollen, wenn in dem Unterschiede derselben, nachdem er auf die Form $c\eta+c_1\eta_1+\ldots+c_{e-1}\eta_{e-1}$ gebracht worden ist, alle Coëfficienten $c, c_1, c_2, \ldots c_{e-1}$ durch q theilbar sind. Nach dieser Erklärung kann man hier, ebenso wie bei den gewöhnlichen Congruenzen, alle Glieder, welche den Modul q als Factor enthalten, weglassen; auch kann man diese Congruenzen ebenso mit einander addiren, subtrahiren, multipliciren und zu Potenzen erheben.

Wir gehen nun von dem bekannten Satze aus, dafs, wenn q eine Primzahl ist, die Coëfficienten $b, b_1, b_2, \ldots b_{q-1}$ des entwickelten Products

$$z(z-1)(z-2)\ldots(z-q+1) = z^q-b_1z^{q-1}+b_2z^{q-2}-\ldots+b_{q-1}z$$

alle durch q theilbar sind, mit Ausnahme des letzten b_{q-1}, welcher durch q dividirt den Rest -1 läfst. Es ist nemlich für jeden Werth des z dieses Product, als Product von q auf einander folgenden ganzen Zahlen, durch q theilbar, also ist auch

$$z^q-b_1z^{q-1}+b_2z^{q-2}-\ldots+b_{q-1}z \equiv 0 \quad \text{Mod. } q,$$

und da nach dem *Fermat*schen Satze $z^q\equiv z$ ist, so ist auch

$$-b_1z^{q-1}+b_2z^{q-2}-\ldots+(b_{q-1}+1)z \equiv 0 \quad \text{Mod. } q.$$

Diese Congruenz vom Grade $q-1$ kann aber nicht q verschiedene Wurzeln haben: also mufs sie identisch erfüllt werden, woraus $b_1\equiv b_2\equiv b_3\equiv\ldots$ $\ldots\equiv b_{q-2}\equiv b_{q-1}+1$ Mod. q folgt.

Nimmt man nun in der obigen Gleichung $z=y-\eta_k$ und läfst die durch q theilbaren Glieder weg, so erhält man die Congruenz

$$(A.)\quad (y-\eta_k)(y-1-\eta_k)(y-2-\eta_k)\ldots(y-q+1-\eta_k)$$
$$\equiv (y-\eta_k)^q-(y-\eta_k) \quad \text{Mod. } q.$$

Bekanntlich sind aber in der binomischen Entwickelung einer $q$ten Potenz, wenn q eine Primzahl ist, alle Glieder, mit Ausnahme des ersten und letzten, durch q theilbar, also ist $(y-\eta_k)^k\equiv y^q-\eta_k^q$. Ferner, ist η_k ein Polynom von f Gliedern, und wenn ein solches zur $q$ten Potenz erhoben wird, so sind die Coëfficienten aller Glieder durch q theilbar, mit Ausnahme der $q$ten Potenzen der einzelnen f Glieder, also ist

$$\eta_k^q \equiv x^{qg^k}+x^{qg^{e+k}}+x^{qg^{2e+k}}+\ldots+x^{qg^{(f-1)e+k}} \quad \text{Mod. } q,$$

und wenn $q\equiv g^r$ Mod. p ist, so hat man hiernach

$$\eta_k^q \equiv \eta_{r+k} \quad \text{Mod. } q,$$

in dem besondern Falle aber, wo $q=p$ ist,

$$\eta_k^p \equiv f \quad \text{Mod. } p.$$

Die Congruenz $(y-\eta_k)^q \equiv y^q - \eta_k^q$ Mod. q geht also allgemein in $(y-\eta_k)^q \equiv y - \eta_{k+r}$ Mod. q über, und für den besondern Fall $q=p$ giebt sie $(y-\eta_k)^p \equiv y - f$ Mod. p.

Giebt man nun hierin dem k nach einander die Werthe $0, 1, 2, \ldots e-1$ und bildet das Product, so erhält man

$$(\varphi(y))^p = (y-f)^e \quad \text{Mod. } p;$$

woraus folgt, dafs $\varphi(y)$ für $y \equiv f$ den Factor p hat, aber für keinen andern Werth des y; oder dafs die Congruenz $\varphi(y) \equiv 0$ Mod. p stets eine reelle Wurzel hat, nemlich $y = f = \frac{p-1}{e}$.

Wir kehren nun zur Untersuchung des allgemeinen Primfactors q zurück, für welchen $q = g^r$ Mod. p ist. Vermöge der Congruenz $(y-\eta_k)^q \equiv y - \eta_{k+r}$ verwandelt sich die Congruenz (*A.*) in folgende:

$$(B.) \quad (y-\eta_k)(y-1-\eta_k)(y-2-\eta_k)\ldots(y-q+1-\eta_k) \equiv \eta_k - \eta_{k+r} \text{ Mod. } q.$$

Es sind die beiden Fälle besonders zu betrachten: erstens wo r durch e theilbar ist, und zweitens, wo dies nicht der Fall ist. Ist r durch e theilbar, so ist $\eta_k = \eta_{k+r}$, also

$$(y-\eta_k)(y-1-\eta_k)(y-2-\eta_k)\ldots(y-q+1-\eta_k) \equiv 0 \quad \text{Mod. } q.$$

Setzt man nun nach einander $k = 0, 1, 2, \ldots e-1$ und bildet das Product aller dieser Congruenzen, so erhält man

$$\varphi(y)\varphi(y-1)\varphi(y-2)\ldots\varphi(y-q+1) \equiv 0 \quad \text{Mod. } q^e.$$

* Es müssen also immer e dieser Factoren durch q theilbar sein, oder auch einige derselben den Factor q mehreremal enthalten, wenn $q \equiv g^r$ Mod. p, und r durch e theilbar ist, dafs heifst, wenn q ein $e$ter Potenzrest der Primzahl q ist. Hieraus erhält man folgenden Lehrsatz: *

„Jede Primzahl, welche ein $e$ter Potenzrest von p ist, ist ein Divisor „der Form $\varphi(y)$; oder auch so: die Congruenz $\varphi(y) \equiv 0$ Mod. q hat, „wenn der Modul q eine Primzahl und zugleich $e$ter Potenzrest von p „ist, immer e reelle Wurzeln, welche in besondern Fällen auch zum „Theil einander gleich werden können."

Der besondere Fall, wo die Perioden eingliedrig, also die imaginären Wurzeln der Gleichung $x^p = 1$ selbst sind, giebt das bekannte Resultat, dafs die Congruenz $x^{p-1} + x^{p-2} + x^{p-3} + \ldots + x + 1 \equiv 0$ Mod. q stets $p-1$ reelle Wurzeln hat, wenn die Primzahl q ein $p-1$ter Potenzrest von p ist, d. h. wenn $q=2mp+1$ ist.

Nachdem gezeigt worden, dafs alle diejenigen Primzahlen, welche $e$te Potenzreste von p sind, Divisoren der Form $\varphi(y)$ sind, ist zweitens zu untersuchen, ob diese Form, aufser diesen genannten und dem Divisor p, noch andere Divisoren haben kann, oder nicht. Es sei also wieder $q\equiv g^r$ Mod. p, aber r nicht durch e theilbar. In diesem Falle giebt die Congruenz (*B.*), wenn nach einander $k=0, 1, 2, \ldots. e-1$ gesetzt und das Product gebildet wird:

$$\varphi(y)\varphi(y-1)\varphi(y-2)\ldots.\varphi(y-q+1)\equiv P \text{ Mod. } q,$$

wo $$P=(\eta-\eta_r)(\eta_1-\eta_{r+1})(\eta_2-\eta_{r+2})\ldots.(\eta_{e-1}-\eta_{r+e-1}) \text{ ist.}$$

P, als symmetrische Function aller Perioden, ist eine ganze Zahl. Für bestimmte p und e kann diese Zahl, auch wenn man alle verschiedenen Werthe des r zuläfst, nur eine endliche, bestimmte und verhältnifsmäfsig sehr geringe Anzahl verschiedener Primfactoren enthalten, und da $\varphi(y)$ keine andern Primfactoren der genannten Art enthalten kann, als diejenigen, welche auch in P vorkommen, so folgt, dafs $\varphi(y)$ nur ausnahmsweise eine stets begrenzte Anzahl solcher Primfactoren enthalten kann, welche nicht $e$te Potenzreste von p sind. Um näher zu untersuchen, in welchen Fällen dergleichen ausnahmsweise Primfactoren des P, und somit auch des $\varphi(y)$, Statt haben können, gebrauchen wir die Congruenz

$$(\eta_k-\eta_{r+k})^q\equiv\eta_{k+r}-\eta_{k+2r} \text{ Mod. } q,$$

deren Richtigkeit nach den oben aufgestellten Principien in die Augen fällt. Werden beide Seiten dieser Congruenz zu wiederholten Malen zur $q$ten Potenz erhoben, so erhält man die allgemeinere Congruenz

$$(C.)\qquad (\eta_k-\eta_{r+k})^{q^h}\equiv\eta_{hr+k}-\eta_{(h+1)r+k} \text{ Mod. } q.$$

Macht man hierin nacheinander $h=0, 1, 2, \ldots. e-1$ und bildet das Product, so erhält man

$$(\eta_k-\eta_{r+k})^{1+q+q^2+\ldots.+q^{e-1}}\equiv(\eta_k-\eta_{r+k})(\eta_{r+k}-\eta_{2r+k})\ldots.(\eta_{(e-1)r+k}-\eta_{er+k}) \text{ Mod. } q.$$

Wenn nun r keinen gemeinschaftlichen Factor mit e hat, so sind die Indices der Perioden $k, r+k, 2r+k, \ldots. (e-1)r+k$, in anderer Ordnung genommen, den Indices $0, 1, 2, \ldots. e-1$ congruent, für den Modul e; das Product rechterhand ist also kein anderes als das Product P, und da P nach der Voraussetzung durch q theilbar sein soll, so hat man

$$(\eta_k-\eta_{r+k})^{1+q+q^2+\ldots.+q^{e-1}}\equiv 0 \text{ Mod. } q.$$

Wenn nun zur Potenz $q-1$ erhoben wird, so ist

$$(\eta_k-\eta_{r+k})^{q^e-1}\equiv 0 \text{ Mod. } q,$$

und, wenn mit $\eta_k-\eta_{r+k}$ multiplicirt wird,

$$(\eta_k-\eta_{k+r})^{q^e}\equiv 0 \text{ Mod. } q,$$

woraus nach der Congruenz (*C.*) folgt:

$$\eta_k-\eta_{r+k}\equiv 0 \text{ Mod. } q;$$

welches unmöglich ist. Das Product P hat also keinen Primfactor q von der Art, dafs $q\equiv g^r$ Mod. p, wo r keinen gemeinschaftlichen Factor mit e hat. Für den Fall also, wo e Primzahl ist, hat man folgenden Lehrsatz:

„Die Form $\varphi(y)$ hat, wenn der Grad derselben e eine Primzahl ist, aufser „dem Divisor p nur solche Divisoren, welche $e$te Potenzreste von p sind."

Für den Fall aber, wo der Grad der Form $\varphi(y)$ keine Primzahl ist, kann man das gefundene Resultat folgendermaafsen aussprechen.

„Die Form $\varphi(y)$ hat aufser dem Divisor p im allgemeinen nur solche „Primzahlen zu Divisoren, welche $e$te Potenzreste von p sind; aufserdem „aber kann sie auch eine endliche bestimmte Anzahl anderer Divisoren „haben, welche, wenn e die von Eins verschiedenen Divisoren $\alpha, \beta, \gamma, \ldots.$ „enthält, αte oder βte oder γte Potenzreste von p sein müssen."

Man kann die Bedingungen, unter welchen $\varphi(y)$ solche besondere Divisoren enthält, die nicht $e$te Potenzreste sind, noch etwas genauer angeben. Wenn nemlich r und e den gröfsten gemeinschaftlichen Factor α enthalten, und $r=r'\alpha$, $e=e'\alpha$ ist, so setze man in der Congruenz (*C.*) nach einander $h=0, 1, 2, \ldots. e'-1$ und bilde das Product der so erhaltenen Congruenzen:

$$\eta_k-\eta_{r+k})^{1+q+q^2+\ldots.+q^{e'-1}}\equiv(\eta_k-\eta_{r+k})(\eta_{r+k}-{}_{2r+k})\ldots.(\eta_{(e'-1)r+k}-\eta_{e'r+k}).$$

Giebt man hierin wieder dem k die Werthe 0, 1, 2, ... $\alpha-1$ und bildet das Product, so wird dieses Product auf der rechten Seite gleich P; wovon man sich sogleich überzeugt, wenn man bemerkt, dafs die Zahlen von der Form $hr+k$ für $h=0, 1, 2, \ldots. e'-1$ und $k=0, 1, 2, \ldots. \alpha-1$, wenn $r=r'\alpha$ und $e=e\alpha$, für den Modul e alle Reste 0, 1, 2, 3, ... $e-1$ geben. Man hat also, da P durch q theilbar sein soll,

$$\{(\eta-\eta_r)(\eta_1-\eta_{r+1})\ldots.(\eta_{\alpha-1}-\eta_{\alpha+r-1})\}^{1+q+q^2+\ldots.+q^{e'-1}}\equiv 0 \text{ Mod. } q.$$

Erhebt man wieder zur Potenz $q-1$ und multiplicirt mit $(\eta-\eta_r)(\eta_1-\eta_{r+1})\ldots$ $\ldots(\eta_{\alpha-1}-\eta_{\alpha+r-1})$, so folgt hieraus

$$\{(\eta-\eta_r)(\eta_1-\eta_{r+1})\ldots.(\eta_{\alpha+1}-\eta_{\alpha+r-1})\}^{q^{e'}}\equiv 0 \text{ Mod. } q,$$

welches vermöge der Congruenz $(C.)$ folgende einfachere Form annimmt:

$$(\eta-\eta_r)(\eta_1-\eta_{r+1})\ldots.(\eta_{\alpha-1}-\eta_{\alpha+r-1})\equiv 0 \quad \text{Mod. } q.$$

Es müssen also schon die ersten α Factoren des Products P den Factor q enthalten, damit P oder $\varphi(y)$ denselben enthalten könne. Man könnte durch diese einschränkende Bedingung auf die Vermuthung kommen, dafs diese ausnahmsweisen Factoren, welche nicht $e$te Potenzreste sind, auch wenn e eine zusammengesetzte Zahl ist, überhaupt gar nicht vorkommen dürften: dafs dieses indefs nicht der Fall ist, vielmehr wirklich dergleichen Factoren Statt haben, kann man an folgendem einfachen Beispiele sehen. Nimmt man $p=109$, $e=6$, so erhält man nach bekannten Methoden:

$$\varphi(y)=y^6+y^5-45y^4-10y^3+135y^2+9y-27,$$

woraus sich leicht für $y=0, 1, 2, 3, 4, 5$ folgende Werthe des $\varphi(y)$ berechnen lassen: $\varphi(0)=-3^3$; $\varphi(1)=2^6$; $\varphi(2)=-173$; $\varphi(3)=-2^6.3^3$; $\varphi(4)=-4871$; $\varphi(5)=-2^6.113$. Die hierin vorkommenden Divisoren 2 und 3 sind keine 6ten Potenzreste von 109; dieselben sind also solche ausnahmsweise Divisoren; und es ist 2 ein cubischer Rest und 3 quadratischer Rest von 109; welches sehr wohl mit dem oben gefundenen Lehrsatze stimmt. Übrigens sind 2 und 3 im gegenwärtigen Falle die einzigen Primfactoren von $\varphi(y)$, welche nicht sechste Potenzreste sind.

Wir zeigen nun noch von zwei wichtigen speciellen Fällen, dafs für sie dergleichen Factoren, welche nicht $e$te Potenzreste von p sind, niemals Statt haben können: nemlich für den Fall, wo die Perioden eingliedrig, und wo sie zweigliedrig sind. Ist $e=p-1$ und $f=1$, so ist $\eta=x$, $\eta_1=x^g$, $\eta_2=x^{g^2}$ etc.; es ist also

$$P=(x-x^g)(x^g-x^{g^2})\ldots.(x^{g^{p-2}}-x^{g^{p-1}}),$$

und wenn man von dem ersten Factor x, vom zweiten x^g, vom dritten x^{g^2}etc., heraushebt, so wird

$$P=(1-x^{g-1})(1-x^{g(g-1)})\ldots.(1-x^{g^{p-2}(g-1)}),$$

und da

$$(z-x^m)(z-x^{mg})\ldots.(z-x^{m.g^{p-2}})=z^{p-1}+z^{p-2}+\ldots.z+1$$

ist, so hat man, wenn $z=1$ und $m=g-1$ genommen wird,

$$P=p.$$

Da nun P keinen andern Factor enthält als p, so folgt hieraus der bekannte Lehrsatz: dafs die Form $y^{p-1}+y^{p-2}+\ldots.+y+1$ aufser dem Divisor p nur solche Divisoren enthalten kann, welche $p-1$te Potenzreste von p sind, deren lineäre Form also $q=2mp+1$ ist.

Wenn ferner die Perioden zweigliedrig sind, also $e=\frac{1}{2}(p-1)$, $f=2$ ist, so ist $\eta=x+x^{-1}$, $\eta_1=x^g+x^{-g}$, $\eta_2=x^{g^2}+x^{-g^2}$ etc., und es wird für diesen Fall bekanntlich

$$\varphi(y)=y^e+y^{e-1}-\frac{e-1}{1}y^{e-2}-\frac{e-2}{1}y^{e-3}+\frac{(e-2)(e-3)}{1\cdot 2}y^{e-4}+\frac{(e-3)(e-4)}{1\cdot 2}y^{e-5}-\dots$$

Ferner ist

$$\eta_k-\eta_{r+k}=x^{g^k}+x^{-g^k}-x^{g^{r+k}}-x^{-g^{r+k}},$$

welcher Ausdruck folgendermaafsen in Factoren zerfällt:

$$\eta_k-\eta_{r+k}=x^{g^k}(1-x^{(g^r-1)g^k})(1-x^{-(g^r+1)g^k}).$$

Giebt man nun dem k die Werthe $0, 1, 2, \dots p-2$ und bildet das Product, so erhält man leicht

$$P^2=p^2, \quad \text{also} \quad P=\pm p.$$

In diesem Falle hat also P ebenfalls keinen andern Divisor als p, woraus folgt, dafs die Form

$$\varphi(y)=y^e+y^{e-1}-\frac{e-1}{1}y^{e-2}-\frac{e-2}{1}y^{e-3}+\dots$$

aufser dem Divisor p nur solche Divisoren hat, welche $\frac{1}{2}(p-1)$te Potenzreste von p sind, also von der Form $2mp\pm 1$.

Dieselbe Methode, welche in dem Vorhergehenden für die Form $\varphi(y)$ angewendet wurde, giebt fast ganz in derselben Weise die Primfactoren einer weit allgemeineren, sehr merkwürdigen Form vom $e$ten Grade mit e unbestimmten Zahlen. Wenn man nemlich folgende zusammengehörige lineäre Functionen der Perioden:

$$\begin{aligned}
F(\eta) &= z\eta+z_1\eta_1+z_2\eta_2+\dots+z_{e-1}\eta_{e-1},\\
F(\eta_1) &= z\eta_1+z_1\eta_2+z_2\eta_3+\dots+z_{e-1}\eta,\\
&\dots\dots\dots\dots\dots\dots\dots\dots\\
F(\eta_{e-1}) &= z\eta_{e-1}+z_1\eta+z_2\eta_1+\dots+z_{e-1}\eta_{e-2}
\end{aligned}$$

mit einander multiplicirt, so wird das Product derselben

$$\Psi=F(\eta)F(\eta_1)F(\eta_2)\dots F(\eta_{e-1})$$

eine homogene Form vom $e$ten Grade der e unbestimmten Zahlen $z, z_1, z_2, \dots z_{e-1}$, mit ganzzahligen Coëfficienten; welche Form, wie wir sogleich zeigen werden, in Beziehung auf ihre Divisoren ganz mit der oben untersuchten specielleren Form $\varphi(y)$ übereinstimmt. Wir beweisen zunächst folgenden Lehrsatz:

„Jede Primzahl q, welche ein $e$ter Potenzrest von p ist, so wie auch p „selbst, ist ein Divisor der Form Ψ."

Untersuchen wir zuerst den Divisor p, so haben wir, nach denselben Principien, welche oben angewendet wurden,

$$F(\eta_k)^p \equiv (z+z_1+z_2+\ldots\ldots+z_{e-1})f. \quad \text{Mod } p,$$

also, wenn die unbestimmten Zahlen z, z_1, z_2, ... z_{e-1} so bestimmt werden, dafs ihre Summe durch p theilbar ist, so wird $F(\eta_k)^p \equiv 0$ Mod. p, und wenn man $k=0, 1, 2, \ldots\ldots e-1$ setzt und diese Congruenzen mit einander multiplicirt, so erhält man $\Psi^p \equiv 0$ Mod. p, also auch $\Psi \equiv 0$ Mod. p; p ist also ein Divisor der Form Ψ.

Um weiter zu beweisen, dafs auch jede Primzahl q, welche ein $e$ter Potenzrest von p ist, immer Divisor von Ψ sei, setze ich die zweite Periode η als ganze rationale Function der ersten Periode η dargestellt; welches bekanntlich immer möglich ist. Es sei also $\eta_1 = \Theta(\eta)$, so ist $\eta_2 = \Theta\Theta(\eta)$, $\eta_3 = \Theta\Theta\Theta(\eta)$ etc. Statt η aber werde eine unbestimmte Gröfse y genommen und folgender Ausdruck gebildet:

$$F(y) \cdot F(\Theta y) \cdot F(\Theta\Theta y) \ldots\ldots F(\Theta^{(e-1)} y) - \Psi,$$

welcher eine ganze rationale Function von y sein wird. Derselbe verschwindet offenbar, wenn $y=\eta$ oder $y=\eta_1$ oder $y=\eta_2$ etc. genommen wird, und mufs deshalb den Factor $(y-\eta)(y-\eta_1)(y-\eta_2) \ldots\ldots (y-\eta_{e-1})$ enthalten, welchen wir oben durch $\varphi(y)$ bezeichnet haben. Es ist demnach

$$F(y) F(\Theta y)(F\Theta\Theta y) \ldots\ldots F(\Theta^{(e-1)} y) - \Psi = \varphi(y) \cdot \Phi,$$

wo auch Φ eine ganze rationale Function von y bedeutet. Wird nun für y irgend eine Wurzel der Congruenz $\varphi(y) = 0$ Mod. q genommen (welche, wenn q ein $e$ter Potenzrest von p ist, immer e reale Wurzeln hat, wie wir oben gezeigt haben), so hat man:

$$F(y) F(\Theta y) F(\Theta\Theta y) \ldots\ldots F(\Theta^{(e-1)} y) \equiv \Psi \quad \text{Mod. } q.$$

Wenn also irgend ein Factor dieses Productes durch q theilbar wird, so wird allemal auch Ψ durch q theilbar. Es lassen sich daher die unbestimmten Zahlen z, z_1, z_2, ... z_{e-1} immer so bestimmen, dafs die Form Ψ den Divisor q erhält, und die Bestimmung derselben ist einfach die, dafs, wenn man in

$$F(\eta) = z\eta + z_1\eta_1 + z_2\eta_2 + \ldots\ldots + z_{e-1}\eta_{e-1}$$

für η, η_1, η_2, ... η_{e-1} nicht mehr die Wurzeln der Gleichung $\varphi(y) = 0$, sondern die Wurzeln der Congruenz $\varphi(y) \equiv 0$ Mod. q, aber in der gehörigen Ordnung genommen, substituirt hat, $F(\eta) \equiv 0$ Mod. q werden mufs.

Die Primfactoren, welche $e$te Potenzreste von p sind, machen auch hier die hauptsächlichsten Divisoren der Form Ψ aus, und aufser diesen haben gewisse andere nur ausnahmsweise Statt, deren Bedingungen wir jetzt näher untersuchen wollen. Es sei q eine Primzahl, von der Art, dafs $q \equiv g^r$ Mod. p und r nicht durch e theilbar ist, so ist

$$F(\eta)^q \equiv F(\eta_r) \text{ Mod. } q.$$

Erhebt man dies nun zu wiederholten Malen zur Potenz q, so erhält man

$$F(\eta)^{q^h} \equiv F(\eta_{hr}) \text{ Mod. } q.$$

Setzt man nach einander $h = 0, 1, 2, 3, \ldots. e-1$ und multiplicirt die so erhaltenen Congruenzen in einander, so wird

$$F(\eta)^{1+q+q^2+\ldots.+q^{e-1}} \equiv F(\eta)F(\eta_r)F(\eta_{2r}) \ldots. F(\eta_{(e-1)r}) \text{ Mod. } q.$$

Wenn nun r und e keinen gemeinschaftlichen Factor haben, so sind die Indices $0, r, 2r, 3r, \ldots. (e-1)r$ den Indices $0, 1, 2, \ldots. e-1$, in anderer Ordnung genommen, congruent für den Modul e, also ist das Product rechterhand gleich Ψ. Soll nun Ψ den Divisor q enthalten, so folgt, dafs

$$F(\eta)^{1+q+q^2+\ldots.+q^{e-1}} \equiv 0 \text{ Mod. } q$$

sein mufs, woraus, ebenso wie oben, $F(\eta) \equiv 0$ Mod. q folgt. Damit aber $F(\eta)$ durch q theilbar sei, müssen alle einzelnen Coëfficienten der Perioden, also die Unbestimmten $z, z_1, z_2, \ldots. z_{e-1}$, durch q theilbar sein. Hieraus erhalten wir folgenden Lehrsatz:

„Wenn die unbestimmten Zahlen der Form Ψ nicht alle einen gemein„schaftlichen Factor haben, so enthält diese Form aufser dem Divisor p „nur solche Primfactoren, welche $e$te Potenzreste von p sind; oder: wenn „e die von Eins verschiedenen Divisoren $\alpha, \beta, \gamma, \ldots.$ enthält, so kann die „Form auch solche Primfactoren enthalten, welche αte oder βte oder γte …. „Potenzreste von p sind."

Daraus geht auch folgender Zusatz hervor:

„Wenn die unbestimmten Zahlen der Form Ψ nicht alle einen gemeinschaft„lichen Factor enthalten, und der Grad e dieser Form ist eine Primzahl, so „enthält dieselbe aufser dem Divisor p nur solche Divisoren, welche $e$te „Potenzreste von p sind."

Ist e nicht eine Primzahl, sondern α ein Divisor von e, so sei $q \equiv g^r$ Mod. p und α der gröfste gemeinschaftliche Factor von r und e, so dafs $r \equiv r'\alpha$ und

15*

$e = e'\alpha$. In diesem Falle hat man, vermöge der Congruenz

$$F(\eta_k^{q^h}) \equiv F(\eta_{rh+k}) \quad \text{Mod. } q,$$

wenn nach einander $h = 0, 1, 2, \ldots. e'-1$ gesetzt wird und sodann für jeden einzelnen dieser Werthe des h die Zahl k die Werthe $0, 1, 2, \ldots. \alpha-1$ bekommt,

$$\{F(\eta)F(\eta_1)F(\eta_2)\ldots.F(\eta_{\alpha-1})\}^{1+q+q^2+\ldots.+q^{e'-1}} \equiv \Psi \quad \text{Mod. } q;$$

denn das Product rechter Hand wird offenbar gleich Ψ, weil $hr+k$ für die angegebenen Werthe des h und k allen den Zahlen $0, 1, 2, \ldots. e-1$ congruent wird, für den Modul e. Soll nun Ψ den Factor q haben, so folgt, dafs

$$\{F(\eta)F(\eta_1)\ldots.F(\eta_{\alpha-1})\}^{1+q+q^2+\ldots.+q^{e'-1}} \equiv 0 \quad \text{Mod. } q$$

sein mufs, woraus man leicht folgert, dafs auch

$$F(\eta)F(\eta_1)F(\eta_2)\ldots.F(\eta_{\alpha-1}) \equiv 0 \quad \text{Mod. } q$$

sein mufs. Damit also Ψ einen Primfactor q habe, welcher nicht $e$ter Potenzrest, sondern nur αter Potenzrest von p ist, wo α ein Divisor von e, mufs allemal schon das Product der ersten α Factoren des Ψ diesen Factor q enthalten.

Zur Theorie der complexen Zahlen

Journal für die reine und angewandte Mathematik 35, 319–326 (1847)

Es ist mir gelungen, die Theorie derjenigen complexen Zahlen, welche aus höheren Wurzeln der Einheit gebildet sind und welche bekanntlich in der Kreistheilung, in der Lehre von den Potenzresten und den Formen höherer Grade eine wichtige Rolle spielen, zu vervollständigen und zu vereinfachen; und zwar durch Einführung einer eigenthümlichen Art imaginärer Divisoren, welche ich *ideale complexe Zahlen* nenne; worüber eine kurze Mittheilung zu machen ich mir erlaube.

Wenn α eine imaginäre Wurzel der Gleichung $\alpha^\lambda = 1$, λ eine Primzahl ist und a, a_1, a_2, etc. ganze Zahlen sind, so ist $f(\alpha) = a + a_1\alpha + a_2\alpha^2 + \ldots + a_{\lambda-1}\alpha^{\lambda-1}$ eine complexe ganze Zahl. Eine solche complexe Zahl kann entweder in Factoren derselben Art zerlegt werden; oder auch nicht. Im ersten Fall ist sie eine zusammengesetzte Zahl: im andern Fall ist sie bisher eine complexe Primzahl genannt worden. Ich habe nun aber bemerkt, dafs, wenn auch $f(\alpha)$ auf keine Weise in complexe Factoren zerlegt werden kann, sie deshalb noch nicht die wahre Natur einer complexen Primzahl hat, weil sie schon gewöhnlich der ersten und wichtigsten Eigenschaft der Primzahlen ermangelt: nämlich, dafs das Product zweier Primzahlen durch keine von ihnen verschiedene Primzahl theilbar ist. Es haben vielmehr solche Zahlen $f(\alpha)$, wenn gleich sie nicht in complexe Factoren zerlegbar sind, dennoch die Natur der zusammengesetzten Zahlen; die Factoren aber sind alsdann nicht wirkliche, sondern *ideale complexe Zahlen.* Der Einführung solcher idealen complexen Zahlen liegt derselbe einfache Gedanke zu Grunde, wie der Einführung der imaginären Formeln in die Algebra und Analysis; namentlich bei der Zerfällung der ganzen rationalen Functionen in ihre einfachsten Factoren, die linearen. Ferner ist es auch dasselbe Bedürfnifs, durch welches genöthigt, *Gaufs* bei den Untersuchungen über die biquadratischen Reste (weil hier alle Primfactoren von der Form $4m+1$ die Natur zusammengesetzter Zahlen zeigen) die complexen Zahlen von der Form $a + b\sqrt{-1}$ zuerst einführte.

42 *

Um nun zu einer festen Definition der wahren (gewöhnlich idealen) Primfactoren der complexen Zahlen zu gelangen, war es nöthig, die unter allen Umständen bleibenden Eigenschaften der Primfactoren complexer Zahlen zu ermitteln, welche von der Zufälligkeit, ob die wirkliche Zerlegung Statt habe, oder nicht, ganz unabhängig wären: ungefähr eben so, wie man, wenn in der Geometrie von der gemeinschaftlichen Sehne zweier Kreise gesprochen wird, auch dann, wenn die Kreise sich nicht schneiden, eine wirkliche Definition dieser idealen gemeinschaftlichen Sehne sucht, welche für alle Lagen der Kreise pafst. Dergleichen bleibende Eigenschaften der complexen Zahlen, welche geschickt sind, als Definitionen der idealen Primfactoren benutzt zu werden, giebt es mehrere, welche im Grunde immer auf dasselbe Resultat führen und von denen ich eine als die einfachste und allgemeinste gewählt habe.

Ist p eine Primzahl von der Form $m\lambda+1$, so läfst sie sich in vielen Fällen als Product von folgenden $\lambda-1$ complexen Factoren darstellen: $p = f(\alpha).f(\alpha^2).f(\alpha^3)\ldots.f(\alpha^{\lambda-1})$; wo aber eine Zerlegung in wirkliche complexe Primfactoren nicht möglich ist: dann sollen die idealen Primfactoren eintreten, um dieselbe zu leisten. Ist $f(\alpha)$ eine wirkliche complexe Zahl und ein Primfactor von p, so hat sie die Eigenschaft, dafs, wenn statt der Wurzel der Gleichung $\alpha^\lambda = 1$ eine bestimmte Congruenzwurzel von $\xi^\lambda \equiv 1$, mod. p, substituirt wird, $f(\xi) \equiv 0$, mod. p, ist. Also auch, wenn in einer complexen Zahl $\Phi(\alpha)$ der Primfactor $f(\alpha)$ enthalten ist, wird $\Phi(\xi) \equiv 0$, mod. p; und umgekehrt: wenn $\Phi(\xi) \equiv 0$, mod. p, und p in $\lambda-1$ complexe Primfactoren zerlegbar ist, enthält $\Phi(\alpha)$ den Primfactor $f(\alpha)$. Die Eigenschaft $\Phi(\xi) \equiv 0$, mod. p ist nun eine solche, welche für sich selbst von der Zerlegbarkeit der Zahl p in $\lambda-1$ Primfactoren gar nicht abhangt; sie kann demnach als Definition benutzt werden, indem bestimmt wird, dafs die complexe Zahl $\Phi(\alpha)$ den idealen Primfactor von p enthält, welcher zu $\alpha = \xi$ gehört, wenn $\Phi(\xi) \equiv 0$, mod. p, ist. Jeder der $\lambda-1$ complexen Primfactoren von p wird so durch eine Congruenzbedingung ersetzt. Dies reicht hin, um zu zeigen, dafs die complexen Primfactoren, sie seien wirklich, oder nur ideal vorhanden, den complexen Zahlen denselben bestimmten Character ertheilen. In der hier gegebenen Weise aber gebrauchen wir die Congruenzbedingungen nicht als Definitionen der idealen Primfactoren, weil diese nicht hinreichend sein würden, mehrere gleiche, in einer complexen Zahl vorkommende ideale Primfactoren vorzustellen, und weil sie, zu beschränkt, nur ideale Primfactoren der realen Primzahlen von der Form $m\lambda-1$ geben würden.

Jeder Primfactor einer complexen Zahl ist immer zugleich auch Primfactor irgend einer realen Primzahl q, und die Beschaffenheit der idealen Primfactoren ist besonders von dem Exponenten abhängig, zu welchem q gehört, für den Modul λ. Derselbe sei f, so dafs $q^f \equiv 1$, mod. λ, und $\lambda - 1 = e \cdot f$. Eine solche Primzahl q läfst sich niemals in mehr als e complexe Primfactoren zerlegen, welche, wenn diese Zerlegung wirklich ausführbar ist, sich als lineare Functionen der e Perioden von je f Gliedern darstellen. Diese Perioden der Wurzeln der Gleichung $\alpha^\lambda = 1$ bezeichne ich durch η, η_1, η_2, ... η_{e-1}; und zwar in der Ordnung, dafs jede in die folgende übergeht, wenn α in α^γ verwandelt wird, wo γ eine primitive Wurzel von λ ist. Bekanntlich sind die Perioden die e Wurzeln einer Gleichung vom $e$ten Grade; und diese, als Con-
* gruenz betrachtet, für den Modul q, hat immer e reale Congruenzwurzeln, welche ich durch u, u_1, u_2, ... u_{e-1} bezeichne und in einer entsprechenden Reihenfolge nehme, wie die Perioden, für welche, aufser der Congruenz vom $e$ten Grade, noch andere leicht zu findende Congruenzen gebraucht werden. Wird nun die aus Perioden gebildete complexe Zahl $c'\eta + c'_1\eta_1 + c'_2\eta_2 + \ldots + c'_{e-1}\eta_{e-1}$ kurz durch $\Phi(\eta)$ bezeichnet, so giebt es unter den Primzahlen q, welche zum Exponenten f gehören, immer solche, die sich auf die Form

$$q = \Phi(\eta)\Phi(\eta_1)\Phi(\eta_2)\ldots\Phi(\eta_{e-1})$$

bringen lassen, in welcher auch die e Factoren niemals eine weitere Zerlegung gestatten. Setzt man statt der Perioden ihre entsprechenden Congruenzwurzeln, wobei sich eine Periode beliebig festsetzen läfst, welche einer bestimmten Congruenzwurzel entsprechen soll, so wird immer einer der e Primfactoren congruent Null, für den Modul q. Enthält nun irgend eine complexe Zahl $f(\alpha)$ den Primfactor $\Phi(\eta)$, so wird sie die Eigenschaft haben, für $\eta = u_k$, $\eta_1 = u_{k+1}$, $\eta_2 = u_{k+2}$, etc. congruent Null zu werden, für den Modul q. Diese Eigenschaft nun (welche eigentlich f besondere Congruenzbedingungen in sich schliefst, deren Entwicklung zu weit führen würde) ist eine bleibende; auch für diejenigen Primzahlen q, welche eine Zerlegung in die e wirklichen complexen Primfactoren nicht gestatten. Sie könnte daher als Definition der complexen Primfactoren benutzt werden, würde aber den Mangel haben, dafs sie die in einer complexen Zahl vorhandenen gleichen idealen Primfactoren nicht ausdrückte.

Die von mir gewählte Definition der idealen complexen Primfactoren, welche im Wesentlichen zwar mit der hier angedeuteten übereinstimmt, aber einfacher und allgemeiner ist, beruht darauf, dafs sich, wie ich besonders beweise, immer eine aus Perioden gebildete complexe Zahl $\psi(\eta)$ finden läfst, von der Art, dafs $\psi(\eta)\psi(\eta_1)\psi(\eta_2)\ldots\psi(\eta_{e-1})$ (welches eine ganze Zahl ist) durch q theilbar sei, aber nicht durch q^2. Diese complexe Zahl $\psi(\eta)$ hat alsdann immer die obige Eigenschaft, dafs sie congruent Null wird, modulo q, wenn statt der Perioden die entsprechenden Congruenzwurzeln gesetzt werden, also $\psi(\eta) \equiv 0$, mod. q, für $\eta = u$, $\eta_1 = u_1$, $\eta_2 = u_2$, etc. Ich setze nun $\psi(\eta_1)\psi(\eta_2)\ldots\psi(\eta_{e-1}) = \Psi(\eta)$ und definire die idealen Primzahlen folgendermaafsen:

Wenn $f(\alpha)$ die Eigenschaft hat, dafs das Product $f(\alpha).\Psi(\eta_r)$ durch q theilbar ist, so soll dies so ausgedrückt werden: Es enthält $f(\alpha)$ den idealen Primfactor von q, welcher zu $u = \eta_r$ gehört. Ferner, wenn $f(\alpha)$ die Eigenschaft hat, dafs $f(\alpha)(\Psi(\eta_r))^\mu$ durch q^μ theilbar ist, aber $f(\alpha)(\Psi(\eta_r))^{\mu+1}$ nicht theilbar durch $q^{\mu+1}$, so soll dies heifsen: Es enthält $f(\alpha)$ den zu $u = \eta_r$ gehörigen idealen Primfactor von q genau μ mal.

Es würde hier zu weit führen, wenn ich den Zusammenhang und die Übereinstimmung dieser Definition mit den oben angedeuteten, welche durch Congruenzbedingungen gegeben werden, entwickeln wollte; ich bemerke nur, dafs die Bedingung: $f(\alpha)\Psi(\eta_r)$ sei durch q theilbar, f verschiedenen Congruenzbedingungen vollkommen gleichbedeutend ist, und dafs die Bedingung: $f(\alpha)(\Psi(\eta_r))^\mu$ sei durch q^μ theilbar, sich allemal durch $\mu.f$ Congruenzbedingungen vollständig ersetzen läfst. Die ganze von mir bereits fertig ausgearbeitete Theorie der idealen complexen Zahlen, deren Hauptsätze ich hier mittheilen will, ist eine Rechtfertigung sowohl der gegebenen Definition, als auch der gewählten Benennung. Diese Hauptsätze sind folgende:

Das Product zweier oder mehrerer complexen Zahlen hat genau dieselben idealen Primfactoren, wie die Factoren zusammengenommen.

Wenn eine complexe Zahl (welche als Product von Factoren auftritt) alle e Primfactoren von q enthält, so ist sie auch durch q selbst theilbar; enthält sie aber irgend einen dieser idealen Primfactoren nicht, so ist sie nicht durch q theilbar.

Wenn eine complexe Zahl (in Form eines Products) alle e idealen Primfactoren von q, enthält, und zwar jeden wenigstens μ mal, so ist sie durch q^μ theilbar.

Wenn $f(\alpha)$ genau m ideale Primfactoren von q enthält, sie mögen verschieden, oder zum Theil, oder sämmtlich gleich sein, so enthält die Norm $Nf(\alpha) = f(\alpha)f(\alpha^2)\ldots f(\alpha^{\lambda-1})$ genau den Factor q^{mf}.

Jede complexe Zahl enthält nur eine endliche, bestimmte Anzahl idealer Primfatoren.

Zwei complexe Zahlen, welche genau dieselben idealen Primfactoren enthalten, unterscheiden sich nur durch eine complexe Einheit, welche als Factor hinzutreten kann.

Eine complexe Zahl ist durch eine andere theilbar, wenn alle idealen Primfactoren des Divisors auch in dem Dividendus enthalten sind; und der Quotient enthält genau den Überschufs der idealen Primfactoren des Dividendus über die des Divisors.

Aus diesen Sätzen geht hervor, dafs die Rechnung mit complexen Zahlen durch Einführung der idealen Primfactoren ganz dieselbe geworden ist, wie die Rechnung mit den ganzen Zahlen und den ganzzahligen realen Primfactoren derselben. Es erledigt sich somit die Klage, welche ich in dem Breslauer Programm zur Jubelfeier der Universität Königsberg S. 18 aussprach: *Maxime dolendum videtur, quod haec numerorum realium virtus, ut in factores primos dissolvi possint, qui pro eodem numero semper iidem sint, non eadem est numerorum complexorum, quae si esset tota haec doctrina, quae magnis adhuc difficultatibus laborat, facile absolvi et ad finem perduci posset. etc.* Auch sieht man, dafs die idealen Primfactoren die innere Natur der complexen Zahlen aufschliefsen, sie gleichsam durchsichtig machen und das innere crystallinische Gefüge derselben zeigen. Ist nämlich eine complexe Zahl nur unter der Form $a + a_1\alpha + a_2\alpha^2 + \ldots + a_{\lambda-1}\alpha^{\lambda-1}$ gegeben, so läfst sich vorläufig wenig über dieselbe aussagen, bis man durch die idealen Primfactoren derselben (welche hier immer durch directe Methoden vollständig gefunden werden können) ihre einfachsten qualitativen Bestimmungen gefunden hat, welche als Grundlagen aller ferneren zahlentheoretischen Untersuchungen dienen.

Die idealen Factoren der complexen Zahlen treten, wie gezeigt, als Factoren von wirklichen complexen Zahlen auf: es müssen deshalb ideale Factoren, mit andern passenden multiplicirt, immer wirkliche complexe Zahlen zu Producten geben. Diese Frage nun, über die Zusammensetzung der idealen Factoren zu wirklichen complexen Zahlen, ist, wie ich an den bereits von mir gefundenen Resultaten zeigen werde, von grofsem Interesse, weil sie mit den wichtigsten Abschnitten der Zahlentheorie in innigem Zusammenhange steht. Die beiden wichtigsten Resultate über diese Frage sind folgende:

Es giebt immer eine endliche, bestimmte Anzahl idealer complexer Multiplicatoren, welche nöthig und hinreichend sind, um alle möglichen idealen complexen Zahlen zu wirklichen zu machen *).

Jede *ideale* complexe Zahl hat die Eigenschaft, dafs eine bestimmte ganze Potenz derselben zu einer *wirklichen* complexen Zahl wird.

Ich gehe in einige nähere Entwickelungen dieser beiden Sätze ein. Zwei ideale complexe Zahlen, welche, mit einer und derselben idealen Zahl multiplicirt, beide zu wirklichen complexen Zahlen machen, nenne ich *äquivalent* oder derselben Classe angehörig, weil diese Untersuchung über die wirklichen und die idealen complexen Zahlen vollständig identisch ist mit der Classification gewisser zusammengehöriger Formen vom $\lambda-1$ten Grade mit $\lambda-1$ Variabeln, über welche *Dirichlet* die Hauptresultate gefunden, aber noch nicht veröffentlicht hat, so dafs ich nicht genau weifs, ob sein Princip der Classification mit diesem, aus der Theorie der complexen Zahlen sich ergebenden genau übereinstimmt. Als besonderer Fall ist die Theorie der Formen vom zweiten Grade mit zwei Variabeln, jedoch nur wenn die Determinante eine Primzahl λ ist, mit in diesen Untersuchungen begriffen, und es stimmt hier unsere Classification mit der *Gaufs*ischen, aber nicht mit der von *Legendre* überein. Auch wirft dieselbe ein helleres Licht auf die *Gaufs*ische Classification der Formen vom zweiten Grade, und auf den wahren Grund der Unterscheidung von *Aequivalentia propria et impropria,* welche, wie nicht zu läugnen, so, wie sie in den *Disquisitiones arihmeticae* auftritt, immer einen Schein des Unpassenden behält. Wenn nämlich dort zwei Formen, wie $ax^2+2bxy+cy^2$ und $ax^2-2bxy+cy^2$, oder $ax^2+2bxy+cy^2$ und $cx^2+2bxy+ay^2$, als verschiedenen Classen angehörend betrachtet werden, während in Wahrheit ein wesentlicher Unterschied derselben nicht aufzufinden ist, und wenn andererseits die *Gaufs*ische Classification dennoch als die der Natur der Sache am meisten entsprechende anerkannt werden mufs: so wird man genöthigt, die sich wirklich nur ganz äufserlich von einander unterscheidenden Formen, wie $ax^2+2bxy+cy^2$ und $ax^2-2bxy+cy^2$, blofs als die Repräsentanten zweier andern, aber wesentlich verschiedenen Begriffe der Zahlentheorie aufzufassen. Diese aber sind in Wahrheit nichts anderes als zwei verschiedene ideale Factoren, welche einer und derselben Zahl angehören. Die ganze Theorie der

*) Ein Beweis dieses wichtigen Satzes, wenn gleich in weit geringerer Allgemeinheit und in ganz anderer Form, findet sich in der Dissertation: *De unitatibus complexis* von *L. Kronecker,* Berlin 1845.

Formen vom zweiten Grade, mit zwei Variabeln, kann nämlich als Theorie der complexen Zahlen von der Form $x+y\sqrt{D}$ aufgefafst werden, und führt dann nothwendig zu idealen complexen Zahlen derselben Art. Diese classificiren sich aber eben so nach den idealen Multiplicatoren, welche nöthig und hinreichend sind, um sie zu wirklichen complexen Zahlen von der Form $x+y\sqrt{D}$ zu machen. Mit der *Gaufs*ischen Classification übereinstimmend, erschliefsen diese so den wahren Grund derselben.

Die allgemeine Untersuchung über die idealen complexen Zahlen hat die gröfste Analogie mit dem bei *Gaufs* sehr schwierig behandelten Abschnitte: *De compositione formarum*, und die Hauptresultate, welche *Gaufs* für die quadratischen Formen pag. 337 sqq. bewiesen hat, finden auch für die Zusammensetzung der allgemeinen idealen complexen Zahlen Statt. Es gehört hier zu jeder Classe idealer Zahlen eine andere Classe, welche, mit dieser multiplicirt, wirkliche complexe Zahlen hervorbringt (die wirklichen complexen Zahlen bilden hier das Analogon der *Classis principalis*). Es sind hier auch Classen, welche, mit sich selbst multiplicirt, wirkliche complexe Zahlen (die *Classis principalis*) geben, also *ancipites;* namentlich ist die *Classis principalis* selbst stets eine *Classis anceps*. Nimmt man eine ideale complexe Zahl $f(\alpha)$ und erhebt sie zu Potenzen, so gelangt man, nach dem zweiten der obigen Sätze, immer zu einer Potenz, welche eine wirkliche complexe Zahl ist; wenn h die kleinste Zahl ist, für welche $(f(\alpha))^h$ eine wirkliche complexe Zahl ist, so gehören $f(\alpha)$, $(f(\alpha))^2$, $(f(\alpha))^3$, ... $f(\alpha)^h$ alle verschiedenen Classen an. Es kann nun geschehen, dafs diese, namentlich bei passender Wahl des $f(\alpha)$, alle vorhandenen Classen erschöpfen: ist es nicht der Fall, so wird leicht bewiesen, dafs die Anzahl aller Classen wenigstens immer ein Vielfaches von h ist. Ich bin vorläufig noch nicht tiefer in dieses Gebiet der Theorie der complexen Zahlen eingedrungen; namentlich habe ich eine Untersuchung der wahren Anzahl der Classen noch nicht unternommen, weil, wie ich durch mündliche Mittheilungen erfahren habe, *Dirichlet*, nach ähnlichen Principien, wie in seinen berühmten Abhandlungen über die quadratischen Formen, diese Anzahl bereits gefunden hat. Ich bemerke nur noch dies Eine über den Character der idealen complexen Zahlen, dafs sie, nach dem zweiten der obigen Sätze, als bestimmte Wurzeln aus wirklichen complexen Zahlen überall angesehen und dargestellt werden können, oder dafs sie immer die Form $\sqrt[h]{(\Phi(\alpha))}$ annehmen, wo $\Phi(\alpha)$ eine wirkliche complexe Zahl ist, und h eine ganze Zahl.

Aus den verschiedenen Anwendungen, welche ich von dieser Theorie der complexen Zahlen schon gemacht habe, hebe ich nur die Anwendung auf die Kreistheilung hervor; als Vervollständigung Dessen, was ich in dem erwähnten Programm bereits mittheilte. Setzt man

$$(\alpha, x) = x + \alpha x^g + \alpha^2 x^{g^2} + \dots + \alpha^{p-2} x^{g^{p-2}},$$

wo $\alpha^\lambda = 1$, $x^p = 1$, $p = m\lambda + 1$, und g eine primitive Wurzel der Primzahl p ist, so ist bekanntlich $(\alpha, x)^\lambda$ eine von x unabhängige, aus den Wurzeln der Gleichung $\alpha^\lambda = 1$ gebildete complexe Zahl. Für diese habe ich in dem erwähnten Programm, unter der Voraussetzung, dafs p sich in $\lambda - 1$ wirkliche complexe Primfactoren zerlegen läfst, deren einer $f(\alpha)$ sei, folgenden Ausdruck gefunden:

$$(\alpha, x)^\lambda = \pm \alpha^h \overset{m_1}{f(\alpha)} . \overset{m_2}{f(\alpha^2)} . \overset{m_3}{f(\alpha^3)} \dots \overset{m_{\lambda-1}}{f(\alpha^{\lambda-1})},$$

in welchem die Potenz-Exponenten m_1, m_2, m_3, etc. so bestimmt sind, dafs allgemein m_k positiv kleiner als λ und $k.m_k \equiv 1$, mod. λ ist. Genau derselbe einfache Ausdruck gilt nun, wie sich leicht beweisen läfst, ganz allgemein, auch wenn $f(\alpha)$, der Primfactor von p, nicht ein wirklicher, sondern nur ein idealer ist. Um aber in letzterem Falle den Ausdruck des $(\alpha, x)^\lambda$ in Form einer wirklichen complexen Zahl zu haben, darf man nur das ideale $f(\alpha)$ als Wurzel aus einer wirklichen complexen Zahl darstellen, oder eine der (wenn auch indirecten) Methoden anwenden, welche dazu dienen, eine wirkliche complexe Zahl herzustellen, deren ideale Primfactoren gegeben sind.

Über die Zerlegung der aus Wurzeln der Einheit gebildeten complexen Zahlen in ihre Primfactoren

Journal für die reine und angewandte Mathematik 35, 327–367 (1847)

In dem vorstehenden Aufsatze, welchen ich als Einleitung zu dem hier folgenden anzusehen bitte, habe ich vor einiger Zeit die Resultate meiner Untersuchungen über die Zerlegung der complexen Zahlen niedergelegt. Die Entwickelung und Begründung derselben soll nun der Gegenstand der gegenwärtigen Abhandlung sein.

§. 1.

Es sei λ eine Primzahl und α eine imaginäre Wurzel der Gleichung $\alpha^\lambda = 1$, so ist die allgemeinste Form der complexen Zahlen, welche wir hier untersuchen:

$$\varphi(\alpha) = a_1\alpha + a_2\alpha^2 + a_3\alpha^3 + \ldots + a_{\lambda-1}\alpha^{\lambda-1};$$

in welchem Ausdruck die Coëfficienten a_1, a_2, a_3, etc. *ganze* Zahlen sind. In besonderen Fällen können alle diejenigen Coëfficienten der complexen Zahl, welche mit den zu denselben Perioden gehörenden Wurzeln der Einheit multiplicirt sind, einander gleich werden; so dafs die complexe Zahl zu einer linearen Function der Perioden wird. Die Untersuchung dieser besondern Art complexer Zahlen, welche andrerseits, da die Perioden auch eingliedrig sein können, die allgemeine complexe Zahl als besondern Fall in sich schliefst, bildet die Grundlage für die Auffindung der Primfactoren aller aus den Wurzeln der Einheit geformten complexen Zahlen; weshalb wir zunächst nur von diesen zu handeln haben.

Besteht die Zahl $\lambda - 1$ aus den beiden Factoren e und f, so dafs $\lambda - 1 = e \cdot f$, und bezeichnet γ eine primitive Wurzel von λ, so kann man die imaginären Wurzeln der Gleichung $\alpha^\lambda = 1$ folgendermafsen in e Perioden von je f Gliedern zusammenordnen:

$$\begin{aligned}
\eta &= \alpha + \alpha^{\gamma^e} + \alpha^{\gamma^{2e}} + \ldots + \alpha^{\gamma^{(f-1)e}}, \\
\eta_1 &= \alpha^{\gamma} + \alpha^{\gamma^{e+1}} + \alpha^{\gamma^{2e+1}} + \ldots + \alpha^{\gamma^{(f-1)e+1}}, \\
\eta_2 &= \alpha^{\gamma^2} + \alpha^{\gamma^{e+2}} + \alpha^{\gamma^{2e+2}} + \ldots + \alpha^{\gamma^{(f-1)e+2}}, \\
&\ldots\ldots\ldots\ldots\ldots\ldots \\
\eta_{e-1} &= \alpha^{\gamma^{e-1}} + \alpha^{\gamma^{2e-1}} + \alpha^{\gamma^{3e-1}} + \ldots + \alpha^{\gamma^{fe-1}}.
\end{aligned}$$

43 *

Das Product zweier Perioden läfst sich bekanntlich als lineare Function der ähnlichen Perioden mit ganzzahligen Coëfficienten darstellen; deshalb sei

$$(1.)\quad \eta.\eta_k = \overset{k}{\mu} f + \overset{k}{m}\eta + \overset{k}{m}_1\eta_1 + \overset{k}{m}_2\eta_2 + \dots + \overset{k}{m}_{e-1}\eta_{e-1},$$

woraus auch die scheinbar allgemeinere Gleichung folgt:

$$(2.)\quad \eta_r\eta_{r+k} = \overset{k}{\mu} f + \overset{k}{m}\eta_r + \overset{k}{m}_1\eta_{r+1} + \overset{k}{m}_2\eta_{r+2} + \dots + \overset{k}{m}_{e-1}\eta_{r+e-1}.$$

Der Coëfficient $\overset{k}{\mu}$ ist immer gleich Null, aufser wenn f gerade und $k=0$, oder f ungerade und zugleich $k=\frac{1}{2}e$ ist; in welchen beiden Fällen $\overset{k}{\mu}=1$ ist. Die übrigen Coëfficienten $\overset{k}{m}_h$ haben sehr einfache Beziehungen zu einander, welche wir kurz entwickeln wollen. Multiplicirt man zwei Perioden von f Gliedern mit einander, so ist die Anzahl aller Glieder des entwickelten Products gleich ff; deshalb mufs, wenn in der Gleichung (1.) die Perioden als Summen von je f Wurzeln aufgefafst werden, die Summe aller Coëfficienten gleich ff, also

$$ff = \overset{k}{\mu} f + \overset{k}{m} f + \overset{k}{m}_1 f + \overset{k}{m}_2 f + \dots + \overset{k}{m}_{e-1} f,$$

mithin

$$(3.)\quad f = \overset{k}{\mu} + \overset{k}{m} + \overset{k}{m}_1 + \overset{k}{m}_2 + \dots + \overset{k}{m}_{e-1}$$

sein. Setzt man ferner in der Gleichung (2.) nach einander $r=0, 1, 2, \dots e-1$ und nimmt die Summe, so erhält man

$$\sum_0^{e-1}{}_r\,\eta_r\eta_{r+k} = ef\overset{k}{\mu} - \overset{k}{m} - \overset{k}{m}_1 - \overset{k}{m}_2 - \dots - \overset{k}{m}_{e-1};$$

also, vermöge (1.), und weil $ef=\lambda-1$ ist:

$$(4.)\quad \sum_0^{e-1}{}_r\,\eta_r\eta_{r+k} = \overset{k}{\mu}\lambda - f.$$

Berücksichtigt man noch den oben angegebenen Werth von $\overset{k}{\mu}$, so zeigt sich, dafs diese Summe immer gleich $-f$ ist; mit Ausnahme der beiden Fälle: erstens, wo f gerade und $k=0$, und zweitens, wo f ungerade und $k=\frac{1}{2}e$ ist; in welchen Fällen diese Summe gleich $\lambda-f$ ist.

Verwandelt man ferner in der Gleichung (2.) k in $e-k$, so wird

$$\eta_r\eta_{r-k} = \overset{e-k}{\mu} f + \overset{e-k}{m}\eta_r + \overset{e-k}{m}_1\eta_{r+1} + \overset{e-k}{m}_2\eta_{r+2} + \dots + \overset{e-k}{m}_{e-1}\eta_{r+e-1}.$$

Setzt man aber in derselben Gleichung (2.) $r-k$ statt r, so ist

$$\eta_r\eta_{r-k} = \overset{k}{\mu} f + \overset{k}{m}\eta_{r-k} + \overset{k}{m}_1\eta_{r-k+1} + \overset{k}{m}_2\eta_{r-k+2} + \dots + \overset{k}{m}_{e-1}\eta_{r-k+e-1}.$$

Durch Vergleichung dieser beiden Ausdrücke erhält man allgemein

$$(5.)\quad \overset{k}{m}_h = \overset{e-k}{m}_{h-k}.$$

Multiplicirt man endlich die Gleichung (2.) mit η_{h+r} und nimmt die Summe für $r=0, 1, 2, \ldots e-1$, so erhält man vermittels der Formel (4.):

$$\sum_0^{e-1}{}_r\, \eta_r \eta_{k+r} \eta_{h+r} = -ff + \lambda \overset{k}{m}_h, \text{ wenn } f \text{ gerade ist und}$$

$$\sum_0^{e-1}{}_r\, \eta_r \eta_{k+r} \eta_{h+r} = -ff + \lambda \overset{k}{m}_{h+\frac{1}{2}e}, \text{ wenn } f \text{ ungerade ist;}$$

und weil sich in den Summen zur Linken h und k vertauschen lassen, so folgt

$$(6.)\quad \begin{cases} \overset{k}{m}_h = \overset{h}{m}_k, \text{ wenn } f \text{ gerade und} \\ \overset{k}{m}_{h+\frac{1}{2}e} = \overset{h}{m}_{k+\frac{1}{2}e}, \text{ wenn } f \text{ ungerade ist.} \end{cases}$$

Die Gleichung (1.), welche die Grundlage jeder Rechnung mit den Perioden ist, drückt eigentlich ein System von e verschiedenen Gleichungen für $k=0, 1, 2, \ldots e-1$ aus. Diese zusammen reichen auch gerade hin, um aus ihnen die e Perioden selbst zu finden. Eliminirt man nämlich alle Perioden, aufser der ersten η, so erhält man für η eine Gleichung vom $e$ten Grade; und zwar genau jene bekannte Gleichung, deren Wurzeln alle e Perioden sind. Läfst man ferner irgend eine der e Gleichungen, welche in (1.) enthalten sind, weg, und betrachtet die erste Periode η als bekannt, die übrigen $e-1$ Perioden als unbekannt, so erhält man ein System von $e-1$ Gleichungen mit $e-1$ Unbekannten; und zwar ein lineares System, aus welchem sich $\eta_1, \eta_2, \ldots \eta_{e-1}$ alle rational durch η ausdrücken lassen. Man erhält so die Ausdrücke der Perioden als rationale Functionen der ersten (d. h. einer beliebigen andern); welche Ausdrücke *Gaufs* auf andere Weise herleitet.

Wir fassen jetzt das System der in (1.) enthaltenen Gleichungen als ein System von Congruenzen auf, für den Modul q; wo q eine Primzahl sein soll, welche der Bedingung $q^f \equiv 1$ mod. λ genügt. Anstatt der Perioden $\eta, \eta_1, \eta_2, \ldots \eta_{e-1}$ setzen wir die unbestimmten ganzen Zahlen $u, u_1, u_2, \ldots u_{e-1}$. Dies giebt folgendes System von Congruenzen:

$$(7.)\quad u u_k \equiv \mu f + \overset{k}{m} u + \overset{k}{m}_1 u_1 + \overset{k}{m}_2 u_2 + \ldots + \overset{k}{m}_{e-1} u_{e-1}, \text{ mod. } q,$$

für $k=0, 1, 2, \ldots e-1$. Es giebt nun immer e wirkliche ganze Zahlen $u, u_1, u_2, \ldots u_{e-1}$, welche diesem Systeme von Congruenzen genugthun:
* denn eliminirt man auch hier alle Unbekannten, die erste u ausgenommen,

so erhält man eine Congruenz vom $e$ten Grade für u, welche mit der Gleichung für η, welche alle Perioden zu Wurzeln hat, genau übereinstimmt.

Diese Gleichung sei

$$(8.)\quad y^e+A_1y^{e-1}+A_2y^{e-2}+\ldots+A_e = Y = 0:$$

so hat die Congruenz $Y\equiv 0$ mod. q (wie ich in der Abhandlung über die Divisoren gewisser Formen der Zahlen, welche aus der Theorie der Kreistheilung entstehen, in diesem Journal Bd. 30. S. 107 bewiesen habe) immer e reale Wurzeln; falls nämlich, wie vorausgesetzt worden, q eine Primzahl von der Art ist, dafs $q^f\equiv 1$, mod. λ. Wenn nun u aus der Congruenz $Y\equiv 0$, mod. q, bestimmt ist, so werden u_1, u_2, u_3, ... u_{e-1} mittels der in (7.) enthaltenen Gleichungen nur noch linear bestimmt; eine Unmöglichkeit kann demnach nirgends eintreten, da überdies der Modul eine Primzahl ist. Man würde auch mit der einzigen Congruenz $Y\equiv 0$ mod. q ausreichen, um alle Zahlen u, u_1, u_2, ... u_{e-1} zu bestimmen, weil sie alle auf gleiche Weise Wurzeln dieser Congruenz sind; aber man würde so nicht finden können, in welcher Ordnung die Wurzeln dieser Congruenz zu nehmen sind, damit sie auch den in (7.) enthaltenen Congruenzen genügen.

Da man jede beliebige der e Congruenzwurzeln von $Y\equiv 0$ mod. q als erste ansehen kann (wonach sich dann nur die Reihenfolge der übrigen zu richten hat, welche cyklisch immer dieselbe bleibt), so folgt, dafs die den Congruenzen, welche in (7.) enthalten sind, genügenden Zahlen u, u_1, u_2, ... u_{e-1}, ebenso auch der scheinbar allgemeineren, der Gleichung (2.) entsprechenden Congruenz

$$(9.)\quad u_r u_{r+k} \equiv \overset{k}{\mu} f+\overset{k}{m} u_r+\overset{k+1}{m_1} u_{r+1}+\overset{k}{m_2} u_{r+2}+\ldots+\overset{k}{m_{e-1}} u_{r-1}, \text{ mod. } q,$$

genügen; für alle Werthe der Zahlen r und k. Jeder der e Perioden η, η_1, η_2, ... η_{e-1} entspricht also eine der Congruenzwurzeln u, u_1, u_2, ... u_{e-1}, in der Art, dafs, wenn man statt der Perioden in den Gleichungen (2.) ihre analogen Congruenzwurzeln setzt, diese Gleichungen zu richtigen Congruenzen für den Modul q werden.

§. 2.

Die Congruenzwurzeln u, u_1, u_2, ... u_{e-1} befolgen, wie wir im vorigen Paragraphen gezeigt haben, in ihrer Multiplication, genau dieselben Gesetze wie die Perioden; und da ohnedies die Addition und Subtraction für die Congruenzwurzeln dieselbe ist wie für die Perioden, so folgt der wichtige Satz: ***Dafs zu jeder Gleichung, unter den Perioden, welche nur Additio-***

nen, Subtractionen und Multiplicationen derselben enthalten, stets eine entsprechende Congruenz sein mufs, welche man dadurch erhält, dafs man nur anstatt der Perioden ihre entsprechenden Congruenzwurzeln setzt. Es gehören ferner zu einer einzigen solchen rationalen Gleichung immer noch $e-1$ conjugirte, welche aus ihr durch Vertauschung der Perioden, mit Beibehaltung der cyklischen Ordnung derselben, gebildet werden, oder, was dasselbe ist, durch gleiche Vergröfserung aller Indices der Perioden; wobei von den Indices, welche gröfser als $e-1$ werden, e, oder Vielfache von e, abzuziehen sind. Dafs nämlich eine solche Veränderung jeder rationalen Gleichung, unter den Perioden, die man auch nur als eine Änderung der Wurzel α der Gleichung $\frac{\alpha^\lambda-1}{\alpha-1}=0$ ansehen kann, stets gestattet ist, folgt sehr leicht aus der Irreductibilität dieser Gleichung. Ebenso gehören auch eigentlich zu einer solchen Gleichung unter den Perioden nicht nur eine, sondern stets e Congruenzen, weil man die Congruenzwurzeln, mit Beibehaltung ihrer cyklischen Ordnung, auf e verschiedene Arten den Perioden zuordnen kann.

Hiernach folgt z. B. aus der Gleichung $\eta+\eta_1+\eta_2+\ldots\ldots+\eta_{e-1}=-1$ die Congruenz

$$(1.)\quad u+u_1+u_2+\ldots\ldots+u_{e-1}\equiv -1 \text{ mod. } q;$$

ferner aus der Gleichung (4.) §. 1. die Congruenz

$$(2.)\quad \sum_0^{e-1}{}_r u_r u_{r+k}\equiv -f \text{ mod. } q;$$

mit Ausnahme der beiden Fälle: erstens, wo f gerade und $k=0$, und zweitens, wo f ungerade und $k=\frac{1}{2}e$ ist; in welchen Fällen diese Summe congruent $\lambda-f$ ist.

Bezeichnet ferner $F(\eta)$ irgend eine ganze rationale Function der Perioden $\eta,\ \eta_1,\ \eta_2,\ \ldots\ldots\ \eta_{e-1}$ mit ganzzahligen Coëfficienten, welche irgend wie als Product von Factoren, oder als Summe solcher Producte erscheinen mag: so kann man dieselbe bekanntlich immer auf die Form

$$(3.)\quad F(\eta)=a\eta+a_1\eta_1+a_2\eta_2+\ldots\ldots+a_{e-1}\eta_{e-1}$$

bringen. Setzt man nun statt der Perioden die Congruenzwurzeln, und zwar so, dafs u_r statt η, u_{r+1} statt η_1, u_{r+2} statt η_2, u. s. w. gesetzt wird, in welchem Falle wir die ganze Zahl, in welche $F(\eta)$ übergeht, durch $F(u_r)$ bezeichnen, so ist

$$(4.)\quad F(u_r)\equiv au_r+a_1u_{r+1}+a_2u_{r+2}+\ldots\ldots+a_{e-1}u_{r-1} \text{ mod. } q,$$

für jeden beliebigen Werth von r. Enthält $F(\eta)$ den realen Factor q, so müssen, nachdem diese complexe Zahl auf die einfache Grundform (3.) gebracht ist, alle Coëfficienten a, a_1, a_2, ... a_{e-1} einzeln durch q theilbar sein; und in diesem Falle mufs deshalb auch $F(u_r)\equiv 0$ mod. q sein, für jeden Werth von r. Stellt man sich ferner nicht nur die complexe Zahl $F(\eta)$, sondern auch alle ihre conjugirten, welche durch cyklische Veränderung der Perioden entstehen, auf die einfache Grundform gebracht vor, so erhält man

$$(5.)\quad \begin{cases} F(\eta) = a\eta + a_1\eta_1 + a_2\eta_2 + \ldots + a_{e-1}\eta_{e-1}, \\ F(\eta_1) = a\eta_1 + a_1\eta_2 + a_2\eta_3 + \ldots + a_{e-1}\eta, \\ F(\eta_2) = a\eta_2 + a_1\eta_3 + a_2\eta_4 + \ldots + a_{e-1}\eta_1, \\ \ldots\ldots\ldots\ldots\ldots\ldots \\ F(\eta_{e-1}) = a\eta_{e-1} + a_1\eta + a_2\eta_1 + \ldots + a_{e-1}\eta_{e-2}. \end{cases}$$

Multiplicirt man diese Gleichungen der Reihe nach mit η_k, η_{k+1}, η_{k+2}, ... u. s. w. und addirt, so erhält man, vermöge der Gleichung (4. §. 1.), wenn f *gerade* ist:

$$(6.)\quad \eta_k F(\eta) + \eta_{k+1}F(\eta_1) + \eta_{k+2}F(\eta_2) + \ldots + \eta_{k-1}F(\eta_{e-1}) = a_k\lambda - (a + a_1 + \ldots + a_{e-1})f;$$

und da die Addition aller Gleichungen (5.)

$$(7.)\quad F(\eta) + F(\eta_1) + F(\eta_2) + \ldots + F(\eta_{e-1}) = -(a + a_1 + \ldots + a_{e-1})$$

giebt, so erhält man:

$$(8.)\quad (\eta_k - f)F(\eta) + (\eta_{k+1} - f)F(\eta_1) + \ldots + (\eta_{k-1} - f)F(\eta_{e-1}) = a_k.\lambda.$$

Wenn f *ungerade* ist, modificirt sich dieses Resultat nur insofern, dafs anstatt $a_k.\lambda$ alsdann $a_{k+\frac{1}{2}e}.\lambda$ herauskommt. Setzt man nun statt der Perioden die Congruenzwurzeln, so ergiebt sich

$$(u_k - f)F(u) + (u_{k+1} - f)F(u_1) + \ldots + (u_{k-1} - f)F(u_{e-1}) \equiv a_k.\lambda, \text{ mod. } q,$$

oder congruent $a_{k+\frac{1}{2}e}.\lambda$, wenn f *ungerade* ist. Wenn nun $F(u_r)\equiv 0$ mod. q ist, für jeden Werth von r, so folgt aus dieser Congruenz, dafs sämmtliche Coëfficienten a, a_1, a_2, ... a_{e-1} durch q theilbar sind, dafs also auch die complexe Zahl $F(\eta)$ den realen Factor q hat. Es sind demnach folgende zwei Sätze bewiesen worden, von welchen einer der umgekehrte des andern ist:

Wenn eine ganze rationale Function der Perioden mit ganzzahligen Coëfficienten durch den realen Factor q theilbar ist, so sind auch alle diejenigen ganzen Zahlen durch q theilbar, welche entstehen, wenn man statt der Perioden ihre entsprechenden Congruenzwurzeln für den Modul q setzt;

und umgekehrt:

Wenn alle die e ganzen Zahlen, welche aus einer ganzen rationalen Function der Perioden entstehen, indem statt der Perioden die Congruenzwurzeln für den Modul q gesetzt werden, durch q theilbar sind, so ist diese Function der Perioden selbst durch q theilbar.

Ist $\varphi(\eta)$ irgend eine aus den Perioden gebildete complexe Zahl, so nennen wir das Product aller conjugirten complexen Zahlen, welches immer eine ganze Zahl ist, *die Norm* der complexen Zahl $\varphi(\eta)$ und bezeichnen dasselbe durch

$$N\varphi(\eta) = \varphi(\eta)\varphi(\eta_1)\varphi(\eta_2)\dots\varphi(\eta_{e-1}).$$

Wendet man nun die sieben gefundenen Sätze auf diese Norm an, so ergiebt sich, *dafs: Wenn die Norm* $N\varphi(\eta)$ *durch* q *theilbar ist, stets auch eine der ganzen Zahlen* $\varphi(u)$, $\varphi(u_1)$, $\varphi(u_2)$, ... $\varphi(u_{e-1})$ *durch* q *theilbar sein mufs; und umgekehrt, dafs, wenn eine dieser Zahlen durch* q *theilbar ist, auch die Norm durch* q *theilbar sein mufs.*

§. 3.

Für die Untersuchung der Primfactoren jeder gegebenen complexen Zahl ist es noch sehr wichtig, zu beweisen, dafs es stets solche complexe, aus Perioden gebildete Zahlen giebt, deren Norm durch q theilbar ist, aber nicht durch q^2; und zugleich zu zeigen, wie diese complexen Zahlen gefunden werden können. Im allgemeinen wird es schon unter den complexen Zahlen $u-\eta$, $u_1-\eta$, $u_2-\eta$, ... $u_{e-1}-\eta$, deren Normen alle durch q theilbar sind, einige geben, welche die verlangte Eigenschaft haben; ja es ist leicht zu beweisen, dafs immer wenigstens eine dieser complexen Zahlen genügt, sobald es unter den Congruenzwurzeln u, u_1, u_2, ... u_{e-1} nur noch eine giebt, welche keiner andern gleich ist. Da dies aber wirklich, namentlich für sehr kleine Werthe von q und grofse Werthe von e, nicht immer der Fall ist, so wenden wir, um diese verlangten complexen Zahlen zu finden, folgende Methode an, welche stets sicher zum Ziele führt.

Wir suchen zunächst eine complexe Zahl $\varphi(\eta)$ von der Art, dafs das Product $\varphi(\eta_r)\varphi(\eta_s)$ durch q theilbar sei, sobald r und s ***verschieden*** sind; aber nicht durch q theilbar, sobald r und s einander ***gleich*** sind. Eine solche Zahl ist immer die folgende:

$$(1.)\qquad \varphi(\eta) = f + u\eta + u_1\eta_1 + u_2\eta_2 + \dots + u_{e-1}\eta_{e-1}.$$

Es ist nämlich

$$\varphi(u_r) \equiv f + uu_r + u_1 u_{r+1} + u_2 u_{r+2} + \dots + u_{e-1} u_{r-1}, \text{ mod. } q,$$

also, vermöge der Congruenz (2. §. 2.),

$$\varphi(u_r) \equiv 0, \text{ mod. } q,$$

mit Ausnahme der beiden Fälle: erstens, wo f *gerade* und $r = 0$, und zweitens, wo f *ungerade* und $r = \frac{1}{2}e$ ist; in welchen Fällen $\varphi(u_r) \equiv \lambda$ ist. Eben so ist $\varphi(u_s) \equiv 0$, oder $\equiv \lambda$, unter denselben Bedingungen. Demnach ist das Product

$$(2.) \quad \varphi(u_r)\varphi(u_s) \equiv 0, \text{ mod. } q,$$

für jeden Werth von r und s; mit Ausnahme der beiden Fälle: erstens, wo $r = s = 0$ und f *gerade*, und zweitens, wo $r = s = \frac{1}{2}e$ und f *ungerade* ist. Läfst man nun die Congruenzwurzeln alle möglichen Werthe durchlaufen, mit Beibehaltung der cyklischen Ordnung, so enthält diese eine Congruenz (2.) eigentlich e Congruenzen, aus welchen mittels des zweiten Lehrsatzes im vorigen Paragraphen sogleich folgt, dafs

$$\varphi(\eta_r)\varphi(\eta_s) \equiv 0, \text{ mod. } q,$$

ist; mit alleiniger Ausnahme des Falles $r = s$; wie es verlangt wurde.

Ich untersuche nun weiter von der complexen Zahl $\lambda - \varphi(\eta) = \psi(\eta)$ die Norm

$$N\psi(\eta) = (\lambda - \varphi(\eta))(\lambda - \varphi(\eta_1))(\lambda - \varphi(\eta_2)) \dots (\lambda - \varphi(\eta_{e-1})).$$

Entwickelt man das Product rechts, und läfst dabei alle durch q theilbaren Theile weg, so erhält man mittels der Congruenz $\varphi(\eta_r)\varphi(\eta_s) \equiv 0$, mod. q, folgende Congruenz:

$$N\psi(\eta) = \lambda^e - \lambda^{e-1}(\varphi(\eta) + \varphi(\eta_1) + \varphi(\eta_2) + \dots + \varphi(\eta_{e-1})), \text{ mod. } q,$$

und da

$$\varphi(\eta) + \varphi(\eta_1) + \varphi(\eta_2) + \dots + \varphi(\eta_{e-1}) \equiv \lambda, \text{ mod. } q,$$

ist, so ist

$$N\psi(\eta) \equiv 0, \text{ mod. } q.$$

Eben so soll jetzt die Theilbarkeit durch q der complexen Zahl

$$\Psi(\eta) = \psi(\eta_1)\psi(\eta_2) \dots \psi(\eta_{e-1}) \text{ oder}$$
$$\Psi(\eta) = (\lambda - \varphi(\eta_1))(\lambda - \varphi(\eta_2)) \dots (\lambda - \varphi(\eta_{e-1}))$$

untersucht werden, welche, mit Ausnahme des ersten Factors, alle Factoren der Norm von $\psi(\eta)$ enthält. Durch Entwicklung und Vernachläfsigung aller durch q theilbaren Theile erhält man hier:

$$\Psi(\eta) \equiv \lambda^{e-1} - \lambda^{e-2}(\varphi(\eta_1) + \varphi(\eta_2) + \dots + \varphi(\eta_{e-1})), \text{ mod. } q,$$

und da

$$\varphi(\eta_1)+\varphi(\eta_2)+\ldots.+\varphi(\eta_{e-1}) \equiv \lambda-\varphi(\eta)$$

ist, so ist

$$\Psi(\eta) \equiv \lambda^{e-1}-\lambda^{e-2}(\lambda-\varphi(\eta)) \equiv \lambda^{e-2}\varphi(\eta), \text{ mod. } q;$$

also ist $\Psi(\eta)$ nicht durch q theilbar. Ich behaupte nun, dafs immer $\psi(\eta)$, oder doch $\psi(\eta)+q$, eine der verlangten complexen Zahlen ist, deren Norm den Factor q einmal enthält, aber nicht mehrmals. Um dies zu beweisen, entwickle ich $N(\psi(\eta)+q)$ nach dem Modul q^2, was

$$N(\psi(\eta)+q) \equiv N\psi(\eta)+q(\Psi(\eta)+\Psi(\eta_1)+\ldots.+\Psi(\eta_{e-1})), \text{ mod. } q^2,$$

giebt. Wenn also wirklich $\psi(\eta)$ nicht eine der verlangten Zahlen, sondern, aufser durch q, auch durch q^2 theilbar ist, so ist

$$N(\psi(\eta)+q) \equiv q(\Psi(\eta)+\Psi(\eta_1)+\ldots.+\Psi(\eta_{e-1})), \text{ mod. } q^2.$$

Multiplicirt man mit $\Psi(\eta)$ und beachtet, dafs $\Psi(\eta)\Psi(\eta_r)$ immer den Factor q enthält, aufser wenn $r=0$, so erhält man

$$\Psi(\eta)N(\psi(\eta)+q) \equiv q(\Psi(\eta))^2, \text{ mod. } q^2;$$

und da $(\Psi(\eta))^2$ nicht mit q aufgeht, so folgt, dafs $N(\psi(\eta)+q)$ nicht durch q^2 theilbar ist. Es giebt also stets complexe Zahlen, deren Normen einen bestimmten realen Primfactor q *nur einmal* enthalten.

§. 4.

Wir wenden uns jetzt zu den allgemeineren complexen Zahlen, welche nicht aus den Perioden, sondern irgend wie aus den Wurzeln der Gleichung $\alpha^\lambda=1$ gebildet sind und welche also auf die Form

$$f(\alpha) = a_1\alpha+a_2\alpha^2+a_3\alpha^3+\ldots.+a_{\lambda-1}\alpha^{\lambda-1}$$

gebracht werden können. Jede complexe Zahl ist ein Factor irgend einer realen ganzen Zahl; namentlich ist sie stets ein Factor der Norm: deshalb ist auch jeder Primfactor einer complexen Zahl zugleich Primfactor einer realen ganzen Zahl; und zwar immer Primfactor der Norm. Aus diesem Grunde haben wir zunächst die Bedingungen zu suchen, unter welchen die Norm einer complexen Zahl einen gegebenen realen Primfactor q enthält. Alle nicht durch λ theilbaren Zahlen können nun bekanntlich nach den verschiedenen Exponenten eingetheilt werden, zu welchen sie gehören, für den Modul λ (S. *Gaufs* Disquisitiones arithm. §. 52.); und diese Eintheilung begründet auch die wesentlich verschiedenen Charactere der Divisoren der Norm.

Es sei demnach q eine Primzahl, welche zum Exponenten von f (einem Divisor von $\lambda-1$) gehört, so dafs $q^f\equiv 1$, mod. λ, aber so, dafs keine niederere

44 *

Potenz von q der Eins congruent sei: so ist (Vergl. die Abhandlung über die Divisoren etc. in diesem Journal Bd. XXX. S. 115)

$$(1.)\quad (f(\alpha))^q \equiv f(\alpha^q),\ \text{mod. } q:$$

also, wenn man wiederholt zur $q$ten Potenz erhebt:

$$(2.)\quad (f(\alpha))^{q^h} \equiv f(\alpha^{q^h}),\ \text{mod. } q,$$

und folglich, wenn $h = 0, 1, 2, \ldots f-1$ gesetzt und multiplicirt wird:

$$(3.)\quad (f(\alpha))^{1+q+q^2+\cdots+q^{f-1}} \equiv f(\alpha)f(\alpha^q)f(\alpha^{q^2})\ldots f(\alpha^{p^{f-1}}),\ \text{mod. } q.$$

Setzt man nun α^{γ^m} statt α, und für q, wo es als Exponent des α vorkommt, seinen congruenten Werth, als Potenz einer primitiven Wurzel γ, welche, da $q^f \equiv 1$, mod. λ eine e^{te} Potenz, also $p \equiv \gamma^{re}$, mod. λ, sein mufs: so erhält man

$$(4.)\quad (f(\alpha^{\gamma^m}))^{1+q+q^2+\cdots+q^{f-1}} \equiv f(\alpha^{\gamma^m})f(\alpha^{\gamma^{re+m}})f(\alpha^{\gamma^{2re+m}})\ldots f(\alpha^{\gamma^{(f-1)re+m}}),\ \text{mod. } q.$$

Es hat r keinen gemeinschaftlichen Factor mit f, weil q nach der Voraussetzung so zum Exponenten f gehört, dafs q^f, aber keine niedrigere Potenz von q, der Eins congruent wird für den Modul λ. Giebt man also dem m nach einander e Werthe, welche alle nach dem Modul e incongruent sind, so erhält man durch Multiplication dieser Congruenzen rechts das Product aller conjugirten complexen Zahlen, d. h. die Norm der complexen Zahl $f(\alpha)$, folglich

$$(5.)\quad (\Pi f(\alpha^{\gamma^m}))^{1+q+q^2+\cdots+q^{f-1}} \equiv Nf(\alpha),\ \text{mod. } q;$$

wo das Productzeichen Π sich auf die e verschiedenen Werthe des m bezieht, welche nur der einen Bedingung unterworfen sind, dafs sie alle incongruent sein müssen, für den Modul e. Wenn nun die Norm $Nf(\alpha)$ durch q theilbar ist, so ist auch

$$(\Pi f(\alpha^{g^m}))^{1+q+q^2+\cdots+q^{f-1}} \equiv 0,\ \text{mod. } q;$$

woraus nothwendig folgt:

$$(6.)\quad \Pi f(\alpha^{g^m}) \equiv 0,\ \text{mod. } q.$$

Wir haben also folgenden Satz:

Wenn** $Nf(\alpha)$ **durch** q **theilbar ist, wo** q **eine zum Exponenten** f **gehörende Primzahl bezeichnet, so müssen von den** $\lambda-1$ **conjugirten complexen Zahlen** $f(\alpha)$, $f(\alpha^g)$, $f(\alpha^{g^2})$, **etc. je** e, **deren Wurzeln nur verschiedenen von den** e **Perioden, zu je** f **Gliedern, angehören, immer ein Product geben, welches durch** q **theilbar ist.

Aufserdem folgt noch als Zusatz:

Wenn die Norm** $Nf(\alpha)$ **durch den zum Exponenten** f **gehörenden

Primfactor q theilbar ist, so mufs sie immer f mal den Factor q enthalten, oder durch q^f theilbar sein.

Um nun weiter die Congruenzbedingungen zu finden, welchen die Coëfficienten der complexen Zahl $f(\alpha)$ genügen müssen, damit die Norm derselben durch q theilbar sei, werden wir dieser complexen Zahl eine Form geben, in welcher statt der einfachen Wurzeln α, α^2, α^3, etc., insoweit es möglich ist, die Perioden von je f Gliedern auftreten. Nach *Gaufs* Disqu. arithm. §. 348, sind alle Wurzeln, welche in einer Periode von f Gliedern vorkommen, zugleich Wurzeln einer Gleichung vom $f$ten Grade von der Form

$$\alpha^f + P_1\alpha^{f-1} + P_2\alpha^{f-2} + \ldots + P_f = 0,$$

deren Coëfficienten ganze und ganzzahlige Functionen der Perioden von je f Gliedern sind. Mittels dieser Gleichung kann man aus dem Ausdrucke

$$f(\alpha) = a_1\alpha + a_2\alpha^2 + a_3\alpha^3 + \ldots + a_{\lambda-1}\alpha^{\lambda-1}$$

alle Potenzen des α, von $\alpha^{\lambda-1}$ bis zu α^f hinab, eliminiren, und erhält dadurch einen Ausdruck von folgender Form:

$$(7.)\quad f(\alpha) = \varphi(\eta) + \alpha\varphi_1(\eta) + \alpha^2\varphi_2(\eta) + \ldots + \alpha^{f-1}\varphi_{f-1}(\eta),$$

wo $\varphi(\eta)$, $\varphi_1(\eta)$, $\varphi_2(\eta)$, etc. aus den Perioden von je f Gliedern gebildete complexe ganze Zahlen sind. Es läfst sich auch eine bestimmte complexe Zahl $f(\alpha)$ nur auf eine einzige Weise auf diese Form bringen.

Der gefundene Satz über die Theilbarkeit der Norm durch den Primfactor q läfst sich nun folgendermafsen ausdrücken:

Wenn $Nf(\alpha)$ durch den zum Exponenten f gehörenden Primfactor q theilbar ist, so mufs das Product der e Factoren

$$(8.)\quad \begin{cases} \left(c f(\alpha) + c_1 f(\alpha^{\gamma^e}) + c_2 f(\alpha^{\gamma^{2e}}) + \ldots + c_{f-1} f(\alpha^{\gamma^{(f-1)e}})\right), \\ \left(\overset{1}{c} f(\alpha^{\gamma}) + \overset{1}{c}_1 f(\alpha^{\gamma^{e+1}}) + \overset{1}{c}_2 f(\alpha^{\gamma^{2e+1}}) + \ldots + \overset{1}{c}_{f-1} f(\alpha^{\gamma^{(f-1)e+1}})\right), \\ \cdot\;\cdot\;\cdot\;\cdot\;\cdot\;\cdot\;\cdot\;\cdot\;\cdot\;\cdot\;\cdot\;\cdot\;\cdot\;\cdot\;\cdot\;\cdot\;\cdot\;\cdot \\ \left(\overset{e-1}{c} f(\alpha^{\gamma^{e-1}}) + \overset{e-1}{c}_1 f(\alpha^{\gamma^{2e-1}}) + \overset{e-1}{c}_2 f(\alpha^{\gamma^{3e-1}}) + \ldots + \overset{e-1}{c}_{f-1} f(\alpha^{\gamma^{fe-1}})\right) \end{cases}$$

stets durch q theilbar sein, für alle beliebigen Werthe der durch c bezeichneten $\lambda-1$ Coëfficienten: denn alle einzelnen Theile dieses entwickelten Products sind dem obigen Satze zufolge durch q theilbar; ganz abgesehen von den Coëfficienten. Setzen wir nun für $f(\alpha)$ den gefundenen Ausdruck:

$$f(\alpha) = \varphi(\eta) + \alpha\varphi_1(\eta) + \alpha^2\varphi_2(\eta) + \ldots + \alpha^{f-1}\varphi_{g-1}(\eta)$$

und der Kürze wegen

$$(9.)\begin{cases} \overset{k}{c}+\overset{k}{c}_1+\overset{k}{c}_2+\dots+\overset{k}{c}_{f-1}=\overset{k}{C}, \\ \alpha\overset{k}{c}+\alpha^{\gamma^e}\overset{k}{c}_1+\alpha^{\gamma^{2e}}\overset{k}{c}_2+\dots+\alpha^{\gamma^{(f-1)e}}\overset{k}{c}_{f-1}=\overset{k}{C}_1, \\ \alpha^2\overset{k}{c}+\alpha^{2\gamma^e}\overset{k}{c}_1+\alpha^{2\gamma^{2e}}\overset{k}{c}_2+\dots+\alpha^{2\gamma^{(f-1)e}}\overset{k}{c}_{f-1}=\overset{k}{C}_2, \\ \dots\dots\dots\dots\dots\dots \\ \alpha^{f-1}\overset{k}{c}+\alpha^{(f-1)\gamma^e}\overset{k}{c}_1+\alpha^{(f-1)\gamma^{2e}}\overset{k}{c}_2+\dots+\alpha^{(f-1)\gamma^{(f-1)e}}\overset{k}{c}_{f-1}=\overset{k}{C}_{f-1}, \end{cases}$$

so verwandelt sich dieses Product in folgendes:

$$(10.)\begin{cases} \left(C\varphi(\eta)+C_1\varphi_1(\eta)+C_2\varphi_2(\eta)+\dots+C_{f-1}\varphi_{f-1}(\eta)\right) \\ \times\left(\overset{1}{C}\varphi(\eta_1)+\overset{1}{C}_1\varphi_1(\eta_1)+\overset{1}{C}_2\varphi_2(\eta_1)+\dots+\overset{1}{C}_{f-1}\varphi_{f-1}(\eta_1)\right) \\ \dots\dots\dots\dots\dots\dots \\ \times\left(\overset{e-1}{C}\varphi(\eta_{e-1})+\overset{e-1}{C}_1\varphi_1(\eta_{e-1})+\overset{e-1}{C}_1\varphi_2(\eta_{e-1})+\dots+\overset{e-1}{C}_{f-1}\varphi_{f-1}\varphi_{f-1}(\eta_{e-1})\right); \end{cases}$$

welches also ebenfalls durch q theilbar sein mufs, und zwar für alle beliebigen Werthe der $\lambda-1$ Gröfsen C: denn da die mit c bezeichneten Gröfsen völlig beliebig waren, so folgt vermöge der Gleichungen (9.) das Gleiche für die Coëfficienten C. Nimmt man diese Coëfficienten unabhängig von α an, und setzt allgemein $\overset{k}{C}_h=C_h$, so geht das Product (10.) in die Norm der nur aus Perioden gebildeten complexen Zahl

$$C\varphi(\eta)+C_1\varphi_1(\eta)+C_2\varphi_2(\eta)+\dots+C_{f-1}\varphi_{(f-1)}(\eta)$$

über; und damit dieselbe den Factor q habe, mufs nach (§. 2.), wenn statt der Perioden die zugehörigen Congruenzwurzeln gesetzt werden,

$$(11.)\quad C\varphi(u_r)+C_1\varphi_1(u_r)+C_2\varphi_2(u_r)+\dots+C_{f-1}\varphi_{f-1}(u_r)\equiv 0,\ \text{mod. } q,$$

sein, für irgend einen Werth von r. Diese Congruenz aber kann, da die Coëfficienten noch völlig beliebig sind, nicht bestehen, ohne dafs die einzelnen Glieder der Null congruent sind, also nicht ohne dafs

$$\varphi(u_r)\equiv 0,\quad \varphi_1(u_r)\equiv 0,\quad \varphi_2(u_r)\equiv 0,\quad \dots\quad \varphi_{f-1}(u_r)\equiv 0,\ \text{mod. } q,$$

ist. Umgekehrt: wenn diese f Congruenzen erfüllt werden, so ist auch

$$f(\alpha)f(\alpha^\gamma)f(\alpha^{\gamma^2})\dots f(\alpha^{\gamma^{e-1}})\equiv 0,\ \text{mod. } q,$$

also $Nf(\alpha)$ durch q^f theilbar. Diese Resultate lassen sich durch folgenden Satz aussprechen:

Wenn $Nf(\alpha)$ durch die Primzahl q theilbar ist, welche zum Exponenten f gehört, für den Modul λ: so müssen die f Congruenzbedingungen

$$\varphi(u_r)\equiv 0,\quad \varphi_1(u_r)\equiv 0,\quad \varphi_2(u_r)\equiv 0,\quad \dots\quad \varphi_{f-1}(u_r)\equiv 0,\ \text{mod. } q,$$

erfüllt werden, für irgend einen Werth von r; welche Congruenzen man dadurch erhält, dafs man f(α) auf die Form

$$f(\alpha) = \varphi(\eta) + \alpha\varphi_1(\eta) + \alpha^2\varphi_2(\eta) + \dots + \alpha^{f-1}\varphi_{f-1}(\eta)$$

bringt und statt der Perioden in $\varphi(\eta)$, $\varphi(\eta_1)$, $\varphi(\eta_2)$, ... *die entsprechenden Congruenzwurzeln setzt. Umgekehrt: wenn die f Congruenzbedingungen erfüllt werden, so ist auch* $Nf(\alpha)$ *durch* q, *(also auch durch* q^f*) theilbar.*

Es giebt noch eine andere, aber weniger brauchbare Art, die Bedingung auszudrücken, dafs $Nf(\alpha)$ durch q theilbar sein soll. Bildet man nämlich das Product

$$f(\alpha)f(\alpha^{\gamma^e})f(\alpha^{\gamma^{2e}})\dots f(\alpha^{\gamma^{(f-1)e}}) = F(\eta),$$

welches, als symmetrische Function der in einer Periode enthaltenen Wurzeln, eine Function der Perioden von je f Gliedern ist, so erhält man

$$Nf(\alpha) = F(\eta)F(\eta_1)F(\eta_2)\dots F(\eta_{e-1}).$$

Damit nun $Nf(\alpha)$ durch q theilbar sei, ist es nothwendig und hinreichend, dafs $F(u_r) \equiv 0$, mod. q, sei, für irgend einen Werth von r. Es ist bemerkenswerth, dafs hier nur eine einzige Congruenzbedingung gefunden wurde, und zwar, in Beziehung auf die Coëfficienten von $f(\alpha)$, vom $f$ten Grade, während sich oben f Congruenzbedingungen vom ersten Grade für die Theilbarkeit der Norm $Nf(\alpha)$ durch die Primzahl q ergaben; und da beide in gleicher Weise hinreichend und nothwendig sind, so ist zu schliefsen, dafs die eine Congruenz vom $f$ten Grade $F(u_r) \equiv 0$, mod. q, genau Dasselbe ausdrückt, wie die f linearen Congruenzen $\varphi(u_r) \equiv 0$, $\varphi_1(u_r) \equiv 0$, ... $\varphi_{f-1}(u_r) \equiv 0$, mod. q, oder dafs jene eine Congruenz nicht erfüllt werden kann, ohne dafs diese f Congruenzen zugleich erfüllt werden.

§. 5.

Wenn für eine complexe Zahl $f(\alpha)$, welche auf die Form

$$f(\alpha) = \varphi(\eta) + \alpha\varphi_1(\eta) + \alpha^2\varphi_2(\eta) + \dots + \alpha^{f-1}\varphi_{f-1}(\eta)$$

gebracht ist, die f Congruenzbedingungen

$$\varphi(u_r) \equiv 0, \quad \varphi_1(u_r) \equiv 0, \quad \varphi_2(u_r) \equiv 0, \quad \dots + \varphi_{f-1}(u_r) \equiv 0, \text{ mod. } q,$$

erfüllt werden, so wollen wir dies künftig kurz so ausdrücken: es sei $f(\alpha) \equiv 0$, mod. q, für $\eta = u_r$. Durch Festsetzung einer einzigen Congruenzwurzel, welche einer Periode entsprechen soll, ist, da die cyklische Ordnung bei beiden stets unverändert bleibt, zugleich für jede Periode die entsprechende Congruenzwurzel bestimmt; oder in der einen Bestimmung für $\eta = u_r$ liegen zugleich die folgenden: $\eta_1 = u_{r+1}$, $\eta_2 = u_{r+2}$, etc.; auch $\eta_{e-r} = u$.

Hiernach ist leicht einzusehen, dafs, wenn $f(\alpha)$ als ein Product von Factoren auftritt, und einer derselben hat die Eigenschaft, congruent Null zu werden für $\eta = u_r$: dafs dann auch das entwickelte Product dieselbe Eigenschaft haben mufs. Ist nämlich

$$(1.) \quad f(\alpha) g(\alpha) = h(\alpha)$$

und $f(\alpha) \equiv 0$, mod. q, für $\eta = u_r$ und man bringt $f(\alpha)$ auf die Form

$$f(\alpha) = \varphi(\eta) + \alpha\varphi_1(\eta) + \alpha^2\varphi_2(\eta) + \ldots + \alpha^{f-1}\varphi_{f-1}(\eta),$$

so sind alle Glieder des Products $f(\alpha)g(\alpha)$ mit einer der complexen Zahlen $\varphi(\eta)$, $\varphi_1(\eta)$, $\varphi_2(\eta)$, etc. multiplicirt; sie werden also alle durch q theilbar, für $\eta = u_r$. Um aber den umgekehrten Satz zu beweisen, nämlich, dafs, wenn $h(\alpha) \equiv 0$, mod. q, für $\eta = u_r$, und $g(\alpha)$ nicht congruent Null ist, für $\eta = u_r$, nothwendig $f(\alpha) \equiv 0$, mod. q, für $\eta = u_r$ sein mufs, setze ich

$$f(\alpha)f(\alpha^{\gamma^e})f(\alpha^{\gamma^{2e}}) \ldots f(\alpha^{\gamma^{(f-1)e}}) = F(\eta),$$
$$g(\alpha)g(\alpha^{\gamma^e})g(\alpha^{\gamma^{2e}}) \ldots g(\alpha^{\gamma^{(f-1)e}}) = G(\eta),$$
$$h(\alpha)h(\alpha^{\gamma^e})h(\alpha^{\gamma^{2e}}) \ldots h(\alpha^{\gamma^{(f-1)e}}) = H(\eta).$$

Dann erhält man aus der Gleichung $f(\alpha)g(\alpha) = h(\alpha)$:

$$(2.) \quad F(\eta)G(\eta) = H(\eta).$$

Wenn nun $h(\alpha) \equiv 0$, mod. q, für $\eta = u_r$ ist, so ist auch $H(\eta) \equiv 0$, mod. q, für $\eta = u_r$, oder (wie wir dies bisher immer kurz bezeichneten) $H(u_r) \equiv 0$, mod. q: also mufs, wenn in (2.) u_r statt η gesetzt wird, auch eine der beiden ganzen Zahlen $F(u_r)$ oder $G(u_r)$ durch q theilbar sein; und wenn nach der Voraussetzung $g(\alpha)$ nicht $\equiv 0$, mod. q, für $\eta = u_r$ ist, so ist auch $G(u_r)$ nicht congruent Null, also $F(u_r) \equiv 0$, mod. q. Die Bedingung $F(u_r) \equiv 0$, mod. q, ist aber, wie zu Ende des vorigen Paragraphen gezeigt wurde, identisch mit der Bedingung $\varphi(u_r) \equiv 0$, $\varphi_1(u_r) \equiv 0$, $\varphi_2(u_r) \equiv 0$, ... $\varphi_{f-1}(u_r) \equiv 0$, mod. q, also auch identisch mit der Bedingung $f(\alpha) \equiv 0$, mod. q, für $\eta = u_r$.

Da nun bewiesen ist, dafs die Bedingung $f(\alpha) \equiv 0$, mod. q, für $\eta = u_r$ stets dieselbe bleibt, es mag $f(\alpha)$ in entwickelter Form auftreten, oder in Form eines Products aus zweien, und folglich auch aus mehreren complexen Zahlen: so folgt von selbst der Satz:

Wenn eine aus Factoren bestehende complexe Zahl durch q theilbar ist, so müssen für alle Werthe $\eta = u$, $\eta = u_1$, $\eta = u_2$, ... $\eta = u_{e-1}$ irgend einige ihrer Factoren congruent Null werden, mod. q.

und umgekehrt:

Wenn für jeden der Werthe $\eta = u$, $\eta = u_1$, $\eta = u_2$, ... $\eta = u_{e-1}$ irgend einer der Factoren einer complexen Zahl congruent Null wird, mod. q: so enthält dieselbe den realen Factor q.

Die f Congruenzbedingungen, welche wir in dem Ausdrucke $f(\alpha) \equiv 0$, mod. q, für $\eta = u_r$ zusammenfassen, lassen sich noch auf eine andere Weise sehr einfach ausdrücken. Ist nämlich $\psi(\eta)$ eine complexe Zahl von der Art, wie wir sie in (§. 3.) fanden, nemlich, dafs die Norm derselben oder das Product $\psi(\eta)\psi(\eta_1)\psi(\eta_2)\ldots\psi(\eta_{e-1})$ durch q theilbar sei, nicht aber durch q^2: so mufs, wie oben gezeigt, $\psi(u_r) \equiv 0$, mod. q, sein, für irgend einen bestimmten Werth von r. Wird nun, wie in (§. 3.),

$$\Psi(\eta) = \psi(\eta_1)\psi(\eta_2)\ldots\psi(\eta_{e-1})$$

gesetzt, so behaupte ich, *dafs die Bedingung $f(\alpha) \equiv 0$, mod. q, für $u = \eta_r$ identisch ist mit der Bedingung $f(\alpha)\Psi(\eta) \equiv 0$, mod. q.* Es ist nämlich $\Psi(\eta)$, weil es die $e-1$ Factoren $\psi(\eta_1)$, $\psi(\eta_2)$, ... $\psi(\eta_{e-1})$ enthält, für alle Werthe $\eta = u$, u_1, u_2, ... u_{e-1}, mit Ausnahme des Werths $\eta = u_r$, durch q theilbar: wenn also $f(\alpha) \equiv 0$, mod. q ist, für $\eta = u_r$, so ist das Product $f(\alpha)\Psi(\eta)$ für alle Werthe $\eta = u$, u_1, u_2, ... u_{e-1}, ohne Ausnahme, congruent Null, mod. q, und enthält also den realen Factor q. Wenn ferner, umgekehrt, $f(\alpha)\Psi(\eta)$ durch q theilbar ist, so müssen die beiden Factoren zusammen für alle Werthe $\eta = u$, u_1, u_2, ... u_{e-1} durch q theilbar sein, und da $\Psi(\eta)$ für $\eta = u_r$ nicht durch q theilbar ist, so mufs nothwendig der andere Factor $f(\alpha) = 0$, mod. q, für $\eta = u_r$ es sein.

§. 6.

Die Coëfficienten einer complexen Zahl von der Form

$$\varphi(\eta) = a\eta + a_1\eta_1 + a_2\eta_2 + \ldots + a_{e-1}\eta_{e-1}$$

lassen sich, wie in (§. 2.) gezeigt, auf unendlich viele verschiedene Arten so bestimmen, dafs die Norm $N\varphi(\eta)$ durch die zum Exponenten f gehörende Primzahl q theilbar, also ein Vielfaches von q wird. In vielen Fällen, aber nicht immer, gelingt es sogar, complexe Zahlen zu finden, deren Norm die reale Primzahl q selbst ist, so dafs $q = \varphi(\eta)\varphi(\eta_1)\varphi(\eta_2)\ldots\varphi(\eta_{e-1})$; die complexe Zahl $\varphi(\eta)$ mufs alsdann, wenn statt der Perioden die entsprechenden Congruenzwurzeln gesetzt werden, durch q theilbar werden, so dafs $\varphi(u_r) \equiv 0$, mod. q, ist, für irgend einen bestimmten Werth von r. Wenn nun die Norm von $\varphi(\eta)$ gleich q ist, so ist $\varphi(\eta)$ ein nicht weiter in Factoren zerlegbarer

Primfactor der realen Zahl q. Wäre nämlich $\varphi(\eta)$ in zwei complexe Factoren zerlegbar, so dafs $\varphi(\eta) = f(\alpha)$ wäre, so müfste zunächst, weil $\varphi(\eta) \equiv 0$, mod. q, für $\eta = u_r$ ist, auch einer der beiden Factoren, zu welchen ich $f(\alpha)$ nehme, congruent Null sein, mod. q, für $\eta = u_r$. Aus $\varphi(\eta) = f(\alpha) g(\alpha)$ folgt aber weiter, wenn man α in α^γ, α^{γ^2}, ... $\alpha^{\gamma^{e-1}}$ verwandelt und die Gleichungen multiplicirt:

$$\varphi(\eta)\varphi(\eta_1)\varphi(\eta_2)\dots\varphi(\eta_{e-1})$$
$$= f(\alpha)f(\alpha^\gamma)f(\alpha^{\gamma^2})\dots f(\alpha^{\gamma^{e-1}})g(\alpha)g(\alpha^\gamma)g(\alpha^{\gamma^2})\dots g(\alpha^{\gamma^{e-1}}).$$

Da nun $f(\alpha) \equiv 0$, für $\eta = u_r$ ist, so ist das Product $f(\alpha)f(\alpha^\gamma)f(\alpha^{\gamma^2})\dots f(\alpha^{\gamma^{e-1}})$ durch q theilbar, also gleich $qF(\alpha)$; und da $\varphi(\eta)\varphi(\eta_1)\varphi(\eta_2)\dots\varphi(\eta_{e-1}) = q$ ist, so erhält man, nach Weglassung des gemeinschaftlichen Factors q:

$$1 = F(\alpha)g(\alpha)g(\alpha^\gamma)g(\alpha^{\gamma^2})\dots g(\alpha^{\gamma^{e-1}}):$$

es müssen also alle Factoren rechts nur complexe Einheiten und mithin mufs namentlich $g(\alpha)$ eine complexe Einheit sein. Wenn man demnach die Zahl $\varphi(\eta)$, deren Norm die Primzahl q ist, auf irgend eine Weise in zwei Factoren zerlegt, so ist einer derselben stets nur eine complexe Einheit, und $\varphi(\eta)$ ist also wirklich eine complexe ***Primzahl.*** Wenn nun irgend eine complexe Zahl $f(\alpha)$ diesen Primfactor $\varphi(\eta)$ der realen Primzahl q als Factor enthält, so dafs $f(\alpha) = \varphi(\eta)g(\alpha)$ ist, so mufs $f(\alpha) \equiv 0$, mod. q, für $\eta = u_r$ sein, weil $\varphi(u_r) \equiv 0$, mod. q, ist. Eine Umkehrung dieses Satzes läfst sich nicht ohne Weiteres aufstellen; denn in vielen Fällen existirt ein solcher Primfactor von q nicht, während die Congruenzbedingung $f(\alpha) \equiv 0$, mod. q, für $\eta = u_r$ wirklich erfüllt wird. Diese Congruenzbedingung aber, als die bleibende und jener Zufälligkeit, ob q sich als Product von e conjugirten complexen Zahlen darstellen lasse, nicht unterworfene Eigenschaft einer complexen Zahl, soll nun als Definition der complexen Primfactoren benutzt werden, welche selbst sodann entweder als wirkliche complexe Zahlen für sich darstellbar sein können, oder auch nicht; in welchem letzteren Falle sie *ideale* Primfactoren genannt werden sollen. Anstatt der Congruenzbedingung selbst aber werden wir den Ausdruck derselben am Ende des vorigen Paragraphen wählen, weil dieser sich am leichtesten auch auf den Fall ausdehnen läfst, wo einer und derselbe Primfactor mehrmals in einer complexen Zahl enthalten ist.

Die allgemeine Definition der realen oder idealen Primfactoren einer gegebenen complexen Zahl ist also folgende:

Es sei $\psi(\eta)$ eine aus den e Perioden von je f Gliedern gebildete complexe Zahl, von der Art, dafs die Norm $\psi(\eta)\psi(\eta_1)\psi(\eta_2)\dots\psi(\eta_{e-1})$

durch die zum Exponenten f gehörende reale Primzahl q theilbar ist, nicht aber durch q^2, so wie, dafs $\psi(u) \equiv 0$, mod. q: dann setze man

$$\Psi(\eta) = \psi(\eta_1)\psi(\eta_2)\ldots\psi(\eta_{e-1}).$$

Wenn nun irgend eine complexe Zahl $f(\alpha)$ die Eigenschaft hat, dafs das Product $f(\alpha)\Psi(\eta_r)$ durch q theilbar ist, so soll dies so ausgedrückt werden: es enthalte $f(\alpha)$ den zu $\eta_r = u$ gehörenden Primfactor von q. Wenn ferner $f(\alpha)$ die Eigenschaft hat, dafs $f(\alpha)(\Psi(\eta_r))^\mu$ durch q^μ theilbar ist, aber $f(\alpha)(\Psi(\eta_r))^{\mu+1}$ nicht theilbar durch $q^{\mu+1}$, so soll dies heifsen: es enthalte $f(\alpha)$ den zu $\eta_r = u$ gehörenden Primfactor von q genau μ mal.

Die Zweckmäfsigkeit dieser Definition kann sich erst aus der auf dieselbe gegründeten Theorie ergeben; in welcher wir zeigen werden, dafs man mit den Eigenschaften der complexen Zahlen, welche wir so eben als Primfactoren derselben definirten, und welche in vielen Fällen auch *wirkliche* Primfactoren derselben geben, genau eben so rechnen kann, wie mit den ganzzahligen Primfactoren der zusammengesetzten ganzen Zahlen. Zunächst aber ist hier noch nachzuweisen, dafs diese Primfactoren von den *besondern* Eigenschaften der zu ihrer Auffindung zu benutzenden complexen Zahl $\Psi(\eta)$ ganz unabhängig sind, oder dafs man immer genau dieselben Primfactoren erhält, auch wenn man statt der complexen Zahl $\Psi(\eta)$ eine andere Zahl von denselben oben angegebenen *allgemeinen* Eigenschaften anwendet.

Es sei also $\Psi'(\eta)$ eine andere complexe Zahl, von der Art, dafs $\Psi'(\eta)\Psi'(\eta_1)\Psi'(\eta_2)\ldots\Psi'(\eta_{e-1})$ durch q, aber nicht durch q^2 theilbar ist, und $\Psi'(u) \equiv 0$, mod. q; auch sei $\Psi'(\eta) = \psi(\eta_1)\psi(\eta_2)\ldots\psi(\eta_{e-1})$: so wird behauptet, dafs, wenn $f(\alpha)(\Psi(\eta_r))^\mu$ durch q^μ theilbar ist, nicht aber $f(\alpha)(\Psi(\eta_r))^{\mu+1}$ durch $q^{\mu+1}$, auch $f(\alpha)(\Psi'(\eta_r))^\mu$ durch q^μ theilbar sein mufs, aber $f(\alpha)(\Psi'(\eta_r))^{\mu+1}$ nicht theilbar durch $q^{\mu+1}$. Der Voraussetzung zufolge hat man

$$f(\alpha)(\Psi(\eta_r))^\mu = q^\mu Q(\alpha),$$

und $Q(\alpha)\Psi(\eta_r)$ ist nicht theilbar durch q, oder, was Dasselbe ist: $Q(\alpha)$ ist nicht congruent Null, mod. q, für $\eta_r = u$. Multiplicirt man nun mit $(\psi(\eta_r))^\mu(\Psi'(\eta_r))^\mu$, so ergiebt sich

$$f(\alpha)(\psi(\eta_r))^\mu(\Psi(\eta_r))^\mu(\Psi'(\eta_r))^\mu = q^\mu Q(\alpha)(\psi(\eta_r))^\mu(\Psi'(\eta_r))^\mu.$$

Nun ist $\psi(\eta_r)\Psi(\eta_r)$, die Norm von $\psi(\eta)$, gleich qP; wo P eine nicht durch q theilbare ganze Zahl ist; ferner ist $\psi(\eta_r)\Psi'(\eta_r) = qR(\eta_r)$, d. h. eine durch q theilbare complexe Zahl. Setzt man diese Ausdrücke in die Gleichung, und

45*

läfst den gemeinschaftlichen Factor q^μ weg, so erhält man

$$f(\alpha)\Psi'(\eta_r))^\mu P^\mu = q^\mu Q(\alpha)(R(\eta_r))^\mu;$$

und da P nicht durch q theilbar ist, so folgt, dafs $f(\alpha)(\Psi(\eta_r))^\mu$ durch q^μ theilbar sein mufs. Multiplicirt man nun noch einmal mit $\Psi'(\eta_r)$, so erhält man

$$f(\alpha)(\Psi'(\eta_r))^{\mu+1}P^\mu = q^\mu Q(\alpha)(R(\eta_r))^\mu\Psi'(\eta_r).$$

Es ist aber $Q(\alpha)(R(\eta_r))^\mu\Psi'(\eta_r)$ nicht theilbar durch q, weil keiner der drei Factoren für $\eta_r = u$ mit q aufgeht: also ist auch $f(\alpha)(\Psi'(\eta_r))^{\mu+1}$ nicht theilbar durch $q^{\mu+1}$. Es ist daher in der That Dasselbe, ob man die complexe Zahl $\Psi'(\eta)$, oder $\Psi'(\eta)$, zur Untersuchung des in $f(\alpha)$ enthaltenen Primfactors anwendet.

Da wir die Bedingung, dafs $f(\alpha)$ den zu $\eta_r = u$ gehörenden Primfactor von q einmal enthalte, durch f Congruenzen ausgedrückt haben, welchen die Coëfficienten der complexen Zahl $f(\alpha)$ genügen müssen, so wollen wir hier auch zeigen, wie die Bedingung, dafs $f(\alpha)$ einen solchen Primfactor mehreremale enthalte, durch Congruenzen auszudrücken sei. Bringt man $f(\alpha)$ wieder auf die Form

$$f(\alpha) = \varphi(\eta) + \alpha\varphi_1(\eta) + \alpha^2\varphi_2(\eta) + \ldots + \alpha^{f-1}\varphi_{f-1}(\eta),$$

so ergiebt sich

$$f(\alpha)\Psi(\eta_r) = \varphi(\eta)\Psi(\eta_r) + \alpha\varphi_1(\eta)\Psi(\eta_r) + \ldots + \alpha^{f-1}\varphi_{f-1}(\eta)\Psi(\eta_r).$$

Wenn nun $f(\alpha)$ den zu $\eta_r = u$ gehörenden Primfactor der zum Exponenten f gehörenden realen Primzahl q einmal enthält, so dafs $f(\alpha)\Psi(\eta_r)$ durch q theilbar ist, so müssen, wie (§. 4.) zeigt, $\varphi(\eta)$, $\varphi_1(\eta)$, ... $\varphi_{f-1}(\eta)$ alle denselben idealen Primfactor enthalten: also müssen auch $\varphi(\eta)\Psi(\eta_r)$, $\varphi_1(\eta)\Psi(\eta_r)$, ... $\varphi_{f-1}(\eta)\Psi(\eta_r)$ alle durch q theilbar sein. Setzt man demnach

$$\varphi(\eta)\Psi(\eta_r) = q\varphi'(\eta),\quad \varphi_1(\eta)\Psi(\eta_r) = q\varphi_1'(\eta),\quad \ldots\quad \varphi_{f-1}(\eta)\Psi(\eta_r) = q\varphi_{f-1}'(\eta),$$

so erhält man

$$f(\alpha)\Psi(\eta_r) = q[\varphi'(\eta) + \alpha\varphi_1'(\eta) + \alpha^2\varphi_2'(\eta) + \ldots + \alpha^{f-1}\varphi_{f-1}'(\eta)].$$

Wenn nun $f(\alpha)$ den zu $\eta_r = u$ gehörenden Primfactor von q zweimal enthält, so ist $f(\alpha)(\Psi(\eta_r))^2$ durch q^2 theilbar; multiplicirt man daher noch einmal mit $\Psi(\eta_r)$, so müssen wieder $\varphi'(\eta)\Psi(\eta_r)$, $\varphi_1'(\eta)\Psi(\eta_r)$, ... $\varphi_{f-1}'(\eta)\Psi(\eta_r)$ alle einzeln durch q theilbar sein, folglich $\varphi(\eta)(\Psi(\eta_r))^2$, $\varphi_1(\eta)(\Psi(\eta_r))^2$, ... $\varphi_{f-1}(\eta)(\Psi(\eta_r))^2$ alle theilbar durch q^2. So fortfahrend, zeigt sich, dafs, wenn $f(\alpha)(\Psi(\eta_r))^\mu$ durch q^μ theilbar ist, allemal auch $\varphi(\eta)(\Psi(\eta_r))^\mu$, $\varphi_1(\eta)(\Psi(\eta_r))^\mu$, ... $\varphi_{f-1}(\eta)(\Psi(\eta_r))^\mu$ einzeln durch q^μ theilbar sein müssen; oder, was Dasselbe ist: wenn $f(\alpha)$ den zu $\eta_r = u$ gehörenden Primfactor von q μmal enthält, so müssen die complexen, aus den Perioden gebildeten Zahlen $\varphi(\eta)$, $\varphi_1(\eta)$, $\varphi_2(\eta)$, ... $\varphi_{f-1}(\eta)$ den nämlichen Primfactor jede μmal enthalten. Entwickelt man nun das Product $\varphi_k(\eta)(\Psi(\eta_r))^\mu$ in linearer Form, so dafs

$$\varphi_k(\eta)(\Psi(\eta_r))^\mu = C\eta + C_1\eta_1 + C_2\eta_2 + \ldots + C_{e-1}\eta_{e-1},$$

so folgt sehr leicht, dafs

$$\varphi_k(\eta)(\Psi(\eta_r))^\mu + \varphi_k(\eta_1)(\Psi(\eta_{r+1}))^\mu + \ldots + \varphi_k(\eta_{e-1})(\Psi(\eta_{r+e-1}))^\mu \\ = -(C + C_1 + C_2 + \ldots + C_{e-1})$$

ist. Multiplicirt man noch einmal mit $(\Psi(\eta_r))^\mu$ und erwägt, dafs $\Psi(\eta_r)\Psi(\eta_s)$ immer durch q theilbar ist, den einen Fall $r = s$ ausgenommen, so erhält man

$$\varphi_k(\eta)(\Psi(\eta_r))^{2\mu} \equiv -(C + C_1 + C_2 + \ldots + C_{e-1})(\Psi(\eta_r))^\mu, \text{ mod. } q^\mu.$$

Hieraus folgt, dafs die Bedingung: $\varphi(\eta_r)(\Psi(\eta_r))^\mu$ sei durch q^μ theilbar, identisch ist mit der Bedingung $C + C_1 + C_2 + \ldots + C_{e-1} \equiv 0$, mod. q^μ. Also die Bedingung, dafs $f(\alpha)$ den zu $\eta_r = u$ gehörenden (idealen) Primfactor von q μmal enthalte, wird durch f Congruenzen für den Modul q^n ausgedrückt, welche unter den Coëfficienten der complexen Zahl $f(\alpha)$ Statt finden müssen. Wir bemerken noch ausdrücklich, dafs diese Congruenzen in Beziehung auf die Coëfficienten von $f(\alpha)$ nur vom ersten Grade sind; denn $\varphi_k(\eta)$ enthält dieselben nur linear, und $(\Psi(\eta_r))^\mu$ enthält sie gar nicht; also enthalten auch C, C_1, C_2, ... C_{e-1} die Coëfficienten nur in linearer Weise.

§. 7.

Wir haben nun dieselben einfachen Sätze, welche für die Rechnnng mit den realen ganzen Primfactoren der ganzen Zahlen gelten, für die in dem vorigen Paragraphen definirten idealen oder wirklichen Primfactoren der complexen Zahlen aufzustellen und zu beweisen. Es werde zunächst folgender Satz bewiesen:

Das entwickelte Product zweier oder mehrerer complexen Zahlen hat genau dieselben Primfactoren wie die Factoren des Products zusammengenommen.

Es seien $f(\alpha)$ und $g(\alpha)$ zwei complexe Factoren, $h(\alpha)$ das entwickelte Product derselben, so dafs $f(\alpha)g(\alpha) = h(\alpha)$. Es enthalte $f(\alpha)$ den zu $\eta_r = u$ gehörenden Primfactor von q genau μmal, $g(\alpha)$ denselben Factor genau νmal: so wird behauptet, dafs $h(\alpha)$ denselben genau $\mu + \nu$mal enthält. Nach der Voraussetzung ist $f(\alpha)(\Psi(\eta_r))^\mu = q^\mu Q(\alpha)$ und $g(\alpha)(\Psi(\eta_r))^\nu = q_\nu R(\alpha)$, und weder $Q(\alpha)\Psi(\eta_r)$ noch $R(\alpha)\Psi(\eta_r)$ ist durch q theilbar, weil sonst $f(\alpha)$ oder $g(\alpha)$ den zu $\eta_r = u$ gehörenden Primfactor von q, $\mu+1$ mal oder $\nu+1$ mal enthalten würden; gegen die Voraussetzung. Hieraus folgt $f(\alpha)g(\alpha)(\Psi(\eta_r))^{\mu+\nu} = q^{\mu+\nu}Q(\alpha)R(\alpha)$, also auch $h(\alpha)(\Psi(\eta_r))^{\mu+\nu} = q^{\mu+\nu}Q(\alpha)R(\alpha)$. Multiplicirt man noch einmal mit $\Psi(\eta_r)$, so ist $h(\alpha)(\Psi(\eta_r))^{\mu+\nu+1}$ auch durch keine höhere Potenz als durch $q^{\mu+\nu}$ theilbar; denn $Q(\alpha)R(\alpha)\Psi(\eta_r)$ ist nicht durch q theilbar, weil für $\eta_r = u$ keiner der drei Factoren dieses Products der Null congruent wird. Was nun hier von einem beliebigen idealen Primfactor bewiesen ist, gilt offenbar für alle; und da man ferner auch jeden der Factoren des Products wieder in Factoren zerfället sich vorstellen kann, so ist klar, dafs der aufgestellte Satz auch eben so für jedes Product beliebig vieler Factoren gilt.

Dafs eine complexe Zahl, welche als Product mehrerer Factoren auftritt, durch q theilbar ist, wenn sie alle e ideale Primfactoren von q enthält, und nicht durch q theilbar, wenn sie irgend einen dieser Primfactoren nicht enthält, ist schon oben in (§. 5.) gezeigt worden; nur ist daselbst noch nicht der Ausdruck *idealer Primfactor,* sondern statt dessen die ihm gleichbedeutende Congruenzbedingung gebraucht worden. Wir erweitern hier den Satz wie folgt:

Wenn eine complexe Zahl $f(\alpha)$ alle idealen Primfactoren von q enthält, und zwar denjenigen, welcher am wenigsten oft darin vorkommt, μ mal, so ist die Zahl durch q^μ theilbar.

Nach der Voraussetzung hat man folgende Congruenzen:

$$f(\alpha)(\Psi(\eta))^\mu \equiv 0, \quad f(\alpha)(\Psi(\eta_1))^\mu \equiv 0, \quad \dots \quad f(\alpha)(\Psi(\eta_{e-1}))^\mu \equiv 0, \quad \text{mod. } q^\mu:$$

also ist auch die Summe.

$$f(\alpha)[(\Psi(\eta))^\mu + (\Psi(\eta_1))^\mu + \dots + (\Psi(\eta_{e-1}))^\mu] \equiv 0, \quad \text{mod. } q^\mu.$$

Der Ausdruck in den Klammern ist aber eine reale ganze Zahl, welche nicht durch q theilbar ist, weil sie sogar, wie leicht zu zeigen, keinen einzigen idealen Primfactor von q enthält: also mufs $f(\alpha)$ durch q^μ theilbar sein; wie es behauptet wurde.

Wenn die complexe Zahl $f(\alpha)$ genau m ideale Primfactoren der realen, zum Exponenten f gehörenden Primzahl q enthält, sie mögen verschieden, oder zum Theil, oder alle dieselben sein: so enthält die Norm $Nf(\alpha)$ den Factor q^{mf}; aber keine höhere Potenz von q.

Die conjugirten complexen Zahlen $f(\alpha)$, $f(\alpha^\gamma)$, $f(\alpha^{\gamma^2})$, ... $f(\alpha^{\gamma^{\lambda-2}})$ enthalten alle gleich viele, und zwar jede genau m ideale Primfactoren von q.

Ändert man nämlich nur die Benennungen der Primfactoren in $f(\alpha^{\gamma^h})$, jeden um eine Stufe, so erhält man die in $f(\alpha^{\gamma^{h+1}})$ enthaltenen Primfactoren. Das Product aller dieser $\lambda-1$ conjugirten Factoren, welches gleich $Nf(\alpha)$ ist, mufs also genau $(\lambda-1)m$, das heifst $e.f.m$ ideale Primfactoren von q enthalten. Ferner ist leicht zu sehen, dafs in diesem Producte alle e verschiedenen Primfactoren von q gleichvielmal vorkommen, weil sonst $Nf(\alpha)$ gar nicht einmal eine ganze Zahl sein könnte. Es mufs also jeder dieser Primfactoren genau mf mal vorkommen, und folglich mufs $Nf(\alpha)$ genau durch q^{fm} theilbar sein.

Hieraus folgt von selbst der wichtige Satz:

Jede gegebene complexe Zahl enthält nur eine endliche bestimmte Anzahl (idealer) Primfactoren.

Umgekehrt aber ist, wenn die idealen Primfactoren einer complexen Zahl bekannt sind, zu untersuchen, ob dieselben nur einer einzigen bestimmten, oder auch verschiedenen complexen Zahlen angehören können. Sind $f(\alpha)$ und $\varphi(\alpha)$ zwei complexe Zahlen, welche genau dieselben (idealen) Primfactoren enthalten, so ist zunächst aus dem so eben bewiesenen Satze klar, dafs die Normen derselben gleich sein müssen, also dafs $Nf(\alpha)=N\varphi(\alpha)$ sein mufs. Es mag nun $f(\alpha)$ sowohl, als $\varphi(\alpha)$, m Primfactoren der zum Exponenten f gehörenden Primzahl q enthalten, m' Primfactoren der zum Exponenten f' gehörenden Primzahl q, m'' Primfactoren der zum Exponenten f'' gehörenden Primzahl q'', etc., so ist:

$$Nf(\alpha) = N\varphi(\alpha) = q^{mf}.q'^{m'f'}.q''^{m''f''}\dots.$$

Da $f(\alpha)$ dieselben (idealen) Primfactoren enthält wie $\varphi(\alpha)$, so mufs auch $f(\alpha)\varphi(\alpha^g)\varphi(\alpha^{g^2})\dots\varphi(\alpha^{g^{\lambda-2}})$ genau dieselben enthalten wie $N\varphi(\alpha)$: also mufs es alle (idealen) Primfactoren von q enthalten, jeden mf mal, alle Primfactoren von q', jeden $m'f'$ mal, alle Primfactoren von q'', jeden $m''f''$ mal u. s. w.; mithin mufs, nach einem oben bewiesenen Satze, $f(\alpha)\varphi(\alpha^g)\varphi(\alpha^{g^2})\dots\varphi(\alpha^{g^{\lambda-2}})$ durch q^{mf} und eben so durch $q'^{m'f'}$, durch $q''^{m''f''}$ u. s. w. theilbar sein, folglich auch durch das Product davon, das heifst durch $N\varphi(\alpha)$. Es ist demnach

$$\frac{f(\alpha)\varphi(\alpha^g)\varphi(\alpha)^{g^2})\dots\varphi(\alpha^{g^{\lambda-2}})}{N\varphi(\alpha)} = \frac{f(\alpha)}{\varphi(\alpha)} = E(\alpha);$$

wo $E(\alpha)$ eine ganze complexe Zahl bedeutet. Hieraus folgt weiter

$$f(\alpha) = \varphi(\alpha)E(\alpha) \quad \text{und} \quad Nf(\alpha) = N\varphi(\alpha).NE(\alpha),$$

und da $N\varphi(\alpha) = Nf(\alpha)$ ist, so folgt

$$NE(\alpha) = 1;$$

also ist $E(\alpha)$ eine complexe Einheit und wir haben folgenden Satz:

Zwei complexe Zahlen, welche genau dieselben (idealen) Primfactoren haben, unterscheiden sich nur durch eine complexe Einheit, welche als Factor hinzutreten kann.

Es möge jetzt ein Divisor $\varphi(\alpha)$ genau dieselben (idealen) Primfactoren haben, welche wir so eben für diese complexe Zahl annahmen; ein Dividendus $f(\alpha)$ aber möge nicht nur dieselben Factoren haben, sondern aufser diesen noch andere, oder auch die nemlichen mehrmals: dann läfst sich leicht zeigen, dafs $f(\alpha)$ durch $\varphi(\alpha)$ theilbar, d. h. dafs der Quotient eine complexe *ganze* Zahl ist. Bringt man nämlich den Quotienten wieder auf die Form

$$\frac{f(\alpha)}{\varphi(\alpha)} = \frac{f(\alpha)\varphi(\alpha^g)\varphi(\alpha^{g^2})\ldots.\varphi(\alpha^{g^{\lambda-2}})}{N\varphi(\alpha)},$$

so hat der Zähler $f(\alpha)\varphi(\alpha^g)\varphi(\alpha^{g^2})\ldots.\varphi(\alpha^{g^{\lambda-2}})$ alle (idealen) Primfactoren, welche der Nenner $N\varphi(\alpha)$ enthält; denn $f(\alpha)$ enthält der Voraussetzung nach alle Primfactoren von $\varphi(\alpha)$. Dieser Zähler enthält also alle Primfactoren von q, und zwar jeden m mal; weshalb er denn durch q^{mf} theilbar sein mufs. Ferner enthält er alle Primfactoren von q', und zwar jeden m' mal, weshalb er durch $q'^{m'f'}$ theilbar sein mufs, u. s. w. Der Zähler enthält also die ganze Zahl $q^{mf}q'^{m'f'}q''^{m''f''}\ldots.$ als Factor, welche gleich $N\varphi(\alpha)$ ist: folglich ist

$$\frac{f(\alpha)}{\varphi(\alpha)} = Q(\alpha),$$

und $Q(\alpha)$ ist eine complexe ganze Zahl. Hieraus folgt weiter $f(\alpha) = \varphi(\alpha)Q(\alpha)$; und da das Product die idealen Primfactoren der beiden Factoren zusammen enthält, so folgt, dafs $Q(\alpha)$ genau alle diejenigen Primfactoren enthalten mufs, welche den Überschufs der Primfactoren des Dividendus $f(\alpha)$ über die des Divisors $\varphi(\alpha)$ ausmachen. Wir erhalten also folgenden Satz:

Eine complexe Zahl ist durch eine andere theilbar, wenn alle (idealen) Primfactoren des Divisors auch im Dividendus enthalten sind, und der Quotient enthält genau den Überschufs der (idealen) Primfactoren des Dividendus über die des Divisors.

Aufser den hier behandelten Primfactoren derjenigen Primzahlen, welche zu Divisoren von $\lambda-1$ für den Modul λ gehören, müssen noch die Primfactoren von λ selbst, besonders erwähnt werden. Da

$$(x-\alpha)(x-\alpha^2)\ldots.(x-\alpha^{\lambda-1}) = x^{\lambda-1}+x^{\lambda-2}+\ldots.+x+1$$

ist, so erhält man, wenn man $x=1$ setzt:

$$(1-\alpha)(1-\alpha^2)(1-\alpha^3)\ldots.(1-\alpha^{\lambda-1}) = \lambda.$$

Es sind demnach $1-\alpha$, $1-\alpha^2$, etc. Factoren von λ, und zwar Primfactoren, weil, wenn $1-\alpha = f(\alpha)\varphi(\alpha)$ gesetzt wird, auch $Nf(\alpha)\cdot N\varphi(\alpha) = \lambda$ sein mufs; woraus folgt, dafs einer der beiden ganzzahligen Factoren $Nf(\alpha)$ oder $N\varphi(\alpha)$ gleich Eins, also dafs eine der beiden complexen Zahlen $f(\alpha)$ oder $\varphi(\alpha)$ eine complexe Einheit sein mufs. Diese Primfactoren von λ haben das Eigenthümliche, dafs sie, wenn man die Einheiten abrechnet, mit welchen sie multiplicirt vorkommen, einander gleich sind; denn es ist $1-\alpha^n = (1-\alpha)(1+\alpha+\alpha^2+\ldots+\alpha^{n-1})$, und $1+\alpha+\alpha^2+\ldots+\alpha^{n-1}$ ist eine complexe Einheit. Es ist also hier niemals die Frage, welche Primfactoren von λ in einer complexen Zahl enthalten sind, sondern nur, wie viele derselben es sind. Ist $f(\alpha)$ irgend eine complexe Zahl, so hat man offenbar

$$(f(\alpha))^\lambda = f(1), \text{ mod. } \lambda.$$

Wenn nun $f(\alpha)$ einen Primfactor von λ enthält, so ist $(f(\alpha))^\lambda$, weil es deren λ hat, durch λ theilbar; mithin mufs auch $f(1)$, d. h. die Summe aller Coëfficienten von $f(\alpha)$, durch λ theilbar sein. Wenn aber diese Bedingung erfüllt ist, so kann man durch mehrmalige Addition oder Subtraction der Gleichung $1+\alpha+\alpha^2+\ldots+\alpha^{\lambda-1} = 0$ diese Summe der Coëfficienten gleich Null machen; und alsdann ist die complexe Zahl offenbar durch $1-\alpha$ theilbar; sogar wenn α eine belibige veränderliche Gröfse bezeichnet. Will man daher wissen, wie viele Primfactoren von λ eine complexe Zahl enthalte, so hat man nur zu untersuchen, wie viele mal nach einander sie sich durch $1-\alpha$ dividiren lasse.

§. 8.

Nachdem in dem vohergehenden Paragraphen die einfachen Rechnungsregeln für die Primfactoren der complexen Zahlen gefunden worden, (welche stets dieselben sind, es mögen die Primfactoren für sich als complexe Zahlen darstellbar sein, oder nur ideale) wollen wir jetzt noch über die Wirklichkeit oder Idealität der durch ihre Primfactoren gegebenen complexen Zahlen einige Untersuchungen anstellen, bei welchen alles hauptsächlich nur auf die eine Hauptfrage ankommen wird: ***Welche ideale Primfactoren setzen sich zu wirklichen complexen Zahlen zusammen; und welche nicht?***

Das Product aller conjugirten idealen Primfactoren einer gegebenen realen Primzahl q ist stets eine wirkliche complexe Zahl, und zwar die Zahl q selbst, welche auch noch eine complexe Einheit zum Factor haben kann. Die Zahl q enthält nämlich wirklich nach unserer Definition alle conjugirten Primfactoren von q, deren Anzahl gleich e ist, wenn q zum Exponenten f gehört, mod. λ, und $ef = \lambda - 1$ ist.

Ist ferner $I(\alpha)$ irgend eine (ideale) complexe Zahl, welche beliebige gegebene Primfactoren enthält, so ist immer die Norm

$$NI(\alpha) = I(\alpha).I(\alpha^{\gamma}).I(\alpha^{\gamma^2})\dots.I(\alpha^{\gamma^{\lambda-1}})$$

eine ***wirkliche*** Zahl, und zwar: Wenn $I(\alpha)$ den zu $u=\eta_r$ gehörenden Primfactor von q, welcher zum Exponenten f gehört, μ mal enthält, ferner den zu $u'=\eta'_{r'}$ gehörenden Primfactor von q', welcher zum Exponenten f' gehört μ' mal, u. s. w., so ist

$$NI(\alpha) = q^{\mu f} q'^{\mu' f'}\dots.;$$

denn in der That enthält $q^{\mu f} q'^{\mu' f'}\dots.$ alle in $I(\alpha)I(\alpha^{\gamma})I(\alpha^{\gamma^2})\dots.I(\alpha^{\gamma^{\lambda-1}})$ vorkommenden Primfactoren, jeden genau eben so oft, als er in diesem Producte vorkommt; und aufser diesen keinen. Dies ist aber auch die einzige allgemeine Art der Zusammensetzung idealer complexer Zahlen zu wirklichen, welche sich von selbst darbietet.

Wenn irgend eine ideale complexe Zahl $I(\alpha)$ gegeben ist, so läfst sich die Aufgabe, andere ideale Zahlen zu finden, welche, mit $I(\alpha)$ multiplicirt, ***wirkliche*** complexe Zahlen geben, stets auf unendlich viele verschiedene Arten lösen. Diese Aufgabe kann nämlich auch so ausgedrückt werden: Wirkliche complexe Zahlen zu finden, welche gegebene (ideale) Primfactoren haben, und welche aufser diesen gegebenen noch irgend andere enthalten dürfen. Die Lösung beruht aber nur darauf, einer gewissen Anzahl Congruenzen vom ersten Grade genugzuthun, welche für die Coëfficienten einer wirklichen complexen Zahl Statt haben müssen, und welche sich niemals widersprechen, sondern stets eine unendliche Anzahl verschiedener Lösungen zulassen. Es giebt deshalb auch immer eine unendliche Menge idealer Zahlen, welche, mit einer gegebenen idealen Zahl multiplicirt, eine wirkliche complexe Zahl erzeugen. Wählt man nun unter allen diesen immer nur diejenigen, deren Normen möglichst klein sind, so findet man das sehr bemerkenswerthe Resultat: ***Dafs stets eine endliche bestimmte Zahl idealer Multiplicatoren hinreicht, um alle idealen complexen Zahlen zu wirklichen zu machen.*** Wir wollen jetzt den Beweis dieses ersten Hauptsatzes geben.

Die ideale complexe Zahl $I(\alpha)$ enthalte wieder, wie oben, den zu $u=\eta_r$ gehörenden Primfactor der zum Exponenten f gehörenden Primzahl q, und zwar μ mal; ferner den zu $u'=\eta'_{r'}$ gehörenden Primfactor der zum Exponenten f' gehörenden Primzahl q', μ' mal u. s. w. und es sei

$$F(\alpha) = a_1\alpha + a_2\alpha^2 + a_3\alpha^3 + \dots. + a_{\lambda-1}\alpha^{\lambda-1}$$

eine wirkliche complexe Zahl, welche alle Primfactoren von $I(\alpha)$, also auch den idealen Factor $I(\alpha)$ selbst enthalten soll. Es sind nun, damit $F(\alpha)$ den zu $u = \eta_r$ gehörenden Primfactor von q μmal enthalte, f Congruenzen für den Modul q^μ zu erfüllen. Dieselben seien

$$\Phi = 0, \quad \Phi_1 = 0, \quad \Phi_2 = 0, \quad \ldots\ldots \quad \Phi_{f-1} = 0, \quad \text{mod. } q^\mu.$$

Ferner sind, damit $F(\alpha)$ den zu $u' = \eta'_{r'}$ gehörenden Primfactor von q', μ'mal enthalte, weiter f' Congruenzen für den Modul $q'^{\mu'}$ zu erfüllen; sie seien

$$\Phi' = 0, \quad \Phi'_1 = 0, \quad \Phi'_2 = 0, \quad \ldots\ldots \quad \Phi'_{f-1} = 0, \quad \text{mod. } q'^{\mu'};$$

und so fort. Alle diese Congruenzen sind auch, wie gezeigt, in Beziehung auf die zu bestimmenden Coëfficienten $a_1, a_2, a_3, \ldots\ldots a_{\lambda-1}$ nur linear.

Wir geben nun allen den $\lambda-1$ Coëfficienten der complexen Zahl $F(\alpha)$ alle die Werthe $0, 1, 2, 3, \ldots\ldots k-1$, so dafs im Ganzen $k^{\lambda-1}$ Werthcombinationen Statt finden, welche eben so viele verschiedene complexe Zahlen geben. Für diese verschiedenen Werthcombinationen kann jede der Gröfsen $\Phi, \Phi_1, \Phi_2, \ldots\ldots \Phi_{f-1}$ nur q^μ verschiedene Reste lassen, für den Modul q^μ; eben so jede der f' Gröfsen $\Phi', \Phi'_1, \Phi'_2, \ldots\ldots \Phi'_{f'-1}$, nur $q'^{\mu'}$ verschiedene Reste, für den Modul $q'^{\mu'}$ etc. Die Anzahl aller verschiedenen Restcombinationen, welche überhaupt Statt haben können, ist also $q^{\mu f} q'^{\mu' f'} \ldots\ldots$ Nimmt man nun k so grofs an, dafs $k^{\lambda-1} > q^{\mu f} q'^{\mu' f'} \ldots\ldots$, also die Anzahl der Werthcombinationen der Coëfficienten gröfser ist als die Anzahl aller Restcombinationen der durch Φ bezeichneten Gröfsen, so müssen immer gewisse Restcombinationen sich wiederholen. Es mögen für die bestimmten Werthe $a_1 = m_1$, $a_2 = m_2$, $a_3 = m_3, \ldots\ldots a_{\lambda-1} = m_{\lambda-1}$ die Gröfsen $\Phi, \Phi_1, \ldots\ldots \Phi_{f-1}$, modulo q^μ, $\Phi', \Phi_1, \ldots\ldots \Phi'_{f'-1}$, modulo $q'^{\mu'}$, etc. alle einzeln dieselben Reste geben, wie für die Werthe $a_1 = n_1$, $a_2 = n_2$, $a_3 = n_3, \ldots\ldots a_{\lambda-1} = n_{\lambda-1}$: so folgt, dafs die Gröfsen $a_1, a_2, a_3, \ldots\ldots a_{\lambda-1}$, weil sie in allen diesen Ausdrücken nur linear vorkommen, alle congruent Null werden müssen, für $a_1 = m_1 - n_1$, $a_2 = m_2 - n_2$, $a_3 = m_3 - n_3, \ldots\ldots a_{\lambda-1} = m_{\lambda-1} - n_{\lambda-1}$. Alle Congruenzen also, welche nöthig und hinreichend sind, damit $F(\alpha)$ die verlangten idealen Primfactoren habe, lassen sich immer befriedigen, wenn man den Coëfficienten nur positive oder negative Werthe giebt, welche, abgesehen vom Vorzeichen, nicht gröfser sind als $k-1$; wo k nur durch die Bedingung $k^{\lambda-1} > q^{\mu f} q'^{\mu' f'} \ldots\ldots$ bestimmt wird.

Da wir nun eine gewisse Grenze der Gröfse gefunden haben, welche die Coëfficienten der complexen Zahl $F(\alpha)$ nicht zu überschreiten brauchen,

46 *

um den Bedingungen der Aufgabe zu genügen, so wollen wir jetzt sehen, wie grofs höchstens die Norm einer solchen complexen Zahl werden könne.

Multiplicirt man die beiden reciproken complexen Zahlen $F(\alpha)$ und $F(\alpha^{-1})$ mit einander, verwandelt sodann α in α^2, α^3, ... $\alpha^{\frac{1}{2}(\lambda-1)}$ und addirt, so erhält man sehr leicht folgende Gleichung:

$$F(\alpha)F(\alpha^{-1})+F(\alpha^2)F(\alpha^{-2})+F(\alpha^3)F(\alpha^{-3})+\ldots.+F(\alpha^{\frac{1}{2}(\lambda-1)})F(\alpha^{\frac{1}{2}(\lambda+1)})$$
$$=\tfrac{1}{2}\lambda(a_1^2+a_2^2+a_3^2+\ldots.+a_{\lambda-1}^2)-\tfrac{1}{2}(a_1+a_2+a_3+\ldots.+a_{\lambda-1})^2.$$

Sind nun die Coëfficienten a_1, a_2, ... $a_{\lambda-1}$ alle, absolut genommen, nicht gröfser als $k-1$, so folgt hieraus:

$$F(\alpha)F(\alpha^{-1})+F(\alpha^2)F(\alpha^{-2})+\ldots.+F(\alpha^{\frac{1}{2}(\lambda-1)})F(\alpha^{\frac{1}{2}(\lambda+1)})\leqq\tfrac{1}{2}\lambda(\lambda-1)(k-1)^2.$$

Aus dem möglich-gröfsten Werthe, welchen die Summe aller dieser stets positiven Gröfsen haben kann, läfst sich aber sehr leicht auf den möglich-gröfsten Werth des Products derselben, welches die Norm $NF(\alpha)$ ist, schliefsen. Ein solches Product positiver Gröfsen, deren Summe gegeben ist, erhält nämlich seinen möglich-gröfsten Werth, wenn die einzelnen Factoren desselben alle einander gleich angenommen werden: im gegenwärtigen Falle also, wenn jeder der Factoren gleich dem $\frac{1}{2}(\lambda-1)$ten Theile der möglich-gröfsten Summe aller gleichen genommen wird; also gleich $\lambda(k-1)^2$. Hierdurch wird das möglich-gröfste Product aller gleich $\lambda^{\frac{1}{2}(\lambda-1)}(k-1)^{\lambda-1}$. Man erhält also

$$NF(\alpha)\leqq\lambda^{\frac{1}{2}(\lambda-1)}(k-1)^{\lambda-1}.$$

Nehmen wir nun die Zahl k so an, dafs, wie es oben verlangt wurde, $k^{\lambda-1}>q^{\mu f}.q'^{\mu' f'}\ldots..$, aber $(k-1)^{\lambda-1}<q^{\mu f}.q'^{\mu' f'}\ldots.$ ist, so ergiebt sich

$$NF(\alpha)<\lambda^{\frac{1}{2}(\lambda-1)}.q^{\mu f}.q'^{\mu' f'}\ldots.$$

Die wirkliche complexe Zahl $F(\alpha)$ enthält nach der Voraussetzung den idealen Factor $I(\alpha)$. Setzt man daher $F(\alpha)=M(\alpha)\cdot I(\alpha)$, so ist $NF(\alpha)=NM(\alpha)\cdot NI(\alpha)$, und da $NI(\alpha)=q^{\mu f}.q'^{\mu' f'}\ldots.$, so ist

$$NM(\alpha)<\lambda^{\frac{1}{2}(\lambda-1)}.$$

Die Multiplicatoren also, welche beliebige ideale complexe Zahlen zu wirklichen machen, lassen sich immer so annehmen, dafs die Normen derselben kleiner sind als die bestimmte Zahl $\lambda^{\frac{1}{2}(\lambda-1)}$; und da die Anzahl solcher idealer Zahlen nur endlich und begrenzt sein kann, so erhalten wir den Beweis folgenden Satzes:

Es giebt immer eine endliche bestimmte Anzahl idealer complexer Multiplicatoren, welche nöthig und hinreichend sind, um alle idealen complexen Zahlen zu wirklichen zu machen.

§. 9.

Auf den im vorigen Paragraph bewiesenen Hauptsatz gründen wir nun eine Classification der idealen complexen Zahlen, indem wir festsetzen:

Alle diejenigen (idealen) complexen Zahlen, welche durch Multiplication mit einer und derselben (idealen) Zahl zu wirklichen complexen Zahlen werden, sollen äquivalent heifsen und zusammen eine Classe ausmachen.

Eine *wirkliche* complexe Zahl, mit einer *idealen* multiplicirt, kann immer nur ein *ideales,* aber niemals ein wirkliches Product geben. Dies ist eine unmittelbare Folge des in (§. 7.) bewiesenen Satzes, welchem zufolge eine wirkliche complexe Zahl durch eine andere wirkliche theilbar ist, wenn alle idealen Primfactoren des Divisors auch im Dividendus enthalten sind und der Quotient alsdann wieder eine wirkliche complexe Zahl ist. Wenn nämlich eine wirkliche Zahl, mit einer idealen multiplicirt, eine wirkliche Zahl zum Product gäbe, so müfste auch eine wirkliche Zahl, durch eine andere wirkliche dividirt, welche keine andern idealen Primfactoren hat als diese, eine ideale Zahl zum Quotienten geben; was dem angeführten Satze widerspricht.

Es seien nun $f(\alpha)$ und $\varphi(\alpha)$ zwei *äquivalente* ideale Zahlen; und zwar sollen beide, mit derselben idealen Zahl $M(\alpha)$ multiplicirt, zu wirklichen werden. Es werde ferner auch $f(\alpha)$, mit $\psi(\alpha)$ multiplicirt, zu einer wirklichen Zahl: so behaupte ich, dafs auch $\varphi(\alpha)$, mit $\psi(\alpha)$ multiplicirt, eine wirkliche Zahl geben mufs. Da nämlich $I(\alpha)\cdot M(\alpha)$ wirklich ist, so folgt, dafs auch $f(\alpha^2)f(\alpha^3)\ldots f(\alpha^{\lambda-1})\cdot M(\alpha^2)\cdot M(\alpha^3)\ldots M(\alpha^{\lambda-1})$ wirklich sein mufs. Wird dies mit der wirklichen Zahl $\varphi(\alpha)M(\alpha)$ multiplicirt, so folgt ferner, dafs $\varphi(\alpha)f(\alpha^2)f(\alpha^3)\ldots f(\alpha^{\lambda-1})\cdot NM(\alpha)$ eine wirkliche Zahl sein mufs; und da der Factor $NM(\alpha)$ wirklich ist, so mufs auch der andere Factor $\varphi(\alpha)f(\alpha^2)f(\alpha^3)\ldots f(\alpha^{\lambda-1})$ wirklich sein. Multiplicirt man endlich noch mit der wirklichen Zahl $f(\alpha)\psi(\alpha)$ und erwägt, dafs $Nf(\alpha)$ wirklich ist, so folgt, dafs $\varphi(\alpha)\psi(\alpha)$ ebenfalls wirklich sein mufs; wie behauptet; also:

Wenn zwei ideale complexe Zahlen äquivalent sind, so macht jeder Multiplicator, durch welchen die eine zu einer wirklichen Zahl wird, auch die andere zu einer wirklichen Zahl.

Aus diesem Satze folgt unmittelbar auch der folgende:

Wenn zwei ideale Zahlen einer und derselben dritten Zahl äquivalent sind, so sind sie auch unter einander äquivalent.

Es müssen nämlich nach dem vorigen Satze diese Zahlen alle drei einen und denselben Multiplicator haben, welcher sie zu wirklichen complexen Zahlen macht.

Durch diese Sätze erlangt der Begriff der *Äquivalenz* idealer complexer Zahlen erst seine wahre Vollkommenheit; indem sich zeigt, dafs die Äquivalenz eine von der zufälligen Wahl der Multiplicatoren unabhängige Beziehung der idealen Zahlen zu einander ist, und dafs auch die Classification nach der Äquivalenz und die endliche Anzahl der verschiedenen Classen nur eine einzige bestimmte ist, für jeden Werth von λ.

In die Classification der idealen complexen Zahlen begreifen wir auch die wirklichen mit ein. Der Ausdruck *ideale complexe Zahl* hat nämlich zwei verschiedene Bedeutungen: eine weitere und eine engere; in der weiteren Bedeutung des Idealen ist das Wirkliche nur ein besonderer Fall desselben; in der engeren ist das Wirkliche der Gegensatz des Idealen: es verhält sich hier das Ideale und Wirkliche, wie das Imaginäre und Reale. Die wirklichen complexen Zahlen müssen unter einander alle als äquivalent betrachtet werden; sie machen also eine Classe für sich aus, welche wir als die erste Classe ansehen und *Hauptclasse* nennen. Für die Anordnung der übrigen Classen lassen sich nicht eher bestimmte Regeln geben, als bis von der Zusammensetzung der verschiedenen Classen äquivalenter Zahlen gehandelt worden ist; über welche wir Folgendes erwähnen:

Äquivalente Zahlen, mit äquivalenten multiplicirt, geben stets äquivalente Producte.

Es seien nämlich $f(\alpha)$ und $f_1(\alpha)$ zwei äquivalente Zahlen; ein Multiplicator, welcher beide zu wirklichen macht, sei $m(\alpha)$; es seien ferner $\varphi(\alpha)$ und $\varphi_1(\alpha)$ zwei andere äquivalente Zahlen, und $M(\alpha)$ der Multiplicator, welcher beide zu wirklichen macht: so ist offenbar $m(\alpha)M(\alpha)$ ein Multiplicator, der sowohl das Product $f(\alpha)\cdot\varphi(\alpha)$, als auch das Product $f_1(\alpha)\cdot\varphi_1(\alpha)$ zu einer wirklichen complexen Zahl macht; mithin sind $f(\alpha)\cdot\varphi(\alpha)$ und $f_1(\alpha)\cdot\varphi_1(\alpha)$ äquivalent. Dieser Satz läfst sich auch folgendermaafsen ausdrücken:

Wenn man irgend zwei ideale Zahlen mit einander multiplicirt, so ist die Classe, welcher das Product angehört, durch die Classen, welchen die Factoren angehören, vollständig bestimmt.

Multiplicirt man irgend eine ideale Zahl mit einer Zahl aus jeder Classe, so erhält man so viele Producte, als Classen vorhanden sind, und es kann keines derselben irgend einem andern äquivalent sein; oder jedes dieser Producte gehört einer andern Classe an. Unter diesen mufs es also auch immer ein Product geben, und zwar ***nur*** eins, welches wirklich ist oder der Hauptclasse angehört; also:

Zu jeder Classe idealer Zahlen gehört eine bestimmte Zahl, welche, mit ihr zusammengesetzt, die Hauptclasse giebt.

Es giebt hier auch noch solche Classen, welche, mit sich selbst zusammengesetzt, die Hauptclasse geben (classes ancipites); namentlich ist die Hauptclasse selbst stets eine solche. Aufser dieser existiren aber für besondere Werthe von λ auch noch andere mit dieser Eigenschaft.

Erhebt man irgend eine ideale Zahl $f(\alpha)$ zu Potenzen, so ist klar, dafs die Reihe idealer Zahlen

$$f(\alpha),\ f(\alpha)^2,\ f(\alpha)^3,\ f(\alpha)^4,\ \text{etc.}$$

äquivalente Zahlen enthalten mufs; denn die Anzahl nicht äquivalenter Zahlen ist eine endliche. Es seien nun $f(\alpha)^r$ und $f(\alpha)^s$ äquivalent, und $m(\alpha)$ sei ein Multiplicator, welcher beide zu wirklichen Zahlen macht, so ist $f(\alpha)^r \cdot m(\alpha)$ wirklich, und auch $f(\alpha)^s \cdot m(\alpha)$ ist wirklich. Ist nun $s > r$, so kann man demselben die Form $f(\alpha)^{s-r} \cdot f(\alpha)^r \cdot m(\alpha)$ geben, und da $f(\alpha)^r m(\alpha)$ wirklich ist, so mufs auch der andere Factor $f(\alpha)^{s-r}$ wirklich sein. Wir erhalten hieraus folgenden wichtigen Satz:

Jede ideale complexe Zahl hat die Eigenschaft, dafs gewisse ganze Potenzen derselben wirkliche complexe Zahlen sind. Alle idealen complexen Zahlen lassen sich daher als Wurzeln aus wirklichen complexen Zahlen darstellen

Es sei nun die $h$te Potenz der idealen Zahl $f(\alpha)$ die niedrigste, für welche $f(\alpha)^h$ zu einer wirklichen complexen Zahl wird, so müssen alle die Zahlen

$$1,\ f(\alpha),\ f(\alpha)^2,\ f(\alpha)^3,\ \ldots\ f(\alpha)^{h-1}$$

verschiedenen Classen angehören; denn wäre $f(\alpha)^r$ äquivalent $f(\alpha)^s$, wo $s > r$ ist, und s und r beide kleiner sind als h, so würde wie oben folgen, dafs $f(\alpha)^{s-r}$ eine wirkliche Zahl sei; gegen die Voraussetzung, da $s - r < h$ ist. Diese h complexen Zahlen können nun entweder alle vorhandenen Classen grade erschöpfen, in welchem Fall die Anzahl aller Classen gleich h ist, oder es können noch andere, nicht äquivalente ideale Zahlen vorhanden sein. Es sei $\varphi(\alpha)$ eine solche ideale Zahl, welche mit keiner der obigen äquivalent ist, so behaupte ich, dafs auch die idealen Zahlen

$$\varphi(\alpha),\ \varphi(\alpha)f(\alpha),\ \varphi(\alpha)f(\alpha)^2,\ \ldots\ \varphi(\alpha)f(\alpha)^{h-1}$$

alle, sowohl unter sich, als auch mit den obigen, nicht äquivalent sind. Wäre nämlich $\varphi(\alpha)f(\alpha)^r$ äquivalent $\varphi(\alpha)f(\alpha)^s$, wo r und s beide kleiner als h sind, so würde auch $f(\alpha)^r$ äquivalent $f(\alpha)^s$ sein; was unmöglich ist; und wäre $\varphi(\alpha)f(\alpha)^r$ äquivalent $f(\alpha)^s$, so würde durch Multiplication mit $f(\alpha)^{h-r}$ folgen, dafs $\varphi(\alpha)$ äquivalent $f(\alpha)^{s+h-r}$ oder äquivalent $f(\alpha)^{s-r}$ sei; welches ebenfalls der über $\varphi(\alpha)$ gemachten Voraussetzung zuwider ist. Es kann nun geschehen, dafs diese idealen Zahlen, zusammen mit den vorigen, alle nicht äquivalenten Zahlen erschöpfen; in welchem Falle die Anzahl aller Classen gleich $2h$ sein würde: es kann aber auch sein, dafs es aufser diesen noch andere nicht äquivalente Zahlen giebt. In diesem letztern Falle beweiset man eben so, dafs eine mit den obigen nicht äquivalente Zahl $\psi(\alpha)$ stets die ganze Gruppe

$$\psi(\alpha),\ \psi(\alpha)f(\alpha),\ \psi(\alpha)f(\alpha)^2,\ \ldots\ \psi(\alpha)f(\alpha)^{h-1}$$

nach sich zieht; welche Zahlen sowohl unter sich, als mit den obigen, nicht äquivalent sind. Sämmtliche nicht äquivalente ideale Zahlen ordnen sich so immer in Gruppen von je h Zahlen; woraus folgt, dafs die vollständige Anzahl derselben immer ein Vielfaches von h sein mufs. Wir sprechen dies Resultat folgendermaafsen aus:

Die vollständige Anzahl aller Classen ist immer ein Vielfaches des niedrigsten Exponenten derjenigen Potenz, zu welcher eine ideale Zahl erhoben werden mufs, um zu einer wirklichen zu werden.

Wenn es nun wirklich eine ***ideale*** Zahl $f(\alpha)$ giebt, von der Art, dafs $f(\alpha)^h$, aber keine niedrigere Potenz von $f(\alpha)$, eine ***wirkliche*** complexe Zahl ist, und dafs $1, f(\alpha), f(\alpha)^2, f(\alpha)^3, \ldots f(\alpha)^{h-1}$ alle nicht äquivalenten idealen Zahlen erschöpfen, so erhält man eine passende Anordnung aller Classen, wenn man die wirklichen zur ersten Classe, die Classe, welche $f(\alpha)$ enthält, zur zweiten, die $f(\alpha)^2$ enthaltende zur dritten nimmt u. s. w. Es kann alsdann die Classe, welcher das Product zweier oder mehrerer Factoren angehört, aus den Classen, welchen die Factoren angehören, unmittelbar bestimmt werden. weil alsdann offenbar die Summe der um Eins verminderten Classenzahlen aller Factoren der um Eins verminderten Classenzahl des Products, für den Modul h, congruent ist. Wenn aber eine solche ideale Zahl $f(\alpha)$ nicht existirt, deren Potenzen alle Classen nicht äquivalenter Zahlen erschöpfen, so kann man nach diesem Principe die Classen nur in Gruppen theilen und die den einzelnen Gruppen angehörenden Classen ordnen; für die Ordnung der Gruppen unter einander ist aber dann ein anderes Princip nöthig.

§. 10.

Wir haben schon oben bemerkt, dafs die idealen Primfactoren stets Primfactoren ganzer realer Primzahlen sind, und dafs die Natur derselben vorzüglich von den Exponenten (Divisoren von $\lambda-1$) abhangt, zu welchen diese realen Primzahlen für den Modul λ gehören. Unter allen diesen sind die idealen Primfactoren der zum Exponenten ***Eins*** gehörenden realen Primzahlen, d. h. der Primzahlen von der Form $2m\lambda+1$, als die einfachsten zu betrachten, welche vor den übrigen sich auszeichnen und darum besondere Beachtung verdienen. Von den nur aus solchen idealen Primfactoren zusammengesetzten idealen Zahlen wollen wir nun zeigen, dafs sie auch allein alle oben angegebenen Classen idealer Zahlen erschöpfen, und dafs die übrigen idealen Zahlen, welche noch andere, zu höhern Exponenten gehörige Primfactoren enthalten, für sich keine neuen Classen geben, sondern nur jenen sich einordnen *). Wir beweisen zu diesem Zwecke zunächst folgenden Satz:

Jeder ideale Primfactor, der zu einem Exponenten gehört, welcher gröfser als Eins ist, ist einer idealen Zahl äquivalent, deren ideale Primfactoren alle zu niedrigeren Exponenten gehören.

Um diesen Satz zu beweisen, gehen wir von der Gleichung aus, deren Wurzeln alle die in einer und derselben Periode von f Gliedern enthaltenen Wurzeln der Gleichung $\alpha^\lambda=1$ sind, und von welcher schon oben (§. 4.) Gebrauch gemacht wurde, nemlich von der Gleichung

$$(1.)\quad \alpha^f+P_1(\eta)\alpha^{f-1}+P_2(\eta)\alpha^{f-2}+\ldots+P_f(\eta)=0,$$

in welcher die Coëfficienten $P_1(\eta)$, $P_2(\eta)$, etc. aus den Perioden von je f Gliedern gebildete ganze complexe Zahlen sind. Es seien ferner $P_1(u)$, $P_2(u)$, $P_3(u)$, etc. diejenigen ganzen Zahlen, welche man erhält, wenn man in jenen statt der Perioden η, η_1, η_2, etc. die entsprechenden Congruenzwurzeln u, u_1, u_2, etc. für den Modul q setzt, der zum Exponenten f gehört. Setzt man jetzt

$$(2.)\quad F(\alpha)=\alpha^f+P_1(u)\alpha^{f-1}+P_2(u)\alpha^{f-2}+\ldots+P_{f-1}(u)\alpha+P_f(u)+q,$$

so erhält man, wenn die Gleichung (1.) subtrahirt wird, und wenn man erwägt, dafs $P_f(\eta)=\pm1$, also $P_f(u)=P_f(\eta)$ ist:

$$(3.)\quad F(\alpha)=(P_1(u)-P_1(\eta))\alpha^{f-1}+(P_2(u)-P_2(\eta))\alpha^{f-2}+\ldots$$
$$\ldots+(P_{f-1}(u)-P_{f-1}(\eta))\alpha+q.$$

*) Statt: Idealer Primfactor einer realen, zum Exponenten f gehörigen Primzahl, werden wir hier und im Folgenden kürzer: *Zum Exponenten f gehöriger idealer Primfactor* schreiben.

Diese complexe Zahl $F(\alpha)$ hat nun, ***Erstens***, keinen idealen Primfactor, welcher zu einem höhern Exponenten gehörte, als zum Exponenten f. Wenn nämlich f' ein gröfserer Divisor von $\lambda-1$ ist, als f, so mufs man nach (§. 4.), um zu sehen ob eine wirkliche complexe Zahl einen zum Exponenten f' gehörenden idealen Primfactor haben könne, oder, was Dasselbe ist, ob die Norm derselben durch eine zum Exponenten f' gehörende reale Primzahl theilbar sein könne, aus der zu untersuchenden Zahl mit Hülfe der Gleichung, deren Wurzeln die sämmtlichen, eine Periode von f Gliedern bildenden Wurzeln sind, alle Potenzen von α, welche höher sind als $\alpha^{f'-1}$, eliminiren; worauf dann alle Glieder einzeln diesen idealen Factor haben müssen. Im gegenwärtigen Falle enthält $F(\alpha)$ an sich schon keine höhern Potenzen von α: es hat also bereits die verlangte Form, welche im Allgemeinen durch Elimination der höhern Potenzen von α hervorzubringen ist. Es müfsten demnach in dem Ausdrucke von $F(\alpha)$, welchen die Gleichung (2.) giebt, wenn $F(\alpha)$ einen zum Exponenten f' gehörenden idealen Primfactor haben sollte, alle Coëfficienten der einzelnen Glieder, also auch der erste Coëfficient ***Eins***, denselben enthalten; was nicht möglich ist.

Zweitens hat $F(\alpha)$ auch keinen idealen Primfactor einer andern, zum Exponenten f gehörenden Primzahl, als der Primzahl q; denn es müfste nach (§. 4.) jeder solcher Primfactor auch ein gemeinschaftlicher Factor aller Glieder der complexen Zahl $F(\alpha)$ in der Form (3.) sein, also namentlich auch ein idealer Primfactor des letzten Gliedes q.

Drittens enthält $F(\alpha)$ auch nur einen einzigen idealen Primfactor von q, nämlich den zu $u=\eta$ gehörenden Factor, und diesen nur einmal. Dafs es den genannten idealen Primfactor wirklich enthält, ist klar, weil $F(\alpha)\equiv 0$, mod. q, ist, für $u=\eta$; und wenn es den Factor mehrmals enthielte, so müfsten in der Form (3.) alle einzelnen Glieder, also auch das letzte Glied q, denselben mehrmals enthalten; welches nicht der Fall ist. Bildet man ferner folgende complexe Zahlen:

$$\begin{array}{ll}
F(\alpha) & = \alpha^f + P_1(u)\alpha^{f-1} + P_2(u)\alpha^{f-2} + \dots + P_f(u) + q, \\
F_1(\alpha) & = \alpha^f + P_1(u_1)\alpha^{f-1} + P_2(u_1)\alpha^{f-2} + \dots + P_f(u_1) + q, \\
F_2(u) & = \alpha^f + P_1(u_2)\alpha^{f-1} + P_2(u_2)\alpha^{f-2} + \dots + P_f(u_2) + q, \\
\dots & \dots \\
F_{e-1}(\alpha) & = \alpha^f + P_1(u_{e-1})\alpha^{f-1} + P_2(u_{e-1})\alpha^{f-2} + \dots + P_f(u_{e-1}) + q,
\end{array}$$

so enthält $F(\alpha)$ den zu $u_1=\eta$ gehörenden idealen Primfactor von q, $F_1(\alpha)$ den zu $u_1=\eta$ gehörenden, $F_2(\alpha)$ den zu $u_2=\eta$ gehörenden Factor u. s. w.;

das Product aller dieser complexen Zahlen ist also, weil es alle idealen Primfactoren von q enthält, durch q theilbar. Enthielte nun aber $F(\alpha)$ aufser dem zu $u = \eta$ gehörenden Primfactor noch einen andern, welcher zu $u_r = \eta$ gehören möge, so müfste das Product dieser complexen Zahlen, auch mit Ausschlufs der Zahl $F_r(\alpha)$, durch q theilbar sein, weil es auch schon ohne $F_r(\alpha)$ alle idealen Primfactoren von q enthielte. Dieses Product würde folgende Form haben:

$$\alpha^{(e-1)f} + C_1\alpha^{(e-1)f-1} + C_2\alpha^{(e-1)f-2} + \ldots + C_{(e-1)f};$$

wo C_1, C_2, etc. ganze Zahlen sind. Es müfsten also wieder alle einzelnen Glieder, folglich auch das erste, dessen Coëfficient Eins ist, durch q theilbar sein; was nicht der Fall ist.

Wir schliefsen hieraus, dafs es immer wirkliche complexe Zahlen giebt, die nur einen einzigen, zum Exponenten f gehörenden idealen Primfactor enthalten, deren übrige Primfactoren aber alle zu niedrigeren Exponenten gehören. Dafs der Fall $f = 1$ eine Ausnahme macht, ist klar. Bezeichnet man nun den einen, zum Exponenten f gehörenden idealen Primfactor, welcher in $F(\alpha)$ enthalten ist, durch $f(\alpha)$, und das Product aller übrigen, welche nur zu niedrigeren Exponenten gehören, durch $\varphi(\alpha)$, so dafs $F(\alpha) = f(\alpha)\cdot\varphi(\alpha)$ eine wirkliche complexe Zahl ist: so folgt, dafs $f(\alpha)$ äquivalent ist mit $\varphi(\alpha^2)\varphi(\alpha^3)\ldots\varphi(\alpha^{\lambda-1})$; denn beide, mit $\varphi(\alpha)$ multiplicirt, geben *wirkliche* Zahlen. Hiermit ist auch der obige Satz vollständig bewiesen.

Aus dem so eben bewiesenen Satze folgt von selbst der allgemeine Satz:

Jede beliebige ideale Zahl ist einer andern äquivalent, deren ideale Primfactoren alle nur zum Exponenten Eins gehören.

Wenn nämlich in $\varphi(\alpha)$, und folglich auch in $\varphi(\alpha^2)\varphi(\alpha^3)\ldots\varphi(\alpha^{\lambda-1})$, noch Primfactoren vorhanden sind, die nicht zum Exponenten Eins gehören, so kann man sie immer durch äquivalente ideale Zahlen ersetzen, deren Primfactoren zu immer niedrigeren Exponenten gehören; und so fortfahrend gelangt man nothwendig dahin, dafs alle Primfactoren der äquivalenten Zahl nur zum Exponenten ***Eins*** gehören. Wenn ferner alle beliebigen Primfactoren solchen idealen Zahlen äquivalent sind, so folgt das Nemliche für alle beliebigen zusammengesetzten idealen Zahlen von selbst.

Anmerkung. Ich kann nicht umhin, hier, wo ich die allgemeine Theorie der Zerlegung der complexen Zahlen, wenn auch unvollendet, verlasse, um in den folgenden Paragraphen noch einige Anwendungen zu geben, auf die grofse Analogie aufmerksam zu machen, welche diese Theorie mit der ***Chemie***

47 *

hat. Der chemischen Verbindung entspricht für die complexen Zahlen die Multiplication; den Elementen, oder eigentlich den Atomgewichten derselben, entsprechen die Primfactoren; und die chemischen Formeln für die Zerlegung der Körper sind genau dieselben, wie die Formeln für die Zerlegung der Zahlen. Auch selbst die idealen Zahlen unserer Theorie finden sich in der Chemie, vielleicht nur allzuoft, als hypothetische Radicale, welche bisher noch nicht dargestellt worden sind, die aber, so wie die idealen Zahlen, in den Zusammensetzungen ihre Wirklichkeit haben. Das Fluor, für sich bisher nicht darstellbar und noch den Elementen zugezählt, kann als Analogon eines idealen Primfactors gelten. Die Idealität in der Chemie verhält sich aber darin wesentlich anders, als die der complexen Zahlen, dafs chemische ideale Stoffe, mit wirklichen verbunden, auch wirkliche Stoffe produciren; was bei den idealen Zahlen nicht der Fall ist. In der Chemie hat man ferner zur Prüfung der in einem unbekannten aufgelöseten Körper enthaltenen Stoffe die Reagentien, welche Niederschläge geben, aus denen die Anwesenheit der verschiedenen Stoffe sich erkennen läfst. Ganz Dasselbe findet für die complexen Zahlen Statt; denn es sind die oben mit Ψ bezeichneten complexen Zahlen ebenso die Reagentien für die idealen Primfactoren, und die reale Primzahl q, welche nach der Multiplication mit einer solchen als Factor aus dem Producte heraustritt, ist genau Dasselbe, wie der unlösliche Niederschlag, der nach Anwendung des Reagens zu Boden fällt. Auch der Begriff der Äquivalenz ist in der Chemie fast derselbe, wie in der Theorie der complexen Zahlen. So wie nämlich dort zwei Gewichtsmengen verschiedener Stoffe äquivalent heifsen, wenn sie sich gegenseitig vertreten können, entweder zum Zwecke des Neutralisirens, oder um Isomorphie hervorzubringen: so sind zwei ideale Zahlen äquivalent, wenn sie für den Zweck, eine andere ideale Zahl zu einer wirklichen zu machen, sich gegenseitig vertreten können. — Diese hier angedeuteten Analogieen sind nicht etwa als blofse Spiele des Witzes zu betrachten, sondern haben ihren guten Grund darin, dafs die Chemie, so wie der hier behandelte Theil der Zahlentheorie, beide denselben Grundbegriff, nämlich den der *Zusammensetzung*, wenn gleich innerhalb verschiedener Sphären des Seins, zu ihrem Principe haben; woraus folgt, dafs auch die diesem verwandten, mit ihm nothwendig gegebenen Begriffe sich in beiden auf ähnliche Weise finden müssen. Die Chemie der natürlichen Stoffe und die hier behandelte Chemie der complexen Zahlen sind beide als Verwirklichungen des Begriffs der Zusammensetzung und der davon abhängigen Begriffs-Sphäre anzusehen: jene als eine physische, mit den Zufälligkeiten der äufsern Existenz verbundene und deshalb reichere, diese als eine mathematische, in ihrer innern Nothwendigkeit vollkommen reine, aber dafür auch ärmere, als jene.

§. 11.

Die hauptsächlichsten Schwierigkeiten der Lehre von der Kreistheilung liegen bekanntlich in gewissen complexen, in derselben auftretenden Zahlen. Es bleibt von der rein theoretischen Seite eine genauere Einsicht in die Natur dieser complexen Zahlen, von der practischen Seite eine leichtere Methode für die Bildung derselben noch zu wünschen. Diese beiden Mängel dürften nun mittels der hier gegebenen Theorie der complexen Zahlen vollständig gehoben werden können.

Ist p eine Primzahl von der Form $p = \mu\lambda + 1$, g eine primitive Wurzel von p, α eine imaginäre Wurzel der Gleichung $\alpha^\lambda = 1$, und x eine imaginäre Wurzel der Gleichung $x^p = 1$, so beruht die hauptsächlichste Aufgabe der Kreistheilung in der Bildung der λten Potenz des Ausdrucks

$$(\alpha, x) = x + \alpha x^g + \alpha^2 x^{g^2} + \ldots + \alpha^{p-2} x^{g^{p-2}},$$

welche Potenz, von x unabhängig, eine aus den Wurzeln α, α^2, α^3, etc. gebildete complexe Zahl ist. Diese complexe Zahl setzt ***Jacobi*** (S. Monatsberichte der Berliner Akademie vom Jahre 1837, wieder abgedruckt in diesem Journal Bd. XXX. S. 166) als Product aus complexen, ebenfalls von x unabhängigen Zahlen zusammen, welche durch die Gleichung

$$\psi_k(\alpha) = \frac{(\alpha, x)(\alpha^k, x)}{(\alpha^{k+1}, x)}$$

bestimmt sind. Es ist nämlich

$$(\alpha, x)^\lambda = p\psi_1(\alpha)\psi_2(\alpha)\psi_3(\alpha) \ldots \psi_{\lambda-2}(\alpha).$$

Nun weiset auch ***Jacobi*** a. a. O. nach, dafs, wenn r eine primitive Wurzel der Gleichung $r^{p-1} = 1$, und g dieselbe primitive Wurzel der Congruenz $g^{p-1} \equiv 1$ ist, welche auch schon in (α, x) vorkommt, und wenn in dem von x unabhängigen Ausdrucke

$$\psi(r) = \frac{(r^{-m}, x)(r^{-n}, x)}{(r^{-m-n}, x)}$$

anstatt der primitiven Wurzel r, die primitive Congruenzwurzel g gesetzt wird, dafs dann

$$\psi(g) \equiv 0, \text{ mod. } p,$$

ist, sobald $m + n > p - 1$, aber m und n für sich kleiner als $p - 1$ und positiv sind. Setzt man nun

$$r^{-m} \equiv \alpha, \quad g^{p-1-m} \equiv u, \text{ mod. } p,$$

$$n \equiv km, \text{ mod. } p-1, \qquad m \equiv h\mu, \text{ mod. } p-1,$$

so ergiebt sich

$$\psi(r) = \psi_k(\alpha^h), \quad \text{also auch} \quad \psi(\alpha) \equiv \psi_k(u^h).$$

Es ist demnach $\psi_k(u^h) \equiv 0$, mod. p, für $m+n > p-1$; welche Bedingung sich nach den festgesetzten Werthen von m und n darauf vereinfacht, dafs die positiven Reste von hk und von h, für den Modul λ, zusammen gröfser sein müssen als λ.

Dieses *Jacobi*sche Resultat giebt unmittelbar die idealen Primfactoren der complexen Zahl $\psi_k(\alpha)$. In unsere Ausdrucksweise übersetzt, heifst es: ***Es enthält*** $\psi_k(\alpha)$ ***den zu*** $\alpha = u^h$ ***gehörenden idealen Primfactor von*** p***, wenn*** h ***und der positive Rest von*** hk***, mod.*** λ***, zusammen gröfser sind als*** λ. Die Anzahl der Werthe von h, welche dieser Bedingung genügen, ist offenbar gleich $\frac{1}{2}(\lambda-1)$; denn von je zwei Werthen von h, deren Summe gleich λ ist, genügt immer einer, und zwar ***nur*** einer von beiden. Es sind also dadurch $\frac{1}{2}(\lambda-1)$ Primfactoren der complexen Zahl $\psi_k(\alpha)$ gegeben, und aufser diesen können keine andern vorhanden sein; denn $\psi_k(\alpha^{-1})$ mufs genau eben so viele Primfactoren enthalten als $\psi_k(\alpha)$, und da $\psi_k(\alpha)\psi_k(\alpha^{-1}) = p$, p aber nur $\lambda-1$ (ideale) Primfactoren enthält, so mufs $\psi_k(\alpha)$ genau $\frac{1}{2}(\lambda-1)$ Primfactoren enthalten, welche die oben gefundenen sind. Bezeichnet $f(\alpha)$ einen idealen Primfactor von p, und zwar den zu $\alpha = u$ gehörenden; bedeutet ferner m^h diejenige positive Zahl, welche kleiner als λ ist und der Congruenz $hm_h \equiv 1$, mod. λ, genügt, so ist

$$\psi_k(\alpha) = \pm \alpha^r \Pi f(\alpha^{m_h});$$

wo das Productzeichen Π sich auf alle diejenigen $\frac{1}{2}(\lambda-1)$ Werthe von h bezieht, für welche h und der positive Rest von hk, mod. λ, zusammen gröfser sind als λ. Nach dem vorletzten Satze in (§. 7.) mufs diesem Producte, welches nur durch seinen idealen Primfactor bestimmt ist, irgend eine complexe Einheit $E(\alpha)$ als Factor beigegeben werden; diese mufs aber vermöge der Gleichung $\psi_k(\alpha)\psi_k(\alpha^{-1}) = p$ so beschaffen sein, dafs $E(\alpha)E(\alpha^{-1}) = 1$ ist; und dieser Gleichung genügt keine andere Einheit, und überhaupt keine andere complexe Zahl als $\pm\alpha^r$, weshalb wir diese Einheit als Factor des Products hinzugefügt haben.

Mit den gefundenen Primfactoren von $\psi_k(\alpha)$ sind nun diejenigen der Potenz $(\alpha, x)^\lambda$ zugleich mit gegeben. Der bestimmte, zu $\alpha = u^h$ gehörige ideale

Primfactor von p kommt nämlich in dem Producte $\psi_1(\alpha)\psi_2(\alpha)\psi_3(\alpha)\ldots.\psi_{\lambda-2}(\alpha)$ gerade so viele mal vor, als es Zahlen k giebt, für welche der positive Rest von kh, wenn h hinzu addirt wird, gröfser wird als λ, also für welche der positive Rest von kh, mod. λ, gröfser ist als $\lambda-h$. Nun hat für $k = 1, 2, 3, \ldots. \lambda-2$ das Product hk alle Reste $1, 2, 3, \ldots. \lambda-1$, mit Ausnahme von $\lambda-h$; die Anzahl derer, die gröfser sind als $\lambda-h$, ist also $h-1$; folglich kommt der zu $\alpha = u^h$ gehörende ideale Primfactor von p, welcher nach der obigen Bezeichnung $f(\alpha^{m_h})$ ist, in diesem Producte $h-1$ mal vor. Man erhält folglich

$$\psi_1(\alpha)\psi_2(\alpha)\ldots.\psi_{\lambda-2}(\alpha) = \pm\alpha^s f(\alpha^{m_2})^1\cdot f(\alpha^{m_3})^2\cdot f(\alpha^{m_4})^3\ldots.f(\alpha^{m_{\lambda-1}})^{\lambda-2},$$

und wenn mit $p = f(\alpha)\cdot f(\alpha^2)\cdot f(\alpha^3)\ldots.f(\alpha^{\lambda-1})$, oder, was Dasselbe ist, mit $p = f(\alpha^{m_1})\cdot f(\alpha^{m_2})\cdot f(\alpha^{m_3})\ldots.f(\alpha^{m_{\lambda-1}})$ multiplicirt wird:

$$(\alpha, x)^\lambda = \pm\alpha^s f(\alpha^{m_1})^1\cdot f(\alpha^{m_2})^2\cdot f(\alpha^{m_3})^3\ldots.f(\alpha^{m_{\lambda-1}})^{\lambda-1};$$

welches nach einer andern Anordnung der Factoren auch so dargestellt werden kann:

$$(\alpha, x)^\lambda = \pm\alpha^s f(\alpha)^{m_1}\cdot f(\alpha^2)^{m_2}\cdot f(\alpha^{\lambda-1})^{m_{\lambda-1}}.$$

(Man vergleiche das Breslauer Programm zur Jubelfeier der Universität Königsberg, in welchem ich diese Darstellung des $(\alpha, x)^\lambda$ zuerst aufgestellt und nach einer andern einfachen Methode bewiesen habe; für den Fall, dafs $f(\alpha)$ ein wirklicher (nicht idealer) complexer Primfactor von p ist.)

Zur Bestimmung des allein noch unbestimmt gebliebenen Factors $\pm\alpha^s$ erwäge man, dafs $(\alpha, x)^\lambda \equiv (1, x^\lambda) \equiv -1$, mod. λ, ist; es mufs also dieser Factor so genommen werden, dafs auch die andere Seite dieser Gleichung congruent -1 werde, für den Modul λ *).

Für die Erkenntnifs der innern Natur dieser complexen Zahlen der Kreistheilung lassen die gegebenen Zerlegungen in die Primfactoren nichts zu wünschen übrig. Was ferner die wirkliche Berechnung derselben für bestimmte Werthe des p und λ betrifft, so reduciren sie die ganze Aufgabe auf den einen Punct: einen complexen Primfactor von p zu finden, welcher immer, entweder als wirkliche complexe Zahl, oder doch als Wurzel aus einer solchen dargestellt werden kann. Im ersten Fall läfst sich ein solcher Primfactor durch indirecte Methoden, welche ich in dem erwähnten Programm ent-

*) Ich verdanke diese Bemerkung einer Privat-Mittheilung des Hrn. Dr. *Eisenstein*.
Anm. d. Verf.

wickelt habe, stets sehr leicht finden; ist er ideal, so mufs man zunächst ermitteln, zu welcher Potenz er erhoben werden mufs, um zu einem wirklichen Factor zu werden. Diese Potenz läfst sich alsdann durch dieselben Methoden finden, und die entsprechende Wurzel daraus giebt den gesuchten idealen Factor. In diesem letzten Falle giebt es übrigens noch andere Mittel zur wirklichen Berechnung von $(\alpha, x)^\lambda$; welche ich jedoch hier übergehe. Hätte man eine vollständige Tafel aller complexen Primfactoren der realen Primzahlen (z. B. bis $p = 997$, als so weit der „Canon arithmeticus" von *Jacobi* reicht), nicht blofs für den Fall, dafs λ eine Primzahl ist, sondern auch wenn es eine Potenz einer Primzahl ist: so hätte man für den ganzen Bereich dieser Tafel auch die Lösung aller Kreistheilungen, ohne alle weitere Rechnung. Einen Anfang für eine solche Tafel der complexen Primfactoren habe ich in dem erwähnten Programme gegeben.

§. 12.

Das Product idealer Primfactoren, durch welches wir die in der Kreistheilung vorkommenden, mit $\psi_k(\alpha)$ bezeichneten complexen Zahlen ausgedrückt haben, nämlich

$$\pm\alpha^r \Pi f(\alpha^{m_h}) = \psi_k(\alpha),$$

wo h diejenigen $\frac{1}{2}(\lambda-1)$ Werthe hat, welche der Bedingung genügen, dafs der kleinste positive Rest von kh, mod. λ, um h vermehrt, gröfser als λ ist, und wo m_h eine der Congruenz hm_h, mod. λ, genügende Zahl bedeutet, hat die merkwürdige Eigenschaft, immer eine *wirkliche* complexe Zahl zu geben; nicht blofs, wenn, wie im vorigen Paragraph, $f(\alpha)$ ein idealer Primfactor der Primzahl p von der Form $\mu\lambda+1$ ist, sondern auch wenn $f(\alpha)$ irgend eine beliebige ideale complexe Zahl ist. Ist nämlich $f(\alpha)$ irgend eine beliebige ideale Zahl, so ist sie nach (§. 10.) stets solchen idealen Zahlen äquivalent, deren ideale Primfactoren nur zum Exponenten *Eins* gehören. Was also in Beziehung auf die Wirklichkeit des obigen Products von einer solchen idealen Zahl gilt, das gilt auch von jeder beliebigen. Ist aber $f(\alpha)$ ein Product aus idealen Primfactoren, welche alle zum Exponenten *Eins* gehören, d. h. welche nur ideale Primfactoren realer Primzahlen von der Form $\mu\lambda+1$ sind, so zerfällt das obige Product in ein Product ähnlicher Producte, welche alle einzeln, dem vorigen Paragraphen zufolge, wirkliche complexe Zahlen sind. Es ist demnach in der That $\Pi f(\alpha^{m_h})$ stets eine wirkliche complexe Zahl, wenn $f(\alpha)$ irgend eine beliebige ideale Zahl bedeutet. Dasselbe gilt nun auch von dem Producte

$$f(\alpha^{m_1})^1 \cdot f(\alpha^{m_2})^2 \cdot f(\alpha^{m_3})^3 \dots f(\alpha^{m_{\lambda-1}})^{\lambda-1} = \boldsymbol{P}(\alpha);$$

denn dieses ist, wie im vorigen Paragraph gezeigt, nur aus den obigen Producten zusammengesetzt.

Es soll nun für $f(\alpha)$ eine aus Perioden von je f Gliedern gebildete ideale Zahl genommen werden, d. h. eine solche, welche, durch $\varphi(\eta)$ bezeichnet, der Bedingung genügt, dafs

$$\varphi(\eta)\varphi(\eta_1)\varphi(\eta_2)\dots\varphi(\eta_{e-1}) = \boldsymbol{N}\varphi(\eta) = \boldsymbol{M}$$

gleich einer realen ganzen Zahl sei. Nach (§. 7.) gestattet auch jede Zahl $\boldsymbol{M}$, welche die Eigenschaft hat, dafs alle ihre Divisoren, zur f^{ten} Potenz erhoben, congruent ***Eins*** sind, für den Modul λ, eine solche ideale Zerlegung in complexe Factoren. Es sei ferner γ eine primitive Wurzel der Congruenz $\gamma^{\lambda-1} \equiv 1$, mod. λ, ferner allgemein γ_r der kleinste positive Rest, welchen γ^r läfst, für den Modul λ, und dann

$$\begin{aligned}
1 + \gamma_e + \gamma_{2e} + \dots + \gamma_{(f-1)e} &= \lambda s_0,\\
\gamma_1 + \gamma_{e+1} + \gamma_{2e+1} + \dots + \gamma_{(f-1)e+1} &= \lambda s_1,\\
\gamma_2 + \gamma_{e+2} + \gamma_{2e+2} + \dots + \gamma_{(f-1)e+2} &= \lambda s_2,\\
\dots\dots\dots\dots\dots\dots\\
\gamma_{e-1} + \gamma_{2e-1} + \gamma_{3e-1} + \dots + \gamma_{fe-1} &= \lambda s_{e-1}:
\end{aligned}$$

so sind $s_0, s_1, s_2, \dots s_{e-1}$ ganze Zahlen, weil diese Summen bekanntlich alle durch λ theilbar sind. Schreibt man nun das Product $\boldsymbol{P}(\alpha)$ in der Form

$$\boldsymbol{P}(\alpha) = f(\alpha)^1 \cdot f(\alpha^{\gamma^{-1}})^{\gamma_1} \cdot f(\alpha^{\gamma^{-2}})^{\gamma_2} \dots f(\alpha^{\gamma^{2-\lambda}})^{\gamma_{\lambda-2}},$$

so erhält man, wenn $\varphi(\eta)$ statt $f(\alpha)$ gesetzt wird,

$$(1.)\quad \boldsymbol{P}(\alpha) = \varphi(\eta)^{\lambda s_0} \cdot \varphi(\eta_{-1})^{\lambda s_1} \cdot \varphi(\eta_{-2})^{\lambda s_2} \dots \varphi(\eta_1)^{\lambda s_{e-1}}.$$

Es ist aber, wie sich leicht zeigen läfst, im gegenwärtigen Fall nicht nur $\boldsymbol{P}(\alpha)$ eine wirkliche complexe Zahl, sondern auch die λ^{te} Wurzel daraus; denn es ist allgemein

$$\frac{\boldsymbol{P}(\alpha)\boldsymbol{P}(\alpha^k)}{\boldsymbol{P}(\alpha^{k+1})} = (\psi_k(\alpha))^\lambda;$$

und wenn $k+1$ einer der Zahlen $\gamma_e, \gamma_{2e}, \dots \gamma_{(f-1)e}$ gleich angenommen wird, so ist $\boldsymbol{P}(\alpha) = \boldsymbol{P}(\alpha^{k+1})$ für $f(\alpha) = \varphi(\eta)$, also auch

$$\boldsymbol{P}(\alpha^k) = (\psi_k(\alpha))^\lambda,$$

d. h. $\boldsymbol{P}(\alpha^k)$, und folglich auch $\boldsymbol{P}(\alpha)$ selbst die λ^{te} Potenz einer ***wirklichen*** complexen Zahl. Deshalb ist denn auch

$$\varphi(\eta)^{s_0}\cdot\varphi(\eta_{-1})^{s_1}\cdot\varphi(\eta_{-2})^{s_2}\ldots\ldots\varphi(\eta_1)^{s_{e-1}}$$

eine ***wirkliche*** complexe Zahl. Es sei nun s_r die kleinste der Zahlen s_0, s_1, s_2, ... s_{e-1}, so ist diese wirkliche complexe Zahl noch durch die wirkliche Zahl $[\varphi(\eta)\varphi(\eta_{-1})\varphi(\eta_{-2})\ldots\ldots\varphi(\eta_1)]^{s_r}$ theilbar, und es ist

$$(2.)\qquad \varphi(\eta)^{s_0-s_r}\cdot\varphi(\eta_{-1})^{s_1-s_r}\cdot\varphi(\eta_{-2})^{s_2-s_r}\ldots\ldots\varphi(\eta_1)^{s_{e-1}-s_r}=\Phi(\eta)$$

gleich einer ***wirklichen*** complexen Zahl. Dieses Resultat ist in dem Fall, wo f (die Anzahl der in einer Periode enthaltenen Wurzeln) eine ***gerade*** Zahl ist, nichtssagend: denn alsdann sind die Zahlen s_0, s_1, s_2, ... s_{e-1} alle einander gleich, und man erhält nichts weiter, als dafs ***Eins*** eine wirkliche Zahl ist. Wenn aber f ungerade ist, so sind die Zahlen s_0, s_1, s_2, ... s_{e-1} nicht alle gleich, sondern es ist nur $s_0+s_{\frac{1}{2}e}=s_1+s_{\frac{1}{2}e+1}=s_2+s_{\frac{1}{2}e+2}=\ldots\ldots=f$. Multiplicirt man also $\Phi(\eta)$ mit der reciproken complexen Zahl, welche $\Phi(\eta_{\frac{1}{2}e})$ ist, so erhält man

$$(3.)\qquad \Phi(\eta)\cdot\Phi(\eta_{\frac{1}{2}e})=M^{f-2s_r}.$$

Dies ist das analoge Resultat zu dem von ***Jacobi*** gefundenen $\psi_k(\alpha)\cdot\psi_k(\alpha^{-1})=p$. Der merkwürdigste specielle Fall ist unstreitig der, wenn $e=2$, $f=\frac{1}{2}(\lambda-1)$ und, da f ungerade sein mufs, λ von der Form $4n+3$ ist. Hier kommen nur die beiden Perioden η und η_1 vor, deren Werthe bekanntlich $\frac{1}{2}(-1\pm\sqrt{-\lambda})$ sind; ferner ist $\Phi(\eta)=x+y\eta$, $\Phi(\eta_{\frac{1}{2}e})=x+y\eta_1$ und $(x+y\eta)(x+y\eta_1)=x^2-xy+\frac{1}{4}(\lambda+1)y^2$; auch ist hier λs_0 gleich der Summe der quadratischen Reste, und λs_1 gleich der Summe der quadratischen Nichtreste. Bezeichnet man diese Summen durch Σa und Σb, so erhält man

$$(4.)\qquad M^{\frac{1}{\lambda}(\Sigma b-\Sigma a)}=x^2-xy+\tfrac{1}{4}(\lambda+1)y^2;$$

das heifst: die Potenz mit dem Exponenten $\frac{1}{\lambda}(\Sigma b-\Sigma a)$ von einer jeden Zahl, welche überhaupt durch quadratische Formen der Determinante $-\lambda$ dargestellt werden kann, ist stets durch die Hauptform $x^2-xy+\frac{1}{4}(\lambda-1)y^2$ darstellbar. Dasselbe kann auch so ausgedrückt werden: Jede Classe quadratischer Formen der Determinante $-\lambda$, wenn sie $\frac{1}{\lambda}(\Sigma b-\Sigma a)$ mal mit sich selbst zusammengesetzt wird, giebt die Hauptclasse. Es ist dies dasselbe Resultat, aus welchem, zusammengehalten mit dem Satze §. 305. der Disqu. arithm. von Gaufs, ***Jacobi*** zuerst vermuthet hat, dafs die Zahl $\frac{1}{\lambda}(\Sigma b-\Sigma a)$ gleich der Anzahl der nicht äquivalenten Classen der quadratischen Formen der Determinante $-\lambda$ sein dürfte.

(S. dieses Journal Bd. IX. S. 189.) Die allgemeinere Formel (3.) giebt zu ähnlichen Vermuthungen über einen Zusammenhang der Zahlen $s_0, s_1, s_2, \dots s_{e-1}$, namentlich der Zahl $f-2s_r$ mit der Anzahl der nicht äquivalenten complexen Zahlen für jeden Werth von λ Anlafs.

Die Wirklichkeit des Products $\boldsymbol{P}(\alpha)$ für alle beliebigen idealen Zahlen $f(\alpha)$ gewährt noch ein Mittel, zu finden, welche Potenz einer idealen Zahl zu einer wirklichen wird. Giebt man nämlich in dem Ausdrucke von $\boldsymbol{P}(\alpha)$ dem α die Werthe α, α^2, α^3, ... $\alpha^{\lambda-1}$, so erhält man $\lambda-1$ Gleichungen, in welchen man $f(\alpha)$, $f(\alpha^2)$, etc. als Unbekannte betrachten kann; und aus diesen Gleichungen, welche für die Logarithmen dieser unbekannten Factoren nur linear sind, kann man, weil sie nicht ganz unabhängig von einander sind, zwar nicht $f(\alpha)$ selbst finden, aber man kann stets den Quotienten $\frac{f(\alpha)}{f(\alpha^{-1})}$, oder vielmehr eine bestimmte Potenz desselben finden, welche, durch $\boldsymbol{P}(\alpha)$, $\boldsymbol{P}(\alpha^2)$, etc. ausgedrückt, eine wirkliche complexe Zahl ist. Ist nun $\varphi(\alpha)$ eine ideale complexe Zahl, von der Art, dafs sie der Bedingung: $\varphi(\alpha)$ äquivalent $f(\alpha^{-1})\varphi(\alpha)$, genügt, für irgend eine Bestimmung der idealen Zahl $f(\alpha)$: so kann man durch Auflösung eines Systems linearer Gleichungen, oder vielmehr nur durch Ausrechnung der Determinante dieses Systems, stets eine Zahl n finden, welche der Bedingung genügt, dafs $(\varphi(\alpha))^n$ eine wirkliche complexe Zahl sei. Die wirkliche Ausrechnung hat folgende bestimmte Werthe des n gegeben: $n=1$ für $\lambda=5$, 7, 11, 13, 17 und 19, $n=3$ für $\lambda=23$, $n=2$ für $\lambda=29$, $n=9$ für $\lambda=31$, $n=37$ für $\lambda=37$, $n=11$ für $\lambda=41$, $n=211$ für $\lambda=43$, $n=5.139$ für $\lambda=47$.

Die vollständige Bestimmung derjenigen Potenzen idealer Zahlen, welche zu wirklichen werden, so wie auch die Bestimmung der Anzahl nicht äquivalenter idealer Zahlen, erfordert aber noch wesentlich andere Principien, als in der gegenwärtigen Abhandlung enthalten sind. Wir verfolgen diese wichtige Frage auch schon deshalb jetzt nicht weiter, weil, wie bereits erwähnt, die Veröffentlichung einer Arbeit von ***Dirichlet*** nahe bevorsteht, in welcher er dieselbe Frage für einen sehr nahe verwandten Gegenstand vollständig gelöset hat.

Breslau, im September 1846.

Über die Vierecke, deren Seiten und Diagonalen rational sind

Journal für die reine und angewandte Mathematik 37, 1–20, Taf. I (1848)

Die Aufgabe: Vierecke zu bilden, deren Seiten und Diagonalen sich durch rationale Zahlen ausdrücken lassen, findet sich schon in den Werken der Inder, namentlich des ***Brahmegupta*** behandelt, welcher nach ***Colebrooke's*** Annahme etwa sechs hundert Jahre nach Christi Geburt gelebt hat. Die hierher gehörenden Sätze des Indischen Mathematikers haben ein sehr mysteriöses Ansehen, so dafs ihr wahrer Sinn nur schwer zu erkennen ist, welchen jedoch ***Chasles*** in der 12ten Note zu seiner Geschichte der Geometrie glücklich enträthselt hat. Es finden sich daselbst auch über den ganzen Abschnitt von den ebenen Figuren viele sehr schätzenswerthe Aufklärungen; die ***Methode*** indessen, von welcher ***Brahmegupta*** Gebrauch gemacht hat, um zu den erwähnten Sätzen über das Viereck zu gelangen, scheint dieser geistvolle Geometer nicht genau erkannt zu haben. Wir wollen hier eine kurze Auseinandersetzung derselben geben.

In dem vierten Abschnitte von ***Colebrooke's*** „Algebra with Arithmetic and Mensuration, translated from the Sanscrit of ***Brahmegupta*** and ***Bhascara***", mit der Überschrift „Plane figure. Triangle and quadrilateral.", welcher in achtzehn Sätzen §. 21. bis §. 38. vom Dreieck und vom Viereck handelt, enthalten die ersten zwölf Sätze nur Regeln zur Berechnung der Stücke dieser Figuren; nämlich der Seiten, des Inhalts, der Perpendikel, der von diesen gebildeten Abschnitte auf der Grundlinie, der Diagonalen und ihrer Theile, und des Radius des umschriebenen Kreises.

Diese Sätze gelten zum Theil in der Allgemeinheit, in welcher sie ausgesprochen sind, zum Theil aber nur unter gewissen, im Texte nicht angegebenen Bedingungen. Namentlich ist, damit sie richtig seien, bei den Vierecken stets die Bedingung hinzuzufügen, dafs sich dieselben einem Kreise einschreiben lassen; bei einigen auch noch die Bedingung, dafs die Diagonalen auf einander senkrecht stehen. Auf diese folgen sodann die sechs Sätze §. 33. bis §. 38., welche mehr arithmetischer Natur sind, da sie von der Bildung von

Dreiecken und Vierecken handeln, deren Stücke sich durch rationale oder ganze Zahlen ausdrücken lassen. Mit diesen Sätzen haben wir es also hier hauptsächlich zu thun; weshalb wir dieselben, so wie sie in der englischen Übersetzung lauten, wörtlich hersetzen.

33. The sum of the squares of two unalike quantities are the sides of an isosceles triangle; twice the product of the same two quantities is the perpendicular, and twice the difference of their squares is the base.

34. The square of an assumed quantity being twice set down and divided by two other assumed quantities and the quotients being severally added to the quantity first put, the moieties of the sums are the sides of a scalene triangle: from the same quotients the two assumed quantities being subtracted, the sum of the moieties of the differences is the base.

35. The square of the side assumed at pleasure, being divided and then lessened by an assumed quantity, the half of the remainder is the upright of an oblong tetragon; and this, added to the same assumed quantity, is the diagonal.

36. Let the diagonals of an oblong be the flanks of a tetragon, having two equal sides. The square of the side of the oblong, being divided by an assumed quantity, and then lessened by it, and divided by two, the quotient increased by the upright of the oblong, is the base, and lessened by it, is the summit.

37. The three equal sides of a tetragon, that has three sides equal, are the squares of the diagonal (of te oblong). The fourth is found by subtracting the square of the upright from thrice the square of the (oblong's) side. If it be greatest, it is the base; if least, it is the summit.

38. The uprights and sides of two rectangular triangles, reciprocally multiplied by the diagonals, are four dissimmilar sides of a trapezium. The greatest is the base; the least is the summit, and the two others are the flanks.

Wir übertragen diese Sätze zunächst in die gewöhnliche mathematische Ausdrucksweise; wobei wir zugleich alles im Texte Weggelassene, was nothwendig zur Sache gehört, vervollständigen.

33. Setzt man jede der beiden gleichen Seiten eines gleichschenkligen Dreiecks gleich a^2+b^2, und die Grundlinie gleich $2(a^2-b^2)$, wo a und b beliebige rationale Zahlen sind, so ist auch die Höhe und der Inhalt dieses Dreiecks rational.

34. Wenn die drei Seiten eines schiefwinkligen Dreiecks folgende Werthe haben: $\frac{1}{2}\left(\frac{a^2}{b}+b\right)$, $\frac{1}{2}\left(\frac{a^2}{c}+c\right)$ und $\frac{1}{2}\left(\frac{a^2}{b}-c\right)+\frac{1}{2}\left(\frac{a^2}{c}-c\right)$, wo a, b und c beliebige rationale Zahlen sind, so sind die Höhen und der Inhalt desselben ebenfalls rational.

35. Wenn eine Seite eines Rechtecks gleich a, die andere gleich $\frac{1}{2}\left(\frac{a^2}{b}-b\right)$ genommen wird, wo a und b rationale Zahlen sind, so ist auch die Diagonal dieses Rechtecks rational.

36. Nimmt man von den zwei parallelen Seiten eines Paralleltrapezes die eine gleich $\frac{1}{2}\left(\frac{a^2}{b}-b\right)+\frac{1}{2}\left(\frac{a^2}{c}-c\right)$, die andere gleich $\frac{1}{2}\left(\frac{a^2}{b}-b\right)-\frac{1}{2}\left(\frac{a^2}{c}-c\right)$, und jede der beiden andern einander gleichen, nicht parallelen Seiten gleich $\frac{1}{2}\left(\frac{a^2}{c}+c\right)$: so erhält man, wenn a, b, c rational sind, ein Paralleltrapez, dessen Seiten, Höhe, Diagonalen und Inhalt rational sind.

37. Ein Paralleltrapez mit drei gleichen Seiten, dessen Seiten, Höhe, Inhalt und beide (einander gleiche) Diagonalen rational sind, erhält man, wenn jede der drei gleichen Seiten gleich $(a^2+b^2)^2$ und die vierte Seite gleich $3(a^2-b^2)^2-4a^2b^2$ angenommen wird; wo a und b rationale Gröfsen bezeichnen.

38. Wenn die vier Seiten eines Vierecks, welches sich einem Kreise einschreiben läfst, die Werthe $(a^2+b^2)(c^2-d^2)$, $(a^2-b^2)(c^2+d^2)$, $2cd(a^2+b^2)$ und $2ab(c^2+d^2)$ haben, wo a, b, c und d beliebige rationale Zahlen bezeichnen, so sind auch die beiden Diagonalen, die Abschnitte derselben, so wie der Inhalt des Vierecks, und der Durchmesser des umschriebenen Kreises rational.

Alle diese Sätze lassen sich auf die einfachste Weise durch blofse Zusammensetzung rechtwinkliger pythagoräischer Dreiecke finden, deren Bildung, auf der Lösung der arithmetischen Aufgabe beruhend: zwei Quadratzahlen zu finden, deren Summe wieder eine Quadratzahl ist, den Indern vollständig bekannt war. Wir wollen nun von den einzelnen Sätzen zeigen, wie sie alle fast unmittelbar aus dieser Quelle fliefsen.

Zur Bildung des gleichschenkligen Dreiecks mit rationalen Seiten und Inhalt, welche im Satze 33. gelehrt wird, werden nur zwei congruente pythagoräische rechtwinklige Dreiecke so an einander gesetzt, dafs zwei an einanderliegende gleiche Katheten die Höhe, die beiden andern die Grundlinie bilden.

1*

Die Bildung des schiefwinkligen Dreiecks mit rationalen Seiten und Inhalt, im Satze 34., geschieht durch die Zusammensetzung zweier verschiedener pythagoräischer rechtwinkliger Dreiecke, wenn in beiden eine Kathete gleich gemacht worden ist.

Das Rechteck im Satze 35. entsteht durch Verdoppelung eines pythagoräischen Dreiecks.

Das Paralleltrapez mit zwei einander gleichen, nicht parallelen Seiten wird gebildet, wenn man zunächst die beiden congruenten pythagoräischen Dreiecke BED und BGD (Taf. I. Fig. 1.) zu einem Rechtecke $EBGD$ zusammensetzt, dessen Seiten und Diagonalen rational sind. Setzt man darauf ein zweites pythagoräisches Dreieck BEA, dessen Kathete einer Rechtecksseite gleich gemacht ist, an das Rechteck an, und schneidet auf der andern Seite das diesem congruente Dreieck CGD von dem Rechtecke ab, so erhält man das Paralleltrapez $ABCD$ mit rationalen Seiten, Diagonalen, Höhe und Inhalt. Ist $BE = a$, $DE = \frac{1}{2}\left(\frac{a^2}{b} - b\right)$, so wird $BD = AC = \frac{1}{2}\left(\frac{a^2}{b} + b\right)$; nimmt man ferner $AE = GC = \frac{1}{2}\left(\frac{a^2}{c} - c\right)$, so wird $AB = CD = \frac{1}{2}\left(\frac{a^2}{c} + c\right)$, also $AD = \left(\frac{a^2}{b} - b\right) + \frac{1}{2}\left(\frac{a^2}{c} - c\right)$ und $BC = \frac{1}{2}\left(\frac{a^2}{b} - b\right) - \frac{1}{2}\left(\frac{a^2}{c} - c\right)$; wie der Satz es vorschreibt.

Das in dem folgenden Satze 37. construirte Paralleltrapez mit drei gleichen Seiten erhält man auf folgende Weise. Man setzt zunächst die beiden congruenten rechtwinkligen pythagoräischen Dreiecke AEB und AED (Fig. 2.) zu einem gleichschenkligen Dreiecke ABD zusammen, dessen Seiten und Höhen rational sind. Es wird also dadurch auch die Höhe BF rational, und eben so werden es die beiden Abschnitte AF und FD der Grundlinie AD. Verbindet man nun das dem Dreiecke BFD congruente Dreieck BGD mit diesem zu einem Rechtecke $BFDG$ und schneidet davon das Dreieck CGD ab, welches mit AFB congruent ist, so hat man das verlangte Paralleltrapez $ABCD$. Ist $AE = 2ab$, $BE = DE = a^2 - b^2$, so ist $BA = AD = DC = a^2 + b^2$; ferner ist in dem Dreiecke ABD: $AE.BD = BF.AD$, also $BF = \frac{AE.BD}{AD} = \frac{4ab(a^2-b^2)}{a^2+b^2}$. Hieraus folgt nach den pythagoräischen Lehrsatze $DF = \frac{2(a^2-b^2)^2}{a^2+b^2}$, also $AF = CG = a^2+b^2 - \frac{2(a^2-b^2)^2}{a^2+b^2}$ und hieraus $DF - AF = \frac{4(a^2-b^2)^2}{a^2+b^2} - a^2 - b^2 = BC$ oder $BC = \frac{3(a^2-b^2)^2 - 4a^2b^2}{a^2+b^2}$. Multiplicirt man noch alle diese Stücke mit a^2+b^2, so hat man die im Satze angegebene Regel.

Der nun folgende letzte Satz 38. ist der merkwürdigste der sechs Sätze des ***Brahmegupta***, weil er ein etwas allgemeineres Viereck zu bilden lehrt, dessen Stücke rational sind. Setzt man zunächst (Fig. 3.) die beiden rechtwinkligen pythagoräischen Dreiecke AEB und CEB mit der in beiden gleich gemachten Kathete BE an einander, sodann an CE das Dreieck CED, welches dem Dreieck AEB durch passende Vervielfältigung und Theilung seiner drei Seiten auf die Weise ähnlich gemacht ist, dafs die der BE entsprechenden Kathete gleich CE ist, und setzt eben so das Dreieck AED, welches dem Dreiecke BEC auf die Weise ähnlich gemacht ist, dafs die der Kathete BE entsprechende Kathete desselben gleich AE ist, an AE an: so ist auch die Kathete DE in diesen beiden neuen Dreiecken dieselbe, und man erhält das verlangte Viereck $ABCD$. Der Punct D kann auch einfach so bestimmt werden, dafs man um das Dreieck ABC einen Kreis beschreibt und die Höhe des Dreiecks AE verlängert, bis sie die Peripherie des Kreises in D schneidet. Nimmt man die beiden rechtwinkligen pythagoräischen Dreiecke, deren Seiten $2ab$, a^2-b^2, a^2+b^2 und $2cd$, c^2-d^2, c^2+d^2 sind, und mnltiplicirt, um zwei Katheten in beiden gleich zu machen, die drei Seiten des ersten mit $2cd$, die des zweiten mit $2ab$, so erhält man, nachdem dieselben zu dem Dreiecke ABC zusammengesetzt worden sind: $AE=2cd(a^2-b^2)$, $BE=4abcd$, $CE=2ab(c^2-d^2)$, $AB=2cd(a^2+b^2)$, $BC=2ab(c^2+d^2)$; ferner wird wegen der Ähnlichkeit der gegenüberliegenden Dreiecke, $DE=(a^2-b^2)(c^2-d^2)$, $CD=(a^2+b^2)(c^2-d^2)$, $DA=(a^2-b^2)(c^2+d^2)$; wie es im Satze verlangt worden ist.

Die sämmtlichen Sätze des ***Brahmegupta*** von der Bildung von Dreiecken und Vierecken mit rationalen Stücken beruhen also hauptsächlich nur auf der Bildung rechtwinkliger Dreiecke mit rationalen Seiten. Die dazu nöthigen geometrischen Kenntnisse beschränken sich, wie man sieht, auf den pythagoräischen Lehrsatz; wozu allenfalls noch der Satz gerechnet werden kann, dafs die vier Abschnitte, in welche zwei Sehnen eines Kreises sich gegenseitig theilen, in gleicher Proportion stehen. Aber selbst mit diesen beschränkten Mitteln hätte ***Brahmegupta*** weit mehr leisten können, da sie nicht nur zur Bildung der besonderen Vierecke des Satzes 38., in welchen die Diagonalen auf einander senkrecht stehen, sondern sogar zur allgemeinsten Lösung der Aufgabe über das dem Kreise einschreibbare Viereck mit rationalen Stücken vollständig ausreichen. Diese allgemeinste Lösung wird nämlich folgendermaafsen gefunden.

Man nehme drei rechtwinklige Dreiecke mit rationalen Seiten, und lege sie, nachdem man eine Kathete in allen gleich gemacht hat, mit dieser Kathete so an- und auf einander, dafs die andern Katheten in einer geraden Linie liegen. Durch Zusammensetzung der drei Dreiecke BFA, BFE und BFC (Fig. 4.) erhält man dann die beiden an einander passenden schiefwinkligen Dreiecke BEA und BEC, welche zusammen das Dreieck ABC bilden. Um dieses Dreieck ABC ziehe man einen Kreis, verlängere BE bis an die Peripherie desselben nach D, ziehe DA und DC: so ist $ABCD$ das verlangte Viereck, dessen Seiten, Diagonalen, Abschnitte der Diagonalen, Radius des umschriebenen Kreises und Inhalt rational sind. Nimmt man für die Seiten der zum Grunde gelegten drei pythagoräischen Dreiecke folgende: *Erstlich*, $2ab$, a^2-b^2, a^2+b^2; *Zweitens*, $2cd$, c^2-d^2, c^2+d^2; *Drittens*, $2ef$, e^2-f^2, e^2+f^2, so erhält man leicht folgende Ausdrücke für die vier Seiten des Vierecks:

$$AB = (a^2+b^2)(e^2+f^2)cd,$$
$$BC = ab(c^2+d^2)(e^2+f^2),$$
$$CD = (a^2+b^2)(ef(c^2-d^2)-cd(e^2-f^2)),$$
$$DA = (c^2+d^2)(ef(a^2-b^2)-ab(e^2-f^2)).$$

Diese Ausdrücke geben, wie sich später zeigen wird, ganz allgemein alle dem Kreise einschreibbaren Vierecke mit rationalen Seiten, Diagonalen und Inhalt. *Brahmegupta* ist bis zu dieser vollständigen Auflösung der von ihm behandelten Aufgabe nicht vorgedrungen, obgleich sie, wie wir sahen, durch dieselben einfachen Mittel als jene beschränkteren möglich war.

Die hier gegebene vollständige Darlegung der sechs Sätze des *Brahmegupta* widerstreitet in einem Hauptpuncte der Ansicht, welche *Chasles* über dieselben und über den ganzen Abschnitt, welchem sie angehören, aufgestellt hat. *Chasles* sieht nämlich die vorhergehenden zwölf Sätze 21. bis 32. als Vorarbeiten an, welche hauptsächlich nur dazu dienen sollen, diese sechs arithmetischen Sätze, und namentlich den letzten derselben, vorzubereiten, und er meint, dafs alle diese vorhergehenden Sätze bei der Lösung der Aufgabe über das Viereck mit rationalen Stücken ihre Anwendung finden, so dafs keiner derselben dieser Aufgabe fremd, oder für sie überflüssig sei; welches, wie wir zeigten, nicht der Fall ist. Eine genaue Entwickelung der übrigen rein geometrischen Sätze dieses Abschnitts würde uns zu weit von unserem Ziele entfernen; dieselbe würde die zu hohen Erwartungen, welche *Chasles* von verloren gegangenen geometrischen Methoden der Inder hegt, gar sehr herabstimmen, und einen neuen Beleg für die Behauptung von *Colebrooke* liefern, dafs sich die Geometrie der Inder in der damaligen Zeit nur auf einer sehr niedern Stufe der Ausbildung befunden habe, während die arithmetischen Kenntnisse derselben viel bedeutender waren. Die geometrische Methode der Inder zur Zeit ***Brahmegupta's*** scheint nemlich nach den vorliegenden Proben hauptsächlich nur in einer sehr unwissenschaftlichen Art von Induction bestanden zu haben, durch welche sie ihre Sätze fanden, die ihnen sodann als Regeln galten und eines Beweises nicht mehr bedürftig schienen. Sie bildeten sich ihre Figuren, namentlich Dreiecke und Vierecke, so, dafs die einzelnen Stücke derselben durch Zahlen, und zwar immer wo möglich durch rationale oder ganze Zahlen ausgedrückt wurden; aus diesen Zahlen hauptsächlich abstrahirten sie ihre allgemeinen Regeln für die Berechnung der Stücke; und wenn dieselben in mehreren vorliegenden Fällen galten, so schrieben sie ihnen ohne Bedenken allgemeine Gültigkeit zu. Hieraus glaube ich, ist es zu erklären, dafs ***Brahmegupta*** viele, nur unter besondern Bedingungen geltende Sätze so giebt, als wären sie allgemein gültig. Dafs namentlieh viele der Sätze über das Viereck nur unter der Bedingung gelten, dafs die Diagonalen auf einander senkrecht stehen, hat seinen Grund wohl darin, dafs ***Brahmegupta*** hauptsächlich nur Vierecke dieser Art zu bilden verstand, und dafs er deshalb hauptsächlich nur von solchen seine Regeln abstrahirte. Diese weggelassenen nothwendigen Bedingungen, welche, wie ***Chasles*** annimmt, immer hinzugedacht, wenn gleich nicht ausgesprochen wurden, sind nach meiner Ansicht, wenigstens grofsentheils, aus wirklicher Unkenntnifs weggelassen; wofür sich als äufserer Beweisgrund auch die Ansicht des indischen Mathematikers ***Bhascara*** anführen läfst, welcher fast sechs hundert Jahre nach ***Brahmegupta*** lebte und, wenn auch kein grofser Geist, doch ein Kenner der ältern Mathematiker seines Vaterlandes war; dieser tadelt ***Brahmegupta*** gerade wegen jener Vernachlässigungen, welche das Viereck betreffen, indem er sagt: „Yet though indeterminate diagonals have been sought as determinate by ***Brahmegupta*** and others," und nennt an einer andern Stelle Den, welcher dieses thut, einen dummen Teufel (blundering devil).

Wir wollen nun unsere eigene Auflösung der allgemeineren Aufgabe entwickeln: Vierecke zu finden, deren Seiten und Diagonalen rational sind; indem wir zunächst beweisen, dafs in jedem solchen Vierecke auch die Abschnitte rational sein müssen, in welche die beiden Diagonalen sich gegenseitig theilen. Ist $ABCD$ (Fig. 5.) das Viereck, dessen Seiten AB, BC, CD, DA, so wie die beiden Diagonalen AC, BD rational sind, und man nennt die Winkel $BAC = u$, $DAC = v$, $AEB = w$, so sind die drei Cosinus $\cos u$, $\cos v$ und $\cos(u+v)$ rationale Zahlen; denn diese drei Winkel u, v, $u+v$ sind Winkel in Dreiecken, deren drei Seiten rational sind, und der Cosinus des Winkels eines Dreiecks ist eine rationale Function der drei Seiten desselben. Hieraus folgt nun nach der Formel $\cos(u+v) = \cos u \cos v - \sin u \sin v$, dafs das Product $\sin u \sin v$ rational ist. Da ferner $\sin v^2 = 1 - \cos v^2$ rational ist, so folgt durch Division mit $\sin v^2$, dafs auch $\frac{\sin u}{\sin v}$ rational ist. Nun geben die Dreiecke AEB und AED die Gleichungen: $BE \sin w = AB \sin u$ und $DE \sin w = DA \sin v$, also $\frac{BE}{DE} = \frac{AB \sin u}{AD \sin v}$; woraus folgt, dafs $\frac{BE}{DE}$ rational sein mufs. Addirt man zu diesem Quotienten *Eins*, so ist auch $\frac{BE+DE}{DE}$ rational, d. h. $\frac{BD}{DE}$ rational, und, weil auch die Diagonal BD rational ist, so folgt, dafs DE rational ist; also auch BE. Derselbe Beweis gilt eben so für die beiden Abschnitte AE und CE der andern Diagonal AE. Wir erhalten daher folgenden

Lehrsatz. *In jedem Vierecke, welches rationale Seiten und Diagonalen hat, sind auch die vier Abschnitte rational, in welche die Diagonalen sich gegenseitig theilen.*

Das gesuchte Viereck ist also immer aus vier Dreiecken mit rationalen Seiten zusammengesetzt, in deren jedem ein Winkel, als Winkel, den die beiden Diagonalen bilden, bestimmt ist. Deshalb lösen wir denn jetzt die einfachere Aufgabe: Dreiecke mit rationalen Seiten zu bilden, welche einen gegebenen Winkel w haben. Dieser Winkel w ist, damit die Aufgabe überhaupt lösbar sei, stets so anzunehmen, dafs der Cosinus desselben eine rationale Zahl ist. Es sei demnach in dem Dreiecke AEB (Fig. 5.) $\cos w = \frac{m}{n}$, $AE = \alpha$, $BE = \beta$ und $AB = a$, so ist

$$1. \quad a^2 = \alpha^2 + \beta^2 - \frac{2m}{n}\alpha\beta.$$

Nimmt man an, dafs die drei Seiten α, β und a *ganze* Zahlen sein sollen, ohne einen, allen dreien gemeinschaftlichen Factor (welches eben so allgemein ist, als die Annahme rationaler Zahlen) und dafs der Bruch $\frac{m}{n}$ in den kleinsten Zahlen ausgedrückt ist, so mufs $2\alpha\beta$ durch n theilbar sein, und es sind nun die beiden Fälle zu unterscheiden, wo *n ungerade* und wo *n gerade* ist. Man untersuche zunächst den ersten dieser Fälle: *n ungerade.* Für diesen ist $\alpha\cdot\beta$ durch n theilbar; α und β müssen also jedes irgend einen der zwei Factoren des n enthalten. Setzt man demnach $n = r\cdot s$, so ist $\alpha = r\cdot\alpha'$, $\beta = s\cdot\beta'$ zu setzen, und folglich

$$2.\qquad a^2 = r^2\alpha'^2 + s^2\beta'^2 - 2m\alpha'\beta'.$$

α' und β' haben nun keinen gemeinschaftlichen Factor, weil sonst α, β, und auch a, denselben haben müfsten; gegen die Voraussetzung: also ist auch wenigstens eine der Zahlen α' und β' ungerade, z. B. β'. Wird mit r^2 multiplicirt, so kann man der Gleichung (2.) auch folgende Form geben:

$$3.\qquad r^2a^2 = (r^2\alpha' - m\beta')^2 + (n^2 - m^2)\beta'^2,$$

oder auch die Form:

$$4.\qquad (ra + r^2\alpha' - m\beta')(ra - r^2\alpha' + m\beta') = (n^2 - m^2)\beta'^2.$$

Die beiden Factoren $ra + r^2\alpha' - m\beta'$ und $ra - r^2\alpha' + m\beta'$ haben nun keinen gemeinschaftlichen Primfactor mit β', weil sonst auch ihre Summe und ihre Differenz, nämlich $2ra$ und $2(r^2\alpha' - m\beta')$, also auch $r^2\alpha'$ und folglich auch a^2, oder a, denselben Primfactor haben müfsten. Zerfället man daher $n^2 - m^2$ in irgend zwei Factoren p und q, so dafs $n^2 - m^2 = pq$ ist, so mufs

$$5.\qquad \begin{cases} ra + r^2\alpha' - m\beta' = py^2, \\ ra - r^2\alpha' + m\beta' = qz^2 \text{ und} \\ \qquad\qquad\quad \beta' = yz \end{cases}$$

sein. Aus diesen drei Gleichungen folgt

$$6.\qquad \frac{2r^2\alpha'}{\beta} = \frac{py}{z} + 2m - \frac{qz}{y},$$

und da $\alpha = r\alpha'$, $\beta = s\beta'$, $rs = n$, also $\frac{2r^2\alpha'}{\beta'} = \frac{2n\alpha}{\beta}$ ist:

$$7.\qquad \frac{2n\alpha}{\beta} = \frac{py}{z} + 2m - \frac{qz}{y}.$$

Setzt man nun $\frac{py}{z} = n\xi$, wo ξ eine beliebige rationale (gebrochene) Zahl bedeutet, so wird $\frac{qz}{y} = \frac{pq}{n\xi} = \frac{n^2 - m^2}{n\xi}$, also, wenn endlich noch $\frac{m}{n} = c$ gesetzt wird,

$$\frac{2\alpha}{\beta} = \xi + 2c - \frac{(1 - c^2)}{\xi} \text{ oder}$$

$$7\,a.\qquad \frac{\alpha}{\beta} = \frac{(\xi + c)^2 - 1}{2\xi} = \frac{(\xi + c + 1)(\xi + c - 1)}{2\xi}.$$

Diese Gleichung (7.) ist nicht allein nothwendig, sondern auch hinreichend für die Construction des verlangten Dreiecks; denn vermöge derselben ist

$$8. \quad a = \sqrt{(\alpha^2+\beta^2-2c\alpha\beta)} = \frac{\beta(\xi^2+1-c^2)}{2\xi}.$$

Wir haben jetzt eigentlich noch den zweiten Fall, wo *n gerade* ist, eben so zu behandeln: da er aber nach derselben Methode genau zu denselben Resultaten (7. und 8.) führt, so wollen wir ihn nicht besonders herschreiben. Der Satz, welchen wir erlangt haben, ist folgender:

Lehrsatz. ***Wenn in einem Dreiecke, dessen Seiten rational sein sollen, ein Winkel so gegeben ist, dafs der Cosinus desselben rational und gleich c ist, so mufs das Verhältnifs der beiden, den Winkel einschliefsenden Seiten sich durch den Ausdruck*** $\frac{(\xi+c)^2-1}{2\xi}$ ***darstellen lassen, in welchem*** ξ ***irgend eine rationale Zahl bedeutet; und umgekehrt: wenn das Verhältnifs dieser beiden rationalen Seiten sich in dieser Form darstellen läfst, so ist auch die dritte Seite rational.***

Um diesen Satz auf das ***Viereck*** anzuwenden, bezeichne man die Theile, in welche die Diagonalen sich gegenseitig theilen, und welche, wie oben gezeigt, rational sein müssen, durch α, β, γ, δ, so dafs (in Fig. 5.) $AE = \alpha$, $BE = \beta$, $CE = \gamma$, $DE = \delta$ ist, und construire das Viereck mit diesen vier Stücken und dem Winkel, den die beiden Diagonalen bilden und dessen Cosinus rational und gleich c ist. Da es ferner nur auf die Verhältnisse dieser vier Abschnitte der Diagonalen ankommt, nicht auf die absoluten Längen, so kann man eine dieser Gröfsen willkürlich annehmen, oder auch, am einfachsten, sie der Einheit gleich setzen. Wird demnach $\beta = 1$ gesetzt, so erhält man dafür, dafs aufser den Stücken α, β, γ, δ auch die vier Seiten des Vierecks ***ABCD*** rational werden, folgende nothwendige und hinreichende Bedingungen:

$$9. \quad \begin{cases} \alpha = \frac{(\xi+c)^2-1}{2\xi}, & \gamma = \frac{(\eta-c)^2-1}{2\eta}, \\ \frac{\delta}{\alpha} = \frac{(x-c)^2-1}{2x}, & \frac{\delta}{\gamma} = \frac{(y+c)^2-1}{2y}. \end{cases}$$

Da die drei Gröfsen α, γ, δ diesen vier Gleichungen genügen müssen, so folgt, dafs die fünf rationalen Zahlen ξ, η, x, y und c nicht ganz beliebig sind, sondern folgender, aus der Elimination des α, γ und δ hervorgehenden Gleichung genügen müssen:

$$10. \quad \left(\frac{(\eta-c)^2-1}{2\eta}\right)\left(\frac{(y+c)^2-1}{2y}\right) = \left(\frac{(\xi+c)^2-1}{2\xi}\right)\left(\frac{(x-c)^2-1}{2x}\right).$$

Umgekehrt giebt aber auch jede rationale Bestimmung dieser fünf Gröfsen, welche der Gleichung (10.) genügt, und von welchen Gröfsen c, als Cosinus eines realen Winkels, kleiner als Eins sein mufs, eine passende Bestimmung der drei Gröfsen α, γ, δ, für welche auch die Seiten des Vierecks rational werden: denn wenn noch der Kürze wegen $1-c^2$ durch k^2 bezeichnet wird, so dafs k den Sinus des Winkels der beiden Diagonalen bezeichnet, der zwar nicht nothwendig *selbst*, aber dessen *Quadrat* stets rational ist, so erhält man nach der Gleichung (8.) folgende Ausdrücke der vier Seiten:

$$11.\quad \begin{cases} AB = \frac{\xi^2+k^2}{2\xi}, & BC = \frac{\eta^2+k^2}{2\eta}, \\ CD = \left(\frac{(\eta-c)^2-1}{2\eta}\right)\left(\frac{y^2+k^2}{2y}\right), & DA = \left(\frac{(\xi+c)^2-1}{2\xi}\right)\left(\frac{x^2+k^2}{2x}\right). \end{cases}$$

Die vollständige Auflösung der Aufgabe, alle die Vierecke zu finden, deren vier Seiten und beide Diagonalen rational sind, liegt demnach allein in der Auflösung der Gleichung (10.) durch rationale Zahlen. Will man aufserdem die zweite Bedingung hinzufügen, dafs auch der *Inhalt* des Vierecks rational sein soll, so macht dies keine besondern Schwierigkeiten; denn dieser Inhalt hat, aus den Inhalten der vier Dreiecke zusammengesetzt, folgenden Ausdruck: $\frac{1}{2}(\alpha\beta+\beta\gamma+\gamma\delta+\delta\alpha)\sin w$; die einzige nothwendige und hinreichende Bedingung, damit auch dieser rational werde, ist also nur die, dafs aufser $\cos w = c$ auch noch $\sin w = k$ rational sei; und diese Bedingung wird auf die allgemeinste Art dadurch befriedigt, dafs man dem c die Form $c = \frac{r^2-1}{r^2+1}$ giebt; wodurch $k = \frac{2r}{r^2+1}$ wird. Von den fünf Gröfsen, welche die Gleichung (10.) enthält, betrachte man nun die drei ξ, η und c als willkürlich anzunehmende rationale Zahlen, und nur x und y als die beiden Unbekannten, welche, wenn jene gegeben sind, allemal so bestimmt werden sollen, dafs sie rational sind und der Gleichung (10.) genügen. Diese Gleichung ist in Beziehung auf jede der Unbekannten vom zweiten Grade und kann, wenn für die Ausdrücke $\frac{(\xi+c)^2-1}{2\xi}$ und $\frac{(\eta-c)^2-1}{2\eta}$ der Kürze wegen die Zeichen α und γ beibehalten werden, folgendermaafsen dargestellt werden:

$$12.\quad \gamma\cdot\left(\frac{(y+c)^2-1}{2y}\right) = \alpha\cdot\left(\frac{(x-c)^2-1}{2x}\right),$$

oder, nach den Unbekannten geordnet, durch:

$$13.\quad \begin{cases} \gamma x y^2 - (\alpha x^2 - 2c(\alpha+\gamma)x - \alpha k^2)y - k^2\gamma x = 0 \text{ und} \\ \alpha y x^2 - (\gamma y^2 + 2c(\alpha+\gamma)x - \gamma k^2)x - k^2\alpha y = 0. \end{cases}$$

2*

Löset man diese Gleichung in Beziehung auf y auf, so erhält man:

$$14.\quad y = \frac{\alpha x^2 - 2c(\alpha+\gamma)x - \alpha k^2 \pm \sqrt{[(\alpha x^2 - 2c(\alpha+\gamma)x - \alpha k^2)^2 + 4k^2\gamma^2 x^2]}}{2\gamma x}.$$

Die ganze Aufgabe reducirt sich also jetzt darauf, die rationalen Werthe von x zu finden, für welche auch die Wurzel

$$\sqrt{[(\alpha x^2 - 2c(\alpha+\gamma)x - \alpha k^2)^2 + 4k^2\gamma^2 x^2]}$$

rational wird. Diese Aufgabe ist bekanntlich schon von *Euler* mehrmals behandelt worden; und zwar zuletzt in einer Abhandlung vom Jahre 1780, welche erst im Jahre 1830 in dem elften Bande der „Mémoires de l'Académie de St. Petersbourg" erschienen ist. Auch hat *Jacobi* in der Abhandlung „De usu theoriae integralium ellipticorum et Abelianorum in analysi Diophantea," im dreizehnten Bande dieses Journals S. 353, die Aufgabe mit Hülfe der elliptischen Functionen gelöset, und dabei bemerkt, dafs diese Methode mit der *Euler*schen im Wesentlichen übereinstimmt. Es mufs, damit nach den genannten Methoden die Lösung der Aufgabe überhaupt gelinge, immer eine bestimmte Anzahl, und wenigstens *eine* der Fundamental-Auflösungen bekannt sein; aus welchen dann eine unendliche Anzahl neuer hergeleitet wird: die Fundamental-Auflösungen aber direct zu finden und durch die aus ihnen herzuleitenden alle möglichen Auflösungen zu erschöpfen, ist ein noch ungelösetes Problem, welches bedeutenden Schwierigkeiten unterworfen zu sein scheint. Wir müssen darum hier darauf verzichten, die vollkommen allgemeinen Ausdrücke für die Stücke des Vierecks mit rationalen Seiten und Diagonalen in entwickelter Form, d. h. so darzustellen, dafs die nöthige Bedingungsgleichung (10.) von selbst erfüllt wird, sind indessen durch die *Euler*'sche Methode in den Stand gesetzt, eine unendliche Anzahl solcher Ausdrücke zu liefern, welche sehr allgemein sind, da sie die drei willkürlichen rationalen Zahlen ξ, η und c enthalten; denn jeder rationale Werth von x, welcher die obige Wurzelgröfse rational macht, giebt eine solche Formel.

Die einfachsten rationalen Werthe von x, welche der Gleichung (10.) so genügen, dafs auch y rational wird, und welche sich von selbst darbieten und also hier als Fundamental-Auflösungen benutzt werden sollen, sind $x=0$, $x=1+c$ und $x=\eta$. Diesen entsprechen nämlich die Werthe $y=0$, $y=1-c$ und $y=\xi$. Der erste dieser Werthe $x=0$ und $y=0$ giebt für sich kein wirkliches Viereck, sondern nur ein solches, dessen eine Winkelspitze im Unendlichen liegt. Ebenso giebt $x=1+c$, $y=1-c$ nur ein Viereck, von welchem zwei Winkelspitzen ineinanderfallen: welches also ein ***Dreieck*** ist.

Der Werth $x=\eta$, $y=\xi$ aber giebt ein ***wirkliches Viereck,*** und zwar das dem Kreise eingeschriebene. Für diese Werthe $x=\eta$ und $y=\xi$ erhält man nämlich aus (9. und 11.) folgende Ausdrücke:

$$15.\quad \alpha=\frac{(\xi+c)^2-1}{2\xi},\quad \beta=1,\quad \gamma=\frac{(\eta-c)^2-1}{2\eta},\quad \delta=\frac{((\xi+c)^2-1)((\eta-c)^2-1)}{4\xi\eta}$$

und

$$16.\quad \begin{cases} AB=\dfrac{\xi^2+k^2}{2\xi}, & BC=\dfrac{\eta^2+k^2}{2\eta}, \\ CD=\dfrac{((\eta-c)^2-1)(\xi^2+k^2)}{4\xi\eta}, & DA=\dfrac{((\xi+c)^2-1)(\eta^2+k^2)}{4\xi\eta}. \end{cases}$$

Diese Formeln enthalten auch den allgemeinsten Ausdruck aller einem Kreise einzuschreibenden Vierecke mit rationalen Seiten und Diagonalen; denn da diese der Bedingung $\alpha\gamma=\beta\delta$ genügen müssen, so folgt, dafs kein anderer Werth von δ möglich ist. Soll aufser den Seiten und Diagonalen auch noch der ***Inhalt*** rational werden, so ist es, wie oben gezeigt, hinreichend, und nothwendig, das c eine rationale Zahl von der Form $\frac{r^2-1}{r^2+1}$ sei; wodurch auch k rational wird. Für einen solchen Werth von c sind auch die hier gegebenen Formeln für das dem Kreise einzuschreibende Viereck, wie sich leicht erkennen läfst, wesentlich identisch mit den Formeln, welche wir oben durch blofse Zusammensetzung rechtwinkliger pythagoräischer Dreiecke nach der Methode ***Brahmegupta's*** fanden.

Als numerische Beispiele zu diesen Formeln für das dem Kreise einzuschreibende Viereck wollen wir $c=\frac{1}{2}$ setzen, also $w=60^0$; ferner $\xi=2$, $\eta=3$: dann ist $\alpha=\frac{3}{8}$, $\beta=1$, $\gamma=\frac{7}{8}$, $\gamma=\frac{21}{64}$, oder in ganzen Zahlen: $\alpha=24$, $\beta=64$, $\gamma=56$, $\delta=21$; die Seiten sind $AB=56$, $BC=104$, $CD=49$, $DA=39$ und die Diagonalen $AC=80$, $BD=85$. Ein anderes Beispiel, auch mit rationalem ***Inhalt,*** ist $c=\frac{3}{5}$, $\xi=\frac{8}{15}$, $\eta=\frac{4}{5}$; hiernach werden die Abschnitte der Diagonalen $\alpha=\frac{4}{15}$, $\beta=1$, $\gamma=\frac{7}{15}$, $\delta=\frac{28}{225}$, oder in ganzen Zahlen: $\alpha=60$, $\beta=225$, $\gamma=105$, $\delta=28$; ferner die Seiten $AB=195$, $BC=300$, $CD=91$, $DA=80$ und die Diagonalen $AC=165$, $BD=253$ und der Inhalt gleich 16698.

Die ***Euler***sche Methode, deren wir uns jetzt bedienen wollen, um aus den drei bekannten Werthen von x andere abzuleiten, die ebenfalls die obige Wurzel rational machen und also rationale Werthe des y geben, beruht hauptsächlich darauf, dafs der Ausdruck unter dem Wurzelzeichen, welcher nach x vom vierten Grade ist, auf die Form P^2+QR gebracht wird, wo

P, Q und R rationale ganze Functionen vom zweiten Grade sind, mit rationalen Coëfficienten. Aus diesen wird sodann die quadratische Gleichung

$$17. \quad Qz^2 + 2Pz - R = 0$$

gebildet, welche auch nach x quadratisch ist und also ebenfalls auf die Form

$$18. \quad Sx^2 + 2Tx - U = 0$$

sich bringen läfst. Wenn nun für irgend einen Werth von x die gegebene Wurzel, welche durch $\sqrt{(P^2+QR)}$ dargestellt ist, rational wird, so wird auch z rational; und zwar erhält es zwei rationale Werthe. Setzt man einen derselben in die Gleichung (18.) in S, T und U, so giebt dieselbe zwei rationale Werthe von x, deren einer der ursprünglich bekannte, der andere neu ist, und welcher ebenfalls die Wurzel $\sqrt{(P^2+QR)}$ rational macht, weil er einen rationalen Werth von z giebt. Legt man nun eben so diesen neuen Werth von x zum Grunde, so findet man auf gleiche Weise wieder einen neuen Werth dazu; und so kann man bis in's Unendliche fortfahren, wenn sich nicht etwa ein Werth, der schon einmal da war, wiederholt; in welchem Fall man nur eine endliche Periode von verschiedenen Werthen erhält.

Die erste für diese Methode nöthige Operation, nämlich eine quadratische Gleichung zu finden, deren Wurzel z sich durch die rational zu machende Wurzelgröfse ausdrücken läfst, und welche auch in Beziehung auf x selbst vom zweiten Grade ist, wurde in unserem Falle auf eine Weise schon ausgeführt; denn die Gleichung (10.), welche, einerseits nach y, andrerseits nach x geordnet, die Formen (13.) annimmt, ist eine Gleichung dieser Art. Grade diese aber leistet nur sehr wenig, denn sie giebt, wie leicht zu sehen, niemals eine unendliche Reihe von Werthen des x, sondern immer nur eine Periode von zweien. Ist nämlich x irgend ein genügender Werth, welcher auch y rational macht, so ergiebt sich als zweiter Werth nur $-\frac{k^2}{x}$, und dieser führt wieder auf den Werth von x zurück. Der Werth $-\frac{k^2}{x}$ giebt aber niemals ein anderes Viereck als x selbst, weil der Ausdruck von δ derselbe bleibt, wenn man x in $-\frac{k^2}{x}$ verwandelt. Eben so bleibt auch α ungeändert, wenn ξ in $-\frac{k^2}{\xi}$, und γ, wenn η in $-\frac{k^2}{\eta}$ verwandelt wird; welche Bemerkung zu beachten nöthig ist, um verschiedene Werthe von x als Werthe zu erkennen, die keine wesentlich verschiedenen Formeln für Vierecke mit rationalen Seiten und Diagonalen geben.

Obschon nun die Gleichungen (13.) selbst, nicht eine unendliche Reihe rationaler Werthe des x geben, für welche auch y rational wäre, so reicht doch eine leichte Änderung hin, sie dazu passend zu machen. Setzt man nämlich $xy = z$, so erhält man:

$$19. \quad \gamma z^2 - (\alpha x^2 - 2c(\alpha+\gamma)x - \alpha k^2)z - k^2\gamma x^2 = 0,$$

oder, nach Potenzen von x geordnet:

$$20. \quad (\alpha z + k^2\gamma)x^2 - 2c(\alpha+\gamma)zx - z(\gamma z + k^2 z) = 0.$$

Sind nun z und z' die beiden Wurzeln der einen, x und x' die beiden Wurzeln der andern Gleichung, so ist bekanntlich

$$21. \quad zz = -k^2 x^2, \qquad z + z' = \frac{\alpha x^2 - 2c(\alpha+\gamma)x - k^2\alpha}{\gamma},$$

$$22. \quad xx' = -\frac{z(\gamma z + k^2\alpha)}{\alpha z + k^2\gamma}, \qquad x + x' = \frac{2c(\alpha+\gamma)z}{\alpha z + k^2\gamma}.$$

Gehen wir jetzt von dem bekannten Werthe 0 von x aus, so finden sich für diesen von z die beiden Werthe $z = 0$ und $z = -\frac{k^2\alpha}{\gamma}$. Der erstere $z = 0$ ist zu verwerfen, weil er nur $x = 0$ zurückgiebt; die beiden Werthe $z = -\frac{k^2\alpha}{\gamma}$ und $x = 0$ aber geben nach der Gleichung (22.):

$$23. \quad x' = \frac{2c\alpha}{\alpha - \gamma};$$

welches ein *neuer* Werth des x ist, der z, und daher auch y, rational macht. Vermittelst einer der Gleichungen (21.) findet man ferner für den zweiten Werth von z, welcher zu diesem Werthe x' gehört:

$$z' = \frac{4c^2\alpha\gamma}{(\alpha-\gamma)^2}$$

und hieraus wieder vermittelst einer der beiden Gleichungen (22.) für den zweiten Werth von x, welcher zu diesem Werthe von z gehört:

$$24. \quad x'' = -\frac{2c\alpha(4c^2\gamma^2 + k^2(\alpha-\gamma)^2)}{(\alpha-\gamma)(4c^2\alpha^2 + k^2(\alpha-\gamma)^2)}.$$

Zu diesem gehört wieder, als neuer Werth von z, folgender:

$$z'' = -\frac{k^2\alpha(4c^2\gamma^2 + k^2(\alpha-\gamma)^2)^2}{(4c^2\alpha^2 + k^2(\alpha-\gamma)^2)^2},$$

welcher wieder folgenden neuen Werth des x giebt:

$$25. \quad x''' = -\frac{4ck^2\alpha(\alpha-\gamma)(4c^2\gamma^2 + k^2(\alpha-\gamma)^2)(2c^2(\alpha^2+\gamma^2) + k^2(\alpha-\gamma)^2)}{(4c^2\alpha^2 + k^2(\alpha-\gamma)^2)(4c^2\alpha\gamma + k^2(\alpha-\gamma)^2)(4c^2\alpha\gamma - k^2(\alpha-\gamma)^2)};$$

und so kann man mit Leichtigkeit die Reihe der Werthe von x, welche jedoch immer complicirter werden, weiter fortsetzen. Die zu ihnen gehörigen Werthe von y erhält man ohne alle Rechnung durch Vertauschung der Buchstaben, weil die Gleichung (10.) oder (13.) ungeändert bleibt, wenn man ξ mit η, c mit $-c$ und x mit y vertauscht, wodurch auch α und γ vertauscht werden. Wir erhalten so die erste Reihe allgemeiner Formeln für Vierecke mit rationalen Seiten und Diagonalen; was sich folgendermaafsen ausdrücken läfst:

Wenn man in dem Vierecke ABCD den Winkel der Diagonalen so bestimmt, dafs dessen Cosinus eine rationale Zahl c ist, und die vier Abschnitte α, β, γ, δ, in welche die Diagonalen sich gegenseitig theilen, so annimmt, dafs

$$26.\quad \alpha = \frac{(\xi+c)^2-1}{2\xi},\quad \beta = 1,\quad \gamma = \frac{(\eta-c)^2-1}{2\eta},$$

$$\delta = \left(\frac{(\xi+c)^2-1}{2\xi}\right)\cdot\left(\frac{(x-c)^2-1}{2x}\right)$$

ist, wo ξ und η beliebige rationale Zahlen sind, x aber irgend einen der Werthe

$$x' = \frac{2c\alpha}{\alpha-\gamma},$$

$$x'' = -\frac{2c\alpha(4c^2\gamma^2+k^2(\alpha-\gamma)^2)}{(\alpha-\gamma)(4c^2\alpha^2+k^2(\alpha-\gamma)^2)},$$

$$x''' = -\frac{4ck^2\alpha(\alpha-\gamma)(4c^2\gamma^2+k^2(\alpha-\gamma)^2)(2c^2(\alpha^2+\gamma^2)+k^2(\alpha-\gamma)^2)}{(4c^2\alpha^2+k^2(\alpha-\gamma)^2)(4c^2\alpha\gamma+k^2(\alpha-\gamma)^2)(4c^2\alpha\gamma-k^2(\alpha-\gamma)^2)}$$

etc. etc.

erhält: so sind, aufser den beiden Diagonalen, auch alle vier Seiten des Vierecks rational. Die Ausdrücke für die vier Seiten sind:

$$27.\quad \begin{cases} AB = \dfrac{\xi^2+k^2}{2\xi}; & BC = \dfrac{\eta^2+k^2}{2\eta}; \\ CD = \left(\dfrac{(\eta+c)^2-1}{2\eta}\right)\cdot\left(\dfrac{y^2+k^2}{2y}\right); & DA = \left(\dfrac{(\xi+c)^2-1}{2\xi}\right)\cdot\left(\dfrac{x^2+k^2}{2x}\right); \end{cases}$$

wo y, dem jedesmaligen Werthe des x entsprechend, einen der folgenden Werthe hat:

$$y' = \frac{2c\gamma}{\alpha-\gamma},$$

$$y'' = -\frac{2c\gamma(4c^2\alpha^2+k^2(\alpha-\gamma)^2)}{(\alpha-\gamma)(4c^2\gamma^2+k^2(\alpha-\gamma)^2)},$$

$$y''' = -\frac{4ck^2\gamma(\alpha-\gamma)(4c^2\alpha^2+k^2(\alpha-\gamma)^2)(2c^2(\alpha^2+\gamma^2)+k^2(\alpha-\gamma)^2)}{(4c^2\gamma^2+k^2(\alpha-\gamma)^2)(4c^2\alpha\gamma+k^2(\alpha-\gamma)^2)(4c^2\alpha\gamma-k^2(\alpha-\gamma)^2)}$$

etc. etc.

Als Zahlenbeispiel sei $c=\frac{1}{2}$, also $w=60^0$, $\xi=\frac{3}{2}$, $\eta=3$. Dann ist $\alpha=1$, $\beta=1$, $\gamma=\frac{7}{8}$; ferner $x'=8$, also $\delta=\frac{221}{64}$, oder in ganzen Zahlen, $\alpha=64$, $\beta=64$, $\gamma=56$, $\delta=221$; ferner $AB=64$, $BC=104$, $CD=199$, $DA=259$ und die Diagonalen sind: $AC=120$, $BD=285$. Ein anderes Beispiel sei $c=\frac{4}{5}$, $\xi=1$, $\eta=3$; was $\alpha=\frac{28}{25}$, $\beta=1$, $\gamma=\frac{16}{25}$, also $x'=\frac{56}{15}$ und $\delta=\frac{1711}{1500}$ giebt, oder in ganzen Zahlen: $\alpha=1680$, $\beta=1500$, $\gamma=960$, $\delta=1711$; ferner sind die Seiten $AB=1020$, $BC=2340$, $CD=1105$ $DA=3217$, die Diagonalen $AC=2640$, $BD=3211$, und der Inhalt ist gleich 2543112.

Sucht man weiter eine neue Reihe von passenden Werthen von x, so bietet sich zunächst der Werth $x=1+c$ als Fundamentalwerth dar, von welchem sich ausgehen läfst. Es findet sich aus demselben vermittels der Gleichungen (19. und 20.) wirklich eine neue Reihe Werthe von x, und aus diesen eine Reihe allgemeiner Formeln für die Vierecke. Die durch dieselben ausgedrückten Vierecke aber lassen sich alle aus den so eben gefundenen ableiten, wenn man blofs überall $-\frac{\alpha\gamma}{\delta}$ statt δ setzt. Dafs diese Verwandlung an jedem Vierecke, dessen Seiten und Diagonalen rational sind, ausgeführt werden könne, ohne dafs Seiten und Diagonalen aufhörten rational zu sein, ist leicht zu zeigen. Beschreibt man nämlich um das Dreieck CDA (Fig. 6.) einen Kreis, welcher BD in D' trifft, so ist $ABCD'$ ebenfalls ein Viereck mit rationalen Seiten und Diagonalen, und es ist $ED'=-\frac{\alpha\gamma}{\delta}$. Da sich die Verwandlung auf alle vier Abschnitte der Diagonalen anwenden läfst, so erhält man aus einem Vierecke mit rationalen Seiten und Diagonalen noch vier andere; und dazu noch zwei, wenn man die Verwandlung an zwei gegenüberliegenden Winkelspitzen ausführt. Alle übrigen Vierecke, welche durch wiederholte Verwandlungen dieser Art entstehen, sind den sechs bezeichneten ähnlich und geben also nichts Neues.

Eine doppelt-unendliche Reihe wirklich brauchbarer Werthe von x geht aus $x=\eta$ hervor. Dieser Werth von x giebt nach (19.) für z die beiden Werthe $z=\xi\eta$ und $z=-\frac{k^2\eta}{\xi}$. Vermittels einer der Gleichungen (22.) erhält man hieraus folgende zwei Werthe von x:

$$28. \quad x'=-\frac{\eta(\xi\eta-2c\xi+k^2)}{\xi\eta+2c\eta+k^2} \quad \text{und} \quad x'=-\frac{k^2\eta(\xi-\eta+2c)}{2c\xi\eta+k^2\xi-k^2\eta}.$$

Jeder dieser Werthe zieht eine unendliche Reihe anderer nach sich, welche wir hier nicht weiter ausführen wollen. Da ferner auch von diesen beiden Werthen der eine aus dem andern entsteht, wenn $-\frac{k^2}{\eta}$ statt η gesetzt wird, so geben beide keine wesentlich von einander verschiedenen Formeln für Vierecke mit rationalen Stücken. Man kann also auch einen dieser Werthe unbeachtet lassen. Der erste giebt nach Gleichung (9.):

$$\delta = -\frac{\alpha\gamma(\xi\eta+(1+c)^2)(\xi\eta+(1-c)^2)}{(\xi\eta-2c\xi+k^2)(\xi\eta+2c\eta+k^2)}.$$

Setzt man aber $-\frac{\alpha\gamma}{\delta}$ statt δ, welches, wie gezeigt wurde, gestattet ist, so erhält man von δ den einfacheren Werth

$$\delta = \frac{(\xi\eta-2c\xi+k^2)(\xi\eta+2c\eta+k^2)}{(\xi\eta+(1+c)^2)(\xi\eta+(1-c)^2)}.$$

Hieraus ergiebt sich folgender Satz:

Wenn in dem Vierecke ABCD der Winkel der Diagonalen so angenommen wird, dafs der Cosinus c desselben rational ist, und man den vier Abschnitten, in welche die Diagonalen sich gegenseitig theilen, die Werthe:

$$29. \quad \alpha = \frac{(\xi+c)^2-1}{2\xi}, \quad \beta = 1, \quad \gamma = \frac{(\eta-c)^2-1}{2\eta},$$
$$\delta = \frac{(\xi\eta-2c\xi+k^2)(\xi\eta+2c\eta+k^2)}{(\xi\eta+(1+c)^2)(\xi\eta+(1-c)^2)}$$

giebt, wo ξ und η beliebige rationale Zahlen sind, so werden aufser den Diagonalen auch die Seiten des Vierecks rational und es ist:

$$30. \quad \begin{cases} AB = \frac{\xi^2+k^2}{2\xi}, \quad BC = \frac{\eta^2+k^2}{2\eta}, \\ CD = \frac{\eta^2(\xi\eta-2c\xi+k^2)^2+k^2(\xi\eta+2c\eta+k^2)^2}{2\eta(\xi\eta+(1+c)^2)(\xi\eta+(1-c)^2)}, \\ DA = \frac{\xi^2(\xi\eta+2c\eta+k^2)^2+k^2(\xi\eta-2c\xi+k^2)^2}{2\xi(\xi\eta+(1+c)^2)(\xi\eta+(1-c)^2)}. \end{cases}$$

Zu einem Zahlenbeispiel sei $c=\frac{1}{2}$, $w=60^0$, $\xi=\frac{3}{2}$, $\eta=3$. Dies giebt nach Wegschaffung der Brüche, $\alpha=456$, $\beta=456$, $\gamma=399$, $\delta=440$; ferner $AB=456$, $BC=741$, $CD=421$, $DA=776$, und für die Diagonalen $AC=855$, $BD=896$. Ein zweites Beispiel $c=\frac{3}{5}$, $\xi=\frac{6}{5}$, $\eta=\frac{12}{5}$ giebt, nach Wegschaffung der Brüche, $\alpha=4522$, $\beta=4845$, $\gamma=2261$, $\delta=3900$, $AB=4199$, $BC=6460$, $CD=3121$, $DA=7538$, $AC=6783$, $BD=8745$ und den Inhalt gleich 23726934.

Von den unendlich vielen ähnlichen allgemeinen Sätzen wollen wir jetzt nur noch einen entwickeln, und zwar einen solchen, der auch auf den Fall anwendbar ist, wo $c=0$, d. h. wo die Diagonalen auf einander senkrecht stehen; denn die bisher aufgestellten allgemeinen Sätze für die unregelmäfsigen, dem Kreise nicht einschreibbaren Vierecke sind grade für diesen Fall nichtssagend.

Setzt man in der Fundamentalgleichung (13.) $y=\frac{\alpha}{\gamma}(\eta-x)z+\xi$, so erhält man folgende verwandelte Gleichung:

31. $$\alpha x(\eta-x)z^2+(\alpha k^2+2(\xi\gamma+c\alpha+c\gamma)x-\alpha x^2)z+\frac{\gamma\xi}{\eta}(k^2+\eta x)=0,$$

welche, nach Potenzen von x geordnet, auch wie folgt dargestellt werden kann:

32. $$\alpha z(1+z)x^2-[\gamma\xi+2(\xi\gamma+c\alpha+c\gamma)z-\alpha z^2]x-\frac{k^2}{\eta}(\gamma\xi+\alpha\eta z)=0.$$

Nimmt man nun $x=\eta$ an, so erhält man für z den Werth

$$z=-\frac{\xi^2(\eta^2+k^2)}{\eta^2(\xi^2+k^2)}.$$

Zu diesem giebt die quadratische Gleichung (32.) von x, aufser dem Werthe η, noch den Werth:

33. $$x=\frac{2\eta(c\xi\eta+ck^2+k^2\xi-k^2\eta)(\xi^2+k^2)}{((\xi+c)^2-1)(\xi-\eta)(\eta^2+k^2)}.$$

Mit Übergehung der unendlichen Reihe neuer Werthe, welche wieder aus diesem folgen, erhalten wir hieraus folgende neue, allgemeine Regel:

Wenn man die vier Abschnitte α, β, γ, δ der Diagonalen eines Vierecks und den Winkel derselben w so bestimmt, dafs

34. $$\cos w=c,\quad \alpha=\frac{(\xi+c)^2-1}{2\xi},\quad \beta=1,\quad \gamma=\frac{(\eta-c)^2-1}{2\eta},$$
$$\delta=\left(\frac{(\xi+c)^2-1}{2\xi}\right)\cdot\left(\frac{(x-c)^2-1}{2x}\right),$$

wo

$$x=\frac{2\eta(c\xi\eta+ck^2+k^2\xi-k^2\eta)(\xi^2+k^2)}{((\xi+c^2)-1)(\xi-\eta)(\eta^2+k^2)}$$

ist, ξ, η und c aber beliebige rationale Zahlen sind, so sind in diesem Viereck, aufser den beiden Diagonalen, auch die vier Seiten rational; und zwar sind sie:

35. $$\begin{cases} AB=\dfrac{\xi^2+k^2}{2\xi}, & BC=\dfrac{\eta^2+k^2}{2\eta}, \\ CD=\left(\dfrac{(\eta-c)^2-1}{2\eta}\right)\cdot\left(\dfrac{y^2+k^2}{2y}\right), & DA=\left(\dfrac{(\xi+c)^2-1}{2\xi}\right)\cdot\left(\dfrac{x^2+k}{2x}\right), \end{cases}$$

3*

wo

$$\gamma = \frac{2\xi(c\xi\eta+ck^2+k^2\xi-k^2\eta)(\eta^2+k^2)}{((\eta-c)^2-1)(\xi-\eta)(\xi^2+k^2)}.$$

Als Zahlenbeispiel sei $w=60^0$, $c=\frac{1}{2}$, $\xi=-\frac{1}{4}$, $\eta=3$. Dies giebt, nach Aufhebung der Brüche, $\alpha=240$, $\beta=128$, $\gamma=112$, $\delta=57$, $AB=208$, $BC=208$, $CD=97$, $DA=273$, $AC=352$ und $BD=185$.

Für den besondern Fall, wo die Diagonalen auf einander senkrecht stehen, also $c=0$ und $k=1$ ist, läfst sich der Satz wie folgt ausdrücken:

Wenn in einem Vierecke, dessen beide Diagonalen auf einander senkrecht stehen, die vier Abschnitte der Diagonalen die Werthe

$$36.\quad \begin{cases} \alpha = \dfrac{\xi^2-1}{2\xi}, \qquad \beta = 1, \qquad \gamma = \dfrac{\eta^2-1}{2\eta}, \\ \delta = \dfrac{(\xi\eta+\xi+\eta-1)(\xi\eta+\xi-\eta+1)(\xi\eta-\xi+\eta+1)(-\xi\eta+\xi+\eta+1)}{8\xi\eta(\xi^2+1)(\eta^2+1)} \end{cases}$$

haben, wo ξ und η beliebige rationale Zahlen sind, so sind auch die Seiten des Vierecks rational; und zwar sind ihre Ausdrücke:

$$37.\quad \begin{cases} AB = \dfrac{\xi^2+1}{2\xi}, \qquad BC = \dfrac{\eta^2+1}{2\eta}, \\ CD = \dfrac{4\xi^2(\eta^2+1)^2+(\eta^2-1)^2(\xi^2+1)^2}{8\xi\eta(\xi^2+1)(\eta^2+1)}, \\ DA = \dfrac{4\eta^2(\xi^2+1)^2+(\xi^2-1)^2(\eta^2+1)^2}{8\xi\eta(\xi^2+1)(\eta^2+1)}. \end{cases}$$

Als Zahlenbeispiel für diesen Fall, wo $w=90^0$ ist, sei $\xi=2$, $\eta=\frac{3}{2}$; dann ist $\alpha=\frac{3}{4}$, $\beta=1$, $\gamma=\frac{5}{12}$, $\delta=\frac{693}{2080}$, oder in ganzen Zahlen: $\alpha=4680$, $\beta=6240$, $\gamma=2600$, $\delta=2079$, $AB=7800$, $BC=6760$, $CD=3329$, $DA=5121$, $AC=7280$, $BD=8319$ und der Inhalt $=30281160$.

Breslau im December 1846.

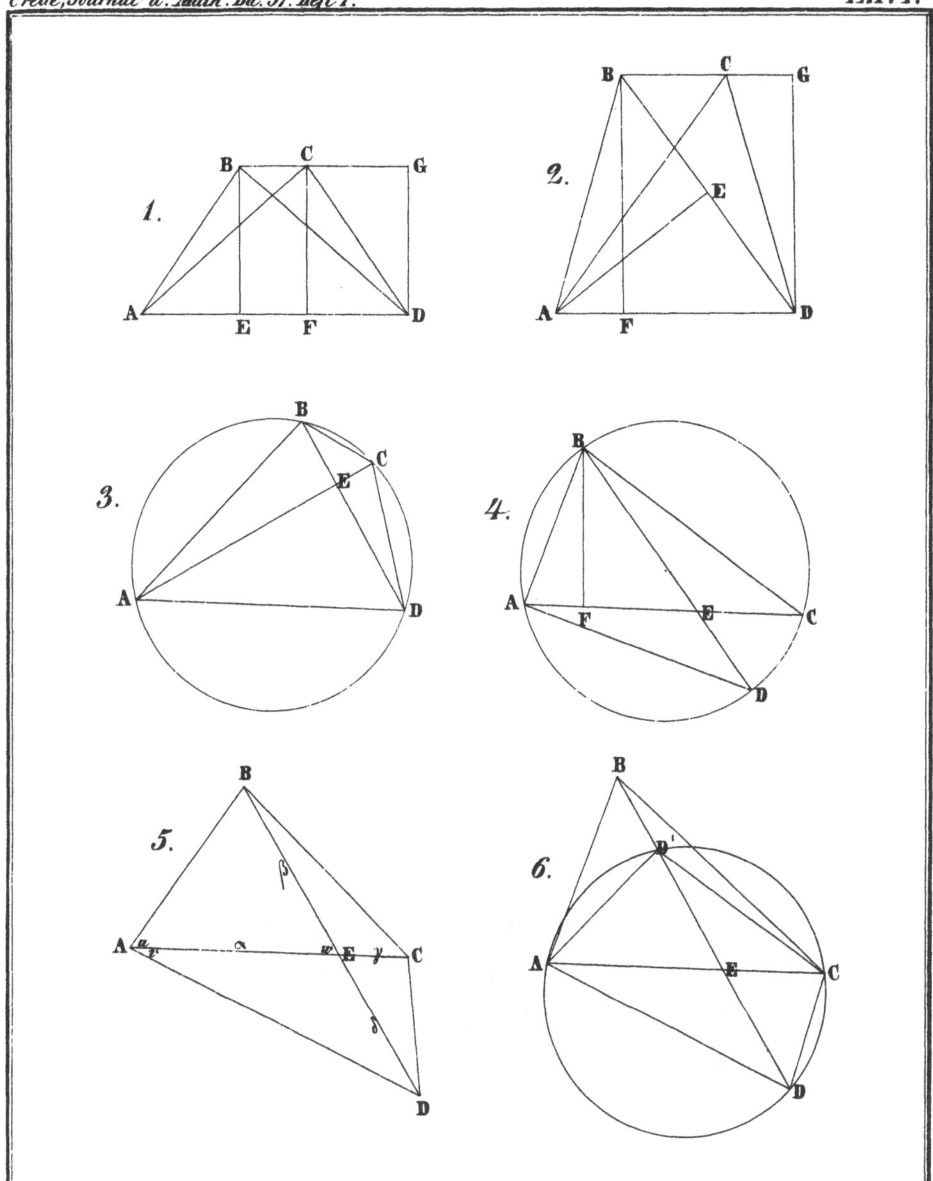
1.
A
B
C
D
E
F
G
2.
A
B
C
D
E
F
G
3.
A
B
C
D
E
4.
A
B
C
D
E
F
5.
A
B
C
D
E
α
β
γ
δ
6.
A
B
C
D
D'
E

Beweis des Fermat'schen Satzes der Unmöglichkeit von $x^\lambda + y^\lambda = z^\lambda$ für eine unendliche Anzahl Primzahlen λ

Monatsberichte der Königlichen Preußischen Akademie der Wissenschaften zu Berlin aus dem Jahre 1847, 132–141, 305–319

Hr. Lejeune Dirichlet theilte folgenden Auszug aus einem an ihn gerichteten Briefe des Hrn. Kummer in Breslau, Correspondenten der Akademie, mit.

Es ist mir neulich gelungen, den Fermatschen Satz der Unmöglichkeit von $x^\lambda - y^\lambda = z^\lambda$ für eine unendliche Anzahl von Primzahlen λ zu beweisen, ich weiſs nur noch nicht recht, für welche; denn der Beweis gründet sich auf zwei Voraussetzungen über die Primzahl λ, zu deren allgemeiner Ergründung eine genauere Erkenntnifs der complexen Einheiten und der Formenanzahlen für die aus λ^{ten} Wurzeln der Einheit gebildeten complexen Zahlen gehört, welche mir jetzt noch nicht zu Gebote steht, Dir aber vielleicht leicht sein wird, weshalb ich mir eben die Freiheit nehme, Dir die Sache mitzutheilen*).

Wenn λ eine Primzahl ist, so haben die aus λ^{ten} Wurzeln der Einheit gebildeten complexen Zahlen, wie ich in einer Abhandlung, die jetzt im Crelleschen Journal gedruckt wird, vollkommen streng bewiesen habe, die Eigenschaft, dafs eine jede nur auf eine einzige Weise in ein Product von Primfactoren (wirklichen oder idealen) zerlegt werden kann. Hieraus folgt unmittelbar der Satz:

Wenn eine complexe Zahl eine Potenz ist und man kann sie in Factoren zerlegen, welche keinen gemeinschaftlichen Theiler haben, so sind diese Factoren für sich ebenfalls solche Potenzen, welche aufserdem nur noch mit complexen Einheiten multiplicirt sein können.

Ich mache nun über die Primzahl λ noch folgende zwei Voraussetzungen:

(*A.*) Es soll λ eine solche Primzahl sein, dafs die Anzahl der nicht äquivalenten Formen, welche zu derselben gehören, nicht durch λ selbst theilbar sei, oder nach meiner Anschauungsweise: dafs die Anzahl aller nicht äquivalenten idealen complexen Zahlen nicht durch λ theilbar sei, oder noch anders ausgesprochen: dafs niemals die λ^{te} Potenz einer idealen complexen Zahl zu einer wirklichen werde.

Ferner:

(*B.*) Es soll λ eine solche Primzahl sein, dafs jede complexe Einheit, welche für den Modul λ einer realen ganzen Zahl congruent wird, nur eine λ^{te} Potenz einer anderen Einheit sei, oder: wenn $\alpha^\lambda = 1$ ist und $E(\alpha)$, $e(\alpha)$ complexe Einheiten bezeichnen, dafs die Congruenz $E(\alpha) \equiv c$, mod. λ (c ganze reale Zahl) nothwendig die Gleichung $E(\alpha) = (e(\alpha))^\lambda$ nach sich zieht.

---

*) Man sehe hierüber die Bemerkung am Ende dieser Mittheilung.

Der umgekehrte Satz dieses letzteren findet, wie sich von selbst versteht, ganz allgemein Statt. Für die Zahlen 3, 5 und 7 finden die gemachten Voraussetzungen wirklich Statt, wie ich streng bewiesen habe. Es ist auch nach meinen bisherigen Arbeiten in dieser Sache höchst wahrscheinlich, dafs sie, wenn nicht für alle Primzahlen, so doch für eine unendliche Anzahl derselben gelten.

Ich beweise zuerst, dafs die Gleichung $x^\lambda - y^\lambda = z^\lambda$ nicht bestehen kann, ohne dafs eine der drei Zahlen x, y, z durch λ theilbar ist. Angenommen, es wäre keine dieser drei Zahlen durch λ theilbar, so würde die Gleichung $x^\lambda - y^\lambda = z^\lambda$ folgende nach sich ziehen:

$$x - \alpha^\varkappa y = E(\alpha^\varkappa) f(\alpha^\varkappa)^\lambda$$

denn die Factoren von $x^\lambda - y^\lambda$, nämlich $x - y$, $x - \alpha y$, $x - \alpha^2 y$ etc. haben keinen gemeinschaftlichen Theiler. Ich mache aus dieser Gleichung eine Congruenz für den Modul λ, wobei ich bemerke, dafs $f(\alpha^\varkappa)^\lambda$ eine w i r k l i c h e complexe Zahl ist, also nach der Voraussetzung (A.) auch $f(\alpha^\varkappa)$ wirklich, und darum $f(\alpha^\varkappa)^\lambda \equiv c$, mod. λ, wo c eine ganze reale Zahl bedeutet. Man erhält so die Congruenz $x - \alpha^\varkappa y \equiv E(\alpha^\varkappa).c$, mod. λ, und wenn $\varkappa = +1$ und $\varkappa = -1$ gesetzt wird,

$$x - \alpha y \equiv E(\alpha).c, \quad x - \alpha^{-1} y \equiv E(\alpha^{-1})c, \text{ mod. } \lambda.$$

Es ist aber nach einer bekannten Eigenschaft der Einheiten

$$E(\alpha^{-1}) = \alpha^r E(\alpha)$$

darum erhält man durch Elimination von $E(\alpha)$ und $E(\alpha^{-1})$ die Congruenz

$$x(\alpha^r - 1) + y(\alpha^{-1} - \alpha^{r+1}) \equiv 0, \text{ mod. } \lambda.$$

Da nun nach der Annahme keine der drei Zahlen x, y, z durch λ theilbar sein soll, so kann diese Congruenz (mit Ausnahme des besonderen Falles wo $\lambda = 3$) nicht anders befriedigt werden, als wenn $r = -1$ genommen wird, wodurch man erhält

$$x + y \equiv 0, \text{ mod. } \lambda.$$

Setzt man die ursprüngliche Gleichung in die Form $x^\lambda - z^\lambda \equiv y^\lambda$ und behandelt ebenso die Factoren von $x^\lambda - z^\lambda$, so erhält man ganz auf dieselbe Weise

$$x + z \equiv 0, \text{ mod. } \lambda,$$

und diese beiden Congruenzen, verbunden mit der Gleichung $x^\lambda = z^\lambda + y^\lambda$, oder mit der aus ihr unmittelbar folgenden Congruenz $x \equiv z + y$, mod. λ, geben $3x \equiv 0$, $3y \equiv 0$, $3z \equiv 0$, mod. λ, welches zeigt, dafs die gemachte Annahme eine sich selbst widersprechende ist, dafs also eine dieser drei Zahlen wirklich durch λ theilbar sein mufs.

Nachdem so dieser Nebenfall abgemacht ist, gehe ich zum Beweise des Hauptfalles über. Anstatt nun hier die Gleichung $x^\lambda - y^\lambda = z^\lambda$ zu nehmen, in welcher z durch λ theilbar sein soll, nehme ich die allgemeinere Gleichung

$$u^\lambda - v^\lambda = E(\alpha)(1-\alpha)^{m\lambda}.\, w^\lambda, \qquad (1)$$

in welcher u, v, w complexe Zahlen sind, ohne gemeinschaftlichen Factor, und $E(\alpha)$ eine complexe Einheit. Ich unterwerfe jedoch die beiden complexen Zahlen u und v der einschränkenden Bedingung, dafs sie folgende Form haben sollen:

$$u = c + (1-\alpha)^{(m-1)\lambda+1}.\, \varphi, \quad v = c + (1-\alpha)^{(m-1)\lambda+1}.\, \psi, \qquad (2)$$

wo c eine reale ganze Zahl ist, φ und ψ beliebige complexe Zahlen. Aufserdem setze ich noch fest, dafs $m > 1$ sein soll.

Die Factoren des $u^\lambda - v^\lambda$, nämlich $u - v$, $u - \alpha v$, $u - \alpha^2 v$, etc. haben nun alle den gemeinschaftlichen Factor $1 - \alpha$, für welchen auch $1 - \alpha^2$, $1 - \alpha^3$, etc. genommen werden kann, aufser diesem aber haben sie keinen, weil sonst u und v denselben haben müfsten. Nach Absonderung der Factoren $1 - \alpha$ müssen diese also selbst λ^{te} Potenzen sein, mit irgend welchen Einheiten multiplicirt. Es ist darum

$$u - v = e(\alpha)(1-\alpha)^{(m-1)\lambda+1}.\, w_,^\lambda \qquad (3)$$

$$u - \alpha^r v = e_r(\alpha)(1-\alpha^r)\, t_r^\lambda, \qquad (4)$$

wo $e(\alpha)$, $e_r(\alpha)$ Einheiten sind, $w_,$, t_r complexe Zahlen, ohne einen gemeinschaftlichen Factor, und zwar wirkliche nach der Voraussetzung (A). Werden für u und v ihre Werthe aus (2) entnommen und in (4) substituirt, so erhält man

$$c + (1-\alpha)^{(m-1)\lambda}.\left(\frac{1-\alpha}{1-\alpha^r}\right)(\varphi - \alpha^r\psi) = e_r(\alpha)\, t_r^\lambda. \qquad (5)$$

Ich mache hieraus eine Congruenz mod. λ, und bemerke, dafs $(1-\alpha)^{\lambda-1}$, also um so mehr $(1-\alpha)^{(m-1)\lambda}$ (da $m > 1$ ist), durch λ theilbar ist. Es wird daher

$$c \equiv e_r(\alpha)\, t_r^\lambda, \text{ mod. } \lambda,$$

4**

und weil die λ^{te} Potenz der wirklichen complexen Zahl t_r einer realen ganzen Zahl b congruent ist, so ist

$$c \equiv e_r(\alpha)\,.\,b, \text{ mod. } \lambda,$$

also $e_r(\alpha)$ selbst ist einer realen ganzen Zahl congruent für den Modul λ, darum muſs nach der Voraussetzung (B) $e_r(\alpha)$ sich als λ^{te} Potenz einer anderen Einheit darstellen lassen, und es wird

$$e_r(\alpha)\,t_r^\lambda = u_,^\lambda.$$

Hiernach nimmt die Gleichung (4) folgende Gestalt an:

$$(6)\qquad u - \alpha^r v = (1-\alpha^r)\,u_,^\lambda,$$

und wenn für r ein anderer Werth s gesetzt wird, ist ebenso

$$(7)\qquad u - \alpha^s v = (1-\alpha^s)\,v_,^\lambda.$$

Aus (6) und (7) hat man

$$u_,^\lambda - v_,^\lambda = \frac{u-\alpha^r v}{1-\alpha^r} - \frac{u-\alpha^s v}{1-\alpha^s} = \frac{(\alpha^r-\alpha^s)(u-v)}{(1-\alpha^r)(1-\alpha^s)},$$

also, wenn für $u - v$ sein Werth aus (3) gesetzt wird:

$$u_,^\lambda - v_,^\lambda = \frac{e(\alpha)\,(\alpha^r-\alpha^s)(1-\alpha)}{(1-\alpha^r)(1-\alpha^s)}\,(1-\alpha)^{(m-1)\lambda}\,.\,w_,^\lambda,$$

und wenn $\dfrac{e(\alpha)\,(\alpha^r-\alpha^s)(1-\alpha)}{(1-\alpha^r)(1-\alpha^s)}$, welches wieder eine Einheit ist, durch $E_,(\alpha)$ bezeichnet wird, so ist:

$$(8)\qquad u_,^\lambda - v_,^\lambda = E_,(\alpha)\,(1-\alpha)^{(m-1)\lambda}\,.\,w_,^\lambda.$$

Dieſs ist eine Gleichung derselben Form als die Gleichung (1), nur mit dem Unterschiede, daſs m um eine Einheit kleiner ist. Damit nun dieselbe Schluſsfolge ebenso wieder für diese Gleichung Statt habe, und diese ebenso eine neue Gleichung derselben Form gewähre, müssen wir nur noch beweisen, daſs auch für sie die beiden Bedingungen

$$u_, = c_, + (1-\alpha)^{(m-2)\lambda+1}\,.\,\varphi_,,\qquad v_, = c_, + (1-\alpha)^{(m-2)\lambda+1}\,.\,\psi_,,$$

welche den Bedingungen (2) entsprechen, mit erfüllt sind. Zu diesem Zwecke nehme ich die Gleichung (5), in welcher $e_r(\alpha)\,t_r^\lambda = u_,^\lambda$ zu setzen ist, und verwandle sie in eine Congruenz für den Modul $(1-\alpha)^{(m-1)\lambda}$, so wird:

$$(9)\qquad c \equiv u_,^\lambda, \text{ mod. } (1-\alpha)^{(m-1)\lambda}.$$

Ich setze nun $u_{,}$ in die Form $u_{,} = a + (1-\alpha)\,\theta$ (wo a real und θ complex ist), welche jede complexe Zahl annehmen kann, hierdurch wird nach dem binomischen Lehrsatze:

$$u_{,}^{\lambda} = a^{\lambda} + \lambda(1-\alpha)\,a^{\lambda-1}\,\theta + \lambda\left(\frac{\lambda-1}{2}\right)(1-\alpha)^2 a^{\lambda-2}\,\theta^2 + \ldots + (1-\alpha)^{\lambda}\theta^{\lambda},$$

und wenn man bedenkt, dafs die Binomial-Coefficienten den Factor λ enthalten, welcher selbst durch $(1-\alpha)^{\lambda-1}$ theilbar ist, so hat man sogleich

$$u_{,}^{\lambda} \equiv a^{\lambda},\ \text{mod.}\ (1-\alpha)^{\lambda}. \qquad (10)$$

Diese Congruenz verbunden mit (9) giebt

$$c \equiv a^{\lambda},\ \text{mod.}\ (1-\alpha)^{\lambda},$$

oder was dasselbe ist:

$$c \equiv a^{\lambda},\ \text{mod.}\ \lambda\,(1-\alpha),\ \text{oder}\ \frac{c-a^{\lambda}}{\lambda} \equiv 0.\ \text{mod.}\ (1-\alpha).$$

Die reale ganze Zahl kann aber nicht durch $1-\alpha$ theilbar sein ohne durch λ theilbar zu sein, darum ist $\frac{c-a^{\lambda}}{\lambda}$ durch λ theilbar, oder $$c \equiv a^{\lambda},\ \text{mod.}\ \lambda^2.$$

Wenn nun c die Eigenschaft hat, einer λ^{ten} Potenz congruent zu sein für den Modul λ^2, so ist es auch, wenn eine beliebig hohe Potenz von λ, z. B. λ^{μ}, als Modul genommen wird, einer λ^{ten} Potenz congruent, darum hat man $c \equiv c_{,}^{\lambda}$ mod. λ^{μ}, für jeden Werth des μ, also auch

$$c \equiv c_{,}^{\lambda},\ \text{mod.}\ (1-\alpha)^{(m-1)\lambda}.$$

Die Congruenz (9) verwandelt sich daher in folgende

$$u_{,}^{\lambda} - c_{,}^{\lambda} \equiv 0,\ \text{mod.}\ (1-\alpha)^{(m-1)\lambda}.$$

Von den Factoren des $u_{,}^{\lambda} - c_{,}^{\lambda}$ (nämlich $u_{,} - c_{,}$, $\alpha u_{,} - c_{,}$, $\alpha^2 u_{,} - c_{,}$ etc.) mufs nun wieder jeder den Factor $1-\alpha$ einmal enthalten, einer aber mufs ihn $(m-1)\,\lambda - \lambda + 1$ mal enthalten, und weil $u_{,}$ insofern unbestimmt ist, dafs man dafür auch $\alpha^{\varkappa} u_{,}$ setzen kann, so kann man als diesen Factor offenbar den ersten $u_{,} - c_{,}$ wählen. Man hat also $u_{,} - c_{,} \equiv 0$ mod. $(1-\alpha)^{(m-2)\lambda+1}$, oder

$$u_{,} = c_{,} + (1-\alpha)^{(m-2)\lambda+1}.\ \phi_{,}. \qquad (11)$$

Dasselbe gilt nun offenbar auch für $v_{,}$, so dafs ebenso

$$v_{,} = c_{,} + (1-\alpha)^{(m-2)\lambda+1}.\ \psi_{,}. \qquad (12)$$

Es bleiben also auch diese beiden Bedingungen für die neue Gleichung (8) erfüllt. Man kann darum ebenso aus dieser wieder eine neue Gleichung derselben Form, mit den entsprechenden beiden Bedingungen ableiten und so fort, bis man endlic weil bei jeder solchen Operation m um eine Einheit abnimmt, dahin gelangt, dafs $m = 1$ ist, wo diese Operation nicht weiter fortgesetzt werden kann. Es bleibt also folgende Gleichung übrig, auf deren Erfüllung die Gleichung (1) immer beruht:

$$(13) \qquad u^\lambda - v^\lambda = E(\alpha)(1-\alpha)^\lambda . w^\lambda .$$

Die Unmöglichkeit dieser Gleichung läfst sich aber sehr einfach dadurch darthun, dafs gezeigt wird: dafs die Form $u^\lambda - v^\lambda$, wenn sie überhaupt den Factor $1-\alpha$ enthält, denselben immer wenigstens $\lambda+1$ mal enthalten mufs. Ich beweise diefs auf folgende Weise: Zunächst mufs jeder der λ Factoren $u-v$, $u-\alpha v$, $u-\alpha^2 v$ den Factor $1-\alpha$ überhaupt enthalten, also namentlich auch $u-v$. Entwickelt man nun die complexen Zahlen u und v nach Potenzen von $1-\alpha$, wobei nur die ersten beiden Glieder in Betracht kommen, so erhält man

$$u = a + b(1-\alpha) + (1-\alpha)^2 \varphi, \quad v = a + c(1-\alpha) + (1-\alpha)^2 \psi,$$

ferner wird $\alpha^r = \big(1-(1-\alpha)\big)^r = 1 - r(1-\alpha) + (1-\alpha)^2 . f$, also

$$u - \alpha^r v = (b - c + ar)(1-\alpha) + (1-\alpha)^2 F$$

für alle Werthe $r = 0, 1, 2 \ldots . \lambda - 1$. Für einen dieser Werthe ist aber offenbar auch $b - c + ar \equiv 0 \bmod. \lambda$, also $b - c + ar$ durch $1-\alpha$ theilbar, also $u - \alpha^r v$ durch $(1-\alpha)^2$ theilbar. Da nun die übrigen $\lambda - 1$ Factoren jeder einmal den Factor $1-\alpha$ enthalten, so folgt, dafs $u^\lambda - v^\lambda$ diesen Factor stets wenigstens $\lambda+1$ mal enthält, oder dafs es durch $(1-\alpha)^{\lambda+1}$ theilbar sein mufs. Die Gleichung (13) (in welcher w keinen Factor $1-\alpha$ enthält) ist also unmöglich und darum ist auch die Gleichung (1) mit ihren beiden Bedingungen (2) unmöglich zu befriedigen.

Die Gleichung $x^\lambda - y^\lambda = z^\lambda$, wo z durch λ theilbar, ist in der Gleichung (1) enthalten, auch finden für dieselbe die beiden Bedingungen (2) mit Statt, darum ist die Gleichung $x^\lambda - y^\lambda = z^\lambda$ überhaupt unmöglich für alle die Primzahlen λ, welche den oben aufgestellten Voraussetzungen (*A.*) und (*B.*) genügen.

Der Fermatsche Satz ist zwar mehr ein Curiosum als ein Hauptpunkt der Wissenschaft, dessen ungeachtet halte ich diese meine Beweisart für bemerkenswerth, da sich dieselbe von den bisher gebrauchten Methoden darin wesentlich unterscheidet, dafs hier nur eine endliche Reihe von Gleichungen abgeleitet wird, welche damit schliefst, dafs die letzte derselben wegen einer einfachen Congruenz-Bedingung unmöglich ist. Die beiden Voraussetzungen (A) und (B) aber scheinen mir viel wissenswerther zu sein, als der Fermatsche Satz selbst, und wenn ich mich nicht sehr täusche, oder vielmehr, wenn die Formenanzahl der complexen Zahlen in ihrem Zusammenhange mit den complexen Einheiten die Analogie der quadratischen Formen befolgt, so bilden diese beiden Voraussetzungen wesentlich nur eine, oder es ist stets eine mit der andern zugleich erfüllt. Sollte sich dies bestätigen, so würde also der hier gegebene Beweis des Fermatschen Satzes nur die eine Voraussetzung (A.) zu seiner Richtigkeit nöthig haben. Ich neige mich zu der Ansicht hin, dafs es wirklich solche Primzahlen λ giebt, deren Formenanzahl durch λ theilbar ist, und ich habe Gründe zu der Vermuthung, dafs z. B. die Zahl $\lambda = 37$ zu denselben gehört. Jedenfalls wäre die Ergründung der beiden Voraussetzungen (A) und (B) und ihres etwaigen Zusammenhanges als Vervollständigung meines Beweises für mich von grofsem Werth, weshalb ich diese Untersuchung Deiner Aufmerksamkeit dringend empfehle.

Breslau, 11. April 1847.

Bemerkung von Hrn. Lejeune Dirichlet zu vorstehender Mittheilung:

Was die zweite der beiden Voraussetzungen betrifft, welche dem scharfsinnigen Beweise des Hrn. Kummer zu Grunde liegen, so läfst sich deren Richtigkeit für jeden besondern Werth von λ mit Hülfe der allgemeinen Theorie der complexen Einheiten prüfen, über welche im Märzbericht von 1846 einige Andeutungen gegeben worden sind und welche in einem der nächsten Hefte des Crelleschen Journals bekannt gemacht werden wird. Nach der erwähnten Theorie läfst sich nämlich für jedes λ der allgemeine Ausdruck aller aus λ^{ten} Wurzeln der Einheit zusammengesetzten complexen Einheiten aufstellen, und die Bildung dieses Ausdrucks bietet keine andere Schwierigkeit dar, als die einer mit wachsendem λ rasch an Complication zunehmenden numerischen Rechnung. Ist dieser Ausdruck, der $\frac{\lambda-3}{2}$ zu unbestimmten Potenzen erhobene Fundamentaleinheiten enthält, bekannt, so läſst sich ohne groſse Mühe bald entscheiden, ob die in der Voraussetzung (B) ausgesprochene Bedingung erfüllt ist, d. h. ob der Ausdruck nur dann nach dem Modul λ einer reellen Zahl congruent werden kann, wenn alle Exponenten durch λ aufgehen.

Die Voraussetzung (A) bezieht sich auf eine Theorie, welche mit der der quadratischen Formen die gröſste Analogie hat. Wie nämlich nicht jede Zahl m, für welche die Congruenz $\xi^2 \equiv D \,(\text{mod. } m)$ möglich ist, immer in der Form $x^2 - Dy^2$ enthalten ist, sondern im Allgemeinen eine beschränkte Anzahl wesentlich verschiedener quadratischer Formen existirt, durch welche sämmtliche Zahlen m dargestellt werden können, so finden ähnliche Beziehungen zwischen höheren Congruenzen und ihnen entsprechenden höheren Formen Statt. Betrachtet man z. B. die Congruenz $\frac{\xi^\lambda - 1}{\xi - 1} \equiv 0 \,(\text{mod. } m)$, hinsichtlich welcher schon Euler die Moduln m, für welche sie möglich ist, vollständig bestimmt hat, so sind auch diese Zahlen m nicht immer von derForm $\phi(\alpha)\,\phi(\alpha^2) \ldots\ldots \phi(\alpha^{\lambda-1})$, wo $\phi(\alpha) = x_0 + \alpha x_1 + \alpha^2 x_2 + \ldots\ldots + \alpha^{\lambda-2} x_{\lambda-2}$, und $x_0, x_1 \ldots\ldots x_{\lambda-2}$ unbestimmte ganze Zahlen bezeichnen, aber es existiren immer Formen in endlicher Anzahl, welche wie die vorige in lineare Factoren zerlegt werden können, und mit dieser vereinigt alle Zahlen m ausdrücken, für welche die Eulersche Congruenz auflösbar ist. Nachdem die Darstellbarkeit aller Einheiten durch $\frac{\lambda-3}{2}$ Fundamentaleinheiten erkannt worden war, was hier das Analogon von der allgemeinen Lösung der Fermatschen Gleichung $x^2 - Dy^2 = 1$ ist, lag der Versuch nahe, die Analogie zwischen den quadratischen und diesen höheren Formen weiter zu verfolgen und namentlich die Anzahl der letztern durch ähnliche Mittel zu bestimmen, durch welche dieselbe Frage in der Theorie der quadratischen Formen früher erledigt worden war. Diese Untersuchung, auf welche sich Hr. Kummer im Eingange seiner Mittheilung bezieht, ist denn auch vor etwa drei Jahren mit Hülfe eines neuen Princips, dessen es bei der Ermittelung der

Formenzahl für den 2ten Grad nicht bedurft hatte, glücklich zu Ende und zu einem Resultate geführt worden, welches durch seine Form merkwürdig scheint und so einfach ist, als man es bei einer Frage, die Formen aller Grade umfafst, nur immer erwarten konnte. Der von mir für die Anzahl der Formen $\lambda - 1^{\text{ten}}$ Grades gefundene Ausdruck, welcher, wie die Analogie mit den quadratischen Formen vorhersehen liefs, die oben erwähnten $\frac{\lambda-3}{2}$ Fundamentaleinheiten enthält, giebt, sobald diese bekannt sind, das Mittel, durch eine ziemlich einfache numerische Rechnung die Voraussetzung (A) zu prüfen. Ob es aber möglich sein werde, aus der Art, wie die Fundamentaleinheiten in den Ausdruck für die Formenzahl eingehn, etwas Allgemeines über den Zusammenhang der beiden Voraussetzungen abzuleiten und die von Hrn. Kummer am Ende seines Briefes ausgesprochene Vermuthung zu prüfen, dafs die erste Voraussetzung die zweite immer involvire, darüber wage ich für jetzt nicht zu entscheiden. Eine solche Entscheidung wird nur das Ergebnifs einer sorgfältigen aus dem eben besprochenen Gesichtspunkt vorzunehmenden Discussion des Ausdrucks für die Formenzahl sein können.

Hierauf legte Hr. Lejeune Dirichlet folgende von Hrn. Kummer, Correspondenten der Akademie, eingegangene Mittheilung vor.

Mein Beweis des Fermatschen Satzes, welchen Hr. Lejeune Dirichlet der Königlichen Akademie der Wissenschaften mitgetheilt hat, gründet sich auf zwei Voraussetzungen, von denen ich damals noch nicht allgemein entscheiden konnte, für welche Primzahlen sie gelten und für welche nicht. Es fehlte mir hierzu namentlich noch der Ausdruck für die Anzahl der nichtäquivalenten Klassen aller idealen complexen Zahlen, oder ihrer zugehörigen Formen, über welche Hr. Dirichlet schon längst eine Abhandlung versprochen hat. Nachdem ich nun seither auf das Erscheinen derselben gewartet hatte, habe ich es unternommen, mit Hülfe einiger von Dirichlet mündlich erhaltenen Andeutungen, den verlangten Ausdruck selbst herzuleiten, und es ist mir nicht nur gelungen denselben zu finden,

9

sondern auch die beiden Voraussetzungen meines Beweises des Fermatschen Satzes aus ihm vollständig zu ergründen. Den Ausdruck für die Anzahl der nicht äquivalenten Klassen idealer complexer Zahlen stelle ich hier nur als Ausgangspunkt auf, und verweise für den Beweis desselben auf die zu erwartende Abhandlung von Dirichlet, welchem diese Untersuchung als unbestrittenes Eigenthum angehört; für die daraus zu ziehenden Folgerungen zur Vervollständigung meines Beweises des Fermatschen Satzes aber werde ich mir erlauben die nöthigen Entwickelungen der Akademie der Wissenschaften in der Kürze mitzutheilen.

Es sei λ eine ungerade Primzahl, α eine imaginäre Wurzel der Gleichung $\alpha^\lambda = 1$, β eine primitive Wurzel der Gleichung $\beta^{\lambda-1} = 1$, und g eine primitive Wurzel der Congruenz $g^{\lambda-1} \equiv 1$, mod. λ; die kleinsten positiven Reste, welche g, g^2, g^3 etc. für den Modul λ lassen, sollen durch g_1, g_2, g_3 etc. bezeichnet werden, auch soll der Kürze wegen $\frac{\lambda - 1}{2} = \mu$ gesetzt werden. Ferner sei

$$\varepsilon_1(\alpha),\ \varepsilon_2(\alpha)\ \ldots\ldots\ \varepsilon_{\mu-1}(\alpha)$$

ein System von Fundamental-Einheiten für die aus λ^{ten} Wurzeln der Einheit gebildeten complexen Zahlen, und Δ die Determinente der Gröſsen:

$$\begin{array}{llll}
l\varepsilon_1(\alpha), & l\varepsilon_2(\alpha), & \ldots\ldots\ldots & l\varepsilon_{\mu-1}(\alpha) \\
l\varepsilon_1(\alpha^g), & l\varepsilon_2(\alpha^g), & \ldots\ldots\ldots & l\varepsilon_{\mu-1}(\alpha^g) \\
\vdots & & & \vdots \\
l\varepsilon_1(\alpha^{g^{\mu-2}}), & l\varepsilon_2(\alpha^{g^{\mu-2}}), & \ldots\ldots & l\varepsilon_{\mu-1}(\alpha^{g^{\mu-2}}).
\end{array}$$

Ferner sei

$$e(\alpha) = \sqrt{\frac{(1-\alpha^g)\,(1-\alpha^{-g})}{(1-\alpha)\,(1-\alpha^{-1})}},$$

welches eine ganze complexe Einheit ist, und sei D die Determinante der Gröſsen:

$$\begin{array}{llll}
le(\alpha), & le(\alpha^g) & \ldots\ldots & le(\alpha^{g^{\mu-2}}) \\
le(\alpha^g), & le(\alpha^{g^2}) & \ldots\ldots & le(\alpha^{g^{\mu-1}}) \\
\vdots & & & \vdots \\
le(\alpha^{g^{\mu-2}}), & le(\alpha^{g^{\mu-1}}) & \ldots\ldots & le(\alpha^{g^{2\mu-4}}).
\end{array}$$

Ferner sei

$$\phi(\beta) = 1 + g_1\beta + g_2\beta^2 + g_3\beta^3 + \ldots + g_{\lambda-2}\beta^{\lambda-2}$$

und

$$P = \phi(\beta).\ \phi(\beta^3)\,\phi(\beta^5)\ldots\phi(\beta^{\lambda-2}).$$

Endlich bezeichne noch H die Anzahl der nichtäquivalenten Klassen aller idealen complexen Zahlen: so ist

$$H = \frac{P}{(2\lambda)^{\mu-1}} \cdot \frac{D}{\Delta}$$

und es ist sowohl $\frac{P}{(2\lambda)^{\mu-1}}$, als auch $\frac{D}{\Delta}$ jedes für sich gleich einer ganzen Zahl. *)

Da nun die erste der beiden Voraussetzungen, auf welche ich meinen Beweis des Fermatschen Satzes gegründet habe, die war, dafs die Anzahl der nichtäquivalenten Klassen aller idealen complexen Zahlen, d. i. H, nicht durch λ theilbar sein soll, so werde ich die beiden Factoren, aus welchen H besteht besonders untersuchen, indem ich mit dem ersten $\frac{P}{(2\lambda)^{\mu-1}}$ anfange. Aus

$$\phi(\beta) = 1 + g_1\beta + g_2\beta^2 + \ldots + g_{\lambda-2}\beta^{\lambda-2}$$

folgt

$$(g\beta - 1)\,\phi(\beta) = gg_{\lambda-2} - 1 + (gg_{\lambda-1} - g_1)\,\beta + (gg_1 - g_2)\beta^2 + gg_{\lambda-3} - g_{\lambda-2})\,\beta^{\lambda-2}$$

und es sind nun, wie klar ist, die Coefficienten aller einzelnen Glieder durch λ theilbar, setzt man also

$$gg_{k-1} - g_k = \lambda b_k$$

---

*) Ich habe nicht allein diese Klassenzahl, sondern auch die entsprechenden für alle nicht aus den einfachen Wurzeln der Gleichung $\alpha^\lambda = 1$, sondern aus den Perioden derselben gebildeten complexen Zahlen vollständig gefunden, und daraus bewiesen, dafs der Factor $\frac{D}{\Delta}$ für sich genau die Anzahl der nicht äquivalenten Klassen aller aus den zweigliedrigen Perioden $\alpha + \alpha^{-1}$, $\alpha^2 + \alpha^{-2}$ etc. gebildeten complexen Zahlen ausdrückt, also auch stets eine ganze Zahl ist.

und

$$b_0 + b_1 \beta + b_2 \beta^2 + \dots\dots + b_{\lambda-2} \beta^{\lambda-2} = \psi(\beta),$$

so hat man

$$(g\beta - 1)\,\phi(\beta) = \lambda\,\psi(\beta),$$

also auch

$$(g^\mu + 1)\,P = \lambda^\mu \psi(\beta)\,\psi(\beta^3)\,\psi(\beta^5) \dots\dots \psi(\beta^{\lambda-2}).$$

$g^\mu + 1$, da $\mu = \frac{\lambda - 1}{2}$, ist bekanntlich durch λ theilbar, man kann also setzen $g^\mu + 1 = \lambda G$ und hat sodann

$$\frac{GP}{\lambda^{\mu-1}} = \psi(\beta)\,\psi(\beta^3)\,\psi(\beta^5) \dots \psi(\beta^{\lambda-2})$$

auch kann man immer die primitive Wurzel g so wählen, dafs $g^\mu + 1$ nicht durch λ^2 theilbar ist, dafs also G nicht weiter durch λ theilbar ist. Es kann also $\frac{P}{(2\lambda)^{\mu-1}}$ nur dann durch λ theilbar sein, wenn $\psi(\beta)\,\psi(\beta^3) \dots\dots \psi(\beta^{\lambda-2})$ durch λ theilbar ist, und umgekehrt. Nun ist aber offenbar, wenn man anstatt der primitiven Gleichungswurzel β der Gleichung $\beta^{\lambda-1} = 1$, die primitive Congruenzwurzel g der Congruenz $g^{\lambda-1} \equiv 1$, mod. λ, setzt

$$\psi(\beta)\,\psi(\beta^3) \dots\dots \psi(\beta^{\lambda-2}) \equiv \psi(g)\,\psi(g^3) \dots \psi(g^{\lambda-2}) \text{ mod. } \lambda.$$

also die Bedingung, $\frac{P}{(2\lambda)^{\mu-1}}$ theilbar durch λ, ist identisch mit der, dafs einer der Factoren des Productes $\psi(g) \,.\, \psi(g^3) \dots\dots \psi(g^{\lambda-2})$ durch λ theilbar ist, also

$$b_0 + b_1 g^{2n-1} + b_2 g^{2(2n-1)} + \dots + b_{\lambda-2} g^{(\lambda-2)(2n-1)} \equiv o, \text{ mod. } \lambda.$$

für irgend einen der Werthe $n = 1, 2, 3, \dots \mu$, welche Congruenz, wenn sie durch g^{2n-1} dividirt wird, und wenn die Exponenten des g zum Theil zu Indices gemacht werden, auch so geschrieben werden kann.

$$b_0 g_{\lambda-2}^{2n-1} + b_1 + b_2 g_1^{2n-1} + b_3 g_2^{2n-1} + \dots + b_{\lambda-2} g_{\lambda-3}^{2n-1} \equiv o, \text{ mod. } \lambda.$$

Aus der Definition der Coefficienten $\lambda b_k = g g_{k-1} - g_k$ folgt nun aber unmittelbar, dafs erstens $b_k = o$ ist, für alle Werthe des g_{k-1}, welche zwischen o und $\frac{\lambda}{g}$ liegen, ferner $b_k = 1$, für alle

Werthe des g_{k-1}, welche zwischen $\frac{\lambda}{g}$ und $\frac{2\lambda}{g}$ liegen, oder allgemein $b_k = s$, für alle Werthe des g_{k-1}, welche zwischen $\frac{s\lambda}{g}$ und $\frac{(s+1)\lambda}{g}$ liegen. Bezeichnet nnn t_s die gröſste in $\frac{s\lambda}{g}$ enthaltene ganze Zahl, so kann man, diejenigen Glieder zusammenfassend, für welche die Coefficienten b_k gleiche Werthe haben, die obige Congruenz folgendermaſsen darstellen:

$$\begin{aligned}
&+1\left((t_1+1)^{2n-1}+(t_1+2)^{2n-1}+\ldots+t_2^{2n-1}\right)\\
&+2\left((t_2+1)^{2n-1}+(t_2+2)^{2n-1}+\ldots+t_3^{2n-1}\right)\\
&\quad\vdots\\
&+(g-1)\left((t_{g-1}+1)^{2n-1}+(t_{g-1}+2)^{2n-1}+\ldots\right.\\
&\qquad\left.+(\lambda-1)^{2n-1}\right)\equiv o, \text{ mod. } \lambda.
\end{aligned}$$

Ich mache jetzt von folgender bekannten ganzen rationalen Function Gebrauch:

$$X(x)=\frac{x^{2n}}{\Pi_{2n}}-\frac{x^{2n-1}}{2\Pi_{2n-1}\Pi_1}+\frac{B_1\,x^{2n-2}}{\Pi_{2n-2}\Pi_2}-\ldots\ldots\frac{(-1)^n B_{n-1}x^2}{\Pi_2\Pi_{2n-2}}$$

in welcher $B_1, B_2 \ldots\ldots B_{n-1}$ die Bernouillischen Zahlen sind und $\Pi r = 1.\ 2.\ 3 \ldots\ldots r$. Diese Function $X(x)$ stellt, wenn x eine ganze Zahl ist, die Summe der Reihe $1^{2n-1}+2^{2n-1}+3^{2n-1}+\ldots+(x-1)^{2n-1}$ dar, dividirt durch Π_{2n-1}, darum läſst sich vermittelst derselben die obige Congruenz folgendermaſsen darstellen:

$$X(t_1+1)+X(t_2+1)+\ldots\ldots+X(t_{g-1}+1)\equiv o, \text{ mod. } \lambda.$$

Da t_s die in $\frac{s\lambda}{g}$ enthaltene gröſste ganze Zahl ist, so kann man setzen

$$t_s=\frac{s\lambda-r_s}{g}$$

wo r_s positiv und kleiner als g ist. Hierdnrch erhält man

$$\begin{aligned}
&X\left(\frac{\lambda-r_1+g}{g}\right)+X\left(\frac{2\lambda-r_2+g}{g}\right)\\
&+\ldots+X\left(\frac{(g-1)\lambda-r_{g-1}+g}{g}\right)\equiv o.
\end{aligned}$$

Läfst man jetzt die Vielfachen von λ weg, und bemerkt, dafs $r_1, r_2, \ldots r_{g-1}$, wenn auch in anderer Ordnung, mit den Zahlen $1, 2, 3, \ldots g-1$ zusammenfallen, so hat man

$$X\left(\frac{1}{g}\right) + X\left(\frac{2}{g}\right) + \ldots + X\left(\frac{g-1}{g}\right) \equiv o, \text{ mod. } \lambda,$$

welche Congruenz zwar Brüche enthält, die aber, weil der Modul λ in keinem der Nenner vorkommt, sogleich auf ganze Zahlen gebracht werden können und darum nicht stören. Die Function $X(x)$ hat unter anderen merkwürdigen Eigenschaften auch die, dafs sie sich in folgende unendliche Reihe entwickeln läfst:

$$X(x) = \frac{(-1)^n B_n}{\Pi_{2n}} - \frac{(-1)^n . 2}{(2\pi)^{2n}}\left(\frac{\cos 2x\pi}{1^{2n}} + \frac{\cos 4x\pi}{2^{2n}} + \frac{\cos 6x\pi}{3^{2n}} + \ldots.\right)$$

gültig in den Grenzen $x = o$ bis $x = 1$. Setzt man in diesem Ausdrucke nach einander $x = o$, $\frac{1}{g}$, $\frac{2}{g}$, $\ldots$ $\frac{g-1}{g}$ und addirt, wobei zu bemerhen ist, dafs $X(o) = o$, so wird im allgemeinen

$$1 + \cos\frac{2k\pi}{g} + \cos\frac{4k\pi}{g} + \ldots. + \cos\frac{2(g-1)k\pi}{g} = o,$$

nur wenn k ein Vielfaches von g ist, wird diese Summe nicht gleich Null, sondern gleich g, darum wird

$$X\left(\frac{1}{g}\right) + X\left(\frac{2}{g}\right) + \ldots. + X\left(\frac{g-1}{g}\right) = \frac{(-1)^n g B_n}{\Pi_{2n}} - \frac{(-1)^n 2g}{(2\pi)^n}\left(\frac{1}{g^{2n}} + \frac{1}{(2g)^{2n}} + \frac{1}{(3g)^{2n}} + \ldots \text{in inf.}\right)$$

es ist aber bekanntlich

$$1 + \frac{1}{2^{2n}} + \frac{1}{3^{2n}} + \ldots \text{in inf.} = \frac{(2\pi)^{2n} B_n}{2\Pi_{2n}},$$

darum wird

$$X\left(\frac{1}{g}\right)+X\left(\frac{2}{g}\right)+\ldots+X\left(\frac{g-1}{g}\right)=\frac{(-1)^n g B_n}{\Pi_{2n}}$$
$$-\frac{(-1)^n g B_n}{g^{2n}\Pi_{2n}}$$

also

$$X\left(\frac{1}{g}\right)+X\left(\frac{2}{g}\right)+\ldots+X\left(\frac{g-1}{g}\right)=\frac{(-1)^n (g^{2n}-1) B_n}{g^{2n-1}\Pi_{2n}}.$$

Die Congruenz also, von deren Erfüllung oder Nichterfüllung es abhängt, ob der erste Factor der Klassenanzahl, nämlich $\frac{P}{(2\lambda)^{\mu-1}}$, durch λ theilbar ist, oder nicht, hat nun folgende Gestalt angenommen:

$$\frac{(g^{2n}-1)B_n}{g^{2n-1}\Pi_{2n}} \equiv o, \text{ mod. } \lambda.$$

Der Nenner, als nicht durch λ theilbar, kann sogleich wegfallen, aufserdem ist aber auch $g^{2n}-1$ nicht dusch λ theilbar für die Werthe $n=1, 2, 3, \ldots \mu-1$, für diese also geht die Bedingungs-Congruenz einfach in

$$B_n \equiv o, \text{ mod. } \lambda,$$

über. Was den Fall $n=\mu$ betrifft, so enthält für diesen $g^{2n}-1$ den Factor λ, und zwar nur einmal, weil $g^{\mu}+1$ ihn nur einmal enthält, aber dafür hat auch B_{μ} den Factor $2\mu+1=\lambda$ im Nenner, wie aus den bekannten Bildungsgesetzen der Bernouillischen Zahlen hervorgeht. Nach Weghebung dieses Factors aus dem Zähler und Nenner sieht man, dafs für den Fall $n=\mu$ die obige Bedingungs-Congruenz niemals erfüllt ist. Wir haben also als Resultat folgenden Satz:

Der erste Factor $\frac{P}{(2\lambda)^{\mu-1}}$ der Klassenanzahl H ist theilbar durch λ, wenn λ eine solche Primzahl ist, welche als Factor des Zählers einer der ersten $\frac{\lambda-3}{2}$ Bernouillischen Zahlen vorkommt, für alle übrigen Primzahlen λ ist dieser Factor nicbt durch λ theilbar.

Es ist nun für den zweiten Factor der Klassenzahl, nämlich $\frac{D}{\Delta}$, die Untersuchnng zu führen, unter welchen Bedingungen derselbe durch λ theilbar ist, und unter welchen nicht. Hierzu ist glücklicherweise die Kenntnifs der in jedem besonderen Falle nur mit äufserster Mühe zu ermittelnden, in Δ enthaltenen, Fundamental-Einheiten nicht nöthig, sondern nur die Definition derselben. Hat man ein System von $\mu - 1$ unabhängigen Einheiten, welche aber im allgemeinen nicht die Fundamental-Einheiten selbst sind, so erhält man aus ihnen alle Einheiten, indem man diese zu Potenzen erhebt, deren Exponenten auch rationale Brüche sein können, und mit einander multiplicirt. Ein solches System unabhängiger Einheiten ist aber das obige

$$e(\alpha),\ e(\alpha^{\gamma}) \ldots\ldots e(\alpha^{\gamma^{\mu-2}}).$$

Setzt man nun

$$\varepsilon_1(\alpha) = e(\alpha)^{r_1^1} \cdot e(\alpha^{\gamma})^{r_2^1} \ldots\ldots e(\alpha^{\gamma^{\mu-2}})^{r_{\mu-1}^1}$$

$$\varepsilon_2(\alpha) = e(\alpha)^{r_1^2} \cdot e(\alpha^{\gamma})^{r_2^2} \ldots\ldots e(\alpha^{\gamma^{\mu-2}})^{r_{\mu-1}^2}$$

$$\vdots$$

$$\varepsilon_{\mu-1}(\alpha) = e(\alpha)^{r_1^{\mu-1}} \cdot e(\alpha^{\gamma})^{r_2^{\mu-1}} \ldots\ldots e(\alpha^{\gamma^{\mu-2}})^{r_{\mu-1}^{\mu-1}},$$

wo die mit zwei Indices versehenen gebrochenen Potenzexponenten r_h^k so zu nehmen sind, dafs $\varepsilon_1(\alpha)$, $\varepsilon_2(\alpha) \ldots\ldots \varepsilon_{\mu-1}(\alpha)$ wirklich zn ganzen complexen Einheiten werden, so sind diese Einheiten Fundamental-Einheiten, wenn die Determinante der gebrochenen Exponenten, nämlich $\Sigma \pm r_1^1\, r_2^2 \ldots r_{\mu-1}^{\mu-1}$, den möglichst kleinsten Werth hat, aber nicht gleich Null ist. Es mögen nun wirklich die Exponenten r_h^k dieser Bedingung gemäfs bestimmt werden, so dafs $\varepsilon_1(\alpha)$, $\varepsilon_2(\alpha) \ldots\ldots \varepsilon_{\mu-1}(\alpha)$ wirkliche Fundamental-Einheiten sind, so hat man, wenn die Logarithmen genommen werden,

$$l\,\varepsilon_k(\alpha) = r_1^k\, l e(\alpha) + r_2^k\, l e(\alpha^{\gamma}) + \ldots + r_{\mu-1}^k\, l e(\alpha^{\gamma^{\mu-2}})$$

und es ist nun nach einem bekannten Satze über Determinanten

$$\Delta = D \,.\, \Sigma \pm r_1^1\, r_2^2 \ldots r_{\mu-1}^{\mu-1}$$

also

$$\frac{D}{\Delta} = \frac{1}{\Sigma \pm r_1^1 \, r_2^2 \dots r_{\mu-1}^{\mu-1}}$$

Bringt man nun die rationalen Brüche r^k_1, $r^k_2, \dots r^k_{\mu\mu 1}$, welche als Exponenten in einer Fundamental-Einheit $\varepsilon_k(\alpha)$ vorkommen, unter einer gemeinschaftlichen, und zwar den möglichst kleinsten Nenner, welcher n_k sei, so kann man allgemein setzen

$$r^k_h = \frac{m^k_h}{n_k}$$

wo m^k_h und n_k ganze Zahlen sind, von der Art, daſs n_k nicht mit allen Zahlen m^k_1, $m^k_2 \dots m^k_{\mu-1}$ einen gemeinschaftlichen Factor habe. Hierdurch wird

$$\frac{D}{\Delta} = \frac{n_1 \, n_2 \dots n_{\mu-1}}{\Sigma \pm m_1^1 \, m_2^2 \dots m_{\mu-1}^{\mu-1}}.$$

Es kann also $\frac{D}{\Delta}$ den Factor λ nur dann enthalten, wenn eine der Zahlen n_1, $n_2 \dots n_{\mu-1}$ durch λ theilbar ist, also nur dann, wenn eine Gleichung von folgender Form Statt hat:

$$\overset{n}{\varepsilon}(\alpha) = \overset{m_1}{e(\alpha)} \, . \, \overset{m_2}{e(\alpha^g)} \dots \overset{m_{\mu-1}}{e(\alpha^{g^{\mu-1}})}$$

in welcher n durch λ theilbar ist, aber m_1, $m_2, \dots m_{\mu-1}$ nicht alle durch λ theilbar sind. Nun ist aber jede λ^{te} Potenz einer ganzen complexen Zahl immer einer realen ganzen Zahl congruent, für den Modul λ, also auch $\varepsilon^n(\alpha) \equiv c$, mod. λ, wo c eine reale ganze Zahl bedeutet, also es muſs sein

$$\overset{m_1}{e(\alpha)} \, . \, \overset{m_2}{e(\alpha^g)} \dots \overset{m_{\mu-1}}{e(\alpha^{g^{\mu-2}})} \equiv c, \text{ mod. } \lambda,$$

wenn $\frac{D}{\Delta}$ den Factor λ enthalten soll, wo m_1, $m_2 \dots m_{\mu-1}$ nicht alle durch λ theilbar sind. Um nun die Bedingung zu erforschen unter welcher eine solche Congruenz Statt haben kann, mache ich daraus die Gleichung

$$e(\alpha)^{m_1}\,.\,e(\alpha^g)^{m_2}\ldots\,e(\alpha^{g^{\mu-2}})^{m_{\mu-1}} = c + \lambda\,\phi(\alpha),$$

welche nach bekannten Principien die für jeden Werth der Variabeln x geltende Gleichung nach sich zieht:

$$e(x)^{m_1}\,.\,e(x^g)^{m_2}\ldots\,e(x^{g^{\mu-2}})^{m_{\mu-1}} = c + \lambda.\phi(x)$$
$$+ (1 + x + x^2 + \ldots + x^{\lambda-1})\,\psi(x).$$

Ich nehme auf beiden Seiten die Differenzialquotienten der Logarithmen, wobei $\frac{d\,e(x)}{dx}$ durch $e'(x)$ bezeichnet wird, multiplicire mit x, und gebe sodann dem x seinen besonderen Werth $x = \alpha$ zurück, so wird

$$m_1\,\frac{\alpha\,e'(\alpha)}{e(\alpha)} + m_2\,g\,\frac{\alpha^g\,e'(\alpha^g)}{e(\alpha^g)} + \ldots + m_{\mu-1}\,g^{\mu-2}\,\frac{\alpha^{g^{\mu-2}}\,e'(\alpha^{g^{\mu-2}})}{e(\alpha^{g^{\mu-2}})}$$
$$= \frac{\lambda\,\alpha\,\phi'(\alpha) + (\alpha + 2\alpha^2 + 3\alpha^3 + \ldots + (\lambda-1)\,\alpha^{\lambda-1})\,\psi(\alpha)}{c + \lambda\,\phi(\alpha)}.$$

Diese Gleichung enthält nur complexe Einheiten in den Nennern, also wesentlich nur ganze complexe Zahlen. Ich verwandle dieselbe wieder in eine Congruenz für den Modul λ, indem ich die Vielfachen von λ weglasse. Dabei setze ich die complexe Zahl $\psi(\alpha)$ in die Form $a + (1-\alpha)\,f(\alpha)$, wo a eine reale ganze Zahl ist, und bemerke, dafs $(1-\alpha)\,(\alpha + 2\alpha^2 + \ldots + (\lambda-1)\,\alpha^{\lambda-1}) = -$ ist. Endlich bestimme ich noch die ganze Zahl M so, dafs $2\,c\,M \equiv a$, mod. λ, ist und setze der Kürze wegen $\frac{2\,\alpha\,e'(\alpha)}{e(\alpha)} = F(\alpha)$, so ist

$$m_1\,F(\alpha) + m_2 g\,F(\alpha^g) + \ldots + m_{\mu-1}\,g^{\mu-2}\,F(\alpha^{g^{\mu-2}})$$
$$\equiv M\,(\alpha + 2\alpha^2 + 3\alpha^3 + \ldots + (\lambda-1)\,\alpha^{\lambda-1}).$$

Ich entwickele jetzt die complexe ganze Zahl

$$F(\alpha) = \frac{2\,\alpha\,e'(\alpha)}{e(\alpha)} = \frac{1+\alpha}{1-\alpha} - \frac{g\,(1+\alpha^g)}{1-\alpha^g}.$$

Aus

$$(1-\alpha)(\alpha+2\alpha^2+3\alpha^3+\ldots+(\lambda-1)\alpha^{\lambda-1})=-\lambda$$

folgt

$$\frac{1}{1-\alpha}=\frac{-(\alpha+2\alpha^2+3\alpha^3+\ldots+(\lambda-1)\alpha^{\lambda-1})}{\lambda}$$

und wenn mit $1+\alpha$ multiplicirt wird:

$$\frac{1+\alpha}{1-\alpha}=\frac{-(\lambda+2\alpha+4\alpha^2+6\alpha^3+\ldots+2(\lambda-1)\alpha^{\lambda-1})}{\lambda},$$

welches anders geordnet auch so dargestellt werden kann:

$$\frac{1+\alpha}{1-\alpha}=\frac{-(\lambda+2\alpha+2g_1\alpha^g+2g_2\alpha^{g^2}+\ldots+2g_{\lambda-2}\alpha^{g^{\lambda-1}})}{\lambda}$$

Wird noch α in α^g verwandelt und mit g multiplicirt, so ist auch

$$\frac{g(1+\alpha^g)}{1-\alpha^g}=\frac{-(g\lambda+2gg_{\lambda-2}\alpha+2g\alpha^g+\ldots\ldots+2gg_{\lambda-3}\alpha^{g^{\lambda-2}})}{\lambda}$$

und wenn diese Gleichung von der vorhergehenden subtrahirt wird, so hat man

$$F(\alpha)=g-1+\frac{2(gg_{\lambda-2}-1)}{\lambda}\alpha+\frac{2(g-g_1)}{\lambda}\alpha^g$$
$$+\ldots+\frac{2(gg_{\lambda-3}-g_{\lambda-2})}{\lambda}\alpha^{g^{\lambda-2}}$$

Dieselben Coefficienten sind aber schon in dem ersten Theile unserer Untersuchung vorgekommen; werden daher auch dieselben Zeichen für sie angewendet, so ist

$$F(\alpha)=g-1+2b_0\alpha+2b_1\alpha^g+2b_2\alpha^{g^2}+\ldots+2b_{\lambda-2}\alpha^{g^{\lambda-2}}$$

wo $b_0, b_1, b_2, \ldots b_{\lambda-2}$ dieselben Coefficienten sind, welche schon oben bei dem ersten Theile der Untersuchung vorgekommen sind, und welche durch die Gleichung $\lambda b_k = gg_{k-1}-g_k$ bestimmt sind. Wird noch $2b_k+1-g=c_k$ gesetzt und bemerkt, dafs

$$c_{k+\mu}=-c_k$$

ist, so nimmt die Entwickelung des $F(\alpha)$ folgende Form an:

$$F(\alpha) = c(\alpha - \alpha^{-1}) + c_1(\alpha^g - \alpha^{-g}) + \ldots + c_{\mu-1}(\alpha^{g^{\mu-1}} - \alpha^{-g^{\mu-1}}).$$

Wird nun diese Entwickelung in der gefundenen Congruenz substituirt, so erhält man aus der Vergleichung der Coefficienten der einzelnen Glieder folgende μ Congruenzen

$$\begin{array}{l} m_1 c - m_2 g c_{\mu-1} - m_3 g^2 c_{\mu-2} - \ldots\ldots - m_{\mu-1} g^{\mu-2} c_2 \equiv M. \\ m_1 c_1 + m_2 g c - m_3 g^2 c_{\mu-1} - \ldots\ldots - m_{\mu-1} g^{\mu-2} c_3 \equiv g M \\ \vdots \\ m_1 c_{\mu-1} + m_2 g c_{\mu-2} + m_3 g^2 c_{\mu-3} + \ldots + m_{\mu-1} g^{\mu-2} c_1 \equiv g^{\mu-1} M. \end{array}$$

Um M zu eliminiren multiplicire ich diese Congruenzen der Reihe nach mit $1, g^{2n-1}, g^{2(2n-1)}, \ldots g^{(\mu-1)(2n-1)}$ und addire, so wird $1 + g^{2n} + g^{4n} + \ldots\ldots + g^{2(\mu-1)n} \equiv o$ mod. λ, für jeden der Werthe $n = 1, 2, 3, \ldots \mu - 1$, die rechte Seite der Congruenz verschwindet also, die linke aber zerfällt von selbst in das Product zweier Factoren, und es wird

$$\begin{array}{l} (m_1 + m_2 g^{2n} + m_3 g^{4n} + \ldots + m_{\mu-1} g^{2(\mu-2)n})\,(c + c_1 g^{2n-1} \\ \quad + c_2 g^{2(2n-1)} + \ldots + c_{\mu-1} g^{(\mu-1)(2n-1)}) \equiv o \end{array}$$

als die Bedingung, welche für alle Werthe $n = 1, 2, \ldots \mu - 1$ erfüllt sein mufs, damit $\frac{D}{\Delta}$ den Factor λ enthalten könne. Wenn nun der Factor $c + c_1 g^{2n-1} + \ldots + c_{\mu-1} g^{(\mu-1)(2n-1)}$ für keinen dieser Werthe des n durch λ theilbar ist, so mufs der andere Factor für alle diese Werthe congruent Null sein, es mufs also folgendes System von Congruenzen Statt haben:

$$\begin{array}{l} m_1 + m_2 g^2 + m_3 g^4 + \ldots\ldots + m_{\mu-1} g^{2(\mu-2)} \equiv o \\ m_1 + m_2 g^4 + m_3 g^8 + \ldots\ldots + m_{\mu-1} g^{4(\mu-2)} \equiv o \\ \vdots \\ m_1 + m_2 g^{2(\mu-1)} + m_3 g^{4(\mu-1)} + \ldots + m_{\mu-1} g^{(\mu-2)(\mu-1)} \equiv o. \end{array}$$

Die Determinante dieses Systems linearer Congruenzen ist nicht congruent Null, denn sie ist bekanntlich gleich $(1 - g^2)(1 - g^4) \ldots\ldots (1 - g^{2(\mu-2)})$, darum müssen alle Gröfsen $m_1, m_2, \ldots\ldots m_{\mu-1}$ einzeln congruent Null sein, welches gegen die Voraussetzung ist. Es mufs also nothwendig der andere Factor der Congruenz für irgend einen der Werthe $n = 1, 2, 3, \ldots \mu - 1$ congruent Null sein, wenn $\frac{D}{\Delta}$ durch λ theilbar sein soll, also

$$c+c_1 g^{2n-1}+c_2 g^{2(2n-1)}+\ldots+c_{\mu-1} g^{(\mu-1)(2n-1)}\equiv o \text{ mod. } \lambda.$$

Vermöge der Gleichung $c_{k+\mu}=-c_k$ und der Congruenz $g^{\mu} \equiv -1$, mod. λ, kann man auch die Anzahl der Glieder verdoppeln, so dafs

$$c+c_1 g^{2n-1}+c_2 g^{2(2n-1)}+\ldots+c_{\lambda-2} g^{(\lambda-2)(2n-1)}\equiv o$$

als die Congruenz genommen werden kann, welche nothwendig erfüllt werden mufs, wenn $\frac{D}{\Delta}$ durch λ theilbar sein soll. Wird zu dieser die Congruenz

$$(g-1)(1+g^{2n-1}+g^{2(2n-1)}+\ldots+g^{(\lambda-2)(2n-1)})\equiv o$$

hinzu addirt und für c_k+g-1 das Zeichen $2b_k$ zurückgesetzt, so hat man endlich

$$b_0+b_1 g^{2n-1}+b_2 g^{2(2n-1)}+\ldots\ldots+b_{\lambda-2} g^{(\lambda-2)(2n-1)}\equiv o$$

und diefs ist die schon oben vollständig untersuchte Congruenz, welche, wie gezeigt worden ist, nur dann Statt hat, wenn λ eine Primzahl ist, welche als Factor des Zählers einer der ersten $\frac{\lambda-3}{2}$ Bernouillischen Zahlen vorkommt. Es kann also der zweite Factor $\frac{D}{\Delta}$ der Formenzahl H nur dann durch λ theilbar sein, wenn auch der erste $\frac{P}{(2\lambda)^{\mu-1}}$ durch λ theilbar ist. Wir fassen nun das gefundene Resultat in folgenden Lehrsatz zusammen:

> Die Anzahl aller nicht äquivalenten Klassen der aus λ^{ten} Wurzeln der Einheit gebildeten idealen complexen Zahlen ist durch λ theilbar, wenn λ in dem Zähler einer der ersten $\frac{\lambda-3}{2}$ Bernouillischen Zahlen als Factor vorkommt; dagegen für alle anderen Primzahlen λ ist diese Klassenanzahl nicht durch λ theilbar.

Es ist somit die erste der beiden Voraussetzungen, auf welchen mein Beweis des Fermatschen Satzes beruht, vollständig ergründet, und ich wende mich nun zu der zweiten Voraussezzung, nämlich dafs keine Einheit einer realen ganzen

Zahl congruent sein soll, für den Modul λ, ohne eine λ^{te} Potenz einer anderen Einheit zu sein.

Es sei

$$E(\alpha) = \pm \alpha^k e(\alpha)^{\frac{m_1}{n}} \cdot e(\alpha^g)^{\frac{m_2}{n}} \ldots . e(\alpha^{g^{\mu-2}})^{\frac{m_{\mu-1}}{n}},$$

welche Form alle möglichen Einheiten umfaſst. Die Bedingung $E(\alpha) \equiv c$, mod. λ, giebt zunächst $E(\alpha^{-1}) \equiv c$, also auch $E(\alpha^{-1}) \equiv E(\alpha)$, woraus $\alpha^{-k} \equiv \alpha^k$ folgt: es muſs darum zunächst $k = o$ sein, und wenn zur n^{ten} Potenz erhoben wird, hat man:

$$E(\alpha)^n = \pm e(\alpha)^{m_1} e(\alpha^g)^{m_2} \ldots . e(\alpha^{g^{\mu-2}})^{m_{\mu-1}}$$

und weil $E(\alpha) \equiv c$ ist:

$$c^n \equiv \pm e(\alpha)^{m_1} . e(\alpha^g)^{m_2} \ldots . e(\alpha^{g^{\mu-2}})^{m_{\mu-1}}, \text{ mod. } \lambda.$$

Wenn nun λ eine Primzahl ist, welche in den Zählern der ersten $\frac{\lambda-3}{2}$ Bernouillischen Zahlen als Factor nicht vorkommt, so kann, wie wir oben bewiesen haben, eine solche Congruenz nicht bestehen, ohne daſs m_1, $m_2 \ldots . m_{\mu-1}$ alle durch λ theilbar sind. In diesem Falle ist darum $E^n(\alpha)$ gleich einer λ^{ten} Potenz einer Einheit, und n nicht theilbar durch λ, weil der Nenner n nicht mit allen Zählern $m_1, m_2, \ldots m_{\mu-1}$ einen gemeinschaftlichen Factor haben soll. Wenn aber $E^n(\alpha)$ gleich einer λ^{ten} Potenz ist, und n nicht durch λ theilbar, so wird leicht geschlossen, daſs auch $E(\alpha)$ eine λ^{te} Potenz sein muſs. Bestimmt man nämlich die beiden Zahlen s und t so, daſs $ns - \lambda t = 1$, und erhebt $E^n(\alpha)$ zur s^{ten} Potenz, so ist auch $E^{ns}(\alpha) = E^{1+\lambda t}(\alpha) = E(\alpha) . E^{\lambda t}(\alpha)$ gleich einer λ^{ten} Potenz einer Einheit, also auch $E(\alpha)$ eine λ^{te} Potenz einer Einheit. Hiermit ist folgender Lehrsatz bewiesen:

Wenn λ eine Primzahl ist, welche in den Zählern der ersten $\frac{\lambda-3}{2}$ Bernouillischen Zahlen als Factor nicht vorkommt, so ist jede aus λ^{ten} Wurzeln der Einheit gebildete complexe Einheit, welche einer realen ganzen Zahl congruent ist, für den Modul λ, eine λ^{te} Potenz einer anderen complexen Einheit.

Es ist also wirklich meine früher ausgesprochene Vermuthung vollkommen begründet, dafs die zweite Voraussetzung für meinen Beweis des Fermatschen Lehrsatzes, nämlich die über die Einheiten mit der ersten über die Klassenanzahl allemal zugleich mit erfüllt ist. Auch bestätigt sich meine Vermuthung wegen der Primzahl 37, für welche die beiden Voraussetzungen wirklich nicht gelten, weil sie in der 16ten Bernouillischen Zahl als Factor des Zählers vorkommt. Übrigens ist 37 auch die kleinste von allen Primzahlen für welche diese Voraussetzungen nicht gelten, für $\lambda = 3, 5, 7, 11, 13, 17, 19, 23, 29, 31, 41, 43$, etc. gelten dieselben, und man kann leicht aus den bekannten Werthen der Bernouillischen Zahlen, welche bis zur 31ten berechnet sind, alle Primzahlen bis $\lambda = 61$ prüfen, ob sie den Voraussetzungen genügen oder nicht. Der Fermatsche Satz, insoweit er bisher streng bewiesen ist heifst nun:

Die Gleichung $x^\lambda - y^\lambda = z^\lambda$, in welcher λ eine ungerade Primzahl ist, die in keiner der ersten $\frac{\lambda - 3}{2}$ Bernouillischen Zahlen als Factor des Zählers vorkommt, ist in ganzen Zahlen unlösbar.

Über meinen Beweis dieses Satzes, wie er im Aprilhefte der Monatsberichte steht, bemerke ich hier noch, dafs ich denselben noch etwas vereinfacht habe, und dafs er sich mit Leichtigkeit auch auf den allgemeineren Fall ausdehnen läfst, wo x, y, z nicht nur reale sondern auch complexe, aus λ^{ten} Wurzeln der Einheit gebildete Zahlen sind.

Breslau d. 16. Sept. 1847.

Extrait d'une lettre de M. Kummer à M. Liouville

Journal de mathématiques pures et appliquées XII, 136 (1847)

« Breslau, le 28 avril 1847.

» ... Engagé par mon ami M. Lejeune-Dirichlet, je prends la liberté de vous envoyer quelques exemplaires d'une Dissertation que j'ai écrite, il y a trois ans, à l'occasion du jubilé séculaire de l'Université de Kœnigsberg, et d'une autre Dissertation d'un de mes amis et disciples, M. Kronecker, jeune géomètre distingué. Dans ces Mémoires, que je vous prie d'accepter en signe de ma profonde estime, vous trouverez des développements sur quelques points de la théorie des nombres complexes composés des racines de l'unité, c'est-à-dire de l'équation $r^n = 1$, qui ont été récemment le sujet de quelques discussions au sein de votre illustre Académie, à l'occasion de l'essai d'une Démonstration du théorème de Fermat, proposé par M. Lamé. Quant à la proposition élémentaire pour ces nombres complexes, *qu'un nombre complexe composé ne peut être décomposé en facteurs premiers que d'une seule manière*, que vous regrettez très-justement dans cette démonstration défectueuse en outre en quelques autres points, je puis vous assurer *qu'elle n'a pas lieu généralement* tant qu'il s'agit de nombres complexes de la forme $\alpha_0 + \alpha_1 r + \alpha_2 r^2 + \ldots + \alpha_{n-1} r^{n-1}$, mais qu'on peut la sauver en introduisant un nouveau genre de nombres complexes que j'ai appelé *nombre complexe idéal*. Les résultats de mes recherches sur cette matière ont été communiqués à l'Académie de Berlin et imprimés dans les Comptes rendus (mars 1846); un Mémoire sur le même sujet paraîtra bientôt dans le Journal de M. Crelle. Les applications de cette théorie à la démonstration du théorème de Fermat m'ont occupé depuis longtemps, et j'ai réussi à faire dépendre l'impossibilité de l'équation $x^n - y^n = z^n$ de deux propriétés du nombre premier n, en sorte qu'il ne reste plus qu'à rechercher si elles appartiennent à tous les nombres premiers. Dans le cas où ces résultats vous paraîtraient dignes de quelque attention, vous les trouverez exposés dans le Compte rendu de l'Académie de Berlin de ce mois. »

Note de M. Liouville. — Le Mémoire de M. Kummer, dont il est d'abord question dans cette Lettre, et qui porte la date de 1844, est écrit en latin, sous ce titre: *De numeris complexis qui radicibus unitatis et numeris integris realibus constant.* Celui de M. Kronecker, qui y fait suite et qui est intitulé : *De unitatibus complexis,* traite spécialement des diviseurs complexes du nombre 1 ; il a paru en 1845, et l'auteur annonce qu'il reprendra la question avec plus de détails dans le Journal de M. Crelle. Le Mémoire de M. Kummer offrant beaucoup d'intérêt et ne paraissant pas avoir été jusqu'ici connu en France, nous le donnerons en entier dans un prochain cahier, à la suite d'un travail de M. Lamé, sur le même sujet, qui est imprimé depuis quelque temps déjà. On devra consulter, en outre, le Journal de M. Crelle, les Comptes rendus de l'Académie de Berlin, et enfin ceux de notre Académie des Sciences qui contiennent des recherches étendues de M. Cauchy. Nous n'avons pas à examiner ici en quoi les auteurs que nous citons s'accordent ou diffèrent, ni quels sont les droits de chacun à l'antériorité de telle ou telle découverte. C'est au temps à fixer la valeur de leurs travaux et à mettre toute chose à sa place.

Bestimmung der Anzahl nicht äquivalenter Classen für die aus λten Wurzeln der Einheit gebildeten complexen Zahlen und die idealen Factoren derselben

Journal für die reine und angewandte Mathematik 40, 93–116 (1850)

Die gegenwärtige Untersuchung über die Anzahl der Classen der aus λ^{ten} Wurzeln der Einheit gebildeten complexen Zahlen und der idealen Factoren derselben wird sich genau an meine früher über diese Art der complexen Zahlen veröffentlichten Arbeiten anschliefsen, namentlich an die Abhandlung No. 16. im 35ten Bande dieses Journals, in welcher ich die idealen complexen Zahlen zuerst eingeführt und genau definirt, auch den Begriff der Äquivalenz für dieselben bestimmt, sie hiernach in Classen getheilt und bewiesen habe, dafs für jede Primzahl λ nur eine endliche bestimmte Anzahl nicht äquivalenter Classen existirt. Die genaue Bestimmung dieser Classen-Anzahl aber habe ich daselbst nicht versucht, weil mir bekannt war, dafs *Lejeune-Dirichlet* diese Arbeit, wenn gleich nicht für die idealen complexen Zahlen, so doch für eine bestimmte Art von homogenen Formen $\lambda-1^{\text{ten}}$ Grades mit $\lambda-1$ unbestimmten Zahlen, welche (nur in der Anschauungsweise von diesen verschieden) wesentlich mit ihnen übereinstimmen, schon seit einiger Zeit ausgeführt hatte. Als ich mich nun aber mit der Anwendung der Theorie der complexen Zahlen auf den allgemeinen Beweis des *Fermat*'schen Satzes, dafs die Gleichung $x^n+y^n=z^n$, wenn $n>2$, durch ganze Zahlen nicht aufzulösen ist, beschäftigte, war mir die Bestimmung dieser Anzahl der Classen nicht äquivalenter idealer complexer Zahlen dazu unentbehrlich und ich entschlofs mich, die Arbeit, welche bei Zugrundelegung der von *Dirichlet* für die Bestimmung der zu einer gegebenen Determinante gehörenden Classen nicht äquivalenter quadratischer Formen gefundenen Principien, auf dem Standpuncte, zu welchem ich die Theorie der complexen idealen Zahlen in der Abhandlung No. 16. 35ten Bandes dieses Journals bereits gebracht hatte, keine principiellen Schwierigkeiten mehr bot, selbst auszuführen. Das Resultat, insoweit ich es zur Ergründung des oben angegebenen *Fermat*'schen Satzes brauchte, habe ich in den Monatsberichten der Berliner Akademie vom September 1847 veröffentlicht, aber erst jetzt, nachdem ich von *Dirichlet*, dem die Priorität hier zugehört, selbst aufgefordert worden bin, will ich in der gegenwärtigen Abhandlung auch die Methode auseinandersetzen, welche mich zu diesem Resultate geführt hat. Hierauf werde ich sodann in einer zweiten Abhandlung, als weitere Vorarbeit für meinen Beweis jenes *Fermat*'schen Satzes, die besondere Untersuchung über die Theilbarkeit dieser Classenzahl durch λ und über gewisse complexe Einheiten, welche ich in den Monatsberichten der Berliner Akademie vom September 1847 kurz entwickelt habe, in etwas veränderter Fassung reproduciren, und endlich, eben so in einer dritten Abhandlung, den auf diese sich stützenden Beweis des *Fermat*'schen Satzes, dessen Grundzüge ich zuerst in einem Briefe an *Lejeune-Dirichlet* mitgetheilt habe, welcher in dem Aprilhefte der Monatsberichte der Berliner Akademie des Jahres 1847 abgedruckt ist.

Es bezeichne q_d eine für den Modul λ zum Exponenten d, einem Divisor von $\lambda-1$, gehörende Primzahl; welche also der Bedingung genügt, dafs $q_d^d \equiv 1$ mod. λ, dafs aber keine niedrigere Potenz von q_d als die d^{te} der Einheit congruent sei, für den Modul λ. Die Primzahl q_d hat unter dieser Voraussetzung allemal genau $\frac{\lambda-1}{d} = \delta$ complexe Primfactoren, welche wirklich, oder ideal sein können; wie es in der Abhandlung No. 16. 35ten Bandes dieses Journals gezeigt worden ist. Eben so sei $q'_{d'}$ eine für den Modul λ zum Exponenten d' gehörende Primzahl, welche darum $\frac{\lambda-1}{d'} = \delta'$ (ideale) complexe Primfactoren enthält, u. s. w. Es stelle ferner $F(\alpha)$, wo $\alpha^\lambda = 1$ ist, irgend eine beliebige ideale oder wirkliche complexe Zahl vor, so hat die Norm derselben $NF(\alpha)$ stets die Form $\lambda^n . q_d^{md} . q'^{m'd'}_{d'} . q''^{m''d''}_{d''} \ldots$, in welcher $q, q', q'', \ldots$ $d, d', d'', \ldots$ die angegebenen Bedeutungen haben und $n, m, m', m'', \ldots$ beliebige, nicht negative ganze Zahlen sind.

Es sei nun

$$1.\quad R = \Sigma \frac{s-1}{(NF(\alpha))^s},$$

wo das Summenzeichen Σ sich auf alle *verschiedene* complexe, ideale und wirkliche Zahlen $F(\alpha)$ bezieht, und wo als verschieden diejenigen gelten, welche verschiedene Primfactoren enthalten, nicht aber diejenigen, welche, dieselben idealen Primfactoren enthaltend, nur mit verschiedenen complexen Einheiten multiplicirt sind. Die unbestimmte Zahl s wird gröfser als Eins angenommen und später der Grenzwerth der Reihe R für $s = 1$ untersucht werden. Alle einzelnen Glieder der Reihe R sind nun von der Form

$$2. \qquad \frac{s-1}{(\lambda^n . q_d^{md} . q'^{m'd'}_{d'} . q''^{m''d''}_{d''} \ldots)^s}$$

und es ist zunächst die Frage zu erörtern: wievielmal ein bestimmtes solches Glied in der Reihe R vorkomme, d. i., wieviele verschiedene ideale oder wirkliche complexe Zahlen $F(\alpha)$ es gebe, deren Norm gleich $\lambda^n . q_d^{md} . q'^{m'd'}_{d'} \ldots$ ist. Dies geschieht sehr einfach auf folgende Weise. Erstens: da die Norm den Factor q_d^{md} enthält, so mufs die complexe Zahl $F(\alpha)$ von den δ idealen Factoren des q_d genau m enthalten, welche entweder verschieden, oder zum Theil, oder alle gleich sein können. Von vorhandenen δ Elementen lassen sich aber m, wenn Wiederholungen gestattet sind, auf

$$\frac{\delta(\delta+1)(\delta+2) \ldots (\delta+m-1)}{1.2.3 \ldots m}$$

verschiedene Arten nehmen. Auf eben so viele verschiedene Arten kann also der Factor der Norm q_d^{md} entstehen. Eben so ergiebt sich, dafs der Factor der Norm $q'^{m'd'}_{d'}$ auf

$$\frac{\delta'(\delta'+1) \ldots (\delta'+m'-1)}{1.2 \ldots m'}$$

verschiedene Arten entstehen kann u. s. f. Der besondere Factor der Norm λ^n aber kann nur auf eine einzige Weise entstehen, nämlich, wenn $F(\alpha)$ von den complexen Primfactoren des λ, welche alle einander gleich sind und durch $1-\alpha$ dargestellt werden, genau n enthält. Hieraus folgt, wenn

$$3. \qquad NF(\alpha) = \lambda^n . q_d^{md} . q'^{m'd'}_{d'} . q''^{m''d''}_{d''} \ldots .$$

ist, dafs es genau

$$\frac{\delta(\delta+1) \ldots (\delta+m-1)}{1.2 \ldots m} . \frac{\delta'(\delta'+1) \ldots (\delta'+m'-1)}{1.2 \ldots m'} . \frac{\delta''(\delta''+1) \ldots (\delta''+m''-1)}{1.2 \ldots m''} \ldots .$$

verschiedene complexe Zahlen $F(\alpha)$ giebt, welche diese bestimmte Norm haben. Es läfst sich also die Reihe R jetzt in folgende Form setzen:

$$4. \qquad R = (s-1)\Sigma \frac{\frac{\delta(\delta+1) \ldots (\delta+m-1)}{1.2 \ldots m} . \frac{\delta'(\delta'+1) \ldots (\delta'+m'-1)}{1.2 \ldots m'} \ldots .}{\lambda^{ns} . q_d^{mds} . q'^{m'd's}_{d'} \ldots .},$$

wo das Summenzeichen sich auf alle verschiedenen Primzahlen $q, q', \ldots .,$

13 *

mit deren jeder auch d und δ, d' und δ' u. s. w. gegeben ist, und auf alle nicht negativen ganzzahligen Werthe von n, m, m' u. s. w. bezieht. Dieser Reihe sieht man auf den ersten Blick an, dafs sie das Product folgender Binomialreihen ist:

$$1+\frac{1}{\lambda^s}+\frac{1}{\lambda^{2s}}+\frac{1}{\lambda^{3s}}+\dots.$$

$$1+\frac{\delta}{1}\cdot\frac{1}{q_d^{ds}}+\frac{\delta(\delta+1)}{1\,.\,2}\cdot\frac{1}{q_d^{2ds}}+\dots.$$

$$1+\frac{\delta'}{1}\cdot\frac{1}{q_{d'}^{d's}}+\frac{\delta'(\delta'+1)}{1\,.\,2}\cdot\frac{1}{q'^{2d's}_{d'}}+\dots.$$

u. s. w.

Summirt man diese einzelnen Binomialreihen, so erhält man R folgendermafsen in Form eines Products dargestellt:

$$5.\qquad R=\frac{s-1}{1-\frac{1}{\lambda^s}}\cdot\Pi\left(\frac{1}{1-\frac{1}{q_d^{ds}}}\right)^{\delta}\cdot\Pi\left(\frac{1}{1-\frac{1}{q'^{d's}_{d'}}}\right)^{\delta'}\dots;$$

wo das erste Productzeichen sich auf alle verschiedenen Primzahlen q_d bezieht, welche zum Exponenten d gehören, für den Modul λ; das zweite auf alle zum Exponenten d' gehörenden Primzahlen $q'_{d'}$ u. s. f., und wo genau so viele besondere Producte vorkommen, als $\lambda-1$ verschiedene Divisoren hat, nämlich d, d', u. s. w. Es sei nun β eine primitive Wurzel der Gleichung $\beta^{\lambda-1}=1$ und r irgend eine Zahl, welche mit $\lambda-1$ den gröfsten gemeinschaftlichen Theiler δ hat, so ist:

$$\left(1-\frac{1}{q_d^{ds}}\right)^{\delta}=\left(1-\frac{1}{q_d^s}\right)\left(1-\frac{\beta^r}{q_d^s}\right)\left(1-\frac{\beta^{2r}}{q_d^s}\right)\dots\left(1-\frac{\beta^{(\lambda-2)r}}{q_d^s}\right),$$

also

$$6.\qquad \Pi\left(\frac{1}{1-\frac{1}{q_d^{ds}}}\right)^{\delta}=\Pi\,\overset{\lambda-2}{\underset{0}{\Pi_k}}\left(\frac{1}{1-\frac{\beta^{kr}}{q_d^s}}\right).$$

Ich gebe nun dem r den Werth Ind. q_d (mod. λ), da Ind. (q_d) und $\lambda-1$ den gröfsten gemeinschaftlichen Factor δ haben; wie es sein soll. Dafs dies wirklich der Fall ist, wird kurz so gezeigt: Setzt man $q_d\equiv g^h$, mod. λ, wo g eine primitive Wurzel von λ bedeutet, so ist $d.h$ durch $\lambda-1$ theilbar; aber kein niedrigeres Vielfache von h, als das $d$fache, ist durch $\lambda-1$ theilbar, und da $\lambda-1=d.\delta$ ist, so haben $h=$ Ind. q_d und $\lambda-1$ den gemeinschaftlichen

Factor δ; aber keinen gröfseren. Werden nun der in Gleichung (6.) gefundene Ausdruck, in welchem $r = \text{Ind.}\, q_d$, mod. λ, ist, und eben so die entsprechenden Ausdrücke für die andern Producte in der Gleichung (5.) substituirt, so nimmt dieselbe, wie leicht zu sehen ist, folgende einfache Gestalt an:

$$7. \qquad R = \frac{s-1}{1-\frac{1}{\lambda^s}} \overset{\lambda-2}{\underset{0}{\Pi_k}} \Pi \frac{1}{1-\frac{\beta^{k\,\text{Ind.}\,q}}{q^s}},$$

wo die Primzahlen q nicht weiter nach den Exponenten zu unterscheiden sind, zu denen sie gehören, für den Modul λ, und das zweite Productzeichen Π sich auf alle Primzahlen q ohne Unterschied bezieht, mit alleiniger Ausnahme der Primzahl λ; denn alle zu den einzelnen Exponenten d, d', d'' u. s. w. gehörenden Primzahlen q_d, $q'_{d'}$, $q''_{d''}$ u. s. w. machen zusammengenommen alle Primzahlen aufser λ aus. Macht man jetzt von der Reihen-Entwicklung

$$\frac{1}{1-\frac{\beta^{k\,\text{Ind.}\,q}}{q^s}} = 1 + \frac{\beta^{k\,\text{Ind.}\,q}}{q^s} + \frac{\beta^{2k\,\text{Ind.}\,q}}{q^{2s}} + \frac{\beta^{3k\,\text{Ind.}\,q}}{q^{3s}} + \dots.$$

Gebrauch und stellt sich alle die unendlich vielen, den verschiedenen Werthen des q entsprechenden, in dieser Form enthaltenen Reihen mit einander multiplicirt vor, wobei man von den bekannten Formeln $\text{Ind.}\,q + \text{Ind.}\,q' \equiv \text{Ind.}\,qq'$, mod. $\lambda-1$, und $n.\text{Ind.}\,q \equiv \text{Ind.}\,q^n$, mod. $\lambda-1$, Gebrauch macht, so übersieht man sehr leicht, dafs

$$\Pi \frac{1}{1-\frac{\beta^{k\,\text{Ind.}\,q}}{q^s}} = \Sigma \frac{\beta^{k\,\text{Ind.}\,n}}{n^s}$$

ist; wo das Summenzeichen Σ sich auf alle ganzzahligen positiven Werthe des n bezieht; mit Ausschlufs der durch λ theilbaren. Demnach hat man

$$8. \qquad R = \frac{s-1}{1-\frac{1}{\lambda^s}} \Sigma \frac{1}{n^s} \cdot \Sigma \frac{\beta^{\text{Ind.}\,n}}{n^s} \cdot \Sigma \frac{\beta^{2\,\text{Ind.}\,n}}{n^s} \dots \Sigma \frac{\beta^{(\lambda-2)\,\text{Ind.}\,n}}{n^s}.$$

Läfst man nun das s abnehmend sich der Grenze Eins unendlich nähern, so erhält man bekanntlich

$$(s-1)\Sigma \frac{1}{n^s} = \frac{\lambda-1}{\lambda}, \quad \text{für } s=1,$$

und die übrigen in dem Ausdrucke (8.) enthaltenen unendlichen Reihen sind bekanntlich für $s=1$ convergent und haben endliche bestimmte Werthe. Deshalb wird

$$9.\quad R = \Sigma\frac{\beta^{\text{Ind.}n}}{n}.\Sigma\frac{\beta^{2\,\text{Ind.}n}}{n}\dots\Sigma\frac{\beta^{(\lambda-2)\,\text{Ind.}n}}{n},\quad \text{für } s=1.$$

Die einzelnen unendlichen Reihen, welche als Factoren dieses Products auftreten, lassen sich durch endliche Ausdrücke summiren, und da diese Summen, welche wesentlich verschieden sind, je nachdem k ungerade oder gerade ist, von *Dirichlet* in der Abhandlung über die arithmetische Reihe gefunden worden sind, so beschränken wir uns hier darauf, nur die Resultate mitzutheilen, welche wir in folgende Form setzen:

$$\Sigma\frac{\beta^{-(2r-1)\,\text{Ind.}n}}{n} = \frac{\pi\sqrt{-1}}{\lambda(\beta^{2r-1},\alpha)}\left(1+g_1\beta^{2r-1}+g_2\beta^{2(2r-1)}+\dots+g_{\lambda-2}\beta^{(\lambda-2)(2r-1)}\right),$$

wo $g_1, g_2, g_3, \dots g_{\lambda-2}$ die kleinsten positiven Reste bezeichnen, welche die Potenzen einer primitiven Wurzel $g, g^2, g^3, \dots g^{\lambda-2}$ geben, für den Modul λ, und wo (β^{2r-1},α) den in der Kreistheilung bekannten Ausdruck

$$(\beta^{2r-1},\alpha) = \alpha+\beta^{2r-1}.\alpha^{g}+\beta^{2(2r-1)}.\alpha^{g^2}+\dots+\beta^{(\lambda-2)(2r-1)}.\alpha^{g^{\lambda-2}}$$

bezeichnet. Ferner

$$\Sigma\frac{\beta^{-2r\,\text{Ind.}n}}{n} = \frac{2\left(l.e(\alpha)+\beta^{2r}l.e(\alpha^{g})+\beta^{4r}l.e(\alpha^{g^2})+\dots+\beta^{(\lambda-3)r}l.e(\alpha^{g^{\frac{1}{2}(\lambda-3)}})\right)}{(1-\beta^{-2r})(\beta^{2r},\alpha)},$$

wo $e(\alpha)$ eine bestimmte complexe Einheit bezeichnet, nämlich

$$e(\alpha) = \sqrt{\left(\frac{(1-\alpha^{g})(1-\alpha^{-g})}{(1-\alpha)(1-\alpha^{-1})}\right)}$$

und

$$(\beta^{2r},\alpha) = \alpha+\beta^{2r}.\alpha^{g}+\beta^{4r}.\alpha^{g^2}+\dots+\beta^{(\lambda-2)2r}.\alpha^{g^{\lambda-2}}.$$

Substituirt man nun diese Summen in dem Ausdrucke (9.) der Reihe R, wobei man der Kürze wegen das Product

$$\Pi_r{}_{1}^{\frac{1}{2}(\lambda-1)}\left(1+g_1.\beta^{2r-1}+g_2.\beta^{2(2r-1)}+\dots+g_{\lambda-2}.\beta^{(\lambda-2)(2r-1)}\right) = P$$

und das Product

$$\Pi_r{}_{1}^{\frac{1}{2}(\lambda-3)}\left(le(\alpha)+\beta^{2r}.le(\alpha^{g})+\dots+\beta^{(\lambda-3)r}.le(\alpha^{g^{\frac{1}{2}(\lambda-3)}})\right) = Q$$

setzt und bemerkt, dafs $(\beta^k,\alpha)(\beta^{-k},\alpha)=\pm\lambda$, $(-1,\alpha)=\pm\sqrt{\pm\lambda}$ und

$$(1-\beta^{-2})(1-\beta^{-4})\dots(1-\beta^{-(\lambda-3)}) = \tfrac{1}{2}(\lambda-1),$$

so hat man

$$10.\quad R = \frac{\pm\pi^{\frac{1}{2}(\lambda-1)}.2^{\frac{1}{2}(\lambda-1)}.P.Q}{(\lambda-1).\lambda^{\lambda-\frac{1}{2}}},\quad \text{für } s=1.$$

Das Product Q läfst noch eine Vereinfachung zu. Nennt man nämlich D die Determinante des Systems folgender $\left(\frac{\lambda-3}{2}\right)^2$ Gröfsen:

$$\begin{array}{llll} le(\alpha), & le(\alpha^g), & \dots\dots & le(\alpha^{g^{\frac{1}{2}(\lambda-5)}}), \\ le(\alpha^g), & le(\alpha^{g^2}), & \dots\dots & le(\alpha^{g^{\frac{1}{2}(\lambda-3)}}), \\ \dots & \dots & \dots & \dots \\ le(\alpha^{g^{\frac{1}{2}(\lambda-5)}}), & le(\alpha^{g^{\frac{1}{2}(\lambda-3)}}), & \dots\dots & le(\alpha^{g^{\frac{1}{2}(\lambda-5)}}), \end{array}$$

so zeigt sich sehr leicht, dafs $Q=\frac{1}{2}(\lambda-1).D$ ist, dafs also R für $s=1$ auch so dargestellt werden kann:

$$11. \quad R = \frac{\pm\,\pi^{\frac{1}{2}(\lambda-1)}\,2^{\frac{1}{2}(\lambda-3)}.P.D}{\lambda^{\lambda-\frac{3}{2}}}, \quad \text{für } s=1.$$

Wir wenden uns jetzt zu dem zweiten Theile unserer Untersuchung, nämlich zu der Summation derselben Reihe R nach einem andern Principe.

Stellt man sich in der Reihe

$$R = \Sigma\frac{s-1}{(NF(\alpha))^s},$$

wo $F(\alpha)$ alle idealen und wirklichen complexen Zahlen umfafst, diese alle in die nicht äquivalenten Classen abgetheilt vor, so dafs die erste Classe, welche alle *wirklichen* complexen Zahlen enthält, durch $f(\alpha)$, die zweite durch $f_1(\alpha)$, die dritte durch $f_2(\alpha)$ u. s. w. bezeichnet wird, so hat man, wenn die Anzahl aller nicht äquivalenten Classen, mit deren Auffindung sich diese ganze Abhandlung beschäftigt, durch H bezeichnet wird:

$$12. \quad R = \Sigma\frac{s-1}{(Nf(\alpha))^s}+\Sigma\frac{s-1}{(Nf_1(\alpha))^s}+\dots\dots+\Sigma\frac{s-1}{(Nf_{H-1}(\alpha))^s}.$$

Es ist nun für jede einzelne dieser Summen die Grenze zu suchen, welche sie für $s=1$ erreicht. Dieser Grenzwerth wird aber, wie wir später beweisen werden, für alle verschiedene Classen, d. h. für alle diese verschiedenen Summen, genau derselbe, und soll darum nur für die erste Summe, welche die erste Classe der complexen Zahlen, also alle *wirklichen* complexen Zahlen umfafst, hier vollständig bestimmt werden.

Alle wirklichen complexen Zahlen sind in der Form

$$13. \quad f(\alpha) = x\alpha+x_1\alpha^g+x_2\alpha^{g^2}+\dots\dots+x_{\lambda-2}\alpha^{g^{\lambda-2}}$$

enthalten, in welcher x, x_1, x_2, ... $x_{\lambda-2}$ beliebige ganze Zahlen sind. Da aber, vermöge der Einheiten, mit welchen sie multiplicirt vorkommen, jede derselben auf unendlich viele verschiedene Arten durch diese Form dargestellt wird, so suchen wir zunächst die Bedingungen, welchen $x, x_1, \ldots\ldots x_{\lambda-2}$ unterworfen werden müssen, damit jede complexe Zahl nur ein einzigesmal durch diese Form dargestellt werde. Zu diesem Zwecke gebrauchen wir ein System von Fundamental-Einheiten, durch deren Potenz-Erhebung und Multiplication mit einander, wenn noch die einfache Einheit $\pm\alpha^m$ als Factor hinzugefügt wird, alle complexen Einheiten erzeugt werden. Diejenigen unserer geehrten Leser, welche mit der Theorie der complexen Einheiten noch nicht vertraut sein sollten, bitten wir, hierüber einen Aufsatz von ***Dirichlet*** in den Monats-Berichten der Berliner Akademie, vom März des Jahres 1846, nachzusehen. Es sei $\varepsilon_1(\alpha), \varepsilon_2(\alpha), \ldots\ldots \varepsilon_{\mu-1}(\alpha)$ ein solches System von Fundamental-Einheiten, und es sei hier, wie auch in dem Folgenden, μ ein abgekürztes Zeichen für $\frac{1}{2}(\lambda-1)$, so umfafst die Form

$$14. \qquad \pm\alpha^m.\varepsilon_1(\alpha)^{m_1}.\varepsilon_2(\alpha)^{m_2}\ldots\ldots\varepsilon_{\mu-1}(\alpha)^{m_{\mu-1}}.\varphi(\alpha) = f(\alpha)$$

alle möglichen Darstellungen einer und derselben complexen Zahl $f(\alpha)$, wenn den Potenz-Exponenten $m, m_1, m_2, \ldots\ldots m_{\mu-1}$ alle möglichen ganzzahligen Werthe gegeben werden. Wir abstrahiren jetzt von dem Factor $\pm\alpha^m$, welcher für sich bewirkt, dafs jede complexe Zahl auf 2λ verschiedene Arten dargestellt wird, und bemerken, dafs $\varepsilon_1(\alpha), \varepsilon_2(\alpha), \ldots\ldots \varepsilon_{\mu-1}(\alpha)$ lauter reale Gröfsen sind, welche auch immer positiv angenommen werden können und sollen: $f(\alpha)$ aber und $\varphi(\alpha)$ können imaginair sein. Wenn nun nur die realen Theile dieser complexen Zahlen in Betracht gezogen werden, welche gleich $\sqrt{(f(\alpha)f(\alpha^{-1}))}$ und $\sqrt{(\varphi(\alpha)\varphi(\alpha^{-1}))}$ sind, so giebt die Gleichung (14.), nachdem von dem Factor $\pm\alpha^m$ abstrahirt worden, wenn auf beiden Seiten die natürlichen Logarithmen genommen werden:

$$m_1 l\varepsilon_1(\alpha) + m_2 l\varepsilon_2(\alpha) + \ldots\ldots + m_{\mu-1} l\varepsilon_{\mu-1}(\alpha) + \tfrac{1}{2}l(r\varphi\alpha\varphi\alpha^{-1}) = \tfrac{1}{2}l(rf\alpha f\alpha^{-1}),$$

wo r eine ganz beliebige Gröfse ist, die wir hier hinzugefügt haben, um sie später zweckmäfsig zu benutzen. Verwandelt man noch α in $\alpha^g, \alpha^{g^2}, \ldots\ldots \alpha^{g^{\mu-2}}$, so erhält man folgendes System von Gleichungen:

$$15. \quad \left\{\begin{array}{l} m_1 l\varepsilon_1(\alpha) + m_2 l\varepsilon_2(\alpha) + \ldots\ldots + m_{\mu-1} l\varepsilon_{\mu-1}(\alpha) + \tfrac{1}{2}l(r\varphi\alpha\varphi\alpha^{-1}) \\ \qquad = \tfrac{1}{2}l(rf\alpha f\alpha^{-1}), \\ m_1 l\varepsilon_1(\alpha^g) + m_2 l\varepsilon_2(\alpha^g) + \ldots\ldots + m_{\mu-1} l\varepsilon_{\mu-1}(\alpha^g) + \tfrac{1}{2}l(r\varphi\alpha^g\varphi\alpha^{-g}) \\ \qquad = \tfrac{1}{2}l(rf\alpha^g f\alpha^{-g}), \\ \ldots\ldots\ldots\ldots\ldots\ldots\ldots\ldots \\ m_1 l\varepsilon_1(\alpha^{g^{\mu-2}}) + m_2 l\varepsilon_2(\alpha^{g^{\mu-2}}) + \ldots\ldots + m_{\mu-1} l\varepsilon_{\mu-1}(\alpha^{g^{\mu-2}}) + \tfrac{1}{2}l(r\varphi\alpha^{g^{\mu-2}}\varphi\alpha^{-g^{\mu-2}}) \\ \qquad = \tfrac{1}{2}l(rf\alpha^{g^{\mu-2}} f\alpha^{-g^{\mu-2}}). \end{array}\right.$$

Die Auflösung dieser Gleichungen, in welchen $m_1, m_2, \ldots m_{\mu-1}$ als Unbekannte angesehen werden sollen, möge ergeben:

$$16.\quad m_k = \frac{\overset{1}{A}_k l(rf\alpha f\alpha^{-1}) + \overset{2}{A}_k l(rf\alpha^g f\alpha^{-g}) + \ldots + \overset{\mu-1}{A_k} l(rf\alpha^{g^{\mu-2}} f\alpha^{-g^{\mu-2}})}{2\Delta}$$
$$- \frac{\overset{1}{A}_k l(r\varphi\alpha\,\varphi\alpha^{-1}) + \overset{2}{A}_k l(r\varphi\alpha^g \varphi\alpha^{-g}) + \ldots + \overset{\mu-1}{A_k} l(r\varphi\alpha^{g^{\mu-2}} \varphi\alpha^{-g^{\mu-2}})}{2\Delta},$$

für $k = 1, 2, 3, \ldots \mu-1$. Unterwirft man jetzt die Gröfsen $f\alpha.f\alpha^{-1}$, $f\alpha^g.f\alpha^{-g}$, ... $f\alpha^{g^{\mu-2}}.f\alpha^{-g^{\mu-2}}$, oder die in ihnen enthaltenen Coëfficienten $x, x_1, \ldots x_{\lambda-2}$ den $\mu-1$ Bedingungen, dafs die Ausdrücke

$$17.\quad z_k = \frac{\overset{1}{A}_k l(rf\alpha f\alpha^{-1}) + \overset{2}{A}_k l(rf\alpha^g f\alpha^{-g}) + \ldots + \overset{\mu-1}{A_k} l(rf\alpha^{g^{\mu-2}} f\alpha^{-g^{\mu-2}})}{2\Delta}$$

für $k = 1, 2, 3, \ldots \mu-1$ alle in den Grenzen 0 und 1 liegen sollen, so giebt es, wie leicht zu sehen, für jede gegebene complexe Zahl $\varphi(\alpha)$ nur ein einziges System von ganzzahligen Werthen der $m_1, m_2, \ldots m_{\mu-1}$, welche diesen Bedingungen und zugleich den in (16.) enthaltenen Gleichungen genügen: also die Einheit, mit welcher eine complexe Zahl behaftet ist, wenn sie diesen $\mu-1$ Bedingungen genügen soll, ist, mit alleiniger Ausnahme des Factors $\pm\alpha^m$, vollständig bestimmt, und da dieser für sich genau 2λ verschiedene Werthe hat, so schliefsen wir, wenn die Bedingungen, dafs $z_1, z_2, \ldots z_{\mu-1}$ alle in den Grenzen 0 und 1 liegen, erfüllt sind, dafs jede gegebene complexe Zahl $f(\alpha)$ genau 2λ verschiedene Darstellungen durch die Form

$$f(\alpha) = x\alpha + x_1\alpha^g + \ldots + x_{\lambda-2}\alpha^{g^{\lambda-2}}$$

zuläfst. Kehrt man nun das System von Gleichungen, welches in der Form (17.) enthalten ist, wieder um, so lassen sich die $\mu-1$ Bedingungen auch so darstellen:

$$18.\quad \begin{cases} l\varepsilon_1(\alpha)z_1 + l\varepsilon_2(\alpha)z_2 + \ldots + l\varepsilon_{\mu-1}(\alpha)z_{\mu-1} = \tfrac{1}{2}l(rf\alpha f\alpha^{-1}) \\ l\varepsilon_1(\alpha^g)z_1 + l\varepsilon_2(\alpha^g)z_2 + \ldots + l\varepsilon_{\mu-1}(\alpha^g)z_{\mu-1} = \tfrac{1}{2}l(rf\alpha^g f\alpha^{-g}) \\ \ldots\ldots\ldots\ldots\ldots\ldots\ldots\ldots \\ l\varepsilon_1(\alpha^{g^{\mu-2}})z_1 + l\varepsilon_2(\alpha^{g^{\mu-2}})z_2 + \ldots + l\varepsilon_{\mu-1}(\alpha^{g^{\mu-2}})z_{\mu-1} = \tfrac{1}{2}l(rf\alpha^{g^{\mu-2}} f\alpha^{-g^{\mu-2}}), \end{cases}$$

und wenn man den Gröfsen $x, x_1, \ldots x_{\lambda-2}$ in der complexen Zahl $f(\alpha) = x\alpha + x_1\alpha^g + \ldots + x_{\lambda-2}\alpha^{g^{\lambda-2}}$ alle ganzzahligen Werthe von $-\infty$ bis $+\infty$ giebt, welche mit diesen Bedingungen verträglich sind, so erhält man alle verschiedenen *wirklichen* complexen Zahlen, jede genau 2λ mal.

Der Grenzwerth der Summe

$$\Sigma\frac{s-1}{(Nfa)^s},\quad \text{für } s=1,$$

ist nun nach den Principien, welche ***Dirichlet*** in seiner berühmten Abhandlung über die arithmetische Reihe entwickelt hat, gleich der Anzahl der Glieder dieser Reihe, für welche $Nf(\alpha) < T$ ist, dividirt durch T, für $T=\infty$. Die Anzahl aller Glieder, welche $Nf(\alpha) < T$ geben, ist aber gleich dem 2λten Theile der Anzahl aller Werth-Systeme der $x, x_1, \ldots x_{\lambda-2}$, für welche die obigen $\mu-1$ Bedingungen gelten, verbunden mit der Bedingung $Nf(\alpha) < T$. Wird jetzt statt der Gröfsen $x, x_1, \ldots x_{\lambda-2}$ überall $\frac{x}{t}, \frac{x_1}{t}, \ldots \frac{x_{\lambda-2}}{t}$ gesetzt, so geht $f(\alpha)$ in $\frac{f(\alpha)}{t}$ und $Nf(\alpha)$ in $\frac{Nf(\alpha)}{t^{\lambda-1}}$ über; und macht man in den $\mu-1$ Bedingungsgleichungen bei (18.) $r=t^2$, so werden dieselben bei dieser Veränderung:

$$19.\quad \begin{cases} l\varepsilon_1(\alpha)z_1 + l\varepsilon_2(\alpha)z_2 + \ldots + l\varepsilon_{\mu-1}(\alpha)z_{\mu-1} = \frac{1}{2}l(f\alpha f\alpha^{-1}) \\ l\varepsilon_1(\alpha^g)z_1 + l\varepsilon_2(\alpha^g)z_2 + \ldots + l\varepsilon_{\mu-1}(\alpha^g)z_{\mu-1} = \frac{1}{2}l(f\alpha^g f\alpha^{-g}) \\ \ldots\ldots\ldots\ldots\ldots\ldots\ldots\ldots \\ l\varepsilon_1(\alpha^{g^{\mu-2}})z_1 + l\varepsilon_2(\alpha^{g^{\mu-2}})z_2 + \ldots + l\varepsilon_{\mu-1}(\alpha^{g^{\mu-2}})z_{\mu-1} = \frac{1}{2}l(f\alpha^{g^{\mu-2}} f\alpha^{-g^{\mu-2}}); \end{cases}$$

wo $z_1, z_2, \ldots z_{\mu-1}$ sämmtlich in den Grenzen 0 und 1 liegende Gröfsen sind. Die neuen Gröfsen $x, x_1, \ldots x_{\lambda-2}$ haben aber nun nicht mehr alle ganzzahligen Werthe, sondern alle um die Differenz t abnehmenden oder zunehmenden Werthe von $-\infty$ bis $+\infty$, welche mit diesen Bedingungen bei (19.) verträglich sind. Setzt man nun $T=\frac{1}{t^{\lambda-1}}$, so wird die obige Bedingung $Nf(\alpha) < T$, für die neuen Werthe der $x, x_1, \ldots x_{\lambda-2}$ zu $Nf(\alpha) < 1$. Endlich wird t unendlich klein genommen, wodurch T unendlich grofs wird, und die Gröfsen $x, x_1, \ldots x_{\lambda-2}$ zu continuirlich veränderlichen Gröfsen werden. Die Anzahl der Werthe, welche x annehmen kann, wird gleich $\frac{1}{t}\int dx$; eben so die Anzahl der Werthe, welche x_1 annimmt, gleich $\frac{1}{t}\int dx_1$, u. s. w.; wo die Integrale in den Grenzen zu nehmen sind, welche die obigen Bedingungen angeben. Die Anzahl aller Werthsysteme wird also

$$\frac{1}{t^{\lambda-1}}\int^{(\lambda-1)} dx\, dx_1 \ldots dx_{\lambda-2}$$

und dieselbe durch $T=\frac{1}{t^{\lambda-1}}$ dividirt, giebt

$$\int^{(\lambda-1)} dx\, dx_1 \ldots dx_{\lambda-2},$$

Dieses $\lambda-1$fache Integral, dividirt durch 2λ, drückt also, nach den oben angeführten Principien von *Dirichlet*, die Summe der Reihe $\Sigma\frac{s-1}{(Nf\alpha)^s}$ aus, für $s=1$, oder es ist:

$$20.\qquad \Sigma\frac{s-1}{(Nf\alpha)^s}=\frac{1}{2\lambda}\int^{(\lambda-1)}dx\,dx_1\ldots dx_{\lambda-2},\quad \text{für } s=1;$$

wo $f(\alpha)$ alle verschiedenen *wirklichen* complexen Zahlen umfafst und das $\lambda-1$fache Integral in Beziehung auf seine $\lambda-1$ Variabeln in denjenigen zwischen $-\infty$ und $+\infty$ liegenden Grenzen zu nehmen ist, welche den Bedingungen (19.) und aufserdem der Bedingung $Nf(\alpha)<1$ genügen.

Um dieses Integral zu finden, führe ich statt der $\lambda-1$ Variabeln x, $x_1, \ldots x_{\lambda-2}$ die neuen Variabeln $u, u_1, u_2, \ldots u_{\lambda-2}$ ein, welche ich durch folgende Gleichungen bestimme:

$$21.\qquad \begin{cases} f(\alpha) = u+u_\mu i & f(\alpha^{-1}) = u-u_\mu i\\ f(\alpha^g) = u_1+u_{\mu+1} i & f(\alpha^{-g}) = u_1-u_{\mu+1} i\\ \cdots\cdots & \cdots\cdots\\ f(\alpha^{g^{\mu-1}}) = u_{\mu-1}+u_{2\mu-1} i & f(\alpha^{-g^{\mu-1}}) = u_{\mu-1}-u_{2\mu-1} i, \end{cases}$$

wo der Kürze wegen $\sqrt{-1}$ durch i bezeichnet ist. Hiernach ist allgemein

$$u_k=\frac{f(\alpha^{g^k})+f(\alpha^{-g^k})}{2},\quad \text{für } k=0,1,2,\ldots \mu-1,$$

$$u_k=\frac{f(\alpha^{g^k})-f(\alpha^{-g^k})}{2i},\quad \text{für } k=\mu,\mu+1,\ldots 2\mu-1,$$

$$\frac{\partial u_k}{\partial x_h}=\frac{\alpha^{g^{k+h}}+\alpha^{-g^{k+h}}}{2},\quad \text{für } k=0,1,2,\ldots \mu-1,$$

$$\frac{\partial u_k}{\partial x_h}=\frac{\alpha^{g^{k+h}}-\alpha^{-g^{k+h}}}{2i},\quad \text{für } k=\mu,\mu+1,\ldots 2\mu-1.$$

Die Transformation des $\lambda-1$fachen Integrals, bei welcher die Variabeln x, $x_1, \ldots x_{\lambda-2}$ in die neuen Variabeln $u, u_1, u_2, \ldots u_{2\mu-1}$ verwandelt werden sollen, wird nun bekanntlich durch die Functional-Determinante

$$22.\qquad \Sigma\pm\frac{\partial u}{\partial x}\cdot\frac{\partial u_1}{\partial x_1}\cdot\frac{\partial u_2}{\partial x_2}\cdots\frac{\partial u_{\lambda-2}}{\partial x_{\lambda-2}}$$

bestimmt, deren Bildung darum zunächst auszuführen ist. Zu diesem Zwecke bilde ich den Ausdruck

$$23.\qquad c_k^h=\frac{\partial u_k}{\partial x}+\beta^h\frac{\partial u_k}{\partial x_1}+\beta^{2h}\frac{\partial u_k}{\partial x_2}+\ldots+\beta^{(\lambda-2)h}\frac{\partial u_k}{\partial x_{\lambda-2}};$$

und suche die Determinante der Gröfsen c_k^h, nämlich

$$\Sigma\pm c_0^0 c_1^1 c_2^2 c_3^3 \ldots c_{\lambda-2}^{\lambda-2},$$

14*

welche, nach einem bekannten Satze über Determinanten, gleich ist dem Producte der gesuchten Functional-Determinante der Gröfsen $\frac{\partial u_k}{\partial x_h}$ in die Determinante der Gröfsen β^{kh}. Diese letztere Determinante ist aber bekanntlich gleich dem Producte aller Differenzen je zweier der Gröfsen $1, \beta, \beta^2, \dots \beta^{\lambda-1}$, dessen Werth nach der Formel

$$(1-\beta)(1-\beta^2)(1-\beta^3)\dots(1-\beta^{\lambda-2}) = \lambda-1$$

gleich $(\lambda-1)^{\mu}$ gefunden wird. Man hat daher

$$24. \quad \Sigma \pm c_0^0 c_1^1 c_2^2 \dots c_{\lambda-2}^{\lambda-2} = (\lambda-1)^{\mu}\, \Sigma \pm \frac{\partial u}{\partial x}\,\frac{\partial u_1}{\partial x_1} \dots \frac{\partial u_{\lambda-2}}{\partial x_{\lambda-2}}.$$

Setzt man nun, um c_k^h zu finden, in dem Ausdrucke (23.) für die partiellen Differentialquotienten $\frac{\partial u_k}{\partial x}, \frac{\partial u_k}{\partial x_1}$, u. s. w. ihre Werthe, so erhält man

$$c_k^h = \frac{\alpha^{g^k}+\alpha^{-g^k}}{2} + \beta^h\Big(\frac{\alpha^{g^{k+1}}+\alpha^{-g^{k+1}}}{2}\Big) + \dots + \beta^{(\lambda-2)h}\Big(\frac{\alpha^{g^{k-1}}+\alpha^{-g^{k-1}}}{2}\Big),$$

wenn $k = 0, 1, 2, \dots \mu-1$; dagegen

$$c_k^h = \frac{\alpha^{g^k}-\alpha^{-g^k}}{2i} + \beta^h\Big(\frac{\alpha^{g^{k+1}}-\alpha^{-g^{k+1}}}{2i}\Big) + \dots + \beta^{(\lambda-2)h}\Big(\frac{\alpha^{g^{k-1}}-\alpha^{-g^{k-1}}}{2i}\Big),$$

wenn $k = \mu, \mu+1, \dots 2\mu-1$ ist;

und diese Werthe lassen sich wieder durch Anwendung des bekannten Ausdrucks der Kreistheilung

$$(\beta, \alpha) = \alpha + \beta\alpha^g + \beta^2\alpha^{g^2} + \dots + \beta^{\lambda-2}\alpha^{g^{\lambda-2}}$$

in folgender Form darstellen:

$$c_k^h = \frac{(\beta^h, \alpha^{g^k}) + (\beta^h, \alpha^{-g^k})}{2}, \quad \text{wenn } k = 0, 1, 2, \dots \mu-1 \text{ und}$$

$$c_k^h = \frac{(\beta^h, \alpha^{g^k}) - (\beta^h, \alpha^{-g^k})}{2i}, \quad \text{wenn } k = \mu, \mu+1, \dots 2\mu-1;$$

und da

$$(\beta^h, \alpha^{g^k}) = \beta^{-kh}(\beta^h, \alpha); \quad (\beta^h, \alpha^{-g^k}) = (-1)^h \beta^{-kh}(\beta^h, \alpha)$$

ist, so nehmen diese Ausdrücke folgende einfache Gestalt an:

$$25. \quad \begin{cases} c_k^h = \Big(\frac{1+(-1)^h}{2}\Big)\beta^{-kh}(\beta^h, \alpha), & \text{für } k = 0, 1, 2, \dots \mu-1, \\ c_k^h = \Big(\frac{1-(-1)^h}{2i}\Big)\beta^{-kh}(\beta^h, \alpha), & \text{für } k = \mu, \mu+1, \dots 2\mu-1. \end{cases}$$

Die Gröfsen c_k^h haben, wie man hieraus sieht, die Eigenschaft, dafs sie gleich Null werden: erstens wenn h ungerade ist und k einen der Werthe 0, 1,

2, ... $\mu-1$ hat; zweitens wenn h gerade ist und k einen der Werthe μ, $\mu+1$, ... $2\mu-1$ hat. Vermöge dieser beiden Eigenschaften zerfällt die Determinante der Gröfsen c_k^h in ein Product zweier Determinanten und man hat

$$26. \quad \Sigma\pm c_0^0 c_1^1 c_2^2 \ldots c_{\lambda-2}^{\lambda-2} = \Sigma\pm c_0^0 c_1^2 c_2^4 \ldots c_{\mu-1}^{2\mu-1} . \Sigma\pm c_\mu^1 c_{\mu+1}^3 c_{\mu+2}^5 \ldots c_{2\mu-1}^{2\mu-1}.$$

Diese beiden Determinanten lassen sich nun mittels der gefundenen Ausdrücke der Gröfsen c_k^h sehr leicht finden. Aus den Gleichungen (25.) folgt nämlich

$$c_k^{2h} = \beta^{-2hk}(\beta^{2h}, \alpha), \quad \text{wenn } k = 0, 1, 2, \ldots \mu-1,$$
$$c_k^{2h+1} = -i\beta^{-(2h+1)k}(\beta^{2h+1}, \alpha), \quad \text{wenn } k = \mu, \mu+1, \ldots 2\mu-1,$$

und da (β^{2h}, α) und $-i(\beta^{2h+1}, \alpha)$ den untern Index k nicht enthalten, so haben alle Glieder der ersten dieser beiden Determinanten den gemeinschaftlichen Factor $(1, \alpha)(\beta^2, \alpha)(\beta^4, \alpha) \ldots (\beta^{\lambda-3}, \alpha)$, und der andere Factor ist die Determinante der Gröfsen β^{-2hk}; eben so haben alle Glieder der zweiten dieser Determinanten den gemeinschaftlichen Factor $(\beta, \alpha), (\beta^3, \alpha) \ldots (\beta^{\lambda-2}, \alpha) . i^\mu$, und als der andere Factor bleibt die Determinante der Gröfsen $\beta^{-(2h+1)k}$. Die beiden Determinanten der Gröfsen β^{-2hk} und $\beta^{-(2h+1)k}$, deren eine gleich dem Producte der Differenzen je zweier der Gröfsen $1, \beta^{-2}, \beta^{-4}, \ldots \beta^{-(\lambda-3)}$, die andere gleich dem Producte je zweier der Gröfsen $\beta^{-1}, \beta^{-3}, \beta^{-5}, \ldots \beta^{-(\lambda-2)}$ ist, sind, wie leicht zu zeigen, einander gleich, und zwar ist jede gleich $\mu^{\frac{1}{2}\mu}$; also hat man

$$\Sigma\pm c_0^0 c_1^2 c_2^4 \ldots c_{\mu-1}^{\lambda-3} = \mu^{\frac{1}{2}\mu}(1, \alpha)(\beta^2, \alpha)(\beta^4, \alpha) \ldots (\beta^{\lambda-3}, \alpha),$$
$$\Sigma\pm c_\mu^1 c_{\mu+1}^3 c_{\mu+2}^5 \ldots c_{2\mu-1}^{\lambda-2} = \mu^{\frac{1}{2}\mu}(\beta, \alpha)(\beta^3, \alpha) \ldots (\beta^{\lambda-2}, \alpha) . i^\mu,$$

mithin vermöge der Gleichung (26.),

$$\Sigma\pm c_0^0 c_1^1 c_2^2 \ldots c_{\lambda-2}^{\lambda-2} = \mu^\mu (1, \alpha)(\beta, \alpha)(\beta^2, \alpha) \ldots (\beta^{\lambda-2}, \alpha),$$

und da

$$(1, \alpha) = -1, \quad (\beta^k, \alpha)(\beta^{-k}, \alpha) = \pm\lambda, \quad (\beta^\mu, \alpha) = (-1, \alpha) = \pm\sqrt{\pm\lambda}$$

ist, so ergiebt sich, je zwei solcher Factoren zusammenfassend:

$$\Sigma\pm c_0^0 c_1^1 c_2^2 \ldots c_{\lambda-2}^{\lambda-2} = \pm\mu^\mu \lambda^{\mu-1} . \sqrt{\pm\lambda} . i^\mu.$$

Das Vorzeichen unter der Wurzel $\sqrt{\pm\lambda}$ ist bekanntlich so zu nehmen, dafs es $+$ ist, wenn λ von der Form $4n+1$, also $\frac{1}{2}(\lambda-1) = \mu$ eine grade Zahl ist, aber $-$, wenn λ von der Form $4n+3$, also $\frac{1}{2}(\lambda-1) = \mu$ eine ungerade Zahl ist. Im ersteren Falle wird $i^\mu = \pm 1$, im zweiten Falle $i^\mu = \pm i$, also ist

$$\Sigma\pm c_0^0 c_1^1 c_2^2 \ldots c_{\lambda-2}^{\lambda-2} = \pm\mu^\mu \lambda^{\mu-\frac{1}{2}};$$

mithin endlich, vermöge Gleichung (24.), weil $\lambda-1=2\mu$ ist:

$$\Sigma\pm\frac{\partial u}{\partial x}\cdot\frac{\partial u_1}{\partial x_1}\cdot\frac{\partial u_2}{\partial x_2}\cdots\frac{\partial u_{\lambda-2}}{\partial x_{\lambda-2}}=\frac{\lambda^{\mu-\frac{1}{2}}}{2^\mu}.$$

Hiernach verwandelt sich das $\lambda-1$fache Integral der Gleichung (20.), dessen Variabeln $x, x_1, \ldots x_{\lambda-2}$ sind, durch Division mit dieser Functional-Determinante in ein anderes, dessen Variabeln $u, u_1, u_2, \ldots u_{\lambda-2}$ sind, und man erhält

$$27.\quad \Sigma\frac{s-1}{(Nf(\alpha))^s}=\frac{2^{\mu-1}}{\lambda^{\mu+\frac{1}{2}}}\int^{(\lambda-1)}du\,du_1\,du_2\ldots du_{\lambda-2}\ (\text{für } s=1).$$

Ich unterwerfe nun dieses Integral wieder einer neuen Transformation, indem ich statt der Variabeln $u, u_1, u_2, \ldots u_{\lambda-2}$ die neuen $v, v_1, v_2, \ldots v_{\mu-1}, w, w_1, w_2, \ldots w_{\mu-1}$ einführe, welche mittels der Gleichungen

$$u_k=e^{v_k}.\cos w_k,\quad u_{k+\mu}=e^{v_k}.\sin w_k$$

bestimmt sind, wo k die Werthe $0, 1, 2, \ldots \mu-1$, hat. Es läfst sich hier die Functional-Determinante, welche die partiellen Differentialquotienten der Variabeln $u, u_1, u_2, \ldots u_{\lambda-2}$ in Beziehung auf die neu einzuführenden Variabeln $v, v_1, \ldots v_{\mu-1}, w, w_1, \ldots w_{\mu-1}$ genommen enthält, leichter finden, als die umgekehrte, weshalb wir diese zur Transformation des Integrals benutzen wollen. Mittels der Ausdrücke

$$\frac{\partial u_k}{\partial v_k}=+e^{v_k}\cos w_k,\quad \frac{\partial u_{k+\mu}}{\partial v_k}=+e^{v_k}\sin w_k,$$

$$\frac{\partial u_k}{\partial w_k}=-e^{v_k}\sin w_k,\quad \frac{\partial u_{k+\mu}}{\partial w_k}=+e^{v_k}\cos w_k,$$

welche für alle Werthe $k=0, 1, 2, \ldots \mu-1$ gelten, und da allgemein $\frac{\partial u_k}{\partial v_h}$, so wie $\frac{\partial u_k}{\partial w_h}$, mit Ausnahme der beiden Fälle $h=k$ und $h+\mu=k$, gleich Null sind, kann man, nach bekannten Sätzen, den Werth dieser Functional-Determinante leicht finden, weshalb wir uns mit der Herleitung derselben nicht aufhalten, sondern nur das Resultat geben, welches

$$\Sigma\pm\frac{\partial u}{\partial v}\frac{\partial u_1}{\partial v_1}\cdots\frac{\partial u_\mu}{\partial w}\frac{\partial u_{\mu+1}}{\partial w_1}\cdots\frac{\partial u_{2\mu-1}}{\partial w_{\mu-1}}=e^{2v}.e^{2v_1}\ldots e^{2v_{\mu-1}}\text{ ist.}$$

Wird nun mittels dieser Functional-Determinante das Integral der Gleichung (27.) transformirt, so erhält man

$$28.\quad \Sigma\frac{s-1}{(Nf\alpha)^s}=\frac{2^{\mu-1}}{\lambda^{\mu+\frac{1}{2}}}\int^{(\lambda-1)}e^{2v}.e^{2v_1}\ldots e^{2v_{\mu-1}}dv\,dv_1\ldots dv_{\mu-1}dw\,dw_1\ldots dw_{\mu-1},$$

für $s=1$. Die einschränkenden Bedingungen, welche die Grenzen der

Bd. 40

Integrationen für die ursprünglichen Variabeln $x, x_1, x_2, \ldots x_{\lambda-2}$ bestimmen, nämlich die in den Gleichungen (19.) enthaltenen, verbunden mit der Bedingung $Nf(\alpha) < 1$, werden nun, weil allgemein $f(\alpha^{g^k}).f(\alpha^{-g^k}) = e^{2v_k}$ ist, für die neuen Variabeln folgendermaſsen dargestellt:

$$29.\quad \begin{cases} l\varepsilon_1(\alpha)z_1 + l\varepsilon_2(\alpha)z_2 + \ldots + l\varepsilon_{\mu-1}(\alpha)z_{\mu-1} = v \\ l\varepsilon_1(\alpha^g)z_1 + l\varepsilon_2(\alpha^g)z_2 + \ldots + l\varepsilon_{\mu-1}(\alpha^g)z_{\mu-1} = v_1 \\ \ldots\ldots\ldots\ldots\ldots\ldots\ldots\ldots \\ l\varepsilon(\alpha^{g^{\mu-2}})z_1 + l\varepsilon_2(\alpha^{g^{\mu-2}})z_2 + \ldots + l\varepsilon_{\mu-1}(\alpha^{g^{\mu-2}})z_{\mu-1} = v_{\mu-2}, \end{cases}$$

wo $z_1, z_2, \ldots z_{\mu-1}$ beliebige, zwischen 0 und 1 liegende Gröſsen sind und

$$e^{2v}.e^{2v_1}.e^{2v_2} \ldots e^{2v_{\mu-1}} < 1 \text{ ist.}$$

Diese Bedingungen enthalten die Gröſsen $w, w_1, w_2, \ldots w_{\mu-1}$ gar nicht, und da

$$30.\quad \operatorname{tang} w_k = \frac{u_{\mu+k}}{u_k}$$

ist und dieser Bruch wegen der Werthe der $x, x_1, \ldots x_{\lambda-1}$, welche nur, μ Bedingungsgleichungen unterworfen, von $-\infty$ bis $+\infty$ variiren können, selbst von $-\infty$ bis $+\infty$ variirt, und nach den Ausdrücken von u_k und $u_{k+\mu}$, welche ebenfalls sowohl positiv als negativ sein können, auch $\sin w_k$ und $\cos w_k$ sowohl positive als negative Werthe annehmen müssen: so folgt, daſs w_k in den Grenzen $-\frac{1}{2}\pi$ bis $+\frac{3}{2}\pi$ zu nehmen ist. Führt man nun alle Integrationen in Beziehung auf die Variabeln $w, w_1, \ldots w_{\mu-1}$ in diesen Grenzen aus, so erhält man

$$31.\quad \Sigma\frac{s-1}{(Nf\alpha)^s} = \frac{2^{2\mu-1}.\pi^\mu}{\lambda^{\mu+\frac{1}{2}}}\int^{(\mu)} e^{2v}e^{2v_1} \ldots e^{2v_{\mu-1}}dv\,dv_1 \ldots da_{\mu-1}.$$

Ich führe nun zunächst weiter die Integration in Bezug auf die bestimmte Variable $v_{\mu-1}$ aus. Da $\int e^{2v_{\mu-1}}dv_{\mu-1} = \frac{1}{2}e^{2v_{\mu-1}}$ ist, und da $e^{2v}.e^{2v_1} \ldots e^{2v_{\mu-1}}$ in den Grenzen 0 und 1 liegen muſs, also $v_{\mu-1}$ in den Grenzen $e^{2v_{\mu-1}} = 0$ bis $e^{2v_{\mu-1}} = e^{-2v}.e^{-2v_1} \ldots e^{-2v_{\mu-2}}$ zu nehmen ist, so wird

$$32.\quad \Sigma\frac{s-1}{(Nf\alpha)^s} = \frac{2^{2\mu-2}.\pi^\mu}{\lambda^{\mu+\frac{1}{2}}}\int^{(\mu-1)} dv\,dv_1 \ldots dv_{\mu-2}.$$

Endlich werden noch statt der Variabeln $v, v_1, \ldots v_{\mu-2}$ die Variabeln $z_1, z_2, \ldots z_{\mu-1}$ eingeführt, welche durch die Gleichungen (29.) bestimmt und für welche alle Integrationen in den Grenzen 0 und 1 auszuführen sind. Hierzu braucht man die Determinante dieses Systems lineärer Gleichungen (29.), welche ich durch Δ bezeichne, durch deren Anwendung man sogleich

$$\Sigma\frac{s-1}{(Nf\alpha)^s} = \frac{2^{2\mu-2}\pi^\mu\Delta}{\lambda^{\mu+\frac{1}{2}}}\int^{(\mu-1)}dz_1dz_2\ldots dz_{\mu-1}$$

erhält; und da alle Integrationen in Bezug auf $z_1, z_2, \ldots z_{\mu-1}$ in den Grenzen 0 und 1 auszuführen sind, so hat dieses $\mu-1$fache Integral selbst den Werth 1 und es ist

$$33.\quad \Sigma\frac{s-1}{(Nf\alpha)^s} = \frac{2^{2\mu-2}.\pi^\mu.\Delta}{\lambda^{\mu+\frac{1}{2}}},\ \text{für } s=1.$$

Es ist nun weiter zu zeigen, dafs auch die übrigen Glieder des Ausdrucks (12.) alle denselben Werth haben, wie dieses erste, für $s=1$; d. h. dafs allgemein

$$\Sigma\frac{s-1}{(Nf_k(\alpha))^s} = \Sigma\frac{s-1}{(Nf(\alpha))^s},\ \text{für } s=1$$

ist, wo $f_k(\alpha)$ alle idealen complexen Zahlen einer bestimmten Classe bezeichnet. Diese Zahlen $f_k(\alpha)$, da sie äquivalent sind, haben alle einen und denselben idealen Multiplicator $\varphi(\alpha)$, mit welchem multiplicirt sie zu *wirklichen* complexen Zahlen werden, und es ist, wenn $\varphi(\alpha)f_k(\alpha) = F_k(\alpha)$ gesetzt wird:

$$34.\quad \Sigma\frac{s-1}{(Nf_k(\alpha))^s} = N\varphi(\alpha)\Sigma\frac{s-1}{(NF_k(\alpha))^s},\ \text{für } s=1;$$

wo $F_k(\alpha)$ alle wirklichen complexen Zahlen bedeutet, welche den bestimmten idealen Factor $\varphi(\alpha)$ haben. Die Untersuchung des Werths der Summe

$$\Sigma\frac{s-1}{(NF_k(\alpha))^s},\ \text{für } s=1,$$

ist der obigen Untersuchung der entsprechenden Summe für die complexen Zahlen $f(\alpha)$ vollkommen gleich; nur dafs hier noch gewisse einschränkende Bedingungen hinzutreten, welche die Coëfficienten der wirklichen complexen Zahl $F_k(\alpha)$ erfüllen müssen, damit $F_k(\alpha)$ den idealen complexen Factor $\varphi(\alpha)$ enthalte. Die Bedingung, dafs eine wirkliche complexe Zahl einen gegebenen idealen Factor enthalte, wird aber, wie wir in der Abhandlung 16. Band 35. dieses Journals gezeigt haben, immer durch eine Anzahl Congruenz-Bedingungen ausgedrückt, welche die Coëfficienten der wirklichen complexen Zahl erfüllen müssen und welche durch die idealen Primfactoren dieses idealen Factors vollkommen bestimmt sind. Es möge nun $\varphi(\alpha)$, in seine idealen Primfactoren zerlegt, einen bestimmten der Primfactoren der zum Exponenten d gehörenden realen Primzahl q_d genau m mal enthalten, ferner einen bestimmten der Primfactoren der zum Exponenten d' gehörenden Primzahl $q'_{d'}$ genau m' mal, u. s. w., so wird, wie wir in der oft erwähnten

Abhandlung gezeigt haben, die Bedingung, dafs $F_k(\alpha)$ den ersten dieser Primfactoren m mal enthalte, genau durch d Congruenzen für den Modul q_d^m ausgedrückt, welche in Beziehung auf die Coëfficienten der wirklichen complexen Zahl $F_k(\alpha) = x\alpha + x_1\alpha^g + x_2\alpha^{g^2} + \ldots + x_{\lambda-2}\alpha^{g^{\lambda-2}}$ lineär sind. Eben so wird die Bedingung, dafs $F_k(\alpha)$ den andern Primfactor m' mal enthalte, durch d' solche Congruenzen für den Modul $q'^{m'}_{d'}$ ausgedrückt; u. s. w. Eine einzige lineäre Congruenz unter den Gröfsen x, x_1, ... $x_{\lambda-2}$ für den Modul q_d^m macht aber, dafs von allen den Werth-Systemen dieser Gröfsen, welche den übrigen Bedingungen genügen, nur der q_d^mte Theil zu nehmen ist, und d solche Congruenz-Bedingungen zusammen machen demgemäfs, dafs nur der q_d^{md}te Theil gelten kann. Eben so bewirken die d' lineären Congruenzen, mod. $q'^{m'}_{d'}$, dafs von den Werthsystemen der x, x_1, ... $x_{\lambda-2}$ nur der $q'^{m'd'}_{d'}$te Theil zu nehmen ist, u. s. w. Der Werth der Summe $\Sigma\frac{s-1}{(NF_k(\alpha))^s}$ ist also genau gleich dem $q_d^{md}.q'^{m'd'}_{d'}\ldots$ ten Theile der Summe $\Sigma\frac{s-1}{(Nf\alpha)^s}$, für $s=1$, wenn $F_k(\alpha)$ alle wirklichen complexen Zahlen bedeutet, welche den idealen Factor $\varphi(\alpha)$ enthalten, und $f(\alpha)$ alle wirklichen complexen Zahlen ohne Ausnahme. Die Gleichung (34.) geht daher in

$$35. \quad \Sigma\frac{s-1}{(Nf_k(\alpha))^s} = \frac{N\varphi(\alpha)}{q_d^{md}.q'^{m'd'}_{d'}\ldots}\Sigma\frac{s-1}{(Nf(\alpha))^s}, \text{ für } s=1,$$

über, und da vermöge der idealen Primfactoren, welche $\varphi(\alpha)$ nach der Voraussetzung enthält, $N\varphi(\alpha) = q_d^{md}.q'^{m'd'}_{d'}\ldots$ ist, so hat man endlich

$$36. \quad \Sigma\frac{s-1}{(Nf_k(\alpha))^s} = \Sigma\frac{s-1}{(Nf(\alpha))^s}; \text{ für } s=1;$$

was zu beweisen war.

Da wir nun den Werth des ersten Gliedes der Gleichung (12.) gefunden und bewiesen haben, dafs alle diese Glieder, deren Anzahl gleich H, der Anzahl aller nicht äquivalenten Classen ist, einander gleich sind, so haben wir folgenden zweiten Ausdruck des Werths der Reihe R, für $s=1$:

$$37. \quad R = \frac{2^{2\mu-2}.\pi^\mu.\Delta.H}{\lambda^{\mu+\frac{1}{2}}}, \text{ für } s=1;$$

und da der erste Ausdruck derselben Reihe nach der Gleichung (11.)

$$R = \pm \frac{\pi^{\frac{1}{2}(\lambda-1)}.2^{\frac{1}{2}(\lambda-3)}.P.D}{\lambda^{\lambda-\frac{3}{2}}} \quad \text{für } s = 1$$

war, so haben wir endlich, durch Gleichsetzung beider Ausdrücke:

$$\frac{2^{2\mu-2}.\pi^{\mu}.\Delta.H}{\lambda^{\mu+\frac{1}{2}}} = \frac{\pi^{\frac{1}{2}(\lambda-1)}.2^{\frac{1}{2}(\lambda-3)}.P.D}{\lambda^{\lambda-\frac{3}{2}}},$$

also, nach gehöriger Reduction, da $\mu = \frac{1}{2}(\lambda-1)$ ist, für den gesuchten Ausdruck der Anzahl aller nichtäquivalenten Classen der aus λten Wurzeln der Einheit gebildeten idealen complexen Zahlen:

$$38. \quad H = \frac{P.D}{(2\lambda)^{\mu-1}.\Delta}.$$

Der bessern Übersicht wegen wiederhôle ich hier noch einmal die Bedeutung der einzelnen in dieser Formel vorkommenden Zeichen. Es ist $\mu = \frac{1}{2}(\lambda-1)$; ferner ist

$$P = \varphi(\beta)\varphi(\beta^3)\varphi(\beta^5)\ldots\varphi(\beta^{2\mu-1}),$$

wenn

$$\varphi(\beta) = 1+g_1\beta+g_2\beta^2+g_3\beta^3+\ldots+g_{\lambda-2}\beta^{\lambda-2},$$

$g_1, g_2, g_3, \ldots g_{\lambda-2}$ die kleinsten positiven Reste sind, welche die Potenzen $g, g^2, g^3, \ldots g^{\lambda-2}$ einer primitiven Wurzel der Primzahl λ für den Modul λ geben, und β eine primitive Wurzel der Gleichung $\beta^{\lambda-1} = 1$ ist. Ferner ist D die Determinante des Systems folgender Gröfsen:

$$\begin{array}{llll} le(\alpha), & le(\alpha^{g}), & \ldots & le(\alpha^{g^{\mu-2}}), \\ le(\alpha^{g}), & le(\alpha^{g^2}), & \ldots & le(\alpha^{g^{\mu-1}}), \\ \cdot & \cdot & \cdot & \cdot \\ le(\alpha^{g^{\mu-2}}), & le(\alpha^{g^{\mu-1}}), & \ldots & le(\alpha^{g^{2\mu-4}}). \end{array}$$

Sodann ist

$$e(\alpha) = \sqrt{\left(\frac{(1-\alpha^{g})(1-\alpha^{-g})}{(1-\alpha)(1-\alpha^{-1})}\right)}$$

und Δ die Determinante des Systems

$$\begin{array}{llll} l\varepsilon_1(\alpha), & l\varepsilon_2(\alpha), & \ldots & l\varepsilon_{\mu-1}(\alpha), \\ l\varepsilon_1(\alpha^{g}), & l\varepsilon_2(\alpha^{g}), & \ldots & l\varepsilon_{\mu-1}(\alpha^{g}), \\ \cdot & \cdot & \cdot & \cdot \\ l\varepsilon_1(\alpha^{g^{\mu-2}}), & l\varepsilon_2(\alpha^{g^{\mu-2}}), & \ldots & l\varepsilon_{\mu-1}(\alpha^{g^{\mu-2}}), \end{array}$$

wo $\varepsilon_1(\alpha), \varepsilon_2(\alpha), \ldots \varepsilon_{\mu-1}(\alpha)$ ein System von Fundamental-Einheiten sind.

Da eine vollständige Theorie der aus λten Wurzeln der Einheit gebildeten complexen Zahlen zugleich mit die Theorie der aus den Perioden dieser Wurzeln gebildeten complexen Zahlen umfassen mufs und in vielen Stücken sich sogar auf diese stützt, so habe ich nicht versäumt, auch die Ausdrücke für die Anzahl der nichtäquivalenten Classen dieser aus Perioden gebildeten idealen complexen Zahlen vollständig auszuarbeiten. Die anzuwendenden Principien, so wie die Ausführung derselben im Einzelnen, sind aber mit den hier entwickelten so übereinstimmend, dafs ich die Entwicklung dieser Ausdrücke hier übergehe und mich darauf beschränke, nur die Resultate den Lesern mitzutheilen.

Ordnet man die λten Wurzeln der Einheit α, α^g, α^{g^2}, ... $\alpha^{g^{\lambda-2}}$ nach den *Gaufsischen* Perioden, und zwar in e Perioden von je f Gliedern, wo $e.f=\lambda-1$ ist, bildet die aus diesen Perioden zusammengesetzten complexen Zahlen und theilt dieselben und ihre sämmtlichen idealen Factoren in alle nichtäquivalenten Classen ein, welche für dieselben bestehen und deren Anzahl gleich H sei: so müssen in dem Ausdrucke des H die beiden Fälle unterschieden werden, wo f *gerade* und wo f *ungerade* ist, und man erhält dann folgende Resultate.

Erstens, wenn f *gerade* ist, d. h. wenn die Perioden aus einer *geraden* Anzahl von Wurzeln der Gleichung $\alpha^\lambda=1$ bestehen, ist

$$39.\qquad H=\frac{D}{\varDelta};$$

wo D die Determinante folgender $(e-1)^2$ Gröfsen:

$$\begin{array}{llll} lE(\alpha), & lE(\alpha^g), & \dots & lE(\alpha^{g^{e-2}}), \\ lE(\alpha^g), & lE(\alpha^{g^2}) & \dots & lE(\alpha^{g^{e-1}}), \\ \cdot & \cdot & \cdot & \cdot \\ lE(\alpha^{g^{e-2}}), & lE(\alpha^{g^{e-1}}), & \dots & lE(\alpha^{g^{2e-4}}) \end{array}$$

und

$$E(\alpha)=\sqrt{\left(\frac{(1-\alpha^g)(1-\alpha^{g^{e+1}})\dots(1-\alpha^{g^{(f-1)e+1}})}{(1-\alpha)(1-\alpha^{g^e})\dots(1-\alpha^{g^{(f-1)e}})}\right)}$$

ist; welches eine aus den e Perioden von je f Gliedern gebildete complexe Einheit und wo $\varDelta$ die Determinante folgender $(e-1)^2$ Gröfsen ist:

15 *

$$\begin{array}{llll} l\varepsilon_1(\alpha), & l\varepsilon_2(\alpha), & \dots & l\varepsilon_{e-1}(\alpha), \\ l\varepsilon_1(\alpha^g), & l\varepsilon_2(\alpha^g), & \dots & l\varepsilon_{e-1}(\alpha^g), \\ \cdot & \cdot & \cdot & \cdot \\ l\varepsilon_1(\alpha^{g^{e-2}}), & l\varepsilon_2(\alpha^{g^{e-2}}), & \dots & l\varepsilon_{e-1}(\alpha^{g^{e-2}}). \end{array}$$

$\varepsilon_1(\alpha)$, $\varepsilon_2(\alpha)$, ... $\varepsilon_{e-1}(\alpha)$ ist ein System von Fundamental-Einheiten, welche aus den e Perioden von je f Gliedern gebildet sind.

Zweitens, wenn f ***ungerade*** ist, d. h. wenn die Perioden aus einer ***ungeraden*** Anzahl von Wurzeln der Gleichung $\alpha^\lambda = 1$ bestehen, so ist

$$40. \quad H = \frac{P.D}{2^{\frac{1}{2}e-1}\Delta},$$

wo D die Determinante folgender $\frac{1}{2}(e-1)^2$ Gröfsen:

$$\begin{array}{llll} lE(\alpha), & lE(\alpha^g), & \dots & lE(\alpha^{g^{\frac{1}{2}e-2}}), \\ lE(\alpha^g), & lE(\alpha^{g^2}), & \dots & lE(\alpha^{g^{\frac{1}{2}e-1}}), \\ \cdot & \cdot & \cdot & \cdot \\ lE(\alpha^{g^{\frac{1}{2}e-2}}), & lE(\alpha^{g^{\frac{1}{2}e-1}}), & \dots & lE(\alpha^{g^{e-4}}) \end{array}$$

und

$$E(\alpha) = \sqrt{\left(\frac{(1-\alpha^g)(1-\alpha^{g^{\frac{1}{2}e+1}}) \dots (1-\alpha^{g^{(2f-1)\frac{1}{2}e+1}})}{(1-\alpha)(1-\alpha^{g^{\frac{1}{2}e}}) \dots (1-\alpha^{g^{(2f-1)\frac{1}{2}e}}}\right)}$$

ist; welches eine aus $\frac{1}{2}e$ Perioden von je $2f$ Gliedern gebildete complexe Einheit ist. Ferner ist Δ die Determinante folgender $(\frac{1}{2}e-1)^2$ Gröfsen:

$$\begin{array}{llll} l\varepsilon_1(\alpha), & l\varepsilon_2(\alpha), & \dots & l\varepsilon_{\frac{1}{2}e-1}(\alpha) \\ l\varepsilon_1(\alpha^g), & l\varepsilon_2(\alpha^g), & \dots & l\varepsilon_{\frac{1}{2}e-1}(\alpha^g) \\ \cdot & \cdot & \cdot & \cdot \\ l\varepsilon_1(\alpha^{g^{\frac{1}{2}e-2}}), & l\varepsilon_2(\alpha^{g^{\frac{1}{2}e-2}}), & \dots & l\varepsilon_{\frac{1}{2}e-1}(\alpha^{g^{\frac{1}{2}e-2}}) \end{array}$$

und $\varepsilon_1(\alpha)$, $\varepsilon_2(\alpha)$, ... $\varepsilon_{\frac{1}{2}e-1}(\alpha)$ ist ein System von Fundamental-Einheiten, welche aus den $\frac{1}{2}e$ Perioden von je $2f$ Gliedern gebildet sind, wobei endlich P folgendes Product bezeichnet:

$$\begin{aligned} P &= \varphi(\beta)\varphi\beta^3)\varphi(\beta^5) \dots \varphi(\beta^{e-1}) \\ \varphi(\beta) &= A + A_1\beta + A_2\beta^2 + \dots + A_{e-1}\beta^{e-1} \\ A &= \frac{1}{\lambda}(1 + g_e + g_{2e} + \dots + A_{(f-1)e}) \\ A_1 &= \frac{1}{\lambda}(g_1 + g_{e+1} + g_{2e+1} + \dots + g_{(f-1)e+1}) \\ &\cdots \\ A_{e-1} &= \frac{1}{\lambda}(g_{e-1} + g_{2e-1} + g_{3e-1} + \dots + g_{fe-1}). \end{aligned}$$

$g_1, g_2, \ldots g_{\lambda-2}$ bezeichnen, wie oben, die kleinsten positiven Reste, welche die Potenzen einer primitiven Wurzel $g, g^2, g^3, \ldots g^{\lambda-2}$ für den Modul λ geben, und β ist eine primitive Wurzel der Gleichung $\beta^e = 1$.

Ich füge noch einige Bemerkungen über die angegebenen Resultate hinzu. Die Anzahl H der nichtäquivalenten Classen aller aus den einfachen Wurzeln $\alpha, \alpha^g, \ldots \alpha^{g^{\lambda-2}}$ gebildeten complexen Zahlen und der idealen Factoren derselben, welche in der Formel (38.) gegeben ist, besteht aus zwei wesentlich verschiedenen Factoren, nämlich dem Factor $\frac{P}{(2\lambda)^{\mu-1}}$ und dem Factor $\frac{D}{\Delta}$; welche beide für sich ganze Zahlen sind. Für den Factor $\frac{D}{\Delta}$ folgt dies unmittelbar aus der Gleichung (39.), welche zeigt, dafs dasselbe $\frac{D}{\Delta}$ für sich die Anzahl aller nichtäquivalenten Classen für die aus den zweigliedrigen Perioden zu bildenden idealen complexen Zahlen darstellt, also nothwendig eine ganze Zahl ist. Um in Beziehung auf den andern Factor des H in der Formel (38.) Dasselbe zu zeigen, verwandele ich die Factoren von der Form $\varphi(\beta^{2n-1})$, aus denen P zusammengesetzt ist, nämlich

$$\varphi(\beta^{2n-1}) = 1 + g_1\beta^{2n-1} + g_2\beta^{2(2n-1)} + \ldots + g_{\lambda-2}\beta^{(\lambda-2)(2n-1)}$$

mit Hülfe der Gleichung $\beta^\mu = -1$ und $g_{k+\mu} = \lambda - g_k$ in die Form

$$\varphi(\beta^{2n-1}) = 2 - \lambda + (2g_1 - \lambda)\beta^{2n-1} + (2g_2 - \lambda)\beta^{2(2n-1)} + \ldots + (2g_{\mu-1} - \lambda)\beta^{(\mu-1)(2n-1)},$$

in welchem Ausdrucke alle Coëfficienten der einzelnen Glieder ungerade Zahlen sind. Es kommt nun darauf an, wie viel mal μ den Factor 2 enthält. Setzt man deshalb $\mu = 2^r.\nu$, wo ν ungerade ist und r auch gleich Null sein kann, und addirt zu dem Ausdrucke des $\varphi(\beta^{2n-1})$ die Gleichung

$$1 - \beta^{2^r(2n-1)} + \beta^{2.2^r(2n-1)} - \ldots + \beta^{(\nu-1)2^r(2n-1)} = 0,$$

so wie dieselbe Gleichung, multiplicirt mit $\beta^{2n-1}, \beta^{2(2n-1)}, \ldots \beta^{(2^r-1)(2n-1)}$, welche für jeden Werth des n Statt hat, mit alleiniger Ausnahme des Falles $2n-1 = \nu$: so werden alle Coëfficienten in $\varphi(\beta^{2n-1})$ zu graden Zahlen; also ist $\varphi(\beta^{2n-1})$, mit Ausnahme des Falles $2n-1 = \nu$, immer durch 2 theilbar, und das Product P, welches aus μ solchen Factoren besteht, ist theilbar durch $2^{\mu-1}$. Um weiter zu zeigen, dafs P auch durch $\lambda^{\mu-1}$ theilbar ist, multiplicire ich $\varphi(\beta^{2n-1})$

mit $g\beta^{2n-1}-1$, wo g eine primitive Wurzel und deshalb $g^{\lambda-1}-1$ durch λ und $g^{\mu}+1$ durch λ theilbar ist. Ich wähle auch, was immer angeht, die primitive Wurzel so, dafs $g^{\mu}+1$ den Factor λ nur *einmal* enthält. Es wird alsdann

$$(g\beta^{2n-1})\varphi(\beta^{2n-1}) = gg_{\lambda-2}-1+(gg_{\lambda-1}-g_1)\beta^{2n-1}+(gg_1-g_2)\beta^{2(2n-1)}+\dots$$
$$+(gg_{\lambda-3}-g_{\lambda-2})\beta^{(\lambda-2)(2n-1)},$$

und es sind nun alle Coëfficienten dieses Ausdrucks durch λ theilbar; wie aus der Congruenz $g_k \equiv g^k$, mod. λ, sogleich erhellt. Setzt man nun in dem durch λ theilbaren Ausdrucke $(g\beta^{2n-1})\varphi(\beta^{2n-1})$ nach einander $n=1, 2, 3, \dots \mu$ und bildet das Product, so erhält man $(g^{\mu}+1)P$ theilbar durch λ^{μ}, und da $g^{\mu}+1$ den Factor λ nur einmal enthält, so mufs P den Factor $\lambda^{\mu-1}$ enthalten. Da wir nun bewiesen haben, dafs P durch $2^{\mu-1}$ und auch durch $\lambda^{\mu-1}$ theilbar ist, so ist wirklich $\frac{P}{(2\lambda)^{\mu-1}}$ eine ganze Zahl; wie behauptet wurde.

In Beziehung auf die Classenzahl der aus Perioden zu bildenden idealen complexen Zahlen tritt ein wesentlicher Unterschied auf, zwischen denen, bei welchen die Perioden eine ungerade oder eine gerade Anzahl von Gliedern enthalten. Wenn nämlich die Perioden eine ungerade Anzahl von Gliedern enthalten, so besteht immer die Classen-Anzahl aus zwei solchen verschiedenen ganzzahligen Factoren, während in den Ausdrücken für die Classenzahl bei den aus Perioden von gerader Gliederzahl gebildeten idealen complexen Zahlen nur der eine dieser beiden Factoren auftritt.

Für den Fall, wo es nur zwei Perioden giebt, deren jede aus $\frac{1}{2}(\lambda-1)$ Gliedern besteht, geben die Formeln (39. und 40.) die Anzahl der Classen für die quadratischen Formen, deren Determinante eine Primzahl gleich λ ist, und zwar die eine für die positive Determinante λ, die andere für dieselbe negative Determinante.

* Eine sehr merkwürdige einfache Beziehung, in welcher die Classenzahlen der aus Perioden gebildeten idealen complexen Zahlen zur Classenzahl der aus den einfachen Wurzeln der Gleichung $\alpha^{\lambda}=1$ gebildeten steht, ist die, dafs jene immer aliquote genaue Theile von dieser sind, da, wo sie ihr nicht völlig gleich sind. Diese Eigenschaft würde sich aus den gefundenen Ausdrücken für die Classen-Anzahl allgemein nur sehr schwer entwickeln lassen:

sie kann aber mit Leichtigkeit unmittelbar aus der Definition der Äquivalenz der idealen complexen Zahlen hergeleitet werden; wobei man nur den einen Satz braucht, dafs die Classen-Anzahl stets eine endliche, bestimmte ist. Es mögen $\eta, \eta_1, \eta_2, \ldots \eta_{e-1}$ e Perioden von je f Gliedern sein, wo $ef = \lambda - 1$ ist und ferner $\varphi(\eta), \varphi_1(\eta), \varphi_2(\eta), \ldots \varphi_{h-1}(\eta)$ die h nichtäquivalenten Classen der aus diesen Perioden gebildeten idealen complexen Zahlen repräsentiren, und zwar so, dafs $\varphi(\eta)$ eine beliebige solche complexe Zahl der ersten Classe, $\varphi_1(\eta)$ eine beliebige der zweiten Classe u. s. w. darstellt. Da nun die aus den Perioden gebildeten idealen complexen Zahlen alle in den aus den einfachen Wurzeln der Gleichung $\alpha^\lambda = 1$ gebildeten mit einbegriffen sind, so müssen die durch $\varphi(\eta), \varphi_1(\eta), \ldots \varphi_{h-1}(\eta)$ repräsentirten Classen alle in den verschiedenen Classen dieser allgemeineren complexen Zahlen vorkommen, deren Anzahl gleich H sei. Es kann nun erstens der Fall eintreten, dafs aufser den h Classen, welche die aus Perioden gebildeten complexen idealen Zahlen geben, für die aus den einfachen Wurzeln der Gleichung $\alpha^\lambda = 1$ gebildeten gar keine andern Classen Statt haben, dafs also $h = H$ ist. Wenn dies aber nicht der Fall ist, so sei $f(\alpha)$ eine ideale complexe Zahl, welche nicht in den h Classen vorkommt, in welchen $\varphi(\eta), \varphi_1(\eta), \ldots \varphi_{h-1}(\eta)$ vorkommen: dann gehören, wie leicht zu zeigen ist, $f(\alpha)\varphi(\eta), f(\alpha)\varphi_1(\eta), f(\alpha)\varphi_2(\eta), \ldots f(\alpha)\varphi_{h-1}(\eta)$ nur solchen Classen an, welche weder unter sich, noch mit den Classen, denen $\varphi(\eta), \varphi_1(\eta), \varphi_2(\eta), \ldots \varphi_{h-1}(\eta)$ angehören, äquivalent sind. Wäre nämlich zunächst $f(\alpha)\varphi_r(\eta)$ äquivalent $f(\alpha)\varphi_s(\eta)$, so multiplicire ich beide mit demselben Multiplicator $F(\alpha)$, welcher bewirkt, dafs $F(\alpha)f(\alpha)$ eine *wirkliche* complexe Zahl ist; woraus dann folgen würde, dafs $\varphi_r(\eta)$ äquivalent $\varphi_s(\eta)$ sein müfste; gegen die Voraussetzung. Wäre ferner $f(\alpha)\varphi_r(\eta)$ äquivalent $\varphi_s(\eta)$, so multiplicire ich beide mit einem solchen Multiplicator $\Phi_r(\eta)$, welcher $\Phi_r(\eta)\varphi_r(\eta)$ zu einer wirklichen complexen Zahl macht; woraus folgen würde, dafs $f(\alpha)$ äquivalent $\Phi_r(\eta)\varphi_r(\eta)$ sei, welches ebenfalls gegen die Voraussetzung ist, weil $f(\alpha)$, wenn es einer aus den Perioden gebildeten idealen complexen Zahl äquivalent wäre, schon in den ersten h Classen, welche durch $\varphi(\eta), \varphi_1(\eta), \ldots \varphi_{h-1}(\eta)$ repräsentirt sind, enthalten sein müfste. Eine einzige, nicht in diesen h Classen enthaltene complexe ideale Zahl macht also, dafs eine neue Gruppe von h Classen hinzukommt. Eben so macht eine einzige nicht in diesen $2h$ Classen enthaltene ideale complexe Zahl, dafs eine ganze dritte Gruppe von h Classen, die weder unter sich, noch mit den vorigen äquivalent sind, hinzukommt: dafs also $3h$ Classen existiren; und so geht dies weiter, bis wirklich alle H Classen erschöpft sind, welche sich, wie hieraus zu ersehen ist, nur durch ein genaues Vielfaches von h erschöpfen lassen. Es ist also immer H ein Vielfaches von h, oder h ein genau aliquoter Theil von H; was zu beweisen war. Eben so wird bewiesen, dafs wenn f' ein Vielfaches von f ist, und zugleich ein Theiler von $\lambda-1$, die Classen-Anzahl der aus Perioden von je f' Gliedern gebildeten idealen complexen Zahlen immer ein genau aliquoter Theil von der Classen-Anzahl für diejenigen ist, welche aus Perioden von je f Gliedern gebildet sind.

Breslau, den 16ten Juni 1849.

Zwei besondere Untersuchungen über die Classen-Anzahl und über die Einheiten der aus λten Wurzeln der Einheit gebildeten complexen Zahlen

Journal für die reine und angewandte Mathematik 40, 117–129 (1850)

Nachdem ich in der vorhergehenden Abhandlung die Classen-Anzahl für die aus λten Wurzeln der Einheit gebildeten complexen Zahlen angegeben habe, will ich jetzt zunächst folgende Frage vollständig beantworten:

Für welche Werthe der Primzahl λ ist die Classen-Anzahl der aus λten Wurzeln der Einheit gebildeten complexen idealen Zahlen durch λ theilbar; und für welche nicht?

Nach der Gleichung (38.) in der vorhergehenden Abhandlung ist diese Classen-Anzahl:

$$1. \quad H = \frac{P}{(2\lambda)^{\mu-1}} \cdot \frac{D}{\Delta},$$

wo die beiden Factoren, aus welchen sie zusammengesetzt ist, nämlich $\frac{P}{(2\lambda)^{\mu-1}}$ und $\frac{D}{\Delta}$, für sich ganze Zahlen sind; es wird also die Untersuchung, ob H durch λ theilbar sei, oder nicht, für diese beiden Factoren besonders zu führen sein. Ich mache den Anfang mit dem Factor $\frac{P}{(2\lambda)^{\mu-1}}$, in welchem

* $$G = \varphi(\beta)\varphi(\beta^3)\varphi(\beta^5)\ \ldots\ \varphi(\beta^{\lambda-2}),$$

$$\varphi(\beta) = 1 + g_1\beta + g_2\beta^2 + g_3\beta^3 + \ \ldots\ + g_{\lambda-2}\beta^{\lambda-2} \text{ ist.}$$

Ich multiplicire $\varphi(\beta)$ mit $g\beta - 1$, so wird

$$(g\beta-1)\varphi(\beta) = gg_{\lambda-2} - 1 + (gg_{\lambda-1} - g_1)\beta + (gg_1 - g_2)\beta^2 + \ \ldots$$
$$+ (gg_{\lambda-3} - g_{\lambda-2})\beta^{\lambda-2},$$

und es sind nun, wie schon in der vorhergehenden Abhandlung gezeigt, die Coëfficienten aller einzelnen Glieder durch λ theilbar. Setzt man also

$$gg_{k-1} - g_k = \lambda b_k$$

und

$$b_0 + b_1\beta + b_2\beta^2 + \ \ldots\ + b_{\lambda-2}\beta^{\lambda-2} = \psi(\beta),$$

so erhält man

$$(g\beta-1)\varphi(\beta) = \lambda\psi(\beta),$$

und hieraus, wenn β in β^3, β^5, ... $\beta^{\lambda-2}$ verwandelt und das Product gebildet wird, bei welchem

$$(g\beta-1)(g\beta^3-1)(g\beta^5-1)\ \dots\ (g\beta^{\lambda-2}-1) = g^\mu+1$$

ist:

$$(g^\mu+1)P = \lambda^\mu\psi(\beta)\psi(\beta^3)\psi(\beta^5)\ \dots\ \psi(\beta^{\lambda-2}).$$

Es ist $g^\mu+1$ (da g eine primitive Wurzel des λ und $\mu=\frac{1}{2}(\lambda-1)$ ist) bekanntlich durch λ theilbar; man kann also $g^\mu+1=\lambda G$ setzen, wo G eine ganze Zahl ist. Man weifs auch, dafs nur in besondern Fällen, für einzelne Werthe der primitiven Wurzel g, der Ausdruck $g^\mu+1$ den Factor λ zweimal, oder wohl gar mehrmal enthalten kann: wählt man aber, was immer möglich ist, die primitive Wurzel g so, dafs dies nicht der Fall ist, so ist G nicht durch λ theilbar. Man erhält also

* 2. $$2^{\mu-1}g\cdot\frac{P}{(2\lambda)^{\mu-1}} = \psi(\beta)\psi(\beta^3)\psi(\beta^5)\ \dots\ \psi(\beta^{\lambda-2}).$$

Es kann demnach $\frac{P}{(2\lambda)^{\mu-1}}$ nur dann durch λ theilbar sein, wenn das Product $\psi(\beta)\psi(\beta^3)\dots\psi(\beta^{\lambda-2})$ durch λ theilbar ist; und umgekehrt, wenn dieses Product durch λ theilbar ist, so ist auch wirklich dieser erste Factor der Classen-Anzahl H durch λ theilbar. Nun ist aber offenbar, wenn man statt der primitiven Wurzel β der Gleichung $\beta^{\lambda-1}=1$ die primitive Congruenzwurzel g der Congruenz $g^{\lambda-1}\equiv 1$, mod. λ, setzt:

$$\psi(\beta)\psi(\beta^3)\ \dots\ \psi(\beta^{\lambda-2}) \equiv \psi(g)\psi(g^3)\ \dots\ \psi(g^{\lambda-2}), \text{ mod. } \lambda;$$

also wenn irgend einer der ganzzahligen Factoren des Products $\psi(g)\psi(g^3)\dots$ $\dots\psi(g^{\lambda-2})$ durch λ theilbar ist, so mufs auch $\frac{P}{(2\lambda)^{\mu-1}}$ durch λ theilbar sein; und umgekekrt, wenn keine der ganzzahligen Gröfsen $\psi(g)$, $\psi(g^3)$, ... $\psi(g^{\lambda-2})$ durch λ theilbar ist, so ist auch $\frac{P}{(2\lambda)^{\mu-1}}$ nicht durch λ theilbar. Es hangt also für diesen ersten Factor der Classen-Anzahl H Alles davon ab, ob die Congruenz

3. $$b_0+b_1g^{2n-1}+b_2g^{2(2n-1)}+\ \dots\ +b_{\lambda-2}g^{(\lambda-2)(2n-1)} \equiv 0, \text{ mod. } \lambda,$$

für irgend einen der Werthe $n=1, 2, \dots \mu$ erfüllt wird; oder für keinen derselben. Dividirt man die Congruenz durch g^{2n-1}, und verwandelt mit Hülfe der Congruenz $g^k\equiv g_k$, mod. λ, die Exponenten 1, 2, 3, ... $\lambda-2$ in Indices, so wird aus dieser Congruenz folgende gleichbedeutende:

$$b_0 g_{\lambda-2}^{2n-1} + b_1 + b_2 g_1^{2n-1} + b_3 g_2^{2n-1} + \dots + b_{\lambda-2} g_{\lambda-3}^{2n-1} \equiv 0, \text{ mod. } \lambda.$$

Aus der Definition der Coëfficienten b_k, nämlich

$$\lambda b_k = g g_{k-1} - g_k,$$

folgt nun aber unmittelbar, dafs erstens $b_k = 0$ ist, für alle Werthe des g_{k-1}, welche zwischen 0 und $\frac{\lambda}{g}$ liegen; ferner $b_k = 1$ für alle Werthe des g_{k-1}, welche zwischen $\frac{\lambda}{g}$ und $\frac{2\lambda}{g}$ liegen, oder allgemein $b_k = s$ für alle Werthe des g_{k-1}, welche zwischen $\frac{s\lambda}{g}$ und $\frac{(s+1)\lambda}{g}$ liegen. Bezeichnet man durch t_s die gröfste in $\frac{s\lambda}{g}$ enthaltene ganze Zahl, so läfst sich, diejenigen Glieder zusammenfassend, für welche die Coëfficienten b_k gleiche Werthe haben, die obige Congruenz folgendermafsen darstellen:

$$4. \quad \begin{cases} +1((t_1+1)^{2n-1} + (t_1+2)^{2n-1} + \dots + t_2^{2n-1}) \\ +2((t_2+1)^{2n-1} + (t_2+2)^{2n-1} + \dots + t_3^{2n-1}) \\ \dots\dots\dots\dots\dots\dots \\ +(g-1)((t_{g-1}+1)^{2n-1} + (t_{g-1}+2)^{2n-1} + \dots + (\lambda-1)^{2n-1}) \equiv 0, \text{ mod. } \lambda; \end{cases}$$

welches wieder leicht auf folgende Form gebracht wird:

$$5. \quad \begin{cases} (t_1+1)^{2n-1} + (t_1+2)^{2n-1} + \dots + (\lambda-1)^{2n-1} \\ +(t_2+1)^{2n-1} + (t_2+2)^{2n-1} + \dots + (\lambda-1)^{2n-1} \\ \dots\dots\dots\dots\dots\dots \\ +(t_{g-1}+1)^{2n-1} + (t_{g-1}+2)^{2n-1} + \dots + (\lambda-1)^{2n-1} \equiv 0, \text{ mod. } \lambda. \end{cases}$$

Subtrahirt man endlich von jeder einzelnen Zeile dieser Congruenz die Congruenz

$$1^{2n-1} + 2^{2n-1} + 3^{2n-1} + \dots + (\lambda-1)^{2n-2} \equiv 0, \text{ mod. } \lambda,$$

und verändert alle Vorzeichen, so ergiebt sich

$$6. \quad \begin{cases} 1^{2n-1} + 2^{2n-1} + 3^{2n-1} + \dots + t_1^{2n-1} \\ + 1^{2n-1} + 2^{2n-1} + 3^{2n-1} + \dots + t_2^{2n-1} \\ \dots\dots\dots\dots\dots\dots \\ + 1^{2n-1} + 2^{2n-1} + 3^{2n-1} + \dots + t_{g-1}^{2n-1} \equiv 0, \text{ mod. } \lambda. \end{cases}$$

Ich mache jetzt von folgender bekannten ganzen rationalen Function Gebrauch:

$$7. \quad X(x) = \frac{x^{2n}}{\Pi 2n} - \frac{x^{2n-1}}{2\Pi 2n-1.\Pi 1} + \frac{B_1 x^{2n-2}}{\Pi 2n-2.\Pi 2} - \dots \frac{(-1)^n B_{n-1} x^2}{\Pi 2.\Pi 2n-2};$$

16 *

in welcher $B_1, B_2, \ldots B_{n-1}$ die *Bernoulli'schen* Zahlen sind und $\Pi r = 1.2.3\ldots r$ ist. Diese Function $X(x)$ drückt, wenn x eine ganze Zahl ist, die Summe der Reihe $1^{2n-1}+2^{2n-1}+3^{2n-1}+\ldots+(x-1)^{2n-1}$ aus, dividirt durch $\Pi(2n-1)$; also läfst sich mittels derselben die Congruenz (6.) folgendermafsen darstellen:

$$8.\quad X(t_1+1)+X(t_2+1)+\ldots+X(t_{g-1}+1)\equiv 0,\ \text{mod.}\ \lambda.$$

Da nun t_s die in $\frac{s\lambda}{g}$ enthaltene gröfste ganze Zahl ist, so kann man

$$t_s=\frac{s\lambda-r_s}{g}$$

setzen, wo r_s positiv und kleiner als g ist. Hierdurch erhält man aus der Congruenz (8.) folgende gleichbedeutende:

$$9.\quad X\Big(\frac{\lambda-r_1+g}{g}\Big)+X\Big(\frac{2\lambda-r_2+g}{g}\Big)+\ldots+X\Big(\frac{(g-1)\lambda-r_{g-1}+g}{g}\Big)\equiv 0,\ \text{mod.}\ \lambda.$$

Die Zahlen $r_1, r_2, \ldots. r_{g-1}$ fallen, wenn auch in anderer Ordnung, mit den Zahlen $1, 2, 3, \ldots. g-1$ zusammen; denn sie liegen alle zwischen 0 und g und sind alle verschieden von einander, weil, wenn $r_k=r_h$ wäre, auch $t_k\equiv t_h$, mod. λ, sein müfste; was unmöglich ist, da t_k und t_h beide kleiner als λ sind. Läfst man nun aus der Congruenz (9.) die Vielfachen des Modul λ weg und setzt statt der Zahlen $r_1, r_2, \ldots. r_{g-1}$ die Zahlen 1, 2, 3, ... $g-1$, so geht diese Congruenz in folgende einfachere über:

$$10.\quad X\Big(\frac{1}{g}\Big)+X\Big(\frac{2}{g}\Big)+\ldots+X\Big(\frac{g-1}{g}\Big)\equiv 0,\ \text{mod.}\ \lambda;$$

welche Congruenz zwar Brüche enthält, die aber, da der Modul λ in keinem der Nenner vorkommt, sogleich auf ganze Zahlen gebracht werden können, und also nicht stören.

Die Function $X(x)$ hat unter andern merkwürdigen Eigenschaften auch die, dafs sie sich in folgende unendliche Reihe entwickeln läfst:

$$11.\quad X(x)=\frac{(-1)^n B_n}{\Pi 2n}-\frac{(-1)^n 2}{(2\pi)^{2n}}\Big(\frac{\cos 2x\pi}{1^{2n}}+\frac{\cos 4x\pi}{2^{2n}}+\frac{\cos 6x\pi}{3^{2n}}+\ldots\Big),$$

gültig in den Grenzen $x=0$ bis $x=1$. Setzt man in diesem Ausdrucke nach einander $x=0, \frac{1}{g}, \frac{2}{g}, \ldots \frac{g-1}{g}$ und addirt (wobei zu bemerken ist, dafs $X(0)=0$ ist, und dafs im allgemeinen auch die Summe

$$1+\cos\frac{2k\pi}{g}+\cos\frac{4k\pi}{g}+\ldots+\cos\frac{2(g-1)k\pi}{g}$$

gleich Null und im Falle, dafs k ein Vielfaches von g ist, gleich g wird): so erhält man

$$X\left(\frac{1}{g}\right)+X\left(\frac{2}{g}\right)+\dots+X\left(\frac{g-1}{g}\right)$$
$$=\frac{(-1)^n g B_n}{\Pi 2n}-\frac{(-1)^n 2g}{(2\pi)^{2n}}\left(\frac{1}{g^{2n}}+\frac{1}{(2g)^{2n}}+\frac{1}{(3g)^{2n}}+\dots \text{ in inf.}\right).$$

Es ist aber bekanntlich

$$1+\frac{1}{2^{2n}}+\frac{1}{3^{2n}}+\dots \text{ in inf.} = \frac{(2\pi)^{2n} B_n}{2\Pi 2n},$$

also wird

$$X\left(\frac{1}{g}\right)+X\left(\frac{2}{g}\right)+\dots+X\left(\frac{g-1}{g}\right)=\frac{(-1)^n g B_n}{\Pi 2n}-\frac{(-1)^n g B_n}{g^{2n}\Pi 2n},$$

und endlich

$$12. \quad X\left(\frac{1}{g}\right)+X\left(\frac{2}{g}\right)+\dots+X\left(\frac{g-1}{g}\right)=\frac{(-1)^n (g^{2n}-1) B_n}{g^{2n-1}\Pi 2n}.$$

Die Congruenz also, von deren Erfüllung oder Nichterfüllung es abhangt, ob der erste Factor der Classen-Anzahl, nämlich $\frac{P}{(2\lambda)^{\mu-1}}$, durch λ theilbar ist, oder nicht, hat nun folgende Gestalt angenommen:

$$13. \quad \frac{(g^{2n}-1) B_n}{g^{2n-1}\Pi 2n} \equiv 0, \text{ mod. } \lambda.$$

Der Nenner, als nicht durch den Modul λ theilbar, kann sogleich wegfallen; aufserdem ist aber auch $g^{2n}-1$ nicht durch λ theilbar, für die Werthe $n=1$, $2, 3, \dots \mu-1$; für diese also geht die Bedingungs-Congruenz einfach in

$$14. \quad B_n \equiv 0, \text{ mod. } \lambda$$

über. Der Fall $n=\mu$ erfordert eine besondere Betrachtung, weil für denselben die Congruenz (13.) die Form $\frac{0}{0}$ annimmt, wegen des in $2^{2\mu}-1$ und ebenfalls im Nenner der μten *Bernoulli*'schen Zahl B_μ enthaltenen Factors λ. Für diesen Fall drücke ich die μte *Bernoulli*'sche Zahl durch die niedrigeren aus, nach der bekannten Formel:

$$B_\mu=\frac{\Pi 2\mu\, B_{\mu-1}}{2^2\Pi 3\,\Pi 2\mu-2}-\frac{\Pi 2\mu\, B_{\mu-2}}{2^4\Pi 5\,\Pi 2\mu-4}+\dots+\frac{(-1)^\mu}{2^{2\mu}(2\mu+1)}+\frac{(-1)^{\mu+1}}{2^{2\mu}}.$$

In dieser Formel enthält nur das einzige, vorletzte Glied den Factor $2\mu+1=\lambda$ im Nenner; alle übrigen Glieder sind Brüche, in deren Nenner λ als Factor nicht vorkommt. Stellt man sich dieselben alle in einem einzigen Bruche $\frac{M}{N}$ zusammengefafst vor, wo N nicht durch λ theilbar ist, so erhält man

$$B_\mu=\frac{(-1)^\mu}{2^{2\mu}\lambda}+\frac{M}{N}.$$

Setzt man weiter $g^{2\mu}-1=\lambda K$, wo k nicht weiter durch λ theilbar ist, weil wir schon oben uns vorgesetzt haben, g immer so zu wählen, dafs $g^{\mu}+1$, und also auch $g^{2\mu}-1$, den Factor λ nur ein einzigesmal enthalte: so giebt die Gleichung (12.) für den Fall $n=\mu$:

$$X\Big(\frac{1}{g}\Big)+X\Big(\frac{2}{g}\Big)+\ldots+X\Big(\frac{g-1}{g}\Big)=\frac{(-1)^{\mu}K}{g^{2\mu-1}\Pi 2\mu}\Big(\frac{(-1)^{\mu}}{2^{2\mu}}+\frac{\lambda M}{N}\Big).$$

Die Congruenz-Bedingung (10.) giebt also in diesem Falle

$$\frac{K}{2^{2\mu}g^{2\mu-1}\Pi 2\mu}\equiv 0,\ \text{mod.}\,\lambda;$$

* und da nun k nicht durch λ theilbar ist, so findet diese Congruenz nicht Statt. Die obige Congruenzbedingung wird also in dem Falle $n=\mu$ niemals erfüllt und es hangt Alles nur davon ab, ob die Congruenz (14.), nämlich $B_n\equiv 0$, mod. λ, für irgend einen der Werthe $n=1,2,3,\ldots\mu-1$ Statt hat, oder nicht.

Wir haben demnach als Resultat folgenden Satz:

Der erste Factor $\frac{P}{(2\lambda)^{\mu-1}}$ der Classen-Anzahl H ist theilbar durch λ, wenn λ eine solche Primzahl ist, welche als Factor des Zählers einer der ersten $\frac{1}{2}(\lambda-3)$ *Bernoulli*'schen Zahlen vorkommt: für alle übrigen Primzahlen λ ist dieser Factor nicht durch λ theilbar.

Es ist jetzt ferner für den zweiten Factor der Classen-Anzahl H, nämlich $\frac{D}{\Delta}$, zu untersuchen, unter welchen Bedingungen derselbe durch λ theilbar sei, und unter welchen nicht. Hierzu ist glücklicherweise die Kenntnifs der in jedem besonderen Falle nur mit äufserster Mühe zu ermittelnden, in Δ enthaltenen Fundamental-Einheiten nicht nöthig, sondern nur die Definition derselben. Hat man ein System von $\mu-1$ *unabhängigen* Einheiten, welche aber im allgemeinen nicht die Fundamental-Einheiten selbst sind, so erhält man aus ihnen alle Einheiten, indem man dieselben zu Potenzen erhebt, deren Exponenten auch rationale Brüche sein können und solche mit einander multiplicirt. Ein solches System unabhängiger Einheiten ist aber das System

$$e(\alpha),\ e(\alpha^{g}),\ e(\alpha^{g^2})\ \ldots\ e(\alpha^{g^{\mu-2}}),$$

welches in D enthalten ist. Setzt man also

$$15.\quad\begin{cases}\varepsilon_1(\alpha)=e(\alpha)^{r_1^1}.\,e(\alpha^{g})^{r_2^1}\ \ldots\ e(\alpha^{g^{\mu-2}})^{r_{\mu-1}^1}\\ \varepsilon_2(\alpha)=e(\alpha)^{r_1^2}.\,e(\alpha^{g})^{r_2^2}\ \ldots\ e(\alpha^{g^{\mu-2}})^{r_{\mu-1}^2}\\ \ldots\ldots\ldots\ldots\ldots\ldots\\ \varepsilon_{\mu-1}(\alpha)=e(\alpha)^{r_1^{\mu-1}}.\,e(\alpha^{g})^{r_2^{\mu-1}}\ \ldots\ e(\alpha^{g^{\mu-2}})^{r_{\mu-1}^{\mu-1}},\end{cases}$$

wo die gebrochenen Potenz-Exponenten r_h^k mit zwei Indices so zu nehmen sind, dafs $\varepsilon_1(\alpha)$, $\varepsilon_2(\alpha)$, ... $\varepsilon_{\mu-1}(\alpha)$ wirklich zu ganzen complexen Einheiten werden: so sind diese Einheiten Fundamental-Einheiten, wenn die Determinante der gebrochenen Exponenten, nämlich $\Sigma \pm r_1^1 r_2^2 \dots r_{\mu-1}^{\mu-1}$, den möglich-kleinsten Werth hat, aber nicht gleich Null ist. Es mögen nun wirklich die Exponenten r_h^k dieser Bedingung gemäfs bestimmt sein, so dafs $\varepsilon_1(\alpha)$, $\varepsilon_2(\alpha)$, ... $\varepsilon_{\mu-1}(\alpha)$ wirkliche Fundamental-Einheiten sind, so hat man, wenn die Logarithmen genommen werden:

$$16. \quad l\varepsilon_k(\alpha) = r_1^k le(\alpha) + r_2^k le(\alpha^g) + \dots + r_{\mu-1}^k le\left(\alpha^{g^{\mu-2}}\right).$$

Da nun $\varDelta$ die Determinante der Gröfsen $l\varepsilon_1(\alpha)$, $l\varepsilon_2(\alpha)$, ... $l\varepsilon_{\mu-1}(\alpha)$ und derer, welche durch Verwandlung des α in α^g, α^{g^2}, ... $\alpha^{g^{\mu-2}}$ aus denselben entstehen, bezeichnet, und eben so D die Determinante der Gröfsen $le(\alpha)$, $le(\alpha^g)$, ... $le\left(\alpha^{g^{\mu-2}}\right)$ und derer, welche aus diesen entstehen, indem man α in α^g, α^{g^2}, ... $\alpha^{g^{\mu-2}}$ verwandelt: so hat man nach einem bekannten Satze über Determinanten:

$$\varDelta = D.\Sigma \pm r_1^1 r_2^2 \dots r_{\mu-1}^{\mu-1};$$

also

$$17. \quad \frac{D}{\varDelta} = \frac{1}{\Sigma \pm r_1^1 r_2^2 \dots r_{\mu-1}^{\mu-1}}.$$

Bringt man nun diejenigen rationalen Brüche r_1^k, r_2^k, ... $r_{\mu-1}^k$, welche als Exponenten in einer und derselben Fundamental-Einheit $\varepsilon_k(\alpha)$ vorkommen, unter einen gemeinschaftlichen, und zwar den möglich-kleinsten Nenner, welcher n_k sein mag: so kann man allgemein

$$r_h^k = \frac{m_h^k}{n_k}$$

setzen, wo m_h^k und n_k ganze Zahlen sind, von der Art, dafs n_k nicht mit allen m_1^k, m_2^k, ... $m_{\mu-1}^k$ einen gemeinschaftlichen Factor hat. Hierdurch wird

$$18. \quad \frac{D}{\varDelta} = \frac{n_1 n_2 \dots n_{\mu-1}}{\Sigma \pm m_1^1 m_2^2 \dots m_{\mu-1}^{\mu-1}}.$$

Es kann also $\frac{D}{\varDelta}$ den Factor λ nur dann enthalten, wenn eine der ganzen Zahlen n_1, n_2, ... $n_{\mu-1}$ durch λ theilbar ist, also nur dann, wenn eine Gleichung von folgender Form Statt hat:

$$19. \quad \varepsilon(\alpha)^n = e(\alpha)^{m_1}.e(\alpha^g)^{m_2} \dots e\left(\alpha^{g^{\mu-2}}\right)^{m_{\mu-1}},$$

in welcher n durch λ theilbar ist, aber $m_1, m_2, \ldots m_{\mu-1}$ nicht alle durch λ theilbar sind. Nun ist aber jede λte Potenz einer ganzen complexen Zahl immer einer realen ganzen Zahl congruent, für den Modul λ, also auch $\varepsilon(\alpha)^n \equiv c$, mod. λ, wo c eine reale ganze Zahl bedeutet: also mufs

$$20. \quad e(\alpha)^{m_1}.e(\alpha^g)^{m_2} \ldots e\left(\alpha^{g^{\mu-2}}\right)^{m_{\mu-1}} \equiv c, \text{ mod. } \lambda,$$

sein, wenn $\frac{D}{\Delta}$ den Factor λ enthalten soll, und wo $m_1, m_2, \ldots m_{\mu-1}$ nicht alle durch λ theilbar sind. Um nun die Bedingung zu erforschen, unter welcher eine solche Congruenz Statt haben kann, mache ich daraus die Gleichung

$$21. \quad e(\alpha)^{m_1}.e(\alpha^g)^{m_2} \ldots e\left(\alpha^{g^{\mu-2}}\right)^{m_{\mu-1}} = c+\lambda\varphi(\alpha),$$

in welcher $\varphi(\alpha)$ eine ganze complexe Zahl bedeutet. Eine solche Gleichung unter ganzen complexen Zahlen, welche für jeden Werth des α gilt, der der Gleichung $1+\alpha+\alpha^2+\ldots+\alpha^{\lambda-1}=0$ genügt, zieht nach bekannten Principien immer eine für jeden beliebigen Werth das x geltende nach sich, wenn man α in x verwandelt und ein Glied von der Form $(1+x+x^2+\ldots+x^{\lambda-1})\psi(x)$ hinzufügt. Demgemäfs erhält man hier die für jeden beliebigen Werth der Variabeln x geltende Gleichung

$$22. \quad e(x)^{m_1}.e(x^g)^{m_2} \ldots e\left(\alpha^{g^{\mu-2}}\right)^{m_{\mu-1}}$$
$$= c+\lambda\varphi(x)+(1+x+x^2+\ldots+x^{\lambda-1})\psi(x).$$

Ich nehme jetzt auf beiden Seiten die Differentialquotienten der Logarithmen, wobei $\frac{de(x)}{dx}$ durch $e'(x)$ bezeichnet wird, multiplicire mit x und gebe sodann dem x seinen besonderen Werth $x=\alpha$ zurück: so wird

$$23. \quad m_1\frac{\alpha e'(\alpha)}{e(\alpha)}+m_2 g\frac{\alpha^g e'(\alpha^g)}{e(\alpha^g)}+\ldots+m_{\mu-1}g^{\mu-2}\frac{\alpha^{g^{\mu-2}}e'(\alpha^{g^{\mu-2}})}{e(\alpha^{g^{\mu-2}})}$$
$$= \frac{\lambda\alpha\varphi'(\alpha)+(\alpha+2\alpha^2+3\alpha^3+\ldots+(\lambda-1)\alpha^{\lambda-1})\psi(\alpha)}{c+\lambda\varphi(\alpha)}.$$

Diese Gleichung enthält nur complexe Einheiten in den Nennern; also wesentlich nur ganze complexe Zahlen. Ich verwandele sie wieder in eine Congruenz für den Modul λ, indem ich die Vielfachen von λ weglasse; dabei bringe ich die complexe Zahl $\psi(\alpha)$ auf die Form $a+(1-\alpha)f(\alpha)$, wo a eine reale ganze Zahl ist, und bemerke, dafs

$$(1-\alpha)(\alpha+2\alpha^2+3\alpha^3+\ldots+(\lambda-1)\alpha^{\lambda-1}) = -\lambda,$$

also

$$(\alpha+2\alpha^2+3\alpha^3+\ldots+(\lambda-1)\alpha^{\lambda-1})\psi(\alpha) \equiv (\alpha+2\alpha^2+3\alpha^3+\ldots+(\lambda-1)\alpha^{\lambda-1})a$$

für den Modul λ ist. Endlich bestimme ich noch die ganze Zahl M so, dafs $2cM \equiv a$, mod. λ, und setze der Kürze wegen

$$\frac{2\alpha e'(\alpha)}{e(\alpha)} = F(\alpha):$$

so ist

$$24. \quad m_1 F(\alpha) + m_2 g F(\alpha^g) + \ldots + m_{\mu-1} g^{\mu-2} F(\alpha^{g^{\mu-2}}) \equiv M(\alpha + 2\alpha^2 + 3\alpha^3 + \ldots + (\lambda-1)\alpha^{\lambda-1})$$

für den Modul λ. Ich entwickele jetzt die complexe ganze Zahl $F(\alpha)$. Es ist

$$e(x) = \sqrt{\left(\frac{(1-x^g)(1-x^{-g})}{(1-x)(1-x^{-1})}\right)},$$

also

$$\frac{e'(x)}{e(x)} = \frac{x^{-1}(1+x)}{2(1-x)} - \frac{g x^{-1}(1+x^g)}{2(1-x^g)},$$

und daher

$$F(\alpha) = \frac{2\alpha e'(\alpha)}{e(\alpha)} = \frac{1+\alpha}{1-\alpha} - \frac{g(1+\alpha^g)}{1-\alpha^g}.$$

Aus der schon oben benutzten Gleichung

$$(1-\alpha)(\alpha + 2\alpha^2 + 3\alpha^3 + \ldots + (\lambda-1)\alpha^{\lambda-1}) = -\lambda$$

folgt nun

$$\frac{1}{1-\alpha} = \frac{-(\alpha + 2\alpha^2 + 3\alpha^3 + \ldots + (\lambda-1)\alpha^{\lambda-1})}{\lambda},$$

und wenn mit $1+\alpha$ multiplicirt wird,

$$\frac{1+\alpha}{1-\alpha} = \frac{-(\lambda + 2\alpha + 4\alpha^2 + 6\alpha^3 + \ldots + 2(\lambda-1)\alpha^{\lambda-1})}{\lambda};$$

welches, wenn man statt der Zahlen 1, 2, 3, ... $\lambda-1$ in den Coëfficienten die mit diesen, wenn auch in anderer Ordnung zusammenfallenden 1, g, g_2, ... $g_{\lambda-2}$, und in den Exponenten statt ihrer die Potenzen 1, g, g^2, ... $g^{\lambda-2}$ setzt, auch so dargestellt werden kann:

$$\frac{1+\alpha}{1-\alpha} = \frac{-(\lambda + 2\alpha + 2g_1\alpha^g + 2g_2\alpha^{g^2} + \ldots + 2g_{\lambda-2}\alpha^{g^{\lambda-2}})}{\lambda}.$$

Wird noch α in α^g verwandelt und mit g multiplicirt, so ist auch

$$\frac{g(1+\alpha^g)}{1-\alpha^g} = \frac{-(g\lambda + 2g\alpha^g + 2gg_1\alpha^{g^2} + \ldots + 2gg_{\lambda-2}\alpha^{g^{\lambda-1}})}{\lambda}.$$

Wird diese Gleichung von der vorhergehenden subtrahirt, so erhält man

$$F(\alpha) = g - 1 + \frac{2(gg_{\lambda-2}-1)}{\lambda}\alpha + \frac{2(g-g_1)}{\lambda}\alpha^g + \ldots + \frac{2(gg_{\lambda-3}-g_{\lambda-2})}{\lambda}\alpha^{g^{\lambda-2}}.$$

Dieselben Coëfficienten sind aber schon in dem ersten Theile unserer Untersuchung vorgekommen, wo wir sie durch b_0, b_1, b_2, ... $b_{\lambda-2}$ bezeichneten. Werden demnach dieselben Zeichen auch hier angewendet, indem allgemein

$$gg_{k-1} - g_k = \lambda b_k$$

gesetzt wird, so ist

$$25. \quad F(\alpha) = g - 1 + 2b_0\alpha + 2b_1\alpha^g + 2b_2\alpha^{g^2} + \dots + 2b_{\lambda-2}\alpha^{g^{\lambda-2}}.$$

Setzt man noch

$$2b_k + 1 - g = c_k$$

und bemerkt, dafs $c_{k+\mu} = -c_k$, weil $b_{k+\mu} + b_k = g - 1$ ist, so nimmt die Entwicklung des $F(\alpha)$ folgende Form an:

$$26. \quad F(\alpha) = c(\alpha - \alpha^{-1}) + c_1(\alpha^g - \alpha^{-g}) + \dots + c_{\mu-1}(\alpha^{g^{\mu-1}} - \alpha^{-g^{\mu-1}}).$$

Werden nun diese Entwicklung und die daraus hervorgehenden von $F(\alpha^g)$, $F(\alpha^{g^2})$, ... $F(\alpha^{g^{\mu-2}})$ in der Congruenz (24.) substituirt, so erhält man aus der Vergleichung der Coëfficienten der einzelnen Glieder folgende μ Congruenzen:

$$27. \quad \begin{cases} m_1 c \;\; - m_2 g c_{\mu-1} - m_3 g^2 c_{\mu-2} - \dots - m_{\mu-1} g^{\mu-2} c_2 \equiv M \\ m_1 c_1 + m_2 g c \;\;\; - m_3 g^2 c_{\mu-1} - \dots - m_{\mu-1} g^{\mu-2} c_3 \equiv gM \\ \dots\dots\dots\dots\dots\dots\dots\dots \\ m_1 c_{\mu-1} + m_2 g c_{\mu-2} + m_3 g^2 c_{\mu-3} + \dots + m_{\mu-1} g^{\mu-2} c_1 \equiv g^{\mu-1} M. \end{cases}$$

Um M zu eliminiren, multiplicire ich diese Congruenzen der Reihe nach mit 1, g^{2n-1}, $g^{2(2n-1)}$, ... $g^{(\mu-1)(2n-1)}$ und addire, so ergiebt sich rechterhand

$$1 + g^{2n} + g^{4n} + \dots + g^{2(\mu-1)n} \equiv 0, \text{ mod. } \lambda,$$

für jeden der Werthe $n = 1, 2, 3, \dots \mu-1$. Die Seite rechts dieser Congruenz verschwindet also für alle jene Werthe des n: die links aber zerfällt von selbst in das Product zweier Factoren und es wird

$$28. \quad (m_1 + m_2 g^{2n} + m_3 g^{4n} + \dots + m_{\mu-1} g^{2(\mu-2)n}) \\ \times (c + c_1 g^{2n-1} + c_2 g^{2(2n-1)} + \dots + c_{\mu-1} g^{(\mu-1)(2n-1)}) \equiv 0;$$

welches die Bedingung ist, die für alle Werthe $n = 1, 2, 2, \dots \mu - 1$ erfüllt werden mufs, damit $\frac{D}{\Delta}$ den Factor λ enthalten könne. Wenn nun der Factor $c + c_1 g^{2n-1} + c_2 g^{2(2n-1)} + \dots + c_{\mu-1} g^{(\mu-1)(2n-1)}$ für keinen der Werthe von n durch λ theilbar ist, so mufs der andere Factor für alle Werthe congruent Null sein. Es mufs also folgendes System von Congruenzen Statt haben:

$$29. \quad \left.\begin{cases} m_1 + m_2 g^2 \;\;\; + m_3 g^4 \;\;\; + \dots + m_{\mu-1} g^{2(\mu-2)} \equiv 0 \\ m_1 + m_2 g^4 \;\;\; + m_3 g^8 \;\;\; + \dots + m_{\mu-1} g^{4(\mu-2)} \equiv 0 \\ \dots\dots\dots\dots\dots\dots\dots\dots \\ m_1 + m_2 g^{2(\mu-1)} + m_3 g^{4(\mu-1)} + \dots + m_{\mu-1} g^{(\mu-2)(2\mu-2)} \equiv 0 \end{cases}\right\} \text{ mod. } \lambda.$$

Die Determinante dieses Systems linearer Congruenzen ist nicht congruent Null, denn sie ist bekanntlich aus lauter Factoren von der Form $g^{2k} - g^{2h}$ zusammengesetzt, in denen h und k kleiner als $\mu - 1$ und von einander verschieden sind: mithin ist keine dieser Congruenzen mit den übrigen identisch, oder schon in denselben enthalten. Diese Congruenzen sind also nicht anders zu erfüllen, als wenn alle die Gröfsen m_1, m_2, ... $m_{\mu-1}$ einzeln congruent Null sind, für den Modul λ; welches gegen die Voraussetzung ist. Es mufs also nothwendig der andere Factor der Congruenz (28.) für irgend einen der Werthe $n = 1, 2, 3, \dots \mu - 1$ congruent Null sein, wenn $\frac{D}{\Delta}$ durch λ theilbar sein soll, mithin mufs man

30. $$c + c_1 g^{2n-1} + c_2 g^{2(2n-1)} + \dots + c_{\mu-1} g^{(\mu-1)(2n-1)} \equiv 0, \text{ mod. } \lambda,$$

haben. Vermöge der Gleichung $c_{k+\mu} = -c_k$ und der Congruenz $g^{k+\mu} \equiv -g^k$, mod. λ, kann man auch die Anzahl der Glieder verdoppeln, so dafs

$$c + c_1 g^{2n-1} + c_2 g^{2(2n-1)} + \dots + c_{\lambda-2} g^{(\lambda-2)(2n-1)} \equiv 0$$

als die Congruenz genommen werden kann, welche nothwendig erfüllt werden mufs, wenn $\frac{D}{\Delta}$ durch λ theilbar sein soll. Wird zu dieser die Congruenz

$$(g-1)(1 + g^{2n-1} + g^{2(2n-1)} + \dots + g^{(\lambda-2)(2n-1)}) \equiv 0$$

addirt und für $c_k + g - 1$ wieder das Zeichen $2b_k$ gesetzt und durch 2 dividirt, so erhält man endlich die Congruenz in der Form

31. $$b_0 + b_1 g^{2n-1} + b_2 g^{2(2n-1)} + \dots + b^{(\lambda-2)(2n-1)} \equiv 0, \text{ mod. } \lambda.$$

Dies ist aber die schon oben vollständig untersuchte Congruenz, welche, wie gezeigt wurde, nur dann Statt hat, wenn λ eine derjenigen Primzahlen ist, die als Factor des Zählers in einer der ersten $\frac{1}{2}(\lambda - 3)$ *Bernoulli*schen Zahlen vorkommen. Das Resultat dieses zweiten Theils unserer Untersuchung ist also folgendes:

Der zweite Factor $\frac{D}{\Delta}$ der Classen-Anzahl H kann nur dann durch λ theilbar sein, wenn auch der erste Factor $\frac{P}{(2\lambda)^{\mu-1}}$ durch λ theilbar ist.

Der umgekehrte Satz von diesem ist nicht mitbewiesen und findet, wie ich vermuthe, auch gar nicht Statt. Fassen wir nun das Resultat mit dem vorigen zusammen, so haben wir folgenden Lehrsatz:

Die Anzahl aller nichtäquivalenten Classen der aus λten Wurzeln der Einheit gebildeten idealen complexen Zahlen ist durch λ theilbar, wenn λ eine solche Primzahl ist, welche als Factor im Zähler einer der ersten $\frac{1}{2}(\lambda - 3)$ *Bernoulli*schen Zahlen vorkommt; dagegen für alle andern Primzahlen λ ist diese Classen-Anzahl nicht durch λ theilbar.

---

17 *

Die zweite Untersuchung, welche ich hier durchführen will, soll die Frage erörtern:

Unter welchen Bedingungen kann eine complexe Einheit einer realen ganzen Zahl congruent sein, für den Modul λ, ohne eine λte Potenz einer andern complexen Einheit zu sein?

Es sei die zu untersuchende Éinheit

$$E(\alpha) = \pm\alpha^k e(\alpha)^{\frac{m_1}{n}}.e(\alpha^g)^{\frac{m_2}{n}} \ldots e(\alpha^{g^{\mu-2}})^{\frac{m_{\mu-1}}{n}};$$

welche Form alle möglichen Einheiten umfafst. Die Bedingung $E(\alpha)\equiv c$, mod. λ, wo c eine reale ganze Zahl ist, giebt auch $E(\alpha^{-1})\equiv c$, also auch $E(\alpha)\equiv E(\alpha^{-1})$; woraus $\alpha^k=\alpha^{-k}$ folgt. Es mufs demnach $\alpha^k=1$ sein, und wenn zur $n$ten Potenz erhoben wird, so ergiebt sich, da $E(\alpha)\equiv c$ ist:

$$c^n \equiv \pm e(\alpha)^{m_1}.e(\alpha^g)^{m_2} \ldots e(\alpha^{g^{\mu-2}})^{m_{\mu-1}}, \text{ mod. } \lambda.$$

Diese Congruenz ist aber mit der obigen, bereits vollständig untersuchten Congruenz (20.) vollkommen identisch, und wir wissen, dafs wenn λ eine Primzahl ist, welche in den Zählern der ersten $\frac{1}{2}(\lambda-3)$ ***Bernoulli****schen* Zahlen als Factor nicht vorkommt, eine solche Congruenz nicht bestehen kann, ohne dafs $m_1, m_2, \ldots m_{\mu-1}$ alle durch λ theilbar sind. In diesem Falle ist also $E(\alpha)^n$ gleich einer λten Potenz einer Einheit, und n ist nicht theilbar durch λ, indem der Nenner n nicht mit allen Zählern $m_1, m_2, \ldots m_{\mu-1}$ einen gemeinschaftlichen Factor haben soll. Wenn aber $E(\alpha)^n$ gleich einer λten Potenz und n nicht durch λ theilbar sein soll, so folgt leicht, dafs auch $E(\alpha)$ eine λte Potenz sein mufs. Bestimmt man nämlich die beiden Zahlen s und t so, dafs $ns-\lambda t=1$ ist und erhebt $E(\alpha)^n$ zur $s$ten Potenz, so ist auch $E(\alpha)^{ns} = E(\alpha)^{\lambda t+1} = E(\alpha)^{\lambda t}.E(\alpha)$ gleich einer λten Potenz einer Einheit. Es ist also folgender Lehrsatz bewiesen:

Wenn λ eine Primzahl ist, welche in keiner der ersten $\frac{1}{2}(\lambda-3)$ ***Bernoulli****schen* Zahlen als Factor des Zählers vorkommt, so ist jede aus λten Wurzeln der Einheit gebildete complexe Einheit, welche einer realen ganzen Zahl congruent ist, für den Modul λ, eine λte Potenz einer andern complexen Einheit.

Hiermit ist die aufgestellte Frage genügend beantwortet; denn, zu untersuchen, in wie weit auch die Umkehrung dieses Satzes richtig sei, liegt aufser unserem Zwecke.

Aus diesen Untersuchungen geht hervor, dafs die aus λten Wurzeln der Einheit gebildeten complexen Zahlen in ihren tiefer liegenden Eigenschaften nicht unwesentliche Unterschiede haben, je nachdem λ eine Primzahl ist, welche in dem Zähler einer der ersten $\frac{1}{2}(\lambda-3)$ *Bernoulli*schen Zahlen als Factor vorkommt, oder nicht; welche Unterschiede ich auch noch bei andern Untersuchungen über diese complexen Zahlen wahrzunehmen Gelegenheit gehabt habe. Aus den bekannten Zahlenwerthen der ersten *Bernoulli*schen Zahlen habe ich alle Primzahlen bis $\lambda=43$ in dieser Beziehung geprüft, und gefunden, dafs unter denselben nur eine ist, welche im Zähler der ersten $\frac{1}{2}(\lambda-3)$ *Bernoulli*schen Zahlen als Factor vorkommt, nämlich $\lambda=37$, da 37 ein Factor des Zählers der 16ten *Bernoulli*schen Zahl ist, während die Primzahlen $\lambda=3, 5, 7, 11, 13, 17, 19, 23, 29, 31, 41, 43$ diese Eigenschaft nicht haben. Überhaupt scheint in der Reihe aller Primzahlen jene besondere Art ziemlich sparsam vertheilt zu sein, so dafs dieselbe füglich nur als Ausnahme zu behandeln sein möchte.

Breslau, den 18ten Juni 1849.

Allgemeiner Beweis des Fermat'schen Satzes, daß die Gleichung $x^\lambda + y^\lambda = z^\lambda$ durch ganze Zahlen unlösbar ist, für alle diejenigen Potenz-Exponenten λ, welche ungerade Primzahlen sind und in den Zählern der ersten $\frac{1}{2}(\lambda-3)$ Bernoulli'schen Zahlen als Factoren nicht vorkommen

Journal für die reine und angewandte Mathematik 40, 130–138 (1850)

In den vorhergehenden Abhandlungen haben wir die Theorie der complexen Zahlen bis zu dem Puncte geführt, dafs mit Hülfe derselben der Beweis dieses Fermatschen Satzes, wenn gleich noch nicht vollkommen allgemein, so doch für alle diejenigen Potenzen, deren Exponenten die in der Überschrift bezeichnete Bedingung erfüllen, leicht und sicher geführt werden kann. Da es hierbei wenig Unterschied macht, ob man x, y, z nur als reale ganze Zahlen, oder, allgemeiner, als complexe, aus λten Wurzeln der Einheit gebildete Zahlen annimmt, so wollen wir den Beweis sogleich für complexe Zahlen geben. Die zu untersuchende Gleichung sei demnach

$$1.\quad u^\lambda + v^\lambda + w^\lambda = 0;$$

wo u, v und w wirkliche complexe Zahlen bezeichnen. Ferner sei λ eine Primzahl, welche in keiner der ersten $\frac{1}{2}(\lambda-3)$ Bernoullischen Zahlen als Factor des Zählers vorkommt. Unter dieser Voraussetzung haben die hier vorkommenden complexen Zahlen nach den in der vorhergehenden Abhandlung bewiesenen Sätzen, erstens, die Eigenschaft, dafs die Anzahl aller nicht-äquivalenten Classen nicht durch λ theilbar ist; woraus wir sogleich die für den folgenden Beweis bemerkenswerthe Folgerung ziehen, dafs hier niemals eine λte Potenz einer *idealen* complexen Zahl zu einer wirklichen werden kann, oder dafs, wenn eine λte Potenz einer complexen Zahl gleich einer *wirklichen* ist, diese complexe Zahl selbst eine wirkliche sein mufs (Man sehe die Abhandlung No. 16. Band 35. S. 356 dieses Journals). Zweitens ist bei dieser Voraussetzung, nach dem letzten Satze der vorhergehenden Abhandlung, jede complexe Einheit, welche für den Modul λ einer realen ganzen Zahl congruent wird, stets eine λte Potenz einer andern Einheit. Dafs die complexen * Zahlen u, v und w so angenommen werden, dafs sie nicht alle drei einen gemeinschaftlichen Factor haben und dafs darum auch nicht zwei derselben einen gemeinschaftlichen Factor haben dürfen, versteht sich von selbst.

Der Beweis der Unmöglichkeit der Gleichung (1.) zerfällt nun in zwei Theile, deren erster den Fall betrifft, wo von den drei complexen Zahlen u, v und w keine den Factor $1-\alpha$ hat, der zweite aber den Fall, wo eine derselben durch $1-\alpha$ theilbar ist.

Es sei erstens in der Gleichung

$$u^\lambda+v^\lambda+w^\lambda = 0$$

keine der complexen Zahlen u, v, w durch $1-\alpha$ theilbar. Da in der gegebenen Gleichung nur die λten Potenzen von u, v, w vorkommen, so kann man diese complexen Zahlen mit beliebigen λten Wurzeln der Einheit multipliciren; man kann also $\alpha^h u$ statt u setzen, wo h eine beliebige ganze Zahl ist, welche sich, wie leicht zu zeigen, immer so bestimmen läfst, dafs $\alpha^h u$ die Form $a+(1-\alpha)^2 P$ erhält; wo a eine reale ganze Zahl und P eine complexe ganze Zahl ist. Dieselbe Form kann auch dem v und w gegeben werden. Es sollen deshalb hier überall für u, v und w folgende Formen angenommen werden:

$$2.\quad \begin{cases} u = a+(1-\alpha)^2 P \\ v = b+(1-\alpha)^2 Q \\ w = c+(1-\alpha)^2 R. \end{cases}$$

Die realen ganzen Zahlen a, b, c sind wegen der Voraussetzung, dafs u, v, w nicht durch $1-\alpha$ theilbar sein sollen, nicht durch λ theilbar. Ich zerlege nun die Form $u^\lambda+v^\lambda$ in ihre Factoren und erhalte so aus der Gleichung $u^\lambda+v^\lambda+w^\lambda=0$ folgende:

$$3.\quad (u+v)(u+\alpha v)(u+\alpha^2 v)\ \ldots\ (u+\alpha^{\lambda-1}v) = -w^\lambda.$$

Diese λ Factoren haben keinen gemeinschaftlichen Theiler: denn hätten $u+\alpha^r v$ und $u+\alpha^s v$ einen solchen, so müfsten auch $(\alpha^r-\alpha^s)u$ und $(\alpha^r-\alpha^s)v$ denselben Theiler haben, und da u und v relative Primzahlen sind, so könnte nur $\alpha^r-\alpha^s$ der gemeinschaftliche Theiler sein. Es ist aber $\alpha^r-\alpha^s$ gleich $1-\alpha$, multiplicirt mit einer complexen Einheit, und dieses kann nicht Theiler eines jener λ Factoren sein, weil sonst, gegen die Annahme, auch w^λ und folglich auch w durch $1-\alpha$ theilbar sein müfste. Da nun alle diese Factoren auf der Seite links der Gleichung (3.) relative Primzahlen sind und ihr

Product gleich einer λten Potenz ist, so müssen sie alle einzeln gleich λten Potenzen gewisser idealen complexen Zahlen sein, multiplicirt mit irgend welchen complexen Einheiten. Es folgt dies unmittelbar, eben so wie für gewöhnliche ganze Zahlen, aus dem in der Abhandlung No. 16. Band. 35. pag. 348 bewiesenen Satze, dafs, abgesehen von den Einheiten, welche als Factoren zutreten können, jede complexe Zahl sich nur auf eine einzige bestimmte Weise als Product ihrer idealen Primfactoren darstellen läfst. Man erhält daher allgemein für alle Werthe $r=0, 1, 2, \ldots \lambda-1$:

$$4.\quad u+\alpha^r v = \alpha^\varrho E_r(\alpha).t_r^\lambda;$$

wo t_r eine complexe Zahl ist, Factor von w, und $\alpha^\varrho E_r(\alpha)$ eine Einheit, von der Art, dafs $E_r(\alpha)=E_r(\alpha^{-1})$ ist. In zwei Factoren, α^ϱ und $E_r(\alpha)$, deren einer nur eine λte Wurzel der Einheit ist, der andere die Eigenschaft hat, bei der Verwandlung des α in α^{-1} ungeändert zu bleiben, läfst sich nämlich, wie bekannt, jede beliebige complexe Einheit zerlegen. Da nach Gleichung (4.) t_r^λ gleich einer wirklichen complexen Zahl ist, so schliefsen wir sogleich, nach Dem, was oben gezeigt, dafs auch t_r selbst eine wirkliche complexe Zahl sein mufs; und da jede λte Potenz einer wirklichen complexen Zahl bekanntlich einer realen ganzen Zahl congruent ist, für den Modul λ, so setze ich $t_r^\lambda \equiv m$, mod. λ; wo m eine reale ganze Zahl ist. Die Gleichung (4.) geht dadurch in die Congruenz

$$5.\quad u+\alpha^r v \equiv \alpha^\varrho.E_r(\alpha)m, \text{ mod. } \lambda,$$

über. Wird nun α in α^{-1} verwandelt, wodurch u in u', v in v', w in w' übergehen mag, so ist

$$6.\quad u'+\alpha^{-r}v' \equiv \alpha^{-\varrho}E_r(\alpha)m, \text{ mod. } \lambda;$$

aus welchen beiden Congruenzen durch Elimination des m

$$7.\quad \alpha^{-\varrho}(u+\alpha^r v) \equiv \alpha^\varrho(u'+\alpha^{-r}v'), \text{ mod. } \lambda,$$

folgt. Nimmt man statt des Moduls λ den Modul $(1-\alpha)^2$, welcher ein Divisor von λ ist, und bemerkt, dafs nach den Gleichungen (2.) $u\equiv a$, $v\equiv b$, $u'\equiv a$, $v'\equiv b$, für den Modul $(1-\alpha)^2$ ist, so erhält man

$$8.\quad \alpha^{-\varrho}(a+\alpha^r b) \equiv \alpha^\varrho(a+\alpha^{-r}b), \text{ mod. } (1-\alpha)^2,$$

und da allgemein $\alpha^h \equiv 1-h(1-\alpha)$, mod. $(1-\alpha)^2$, ist, so geht diese Congruenz in die folgende über:

$$2(a+b)\varrho \equiv 2br, \text{ mod. } (1-\alpha).$$

Da nun reale ganze Zahlen, welche durch $1-\alpha$ theilbar sind, auch durch λ

theilbar sein müssen, so ist

$$9.\quad (a+b)\varrho \equiv br,\ \text{mod. } \lambda.$$

Nennt man nun k diejenige ganze Zahl, welche der Congruenz

$$10.\quad (a+b)k \equiv b,\ \text{mod. } \lambda,$$

genügt, so ist k von r unabhängig und $\varrho\equiv k.r$, also giebt die Congruenz (7.)

$$11.\quad \alpha^{-kr}(u+\alpha^r v) \equiv \alpha^{+kr}(u'+\alpha^{-r}v'),\ \text{mod. } \lambda.$$

Für den besondern Fall $r=0$ hat man, da $a+b$ nicht $\equiv 0$, mod. λ, sein kann, aus der Congruenz (9.): $\varrho\equiv 0$, mod. λ, also

$$12.\quad u+v \equiv u'+v',\ \text{mod. } \lambda,$$

und da u, v, w in der gegebenen Gleichung $u^\lambda+v^\lambda+w^\lambda=0$ beliebig vertauscht werden können, so ist auch

$$13.\quad \left.\begin{aligned} u+w &\equiv u'+w' \\ v+w &\equiv v'+w' \end{aligned}\right\}\ \text{mod. } \lambda,$$

und aus diesen Congruenzen folgen die drei einfacheren:

$$14.\quad \left.\begin{aligned} u &\equiv u' \\ v &\equiv v' \\ w &\equiv w' \end{aligned}\right\}\ \text{mod. } \lambda.$$

Hiernach verwandelt sich die für jeden beliebigen Werth von r geltende Congruenz (11.) in folgende:

$$15.\quad \alpha^{-kr}(u+\alpha^r v) \equiv \alpha^{kr}(u+\alpha^{-r}v),\ \text{mod. } \lambda,$$

oder in

$$u(\alpha^{kr}-\alpha^{-kr})+v(\alpha^{(k-1)r}-\alpha^{-(k-1)r}) \equiv 0,\ \text{mod. } \lambda.$$

Ich setze $r=1$ und $r=2$, und erhalte dadurch:

$$16.\quad \left.\begin{aligned} u(\alpha^k-\alpha^{-k})+v(\alpha^{(k-1)}-\alpha^{-(k-1)}) &\equiv 0, \\ u(\alpha^{2k}-\alpha^{-2k})+v(\alpha^{2(k-1)}-\alpha^{-2(k-1)}) &\equiv 0, \end{aligned}\right\}\ \text{mod. } \lambda,$$

und wenn die erste dieser beiden Congruenzen mit $\alpha^k+\alpha^{-k}$ multiplicirt und die zweite davon abgezogen wird, so ist nach Weghebung des gemeinschaftlichen Factors v, welcher nicht durch $1-\alpha$ theilbar, also zu λ relative Primzahl ist,

$$(\alpha^k+\alpha^{-k})(\alpha^{(k-1)}-\alpha^{-(k-1)})+(\alpha^{2(k-1)}-\alpha^{-2(k-1)}) \equiv 0,\ \text{mod. } \lambda,$$

also

$$(\alpha^{k-1}-\alpha^{-(k-1)})(\alpha^k+\alpha^{-k}-\alpha^{k-1}-\alpha^{-(k-1)}) \equiv 0,\ \text{mod. } \lambda,$$

folglich

$$17.\quad (\alpha^{k-1}-\alpha^{-(k-1)})(\alpha^{-k}-\alpha^{k-1})(1-\alpha) \equiv 0,\ \text{mod. } \lambda.$$

Wenn nun keiner dieser drei Factoren für sich gleich Null ist, so enthält das Product derselben den Factor $1-\alpha$ dreimal: es müfste denselben aber ebensovielmal enthalten als λ, also $\lambda - 1$ mal, damit die Congruenz wirklich Statt habe. Mit Ausschlufs des einzigen Falles $\lambda = 3$ kann also diese Congruenz (17.) nicht Statt finden, wenn nicht

$$18. \quad \begin{cases} \text{entweder} & \alpha^{k-1} - \alpha^{-(k-1)} = 0, \\ \text{oder} & \alpha^{-k} - \alpha^{k-1} = 0 \end{cases}$$

ist. Es mufs also entweder $k \equiv 1$, oder $2k \equiv 1$, mod. λ, sein. Der erste Fall $k \equiv 1$ würde aber der Congruenz (10.) zufolge $a \equiv 0$, mod. λ, geben, und kann deshalb nicht Statt haben. Der zweite Fall $2k \equiv 1$ giebt der Congruenz (10.) zufolge $a \equiv b$, mod. λ, woraus durch blofse Vertauschung der Buchstaben folgt, dafs auch $a \equiv c$ und $b \equiv c$, mod. λ, sein mufs. Aus der Gleichung $u^\lambda + v^\lambda + w^\lambda = 0$ folgt aber, nach den bei (2.) angenommenen Ausdrücken von u, v und w, dafs auch $a^\lambda + b^\lambda + c^\lambda \equiv 0$, mod. λ, sein mufs, also auch $a + b + c \equiv 0$, mod. λ, und da a, b und c congruent sind, endlich $3a \equiv 0$, mod. λ, welches, mit Ausnahme des schon oben ausgeschlossenen Falles $\lambda = 3$, ebenfalls unmöglich ist, weil nach der Voraussetzung u den Factor $1-\alpha$ nicht enthalten und also auch a nicht durch λ theilbar sein darf. Hiermit ist nun der erste Theil des Beweises vollständig gegeben, indem gezeigt worden ist, dafs die Gleichung $u^\lambda + v^\lambda + w^\lambda = 0$, wenn keine der complexen Zahlen u, v und w den Factor $1-\alpha$ enthält, immer eine unmögliche Congruenz für den Modul λ nach sich zieht; mit Ausnahme des Falles $\lambda = 3$, welchen wir hier nicht besonders betrachten wollen.

Es sei zweitens in der Gleichung $u^\lambda + v^\lambda + w^\lambda = 0$ eine der drei Zahlen u, v, w durch $1-\alpha$ theilbar; zu welcher w genommen werden soll. Dieselbe kann den Factor $1-\alpha$ auch mehrmals enthalten. Setzt man daher $(1-\alpha)^m w$ statt w, so dafs nun w den Factor $1-\alpha$ nicht weiter enthält, so ist die zu untersuchende Gleichung:

$$u^\lambda + v^\lambda + (1-\alpha)^{m\lambda} . w^\lambda = 0.$$

Statt dieser aber setze ich die etwas allgemeinere

$$19. \quad u^\lambda + v^\lambda = E(\alpha)(1-\alpha)^{m\lambda} . w^\lambda,$$

in welcher $E(\alpha)$ eine beliebige complexe Einheit bezeichnet. Durch Zerlegung des Ausdrucks $u^\lambda + v^\lambda$ in Factoren erhält man

$$20. \quad (u+v)(u+\alpha v)(u+\alpha^2 v) \dots (u+\alpha^{\lambda-1} v) = E(\alpha)(1-\alpha)^{m\lambda} . w^\lambda.$$

Die λ Factoren $u+v$, $u+\alpha v$, u. s. w. haben hier alle den gemeinschaftlichen gröfsten Factor $1-\alpha$; aufser diesem haben je zwei derselben keinen gemeinschaftlichen Theiler. Nimmt man nämlich für u und v wieder, wie oben, die Formen $u = a + (1-\alpha)^2 P$ und $v = b + (1-\alpha)^2 Q$ an, so erhält man

$$21. \quad u + \alpha^r v = a + b - rb(1-\alpha), \text{ mod. } (1-\alpha)^2;$$

es mufs aber $u + \alpha^r v$ wenigstens für einen Werth von r durch $1-\alpha$ theilbar sein, weil das Product aller dieser Factoren durch $(1-\alpha)^{m\lambda}$ theilbar ist: also mufs $a+b$ durch $1-\alpha$, folglich auch durch λ theilbar sein, und die Congruenz (21.) verwandelt sich in

$$22. \quad u + \alpha^r v \equiv rb(1-\alpha), \text{ mod. } (1-\alpha)^2;$$

woraus zunächst folgt, dafs für jeden Werth von r, $u + \alpha^r v$ den Factor $1-\alpha$ enthalten mufs, statt dessen auch der Factor $1-\alpha^r$ genommen werden kann, welcher sich von diesem nur durch eine complexe Einheit unterscheidet, die als Factor hinzutritt: ferner, dafs $u + \alpha^r v$ diesen Factor $1-\alpha^r$ oder $1-\alpha$ nur einmal enthalten kann, mit Ausnahme des Falles $r = 0$. Die Gröfse $u+v$ aber enthält wirklich den Factor $1-\alpha$ mehrmals, und zwar vermöge der Gleichung (20.) genau $m\lambda - \lambda + 1$ mal, indem die übrigen $\lambda - 1$ Factoren ihn jeder einmal enthalten und das Product aller $m\lambda$ mal. Setzt man nun

$$23. \quad u + v = (1-\alpha)^{m\lambda - \lambda + 1} . \varphi$$

und

$$24. \quad u + \alpha^r v = (1-\alpha^r) \varphi_r,$$

so geht die Gleichung (20.) in folgende über:

$$25. \quad \varphi . \varphi_1 . \varphi_2 \ldots \varphi_{\lambda-1} = E(\alpha) w^\lambda$$

und es müssen nun die Factoren φ, φ_1, φ_2, ... $\varphi_{\lambda-1}$, welche unter sich alle relative Primzahlen sind und deren Product gleich einer mit einer Einheit multiplicirten λten Potenz ist, alle einzeln ebenfalls solche mit Einheiten multiplicirte λte Potenzen sein. Demnach kann man setzen:

$$\varphi = e(\alpha) w_1^\lambda \text{ und } \varphi_r = e_r(\alpha) t_r^\lambda,$$

woraus

$$26. \quad u + v = e(\alpha)(1-\alpha)^{m\lambda - \lambda + 1} . w_1^\lambda$$

und

$$27. \quad u + \alpha^r v = e_r(\alpha)(1-\alpha^r) t_r^\lambda$$

für alle Werthe $r = 1, 2, 3, \ldots \lambda - 1$ folgt. Die complexen Zahlen w_1 und t_r sind hier ebenfalls nur wirkliche complexe Zahlen, weil die λten Potenzen

18 *

derselben wirkliche complexe Zahlen sind. Giebt man dem r einen andern Werth s, so erhält man ebenfalls

$$28.\quad u+\alpha^s v = e_s(\alpha)(1-\alpha^s).t_s^\lambda,$$

und wenn man aus diesen drei Gleichungen u und v eliminirt, so erhält man

$$29.\quad e_r(\alpha)t_r^\lambda - e_s(\alpha)t_s^\lambda = \frac{e(\alpha^r-\alpha^s)(1-\alpha)}{(1-\alpha^r)(1-\alpha^s)}(1-\alpha)^{(m-1)\lambda}.w_1^\lambda.$$

Dividirt man durch $e_r(\alpha)$ und setzt

$$\frac{-e_s(\alpha)}{e_r(\alpha)} = \varepsilon(\alpha) \text{ und}$$

$$\frac{e(\alpha)(\alpha^r-\alpha^s)(1-\alpha)}{e_r(\alpha)(1-\alpha^r)(1-\alpha^s)} = E_1(\alpha),$$

wo $\varepsilon(\alpha)$ und $E_1(\alpha)$ ebenfalls nur complexe Einheiten sind, so ergiebt sich

$$30.\quad t_r^\lambda+\varepsilon(\alpha)t_s^\lambda = E_1(\alpha)(1-\alpha)^{(m-1)\lambda}.w_1^\lambda.$$

Ist nun $m>1$, so ist bekanntlich $(1-\alpha)^{(m-1)\lambda}\equiv 0$, mod. λ. Ferner, da t_r und t_s wirkliche complexe Zahlen sind, so müssen die λten Potenzen derselben realen ganzen Zahlen congruent sein für den Modul λ, also mufs $t_r^\lambda\equiv c$ und $t_s^\lambda\equiv k$, mod. λ, sein. Die Gleichung (30.) giebt daher folgende Congruenz:

$$c+\varepsilon(\alpha)k \equiv 0, \text{ mod. } \lambda,$$

aus welcher folgt, dafs die Einheit $\varepsilon(\alpha)$ einer realen ganzen Zahl congruent ist, für den Modul λ, dafs also $\varepsilon(\alpha)$ eine λte Potenz einer andern Einheit sein mufs, mithin $\varepsilon(\alpha)=\varepsilon_1(\alpha)^\lambda$. Setzt man nun $\varepsilon_1(\alpha)t_s=v_1$ und statt t_r das Zeichen u_1, so geht die Gleichung (30.) in folgende über:

$$31.\quad u_1^\lambda+v_1^\lambda = E_1(\alpha)(1-\alpha)^{(m-1)\lambda}.w_1^\lambda.$$

Diese Gleichung ist aber der Form nach der Gleichung (19.), aus welcher sie abgeleitet ist, vollkommen gleich und unterscheidet sich von ihr nur dadurch, dafs m um eine Einheit kleiner ist. Wendet man also auf die Gleichung (31.) dieselbe Methode an, so erhält man aus ihr wieder eine Gleichung von derselben Form, in welcher m um zwei Einheiten kleiner ist, als in der Gleichung (19.) u. s. w. Durch Wiederholung dieses Verfahrens gelangt man stets zu einer Gleichung von derselben Form wie (19.), in welcher $m=1$ ist: auf diese aber ist sodann die Methode, welche, wie wir oben ausdrücklich bemerkt haben, $m>1$ voraussetzt, nicht weiter anwendbar. Man erhält also eine Gleichung von der Form

$$32.\quad u^\lambda+v^\lambda = E(\alpha).(1-\alpha)^\lambda.w^\lambda.$$

Die Unmöglichkeit dieser Gleichung läfst sich einfach dadurch beweisen, dafs gezeigt wird: die Form $u^\lambda+v^\lambda$, wenn sie überhaupt den Factor $1-\alpha$ enthält, müsse denselben wenigstens $\lambda+1$ mal enthalten. Um dies zu beweisen, setze ich für u und v wieder wie oben die Formen

$$u = a+(1-\alpha)^2 P, \quad v = b+(1-\alpha)^2 Q,$$

so ergiebt sich wieder

$$33. \quad u+\alpha^r v \equiv a+b-rb(1-\alpha), \text{ mod. } (1-\alpha)^2.$$

Da nun $u^\lambda+v^\lambda$ durch $1-\alpha$ theilbar ist und deshalb auch wenigstens einer der Factoren dieses Ausdrucks, welche alle die Form $u+\alpha^r v$ haben, durch $1-\alpha$ theilbar sein mufs, so folgt, dafs $a+b$ durch $1-\alpha$ und deshalb auch durch λ theilbar ist. Die Congruenz (33.) geht demnach, eben so wie oben, in folgende über:

$$34. \quad u+\alpha^r v \equiv rb(1-\alpha), \text{ mod. } (1-\alpha)^2.$$

Für $r=0$ ist insbesondere

$$35. \quad u+v \equiv 0, \text{ mod. } (1-\alpha)^2.$$

Es sind also alle die Factoren der Form

$$u^\lambda+v^\lambda = (u+v)(u+\alpha v)(u+\alpha^2 v) \ldots (u+\alpha^{\lambda-1}v)$$

durch $1-\alpha$ theilbar; der Factor $u+v$ aber ist durch $(1-\alpha)^2$ theilbar. Die Anzahl aller in $u^\lambda+v^\lambda$ enthaltenen Factoren $1-\alpha$ ist demnach mindestens gleich $\lambda+1$; was zu beweisen war. Die Gleichung (32.), in welcher w nicht durch $1-\alpha$ theilbar ist, enthält also den Widerspruch in sich, dafs die Seite derselben links durch $(1-\alpha)^{\lambda+1}$ theilbar ist, die rechts aber nicht. Diese Gleichung ist also eine unmögliche, und darum ist auch die Gleichung (19.), aus welcher sie abgeleitet wurde, unmöglich: d. h. eine solche, welche durch complexe ganze Zahlen auf keine Weise erfüllt werden kann.

Die Gleichung $u^\lambda+v^\lambda+w^\lambda=0$ ist also in beiden Fällen unmöglich, sowohl wenn keine der complexen Zahlen u, v, w durch $1-\alpha$ theilbar ist, als auch wenn eine derselben durch $1-\alpha$ theilbar angenommen wird. Der ***Fermat***sche Satz ist demnach nicht nur für reale ganze Zahlen, sondern sogar für complexe, aus λten Wurzeln der Einheit gebildete ganze Zahlen bewiesen, für alle diejenigen Potenzen, deren Exponenten λ Primzahlen sind und welche die Bedingung erfüllen, dafs sie in keiner der ersten $\frac{1}{2}(\lambda-3)$ ***Bernoulli***schen

Zahlen als Factoren des Zählers vorkommen. Da $\lambda=5$, 7, 11, 13, 17, 19, 23, 29, 31, 41, 43 diese Bedingung erfüllen, so ist namentlich für alle diese der *Fermat*sche Satz bewiesen. Für $\lambda=37$ aber, wird die angegebene Bedingung nicht erfüllt: also ist auch der *Fermat*sche Satz für 37te Potenzen nicht bewiesen. Meine gegenwärtigen Kenntnisse der Theorie der complexen Zahlen haben mir auch noch nicht die Mittel gewährt, für $\lambda=37$ und für die übrigen Primzahlen, welche der angegebenen Bedingung nicht genügen, die Nicht-Auflösbarkeit oder Auflösbarkeit der *Fermat*schen Gleichung zu ergründen.

Breslau, den 19ten Juni 1849.

Allgemeine Reciprocitätsgesetze für beliebig hohe Potenzreste

Monatsberichte der Königlichen Preußischen Akademie der Wissenschaften zu Berlin aus dem Jahre 1850, 154–165

Hr. Dirichlet trug den Inhalt des folgenden, von Herrn Kummer in Breslau, Correspondenten der Klasse, eingesandten Aufsatzes vor:

Bei meinen Untersuchungen über die Theorie der complexen Zahlen und den Anwendungen derselben auf den Beweis des

Fermatschen Lehrsatzes, welchen ich der Akademie der Wissenschaften vor drei Jahren mitzutheilen die Ehre gehabt habe, ist es mir gelungen die allgemeinen Reciprocitätsgesetze für beliebig hohe Potenzreste zu entdecken, welche nach dem gegenwärtigen Stande der Zahlentheorie als die Hauptaufgabe und die Spitze dieser Wissenschaft anzusehen sind. Durch besondere mit Hülfe des *Canon arithmeticus* angelegte Tabellen habe ich sodann die Richtigkeit der gefundenen Gesetze für fünfte, siebente und drei und zwanzigste Potenzreste, für ziemlich ausgedehnte Grenzen der complexen Primzahlen, nachgewiesen. Ich unterliefs damals die öffentliche Bekanntmachung der von mir gefundenen Gesetze, weil ich von der Auffindung eines allgemeinen Beweises derselben nicht weit entfernt zu sein glaubte und theilte sie nur an Hrn. Lejeune-Dirichlet, in einem Schreiben vom 20. Januar 1848, und durch diesen an Hrn. Jacobi mit, zugleich mit den berechneten Tafeln, welche ihre Richtigkeit nachwiesen, so weit diefs eine ziemlich umfangreiche Induction vermag. Meine Hoffnung die vollständigen Beweise zu finden, ist seitdem nicht erfüllt worden, theils wegen anderweitiger Beschäftigungen, welche mich von diesen Studien abgezogen haben, vorzüglich aber wohl wegen der in der Sache liegenden Schwierigkeiten, welche zu überwinden ich nicht vermocht habe. Da aber die Kenntnifs dieser Gesetze einen nicht unbedeutenden Schritt zur Auffindung eines Beweises derselben zu bilden scheint, so habe ich mich entschlossen, im Interesse der Wissenschaft durch Mittheilung an Eine Hohe Akademie der Wissenschaften sie zu einem Gemeingute aller Mathematiker zu machen, welche an der Fortbildung der Zahlentheorie arbeiten. Aufser dem Reciprocitätsgesetze für zwei beliebige complexe Primzahlen habe ich auch die Gesetze über den Charakter der complexen Einheiten, so wie auch der Zahl, welche der Exponent derjenigen Potenz ist, in Beziehung auf welche der Charakter betrachtet wird, und der complexen Primfaktoren dieser Zahl durch allgemeine Methoden nicht allein gefunden, sondern auch in vollkommener Strenge hergeleitet. Diese die Ergänzungsfälle des Reciprocitätsgesetzes ergründenden Methoden, mit den Hauptresultaten, werde ich mir ebenfalls erlauben in der Kürze darzustellen.

Ich betrachte in dem Folgenden die Potenzreste nur in Beziehung auf solche Potenzen, deren Exponenten ungrade Primzahlen sind. Es bezeichne daher λ eine ungrade Primzahl und α eine λ^{te} Wurzel der Einheit, d. h. eine imaginäre Wurzel der Gleichung $\alpha^\lambda = 1$. Die Reciprocitätsgesetze für die λ^{ten} Potenzreste und Nichtreste werden am einfachsten für die aus λ^{ten} Wurzeln der Einheit gebildeten complexen Primzahlen dargestellt, welche zum Theil wirklich zum Theil aber ideal sind. Das dem Legendreschen Zeichen für quadratische Reste analog gebildete Zeichen:

$$\left(\frac{\phi(\alpha)}{f(\alpha)}\right) \equiv \phi(\alpha)^{\frac{Nf(\alpha)-1}{\lambda}} \equiv \alpha^{\varkappa}, \qquad \text{Mod.}\, f(\alpha),$$

in welchem $Nf(\alpha)$ die Norm von $f(\alpha)$ bedeutet, hat aber nur dann einen bestimmten Sinn, wenn $\phi(\alpha)$ eine wirkliche complexe Zahl ist, $f(\alpha)$ aber, welches hier nur als Modul und in $Nf(\alpha)$ vorkommt, kann ebensogut eine ideale als wirkliche Primzahl sein. Damit nun dieses Zeichen, auch wenn $\phi(\alpha)$ ideal ist, eine bestimmte Bedeutung habe, erhebe ich $\phi(\alpha)$ zu derjenigen Potenz, welche diese ideale Zahl zu einer wirklichen macht, diefs sei die h^{te} Potenz, und ich definire, es soll sein:

$$\left(\frac{\phi(\alpha)}{f(\alpha)}\right) \equiv \alpha^{\varkappa}, \qquad \text{Mod.}\, f(\alpha),$$

$$\text{wenn } \left(\phi(\alpha)^h\right)^{\frac{Nf(\alpha)-1}{\lambda}} \equiv \alpha^{\varkappa'}, \text{ Mod.}\, f(\alpha), \text{ und } \varkappa' \equiv h\varkappa, \text{ Mod. } \lambda.$$

Diese Zurückführung des Charakters der idealen Zahl $\phi(\alpha)$ auf den der wirklichen Zahl $\phi(\alpha)^h$ setzt nothwendig voraus, dafs h nicht durch λ theilbar ist, denn wenn diefs der Fall wäre, so würde $\varkappa$ aus der Congruenz $\varkappa' \equiv h\varkappa$, Mod. λ, sich nicht bestimmen lassen. Demgemäfs werde ich hier das Reciprocitätsgesetz nur für diejenigen λ^{ten} Potenzreste aufstellen, für welche λ die Eigenschaft hat, dafs keine λ^{te} Potenz einer aus λ^{ten} Wurzeln der Einheit gebildeten complexen idealen Zahl zu einer wirklichen wird. Diese Bedingung fällt, wie ich in meinem Beweise des Fermatschen Satzes im Octoberhefte des Jahrganges 1847 der Monatsberichte gezeigt habe, mit der zusammen, dafs die Anzahl aller nicht äquivalenten Klassen der aus λ^{ten} Wurzeln der Einheit gebildeten idealen complexen Zahlen nicht durch λ theilbar sei, und ist, wie ich ebendaselbst gezeigt habe, identisch mit der Bedingung, dafs λ als Faktor einer der ersten $\frac{\lambda-3}{2}$ Bernouillischen Zahlen nicht vorkomme. Ich schliefse also hier bei der Aufstellung des Reciprocitätsgesetzes genau dieselben Ausnahms-Zahlen λ aus, auf welche auch mein Beweis des Fermatschen Lehrsatzes sich nicht erstreckt.

Die hauptsächlichste Schwierigkeit für die Auffindung des allgemeinen Reciprocitätsgesetzes lag nun darin, dafs die einzelnen complexen Primzahlen, vermöge der Einheiten, mit welchen sie behaftet sind, in unendlich vielen verschiedenen Formen sich darstellen lassen und dafs für diese verschiedenen Formen auch die Charaktere derselben in Beziehung auf λ^{te} Potenzen sich ändern. Es war also hier diejenige primäre Form der complexen Zahlen zu suchen, für welche das Reciprocitätsgesetz seinen einfachsten Ausdruck gewinnt. Bei diesem Geschäfte leitete mich der einfache Gedanke, dafs diejenigen complexen Zahlen, welche nicht willkürlich gemacht, sondern in ganz bestimmten Formen entstanden wären, namentlich die complexen Zahlen, welche die Kreistheilung erzeugt, dieser bevorzugten primären Form theilhaftig sein möchten. In diesem Sinne nahm ich mir die complexe Zahl $\psi(\alpha)$, nach Jacobi's Bezeichnung, als Muster und erweiterte die Grundeigenschaft derselben, nach welcher sie mit ihrer reciproken multiplicirt einer ganzen Zahl gleich wird, dahin, dafs die complexe Zahl $\phi(\alpha)$ in ihrer primären Form die Eigenschaft haben sollte, mit ihrer reciproken $\phi(\alpha^{-1})$ multiplicirt, einer gewöhnlichen ganzen Zahl congruent zu sein, für den Modul λ, dafs also, wenn c eine reale ganze Zahl bedeutet,

$$\phi(\alpha)\,\phi(\alpha^{-1}) \equiv c, \qquad \text{Mod. } \lambda.$$

In der That kann man jede gegebene complexe wirkliche Zahl durch Multiplication mit Einheiten so zubereiten, dafs sie dieser Bedingung genügt, ausgenommen in dem schon oben ausgeschlossenen Falle, wo λ eine der Ausnahmszahlen ist, welche als Faktor einer der ersten $\frac{\lambda-3}{2}$ Bernouillischen Zahlen vorkommt. Der Beweis dieser Behauptung läfst sich nach der pag. 314 Jahrgang 1847 der Monatsberichte von mir gebrauchten Methode leicht führen und gewährt zugleich ein einfaches, leicht ausführbares Verfahren, in jedem besonderen Falle die passenden Einheiten zu finden, mit welchen eine gegebene wirkliche complexe Zahl multiplicirt werden mufs, um der Bedingung $\varphi(\alpha)\,\varphi(\alpha^{-1}) \equiv c$, Mod. λ, zu genügen.

Diese Bedingung reicht aber zur Bestimmung der primären complexen Zahl noch nicht aus, denn es kann eine und dieselbe complexe Zahl auf verschiedene Weisen in die dieser Bedingung genügende Form gebracht werden, und zwar so, dafs diese verschiedenen Formen auch verschiedene Charaktere derselben in Beziehung auf λ^{te} Potenzreste ergeben. Ich nehme darum noch eine zweite Eigenthümlichkeit der complexen Zahlen der Kreistheilung $\psi(\alpha)$ zur allgemeinen Bestimmung der primären Form hinzu, nämlich die Eigenschaft der Coëfficienten von

$$\psi(\alpha) = a + a_1\alpha + a_2\alpha^2 + \ldots + a_{\lambda-1}\alpha^{\lambda-1},$$

nach welcher

$$1a_1 + 2a_2 + 3a_3 + \ldots + (\lambda-1)a_{\lambda-1} \equiv 0, \quad \text{Mod. } \lambda,$$

welche auch so ausgedrückt werden kann, dafs $\psi(\alpha)$ einer realen ganzen Zahl congruent ist für den Modul $(1-\alpha)^2$. Als die zweite Bedingung, welcher jede complexe Zahl genügen soll, um eine primäre zu sein, nehme ich darum diese: $\varphi(\alpha) \equiv b$, Mod. $(1-\alpha)^2$. In die dieser Bedingung genügende Form läfst sich jede gegebene complexe Zahl durch Multiplikation mit einer passenden Potenz von α, also durch Multiplikation mit einer der einfachen Einheiten bringen.

Als vollständige Definition der primären Form der complexen Zahlen, welche aus λ^{ten} Wurzeln der Einheit gebildet sind, wo λ eine ungrade Primzahl ist, welche als Faktor einer der $\frac{\lambda-3}{2}$ ersten Bernouillischen Zahlen nicht vorkommt, nehme ich also die folgende:

Eine wirkliche complexe Zahl $\varphi(\alpha)$ soll eine primäre heifsen, wenn sie den beiden Bedingungen genügt,

$$\text{dafs } \varphi(\alpha)\,\varphi(\alpha^{-1}) \equiv c, \qquad \text{Mod. } \lambda,$$
$$\text{und } \varphi(\alpha) \equiv b, \qquad \text{Mod. } (1-\alpha)^2,$$

wo b und c reale ganze Zahlen bedeuten.

Nach dieser Definition kann zwar ebenfalls noch jede gegebene complexe Zahl auf unendlich viele verschiedene Weisen in die primäre Form gebracht werden, aber in allen diesen primären Formen hat sie wirklich nur einen bestimmten Charakter in Beziehung auf λ^{te} Potenzreste. Die verschiedenen Formen der primären complexen Zahl unterscheiden sich nämlich nur dadurch von einander, dafs eine λ^{te} Potenz einer beliebigen Einheit als Faktor hinzutreten kann, welches in Beziehung auf λ^{te} Potenzreste keinen Unterschied macht. Der Beweis dieser Behauptung folgt sehr leicht aus dem pag. 318 der Monatsberichte von 1847 von mir bewiesenen Satze: Wenn λ eine Primzahl ist, welche in den Zählern der ersten $\frac{\lambda-3}{2}$ Bernouillischen Zahlen als Faktor nicht vorkommt, so ist jede aus λ^{ten} Wurzeln der Einheit gebildete complexe Einheit, welche einer realen ganzen Zahl congruent ist, für den Modul λ, eine λ^{te} Potenz einer anderen complexen Einheit.

Wenn nun der Sinn des Zeichens $\left(\frac{\phi(\alpha)}{f(\alpha)}\right)$ für wirkliche oder ideale complexe Primzahlen $\phi(\alpha)$ und $f(\alpha)$ so fixirt wird wie er oben angegeben worden ist, und wenn die so eben aufgestellte Definition der primären complexen Zahl zu Grunde gelegt wird, so läfst sich das allgemeine Reciprocitätsgesetz in Beziehung auf diejenigen λ^{ten} Potenzreste, für welche λ nicht eine von den in den ersten $\frac{\lambda-3}{2}$ Bernouillischen Zahlen als Faktor vorkommenden Ausnahms-Zahlen ist, wie ich dasselbe gefunden und durch Induction innerhalb ziemlich ausgedehnter Grenzen als richtig nachgewiesen habe, ganz einfach durch die Gleichung

$$\left(\frac{\phi(\alpha)}{f(\alpha)}\right) = \left(\frac{f(\alpha)}{\phi(\alpha)}\right)$$

darstellen, wenn nämlich die complexen Primzahlen $\phi(\alpha)$ und $f(\alpha)$ in der primären Form genommen werden, oder im Falle dafs diese ideal sind, wenn die zu wirklichen complexen Zahlen werdenden Potenzen derselben in der primären Form genommen werden.

Für $\lambda = 3$ hat Jacobi dieses Reciprocitätsgesetz durch die Kreistheilung vollständig bewiesen. Für höhere Werthe des λ giebt die Kreistheilung immer eine brauchbare, aber nicht für sich allein ausreichende Reciprocitäts-Gleichung, welche ich bei meinem zur Prüfung des eben ausgesprochenen Gesetzes angestellten Inductions-Verfahren mit benutzt habe. Ich habe so durch berechnete Tafeln nachgewiesen, dafs dieses Reciprocitätsgesetz für fünfte Potenzreste für je zwei aller derjenigen complexen Primzahlen gültig ist, welche Primfactoren der realen ganzen Zahlen bis 200 sind. Für die siebenten Potenzreste habe ich die Gültigkeit desselben Gesetzes nachgewiesen für je zwei aller complexen Primfactoren der realen ganzen Primzahlen von der Form $7n+1$, welche nicht gröfser als 200 sind. Dagegen habe ich mich von der Richtigkeit desselben für die Primfactoren, welche realen ganzen Primzahlen der Formen $7n+1$ und $7n+3$, $7n+5$ und $7n+2$, $7n+4$ angehören, nur so weit überzeugt, als die Kreistheilungsgleichung reicht, weil da, wo die complexen Zahlen nicht mehr realen ganzen Zahlen congruent sind, der *Canon arithmeticus* nicht mehr mit Erfolg anzuwenden ist und die Rechnungen zu zeitraubend werden. Als Beispiel für höhere Potenzen habe ich $\lambda = 23$ gewählt, weil für diesen Werth zuerst die wirklichen complexen Primzahlen nicht mehr ausreichen und darum ideale Primzahlen anzuwenden sind. Meine Induction erstreckt sich hier auf 44 complexe Primzahlen, nämlich auf die 22 idealen Primfactoren der realen ganzen Primzahl 47, deren dritte Potenzen zu wirklichen complexen Zahlen werden, und auf die 22 wirklichen complexen Primfactoren der realen Primzahl 599. Bei allen möglichen 946 Verbindungen dieser 44 complexen Primzahlen habe ich das obige Reciprocitätsgesetz bestätigt gefunden.

Durch alle diese Resultate meiner Induction glaube ich zu der vorläufigen Annahme berechtigt zu sein, dafs das von mir gefundene und aufgestellte Gesetz richtig ist, und dafs für alle diejenigen Potenzen, auf welche es sich bezieht, es von jetzt an sich nicht weiter um die Auffindung, sondern nur noch um den Beweis des Reciprocitätsgesetzes handelt. Aufserdem gehören zur Theorie der λ^{ten} Potenzreste wesentlich noch diejenigen Sätze, welche sich auf den Charakter beziehen, den die Zahl λ selbst, so wie ihre Primfactoren von der Form $1-\alpha^{n}$ und ausserdem die complexen Einheiten haben. Diese Ergänzungssätze habe ich nicht allein gefunden, sondern auch vollständig ergründet und bewiesen, und ich werde mir erlauben, dieselben in der Kürze hier zu entwickeln.

Es sei p eine reale Primzahl von der Form $\nu\lambda+1$, λ eine ungrade Primzahl, g primitive Wurzel der Congruenz $g^{p-1}\equiv 1$, Mod. p, γ primitive Wurzel der Congruenz $\gamma^{\lambda-1}\equiv 1$, Mod. λ,

x eine imaginäre Wurzel der Gleichung $x^p = 1$, α eine imaginäre Wurzel der Gleichung $\alpha^\lambda = 1$. Ferner seien $\eta, \eta_1, \eta_2, \ldots$ $\ldots \eta_{\lambda-1}$ die λ Perioden, welche aus je ν Wurzeln der Gleichung $x^p = 1$ gebildet sind, wo $\nu = \frac{p-1}{\lambda}$, so dafs

$$\begin{aligned}
\eta \;\; &= x + x^{g^\lambda} + \ldots\ldots\ldots\ldots + x^{g^{p-1-\lambda}} \\
\eta_1 \;\; &= x^g + x^{g^{\lambda+1}} + \ldots\ldots\ldots + x^{g^{p-\lambda}} \\
&\vdots \\
\eta_{\lambda-1} &= x^{g^{\lambda-1}} + x^{g^{2\lambda-1}} + \ldots + x^{g^{p-2}}
\end{aligned}$$

Aufserdem sei $f(\alpha)$ ein complexer Primfactor von p und Ind. y der Index von y für die primitive Wurzel g und für den Modul p, wenn y real ist, aber für den Modul $f(\alpha)$, wenn y eine complexe Zahl ist. Das Product zweier Perioden läfst sich bekanntlich als lineäre Function aller Perioden so darstellen, dafs

$$\begin{aligned}
\eta^2 &= \nu + m\eta + m_1\eta_1 + \ldots\ldots + m_{\lambda-1}\eta_{\lambda-1} \\
\eta\eta_\varkappa &= \quad\;\; \overset{\varkappa}{m}\eta + \overset{\varkappa}{m}_1\eta_1 + \ldots\ldots + \overset{\varkappa}{m}_{\lambda-1}\eta_{\lambda-1}
\end{aligned}$$

wo allgemein $\overset{\varkappa}{m}_h$ der Anzahl der Werthe des r aus der Reihe $r = 0, 1, 2, \ldots \nu - 1$ gleich ist, welche der Congruenz

* $$g^{r\lambda+\varkappa} + 1 \equiv g^h, \qquad \text{Mod. } p,$$

genügen, oder, was dasselbe ist, der Congruenz

$$\text{Ind. } (g^{r\lambda+\varkappa} + 1) \equiv h$$

für den Modul $p-1$, also auch für den Modul λ. Aus dieser allgemeinen Bestimmung der Zahlen $\overset{\varkappa}{m}_h$ erkennt man sogleich die Richtigkeit folgender Congruenz:

$$\begin{aligned}
&\text{Ind. } (g^\varkappa + 1) + \text{Ind. } (g^{\varkappa+\lambda} + 1) + \ldots\ldots \text{Ind. } (g^{\varkappa+(\nu-1)\lambda} + 1) \\
&\equiv 1\overset{\varkappa}{m}_1 + 2\overset{\varkappa}{m}_2 + 3\overset{\varkappa}{m}_3 + \ldots\ldots + (\lambda-1)\overset{\varkappa}{m}_{\lambda-1}, \qquad \text{Mod. } \lambda.
\end{aligned}$$

Nun ist aber, wie leicht zu zeigen, die Congruenz

$$z^\nu - 1 \equiv (z-1)(z-g^\lambda)(z-g^{2\lambda})\ldots(z-g^{(\nu-1)\lambda}), \quad \text{Mod. } p,$$

für alle beliebigen ganzzahligen Werthe des z erfüllt; setzt man daher in derselben $z \equiv -g^{p-1-\varkappa}$, Mod. p, und multiplicirt auf beiden Seiten mit $g^{\nu\varkappa}$, so wird

$$1 - g^{\nu\varkappa} \equiv (1+g^\varkappa)(1+g^{\varkappa+\lambda})\ldots(1+g^{\varkappa+(\nu-1)\lambda}), \quad \text{Mod. } p,$$

also auch

$$\text{Ind.}\,(1-g^{\nu\varkappa}) \equiv \text{Ind.}\,(1+g^{\varkappa}) + \text{Ind.}\,(1+g^{\varkappa+\lambda}) + \ldots\ldots + \text{Ind.}\,(1+g^{\varkappa+(\nu-1)\lambda})$$

für den Modul λ, also nach der obigen Gleichung

$$\text{Ind.}\,(1-g^{\nu\varkappa}) \equiv 1\overset{\varkappa}{m}_1 + 2\overset{\varkappa}{m}_2 + \ldots\ldots + (\lambda-1)\overset{\varkappa}{m}_{\lambda-1}, \quad \text{Mod. } \lambda.$$

Wenn nun $f(\alpha)$ ein complexer Primfactor des p ist und zwar derjenige der $\lambda-1$ conjugirten, welcher congruent Null wird für den Modul p, wenn g^{ν} statt α gesetzt wird: so kann anstatt Ind. $(1-g^{\nu\varkappa})$ gesetzt werden Ind. $(1-\alpha^{\varkappa})$, wo aber jetzt das Zeichen Ind. nicht mehr den Index in Beziehung auf p, sondern in Beziehung auf den Modul $f(\alpha)$ bedeutet. Es ist daher

$$\text{Ind.}\,(1-\alpha^{\varkappa}) \equiv 1\overset{\varkappa}{m}_1 + 2\overset{\varkappa}{m}_2 + \ldots\ldots + (\lambda-1)\overset{\varkappa}{m}_{\lambda-1}, \quad \text{Mod. } \lambda.$$

Diese Congruenz giebt den Charakter eines Primfactors $1-\alpha^{\varkappa}$ der Zahl λ in Beziehung auf λ^{te} Potenzreste, ausgedrückt durch die Zahlen $\overset{\varkappa}{m}_h$, welche als bekannte anzusehen sind, da die Kreistheilung dieselben vollständig und auch so einfach bestimmt, als nur überhaupt die Zahlen der Kreistheilung sich bestimmen lassen.

Setzt man in diesem Ausdrucke des Ind. $(1-\alpha^{\varkappa})$ nach einander $\varkappa = 1, 2, 3, \ldots \lambda-1$, und addirt, indem man von der bekannten Formel

$$\overset{1}{m}_h + \overset{2}{m}_h + \ldots\ldots + \overset{\lambda-1}{m}_h = \nu$$

Gebrauch macht, so erhält man den Charakter der Zahl λ selbst, durch folgende Congruenz ausgedrückt:

$$\text{Ind.}\,(\lambda) \equiv -(m_1 + 2m_2 + 3m_3 + \ldots\ldots + (\lambda-1)m_{\lambda-1}), \quad \text{Mod. } \lambda.$$

Nimmt man endlich die Einheit

$$\sqrt[2]{\frac{(1-\alpha^{\gamma})\,(1-\alpha^{-\gamma})}{(1-\alpha)\;\,(1-\alpha^{-1})}} = \frac{\alpha^{\frac{1-\gamma}{2}}(1-\alpha^{\gamma})}{1-\alpha} = e(\alpha),$$

welche ich gewöhnlich mit dem Namen Kreistheilungs-Einheit der complexen Zahlen zu benennen pflege, so erhält man aus dem Ausdrucke des Ind. $(1-\alpha^{\varkappa})$ sogleich

$$\text{Ind.}\,e(\alpha^{\varkappa}) \equiv \frac{\varkappa(1-\gamma)\,(p-1)}{2\lambda} + 1\overset{\varkappa\gamma}{m}_1 + 2\overset{\varkappa\gamma}{m}_2 + \ldots + (\lambda-1)\overset{\varkappa\gamma}{m}_{\lambda-1} - 1\overset{\varkappa}{m}_1 - 2\overset{\varkappa}{m}_2 - \ldots - (\lambda-1)\overset{\varkappa}{m}_{\lambda-1}$$

für den Modul λ.

Ein System conjugirter Kreistheilungs-Einheiten ist ein unabhängiges System, welches die Eigenschaft hat, dafs überhaupt jede Einheit sich als ein Product von Potenzen der Kreistheilungs-Einheiten darstellen läfst, und zwar so, dafs die Potenzexponenten nur ganze Zahlen oder rationale Brüche sind. Diese Brüche aber können, wie ich in der mehrmals erwähnten Abhandlung bewiesen habe, in ihren Nennern den Faktor λ nur dann enthalten, wenn die Klassenanzahl der idealen complexen Zahlen durch λ theilbar ist, also nur dann, wenn λ in einer der ersten $\frac{\lambda-3}{2}$ Bernouillischen Zahlen als Faktor vorkommt. Schliefst man diesen Fall auch hier aus, so sind mit Ind. $e(\alpha^{\varkappa})$ zugleich die Indices aller complexen Einheiten bekannt, denn wenn $E(\alpha)$ eine beliebige Einheit ist, so kann man sie stets in die Form setzen

$$E(\alpha) = e(\alpha)^{n} \,.\, e(\alpha^{\gamma})^{n_1} . \, e\left(\alpha^{\gamma^2}\right)^{n_2} \ldots . \, e\left(\alpha^{\gamma^{\frac{\lambda-3}{2}}}\right)^{n_{\frac{\lambda-3}{2}}},$$

man hat daher

$$\text{Ind.}\, E(\alpha) \equiv n \,\text{Ind.}\, e(\alpha) + n_1 \,\text{Ind.}\, e(\alpha^{\gamma}) + \ldots + n_{\frac{\lambda-3}{2}} \,\text{Ind.}\, e\left(\alpha^{\gamma^{\frac{\lambda-3}{2}}}\right)$$

für den Modul λ, wo man die rationalen Brüche n, n_1, $n_2, \ldots$ $\ldots n_{\frac{\lambda-3}{2}}$, in deren Nennern λ nicht enthalten ist, durch die ganzen Zahlen ersetzen kann, denen sie congruent sind.

Die Zahlen $\overset{\varkappa}{m}_h$, durch welche hier die Indices von $1-\alpha^{\varkappa}$, λ und der Einheiten ausgedrückt sind, lassen sich auf mannigfache Weisen auf andere in der Lehre von der Kreistheilung vorkommende Zahlen reduciren, namentlich auch auf die Coëfficienten der Jacobischen complexen Zahlen ψ, welche wir schon oben erwähnt haben. Nimmt man nämlich

$$F(\alpha, x) = x + \alpha x^{g} + \alpha^2 x^{g^2} + \ldots + \alpha^{p-2} x^{g^{p-2}},$$

und

$$\frac{F(\alpha, x)\, F(\alpha^{r}, x)}{F(\alpha^{r+1}, x)} = \psi_r(\alpha) = \overset{r}{a} + \overset{r}{a}_1 \alpha + \overset{r}{a}_2 \alpha^2 + \ldots + \overset{r}{a}_{\lambda-1} \alpha^{\lambda-1},$$

so ist allgemein

$$\overset{\varkappa}{m}_h = \frac{1}{\lambda} \left(\sum_{1}^{\lambda-2} {}_{s}\, \overset{s}{a}_{h+s\varkappa} - \frac{(\lambda-1)(\lambda-3)}{2} + 1 \right)$$

für die besonderen Fälle $h = \varkappa$ und $\varkappa = 0$ aber ist

$$\overset{h}{m}_h = \frac{1}{\lambda}\left(\sum_1^{\lambda-2}{}_s^s\, a_{h(1+s)} - \frac{(\lambda-1)(\lambda-3)}{2}\right),$$

$$\overset{0}{m}_h = \frac{1}{\lambda}\left(\sum_1^{\lambda-2}{}_s^s\, a_h - \frac{(\lambda-1)(\lambda-3)}{2}\right).$$

Ich will diese und ähnliche Verwandlungen, welche man mit den gefundenen Ausdrücken von Ind. $(1-\alpha^{x})$, Ind. (λ) und Ind. $e(\alpha^{x})$ vornehmen kann, hier nicht ausführlich entwickeln, sondern nur die merkwürdigsten Resultate mittheilen, welche man auf diese Weise für die Indices der Kreistheilungs-Einheiten erhält. Diese stellen sich am einfachsten dar für die auch in anderen Beziehungen sehr bemerkenswerthe zusammengesetzte Kreistheilungs-Einheit

$$E_n(\alpha) = e(\alpha)\,.\,e(\alpha^{\gamma})^{\gamma^{-2n}}\,.\,e(\alpha^{\gamma^2})^{\gamma^{-4n}}\ldots\ldots e\left(\alpha^{\gamma^{\frac{\lambda-3}{2}}}\right)^{\gamma^{-(\lambda-3)n}}$$

von welcher man, wenn man will, sogleich auch wieder zur einfachen Einheit $e(\alpha)$ übergehen kann, vermittelst der Formel:

$$\text{Ind.}\, e\left(\alpha^{\gamma^h}\right) \equiv -2\left(\gamma^{2h}\,\text{Ind.}\, E_1(\alpha) + \gamma^{4h}\,\text{Ind.}\, E_2(\alpha) + \ldots\ldots\ldots \right.$$
$$\left. \ldots\ldots + \gamma^{(\lambda-3)h}\,\text{Ind.}\, E_{\frac{\lambda-3}{2}}(\alpha)\right)$$

Der Index dieser Einheit $E_n(\alpha)$ wird durch folgende sehr merkwürdige Congruenz bestimmt:

$$\text{Ind.}\, E_n(\alpha) \equiv \frac{(\gamma^{2n}-1)\, d_0^{\lambda-2n}\, l\psi_r(e^{v})}{2\left(1+r^{\lambda-2n}-(1+r)^{\lambda-2n}\right) dv^{\lambda-2n}},$$

wo $l\psi_r(e^{v})$ den natürlichen Logarithmen der complexen Kreistheilungszahl $\psi_r(\alpha)$ bedeutet, in welcher anstatt α die variable Exponentialgröfse e^{v} zu setzen und nach der $\lambda - 2n$ maligen Differenziation $v = 0$ zu nehmen ist.

Noch einfacher wird der analoge Ausdruck des Ind. $E_n(\alpha)$ durch

$$F(\alpha, x) = x + \alpha x^{g} + \alpha^2 x^{g^2} + \ldots + \alpha^{p-2} x^{g^{p-2}}$$

worin ebenfalls α in e^{v} zu verwandeln ist, nämlich

$$\text{Ind.}\, E_n(\alpha) \equiv \frac{\gamma^{2n}-1}{2}\,.\,\frac{d_0^{\lambda-2n}\, lF(e^{v}, x)}{dv^{\lambda-2n}}$$

Die p^{te} Wurzel der Einheit x verschwindet aus diesem Ausdrucke nach Ausführung der Differenziation, wenn $v = 0$ gesetzt wird, von selbst.

Endlich füge ich noch einen Ausdruck hinzu, welcher den Ind. $E_n(\alpha)$ giebt, ausgedrückt durch die Coëfficienten der complexen Primzahl $f(\alpha)$, in Beziehung auf welche als den Modul das Zeichen Ind. zu nehmen ist, oder wenn $f(\alpha)$ ideal ist, durch die Coëfficienten derjenigen Potenz dieser idealen Zahl, welche eine wirkliche complexe Zahl ist. Diese Potenz sei die h^{te}, so dafs

$$f(\alpha)^h = b + b_1\alpha + b_2\alpha^2 + \ldots + b_{\lambda-1}\alpha^{\lambda-1}$$

eine wirkliche complexe Zahl ist, welche überdiefs so gewählt werden soll, dafs

$$b_1 + 2b_2 + 3b_3 + \ldots\ldots + (\lambda - 1)b_{\lambda-1} \equiv 1 \quad \text{Mod. } \lambda.$$

Man hat so den folgenden Ausdruck:

$$\text{Ind. } E_n(\alpha) \equiv \frac{(-1)^n B_n(\gamma^{2n} - 1)}{4nh} \cdot \frac{d^{\lambda-2n}\, l\left(f(e^v)^h\right)}{dv^{\lambda-2n}},$$

wo B_n die n^{te} Bernouillische Zahl bezeichnet. Bringt man ausserdem $f(\alpha)^h$ in die primäre Form, wie dieselbe oben vollständig definirt worden ist, so erhält man Ind. $E_n(\alpha)$ durch die Coëfficienten von $f(\alpha)^h$ unmittelbar ausgedrückt durch die Congruenz

$$\text{Ind. } E_n(\alpha) \equiv \frac{(-1)^n B_n(\gamma^{2n} - 1) \sum\limits_0^{\lambda-1}{}_s \sum\limits_0^{\lambda-1} s^{\lambda-2n-1}(s+t)\, b_s\, b_{s+t}}{4nh\,(b + b_1 + b_2 + \ldots + b_{\lambda-1})^2}$$

Ich habe für die Herleitung dieser und ähnlicher Ausdrücke der Indices der Einheiten noch eine andere von der hier kurz entwickelten wesentlich verschiedene Methode gefunden, welche ich aber, zugleich mit einigen für den allgemeinen Beweis des oben aufgestellten Reciprocitätsgesetzes nicht unwichtigen Vorarbeiten, ein andermal zu veröffentlichen gedenke.

Breslau, d. 14. Mai 1850. **Kummer.**

Über eine allgemeine Eigenschaft der rationalen Entwicklungscoefficienten einer bestimmten Gattung analytischer Functionen

Journal für die reine und angewandte Mathematik 41, 368–372 (1851)

Als ich vor einiger Zeit eine bestimmte Aufgabe der Zahlentheorie nach zwei verschiedenen Methoden lösete, erhielt ich als Resultat der Vergleichung beider Auflösungen eine bisher unbekannte Eigenschaft der *Bernoullischen Zahlen,* welche durch folgende Congruenz ausgedrückt wird:

$$\frac{B_\nu}{\nu} \equiv (-1)^{\frac{1}{2}(\lambda-1)} \frac{B_{\nu+\frac{1}{2}(\lambda-1)}}{\nu+\frac{1}{2}(\lambda-1)}, \text{ Mod. } \lambda;$$

wo B_ν die νte *Bernoulli*sche Zahl, λ eine beliebige ungerade Primzahl und ν eine beliebige, nur der einen Bedingung unterworfene ganze Zahl ist, dafs sie nicht ein Vielfaches von $\frac{1}{2}(\lambda-1)$ sei. Als ich jetzt einen directen Beweis dieses Satzes suchte, fand ich, dafs diese Eigenschaft der *Bernoulli*schen Zahlen nur ein besonderer Fall von allgemeinen Eigenschaften der rationalen Entwickelungs-Coëfficienten einer sehr ausgebreiteten Gattung analytischer Functionen ist. Das von mir gefundene allgemeine Resultat läfst sich folgendermaafsen in Form eines Lehrsatzes aussprechen:

Lehrsatz. Wenn $\varphi(x)$ eine Function von x ist, welche in eine nach aufsteigenden ganzen Potenzen der Gröfse $z = e^{rx} - e^{sx}$ geordnete Reihe sich so entwickeln läfst, dafs die Coëfficienten dieser Entwickelung rationale Zahlen sind, in deren Nennern die Primzahl λ als Factor nicht vorkommt, und wobei r und s rationale Zahlen sind, deren Nenner ebenfalls den Primfactor λ nicht enthalten: wenn ferner dieselbe Function $\varphi(x)$, nach Potenzen von x entwickelt, folgende Reihe giebt:

$$\varphi(x) = A + \frac{A_1 x}{1} + \frac{A_2 x^2}{1.2} + \frac{A_3 x^3}{1.2.3} + \cdots,$$

so findet unter den Coëfficienten dieser Entwickelung folgende Congruenz Statt:

$$A_m - \frac{n}{1} A_{m+\lambda-1} + \frac{n(n-1)}{1.2} A_{m+2\lambda-2} - \cdots (-1)^n A_{m+n\lambda-n} \equiv 0,$$

für den Modul λ^n. Die Zahlen m und n sind beliebige ganze Zahlen und nur der einen Bedingung unterworfen, dafs $m \geqq n$ ist. Die besonderen Fälle für $n = 1, 2, 3$ sind:

$$A_m - A_{m+\lambda-1} \equiv 0, \text{ Mod. } \lambda, \text{ wenn } m \geqq 1,$$
$$A_m - 2A_{m+\lambda-1} + A_{m+2\lambda-2} \equiv 0, \text{ Mod. } \lambda^2, \text{ wenn } m \geqq 2,$$
$$A_m - 3A_{m+\lambda-1} + 3A_{m+2\lambda-2} - A_{m+3\lambda-3} \equiv 0, \text{ Mod. } \lambda^3, \text{ wenn } m \geqq 3,$$

u. s. w.

Beweis. Nach der Voraussetzung des Lehrsatzes ist

$$\varphi(x) = a + a_1(e^{rx} - e^{sx}) + a_2(e^{rx} - e^{sx})^2 + a_3(e^{rx} - e^{sx})^3 + \cdots,$$

oder, wenn man das Summenzeichen Σ anwendet,

$$\varphi(x) = \Sigma a_k(e^{rx} - e^{sx})^k,$$

wo a_k, r und s rationale Zahlen sind, deren Nenner den Factor λ nicht enthalten. Wird nun $(e^{rx} - e^{sx})^k$ nach dem binomischen Lehrsatze entwickelt, wobei der $h$te Binomial-Coëfficient der $k$ten Potenz einfach durch $(k)_h$ bezeichnet werden möge, so ist

$$(e^{rx} - e^{sx})^k = \Sigma(-1)^h (k)_h e^{(r(k-h)+sh)x},$$

also

$$\varphi(x) = \Sigma\Sigma(-1)^h (k)_h a_k e^{(r(k-h)+sh)x}.$$

Wird der $m$te Differentialquotient dieses Ausdrucks des $\varphi(x)$ genommen, für den Werth $x = 0$, welcher nach dem *Maclaurin*schen Satze gleich dem Entwickelungs-Coëfficienten A_m ist, so ergiebt sich

$$A_m = \Sigma\Sigma(-1)^h (k)_h a_k (r(k-h) + sh)^m.$$

Wenn man nun m in $m + \lambda - 1$, $m + 2\lambda - 2$, ... $m + n\lambda - n$ verwandelt, so erhält man mittels dieser Ausdrücke der Coëfficienten A_m, $A_{m+\lambda-1}$, $A_{m+2\lambda-2}$ etc. die Gleichung

$$A_m - \frac{n}{1} A_{m+\lambda-1} + \frac{n(n-1)}{1.2} A_{m+2\lambda-1} - \cdots (-1)^n A_{m+n\lambda-n} =$$
$$\Sigma\Sigma(-1)^h (k)_h a_k (r(k-h) + sh)^m (1 - (r(k-h) + sh)^{\lambda-1})^n.$$

Der Ausdruck $1 - (r(k-h) + sh)^{\lambda-1}$ ist nach dem *Fermat*schen Satze durch λ theilbar; mit alleiniger Ausnahme des Falles, wo $r(k-h) + sh$ durch λ theilbar ist. Hieraus folgt, dafs, wenn $m \geqq n$ angenommen wird, immer der eine oder der andere der beiden Factoren des Products

$$(r(k-h) + sh)^m (1 - (r(k-h) + sh)^{\lambda-1})$$

durch λ^n theilbar ist, dafs also

$$A_m - \frac{n}{1} A_{m+\lambda-1} + \frac{n(n-1)}{1.2} A_{m+2\lambda-2} - \cdots (-1)^n A_{m+n\lambda-n} \equiv 0, \text{ Mod. } \lambda^n,$$

ist. W. z. b. w.

Um den gefundenen und bewiesenen allgemeinen Satz zunächst auf die *Bernoulli*schen Zahlen anzuwenden, mache ich von der bekannten Reihen-Entwickelung

$$\frac{1}{e^x-1} = \frac{1}{x} - \tfrac{1}{2} + \frac{B_1 x}{1.2} - \frac{B_2 x^3}{1.2.3.4} + \frac{B_3 x^5}{1.2.3.4.5.6} - \cdots$$

Gebrauch, in welcher B_1, B_2 B_3, ... die *Bernoulli*schen Zahlen bedeuten. Ich setze γx statt x, multiplicire mit γ und ziehe die unveränderte Gleichung von dieser neuen ab, so ergiebt sich

$$\frac{\gamma}{e^{\gamma x}-1} - \frac{1}{e^x-1} = -\tfrac{1}{2}(\gamma-1) + \frac{B_1(\gamma^2-1)x}{1.2} - \frac{B_2(\gamma^4-1)x^3}{1.2.3.4} + \cdots$$

Wird nun

$$\frac{\gamma}{e^{\gamma x}-1} - \frac{1}{e^x-1} = \varphi(x) \text{ und } e^x-1 = z$$

gesetzt, so erhält man

$$\varphi(x) = \frac{\gamma}{(1+z)^\gamma-1} - \frac{1}{z},$$

welches, wie leicht zu sehen, in eine nach positiven ganzen Potenzen von $z = e^x-1$ geordnete Reihe sich entwickeln läfst, und zwar so, dafs die Coëfficienten rational sind und nur Potenzen von γ, also, wenn γ nicht durch λ theilbar ist, keinen Factor λ in ihren Nennern enthalten. Die Function $\varphi(x)$ ist also eine solche, wie der Lehrsatz sie verlangt; und zwar, wenn in demselben $r=1$ und $s=0$ angenommen wird.

Ferner ist für den vorliegenden Fall

$$A_2 = 0, \quad A_4 = 0, \quad A_6 = 0 \text{ etc.},$$

$$A_1 = +\tfrac{1}{2}(B_1(\gamma^2-1)), \quad A_3 = -\tfrac{1}{4}(B_2(\gamma^4-1)), \quad A_5 = +\tfrac{1}{6}(B_3(\gamma^6-1)) \text{ etc.},$$

also allgemein

$$A_{2\nu} = 0, \quad A_{2\nu-1} = -\frac{(-1)^\nu B_\nu(\gamma^{2\nu}-1)}{2\nu}.$$

Die erste der specielleren Congruenzen des Lehrsatzes, nämlich $A_m - A_{m+\lambda-1} \equiv 0$, Mod. λ, giebt daher, wenn $m = 2\nu-1$ angenommen wird:

$$-\frac{(-1)^\nu B_\nu(\gamma^{2\nu}-1)}{2\nu} \equiv -\frac{(-1)^{\nu+\frac{1}{2}(\lambda-1)} B_{\nu+\frac{1}{2}(\lambda-1)}(\gamma^{2\nu+\lambda-1}-1)}{2\nu+\lambda-1}, \text{ Mod. } \lambda.$$

Da nach dem *Fermat*schen Satze $\gamma^{2\nu+\lambda-1}-1\equiv\gamma^{2\nu}-1$, Mod. λ ist, so kann $\gamma^{2\nu}-1$ als gemeinschaftlicher Factor weggehoben werden, sobald diese Gröfse nicht durch λ theilbar ist. Man wähle nun die bisher unbestimmt gelassene Zahl γ so, dafs sie eine primitive Wurzel der Primzahl λ ist, dafs also $\gamma^{2\nu}-1$ nur dann durch λ theilbar ist, wenn ν ein Vielfaches von $\frac{1}{2}(\lambda-1)$ ist, welche Werthe des ν ich ausschliefse. Alsdann kann $\gamma^{2\nu}-1$ als gemeinschaftlicher Factor der Congruenz hinweggehoben werden und die gefundene Congruenz geht in folgende über:

$$\frac{B_\nu}{\nu}\equiv(-1)^{\frac{1}{2}(\lambda-1)}\frac{B_{\nu+\frac{1}{2}(\lambda-1)}}{\nu+\frac{1}{2}(\lambda-1)},\quad \text{Mod.}\,\lambda,$$

die für alle ganzzahligen positiven Werthe des ν gültig ist; mit Ausschlufs derer, welche Vielfache von $\frac{1}{2}(\lambda-1)$ sind. Eben so giebt die Congruenz $A_m-2A_{m+\lambda-1}+A_{m+2\lambda-2}\equiv 0$, Mod. λ^2, für den vorliegenden Fall:

$$-\frac{(-1)^\nu B_\nu(\gamma^{2\nu}-1)}{2\nu}+\frac{2(-1)^{\nu+\frac{1}{2}(\lambda-1)}B_{\nu+\frac{1}{2}(\lambda-1)}(\gamma^{2\nu+\lambda-1}-1)}{2\nu+\lambda-1}$$
$$-\frac{(-1)^\nu B_{\nu+\lambda-1}(\gamma^{2\nu+2\lambda-2}-1)}{2\nu+2\lambda-2}\equiv 0,$$

für den Modul λ^2; was mittels des *Fermat*schen Satzes und mit Hülfe der vorhergehenden Congruenz sich auf folgende einfachere Gestalt bringen läfst:

$$\frac{B_\nu}{\nu}-\frac{(-1)^{\frac{1}{2}(\lambda-1)}2B_{\nu+\frac{1}{2}(\lambda-1)}}{\nu+\frac{1}{2}(\lambda-1)}+\frac{B_{\nu+\lambda-1}}{\nu+\lambda-1}\equiv 0,\quad \text{Mod.}\,\lambda^2.$$

Die Zahl ν mufs in dieser Congruenz, aufser dafs sie nicht ein Vielfaches von $\frac{1}{2}(\lambda-1)$ sein darf, auch gröfser als 1 sein; wie aus der Bedingung $m\geqq 2$, also $2\nu-1\geqq 2$ hervorgeht. Die Congruenz $A_m-3A_{m+\lambda-1}+3A_{m+2\lambda-2}-A_{m+3\lambda-3}\equiv 0$, Mod. λ^3, giebt auf gleiche Weise

$$\frac{B_\nu}{\nu}-\frac{(-1)^{\frac{1}{2}(\lambda-1)}3B_{\nu+\frac{1}{2}(\lambda-1)}}{\nu+\frac{1}{2}(\lambda-1)}+\frac{3B_{\nu+\lambda-1}}{\nu+\lambda-1}-\frac{(-1)^{\frac{1}{2}(\lambda-1)}B_{\nu+\frac{3}{2}(\lambda-1)}}{\nu+\frac{3}{2}(\lambda-1)}\equiv 0,\quad \text{Mod.}\,\lambda^3,$$

wo ν gröfser als 1 sein mufs und nicht ein Vielfaches von $\frac{1}{2}(\lambda-1)$ sein darf. Auf diese Weise fortfahrend erhält man weiter die entsprechenden Congruenzen für die *Bernoulli*schen Zahlen für die Moduln λ^4, λ^5 u. s. w.; und allgemein

$$\frac{B_\nu}{\nu}-(-1)^\mu\frac{n}{1}\frac{B_{\nu+\mu}}{\nu+\mu}+\frac{n(n-1)}{1.2}\frac{B_{\nu+2\mu}}{\nu+2\nu}-(-1)^\mu\frac{n(n-1)(n-2)}{1.2.3}\frac{B_{\nu+3\mu}}{\nu+3\mu}+\cdots\equiv 0$$

für den Modul λ^n, wo $\nu\geqq\frac{1}{2}(n+1)$ sein mufs und kein Vielfaches von $\frac{1}{2}(\lambda-1)$ sein darf, und wo der Kürze wegen $\frac{1}{2}(\lambda-1)$ durch den Buchstaben μ bezeichnet ist.

Zu einem zweiten Beispiel der Anwendung des allgemeinen Lehrsatzes nehme ich die Coëfficienten der Secantenreihe

$$\sec(v) = 1+\frac{C_1 v^2}{1.2}+\frac{C_2 v^4}{1.2.3.4}+\frac{C_3 v^6}{1.2.3.4.5.6}+\cdots,$$

wo $C_1=1$, $C_2=5$, $C_3=61$, $C_4=1385$ u. s. w. ist. Setzt man $v=x\sqrt{-1}$, so erhält man

$$\frac{2}{e^x+e^{-x}} = 1-\frac{C_1 x^2}{1.2}+\frac{C_2 x^4}{1.2.3.4}-\frac{C_3 x^6}{1.2.3.4.5.6}+\cdots$$

Nimmt man nun

$$\varphi(x) = \frac{2}{e^x+e^{-x}}$$

und setzt $e^{\frac{1}{2}x}-e^{-\frac{1}{2}x}=z$, so wird $e^x+e^{-x}=z^2+2$, also

$$\varphi(x) = \frac{1}{1+\frac{1}{2}z^2}.$$

Hieraus ist klar, dafs sich $\varphi(x)$ in eine nach Potenzen von $z=e^{\frac{1}{2}x}-e^{-\frac{1}{2}x}$ geordnete Reihe entwickeln läfst, deren rationale Coëfficienten in ihren Nennern nur Potenzen der Zahl 2, also keinen Factor λ enthalten. Nimmt man daher in dem allgemeinen Satze $r=\frac{1}{2}$, $s=-\frac{1}{2}$, so findet sich, dafs $\varphi(x)$ eine Function von x ist, von der Form, wie der Lehrsatz sie verlangt. Aufserdem ergiebt sich durch Vergleichung des vorliegenden Falles mit dem allgemeinen Satze:

$$A_{2\nu-1} = 0 \quad \text{und} \quad A_{2\nu} = (-1)^\nu C_\nu.$$

Die daselbst gefundenen Congruenzen geben also für die Secanten-Coëfficienten

$$C_\nu-(-1)^{\frac{1}{2}(\lambda-1)}C_{\nu+\frac{1}{2}(\lambda-1)} \equiv 0, \text{ Mod. } \lambda, \text{ wenn } \nu \geqq 1,$$

$$C_\nu-(-1)^{\frac{1}{2}(\lambda-1)}2C_{\nu+\frac{1}{2}(\lambda-1)}+C_{\nu+\lambda-1} \equiv 0, \text{ Mod. } \lambda^2, \text{ wenn } \nu \geqq 0,$$

$$C_\nu-(-1)^{\frac{1}{2}(\lambda-1)}3C_{\nu+\frac{1}{2}(\lambda-1)}+3C_{\nu+\lambda-1}-(-1)^{\frac{1}{2}(\lambda-1)}C_{\nu+\frac{3}{2}(\lambda-1)} \equiv 0, \text{ Mod. } \lambda^3, \text{ wenn } \nu \geqq 2$$

und allgemein

$$C_\nu-(-1)^\mu\frac{n}{1}C_{\nu+\mu}+\frac{n(n-1)}{1.2}C_{\nu+2\mu}-(-1)^\mu\frac{n(n-1)(n-2)}{1.2.3}C_{\nu+3\mu}+\cdots \cdots \pm C_{\nu+n\mu} = 0,$$

für den Modul λ^n, wenn $\nu \geqq \frac{1}{2}n$ ist, und wo der Kürze wegen $\frac{1}{2}(\lambda-1)$ durch den Buchstaben μ bezeichnet ist.

Breslau, den 30ten Juli 1850.

Mémoire sur la théorie des nombres complexes composés de racines de l'unité et de nombres entiers

Journal de mathématiques pures et appliquées XVI, 377–498 (1851)

Dans l'état actuel de la science, on entend généralement par nombre complexe une fonction entière, à coefficients entiers, des racines irrationnelles d'une ou de plusieurs équations algébriques dont les coefficients sont également des nombres entiers. Le produit de tous les nombres complexes qu'on déduit d'un d'entre eux en changeant les racines des équations qu'il contient, étant une fonction symétrique de ces racines, sera toujours délivré de toute irrationnalité. Ce produit, qu'on appelle la *norme* du nombre complexe, sera donc un nombre entier; et, par conséquent, tout nombre complexe sera facteur irrationnel d'un nombre entier. De même, si l'on prend les coefficients du nombre complexe pour des indéterminés, la norme représentera une forme homogène d'un certain degré, du genre de celles qui sont décomposables en facteurs linéaires. La théorie des nombres complexes revient, au fond, à la théorie de ces formes, et, à cet égard, elle fait partie d'une des plus belles branches de l'Arithmétique supérieure. C'est sous ce point de vue que M. Lejeune-Dirichlet a fait des recherches très-générales sur les formes de degrés quelconques qui dépendent des normes des nombres complexes. Il a jeté les fondements de cette théorie en découvrant les propriétés générales de ces formes; mais, malheureusement, il n'en a publié jusqu'à présent que quelques-uns des résultats principaux, en ne donnant que des notions générales sur les principes nouveaux dont il s'est servi pour y parvenir. D'un autre côté, la théorie des nombres complexes peut être considérée comme la théorie de la décomposition des nombres en facteurs irrationnels, et c'est sous ce point de vue qu'elle a un grand intérêt, aussi bien en elle-même que pour les applications nombreuses et importantes qu'on en a faites dans plusieurs questions relatives à l'Arithmétique et à l'Algèbre supérieure.

Je ne traiterai, dans ce Mémoire, que les nombres complexes dont les irrationnalités sont les racines imaginaires de l'unité ou de l'équation binôme

$$\alpha^\lambda = 1,$$

genre spécial de nombres complexes, mais duquel l'importance pour la théorie générale est comparable à celle qu'il faut attribuer à la solution de l'équation binôme pour les équations algébriques les plus générales. La théorie de ces nombres complexes a été depuis longtemps le sujet de mes recherches, que j'ai publiées dans les *Comptes rendus de l'Académie de Berlin* et dans le Journal de M. Crelle [*]. En reprenant ici cette matière, j'ai en vue de compléter et de réunir la substance principale de ces divers Mémoires pour en former un Traité entier et continu qui puisse servir de base sûre à des recherches ultérieures dans cette partie de la théorie des nombres. J'ajouterai aussi deux applications des nombres complexes, dont l'une se rapporte à la théorie de la division du cercle, l'autre à la démonstration du dernier théorème de Fermat.

§ I.

Définitions et théorèmes préliminaires.

Les nombres complexes dont nous nous occuperons dans ce Mémoire sont des fonctions rationnelles et entières, à coefficients entiers, d'une racine imaginaire de l'équation

$$\alpha^\lambda = 1,$$

λ étant un nombre premier impair. A l'aide de l'équation

$$1 + \alpha + \alpha^2 + \ldots + \alpha^{\lambda-1} = 0,$$

[*] Rappelons aussi les premiers essais, déjà très-intéressants, que M. Kummer avait consignés dans un Mémoire imprimé d'abord à Breslau, en 1844, et inséré depuis au tome XII du présent Journal. (J. LIOUVILLE.)

qui contient toutes les racines imaginaires de l'équation $\alpha^\lambda = 1$, un nombre complexe $f(\alpha)$ est toujours réductible à la forme

$$f(\alpha) = a + a_1\alpha + a_2\alpha^2 + \ldots + a_{\lambda-2}\alpha^{\lambda-2},$$

$a, a_1, a_2, \ldots, a_{\lambda-2}$ étant des nombres entiers. On démontre aisément que cette réduction ne pourrait être effectuée que d'une seule manière; car, si l'on avait

$$f(\alpha) = a + a_1\alpha + a_2\alpha^2 + \ldots + a_{\lambda-2}\alpha^{\lambda-2}$$

et

$$f(\alpha) = b + b_1\alpha + b_2\alpha^2 + \ldots + b_{\lambda-2}\alpha^{\lambda-2},$$

on en conclurait

$$a - b + (a_1 - b_1)\alpha + (a_2 - b_2)\alpha^2 + \ldots + (a_{\lambda-2} - b_{\lambda-2})\alpha^{\lambda-2} = 0,$$

équation qui ne peut pas subsister à cause de l'irréductibilité de l'équation

$$1 + \alpha + \alpha^2 + \ldots + \alpha^{\lambda-1} = 0.$$

En prenant pour α successivement toutes les racines différentes, que nous représenterons comme puissances de l'une d'elles par α, α^2, $\alpha^3, \ldots, \alpha^{\lambda-1}$, on obtient les $\lambda - 1$ nombres complexes

$$f(\alpha),\quad f(\alpha^2),\quad f(\alpha^3), \ldots,\quad f(\alpha^{\lambda-1}),$$

que nous appellerons *nombres complexes conjugués*. Deux nombres conjugués, tels que $f(\alpha^n)$ et $f(\alpha^{-n})$, dont les racines sont réciproques, seront appelés *nombres complexes réciproques*.

Le produit de tous les nombres conjugués

$$f(\alpha),\quad f(\alpha^2),\quad f(\alpha^3), \ldots,\quad f(\alpha^{\lambda-1}),$$

étant une fonction symétrique de toutes les racines de l'équation

$$1 + \alpha + \alpha^2 + \ldots + \alpha^{\lambda-1} = 0,$$

se réduit à un nombre rationnel et entier qui, suivant M. Lejeune-Dirichlet, sera appelé la *norme* du nombre complexe $f(\alpha)$, et qui sera

48..

désigné par la lettre N, en sorte qu'on ait

$$\mathrm{N}f(\alpha)=f(\alpha).f(\alpha^2).f(\alpha^3)\ldots f(\alpha^{\lambda-1}).$$

Il suit immédiatement de la définition :

1°. Que les nombres conjugués ont tous la même norme

$$\mathrm{N}f(\alpha^k)=\mathrm{N}f(\alpha);$$

2°. Que la norme du produit de deux ou de plusieurs nombres complexes est égale au produit des normes des facteurs

$$\mathrm{N}[f(\alpha).\varphi(\alpha)]=\mathrm{N}f(\alpha).\mathrm{N}\varphi(\alpha).$$

La norme du nombre complexe

$$f(\alpha)=a+a_1\alpha+a_2\alpha^2+\ldots+a_{\lambda-2}\alpha^{\lambda-2}$$

est une fonction homogène du degré $\lambda-1$ des $\lambda-1$ nombres indéterminés a, a_1, $a_2,\ldots$, $a_{\lambda-2}$. Donc la théorie des nombres complexes est, au fond, la même que la théorie de ces formes, et, pour cette raison, elle donne lieu à des questions semblables à celles qu'on connaît de la théorie des formes quadratiques. Le développement effectif de la norme comme forme du degré $\lambda-1$ étant très-pénible, nous pourrons nous en dispenser en la représentant toujours comme produit de facteurs conjugués; d'ailleurs la discussion de ces formes nous paraît moins simple que celle des nombres complexes eux-mêmes, qui en sont les facteurs, les éléments, pour ainsi dire, et dont l'analogie avec les nombres entiers est frappante, comme on le verra dans la suite.

L'addition et la soustraction des nombres complexes n'offrent aucune difficulté, car les deux nombres complexes

$$f(\alpha)=a+a_1\alpha+a_2\alpha^2+\ldots+a_{\lambda-2}\alpha^{\lambda-2}$$

et

$$\varphi(\alpha)=b+b_1\alpha+b_2\alpha^2+\ldots+b_{\lambda-2}\alpha^{\lambda-2},$$

donnent immédiatement la somme et la différence

$$f(\alpha)\pm\varphi(\alpha)=a\pm b+(a_1\pm b_1)\alpha+(a_2\pm b_2)\alpha^2+\ldots$$
$$+(a_{\lambda-2}\pm b_{\lambda-2})\alpha^{\lambda-2}.$$

Le résultat de la multiplication des nombres $f(\alpha)$ et $\varphi(\alpha)$ sera toujours un nombre $\psi(\alpha)$ de la forme

$$\psi(\alpha) = c + c_1 \alpha + c_2 \alpha^2 + \ldots + c_{\lambda-1} \alpha^{\lambda-1},$$

qui, au moyen de l'équation

$$1 + \alpha + \alpha^2 + \ldots + \alpha^{\lambda-1} = 0,$$

se réduira à

$$\psi(\alpha) = c - c_{\lambda-1} + (c_1 - c_{\lambda-1})\alpha + (c_2 - c_{\lambda-1})\alpha^2 + \ldots \\ + (c_{\lambda-2} - c_{\lambda-1})\alpha^{\lambda-2},$$

et l'on aura

$$\begin{aligned} c \, &= ab + a_{\lambda-2} b_2 + a_{\lambda-3} b_3 + \ldots + a_2 b_{\lambda-2}, \\ c_1 &= a_1 b + ab_1 + a_{\lambda-2} b_3 + \ldots + a_3 b_{\lambda-2}, \\ c_2 &= a_2 b + a_1 b_1 + ab_2 + \ldots + a_4 b_{\lambda-2}, \\ &\ldots\ldots\ldots\ldots\ldots\ldots\ldots\ldots\ldots \\ c_{\lambda-1} &= a_{\lambda-2} b_1 + a_{\lambda-3} b_2 + \ldots + a_1 b_{\lambda-2}. \end{aligned}$$

L'addition de ces équations donne

$$c + c_1 + c_2 + \ldots + c_{\lambda-1} \\ = (a + a_1 + a_2 + \ldots + a_{\lambda-2})(b + b_1 + b_2 + \ldots + b_{\lambda-2}),$$

et la somme des coefficients du produit $\psi(\alpha)$ dans sa forme réduite, étant égale à

$$c + c_1 + c_2 + \ldots + c_{\lambda-1} - \lambda c_{\lambda-1},$$

on en conclut ce théorème :

La somme des coefficients du produit de deux nombres complexes est congrue au produit des sommes des coefficients des facteurs pour le module λ.

Il est clair que ce théorème subsiste également pour un nombre quelconque de facteurs. En l'appliquant aux facteurs conjugués dont se compose la norme du nombre

$$f(\alpha) = a + a_1 \alpha + a_2 \alpha^2 + \ldots + a_{\lambda-2} \alpha^{\lambda-2},$$

pour lesquels les sommes des coefficients sont toutes égales, on aura

$$\mathrm{N} f(\alpha) \equiv (a + a_1 + a_2 + \ldots + a_{\lambda-2})^{\lambda-1} \pmod{\lambda}.$$

On en conclut :

Pour que la norme d'un nombre complexe soit divisible par λ, il faut et il suffit que la somme des coefficients de ce nombre soit divisible par λ.

Mais si cette somme n'est pas divisible par λ, on a, en vertu du théorème de Fermat,

$$(a + a_1 + a_2 + \ldots + a_{\lambda-2})^{\lambda-1} \equiv 1 \quad (\text{mod. } \lambda),$$

et, par conséquent,

$$\mathrm{N}f(\alpha) \equiv 1 \quad (\text{mod. } \lambda).$$

Il suit de là cet autre théorème :

La norme de tout nombre complexe dont la somme des coefficients n'est pas divisible par λ, est de la forme linéaire $m\lambda + 1$.

La division des nombres complexes se réduit immédiatement au cas où le diviseur est un nombre entier non complexe. En effet, $\varphi(\alpha)$ étant le dividende et $f(\alpha)$ le diviseur, on les multipliera tous les deux par $f(\alpha^2).f(\alpha^3)\ldots f(\alpha^{\lambda-1})$ et l'on aura le diviseur $\mathrm{N}f(\alpha)$, qui sera entier. Pour que $\varphi(\alpha)$ soit divisible par $f(\alpha)$, il faut que le quotient soit égal à un nombre entier complexe $\psi(\alpha)$. L'équation

$$\frac{\varphi(\alpha)}{f(\alpha)} = \psi(\alpha)$$

donne

$$\frac{\varphi(\alpha)f(\alpha^2)f(\alpha^3)\ldots f(\alpha^{\lambda-1})}{\mathrm{N}f(\alpha)} = \psi(\alpha),$$

d'où l'on conclut que, dans le produit

$$\varphi(\alpha).f(\alpha^2).f(\alpha^3)\ldots f(\alpha^{\lambda-1}),$$

développé et réduit à la forme

$$c + c_1\alpha + c_2\alpha^2 + \ldots + c_{\lambda-2}\alpha^{\lambda-2},$$

tous les coefficients c, c_1, $c_2, \ldots$, $c_{\lambda-2}$ doivent être divisibles par $\mathrm{N}f(\alpha)$, et cette condition étant remplie, on aura effectivement $\varphi(\alpha)$ divisible par $f(\alpha)$.

§ II.

Théorie des unités complexes.

Les nombres complexes dont la norme est égale à l'unité sont appelés *unités complexes*. Ces unités, toujours infinies en nombre, excepté le seul cas de $\lambda = 3$, jouent un rôle principal dans toutes les questions sur les nombres complexes, et c'est pour cette raison que la discussion des unités doit être mise à la tête de cette théorie.

D'abord, il est visible que

$$\pm 1, \quad \pm \alpha, \quad \pm \alpha^2, \ldots, \quad \pm \alpha^{\lambda-1}$$

satisfont à la définition des unités; nous les appellerons des *unités simples*. De plus, il est aisé de voir que le nombre complexe

$$1 + \alpha + \alpha^2 + \ldots + \alpha^{r-1} = \frac{1-\alpha^r}{1-\alpha},$$

r désignant un nombre entier quelconque, non divisible par λ, a pour norme la fraction

$$\frac{(1-\alpha^r)(1-\alpha^{2r})(1-\alpha^{3r})\ldots\left[1-\alpha^{(\lambda-1)r}\right]}{(1-\alpha)(1-\alpha^2)(1-\alpha^3)\ldots(1-\alpha^{\lambda-1})}$$

qui se réduit à l'unité. Donc $\frac{1-\alpha^r}{1-\alpha}$ sera aussi une unité complexe, et, puisqu'une puissance entière quelconque et le produit de plusieurs unités sont toujours une unité, il est clair que

$$\pm \alpha^k \left(\frac{1-\alpha^r}{1-\alpha}\right)^m \left(\frac{1-\alpha^s}{1-\alpha}\right)^n \left(\frac{1-\alpha^t}{1-\alpha}\right)^p \ldots$$

sera une unité pour toutes les valeurs entières des exposants m, n, p, etc. On a ainsi, en général, une infinité d'unités différentes. Mais ce qui constitue dans cette théorie la question la plus importante et en même temps la plus délicate, c'est la représentation de *toutes* les unités sous la forme la plus simple.

Bd. XVI (1851)

Nous allons aborder cette question par la démonstration du théorème :

Chaque unité complexe, divisée par sa réciproque, donne pour quotient une unité simple.

C'est-à-dire $E(\alpha)$ étant une unité quelconque, on a toujours

$$\frac{E(\alpha)}{E(\alpha^{-1})} = \pm \alpha^k.$$

D'abord il est visible que ce quotient est un entier complexe. Nous pourrons donc poser

$$\frac{E(\alpha)}{E(\alpha^{-1})} = \varphi(\alpha) = a + a_1 \alpha + a_2 \alpha^2 + \ldots + a_{\lambda-1} \alpha^{\lambda-1},$$

et nous aurons

$$\varphi(\alpha)\varphi(\alpha^{-1}) = 1.$$

En effectuant la multiplication, on trouve

$$\begin{aligned}
\varphi(\alpha)\varphi(\alpha^{-1}) &= A + A_1 \alpha + A_2 \alpha^2 + \ldots + A_{\lambda-1} \alpha^{\lambda-1},\\
A &= a^2 + a_1^2 + a_2^2 + \ldots + a_{\lambda-1}^2,\\
A_1 &= aa_1 + a_1 a_2 + a_2 a_3 + \ldots + a_{\lambda-1} a,\\
A_2 &= aa_2 + a_1 a_3 + a_2 a_4 + \ldots + a_{\lambda-1} a_1,\\
&\ldots\ldots\ldots\ldots\ldots\ldots\ldots\ldots\\
A_{\lambda-1} &= aa_{\lambda-1} + a_1 a + a_2 a_1 + \ldots + a_{\lambda-1} a_{\lambda-2},
\end{aligned}$$

et, en prenant la somme de ces coefficients,

$$A + A_1 + A_2 + \ldots + A_{\lambda-1} = (a + a_1 + a_2 + \ldots + a_{\lambda-1})^2.$$

De l'équation

$$\varphi(\alpha)\varphi(\alpha^{-1}) = A + A_1 \alpha + A_2 \alpha^2 + \ldots + A_{\lambda-1} \alpha^{\lambda-1} = 1$$

il suit aussi

$$A_1 = A_2 = A_3 = \ldots = A_{\lambda-1} \quad \text{et} \quad A = A_1 + 1,$$

et de là

$$1 + \lambda A_1 = (a + a_1 + a_2 + \ldots + a_{\lambda-1})^2$$

ou

$$\pm 1 \equiv a + a_1 + a_2 + \ldots + a_{\lambda-1} \pmod{\lambda};$$

et, puisque le nombre complexe $\varphi(\alpha)$ ne change pas de valeur quand on augmente ou diminue tous ses coefficients d'une même quantité, il est évident qu'on peut prendre

$$a + a_1 + a_2 + \ldots + a_{\lambda-1} = \pm 1,$$

et de là on a

$$A_1 = 0, \quad A_2 = 0, \quad A_3 = 0, \ldots, \quad A_{\lambda-1} = 0,$$
$$A = a^2 + a_1^2 + a_2^2 + \ldots + a_{\lambda-1}^2 = 1.$$

Mais la somme des carrés des nombres entiers $a, a_1, a_2, \ldots, a_{\lambda-1}$ ne pourra jamais être égale à l'unité, à moins qu'un quelconque d'entre eux ne soit égal à l'unité et tous les autres égaux à zéro. On a donc effectivement

$$\frac{E(\alpha)}{E(\alpha^{-1})} = \varphi(\alpha) = \pm \alpha^k.$$

On conclut facilement de ce théorème, que toutes les unités complexes peuvent être décomposées en deux facteurs, dont l'un soit une unité simple α^h, et l'autre une fonction des périodes à deux termes $\alpha + \alpha^{-1}$, $\alpha^2 + \alpha^{-2}$, $\alpha^3 + \alpha^{-3}$, etc., de sorte qu'on a toujours

$$E(\alpha) = \alpha^h[c + c_1(\alpha + \alpha^{-1}) + c_2(\alpha^2 + \alpha^{-2}) + \ldots + c_\mu(\alpha^\mu + \alpha^{-\mu})],$$

où nous avons fait, pour abréger,

$$\frac{\lambda - 1}{2} = \mu.$$

Ainsi, pour le cas de $\lambda = 3$, on a

$$E(\alpha) = \alpha^h[c + c_1(\alpha + \alpha^{-1})],$$

et, en réduisant au moyen de l'équation $1 + \alpha + \alpha^2 = 0$,

$$E(\alpha) = \alpha^h(c - c_1),$$

d'où

$$c - c_1 = \pm 1;$$

donc, dans ce cas, il n'y a pas d'autres unités que

$$\pm 1, \quad \pm \alpha, \quad \pm \alpha^2.$$

Pour le cas de $\lambda = 5$, on a

$$\mathrm{E}(\alpha) = \alpha^h[c + c_1(\alpha + \alpha^{-1}) + c_2(\alpha^2 + \alpha^{-2})],$$

et, en réduisant et mettant pour abréger $c - c_2 = t$, $c_1 - c_2 = u$,

$$\mathrm{E}(\alpha) = \alpha^h[t + (\alpha + \alpha^4)u],$$

et de là

$$\mathrm{E}(\alpha)\mathrm{E}(\alpha^2) = \alpha^{3h}(t^2 - tu - u^2).$$

On en conclut que la forme quadratique $t^2 - tu - u^2$ doit être égale à l'unité, et connaissant la solution générale de l'équation

$$t^2 - tu - u^2 = 1,$$

on en tire toutes les valeurs des unités complexes pour le cas de $\lambda = 5$, lesquelles peuvent être représentées sous la forme simple

$$\mathrm{E}(\alpha) = \pm \alpha^h(\alpha + \alpha^4)^m,$$

m étant un entier quelconque positif ou négatif.

L'exemple donné pour $\lambda = 5$ suffit pour faire voir ce qu'il faut chercher pour une valeur quelconque du nombre premier λ. Mais, en allant plus loin, on est bientôt arrêté par de grandes difficultés qu'on ne pourra guère surmonter qu'à l'aide de principes nouveaux. Ces principes, dont nous ferons usage dans la théorie des unités complexes, sont dus à M. Lejeune-Dirichlet qui les a signalés dans une Note insérée dans les *Comptes rendus de l'Académie de Berlin* du 30 mars de l'année 1846. Nous reproduirons aussi plusieurs des beaux résultats trouvés par M. Kronecker, et publiés en 1845 dans un Mémoire sur les unités complexes.

Prenons un système de $\frac{\lambda - 3}{2}$ unités complexes

$$c_1(\alpha), \quad c_2(\alpha), \quad c_3(\alpha), \ldots, \quad c_{\mu-1}(\alpha),$$

où

$$\mu = \frac{\lambda - 1}{2},$$

qui soient toutes dégagées du facteur $\pm \alpha^h$, en sorte qu'elles ne contiennent que les périodes à deux termes $\alpha + \alpha^{-1}$, $\alpha^2 + \alpha^{-2}$, etc. Ces unités seront évidemment des quantités réelles que nous prendrons toujours comme positives. Cela posé, nous en composons la forme

$$c_1(\alpha)^{m_1} . c_2(\alpha)^{m_2} . c_3(\alpha)^{m_3} \ldots c_{\mu-1}(\alpha)^{m_{\mu-1}},$$

où les exposants m_1, m_2, m_3, etc., sont entiers.

Si cette forme a la propriété que, pour des valeurs différentes des exposants entiers m_1, m_2, m_3, etc., elle ne donne que des unités réellement différentes, ou, ce qui est au fond la même chose, qu'elle ne donne jamais l'unité ordinaire 1 tant que ces exposants ne sont pas tous égaux à zéro, le système des unités

$$c_1(\alpha), \quad c_2(\alpha), \quad c_3(\alpha), \ldots, \quad c_{\mu-1}(\alpha),$$

selon M. Dirichlet, est appelé *un système indépendant*.

La condition du système indépendant que nous venons d'énoncer revient à ce que le déterminant D des quantités logarithmiques

$$\begin{array}{llll}
\mathrm{l}\, c_1(\alpha), & \mathrm{l}\, c_2(\alpha), & \mathrm{l}\, c_3(\alpha), \ldots, & \mathrm{l}\, c_{\mu-1}(\alpha), \\
\mathrm{l}\, c_1(\alpha^{\gamma}), & \mathrm{l}\, c_2(\alpha^{\gamma}), & \mathrm{l}\, c_3(\alpha^{\gamma}), \ldots, & \mathrm{l}\, c_{\mu-1}(\alpha^{\gamma}), \\
\mathrm{l}\, c_1(\alpha^{\gamma^2}), & \mathrm{l}\, c_2(\alpha^{\gamma^2}), & \mathrm{l}\, c_3(\alpha^{\gamma^2}), \ldots, & \mathrm{l}\, c_{\mu-1}(\alpha^{\gamma^2}), \\
\ldots & \ldots & \ldots & \ldots \\
\mathrm{l}\, c_1\left(\alpha^{\gamma^{\mu-2}}\right), & \mathrm{l}\, c_2\left(\alpha^{\gamma^{\mu-2}}\right), & \mathrm{l}\, c_3\left(\alpha^{\gamma^{\mu-2}}\right), \ldots, & \mathrm{l}\, c_{\mu-1}\left(\alpha^{\gamma^{\mu-2}}\right),
\end{array}$$

ne soit pas égal à zéro. (La lettre γ désigne ici, comme dans les formules suivantes, une racine primitive de la congruence $\gamma^{\lambda-1} \equiv 1 \pmod{\lambda}$, et l'on fait toujours $\mu = \frac{\lambda-1}{2}$).

Pour le prouver, posons

$$c_1(\alpha)^{m_1} c_2(\alpha)^{m_2} \ldots c_{\mu-1}(\alpha)^{m_{\mu-1}} = c_1(\alpha)^{n_1} c_2(\alpha)^{n_2} \ldots c_{\mu-1}(\sigma)^{n_{\mu-1}};$$

49..

en divisant par le second membre de cette équation, et faisant

$$m_1 - n_1 = x_1, \quad m_2 - n_2 = x_2, \ldots,$$

on aurait

$$c_1(\alpha)^{x_1} c_2(\alpha)^{x_2} \ldots c_{\mu-1}(\alpha)^{x_{\mu-1}} = 1,$$

et, en changeant la racine α en $\alpha^\gamma, \alpha^{\gamma^2}, \ldots, \alpha^{\gamma^{\mu-1}}$, on aurait de même

$$c_1(\alpha^\gamma)^{x_1} . c_2(\alpha^\gamma)^{x_2} \ldots c_{\mu-1}(\alpha^\gamma)^{x_{\mu-1}} = 1,$$
$$c_1(\alpha^{\gamma^2})^{x_1} . c_2(\alpha^{\gamma^2})^{x_2} \ldots c_{\mu-1}(\alpha^{\gamma^2})^{x_{\mu-1}} = 1,$$
$$\cdots\cdots\cdots\cdots\cdots\cdots$$
$$c_1\left(\alpha^{\gamma^{\mu-1}}\right)^{x_1} . c_2\left(\alpha^{\gamma^{\mu-1}}\right)^{x_2} \ldots c_{\mu-1}\left(\alpha^{\gamma^{\mu-1}}\right)^{x_{\mu-1}} = 1,$$

et, en prenant les logarithmes,

$$x_1 \,\mathrm{l}\, c_1(\alpha) + x_2 \,\mathrm{l}\, c_2(\alpha) + \ldots + x_{\mu-1} \,\mathrm{l}\, c_{\mu-1}(\alpha) = 0,$$
$$x_1 \,\mathrm{l}\, c_1(\alpha^\gamma) + x_2 \,\mathrm{l}\, c_2(\alpha^\gamma) + \ldots + x_{\mu-1} \,\mathrm{l}\, c_{\mu-1}(\alpha^\gamma) = 0,$$
$$\cdots\cdots\cdots\cdots\cdots\cdots$$
$$x_1 \,\mathrm{l}\, c_1\left(\alpha^{\gamma^{\mu-1}}\right) + x_2 \,\mathrm{l}\, c_2\left(\alpha^{\gamma^{\mu-1}}\right) + \ldots + x_{\mu-1} \,\mathrm{l}\, c_{\mu-1}\left(\alpha^{\gamma^{\mu-1}}\right) = 0.$$

En ajoutant ces équations et observant qu'on a

$$c_k(\alpha) . c_k\left(\alpha^\gamma\right) . c_k\left(\alpha^{\gamma^2}\right) \ldots c_k\left(\alpha^{\gamma^{\mu-1}}\right) = 1,$$

et de là

$$\mathrm{l}\, c_k(\alpha) + \mathrm{l}\, c_k\left(\alpha^\gamma\right) + \mathrm{l}\, c_k\left(\alpha^{\gamma^2}\right) + \ldots + \mathrm{l}\, c_k\left(\alpha^{\gamma^{\mu-1}}\right) = 0,$$

on obtient le résultat identique

$$0 = 0.$$

On voit par là qu'une de ces μ équations, par exemple la dernière, peut être rejetée comme déjà contenue dans les autres; on n'aura donc que $\mu - 1$ équations différentes à un même nombre d'inconnues $x_1, x_2, \ldots, x_{\mu-1}$. On sait que, si le déterminant de ce système n'est pas égal à zéro, on n'a que la seule solution

$$x_1 = 0, \quad x_2 = 0, \ldots, \quad x_{\mu-1} = 0,$$

ce qui donne

$$m_1 = n_1, \quad m_2 = n_2, \ldots, \quad m_{\mu-1} = n_{\mu-1};$$

mais on sait aussi que le système a une infinité de solutions différentes si le déterminant est égal à zéro. Donc le système des unités

$$c_1(\alpha), \quad c_2(\alpha), \ldots, \quad c_{\mu-1}(\alpha)$$

sera un système indépendant si le déterminant D n'est pas égal à zéro, et il ne sera jamais un système indépendant si D = o.

C'est un point principal de la théorie générale des unités complexes de prouver qu'il est effectivement des systèmes indépendants de $\mu - 1$ unités. Pour ce but, nous proposons le système de $\mu - 1$ unités conjuguées

$$c(\alpha), \quad c\left(\alpha^{\gamma}\right), \quad c\left(\alpha^{\gamma^2}\right), \ldots, \quad c\left(\alpha^{\gamma^{\mu-2}}\right),$$

dans lequel $c(\alpha)$ signifie l'unité spéciale

$$c(\alpha) = \sqrt{\frac{(1-\alpha^{\gamma})(1-\alpha^{-\gamma})}{(1-\alpha)(1-\alpha^{-1})}} = \pm \frac{\alpha^{\frac{(\lambda-1)(\gamma-1)}{2}}(1-\alpha^{\gamma})}{1-\alpha}.$$

D'après le système exposé ci-dessus, ce système sera indépendant si le déterminant D des quantités

$$\begin{array}{llll}
\mathrm{l}\, c(\alpha), & \mathrm{l}\, c\left(\alpha^{\gamma}\right), & \mathrm{l}\, c\left(\alpha^{\gamma^2}\right), \ldots, & \mathrm{l}\, c\left(\alpha^{\gamma^{\mu-2}}\right), \\
\mathrm{l}\, c\left(\alpha^{\gamma}\right), & \mathrm{l}\, c\left(\alpha^{\gamma^2}\right), & \mathrm{l}\, c\left(\alpha^{\gamma^3}\right), \ldots, & \mathrm{l}\, c\left(\alpha^{\gamma^{\mu-1}}\right), \\
\cdots & \cdots & \cdots & \cdots \\
\mathrm{l}\, c\left(\alpha^{\gamma^{\mu-2}}\right), & \mathrm{l}\, c\left(\alpha^{\gamma^{\mu-1}}\right), & \mathrm{l}\, c\left(\alpha^{\gamma^{\mu}}\right), \ldots, & \mathrm{l}\, c\left(\alpha^{\gamma^{2\mu-4}}\right),
\end{array}$$

n'est pas égal à zéro. Nous tirerons ce déterminant de la résolution effective du système des équations linéaires

$$\begin{aligned}
&x\,\mathrm{l}\, c(\alpha) + x_1\,\mathrm{l}\, c\left(\alpha^{\gamma}\right) + \ldots + x_{\mu-2}\,\mathrm{l}\, c\left(\alpha^{\gamma^{\mu-2}}\right) = \mathrm{A}, \\
&x\,\mathrm{l}\, c\left(\alpha^{\gamma}\right) + x_1\,\mathrm{l}\, c\left(\alpha^{\gamma^2}\right) + \ldots + x_{\mu-2}\,\mathrm{l}\, c\left(\alpha^{\gamma^{\mu-1}}\right) = \mathrm{A}_1, \\
&\cdots\cdots\cdots\cdots\cdots\cdots\cdots\cdots \\
&x\,\mathrm{l}\, c\left(\alpha^{\gamma^{\mu-2}}\right) + x_1\,\mathrm{l}\, c\left(\alpha^{\gamma^{\mu-1}}\right) + \ldots + x_{\mu-2}\,\mathrm{l}\, c\left(\alpha^{\gamma^{2\mu-4}}\right) = \mathrm{A}_{\mu-2}.
\end{aligned}$$

En les ajoutant et faisant

$$A + A_1 + \ldots + A_{\mu-2} = - A_{\mu-1},$$

au moyen de l'équation

$$\mathrm{l}\,c(\alpha) + \mathrm{l}\,c\left(\alpha^{\gamma}\right) + \mathrm{l}\,c\left(\alpha^{\gamma^2}\right) + \ldots + \mathrm{l}\,c\left(\alpha^{\gamma^{\mu-1}}\right) = 0,$$

on en déduit encore l'équation suivante qu'on peut regarder comme complémentaire,

$$x\,\mathrm{l}\,c\left(\alpha^{\gamma^{\mu-1}}\right) + x_1\,\mathrm{l}\,c(\alpha) + \ldots + x_{\mu-2}\,\mathrm{l}\,c\left(\alpha^{\gamma^{\mu-3}}\right) = A_{\mu-1}.$$

En multipliant ces équations par 1, β^{2k}, $\beta^{4k}, \ldots, \beta^{2(\mu-1)k}$ respectivement, β étant une racine primitive de l'équation

$$\beta^{\lambda-1} = 1,$$

et ajoutant, on reconnaît facilement que le premier membre de cette somme se décompose en deux facteurs, et il en résulte

$$\begin{gathered}\left(x + \beta^{-2k} x_1 + \beta^{-4k} x_2 + \ldots + \beta^{-2(\mu-2)k} x_{\mu-2}\right) \\ \times\left[\mathrm{l}\,c(\alpha) + \beta^{2k}\,\mathrm{l}\,c\left(\alpha^{\gamma}\right) + \beta^{4k}\,\mathrm{l}\,c\left(\alpha^{\gamma^2}\right) + \ldots + \beta^{2(\mu-1)k}\,\mathrm{l}\,c\left(\alpha^{\gamma^{\mu-1}}\right)\right] \\ = A + \beta^{2k} A_1 + \beta^{4k} A_2 + \ldots + \beta^{2(\mu-1)k} A_{\mu-1};\end{gathered}$$

de là, en posant, pour abréger,

$$\begin{gathered}\mathrm{l}\,c(\alpha) + \beta^{2k}\,\mathrm{l}\,c\left(\alpha^{\gamma}\right) + \beta^{4k}\,\mathrm{l}\,c\left(\alpha^{\gamma^2}\right) + \ldots \\ + \beta^{2(\mu-1)k}\,\mathrm{l}\,c\left(\alpha^{\gamma^{\mu-1}}\right) = \mathrm{L}\left(\beta^{2k}\right)\end{gathered}$$

et

$$A + \beta^{2k} A_1 + \beta^{4k} A_2 + \ldots + \beta^{2(\mu-1)k} A_{\mu-1} = \psi\left(\beta^{2k}\right),$$

on a

$$x + \beta^{-2k} x_1 + \beta^{-4k} x_2 + \ldots + \beta^{-2(\mu-2)k} x_{\mu-2} = \frac{\psi\left(\beta^{2k}\right)}{\mathrm{L}\left(\beta^{2k}\right)},$$

et de là, en multipliant par $\beta^{2kh} - \beta^{-2k}$, prenant

$$k = 1,\ 2,\ 3, \ldots, \mu - 1,$$

et ajoutant, on trouve la solution complète du système proposé des équations linéaires

$$\mu x_h = \frac{(\beta^{+2h} - \bar{\beta}^{2})\psi(\beta^2)}{\mathrm{L}(\beta^2)} + \frac{(\beta^{4h} - \beta^{-4})\psi(\beta^4)}{\mathrm{L}(\beta^4)} + \ldots + \frac{[\beta^{2(\mu-1)h} - \beta^{-2(\mu-1)}]\psi[\beta^{2(\mu-1)}]}{\mathrm{L}[\beta^{2(\mu-1)}]}.$$

On en conclut que le dénominateur commun de toutes les inconnues x, $x_1, \ldots, x_{\mu-2}$ est donné par le produit

$$\mathrm{L}(\beta^2).\mathrm{L}(\beta^4)\ldots\mathrm{L}(\beta^{2\mu-2}),$$

lequel, dégagé du facteur étranger μ, donne le déterminant cherché

$$\mathrm{D} = \frac{\mathrm{L}(\beta^2).\mathrm{L}(\beta^4).\mathrm{L}(\beta^6)\ldots\mathrm{L}(\beta^{2\mu-2})}{\mu}$$

Ainsi la démonstration de l'indépendance du système proposé des unités conjuguées

$$c(\alpha),\quad c\left(\alpha^{\gamma}\right),\ldots,\quad c\left(\alpha^{\gamma^{\mu-2}}\right)$$

est réduite à prouver que l'expression

$$\mathrm{L}(\beta^{2k}) = \mathrm{l}\,c(\alpha) + \beta^{2k}\,\mathrm{l}\,c\left(\alpha^{\gamma}\right) + \beta^{4k}\,\mathrm{l}\,c\left(\alpha^{\gamma^2}\right) + \ldots + \beta^{2k(\mu-1)}\,\mathrm{l}\,c\left(\alpha^{\gamma^{\mu-1}}\right)$$

n'est égale à zéro pour aucune des valeurs

$$k = 1,\ 2,\ 3,\ldots,\ \mu - 1.$$

Pour cela, nous renvoyons le lecteur au célèbre Mémoire de M. Lejeune-Dirichlet, inséré dans les *Actes de l'Académie de Berlin* de l'année 1837, dans lequel il a démontré le premier que toute série arithmétique dont le premier membre et la différence sont sans facteur commun, contient une infinité de nombres premiers [*]. La méthode ingénieuse de ce grand géomètre repose de même sur cette proposition dont il a donné une démonstration rigoureuse à l'endroit cité. Il est donc prouvé que le système des $\mu - 1$ unités conjuguées

$$c(\alpha),\quad c\left(\alpha^{\gamma}\right),\quad c\left(\alpha^{\gamma^2}\right),\ldots,\quad c\left(\alpha^{\gamma^{\mu-2}}\right)$$

---

[*] Une traduction française de ce Mémoire, par M. Terquem, a paru au tome IV du présent Journal. (J. Liouville.)

est un système indépendant, et, par conséquent, toutes les propriétés générales des systèmes indépendants que nous allons expliquer, conviendront toujours à ce système remarquable.

Revenons au système général des $\mu - 1$ unités

$$c_1(\alpha),\quad c_2(\alpha),\quad c_3(\alpha),\ldots,\quad c_{\mu-1}(\alpha),$$

et supposons toujours qu'il soit indépendant, c'est-à-dire que le déterminant D des logarithmes de ces unités et de leurs conjuguées ne soit pas égal à zéro. Alors nous savons que toutes les unités contenues dans la forme

$$c_1(\alpha)^{m_1}\cdot c_2(\alpha)^{m_2}\cdot c_3(\alpha)^{m_3}\ldots c_{\mu-1}(\alpha)^{m_{\mu-1}}$$

sont différentes entre elles; mais, en général, pour toutes les valeurs entières positives et négatives des nombres m_1, m_2,..., $m_{\mu-1}$, cette forme, multipliée par l'unité simple $\pm\alpha^k$, ne contient pas toutes les unités possibles. Pour remédier à ce défaut, nous ferons abstraction de la restriction que les exposants doivent être des nombres entiers, et nous essayerons d'exprimer toutes les unités, dégagées des unités simples $\pm\alpha^k$, par la forme

$$\mathrm{E}(\alpha) = c_1(\alpha)^{x_1}\cdot c_2(\alpha)^{x_2}\cdot c_3(\alpha)^{x_3}\ldots c_{\mu-1}(\alpha)^{x_{\mu-1}},$$

dans laquelle $x_1, x_2, x_3,\ldots, x_{\mu-1}$ désignent des quantités numériques quelconques. En prenant les logarithmes des deux membres de l'équation proposée et changeant α en α^γ, α^{γ^2},..., $\alpha^{\gamma^{\mu-1}}$, on obtient ce système d'équations linéaires

$$x_1 \,\mathrm{l}\, c_1(\alpha) + x_2 \,\mathrm{l}\, c_2(\alpha) + \ldots + x_{\mu-1} \,\mathrm{l}\, c_{\mu-1}(\alpha) = \mathrm{l}\,\mathrm{E}(\alpha),$$

$$x_1 \,\mathrm{l}\, c_1\left(\alpha^\gamma\right) + x_2 \,\mathrm{l}\, c_2\left(\alpha^\gamma\right) + \ldots + x_{\mu-1} \,\mathrm{l}\, c_{\mu-1}\left(\alpha^\gamma\right) = \mathrm{l}\,\mathrm{E}\left(\alpha^\gamma\right),$$

. . . .

$$x_1 \,\mathrm{l}\, c_1\left(\alpha^{\gamma^{\mu-1}}\right) + x_2 \,\mathrm{l}\, c_2\left(\alpha^{\gamma^{\mu-1}}\right) + \ldots + x_{\mu-1} \,\mathrm{l}\, c_{\mu-1}\left(\alpha^{\gamma^{\mu-1}}\right) = \mathrm{l}\,\mathrm{E}\left(\alpha^{\gamma^{\mu-1}}\right),$$

dont l'une, par exemple la dernière, peut être rejetée parce qu'elle est déjà contenue dans les autres. La résolution de ces équations linéaires, dont le déterminant par l'hypothèse n'est pas égal à zéro, donne toujours des valeurs finies et déterminées des inconnues x_1, $x_2, \ldots, x_{\mu-1}$ qui sont indépendantes de la racine α, parce que ce système reste le même quand on y change α en α^γ, α^{γ^2}, etc. Donc :

Toutes les unités complexes sans exception peuvent être représentées comme produits des puissances de $\mu - 1$ unités indépendantes, jointes aux unités simples $\pm \alpha^h$, en admettant des exposants numériques quelconques de ces puissances.

Si, dans l'expression

$$E(\alpha) = c_1(\alpha)^{x_1} \cdot c_2(\alpha)^{x_2} \cdot c_3(\alpha)^{x_3} \ldots c_{\mu-1}(\alpha)^{x_{\mu-1}},$$

on sépare les plus grands entiers contenus dans les exposants, et qu'on pose

$$x_1 = m_1 + \delta_1, \quad x_2 = m_2 + \delta_2, \ldots, \quad x_{\mu-1} = m_{\mu-1} + \delta_{\mu-1},$$

où $\delta_1, \delta_2, \ldots, \delta_{\mu-1}$ sont renfermés entre les limites 0 et 1, on a

$$E(\alpha) = c_1(\alpha)^{m_1} \cdot c_2(\alpha)^{m_2} \ldots c_{\mu-1}(\alpha)^{m_{\mu-1}} \cdot c_1(\alpha)^{\delta_1} \cdot c_2(\alpha)^{\delta_2} \ldots c_{\mu-1}(\alpha)^{\delta_{\mu-1}}.$$

Le second facteur $c_1(\alpha)^{\delta_1} \cdot c_2(\alpha)^{\delta_2} \ldots c_{\mu-1}(\alpha)^{\delta_{\mu-1}}$, que nous désignons, pour abréger, par $F(\alpha)$, doit être une unité entière complexe aussi bien que $E(\alpha)$. Soit donc

$$F(\alpha) = a(\alpha + \alpha^{-1}) + a_1\left(\alpha^\gamma + \alpha^{-\gamma}\right) + \ldots + a_{\mu-1}\left(\alpha^{\gamma^{\mu-1}} + \alpha^{-\gamma^{\mu-1}}\right),$$

et de là

$$F\left(\alpha^\gamma\right) = a\left(\alpha^\gamma + \alpha^{-\gamma}\right) + a_1\left(\alpha^{\gamma^2} + \alpha^{-\gamma^2}\right) + \ldots + a_{\mu-1}(\alpha + a^{-1}),$$
$$F\left(\alpha^{\gamma^2}\right) = a\left(\alpha^{\gamma^2} + \alpha^{-\gamma^2}\right) + a_1\left(\alpha^{\gamma^3} + \alpha^{-\gamma^3}\right) + \ldots + a_{\mu-1}\left(\alpha^\gamma + \alpha^{-\gamma}\right),$$
. . . .
$$F\left(\alpha^{\gamma^{\mu-1}}\right) = a\left(\alpha^{\gamma^{\mu-1}} + \alpha^{-\gamma^{\mu-1}}\right) + a_1(\alpha + \alpha^{-1}) + \ldots + a_{\mu-1}\left(\alpha^{\gamma^{\mu-2}} + \alpha^{-\gamma^{\mu-2}}\right),$$

En résolvant ce système d'équations par rapport aux coefficients a,

$a_1, a_2, \ldots, a_{\mu-1}$, on trouve facilement

$$\begin{aligned}
-\lambda a &= (2-\alpha-\alpha^{-1})\,\mathrm{F}(\alpha) + \left(2-\alpha^{\gamma}-\alpha^{-\gamma}\right)\mathrm{F}\left(\alpha^{\gamma}\right)+\ldots \\
&\qquad + \left(2-\alpha^{\gamma^{\mu-1}}-\alpha^{-\gamma^{\mu-1}}\right)\mathrm{F}\left(\alpha^{\gamma^{\mu-1}}\right), \\
-\lambda a_1 &= \left(2-\alpha^{\gamma}-\alpha^{-\gamma}\right)\mathrm{F}(\alpha) + \left(2-\alpha^{\gamma^2}-\alpha^{-\gamma^2}\right)\mathrm{F}\left(\alpha^{\gamma}\right)+\ldots \\
&\qquad + (2-\alpha-\alpha^{-1})\,\mathrm{F}\left(\alpha^{\gamma^{\mu-1}}\right), \\
&\ldots\ldots\ldots\ldots\ldots\ldots\ldots\ldots\ldots\ldots \\
-\lambda a_{\mu-1} &= \left(2-\alpha^{\gamma^{\mu-1}}-\alpha^{-\gamma^{\mu-1}}\right)\mathrm{F}(\alpha) + (2-\alpha-\alpha^{-1})\,\mathrm{F}\left(\alpha^{\gamma}\right)+\ldots \\
&\qquad + \left(2-\alpha^{\gamma^{\mu-2}}-\alpha^{-\gamma^{\mu-2}}\right)\mathrm{F}\left(\alpha^{\gamma^{\mu-1}}\right).
\end{aligned}$$

Les quantités $2-\alpha-\alpha^{-1}$, $2-\alpha^{\gamma}-\alpha^{-\gamma}$, etc., sont toutes positives et contenues entre les limites 0 et 4, car on a

$$\alpha+\alpha^{-1} = 2\cos\frac{2k\pi}{\lambda}$$

et

$$2-\alpha-\alpha^{-1} = \left(2\sin\frac{k\pi}{\lambda}\right)^2.$$

En considérant que les exposants $\delta_1, \delta_2, \ldots, \delta_{\mu-1}$ sont renfermés entre les limites 0 et 1, on voit aussi que, pour chaque système donné d'unités indépendantes, les quantités $\mathrm{F}(\alpha)$, $\mathrm{F}\left(\alpha^{\gamma}\right)$, etc., ne peuvent surpasser certaines limites qu'on peut fixer en chaque cas particulier. Il suit de là que les coefficients $a, a_1, a_2, \ldots, a_{\mu-1}$ sont tous négatifs et qu'ils sont contenus entre des limites fixes; et, en considérant encore que ces coefficients sont des nombres entiers, on conclut qu'il n'y a qu'un nombre fini et limité de valeurs convenables de ces coefficients, et, par conséquent, qu'il n'y a qu'un nombre fini d'unités, telles que $\mathrm{F}(\alpha)$.

Le résultat trouvé peut être énoncé comme il suit :

$c_1(\alpha), c_2(\alpha), \ldots, c_{\mu-1}(\alpha)$ étant $\mu-1$ unités indépendantes et m_1, $m_2, m_3, \ldots, m_{\mu-1}$ tous les nombres entiers positifs et négatifs, il y a tou-

jours un nombre fini d'unités qui, multipliées par les unités contenues dans la forme

$$\pm \alpha^h c_1(\alpha)^{m_1} . c_2(\alpha)^{m_2} \ldots c_{\mu-1}(\alpha)^{m_{\mu-1}},$$

produisent toutes les unités possibles.

Revenons à présent à la forme

$$E(\alpha) = c_1(\alpha)^{x_1} . c_2(\alpha)^{x_2} . c_3(\alpha)^{x_3} \ldots c_{\mu-1}(\alpha)^{x_{\mu-1}},$$

et supposons que les exposants $x_1, x_2, \ldots, x_{\mu-1}$ soient déterminés de manière à rendre $E(\alpha)$ une unité entière complexe. Élevons à une puissance entière indéterminée n, et séparons les plus grands entiers de

$$nx_1 = m_1 + \delta_1, \quad nx_2 = m_2 + \delta_2, \ldots, \quad nx_{\mu-1} = m_{\mu-1} + \delta_{\mu-1},$$

de manière que $\delta_1, \delta_2, \ldots, \delta_{\mu-1}$ soient tous entre les limites 0 et 1; nous aurons ainsi

$$E(\alpha)^n = c_1(\alpha)^{m_1} . c_2(\alpha)^{m_2} . c_{\mu-1}(\alpha)^{m_{\mu-1}} . c_1(\alpha)^{\delta_1} . c_2(\alpha)^{\delta_2} \ldots c_{\mu-1}(\alpha)^{\delta_{\mu-1}}.$$

Maintenant, en donnant à l'exposant n les valeurs 1, 2, 3, 4,..., et ainsi de suite, les nombres entiers $m_1, m_2, \ldots, m_{\mu-1}$ et les fractions $\delta_1, \delta_2, \ldots, \delta_{\mu-1}$ changeront de valeur. Mais, parce qu'il est démontré que la forme

$$c_1(\alpha)^{\delta_1} . c_2(\alpha)^{\delta_2} \ldots c_{\mu-1}(\alpha)^{\delta_{\mu-1}},$$

$\delta_1, \delta_2, \ldots, \delta_{\mu-1}$ étant toujours entre 0 et 1, ne peut contenir qu'un nombre fini d'unités différentes, il s'ensuit que ce second facteur se reproduira nécessairement et qu'il restera le même pour une certaine suite de valeurs de l'exposant n. Soient n et n' deux exposants auxquels appartient le même second facteur, et soient $m'_1, m'_2, \ldots, m'_{\mu-1}$ des exposants entiers dans l'expression de $E(\alpha)^{n'}$, on aura

$$E(\alpha)^n = c_1(\alpha)^{m_1} . c_2(\alpha)^{m_2} \ldots c_{\mu-1}(\alpha)^{m_{\mu-1}} . c_1(\alpha)^{\delta_1} . c_2(\alpha)^{\delta_2} \ldots c_{\mu-1}(\alpha)^{\delta_{\mu-1}},$$

$$E(\alpha)^{n'} = c_1(\alpha)^{m'_1} . c_2(\alpha)^{m'_2} \ldots c_{\mu-1}(\alpha)^{m'_{\mu-1}} . c_1(\alpha)^{\delta_1} . c_2(\alpha)^{\delta_2} \ldots c_{\mu-1}(\alpha)^{\delta_{\mu-1}},$$

50..

et, en divisant,

$$\mathrm{E}(\alpha)^{n-n'} = c_1(\alpha)^{m_1-m'_1}\,.\,c_2(\alpha)^{m_2-m'_2}\ldots c_{\mu-1}(\alpha)^{m_{\mu-1}-m'_{\mu-1}}.$$

Nous en concluons le théorème suivant :

Il y a toujours une certaine puissance entière de toute unité complexe telle, que cette puissance de l'unité complexe puisse être exprimée par le produit de puissances entières d'un système donné de $\mu - 1$ *unités indépendantes.*

Le même résultat peut aussi s'énoncer comme il suit :

La forme

$$\pm\,\alpha^h\, c_1(\alpha)^{x_1}\,.\,c_2(\alpha)^{x_2}\,.\,c_3(\alpha)^{x_3}\ldots c_{\mu-1}(\alpha)^{x_{\mu-1}}$$

ne contient des unités entières complexes que pour des valeurs rationnelles des exposants $x_1, x_2, \ldots, x_{\mu-1}$ *dont les dénominateurs ne surpassent pas une certaine limite fixe, mais pour de telles valeurs elle représente toutes les unités possibles.*

Ainsi, par exemple, le système des unités

$$c(\alpha),\quad c\left(\alpha^{\gamma}\right),\quad \left(c^{\gamma^2}\right),\ldots,\quad c\left(\alpha^{\gamma^{\mu-1}}\right),$$

où

$$c(\alpha) = \sqrt{\frac{(1-\alpha^{\gamma})(1-\alpha^{-\gamma})}{(1-\alpha)(1-\alpha^{-1})}},$$

que nous avons démontré être indépendant, suffit pour représenter toutes les unités complexes. Mais, comme cette représentation a l'inconvénient qu'elle exige, en général, des puissances fractionnaires ou des radicaux, et qu'elle ne donne des unités rationnelles et entières que pour des systèmes de valeurs de ces exposants fractionnaires qu'on ne saurait assigner à priori, nous allons en déduire un autre système d'unités indépendantes tel, que le produit de ses puissances *entières* contienne toutes les unités possibles.

Prenons un système de $\mu - 1$ unités indépendantes

$$\varepsilon_1(\alpha),\quad \varepsilon_2(\alpha),\quad \varepsilon_3(\alpha),\ldots,\quad \varepsilon_{\mu-1}(\alpha).$$

D'après le théorème que nous venons de démontrer, certaines puissances entières de ces unités pourront être exprimées comme puissances entières des unités

$$c(\alpha),\quad c\left(\alpha^{\gamma}\right),\quad c\left(\alpha^{\gamma^2}\right),\ldots,\quad c\left(\alpha^{\gamma^{\mu-1}}\right);$$

nous pouvons donc poser

$$\varepsilon_1(\alpha)^{n_1} = c(\alpha)^{r_1^1}.c\left(\alpha^{\gamma}\right)^{r_2^1}\ldots c\left(\alpha^{\gamma^{\mu-2}}\right)^{r_{\mu-1}^1},$$

$$\varepsilon_2(\alpha)^{n_2} = c(\alpha)^{r_1^2}.c\left(\alpha^{\gamma}\right)^{r_2^2}\ldots c\left(\alpha^{\gamma^{\mu-2}}\right)^{r_{\mu-1}^2},$$

. . . .

$$\varepsilon_{\mu-1}(\alpha)^{n_{\mu-1}} = c(\alpha)^{r_1^{\mu-1}}.c\left(\alpha^{\gamma}\right)^{r_2^{\mu-1}}\ldots c\left(\alpha^{\gamma^{\mu-2}}\right)^{r_{\mu-1}^{\mu-1}},$$

les exposants $n_1, n_2,\ldots, n_{\mu-1}$ étant des nombres entiers qui ne surpassent pas des limites finies et déterminées, et les exposants r_1^1, r_2^1, etc., étant également des nombres entiers. En prenant les logarithmes de ces équations, on a

$$n_1 \,\mathrm{l}\,\varepsilon_1(\alpha) = r_1^1 \,\mathrm{l}\, c(\alpha) + r_2^1 \,\mathrm{l}\, c\left(\alpha^{\gamma}\right) + \ldots + r_{\mu-1}^1 \,\mathrm{l}\, c\left(\alpha^{\gamma^{\mu-2}}\right),$$

$$n_2 \,\mathrm{l}\,\varepsilon_2(\alpha) = r_1^2 \,\mathrm{l}\, c(\alpha) + r_2^2 \,\mathrm{l}\, c\left(\alpha^{\gamma}\right) + \ldots + r_{\mu-1}^2 \,\mathrm{l}\, c\left(\alpha^{\gamma^{\mu-2}}\right),$$

. . . .

$$n_{\mu-1} \,\mathrm{l}\,\varepsilon_{\mu-1}(\alpha) = r_1^{\mu-1} \,\mathrm{l}\, c(\alpha) + r_2^{\mu-1} \,\mathrm{l}\, c\left(\alpha^{\gamma}\right) + \ldots + r_{\mu-1}^{\mu-1} \,\mathrm{l}\, c\left(\alpha^{\gamma^{\mu-2}}\right).$$

Désignons par la lettre Δ le déterminant des quantités

$$\begin{array}{lll} \mathrm{l}\,\varepsilon_1(\alpha), & \mathrm{l}\,\varepsilon_2(\alpha),\ldots, & \mathrm{l}\,\varepsilon_{\mu-1}(\alpha), \\ \mathrm{l}\,\varepsilon_1\left(\alpha^{\gamma}\right), & \mathrm{l}\,\varepsilon_2\left(\alpha^{\gamma}\right),\ldots, & \mathrm{l}\,\varepsilon_{\mu-1}\left(\alpha^{\gamma}\right), \\ \ldots & \ldots & \ldots \\ \mathrm{l}\,\varepsilon_1\left(\alpha^{\gamma^{\mu-2}}\right), & \mathrm{l}\,\varepsilon_2\left(\alpha^{\gamma^{\mu-2}}\right),\ldots, & \mathrm{l}\,\varepsilon_{\mu-1}\left(\alpha^{\gamma^{\mu-2}}\right). \end{array}$$

Soit de même D le déterminant des quantités

$$\begin{array}{lll} \mathrm{l}\, c(\alpha), & \mathrm{l}\, c\left(\alpha^{\gamma}\right),\ldots, & \mathrm{l}\, c\left(\alpha^{\gamma^{\mu-2}}\right), \\ \mathrm{l}\, c\left(\alpha^{\gamma}\right), & \mathrm{l}\, c\left(\alpha^{\gamma^2}\right),\ldots, & \mathrm{l}\, c\left(\alpha^{\gamma^{\mu-1}}\right), \\ \ldots & \ldots & \ldots \\ \mathrm{l}\, c\left(\alpha^{\gamma^{\mu-2}}\right), & \mathrm{l}\, c\left(\alpha^{\gamma^{\mu-1}}\right),\ldots, & \mathrm{l}\, c\left(\alpha^{\gamma^{\mu-4}}\right), \end{array}$$

et R le déterminant du système des nombres

$$\begin{array}{lll} r_1^1, & r_2^1,\ldots, & r_{\mu-1}^1, \\ r_1^2, & r_2^2,\ldots, & r_{\mu-1}^2, \\ \ldots & \ldots & \ldots \\ r_1^{\mu-1}, & r_2^{\mu-1},\ldots, & r_{\mu-1}^{\mu-1}. \end{array}$$

En examinant les expressions des quantités $\mathrm{l}\,\varepsilon_1(\alpha)$, $\mathrm{l}\,\varepsilon_2(\alpha)$, etc., et celles qu'on en tire par le changement de la racine α en α^{γ}, α^{γ^2}, etc., on reconnaît facilement que le déterminant Δ se compose des deux déterminants D et R; en effet, on a

$$\Delta = \frac{\mathrm{R}.\mathrm{D}}{n_1.n_2\ldots n_{\mu-1}}.$$

En excluant tous les systèmes de valeurs des entiers r_k^h, qui donneraient

$$\mathrm{R} = 0,$$

et, par conséquent aussi,

$$\Delta = 0,$$

la plus petite valeur de R, qui est un nombre entier, sera

$$\mathrm{R} = 1.$$

De plus, nous savons que le déterminant D a une valeur finie et déterminée différente de zéro, et que les nombres $n_1, n_2, \ldots, n_{\mu-1}$ ne surpassent pas une limite fixe; nous en concluons qu'il y aura toujours une valeur *minimum* finie et déterminée de Δ, c'est-à-dire que, parmi les systèmes en nombres infinis de $\mu - 1$ unités indépendantes, il n'y en aura jamais aucun dont le déterminant Δ soit inférieur à une certaine limite finie.

Nous appelons *système fondamental* tout système indépendant de $\mu - 1$ unités pour lequel le déterminant Δ a cette valeur du *minimum* dont nous venons de prouver l'existence. Supposons aussi que les unités

$$\varepsilon_1(\alpha), \quad \varepsilon_2(\alpha), \ldots, \quad \varepsilon_{\mu-1}(\alpha)$$

représentent un tel système fondamental dont le déterminant Δ ait la valeur la moindre possible. Cela posé, nous allons démontrer que la forme

$$\pm \alpha^h \varepsilon_1(\alpha)^{m_1} . \varepsilon_2(\alpha)^{m_2} . \varepsilon_3(\alpha)^{m_3} \ldots \varepsilon_{\mu-1}(\alpha)^{m_{\mu-1}},$$

pour des valeurs *entières* des exposants m_1, m_2, $m_3, \ldots$, $m_{\mu-1}$, donne toutes les unités sans exception. En effet, imaginons qu'il y ait une unité non comprise dans cette forme, nous savons qu'elle pourrait toujours être représentée par la forme

$$\pm \alpha^h \varepsilon_1(\alpha)^{x_1} . \varepsilon_2(\alpha)^{x_2} . \varepsilon_3(\alpha)^{x_3} \ldots \varepsilon_{\mu-1}(\alpha)^{x_{\mu-1}},$$

pour des valeurs fractionnaires de x_1, $x_2, \ldots$, $x_{\mu-1}$; séparant donc les plus grands entiers contenus dans ces exposants, et faisant comme ci-dessus,

$$x_1 = m_1 + \delta_1, \quad x_2 = m_2 + \delta_2, \ldots, \quad x_{\mu-1} = m_{\mu-1} + \delta_{\mu-1},$$

où m_1, $m_2, \ldots$, $m_{\mu-1}$ sont entiers et δ_1, $\delta_2, \ldots$, $\delta_{\mu-1}$ contenus entre 0 et 1, cette unité complexe prendrait la forme

$$\pm \alpha^h \varepsilon_1(\alpha)^{m_1} . \varepsilon_2(\alpha)^{m_2} \ldots \varepsilon_{\mu-1}(\alpha)^{m_{\mu-1}} . \varepsilon_1(\alpha)^{\delta_1} . \varepsilon_2(\alpha)^{\delta_2} \ldots \varepsilon_{\mu-1}(\alpha)^{\delta_{\mu-1}},$$

et de là il suivrait que

$$\varepsilon_1(\alpha)^{\delta_1} . \varepsilon_2(\alpha)^{\delta_2} \ldots \varepsilon_{\mu-1}(\alpha)^{\delta_{\mu-1}} = \mathrm{E}(\alpha)$$

serait encore une unité entière complexe. Les quantités δ_1, $\delta_2, \ldots$, $\delta_{\mu-1}$ qui sont moindres que l'unité ne pourraient pas toutes être égales à zéro; supposons donc que δ_1 soit différent de zéro et prenons le nouveau système de $\mu - 1$ unités

$$E(\alpha), \quad \varepsilon_2(\alpha), \quad \varepsilon_3(\alpha), \ldots, \quad \varepsilon_{\mu-1}(\alpha).$$

Le déterminant des logarithmes de ce système, que nous désignerons par Δ',

$$\begin{array}{llll} \mathrm{l}\,E(\alpha), & \mathrm{l}\,\varepsilon_2(\alpha), & \mathrm{l}\,\varepsilon_3(\alpha),\ldots, & \mathrm{l}\,\varepsilon_{\mu-1}(\alpha), \\ \mathrm{l}\,E\left(\alpha^{\gamma}\right), & \mathrm{l}\,\varepsilon_2\left(\alpha^{\gamma}\right), & \mathrm{l}\,\varepsilon_3\left(\alpha^{\gamma}\right),\ldots, & \mathrm{l}\,\varepsilon_{\mu-1}\left(\alpha^{\gamma}\right), \\ \ldots & \ldots & \ldots & \ldots \\ \mathrm{l}\,E\left(\alpha^{\gamma^{\mu-2}}\right), & \mathrm{l}\,\varepsilon_2\left(\alpha^{\gamma^{\mu-2}}\right), & \mathrm{l}\,\varepsilon_3\left(\alpha^{\gamma^{\mu-2}}\right),\ldots, & \mathrm{l}\,\varepsilon_{\mu-1}\left(\alpha^{\gamma^{\mu-2}}\right), \end{array}$$

se réduit immédiatement au déterminant Δ; car, au moyen de l'équation

$$\mathrm{l}\,E(\alpha) = \delta_1 \mathrm{l}\,\varepsilon_1(\alpha) + \delta_2 \mathrm{l}\,\varepsilon_2(\alpha) + \ldots + \delta_{\mu-1} \mathrm{l}\,\varepsilon_{\mu-1}(\alpha),$$

et de celles qu'on en déduit en changeant α en $\alpha^{\gamma}, \alpha^{\gamma^2}, \ldots, \alpha^{\gamma^{\mu-1}}$, on trouve

$$\Delta' = \delta_1 \Delta.$$

Mais, δ_1 étant moindre que 1 et différent de zéro, on en conclut qu'il y aurait un système de $\mu - 1$ unités indépendantes à qui appartiendrait un déterminant Δ' moindre que Δ, c'est-à-dire moindre que le plus petit de tous, ce qui serait absurde. Nous avons donc ce théorème important :

Il existe toujours des systèmes de $\mu - 1$ unités fondamentales telles, qu'en les élevant à des puissances entières, multipliant et joignant le facteur $\pm \alpha^h$, on produit toutes les unités possibles, et qu'en prenant des combinaisons différentes des exposants on ne produira que des unités vraiment différentes.

Le calcul effectif de systèmes d'unités fondamentales étant toujours très-pénible, nous ne nous arrêterons pas à expliquer les méthodes propres à ce but; mais nous terminerons cet abrégé des propriétés principales des unités complexes en démontrant qu'il y a toujours une infinité de systèmes différents d'unités fondamentales, et en faisant voir le rapport qui existe entre eux.

Pour cela, nous prenons les $\mu - 1$ unités

$$E_1(\alpha),\quad E_2(\alpha),\quad E_3(\alpha),\ldots,\quad E_{\mu-1}(\alpha),$$

qui s'expriment par les unités fondamentales

$$\varepsilon_1(\alpha),\quad \varepsilon_2(\alpha),\ldots,\quad \varepsilon_{\mu-1}(\alpha),$$

comme il suit :

$$E_1(\alpha) = \varepsilon_1(\alpha)^{r_1^1} . \varepsilon_2(\alpha)^{r_2^1} \ldots \varepsilon_{\mu-1}(\alpha)^{r_{\mu-1}^1},$$
$$E_2(\alpha) = \varepsilon_1(\alpha)^{r_1^2} . \varepsilon_2(\alpha)^{r_2^2} \ldots \varepsilon_{\mu-1}(\alpha)^{r_{\mu-1}^2},$$
$$\ldots\ldots\ldots\ldots\ldots\ldots\ldots\ldots\ldots$$
$$E_{\mu-1}(\alpha) = \varepsilon_1(\alpha)^{r_1^{\mu-1}} . \varepsilon_2(\alpha)^{r_2^{\mu-1}} \ldots \varepsilon_{\mu-1}(\alpha)^{r_{\mu-1}^{\mu-1}},$$

où tous les exposants r_k^h sont entiers. En prenant les logarithmes, on a

$$\mathrm{l}\,E_1(\alpha) = r_1^1\,\mathrm{l}\,\varepsilon_1(\alpha) + r_2^1\,\mathrm{l}\,\varepsilon_2(\alpha) + \ldots + r_{\mu-1}^1\,\mathrm{l}\,\varepsilon_{\mu-1}(\alpha),$$
$$\mathrm{l}\,E_2(\alpha) = r_1^2\,\mathrm{l}\,\varepsilon_1(\alpha) + r_2^2\,\mathrm{l}\,\varepsilon_2(\alpha) + \ldots + r_{\mu-1}^2\,\mathrm{l}\,\varepsilon_{\mu-1}(\alpha),$$
$$\ldots\ldots\ldots\ldots\ldots\ldots\ldots\ldots\ldots$$
$$\mathrm{l}\,E_{\mu-1}(\alpha) = r_1^{\mu-1}\,\mathrm{l}\,\varepsilon_1(\alpha) + r_2^{\mu-1}\,\mathrm{l}\,\varepsilon_2(\alpha) + \ldots + r_{\mu-1}^{\mu-1}\,\mathrm{l}\,\varepsilon_{\mu-1}(\alpha).$$

Donc, en désignant, comme ci-dessus, par R le déterminant des entiers r_1^1, r_2^1, etc., par Δ le déterminant du système des unités fondamentales

$$\varepsilon_1(\alpha),\quad \varepsilon_2(\alpha),\ldots,\quad \varepsilon_{\mu-1}(\alpha),$$

et, de plus, par Δ_1 le déterminant analogue des logarithmes du système $E_1(\alpha)$, $E_2(\alpha)$, etc., on aura, par la même raison que ci-dessus,

$$\Delta_1 = R\Delta.$$

Ainsi, toutes les fois que le déterminant R est égal à l'unité, on aura

$$\Delta_1 = \Delta,$$

et, par conséquent, le système des unités

$$E_1(\alpha),\quad E_2(\alpha),\ldots,\quad E_{\mu-1}(\alpha)$$

sera un système fondamental. Ainsi, d'un seul système fondamental on déduira l'infinité de tous les autres sans exception.

En observant que le déterminant R est un nombre entier, on voit aussi que le quotient qu'on obtient en divisant le déterminant d'un système indépendant quelconque par le déterminant du système fondamental est un nombre entier.

§ III.

Des périodes des racines de l'équation $1 + \alpha + \alpha^2 + \ldots + \alpha^{\lambda - 1} = 0$ *et de leur correspondance avec les racines de congruences analogues.*

Après avoir traité le cas où la norme d'un nombre complexe est égale à l'unité, nous passons à la discussion générale des nombres complexes dont les normes sont des entiers quelconques. Mais pour aborder la question dans toute la généralité qu'elle exige, nous ne nous bornerons pas au cas où les nombres complexes sont composés des simples racines de l'équation

$$1 + \alpha + \alpha^2 + \ldots + \alpha^{\lambda - 1} = 0,$$

mais nous admettrons aussi qu'ils contiennent les périodes de ces racines. Au premier aspect, les nombres complexes composés des périodes paraissent être moins généraux que ceux qui contiennent les simples racines; mais, en considérant que les périodes qui ne consistent qu'en un seul terme sont les simples racines, on voit que la discussion des nombres complexes composés des périodes embrassera tous les nombres complexes, tels que nous les avons proposés au commencement.

Nous commençons par exposer les principes du calcul des périodes dont nous ferons usage dans la suite de ce Mémoire. Soient e et f deux facteurs du nombre $\lambda - 1$, en sorte qu'on ait

$$\lambda - 1 = e.f;$$

soit, comme ci-dessus, γ une racine primitive de la congruence

$$\gamma^{\lambda - 1} \equiv 1 \pmod{\lambda}.$$

Cela posé, les racines imaginaires de l'équation

$$\alpha^\lambda = 1$$

pourront être rangées en e périodes à f termes, comme il suit :

$$\eta = \alpha + \alpha^{\gamma^e} + \alpha^{\gamma^{2e}} + \ldots + \alpha^{\gamma^{(f-1)e}},$$

$$\eta_1 = \alpha^{\gamma} + \alpha^{\gamma^{e+1}} + \alpha^{\gamma^{2e+1}} + \ldots + \alpha^{\gamma^{(f-1)e+1}},$$

$$\eta_2 = \alpha^{\gamma^2} + \alpha^{\gamma^{e+2}} + \alpha^{\gamma^{2e+2}} + \ldots + \alpha^{\gamma^{(f-1)e+2}},$$

. . . .

$$\eta_{e-1} = \alpha^{\gamma^{e-1}} + \alpha^{\gamma^{2e-1}} + \alpha^{\gamma^{3e-1}} + \ldots + \alpha^{\gamma^{fe-1}}$$

Ces périodes forment un cycle tel, qu'en continuant la série

$$\eta, \quad \eta_1, \quad \eta_2, \quad \eta_{e-1},$$

on aura

$$\eta_e = \eta, \quad \eta_{e+1} = \eta_1,$$

et généralement

$$\eta_{ke+h} = \eta_h.$$

Le changement de α en α^{γ^k} ne changera pas l'ordre cyclique des périodes, car au lieu de la suite

$$\eta, \quad \eta_1, \quad \eta_2, \ldots, \quad \eta_{e-1},$$

on aura respectivement

$$\eta_k, \quad \eta_{k+1}, \quad \eta_{k+2}, \ldots, \quad \eta_{k-1}.$$

La somme de toutes les périodes étant la même que la somme de toutes les racines de l'équation

$$1 + \alpha + \alpha^2 + \ldots + \alpha^{\lambda-1} = 0$$

est égale à l'unité négative

$$\eta + \eta_1 + \eta_2 + \ldots + \eta_{e-1} = -1.$$

Le produit de deux périodes est toujours une fonction linéaire de toutes les périodes du même rang. En exécutant la multiplication des

51..

deux périodes η et η_k, on trouve

$$\eta\eta_k = n^k f + m^k\eta + m_1^k\eta_1 + m_2^k\eta_2 + \ldots + m_{e-1}^k\eta_{e-1}.$$

En changeant les périodes η en η_r, η_1 en η_{r+1}, η_2 en η_{r+2}, etc., on a aussi

$$\eta_r\eta_{r+k} = n^k f + m^k\eta_r + m_1^k\eta_{r+1} + m_2^k\eta_{r+2} + \ldots + m_{e-1}^k\eta_{r-1},$$

et, en faisant

$$k = 0,\ 1,\ 2,\ldots,\ e-1,$$

on a le système d'équations

$$(\mathrm{A})\quad\left\{\begin{aligned}
\eta^2 &= nf + m\eta + m_1\eta_1 + m_2\eta_2 + \ldots + m_{e-1}\eta_{e-1},\\
\eta\eta_1 &= n^1 f + m^1\eta + m_1^1\eta_1 + m_2^1\eta_2 + \ldots + m_{e-1}^1\eta_{e-1},\\
\eta\eta_2 &= n^2 f + m^2\eta + m_1^2\eta_1 + m_2^2\eta_2 + \ldots + m_{e-1}^2\eta_{e-1},\\
&\ldots\ldots\ldots\ldots\ldots\ldots\ldots\ldots\ldots\ldots\ldots\\
\eta\eta_{e-1} &= n^{e-1} f + m^{e-1}\eta + m_1^{e-1}\eta_1 + m_2^{e-1}\eta_2 + \ldots + m_{e-1}^{e-1}\eta_{e-1}.
\end{aligned}\right.$$

Dans ces équations fondamentales pour le calcul des périodes, les coefficients n^k sont toujours égaux à zéro, excepté : 1° pour f pair et $k = 0$; 2° pour f impair et $k = \frac{1}{2}e$; on a pour ces deux cas $n^k = 1$. Les autres coefficients m_h^k sont des nombres entiers, tels que m_h^k sera égal au nombre des valeurs positives et moindres que f (zéro y compris) de x et y qui satisfont à la conguence

$$\gamma^{k+xe} + 1 \equiv \gamma^{h+ye} \pmod{\lambda}.$$

En regardant les simples racines α, α^γ, α^{γ^2}, etc., dont chaque période contient un nombre f, on voit que le produit de deux périodes en contiendra f^2, et de là il est aisé de conclure qu'on aura toujours

$$n^k + m^k + m_1^k + m_2^k + \ldots + m_{e-1}^k = f.$$

Si, dans la formule qui donne le produit des deux périodes η_r, η_{r+k}, on fait successivement

$$r = 0,\ 1,\ 2,\ldots,\ e-1.$$

et qu'on ajoute, on trouve

$$\begin{aligned}\eta\eta_k + \eta_1\eta_{k+1} + \eta_2\eta_{k+2} + \ldots + \eta_{e-1}\eta_{k-1} \\ = n^k ef - m^k - m_1^k - m_2^k - \ldots - m_{e-1}^k,\end{aligned}$$

et de là

$$\eta\eta_k + \eta_1\eta_{k+1} + \eta_2\eta_{k+2} + \ldots + \eta_{e-1}\eta_{k-1} = n^k\lambda - f.$$

Donc, d'après la valeur donnée de n^k, cette somme est égale à $-f$, excepté les cas : 1° f pair et $k = 0$; 2° f impair et $k = \frac{1}{2}e$, où cette somme est égale à $\lambda - f$.

En multipliant l'expression donnée du produit de deux périodes η_r, η_{r+k} par η_{r+h}, prenant ensuite

$$r = 0,\ 1,\ 2,\ldots,\ e-1,$$

ajoutant et réduisant à l'aide des formules données ci-dessus, on trouve

$$\begin{aligned}\eta\eta_k\eta_h + \eta_1\eta_{k+1}\eta_{h+1} + \eta_2\eta_{k+2}\eta_{h+2} + \ldots + \eta_{e-1}\eta_{k-1}\eta_{h-1} \\ = -f^2 + \lambda m_h^k,\end{aligned}$$

si f est un nombre pair;

$$\begin{aligned}\eta\eta_k\eta_h + \eta_1\eta_{k+1}\eta_{h+1} + \eta_2\eta_{k+2}\eta_{h+2} + \ldots + \eta_{e-1}\eta_{k-1}\eta_{h-1} \\ = -f^2 + \lambda m_{h+\frac{1}{2}e}^k,\end{aligned}$$

si f est un nombre impair.

Parce que, dans ces formules, les lettres k et h peuvent être échangées entre elles, on en tire cette propriété remarquable des coefficients du système des équations (A),

$$m_h^k = m_k^h,$$

si f est pair;

$$m_{h+\frac{1}{2}e}^k = m_{k+\frac{1}{2}e}^h,$$

si f est impair.

Une autre relation du même genre provient de l'expression du produit $\eta_r\eta_{r+k}$, en y changeant k en $e-k$ et r en $r+k$, et comparant avec l'expression primitive, on trouve

$$m_h^k = m_{h-k}^{e-k}.$$

Dans le système des équations (A) on peut considérer comme inconnues les e périodes, et ces équations suffiront toujours pour les trouver complétement. En effet, l'élimination des périodes

$$\eta_1, \quad \eta_2, \quad \eta_3, \ldots, \quad \eta_{e-1}$$

donnera une équation du degré e à coefficients entiers pour déterminer la première période η, et puisque, évidemment, toute période peut être regardée comme première, cette équation aura nécessairement comme racines toutes les e périodes

$$\eta, \quad \eta_1, \quad \eta_2, \ldots, \quad \eta_{e-1}.$$

On aura ainsi l'équation

$$y^e - A_1 y^{e-1} + A_2 y^{e-2} - A_3 y^{e-3} + \ldots \pm A_e = 0,$$

dont les racines sont toutes les e périodes et dont les coefficients A_1, $A_2, \ldots, A_e$ sont des nombres entiers. De plus, si dans le système (A) on prend la première période η comme connue et les autres comme inconnues, toutes ces équations ne sont que linéaires par rapport aux inconnues $\eta_1, \eta_2, \ldots, \eta_{e-1}$; et de là, en résolvant ces équations, après avoir rejeté une quelconque d'entre elles comme snperflue, on trouvera les périodes $\eta_1, \eta_2, \ldots, \eta_{e-1}$ exprimées rationnellement par la première η. Le résultat de la résolution des équations (A) pourra toujours être réduit à la forme

$$D\eta_k = B + B_1\eta + B_2\eta^2 + B_3\eta^3 + \ldots + B_{e-1}\eta^{e-1},$$

où $B, B_1, B_2, \ldots, B_{e-1}$ et D sont entiers.

Nous ajoutons encore le théorème important, que toute fonction rationnelle et entière des périodes peut être représentée comme fonction *linéaire* de ces périodes. En effet, au moyen des équations (A) on a le produit de deux périodes quelconques exprimé comme fonction linéaire de toutes les périodes; il suit de là qu'en répétant cette réduction, on parviendra à réduire à la forme linéaire les produits de plusieurs périodes, et, par conséquent aussi, toute fonction entière et rationnelle des périodes. De plus, il est aisé de démontrer qu'une telle fonction ne peut être ramenée à la forme

$$a\eta + a_1\eta_1 + a_2\eta_2 + \ldots + a_{e-1}\eta_{e-1}$$

que d'une seule manière; car, si l'on avait aussi

$$b\eta + b_1\eta_1 + b_2\eta_2 + \ldots + b_{e-1}\eta_{e-1}$$

égal à la même fonction rationnelle, il s'ensuivrait

$$a\eta + a_1\eta_1 + \ldots + a_{e-1}\eta_{e-1} = b\eta + b_1\eta_1 + \ldots + b_{e-1}\eta_{e-1},$$

et de là

$$(a - b)\eta + (a_1 - b_1)\eta_1 + \ldots + (a_{e-1} - b_{e-1})\eta_{e-1} = 0,$$

en exprimant les périodes par les racines de l'équation

$$1 + \alpha + \alpha^2 + \ldots + \alpha^{\lambda-1} = 0,$$

et divisant par α, on aurait une équation à coefficients entiers du degré $\lambda-2$ dont la racine serait α, ce qu'on sait être impossible.

Une propriété très-importante de toutes les équations rationnelles entre les périodes

$$\eta,\quad \eta_1,\quad \eta_2,\ldots,\quad \eta_{e-1},$$

c'est de donner toujours des solutions réelles lorsqu'on les envisage comme congruences pour une certaine classe de modules, en sorte qu'à chaque période corresponde un certain nombre entier. Cette correspondance intime entre les périodes comme racines des équations proposées et les racines des congruences analogues nous servira de base pour la recherche des facteurs, et surtout des facteurs premiers des nombres complexes; c'est pour cette raison que nous en donnerons ici les développements nécessaires.

D'abord nous expliquerons une amplification de la définition des congruences dont nous ferons usage dans la suite de ce Mémoire, qui consiste en ce que nous y admettons aussi les périodes

$$\eta,\quad \eta_1,\quad \eta_2,\ldots,\quad \eta_{e-1}.$$

Le sens de telles congruences est fixé comme il suit:

Deux fonctions rationnelles et entières des périodes à coefficients entiers sont censées congrues par rapport à un module entier donné si, dans leur différence réduite à la forme linéaire

$$a\eta + a_1\eta_1 + \ldots + a_{e-1}\eta_{e-1},$$

tous les coefficients a, a_1, $a_2,\ldots,$ a_{e-1} *sont divisibles par le module.*

Une congruence contenant les périodes à f termes équivaut donc toujours à e congruences pour des nombres entiers, qui sont complétement déterminées parce qu'une fonction rationnelle et entière des périodes n'est réductible à cette forme linéaire que d'une seule manière. Selon cette définition, on pourra opérer sur les congruences qui contiennent les périodes de la même manière que sur les congruences ordinaires.

Cela posé, nous partons de la proposition connue que, dans le produit développé de q facteurs,

$$z(z-1)(z-2)\ldots(z-q+1) = z^q - b_1 z^{q-1} + b_2 z^{q-2} - \ldots + b_{q-1} z,$$

où q est un nombre premier, tous les coefficients $b_1, b_2, b_3, \ldots, b_{q-2}$ sont divisibles par q, excepté le dernier b_{q-1} qui donne le reste -1. En posant donc

$$z = y - \eta_k,$$

ou y est un nombre entier, et négligeant les termes divisibles par le module q, on aura la congruence

$$(y-\eta_k)(y-1-\eta_k)(y-2-\eta_k)\ldots(y-q+1-\eta_k)$$
$$\equiv (y-\eta_k)^q - (y-\eta_k) \pmod{q}.$$

Observons aussi que, dans le développement de la puissance du binôme $(y-\eta_k)^q$, q étant un nombre premier, tous les termes sont divisibles par q, excepté le premier et le dernier, d'où

$$(y-\eta_k)^q \equiv y^q - \eta_k^q \pmod{q}.$$

De même, en élevant le polynôme

$$\eta_k = \alpha^{\gamma^k} + \alpha^{\gamma^{k+e}} + \alpha^{\gamma^{k+2e}} + \ldots + \alpha^{\gamma^{k+(f-1)e}}$$

a la puissance q, et rejetant tous les termes divisibles par q, on aura

$$\eta_k^q \equiv \alpha^{\gamma^{\,q^k}} + \alpha^{\gamma q^{k+e}} + \alpha^{\gamma q^{k+2e}} + \ldots + \alpha^{\gamma q^{k+(f-1)e}} \pmod{q};$$

et de là, en posant $q \equiv \gamma^r \pmod{\lambda}$,

$$\eta_k^q \equiv \eta_{k+r} \pmod{q}.$$

Enfin on a, en vertu du théorème de Fermat,

$$y^q \equiv y \pmod{q},$$

d'où

$$(y-\eta_k)(y-1-\eta_k)(y-2-\eta_k)\ldots(y-q+1-\eta_k)$$
$$\equiv \eta_k - \eta_{k+r} \pmod{q}.$$

Supposons maintenant que r soit un multiple de e, en sorte qu'on ait

$$q \equiv \gamma^{re} \pmod{\lambda},$$

ou, ce qui revient au même,

$$q^f \equiv 1 \pmod{\lambda};$$

alors, le second membre de la congruence se réduisant à zéro, on a

$$(y-\eta_k)(y-1-\eta_k)(y-2-\eta_k)\ldots(y-q+1-\eta_k) \equiv 0 \pmod{q}.$$

En multipliant les e congruences contenues dans celle-ci pour les valeurs de

$$k = 0,\ 1,\ 2,\ldots,\ e-1,$$

et en désignant, pour abréger, par $\varphi(y)$ le produit

$$(y-\eta)(y-\eta_1)(y-\eta_2)\ldots(y-\eta_{e-1}),$$

on aura cette congruence pour le module q^e,

$$\varphi(y)\varphi(y-1)\varphi(y-2)\ldots\varphi(y-q+1) \equiv 0 \pmod{q^e}.$$

On en conclut aisément que la congruence

$$\varphi(y) \equiv 0 \pmod{q}$$

aura toujours e racines réelles; en observant encore que le produit désigné par $\varphi(y)$ est exactement le premier membre de l'équation

$$y^e - A_1 y^{e-1} + A_2 y^{e-2} - \ldots \pm A_e = 0,$$

dont les racines sont les e périodes, on aura ce résultat :

L'équation du degré e, dont les racines sont les périodes à f termes, prise pour congruence par rapport à un module q, nombre premier, qui satisfait à la condition

$$q^f \equiv 1 \pmod{\lambda},$$

a toujours e racines réelles.

Cela étant, nous établissons ce système de congruences parfaitement analogues au système des équations (A) :

$$(\mathrm{B})\left\{\begin{array}{l} uu \equiv nf + mu + m_1 u_1 + \ldots + m_{e-1} u_{e-1} \\ uu_1 \equiv n^1 f + m^1 u + m_1^1 u_1 + \ldots + m_{e-1}^1 u_{e-1} \\ uu_2 \equiv n^2 f + m^2 u + m_1^2 u_1 + \ldots + m_{e-1}^2 u_{e-1} \\ \cdots\cdots\cdots\cdots\cdots\cdots\cdots\cdots \\ uu_{e-1} \equiv n^{e-1} f + m^{e-1} u + m_1^{e-1} u + \ldots + m_{e-1}^{e-1} u_{e-1} \end{array}\right\} \pmod{q},$$

où le module q soit un nombre premier, tel qu'on ait

$$q^f \equiv 1 \pmod{\lambda}.$$

* L'élimination des $e - 1$ des inconnus

$$u, \quad u_1, \quad u_2, \ldots, \quad u_{e-1}$$

donne nécessairement une congruence du degré e qui sera parfaitement conforme à l'équation dont les e racines sont les périodes, savoir,

$$y^e - A_1 y^{e-1} + A_2 y^{e-2} - \ldots \pm A_e \equiv 0 \pmod{q},$$

et, puisque cette congruence a toutes ses racines réelles, on en tirera les valeurs réelles des inconnus

$$u, \quad u_1, \quad u_2, \ldots, \quad u_{e-1}.$$

Ainsi, le système des congruences (B) est toujours résoluble par des nombres entiers réels. Observons aussi que, pour le premier des inconnus u, on peut choisir une quelconque des racines de cette congruence; mais celle-ci une fois déterminée, on ne pourra plus choisir arbitrairement la seconde, la troisième et les autres, puisque l'ordre de ces racines dépend des congruences (B).

La conformité des équations (A) avec les congruences (B) nous conduit à la conclusion, que toutes les réductions et transformations des fonctions entières des périodes qu'on peut effectuer au moyen des équations (A), seront absolument les mêmes pour les nombres entiers qu'on obtient en remplaçant les périodes

$$\eta, \quad \eta_1, \quad \eta_2, \ldots, \quad \eta_{e-1}$$

par les nombres

$$u, \quad u_1, \quad u_2, \ldots, \quad u_{e-1},$$

et, puisque toutes les équations rationnelles et entières entre les périodes se réduisent à des identités simples, il en sera toujours de même des congruences analogues. Nous avons donc le théorème suivant :

Toute équation qui ne contient que des fonctions rationnelles et entières des périodes donne immédiatement une congruence quand on y remplace les périodes par les nombres entiers correspondants qui satisfont aux congruences (B).

Par exemple, des équations données ci-dessus on tire les congruences

$$u + u_1 + u_2 + \ldots + u_{e-1} \equiv -1 \quad (\text{mod. } q),$$

et

$$uu_k + u_1 u_{k+1} + u_2 u_{k+2} + \ldots + u_{e-1} u_{k-1} \equiv -f, \quad (\text{mod. } q),$$

excepté les cas où f est pair et $k = 0$, ou bien pour f impair et $k = \frac{1}{2}e$, où cette somme est congrue à $\lambda - f$.

§ IV.

Des facteurs premiers de la norme d'un nombre complexe quelconque.

Une fonction rationnelle et entière des périodes à coefficients entiers, c'est-à-dire un nombre complexe contenant les périodes, peut être considéré comme fonction d'une seule période, par exemple de la première η, et, pour cette raison, nous la désignerons simplement par $F(\eta)$. De même, le nombre entier qui en résulte si l'on remplace les périodes

$$\eta, \quad \eta_1, \quad \eta_2, \ldots, \quad \eta_{e-1}$$

par les racines de congruences analogues

$$u, \quad u_1, \quad u_2, \ldots, \quad u_{e-1},$$

52..

sera désigné par $F(u)$, ou, si l'on prend les périodes

$$\eta, \quad \eta_1, \quad \eta_2, \ldots, \quad \eta_{e-1}$$

respectivement correspondantes aux racines des congruences

$$u_r, \quad u_{r+1}, \quad u_{r+2}, \ldots, \quad u_{r-1},$$

le nombre entier correspondant à $F(\eta)$ sera désigné par $F(u_r)$.

Pour qu'un nombre complexe $F(\eta)$, contenant les périodes seules, soit divisible par un entier non complexe q, il est nécessaire et il suffit que, dans la forme linéaire de ce nombre

$$F(\eta) = a\eta + a_1\eta_1 + a_2\eta_2 + \ldots + a_{e-1}\eta_{e-1},$$

tous les coefficients $a, a_1, a_2, \ldots, a_{e-1}$ aient le facteur commun q. Mais cette méthode de trouver les facteurs entiers non complexes des nombres complexes n'est pas bien applicable au cas d'un nombre complexe donné comme produit de plusieurs facteurs dont la multiplication effective serait très-pénible Pour ce but, nous ferons usage de ce théorème :

Si une fonction entière rationnelle des périodes à coefficients entiers est divisible par le facteur premier non complexe q, qui satisfait à la congruence

$$q^f \equiv 1 \pmod{\lambda},$$

tous les nombres entiers qu'on déduit de ce nombre complexe en remplaçant les périodes par des racines de congruences analogues, seront divisibles par q. Réciproquement, si tous ces e nombres entiers sont divisibles par q, la fonction des périodes dont ils dérivent le sera aussi.

Au moyen des signes adoptés, ce théorème s'exprime de la manière suivante :

Si l'on a

$$F(\eta) \equiv 0 \pmod{q},$$

où q satisfait à la congruence

$$q^f \equiv 1 \pmod{\lambda},$$

il s'ensuit

$$F(u) \equiv 0, \quad F(u_1) \equiv 0, \quad F(u_2) \equiv 0, \ldots, \quad F(u_{e-1}) \equiv 0 \pmod{q};$$

et, réciproquement, si

$$F(u) \equiv 0, \quad F(u_1) \equiv 0, \ldots, \quad F(u_{e-1}) \equiv 0 \pmod{q},$$

il s'ensuit

$$F(\eta) \equiv 0 \pmod{q}.$$

En effet, supposons $F(\eta)$ réduit à la forme linéaire

$$F(\eta) = a\eta + a_1\eta_1 + a_2\eta_2 + \ldots + a_{e-1}\eta_{e-1}$$

en changeant le terme initial des périodes, nous aurons en même temps

$$F(\eta_1) = a\eta_1 + a_1\eta_2 + a_2\eta_3 + \ldots + a_{e-1}\eta,$$
$$F(\eta_2) = a\eta_2 + a_1\eta_3 + a_2\eta_4 + \ldots + a_{e-1}\eta_1,$$
$$\cdots\cdots\cdots\cdots\cdots\cdots\cdots\cdots$$
$$F(\eta_{e-1}) = a\eta_{e-1} + a_1\eta + a_2\eta_1 + \ldots + a_{e-1}\eta_{e-2}.$$

La résolution de ces e équations, linéaires par rapport aux coefficients, donne sans difficulté

$$\lambda a_k = (\eta_k - f)F(\eta) + (\eta_{k+1} - f)F(\eta_1) + \ldots + (\eta_{k-1} - f)F(\eta_{e-1}),$$

si f est pair, et

$$\lambda a_{k+\frac{e}{2}} = (\eta_k - f)F(\eta) + (\eta_{k+1} - f)F(\eta_1) + \ldots + (\eta_{k-1} - f)F(\eta_{e-1}),$$

si f est impair.

Maintenant, si l'on remplace les périodes par les racines des congruences, on a

$$F(u_r) \equiv a u_r + a_1 u_{r+1} + a_2 u_{r+2} + \ldots + a_{e-1} u_{r-1} \pmod{q},$$

et puisque, $F(\eta)$ étant divisible par q, il en sera de même des coefficients $a, a_1, a_2, \ldots, a_{e-1}$, on en conclut que le nombre $F(u_r)$ sera divisible par q pour toutes les valeurs

$$r = 0, 1, 2, \ldots, e-1.$$

De même, on a

$$\lambda a_k \equiv (u_k - f)F(u) + (u_{k+1} - f)F(u_1) + \ldots$$
$$+ (u_{k-1} - f)F(u_{e-1}) \pmod{\lambda},$$

*

si f est pair, et la même expression est congrue à $\lambda a_{k+\frac{e}{2}}$ si f est impair; donc, si l'on a

$$F(u)\equiv 0,\quad F(u_1)\equiv 0,\ldots,\quad F(u_{e-1})\equiv 0 \quad (\text{mod. } q),$$

on aura aussi

$$a_k\equiv 0 \quad (\text{mod } q),$$

pour toutes les valeurs de

$$k=0,\ 1,\ 2,\ldots,\ e-1,$$

et, par conséquent,

$$F(\eta)\equiv 0 \quad (\text{mod. } q),$$

ce qu'il fallait démontrer.

La norme d'un nombre complexe $F(\eta)$ qui ne contient pas les simples racines α, α^{γ}, α^{γ^2}, etc., mais seulement les périodes à f termes de ces racines, prise par rapport aux périodes, sera le produit des e facteurs

$$F(\eta),\quad F(\eta_1),\quad F(\eta_2),\ldots,\quad F(\eta_{e-1}),$$

qui est toujours un nombre entier. La norme du même nombre $F(\eta)$, prise par rapport aux diverses valeurs de la racine α, contenue dans les périodes, sera composée de $\lambda-1=e.f$ facteurs, dont f à f seront égaux; elle sera donc la $f^{\text{ième}}$ puissance de la norme, prise par rapport aux périodes. Nous ne craignons point d'embarras en désignant la norme, prise par rapport aux périodes, par la lettre N, de même que la norme, prise par rapport aux racines de l'équation

$$1+\alpha+\alpha^2+\ldots+\alpha^{\lambda-1}=0.$$

Donc, en mettant

$$\text{N}F(\eta)=F(\eta).F(\eta_1).F(\eta_2)\ldots F(\eta_{e-1}),$$

et en substituant aux périodes les racines des congruences, nous avons

$$\text{N}F(\eta)\equiv F(u).F(u_1).F(u_2)\ldots F(u_{e-1}) \quad (\text{mod. } q).$$

Cette congruence donne immédiatement le théorème suivant :

Si la norme d'un nombre complexe $F(\eta)$, *composé des périodes seules, est divisible par q, un des nombres*

$$F(u),\quad F(u_1),\quad F(u_2),\ldots,\quad F(u_{e-1})$$

sera de même divisible par q; et réciproquement, si l'un de ces nombres est divisible par q, q sera nécessairement facteur de la norme.

Nous allons maintenant discuter les conditions nécessaires pour que la norme d'un nombre complexe quelconque, composé des racines de l'équation

$$1+\alpha+\alpha^2+\ldots+\alpha^{\lambda-1}=0,$$

ait un facteur premier donné. Quant au nombre λ, nous avons trouvé cette condition dans le § Ier de ce Mémoire, savoir, que la somme des coefficients d'un nombre complexe doit être divisible par λ pour que la norme ait le facteur λ, et réciproquement. Tous les autres nombres premiers peuvent être rangés selon les exposants auxquels ils appartiennent pour le module λ (*voyez* M. Gauss, *Disquisitiones arithmeticæ,* § LII), et cette classification convient parfaitement aux divers diviseurs de la norme dont les caractères sont intimement liés aux plus petits exposants de ses puissances qui sont congrues à l'unité pour le module λ. Nous supposons donc, comme ci-dessus, que q soit un nombre premier tel que

$$q^f\equiv 1 \pmod{\lambda};$$

mais, désormais, nous ajoutons la condition que q appartienne à l'exposant f, en sorte qu'aucune puissance de q inférieure à la $f^{\text{ième}}$ ne soit congrue à l'unité. La puissance d'une racine primitive γ, congrue à un nombre q qui appartient à l'exposant f, sera

$$q\equiv\gamma^{re} \pmod{\lambda},$$

où r n'a aucun facteur commun avec f; car, si l'on prend

$$q\equiv\gamma^k,$$

la condition

$$q^f\equiv 1$$

donne

$$\gamma^{kf}\equiv 1;$$

il suit de là

$$kf\equiv 0 \pmod{\lambda-1},$$

et puisque

$$\lambda - 1 = ef,$$

on a

$$k \equiv 0 \pmod{e},$$

ou

$$k = re$$

et

$$q \equiv \gamma^{re} \pmod{\lambda}.$$

On voit aussi qu'à cause de la condition que q *appartienne* à l'exposant f, le nombre r doit être premier à f; car, si r et f avaient un facteur commun n, on pourrait poser

$$r = nr' \quad \text{et} \quad f = nf',$$

on aurait ainsi

$$q^{f'} \equiv \gamma^{nr'ef'} \equiv \gamma^{r'(\lambda-1)} \equiv 1 \pmod{\lambda},$$

et la puissance $q^{f'}$, inférieure à q^f, serait congrue à l'unité, ce qui est contre l'hypothèse.

Cela posé, nous élevons le nombre complexe

$$f(\alpha) = a + a_1\alpha + a_2\alpha^2 + a_3\alpha^3 + \ldots + a_{\lambda-2}\alpha^{\lambda-2}$$

à la puissance q. En rejetant tous les termes divisibles par q, nous aurons

$$f(\alpha)^q \equiv a^q + a_1^q\alpha^q + a_2^q\alpha^{2q} + \ldots + a_{\lambda-2}^q\alpha^{(\lambda-2)q} \pmod{q}.$$

Donc, en observant que

$$a_k^q \equiv a_k \pmod{q},$$

on a la congruence

$$f(\alpha)^q \equiv f(\alpha^q) \pmod{q},$$

et, en répétant la même opération, on a plus généralement

$$f(\alpha)^{q^h} \equiv f(\alpha^{q^h}) \pmod{q}.$$

Mettant à présent

$$h = 0,\ 1,\ 2, \ldots,\ f-1,$$

et multipliant toutes ces congruences, on a

$$f(\alpha)^{1+q+q^2+\ldots+q^{f-1}} \equiv f(\alpha)\,f(\alpha^q)\,f(\alpha^{q^2})\ldots f(\alpha^{q^{f-1}}) \pmod{q},$$

et, en prenant

$$q \equiv \gamma^{re} \pmod{\lambda},$$

et changeant α en α^{γ^m}, cette congruence devient

$$f\left(\alpha^{\gamma^m}\right)^{1+q+q^2+\ldots+q^{f-1}} \equiv f\left(\alpha^{\gamma^m}\right).f\left(\alpha^{\gamma^{m+re}}\right).f\left(\alpha^{\gamma^{m+2re}}\right)\ldots f\left(\alpha^{\gamma^{m+(f-1)re}}\right).$$

Prenons maintenant, pour m, e valeurs incongrues par rapport au module e, et multiplions ces e congruences; nous aurons

$$\left[\Pi f\left(\alpha^{\gamma^m}\right)\right]^{1+q+q^2+\ldots+q^{f-1}} \equiv \mathrm{N} f(\alpha) \pmod{q},$$

où le signe du produit Π s'étend à toutes les valeurs du nombre m qui sont incongrues par rapport au module e. D'où, en supposant $\mathrm{N} f(\alpha)$ divisible par q,

$$\left[\Pi f\left(\alpha^{\gamma^m}\right)\right]^{1+q+q^2+\ldots+q^{f-1}} \equiv 0 \pmod{q}.$$

En élevant à la puissance $q-1$, multipliant par $\Pi f\left(\alpha^{\gamma^m}\right)$ et observant que, pour tout nombre complexe, l'on a

$$\varphi(\alpha)^{q^f} \equiv \varphi(\alpha) \pmod{q},$$

on obtient enfin

$$\Pi f\left(\alpha^{\gamma^m}\right) \equiv 0 \pmod{q},$$

et de là le théorème suivant :

Si la norme d'un nombre complexe $f(\alpha)$ est divisible par q (nombre premier qui appartient à l'exposant f), il faut que le produit de e à e des $\lambda - 1$ nombres conjugués représentés par $f\left(\alpha^{\gamma^m}\right)$, pour lesquels m n'ait que des valeurs incongrues par rapport au module e, soit divisible par q.

De ce théorème il suit aussi, comme corollaire :

Si la norme d'un nombre complexe $f(\alpha)$ est divisible par un nombre premier q qui appartient à l'exposant f, il faut qu'elle contienne ce facteur f fois, de manière qu'elle soit toujours divisible par q^f.

Cherchons maintenant les conditions qui doivent avoir lieu entre les coefficients du nombre complexe $f(\alpha)$ pour que sa norme soit divisible par q. On sait que toutes les f racines de l'équation

$$1 + \alpha + \alpha^2 + \ldots + \alpha^{\lambda-1} = 0,$$

qui sont contenues dans une seule période à f termes, sont en même temps les racines d'une équation du degré f de la forme

$$\alpha^f + P_1 \alpha^{f-1} + P_2 \alpha^{f-2} + \ldots + P_f = 0,$$

dont les coefficients P_1, P_2, etc., sont des fonctions entières et rationnelles des périodes

$$\eta, \quad \eta_1, \quad \eta_2, \ldots, \quad \eta_{e-1}.$$

Au moyen de cette équation, on pourra éliminer toutes les puissances de α supérieures à α^{f-1} de l'expression du nombre complexe

$$f(\alpha) = a + a_1 \alpha + a_2 \alpha^2 + a_3 \alpha^3 + \ldots + a_{\lambda-2} \alpha^{\lambda-2};$$

donc on aura cette forme

$$f(\alpha) = \varphi(\eta) + \alpha \varphi_1(\eta) + \alpha^2 \varphi_2(\eta) + \ldots + \alpha^{f-1} \varphi_{f-1}(\eta),$$

où

$$\varphi(\eta), \quad \varphi_1(\eta), \quad \varphi_2(\eta), \ldots, \quad \varphi_{f-1}(\eta)$$

désignent des nombres entiers complexes contenant les périodes seules. On voit aussi qu'un nombre complexe donné n'aura jamais deux représentations différentes par cette forme.

Cela posé, nous considérons le produit des e facteurs :

$$\left[c f(\alpha) + c_1 f\left(\alpha^{\gamma^e}\right) + c_2 f\left(\alpha^{\gamma^{2e}}\right) + \ldots + c_{f-1} f\left(\alpha^{\gamma^{(f-1)e}}\right)\right]$$

$$\times\left[c f\left(\alpha^{\gamma}\right) + c_1 f\left(\alpha^{\gamma^{e+1}}\right) + c_2 f\left(\alpha^{\gamma^{2e+1}}\right) + \ldots + c_{f-1} f\left(\alpha^{\gamma^{(f-1)e+1}}\right)\right]$$

. . . .

$$\times\left[c f\left(\alpha^{\gamma^{e-1}}\right) + c_1 f\left(\alpha^{\gamma^{2e-1}}\right) + c_2 f\left(\alpha^{\gamma^{3e-1}}\right) + \ldots + c_{f-1} f\left(\alpha^{\gamma^{fe-1}}\right)\right].$$

En effectuant la multiplication on voit que tous les termes contiendront un de ces produits $\Pi f\left(\alpha^{\gamma^m}\right)$ où m a e valeurs incongrues par rapport au module e; donc, en vertu du théorème que nous venons de démontrer, le produit proposé sera divisible par q en même temps que $\mathrm{N}f(\alpha)$, quelles que soient les quantités

$$c, \quad c_1, \quad c_2, \ldots, \quad c_{f-1}.$$

Prenons pour $f(\alpha)$ la forme

$$f(\alpha) = \varphi(\eta) + \alpha\varphi_1(\eta) + \alpha^2\varphi_2(\eta) + \ldots + \alpha^{f-1}\varphi_{f-1}(\eta),$$

et pour $f\left(\alpha^\gamma\right)$, $f\left(\alpha^{\gamma^2}\right)$, etc., les expressions qu'on déduit de celle-ci en changeant α en α^γ, α^{γ^2}, etc. Posons de plus

$$c + c_1 + c_2 + \ldots + c_{f-1} = \mathrm{C},$$

$$\alpha c + \alpha^{\gamma^e} c_1 + \alpha^{\gamma^{2e}} c_2 + \ldots + \alpha^{\gamma^{(f-1)e}} c_{f-1} = \mathrm{C}_1,$$

$$\alpha^2 c + \alpha^{2\gamma^e} c_2 + \alpha^{2\gamma^{2e}} c_2 + \ldots + \alpha^{2\gamma^{(f-1)e}} c_{f-1} = \mathrm{C}_2,$$

$$\ldots\ldots\ldots\ldots\ldots\ldots\ldots\ldots\ldots\ldots\ldots$$

$$\alpha^{f-1} c + \alpha^{(f-1)\gamma^e} c_1 + \alpha^{(f-1)\gamma^{2e}} c_2 + \ldots + \alpha^{(f-1)\gamma^{(f-1)e}} c_{f-1} = \mathrm{C}_{f-1},$$

nous aurons cette transformation du produit proposé,

$$\begin{aligned}
&[\mathrm{C}\varphi(\eta) + \mathrm{C}_1\varphi_1(\eta) + \mathrm{C}_2\varphi_2(\eta) + \ldots + \mathrm{C}_{f-1}\varphi_{f-1}(\eta)] \\
\times &[\mathrm{C}\varphi(\eta_1) + \mathrm{C}_1\varphi_1(\eta_1) + \mathrm{C}_2\varphi_2(\eta_1) + \ldots + \mathrm{C}_{f-1}\varphi_{f-1}(\eta_1)] \\
&\ldots\ldots\ldots\ldots\ldots\ldots\ldots\ldots\ldots\ldots\ldots \\
\times &[\mathrm{C}\varphi(\eta_{e-1}) + \mathrm{C}_1\varphi_1(\eta_{e-1}) + \mathrm{C}_2\varphi_2(\eta_{e-1}) + \ldots + \mathrm{C}_{f-1}\varphi_{f-1}(\eta_{e-1})],
\end{aligned}$$

qui sera divisible par q en même temps que $\mathrm{N}f(\alpha)$ pour toutes les valeurs des quantités

$$\mathrm{C}, \quad \mathrm{C}_1, \quad \mathrm{C}_2, \ldots, \quad \mathrm{C}_{f-1},$$

qui sont tout à fait arbitraires aussi bien que

$$c, \quad c_1, \quad c_2, \ldots, \quad c_{f-1}.$$

Ce produit étant une norme prise par rapport aux périodes, pour qu'il soit divisible par q il faut et il suffit qu'on ait

$$\mathrm{C}\varphi(u_r) + \mathrm{C}_1\varphi_1(u_r) + \mathrm{C}_2\varphi_2(u_r) + \ldots + \mathrm{C}_{f-1}\varphi_{f-1}(u_r) \equiv 0 \pmod{q}$$

53..

pour quelqu'une des valeurs de

$$r = 0,\ 1,\ 2, \ldots,\ e-1,$$

et parce que les coefficients

$$C,\quad C_1,\quad C_2, \ldots,\quad C_{f-1}$$

sont arbitraires, il faut qu'on ait séparément les f congruences

$$\varphi(u_r) \equiv 0,\quad \varphi_1(u_r) \equiv 0, \ldots,\quad \varphi_{f-1}(u_r) \equiv 0 \pmod{q}.$$

On a donc ce théorème:

Si la norme d'un nombre complexe $f(\alpha)$ est divisible par un nombre premier q, qui appartient à l'exposant f pour le module λ, il faut que les f congruences

$$\varphi(u_r) \equiv 0,\quad \varphi_1(u_r) \equiv 0,\quad \varphi_2(u_r) \equiv 0, \ldots,\quad \varphi_{f-1}(u_r) \equiv 0 \pmod{q},$$

qu'on obtient en mettant le nombre $f(\alpha)$ sous la forme

$$f(\alpha) = \varphi(\eta) + \alpha\varphi_1(\eta) + \alpha^2\varphi_2(\eta) + \ldots + \alpha^{f-1}\varphi_{f-1}(\eta),$$

et remplaçant les e périodes à f termes par les racines de congruences correspondantes, aient lieu pour une certaine valeur de r. Réciproquement, si ces f congruences ont lieu, la norme de $f(\alpha)$ sera divisible par q.

On peut aussi déterminer la condition pour que la norme d'un nombre complexe $f(\alpha)$ soit divisible par q, en faisant

$$f(\alpha) \cdot f\left(\alpha^{\gamma^e}\right) \cdot f\left(\alpha^{\gamma^{2e}}\right) \ldots f\left(\alpha^{\gamma^{(f-1)e}}\right) = F(\eta).$$

Ce produit étant une fonction symétrique de toutes les racines contenues dans l'une des périodes ne sera qu'une fonction des périodes, à cause de quoi nous l'avons désigné par $F(\eta)$. Il suit de là

$$Nf(\alpha) = F(\eta) \cdot F(\eta_1) \cdot F(\eta_2) \ldots F(\eta_{e-1}),$$

et, par conséquent, la condition nécessaire et suffisante pour que $Nf(\alpha)$ soit divisible par q revient simplement à ce que $F(u_r)$ soit divisible par q pour une certaine valeur de r. Ainsi, la même condition pour laquelle nous avons trouvé les f congruences

$$\varphi(u_r)\equiv 0,\quad \varphi_1(u_r)\equiv 0,\quad \text{etc.},$$

linéaires par rapport aux coefficients du nombre complexe $f(\alpha)$, est exprimée par la seule congruence du degré f,

$$F(u_r)\equiv 0 \pmod{q}.$$

Si pour un nombre complexe quelconque

$$f(\alpha)=\varphi(\eta)+\alpha\varphi_1(\eta)+\alpha^2\varphi_2(\eta)+\ldots+\alpha^{f-1}\varphi_{f-1}(\eta),$$

les f congruences

$$\varphi(u_r)\equiv 0,\quad \varphi_1(u_r)\equiv 0,\ldots,\quad \varphi_{f-1}(u_r)\equiv 0 \pmod{q}$$

ont lieu, nous dirons simplement que ce nombre complexe $f(\alpha)$ est congru à zéro, par rapport au module q, pour

$$\eta=u_r,\quad \eta_1=u_{r+1},\quad \eta_2=u_{r+2},\ldots,\quad \eta_{e-1}=u_{r-1},$$

où il suffira aussi d'écrire seulement la condition

$$\eta=u_r$$

qui entraîne les autres.

Cela posé, il est évident qu'un des facteurs d'un produit de deux ou de plusieurs nombres complexes étant congru à zéro pour $\eta=u_r$, il en sera de même du produit développé; mais le théorème réciproque exige une démonstration particulière. Soit donc

$$f(\alpha).\varphi(\alpha)=\chi(\alpha),$$

et, par hypothèse,

$$\chi(\alpha)\equiv 0 \pmod{q}\quad \text{pour}\quad \eta=u_r;$$

nous prouverons que l'un des deux facteurs $f(\alpha)$ ou $\varphi(\alpha)$ doit être aussi congru à zéro pour $\eta=u_r$. En effet, en posant

$$f(\alpha).f\left(\alpha^{\gamma^e}\right).f\left(\alpha^{\gamma^{2e}}\right)\ldots f\left(\alpha^{\gamma^{(f-1)e}}\right)=F(\eta),$$

$$\varphi(\alpha).\varphi\left(\alpha^{\gamma^e}\right).\varphi\left(\alpha^{\gamma^{2e}}\right)\ldots \varphi\left(\alpha^{\gamma^{(f-1)e}}\right)=\Phi(\eta),$$

$$\chi(\alpha).\chi\left(\alpha^{\gamma^e}\right).\chi\left(\alpha^{\gamma^{2e}}\right)\ldots \chi\left(\alpha^{\gamma^{(f-1)e}}\right)=X(\eta),$$

on aura

$$F(\eta)\,\Phi(\eta) = X(\eta),$$

et, puisque par l'hypothèse $\chi(\alpha)$ est congru à zéro pour $\eta = u_r$, on aura aussi

$$X(u_r) \equiv 0 \pmod{q}.$$

Il suit de là que l'un des nombres $F(u_r)$ ou $\Phi(u_r)$ doit de même être congru à zéro, et, en remplaçant la congruence

$$F(u_r) \equiv 0 \pmod{q}$$

par les f congruences équivalentes, qu'on est convenu d'exprimer par

$$f(\alpha) \equiv 0 \pmod{q} \quad \text{pour} \quad \eta = u_r,$$

on en conclut ce théorème :

La condition

$$f(\alpha) \equiv 0 \pmod{q}$$

pour $\eta = u_r$ est toujours la même pour un nombre complexe, soit qu'il consiste de facteurs, soit qu'il ait la forme développée.

De là découle aussi cet autre théorème, qu'on peut regarder comme une généralisation du premier théorème de ce paragraphe, parce qu'il donne, pour les nombres complexes quelconques $f(\alpha)$, la même condition que cet autre pour les nombres complexes des périodes :

Pour qu'un nombre complexe quelconque, représenté comme produit de plusieurs facteurs, soit divisible par q, il faut et il suffit que, pour toutes les substitutions

$$\eta = u_r, \quad \eta = u_1, \quad \eta = u_2, \ldots, \quad \eta = u_{e-1},$$

un de ses facteurs soit congru à zéro pour le module q.

Nous terminerons cette discussion des facteurs premiers de la norme d'un nombre complexe quelconque, en montrant encore une autre manière très-simple et très-utile d'exprimer les f congruences contenues dans l'énoncé

$$f(\alpha) \equiv 0 \pmod{q} \quad \text{pour} \quad \eta = u_r.$$

Pour cela, nous avons besoin d'un nombre complexe contenant les périodes seules dont la norme, prise par rapport aux périodes, soit divisible par q sans être divisible par q^2. De tels nombres existent toujours, et à l'ordinaire ils s'offrent d'eux-mêmes; par exemple, parmi les simples nombres

$$u - \eta, \quad u_1 - \eta, \quad u_2 - \eta, \ldots, \quad u_{e-1} - \eta,$$

dont les normes sont toutes divisibles par q, il y en aura presque toujours plusieurs qui satisferont à la condition requise; mais dans certains cas il se pourrait que les normes de tous ces nombres fussent aussi divisibles par q^2. Alors on fera usage du nombre complexe

$$\psi(\eta) = \lambda - f - u\eta - u_1\eta_1 - u_2\eta_2 - \ldots - u_{e-1}\eta_{e-1}.$$

En remplaçant les périodes par les racines des congruences, on a toujours

$$\psi(u_r) \equiv \lambda \pmod{q},$$

à l'exception des deux cas: 1° de f pair et $r = 0$, où l'on a

$$\psi(u) \equiv 0;$$

2° de f impair et $r = \frac{1}{2}e$, où l'on a

$$\psi\left(u_{\frac{e}{2}}\right) \equiv 0.$$

De là, en appliquant le premier et le second théorème de ce paragraphe, on conclut que $N\psi(\eta)$ est toujours divisible par q, mais que le produit

$$\Psi(\eta) = \psi(\eta_1) \cdot \psi(\eta_2) \cdot \psi(\eta_3) \ldots \psi(\eta_{e-1})$$

n'est pas divisible par q. Observons aussi que le produit $\Psi(\eta_r)\Psi(\eta_s)$, tant que r et s ne sont pas égaux, contient toujours tous les facteurs de la norme $N\psi(\eta)$, et que, par conséquent, il est divisible par q. Cela étant, développons la norme de $\psi(\eta) + q$, en rejetant tous les termes divisibles par q^2, ce qui donne

$$N[\psi(\eta) + q]$$
$$\equiv N\psi(\eta) + q[\Psi(\eta) + \Psi(\eta_1) + \ldots + \Psi(\eta_{e-1})] \pmod{q^2};$$

multipliant encore par $\Psi(\eta)$ et rejetant les termes divisibles par q^2, nous aurons

$$\Psi(\eta)\,\mathrm{N}[\psi(\eta)+q] \equiv \Psi(\eta)\,\mathrm{N}\psi(\eta) + q[\Psi(\eta)]^2 \quad (\mathrm{mod.}\ q^2).$$

Maintenant, si $\mathrm{N}\psi(\eta)$ n'est pas divisible par q^2, mais seulement par q, $\psi(\eta)$ est un nombre complexe tel qu'on le désire. Mais s'il arrivait que $\mathrm{N}\psi(\eta)$ contînt le facteur q^2, on aurait

$$\Psi(\eta)\,\mathrm{N}[\psi(\eta)+q] \equiv q[\Psi(\eta)]^2 \quad (\mathrm{mod.}\ q^2):$$

on voit que la norme du nombre complexe $\psi(\eta)+q$ ne pourrait être divisible par q^2. Ainsi, l'un des deux nombres complexes $\psi(\eta)$ ou $\psi(\eta)+q$ satisfait toujours à la condition exigée.

Soit donc $\psi(\eta)$ un nombre complexe, tel qu'on ait

$$\mathrm{N}\psi(\eta) \equiv 0 \quad (\mathrm{mod.}\ q),$$

sans que

$$\mathrm{N}\psi(\eta) \equiv 0 \quad (\mathrm{mod.}\ q^2);$$

soit aussi

$$\psi(u) \equiv 0 \quad (\mathrm{mod.}\ q),$$

et posons, pour abréger,

$$\psi(\eta_1).\psi(\eta_2).\psi(\eta_3)\ldots\psi(\eta_{e-1}) = \Psi(\eta),$$

je dis que *les f congruences comprises dans l'énoncé*

$$f(\alpha) \equiv 0 \quad (\mathrm{mod.}\ q)$$

pour $\eta = u_r$ sont équivalentes à la congruence

$$f(\alpha)\Psi(\eta_{e-r}) \equiv 0 \quad (\mathrm{mod.}\ q).$$

En effet, d'après les principes établis dans le § III, on conclut de l'expression de $\Psi(\eta)$,

$$\Psi(u_r) \equiv \psi(u_{r+1}).\psi(u_{r+2})\ldots\psi(u_{r-1}) \quad (\mathrm{mod.}\ q);$$

et puisqu'on a

$$\psi(u) \equiv 0 \quad (\mathrm{mod.}\ q),$$

on voit qu'on aura toujours

$$\Psi(u_r) \equiv 0,$$

excepté le seul cas $r=0$. Or nous savons que, pour avoir

$$f(\alpha)\Psi(\eta_{e-r}) \equiv 0 \quad (\mathrm{mod.}\ q),$$

il faut et il suffit qu'on ait pour chacune des suppositions

$$\eta = u, \quad \eta = u_1, \quad \eta = u_2, \ldots, \quad \eta = u_{e-1},$$

ou

$$f(\alpha) \equiv 0 \quad (\text{mod. } q),$$

ou

$$\Psi(\eta_{e-r}) \equiv 0 \quad (\text{mod. } q);$$

et puisque, pour $\eta = u_r$, le second facteur, qui devient $\Psi(u)$, n'est pas congru à zéro, il s'ensuit nécessairement

$$f(\alpha) \equiv 0$$

pour $\eta = u_r$. Réciproquement, si l'on a

$$f(\alpha) \equiv 0 \quad (\text{mod. } q)$$

pour $\eta = u_r$, le produit $f(\alpha)\,\Psi(\eta_{e-r})$ est divisible par q, parce que pour $\eta = u_r$ le premier facteur $f(\alpha)$, et pour toutes les autres substitutions

$$\eta = u, \quad \eta = u_1, \ldots, \quad \eta = u_{r-1}, \quad \eta = u_{r+1}, \ldots, \quad \eta = n_{e-1},$$

le second facteur $\Psi(\eta_{e-r})$ est congru à zéro.

§ V.

Définition et propriétés générales des facteurs idéaux d'un nombre complexe.

Au moyen des résultats trouvés dans le paragraphe précédent, nous serons en état d'assigner tous les nombres complexes dont les normes ont un facteur premier donné; voyons maintenant si, parmi tous ces nombres complexes, il y en a un ou plusieurs dont les normes soient *égales* à ce nombre premier même. D'abord nous savons que quand la norme d'un nombre complexe $f(\alpha)$, prise par rapport aux racines α, est divisible par le nombre premier q qui appartient à l'exposant f, elle ne peut pas être égale à q, à moins qu'on n'ait

$$f = 1,$$

c'est-à-dire à moins que q, appartenant à l'exposant 1, n'ait la forme linéaire

$$q = m\lambda + 1.$$

Cela tient à ce que toute norme est de la forme linéaire $m\lambda + 1$ ou à ce que toute norme divisible par q doit avoir le facteur q^f. Mais parmi les nombres complexes contenant les périodes à f termes, il pourra y en avoir dont les normes prises par rapport aux périodes soient égales au nombre premier q. Donc, il se pourra qu'un nombre premier q, appartenant à l'exposant f, soit décomposable en e facteurs conjugués contenant les périodes à f termes, en sorte qu'on ait

$$\varphi(\eta).\varphi(\eta_1).\varphi(\eta_2)\ldots\varphi(\eta_{e-1}) = \mathrm{N}\varphi(\eta) = q,$$

et en particulier, pour le cas de $f = 1$, il se pourra qu'un nombre premier p de la forme $m\lambda + 1$ soit égal à la norme d'un nombre complexe $f(\alpha)$, en sorte qu'on ait

$$f(\alpha).f(\alpha^2).f(\alpha^3)\ldots f(\alpha^{\lambda-1}) = \mathrm{N}f(\alpha) = p.$$

Mais il s'en faut de beaucoup que chaque nombre premier appartenant à l'exposant f soit décomposable en e facteurs premiers complexes. On distinguera plutôt deux classes de tous les nombres premiers, les uns qui peuvent être représentés comme normes, ainsi que nous venons de l'expliquer, les autres qui n'admettent pas une telle décomposition en facteurs conjugués.

En supposant toujours q premier et appartenant à l'exposant f, sans exclure le cas de $f = 1$, nous aurons pour un nombre q de la première classe,

$$\varphi(\eta).\varphi(\eta_1).\varphi(\eta_2)\ldots\varphi(\eta_{e-1}) = \mathrm{N}\varphi(\eta) = q.$$

Le nombre complexe $\varphi(\eta)$, dont la norme est un nombre premier, n'admet aucune décomposition ultérieure en deux facteurs, sans que l'un de ces facteurs soit une unité complexe. En effet, si l'on avait

$$\varphi(\eta) = \mathrm{F}(\alpha).\mathrm{G}(\alpha),$$

en prenant la norme par rapport aux racines α, on aurait

$$\mathrm{N}\varphi(\eta) = q^f = \mathrm{NF}(\alpha).\mathrm{NG}(\alpha),$$

et puisque l'une de ces deux normes, par exemple $NF(\alpha)$ divisible par q, doit être divisible par q^f, on aura nécessairement

$$NF(\alpha) = q^f, \quad \text{d'où} \quad NG(\alpha) = 1,$$

ou bien $G(\alpha)$ sera une unité complexe. Pour cette raison, tout nombre complexe $\varphi(\eta)$, dont la norme est un nombre premier, sera appelé *nombre premier complexe*. Ainsi, tout nombre premier non complexe q de notre première classe se compose de e facteurs *premiers complexes*.

Pour distinguer les facteurs premiers du nombre q, nous ferons usage des nombres

$$u, \quad u_1, \quad u_2, \ldots, \quad u_{e-1}$$

correspondants aux périodes. En les substituant, on aura pour une certaine valeur de r,

$$\varphi(u_r) \equiv 0 \pmod{q},$$

condition nécessaire pour que la norme de $\varphi(\eta)$ soit divisible par q; on voit par là que, pour tous les divers facteurs premiers de q, on a

$$\begin{aligned} \varphi(\eta) &\equiv 0 \quad \text{pour} \quad \eta = u_r, \\ \varphi(\eta_1) &\equiv 0 \quad \text{pour} \quad \eta = u_{r-1}, \\ \varphi(\eta_2) &\equiv 0 \quad \text{pour} \quad \eta = u_{r-2}, \end{aligned}$$

et généralement

$$\varphi(\eta_k) \equiv 0 \quad \text{pour} \quad \eta = u_{r-k}.$$

Nous distinguerons donc les e facteurs premiers conjugués du nombre q selon les racines de congruences qui doivent être choisies comme correspondantes aux périodes pour que chacun de ces facteurs premiers complexes soit congru à zéro pour le module q.

Lorsqu'on multiplie un nombre premier complexe $\varphi(\eta)$, pour lequel on a

$$\varphi(\eta) \equiv 0 \pmod{q} \quad \text{pour} \quad \eta = u_r,$$

par un nombre complexe quelconque $f(\alpha)$, le produit développé

$$\varphi(\eta) f(\alpha) = F(\alpha)$$

satisfera toujours à

$$F(\alpha) \equiv 0 \quad \text{pour} \quad \eta = u_r.$$

54..

Réciproquement, si un nombre complexe $\mathcal{F}(\alpha)$ satisfait à

$$F(\alpha) \equiv 0 \pmod{q} \quad \text{pour} \quad \eta = u_r,$$

je dis qu'il contiendra nécessairement le facteur premier $\varphi(\eta)$. En effet, le nombre $\varphi(\eta_r)$, dont la norme est égale à q et qui est congru à zéro pour $\eta = u$, est un de ces nombres complexes que nous avons désignés par $\psi(\eta)$ dans le paragraphe précédent, à l'aide desquels la condition

$$F(\alpha) \equiv 0 \pmod{q} \quad \text{pour} \quad \eta = u_r$$

s'exprime par la simple congruence

$$F(\alpha)\Psi(\eta_{e-r}) \equiv 0 \pmod{q}.$$

Prenant donc

$$\varphi(\eta_r) = \psi(\eta)$$

et

$$\Psi(\eta) = \varphi(\eta_{r+1}).\varphi(\eta_{r+2})\ldots\varphi(\eta_{r-1}),$$

et, par conséquent,

$$\Psi(\eta_{e-r}) = \varphi(\eta_1).\varphi(\eta_2).\varphi(\eta_3)\ldots\varphi(\eta_{e-1}),$$

de l'hypothèse

$$F(\alpha) \equiv 0 \pmod{q} \quad \text{pour} \quad \eta = u_r,$$

on conclut

$$F(\alpha).\varphi(\eta_1).\varphi(\eta_2)\ldots\varphi(\eta_{e-1}) \equiv 0 \pmod{q};$$

on pourra donc poser

$$F(\alpha).\varphi(\eta_1).\varphi(\eta_2)\ldots\varphi(\eta_{e-1}) = q f(\alpha);$$

enfin, en multipliant par $\varphi(\eta)$ et divisant par

$$\varphi(\eta).\varphi(\eta_1).\varphi(\eta_2)\ldots\varphi(\eta_{e-1}) = q,$$

on aura

$$F(\alpha) = \varphi(\eta) f(\alpha).$$

Donc, $F(\alpha)$ a effectivement le facteur $\varphi(\eta)$. Nous voyons par là que la condition

$$F(\alpha) \equiv 0 \pmod{q} \quad \text{pour} \quad \eta = u_r$$

définit complétement un facteur premier complexe $\varphi(\eta)$, contenu dans le nombre $F(\alpha)$ toutes les fois que le nombre premier $\varphi(\eta)$ existe réellement, c'est-à-dire que q est décomposable en e facteurs premiers conjugués.

Supposons à présent que q soit un nombre premier qui n'admet pas de décomposition en e facteurs conjugués, ou qui ne saurait être représenté comme norme d'un nombre complexe contenant les périodes à f termes. Alors on ne pourra plus isoler un facteur premier d'un nombre complexe $f(\alpha)$ satisfaisant à la condition

$$f(\alpha) \equiv 0 \pmod{q} \quad \text{pour} \quad \eta = u_r,$$

mais nous dirons néanmoins que ce nombre complexe contient un facteur premier complexe du nombre q, que nous appellerons *facteur premier idéal* du nombre premier q; et puisque nous avons vu que les e facteurs premiers du nombre q, s'ils existent effectivement, sont distingués en ce que chacun d'entre eux devient congru à zéro pour une certaine substitution des racines de congruences au lieu des périodes, nous désignerons ce facteur premier idéal de q comme *appartenant à la substitution* $\eta = u_r$. Un tel facteur idéal, qui ne subsiste pas par soi-même, mais qui a une existence effective dans la combinaison avec les autres facteurs du nombre complexe, dans lequel il est contenu, sera donc défini complétement comme il suit :

Si un nombre complexe $f(\alpha)$ satisfait à la condition

$$f(\alpha) \equiv 0 \pmod{q} \quad \textit{pour} \quad \eta = u_r,$$

nous dirons qu'il contient le facteur premier idéal du nombre premier q qui appartient à la substitution $\eta = u_r$.

Quoiqu'il n'y ait rien d'obscur dans la définition, nous allons expliquer rapidement pourquoi nous avons désigné cette condition de congruence de $f(\alpha)$ par le nom de *facteur idéal* de ce nombre complexe, tout en observant que les développements seuls de la théorie pourront justifier une telle innovation. A cause de l'analogie frappante qui règne entre les facteurs premiers idéaux et les facteurs complexes ordinaires, la dénomination servira à simplifier extrêmement les énoncés des théorèmes. Mais cela ne constitue pas l'avantage principal de la théorie nouvelle; nous croyons surtout que les facteurs idéaux rendent visible, pour ainsi dire, la constitution intérieure des nombres, en sorte que leurs propriétés essentielles soient mises dans leur jour. Un nombre complexe satisfaisant à plusieurs des conditions que nous regardons comme facteurs idéaux de ce nombre, quoiqu'il ne soit pas décomposable en facteurs complexes, se comporte tout à fait comme un nombre composé, et c'est pour cela que nous le considérons en quelque sorte comme un produit de facteurs. L'Algèbre, l'Arithmétique et la Géométrie offrent des analogies nombreuses à notre théorie. On décompose, par exemple, les fonctions rationnelles et entières d'une seule variable en facteurs linéaires, quoique ces facteurs isolés n'existent qu'en des cas particuliers; c'est pour ce but qu'on a créé les quantités imaginaires. En Géométrie, on parle d'une droite passant par les points d'intersection de deux cercles, quand même les points d'intersection n'existent pas. Dans cet exemple, la propriété permanente que les tangentes menées d'un point quelconque de cette ligne aux deux cercles, sont égales entre elles, est l'analogie de la propriété permanente du nombre complexe

$$f(\alpha) \equiv 0 \pmod{q} \quad \text{pour} \quad \eta = u_r,$$

et la propriété accidentelle de cette ligne, de passer par les points d'intersection des deux cercles, est de même analogue à la propriété accidentelle du nombre complexe $f(\alpha)$, d'avoir un facteur premier existant $\varphi(\eta)$. Enfin, l'idée de considérer des facteurs idéaux des nombres complexes est, au fond, la même que celle qui a procréé les nombres complexes eux-mêmes. En effet, on sait que M. Gauss, en observant que, dans la recherche des lois de réciprocité entre les résidus biquadratiques, les nombres premiers de la forme $4n+1$ se comportaient comme nombres composés, les a décomposés en facteurs imaginaires de la forme $a+b\sqrt{-1}$, et qu'il a jeté par là les fondements de la théorie générale des nombres complexes.

Nous avons fixé le sens d'un facteur premier idéal d'un nombre complexe $f(\alpha)$ par la condition

$$f(\alpha) \equiv 0 \pmod{q} \quad \text{pour} \quad \eta = u_r,$$

qui en elle-même n'est qu'une manière concise d'exprimer que les f

congruences

$$\varphi(u_r) \equiv 0, \quad \varphi_1(u_r) \equiv 0, \quad \varphi_2(u_r) \equiv 0, \ldots, \quad \varphi_{f-1}(u_r) \equiv 0 \pmod{q}$$

ont lieu en même temps, congruences qu'on obtient en mettant le nombre $f(\alpha)$ sous la forme

$$f(\alpha) = \varphi(\eta) + \alpha\varphi_1(\eta) + \alpha^2\varphi_2(\eta) + \ldots + \alpha^{f-1}\varphi_{f-1}(\eta),$$

et remplaçant les périodes par les racines des congruences correspondantes. Nous savons aussi que ces f congruences sont comprises dans cette seule,

$$f(\alpha)\Psi(\eta_{e-r}) \equiv 0 \pmod{q},$$

où $\Psi(\eta)$ désigne le produit de $e-1$ facteurs

$$\Psi(\eta) = \psi(\eta_1).\psi(\eta_2)\ldots\psi(\eta_{e-1}),$$

$\psi(\eta)$ étant un nombre complexe qui donne

$$\psi(u) \equiv 0 \pmod{q},$$

et dont la norme, divisible par q, ne soit pas divisible par q^2. Nous pouvons donc énoncer la définition des facteurs premiers idéaux comme il suit :

Si le nombre complexe $f(\alpha)$ satisfait à la congruence

$$f(\alpha)\Psi(\eta_{e-r}) \equiv 0 \pmod{q},$$

où $\Psi(\eta)$ désigne le produit des $e-1$ facteurs

$$\Psi(\eta) = \psi(\eta_1).\psi(\eta_2)\ldots\psi(\eta_{e-1}),$$

et $\psi(\eta)$ un nombre complexe tel qu'on ait

$$\psi(u) \equiv 0 \pmod{q},$$

et dont la norme, divisible par q, ne soit pas divisible par q^2, nous dirons que $f(\alpha)$ contient le facteur premier idéal du nombre q, qui appartient à la substitution $\eta = u_r$.

Pour introduire la multiplicité d'un facteur premier idéal, nous généralisons la définition précédente comme il suit :

Le nombre complexe $f(\alpha)$ est censé contenir exactement n fois *le facteur premier idéal du nombre q, qui appartient à la substitution $\eta = u_r$, lorsqu'il satisfait à la congruence*

$$f(\alpha)[\Psi(\eta_{e-r})]^n \equiv 0 \pmod{q^n},$$

sans que la congruence

$$f(\alpha)[\Psi(\eta_{e-r})]^{n+1} \equiv 0 \pmod{q^{n+1}}$$

ait lieu.

Nous remarquons aussi que la notion du nombre ou facteur complexe idéal sera employée aussi bien dans le sens plus large où les nombres complexes *existants,* comme cas particuliers, sont compris parmi les nombres complexes *idéaux,* que dans le sens plus étroit où les nombres *idéaux* signifient le contraire des nombres complexes *existants,* de même que, dans l'Algèbre, le mot *imaginaire* est employé dans ce double sens.

La condition qu'un nombre complexe contienne un facteur premier idéal multiple s'exprime encore par des congruences linéaires par rapport aux coefficients de ce nombre complexe. En mettant $f(\alpha)$ sous la forme

$$f(\alpha) = \varphi(\eta) + \alpha\varphi_1(\eta) + \alpha^2\varphi_2(\eta) + \ldots + \alpha^{f-1}\varphi_{f-1}(\eta),$$

et supposant que $f(\alpha)$ contient n fois le facteur premier idéal de q, appartenant à la substitution $\eta = u_r$, on aura

$$[\Psi(\eta_{e-r})]^n[\varphi(\eta) + \alpha\varphi_1(\eta) + \alpha^2\varphi_2(\eta) + \ldots + \alpha^{f-1}\varphi_{f-1}(\eta)] \equiv 0$$

pour le module q^n. On en conclut aisément que les f congruences contenues dans la formule

$$[\Psi(\eta_{e-r})]^n\varphi_k(\eta) \equiv 0 \pmod{q^n},$$

doivent avoir lieu séparément pour

$$k = 0, 1, 2, \ldots, f-1.$$

En réduisant à la forme linéaire, on aura

$$[\Psi(\eta_{e-r})]^n\varphi_k(\eta) = C\eta + C_1\eta_1 + C_2\eta_2 + \ldots + C_{e-1}\eta_{e-1},$$

ce qui donne

$$(\Psi\eta_{e-r})^n\varphi_k(\eta) + (\Psi\eta_{e-r})^n\varphi_k(\eta_1) + \ldots + (\Psi\eta_{e-r-1})^n\varphi_k(\eta_{e-1})$$
$$= -(C + C_1 + C_2 + \ldots + C_{e-1}).$$

Multipliant encore par $[\Psi(\eta_{e-r})]^n$, et observant que le produit $\Psi(\eta_r)\Psi(\eta_s)$ est toujours divisible par q, excepté le seul cas de $r=s$, on aura la congruence

$$(\Psi\eta_{e-r})^{2n}\varphi_k(\eta) \equiv -(C+C_1+C_2+\ldots+C_{e-1})(\Psi\eta_{e-r})^n \quad (\text{mod. } q^n).$$

Il suit de là que la condition

$$(\Psi\eta_{e-r})^n\varphi_k(\eta)\equiv 0 \quad (\text{mod. } q^n)$$

est équivalente à la congruence

$$C+C_1+C_2+\ldots+C_{e-1}\equiv 0 \quad (\text{mod. } q^n),$$

qui est linéaire par rapport aux coefficients de $f(\alpha)$. Ainsi, la condition qu'un nombre complexe contienne n fois un facteur premier de q est exprimée par f congruences linéaires par rapport aux coefficients du nombre $f(\alpha)$ pour le module q^n.

Nous allons maintenant démontrer les théorèmes élémentaires pour le calcul des facteurs idéaux, qui seront entièrement les mêmes que pour les facteurs existants. Pour cela, nous proposons d'abord ce théorème :

Le produit de deux ou de plusieurs nombres complexes a précisément les mêmes facteurs premiers idéaux ou existants que les facteurs pris ensemble.

Soient $f(\alpha)$ et $g(\alpha)$ deux nombres complexes et $h(\alpha)$ leur produit, en sorte qu'on ait

$$f(\alpha).g(\alpha)=h(\alpha);$$

si le facteur premier idéal de q, qui appartient à la substitution $\eta=u_r$, est contenu précisément n fois dans $f(\alpha)$ et ν fois dans $g(\alpha)$, je dis qu'il sera contenu $n+\nu$ fois précisément dans $h(\alpha)$. D'après l'hypothèse, on a

$$f(\alpha)[\Psi(\eta_{e-r})]^n = q^n F(\alpha)$$

et

$$g(\alpha)[\Psi(\eta_{e-r})]^\nu = q^\nu G(\alpha),$$

où $F(\alpha)$ et $G(\alpha)$ ne sont pas congrus à zéro pour $\eta=u_r$, puisqu'on aurait alors

$$f(\alpha)[\Psi(\eta_{e-r})]^{n+1} \equiv 0 \quad (\text{mod. } q^{n+1})$$

et

$$g(\alpha)[\Psi(\eta_{e-r})]^{\nu+1} \equiv 0 \quad (\text{mod. } q^{\nu+1}).$$

De ces deux équations il suit

$$f(\alpha).g(\alpha)[\Psi(\eta_{e-r})]^{n+\nu} = h(\alpha)[\Psi(\eta_{e-r})]^{n+\nu} = q^{n+\nu}\mathrm{F}(\alpha).\mathrm{G}(\alpha).$$

En mulipliant de nouveau par $\Psi(\eta_{e-r})$, on a aussi

$$h(\alpha)[\Psi(\eta_{e-r})]^{n+\nu+1} = q^{n+\nu}\,\mathrm{F}(\alpha).\mathrm{G}(\alpha).\Psi(\eta_{e-r}),$$

et puisqu'aucun des trois facteurs du produit $\mathrm{F}(\alpha).\mathrm{G}(\alpha).\Psi(\eta_{e-r})$ n'est congru à zéro pour $\eta = u_r$, ce produit n'est pas divisible par q. On a donc

$$h(\alpha)[\Psi(\eta_{e-r})]^{n+\nu} \equiv 0 \quad (\text{mod. } q^{n+\nu}),$$

sans avoir

$$h(\alpha)[\Psi(\eta_{e-r})]^{n+\nu+1} \equiv 0 \quad (\text{mod. } q^{n+\nu+1}),$$

c'est-à-dire le produit $h(\alpha)$ contient le facteur premier idéal de q, qui appartient à la substitution $\eta = u_r$, $n + \nu$ fois précisément. Ce résultat donne sans peine la démonstration du théorème proposé, car tout ce que nous avons démontré pour le facteur premier idéal de q, qui appartient à la substitution $\eta = u_r$, subsiste de même pour tous les facteurs premiers idéaux, et puisqu'on peut regarder les deux facteurs eux-mêmes comme composés de facteurs, et ainsi de suite, on voit qu'il subsiste également pour un produit d'un nombre quelconque de facteurs.

En faisant usage de la dénomination nouvelle de facteurs premiers idéaux, le théorème du paragraphe précédent, relatif à la condition qu'un nombre complexe $f(\alpha)$ soit divisible par q, pourra être énoncé comme il suit :

Pour qu'un nombre complexe quelconque, représenté comme produit *de plusieurs facteurs, soit divisible par q, il faut et il suffit qu'il contienne tous les e facteurs premiers idéaux du nombre q* [*].

En supposant que dans le nombre $f(\alpha)$ chacun des facteurs premiers idéaux de q soit contenu n fois au moins, on aura les congruences

$$\left.\begin{array}{l} f(\alpha)[\Psi(\eta)]^n \equiv 0, \\ f(\alpha)[\Psi(\eta_1)]^n \equiv 0,\ldots, \\ f(\alpha)[\Psi(\eta_{e-r})]^n \equiv 0, \end{array}\right\} \pmod{q^n},$$

et, en ajoutant ces congruences,

$$f(\alpha)\left([\Psi(\eta)]^n + [\Psi(\eta_1)]^n + \ldots + [\Psi(\eta_{e-1})]^n\right) \equiv 0 \pmod{q^n}.$$

Le second facteur de ce produit, comme fonction symétrique de toutes les périodes, sera un nombre entier non complexe; de plus, il n'est pas divisible par q, puisqu'aucun des facteurs idéaux de q n'y est contenu; il faut donc que le premier facteur soit divisible par q^n, ce qui donne cette généralisation du théorème précédent :

Un nombre complexe, contenant tous les facteurs premiers idéaux de q, chacun n fois au moins, est divisible par q^n.

Supposons maintenant que $f(\alpha)$ contienne n facteurs premiers idéaux de q, soit tous différents ou non, alors chacun des nombres

---

[*] Remarquons ici qu'au moyen de ce théorème on pourra reconnaître le motif de l'emploi des multiplicateurs désignés par $\Psi(\eta)$, qui consistent des $e-1$ facteurs

$$\psi(\eta_1)\cdot\psi(\eta_2)\ldots\psi(\eta_{e-1}).$$

En effet, la condition

$$\psi(u) \equiv 0 \pmod{q}$$

fait voir que $\psi(\eta)$ contient le facteur premier idéal de q, qui appartient à la substitution $\eta = u$, et la condition que $\mathrm{N}\psi(\eta)$ ne soit pas divisible par q^2 exclut tous les autres facteurs premiers idéaux du nombre premier q qui pourraient troubler le résultat. Il s'ensuit que $\Psi(\eta_{e-r})$ contient tous les facteurs idéaux de q, à l'exception d'un seul, qui appartient à la substitution $\eta = u_r$. En multipliant donc par $\Psi(\eta_{e-r})$ un nombre donné $f(\alpha)$, contenant le facteur idéal de q, qui appartient à la substitution $\eta = u_r$, on aura un produit divisible par q, parce qu'il contient tous les facteurs idéaux de q; mais si le nombre $f(\alpha)$ ne contient pas ce facteur idéal, le produit ne sera jamais divisible par q, puisqu'un des facteurs idéaux du nombre q n'y est pas contenu.

55..

conjugués

$$f(\alpha),\quad f\left(\alpha^{\gamma}\right),\quad f\left(\alpha^{\gamma^2}\right),\ldots,\quad f\left(\alpha^{\gamma^{\lambda-2}}\right)$$

contiendra de même n facteurs idéaux de q, et le produit de ces $\lambda - 1$ nombres conjugués, c'est-à-dire la norme $\mathrm{N}f(\alpha)$, en contiendra $(\lambda - 1)n$. De plus, il est aisé de voir que tous ces facteurs idéaux dans la norme $\mathrm{N}f(\alpha)$ se trouveront en nombre égal; donc chacun d'entre eux s'y trouvera $\frac{n(\lambda-1)}{e}$ fois, ce qui est nf fois précisément. Donc, au moyen de la proposition précédente, on a le théorème très-important :

Si le nombre complexe $f(\alpha)$ contient n facteurs premiers idéaux du nombre q (appartenant à l'exposant f), soit que tous ces facteurs soient différents ou non, la norme $\mathrm{N}f(\alpha)$ contiendra toujours le facteur q^{nf}, mais elle ne contiendra jamais une puissance plus élevée de q.

Le nombre premier λ, le seul qui n'est pas contenu dans les nombres désignés par q, se décompose en $\lambda - 1$ facteurs premiers conjugués de la manière suivante :

$$(1-\alpha)(1-\alpha^2)(1-\alpha^3)\ldots(1-\alpha^{\lambda-1}) = \mathrm{N}(1-\alpha) = \lambda.$$

Pour s'assurer que $1 - \alpha^k$ est un nombre premier, on fera

$$1 - \alpha^k = f(\alpha).\varphi(\alpha);$$

d'où, en prenant les normes, on aura

$$\lambda = \mathrm{N}f(\alpha).\mathrm{N}\varphi(\alpha).$$

On en conclut que l'un des deux facteurs $\mathrm{N}f(\alpha)$ et $\mathrm{N}\varphi(\alpha)$ sera nécessairement égal à λ, l'autre égal à l'unité; donc, $1 - \alpha^k$ ne pouvant être décomposé en deux facteurs, sans que l'un d'eux soit une unité complexe, satisfait à la condition d'un nombre premier complexe. Les facteurs premiers de λ sont distingués des facteurs premiers complexes de tous les autres nombres premiers, parce qu'étant dégagés des unités complexes, ils sont tous égaux entre eux. En effet, on a

$$1 - \alpha^k = (1-\alpha).(1 + \alpha + \alpha^2 + \ldots + \alpha^{k-1}),$$

et $1 + \alpha + \alpha^2 + \ldots + \alpha^{k-1}$ n'est qu'une unité complexe. Pour cette raison, il ne s'agira jamais de rechercher *quels* facteurs de λ, mais seulement *combien* de ces facteurs sont contenus dans un nombre complexe donné, et, pour cela, on n'aura qu'à chercher combien de fois la norme du nombre donné contient le facteur λ; car le nombre des facteurs premiers $1 - \alpha$, contenus dans un nombre complexe, est évidemment le même que le nombre des facteurs λ contenus dans sa norme.

Du théorème précédent, et de ce que nous avons remarqué des facteurs premiers du nombre λ, on conclut :

La norme d'un nombre complexe $f(\alpha)$ est toujours de la forme

$$\mathrm{N}f(\alpha) = \lambda^n q^{mf}.q'^{m'f'}.q''^{m''f''}\ldots,$$

où q, q', q'', etc., sont des nombres premiers appartenant respectivement aux exposants f, f', f'', etc., et n, m, m', m'', etc., sont des entiers positifs ou zéro.

En observant que la norme $\mathrm{N}f(\alpha)$ ne contient qu'un nombre fini de facteurs premiers λ, q, q', $q''\ldots$, on voit que le nombre $f(\alpha)$ lui-même ne peut contenir qu'un nombre fini de facteurs premiers idéaux, qui seront aussi parfaitement déterminés par les définitions établies ci-dessus. Il en résulte ce théorème important :

Tout nombre complexe donné ne contient qu'un nombre fini de facteurs premiers idéaux parfaitement déterminés.

C'est surtout ce théorème qui justifie la notion des facteurs premiers idéaux, en montrant l'identité des règles du calcul des facteurs idéaux et des facteurs premiers de l'Arithmétique élémentaire, où le théorème analogue : Que tout nombre composé ne peut être décomposé en facteurs premiers que d'une seule manière, joue le même rôle principal.

Le théorème réciproque pourra s'énoncer comme il suit :

Deux nombres complexes, contenant les mêmes facteurs premiers idéaux, ne diffèrent que par des unités complexes, par lesquelles ils peuvent être multipliés.

En effet, supposons que $f(\alpha)$ et $\varphi(\alpha)$ aient les mêmes facteurs premiers idéaux, et soit

$$Nf(\alpha) = N\varphi(\alpha) = \lambda^n . q^{mf} . q'^{m'f'} \dots;$$

alors le produit

$$f(\alpha).\varphi(\alpha^\gamma).\varphi(\alpha^{\gamma^2})\dots\varphi(\alpha^{\gamma^{\lambda-2}})$$

contiendra nécessairement tous les facteurs premiers idéaux de $N\varphi(\alpha)$: il contiendra $n(\lambda-1)$ fois le facteur premier $1-\alpha$, et, par conséquent, il sera divisible par λ^n; il contiendra aussi tous les facteurs premiers idéaux de q chacun mf fois, tous les facteurs premiers idéaux de q', chacun $m'f'$ fois, et ainsi de suite. Mais puisqu'un nombre complexe, contenant tous les facteurs premiers de q, n fois chacun, est divisible par q^n, il s'ensuit que le produit

$$f(\alpha).\varphi(\alpha^\gamma).\varphi(\alpha^{\gamma^2})\dots\varphi(\alpha^{\gamma^{\lambda-2}})$$

est divisible par q^{mf}, $q'^{m'f'}$, etc., il sera donc divisible par $N\varphi(\alpha)$. De là nous avons

$$\frac{f(\alpha).\varphi(\alpha^\gamma)\,\varphi(\alpha^{\gamma^2})\dots\varphi(\alpha^{\gamma^{\lambda-2}})}{N\varphi(\alpha)} = \frac{f(\alpha)}{\varphi(\alpha)} = E(\alpha),$$

$E(\alpha)$ étant un entier complexe, et il s'ensuit

$$f(\alpha) = \varphi(\alpha)\,E(\alpha) \quad \text{et} \quad Nf(\alpha) = N\varphi(\alpha)\,NE(\alpha);$$

d'où

$$NE(\alpha) = 1. \qquad C.\ Q.\ F.\ D.$$

Pour qu'un nombre complexe $f(\alpha)$ soit divisible par $\varphi(\alpha)$, il faut et il suffit que tous les facteurs premiers idéaux du diviseur $\varphi(\alpha)$ soient contenus dans le dividende $f(\alpha)$.

D'abord il est clair que $f(\alpha)$ ne sera jamais divisible par $\varphi(\alpha)$, à moins qu'il ne contienne tous les facteurs premiers idéaux de $\varphi(\alpha)$; car, si l'on a

$$\frac{f(\alpha)}{\varphi(\alpha)} = Q(\alpha),$$

$Q(\alpha)$ étant un entier, il s'ensuit

$$f(\alpha) = Q(\alpha).\varphi(\alpha),$$

et, puisque le produit développé $f(\alpha)$ a tous les facteurs premiers idéaux des facteurs $Q(\alpha)$ et $\varphi(\alpha)$, pris ensemble, il faut que $f(\alpha)$ contienne tous les facteurs idéaux de $\varphi(\alpha)$. Pour démontrer que cette condition est suffisante, nous observons que, par la même raison que dans la démonstration du théorème précédent, on aura

$$f(\alpha)\varphi\left(\alpha^{\gamma}\right).\varphi\left(\alpha^{\gamma^2}\right)\ldots\varphi\left(\alpha^{\gamma^{\lambda-2}}\right)$$

divisible par $N\varphi(\alpha)$; d'où

$$\frac{f(\alpha).\varphi\left(\alpha^{\gamma}\right).\varphi\left(\alpha^{\gamma^2}\right)\ldots\varphi\left(\alpha^{\gamma^{\lambda-2}}\right)}{N\varphi(\alpha)}=\frac{f(\alpha)}{\varphi(\alpha)}=Q(\alpha),$$

$Q(\alpha)$ étant un entier.

Pour les applications que nous ferons dans la suite, nous ajoutons encore ce théorème, qui découle immédiatement du théorème principal :

Lorsqu'une puissance d'un nombre complexe est décomposée en plusieurs facteurs premiers entre eux, ces facteurs seront séparément des puissances semblables, multipliées par des unités complexes.

§ VI.

De la composition des nombres complexes idéaux.

Puisque les nombres complexes idéaux, comme facteurs des nombres complexes, jouent le même rôle que les facteurs existants, nous les désignerons désormais de la même manière que ceux-ci, par $f(\alpha)$, $\varphi(\alpha)$, etc., en sorte que $f(\alpha)$ par exemple, sera un nombre complexe, satisfaisant à un certain nombre déterminé de conditions caractéristiques pour les facteurs premiers idéaux, abstraction faite de l'existence du nombre $f(\alpha)$.

Cela posé, la norme d'un nombre complexe idéal sera toujours un nombre complexe existant; car, en supposant que $f(\alpha)$ contient m facteurs idéaux du nombre q'_{1}, appartenant à l'exposant f, différents ou non, de même m' facteurs idéaux du nombre q', appartenant à l'exposant f', etc., la norme $Nf(\alpha)$ contenant tous les facteurs idéaux de q, mf fois chacun, sera divisible par q^{mf}, de même elle sera divisible par $q'^{m'f'}$, etc., et, puisque la norme ne contient pas d'autres facteurs idéaux, elle sera égale à $q^{mf}.q'^{m'f'}\ldots$, à une unité près.

Le problème de trouver un nombre complexe existant $\mathrm{F}(\alpha)$, qui eût le facteur idéal $f(\alpha)$, aura toujours une infinité de solutions différentes, c'est-à-dire il y a toujours une infinité de nombres idéaux qui, multipliés par le nombre idéal déterminé $f(\alpha)$, produisent des nombres complexes existants. En choisissant ces multiplicateurs idéaux de manière que la norme du produit soit aussi petite que possible, on parviendra au résultat remarquable, qu'un nombre fini et déterminé de multiplicateurs idéaux suffit à rendre existante l'infinité de tous les nombres idéaux.

Soient, comme ci-dessus, q un nombre premier appartenant à l'exposant f, q' un nombre premier appartenant à l'exposant f', etc., soit $f(\alpha)$ un nombre complexe idéal qui contienne m fois le facteur premier idéal de q, appartenant à la substitution $\eta = u_r$, de plus m' fois le facteur premier idéal de q', appartenant à la substitution $\eta' = u^{r'}$, etc., soit enfin

$$\mathrm{F}(\alpha) = x_1\alpha + x_2\alpha^2 + x_3\alpha^3 + \ldots + x_{\lambda-1}\alpha^{\lambda-1}$$

un nombre complexe existant, qui contienne tous les facteurs premiers idéaux de $f(\alpha)$, c'est-à-dire le nombre idéal $f(\alpha)$ lui-même. La condition que $\mathrm{F}(\alpha)$ contienne m fois le facteur premier de q, appartenant à la substitution $\eta = u_r$, est exprimée par une congruence de la forme

$$[\Psi(\eta_{e-r})]^m\,\mathrm{F}(\alpha) \equiv 0 \pmod{q^m},$$

équivalente à f congruences linéaires par rapport aux coefficients de $\mathrm{F}(\alpha)$, pour le même module q^m. Soient donc

$$\Phi \equiv 0,\quad \Phi_1 \equiv 0,\ldots,\quad \Phi_{f-1} \equiv 0 \pmod{q^m},$$

ces f congruences linéaires. Soient de même

$$\Phi' \equiv 0,\quad \Phi'_1 \equiv 0,\ldots,\quad \Phi'_{f'-1} \equiv 0 \pmod{q'^{m'}},$$

les f' congruences nécessaires et suffisantes, pour que $\mathrm{F}(\alpha)$ contienne

m' fois le facteur premier de q', appartenant à la substitution $\eta' = u'_{r'}$, et ainsi de suite. Cela posé, si l'on donne aux $\lambda - 1$ coefficients $x_1, x_2, x_3, \ldots, x_{\lambda-1}$, toutes les valeurs $0, 1, 2, \ldots, k-1$, en sorte qu'on ait $k^{\lambda-1}$ combinaisons diverses de valeurs de ces coefficients, on voit que chacune des f quantités $\Phi, \Phi_1, \ldots, \Phi_{f-1}$ ne pourra donner que q^m restes différents par rapport au module q^m; de même chacune des f' quantités $\Phi', \Phi'_1, \ldots, \Phi'_{f'-1}$, ne donnera que $q'^{m'}$ restes différents, pour le module $q'^{m'}$, etc. Le nombre de toutes les combinaisons différentes des restes des $f + f' + \ldots$ quantités désignées par Φ, sera donc égal à $q^{mf}.q'^{m'f'}\ldots$. Maintenant si l'on prend le nombre k assez grand pour que le nombre des combinaisons de tous les restes possibles soit inférieur au nombre des combinaisons des valeurs des coefficients, c'est-à-dire $q^{mf}.q'^{m'f'}\ldots < k^{\lambda-1}$, on voit que toutes les combinaisons différentes des valeurs des coefficients ne pourront pas appartenir à des combinaisons différentes des restes, mais que ces combinaisons des restes se reproduiront nécessairement. Soit donc

$$x_1 = a_1, \quad x_2 = a_2, \quad x_3 = a_3, \ldots, \quad x_{\lambda-1} = a_{\lambda-1},$$

une combinaison de valeurs des coefficients, pour lesquelles toutes les quantités désignées par Φ donnent les mêmes restes que pour la combinaison

$$x_1 = b_1, \quad x_2 = b_2, \quad x_3 = b_3, \ldots, \quad x_{\lambda-1} = b_{\lambda-1};$$

par la soustraction des résultats de ces deux substitutions, les restes égaux se détruisent, et puisque les quantités désignées par Φ sont linéaires par rapport aux coefficients $x_1, x_2, x_3, \ldots, x_{\lambda-1}$, on voit que la combinaison des valeurs

$$x_1 = a_1 - b_1, \ x_2 = a_2 - b_2, \ x_3 = a_3 - b_3, \ldots, \ x_{\lambda-1} = a_{\lambda-1} - b_{\lambda-1},$$

satisfera à toutes les congruences nécessaires pour que $F(\alpha)$ contienne le facteur idéal $f(\alpha)$. Ainsi, parmi les nombres complexes dont les coefficients entiers positifs ou négatifs ne surpassent pas la valeur $k-1$, où k satisfait à la condition $k^{\lambda-1} > Nf(\alpha)$, il y en aura toujours qui contiennent le facteur idéal $f(\alpha)$.

En développant le produit des deux nombres réciproques $F(\alpha)$ et $F(\alpha^{-1})$, puis changeant α en α^2, $\alpha^3, \ldots, \alpha^{\frac{\lambda-1}{2}}$, et ajoutant, on obtient

$$F(\alpha).F(\alpha^{-1}) + F(\alpha^2).F(\alpha^{-2}) + \ldots + F\left(\alpha^{\frac{\lambda-1}{2}}\right).F\left(\alpha^{-\frac{\lambda-1}{2}}\right)$$
$$= \tfrac{1}{2}\lambda\left(x_1^2 + x_2^2 + x_3^2 + \ldots + x_{\lambda-1}^2\right) - \tfrac{1}{2}\left(x_1 + x_2 + x_3 + \ldots + x_{\lambda-1}\right)^2;$$

et puisque les coefficients $x_1, x_2, x_3, \ldots, x_{\lambda-1}$, abstraction faite des signes, ne sont pas supérieurs à $k-1$, on en déduit facilement l'inégalité

$$F(\alpha).F(\alpha^{-1}) + F(\alpha^2).F(\alpha^{-2}) + \ldots$$
$$+ F\left(\alpha^{\frac{\lambda-1}{2}}\right).F\left(\alpha^{-\frac{\lambda-1}{2}}\right) \leqq \tfrac{1}{2}\lambda(\lambda-1).(k-1)^2,$$

et en faisant usage de la proposition connue, que le produit de n quantités positives est plus petit que la $n^{ième}$ puissance de la moyenne arithmétique de ces quantités, on en conclut sans difficulté

$$NF(\alpha) \leqq \lambda^{\frac{\lambda-1}{2}}(k-1)^{\lambda-1}.$$

Le nombre k, qui doit satisfaire à la condition $k^{\lambda-1} > Nf(\alpha)$ pourra toujours être pris de manière à rendre $(k-1)^{\lambda-1} < Nf(\alpha)$; par cette supposition on a

$$NF(\alpha) < \lambda^{\frac{\lambda-1}{2}} Nf(\alpha).$$

En représentant actuellement le nombre complexe $F(\alpha)$ comme produit des deux facteurs idéaux $\varphi(\alpha)$ et $f(\alpha)$, on aura

$$F(\alpha) = \varphi(\alpha).f(\alpha),$$

d'où

$$NF(\alpha) = N\varphi(\alpha).Nf(\alpha),$$

et, au moyen de l'inégalité trouvée, il vient

$$N\varphi(\alpha) < \lambda^{\frac{\lambda-1}{2}}.$$

Ainsi, les multiplicateurs idéaux, qui, composés avec tous les nombres complexes idéaux, donnent des nombres complexes existants, peuvent toujours être choisis tels, que leurs normes soient plus petites que $\lambda^{\frac{\lambda-1}{2}}$; et puisque le nombre des facteurs idéaux dont les normes ne surpassent pas cette limite fixe est limité, nous avons ce théorème :

Il y a toujours un nombre fini de multiplicateurs idéaux, qui, composés avec les nombres idéaux, en nombre infini, les rendent tous nombres complexes existants.

Les multiplicateurs qui rendent existants les nombres complexes idéaux, avec lesquels ils sont composés, donnent lieu à la classification des nombres complexes idéaux, que nous établissons par cette définition :

Tous les nombres complexes idéaux qui donnent des produits existants lorsqu'on les multiplie par un même nombre idéal, seront appelés NOMBRES IDÉAUX ÉQUIVALENTS, *et ils seront attribués à une même* CLASSE *des nombres complexes idéaux.*

Nous comprenons dans cette classification les nombres existants eux-mêmes, qui constituent une de ces classes que nous appellerons *la classe principale.*

D'abord nous observons qu'un nombre idéal, multiplié par un nombre existant, ne donnera jamais un produit existant, puisqu'alors le quotient de deux nombres existants serait un nombre idéal.

Soient à présent $f(\alpha)$ et $\varphi(\alpha)$ deux nombres idéaux équivalents, $\psi(\alpha)$ le multiplicateur qui les rend existants, soit aussi $\chi(\alpha)$ un multiplicateur de $f(\alpha)$, tel que $\chi(\alpha).f(\alpha)$ soit un nombre existant; je dis que $\chi(\alpha).\varphi(\alpha)$ sera de même un nombre existant. En effet, puisque $\psi(\alpha).f(\alpha)$ est un nombre existant, il en sera de même du produit

$$\psi(\alpha^2).\psi(\alpha^3)\ldots\psi(\alpha^{\lambda-1}).f(\alpha^2).f(\alpha^3)\ldots f(\alpha^{\lambda-1}).$$

Multipliant donc par les deux nombres existants $\psi(\alpha).\varphi(\alpha)$ et $\chi(\alpha).f(\alpha)$, on aura le nombre existant

$$\varphi(\alpha).\chi(\alpha).\mathrm{N}\psi(\alpha).\mathrm{N}f(\alpha),$$

56..

et, en supprimant le facteur existant $N\psi(\alpha).Nf(\alpha)$, on voit que $\varphi(\alpha).\chi(\alpha)$ est un nombre existant. Nous en concluons :

Les classes des nombres équivalents sont toujours les mêmes, pour tous les multiplicateurs qu'on pourra choisir.

Soit présentement $f(\alpha)$ équivalent à $\psi(\alpha)$, et de même $\varphi(\alpha)$ équivalent à $\psi(\alpha)$, je dis que $f(\alpha)$ sera équivalent à $\varphi(\alpha)$. En effet, tout multiplicateur de $\psi(\alpha)$, qui donne un produit existant, sera de même multiplicateur de $f(\alpha)$ et de $\varphi(\alpha)$. Donc :

Deux nombres idéaux, équivalents à un même troisième nombre idéal, sont équivalents entre eux.

En rapprochant ces résultats du premier théorème de ce paragraphe, on conclut :

Les classes diverses dans lesquelles tous les nombres idéaux se distribuent, sont en nombre fini et complétement déterminées, en sorte qu'un certain nombre idéal n'appartient qu'à une seule classe.

Soient $f(\alpha)$ et $f_1(\alpha)$ deux nombres idéaux équivalents, $\psi(\alpha)$ le multiplicateur qui les rend existants; soient de même $\varphi(\alpha)$ et $\varphi_1(\alpha)$ deux nombres équivalents, et $\chi(\alpha)$ leur multiplicateur; $\psi(\alpha).\chi(\alpha)$ sera évidemment un multiplicateur qui rend existant le produit $f(\alpha).\varphi(\alpha)$, de même que le produit $f_1(\alpha).\varphi_1(\alpha)$. Ces produits seront donc équivalents. De là ce théorème :

Des nombres équivalents, multipliés par des nombres équivalents, donnent toujours des produits équivalents.

Il suit de là que, lorsqu'on multiplie les nombres idéaux d'une certaine classe avec les nombres idéaux d'une autre classe, tous ces produits appartiendront à une même classe déterminée. Ainsi :

La classe du produit de deux nombres idéaux est complétement déterminée par les classes des facteurs.

Lorsqu'on multiplie un nombre idéal donné par un des nombres de chaque classe, on aura autant de produits qu'on a de classes, et l'on s'assure aisément que tous ces produits appartiennent à des classes différentes. Parmi ces produits, il y en aura nécessairement un, et il n'y en aura qu'un, qui appartient à la classe principale. Donc :

Il correspond à chaque classe une certaine classe, en sorte que ces deux classes composées produisent la classe principale, c'est-à-dire que le produit de deux nombres quelconques de ces deux classes est un nombre complexe existant.

Les classes qui produisent la classe principale, lorsqu'elles sont composées avec elles-mêmes, seront appelées *classes ambiguës*. La classe principale est toujours une classe ambiguë ; d'autres classes ambiguës n'existent que pour des valeurs particulières du nombre λ.

Considérons maintenant les diverses puissances entières d'un nombre idéal $f(\alpha)$. Dans la série

$$f(\alpha),\quad f(\alpha)^2,\quad f(\alpha)^3,\quad f(\alpha)^4,\ldots,$$

il y aura nécessairement des nombres idéaux équivalents, car le nombre de ceux qui appartiennent à des classes diverses est fini, aussi bien que le nombre des classes mêmes.

Soient donc $f(\alpha)^r$ et $f(\alpha)^s$ deux nombres idéaux équivalents de cette série où $s > r$; soit aussi $\Psi(\alpha)$ un multiplicateur qui rend $f(\alpha)^r\,\Psi(\alpha)$ et $f(\alpha)^s\,\Psi(\alpha)$ nombres existants ; le quotient de ces deux nombres, ou bien $f(\alpha)^{s-r}$, sera de même un nombre existant. Nous avons donc ce théorème important :

Tout nombre complexe idéal, élevé à une certaine puissance entière, donne un nombre complexe existant.

Ou bien :

Tout nombre complexe idéal peut être représenté comme une certaine racine d'un nombre complexe existant.

Si l'exposant h est le plus petit pour lequel $f(\alpha)^h$ est un nombre existant, les nombres de la série de h termes

$$1,\ f(\alpha),\quad f(\alpha)^2,\quad f(\alpha)^3,\ldots,\quad f(\alpha)^{h-1}$$

appartiendront à des classes diverses ; car, si l'on avait $f(\alpha)^r$ équivalent à $f(\alpha)^s$, r et s étant positifs et moindres que h, on en conclurait, comme ci-dessus, que $f(\alpha)^{s-r}$ serait existant. Il y aurait donc un exposant

$s-r$, moindre que h, pour lequel la puissance du nombre idéal $f(\alpha)$ serait un nombre existant, ce qui est contre l'hypothèse. Il se peut maintenant que les h classes, représentées par les nombres idéaux de la série proposée de h termes, embrassent toutes les classes des nombres idéaux; alors le nombre de toutes les classes diverses sera égal à h. Si cela n'a pas lieu, on choisira arbitrairement un nombre idéal $\varphi(\alpha)$, non équivalent aux h nombres idéaux proposés, et l'on en formera ce second groupe de h nombres idéaux

$$\varphi(\alpha),\quad \varphi(\alpha).f(\alpha),\quad \varphi(\alpha).f(\alpha)^2,\ldots,\quad \varphi(\alpha).f(\alpha)^{h-1}$$

Les h termes de ce groupe ne sont pas équivalents, ni entre eux, ni aux termes du premier groupe. En effet, si l'on avait $\varphi(\alpha).f(\alpha)^r$ équivalent à $\varphi(\alpha).f(\alpha)^s$, r et s étant moindres que h, il s'ensuivrait $f(\alpha)^r$ équivalent à $f(\alpha)^s$, contre l'hypothèse, et, si l'on avait $\varphi(\alpha).f(\alpha)^r$ équivalent à $f(\alpha)^s$, en multipliant par $f(\alpha)^{h-r}$, on aurait aussi $\varphi(\alpha).f(\alpha)^h$ équivalent à $f(\alpha)^{h-r+s}$, ou bien $\varphi(\alpha)$ équivalent à $f(\alpha)^{s-r}$, ce qui est aussi contre l'hypothèse. Maintenant, il se pourra que les $2h$ classes de nombres idéaux, représentées par le premier et le second groupe, embrassent toutes les classes des nombres idéaux. Dans ce cas, le nombre des classes sera égal à $2h$. Mais si cela n'a pas lieu, on prendra arbitrairement un nombre idéal $\psi(\alpha)$, qui ne soit pas équivalent à aucun nombre de ces deux groupes, et l'on en composera ce troisième groupe de h termes,

$$\psi(\alpha),\quad \psi(\alpha).f(\alpha),\quad \psi(\alpha).f(\alpha)^2,\ldots,\quad \psi(\alpha).f(\alpha)^{h-1}.$$

On démontre aisément que ces nombres idéaux ne sont pas équivalents ni entre eux ni aux termes du premier et du second groupe, et l'on en conclut, si ces trois groupes embrassent toutes les classes, que le nombre des classes sera égal à $3h$. Si cela n'a pas lieu, on pourra continuer de la même manière, et l'on voit clairement qu'on parviendra ainsi à ranger toutes les classes des nombres idéaux en groupes de h termes. De là ce théorème :

Le nombre total de toutes les classes des nombres idéaux est un multiple du plus petit exposant de la puissance d'un nombre idéal quelconque qui devient un nombre complexe existant.

Toutes les autres puissances du nombre idéal $f(\alpha)$, qui donnent des nombres complexes existants, seront nécessairement contenues dans la forme $f(\alpha)^{mh}$, d'où l'on conclut ce corollaire :

Si une puissance d'un nombre complexe idéal donne un nombre complexe existant, il faut que l'exposant de cette puissance ait un facteur commun avec le nombre des classes de tous les nombres complexes idéaux.

Il se présente ici la question délicate, si le nombre idéal $f(\alpha)$ pourra toujours être choisi tel, que les nombres idéaux d'un seul groupe

$$1, \quad f(\alpha), \quad f(\alpha)^2, \quad f(\alpha)^3, \ldots, \quad f(\alpha)^{h-1}$$

embrassent toutes les classes des nombres idéaux. Il nous paraît que cela n'a pas lieu dans tous les cas, mais qu'il y a effectivement des nombres premiers λ pour lesquels les puissances d'un seul nombre idéal ne pourront jamais représenter toutes les classes des nombres complexes idéaux. En laissant cela aux recherches plus approfondies des géomètres, nous remarquons que la question analogue pour la composition des formes quadratiques a été traitée par M. Gauss dans la cinquième section, n° 306, des *Disquisitiones Arithmeticæ*.

Remarque. Qu'il me soit permis de signaler ici en peu de mots l'analogie de cette théorie de la composition des nombres idéaux avec les principes fondamentaux de la chimie. La composition des nombres complexes peut être envisagée comme l'analogue de la combinaison chimique ; les facteurs premiers correspondent aux éléments, ou plutôt aux équivalents de ces éléments. Les nombres complexes idéaux sont comparables aux radicaux hypothétiques qui n'existent pas par eux-mêmes, mais seulement dans les combinaisons ; le fluor, en particulier, comme élément qu'on ne sait pas représenter isolément, peut être comparé à un facteur premier idéal. La notion de l'équivalence des nombres idéaux est, au fond, la même que celle de l'équivalence chimique ; car, ainsi que des quantités pondérales équivalentes des matières naturelles peuvent être substituées les unes aux autres pour rendre des sels neutres ou des corps isomorphes, de même les nombres idéaux, remplacés par les facteurs équivalents, ne produisent que des nombres idéaux de la même classe. En comparant les méthodes de l'analyse chimique à celles de la décomposition des nombres complexes, on trouve encore des analogies surprenantes. Car, de même que les réactifs chimiques, joints à un corps en dissolution, donnent des précipités au moyen desquels on reconnaît les éléments contenus dans le corps proposé, de même les nombres que nous avons désignés par $\Psi(\eta)$ comme réactifs des nombres complexes, font connaître les facteurs premiers contenus dans les nombres complexes en mettant en évidence un facteur premier q analogue au précipité chimique. Toutes ces analogies qu'on pourra poursuivre et augmenter à volonté, ne proviennent pas d'un jeu d'esprit oisif, mais elles sont bien fondées en ce que les mêmes idées fondamentales de la composition et décomposition des éléments règnent aussi bien dans la chimie des matières naturelles que dans celle des nombres complexes.

§ VII.

Application à la théorie de la division du cercle.

On sait que toutes les difficultés de la théorie de la division du cercle, c'est-à-dire de la solution algébrique de l'équation binôme $x^p = 1$, se réduisent à la recherche de certains nombres complexes du genre de ceux qui ont été discutés dans ce qui précède. Sous le point de vue théorique, on désire la connaissance approfondie de la constitution intérieure de ces nombres complexes, et pour la pratique, il reste encore à réduire le calcul de ces nombres à la solution d'un seul problème élémentaire et général. C'est ce que nous compléterons au moyen de notre théorie des nombres complexes.

Soient p un nombre premier de la forme $m\lambda + 1$, g une racine primitive de la congruence $g^{p-1} \equiv 1 \pmod{p}$, x une racine imaginaire de l'équation $x^p = 1$, et soient, comme ci-dessus, α une racine imaginaire de l'équation $\alpha^\lambda = 1$, λ un nombre premier. Cela posé, la partie la plus difficile du problème de la résolution algébrique de l'équation $x^p = 1$ tient à la recherche de la puissance $\lambda^{\text{ième}}$ de l'expression résolvante proposée par Lagrange,

$$x + \alpha x^g + \alpha^2 x^{g^2} + \alpha^3 x^{g^3} + \ldots + \alpha^{p-2} x^{g^{p-2}},$$

que nous désignons simplement par (α, x). Cette puissance, indépendante de x, n'est qu'un nombre complexe composé des racines α, α^2, $\alpha^3, \ldots, \alpha^{\lambda-1}$. En faisant usage de l'expression

$$\frac{(\alpha, x)\,(\alpha^k, x)}{(\alpha^{k+1}, x)} = \psi_k(\alpha),$$

qui est de même indépendante de x, et en observant que

$$(\alpha, x).(\alpha^{-1}, x) = p,$$

on aura

$$(\alpha, x)^\lambda = p\psi_1(\alpha).\psi_2(\alpha).\psi_2(\alpha)\ldots\psi_{\lambda-2}(\alpha);$$

d'où l'on voit que tout se réduit à trouver les nombres complexes désignés par $\psi_k(\alpha)$, pour

$$k = 1, 2, 3, \ldots, \lambda - 2.$$

Nous déterminerons ces nombres complexes en définissant les facteurs premiers idéaux qu'ils contiennent, ce qui se fera au moyen de quelques résultats trouvés en même temps par M. Cauchy et par M. Jacobi, et publiés dans les *Mémoires de l'Académie des Sciences de Paris,* de l'année 1840, et dans les *Comptes rendus mensuels de l'Académie de Berlin,* de l'année 1837. Ces deux grands géomètres ont trouvé qu'en désignant par r une racine primitive de l'équation $r^{p-1} = 1$, et faisant

$$x + rx^g + r^2 x^{g^2} + \ldots + r^{p-2} x^{g^{p-2}} = (r, x),$$

$$\frac{(r^{-m}, x).(r^{-n}, x)}{(r^{-m-n}, x)} = \psi(r),$$

la substitution de la racine primitive g de la congruence

$$g^{p-1} \equiv 1 \quad (\text{mod. } p),$$

au lieu de la racine r de l'équation $r^{p-1} = 1$, donne

$$\psi(g) \equiv -\frac{\Pi(m+n)}{\Pi(m)\Pi(n)} \quad (\text{mod. } p),$$

$\Pi(n)$ désignant le produit $1.2.3\ldots n$, d'où il suit

$$\psi(g) \equiv 0 \quad (\text{mod. } p), \quad \text{si} \quad m < p-1, \quad n < p-1, \quad m+n > p-1.$$

Pour en faire l'application au nombre complexe $\psi_k(\alpha)$, nous ferons

$$m = \frac{h(p-1)}{\lambda}, \quad n = \frac{i(p-1)}{\lambda}, \quad r^{-\frac{p-1}{\lambda}} = \alpha,$$

ce qui donne

$$\psi(r) = \frac{(\alpha^h, x).(\alpha^i, x)}{(\alpha^{h+i}, x)};$$

de là, en faisant $i \equiv hk \pmod{\lambda}$,

$$\psi(r) = \frac{(\alpha^h, x).(\alpha^{hk}, x)}{(\alpha^{h(k+1)}, x)} = \psi_k(\alpha^h).$$

Substituant enfin la racine primitive g au lieu de r, et faisant, pour abréger,

$$g^{\frac{p-1}{\lambda}} \equiv u \pmod{p},$$

nous aurons

$$\psi_k(u^{-h}) \equiv 0 \pmod{p} \quad \text{si} \quad h < \lambda, \quad i < \lambda, \quad h + i > \lambda.$$

En considérant les racines α, α^γ, $\alpha^{\gamma^2}, \ldots, \alpha^{\gamma^{\lambda-2}}$ comme périodes monômes, on aura les racines des congruences correspondantes u, u^γ, $u^{\gamma^2}, \ldots, u^{\gamma^{\lambda-2}}$; donc le résultat trouvé pourra s'énoncer comme il suit :

Si la somme du nombre h, positif et moindre que λ, et du plus petit reste positif de hk (mod. λ), *est plus grande que λ, le nombre complexe $\psi_k(\alpha)$ contient le facteur premier idéal de p, qui appartient à la substitution $\alpha = u^{-h}$.*

Le nombre des valeurs de h, qui satisfont à cette condition, est égal à $\frac{\lambda-1}{2}$; car on voit, sans difficulté, que de deux valeurs de h, dont la somme est égale à λ, l'une satisfait toujours, mais qu'elles ne satisfont jamais toutes deux en même temps. Le théorème trouvé fait donc connaître $\frac{\lambda-1}{2}$ facteurs premiers idéaux contenus dans le nombre complexe $\psi_k(\alpha)$, et l'on s'assure facilement qu'il n'y a pas d'autres facteurs premiers dans $\psi_k(\alpha)$. En effet, par l'équation connue

$$\psi_k(\alpha) . \psi_k(\alpha^{-1}) = p,$$

on voit que $\psi_k(\alpha)$ et $\psi_k(\alpha^{-1})$, pris ensemble, contiennent tous les $\lambda - 1$

facteurs premiers de p, et, puisque $\psi_k(\alpha^{-1})$ contient nécessairement autant de facteurs premiers que $\psi_k(\alpha)$, ce nombre complexe contiendra exactement les $\frac{\lambda-1}{2}$ facteurs premiers définis dans le théorème.

En désignant par $f(\alpha)$ le facteur premier (idéal) de p, appartenant à la substitution $\alpha = u$, et par $\left|\frac{k}{h}\right|$ la racine positive, moindre que λ, de la congruence $hx \equiv k$ (mod. λ), nous aurons

$$\psi_k(\alpha) = \mathrm{E}(\alpha)\,\Pi f(\alpha^{-h}),$$

où le signe du produit Π s'étend sur toutes les valeurs de h, moindres que λ, qui satisfont à la condition $\left|\frac{1}{h}\right| + \left|\frac{k}{h}\right| > \lambda$. L'unité complexe $\mathrm{E}(\alpha)$, qu'il faut toujours ajouter au produit de tous les facteurs premiers d'un nombre complexe donné, se réduit, dans ce cas, à l'unité simple $\pm\alpha^r$; car, moyennant la formule

$$\psi_k(\alpha)\,\psi_k(\alpha^{-1}) = p,$$

on a

$$\mathrm{E}(\alpha)\,\mathrm{E}(\alpha^{-1}) = 1,$$

équation qui n'est jamais satisfaite que par les unités simples de la forme $\pm\alpha^r$. L'expression de $\psi_k(\alpha)$ devient donc

$$\psi_k(\alpha) = \pm\alpha^r\,\Pi f(\alpha^{-h}),$$

pour toutes les valeurs positives de $h < \lambda$, qui satisfont à la condition $\left|\frac{1}{h}\right| + \left|\frac{k}{h}\right| > \lambda$. Pour déterminer complétement le signe et l'exposant r de l'unité simple $\pm\alpha^r$, on fera usage de la propriété connue de ces nombres complexes que

$$\psi_k(\alpha) \equiv -1 \quad [\text{mod.}\ (1-\alpha)^2].$$

Des facteurs premiers trouvés de $\psi_k(\alpha)$, on déduit ceux de la puissance

$$(\alpha, x)^{\lambda} = p\,\psi_1(\alpha)\,.\,\psi_2(\alpha)\,.\,\psi_3(\alpha)\ldots\psi_{\lambda-2}(\alpha).$$

Le facteur premier idéal déterminé $f(\alpha^{-h})$ se trouvera dans le produit

57..

$\psi_1(\alpha).\psi_2(\alpha)..\psi_3(\alpha)\ldots\psi_{\lambda-2}(\alpha)$ autant de fois précisément qu'il y a de nombres k, pour lesquels on a

$$\left|\frac{1}{h}\right| + \left|\frac{k}{h}\right| > \lambda,$$

et, puisque $\left|\frac{k}{h}\right|$, pour $k = 1, 2, 3, \ldots, \lambda - 2$, a évidemment toutes les valeurs $\left|\frac{k}{h}\right| = 1, 2, 3, \ldots, \lambda - 1$, excepté la seule $\left|\frac{\lambda - 1}{h}\right| = \lambda - \left|\frac{1}{h}\right|$, le nombre des valeurs de k, qui donnent

$$\left|\frac{1}{h}\right| + \left|\frac{k}{k}\right| > \lambda, \quad \text{ou} \quad \left|\frac{k}{h}\right| > \lambda - \left|\frac{1}{h}\right|,$$

c'est-à-dire le nombre des facteurs $f(\alpha^{-h})$ contenus dans le produit $\psi_1(\alpha).\psi_2(\alpha)\ldots\psi_{\lambda-2}(\alpha)$, sera égal à $\left|\frac{1}{h}\right| - 1$. En y ajoutant un facteur $f(\alpha^{-h})$, contenu dans p, on voit que le facteur premier $f(\alpha^{-h})$ est contenu $\left|\frac{1}{h}\right|$ fois précisément dans la puissance $(\alpha, x)^\lambda$. Ainsi, en prenant

$$h = 1, 2, 3, \ldots, \lambda - 1,$$

on a l'expression

$$(\alpha, x)^\lambda = \pm \alpha^s f(\alpha^{-1})^{\left|\frac{1}{1}\right|} . f(\alpha^{-2})^{\left|\frac{1}{2}\right|} . f(\alpha^{-3})^{\left|\frac{1}{3}\right|} \ldots f[\alpha^{-(\lambda-1)}]^{\left|\frac{1}{\lambda-1}\right|},$$

ou, ce qui est la même chose,

$$(\alpha, x)^\lambda = \pm \alpha^s f(\alpha)^{\left|\frac{1}{\lambda-1}\right|} . f(\alpha^2)^{\left|\frac{1}{\lambda-2}\right|} . f(\alpha^3)^{\left|\frac{1}{\lambda-3}\right|} \ldots f(\alpha^{\lambda-1})^{\left|\frac{1}{1}\right|}.$$

Ici l'unité simple $\pm \alpha^s$ sera, en chaque cas, déterminée complétement par la condition connue

$$(\alpha, x)^\lambda \equiv -1 \pmod{\lambda}.$$

La décomposition en facteurs premiers donne en même temps la connaissance parfaite des nombres complexes qui se présentent dans la théorie de la division du cercle, et le moyen le plus simple pour les calculer; car on voit que tout se réduit au seul problème de trouver un facteur premier complexe du nombre p, qui pourra être représenté comme nombre complexe entier, s'il existe par soi-même, ou comme racine d'un certain degré d'un nombre complexe existant, s'il est idéal. La recherche de ces facteurs premiers se fait assez facilement au moyen de méthodes indirectes qui s'offrent d'elles-mêmes. Il ne serait pas trop pénible, et il serait très-utile, pour cette théorie, de construire une Table de tous les facteurs premiers existants et idéaux des nombres premiers du premier millier, laquelle donnerait tous les nombres nécessaires pour la résolution algébrique de l'équation $x^p = 1$, pour tous les nombres premiers p contenus dans les mêmes limites. Une partie de cette Table a été donnée dans la dissertation *de numeris complexis*, etc., publiée en 1834, et réimprimée dans ce Journal en 1837.

Pour faire mieux apprécier l'usage des formules données dans ce paragraphe, nous rappelons la méthode ingénieuse dont M. Gauss s'est servi pour trouver l'équation du troisième degré, dont les racines sont les trois périodes à $\frac{p-1}{3}$ termes (voir *Disq. Arithm.*, sect. VII, n° 358), qui revient à ce que le nombre $4p$ doit être mis sous la forme du second degré

$$4p = M^2 + 27N^2.$$

Notre méthode donne la même réduction pour le problème général, où il s'agit de λ périodes à $\frac{p-1}{\lambda}$ termes; car le problème de trouver un facteur premier complexe d'un nombre premier p est essentiellement le même que de représenter le nombre p comme une certaine forme du degré $\lambda - 1$ et du même nombre d'indéterminés; on voit aussi que les coefficients du nombre premier complexe, ou, ce qui revient au même, les valeurs des indéterminés de la forme du degré $\lambda - 1$, jouent le même rôle dans le problème général, que les deux nombres M et N de la forme

$$4p = M^2 + 27N^2,$$

pour le cas de $\lambda = 3$.

§ VIII.

Recherche du nombre des classes diverses des nombres complexes idéaux.

Les principes nouveaux dont M. Lejeune-Dirichlet s'est servi dans la recherche du nombre des formes quadratiques qui répondent à un déterminant donné, joints aux résultats exposés dans ce qui précède, suffiront pour achever la recherche analogue du nombre des classes des nombres complexes idéaux. Pour en faire l'application, nous considérons la série infinie

$$R = \sum \frac{s-1}{[\mathrm{NF}(\alpha)]^s},$$

où le signe sommatoire $\sum$ doit être pris par rapport à tous les nombres complexes différents, existants ou idéaux, et où l'on compte pour différents deux nombres complexes qui ne sont pas composés des mêmes facteurs premiers idéaux. Pour que la série proposée soit convergente, le nombre s est supposé positif et plus grand que l'unité, mais on le fera décroître ci-après jusqu'à la limite 1.

La norme d'un nombre complexe, existant ou idéal, étant toujours de la forme

$$\mathrm{NF}(\alpha) = \lambda^n . q^{mf} . q'^{m'f'} . q''^{m''f''} \ldots,$$

où q, q', q'',..., sont des nombres premiers qui appartiennent respectivement aux exposants f, f', f'',..., nous pourrons introduire cette expression dans la série R; mais puisque la même norme convient toujours à plusieurs nombres complexes différents, nous rechercherons d'abord combien de nombres complexes donneront la même norme, c'est-à-dire combien de fois le terme

$$\frac{s-1}{(\lambda^n . q^{mf} . q'^{m'f'} . q''^{m''f''} \ldots)^s}$$

se trouvera dans la série R. On conclut du facteur q^{mf} contenu dans la norme, que le nombre complexe $\mathrm{F}(\alpha)$ contiendra nécessairement m facteurs premiers de q, différents ou non, et parce qu'il y a $\frac{\lambda-1}{f} = e$

facteurs premiers de q, on voit qu'en prenant toutes les combinaisons m à m des e facteurs premiers idéaux, sans exclure les facteurs premiers égaux, on aura toutes les manières de produire le facteur q^{mf} de la norme. Cela se fera, comme on sait, de

$$\frac{e(e+1).(e+2)\ldots(e+m-1)}{1.2.3\ldots m}$$

manières différentes. De même le facteur de la norme $q'^{m'f'}$ pourra être produit de

$$\frac{e'(e'+1).(e'+2)\ldots(e'+m'-1)}{1.2.3\ldots m'}$$

manières différentes, si

$$e' = \frac{\lambda - 1}{f'},$$

et ainsi de suite. Observons encore que le facteur λ^n ne sera produit que d'une seule manière, savoir par le facteur $(1-\alpha)^n$, qui doit être contenu dans $F(\alpha)$. En combinant ces résultats particuliers, on voit que le terme

$$\frac{s-1}{[NF(\alpha)]^s} = \frac{s-1}{(\lambda^n.q^{mf}.q'^{m'f'}.q''^{m''f''}\ldots)^s},$$

pour des valeurs déterminées des nombres premiers q, q', q'',... et des nombres n, m, m', m'',..., sera contenu

$$\frac{e(e+1)\ldots(e+m-1)}{1.2\ldots m}\cdot\frac{e'(e'+1)\ldots(e'+m'-1)}{1.2\ldots m'}\cdot\frac{e''(e''+1)\ldots(e''+m''-1)}{1.2\ldots m''}\ldots$$

fois précisément dans la série R. Donc cette série pourra être représentée de cette manière,

$$R = (s-1)\sum\frac{\dfrac{e(e+1)\ldots(e+m-1)}{1.2\ldots m}\cdot\dfrac{e'(e'+1)\ldots(e'+m'-1)}{1.2\ldots m'}\ldots}{\lambda^{ns}.q^{mfs}.q'^{m'f's}\ldots},$$

où le signe sommatoire est pris par rapport à toutes les valeurs entières positives, zéro y compris, des nombres n, m, m', m''.... Toutes ces sommations s'effectuent sans difficulté au moyen des séries binômiales

$$1+\frac{1}{\lambda^s}+\frac{1}{\lambda^{2s}}+\frac{1}{\lambda^{3s}}+\ldots=\frac{1}{1-\frac{1}{\lambda^s}},$$

$$1+\frac{e}{1}\frac{1}{q^{fs}}+\frac{e(e+1)}{1.2}\cdot\frac{1}{q^{2fs}}+\ldots=\left(\frac{1}{1-\frac{1}{q^{fs}}}\right)^e,$$

$$1+\frac{e'}{1}\frac{1}{q'^{f's}}+\frac{e'(e'+1)}{1.2}\cdot\frac{1}{q'^{2f's}}+\ldots=\left(\frac{1}{1-\frac{1}{q'^{f's}}}\right)^{e'},$$

. . . .

On aura donc la série R exprimée comme produit d'un nombre infini de facteurs

$$R=\frac{s-1}{1-\frac{1}{\lambda^s}}\left(\frac{1}{1-\frac{1}{q^{fs}}}\right)^e\cdot\left(\frac{1}{1-\frac{1}{q'^{f's}}}\right)^{e'}\cdots$$

En séparant les nombres premiers q, suivant les exposants auxquels ils appartiennent, on aura autant de produits infinis qu'il y a de diviseurs différents de $\lambda-1$. Le produit total donnera

$$R=\frac{s-1}{1-\frac{1}{\lambda^s}}\Pi\left(\frac{1}{1-\frac{1}{q^{fs}}}\right)^e\cdot\Pi\left(\frac{1}{1-\frac{1}{q'^{f's}}}\right)^{e'}\cdots,$$

où le premier signe du produit Π se rapporte à tous les nombres premiers q, qui appartiennent à l'exposant f, le second à ceux qui appartiennent à l'exposant f', et ainsi de suite.

Soient à présent β une racine primitive de l'équation $\beta^{\lambda-1}=1$, r un nombre entier, tel que le plus grand diviseur commun de r et $\lambda-1$ est égal à e; on aura évidemment

$$1-\frac{1}{q^{fs}}=\left(1-\frac{1}{q^s}\right)\cdot\left(1-\frac{\beta^r}{q^s}\right)\cdot\left(1-\frac{\beta^{2r}}{q^s}\right)\cdots\left(1-\frac{\beta^{(f-1)r}}{q^s}\right).$$

En élevant à la puissance $-e$, et observant qu'on a généralement

$$\beta^{kr}=\beta^{(k+f)r}=\beta^{(k+2f)r},\ldots,$$

on conclut de là que

$$\left(\frac{1}{1-\frac{1}{q^{fs}}}\right)^e = \left(\frac{1}{1-\frac{1}{q^s}}\right)\cdot\left(\frac{1}{1-\frac{\beta^r}{q^s}}\right)\cdot\left(\frac{1}{1-\frac{\beta^{2r}}{q^s}}\right)\cdots\left(\frac{1}{1-\frac{\beta^{(\lambda-2)r}}{q^s}}\right).$$

Le nombre r, dont le plus grand diviseur commun avec $\lambda - 1$ doit être égal à e, pourra toujours être pris égal à l'indice de q; car, en posant

$$r \equiv \operatorname{ind} q \quad [\text{mod.}\ (\lambda - 1)],$$

on a

$$q \equiv \gamma^r \quad (\text{mod.}\ \lambda),$$

et puisque q appartient à l'exposant f, on aura

$$q^f \equiv \gamma^{rf} \equiv 1,$$

d'où l'on conclut que $r = \operatorname{ind} q$ sera un multiple de e; mais s'il y avait un facteur plus grand de $r = \operatorname{ind} q$ et $\lambda - 1$, f ne serait pas le plus petit exposant pour lequel on a $q^f \equiv 1$ (mod. λ). La substitution de la valeur convenable $r = \operatorname{ind} q$ donne

$$\left(\frac{1}{1-\frac{1}{q^{fs}}}\right)^e = \left(\frac{1}{1-\frac{1}{q^s}}\right)\cdot\left(\frac{1}{1-\frac{\beta^{\operatorname{ind} q}}{q^s}}\right)\cdot\left(\frac{1}{1-\frac{\beta^{2\operatorname{ind} q}}{q^s}}\right)\cdots\left(\frac{1}{1-\frac{\beta^{(\lambda-2)\operatorname{ind} q}}{q^s}}\right).$$

Les nombres f et e n'étant plus contenus dans cette expression, on voit que le même résultat subsiste non-seulement pour les nombres premiers q, qui appartiennent à l'exposant f, mais aussi pour tous les autres q', q'', etc., qui appartiennent aux exposants f', f'', etc. Donc, si l'on fait subir les mêmes transformations à tous les autres facteurs du produit trouvé, on aura cette nouvelle expression de R,

$$R = \frac{s-1}{1-\frac{1}{\lambda^s}}\,\Pi\left(\frac{1}{1-\frac{1}{q^s}}\right).\Pi\left(\frac{1}{1-\frac{\beta^{\operatorname{ind} q}}{q^s}}\right).\Pi\left(\frac{1}{1-\frac{\beta^{2\operatorname{ind} q}}{q^s}}\right)\ldots\Pi\left(\frac{1}{1-\frac{\beta^{(\lambda-2)\operatorname{ind} q}}{q^s}}\right),$$

où tous les $\lambda - 1$ signes des produits s'étendent sur tous les nombres premiers, excepté le seul λ.

Les produits infinis de cette formule sont connus par le célèbre Mémoire de M. Lejeune-Dirichlet sur la Progression arithmétique, lu à l'Académie de Berlin le 27 juillet 1837. Nous ne nous arrêterons pas à développer ici leur transformation en séries infinies très-simples, et la sommation de ces séries, pour le cas de $s = 1$, qu'on trouve dans le premier et dans le quatrième paragraphe du Mémoire cité. En se servant d'une méthode élégante due à Euler, M. Dirichlet a trouvé

$$\Pi \frac{1}{1 - \frac{\beta^{k \operatorname{ind} q}}{q^s}} = \sum \frac{\beta^{k \operatorname{ind} n}}{n^s},$$

où le signe sommatoire s'étend à tous les nombres entiers positifs n, non divisibles par λ. La substitution des séries au lieu des produits donne

$$R = \frac{s-1}{1 - \frac{1}{\lambda^s}} \sum \frac{1}{n^s} \cdot \sum \frac{\beta^{\operatorname{ind} n}}{n^s} \cdot \sum \frac{\beta^{2 \operatorname{ind} n}}{n^s} \cdots \sum \frac{\beta^{(\lambda-2) \operatorname{ind} n}}{n^s}.$$

Maintenant, si l'on fait converger indéfiniment s vers la limite $s = 1$, on aura, d'après un théorème de M. Dirichlet,

$$(s-1) \sum \frac{1}{n^s} = 1 - \frac{1}{\lambda} \quad \text{pour} \quad s = 1,$$

et puisque les autres séries contenues dans l'expression de R ne cessent pas d'être convergentes pour $s = 1$, on a

$$R = \sum \frac{\beta^{\operatorname{ind} n}}{n} \cdot \sum \frac{\beta^{2 \operatorname{ind} n}}{n} \cdots \sum \frac{\beta^{(\lambda-2) \operatorname{ind} n}}{n} \quad \text{pour} \quad s = 1.$$

Les sommes finies de ces séries, trouvées par M. Dirichlet au § IV du Mémoire cité, peuvent être représentées de la manière suivante ·

1°. Si k est impair,

$$\sum \frac{\beta^{-k \operatorname{ind} n}}{n} = \frac{\pi \sqrt{-1}}{\lambda (\beta^k, \alpha)} \left[1 + \gamma_1 \beta^k + \gamma_2 + \beta^{2k} + \gamma_3 \beta^{3k} \ldots + \gamma_{\lambda-2} \beta^{(\lambda-2)k} \right],$$

où $\gamma_1, \gamma_2, \gamma_3, \ldots, \gamma_{\lambda-2}$ désignent les plus petits restes positifs des puissances correspondantes de la racine primitive $\gamma, \gamma^2, \gamma^3, \ldots, \gamma^{\lambda-2}$, pour le module λ, et (β^k, α) l'expression connue de la division du cercle

$$(\beta^k, \alpha) = \alpha + \beta^k \alpha^\gamma + \beta^{2k} \alpha^{\gamma^2} + \beta^{3k} \alpha^{\gamma^3} + \ldots + \beta^{(\lambda-2)k} \alpha^{\gamma^{(\lambda-2)}};$$

2°. Si k est pair,

$$\sum \frac{\beta^{-k \operatorname{ind} n}}{n} = \frac{2\left[\operatorname{l}e(\alpha) + \beta^{k}\operatorname{l}e(\alpha^{\gamma}) + \beta^{2k}\operatorname{l}e(\alpha^{\gamma^2}) + \ldots + \beta^{(\mu-1)k}\operatorname{l}e\left(\alpha^{\gamma^{\mu-1}}\right)\right]}{(1-\beta^{-k}).(\beta^{k}, \alpha)},$$

où l'on a mis, pour abréger,

$$\mu = \frac{\lambda - 1}{2},$$

et où $e(\alpha)$ désigne l'unité complexe

$$e(\alpha) = \sqrt{\frac{(1-\alpha^{\gamma}).(1-\alpha^{-\gamma})}{(1-\alpha).(1-\alpha^{-1})}} = \pm \frac{\alpha^{\mu(\gamma-1)}(1-\alpha^{\gamma})}{1-\alpha},$$

la même qui nous a donné un système indépendant d'unités dans le deuxième paragraphe de ce Mémoire.

Substituons ces sommes dans l'expression trouvée de R, en faisant, pour abréger,

$$1 + \gamma_1 \beta + \gamma_2 \beta^2 + \gamma_3 \beta^3 + \ldots + \gamma_{\lambda-2} \beta^{\lambda-2} = \varphi(\beta),$$

$$\varphi(\beta).\varphi(\beta^3).\varphi(\beta^5)\ldots\varphi(\beta^{\lambda-2}) = \mathrm{P},$$

et comme ci-dessus, dans le § II,

$$\operatorname{l}e(\alpha) + \beta^2 \operatorname{l}e(\alpha^{\gamma}) + \beta^4 \operatorname{l}e(\alpha^{\gamma^2}) + \ldots + \beta^{2\mu-2}\operatorname{l}e\left(\alpha^{\gamma^{\mu-1}}\right) = \mathrm{L}(\beta^2),$$

nous aurons

$$\mathrm{R} = \frac{2^{\mu-1}\pi^{\mu}(-1)^{\frac{\mu}{2}}\,\mathrm{P}\,\mathrm{L}(\beta^2).\mathrm{L}(\beta^4).\mathrm{L}(\beta^6)\ldots\mathrm{L}(\beta^{2\mu-2})}{\lambda^{\mu}(\beta,\alpha).(\beta^3,\alpha)\ldots(\beta^{\lambda-2},\alpha).(1-\beta^{-2}).(1-\beta^{-4})\ldots(1-\beta^{-(2\mu-2)})}.$$

Or nous avons trouvé, dans le § II, qu'en désignant par D le déterminant des quantités

$$\begin{array}{llll}
\operatorname{l}e(\alpha), & \operatorname{l}e(\alpha^{\gamma}), \ldots, & \operatorname{l}e\left(\alpha^{\gamma^{\mu-2}}\right), \\
\operatorname{l}e(\alpha^{\gamma}), & \operatorname{l}e(\alpha^{\gamma^2}), \ldots, & \operatorname{l}e\left(\alpha^{\gamma^{\mu-1}}\right), \\
\ldots & \ldots & \ldots \\
\operatorname{l}e\left(\alpha^{\gamma^{\mu-2}}\right), & \operatorname{l}e\left(\alpha^{\gamma^{\mu-1}}\right), \ldots, & \operatorname{l}e\left(\alpha^{\gamma^{\mu-4}}\right),
\end{array}$$

58..

on a

$$L(\beta^2).L(\beta^4).L(\beta^6)\ldots L(\beta^{2-\mu-2}) = \mu D;$$

de plus, les équations

$$(\beta^k, \alpha).(\beta^{-k}, \alpha) = \pm\lambda \quad \text{et} \quad (\beta^\mu, \alpha) = (-1, \alpha) = \pm\sqrt{(-1)^{\mu\lambda}}$$

donnent

$$(\beta, \alpha).(\beta^2, \alpha)\ldots(\beta^{\lambda-2}, \alpha) = \pm\lambda^{\mu+\frac{1}{2}}\sqrt{(-1)^\mu},$$

et l'on a

$$(1-\beta^{-2}).(1-\beta^{-4})\ldots(1-\beta^{2-\mu-2}) = \mu:$$

donc on aura enfin l'expression simplifiée

$$R = \frac{2^{\mu-1}\pi^\mu PD}{\lambda^{2\mu-\frac{1}{2}}} \quad \text{pour} \quad s = 1.$$

Après avoir trouvé cette somme de la série R, pour le cas limite $s = 1$, nous venons à la seconde partie de notre recherche, dont le but sera de trouver une nouvelle expression de la même série R, afin que la comparaison de ces deux résultats nous donne le nombre cherché des classes des nombres idéaux.

Dans la série proposée

$$R = \sum \frac{s-1}{[NF(\alpha)]^s},$$

où $F(\alpha)$ représente tous les nombres complexes différents, existants et idéaux, nous séparons les termes suivant les classes diverses auxquelles appartiennent les nombres complexes $F(\alpha)$. En désignant par $f(\alpha)$ tous les nombres de la première classe, c'est-à-dire les nombres complexes existants, par $f_1(\alpha)$ les nombres idéaux de la deuxième classe, par $f_2(\alpha)$ ceux de la troisième classe, et ainsi de suite; et en nommant H le nombre de toutes les classes diverses, nous avons

$$R = \sum \frac{s-1}{[Nf(\alpha)]^s} + \sum \frac{s-1}{[Nf_1(\alpha)]^s} + \ldots + \sum \frac{s-1}{[Nf_{H-1}(\alpha)]^s}:$$

il s'agira donc de trouver les sommes de ces H séries particulières, pour le cas limite $s = 1$. Nous démontrerons ci-après que toutes ces sommes, prises par rapport aux diverses classes des nombres idéaux, sont égales entre elles; c'est pour cette raison que nous nous bornerons ici à trouver seulement la première somme, qui ne contient que les nombres complexes existants.

Tous les nombres complexes existants sont contenus dans la forme

$$f(\alpha) = x\alpha + x_1\alpha^{\gamma} + x_2\alpha^{\gamma^2} + \ldots + x_{\lambda-2}\alpha^{\gamma^{\lambda-2}},$$

dans laquelle les coefficients

$$x, \quad x_1, \quad x_2, \ldots, \quad x_{\lambda-2}$$

représentent tous les nombres entiers; mais pour toutes ces valeurs différentes des coefficients, on n'obtiendra pas des nombres complexes différents, car on sait qu'au moyen des unités complexes, le même nombre complexe peut être représenté, par la forme proposée, d'une infinité de manières différentes. Il s'agira donc d'abord de trouver les conditions auxquelles les coefficients

$$x, \quad x_1, \quad x_2, \ldots, \quad x_{\lambda-2}$$

doivent être assujettis, pour que $f(\alpha)$ ne soit qu'une seule fois contenu dans cette forme. A cet effet, j'observe que $f'(\alpha)$ étant une expression déterminée du nombre complexe dont il s'agit, on aura toutes les autres expressions du même nombre, en multipliant celui-ci par une unité quelconque; et puisque toutes les unités sont contenues dans la forme

$$\varepsilon(\alpha) = \pm\,\alpha^n \varepsilon_1(\alpha)^{m_1} \varepsilon_2(\alpha)^{m_2} \ldots \varepsilon_{\mu-1}(\alpha)^{m_{\mu-1}},$$

où $\varepsilon_1(\alpha)$, $\varepsilon_2(\alpha), \ldots, \varepsilon_{\mu-1}(\alpha)$ sont $\mu - 1$ unités fondamentales, la forme

$$f(\alpha) = \pm\,\alpha^n \varepsilon_1(\alpha)^{m_1} . \varepsilon_2(\alpha)^{m_2} \ldots \varepsilon_{\mu-1}(\alpha)^{m_{\mu-1}} f'(\alpha),$$

pour toutes les valeurs entières des exposants n, m_1, $m_2, \ldots, m_{\mu-1}$, contiendra toutes les expressions possibles du même nombre complexe. Lorsqu'on y change α en α^{-1}, on obtient, en multipliant,

$$f(\alpha).f(\alpha^{-1}) = \varepsilon_1(\alpha)^{2m_1} . \varepsilon_2(\alpha)^{2m_2} \ldots \varepsilon_{\mu-1}(\alpha)^{2m_{\mu-1}} f'(\alpha).f'(\alpha^{-1}).$$

En passant aux logarithmes, et changeant α en α^{γ}, $\alpha^{\gamma^2}, \ldots, \alpha^{\gamma^{\mu-2}}$

on a ce système d'équations,

$$\tfrac{1}{2}\,\mathrm{l}\,[rf(\alpha).f(\alpha^{-1})] = m_1\,\mathrm{l}\varepsilon_1(\alpha) + m_2\,\mathrm{l}\varepsilon_2(\alpha)$$
$$+ m_{\mu-1}\,\mathrm{l}\varepsilon_{\mu-1}(\alpha) + \tfrac{1}{2}\,\mathrm{l}\,[rf'(\alpha).f'(\alpha^{-1})],$$

$$\tfrac{1}{2}\,\mathrm{l}\,[rf(\alpha^{\gamma}).f(\alpha^{-\gamma})] = m_1\,\mathrm{l}\varepsilon_1(\alpha^{\gamma}) + m_2\,\mathrm{l}\varepsilon_2(\alpha^{\gamma}) + \ldots$$
$$+ m_{\mu-1}\,\mathrm{l}\varepsilon_{\mu-1}(\alpha^{\gamma}) + \tfrac{1}{2}\,\mathrm{l}\,[rf'(\alpha^{\gamma}).f'(\alpha^{-\gamma})],$$

. . . .

$$\tfrac{1}{2}\,\mathrm{l}\left[rf\left(\alpha^{\gamma^{\mu-2}}\right).f\left(\alpha^{-\gamma^{\mu-2}}\right)\right] = m_1\,\mathrm{l}\varepsilon_1\left(\alpha^{\gamma^{\mu-2}}\right) + m_2\,\mathrm{l}\varepsilon_2\left(\alpha^{\gamma^{\mu-2}}\right) + \ldots$$
$$+ m_{\mu-1}\,\mathrm{l}\varepsilon^{\mu-1}\left(\alpha^{\gamma^{\mu-2}}\right) + \tfrac{1}{2}\,\mathrm{l}\left[rf'\left(\alpha^{\gamma^{\mu-2}}\right).f'\left(\alpha^{-\gamma^{\mu-2}}\right)\right],$$

où le nombre r est tout à fait arbitraire.

Mettons à présent

$$\tfrac{1}{2}\,\mathrm{l}\,[rf'(\alpha).f'(\alpha^{-1})] = y_1\,\mathrm{l}\varepsilon_1(\alpha) + y_2\,\mathrm{l}\varepsilon_2(\alpha) + \ldots + y_{\mu-1}\,\mathrm{l}\varepsilon_{\mu-1}(\alpha),$$

$$\tfrac{1}{2}\,\mathrm{l}\,[rf'(\alpha^{\gamma}).f'(\alpha^{-\gamma})] = y_1\,\mathrm{l}\varepsilon_1(\alpha^{\gamma}) + y_2\,\mathrm{l}\varepsilon_2(\alpha^{\gamma}) + \ldots + y_{\mu-1}\,\mathrm{l}\varepsilon_{\mu-1}(\alpha^{\gamma}),$$

. . . .

$$\tfrac{1}{2}\,\mathrm{l}\left[rf'\left(\alpha^{\gamma^{\mu-2}}\right).f'\left(\alpha^{\gamma^{\mu-2}}\right)\right] = y_1\,\mathrm{l}\varepsilon_1\left(\alpha^{\gamma^{\mu-2}}\right) + y_2\,\mathrm{l}\varepsilon_2\left(\alpha^{\gamma^{\mu-2}}\right) + \ldots$$
$$+ y_{\mu-1}\,\mathrm{l}\varepsilon_{\mu-1}\left(\alpha^{\gamma^{\mu-2}}\right);$$

les inconnues $y_1, y_2, \ldots, y_{\mu-1}$ de ce système d'équations linéaires auront toujours des valeurs finies et déterminées, parce que le déterminant du système, ou bien le déterminant Δ du § II de ce Mémoire, n'est pas égal à zéro. Ce système d'équations, ajouté au précédent, donne aussi

$$\tfrac{1}{2}\,\mathrm{l}\,[rf(\alpha).f(\alpha^{-1})] = (m_1 + y_1)\,\mathrm{l}\varepsilon_1(\alpha) + (m_2 + y_2)\,\mathrm{l}\varepsilon_2(\alpha)$$
$$+ (m_{\mu-1} + y_{\mu-1})\,\mathrm{l}\varepsilon_{\mu-1}(\alpha),$$

$$\tfrac{1}{2}\,\mathrm{l}\,[rf(\alpha^{\gamma}).f(\alpha^{-\gamma})] = (m_1 + y_1)\,\mathrm{l}\varepsilon_1(\alpha^{\gamma}) + (m_2 + y_2)\,\mathrm{l}\varepsilon_2(\alpha^{\gamma})$$
$$+ (m_{\mu-1} + y_{\mu-1})\,\mathrm{l}\varepsilon_{\mu-1}(\alpha^{\gamma}),$$

. . . .

$$\tfrac{1}{2}\,\mathrm{l}\left[rf\left(\alpha^{\gamma^{\mu-2}}\right).f\left(\alpha^{-\gamma^{\mu-2}}\right)\right] = (m_1 + y_1)\,\mathrm{l}\varepsilon_1\left(\alpha^{\gamma^{\mu-2}}\right)$$
$$+ (m_2 + y_2)\,\mathrm{l}\varepsilon_2\left(\alpha^{\gamma^{\mu-2}}\right) + \ldots + (m_{\mu-1} + y_{\mu-1})\,\mathrm{l}\varepsilon_{\mu-1}\left(\alpha^{\gamma^{\mu-2}}\right).$$

On voit par là que les expressions de

$$\tfrac{1}{2}\,\mathrm{l}\,[rf'(\alpha).f'(\alpha^{-1})] \quad \text{et} \quad \tfrac{1}{2}\,\mathrm{l}\,r\,[rf(\alpha).f(\alpha^{-1})],$$

et celles qu'on en déduit par le changement de α en α^{γ}, α^{γ^2},..., $\alpha^{\gamma^{\mu-2}}$, ne diffèrent qu'en ce que les quantités $y_1, y_2, \ldots, y_{\mu-1}$ sont augmentées des nombres entiers positifs ou négatifs $m_1, m_2, \ldots, m_{\mu-1}$. On pourra donc toujours choisir convenablement les nombres entiers $m_1, m_2, \ldots, m_{\mu-1}$, de manière que les quantités $m_1 + y_1$, $m_2 + y_2, \ldots, m_{\mu-1} + y_{\mu-1}$ soient toutes contenues entre les limites o et 1, et par cette condition, les nombres entiers $m_1, m_2, \ldots, m_{\mu-1}$ seront parfaitement déterminés. Mettant donc, pour abréger,

$$m_1 + y_1 = z_1, \quad m_2 + y_2 = z_2, \ldots, m_{\mu-1} + y_{\mu-1} = z_{\mu-1},$$

nous concluons que parmi tous les nombres complexes contenus dans la forme

$$f(\alpha) = \varepsilon_1(\alpha)^{m_1}.\,\varepsilon_2(\alpha)^{m_2} \ldots \varepsilon_{\mu-1}(\alpha)^{m_{\mu-1}} f'(\alpha),$$

il y aura toujours un, et il n'y aura jamais qu'un seul qui satisfait aux conditions

$$\tfrac{1}{2}\,\mathrm{l}[rf(\alpha)f(\alpha^{-1})] = z_1\,\mathrm{l}\,\varepsilon_1(\alpha) + z_2\,\mathrm{l}\,\varepsilon_2(\alpha) + \ldots + z_{\mu-1}\,\mathrm{l}\,\varepsilon_{\mu-1}(\alpha),$$

$$\tfrac{1}{2}\,\mathrm{l}\left[rf\left(\alpha^{\gamma}\right)f\left(\alpha^{-\gamma}\right)\right] = z_1\,\mathrm{l}\,\varepsilon_1\left(\alpha^{\gamma}\right) + z_2\,\mathrm{l}\,\varepsilon_2\left(\alpha^{\gamma}\right) + \ldots + z_{\mu-1}\,\mathrm{l}\,\varepsilon_{\mu-1}\left(\alpha^{\gamma}\right),$$

. . . .

$$\tfrac{1}{2}\,\mathrm{l}\left[rf\left(\alpha^{\gamma^{\mu-2}}\right)f\left(\alpha^{-\gamma^{\mu-2}}\right)\right] = z_1\,\mathrm{l}\,\varepsilon_1\left(\alpha^{\gamma^{\mu-2}}\right) + z_2\,\mathrm{l}\,\varepsilon_2\left(\alpha^{\gamma^{\mu-2}}\right) + \ldots + z_{\mu-1}\,\mathrm{l}\,\varepsilon_{\mu-1}\left(\alpha^{\gamma^{\mu-2}}\right),$$

$z_1, z_2, \ldots, z^{\mu-1}$ désignant des quantités contenues entre les limites o et 1. A cause du facteur $\pm\,\alpha^n$, qu'il faut encore ajouter pour avoir toutes les expressions de $f(\alpha)$, facteur qui donne lieu à 2λ valeurs différentes, il y aura toujours 2λ représentations du nombre $f(\alpha)$, pour lesquelles les conditions posées seront satisfaites. Il serait facile d'ajouter une nouvelle condition des coefficients, en vertu de laquelle le nombre $f(\alpha)$ n'aurait qu'une seule représentation; mais il suffira de savoir que le nombre des représentations de $f(\alpha)$ est exactement égal à 2λ.

D'après les principes établis par M. Lejeune-Dirichlet dans ses recherches sur diverses applications de l'analyse infinitésimale à la théorie des nombres, on obtient la valeur de la somme

$$\sum \frac{s-1}{[\mathrm{N}f(\alpha)]^s}, \quad \text{pour} \quad s = 1,$$

égale au quotient du nombre des termes de cette série, pour lesquels $\mathrm{N}f(\alpha)$ est inférieur à T, divisé par T, lorsqu'on y fait croître la quantité T à l'infini. Nous obtiendrons le nombre des termes pour lesquels on a $\mathrm{N}f(\alpha) < \mathrm{T}$, en recherchant combien il y a de systèmes de valeurs des coefficients $x, x_1, \ldots, x_{\lambda-2}$, compatibles aux $\mu - 1$ conditions posées jointes à la condition $\mathrm{N}f(\alpha) < \mathrm{T}$, et puis divisant ce nombre par 2λ. Pour cela, nous changeons les quantités $x, x_1, \ldots, x_{\lambda-2}$ en $\frac{x}{t}, \frac{x_1}{t}, \ldots, \frac{x_{\lambda-2}}{t}$; d'où $f(\alpha)$ se change en $\frac{f(\alpha)}{t}$, $\mathrm{N}f(\alpha)$ en $\frac{\mathrm{N}f(\alpha)}{t^{\lambda-1}}$; alors les nouvelles quantités $x, x_1, \ldots, x_{\lambda-1}$ n'obtiendront plus toutes les valeurs entières compatibles aux conditions posées, mais les valeurs d'une série arithmétique doublement infinie, dont la différence est égale à t. Les $\mu - 1$ conditions pour les nouvelles quantités $x, x_1, \ldots, x_{\lambda-2}$ ne sont altérées qu'en ce que, au lieu de r on aura $\frac{r}{t^2}$, et puisque r est tout à fait arbitraire, on prendra $r = t^2$, ce qui revient à faire $r = 1$ dans les conditions données ci-dessus. De plus, en prenant $\mathrm{T} = \frac{1}{t^{\lambda-1}}$, au lieu de $\mathrm{N}f(\alpha) < \mathrm{T}$, on a pour les nouveaux coefficients du nombre $f(\alpha)$ la condition $\mathrm{N}f(\alpha) < 1$. Maintenant, si l'on prend t infiniment petit, les quantités $x, x_1, \ldots, x_{\lambda-2}$ seront des variables continues, le nombre des valeurs de x pourra être représenté sous la forme $\frac{1}{t}\int dx$, le nombre des valeurs de x_1 sous la forme $\frac{1}{t}\int dx_1$, etc.; donc le nombre de toutes les combinaisons des valeurs de $x, x_1, \ldots, x_{\lambda-1}$ sera donné par l'intégrale multiple

$$\frac{1}{t^{\lambda-1}}\int^{(\lambda-1)} dx.dx_1.dx_2\ldots dx_{\lambda-2},$$

ce qui donne, en divisant par $T = \frac{1}{t^{\lambda-1}}$,

$$\int^{(\lambda-1)} dx . dx_1 . dx_2 ... dx_{\lambda-2}.$$

Nous en concluons que la valeur de la somme proposée, pour le cas de $s = 1$, est donnée par la formule

$$\sum \frac{s-1}{[\mathrm{N}f(\alpha)]^s} = \frac{1}{2\lambda} \int^{(\lambda-1)} dx . dx_1 . dx_2 ... dx_{\lambda-2},$$

où les intégrations s'étendent sur toutes les valeurs des variables x, $x_1, ..., x_{\lambda-2}$ entre les limites $-\infty$ et $+\infty$, qui satisfont à la condition

* $\mathrm{N}f(\alpha) > 1$ et aux équations

$$\tfrac{1}{2} \mathrm{l}[rf(\alpha).f(\alpha^{-1})] = z_1 \mathrm{l}\varepsilon_1(\alpha) + z_2 \mathrm{l}\varepsilon_2(\alpha) + ... + z_{\mu-1} \mathrm{l}\varepsilon_{\mu-1}(\alpha),$$

$$\tfrac{1}{2} \mathrm{l}\left[rf\left(\alpha^{\gamma}\right).f\left(\alpha^{-\gamma}\right)\right] = z_1 \mathrm{l}\varepsilon_1\left(\alpha^{\gamma}\right) + z_2 \mathrm{l}\varepsilon_2\left(\alpha^{\gamma}\right) + ... + z_{\mu-1} \mathrm{l}\varepsilon_{\mu-1}\left(\alpha^{\gamma}\right),$$

. . . .

$$\tfrac{1}{2} \mathrm{l}\left[rf\left(\alpha^{\gamma^{\mu-2}}\right).f\left(\alpha^{-\gamma^{\mu-2}}\right)\right] = z_1 \mathrm{l}\varepsilon_1\left(\alpha^{\gamma^{\mu-2}}\right) + z_2 \mathrm{l}\varepsilon_2\left(\alpha^{\gamma^{\mu-2}}\right) + ... + z_{\mu-1} \mathrm{l}\varepsilon_{\mu-1}\left(\alpha^{\gamma^{\mu-2}}\right),$$

où les quantités $z_1, z_2, ..., z_{\mu-1}$ sont contenues entre les limites 0 et 1.

Pour déterminer la valeur de l'intégrale

$$\int^{(\lambda-1)} dx . dx_1 . dx_2 ... dx_{\lambda-2} = \mathrm{I},$$

nous y appliquerons quelques transformations successives, qui seront effectuées d'après les règles connues pour la transformation des intégrales multiples.

D'abord nous introduirons les nouvelles variables $y, y_1, y_2, ..., y_{\lambda-2}$, liées aux variables $x, x_1, x_2, ..., x_{\lambda-2}$ par les équations

$$y = f(\alpha), \quad y_1 = f\left(\alpha^{\gamma}\right), \quad y_2 = f\left(\alpha^{\gamma^2}\right), ..., \quad y_{\lambda-2} = f\left(\alpha^{\gamma^{\lambda-2}}\right).$$

Les différentielles partielles des nouvelles variables, prises par rapport aux variables x, $x_1, \ldots, x^{\lambda-2}$, sont données par la formule

$$\frac{dy_k}{dx_h} = \alpha^{\gamma^{k+h}},$$

d'où il suit que le déterminant

$$\sum\left(\pm\frac{dy}{dx}\cdot\frac{dy_1}{dx_1}\cdots\frac{dy_{\lambda-2}}{dx_{\lambda-2}}\right)$$

sera le déterminant des quantités

$$\begin{array}{llll} \alpha, & \alpha\gamma, & \alpha^{\gamma^2},\ldots, & \alpha^{\gamma^{\lambda-2}}, \\ \alpha^{\gamma}, & \alpha^{\gamma^2}, & \alpha^{\gamma^3},\ldots, & \alpha, \\ \ldots & \ldots & \ldots & \ldots \\ \alpha^{\gamma^{\lambda-2}}, & \alpha, & \alpha^{\gamma},\ldots, & \alpha^{\gamma^{\lambda-3}}, \end{array}$$

qu'on sait être égal au produit des différences de toutes ces quantités prises deux à deux,

$$\begin{array}{r} \left(\alpha-\alpha^{\gamma}\right).\left(\alpha-\alpha^{\gamma^2}\right)\ldots\left(\alpha-\alpha^{\gamma^{\lambda-2}}\right), \\ \left(\alpha^{\gamma}-\alpha^{\gamma^2}\right)\ldots\left(\alpha^{\gamma}-\alpha^{\gamma^{\lambda-2}}\right), \\ \ldots\ldots\ldots\ldots \\ \left(\alpha^{\gamma^{\lambda-3}}-\alpha^{\gamma^{\lambda-2}}\right). \end{array}$$

Ce produit, qui contient $\frac{(\lambda-1)(\lambda-2)}{2}$ facteurs premiers de λ, est égal à la puissance $\lambda^{\frac{\lambda-2}{2}}$, d'où résulte

$$\sum\left(\pm\frac{dy}{dx}\cdot\frac{dy_1}{dx_1}\cdots\frac{dy_{\lambda-2}}{dx_{\lambda-2}} = \lambda^{\frac{\lambda-2}{2}}\right);$$

et puisqu'on obtient l'intégrale multiple transformée en divisant par ce déterminant, on aura la première transformée

$$\mathrm{I} = \frac{1}{\lambda^{\frac{\lambda-2}{2}}}\int^{(\lambda-1)} dy\, dy_1 \ldots\, dy_{\lambda-2}.$$

Transformons de nouveau par les substitutions

$$\mathrm{l}y = v + \omega\sqrt{-1},\ \mathrm{l}y_1 = v_1 + \omega_1\sqrt{-1},\ldots,\ \mathrm{l}y_{\mu-1} = v_{\mu-1} + \omega_{\mu-1}\sqrt{-1},$$

$$\mathrm{l}y_\mu = v - \omega\sqrt{-1},\quad \mathrm{l}y_{\mu+1} = v_1 - \omega_1\sqrt{-1},\ldots,$$

$$\mathrm{l}y_{2\mu+1} = v_{\mu-1} - \omega_{\mu-1}\sqrt{-1},$$

nous aurons en général

$$\frac{dy_k}{y_k} = dv_k + \sqrt{-1}\,d\omega_k,\quad \frac{dy_{k+\mu}}{y_{k+\mu}} = dv_k - \sqrt{-1}\,d\omega_k.$$

Alors au lieu de $dy_k\,dy_{k+\mu}$ dans l'intégrale multiple, il faudra substituer l'expression

$$\left(\frac{dy_k}{dv_k}\cdot\frac{dy_{k+\mu}}{d\omega_k} - \frac{dy_k}{d\omega_k}\cdot\frac{dy_{k+\mu}}{dv_k}\right)dv_k\,d\omega_k,$$

qui se réduit à

$$-2\sqrt{-1}\ e^{2v_k}\,dv_k.d\omega_k,$$

et l'on aura l'intégrale transformée

$$\mathrm{I} = \frac{2^\mu}{\lambda^{\frac{\lambda-2}{2}}}\int^{(\lambda-1)} e^{2v}.e^{2v_1}\ldots e^{2v_{\mu-1}}\,dv\,dv_1\ldots dv_{\mu-1}\,d\omega\,d\omega_1\ldots d\omega_{\mu-1}.$$

Ayant posé

$$f\left(\alpha^{\gamma^k}\right) = y_k = e^{v_k+\omega_k\sqrt{-1}},\quad f\left(\alpha^{-\gamma^k}\right) = y_{k+\mu} = e^{v_k-\omega_k\sqrt{-1}},$$

nous avons

$$f\left(\alpha^{\gamma^k}\right).f\left(\alpha^{-\gamma^k}\right) = e^{2v_k},\quad v_k = \tfrac{1}{2}\mathrm{l}\left[f\left(\alpha^{\gamma^k}\right).f\left(\alpha^{-\gamma^k}\right)\right];$$

les limites des nouvelles variables seront donc déterminées par les équations

$$v = z_1\,\mathrm{l}\,\varepsilon_1(\alpha) + z_2\,\mathrm{l}\,\varepsilon_2(\alpha) + \ldots + z_{\mu-1}\,\mathrm{l}\,\varepsilon_{\mu-1}(\alpha),$$

$$v_1 = z_1\,\mathrm{l}\,\varepsilon_1\left(\alpha^\gamma\right) + z_2\,\mathrm{l}\,\varepsilon_2\left(\alpha^\gamma\right) + \ldots + z_{\mu-1}\,\mathrm{l}\,\varepsilon_{\mu-1}\left(\alpha^\gamma\right),$$

. . . .

$$v_{\mu-2} = z_1\,\mathrm{l}\,\varepsilon_1\left(\alpha^{\gamma^{\mu-2}}\right) + z_2\,\mathrm{l}\,\varepsilon_2\left(\alpha^{\gamma^{\mu-2}}\right) + \ldots + z_{\mu-1}\,\mathrm{l}\,\varepsilon_{\mu-1}\left(\alpha^{\gamma^{\mu-2}}\right),$$

59..

où $z_1, z_2, \ldots, z_{\mu-1}$ sont contenus entre les limites 0 et 1. La condition $\mathrm{N} f(\alpha) < 1$ donne, en outre,

$$e^{2v} . e^{2v_1} \ldots e^{2v_{\mu-1}}$$

entre les limites 0 et 1.

Les μ variables $\omega, \omega_1, \ldots, \omega_{\mu-1}$, qui ne sont pas limitées par ces conditions, sont données comme fonctions des variables primitives $x, x_1, \ldots, x_{\lambda-2}$ par l'équation

$$\operatorname{tang}(\omega_k) = \frac{f\left(\alpha^{\gamma^k}\right) - f\left(\alpha^{-\gamma^k}\right)}{\sqrt{-1}\left[f\left(\alpha^{\gamma^k}\right) + f\left(\alpha^{-\gamma^k}\right)\right]},$$

expression où les quantités imaginaires disparaissent d'elles-mêmes; et, puisque les quantités $x, x_1, \ldots, x_{\lambda-2}$ peuvent varier entre les limites $-\infty$ et $+\infty$, on voit que la même chose aura lieu pour $\operatorname{tang}(\omega_k)$: et, en considérant encore que $\sin \omega_k$ et $\cos \omega_k$ peuvent avoir toutes les valeurs entre les limites -1 et $+1$, on conclut que les limites des variables $\omega, \omega_1, \omega_2, \ldots, \omega_{\mu-1}$ seront $-\frac{\pi}{2}$ et $+\frac{3\pi}{2}$, ou tout autre intervalle de 2π. En effectuant les intégrations par rapport à ces variables, entre les limites $-\frac{\pi}{2}$ et $+\frac{3\pi}{2}$, on a

$$\mathrm{I} = \frac{2^{2\mu}\pi^\mu}{\lambda^{\frac{\lambda-2}{2}}} \int^{(\mu)} e^{2v} . e^{2v_1} \ldots e^{2v_{\mu-1}} \, dv . dv_1 \ldots dv_{\mu-1}.$$

Intégrons à présent par rapport à la variable $v_{\mu-1}$, nous aurons

$$\int e^{2v_{\mu-1}} dv_{\mu-1} = \frac{1}{2} e^{2v_{\mu-1}},$$

les limites de cette intégration étant données par la condition $e^{2v} . e^{2v_1} \ldots e^{2v_{\mu-1}} \gtrless {0 \atop 1}$, et celles de $e^{2v_{\mu-1}}$ étant 0 et $e^{-2v} . e^{-2v_1} \ldots e^{2v_{\mu-2}}$. L'intégration entre ces limites donne le résultat très-simple

$$\frac{2^{2\mu-1}\pi^\mu}{\lambda^{\frac{\lambda-2}{2}}} \int^{(\mu-1)} dv . dv_1 \ldots dv_{\mu-2}.$$

Il nous reste encore à effectuer la dernière transformation de cette intégrale, qui consiste dans la substitution des variables $z_1, z_2, \ldots, z_{\mu-1}$, données par les équations

$$v = z_1 \,\mathrm{l}\varepsilon_1(\alpha) + z_2 \,\mathrm{l}\varepsilon_2(\alpha) + \ldots + z_{\mu-1} \,\mathrm{l}\varepsilon_{\mu-1}(\alpha),$$

$$v_1 = z_1 \,\mathrm{l}\varepsilon_1(\alpha^\gamma) + z_2 \,\mathrm{l}\varepsilon_2(\alpha^\gamma) + \ldots + z_{\mu-1} \,\mathrm{l}\varepsilon_{\mu-1}(\alpha^\gamma),$$

. . . .

$$v_{\mu-2} = z_1 \,\mathrm{l}\varepsilon_1\left(\alpha^{\gamma^{\mu-2}}\right) + z_2 \,\mathrm{l}\varepsilon_2\left(\alpha^{\gamma^{\mu-2}}\right) + \ldots + z_{\mu-1} \,\mathrm{l}\varepsilon_{\mu-1}\left(\alpha^{\gamma^{\mu-2}}\right),$$

au lieu des variables $v, v_1, \ldots, v_{\mu-2}$. En désignant, comme ci-dessus, par Δ le déterminant de ce système d'équations linéaires, nous avons tout de suite l'intégrale transformée, et, parce que toutes les intégrations par rapport aux variables $z_1, z_2, \ldots, z_{\mu-1}$, doivent être prises entre les limites 0 et 1, la nouvelle intégrale est évidemment égale à l'unité. Donc la valeur cherchée de l'intégrale I devient

$$\mathrm{I} = \frac{2^{2\mu-1}\pi^\mu \Delta}{\lambda^{\frac{\lambda-2}{2}}}.$$

De là résulte cette expression de la somme proposée

$$\sum \frac{s-1}{[\mathrm{N}f(\alpha)]^s} = \frac{2^{2\mu-2}\pi^\mu \Delta}{\lambda^{\frac{\lambda}{2}}}, \quad \text{pour} \quad s = 1.$$

Nous allons maintenant démontrer que toutes les autres sommes partielles de l'expression

$$\sum \frac{s-1}{[\mathrm{NF}(\alpha)]^s} = \sum \frac{s-1}{[\mathrm{N}f(\alpha)]^s} + \sum \frac{s-1}{[\mathrm{N}f_1(\alpha)]^s} + \ldots \sum \frac{s-1}{\left[\mathrm{N}f_{\mathrm{H}-1}(\alpha)\right]^s}$$

ont les mêmes valeurs que la première, que nous venons de trouver. Pour cela, soit $\varphi(\alpha)$ un multiplicateur idéal qui, joint aux nombres idéaux de la classe $f_k(\alpha)$, les rend tous existants, et posons

$$\varphi(\alpha) f_k(\alpha) = \mathrm{F}_k(\alpha);$$

nous aurons

$$\sum \frac{s-1}{[\mathrm{N}f_k(\alpha)]^s} = \mathrm{N}\varphi(\alpha) \sum \frac{s-1}{[\mathrm{NF}_k(\alpha)]^s}, \quad \text{pour} \quad s = 1,$$

où $F_k(\alpha)$ désigne tous les nombres complexes existants qui ont le facteur idéal $\varphi(\alpha)$. La recherche de la valeur de la somme

$$\sum \frac{s-1}{[NF_k(\alpha)]^s}, \quad \text{pour } s=1,$$

se fera de la même manière que celle de la somme trouvée ci-dessus, avec la seule différence qu'il y aura quelques conditions de plus pour les coefficients des nombres $F_k(\alpha)$, pour exprimer que $F_k(\alpha)$ doit contenir le facteur idéal $\varphi(\alpha)$. Supposons que $\varphi(\alpha)$ contient un certain facteur premier idéal de q, m fois, de plus un facteur premier idéal de q', m' fois, etc., où q, q', etc., appartiennent respectivement aux exposants f, f', etc.; la condition que $F_k(\alpha)$ soit divisible par $\varphi(\alpha)$ sera exprimée par f congruences linéaires pour les coefficients de $F_k(\alpha)$ par rapport au module q^m, de plus par f' congruences linéaires pour le module $q'^{m'}$, etc. Or il est clair qu'étant donné f congruences linéaires par rapport aux coefficients du nombre $F_k(\alpha)$, pour le module q^m, le nombre de tous les systèmes des valeurs convenables de ces coefficients se réduira à la q^{mf} ième partie; de même, les f' congruences pour le module $q'^{m'}$ réduiront ce nombre de valeurs à la $q'^{m'f'}$ ième partie, et ainsi de suite. On en conclut que la valeur de la somme proposée sera égale à la $q^{mf}.q'^{m'f'}\ldots$ ième partie de la valeur de la somme trouvée ci-dessus, et, puisque le produit $q^{mf}.q'^{m'f'}\ldots$ est égal à la norme $N\varphi(\alpha)$, on aura

$$\sum \frac{s-1}{[NF_k(\alpha)]^s} = \frac{1}{N\varphi(\alpha)} \sum \frac{s-1}{[Nf(\alpha)]^s}, \quad \text{pour} \quad s=1;$$

enfin cette équation, jointe à la précédente, donne

$$\sum \frac{s-1}{[Nf_k(\alpha)]^s} = \sum \frac{s-1}{[Nf(\alpha)]^s}, \quad \text{pour} \quad s=1,$$

ce qu'il fallait démontrer.

Le nombre des sommes partielles qui donnent la somme totale R étant égal à H, nous avons

$$R = H \sum \frac{s-1}{[Nf(\alpha)]^s}, \quad \text{pour} \quad s=1;$$

d'où, en substituant la valeur trouvée de cette somme, nous avons la nouvelle expression

$$R = \frac{2^{2\mu-2}\pi^{\mu}\Delta H}{\lambda^{\frac{\lambda}{2}}};$$

ce qui, comparé à la valeur de la même série R, trouvée dans la première partie de cette recherche,

$$R = \frac{2^{\mu-1}\pi^{\mu}P.D}{\lambda^{2\mu-\frac{1}{2}}},$$

donne l'expression cherchée du nombre des classes des nombres complexes idéaux

$$H = \frac{P.D}{(2\lambda)^{\mu-1}.\Delta}.$$

Je répète que dans cette formule la lettre P désigne le produit

$$P = \varphi(\beta).\varphi(\beta^3).\varphi(\beta^5)\ldots\varphi(\beta^{\lambda-2}),$$

où β est une racine primitive de l'équation $\beta^{\lambda-1} = 1$, et où

$$\varphi(\beta) = 1 + \gamma_1\beta + \gamma_2\beta^2 + \gamma_3\beta^3 + \ldots + \gamma_{\lambda-2}\beta^{\lambda-2},$$

$\gamma_1, \gamma_2, \gamma_3, \ldots, \gamma_{\lambda-2}$ étant les plus petits restes positifs des puissances γ, $\gamma^2, \gamma^3, \ldots, \gamma^{\lambda-2}$ d'une racine primitive γ pour le module λ. La lettre μ n'est qu'un signe abrégé de $\frac{\lambda-1}{2}$. D désigne le déterminant des quantités

$$\begin{array}{llll} 1e(\alpha), & 1e(\alpha^{\gamma}),\ldots, & 1e(\alpha^{\gamma^{\mu-2}}), \\ 1e(\alpha^{\gamma}), & 1e(\alpha^{\gamma^2}),\ldots, & 1e(\alpha^{\gamma^{\mu-1}}), \\ \ldots & \ldots & \ldots \\ 1e(\alpha^{\gamma^{\mu-2}}), & 1e(\alpha^{\gamma^{\mu-1}}),\ldots, & 1e(\alpha^{\gamma^{\mu-4}}), \end{array}$$

où

$$e(\alpha) = \sqrt{\frac{(1-\alpha^{\gamma}).(1-\alpha^{-\gamma})}{(1-\alpha)\,(1-\alpha^{-1})}} = \pm\frac{\alpha^{\mu(\gamma-1)}(1-\alpha^{\gamma})}{1-\alpha}.$$

Δ désigne le déterminant des quantités

$$\begin{array}{llll}
\mathrm{l}\,\varepsilon_1(\alpha), & \mathrm{l}\,\varepsilon_2(\alpha),\ldots, & \mathrm{l}\,\varepsilon_{\mu-1}(\alpha), \\
\mathrm{l}\,\varepsilon_1(\alpha^\gamma), & l\,\varepsilon(\alpha^\gamma),\ldots, & \mathrm{l}\,\varepsilon_{\mu-1}(\alpha^\gamma), \\
\ldots & \ldots & \ldots \\
\mathrm{l}\,\varepsilon_1\left(\alpha^{\gamma^{\mu-2}}\right), & \mathrm{l}\,\varepsilon_2\left(\alpha^{\gamma^{\mu-2}}\right),\ldots, & \mathrm{l}\,\varepsilon_{\mu-1}\left(\alpha^{\gamma^{\mu-2}}\right),
\end{array}$$

où $\varepsilon_1(\alpha)$, $\varepsilon_2(\alpha),\ldots,\varepsilon_{\mu-1}(\alpha)$ représentent un système de $\mu-1$ unites fondamentales.

Nous terminerons ce paragraphe en faisant quelques remarques sur la formule trouvée pour le nombre H des classes des nombres complexes idéaux. D'abord nous observons que le nombre H consiste de deux facteurs entiers bien distincts, le premier égal à $\frac{P}{(2\lambda)^{\mu-1}}$, le second égal à $\frac{D}{\Delta}$. On démontre aisément que le premier de ces facteurs est un nombre entier, en faisant voir que le produit P contient $(\mu-1)$ fois le facteur 2, et autant de fois le facteur λ; pour le second facteur, cela découle immédiatement de l'observation faite à la fin du § II de ce Mémoire. Pour le cas de $\lambda=5$, nous avons trouvé que l'unité $e(\alpha)$ est une unité fondamentale, d'où il suit $D=\Delta$; donc on a

$$\frac{D}{\Delta}=1, \quad \text{pour} \quad \lambda=5.$$

Pour le cas de $\lambda=7$, on trouve sans difficulté que $e(\alpha)$ et $e(\alpha^\gamma)$ sont encore deux unités fondamentales; donc on a de même

$$D=\Delta \quad \text{et} \quad \frac{D}{\Delta}=1, \quad \text{pour} \quad \lambda=7.$$

Mais si λ est plus grand, le calcul effectif de ce second facteur est trespénible, parce qu'il exige qu'on recherche d'abord un système d'unités fondamentales. Le calcul du premier facteur $\frac{P}{(2\lambda)^{\mu-1}}$, que nous désignerons simplement comme fonction de λ par $P'(\lambda)$, n'offre pas cette difficulté; je l'ai calculé pour toutes les valeurs du nombre premier λ

de la première centaine. Voici les résultats de ce calcul :

$$
\begin{array}{ll}
P'(3) = 1, & P'(43) = 211, \\
P'(5) = 1, & P'(47) = 695 = 5.139, \\
P'(7) = 1, & P'(53) = 4889, \\
P'(11) = 1, & P'(59) = 41241 = 3.59.233, \\
P'(13) = 1, & P'(61) = 76301 = 41.1861, \\
P'(17) = 1, & P'(67) = 853513 = 67.12739, \\
P'(19) = 1, & P'(71) = 5472271 = 7.7.29.3851, \\
P'(23) = 3, & P'(73) = 11957417 = 89.134353, \\
P'(29) = 8, & P'(79) = 60087849 = 5.53.377911, \\
P'(31) = 9, & P'(83) = 838216959 = 3.279405653, \\
P'(37) = 37, & P'(89) = 13379363737 = 113.118401449, \\
P'(41) = 121, & P'(97) = 411322823001 = 3457.118982593.
\end{array}
$$

On voit que ces nombres vont en croissant avec une rapidité extraordinaire. La loi asymptotique des valeurs de ce premier facteur du nombre des classes H est exprimée par la formule

$$P'(\lambda) = \frac{P}{(2\lambda)^{\mu-1}} = \frac{\lambda^{\frac{\lambda+3}{4}}}{2^{\frac{\lambda-3}{2}} . \pi^{\frac{\lambda-1}{2}}},$$

dont je me réserve la démonstration et les développements ultérieurs à une autre occasion.

§ IX.

Deux recherches spéciales concernant le nombre des classes des nombres complexes idéaux et les unités complexes.

En examinant les valeurs du premier facteur du nombre H, données dans le paragraphe précédent, on voit que, pour les trois nombres premiers de la première centaine

$$\lambda = 37, \quad \lambda = 59 \quad \text{et} \quad \lambda = 67,$$

le nombre H est divisible par λ. Pour de telles valeurs de λ, les nombres complexes sont doués de quelques propriétés singulières, de manière que ces valeurs de λ se présentent dans plusieurs recherches générales comme cas d'exception qui exigent des méthodes particulières. C'est pour cette raison que nous nous proposons ici le problème *de trouver tous les nombres premiers λ pour lesquels le nombre* H *des classes des nombres complexes idéaux est divisible par λ.*

Nous commençons par discuter le premier facteur $\frac{P}{(2\lambda)^{\mu-1}}$ du nombre

$$H = \frac{P}{(2\lambda)^{\mu-1}} \cdot \frac{D}{\Delta},$$

où nous avons posé

$$P = \varphi(\beta).\varphi(\beta^3).\varphi(\beta^5)\ldots\varphi(\beta^{\lambda-2}),$$

$$\varphi(\beta) = 1 + \gamma_1\beta + \gamma_2\beta^2 + \gamma_3\beta^3 + \ldots + \gamma_{\lambda-2}\beta^{\lambda-2}.$$

En multipliant par $\gamma\beta - 1$, on a la formule

$$(\gamma\beta - 1)\varphi(\beta) = (\gamma\gamma_{\lambda-2} - 1) + (\gamma - \gamma_1)\beta + (\gamma\gamma_1 - \gamma_2)\beta^2 + \ldots$$
$$+ (\gamma\gamma_{\lambda-3} - \gamma_{\lambda-2})\beta^{\lambda-2},$$

dans laquelle tous les coefficients sont divisibles par λ; car, en substituant

$$\gamma_{k-1} \equiv \gamma^{k-1} \quad \text{et} \quad \gamma_k \equiv \gamma^k \quad (\text{mod.}\ \lambda),$$

on a

$$\gamma\gamma_{k-1} - \gamma_k \equiv \gamma\gamma^{k-1} - \gamma^k \equiv 0 \quad (\text{mod.}\ \lambda).$$

Donc, en posant

$$\gamma\gamma_{k-1} - \gamma_k = \lambda b_k$$

et

$$b_0 + b_1\beta + b_2\beta^2 + \ldots + b_{\lambda-2}\beta^{\lambda-2} = \psi(\beta),$$

l'équation précédente donne

$$(\gamma\beta - 1)\varphi(\beta) = \lambda\psi(\beta).$$

En changeant β en β^3, β^5,..., $\beta^{\lambda-2}$ et multipliant, on trouve

$$(\gamma\beta - 1).(\gamma\beta^3 - 1)\ldots(\gamma\beta^{\lambda-2} - 1)P = \lambda^\mu\psi(\beta).\psi(\beta^3)\ldots\psi(\beta^{\lambda-2}),$$

et, comme on a

$$(\gamma\beta - 1).(\gamma\beta^3 - 1).(\gamma\beta^5 - 1)\ldots(\gamma\beta^{\lambda-2} - 1) = \gamma^\mu + 1,$$

cette équation devient

$$(\gamma^\mu + 1)\mathrm{P} = \lambda^\mu \psi(\beta).\psi(\beta^3)\ldots\psi(\beta^{\lambda-2}).$$

Le nombre $\gamma^\mu + 1$, γ étant une racine primitive et $\mu = \frac{\lambda - 1}{2}$, est toujours divisible par λ, on pourra donc poser $\gamma^\mu + 1 = \lambda\mathrm{G}$, où G est un entier; on pourra aussi toujours choisir la racine primitive γ, telle que $\gamma^\mu + 1$ ne soit pas divisible par λ^2, c'est-à-dire que G ne soit pas divisible par λ. Cela étant, on aura

$$2^{\mu-1}\mathrm{G}.\frac{\mathrm{P}}{(2\lambda)^{\mu-1}} = \psi(\beta).\psi(\beta^3)\ldots\psi(\beta^{\lambda-2}),$$

et, puisqu'en substituant la racine primitive γ de la congruence

$$\gamma^{\lambda-1} \equiv 1 \quad (\text{mod. } \lambda)$$

à la racine de l'équation

$$\beta^{\lambda-1} = 1,$$

on a évidemment

$$\psi(\beta).\psi(\beta^3)\ldots\psi(\beta^{\lambda-2}) \equiv \psi(\gamma).\psi(\gamma^3)\ldots\psi(\gamma^{\lambda-2}) \quad (\text{mod. } \lambda),$$

cette équation donne la congruence

$$2^{\mu-1}\mathrm{G}.\frac{\mathrm{P}}{(2\lambda)^{\mu-1}} \equiv \psi(\gamma).\psi(\gamma^3)\ldots\psi(\gamma^{\lambda-2}) \quad (\text{mod. } \lambda).$$

Il suit de là que le premier facteur $\frac{\mathrm{P}}{(2\lambda)^{\mu-1}}$ du nombre des classes H sera ou ne sera pas divisible par λ, selon que parmi les facteurs du produit $\psi(\gamma).\psi(\gamma^3)\ldots\psi(\gamma^{\lambda-2})$ il y en a un ou il n'y en a aucun divisible par λ. Ainsi tout se réduit à examiner si la congruence

$$\begin{aligned}\psi(\gamma^{2n-1}) = b_0 + b_1\gamma^{2n-1} + b_2\gamma^{2(2n-1)} + \ldots \\ + b_{\lambda-2}\gamma^{(\lambda-2)(2n-1)} \equiv 0 \quad (\text{mod. } \lambda)\end{aligned}$$

est ou n'est pas remplie pour une quelconque des valeurs de $n = 1, 2, 3, \ldots, \mu$.

60..

En divisant par γ^{2n-1}, puis changeant les exposants $1, 2, 3, \ldots, \lambda-1$ en indice, au moyen de la congruence $\gamma^k \equiv \gamma_k \pmod{\lambda}$, on aura la congruence équivalente

$$b_0 \gamma_{\lambda-2}^{2n-1} + b_1 + b_2 \gamma_1^{2n-1} + b_3 \gamma_2^{2n-1} + \ldots + b_{\lambda-2} \gamma_{\lambda-3}^{2n-1} \equiv 0 \quad (\text{mod. } \lambda).$$

Or il suit de la définition même des coefficients $\lambda b_k = \gamma\gamma_{k-1} - \gamma_k$ qu'on a $b_k = 0$, pour toutes les valeurs de γ_{k-1} comprises entre les limites 0 et $\frac{\lambda}{\gamma}$, $b_k = 1$ pour toutes les valeurs de γ_{k-1} comprises entre les limites $\frac{\lambda}{\gamma}$ et $\frac{2\lambda}{\gamma}$, et généralement $b_k = s$ pour toutes les valeurs comprises entre les limites $\frac{s\lambda}{\gamma}$ et $\frac{(s+1)\lambda}{\gamma}$. Donc, en comprenant tous les termes de cette congruence, pour lesquels les coefficients b_k ont la même valeur, et en désignant généralement par t_s le plus grand entier contenu dans $\frac{s\lambda}{\gamma}$, on pourra la mettre sous cette forme,

$$\begin{gathered} 1[(t_1+1)^{2n-1} + (t_1+2)^{2n-1} + \ldots + t_2^{2n-1}], \\ + 2[(t_2+1)^{2n-1} + (t_2+2)^{2n-1} + \ldots + t_3^{2n-1}], \\ \cdots\cdots\cdots\cdots\cdots\cdots \\ + (\gamma-1)[(t_{\gamma-1}+1)^{2n-1} + (t_{\gamma-1}+2)^{2n-1} + \ldots \\ + (\lambda-1)^{2n-1}] \equiv 0 \quad (\text{mod. } \lambda). \end{gathered}$$

Nous nous servirons ici de la fonction rationnelle et entière de la variable x,

$$\chi(x) = \frac{x^{2n}}{\Pi_{2n}} - \frac{x^{2n-1}}{2\Pi_{2n-1}} + \frac{B_1 x^{2n}}{\Pi_{2n-2}\Pi_2} - \frac{B_2 x^{2n-2}}{\Pi_{2n-4}\Pi_4} + \ldots + \frac{(-1)^n B_{n-1} x^2}{\Pi_2 \Pi_{2n-2}},$$

dans laquelle $B_1, B_2, \ldots, B_{n-1}$ désignent les nombres bernoulliens, et Π_r le produit $1.2.3\ldots r$. Dans le cas où x est un nombre entier positif, la fonction $\chi(x)$ exprime la série des puissances $(2n-1)^{\text{ièmes}}$ des nombres naturels, car on a la formule connue

$$1^{2n-1} + 2^{2n-1} + 3^{2n-1} + \ldots + (x-1)^{2n-1} = \Pi_{2n-1}\chi(x).$$

Au moyen de cette formule, la congruence proposée se représente de la manière suivante :

$$\begin{aligned}&\chi(t_2+1)-\chi(t_1+1)\\ &+2[\chi(t_3+1)-\chi(t_2+1)]+3[\chi(t_4+1)-\chi(t_3+1)]+\ldots\\ &+(\gamma-1)[\chi(\lambda)-\chi(t_{\gamma-1}+1)]\equiv 0 \pmod{\lambda},\end{aligned}$$

ou, plus simplement,

$$\chi(t_1+1)+\chi(t_2+1)+\chi(t_3+1)+\ldots+\chi(t_{\gamma-1}+1)\equiv 0 \pmod{\lambda}.$$

Le nombre t_s étant le plus grand entier contenu dans la fraction $\frac{s\lambda}{\gamma}$, pourra être mis sous la forme

$$t_s=\frac{s\lambda-r_s}{\gamma},$$

où r_s sera un nombre positif, moindre que γ, et la congruence proposée prendra la forme

$$\begin{aligned}&\chi\left(\frac{\lambda-r_1+\gamma}{\gamma}\right)+\chi\left(\frac{2\lambda-r_2+\gamma}{\gamma}\right)+\ldots\\ &\qquad+\chi\left(\frac{(\gamma-1)\lambda-r_{\gamma-1}+\gamma}{\gamma}\right)\equiv 0 \pmod{\lambda};\end{aligned}$$

d'où, en négligeant les multiples du module λ, on aura la congruence simplifiée

$$\chi\left(\frac{\gamma-r_1}{\gamma}\right)+\chi\left(\frac{\gamma-r_2}{\gamma}\right)+\ldots+\chi\left(\frac{\gamma-r_{\gamma-1}}{\gamma}\right)\equiv 0 \pmod{\lambda},$$

dans laquelle on pourra faire disparaître les fractions en multipliant par les dénominateurs, dont aucun ne contient le facteur λ. Les nombres $r_1, r_2,\ldots, r_{\gamma-1}$ coïncident avec les nombres $1, 2, 3,\ldots, \gamma-1$, pris dans un ordre convenable; car on sait qu'ils sont tous positifs, moindres que γ, et l'on démontre aisément qu'ils sont tous différents entre eux. Donc, en mettant $1, 2,\ldots, \gamma-1$, au lieu de $r_1, r_2,\ldots, r_{\gamma-1}$, on obtient la congruence

$$\chi\left(\frac{1}{\gamma}\right)+\chi\left(\frac{2}{\gamma}\right)+\chi\left(\frac{3}{\gamma}\right)+\ldots+\chi\left(\frac{\gamma-1}{\gamma}\right)\equiv 0 \pmod{\lambda}.$$

Pour obtenir l'expression la plus simple de cette congruence, nous développerons ici une formule nouvelle, contenant une propriété curieuse de la fonction $\chi(x)$. Pour cela, nous partons du développement en série, ordonnée suivant les cosinus des multiples de $2x\pi$,

$$\chi(x) = \frac{(-1)^n B_n}{\Pi_{2n}} - \frac{(-1)^n 2}{(2\pi)^{2n}}\left(\frac{\cos 2x\pi}{1^{2n}} + \frac{\cos 4x\pi}{2^{2n}} + \frac{\cos 6x\pi}{3^{2n}} + \ldots\right),$$

dans lequel x doit être pris entre les limites 0 et 1. En prenant successivement $x = 0, \frac{1}{\gamma}, \frac{2}{\gamma}, \ldots, \frac{\gamma - 1}{\gamma}$, et ajoutant ces équations, en observant que la série

$$1 + \cos\frac{2k\pi}{\gamma} + \cos\frac{4k\pi}{\gamma} + \ldots + \cos\frac{2(\gamma-1)k\pi}{\gamma}$$

est égale à γ, si k est un multiple de γ, et qu'elle est égale à zéro dans tous les autres cas, on obtient

$$\chi\left(\frac{1}{\gamma}\right) - \chi\left(\frac{2}{\gamma}\right) + \chi\left(\frac{3}{\gamma}\right) + \ldots + \chi\left(\frac{\gamma-1}{\gamma}\right)$$
$$= \frac{(-1)^n \gamma B_n}{\Pi_{2n}} - \frac{(-1)^n 2\gamma}{(2\pi)^{2n}}\left[\frac{1}{\gamma^{2n}} + \frac{1}{(2\gamma)^{2n}} + \frac{1}{(3\gamma)^{2n}} + \ldots\right],$$

et puisqu'on a

$$1 + \frac{1}{2^{2n}} + \frac{1}{3^{2n}} + \ldots = \frac{(2\pi)^{2n} B_n}{2\Pi_{2n}},$$

l'équation précédente se réduit à

$$\chi\left(\frac{1}{\gamma}\right) + \chi\left(\frac{2}{\gamma}\right) + \ldots + \chi\left(\frac{\gamma-1}{\gamma}\right) = \frac{(-1)^n(\gamma^{2n}-1)B_n}{\gamma^{2n-1}\Pi_{2n}}.$$

A cause de cette propriété de la fonction $\chi(x)$, la congruence de laquelle dépend la divisibilité par λ du premier facteur du nombre des classes H, prend la forme

$$\frac{(\gamma^{2n}-1)B_n}{\gamma^{2n-1}\Pi_{2n}} \equiv 0 \pmod{\lambda}.$$

Le dénominateur $\gamma^{2n-1}\Pi_{2n}$ peut être rejeté, parce qu'il ne contient pas le facteur λ; de même, le facteur $\gamma^{2n} - 1$, qui n'est pas divisible par λ pour les valeurs de $n = 1, 2, 3, \ldots, \mu - 1$, pourra être négligé dans tous ces cas, en sorte qu'on a simplement

$$B_n \equiv 0 \pmod{\lambda}.$$

Le cas de $n = \mu$, où $\gamma^{2\mu} - 1$ est divisible par λ, en sorte que B_μ contient le facteur λ en dénominateur, exige une discussion particulière. D'après une formule connue, le nombre bernoullien B_μ s'exprime au moyen des nombres bernoulliens précédents, de la manière suivante :

$$B_\mu = \frac{\Pi_{2\mu} B_{\mu-1}}{2^2 \Pi_3 \Pi_{2\mu-2}} - \frac{\Pi_{2\mu} B_{\mu-2}}{2^4 \Pi_5 \Pi_{2\mu-4}} + \ldots + \frac{(-1)^\mu}{2^{2\mu}(2\mu+1)} + \frac{(-1)^{\mu+1}}{2^{2\mu}},$$

où l'avant-dernier terme est le seul qui contienne le facteur $2\mu + 1 = \lambda$ en dénominateur. On aura donc

$$B_\mu = \frac{(-1)^\mu}{2^{2\mu}.\lambda} + \frac{M}{N},$$

où N n'est pas divisible par λ, et en multipliant par

$$\gamma^{2\mu} - 1 = (\gamma^\mu - 1)(\gamma^\mu + 1) = (\gamma^\mu - 1)\lambda G,$$

on obtient l'équation

$$\chi\left(\frac{1}{\gamma}\right) + \chi\left(\frac{2}{\gamma}\right) \cdots \chi\left(\frac{\gamma-1}{\gamma}\right) = \frac{(-1)^\mu G(\gamma^\mu - 1)}{\gamma^{2\mu-1}\Pi_{2\mu}} \left[\frac{(-1)^\mu}{2^{2\mu}} + \frac{\lambda M}{N}\right].$$

La congruence de condition, posée ci-dessus, devient donc, dans le cas de $n = \mu$,

$$\frac{G(\gamma^\mu - 1)}{2^{2\mu}\gamma^{2\mu-1}\Pi_{2\mu}} \equiv 0 \pmod{\lambda};$$

d'où, en ayant égard à l'hypothèse que G n'est pas divisible par λ, on voit qu'elle n'est satisfaite pour aucune valeur du nombre premier λ. Ainsi la divisibilité par λ du premier facteur du nombre H dépend de ce que la congruence $B_n \equiv 0 \pmod{\lambda}$ ait lieu pour une quelconque des valeurs de $n = 1, 2, 3, \ldots, \mu - 1$. Le résultat trouvé donne le théorème :

La condition nécessaire et suffisante pour que le premier facteur $\frac{P}{(2\lambda)^{\mu-1}}$ *du nombre des classes* H *soit divisible par* λ, *consiste en ce qu'un quelconque des* $\frac{\lambda-3}{2}$ *premiers nombres bernoulliens soit divisible par* λ.

Passons à la discussion du second facteur $\frac{\mathrm{D}}{\Delta}$ du nombre H. En vertu d'un théorème démontré dans le § II de ce Mémoire, une certaine puissance de toute unité complexe s'exprime comme produit de puissances entières du système indépendant $e(\alpha)$, $e(\alpha^{\gamma}),\ldots, e(\alpha^{\gamma^{\mu-2}})$; nous pouvons donc poser, comme à l'endroit cité,

$$\varepsilon_1(\alpha)^{n_1} = e(\alpha^{\gamma})^{r'_1}.e(\alpha^{\gamma})^{r'_2}\ldots e(\alpha^{\gamma^{\mu-2}})^{r'_{\mu-1}},$$

$$\varepsilon_2(\alpha)^{n_2} = e(\alpha)^{r^2_1}.e(\alpha^{\gamma})^{r^2_2}\ldots e(\alpha^{\gamma^{\mu-2}})^{r^2_{\mu-1}},$$

. . . .

$$\varepsilon_{\mu-1}(\alpha)^{n_{\mu-1}} = e(\alpha)^{r^{\mu-1}_1}.e(\alpha^{\gamma})^{r^{\mu-1}_2}\ldots e(\alpha^{\gamma^{\mu-2}})^{r^{\mu-1}_{\mu-1}},$$

en supposant que $\varepsilon_1(\alpha)$, $\varepsilon_2(\alpha),\ldots, \varepsilon_{\mu-1}(\alpha)$ désignent un système d'unités fondamentales, et que l'exposant n_h n'ait aucun facteur commun avec tous les exposants r^h_1, $r^h_2,\ldots, r^h_{\mu-1}$. Nous obtiendrons de là, comme dans le § II,

$$\frac{\mathrm{D}}{\Delta} = \frac{n_1.n_2\ldots n_{\mu-1}}{\mathrm{R}},$$

d'où il suit que $\frac{\mathrm{D}}{\Delta}$ ne pourra être divisible par λ, à moins qu'un des nombres n_1, $n_2,\ldots, n_{\mu-1}$ ne soit divisible par λ, c'est-à-dire qu'on ait une équation de la forme

$$\varepsilon(\alpha)^{n} = e(\alpha)^{r_1}.e(\alpha^{\gamma})^{r_2}\ldots e(\alpha^{\gamma^{\mu-2}})^{r_{\mu-1}},$$

dans laquelle n soit divisible par λ, sans que les exposants r_1, $r_2,\ldots, r_{\mu-1}$ soient tous divisibles par λ.

Or, on démontre aisément que la $\lambda^{ième}$ puissance de tout nombre complexe existant est congrue à un nombre entier non complexe pour le module λ. En effet, en élevant à la $\lambda^{ième}$ puissance le polynôme

* $$f(\alpha) = a + a_1\alpha + a_2\alpha^2 + \ldots + a_{\lambda-2}\alpha_{\lambda-2}$$

et négligeant tous les termes divisibles par λ, il ne restera que les $\lambda^{ièmes}$

puissances des termes du polynôme, et l'on aura généralement

$$f(\alpha)^\lambda \equiv a^\lambda + a_1^\lambda + a_2^\lambda + \ldots + a_{\lambda-2}^\lambda$$

$$\equiv a + a_1 + a_2 + \ldots + a_{\lambda-2} \quad (\mathrm{mod.}\ \lambda).$$

La puissance $\varepsilon(\alpha)^n$, n étant divisible par λ, sera donc congrue à un nombre entier non complexe c. En le substituant dans l'équation précédente, on aura une condition nécessaire pour que le facteur $\frac{\mathrm{D}}{\Delta}$ soit divisible par λ, exprimée par la congruence

$$e(\alpha)^{r_1}.e(\alpha^\gamma)^{r_2}\ldots e\left(\alpha^{\gamma^{\mu-2}}\right)^{r_{\mu-1}} \equiv c \quad (\mathrm{mod.}\ \lambda),$$

laquelle doit avoir lieu pour des valeurs entières des exposants r_1, $r_2, \ldots$, $r_{\mu-1}$, qui ne sont pas tous divisibles par λ.

La recherche des valeurs du nombre premier λ pour lesquelles cette congruence puisse avoir lieu se fait au moyen d'une méthode générale, dont le point principal consiste en ce que, dans une équation contenant des nombres complexes, on introduit une variable continue x, au lieu de la racine α. Étant donnée une équation

$$f(\alpha) = \varphi(\alpha) \quad \text{ou} \quad f(\alpha) - \varphi(\alpha) = 0,$$

qui aura lieu pour toutes les valeurs de α qui satisfont à l'équation

$$1 + \alpha + \alpha^2 + \ldots + \alpha^{\lambda-1} = 0,$$

la fonction rationnelle et entière de la variable x, $f(x) - \varphi(x)$ sera égale à zéro pour toutes les valeurs de $x = \alpha$, $x = \alpha^2, \ldots$, $x = \alpha^{\lambda-1}$, d'où l'on conclut qu'elle sera divisible par

$$(x-\alpha).(x-\alpha^2)\ldots(x-\alpha^{\lambda-1}) = 1 + x + x^2 + \ldots + x^{\lambda-1};$$

on pourra donc poser

$$f(x) - \varphi(x) = (1 + x + x^2 + \ldots + x^{\lambda-1})\psi(x),$$

ou

$$f(x) = \varphi(x) + (1 + x + x^2 + \ldots + x^{\lambda-1})\psi(x),$$

$\psi(x)$ étant une fonction rationnelle et entière de la variable x.

Pour faire l'application à la congruence proposée, nous la réduirons d'abord à l'équation

$$e(\alpha)^{r_1}.e(\alpha^\gamma)^{r_2}\ldots e\left(\alpha^{\gamma^{\mu-2}}\right)^{r_{\mu-1}} \equiv c+\lambda\varphi(\alpha),$$

dans laquelle $\varphi(\alpha)$ désigne un nombre entier complexe. En substituant la variable x au lieu de la racine α, on aura

$$e(x)^{r_1}.e(x^\gamma)^{r_2}\ldots e\left(x^{\gamma^{\mu-2}}\right)^{r_{\mu-1}}$$
$$\equiv c+\lambda\varphi(x)+(1+x+x^2+\ldots+x^{\lambda-1})\psi(x).$$

Si maintenant on prend les différentielles des logarithmes des deux membres de cette équation, et qu'après avoir multiplié par x, on restitue la valeur particulière $x=\alpha$, on en tire l'équation

$$r_1\frac{\alpha e'(\alpha)}{e(\alpha)}+r_2\gamma\frac{\alpha^\gamma e'(\alpha^\gamma)}{e(\alpha)}+\ldots+r_{\mu-1}\gamma^{\mu-2}\frac{\alpha^{\gamma^{\mu-2}}e'\left(\alpha^{\gamma^{\mu-2}}\right)}{e\left(\alpha^{\gamma^{\mu-2}}\right)}$$
$$=\frac{\lambda\alpha\varphi'(\alpha)+[\alpha+2\alpha^2+3\alpha^3+\ldots+(\lambda-1)\alpha^{\lambda-1}]\psi(\alpha)}{c+\lambda\varphi(\alpha)},$$

où, suivant la notation de Lagrange, $e'(x)$ et $\varphi'(x)$ désignent les premières dérivées de $e(x)$ et $\varphi(x)$. De là, en rejetant les multiples de λ, et faisant, pour abréger,

$$\frac{2e'(\alpha)}{e(\alpha)}=\mathrm{F}(\alpha),$$

on aura la congruence

$$r_1\mathrm{F}(\alpha)+r_2\gamma\mathrm{F}(\alpha^\gamma)+\ldots+r_{\mu-1}\gamma^{\mu-2}\mathrm{F}\left(\alpha^{\gamma^{\mu-2}}\right)$$
$$\equiv\frac{2}{c}[\alpha+2\alpha^2+3\alpha^3+\ldots+(\lambda-1)\alpha^{\lambda-1}]\psi(\alpha)\quad(\text{mod. }\lambda).$$

En mettant le nombre complexe $\psi(\alpha)$ sous la forme

$$\psi(\alpha)=a+(1-\alpha)\psi_1(\alpha),$$

où a est un entier non complexe, et observant qu'on a

$$(1-\alpha)[\alpha+2\alpha^2+3\alpha^3+\ldots+(\lambda-1)\alpha^{\lambda-1}]=-\lambda,$$

puis faisant enfin

$$\mathrm{M}\equiv\frac{2a}{c}\quad(\text{mod. }\lambda),$$

on aura

$$r_1 F(\alpha) + r_2 \gamma F(\alpha^\gamma) + \ldots + r_{\mu-1}\gamma^{\mu-2} F\left(\alpha^{\gamma^{\mu-2}}\right)$$
$$\equiv M[\alpha + 2\alpha^2 + \ldots + (\lambda-1)\alpha^{\lambda-1}].$$

Développons actuellement le nombre entier complexe $F(\alpha)$. En prenant la différentielle logarithmique de l'expression

$$e(x) = \sqrt{\frac{(1-x^\gamma)(1-x^{-\gamma})}{(1-x)(1-x^{-1})}},$$

on obtient

$$\frac{2xe'(x)}{e(x)} = \frac{1+x}{1-x} - \frac{\gamma(1+x^\gamma)}{1-x^\gamma};$$

d'où, en restituant la valeur particulière $x = \alpha$,

$$F(\alpha) = \frac{1+\alpha}{1-\alpha} - \frac{\gamma(1+\alpha^\gamma)}{1-\alpha^\gamma}.$$

Le développement ultérieur se fait au moyen de l'équation

$$(1-\alpha)[\alpha + 2\alpha^2 + 3\alpha^3 + \ldots + (\lambda-1)\alpha^{\lambda-1}] = -\lambda,$$

de laquelle on tire

$$\frac{1}{1-\alpha} = -\frac{[\alpha + 2\alpha^2 + 3\alpha^3 + \ldots + (\lambda-1)\alpha^{\lambda-1}]}{\lambda},$$

et en multipliant par $1+\alpha$,

$$\frac{1+\alpha}{1-\alpha} = -\frac{[\lambda + 2\alpha + 4\alpha^2 + 6\alpha^3 + \ldots + 2(\lambda-1)\alpha^{\lambda-1}]}{\lambda}.$$

Lorsqu'on y remplace les coefficients $1, 2, 3, \ldots, \lambda-1$ par les nombres $1, \gamma_1, \gamma_2, \ldots, \gamma_{\lambda-2}$, qui, abstraction faite de l'ordre, sont absolument les mêmes, et qu'en même temps, au lieu des exposants correspondants, on met $1, \gamma, \gamma^2, \ldots, \gamma^{\lambda-2}$, cette équation prend la forme

$$\frac{1+\alpha}{1-\alpha} = -\frac{\left(\lambda + 2\alpha + 2\gamma_1\alpha^\gamma + 2\gamma_2\alpha^{\gamma^2} + \ldots + 2\gamma_{\lambda-2}\alpha^{\gamma^{\lambda-2}}\right)}{\lambda}.$$

61..

Si l'on change α en α^γ, et qu'on multiplie par γ, on aura aussi

$$\frac{\gamma(1-\alpha^\gamma)}{1-\alpha^\gamma} = -\frac{\left(\gamma + 2\gamma\gamma_{\lambda-2}\alpha + 2\gamma\alpha^\gamma + 2\gamma\gamma_1\alpha^{\gamma^2} + \ldots + 2\gamma\gamma_{\lambda-3}\alpha^{\gamma^{\lambda-2}}\right)}{\lambda};$$

enfin, en soustrayant cette équation de la précédente, on obtient le développement

$$F(\alpha) = \gamma - 1 + \frac{2(\gamma\gamma_{\lambda-2}-1)}{\lambda}\alpha + \frac{2(\gamma-\gamma_1)}{\lambda}\alpha^\gamma + \ldots$$
$$+ \frac{2(\gamma\gamma_{\lambda-3}-\gamma_{\lambda-2})}{\lambda}\alpha^{\gamma^{\lambda-2}},$$

dans lequel les coefficients sont les mêmes que ceux que nous avons désignés par b_k, dans la première partie de cette recherche. En adoptant les mêmes signes, nous avons

$$F(\alpha) = \gamma - 1 + 2b_0\alpha + 2b_1\alpha^\gamma + 2b_3\alpha^{\gamma^2} + \ldots + 2b_{\lambda-2}\alpha^{\gamma^{\lambda-2}}.$$

Substituons le développement trouvé de $F(\alpha)$, de même que ceux de $F(\alpha^\gamma)$, $F(\alpha^{\gamma^2})$, etc., qu'on en déduit par le changement de α en α^γ, α^{γ^2}, etc., dans la congruence proposée, dont le second membre pourra être mis dans la forme

$$M\left(\alpha + \gamma\alpha^\gamma + \gamma^2\alpha^{2\gamma^2} + \ldots + \gamma^{\lambda-2}\alpha^{\gamma^{\lambda-2}}\right).$$

Comparons séparément les coefficients des termes multipliés par α, $\alpha^\gamma, \alpha^{\gamma^2}, \ldots, \alpha^{\gamma^{\lambda-2}}$, et nous aurons ce système de congruences :

$$\left.\begin{array}{l}
r_1 b_0 + \gamma r_2 b_{\lambda-2} + \gamma^2 r_3 b_{\lambda-3} + \ldots + \gamma^{\mu-2} r_{\mu-1} b_{\lambda-\mu+1} - R \equiv M \\
r_1 b_1 + \gamma r_2 b_0 + \gamma^2 r_3 b_{\lambda-2} + \ldots + \gamma^{\mu-2} r_{\mu-1} b_{\lambda-\mu+2} - R \equiv \gamma M \\
r_1 b_2 + \gamma r_2 b_1 + \gamma^2 r_3 b_0 + \ldots + \gamma^{\mu-2} r_{\mu-1} b_{\lambda-\mu+3} - R \equiv \gamma^2 M \\
\ldots\ldots\ldots\ldots\ldots\ldots\ldots\ldots\ldots\ldots\ldots\ldots \\
r_1 b_{\lambda-2} + \gamma r_2 b_{\lambda-3} + \gamma^2 r_3 b_{\lambda-4} + \ldots + \gamma^{\mu-2} r_{\mu-1} b_{\lambda-\mu} - R \equiv \gamma^{\lambda-2} M
\end{array}\right\} \pmod{\lambda}$$

où nous avons fait, pour abréger,

$$r_1 + \gamma r_2 + \gamma^2 r_3 + \ldots + \gamma^{\mu-2} r_{\mu-1} = R.$$

Ces congruences, multipliées respectivement par $1, \gamma^{2n-1}, \gamma^{2(2n-1)},\ldots,$ $\gamma^{(\lambda-2)(2n-1)}$, et ajoutées, donnent une somme qu'on décompose aisément en deux facteurs

$$\left(\begin{array}{l} r_1 + \gamma^{2n} r_2 + \gamma^{4n} r_3 + \ldots \\ + \gamma^{2(\mu-2)n} r_{\mu-1} \end{array}\right) . \left(\begin{array}{l} b_0 + b_1 \gamma^{2n-1} + b_2 \gamma^{2(2n-1)} + \ldots \\ + b_{\lambda-2} \gamma^{(\lambda-2)(2n-1)} \end{array}\right) \equiv 0,$$

pour le module λ et pour les valeurs de $n = 1, 2, 3,\ldots, \mu - 1$. Il suit de cette congruence, que dans les cas où le second facteur

$$b_0 + b_1 \gamma^{2n-1} + b_2 \gamma^{2(2n-1)} + \ldots + b_{\lambda-2} \gamma^{(\lambda-2)(2n-1)}$$

n'est divisible par λ pour aucune des valeurs de $n = 1, 2, 3, \ldots,$ $\mu - 1$, le premier facteur sera nécessairement congru à zéro pour toutes ces valeurs de n. Dans ce cas, on aura le système des congruences :

$$\left.\begin{array}{l} r_1 + \gamma^2 r_2 + \gamma^4 r_3 + \ldots + \gamma^{2(\mu-2)} r_{\mu-1} \equiv 0, \\ r_1 + \gamma^4 r_2 + \gamma^8 r_3 + \ldots + \gamma^{4(\mu-2)} r_{\mu-1} \equiv 0, \\ \ldots\ldots\ldots\ldots\ldots\ldots\ldots\ldots \\ r_1 + \gamma^{2(\mu-1)} r_2 + \gamma^{4(\mu-1)} r_3 + \ldots + \gamma^{(\mu-2)(\mu-1)} r_{\mu-1} \equiv 0, \end{array}\right\} (\text{mod. } \lambda).$$

Le déterminant de ce système de congruences linéaires n'est pas congru à zéro, car on sait qu'il est composé de facteurs de la forme $\gamma^{2k} - \gamma^{2h}$, où k et h sont différents entre eux et moindres que $\mu - 1$. Par cette raison, il n'y a pas d'autres valeurs des inconnues de ce système, que

$$r_1 \equiv 0, \quad r_2 \equiv 0, \quad r_3 \equiv 0,\ldots, \quad r_{\mu-1} \equiv 0 \quad (\text{mod. } \lambda),$$

et puisque ces exposants ne doivent pas être tous divisibles par λ, on en conclut que, dans le cas où

$$b_0 + b_1 \gamma^{2n-1} + b_2 \gamma^{2(2n-1)} + \ldots + b_{\lambda-2} \gamma^{(\lambda-2)(2n-1)}$$

n'est divisible par λ pour aucune des valeurs $n = 1, 2, 3,\ldots, \mu - 1$, le second facteur $\frac{D}{\Delta}$ du nombre des classes H ne sera pas divisible par λ. La condition

$$b_0 + b_1 \gamma^{2n-1} + b_2 \gamma^{2(2n-1)} + \ldots + b_{\lambda-2} \gamma^{(\lambda-2)(2n-1)} \equiv 0 \quad (\text{mod. } \lambda),$$

nécessaire pour que le facteur $\frac{D}{\Delta}$ soit divisible par λ, étant la même que nous avons trouvée comme condition nécessaire et suffisante pour que le premier facteur $\frac{P}{(2\lambda)^{\mu-1}}$ soit divisible par λ, on voit que le second facteur du nombre des classes H ne sera jamais divisible par λ, à moins que le premier facteur ne le soit en même temps. Nous voilà donc arrivés au résultat suivant :

Pour que le nombre H *des classes diverses des nombres complexes idéaux soit divisible par* λ, *il faut et il suffit que le nombre premier* λ *soit facteur du numérateur d'un quelconque des* $\frac{\lambda-3}{2}$ *premiers nombres bernoulliens.*

Nous passons au second problème de ce paragraphe, qui consiste dans la recherche des valeurs du nombre premier λ pour lesquelles une unité complexe puisse être congrue à un nombre entier non complexe, sans être une puissance du degré λ.

Soit $E(\alpha)$ une unité complexe qui satisfasse à la condition

$$E(\alpha) \equiv c \pmod{\lambda},$$

où c est un entier non complexe. Cette unité, exprimée par le système des unités indépendantes

$$e(\alpha), \quad e(\alpha^{\gamma}), \ldots, \quad e(\alpha^{\gamma^{\mu-2}}),$$

prendra la forme

$$E(\alpha) = \pm \alpha^k e(\alpha)^{\frac{r_1}{n}} \cdot e(\alpha^{\gamma})^{\frac{r_2}{n}} \ldots e(\alpha^{\gamma^{\mu-2}})^{\frac{r_{\mu-1}}{n}},$$

où les exposants rationnels $\frac{r_1}{n}, \frac{r_2}{n}, \ldots, \frac{r_{\mu-1}}{n}$ sont réduits au même dénominateur le plus petit possible, en sorte qu'il n'y ait pas de facteur commun de tous les numérateurs avec le dénominateur n. En changeant α en α^{-1} dans la congruence posée, on a aussi

$$E(\alpha^{-1}) \equiv c \pmod{\lambda},$$

d'où

$$E(\alpha) \equiv E(\alpha^{-1}) \pmod{\lambda},$$

et, par conséquent,

$$\alpha^k = \alpha^{-k}, \quad \alpha^k = 1, \quad k \equiv 0 \quad (\text{mod. } \lambda).$$

En substituant cette valeur de $\alpha^k = 1$, élevant à la puissance n et prenant $E(\alpha) \equiv c$, on aura

$$\pm c^n \equiv e(\alpha)^{r_1} . e\left(\alpha^\gamma\right)^{r_2} \ldots e\left(\alpha^{\gamma^{\mu-2}}\right)^{r_{\mu-1}} \quad (\text{mod. } \lambda,$$

Voilà la même congruence que nous avons discutée dans la recherche précédente. Nous y avons trouvé qu'à l'exception des cas où λ se trouve comme facteur du numérateur d'un des $\frac{\lambda-3}{2}$ premiers nombres bernoulliens, cette congruence exige nécessairement que tous les exposants soient divisibles par λ. Alors $E(\alpha)^n$ sera égal à une puissance du degré λ; on pourra donc poser

$$E(\alpha)^n = F(\alpha)^\lambda,$$

et comme tous les nombres $r_1, r_2, \ldots, r_{\mu-1}$ n'ont aucun facteur commun avec n, cet exposant n ne sera pas divisible par λ. Par cette raison, on pourra toujours trouver deux nombres entiers s et t, tels qu'on ait

$$ns - \lambda t = 1 \quad \text{ou} \quad ns = \lambda t + 1,$$

et, en élevant l'équation précédente à la puissance s, on aura

$$E(\alpha)^{ns} = E(\alpha)^{\lambda t + 1} = F(\alpha)^{\lambda s},$$

d'où, en divisant par $E(\alpha)^{\lambda t}$,

$$E(\alpha) = \left[\frac{F(\alpha)^s}{E(\alpha)^t}\right]^\lambda.$$

Ainsi $E(\alpha)$ est une $\lambda^{\text{ième}}$ puissance, et nous avons le théorème:

Si le nombre premier λ n'est contenu dans aucun des $\frac{\lambda-3}{2}$ premiers nombres bernoulliens comme facteur du numérateur, toute unité complexe, congrue à un nombre complexe pour le module λ, sera une $\lambda^{\text{ième}}$ puissance d'une autre unité complexe.

§ X.

Application à la démonstration du dernier théorème de Fermat.

Dans tout ce qui précède, la théorie des nombres complexes est avancée à un point où la démonstration du fameux théorème de Fermat se fait avec facilité pour toutes les puissances dont les exposants satisfont à la condition qui se trouve dans les théorèmes du paragraphe précédent : que le nombre λ ne soit pas contenu comme facteur dans un des premiers $\frac{\lambda-3}{2}$ nombres bernoulliens. Comme la démonstration pour les nombres complexes est aussi facile que pour les nombres entiers non complexes, nous supposerons que, dans l'équation

$$u^\lambda + v^\lambda + w^\lambda = 0,$$

u, v, w soient des nombres complexes existants de la forme

$$a + a_1\alpha + a_2\alpha^2 + \ldots + a_{\lambda-2}\alpha^{\lambda-2},$$

sans aucun facteur commun deux à deux. Nous supposerons aussi que λ soit un nombre premier qui ne se trouve pas comme facteur du numérateur d'un des $\frac{\lambda-3}{2}$ premiers nombres bernoulliens.

Cela posé, nous savons qu'en vertu d'un théorème du paragraphe précédent, le nombre H des classes des nombres complexes idéaux n'est pas divisible par λ. De plus, nous avons trouvé dans le § VI, qu'en supposant $f(\alpha)$ idéal, $f(\alpha)^\lambda$ ne pourra être existant à moins que λ et H n'aient un facteur commun différent de l'unité, ce qui n'a pas lieu. Donc, si λ est un nombre tel que nous l'avons supposé, *la $\lambda^{\text{ième}}$ puissance d'un nombre complexe idéal ne sera pas égale à un nombre complexe existant, et de ce que $f(\alpha)^\lambda$ est un nombre complexe existant, il suivra nécessairement que $f(\alpha)$ est un nombre existant.*

Rappelons encore expressément le théorème démontré à la fin du paragraphe précédent, que pour les valeurs de λ, que nous considérons ici, *chaque unité complexe qui est congrue à un nombre non complexe pour le module λ est égale à une $\lambda^{\text{ième}}$ puissance d'une autre unité.*

La démonstration de l'impossibilité de l'équation proposée consistera en deux parties distinctes, dont la première traitera le cas où aucun des trois nombres u, v, w n'est divisible par $1-\alpha$, l'autre le cas où un de ces nombres aura le facteur $1-\alpha$.

Soit donc proposée en premier lieu l'équation

$$u^\lambda + v^\lambda + w^\lambda = 0,$$

dans laquelle aucun des trois nombres complexes u, v, w ne soit divisible par $1-\alpha$. Observons d'abord que les nombres u, v, w pourront être multipliés par des unités simples de la forme α^k sans que l'équation proposée change de forme. En posant donc $\alpha^k u$ au lieu de u, on pourra toujours déterminer le nombre k tel, que $\alpha^k u$ prenne la forme

$$a + (1-\alpha)^2 P,$$

dans laquelle a est un entier non complexe et P un entier complexe. En effet, le nombre complexe u, ordonné suivant les puissances $1-\alpha$, prendra la forme

$$u = A + A_1(1-\alpha) + A_2(1-\alpha)^2 + \ldots + A_{\lambda-2}\alpha^{\lambda-2};$$

de même, en prenant

$$\alpha = 1-(1-\alpha),$$

on aura

$$\alpha^k = 1 - k(1-\alpha) + \frac{k(k-1)}{1.2}(1-\alpha)^2 - \ldots,$$

d'où, en multipliant et comprenant dans un seul tous les termes multipliés par $(1-\alpha)^2$,

$$\alpha^k u = A + (A_1 - Ak)(1-\alpha) + (1-\alpha)^2 P.$$

Enfin, en prenant le nombre k tel qu'on ait

$$Ak \equiv A_1 \pmod{\lambda},$$

ce qui est toujours possible, puisque A n'est pas divisible par λ, on voit que $\alpha^k u$ prend la forme proposée. En opérant de même sur les nombres v et w, on pourra mettre

$$u = a + (1 - \alpha)^2 P,$$
$$v = b + (1 - \alpha)^2 Q,$$
$$w = c + (1 - \alpha)^2 R,$$

où les nombres a, b, c ne sont pas divisibles par λ, à cause de la supposition que u, v, w ne sont pas divisibles par $1 - \alpha$.

Maintenant, en décomposant la forme $u^\lambda + v^\lambda$, on obtient l'équation

$$(u + v).(u + \alpha v).(u + \alpha^2 v)\ldots(u + \alpha^{\lambda-1} v) = -w^\lambda,$$

et je dis que deux quelconques de ces facteurs sont premiers entre eux. En effet, si $u + \alpha^r v$ et $u + \alpha^s v$ avaient un facteur commun, il est clair que $(\alpha^r - \alpha^s)u$ et $(\alpha^r - \alpha^s)v$ auraient le même facteur commun, et comme u et v sont premiers entre eux et $\alpha^r - \alpha^s$, ou, ce qui est au fond le même, $1 - \alpha$ ne peut diviser un de ces facteurs sans diviser également le nombre w, on voit que ces facteurs sont premiers entre eux.

Cela étant, mettons à profit le théorème démontré à la fin du § V, que lorsqu'une puissance d'un nombre complexe est décomposée en facteurs premiers entre eux, ces facteurs doivent être séparément des puissances semblables multipliées par des unités complexes. Nous en concluons qu'on doit avoir généralement pour toutes les valeurs de $r = 0, 1, 2, \ldots, \lambda - 1$,

$$u + \alpha^r v = \alpha^\rho E_r(\alpha)\, t_r^\lambda,$$

t_r étant un nombre complexe, facteur de w, et $\alpha^\rho E_r(\alpha)$ une unité complexe quelconque, dans laquelle $E_r(\alpha)$ soit égal à $E_r(\alpha^{-1})$. On se rappellera que toute unité complexe consiste de deux facteurs tels que nous venons de les établir. Dans cette équation, t_r^λ est donné comme nombre complexe existant; d'où nous concluons que t_r sera aussi un nombre existant; la $\lambda^{ième}$ puissance t_r^λ sera donc congrue à un nombre entier non complexe pour le module λ, et l'on aura la congruence

$$u + \alpha^r v \equiv \alpha^\rho E_r(\alpha). m \pmod{\lambda}.$$

En désignant par u', v', w' les nombres complexes réciproques de u, v, w, qu'on en déduit en changeant α en α^{-1}, on aura de même

$$u' + \alpha^{-r} v' \equiv \alpha^{-\rho} E_r(\alpha) . m \quad (\text{mod. } \lambda);$$

d'où, en éliminant $E_r(\alpha).m$,

$$\alpha^{-\rho}(u + \alpha^r v) \equiv \alpha^{\rho}(u' + \alpha^{-r} v') \quad (\text{mod. } \lambda).$$

Cette congruence, prise par rapport au module $(1 - \alpha)^2$, diviseur du module λ, donne

$$\alpha^{-\rho}(a + \alpha^r b) \equiv \alpha^{\rho}(a + \alpha^{-r} b) \quad [\text{mod. } (1 - \alpha)^2],$$

car on a, en vertu des expressions de u, v, w données ci-dessus,

$$u \equiv a, \quad v \equiv b, \quad u' \equiv a, \quad v' \equiv b \quad [\text{mod. } (1 - \alpha)^2]:$$

et, comme on a généralement

$$\alpha^h \equiv 1 - h(1 - \alpha) \quad [\text{mod. } (1 - \alpha)^2],$$

cette congruence donne

$$(\alpha + b)\rho \equiv br \quad [\text{mod. } (1 - \alpha)].$$

Or on sait qu'un nombre entier non complexe divisible par $1 - \alpha$ sera toujours divisible par λ; d'où il suit que deux entiers non complexes congrus pour le module $1 - \alpha$ seront aussi congrus pour le module λ. On aura donc

$$(a + b)\rho \equiv br \quad (\text{mod. } \lambda),$$

ce qui fait voir que $a + b$ n'est pas divisible par λ. En déterminant le nombre k par la congruence

$$(a + b)k \equiv b \quad (\text{mod. } \lambda),$$

on aura

$$\rho \equiv kr \quad (\text{mod. } \lambda),$$

et, en substituant cette valeur de ρ dans la congruence plus générale, donnée ci-dessus, on obtient la congruence

$$\alpha^{-kr}(u + \alpha^r v) \equiv \alpha^{kr}(u' + \alpha^{-r} v') \quad (\text{mod. } \lambda),$$

qui subsiste pour toutes les valeurs de $r = 0, 1, 2, \ldots, \lambda - 1$. Lors-

62..

qu'on y prend $r=0$, on trouve

$$u+v\equiv u'+v' \pmod{\lambda},$$

et, puisque les lettres u, v, w, et en même temps les lettres u', v', w' peuvent être échangées entre elles, on aura aussi

$$\left.\begin{array}{l} u+w\equiv u'+w' \\ v+w\equiv v'+w' \end{array}\right\} \pmod{\lambda};$$

enfin, de ces trois congruences, on déduit les congruences plus simples

$$\left.\begin{array}{l} u\equiv u' \\ v\equiv v' \\ w\equiv w' \end{array}\right\} \pmod{\lambda}.$$

En substituant ces valeurs congrues de u, v, w, au lieu de u', v', w', on aura

$$\alpha^{-kr}(u+\alpha^r v)\equiv\alpha^{kr}(u+\alpha^{-r}v) \pmod{\lambda},$$

ou, ce qui est le même

$$(\alpha^{-kr}-\alpha^{kr})\,u+[\alpha^{-(k-1)r}-\alpha^{(k-r)r}]\,v\equiv 0 \pmod{\lambda}.$$

Maintenant, si l'on prend les deux valeurs déterminées $r=1$ et $r=2$, on aura

$$\left.\begin{array}{r} (\alpha^{-k}-\alpha^{k})\,u+[\alpha^{-(k-1)}-\alpha^{k-1}]\,v\equiv 0 \\ (\alpha^{-2k}-\alpha^{2k})\,u+[\alpha^{-2(k-1)}-\alpha^{2(k-1)}]\,v\equiv 0 \end{array}\right\} \pmod{\lambda};$$

d'où, en multipliant la première par $\alpha^{-k}+\alpha^{k}$, soustrayant la seconde et divisant par v, qui est premier par rapport au module, on obtient la congruence

$$(\alpha^{-k}+\alpha^{k})[\alpha^{-k(-1)}-\alpha^{k-1}]-[\alpha^{-2(k-1)}-\alpha^{2(k-1)}]\equiv 0 \pmod{\lambda},$$

qu'on peut mettre sous la forme

$$(\alpha^{k-1}-\alpha^{-k+1})(\alpha^{-k}-\alpha^{k-1})(\alpha-1)\equiv 0 \pmod{\lambda}.$$

Maintenant, si aucun de ces trois facteurs n'est égal à zéro, leur produit contient le facteur $1-\alpha$ trois fois; mais, pour être divisible par λ, il fallait qu'il contînt ce facteur $(\lambda-1)$ fois. Donc, excepté le seul cas $\lambda = 3$, que nous excluons de notre discussion, cette congruence ne peut subsister, à moins que des deux facteurs $\alpha^{k-1} - \alpha^{-k+1}$ et $\alpha^{-k} - \alpha^{k-1}$, l'un ou l'autre soit égal à zéro, ce qui revient à

$$k \equiv 1 \quad \text{ou} \quad 2k \equiv 1 \quad (\text{mod. } \lambda).$$

Le premier de ces deux cas ne peut avoir lieu; car, d'après la congruence $(a + b)k \equiv b$, il s'ensuivrait

$$a \equiv 0 \quad (\text{mod. } \lambda),$$

ce qui est contre l'hypothèse. Le second cas $2k \equiv 1$ donne, en vertu de la même congruence,

$$a \equiv b \quad (\text{mod. } \lambda),$$

comme conséquence nécessaire de l'équation

$$u^\lambda + v^\lambda + w^\lambda = 0,$$

en supposant qu'aucun des trois nombres u, v, w ne soit divisible par $1 - \alpha$. En échangeant les nombres u, v, w, ce qui change les nombres a, b, c en même temps, on obtient

$$a \equiv c \quad \text{et} \quad b \equiv c \quad (\text{mod. } \lambda).$$

Or, en développant la $\lambda^{\text{ième}}$ puissance du binôme

$$u = a + (1 - \alpha)^2 \text{P},$$

et en négligeant tous les termes divisibles par λ, on obtient

$$u^\lambda \equiv a^\lambda \equiv a \quad (\text{mod. } \lambda).$$

De la même manière, on aura aussi

$$\left.\begin{array}{l} v^\lambda \equiv b^\lambda \equiv b \\ w^\lambda \equiv c^\lambda \equiv c \end{array}\right\} \quad (\text{mod. } \lambda);$$

et, en ajoutant,

$$u^\lambda + v^\lambda + w^\lambda \equiv a + b + c \quad (\text{mod. } \lambda).$$

Il faut donc qu'on ait

$$a + b + c \equiv 0 \quad (\text{mod. } \lambda);$$

et, comme ces trois nombres a, b, c sont congrus entre eux, on aura enfin les congruences

$$3a \equiv 0, \quad 3b \equiv 0, \quad 3c \equiv 0 \quad (\text{mod. } \lambda),$$

qui ne peuvent subsister que dans le seul cas $\lambda = 3$.

Donc l'équation proposée

$$u^\lambda + v^\lambda + w^\lambda = 0$$

ne peut être satisfaite par des nombres complexes u, v, w dont aucun n'a le facteur $1 - \alpha$.

Passons maintenant à la seconde partie de notre démonstration, dans laquelle nous supposons qu'un des trois nombres u, v, w soit divisible par $1 - \alpha$. Soit w ce nombre divisible par $1 - \alpha$, qui pourra contenir ce facteur plusieurs fois, et mettons $(1 - \alpha)^m w$ au lieu de w, en sorte que l'équation proposée soit

$$u^\lambda + v^\lambda + (1 - \alpha)^{m\lambda} w^\lambda = 0.$$

Mais nous préférons discuter l'équation plus générale

$$u^\lambda + v^\lambda = E(\alpha)(1 - \alpha)^{m\lambda} w^\lambda,$$

dans laquelle les nombres complexes u, v, w, premiers entre eux, ne sont pas divisibles par $1 - \alpha$, et où $E(\alpha)$ désigne une unité complexe quelconque.

En décomposant l'expression $u^\lambda + v^\lambda$, on a

$$(u + v).(u + \alpha v).(u + \alpha^2 v)\ldots(u + \alpha^{\lambda-1} v) = E(\alpha)(1 - \alpha)^{m\lambda} w^\lambda.$$

Ici les facteurs de ce produit ont tous le facteur commun $1 - \alpha$; car, en faisant, comme ci-dessus,

$$u = a + (1 - \alpha)^2 P,$$
$$v = b + (1 - \alpha)^2 Q,$$

on aura

$$u + \alpha^r v \equiv a + b - rb(1 - \alpha) \quad [\text{mod. } (1 - \alpha)^2]:$$

et comme le facteur $u + \alpha^r v$ doit être divisible par $1 - \alpha$, pour une certaine valeur de r du moins, on voit que $a + b$ doit être divisible par $1-\alpha$, et conséquemment par λ. Cette congruence devient donc

$$u+\alpha^r v \equiv -rb(1-\alpha) \quad [\text{mod.}\,(1-\alpha)^2];$$

d'où l'on voit que tous ces facteurs ont le facteur commun $1-\alpha$, et qu'ils ne le contiennent qu'une seule fois, excepté le cas de $r=0$, où l'on a

$$u+v\equiv 0 \quad [\text{mod.}\,(1-\alpha)^2].$$

Cela étant, il suit immédiatement de l'équation précédente, que $u+v$ contient $(m\lambda-\lambda+1)$ fois précisément le facteur $(1-\alpha)$; on pourra donc poser

$$u+v=(1-\alpha)^{m\lambda-\lambda+1}.\varphi$$

et

$$u+\alpha^r v=(1-\alpha^r)\varphi_r,$$

et, en substituant ces expressions dans l'équation où $u^\lambda+v^\lambda$ est décomposé en λ facteurs, on aura

$$\varphi.\varphi_1.\varphi_2\ldots\varphi_{\lambda-1}=\mathrm{E}(\alpha)\,w^\lambda.$$

Les nombres complexes $\varphi, \varphi_1, \varphi_2,\ldots, \varphi_{\lambda-1}$ sont premiers entre eux; car tout facteur commun de φ_r et φ_s, ou de $u+\alpha^r v$ et $u+\alpha^s v$, diviserait également les nombres $(\alpha^r-\alpha^s)u$ et $(\alpha^r-\alpha^s)v$, dont le plus grand diviseur commun est $1-\alpha$. De là on conclut, par la même raison que dans la première partie de notre démonstration, que les facteurs $\varphi, \varphi_1, \varphi_2,\ldots, \varphi_{\lambda-1}$, premiers entre eux, dont le produit est égal à une $\lambda^{ième}$ puissance multipliée par une unité, seront séparément des puissances du degré λ multipliées par des unités. Donc on peut poser

$$\varphi=e_0(\alpha)w_1^\lambda,$$

$$\varphi_r=e_r(\alpha)\,t_r^\lambda,$$

ce qui donne

$$u+v=e_0(\alpha)(1-\alpha)^{m\lambda-\lambda+1}.w_1^\lambda,$$

$$u+\alpha^k v=e_r(\alpha)(1-\alpha^r)\,t_r^\lambda,$$

pour toutes les valeurs de $r = 1, 2, 3, \ldots, (\lambda - 1)$. En changeant la valeur de r en s, on aura aussi

$$u + \alpha^s v = e_s(\alpha)(1 - \alpha^s) t_s^\lambda,$$

et, en éliminant u et v de ces trois équations, on trouve

$$t_r^\lambda - \frac{e_s(\alpha)}{e_r(\alpha)} t_s^\lambda = \frac{e_0(\alpha)(\alpha^r - \alpha^s)(1-\alpha)}{e_r(\alpha)(1-\alpha^r)(1-\alpha^s)} (1 - \alpha)^{(m-1)\lambda} w_1^\lambda;$$

lorsqu'on y fait, pour abréger,

$$\frac{e_s(\alpha)}{e_r(\alpha)} = -\varepsilon(\alpha),$$

$$\frac{e_0(\alpha)(\alpha^r - \alpha^s)(1-\alpha)}{e_r(\alpha)(1-\alpha^r)(1-\alpha^s)} = E_1(\alpha),$$

où $\varepsilon(\alpha)$ et $E_1(\alpha)$ sont aussi unités complexes, on aura

$$t_r^\lambda + \varepsilon(\alpha) t_s^\lambda = E_1(\alpha)(1 - \alpha^s)^{(m-1)\lambda} w_1^\lambda.$$

Les nombres complexes t_r, t_s et w_1, dont les $\lambda^{\text{ièmes}}$ puissances sont données comme nombres complexes existants, doivent être nombres existants eux-mêmes; les $\lambda^{\text{ièmes}}$ puissances de ces nombres existants seront donc congrues à des entiers non complexes pour le module λ, et l'on aura

$$t_r^\lambda \equiv k, \quad t_s^\lambda \equiv k' \pmod{\lambda},$$

et, comme la puissance $(1-\alpha)^{(m-1)\lambda}$ est divisible par λ, si le nombre entier m est plus grand que l'unité, cette équation donne la congruence

$$k + \varepsilon(\alpha) k' \equiv 0 \pmod{\lambda}.$$

En déterminant le nombre c par la congruence

$$k + ck' \equiv 0 \pmod{\lambda},$$

on aura

$$\varepsilon(\alpha) \equiv c \pmod{\lambda},$$

et l'unité $\varepsilon(\alpha)$, congrue à un nombre entier non complexe pour le module λ, doit être égale à une $\lambda^{\text{ième}}$ puissance d'une autre unité. Donc, en posant

$$\varepsilon(\alpha) = \varepsilon_1(\alpha)^\lambda, \quad \varepsilon_1(\alpha)t_s = v_1 \quad \text{et} \quad t_r = u_1,$$

l'équation précédente devient

$$u_1^\lambda + v_1^\lambda = \mathrm{E}_1(\alpha)(1-\alpha)^{(m-1)\lambda} w_1.$$

Voilà une équation qui ne diffère de l'équation proposée, dont elle dérive, qu'en ce que le nombre m est diminué d'une unité. En appliquant la même méthode à la nouvelle équation, on obtiendra une équation entièrement semblable, dans laquelle le nombre m sera diminué de deux unités; et, en continuant cette réduction, on parviendra nécessairement à une équation semblable, dans laquelle m est réduit à la valeur $m = 1$. La même méthode de réduction, dans laquelle m était supposé plus grand que l'unité, n'est plus applicable à l'équation

$$u^\lambda + v^\lambda = \mathrm{E}(\alpha)(1-\alpha)^\lambda w^\lambda,$$

mais il est facile de démontrer que cette équation ne peut jamais avoir lieu. Pour cela, il suffit de démontrer que *la forme* $u^\lambda + v^\lambda$, *étant divisible par* $1-\alpha$, *doit nécessairement contenir ce facteur* $(\lambda+1)$ *fois au moins.*

En effet, en prenant, comme ci-dessus,

$$u = a + (1-\alpha)^2 \mathrm{P}, \quad v = b + (1-\alpha)^2 \mathrm{Q},$$

on aura

$$u + \alpha^r v \equiv a + b - rb(1-\alpha) \quad [\text{mod. } (1-\alpha)^2],$$

et comme, pour une certaine valeur de r, le facteur $u + \alpha^r v$ de la forme $v^\lambda + v^\lambda$ doit être divisible par $1-\alpha$, on aura nécessairement $a+b$ divisible par $1-\alpha$, et conséquemment divisible par λ, ce qui donne

$$u + \alpha^r v \equiv -rb(1-\alpha) \quad [\text{mod. } (1-\alpha)^2],$$

et, pour le cas $r = 0$,

$$u + v \equiv 0 \quad [\text{mod. } (1-\alpha)^2].$$

On voit donc que tous les facteurs du produit

$$u^\lambda + v^\lambda = (u+v).(u+\alpha v).(u+\alpha^2 v)\ldots(u+\alpha^{\lambda-1} v)$$

sont divisibles par $1-\alpha$, et que le premier facteur $u+v$ est divisible par $(1-\alpha)^2$. Ainsi le nombre des facteurs $1-\alpha$ contenus dans la forme $u^\lambda + v^\lambda$ sera égal à $\lambda+1$ au moins.

Il suit de là que l'équation

$$u^\lambda + v^\lambda = \mathrm{E}(\alpha).(1-\alpha)^\lambda w^\lambda,$$

dans laquelle la forme $u^\lambda + v^\lambda$, divisible par $(1-\alpha)^\lambda$, n'est pas divisible par $(1-\alpha)^{\lambda+1}$, ne pourra jamais subsister, et nous en concluons que l'équation

$$u^\lambda + v^\lambda = \mathrm{E}(\alpha).(1-\alpha)^{m\lambda}.w^\lambda$$

dont elle dérive est aussi impossible.

Le théorème de Fermat, en tant qu'il est rigoureusement démontré dans ce qui précède, pourra donc s'énoncer comme il suit :

Si λ est un nombre premier, qui n'est contenu dans aucun des $\frac{\lambda-3}{2}$ premiers nombres bernoulliens comme facteur du numérateur, l'équation

$$u^\lambda + v^\lambda + w^\lambda = 0$$

ne peut avoir lieu, ni pour des valeurs entières non complexes des
* *nombres u, v, w, ni pour des valeurs complexes de ces nombres formées avec les racines de l'équation $\alpha^\lambda = 1$.*

D'après les valeurs calculées du premier facteur du nombre des classes H, que nous avons données à la fin du § VIII de ce Mémoire, dans la première centaine, il n'y a que les trois nombres premiers $\lambda = 37$, $\lambda = 59$ et $\lambda = 67$, qui sont exceptés de notre démonstration, parce qu'ils se trouvent respectivement comme facteurs du seizième, du vingt-deuxième et du vingt-neuvième nombre bernoullien. Pour de telles puissances qui se présentent comme cas exceptionnels de notre démonstration, il reste encore à démontrer que le théorème de Fermat a lieu en vérité, ou à trouver les solutions de l'équation

$$u^\lambda + v^\lambda + w^\lambda = 0.$$

Über die Ergänzungssätze zu den allgemeinen Reciprocitätsgesetzen

Journal für die reine und angewandte Mathematik 44, 93-146 (1852)

Das Reciprocitätsgesetz für die quadratischen Reste erstreckt sich bekanntlich nicht auf die Primzahl 2, für welche ein besonderer Satz lehrt, ob sie quadratischer Rest einer gegebenen Primzahl sei, oder Nichtrest. Eben so ist von dem Reciprocitätsgesetze für die cubischen Reste die Primzahl 3 ausgenommen; nebst den beiden complexen Primfactoren derselben $1-\alpha$, $1-\alpha^2$, wo α eine dritte Wurzel der Einheit bezeichnet. Für die höheren Potenzreste, welche hier nur für den Fall in Betracht kommen sollen, wo der Potenz-Exponent λ eine ***Primzahl*** ist, sind ebenfalls, aufser dem allgemeinen Reciprocitätsgesetze, Ergänzungssätze nöthig, welche entscheiden, ob der Potenz-Exponent λ und die aus λten Wurzeln der Einheit gebildeten Primfactoren desselben, $1-\alpha$, $1-\alpha^2$, ... $1-\alpha^{\lambda-1}$, für eine gegebene complexe Primzahl λte Potenzreste sind, oder zu welcher Classe der Nichtreste sie gehören. Auch kommen für diese höheren Potenzreste noch die complexen Einheiten hinzu, für welche die Untersuchung noch besonders anzustellen ist. Die Aufgabe, deren vollständige Lösung ich in dem Folgenden geben werde, besteht also darin, die Werthe der auf λte Potenzreste sich beziehenden Symbole

$$\left(\frac{\lambda}{f(\alpha)}\right), \quad \left(\frac{1-\alpha^k}{f(\alpha)}\right) \quad \text{und} \quad \left(\frac{\varepsilon(\alpha)}{f(\alpha)}\right)$$

zu finden, wo α eine λte Wurzel der Einheit, $\varepsilon(\alpha)$ eine beliebige complexe Einheit und $f(\alpha)$ eine complexe Primzahl bedeutet; welche eine wirkliche oder eine ideale sein kann. Die Bedeutung des Symbols $\left(\frac{\varphi(\alpha)}{f(\alpha)}\right)$ ist durch die Congruenz

$$\left(\frac{\varphi(\alpha)}{f(\alpha)}\right) \equiv \varphi(\alpha)^{\frac{Nf(\alpha)-1}{\lambda}} \equiv \alpha^i, \qquad \text{Mod.}\, f(\alpha),$$

definirt, in welcher $\boldsymbol{N}f(\alpha)$ die ***Norm*** von $f(\alpha)$ bedeutet. Für den Fall, dafs $\varphi(\alpha)$ ***ideal*** ist, welcher jedoch vorläufig, da $\varphi(\alpha)$ nur einen der drei Werthe λ, $1-\alpha^k$ oder $\varepsilon(\alpha)$ haben soll, nicht in Betracht kommt, wird diese Definition dahin erweitert, dafs man statt $\varphi(\alpha)$ diejenige Potenz von $\varphi(\alpha)$ nimmt, welche zu einer wirklichen complexen Zahl wird, wenn die ***H***te Potenz es ist. So hat das Symbol $\left(\frac{\varphi(\alpha)^H}{f(\alpha)}\right)$ nach der obigen Definition einen ganz bestimmten Sinn, und es ist sodann $\left(\frac{\varphi(\alpha)}{f(\alpha)}\right)$ nach der Gleichung

$$\left(\frac{\varphi(\alpha)^H}{f(\alpha)}\right) = \left(\frac{\varphi(\alpha)}{f(\alpha)}\right)^H$$

zu definiren; welche immer einen bestimmten Werth dafür giebt, wenn H nicht durch λ theilbar ist. Für diesen Fall aber, welcher, wie ich anderweit gezeigt habe, nur dann vorkommen kann, wenn λ eine von den Ausnahmszahlen ist, für welche eine der ersten $\frac{1}{2}(\lambda-3)$ ***Bernoulli***schen Zahlen in ihrem Zähler λ selbst als Factor enthält, ist die gegebene Definition unzureichend. Der Exponent derjenigen Potenz von α, welcher $\left(\frac{\varphi(\alpha)}{f(\alpha)}\right)$ gleich ist, heifst der ***Index*** von $\varphi(\alpha)$ für den Mod. $f(\alpha)$ und soll in dem Folgenden durch Ind. $\varphi(\alpha)$ bezeichnet werden, so dafs die vorliegende Aufgabe auch so ausgedrückt werden kann: Die Werthe von

$$\text{Ind.}\,\lambda, \quad \text{Ind.}\,(1-\alpha^k) \quad \text{und} \quad \text{Ind.}\,\varepsilon(\alpha)$$

zu finden, für den Mod. $f(\alpha)$.

Es werden hierbei zwei Fälle zu unterscheiden und besonders zu behandeln sein; nämlich ***erstens*** der Fall, wo $f(\alpha)$ eine complexe Primzahl ist, deren Norm $Nf(\alpha) = p$ eine Primzahl von der Form $\nu\lambda+1 = p$ ist, in welchem Falle ich $f(\alpha)$ eine zum Exponenten ***Eins*** gehörende ***complexe*** Primzahl nenne; und ***zweitens*** der Fall, wo $f(\alpha)$ eine zu einem beliebigen andern Exponenten gehörende complexe Primzahl ist, d. h., wo $Nf(\alpha) = q^t$ und q eine für den Modul λ zum Exponenten t gehörende ***nichtcomplexe*** Primzahl ist. In dem ersten Falle, für welchen ich eine einfache Methode und die Hauptresultate im vorigen Jahre der Königlichen Akademie der Wissenschaften zu Berlin mitgetheilt habe (M. s. die Monatsberichte vom Mai 1850), ist die Lehre von der ***Kreistheilung*** allein ausreichend, um die vorliegende Aufgabe zu lösen; in dem zweiten Falle aber ist noch eine, auch in anderer Beziehung nicht unwichtige Erweiterung oder Verallgemeinerung der Theorie der Kreistheilung nöthig; welche in dem Folgenden ebenfalls entwickelt werden soll. Endlich gedenke ich hier auch noch eine Anwendung der gefundenen Resultate anf das ***allgemeine*** Reciprocitätsgesetz zu geben, bestehend in der Lösung der Aufgabe: Wenn das Reciprocitätsgesetz zwischen zwei complexen

Primzahlen gegeben ist, für besondere Bedingungen, welche die Einheiten bestimmen, die als Factoren dieser complexen Zahlen vorhanden sein können: daraus das Reciprocitätsgesetz für dieselben complexen Primfactoren zu finden, welche jenen Bedingungen nicht weiter unterliegen.

§. 1.

Entwickelung der Indices von λ, $1-\alpha^k$ und $\varepsilon(\alpha)$ für den Fall, dafs der Modul, auf welchen sich die Indices beziehen, ein zum Exponenten *Eins* gehörender complexer Primfactor ist.

Es sei $f(\alpha)$ ein complexer, idealer oder wirklicher Primfactor der realen Primzahl p, von der Form $\nu\lambda+1$; λ sei eine *ungerade* Primzahl; g eine primitive Wurzel der Congruenz $g^{p-1}\equiv 1$, Mod. p; γ eine primitive Wurzel der Congruenz $\gamma^{\lambda-1}\equiv 1$, Mod. λ; x eine imaginäre Wurzel der Gleichung $x^p=1$, und α eine imaginäre Wurzel der Gleichung $\alpha^\lambda=1$. Ferner seien $\eta, \eta_1, \eta_2, \ldots \eta_{\lambda-1}$ die λ Perioden, welche aus je $\nu=\frac{p-1}{\lambda}$ Wurzeln der Gleichung $x^p=1$ gebildet werden können, so dafs

$$\eta = x+x^{g^\lambda}+x^{g^{2\lambda}}+\cdots+x^{g^{(\nu-1)\lambda}},$$
$$\eta_1 = x^g+x^{g^{\lambda+1}}+x^{g^{2\lambda+1}}+\cdots+x^{g^{(\nu-1)\lambda+1}} \text{ ist,}$$

u. s. w.

Die Producte je zweier Perioden lassen sich bekanntlich immer als lineäre Functionen aller ähnlichen Perioden darstellen. Diese Ausdrücke nehmen in dem gegenwärtigen Falle, wo λ *ungrade* ist, folgende Gestalt an:

$$\eta^2 = \nu+m\eta+m_1\eta_1+m_2\eta_2+\cdots+m_{\lambda-1}\eta_{\lambda-1},$$
$$\eta\eta_1 = \overset{1}{m}\eta+\overset{1}{m}_1\eta_1+\overset{1}{m}_2\eta_2+\cdots+\overset{1}{m}_{\lambda-1}\eta_{\lambda-1},$$
$$\eta\eta_2 = \overset{2}{m}\eta+\overset{2}{m}_1\eta_1+\overset{2}{m}_2\eta_2+\cdots+\overset{2}{m}_{\lambda-1}\eta_{\lambda-1},$$
$$\cdots\cdots\cdots\cdots\cdots\cdots\cdots$$
$$\eta\eta_{\lambda-1} = \overset{\lambda-1}{m}\eta+\overset{\lambda-1}{m}_1\eta_1+\overset{\lambda-1}{m}_2\eta_2+\cdots+\overset{\lambda-1}{m}_{\lambda-1}\eta_{\lambda-1},$$

in welchen allgemein der Coëfficient $\overset{k}{m}_h$ gleich ist der Anzahl der Werth-
* Verbindungen des r und s, aus den Zahlen $r=0, 1, 2, \ldots r-1$ und
* $s=0, 1, 2, \ldots s-1$, welche der Congruenz

$$g^{k+r\lambda}+1 \equiv g^{h+s\lambda}, \quad \text{Mod. } p,$$

genügen, oder, was Dasselbe ist, gleich der Anzahl der Werthe des r aus

13*

* der Reihe $r = 0, 1, 2, \dots r-1$, welche der Congruenz

$$\text{Ind.}(g^{k+r\lambda}+1) \equiv h, \quad \text{Mod. } \lambda,$$

genügen, wenn das Zeichen des Index (Ind.) sich auf die Primzahl p und deren primitive Wurzel g bezieht. Ich stelle hier, wegen des in dem Folgenden davon zu machenden Gebrauchs, die Haupt-Eigenschaften dieser Zahlen $\overset{k}{m}_h$, welche ich in der Abhandlung über die Zerlegung der aus Wurzeln der Einheit gebildeten complexen Zahlen in ihre Primfactoren (Bd. 35, S. 327 dieses Journals) hergeleitet habe, für den vorliegenden Fall, wo λ, die Anzahl der Perioden, *ungerade* ist, noch einmal zusammen, nämlich:

$$\overset{k+r\lambda}{m}_h = \overset{k}{m}_h, \qquad \overset{k}{m}_{h+s\lambda} = \overset{k}{m}_h,$$

$$\overset{k}{m}_h = \overset{h}{m}_k, \qquad \overset{k}{m}_h = \overset{\lambda-k}{m}_{h-k},$$

$$\overset{k}{m} + \overset{k}{m}_1 + \overset{k}{m}_2 + \cdots + \overset{k}{m}_{\lambda-1} = \nu;$$

aber für $k = 0$,

$$m + m_1 + m_2 + \cdots + m_{\lambda-1} = \nu - 1.$$

Betrachtet man nun die Summe

$$\Sigma\,\text{Ind.}(g^{k+r\lambda}+1)$$

für alle Werthe $0, 1, 2, \dots r-1$ des r, so finden sich in derselben $\overset{k}{m}_0$ Glieder, welche congruent Null werden für den Modul λ; ferner giebt es $\overset{k}{m}_1$ Glieder, welche congruent 1 werden, $\overset{k}{m}_2$ Glieder, welche congruent 2 werden u. s. w.; woraus

$$\Sigma\,\text{Ind.}(g^{k+r\lambda}+1) \equiv 1\overset{k}{m}_1 + 2\overset{k}{m}_2 + 3\overset{k}{m}_3 + \cdots + (\lambda-1)\overset{k}{m}_{\lambda-1}, \quad \text{Mod. } \lambda,$$

folgt. Nun wird aber, wie leicht zu zeigen, für $\nu = \frac{p-1}{\lambda}$ die Congruenz

$$z^\nu - 1 \equiv (z-1)(z-g^\lambda)(z-g^{2\lambda}) \dots (z-g^{(\nu-1)\lambda}), \quad \text{Mod. } p,$$

für alle beliebigen Werthe des z identisch erfüllt. Setzt man daher in derselben $z \equiv -g^{p-1-k}$, Mod. p, und multiplicirt auf beiden Seiten mit $g^{\nu k}$, so wird

$$1 - g^{\nu k} \equiv (g^k+1)(g^{k+\lambda}+1)(g^{k+2\lambda}+1) \dots (g^{k+(\nu-1)\lambda}+1), \quad \text{Mod. } p,$$

also, wenn auf beiden Seiten die Indices genommen werden:

$$\text{Ind.}(1-g^{\nu k}) \equiv \Sigma\,\text{Ind.}(g^{k+r\lambda}+1), \quad \text{Mod. } \lambda,$$

und diese Congruenz, mit der obigen Congruenz verglichen, giebt:

$$\text{Ind.}(1-g^{\nu k}) \equiv 1\overset{k}{m}_1 + 2\overset{k}{m}_2 + 3\overset{k}{m}_3 + \cdots + (\lambda-1)\overset{k}{m}_{\lambda-1}, \quad \text{Mod. } \lambda.$$

Ist nun $f(\alpha)$ ein complexer Primfactor des p, und zwar der zu $\alpha = g^\nu$ gehörige, so ist $g^\nu \equiv \alpha$, Mod. $f(\alpha)$. Ferner ist nach der oben gegebenen

Definition des Zeichens Ind., welches sich auf den Mod. $f(\alpha)$ bezieht,

$$\varphi(\alpha)^{\frac{Nf(\alpha)-1}{\lambda}} \equiv \alpha^{\mathrm{Ind.}\,\varphi(\alpha)}, \quad \mathrm{Mod.}\, f(\alpha),$$

oder weil $Nf(\alpha)=p$, $\frac{p-1}{\lambda}=\nu$ und $\alpha\equiv g^{\nu}$, Mod. $f(\alpha)$, ist:

$$\varphi(g^{\nu})^{\nu} \equiv g^{\nu\,\mathrm{Ind.}\,\varphi(\alpha)},$$

für den Modul $f(\alpha)$ und, da diese Congruenz nur nichtcomplexe Zahlen enthält, auch für den Mod. p. Hieraus folgt, wenn $\varphi(\alpha)=1-\alpha^{k}$ gesetzt wird:

$$\nu\,\mathrm{Ind.}(1-g^{\nu k}) \equiv \nu\,\mathrm{Ind.}(1-\alpha^{k}), \quad \mathrm{Mod.}\, p-1,$$

also, wenn man durch ν dividirt,

$$\mathrm{Ind.}(1-g^{\nu k}) \equiv \mathrm{Ind.}(1-\alpha^{k}), \quad \mathrm{Mod.}\, \lambda.$$

Demnach hat man auch

$$\mathrm{Ind.}(1-\alpha^{k}) \equiv 1\overset{k}{m}_1+2\overset{k}{m}_2+3\overset{k}{m}_3+\cdots+(\lambda-1)\overset{k}{m}_{\lambda-1}, \quad \mathrm{Mod.}\, \lambda;$$

welche Congruenz eine sehr einfache Lösung einer der obigen Aufgaben darstellt. Aus diesem Ausdrucke läfst sich sehr leicht der entsprechende Ausdruck des Ind. (λ) ableiten, indem man nach einander $k=1, 2, 3, \ldots \lambda-1$ setzt; wobei zu beachten ist, dafs

$$(1-\alpha)(1-\alpha^{2})(1-\alpha^{3}) \ldots (1-\alpha^{\lambda-1}) = \lambda$$

und

$$\overset{1}{m}_h+\overset{2}{m}_h+\overset{}{m}_h+\cdots\cdots+\overset{\lambda-1}{m}_h = \nu.$$

So erhält man

$$\mathrm{Ind.}(\lambda) \equiv -(m_1+2m_2+3m_3+\cdots+(\lambda-1)m_{\lambda-1}), \quad \mathrm{Mod.}\, \lambda.$$

Endlich ergeben sich aus derselben Quelle auch die Indices der complexen Einheiten. Nimmt man nämlich die Einheit

$$e(\alpha) = \sqrt[2]{\left(\frac{(1-\alpha^{\gamma})(1-\alpha^{-\gamma})}{(1-\alpha)(1-\alpha^{-1})}\right.} = \pm\frac{\alpha^{\frac{1}{2}(1-\gamma)}(1-\alpha^{\gamma})}{1-\alpha},$$

welche ich ***Kreistheilungs-Einheit*** nenne, so hat man

$$\mathrm{Ind.}\, e(\alpha^{k}) \equiv \tfrac{1}{2}k(1-\gamma)\,\mathrm{Ind.}\,\alpha+\mathrm{Ind.}(1-\alpha^{k\gamma})-\mathrm{Ind.}(1-\alpha^{k}), \quad \mathrm{Mod.}\, \lambda,$$

also

$$\mathrm{Ind.}\, e(\alpha^{k}) \equiv \tfrac{1}{2}k(1-\gamma)\nu+1\overset{k\gamma}{m}_1+2\overset{k\gamma}{m}_2+3\overset{k\gamma}{m}_3+\cdots+(\lambda-1)\overset{k\gamma}{m}_{\lambda-1} \qquad \mathrm{Mod.}\, \lambda.$$
$$-1\overset{k}{m}_1-2\overset{k}{m}_2-3\overset{k}{m}_3-\cdots-(\lambda-)\overset{k}{m}_{\lambda-1},$$

Ein System conjugirter Kreistheilungs-Einheiten ist ein unabhängiges System von Einheiten, welches die Eigenschaft hat, dafs überhaupt jede Einheit ohne Ausnahme sich als ein Product von Potenzen der conjugirten Kreistheilungs-Einheiten darstellen läfst, und zwar so, dafs die Potenz-Exponenten nur rationale Brüche sind; d. h., wenn $\varepsilon(\alpha)$ eine beliebige complexe Einheit ist, so hat man immer

$$\varepsilon(\alpha) = \pm\alpha^k e(\alpha)^n . e(\alpha^\gamma)^{n_1} . e(\alpha^{\gamma^2})^{n_2} \ldots e(\alpha^{\gamma^{\mu-1}})^{n_{\mu-1}},$$

wo zur Abkürzung $\frac{1}{2}(\lambda-1)=\mu$ gesetzt ist, und wo $n, n_1, n_2, \ldots n_{\mu-1}$ rationale Brüche sind. Man erhält daher auch den Index jeder beliebigen Einheit $\varepsilon(\alpha)$ durch die Indices der Kreistheilungs-Einheiten ausgedrückt, nämlich:

$$\text{Ind.}\,\varepsilon(\alpha) \equiv k\nu + n\,\text{Ind.}\,e(\alpha) + n_1\,\text{Ind.}\,e(\alpha^\gamma) + \cdots + n_{\mu-1}\,\text{Ind.}\,e(\alpha^{\gamma^{\mu-1}}), \quad \text{Mod.}\,\lambda.$$

In den Nennern der rationalen Brüche $n, n_1, \ldots n_{\mu-1}$ kann, wie ich in der Abhandlung (Bd. 40. S. 117 dieses Journals) gezeigt habe, der Factor λ nur dann vorkommen, wenn die Classen-Anzahl aller idealen complexen Zahlen durch λ theilbar ist, also nur dann, wenn λ in einer der ersten $\frac{1}{2}(\lambda-3)$ *Bernoulli*schen Zahlen als Factor des Zählers auftritt. Schliefst man diese Ausnahmszahlen λ auch hier aus, so sind mit den Indices der Kreistheilungs-Einheiten $e(\alpha^k)$ zugleich auch die Indices aller möglichen Einheiten gegeben; denn man kann alle in dem Ausdrucke des Ind. $\varepsilon(\alpha)$ vorkommenden Brüche $n, n_1, \ldots n_{\mu-1}$ durch die ganzen Zahlen ersetzen, denen sie congruent sind für den Modul λ.

Hiermit ist also die erste Lösung der Aufgabe, die Indices der Zahlen $1-\alpha^k$ und λ, und der complexen Einheiten zu finden, in sehr einfacher Weise gegeben; für den Fall, dafs die complexe Primzahl, auf welche die Indices sich beziehen, eine solche ist, die für den Modul λ zum Exponenten ***Eins*** gehört. Die Zahlen $\overset{k}{m}_h$, mittels welcher die Lösung dieser Aufgabe gegeben ist, lassen sich auf mannichfache Weise auf andere in der Lehre von der Kreistheilung vorkommende Zahlen reduciren, namentlich auf die Coëfficienten der complexen Zahlen $\psi_r(\alpha)$, welche, wenn man

$$F(\alpha, x) = x + \alpha x^g + \alpha^2 x^{g^2} + \cdots + \alpha^{p-2} x^{g^{p-2}}$$

setzt, durch die Gleichung

$$\frac{F(\alpha, x)F(\alpha^r, x)}{F(\alpha^{r+1}, x)} = \psi_r(\alpha) = \overset{r}{a} + \overset{r}{a}_1\alpha + \overset{r}{a}_2\alpha^2 + \cdots + \overset{r}{a}_{\lambda-1}\alpha^{\lambda-1}$$

bestimmt werden. Durch dieselben können die Zahlen $\overset{k}{m}_h$ in folgender Art ausgedrückt werden:

$$\lambda \overset{k}{m}_h = \overset{\lambda-2}{\Sigma_r}\, \overset{r}{a}_{h+kr} - (\lambda-3)\nu + 1;$$

wo in den besondern Fällen $h=k$, $h=0$ oder $k=0$ die ***Eins*** auf der rechten Seite wegfallen mufs. Da ferner, wie ich in früheren Abhandlungen gezeigt habe, alle in der Kreistheilung vorkommenden Zahlen durch die Coëfficienten eines complexen Primfactors von p sich ausdrücken lassen, so wird man auch die Indices von λ, $1-\alpha^k$ und $\varepsilon(\alpha)$ durch die Coëfficienten von $f(\alpha)$ selbst ausdrücken können. Ich will jedoch nicht alle diese Umformungen der gefundenen Formeln hier wirklich ausführen. Für Ind. λ und Ind. $(1-\alpha^k)$ sind auch die gefundenen Ausdrücke so einfach uud elegant, dafs sie vollständig genügen können; für die Indices der Einheiten dagegen giebt es noch einige merkwürdige und einfache Ausdrücke, welche ich entwickeln will, weil sie auch für andere Untersuchungen wichtig sind. Sie gestalten sich am einfachsten für die zusammengesetzte Kreistheilungs-Einheit

$$E_n(\alpha) = e(\alpha)e(\alpha^\gamma)^{\gamma^{-2n}}.e(\alpha^{\gamma^2})^{\gamma^{-4n}} \dots e(\alpha^{\gamma^{\mu-1}})^{\gamma^{-2(\mu-1)n}},$$

welche überhaupt in mehrfacher Beziehung vor den übrigen Einheiten bevorzugt ist. Für diese Einheit hat man

$$\text{Ind.}\,E_n(\alpha) \equiv \text{Ind.}\,e(\alpha)+\gamma^{-2n}\text{Ind.}\,e(\alpha^\gamma)+\cdots+\gamma^{-2(\mu-1)n}\text{Ind.}\,e(\alpha^{\gamma^{\mu-1}}), \text{ Mod. } \lambda,$$

und man kann umgekehrt Ind. $e(\alpha^{\gamma^h})$ mittels der aus jener leicht abzuleitenden Formel

$$\text{Ind.}\,e(\alpha^{\gamma^h}) \equiv -2(\gamma^{2h}\text{Ind.}\,E_1(\alpha)+\gamma^{4h}\text{Ind.}\,E_2(\alpha)+\cdots$$
$$\cdots+\gamma^{2(\mu-1)h}\text{Ind.}\,E_{\mu-1}(\alpha)), \text{ Mod. } \lambda,$$

durch die Indices von $E_1(\alpha)$, $E_2(\alpha)$, ... $E_{\mu-1}(\alpha)$ ausdrücken.

Substituirt man in diesem Ausdrucke des Ind. $E_n(\alpha)$ für $e(\alpha)$, für $e(\alpha^\gamma)$ u. s. w. ihre Werthe nach dem oben gegebenen Ausdrucke dieser Kreistheilungs-Einheiten, so erhält man nach einigen leichten Reductionen:

* $$2\,\text{Ind.}\,E_n(\alpha) \equiv (\gamma^{2n})(\text{Ind.}(1-\alpha)+\gamma^{-2n}\text{Ind.}(1-\alpha^\gamma)+\cdots+\gamma^{-(\lambda-2)2n}\text{Ind.}(1-\alpha^{\gamma^{\lambda-2}})),$$

und wenn $\gamma^h \equiv k$, Mod. λ, gesetzt wird, was $\gamma^{-2hn} \equiv k^{\lambda-1-2n}$ giebt, so erhält man, mit Anwendung des Summenzeichens:

$$2\,\text{Ind.}\,E_n(\alpha) \equiv (\gamma^{2n}-1)\Sigma k^{\lambda-2n-1}\text{Ind.}(1-\alpha^k),$$

für $k=1, 2, 3, \dots \lambda-1$; wozu, wenn $\lambda-2n-1$, wie hier angenommen werden soll, *positiv* ist, auch der Werth $k=0$ hinzugethan werden kann. Da nun oben

$$\text{Ind.}(1-\alpha^k) \equiv 1\overset{k}{m}_1+2\overset{k}{m}_2+\cdots+(\lambda-1)\overset{k}{m}_{\lambda-1} \equiv \sum_{h=0}^{\lambda-1} h\overset{k}{m}_h$$

gefunden wurde, so hat man

$$2\,\text{Ind.}\,E_n(\alpha) \equiv (\gamma^{2n}-1)\sum_{k=0}^{\lambda-1}\sum_{h=0}^{\lambda-1} hk^{\lambda-2n-1}\overset{k}{m}_h.$$

In diesen Ausdruck sollen nun statt der Zahlen $\overset{k}{m}_h$ die *Perioden* η, η_1, η_2, ... $\eta_{\lambda-1}$ eingeführt werden. Dies geschieht mit Hülfe der Gleichung

$$\eta\eta_k = \overset{k}{m}\eta + \overset{k}{m}_1\eta_1 + \overset{k}{m}_2\eta_2 + \cdots + \overset{k}{m}_{\lambda-1}\eta_{\lambda-1} = \sum_0^{\lambda-1}{}_h \overset{k}{m}_h\eta_h,$$

aus welcher

$$\eta_i\eta_{i+k} = \sum_0^{\lambda-1}{}_h \overset{k}{m}_h\eta_{i+h}$$

folgt. Multiplicirt man mit i und nimmt die Summe für $i=0, 1, 2, 3, \ldots \lambda-1$, so ergiebt sich

$$\sum_0^{\lambda-1}{}_i i\eta_i\eta_{i+k} = \sum_0^{\lambda-1}{}_i \sum_0^{\lambda-1}{}_h i\overset{k}{m}_h\eta_{i+h},$$

oder, was Dasselbe ist,

$$\sum_0^{\lambda-1}{}_i i\eta_i\eta_{i+k} = \sum_0^{\lambda-1}{}_i \sum_0^{\lambda-1}{}_h (i+h)\overset{k}{m}_h\eta_{i+h} - \sum_0^{\lambda-1}{}_i \sum_0^{\lambda-1}{}_h h\overset{k}{m}_h\eta_{i+h}$$

und, da $\Sigma_i\eta_{i+h}=-1$, $\sum_0^{\lambda-1}{}_i(i+h)\eta_{i+h}\equiv\sum_0^{\lambda-1}{}_i i\eta_i$, Mod. λ, und $\sum_0^{\lambda-1}{}_h \overset{k}{m}_h=\nu$ ist, so erhält man

$$\sum_0^{\lambda-1}{}_i i\eta_i\eta_{i+k} \equiv \nu\sum_0^{\lambda-1}{}_i i\eta_i + \sum_0^{\lambda-1}{}_h h\overset{k}{m}_h, \text{ Mod. } \lambda.$$

Multiplicirt man jetzt mit $k^{\lambda-2n-1}$, nimmt die Summe für $k=0, 1, 2, \ldots \lambda-1$, und beachtet, dafs $\Sigma k^{\lambda-2n-1}\equiv 0$, Mod. λ ist, für $k=0, 1, 2, \ldots \lambda-1$, so erhält man

$$\sum_0^{\lambda-1}{}_k \sum_0^{\lambda-1}{}_i ik^{\lambda-2n-1}.\eta_i\eta_{i+k} \equiv \sum_0^{\lambda-1}{}_k \sum_0^{\lambda-1}{}_h hk^{\lambda-2n-1}.\overset{k}{m}_h, \text{ Mod. } \lambda.$$

Der Ausdruck des zweiten Ind. $E_n(\alpha)$ verwandelt sich demnach in den folgenden:

$$2\,\text{Ind.}\, E_n(\alpha) \equiv (\gamma^{2n}-1)\sum_0^{\lambda-1}{}_k \sum_0^{\lambda-1}{}_i ik^{\lambda-2n-1}\eta_i\eta_{i+k}, \text{ Mod. } \lambda.$$

Anstatt dem k in dieser Doppelsumme die Werthe $0, 1, 2, \ldots \lambda-1$ zu geben, kann man ihm auch die Werthe $-i$, $-i+1$, $-i+2$, ... $-i+\lambda-1$ zutheilen, ohne dafs in Beziehung auf den Modul λ diese Summe ihren Werth ändert. Setzt man also $k-i$ statt i, so erhält man

$$2\,\text{Ind.}\, E_n(\alpha) \equiv (\gamma^{2n}-1)\sum_0^{\lambda-1}{}_k \sum_0^{\lambda-1}{}_i i(k-i)^{\lambda-2n-1}.\eta_i\eta_k.$$

Wird jetzt $(k-i)^{\lambda-2n-1}=(i-k)^{\lambda-2n-1}$ nach dem binomischen Lehrsatze entwickelt, so trennen sich in den einzelnen Gliedern die in Beziehung auf i zu nehmenden Summen von den in Beziehung auf k zu nehmenden, und wenn der Kürze wegen

$$\sum_0^{\lambda-1}{}_k k^r \eta_k = \sum_0^{\lambda-2}{}_i i^r \eta_i = D_r$$

gesetzt und für einen Augenblick $\lambda-2n-1$ einfach durch m bezeichnet wird, so erhält man

$$2\,\text{Ind.}\,E_n(\alpha) \equiv (\gamma^{2n}-1)\Big(D_{m+1}D_0 - \frac{m}{1}D_m D_1 + \frac{m(m-1)}{1.2}D_{m-1}D_2 - \cdots\Big).$$

Um die durch D_0, D_1, D_2, ... bezeichneten Gröfsen noch auf andere Weise zu bestimmen, gebe ich dem Ausdruck der Kreistheilung

$$F(\alpha, x) = x + \alpha x^g + \alpha^2 x^{g^2} + \cdots + \alpha^{p-2} x^{g^{p-2}}$$

die Form

$$F(\alpha, x) = \eta + \alpha\eta_1 + \alpha^2\eta_2 + \cdots + \alpha^{\lambda-1}\eta_{\lambda-1}$$

* und verwandle α in ε^v, wo v eine ***continuirliche Variable*** bedeutet. Hiernach wird

$$\frac{d_0^r F(e^v, x)}{dv^r} \equiv 1^r\eta_2 + 2^r\eta_2 + 3^r\eta_3 + \cdots + (\lambda-1)^r\eta_{\lambda-1} \equiv D_r;$$

wo die dem Zeichen des Differentials d zugefügte Null bedeutet, dafs nach der Differentiation $v=0$ zu setzen ist. Verwandelt man v in $-v$, so erhält man eben so:

$$\frac{d_0^r F(e^{-v}, x)}{dv^r} \equiv (-1)^r D^r.$$

Aus diesen Werthen der Gröfsen D_r, als Differentialquotienten, folgt nach einem bekannten Satze der Differentialrechnung über die Differentiation eines Products zweier Factoren:

$$D_{m+1}D_0 - \frac{m}{1}D_m D_1 + \frac{m(m-1)}{1.2}D_{m-1}D_2 - \cdots = \frac{d_0^m\Big(\dfrac{dF(e^v, x)}{dv}\cdot F(e^{-v}, x)\Big)}{dv^m}.$$

Aus der bekannten Gleichung der Kreistheilung

$$F(\alpha, x)F(\alpha^{-1}, x) = p$$

erhält man aber nach bekannten Principien die für jeden Werth der Variabeln v geltende Gleichung

$$F(e^v, x)F(e^{-v}, x) = p + V.W,$$

in welcher $V = 1 + e^v + e^{2v} + \cdots + e^{(\lambda-1)v}$ und W irgend eine ganze rationale Function von e^v ist, deren Coëfficienten ganze, die Wurzel x enthaltende complexe Zahlen sind; folglich ist auch

$$F(e^{-v}, x) = \frac{p}{F(e^v, x)} + \frac{VW}{F(e^v, x)}$$

und

$$\frac{dF(e^v,\,x)}{dv}F(e^{-v},\,x) = \frac{dlF(e^v,\,x)}{dx}(p+VW).$$

Differentiirt man diesen Ausdruck m mal nach einander und setzt $v=0$, indem man bemerkt, dafs für $v=0$, V und alle Differentialquotienten davon, bis zum $\lambda-2$ten einschliefslich, durch λ theilbar sind, da $1^r+2^r+3^r+\cdots$ $\cdots+(\lambda-1)^r\equiv 0$, Mod. λ, für $r=1, 2, 3, \ldots \lambda-2$ ist, und indem man ferner bemerkt, dafs $p\equiv 1$, Mod. λ ist, so erhält man:

$$\frac{d_0^m\left(\frac{dF(e^v,\,x)}{dx}F(e^{-v},\,x)\right)}{dv^m} \equiv \frac{d_0^{m+1}lF(e^v,\,x)}{dv^{m+1}},\ \text{Mod. } \lambda,$$

also

$$D_{m+1}D_0-\frac{m}{1}D_mD_1+\frac{m(m-1)}{1.2}D_{m-1}D_2-\cdots \equiv \frac{d_0^{m+1}lF(e^v,\,x)}{dv^{m+1}},\ \text{Mod. } \lambda.$$

Macht man von diesem einfachen Ausdrucke Gebrauch, und setzt zugleich für m wieder seinen Werth $\lambda-2n-1$, so erhält man folgenden merkwürdigen Ausdruck des Index von $E_n(\alpha)$:

$$\text{Ind. } E_n(\alpha) \equiv \tfrac{1}{2}(\gamma^{2n}-1)\frac{d_0^{\lambda-2n}lF(e^v,\,x)}{dv^{\lambda-2n}},\ \text{Mod. } \lambda.$$

Die in diesem Ausdrucke vorkommende Wurzel x ist in demselben nur scheinbar enthalten, weil sie, wenn nach Ausführung der Differentiation $v=0$ gesetzt wird, und alle Glieder, die den Factor λ haben, weggelassen werden, von selbst mit herausfällt.

Aus dieser Darstellung des Ind. $E_n(\alpha)$ läfst sich sehr leicht eine andere, ebenfalls sehr bemerkenswerthe Darstellung desselben erlangen, in welcher der von α und x abhängige Ausdruck $F(\alpha,\,x)$ durch die nur von α allein abhängige, oben definirte complexe Zahl $\psi_r(\alpha)$ ersetzt wird. Aus der Gleichung

$$F(\alpha,\,x)F(\alpha^r,\,x) = \psi_r(\alpha)F(\alpha^{r+1},\,x)$$

folgt nämlich, wenn α in e^v verwandelt wird, nach denselben Principien wie oben, die für jeden beliebigen Werth der Variabel v geltende Gleichung

$$F(e^v,\,x)F(e^{rv},\,x) = \psi_r(e^v)F(e^{(r+1)v},\,x)+V.W.$$

Und vermöge der Eigenschaft des $V=1+e^v+e^{2v}+\cdots+e^{(\lambda-1)v}$, dafs V selbst, so wie seine Differentialquotienten, bis zum $(\lambda-2)$ten einschliefslich, für den Werth $v=0$ congruent Null sind, nach dem Modul λ, erhält man aus dieser Gleichung

$$\frac{d_0^{\lambda-2n}lF(e^v,\,x)}{dv^{\lambda-2n}}+\frac{d_0^{\lambda-2n}lF(e^{rv},\,x)}{dv^{\lambda-2n}} \equiv \frac{d_0^{\lambda-2n}lF(e^{(r+1)v},\,x)}{dv^{\lambda-2n}}+\frac{d_0^{\lambda-2n}l\psi_r(e^v)}{dv^{\lambda-2n}},$$

oder, vereinfacht:

$$(1+r^{\lambda-2n}-(r+1)^{\lambda-2n})\frac{d_0^{\lambda-2n}lF(e^v,x)}{dv^{\lambda-2n}} \equiv \frac{d_0^{\lambda-2n}l\psi_r(e^v)}{dv^{\lambda-2n}}, \text{ Mod. } \lambda.$$

Durch diese Congruenz verwandelt sich der obige Ausdruck des Ind. $E_n(\alpha)$ in folgenden:

$$\text{Ind. } E_n(\alpha) \equiv \frac{\gamma^{2n}-1}{2(1+r^{\lambda-2n}-(r+1)^{\lambda-2n})}\cdot\frac{d_0^{\lambda-2n}l\psi_r(\alpha)}{dv^{\lambda-2n}}, \text{ Mod. } \lambda.$$

Endlich leite ich hieraus noch einen andern Ausdruck des Ind. $E_n(\alpha)$ ab, welcher deshalb der wichtigste von allen zu sein scheint, weil er diesen Index der Einheit $E_n(\alpha)$ durch die complexe Primzahl $f(\alpha)$ selbst ausdrückt, in Beziehung auf welche das Zeichen Ind. zu nehmen ist. Hierzu wird die Zerfällung der complexen Zahl $\psi_r(\alpha)$ in ihre complexen idealen oder wirklichen Primfactoren gebraucht, welche ich im (35ten Bande dieses Journals, S. 362) gegeben habe und welche sich in folgender Art darstellen läfst:

$$\psi_r(\alpha) = \pm\alpha^k\Pi f(\alpha^{\gamma^h});$$

wo das Productenzeichen Π sich auf alle diejenigen Werthe des h erstreckt, welche nicht negativ, aber kleiner als $\lambda-1$ sind, und dabei der Bedingung genügen, dafs

$$\gamma_{\mu-h}+\gamma_{\mu-h+\text{Ind.}r} > \lambda$$

ist; wo das Zeichen γ_k den kleinsten positiven Werth von γ^k für den Modul λ, Ind. r den Index des r für die primitive Wurzel γ und den Modul λ bezeichnet und $\mu=\frac{1}{2}(\lambda-1)$ ist. Um die in diesem Producte enthaltenen Primfactoren, welche im Allgemeinen ideal sein werden, zu wirklichen complexen Zahlen zu machen, erhebe ich beide Seiten dieser Gleichung zur Hten Potenz und nehme H so an, dafs $f(\alpha)^H$ eine wirkliche complexe Zahl ist. Diese complexe Zahl soll auch durch Multiplication mit einer passenden einfachen Einheit α^m so zubereitet angenommen werden, dafs sie für den Modul $(1-\alpha)^2$ einer nicht complexen ganzen Zahl congruent ist. Alsdann fällt in dem obigen Ausdrucke des $\psi_r(\alpha)$ die einfache Einheit α^k weg, und man hat

$$\psi_r(\alpha)^H = \pm\Pi f(\alpha^{\gamma^h})^H.$$

Diese Gleichung wird wieder, wie oben, in folgende, für alle beliebigen Werthe der Variabel v geltende Gleichung verwandelt:

$$\psi_r(e^v)^H = \Pi f(e^{v\gamma^h})^H + V.W,$$

und giebt alsdann in derselben Weise:

$$H.\frac{d_0^{\lambda-2n}l\psi_r(e^v)}{dv^{\lambda-2n}} = \Sigma_h\frac{d_0^{\lambda-2n}l\left(f(e^{v\gamma^h})^H\right)}{dv^{\lambda-2n}},$$

14 *

oder vereinfacht:,

$$H\cdot\frac{d_0^{\lambda-2n}l\psi_r(e^v)}{dv^{\lambda-2n}} = \frac{d_0^{\lambda-2n}l(f(e^v)^H)}{dv^{\lambda-2n}}\Sigma_h\gamma^{(\lambda-2n)h}.$$

Es ist nun die Summe $\Sigma_h\gamma^{(\lambda-2n)h}$ zu suchen, in welcher dem h alle diejenigen Werthe von $h=0$ bis $h=\lambda-2$ zu geben sind, welche der Bedingung genügen, dafs $\gamma_{\mu-h}+\gamma_{\mu-h+\text{Ind.}r}>\lambda$ ist. Zu diesem Zwecke bemerke ich, dafs, wie leicht zu beweisen, der Ausdruck

$$\frac{1}{\lambda}(\gamma_{\mu-i}+\gamma_{\mu-i+\text{Ind.}r}-\gamma_{\mu-i+\text{Ind.}(r+1)})$$

in allen den Fällen, wo i einen der mit h bezeichneten Werthe hat, für welche $\gamma_{\mu-1}+\gamma_{\mu-1+\text{Ind.}r}>\lambda$ ist, gleich ***Eins*** ist; während derselbe für alle andern Werthe des i gleich Null ist. Hieraus folgt

$$\Sigma_h\gamma^{(\lambda-2n)h} = \Sigma_i\frac{1}{\lambda}(\gamma_{\mu-i}+\gamma_{\mu-i+\text{Ind.}r}-\gamma_{\mu-i+\text{Ind.}(r+1)})\gamma^{(\lambda-2n)i};$$

wo dem h nur die oben angegebenen Werthe, aber dem i alle Werthe $0, 1, 2, \dots \lambda-2$ zu geben sind. Um bequeme Congruenzen für den Modul λ^2 anwenden zu können, welches wegen des als Divisor hier vorkommenden λ nöthig ist, setze ich $\gamma^{\lambda(\lambda-2n)i}$ statt $\gamma^{(\lambda-2n)i}$; was in Beziehung auf den Modul λ keinen Unterschied macht. Ich bezeichne ferner die zu suchende Summe $\Sigma_h\gamma^{(\lambda-2n)h}$ mit dem Buchstaben T, multiplicire mit λ, und erhalte so folgende Congruenz für den Modul λ^2:

$$\lambda T \equiv \Sigma(\gamma_{\mu-i}+\gamma_{\mu-i+\text{Ind.}r}-\gamma_{\mu-i+\text{Ind.}(r+1)})\gamma^{\lambda(\lambda-2n)i},\ \text{Mod.}\ \lambda^2,$$

oder

$$\lambda T \equiv \Sigma\gamma_{\mu-i}\gamma^{\lambda(\lambda-2n)i}+\Sigma\gamma_{\mu+i+\text{Ind.}r}\gamma^{\lambda(\lambda-2n)i}-\Sigma\gamma_{\mu-i+\text{Ind.}(r+1)}\gamma^{\lambda(\lambda-2n)i}.$$

Ich betrachte nun zuerst nur die zweite dieser drei Summen, nämlich die Summe

$$\sum_0^{\lambda-2}{}_i\gamma_{\mu-i+\text{Ind.}r}\gamma^{\lambda(\lambda-2n)i},$$

aus welcher die erste hervorgeht, wenn $r=1$ gesetzt, und die dritte, wenn r in $r+1$ verwandelt wird. Ich verwandle i in $i+\mu+\text{Ind.}r$ und bemerke, dafs nach dieser Verwandlung dem i in der Summe immer noch dieselben Werthe $0, 1, 2, \dots \lambda-2$ zukommen; was daraus folgt, dafs in Beziehung auf den Modul λ^2 die Glieder dieser Summe einen Cyclus bilden. Diese Summe ist daher für den Modul λ^2 der folgenden congruent:

$$\sum_0^{\lambda-2}{}_i\gamma_{-i}\gamma^{\lambda(k-2n)(i+\mu+\text{Ind.}r)},$$

und da $\gamma^{\lambda(\lambda-2n)\mu}\equiv-1$, $\gamma^{\lambda(\lambda-2n)\text{Ind.}r}\equiv r^{\lambda(\lambda-2n)}$ für den Modul λ^2 ist, so wird dieselbe in folgende Form gebracht:

$$-r^{\lambda(\lambda-2n)}\sum_{0}^{\lambda-2}{}_i\gamma_{-i}\gamma^{\lambda(\lambda-2n)i}.$$

Macht man von diesem Werthe Gebrauch, und von den entsprechenden, welche man erhält, wenn $r=1$ und $r+1$ statt r gesetzt wird, so erhält man folgenden Ausdruck der Summe T:

$$\lambda T \equiv -(1+r^{\lambda(\lambda-2n)}-(r+1)^{\lambda(\lambda-2n)})\sum_{0}^{\lambda-2}{}_i\gamma_{-i}\gamma^{\lambda(\lambda-2n)i}, \text{ Mod. } \lambda^2.$$

Man setze jetzt $\gamma_{-i}=k$, so wird $\gamma^{-i}\equiv k$ und $\gamma^i\equiv k^{\lambda-2}$, für den Modul λ; woraus $\gamma^{i\lambda}\equiv k^{\lambda(\lambda-2)}$, Mod. λ^2, also

$$\sum_{0}^{\lambda-2}{}_i\gamma_{-i}\gamma^{\lambda(\lambda-2n)i} \equiv \sum_{0}^{\lambda-1}{}_k k^{\lambda(\lambda-2n)(\lambda-2)+1}, \text{ Mod. } \lambda^2$$

folgt, oder, vereinfacht, mit Hülfe der Congruenz $k^{\lambda(\lambda-1)}\equiv 1$, Mod. λ^2:

$$\sum_{0}^{\lambda-2}{}_i\gamma_{-i}\gamma^{\lambda(\lambda-2n)i} \equiv \sum_{0}^{\lambda-1}{}_k k^{\lambda(2n-1)+1}, \text{ Mod. } \lambda^2.$$

Nun folgt aber aus den bekannten Summen-Ausdrücken der Potenzen der natürlichen Zahlen, dafs

$$\sum_{0}^{\lambda-1} k^{\lambda(2n-1)+1} \equiv (-1)^{m-1}B_m\lambda, \text{ Mod. } \lambda^2$$

ist; wo B_m die $m$te *Bernoulli*sche Zahl bezeichnet und wo der Kürze wegen auf einen Augenblick für $\frac{1}{2}(\lambda(2n-1)+1)$ das einfache Zeichen m gesetzt ist. Demnach ist

$$\lambda T = (-1)^m(1+r^{\lambda(\lambda-2n)}-(r+1)^{\lambda(\lambda-2n)})B_m\lambda, \text{ Mod. } \lambda^2.$$

Und wenn durch λ dividirt wird und für $r^{\lambda(\lambda-2n)}$ und $(r+1)^{\lambda(\lambda-2n)}$ die nach dem Modul λ congruenten einfacheren Potenzen $r^{\lambda-2n}$ und $(r+1)^{\lambda-2n}$ gesetzt werden, so ist

$$T \equiv (-1)^m(1+r^{\lambda-2n}-(r+1)^{\lambda-2n})B_m, \text{ Mod. } \lambda.$$

Die hierin vorkommende *Bernoulli*sche Zahl B_m kann noch durch Reduction auf eine möglichst niedrige *Bernoulli*sche Zahl vereinfacht werden; mittels der von mir in einer kleinen Abhandlung (Bd. 41, S. 371 dieses Journals) bewiesenen Eigenschaft der *Bernoulli*schen Zahlen, dafs

$$\frac{B_n}{n} \equiv \frac{(-1)^\mu B_{n+\mu}}{n+\mu}, \text{ Mod. } \lambda$$

ist, für $\mu=\frac{1}{2}(\lambda-1)$ und für alle Werthe des n, welche nicht Vielfache von μ sind. Diese Congruenz giebt unmittelbar die allgemeinere:

$$\frac{B_n}{n} \equiv \frac{(-1)^{s\mu} B_{n+s\mu}}{n+s\mu}, \text{ Mod. } \lambda,$$

welche für alle ganzzahligen Werthe von s gültig ist. Setzt man in dieser $s=2n-1$, so erhält man für $m=\frac{1}{2}(\lambda(2n-1)+1)$:

$$B_m \equiv (-1)^\mu \frac{B_n}{2n}, \text{ Mod. } \lambda.$$

Wird dieser einfachere Werth des B_m angewendet und erwogen, dafs $(-1)^{m+\mu}=(-1)^n$ ist, so erhält man

$$T \equiv (-1)^n(1+r^{\lambda-2n}-(r+1)^{\lambda-2n})\frac{B_n}{2n}, \text{ Mod. } \lambda,$$

und demnach

$$H\frac{d_0^{\lambda-2n} l\psi_r(e^v)}{dv^{\lambda-2n}} \equiv (-1)^n(1+r^{\lambda-2n}-(r+1)^{\lambda-2n})\frac{B_n}{2n}\frac{d_0^{\lambda-2n} l(f(e^v)^H)}{dv^{\lambda-2n}}, \text{ Mod. } \lambda.$$

Vermöge dieser Congruenz verwandelt sich der zuletzt gegebene Ausdruck des Ind. $E_n(\alpha)$ in folgenden:

$$\text{Ind.} E_n(\alpha) \equiv (-1)^n(\gamma^{2n}-1)\frac{B_n}{4nH}\frac{d_0^{\lambda-2n} l(f(e^v)^H)}{dv^{\lambda-2n}}, \text{ Mod. } \lambda.$$

Dieser Ausdruck, in welchem die Zahlen der Kreistheilung nicht weiter vorkommen, welcher vielmehr den Index der Einheit $E_n(\alpha)$ durch die complexe Primzahl $f(\alpha)$ selbst darstellt, in Beziehung auf welche das Zeichen Index zu nehmen ist, wird in dem Falle, wo H ein Vielfaches von λ ist, nichtssagend. Es sind also auch bei der Anwendung dieser Formel diejenigen Ausnahmewerthe des λ auszuschliefsen, für welche λ ein Factor des Zählers einer der ersten $\frac{1}{2}(\lambda-3)$ ***Bernoulli***schen Zahlen ist. Für den Fall, dafs $f(\alpha)$ eine wirkliche complexe Primzahl ist, hat man einfach $H=1$ zu setzen. Man kann aber auch für den Fall, dafs $f(\alpha)$ eine ideale complexe Primzahl ist, den Ausdruck des Ind. $E_n(\alpha)$ einfach durch

$$\text{Ind.} E_n(\alpha) \equiv (-1)^n(\gamma^{2n}-1)\frac{B_n}{4n}\frac{d_0^{\lambda-2n} lf(e^v)}{dv^{\lambda-2n}}, \text{ Mod. } \lambda$$

darstellen, wenn man ein für allemal festsetzt, dafs für $f(\alpha)$, falls es ideal ist, sein Ausdruck als Hte Wurzel aus der wirklichen complexen Zahl $f(\alpha)^H$ genommen werden soll.

§. 2.

Eine Erweiterung der Theorie der Kreistheilung.

Nachdem in dem vorhergehenden Paragraphen die Indices von λ, $1-\alpha^k$ und $e(\alpha)$ oder $E_n(\alpha)$ gefunden worden sind, für welche der Modul $f(\alpha)$ ein Primfactor einer nichtcomplexen ganzen Primzahl p von der Form $\nu\lambda+1$ ist, mufs noch dieselbe Aufgabe für den allgemeinern Fall gelöset werden, dafs der Modul $f(\alpha)$ eine zu irgend einem Exponenten t (Divisor von $\lambda-1$) gehörende complexe Primzahl ist. Die Lösung dieser Aufgabe erfordert als Vorarbeit eine Erweiterung der Theorie der ***Kreistheilung***, welche ich hier kurz entwickeln will.

Es sei also $f(\alpha)$ ein idealer complexer Primfactor der nichtcomplexen Primzahl q, welche zum Exponenten t (einem Divisor von $\lambda-1$) gehört, für den Modul λ, so dafs $q^t \equiv 1$, Mod. λ, dafs aber keine niedrigere Potenz von q der Einheit congruent ist, für den Modul λ. Wird eine solche complexe Primzahl $f(\alpha)$ zum Modul genommen, so ist ein vollständiges Resten-System dieses Moduls in der Form

$$a+a_1\alpha+a_2\alpha^2+\cdots+a_{t-1}\alpha^{t-1}$$

enthalten, in welcher alle die Zahlen $a, a_1, a_2, \ldots a_{t-1}$ alle ganzzahligen Werthe von 0 bis $q-1$ einschliefslich erhalten können. Die Anzahl aller verschiedenen Reste ist also gleich q^t; nämlich gleich der Anzahl aller Verbindungen der q Zahlen a mit den q Zahlen a_1, mit den q Zahlen a_2, u. s. w. Die Richtigkeit dieser Behauptung ergiebt sich daraus, dafs, ***erstens***, jeder wirklichen complexen Zahl $F(\alpha)$ die Form

$$F(\alpha) = \varphi(\eta)+\alpha\varphi_1(\eta)+\alpha^2\varphi_2(\eta)+\cdots+\alpha^{t-1}\varphi_{t-1}(\eta)$$

gegeben werden kann, wo $\varphi(\eta)$, $\varphi_1(\eta)$, u. s. w. complexe Zahlen sind, welche nur die aus je t Gliedern bestehenden Perioden der Wurzeln $\alpha, \alpha^2, \ldots \alpha^{\lambda-1}$ enthalten (M. s. meine Abhandlung Bd. 35. S. 337 dieses Journals); ***zweitens*** daraus, dafs in Beziehung auf einen solchen Modul $f(\alpha)$ diese Perioden, und also auch die complexen Zahlen $\varphi(\eta)$, $\varphi_1(\eta)$, u. s. w., stets nichtcomplexen ganzen Zahlen congruent sind; und endlich, ***drittens***, daraus, dafs eine complexe Zahl von der Form

$$(a-b)+(a_1-b_1)\alpha+(a_2-b_2)\alpha^2+\cdots+(a_{t-1}-b_{t-1})\alpha^{t-1}$$

nicht durch $f(\alpha)$ theilbar sein kann, ohne dafs zugleich $a\equiv b$, $a_1\equiv b_1$, $a_2\equiv b_2, \ldots a_{t-1}\equiv b_{t-1}$ ist, für den Modul q.

Es giebt ferner für den Modul $f(\alpha)$ stets primitive Wurzeln $g(\alpha)$, in der Art, dafs die verschiedenen Potenzen einer solchen primitiven Wurzel

$$(1.)\quad g(\alpha),\ g(\alpha)^2,\ g(\alpha)^3,\ \ldots\ g(\alpha)^{q^t-2},$$

alle verschiedenen Reste für den Modul $f(\alpha)$, mit alleiniger Ausnahme des Restes 0, vollständig erschöpfen. Der Beweis dieser Behauptung kann eben so leicht und nach derselben Methode gegeben werden, nach welcher man in den Elementen der Zahlentheorie die Existenz der primitiven Wurzeln für die gewöhnlichen Primzahlen beweiset; weshalb ich denselben übergehe.

Wenn $F(\alpha)$ eine beliebige, jedoch nicht durch den Modul $f(\alpha)$ theilbare complexe Zahl bezeichnet, so ist

$$F(\alpha)^{q^t-1} \equiv 1, \quad \text{Mod.} f(\alpha).$$

Wenn ferner $g(\alpha)$ eine primitive Wurzel ist, so hat man

$$g(\alpha)^{\frac{1}{2}(q^t-1)} \equiv -1, \quad g(\alpha)^{\frac{q^t-1}{\lambda}} \equiv \alpha^k, \quad \text{Mod.} f(\alpha);$$

wo k niemals congruent Null ist, für den Modul λ. Es kann auch die primitive Wurzel $g(\alpha)$, welche in dem Folgenden einfach durch g bezeichnet werden soll, immer so angenommen werden, dafs $k = 1$ ist und dafs man also

$$g^{\frac{q^t-1}{\lambda}} \equiv \alpha, \quad \text{Mod.} f(\alpha)$$

hat. Wir wollen ein für allemal festsetzen, dafs die primitive Wurzel g *immer* dieser Bedingung gemäfs angenommen werden soll. Die Beweise der in diesen Congruenzen enthaltenen Sätze, welche nach den in der Theorie der complexen Zahlen gebräuchlichen Methoden leicht sind, glaube ich ebenfalls hier übergehen zu können. Zu bemerken ist noch, dafs die primitive Wurzel g niemals eine nichtcomplexe Zahl sein kann, aufser in dem Falle $t = 1$, für welchen die primitiven Wurzeln des Modul $f(\alpha)$ genau dieselben sind, wie die primitiven Wurzeln der gewöhnlichen Primzahl $p = Nf(\alpha)$.

Ist $g^i \equiv \varphi(\alpha)$, Mod. $f(\alpha)$, so soll i hier ebenfalls der Index von $\varphi(\alpha)$ für den Mod. $f(\alpha)$ heifsen und durch Ind. $\varphi(\alpha)$ bezeichnet werden. Es gelten dann für die Indices ebenfalls die Gleichungen

$$\text{Ind.}\varphi(\alpha) + \text{Ind.}F(\alpha) \equiv \text{Ind.}(\varphi(\alpha)F(\alpha)), \quad \text{Mod.}\, q^t-1 \quad \text{und}$$
$$n\,\text{Ind.}\varphi(\alpha) \equiv \text{Ind.}(\varphi(\alpha)^n), \quad \text{Mod.}\, q^t-1.$$

Nachdem Dieses vorbereitet ist, gehe ich zu dem Hauptgegenstande dieses Paragraphen über, nämlich zu der Erweiterung der Theorie der ***Kreistheilung;*** und zwar knüpfe ich dieselbe an die complexe Zahl $\psi_r(\alpha)$ der Kreistheilung an, welche schon oben definirt wurde und für welche die Lehre von der Kreistheilung folgende sechs Ausdrücke giebt:

$$\psi_r(\alpha) = \Sigma\alpha^{h-(r+1)\,\text{Ind.}(g^h+1)},$$
$$\psi_r(\alpha) = \Sigma\alpha^{h+r\,\text{Ind.}(g^h+1)},$$
$$\psi_r(\alpha) = \Sigma\alpha^{rh-(r+1)\,\text{Ind.}(g^h+1)},$$
$$\psi_r(\alpha) = \Sigma\alpha^{-(r+1)h+r\,\text{Ind.}(g^h+1)},$$
$$\psi_r(\alpha) = \Sigma\alpha^{rh+\text{Ind.}(g^h+1)},$$
$$\psi_r(\alpha) = \Sigma\alpha^{-(r+1)h+\text{Ind.}(g^h+1)};$$

wo die Summenzeichen sich auf die Werthe $h=0, 1, 2, \dots p-2$ erstrecken, mit Ausschlufs des Werths $h=\frac{1}{2}(p-1)$, für welchen $g^h+1\equiv 0$, Mod. p, sein würde.

Die hier auszuführende Erweiterung der Theorie der Kreistheilung soll nun darin bestehen, dafs in diesen Ausdrücken des $\psi_r(\alpha)$ die primitive Wurzel g der Primzahl p als primitive Wurzel für den complexen Modul $f(\alpha)$ aufgefafst und demgemäfs auch das Zeichen Ind. auf denselben Modul $f(\alpha)$ bezogen werden soll. Die Summen werden dann in Beziehung auf die Werthe $h=0, 1, 2, 3, \dots q^t-2$ zu nehmen sein, mit Ausschlufs des Werths $h=\frac{1}{2}(q^t-1)$, für welchen $g^h+1\equiv 0$, Mod. $f(\alpha)$ sein und folglich Ind. (g^h+1) keinen Sinn haben würde. Die complexen Zahlen $\psi_r(\alpha)$, in diesem neuen Sinne genommen, sollen zur Unterscheidung von denen der Kreistheilung durch $\Psi_r(\alpha)$ bezeichnet werden. Die obigen sechs verschiedenen Ausdrücke von $\psi_r(\alpha)$ stimmen auch bei dieser neuen Auffassung der Zeichen g und Ind. ganz miteinander überein; in der Art, dafs sie alle nur verschiedene Ausdrücke einer und derselben complexen Zahl sind. Es wird also hinreichen, nur einen dieser Ausdrücke zu untersuchen, zu welchem ich den letzten wähle. Ich setze also

$$\Psi_r(\alpha) = \Sigma\alpha^{-(r+1)h+\text{Ind.}(g^h+1)};$$

wo das Summenzeichen sich auf die Werthe $h=0, 1, 2, 3, \dots q^t-2$ erstreckt; mit Ausschlufs von $h=\frac{1}{2}(q^t-1)$, und wo g eine primitive Wurzel für den Modul $f(\alpha)$ bedeutet, auch das Zeichen Ind. sich auf denselben Modul und dieselbe primitive Wurzel bezieht. Der Modul $f(\alpha)$ selbst ist ein complexer idealer Primfactor der zum Exponenten t gehörenden gewöhnlichen Primzahl q.

Aus dieser Definition der complexen Zahl $\Psi_r(\alpha)$ folgt zunächst unmittelbar

$$\Psi_{r+\lambda}(\alpha) = \Psi_r(\alpha),\quad \Psi_v(\alpha) = -1,\quad \Psi_{\lambda-1}(\alpha) = -1.$$

Setzt man ferner α^q statt α, so erhält man

$$\Psi_r(\alpha^q) = \Sigma\alpha^{-(r+1)qh+q\,\text{Ind.}(g^h+1)}.$$

Es ist aber $(g^h+1)^q\equiv g^{hq}+1$, für den Modul q, und also auch für den Modul $f(\alpha)$, welcher ein Factor von q ist; demnach ist auch $q\,\text{Ind.}(g^h+1)\equiv \text{Ind.}(g^{qh}+1)$, Mod. (q^t-1). Setzt man Dieses in den obigen Ausdruck des $\Psi_r(\alpha^q)$ und $k\equiv qh$, Mod. (q^t-1), so entsprechen den Werthen $h=0, 1, 2, \dots$ $\dots q^t-2$, mit Ausschlufs von $h=\frac{1}{2}(q^t-1)$, genau die Werthe $k=0, 1, 2, 3, \dots q^t-2$, mit Ausschlufs von $k=\frac{1}{2}(q^t-1)$, wenn auch in anderer Ordnung genommen, und darum ist

$$\Psi_r(\alpha^q) = \Sigma\alpha^{-(r+1)k+\text{Ind.}(g^k+1)},$$

also

$$\Psi_r(\alpha^q) = \Psi_r(\alpha);$$

aus welcher Gleichung auch sogleich die allgemeinere

$$\Psi_r(\alpha^{q^h}) = \Psi_r(\alpha)$$

für jeden beliebigen Werth des h folgt. Man sieht hieraus, dafs die complexe Zahl $\Psi_r(\alpha)$ nicht die einzelnen Wurzeln $\alpha, \alpha^2, \alpha^3, \ldots \alpha^{\lambda-1}$ in sich enthält, sondern nur die ***Perioden*** derselben, welche je t Glieder umfassen. Dies ist die ***erste wesentliche Grund-Eigenschaft*** der complexen Zahl $\Psi_r(\alpha)$.

Multiplicirt man die beiden reciproken complexen Zahlen $\Psi_r(\alpha)$ und $\Psi_r(\alpha^{-1})$ miteinander, so wird das Product derselben durch die Doppelsumme

$$\Psi_r(\alpha)\Psi_r(\alpha^{-1}) = \Sigma\Sigma\alpha^{-(r+1)(h-k)+\text{Ind.}(g^h+1)-\text{Ind.}(g^k+1)}$$

ausgedrückt, für $h=0, 1, 2, 3, \ldots q^t-2$, $k=0, 1, 2, \ldots q^t-2$; mit Ausschlufs von $h=\frac{1}{2}(q^t-1)$ und $k=\frac{1}{2}(q^t-1)$. Ich scheide aus dieser Doppelsumme zunächst alle die Werth-Verbindungen von h und k aus, für welche $h=k$ ist. Ihrer finden offenbar q^t-2 Statt, und es wird für dieselben die Potenz des α unter dem doppelten Summenzeichen nur gleich ***Eins.*** Man hat also

$$\Psi_r(\alpha)\Psi_r(\alpha^{-1}) = q^t-2+\Sigma\Sigma\alpha^{-(r+1)(h-k)+\text{Ind.}(g^h+1)-\text{Ind.}(g^k+1)}$$

für alle ganzzahligen Werthe des h und k von 0 bis q^t-2 einschliefslich; mit Ausschlufs der Werthe $\frac{1}{2}(q^t-1)$ und der Werth-Verbindungen, für welche $h=k$ ist. Man transformire nun diese Doppelsumme, indem man

$$g^{k-h} \equiv g^{k'}, \quad \frac{g^k+1}{g^h+1} \equiv g^{h'}, \quad \text{Mod.}\, f(\alpha),$$

setzt, so erhält man

$$\Psi_r(\alpha)\Psi_r(\alpha^{-1}) = q^t-2+\Sigma\Sigma\alpha^{(r+1)k'-h'}.$$

Die Summenzeichen beziehen sich hier auf die verschiedenen Werthe des k' und h', und es sind denselben alle Werthe von 1 bis q^t-2 einschliefslich zu geben; mit Ausschlufs derjenigen Werth-Verbindungen, für welche $h'=k'$ ist. Vermöge der Congruenzen, welche die neuen Unbestimmten k' und h' durch die alten k und h ausdrücken, entspricht nämlich jeder einzelnen Verbindung der Werthe von h und k eine, und nur eine Werth-Verbindung der angegebenen Werthe von h' und k'; und eben so, umgekehrt: jeder beliebigen Werth-Verbindung des h' und k' entspricht unter den angegebenen Bedingungen eine, und nicht mehr als eine der Werth-Verbindungen des h und k. Ich nehme jetzt mit dieser Doppelsumme noch die kleine Änderung vor, dafs ich sie mit auf diejenigen Werth-Verbindungen erstrecke, welche $h' = k'$ geben. Dies wird ohne Beeinträchtigung der Richtigkeit der obigen Gleichung geschehen können, wenn die Summe aller, der Werth-Verbindung $h' = k'$ entsprechenden Glieder, nämlich $\Sigma\alpha^{rk'}$, in Abzug gebracht wird. Es ist demnach

$$\Psi_r(\alpha)\Psi_r(\alpha^{-1}) = q^t - 2 - \Sigma\alpha^{rk'} + \Sigma\Sigma\alpha^{(r+1)k'-h'},$$

für $h' = 1, 2, 3, \ldots q^t - 2$, $k' = 1, 2, 3, \ldots q^t - 2$. Die Werthe des h' und k' sind nun gänzlich unabhängig von einander, und man kann die Summation in Beziehung auf beide einzeln ausführen.

Summirt man zuerst in Beziehung auf h', so erhält man

$$\Sigma\alpha^{-h'} = \alpha^{-1} + \alpha^{-2} + \alpha^{-3} + \cdots + \alpha^{-(q^t-2)} = -1,$$

also

$$\Psi_r(\alpha)\Psi_r(\alpha^{-1}) = q^t - 2 - \Sigma\alpha^{rk'} - \Sigma\alpha^{(r+1)k'}.$$

Schliefst man die beiden besonderen Fälle aus, wo $r \equiv 0$, oder $r + 1 \equiv 0$, Mod. λ, so ist wieder jede dieser beiden Summen gleich -1, und man hat als Endresultat:

$$\Psi_r(\alpha)\Psi_r(\alpha^{-1}) = q^t;$$

welches die *zweite Grund-Eigenschaft* der complexen Zahl $\Psi_r(\alpha)$ ist, deren Analogie mit der entsprechenden Eigenschaft der complexen Zahlen der Kreistheilung $\psi_r(\alpha)$ auf der Hand liegt.

In den beiden besondern Fällen, $r \equiv 0$ oder $r + 1 \equiv 0$, Mod. λ wird allemal eine der beiden Summen $\Sigma\alpha^{k'r}$ oder $\Sigma\alpha^{(r+1)k'}$, weil sie alsdann aus lauter *Einsen* besteht, gleich $q^t - 2$, die andere aber gleich -1. Man erhält demnach

$$\Psi_0(\alpha)\Psi_0(\alpha^{-1}) = 1, \qquad \Psi_{-1}(\alpha)\Psi_{-1}(\alpha^{-1}) = 1;$$

welches mit den oben angegebenen, aus der Definition unmittelbar folgenden Werthen von $\Psi_0(\alpha) = -1$ und $\Psi_{-1}(\alpha) = -1$ übereinstimmt, aber nichts Neues lehrt.

Nach der *ersten* Grund-Eigenschaft der hier behandelten complexen Zahlen ist:

$$\Psi_r(\alpha^{q^h}) = \Psi_r(\alpha).$$

Wenn nun t eine *gerade* Zahl ist, so ist, weil q zum Exponenten t gehört, $q^t \equiv 1$ und $q^{\frac{1}{2}t} \equiv -1$, Mod. λ. Es ergiebt sich daher, wenn $h = \frac{1}{2}t$ gesetzt wird:

$$\Psi_r(\alpha^{-1}) = \Psi_r(\alpha);$$

15 *

woraus, vermöge der *zweiten* Grund-Eigenschaft dieser complexen Zahlen,

* $$\Psi_r(\alpha) = q^{\frac{1}{2}t}$$

folgt. Für den Fall, dafs t eine *gerade* Zahl ist, ist also die Zahl $\Psi_r(\alpha)$ in Wahrheit nicht eine complexe Zahl, sondern einer Potenz der gewöhnlichen Primzahl q gleich. Das Haupt-Interesse der gegenwärtigen Untersuchung liegt demnach nur in dem Falle, wo t *ungerade* ist, in welchem $\Psi_r(\alpha)$ nicht nur scheinbar, sondern auch *wirklich* eine complexe Zahl ist.

Zur vollständigen Erkenntnifs der complexen Zahlen $\Psi_r(\alpha)$, namentlich für den Fall wo t *ungerade* ist, gehört noch die Zerlegung derselben in ihre complexen, idealen oder wirklichen Primfactoren, welche, weil nach der *zweiten* Grund-Eigenschaft derselben das Product zweier reciproker Zahlen einer Potenz von q gleich ist, nur die idealen Primfactoren des q sein können; also nur die Primfactoren $f(\alpha)$, $f(\alpha^\gamma)$, $f(\alpha^{\gamma^2})$, ... u. s. w. Von diesen sind bekanntlich, da q zum Exponenten t gehört, je t einander gleich, so dafs nur $\frac{\lambda-t}{t}$ wesentlich verschiedene Primfactoren vorhanden sind; als welche, wenn $\frac{\lambda-1}{t} = \tau$ gesetzt wird (welche Bedeutung der Buchstabe τ in dem Folgenden überall beibehalten soll), die Primfactoren

$$f(\alpha),\ f(\alpha^\gamma),\ f(\alpha^{\gamma^2}),\ \ldots f(\alpha^{\gamma^{\tau-1}})$$

genommen werden können. Man hat daher

$$\Psi_r(\alpha) = \boldsymbol{E}(\alpha) f(\alpha)^m . f(\alpha^\gamma)^{m_1} \ldots f(\alpha^{\gamma^{\tau-1}})^{m_{\tau-1}},$$

wo $\boldsymbol{E}(\alpha)$ eine ***Einheit*** ist und m, m_1, ... $m_{\tau-1}$ die noch zu bestimmenden Exponenten sind, welche angeben, wievielmal jeder der verschiedenen Primfactoren des q in $\Psi_r(\alpha)$ enthalten ist.

Mittels der zweiten Grund-Eigenschaft der complexen Zahl $\Psi_r(\alpha)$ kann nun sogleich die eine Hälfte dieser Exponenten durch die andere Hälfte ausgedrückt werden. Wenn nämlich, wie hier vorausgesetzt wird, t eine *ungerade* Zahl ist, so ist τ eine *gerade* Zahl, weil $\lambda-1 = t\tau$ ist. Verwandelt man α in α^{-1} und erwägt, dafs $f(\alpha^{\gamma^h}) = f(\alpha^{\gamma^{h+\tau}}) = f(\alpha^{\gamma^{h+2\tau}})$ u. s. w. ist, so erhält man

$$\Psi_r(\alpha^{-1}) = \boldsymbol{E}(\alpha^{-1}) . f(\alpha)^{m_{\frac{1}{2}\tau}} . f(\alpha^\gamma)^{m_{\frac{1}{2}\tau+1}} \ldots f(\alpha^{\gamma^{\tau-1}})^{m_{\frac{1}{2}\tau-1}}.$$

Multiplicirt man diesen Ausdruck mit dem vorigen, so erhält man, vermöge der Gleichung $\Psi_r(\alpha)\Psi_r(\alpha^{-1}) = q^t$, für die Exponenten m, m_1, m_2, ... $m_{\tau-1}$ folgende Gleichungen:

$$m + m_{\frac{1}{2}\tau} = t,\quad m_1 + m_{\frac{1}{2}\tau+1} = t,\quad \ldots\quad m_{\frac{1}{2}\tau-1} + m_{\tau-1} = t.$$

Also ist die Summe je zweier dieser Exponenten, deren Stellenzeiger um $\frac{1}{2}\tau$ unterschieden sind, immer gleich t.

Die vollständige Bestimmung dieser Exponenten ist etwas mühsam, führt aber, wie sich zeigen wird, zu einem ziemlich einfachen und sehr eleganten Resultate. Es müssen dazu Congruenzen angewendet werden, deren Modul eine Potenz der complexen Primzahl $f(\alpha)$ ist, und es kommt dabei hauptsächlich darauf an, die Wurzel α durch eine Potenz der primitiven Wurzel g zu ersetzen, welcher sie in Beziehung auf den Modul $f(\alpha)^n$ congruent ist. Dies geschieht auf folgende Weise. Wenn der Kürze wegen $\nu = \frac{q^t - 1}{\lambda}$ gesetzt wird (welche Bedeutung der Buchstabe ν hier überall beibehalten soll), so ist, wie oben gezeigt worden:

$$g^\nu \equiv \alpha, \text{ Mod.} f(\alpha),$$

oder, wenn man diese Congruenz als Gleichung schreibt:

$$g^\nu = \alpha + wf(\alpha).$$

Erhebt man dies zur Potenz q, so ist

$$g^{\nu q} = \alpha^q + q\alpha^{q-1} wf(\alpha) + \frac{q(q-1)}{1.2}\alpha^{q-2} w^2 f(\alpha)^2 + \cdots$$

und, da q den Primfactor $f(\alpha)$ einmal enthält:

$$g^{\nu q} \equiv \alpha^q, \text{ Mod.} f(\alpha)^2.$$

Auf dieselbe Weise weiter schliefsend, findet man, dafs allgemein für jeden Werth des k die Congruenz

$$g^{\nu q^k} \equiv \alpha^{q^k}, \text{ Mod.} f(\alpha)^{k+1},$$

Statt hat; und wenn für k ein Vielfaches von t genommen wird, da $q^{nt} \equiv 1$, Mod. λ ist, so ergiebt sich

$$g^{\nu q^{nt}} \equiv \alpha, \text{ Mod.} f(\alpha)^{nt+1};$$

wo n eine beliebige positive ganze Zahl ist.

Nachdem Dieses vorbereitet worden, verwandle ich in dem ursprünglichen Ausdrucke, durch welchen $\Psi_r(\alpha)$ definirt wurde, α in $\alpha^{\gamma^{-i}}$. Dann ist

$$\Psi_r\left(\alpha^{\gamma^{-i}}\right) = \Sigma\alpha^{-(r+1)\gamma^{-i}h + \gamma^{-i}\,\text{Ind.}(g^h+1)}.$$

Man setze ferner $r+1 \equiv \gamma^\varrho$, Mod. λ, so wird $(r+1)\gamma^{-i} \equiv \gamma^{\varrho-i}$, und wenn wieder γ_k den kleinsten positiven Rest von γ^k für den Modul λ bezeichnet, so ist $(r+1)\gamma^{-i} \equiv \gamma_{\varrho-i}$, $\gamma^{-i} \equiv \gamma_{-i}$, Mod. λ. Endlich setze man noch Ind. $(g^{hq^{nt}} + 1)$ statt Ind. $(g^h + 1)$; welches demselben vollkommen gleichbedeutend ist, indem $g^{q^{nt}} \equiv g$ ist, für den Modul $f(\alpha)$. Dann erhält man

$$\Psi_r\left(\alpha^{\gamma^{-i}}\right) = \Sigma\alpha^{-\gamma_{\varrho-i}h + \gamma_{-i}\,\text{Ind.}(g^{hq^{nt}}+1)}$$

Wird nun die Congruenz $g^{\nu q^{nt}} \equiv \alpha$, Mod. $f(\alpha)^{nt+1}$ angewendet, so verwandelt sich diese Gleichung in folgende Congruenz:

$$\Psi_r(\alpha^{\gamma^{-i}}) \equiv \Sigma g^{-\gamma_{\varrho-i}\nu q^{nt}h} \cdot g^{\gamma_{-i}\nu q^{nt}\,\mathrm{Ind.}(g^{hq^{nt}}+1)}, \text{ Mod. } f(\alpha)^{nt+1}.$$

Es ist aber

$$g^{\mathrm{Ind.}(g^{hq^{nt}}+1)} \equiv g^{hq^{nt}}+1, \text{ Mod. } f(\alpha),$$

woraus

$$g^{q^{nt}\,\mathrm{Ind.}(g^{hq^{nt}}+1)} \equiv (g^{hq^{nt}}+1)^{q^{nt}}, \text{ Mod. } f(\alpha)^{nt+1}.$$

folgt, also

$$\Psi_r(\alpha^{\gamma^{-i}}) \equiv \Sigma g^{-\gamma_{\varrho-i}\nu q^{nt}h}(g^{hq^{nt}}+1)^{\gamma_{-i}\nu q^{nt}}, \text{ Mod. } f(\alpha)^{nt+1}.$$

Das Summenzeichen erstreckt sich auf die Werthe $0, 1, 2, \ldots q^t-2$ von h und es ist hier nicht weiter nöthig, den Werth $h=\frac{1}{2}(q^t-1)$ auszuschliefsen, weil für denselben der Ausdruck unter dem Summenzeichen congruent Null ist, für den Mod. $f(\alpha)^{nt+1}$.

Nun ist die unter dem Summenzeichen stehende Potenz des Binoms $g^{hq^{nt}}+1$ nach dem binomischen Lehrsatze zu entwickeln. Wird dabei durch $\Pi(m)$ das Product der Zahlen $1.2.3.4\ldots m$ bezeichnet, so erhält man die daraus hervorgehende Doppelsumme in folgender Form:

$$\Psi_1(\alpha^{\gamma^{-i}}) \equiv \Sigma\Sigma \frac{\Pi(\gamma_{-i}\nu q^{nt})}{\Pi(k)\Pi(\gamma_{-i}\nu q^{nt}-k)} g^{(-\gamma_{\varrho-i}\nu+k)hq^{nt}}, \text{ Mod. } f(\alpha)^{nt+1},$$

für $h=0, 1, 2, 3, \ldots q^t-2$ und $k=0, 1, 2, 3, \ldots \gamma_{-i}\nu q^{nt}$. Setzt man für einen Augenblick $-\gamma_{\varrho-i}\nu q^{nt}+k=l$ und führt die Summation in Beziehung auf h aus, so ist allgemein

$$\Sigma g^{lhq^{nt}} = \frac{1-g^{l(q^t-1)q^{nt}}}{1-g^{lq^{nt}}},$$

also immer congruent Null für den Modul $f(\alpha)^{nt+1}$; mit Ausnahme der Fälle, wo l ein Vielfaches von q^t-1 ist. Es sei also

$$l = -\gamma_{\varrho-i}\nu q^{nt}+k = s(q^t-1) = s\lambda\nu,$$

so ist

$$\Psi_r(\alpha^{\gamma^{-i}}) \equiv \Sigma \frac{(q^t-1)\Pi(\gamma_{-i}\nu q^{nt})}{\Pi(\gamma_{\varrho-i}\nu+s\lambda\nu)\Pi(\gamma_{-i}\nu q^{nt}-\gamma_{\varrho-i}\nu-s\lambda\nu)}, \text{ Mod. } f(\alpha)^{\lambda n+1};$$

wo sich das Summenzeichen auf alle ganzzahligen Werthe von s bezieht, für welche weder $\gamma_{\varrho-i}\nu+s\lambda\nu$, noch $\gamma_{-i}\nu q^{nt}-\gamma_{\varrho-i}\nu-s\lambda\nu$ negativ wird. Wird endlich noch $s=\frac{\gamma_{\varrho-i}(q^{nt}-1)}{\lambda}-z$ gesetzt und aufserdem α in α^{γ^i} verwandelt, so ergiebt sich

$$\Psi_r(\alpha) \equiv \Sigma \frac{(q^t-1)\Pi(\gamma_{-i}\nu q^{nt})}{\Pi(\gamma_{\varrho-i}\nu q^{nt}-\lambda\nu z)\Pi((\gamma_{-i}-\gamma_{\varrho-i})\nu q^{nt}+\lambda\nu z)}$$

für den Modul $f(\alpha^{\gamma^i})^{n\lambda+1}$; wo das Summenzeichen auf alle ganzzahligen Werthe des z sich bezieht, für welche weder $\gamma_{\varrho-i}q^{nt}-\lambda z$, noch $(\gamma_{-i}-\gamma_{\varrho-i})q^{nt}+\lambda z$ negativ wird.

Durch diese Congruenz, in welcher n beliebig grofs angenommen werden kann, wird die Untersuchung, wievielmal $\Psi_r(\alpha)$ den complexen Primfactor $f(\alpha^{\gamma^i})$ enthält, darauf zurückgeführt, zu untersuchen, wievielmal diese Summe, welcher $\Psi_r(\alpha)$ congruent und welche eine nichtcomplexe Zahl ist, den Factor q enthält; denn es ist klar, dafs, wenn n hinreichend grofs angenommen wird, die Summe den Factor q genau ebensovielmal enthalten mufs, als $\Psi_r(\alpha)$ den Factor $f(\alpha^{\gamma^i})$ enthält, welcher ein Primfactor von q und dessen $(n\lambda+1)$te Potenz der Modul dieser Congruenz ist.

Zu diesem Zwecke werde ich zunächst beweisen, dafs, unter der Voraussetzung, $\gamma_{-i}-\gamma_{\varrho-i}$ sei positiv, dasjenige Glied dieser Summe, welches dem Werthe $z=0$ entspricht, von allen die niedrigste Potenz von q als Factor enthält; in der Art, dafs jedes beliebige andere Glied als dieses, eine höhere Potenz von q als Factor enthalten mufs. Demgemäfs wird die Untersuchung, wievielmal die ganze Summe den Factor q enthält, darauf reducirt sein, zu untersuchen, wievielmal das eine dem Werthe $z=0$ entsprechende Glied der Summe den Factor q enthält.

Ich bediene mich hier des folgenden, wenn nicht bekannten, so doch leicht zu beweisenden Satzes:

Lehrsatz: Wenn q eine Primzahl ist und die Zahl A auf die Form

$$A = a+a_1q+a_2q_2+ \cdots +a_{k-1}q^{k-1}$$

gebracht wird, in der Art, dafs die Zahlen $a, a_1, a_2, \ldots a_{k-1}$ nicht negativ und alle kleiner als q sind (welches nichts anderes ist, als die Zahl A nach $q$theiligem Zahlensysteme geschrieben), so ist die Anzahl der in dem Producte

$$1.2.3.4 \ldots A = \Pi(A)$$

enthaltenen Factoren q gleich

$$\frac{A-(a+a_1+a_2+\cdots+a_{k-1})}{q-1}.$$

Mittels dieses Satzes läfst sich nun folgendermaafsen untersuchen, wievielmal der Factor q in dem Binomialcoëfficienten $\frac{\Pi(A+B)}{\Pi(A)\,\Pi(B)}$ enthalten ist. Man giebt beiden Zahlen A und B diese Form, nämlich

$$A = a+a_1q+a_2q^2+\cdots+a_{k-1}q^{k-1} \quad \text{und}$$
$$B = b+b_1q+b_2q^2+\cdots+b_{k-1}q^{k-1},$$

so dafs die Zahlen $a, a_1, a_2, \ldots a_{k-1}, b, b_1, b_2, \ldots b_{k-1}$ alle kleiner als q und nicht negativ sind, und bildet folgende Reihe von Gleichungen:

$$\begin{aligned} a+b &= \varepsilon q+c, \\ \varepsilon+a_1+b_1 &= \varepsilon_1 q+c_1, \\ \varepsilon_1+a_2+b_2 &= \varepsilon_2 q+c_2, \\ &\cdots\cdots\cdots \\ \varepsilon_{k-2}+a_{k-1}+b_{k-1} &= \varepsilon_{k-1}q+c_{k-1}, \end{aligned}$$

in welchen die Zahlen $c, c_1, c_2, \ldots c_{k-1}$ derselben Bedingung unterworfen sein sollen, dafs sie nicht negativ und kleiner als q sind, und in welchen demgemäfs die Zahlen $\varepsilon, \varepsilon_1, \varepsilon_2, \ldots \varepsilon_{k-1}$ nur die Werthe ***Null*** oder ***Eins*** erhalten können. Addirt man alle diese Gleichungen, nachdem die erste mit 1, die zweite mit q, die dritte mit q^2, u. s. w. multiplicirt worden, so erhält man

$$A+B = c+c_1q+c_2q^2+\cdots+c_{k-1}q^{k-1}+\varepsilon_{k-1}q^k.$$

Nennt man N die Anzahl, wievielmal der Factor q in dem Binomialcoëfficienten $\frac{\Pi(A+B)}{\Pi(A)\Pi(B)}$ enthalten ist, so ergiebt sich nach dem obigen Satze:

$$N = \frac{A+B-(c+c_1+\cdots+c_{k-1}+\varepsilon_{k-1})}{q-1} - \frac{(A-(a+a_1+\cdots+a_{k-1}))}{q-1} - \frac{(B-(b+b_1+\cdots+b_{k-1}))}{q-1},$$

oder, vereinfacht,

$$N = \frac{a+a_1+\cdots+a_{k-1}+b+b_1+\cdots+b_{k-1}-c-c_1-\cdots-c_{k-1}}{q-1},$$

und da durch Addition des aufgestellten Systems von Gleichungen

$$a+a_1+\cdots+a_{k-1}+b+b_1+\cdots+b_{k-1}-c-c_1-\cdots-c_{k-1}$$
$$= (q-1)(\varepsilon+\varepsilon_1+\varepsilon_2+\cdots+\varepsilon_{k-1})$$

ist, so erhält man

$$N = \varepsilon+\varepsilon_1+\varepsilon_2+\cdots+\varepsilon_{k-1}.$$

Um die Anwendung auf die vorliegende Summe zu machen, welche dem $\Psi_r(\alpha)$ congruent ist und deren einzelne Glieder Binomialcoëfficienten sind, setze man

$$A = \gamma_{\varrho-i}\nu q^{nt}-\lambda\nu z, \qquad B = (\gamma_{-i}-\gamma_{\varrho-i})\nu q^{nt}+\lambda\nu z,$$
$$A+B = \gamma_{-i}\nu q^{nt},$$

so wird

$$\gamma_{-i}\nu q^{nt} = c+c_1q+c_2q^2+\cdots+c_{k-1}q^{k-1}+\varepsilon_{k-1}q^k,$$

woraus zu sehen, dafs die Zahlen $c, c_1, c_2, \ldots$, bis c_{nt-1} einschliefslich, gleich *Null* sein müssen. Ist nun aufserdem auch $a = a_1 = a_2 = \ldots = a_{nt-1} = 0$ und $b = b_1 = b_2 = \ldots = b_{nt-1} = 0$, so folgt aus dem die Zahlen a, b, c und ε verbindenden Systeme von Gleichungen, dafs auch die Zahlen $\varepsilon, \varepsilon_1, \varepsilon_2, \ldots \varepsilon_{nt-1}$ gleich *Null* sind. Auch werden durch diese Bestimmungen die Werthe der übrigen Zahlen $\varepsilon_{nt}, \varepsilon_{nt+1}, \ldots \varepsilon_{k-1}$ im Allgemeinen gar nicht verändert, und nur in dem ganz besondern Falle, wo in der Gleichung

$$\varepsilon_{nt-1} + a_{nt} + b_{nt} = \varepsilon_{nt} q + c_{nt}$$

$a_{nt} + b_{nt} = q - 1$ wäre, würde ε_{nt} von ε_{nt-1} abhängig sein, indem, wenn ε_{nt-1} nicht gleich *Null*, sondern gleich ***Eins*** wäre, auch ε_{nt} den Werth ***Eins*** statt *Null* erhalten würde. Durch andere als die angenommenen Bestimmungen der Werthe $\varepsilon, \varepsilon_1, \varepsilon_2, \ldots \varepsilon_{nt-1}$ können also die Werthe der folgenden Zahlen in einem besondern Falle zwar *erhöht*, niemals aber *erniedrigt* werden. Für diejenigen Glieder der zu untersuchenden Summe von Binomialcoëfficienten, für welche die Bedingungen $a = a_1 = a_2 = \cdots = a_{nt-1} = 0$ und $b = b_1 = b_2 = \cdots = b_{nt-1} = 0$ erfüllt werden, mufs demnach nothwendig die Totalsumme aller ε, nämlich $\varepsilon + \varepsilon_1 + \cdots + \varepsilon_{k-1}$, kleiner sein als für alle andern; d. h. die Anzahl der in solchen Gliedern enthaltenen Factoren q mufs nothwendig kleiner sein, als die Anzahl der in einem der andern Glieder dieser Summe enthaltenen Factoren q.

Vermöge der Bedingungen $a = a_1 = \cdots = a_{nt-1} = 0$ und $b = b_1 = \cdots$ $\cdots = b_{nt-1} = 0$ ist

$$\gamma_{\varrho-i}\nu q^{nt} - \lambda\nu z = q^{nt}(a_{nt} + a_{nt+1}q + \cdots + a_{k-1}q^{k-nt-1}) \quad \text{und}$$
$$(\gamma - \gamma_{\varrho-i})\nu q^{nt} + \lambda\nu z = q^{nt}(b_{nt} + b_{nt+1}q + \cdots + b_{k-1}q^{k-nt-1});$$

woraus folgt, dafs z durch q^{nt} theilbar sein mufs. Setzt man daher $z = q^{nt}z'$, und beachtet, dafs $\gamma_{\varrho-i} < \lambda$ und nach der obigen Voraussetzung auch $\gamma_{-i} - \gamma_{\varrho-1}$ positiv und $< \lambda$ ist, so zeigt sich aus den Bedingungen, dafs $\gamma_{\varrho-i}q^{nt} - \lambda z$ und $(\gamma_{-i} - \gamma_{\varrho-i})q^{nt} + \lambda z$ nicht negativ werden dürfen, dafs $z' = 0$, also auch $z = 0$ der einzige Werth von z ist, welcher diesen Bedingungen genügt. Da ferner jeder andere Theil der zu untersuchenden Summe den Factor q öfter enthält, als der dem Werthe $z = 0$ angehörende, so folgt, dafs die ganze Summe den Factor q genau eben so oft enthalten mufs, als dieser eine Theil derselben, und dafs also endlich die complexe Zahl $\Psi_r(\alpha)$ den complexen Factor $f(\alpha^{\gamma^i})$ genau eben so vielemal enthält, als der Binomialcoëfficient

$$\frac{\Pi(\gamma_{-i}\nu q^{nt})}{\Pi(\gamma_{\varrho-i}\nu q^{nt})\,\Pi(\gamma_{-i}-\gamma_{\varrho-i})\nu q^{nt}}$$

den Factor q.

Um diese Anzahl mittels des obigen Satzes zu finden, bringe ich die Zahl $\gamma_h\nu$, in welcher $\nu=\frac{q^t-1}{\lambda}$ ist, auf die Form

$$\gamma_h\nu = d+d_1q+d_2q^2+\cdots+d_{t-1}q^{t-1};$$

wo $d, d_1, \ldots d_{t-1}$ alle kleiner als q und nicht negativ sind. Alsdann ist die Anzahl der in dem Producte $\Pi(\gamma_h\nu q^{nt})$ enthaltenen Factoren q gleich

$$\frac{\gamma_h\nu q^{nt}-(d+d_1+d_2+\cdots+d_{t-1})}{q-1}.$$

Multiplicirt man den Ausdruck des $\gamma_h\nu$ mit λ, so erhält man

$$\gamma_hq^t = \gamma_h+\lambda d+\lambda d_1q+\lambda d_2q^2+\cdots+\lambda d_{t-1}q^{t-1}.$$

Hieraus folgt zunächst, dafs $\gamma_h+\lambda d$ durch q theilbar ist. Setzt man also $\gamma_h+\lambda d=\delta q$, so folgt weiter, dafs auch $\delta+\lambda d_1$ durch q theilbar sein mufs u. s. w. Man erhält also folgende Reihe von Gleichungen:

$$\begin{aligned}\gamma_h &+ \lambda d = \delta q,\\ \delta &+ \lambda d_1 = \delta_1 q,\\ \delta_1 &+ \lambda d_2 = \delta_2 q,\\ &\cdots\cdots\\ \delta_{t-2} &+ \lambda d_{t-1} = \delta_{t-1}q;\end{aligned}$$

in welchen die Zahlen $\delta, \delta_1, \delta_2, \ldots \delta_{t-1}$ alle kleiner als λ und positiv sein müssen. Macht man aus diesen Gleichungen Congruenzen für den Modul λ, und ersetzt q durch eine Potenz der primitiven Wurzel γ (welche, weil q zum Exponenten t gehört und $\tau=\frac{\lambda-1}{t}$ ist, nothwendig $q\equiv\gamma^{m\tau}$ sein mufs, wo m eine relative Primzahl zu t ist): so erhält man

$$\gamma^h\equiv\delta\gamma^{m\tau},\quad \delta\equiv\delta_1\gamma^{m\tau},\ \ldots\ \delta_{t-2}\equiv\delta_{t-1}\gamma^{m\tau},$$

also

$$\delta\equiv\gamma^{h+m\tau},\ \delta_1\equiv\gamma^{h-2m\tau},\ \ldots\ \delta_{t-1}\equiv\gamma^h,$$

und da alle diese Zahlen positiv und kleiner als λ sind:

$$\delta=\gamma_{h-m\tau},\ \delta_1=\gamma_{h-2m\tau},\ \ldots\ \delta_{t-1}=\gamma_h.$$

Addirt man alle die Gleichungen, durch welche die Zahlen $d, d_1, \ldots d_{t-1}$ mittels der Zahlen $\delta, \delta_1, \ldots \delta_{t-1}$ bestimmt werden, und setzt für die letzteren ihre gefundenen Werthe, so erhält man

$$\lambda(d+d_1+\cdots+d_{t-1}) = (q-1)(\gamma_{h-m\tau}+\gamma_{h-2m\tau}+\cdots+\gamma_h).$$

Die Reihe der eine Periode bildenden Reste

$$\gamma_{h-m\tau}+\gamma_{h-2m\tau}+\gamma_{h-3m\tau}+\cdots+\gamma_{h-t\tau}$$

ist, weil m relative Primzahl zu t ist, wenn man von der bestimmten Ordnung der Glieder absieht, dieselbe wie

$$\gamma_h+\gamma_{h+\tau}+\gamma_{h+2\tau}+\cdots+\gamma_{h+(t-1)\tau}.$$

Diese ist, wie man weifs, immer durch λ theilbar, aufser für $t=1$, weshalb sie durch λS_h bezeichnet werden soll, so dafs

$$\lambda S_h = \gamma_h+\gamma_{h+\tau}+\gamma_{h+2\tau}+\cdots+\gamma_{h+(t-1)\tau}$$

ist. Demzufolge giebt die vorige Gleichung:

$$d+d_1+d_2+\cdots+d_{t-1} = (q-1)S_h$$

und man erhält die Anzahl der in dem Producte $\Pi(\gamma_h\nu q^{nt})$ enthaltenen Factoren q durch die Zahl

$$\frac{\gamma_h\nu q^{nt}}{q-1}-S_h$$

ausgedrückt. Setzt man nun $\gamma_{-i}-\gamma_{\varrho-i}$, welches nach der Voraussetzung positiv und nothwendig kleiner als λ ist, gleich γ_σ, und nimmt sodann nach einander $h=-i$, $h=\varrho-i$ und $h=\sigma$, so erhält man für die Anzahl der in dem Binomialcoëfficienten

$$\frac{\Pi(\gamma_{-i}\nu q^{nt})}{\Pi(\gamma_{\varrho-i}\nu q^{nt}\,\Pi(\gamma_{-i}-\gamma_{\varrho-i})\nu q^{nt}}$$

enthaltenen Factoren q:

$$S_{\varrho-i}+S_\sigma-S_{-i}.$$

Genau eben so vielemal enthält also, nach Dem was oben bewiesen wurde, die complexe Zahl $\Psi_r(\alpha)$ den complexen Primfactor $f(\alpha^{\gamma^i})$; das heifst: in dem Ausdrucke

$$\Psi_r(\alpha) = E(\alpha)f(\alpha)^m f(\alpha^\gamma)^{m_1}\ldots f(\alpha^{\gamma^{\tau-1}})^{m_{\tau-1}}$$

werden diejenigen Exponenten m_i, für welche $\gamma_{-i}-\gamma_{\varrho-i}$ eine positive Zahl ist, durch die Gleichung

$$m_i = S_{\varrho-i}+S_\sigma-S_{-i}$$

bestimmt.

Aus der Gleichung $\gamma_{-i}-\gamma_{\varrho-i}=\gamma_\sigma$ folgt die Congruenz $\gamma^{-i}-\gamma^{\varrho-i}\equiv\gamma^\sigma$ oder $1-\gamma^\varrho\equiv\gamma^{\sigma+i}$, Mod. λ, und da $\gamma^\varrho\equiv r+1$ ist, so folgt weiter $-r\equiv\gamma^{\sigma+i}$ oder $r\equiv\gamma^{\sigma+i-\mu}$. Es ist also $\sigma+i-\mu\equiv\text{Ind.}\,r$ oder $\sigma\equiv\mu-i+\text{Ind.}\,r$, Mod. $\lambda-1$, wenn das Zeichen Ind., wie oben, sich auf den Mod. λ und dessen primitive Wurzel γ bezieht; und da eben so $\varrho\equiv\text{Ind.}(r+1)$ ist, so ist

$$m_i = S_{-i+\text{Ind.}(r+1)}+S_{-i+\mu+\text{Ind.}\,r}-S_{-i}.$$

16 *

Die mit S_h bezeichneten Zahlen haben aber, wie bekannt, die Eigenschaft, dafs $S_{h+\mu} = t - S_h$ ist. Also kann dem Ausdruck des m_i auch eine der folgenden beiden Formen gegeben werden:

$$m_i = t - S_{-i} - S_{-i+\text{Ind.}\,r} + S_{-i+\text{Ind.}(r+1)},$$

oder

$$m_i = S_{\mu-i} + S_{\mu-i+\text{Ind.}\,r} - S_{\mu-i+\text{Ind.}(r+1)}.$$

Es ist jetzt auch nicht schwer zu zeigen, dafs die Bestimmung der Exponenten m_i, welche bis jetzt nur für diejenigen Werthe des i bewiesen wurde, für welche $\gamma_{-i} - \gamma_{\varrho-i}$ positiv ist, ebensowohl für alle andern Werthe von i gültig bleibt.

Aus der Gleichung $\gamma_{\mu+h} = \lambda - \gamma_h$ folgt, dafs

$$\gamma_{-i} - \gamma_{\varrho-i} = -(\gamma_{-i-\mu} - \gamma_{\varrho-i-\mu}).$$

Wenn also i einen Werth hat, welcher der Bedingung, dafs $\gamma_{-i} - \gamma_{\varrho-i}$ positiv sein soll, nicht genügt, so genügt dagegen allemal der Werth $i + \mu$ dieser Bedingung. Ferner kann das oben gefundene Resultat, nach welchem $m_i + m_{\frac{1}{2}\tau} = t$, weil $m_i = m_{i+\tau} = m_{i+2\tau}$ u. s. w. und $t.\frac{1}{2}\tau = \mu = \frac{1}{2}(\lambda - 1)$ ist, auch so dargestellt werden, dafs $m_i + m_{i+\mu} = t$ ist. Es sei nun i eine Zahl, welche der obigen Bedingung nicht genügt, so ist dagegen $\mu + i$ eine solche Zahl, für welche der gefundene Ausdruck für $m_{i+\mu}$ richtig ist. Setzt man diesen aber in die Gleichung $m_i = t - m_{i+\mu}$, so findet man, vermöge der Gleichung $S_{h+\mu} = t - S_h$, für m_i wieder denselben Ausdruck; welcher also für alle Werthe des i richtig ist

Die Zerlegung der complexen Zahl $\Psi_r(\alpha)$ in ihre complexen Primfactoren, welche nur die complexen Primfactoren des q sind, wird demnach durch folgende Formel ausgedrückt:

$$\Psi_r(\alpha) = E(\alpha) f(\alpha)^m f(\alpha^\gamma)^{m_1} f(\alpha^{\gamma^2})^{m_2} \ldots f(\alpha^{\gamma^{\tau-1}})^{m_{\tau-1}},$$

in welcher $E(\alpha)$ eine complexe Einheit, $f(\alpha)$ ein complexer idealer Primfactor der für den Modul λ zum Exponenten t gehörenden Primzahl q ist; ferner $\tau = \frac{\lambda - 1}{t}$ und

$$m_i = S_{\mu-i} + S_{\mu-i+\text{Ind.}\,r} - S_{\mu-i+\text{Ind.}(r+1)} \quad \text{und}$$

$$\lambda S_h = \gamma_h + \gamma_{h+\tau} + \gamma_{h+2\tau} + \cdots + \gamma_{h+(t-1)\tau}.$$

Die complexe Einheit $E(\alpha)$, welche in diesem Ausdrucke als Factor vorkommt, ist nothwendig unbestimmt, so lange über den complexen Primfactor $f(\alpha)$, welcher im Allgemeinen ideal sein wird, keine nähere Bestimmung gemacht ist. Stellt man sich aber diesen Ausdruck zur Hten Potenz erhoben vor, wo H der Exponent derjenigen Potenz sein soll, für welche $f(\alpha)^H$ zu einer wirklichen complexen Zahl wird, so kann man dieselbe immer so darstellen, dafs sie nicht die einzelnen Wurzeln $\alpha, \alpha^2, \dots \alpha^{\lambda-1}$, sondern nur die Perioden derselben enthält; nämlich die τ Perioden von je t Gliedern. Es mufs alsdann, weil, wie oben bewiesen wurde, $\Psi_r(\alpha)$ auch nur diese Perioden enthält, die Einheit $E(\alpha)$ ebenfalls nur aus diesen Perioden zusammengesetzt sein. Aus der Gleichung $\Psi_r(\alpha)\Psi_r(\alpha^{-1}) = q^t$ folgt ferner, dafs $E\alpha)E(\alpha^{-1}) = 1$ sein mufs; woraus nach einer bekannten Eigenschaft aller Einheiten folgt, dafs $E(\alpha)$ nur gleich $\pm\alpha^k$ sein kann; und da $E(\alpha)$ eine solche einzelne Wurzel α^k nicht enthalten kann, so folgt endlich, dafs $E(\alpha) = \pm 1$ ist; nämlich, wenn $f(\alpha)^H$ als eine, nur die Perioden enthaltende wirkliche complexe Zahl dargestellt angenommen wird. Diese Zerlegung der complexen Zahl $\Psi_r(\alpha)$ soll als die ***dritte Grund-Eigenschaft*** derselben bezeichnet werden.

Schliefslich bemerke ich noch, dafs dieser Ausdruck der complexen Zahl $\Psi_r(\alpha)$ durch ihre Primfactoren für den Fall $t = 1$ den entsprechenden Ausdruck der Kreistheilungszahl $\psi_r(\alpha)$ giebt, welchen ich in dem vorhergehenden Paragraphen dieser Abhandlung angewendet und zuerst in dem Programm der Universität ***Breslau*** zur Jubelfeier der Universität ***Königsberg*** im Jahre 1844 und später in diesem Journale (Band. 35, S. 362) gegeben habe. An demselben Orte (S. 364) wird man auch schon eine Anwendung der complexen Zahl $\Psi_r(\alpha)$ finden, von welcher ich damals nur die Zerlegung in die Primfactoren kannte, und wufste, dafs sie eine wirkliche complexe Zahl sei, für welche aber der in der gegenwärtigen Untersuchung zum Grunde gelegte, der complexen Kreistheilungszahl $\psi_r(\alpha)$ vollständig entsprechende Ausdruck

$$\Psi_r(\alpha) = \Sigma\alpha^{-(r+1)h + \text{Ind.}(g^h+1)}$$

mir damals noch unbekannt war.

§. 3.

Entwickelung der Indices der complexen Einheiten und der Zahlen λ und $1-\alpha^k$, für welche der Modul eine zu einem beliebigen Exponenten gehörende complexe Primzahl ist.

Mittels der in dem vorigen Paragraphen entwickelten Eigenschaften der complexen Zahl $\Psi_r(\alpha)$ können nun die Indices der complexen Einheiten, so wie der Zahlen λ und $1-\alpha^k$, auch für den Fall gefunden werden, wo der Modul $f(\alpha)$, auf welchen sich die Indices beziehen, ein ***complexer***

idealer Primfactor einer zum Exponenten t gehörenden ***nichtcomplexen*** Primzahl q ist. Die Ausdrücke derselben sind den in dem ersten Paragraphen für den speciellen Fall $t=1$ gefundenen, namentlich den beiden letzten derselben, vollständig entsprechend. Ich untersuche also sogleich den $(\lambda-2n)$ten Differentialquotienten des Logarithmen der complexen Zahl $\Psi_r(\alpha)$, in welcher die Wurzel α durch die variable Exponentialgröfse e^v zu ersetzen ist, für den Werth $v=0$.

Es sei demnach

$$\Psi_r(e^v) = \Sigma e^{v(-(r+1)h+\text{Ind.}(g^h+1))}$$

für $h=0, 1, 2, \ldots q^t=2$, mit Ausschlufs von $h=\frac{1}{2}(q^t-1)$. Dann ist zunächst

$$\frac{d^{\lambda-2n}\, l\Psi_r(e^v)}{dv^{\lambda-2n}} = \frac{d^{\lambda-2n-1}\left(\frac{d\Psi_r(e^v)}{dv}\cdot\frac{1}{\Psi_r(e^v)}\right)}{dv^{\lambda-2n-1}},$$

und wenn man zur Abkürzung für einen Augenblick $\lambda-2n-1=m$ und $\frac{1}{\Psi_r(e^v)}=U$ setzt, so folgt hieraus nach den Regeln der Differentiation eines Products zweier Factoren:

$$\frac{d^{\lambda-2n}\, l\Psi_r(e^v)}{dv^{\lambda-2n}} = \frac{d^{m+1}\Psi_r(e^v)}{dv^{m+1}}U + \frac{m}{1}\frac{d^m\Psi_r(e^v)}{dv^m}\frac{dU}{dv} + \cdots$$

Macht man jetzt Gebrauch von der *zweiten* Grund-Eigenschaft der complexen Zahl $\Psi_r(\alpha)$, nach welcher $\Psi_r(\alpha)\,\Psi_r(\alpha^{-1})=q^t$ ist, so erhält man folgende, für jeden beliebigen Werth der Variabel v geltende Gleichung:

$$\Psi_r(e^v)\,\Psi_r(e^{-v}) = q^t + V.W;$$

* wo $V=1+e^v+e^{2v}+\cdots e^{(\lambda-1)v}$ und W eine ganze rationale Function von e^v ist. Hieraus folgt

$$\Psi_r(e^{-v}) = \frac{q^t+VW}{\Psi_r(e^v)} = (q^t+VW)\,U,$$

und wenn man den $i$ten Differentialquotienten nimmt,

$$\frac{d^i\Psi_r(e^{-v})}{dv^i} = \frac{d^iU}{dv^i}(q^t+VW) + \frac{i}{1}\frac{d^{i-1}U}{dv^{i-1}}\frac{d(VW)}{dv} + \cdots$$

Setzt man nun $v=0$ und macht aus dieser Gleichung eine Congruenz für den Modul λ, so verschwinden V und alle Differentialquotienten des V, bis zum $(\lambda-2)$ten einschliefslich, weil sie durch λ theilbar sind; und wenn man noch beachtet, dafs $q^t\equiv 1$, Mod. λ ist, so erhält man

$$\frac{d_0^i\Psi_r(e^{-v})}{dv^i} \equiv \frac{d_0^iU}{dv^i},\ \text{Mod. } \lambda,$$

für alle Werthe 0, 1, 2, ... $\lambda-2$ von i, oder, was Dasselbe ist:

$$\frac{d_0^i U}{dv^i} \equiv (-1)^i \frac{d_0^i \Psi_r(e^v)}{dv^i} \equiv (-1)^i \mathrm{D}_i, \quad \text{Mod. } \lambda;$$

wenn der Kürze wegen der $i$te Differentialquotient von $\Psi_r(e^v)$ für $v=0$ durch Di bezeichnet wird. Der obige Ausdruck des $(\lambda-2n)$ten Differentialquotienten von $l\Psi_r(e^v)$ nimmt daher für $v=0$ folgende Gestalt an:

$$\frac{d_0^{\lambda-2n} l\Psi_r(e^v)}{dv^{\lambda-2n}} \equiv \mathrm{D}_{m+1}\mathrm{D}_0 - \frac{m}{1}\mathrm{D}_m\mathrm{D}_1 + \frac{m(m-1)}{1.2}\mathrm{D}_{m-1}\mathrm{D}_2 - \cdots, \quad \text{Mod. } \lambda.$$

Durch Differentiation des oben gegebenen Ausdrucks für $\Psi_r(e^v)$ erhält man nun für $v=0$:

$$\mathrm{D}^i = \Sigma(-(r+1)h + \text{Ind.}(g^h+1))^i$$

für $h=0, 1, 2, \ldots q^t-2$, mit Ausschlufs von $h=\frac{1}{2}(q^t-1)$. Setzt man, um abzukürzen, für einen Augenblick

$$-(r+1)h + \text{Ind.}(g^h+1) = C_h \quad \text{und} \quad -(r+1)k + \text{Ind.}(g^k+1) = C_k,$$

so ergiebt sich

$$\frac{d_0^{\lambda-2n} l\Psi_r(e^v)}{dv^{\lambda-2n}} \equiv \Sigma\Sigma\left(C_h^{m+1}C_k^0 - \frac{m}{1}C_h^m C_k + \frac{m(m-1)}{1.2}C_h^{m-1}C_k^2 - \cdots\right);$$

wo die Doppelsumme auf alle ganzzahligen Werthe des h und k, von 0 bis q^t-2, sich erstreckt; mit Ausschlufs des Werths $h=\frac{1}{2}(q^t-1)$ und $k=\frac{1}{2}(q^t-1)$. Summirt man nun die Binomialreihe unter den beiden Summenzeichen, so erhält man

$$\frac{d_0^{\lambda-2n} l\Psi_r(e^v)}{dv^{\lambda-2n}} \equiv \Sigma\Sigma C_h(C_h - C_k)^m,$$

und wenn für C_h und C_k und für m wieder ihre Werthe gesetzt werden:

$$\frac{d_0^{\lambda-2n} l\Psi_r(e^v)}{dv^{\lambda-2n}} \equiv \Sigma\Sigma(-(r+1)h + \text{Ind.}(g^h+1))$$

* $$\times(-(r+1)(h-k) + \text{Ind.}(g^h+1) + \text{Ind.}(g^k+1))^{\lambda-2n-1},$$

für $h=0, 1, 2, \ldots q^t-2$ und $k=0, 1, 2, \ldots q^t-2$, mit Ausschlufs von $h=\frac{1}{2}(q^t-1)$, $k=\frac{1}{2}(q^t-1)$.

Man transformire nun diese Doppelsumme, in welcher alle Glieder, die gleichen Werthen von h und k angehören, als congruent Null, von selbst wegfallen, durch dieselbe Substitution, welche oben im zweiten Paragraphen zur Transformation einer ähnlichen Doppelsumme angewendet worden ist, nämlich:

$$g^{k-h} \equiv g^{k'}, \quad \frac{g^k+1}{g^h+1} \equiv g^{h'}, \quad \text{Mod. } f(\alpha),$$

so erhält man diesen $(\lambda-2n)$ten Differentialquotienten durch die Doppelsumme

* $$\Sigma\Sigma(-(r+1)\,\text{Ind.}(g^{h'}-1) + \text{Ind.}(g^{k'}-1) + r\,\text{Ind.}(g^{k'}-g^{h'}))\,((r+1)k'-h)^{\lambda-2n-1}$$

ausgedrückt, wo h' und k' alle Werthe $1, 2, 3, \ldots q^t-2$ bekommen müssen, die Werthverbindung $h'=k'$ aber ausgeschlossen ist. Man zerlege diese Doppelsumme in folgende drei Theile:

$$-(r+1)\Sigma\Sigma \text{Ind.}(g^{h'}-1)\ \left((r+1)k'-h'\right)^{\lambda-2n-1},$$
$$\Sigma\Sigma \text{Ind.}(g^{k'}-1)\ \left((r+1)k'-h'\right)^{\lambda-2n-1},$$
$$r\Sigma\Sigma \text{Ind.}(g^{k'}-g^{h'})\left((r+1)k'-h'\right)^{\lambda-2n-1},$$

und suche die Werthe dieser drei Theile einzeln; mit dem ersten Theile anfangend. Um in demselben die Werthverbindung $k'=h'$ nicht ferner ausschliefsen zu müssen, füge man den dieser Werthverbindung entsprechenden Theil mit entgegengesetztem Vorzeichen hinzu. Der erste Theil ist

$$+r^{\lambda-2n-1}(r+1)\Sigma h'^{\lambda-2n-1}\text{Ind.}(g^{h'}-1)-(r+1)\Sigma\Sigma\text{Ind.}(g^{h'}-1)\left((r+1)k'-h'\right)^{\lambda-2n-1};$$

wo jetzt den k' und h' alle Werthe von 1 bis q^t-2 einschliefslich, ohne Ausnahme, zu geben sind. Ich führe nun die Summation in Beziehung auf k' aus, indem zu bemerken ist, dafs $q^t-1\equiv 0$, Mod. λ und dafs die Summe $\Sigma k'^i$ ebenfalls congruent Null ist, für jeden der Werthe $1, 2, 3, \ldots \lambda-2$ von i. Dies giebt zunächst

$$\Sigma\left((r+1)k'-h'\right)^{\lambda-2n-1}\equiv(q^t-2)h'^{\lambda-2n-1}\equiv -h'^{\lambda-2n-1}.$$

Der erste Theil ist vermöge Dessen dem folgenden Ausdrucke congruent:

$$(r+1)(1+r^{\lambda-2n-1})\Sigma h'^{\lambda-2n-1}\text{Ind.}(g^{h'}-1).$$

Genau auf dieselbe Weise findet man den zweiten Theil congruent

$$-\left(r^{\lambda-2n-1}+(r+1)^{\lambda-2n-1}\right)\Sigma h'^{\lambda-2n-1}\text{Ind.}(g^{h'}-1).$$

Um auch den dritten Theil auf eine ähnliche einfache Summe zu reduciren, dehne man die Werthe von h' und k' auch auf den Werth $h'=0$ und $k'=0$ aus; welches ohne Weiteres geschehen kann, wenn man die diesen Werthen entsprechenden Glieder

$$r\Sigma k'^{\lambda-2n-1}\text{Ind.}(g^{k'}-1) \quad \text{und} \quad r(r+1)^{\lambda-2n-1}\Sigma h'^{\lambda-2n-1}\text{Ind.}(g^{h'}-1)$$

subtrahirt.

Ich verwandle nun diese Doppelsumme, in welcher jetzt h' und k' alle Werthe von 0 bis q^t-2 einschliefslich haben (jedoch mit Ausschlufs der Werthverbindung $h'=k'$), indem ich $h'+k'$ statt h' setze; wodurch sie folgende Form annimmt:

$$r\Sigma\Sigma\left(\text{Ind.}(g^{h'}-1)+k'\right)(rk'-h')^{\lambda-2n-1},$$

für $k'=0, 1, 2, 3, \ldots q^t-2$ und $h'=1, 2, 3, \ldots q^t-2$. Die Summation in Beziehung auf k' läfst sich wieder ohne Schwierigkeit ausführen und giebt das Resultat, dafs die Doppelsumme einfach congruent Null ist, für den Modul λ. Der dritte Theil ist den beiden einfachen Summen, welche zu subtrahiren waren, deshalb congruent, damit die Werthe von h' und k' auch auf $h'=0$ und $k'=0$ ausgedehnt werden konnten. Diese zusammengefafst, geben den dritten Theil congruent

$$-r(1+(r+1)^{\lambda-2n-1})\Sigma h'^{\lambda-2n-1}\,\text{Ind.}\,(g^{h'}-1).$$

Fafst man alle drei Theile wieder in einen zusammen, so erhält man nach einigen leichten Reductionen:

$$\frac{d_0^{\lambda-2n}\, l\Psi_r(e^v)}{dv^{\lambda-2n}} = (1+r^{\lambda-2n}-(r+1)^{\lambda-2n})\Sigma h'^{\lambda-2n-1}\,\text{Ind.}\,(g^{h'}-1),$$

für $h'=1, 2, 3, \dots q^t-2$.

Die hierin enthaltene einfache Summe kann nun auf folgende Weise leicht als Index der complexen Einheit $E_n(\alpha)$ dargestellt werden. Zuerst werden alle Glieder, für welche h' durch λ theilbar ist, als congruent Null, weggelassen, und dann wird $h'=i+\lambda k$ gesetzt und es werden dem i die Werthe $1, 2, 3, \dots \lambda-1$, dem k die Werthe $0, 1, 2, \dots \nu-1$ gegeben (wo $\nu=\frac{q^t-1}{\lambda}$ ist). Dann ist

$$\Sigma h'^{\lambda-2n-1}\,\text{Ind.}\,(g^{h'}-1) \equiv \Sigma\Sigma i^{\lambda-2n-1}\,\text{Ind.}\,(g^{i+k\lambda}-1),$$

für $i=1, 2, 3, \dots \lambda-1$, $k=0, 1, 2, \dots \nu-1$. Nun ist aber, wie leicht zu beweisen,

$$(g^i-1)(g^{i+\lambda}-1)(g^{i+2\lambda}-1)\dots(g^{i+(\nu-1)\lambda}-1) \equiv 1-g^{\nu i}, \text{ Mod. } f(\alpha),$$

also auch

$$\Sigma\,\text{Ind.}\,(g^{i+k\lambda}-1) \equiv \text{Ind.}\,(1-g^{\nu i}) \equiv \text{Ind.}\,(1-\alpha^i), \text{ Mod. } \lambda,$$

indem $g^\nu\equiv\alpha$, Mod. $f(\alpha)$, ist. Die Ausführung der Summation in Beziehung auf k giebt daher

$$\Sigma h'^{\lambda-2n-1}\,\text{Ind.}\,(g^{h'}-1) = \Sigma i^{\lambda-2n-1}\,\text{Ind.}\,(1-\alpha^i),$$

für $i=1, 2, 3, \dots \lambda-1$. Setzt man $i\gamma$ statt i, multiplicirt mit γ^{2n} und subtrahirt von der so erhaltenen Congruenz die unveränderte, so erhält man

$$(\gamma^{2n}-1)\Sigma h'^{\lambda-2n-1}\,\text{Ind.}\,(g^{h'}-1) \equiv \Sigma i^{\lambda-2n-1}\,\text{Ind.}\left(\frac{1-\alpha^{\gamma i}}{1-\alpha^i}\right), \text{ Mod. } \lambda.$$

Verwandelt man endlich i in γ^h, so zeigt sich sogleich, dafs die Summe rechts in dieser Congruenz dem doppelten Index der Einheit

$$E_n(\alpha) = e(\alpha).e(\alpha^\gamma)^{\gamma^{-2n}}.e(\alpha^{\gamma^2})^{\gamma^{-4n}} \dots e(\alpha^{\gamma^{\mu-1}})^{\gamma^{-2(\mu-1)n}}$$

congruent, also

$$(\gamma^{-2n}-1)\Sigma h'^{\lambda-2n-1}\,\text{Ind.}\,(g^{h'}-1) \equiv 2\,\text{Ind.}\,E_n(\alpha)$$

ist. Dadurch hat man nun folgenden Ausdruck für den Index dieser Einheit:

$$\text{Ind.}\,E_n(\alpha) \equiv \frac{(\gamma^{2n}-1)}{2(1+r^{\lambda-2n}-(r+1)^{\lambda-2n})}\cdot\frac{d_0^{\lambda-2n}\, l\Psi_r(e^v)}{dv^{\lambda-2n}}, \text{ Mod. } \lambda.$$

Diese Formel, welche der entsprechenden, im ersten Paragraphen für den besondern Fall $t = 1$ bewiesenen ganz gleich ist, giebt die Indices der Einheiten $E_1(\alpha)$, $E_2(\alpha)$, u. s. w.; also auch die Indices aller Einheiten, welche durch diese sich darstellen lassen. Es ist aber hier nicht nöthig, die Indices aller dieser Einheiten $E_1(\alpha)$, $E_2(\alpha)$, ... $E_{\mu-1}(\alpha)$ nach der Formel besonders zu berechnen, denn es läfst sich leicht ein- für allemal zeigen, dafs für alle Werthe von n, für welche $\lambda - 2n$ nicht durch t theilbar ist, der Index Ind. $E_n(\alpha)$ nur congruent Null ist. Es kann Dies auf folgende Weise aus dem gefundenen Ausdrucke des Ind. $E_n(\alpha)$ selbst abgeleitet werden. Da, vermöge der ersten Grund-Eigenschaft der complexen Zahl $\Psi_r(\alpha)$, dieselbe nur die τ Perioden von je t Gliedern, nicht aber die Wurzeln $\alpha, \alpha^2, \dots \alpha^{\lambda-1}$ einzeln enthält, so ist

$$\Psi_r(\alpha) = \Psi_r(\alpha^{\gamma^\tau}),$$

woraus nach den schon mehrmals angewendeten Principien

$$\Psi_r(e^v) = \Psi_r(e^{v\gamma^\tau}) + V.W$$

und weiter

$$\frac{d_0^{\lambda-2n}\, l\Psi_r(e^v)}{dv^{\lambda-2n}} \equiv \frac{d_0^{\lambda-2n}\, l\Psi_r(e^{v\gamma^\tau})}{dv^{\lambda-2n}},$$

oder

$$\frac{d_0^{\lambda-2n}\, l\Psi_r(e^v)}{dv^{\lambda-2n}} \equiv \gamma^{(\lambda-2n)\tau} \frac{d_0^{\lambda-2n}\, l\Psi_r(e^v)}{dv^{\lambda-2n}}$$

folgt. Es mufs also nothwendig, entweder

$$\gamma^{(\lambda-2n)\tau} - 1 \equiv 0, \quad \text{oder} \quad \frac{d_0^{\lambda-2n}\, l\Psi_r(e^v)}{dv^{\lambda-2n}} \equiv 0$$

sein, für den Mod. λ. In dem letztern Falle ist aber Ind. $E_n(\alpha) \equiv 0$, Mod. λ, und nur wenn $\gamma^{(\lambda-2n)\tau} \equiv 1$, Mod. λ, ist, kann Ind. $E_n(\alpha)$ einen von Null verschiedenen Werth haben; also nur dann, wenn $(\lambda - 2n)\tau$ ein Vielfaches von $\lambda - 1 = t\tau$ oder, was Dasselbe ist, wenn $\lambda - 2n$ ein Vielfaches von t ist; wie behauptet wurde. Dasselbe läfst sich übrigens auch ohne Hülfe der allgemeinen Formel für Ind. $E_n(\alpha)$ auf elementare Art leicht beweisen.

Aus dem gefundenen Ausdruck des Ind. $E_n(\alpha)$ wird nun ein anderer, welcher nicht die complexe Zahl $\Psi_r(\alpha)$, sondern den Primfactor $f(\alpha)$ selbst enthält, mittels der dritten Grund-Eigenschaft der $\Psi_r(\alpha)$, welche ihre Zerlegung in die Primfactoren giebt, auf folgende Weise hergeleitet. Wenn der im Allgemeinen ideale Primfactor $f(\alpha)$ als Hte Wurzel aus der Potenz $f(\alpha)^H$ dargestellt und H so angenommen wird, dafs $f(\alpha)^H$ eine wirkliche complexe Zahl ist, so kann, wie schon oben bemerkt, diese wirkliche complexe Zahl $f(\alpha)^H$

immer so angenommen werden, dafs sie nur die τ Perioden von je t Gliedern enthält. Bei dieser Annahme hat man

$$\Psi_r(\alpha) = \pm f(\alpha)^m f(\alpha^\gamma)^{m_1} f(\alpha^{\gamma^2})^{m_2} \ldots f(\alpha^{\gamma^{\tau-1}})^{m_{\tau-1}} \text{ und}$$

$$m_i = S_{\mu-i} + S_{\mu-i+\text{Ind.}\,r} - S_{\mu-i+\text{Ind.}\,(r+1)}.$$

Daraus folgt sogleich

$$\frac{d_0^{\lambda-2n}\, l\Psi_r(e^v)}{dv^{\lambda-2n}} \equiv m\frac{d_0^{\lambda-2n}\, lf(e^v)}{dv^{\lambda-2n}} + m_1\frac{d_0^{\lambda-2n}\, lf(e^{\gamma v})}{dv^{\lambda-2n}} + \cdots + m_{\tau-1}\frac{d_0^{\lambda-2n}\, lf(e^{\gamma^{\tau-1}v})}{dv^{\lambda-2n}},$$

oder, vereinfacht:

$$\frac{d_0^{\lambda-2n}\, l\Psi_r(e^v)}{dv^{\lambda-2n}} \equiv (m + \gamma^{\lambda-2n} m_1 + \gamma^{2(\lambda-2n)} m_2 + \cdots + \gamma^{(\tau-1)(\lambda-2n)} m_{\tau-1})\frac{d_0^{\lambda-2n}\, lf(e^v)}{dv^{\lambda-2n}}.$$

Es ist nun die rechts in dieser Congruenz vorkommende Summe zu suchen, welche durch den Buchstaben R bezeichnet werden soll, so dafs

$$R = m + \gamma^{\lambda-2n} m_1 + \gamma^{2(\lambda-2n)} m_2 + \cdots + \gamma^{(\tau-1)(\lambda-2n)} m_{\tau-1}$$

ist. Es wird auch hinreichen, diesen Werth von R nur für diejenigen Werthe von n zu suchen, für welche $\lambda-2n$ ein Vielfaches von t ist; denn wenn dies nicht der Fall ist, so giebt die Congruenz, nach Dem was oben gezeigt, nur identisch $0 \equiv 0$, Mod. λ. Wird mehrerer Einfachheit wegen das Summenzeichen angewendet und für m_i sein Werth gesetzt, so ist

$$R = \Sigma\gamma^{i(\lambda-2n)}(S_{\mu-i} + S_{\mu-i+\text{ind.}\,r} - S_{\mu-i+\text{ind.}\,(r+1)})$$

für $i = 0, 1, 2, \ldots \tau-1$. Nun sollen hierin für die drei Gröfsen $S_{\mu-i}$, $S_{\mu-i+\text{ind.}\,r}$ und $S_{\mu-i+\text{Ind.}\,(r+1)}$ ihre Werthe nach der Formel

$$S_h = \frac{1}{\lambda}(\gamma_h + \gamma_{h+\tau} + \gamma_{h+2\tau} + \cdots + \gamma_{h+(t-1)\tau}) = \frac{1}{\lambda}\Sigma\gamma_{h+k\tau}$$

gesetzt werden. Alsdann sind, wegen des als Nenner vorkommenden λ, wieder Congruenzen für den Modul λ^2 zur Bestimmung von R nöthig, und deshalb soll $\gamma^{i\lambda(\lambda-2n)}$ statt $\gamma^{i(\lambda-2n)}$ genommen werden, welches demselben für den Mod. λ congruent ist. Wird sodann mit λ multiplicirt, so erhält man folgende Congruenz:

$$\lambda R \equiv \Sigma\Sigma\gamma^{i\lambda(\lambda-2n)}(\gamma_{\mu-i+k\tau} + \gamma_{\mu-i+k\tau+\text{ind.}\,r} - \gamma_{\mu-i+k\tau+\text{ind.}\,(r+1)}),$$

für den Mod. λ^2, wo die beiden Summenzeichen auf die Werthe $i = 0, 1, 2, \ldots \ldots \tau-1$ und $k = 0, 1, 2, \ldots t-1$ zu beziehen sind. Setzt man $i + k\tau$ statt i und beachtet, dafs bei dieser Veränderung $\gamma^{i\lambda(\lambda-2n)}$ in Beziehung auf den Mod. λ^2 unverändert bleibt, weil $\lambda-2n$ nach der Voraussetzung durch t theilbar und $t\tau = \lambda-1$ ist, so verwandelt sich die Doppelsumme in die einfache:

$$\lambda R \equiv \Sigma\gamma^{i\lambda(\lambda-2n)}(\gamma_{\mu-i} + \gamma_{\mu-i+\text{ind.}\,r} - \gamma_{\mu-i+\text{ind.}\,(r+1)}), \text{ Mod. } \lambda^2,$$

für $i = 0, 1, 2, \ldots \lambda-2$.

17*

Vergleicht man nun dieses Resultat mit dem in der ähnlichen Untersuchung im ersten Paragraphen dieser Abhandlung gefundenen Ausdrucke der dort mit T bezeichneten Summe, so findet sich $\lambda R \equiv \lambda T$, Mod. λ^2, also $R \equiv T$, Mod. λ. Macht man demnach von dem oben gefundenen Werthe von T Gebrauch, so erhält man

$$R \equiv (-1)^n (1 + r^{\lambda-2n} - (r+1)^{\lambda-2n}) \frac{B_n}{2n}, \text{ Mod. } \lambda,$$

also

$$\frac{d_0^{\lambda-2n} l\Psi_r(e^v)}{dv^{\lambda-2n}} \equiv (-1)^n (1 + r^{\lambda-2n} - (r+1)^{\lambda-2n}) \frac{B_n}{2n} \cdot \frac{d_0^{\lambda-2n} lf(e^v)}{dv^{\lambda-2n}}.$$

Der gefundene Ausdruck des Ind. $E_n(\alpha)$ giebt demnach folgenden neuen Ausdruck des Index dieser Einheit:

$$\text{Ind.} E_n(\alpha) \equiv \frac{(-1)^n (\gamma^{2n} - 1) B^n}{4n} \cdot \frac{d_0^{\lambda-2n} lf(e^v)}{dv^{\lambda-2n}}, \text{ Mod. } \lambda.$$

Die Vergleichung dieses Ausdrucks mit dem entsprechenden, für den besondern Fall $t = 1$ im ersten Paragraphen gefundenen, zeigt, dafs er genau derselbe ist, wie jener, dafs also in allen Fällen, ohne Unterschied, d. h. für alle complexen idealen Primfactoren $f(\alpha)$ als Moduln, derselbe Ausdruck gilt.

Es bleibt noch übrig, auch für die zu einem beliebigen Exponenten t gehörigen Primfactoren $f(\alpha)$, als Moduln, den Index von λ und die Indices der Primfactoren des λ von der Form $1 - \alpha^k$ zu suchen. Zunächst läfst sich leicht zeigen, dafs Ind. (λ) allemal congruent Null ist, wenn t nicht gleich Eins ist; also für alle Moduln, mit Ausnahme der in dem ersten Paragraphen behandelten. Für diese Moduln ist nämlich nicht blofs Ind. (λ), sondern überhaupt der Index jeder nichtcomplexen ganzen Zahl congruent Null. Denn aus der Congruenz

$$c^{\frac{q^t-1}{\lambda}} \equiv \alpha^{\text{Ind.} c}, \text{ Mod. } f(\alpha),$$

in welcher c eine beliebige (jedoch nicht durch $f(\alpha)$ theilbare) nichtcomplexe Zahl bedeutet, folgt, wenn α in α^{γ^τ} verwandelt wird, weil $f(\alpha) = f(\alpha^{\gamma^\tau})$ ist:

$$c^{\frac{q^t-1}{\lambda}} \equiv \alpha^{\gamma^\tau \text{Ind.} c}, \text{ Mod. } f(\alpha),$$

also $\alpha^{\text{Ind.} c} \equiv \alpha^{\gamma^\tau \text{Ind.} c}$, Mod. $f(\alpha)$, mithin auch Ind. $c \equiv \gamma^\tau$ Ind. c, Mod. λ, und Ind. $(c) \equiv 0$, Mod. λ; mit Ausnahme des einzigen Falles, wo $\gamma^\tau \equiv 1$, also $\tau = \lambda - 1$ und $t = 1$ ist.

Um auch Ind. $(1 - \alpha^k)$ für die in diesem Paragraphen behandelte Moduln zu finden, nehme man

$$\lambda = (1-\alpha)(1-\alpha^{\gamma})(1-\alpha^{\gamma^2}) \ldots (1-\alpha^{\gamma^{\lambda-2}})$$

an, und gebe dieser Gleichung die Form

$$\lambda = (1-\alpha)^{\lambda-1}\left(\frac{1-\alpha^{\gamma}}{1-\alpha}\right)\left(\frac{1-\alpha^{\gamma^2}}{1-\alpha}\right) \ldots \left(\frac{1-\alpha^{\gamma^{\lambda-2}}}{1-\alpha}\right).$$

Man drücke die Einheiten, welche hierin vorkommen, alle durch die Kreistheilungs-Einheiten $e(\alpha)$, $e(\alpha^{\gamma})$, u. s. w, aus, so erhält man

$$\lambda = (1-\alpha)^{\lambda-1}\alpha^{-\mu}.e(\alpha)^{\lambda-1}.e(\alpha^{\gamma})^{\lambda-2}.e(\alpha^{\gamma^2})^{\lambda-3} \ldots e(\alpha^{\gamma^{\lambda-2}})^{1},$$

und wenn α in α^k verwandelt wird:

$$\lambda = (1-\alpha^k)^{\lambda-1}\alpha^{-\mu k}.e(\alpha^k)^{\lambda-1}.e(\alpha^{k\gamma})^{\lambda-2}.e(\alpha^{k\gamma^2})^{\lambda-3} \ldots e(\alpha^{k\gamma^{\lambda-2}})^{1}.$$

Man nehme nun auf beiden Seiten die Indices für den Modul $f(\alpha)$, so ergiebt sich, weil $\text{Ind.}(\lambda) \equiv 0$ ist:

$$0 \equiv -\text{Ind.}(1-\alpha^k) - \mu k\nu - \text{Ind.}\,e(\alpha^k) - 2\,\text{Ind.}\,e(\alpha^{k\gamma}) - \cdots - (\lambda-1)\,\text{Ind.}\,e(\alpha^{k\gamma^{\lambda-2}}),$$

wo $\lambda\nu = q^t - 1$, oder, mit Anwendung des Summenzeichens,

$$\text{Ind.}(1-\alpha^k) \equiv -\mu k\nu - \Sigma h\,\text{Ind.}\,e(\alpha^{k\gamma^{h-1}})$$

für $h = 1, 2, 3, \ldots \lambda-1$ ist. Man drücke ferner $\text{Ind.}\,e(\alpha^{k\gamma^{h-1}})$ durch die Indices der Einheiten $E_1(\alpha)$, $E_2(\alpha)$, ... $E_{\mu-1}(\alpha)$ mittels der Formel

$$\text{Ind.}\,e(\alpha^{\gamma^{h-1}}) \equiv -2\Sigma\gamma^{2n(h-1)}\,\text{Ind.}\,E_n(\alpha)$$

aus, für $n = 1, 2, 3, \ldots \mu-1$, so wird

$$\text{Ind.}(1-\alpha^k) \equiv -\mu k\nu + 2\Sigma\Sigma h\gamma^{2n(h-1)}\,\text{Ind.}\,E_n(\alpha^k),$$

für $h = 1, 2, 3, \ldots \lambda-1$ und $n = 1, 2, 3, \ldots \mu-1$. Die Summation in Beziehung auf h wird nach der leicht zu beweisenden Formel

$$\Sigma h\gamma^{2n(h-1)} \equiv \frac{1}{1-\gamma^{2n}}$$

ausgeführt und giebt

$$\text{Ind.}(1-\alpha^k) \equiv -\mu k\nu + 2\Sigma\frac{\text{Ind.}\,E_n(\alpha^k)}{1-\gamma^{2n}},$$

für $n = 1, 2, 3, \ldots \mu-1$. Endlich kann man noch von der Eigenschaft der Einheit $E_n(\alpha)$ Gebrauch machen, nach welcher

$$\text{Ind.}\,E_n(\alpha^k) \equiv k^{2n}\,\text{Ind.}\,E_n(\alpha), \quad \text{Mod.}\ \lambda$$

ist. Vermöge derselben erhält man

$$\text{Ind.}(1-\alpha^k) \equiv -\mu k\nu + 2\Sigma\frac{k^{2n}\,\text{Ind.}\,E_n(\alpha)}{1-\gamma^{2n}}, \quad \text{Mod.}\ \lambda,$$

für $n = 1, 2, 3, \ldots \mu-1$. Dieser Ausdruck, welcher darauf sich stützt, daſs $\text{Ind.}(\lambda) \equiv 0$ ist, gilt aber nicht für den Fall $t = 1$, für welchen die in dem ersten Paragraphen gegebenen Ausdrücke von Ind. (λ) und Ind. $(1-\alpha^k)$ zu nehmen sind. Es könnten auch für Ind. $E_n(\alpha)$ die gefundenen Ausdrücke hier eingesetzt werden; da aber dadurch keine besonders eleganten Formeln entstehen, so mag es hinreichen, den Ind. $(1-\alpha^k)$ durch die oben gefundenen Indices der Einheiten $E_1(\alpha)$, $E_2(\alpha)$, ... $E_{\mu-1}(\alpha)$ ausgedrückt zu haben.

§. 4.

Über die Anwendung logarithmischer Ausdrücke der complexen Zahlen auf Congruenzen, deren Modul λ oder eine Potenz von λ ist.

Die Anwendung der in dem vorhergehenden Paragraphen gefundenen Resultate auf die allgemeinen Reciprocitätsgesetze, welche ich in dem Folgenden zu geben gedenke, erfordert, dafs ich zunächst eine allgemeine Methode mittheile, nach welcher Congruenzen für den Modul λ, oder eine Potenz von λ, namentlich wenn sie Producte mehrerer complexen Zahlen und Potenzen derselben enthalten, auf die einfachste Weise eben so behandelt werden können, wie in der Analysis dergleichen Ausdrücke mit Hülfe der Logarithmen. Eine Andeutung einer solchen Methode, welche ich in der Form, wie ich sie hier entwickeln werde, schon seit mehreren Jahren vielfach angewendet, aber bisher noch nicht veröffentlicht habe, findet sich in einer Abhandlung des Herrn *Eisenstein* (im 39ten Bande dieses Journals S. 351), auf welche Abhandlung ich in dem Folgenden noch einmal zurückkommen werde.

Ich stelle für den *Logarithmen* einer complexen Zahl, in Beziehung auf den Modul λ^{n+1} genommen, folgende Definition auf:

Wenn $\varphi(\alpha)$ eine complexe Zahl ist, welche den Factor $1-\alpha$ nicht enthält, so dafs auch $\varphi(1)$ nicht $\equiv 0$, Mod. λ ist, so soll unter dem Ausdrucke

$$l\Big(\frac{\varphi(\alpha)}{\varphi(1)}\Big) \text{ für den Modul } \lambda^{n+1}$$

die endliche Anzahl von Gliedern der Reihe

$$\frac{\varphi(\alpha)-\varphi(1)}{\varphi(1)} - \tfrac{1}{2}\Big(\frac{\varphi(\alpha)-\varphi(1)}{\varphi(1)}\Big)^2 + \tfrac{1}{3}\Big(\frac{\varphi(\alpha)-\varphi(1)}{\varphi(1)}\Big)^3 - \dots$$

verstanden werden, welche nicht congruent Null sind für den Modul λ^{n+1}.

Dafs die Glieder dieser Reihe, von einem bestimmten an, wirklich alle congruent Null werden, für den Modul λ^{n+1}, erhellet daraus, dafs $\varphi(\alpha)-\varphi(1)$ den Factor $1-\alpha$ enthält, und dafs $(1-\alpha)^{\lambda-1}$ durch λ theilbar ist. Wenn nämlich in dem allgemeinen Gliede der Reihe

$$\pm\frac{1}{k}\Big(\frac{\varphi(\alpha)-\varphi(1)}{\varphi(1)}\Big)^k$$

k nicht durch λ theilbar ist, so ist Dasselbe allemal congruent Null für den Modul λ^{n+1}, sobald $k > (\lambda-1)(n+1)$ ist. Wenn aber k den Factor λ enthält, und man setzt $k = c\lambda^a$, wo c nicht weiter durch λ theilbar sein soll, so müssen noch die a Factoren λ des Nenners durch $(\lambda-1)a$ Factoren $1-\alpha$ des Zählers compensirt werden. Wenn daher $k > (\lambda-1)(n+a+1)$ angenommen wird, so sind alle Glieder congruent Null für den Modul λ^{n+1}. In Beziehung auf diesen Modul bricht also diese Reihe nothwendig ab. Die Logarithmen der complexen Zahlen für den Modul λ^{n+1} sind demnach nur endliche Ausdrücke, d. h. sie sind selbst nur gewöhnliche complexe Zahlen. Auch geht aus der Definition hervor, dafs $l\left(\frac{\varphi(\alpha)}{\varphi(1)}\right)$ für den Modul λ^{n+1} nur einen einzigen bestimmten Werth hat, wenn nämlich aufser $\varphi(\alpha)$ auch $\varphi(1)$ nur einen bestimmten Werth hat. Aus den bekannten Eigenschaften der Reihe, durch welche $l\left(\frac{\varphi(\alpha)}{\varphi(1)}\right)$ dargestellt wird, erhält man ferner sogleich für die Grund-Eigenschaften dieser Ausdrücke:

$$\left.\begin{aligned} l\left(\frac{f(\alpha)}{f(1)}\right)+l\left(\frac{\varphi(\alpha)}{\varphi(1)}\right) &\equiv l\left(\frac{f(\alpha)\varphi(\alpha)}{f(1)\varphi(1)}\right) \\ ml\left(\frac{\varphi(\alpha)}{\varphi(1)}\right) &\equiv l\left(\frac{\varphi(\alpha)}{\varphi(1)}\right)^m \end{aligned}\right\} \ldots \text{Mod. } \lambda^{n+1}.$$

Es kommt nun darauf an, für $l\left(\frac{\varphi(\alpha)}{\varphi(1)}\right)$, Mod. λ^{n+1}, den einfachsten Ausdruck zu finden. Zu diesem Ende werde die complexe Zahl $\varphi(\alpha)-\varphi(1)$, welche durch $1-\alpha$ theilbar ist, auf folgende Form gebracht:

* $$\varphi(\alpha)-\varphi(1) = A_1(1-\alpha)+A_2(1-\alpha)^2+\cdots+A_{\lambda-1}\alpha^{\lambda-1}.$$

Stellt man sich diesen Werth in die obige Reihe gesetzt vor, welche nur so weit fortgeführt anzunehmen ist, als die Glieder nicht von selbst wegfallen, weil sie congruent Null werden, und dann die einzelnen Glieder alle entwickelt, nach Potenzen von $1-\alpha$, so erhält man eine Reihe von folgender Form:

$$l\left(\frac{\varphi(\alpha)}{\varphi(1)}\right) \equiv \frac{C_1(1-\alpha)}{1\,.\,\varphi(1)}+\frac{C_2(1-\alpha)^2}{2\,.\,\varphi(1)^2}+\frac{C_3(1-\alpha)^3}{3\,.\,\varphi(1)^3}+\cdots, \text{ Mod. } \lambda^{n+1},$$

welche man ebenfalls nur so weit fortzusetzen braucht, als noch Glieder sich finden, welche nicht congruent Null sind. Wendet man das Summenzeichen an und verwandelt α in $\alpha\gamma^h$, so erhält man

$$l\left(\frac{\varphi(\alpha\gamma^h)}{\varphi(1)}\right) \equiv \Sigma\frac{C_i(1-\alpha\gamma^h)^i}{i\varphi(1)^i}, \text{ Mod. } \lambda^{n+1},$$

und wenn man mit $\gamma^{-hk\lambda^n}$ multiplicirt und die Summe für $h = 0, 1, 2, \ldots \lambda-2$

nimmt, so ergiebt sich

$$\sum_{0}^{\lambda-2}{}_h \gamma^{-hk\lambda^n} l\left(\frac{\varphi(\alpha^{\gamma^h})}{\varphi(1)}\right) = \sum_{0}^{\lambda-2}{}_h \sum_i \frac{C_i \gamma^{-hk\lambda^n}(1-\alpha^{\gamma^h})^i}{i\varphi(1)^i}.$$

Wird nun die $i$te Potenz von $1-\alpha^{\gamma^h}$ nach dem binomischen Lehrsatze entwickelt, so erhält man

$$(1-\alpha^{\gamma^h})^i = \sum_{0}^{i}{}_s \frac{(-1)_s \Pi(i)\alpha^{s\gamma^h}}{\Pi(s)\Pi(i-s)},$$

folglich

$$\sum_{0}^{\lambda-2}{}_h \frac{\gamma^{-hk\lambda^n}(1-\alpha^{\gamma^h})^i}{i} = \sum_{0}^{\lambda-2}{}_h \sum_{0}^{i}{}_s \frac{(-1)_s \Pi(i)\gamma^{-hk\lambda^n}\alpha^{s\gamma^h}}{i\Pi(s)\Pi(i-s)}.$$

Nun führe ich folgende für die hier zu behandelnden logarithmischen Ausdrücke besonders wichtige complexe Zahl ein:

$$X_k(\alpha) = \alpha + \gamma^{-k\lambda^n}\alpha^{\gamma} + \gamma^{-2k\lambda^n}\alpha^{\gamma^2} + \cdots + \gamma^{-(\lambda-2)k\lambda^n}\alpha^{\gamma^{\lambda-2}}.$$

Diese durch $X_k(\alpha)$ bezeichneten complexen Zahlen haben mehrere besondere Eigenschaften, welche denen des entsprechenden Ausdrucks der Kreistheilung

$$F(\beta, \alpha) = \alpha + \beta\alpha^{\gamma} + \beta^2\alpha^{\gamma^2} + \cdots + \beta^{\lambda-2}\alpha^{\gamma^{\lambda-2}},$$

wo α eine Wurzel der Gleichung $\alpha^{\lambda} = 1$ und β eine Wurzel der Gleichung $\beta^{\lambda-1} = 1$ ist, ganz analog sind. Für den gegenwärtigen Zweck aber soll nur von der einen Eigenschaft derselben Gebrauch gemacht werden, nämlich von

$$X_k(\alpha^m) \equiv m^{k\lambda^n} X_k(\alpha), \quad \text{Mod. } \lambda^{n+1};$$

welche Eigenschaft aus der Definition selbst leicht zu beweisen ist, und als deren specieller Fall, für $m = 1$, noch der Congruenz

$$X_k(\alpha^{-1}) \equiv (-1)^k X_k(\alpha), \quad \text{Mod. } \lambda^{n+1},$$

besonders erwähnt werden mag.

Durch Einführung des Zeichens $X_k(\alpha)$ wird die obige Gleichung folgendermafsen dargestellt:

$$\sum_{0}^{\lambda-2}{}_h \frac{\gamma^{-hk\lambda^n}(1-\alpha^{\gamma^h})^i}{i} = \sum_{0}^{i}{}_s \frac{(-1)^s \Pi(i)}{i\Pi(s)\Pi(i-s)} X_k(\alpha^s).$$

Man zerlege nun die Summe rechter Hand in λ Partialsummen, indem man die Fälle sondert, wo $s \equiv 0$, $s \equiv 1$, $s \equiv 2$, ... $s \equiv \lambda - 1$ ist, für den Modul λ. Dabei bezeichne man durch P_t folgende Summe von Binomialcoëfficienten:

$$P_t = (-1)^t\left(\frac{\Pi(i)}{\Pi(t)\Pi(i-t)} - \frac{\Pi(i)}{\Pi(t+\lambda)\Pi(i-t-\lambda)} + \frac{\Pi(i)}{\Pi(t+2\lambda)\Pi(i-t-2\lambda)} - \cdots\right).$$

Alsdann ist

$$\sum_{0}^{\lambda-2}{}_h \frac{\gamma^{-hk\lambda^n}(1-\alpha^{\gamma^h})^i}{i} = \frac{1}{i}\left(P_0 X_k(1) + P_1 X_k(\alpha) + P_2 X_k(\alpha^2) + \cdots + P_{\lambda-1} X_k(\alpha^{\lambda-1})\right).$$

Um aus dieser Gleichung eine Congruenz für den Modul λ^{n+1} machen zu können, werde ich zunächst beweisen, dafs, auch wenn i den Factor λ enthält, dennoch $\frac{P_t}{i}$ denselben niemals im Nenner enthalten kann; oder, dafs P_t diesen Factor λ immer eben so oft enthält als i. Es sei wieder $i = c\lambda^\alpha$, so sind in allen den Fällen, wo t nicht $= 0$ ist, alle einzelnen in P_t enthaltenen Binomialcoëfficienten durch λ^α theilbar; welches mit Hülfe des in (§. 2) angewendeten Satzes über die Anzahl der in dem Producte $\Pi(A)$ enthaltenen Primfactoren von einer bestimmten Art, leicht bewiesen werden kann. Um aber für den Fall $t = 0$ zu zeigen, dafs P_0 durch λ^α theilbar ist, bemerke ich, dafs $P_0 + P_1 + P_2 + \cdots + P_{\lambda-1}$ gleich der Summe aller Binomialcoëfficienten von der $i$ten Potenz mit abwechselnden Vorzeichen, also gleich $(1-1)^i = 0$ ist. Wenn daher, wie gezeigt worden, $P_1, P_2, \ldots P_{\lambda-1}$ alle den Factor λ^α enthalten, so folgt, dafs P_0 ebenfalls durch diesen Factor theilbar sein mufs. Also $\frac{P_t}{i}$, aus dessen Nenner jeder Factor λ sich gegen die Zähler hinweghebt, kann in Beziehung auf den Mod. λ^{n+1} als ganze Zahl betrachtet werden. Wendet man nun die Congruenz

* $$X_k(\alpha) \equiv m^{k\lambda^n} X_k(\alpha), \quad \text{Mod. } \lambda^{n+1}$$

auf die obige Gleichung an, und beachtet, dafs $X_k(1) \equiv 0$, Mod. λ^{n+1} ist, so erhält man

$$\sum_0^{\lambda-2}{}_h \frac{\gamma^{-hk\lambda^n}(1-\alpha\gamma^h)^i}{i} \equiv \frac{1}{i}(1^{k\lambda^n} P_1 + 2^{k\lambda^n} P_2 + \cdots + (\lambda-1)^{k\lambda^n} P_{\lambda-1}) X_k(\alpha),$$

für den Mod. λ^{n+1}; welche Congruenz auch in dem Falle richtig bleibt, wenn i durch λ theilbar ist.

Man entwickle jetzt die Potenz $(1-e^v)^i$ nach dem binomischen Lehrsatze und nehme den $k\lambda^n$ten Differentialquotienten für $v = 0$, so ist

$$\frac{d_0^{k\lambda^n}(1-e^v)^i}{i\,dv^{k\lambda^n}} = \frac{1}{i}\sum_0^i{}_s \frac{(-1)^s \Pi(i)\, s^{k\lambda^n}}{\Pi(s)\,\Pi(i-s)}.$$

Man zerlege ferner diese Summe genau eben so in Partialsummen, wie die obige, indem man von der Congruenz

$$(s+m\lambda)^{k\lambda^n} \equiv s^{k\lambda^n}, \quad \text{Mod. } \lambda^{n+1}$$

Gebrauch macht, so wird

$$\frac{d_0^{k\lambda^n}(1-e^v)^i}{dv^{k\lambda^n}} \equiv \frac{1}{i}(1^{k\lambda^n} P_1 + 2^{k\lambda^n} P_2 + \cdots + (\lambda-1^{k\lambda^n} P_{\lambda-1}), \quad \text{Mod. } \lambda^{n+1},$$

also

$$\sum_{0}^{\lambda-2}{}_h \frac{\gamma^{-hk\lambda^n}(1-\alpha\gamma^h)^i}{i} \equiv \frac{d_0^{k\lambda^n}(1-e^v)^i}{dv^{k\lambda^n}}\cdot X_k(\alpha), \text{ Mod. } \lambda^{n+1},$$

und demzufolge auch

$$\sum_{0}^{\lambda-2}{}_h \gamma^{-hk\lambda^n} l\left(\frac{\varphi(\alpha\gamma^h)}{\varphi(1)}\right) \equiv \Sigma_i \frac{C_i d_0^{k\lambda^n}(1-e^v)^i}{i\varphi(1)^i dv^{k\lambda^n}}\cdot X_k(\alpha).$$

Nun ist aber die Summe

$$\Sigma \frac{C_i(1-e^v)^i}{i\varphi(1)^i}$$

nichts anderes, als die nach Potenzen von $1-e^v$ geordnete Entwickelung des Logarithmus von $\frac{\varphi(e^v)}{\varphi(1)}$, also

$$\Sigma \frac{C_i(1-e^v)^i}{i\varphi(1)^i} = l\left(\frac{\varphi(e^v)}{\varphi(1)}\right) = l\varphi(e^v) - l\varphi(1);$$

folglich ist, wenn hiervon der $k\lambda^n$te Differentialquotient genommen und $v=0$ gesetzt wird:

$$\Sigma \frac{C_i d_0^{k\lambda^n}(1-e^v)^i}{i\varphi(1)^i dv^{k\lambda^n}} = \frac{d_0^{k\lambda^n} l\varphi(e^v)}{dv^{k\lambda^n}},$$

also

$$\sum^{\lambda-2}{}_k \gamma^{-hk\lambda^n} l\left(\frac{\varphi(\alpha\gamma^h)}{\varphi(1)}\right) \equiv \frac{d_0^{k\lambda^n} l\varphi(e^v)}{dv^{k\lambda^n}}\cdot X_k(\alpha), \text{ Mod. } \lambda^{n+1}.$$

Setzt man endlich $k=1, 2, 3, \ldots \lambda-2$ und bildet die Summe, wobei der Werth $h=0$ von den übrigen zu trennen ist, so erhält man folgenden Ausdruck des Logarithmen einer complexen Zahl in Beziehung auf den Mod. λ^{n+1}:

$$(\lambda-1)l\left(\frac{\varphi(\alpha)}{\varphi(1)}\right) \equiv l\left(\frac{N\varphi(\alpha)}{\varphi(1)^{\lambda-1}}\right) + \frac{d_0^{\lambda^n} l\varphi(e^v)}{dv^{\lambda^n}} X_1(\alpha) + \frac{d_0^{2\lambda^n} l\varphi(e^v)}{dv^{2\lambda^n}} X_2(\alpha) + \cdots$$
$$\cdots + \frac{d_0^{(\lambda-2)\lambda^n} l\varphi(e^v)}{dv^{(\lambda-2)\lambda^n}} X_{\lambda-2}(\alpha), \text{ Mod. } \lambda^{n+1}.$$

Der Fall $n=0$, als der in den Anwendungen am häufigsten vorkommende, verdient noch eine besondere Berücksichtigung. Für denselben hat man, weil $N\varphi(\alpha)\equiv 1$ und $\varphi(1)^{\lambda-1}\equiv 1$, Mod. λ ist,

$$\sum_{0}^{\lambda-2}{}_h \gamma^{-hk} l\left(\frac{\varphi(\alpha\gamma^h)}{\varphi(1)}\right) \equiv \frac{d_0^k l\varphi(e^v)}{dv^k}\cdot X_k(\alpha),$$

und

$$-l\left(\frac{\varphi(\alpha)}{\varphi(1)}\right) \equiv \frac{d_0 l\varphi(e^v)}{dv} X_1(\alpha) + \frac{d_0^2 l\varphi(e^v)}{dv^2} X_2(\alpha) + \cdots + \frac{d_0^{\lambda-2} l\varphi(e^v)}{dv^{\lambda-2}} X_{\lambda-2}(\alpha), \text{ Mod. } \lambda,$$

und es ist in diesem Falle

$$X_k(\alpha) = \alpha + \gamma^{-k}\alpha^{\gamma} + \gamma^{-2k}\alpha^{\gamma^2} + \cdots + \gamma^{-(\lambda-2)k}\alpha^{\gamma^{\lambda-2}};$$

wofür man auch den congruenten Ausdruck

$$X_k(\alpha) \equiv \alpha + 2^{\lambda-1-k}\alpha^2 + 3^{\lambda-1-k}\alpha^3 + \cdots + (\lambda-1)^{\lambda-1-k}\alpha^{\lambda-1}$$

setzen kann.

Um nun diese Logarithmen der complexen Zahlen und die für dieselben gefundenen Entwickelungen, in der Rechnung, da wo es sich um Congruenzen für den Mod. λ^{n+1} oder λ handelt, überall mit Sicherheit anwenden zu können, sind noch zwei Sätze nöthig, welche hier aufgestellt und bewiesen werden sollen. Nemlich:

Lehrsatz. Wenn zwei complexe Zahlen congruent sind, für den Modul λ^{n+1}, und wenn auch die beiden ganzen Zahlen, welche man erhält, wenn man in denselben $\alpha = 1$ setzt, congruent sind, für denselben Modul: so sind auch die Logarithmen dieser complexen Zahlen für diesen Modul congruent.

Die beiden complexen Zahlen seien $f(\alpha)$ und $\varphi(\alpha)$, so ist nach der Voraussetzung $f(\alpha) \equiv \varphi(\alpha)$ und $f(1) \equiv \varphi(1)$, Mod. λ^{n+1}. Setzt man der Kürze wegen

$$\frac{f(\alpha)-f(1)}{f(1)} = x, \qquad \frac{\varphi(\alpha)-\varphi(1)}{\varphi(1)} = y,$$

so erhält man

$$l\left(\frac{f(\alpha)}{f(1)}\right) \equiv x - \tfrac{1}{2}x^2 + \tfrac{1}{3}x^3 - \tfrac{1}{4}x^4 + \cdots,$$

$$l\left(\frac{\varphi(\alpha)}{\varphi(1)}\right) \equiv y - \tfrac{1}{2}y^2 + \tfrac{1}{3}y^3 - \tfrac{1}{4}y^4 + \cdots$$

für den Mod. λ^{n+1}; welche Reihen von selbst abbrechen, weil, von einem bestimmten Gliede an, alle folgenden congruent Null werden; die man aber auch über diese Grenze hinaus noch fortgesetzt annehmen kann, so weit man will. Aus den beiden Voraussetzungen des Satzes folgt $x \equiv y$, Mod. λ^{n+1}; woraus sogleich erhellet, dafs auch $\frac{1}{k}x^k \equiv \frac{1}{k}y^k$, Mod. λ^{n+1} ist, für alle diejenigen Werthe von k, welche nicht Vielfache von λ sind. Um zu zeigen, dafs eben Das auch für alle durch λ theilbaren Werthe von k Statt findet, also allgemein für $k = c\lambda^a$, wo c nicht weiter durch λ theilbar ist, setze ich $y = x + \lambda^{n+1}z$. Dann giebt die binomische Entwickelung:

$$y^{c\lambda^a} \equiv x^{c\lambda^a} + c\lambda^{n+a+1}x^{c\lambda^a-1}z + \cdots,$$

18*

also

$$y^{c\lambda^a} \equiv x^{c\lambda^a}, \text{ Mod. } \lambda^{n+a+1},$$

und hieraus folgt

$$\frac{y^{c\lambda^a}}{c\lambda^a} \equiv \frac{x^{c\lambda^a}}{c\lambda^a}, \text{ Mod. } \lambda^{n+1}.$$

Da nun alle einzelnen Glieder der einen Entwicklung den entsprechenden der andern Entwicklung congruent sind, so ist nothwendig

$$l\left(\frac{f(\alpha)}{f(1)}\right) \equiv l\left(\frac{\varphi(\alpha)}{\varphi(1)}\right), \text{ Mod. } \lambda^{n+1};$$

was zu beweisen war.

Der zweite, jetzt zu beweisende Satz (die Umkehrung des ersten) ist nur unter Hinzufügung einer neuen Bedingung richtig und lautet so:

Lehrsatz. Wenn die Logarithmen zweier complexen Zahlen, nemlich $l\left(\frac{f(\alpha)}{f(1)}\right)$ und $l\left(\frac{\varphi(\alpha)}{\varphi(1)}\right)$ congruent sind für den Modul λ^{n+1}, und auch $f(1) \equiv \varphi(1)$, Mod. λ^{n+1} ist, und wenn überdies die complexen Zahlen $f(\alpha)$ und
* $(\varphi\alpha)$ so beschaffen sind, dafs $f(\alpha)-f(1)$ und $\varphi(\alpha)-\varphi(1)$ beide durch $(1-\alpha)^2$ theilbar sind: so sind die complexen Zahlen $f(\alpha)$ und $\varphi(\alpha)$ selbst congruent für den Mod. λ^{n+1}.

Es sei wieder

$$\frac{f(\alpha)-f(1)}{f(1)} = x, \qquad \frac{\varphi(\alpha)-\varphi(1)}{\varphi(1)} = y,$$

und

$$l\left(\frac{f(\alpha)}{f(1)}\right) = u, \qquad l\left(\frac{\varphi(\alpha)}{\varphi(1)}\right) = v,$$

so ist

$$u \equiv x - \tfrac{1}{2}x^2 + \tfrac{1}{3}x^3 - \tfrac{1}{4}x^4 + \cdots, \text{ Mod. } \lambda^{n+1}.$$

Es läfst sich nun diese Reihe nach den gewöhnlichen Methoden umkehren, so dafs x in eine nach Potenzen von u geordnete Reihe entwickelt wird. Aus den ersten N Gliedern der gegebenen Reihe werden so die ersten N Glieder der umgekehrten Reihe bestimmt, und wenn man N grofs genug annimmt, so dafs in beiden Reihen, der gegebenen sowohl, als der umgekehrten, über das Nte Glied hinaus nur noch solche Glieder liegen, welche für den Mod. λ^{n+1} congruent Null sind, so hat man den vollständigen Ausdruck des x durch u in Beziehung auf den Mod. λ^{n+1}. Das Resultat der Umkehrung dieser Reihe ist aus der Analysis bekannt. Es ist

$$x \equiv \frac{u}{1} + \frac{u^2}{1.2} + \frac{u^3}{1.2.3} + \frac{u^4}{1.2.3.4} + \cdots$$

Nun ist zu beweisen, dafs auch in dieser Reihe die Anzahl der ersten

Glieder N immer so grofs angenommen werden kann, dafs alle folgenden Glieder congruent Null werden. Zu diesem Zwecke untersuche ich das allgemeine Glied der Reihe, nämlich

$$\frac{u^k}{\Pi(k)}.$$

Aus dem im zweiten Paragraphen mitgetheilten Satze über die Anzahl, wievielmal ein bestimmter Primfactor in dem Producte $\Pi(A)$ enthalten ist, folgt zunächst, dafs die Anzahl der in dem Producte $\Pi(k)$ enthaltenen Factoren λ stets kleiner ist als $\frac{k}{\lambda-1}$. Ferner enthält, nach der Voraussetzung, $f(\alpha)-f(1)$, also auch x, den Factor $(1-\alpha)^2$; woraus sogleich folgt, dafs auch u den Factor $(1-\alpha)^2$ enthalten mufs. Der Zähler u^k enthält demnach den Factor $1-\alpha$ genau $2k$ mal, und wenn man erwägt, dafs ein Factor λ genau $\lambda-1$ Factoren $1-\alpha$ giebt, so weifs man, dafs der Nenner $\Pi(k)$ stets weniger als k Factoren $1-\alpha$ enthält. Nach Aufhebung der Factoren λ im Nenner, gegen die im Zähler, bleiben folglich im Zähler mehr als k Factoren $1-\alpha$ stehen, und wenn daher $k>(n+1)(\lambda-1)$ angenommen wird, so enthält $\frac{u^k}{\Pi(k)}$ den Factor $(1-\alpha)^{(\lambda-1)(n+1)}$, d. h. den Factor λ^{n+1}. Von einem solchen Gliede an sind demnach alle folgenden $\equiv 0$. Ohne die Bedingung: $f(\alpha)-f(1)$ theilbar durch $(1-\alpha)^2$, würde aber, wie sich hieraus zeigt, die Reihe nach dem Mod. λ^{n+1} keinen endlichen Ausdruck geben.

Ganz auf dieselbe Weise ist nun auch

$$y \equiv \frac{v}{1}+\frac{v^2}{1.2}+\frac{v^3}{1.2.3}+\frac{v^4}{1.2.3.4}+\cdots, \quad \text{Mod. } \lambda^{n+1}.$$

Nach der Voraussetzung des zu beweisenden Satzes ist aber $u\equiv v$, Mod. λ^{n+1}. Schreibt man diese Congruenz, als Gleichung, in die Form $u=v+\lambda^{n+1}w$, so findet sich durch Entwicklung der Potenz des Binoms:

$$\frac{u^k}{\Pi(k)}=\frac{(v+\lambda^{n+1}w)^k}{\Pi(k)}=\frac{v^k}{\Pi(k)}+\frac{\lambda^{n+1}v^{k-1}w}{\Pi(1)\Pi(k-1)}+\frac{\lambda^{2n+2}v^{k-2}w^2}{\Pi(2)\Pi(k-2)}+\cdots$$

Nun heben sich, nach Dem was so eben gezeigt, alle in $\Pi(k-1)$ enthaltenen Factoren λ gegen die in v^{k-1} enthaltenen vollständig hinweg; eben so die in $\Pi(k-2)$ enthaltenen gegen die in v^{k-2} enthaltenen u. s. w. Für den Modul λ^{n+1} bleibt also von dieser binomischen Entwicklung nur das erste Glied stehen; alle übrigen verschwinden, als Vielfache von λ^{n+1}. Folglich ist

$$\frac{u^k}{\Pi(k)}\equiv\frac{v^k}{\Pi(k)}, \quad \text{Mod. } \lambda^{n+1}.$$

Da also alle Glieder der Entwicklung von x einzeln denen der Entwicklung von y congruent sind, so ist nothwendig $x \equiv y$, Mod. λ^{n+1}, d. h.

$$\frac{f(\alpha)-f(1)}{f(1)} \equiv \frac{\varphi(\alpha)-\varphi(1)}{\varphi(1)}, \quad \text{Mod. } \lambda^{n+1};$$

und da ferner nach der Voraussetzung auch $f(1) \equiv \varphi(1)$ ist, so ist

$$f(\alpha) \equiv \varphi(\alpha), \quad \text{Mod. } \lambda^{n+1};$$

was zu beweisen war.

Um den Gebrauch der hier entwickelten Methode an einem Beispiele zu zeigen, wende ich sie zur Lösung folgender in der Theorie der complexen Zahlen sehr wichtigen Aufgabe an.

Aufgabe. Eine gegebene *complexe* Zahl durch Multiplication mit Einheiten in eine solche Form zu bringen, dafs sie, mit ihrer reciproken multiplicirt, ein Product giebt, welches einer *nichtcomplexen* ganzen Zahl congruent ist, für den Mod. λ.

Die Aufgabe ist, wie sich ergeben wird, immer lösbar, aufser wenn λ eine von den Ausnahmezahlen ist, d. h. wenn λ in einer der ersten $\frac{1}{2}(\lambda-3)$ *Bernoulli*schen Zahlen als Factor des Zählers vorkommt. Ich wende zur Lösung der Aufgabe wieder die Einheiten von der Form

$$E_n(\alpha) = e(\alpha).e(\alpha^{\gamma})^{\gamma^{-2n}}.e(\alpha^{\gamma^2})^{\gamma^{-4n}} \dots e(\alpha^{\gamma^{\mu-1}})^{\gamma^{-2(\mu-1)n}}$$

an; nemlich diejenigen, für welche die Indices sich am einfachsten darstellten.

Durch Anwendung der Logarithmen erhält man

$$l\left(\frac{E_n(\alpha)}{E(1)}\right) \equiv l\left(\frac{e(\alpha)}{e(1)}\right) + \gamma^{-2n} l\left(\frac{e(\alpha^{\gamma})}{e(1)}\right) + \dots \gamma^{-2(\mu-1)n} l\left(\frac{e(\alpha^{\gamma^{\mu-1}})}{e(1)}\right),$$

oder, durch Anwendung des Summenzeichens:

$$l\left(\frac{E_n(\alpha)}{F_n(1)}\right) \equiv \sum_{0}^{\mu-1}{}_h \gamma^{-2hn} l\left(\frac{e(\alpha^{\gamma^h})}{e(1)}\right), \quad \text{Mod. } \lambda.$$

Giebt man dem h weiter die Werthe $h=\mu, \mu+1, \mu+2, \dots \lambda-2$, so kehren dieselben Glieder wieder, für den Mod. λ. Wenn daher die Summe für alle Werthe $h=0, 1, 2, \dots \lambda-2$ genommen wird, so verdoppelt sie sich und man erhält

$$2l\left(\frac{E_n(\alpha)}{F_n(1)}\right) \equiv \Sigma \gamma^{-2hn} l\left(\frac{e(\alpha^{\gamma^h})}{e(1)}\right), \quad \text{Mod. } \lambda.$$

Setzt man nun in der einen der für den Mod. λ in diesem Paragraphen gegebenen allgemeinen Formeln $\varphi(\alpha)=e(\alpha)$, so ergiebt sich

$$2l\left(\frac{E_n(\alpha)}{E_n(1)}\right) \equiv \frac{d_0^{2n}\, le(e^v)}{dv^{2n}} X_{2n}(\alpha), \quad \text{Mod. } \lambda.$$

Nun ist

$$e(\alpha) = \pm \frac{\alpha^{\frac{1}{2}(1-\gamma)}(1-\alpha^{\gamma})}{1-\alpha},$$

also

$$e(e^{v}) = \pm \frac{e^{\frac{1}{2}v(1-\gamma)}(1-e^{\gamma v})}{1-e^{v}} = \pm \frac{e^{\frac{1}{2}\gamma v}-e^{-\frac{1}{2}\gamma v}}{e^{\frac{1}{2}v}-e^{-\frac{1}{2}v}}.$$

Die nach Potenzen von v geordnete Reihen-Entwicklung des Logarithmen dieser Gröfse ist bekanntlich

$$le(e^{v}) = l\gamma + \frac{(\gamma^2-1)B_1 v^2}{1.2.2} - \frac{(\gamma^4-1)B_2 v^4}{1.2.3.4.4} + \cdots,$$

wo B_1, B_2, u. s. w. die *Bernoulli*schen Zahlen sind. Hieraus erhält man unmittelbar den $2n$ten Differentialquotienten für den Werth $v=0$, nämlich

$$\frac{d_0^{2n}\, le(e^{v})}{dv^{2n}} = (-1)^{n+1}\frac{(\gamma^{2n}-1)B_n}{2n},$$

also

$$l\left(\frac{E_n(\alpha)}{E_n(1)}\right) \equiv (-1)^{n+1}\frac{(\gamma^{2n}-1)B_n}{4n} X_{2n}(\alpha), \quad \text{Mod } \lambda.$$

Es sei nun $F(\alpha)$ eine beliebige, nicht durch $1-\alpha$ theilbare complexe Zahl, welche auch so angenommen werden soll, dafs $F(\alpha)-F(1)$ durch $(1-\alpha)^2$ theilbar ist, in welche Form sich jede solche complexe Zahl durch Multiplication mit einer passenden einfachen Einheit α^k bringen läfst: so hat man für den Logarithmen derselben folgende Entwicklung:

$$-l\left(\frac{F(\alpha)}{F(1)}\right) \equiv \frac{d_0 lF(e^{v})}{dv} X_1(\alpha) + \frac{d_0^2 lF(e^{v})}{dv^2} X_2(\alpha) \cdots \frac{d_0^{\lambda-2} lF(e^{v})}{dv^{\lambda-2}} X_{\lambda-2}(\alpha), \quad \text{Mod. } \lambda.$$

Nun möge die Zahl M_n durch folgende Congruenz bestimmt werden:

$$\frac{(-1)^{n+1}(\gamma^{2n}-1)B_n}{4n} M_n \equiv \frac{d_0^{2n}\, lF(e^{v})}{dv^{2n}}, \quad \text{Mod. } \lambda.$$

Der Werth von M_n wird durch diese Congruenz immer vollständig bestimmt sein, wenn nicht etwa der Coëfficient des M_n in derselben durch λ theilbar ist. Hat n nur einen der Werthe 1, 2, 3, ... $\mu-1$, so ist $\gamma^{2n}-1$ niemals durch λ theilbar. Es kommt also nur darauf an, dafs die *Bernoulli*sche Zahl B_n für einen der Werthe $n=1, 2, 3, \ldots \mu-1$ nicht durch λ theilbar sei. Ich setze deshalb voraus, dafs λ nicht eine von diesen Ausnahmezahlen sei, welche auch schon bei mehreren der vorhergehenden Untersuchungen ausgeschlossen werden mufsten, und für welche im Allgemeinen die vorliegende Aufgabe immer unlösbar ist. Verbindet man die für M_n aufgestellte Congruenz mit dem gefundenen Ausdrucke des Logarithmus der Einheit $E_n(\alpha)$, so erhält man

$$M_n l\left(\frac{E_n(\alpha)}{E_n(1)}\right) \equiv \frac{d_0^{2n} lF(e^v)}{dv^{2n}} \cdot X_{2n}(\alpha), \quad \text{Mod. } \lambda,$$

und wenn man in der Entwicklung des $-l\left(\frac{F(\alpha)}{F(1)}\right)$ alle Glieder mit geradem Stellenzeiger durch die für $n = 1, 2, 3, \ldots \mu - 1$ in dieser Congruenz gegebenen Ausdrücke ersetzt, so ergiebt sich:

$$-l\left(\frac{F(\alpha)}{F(1)}\right) \equiv M_1 l\left(\frac{E_1(\alpha)}{E_1(1)}\right) + M_2 l\left(\frac{E_2(\alpha)}{E_2(1)}\right) + \cdots + M_{\mu-1} l\left(\frac{E_{\mu-1}(\alpha)}{E_{\mu-1}(1)}\right)$$
$$+ \frac{d_0 lF(e^v)}{dv} X_1(\alpha) + \frac{d_0^3 lF(e^v)}{dv^3} X_3(\alpha) + \cdots + \frac{d_0^{\lambda-2} lF(e^v)}{dv^{\lambda-2}} X_{\lambda-2}(\alpha).$$

Es sei nun

$$F(\alpha).E_1(\alpha)^{M_1}.E_2(\alpha)^{M_2} \ldots E_{\mu-1}(\alpha)^{M_{\mu-1}} = F_1(\alpha),$$

so ist vermöge dieser Congruenz:

$$-l\left(\frac{F_1(\alpha)}{F_1(1)}\right) \equiv \frac{d_0 lF(e^v)}{dv} X_1(\alpha) + \frac{d_0^3 lF(e^v)}{dv^3} X_3(\alpha) \cdots + \frac{d_0^{\lambda-2} lF(e^v)}{dv^{\lambda-3}} X_{\lambda-2}(\alpha),$$

für den Mod. λ. Verwandelt man α in α^{-1}, so wird für alle ungeraden Werthe von k:

$$X_k(\alpha^{-1}) \equiv -X_k(\alpha).$$

Durch diese Änderung wird also lediglich das Vorzeichen von $l\left(\frac{F_1(\alpha)}{F_1(1)}\right)$ umgekehrt, und man erhält

$$l\left(\frac{F_1(\alpha)}{F_1(1)}\right) \equiv -l\left(\frac{F_1(\alpha^{-1})}{F_1(1)}\right) \quad \text{oder} \quad l\left(\frac{F_1(\alpha)F_1(\alpha^{-1})}{F_1(1)E_1(1)}\right) \equiv 0,$$

und hieraus folgt endlich

$$F_1(\alpha)F_1(\alpha^{-1}) \equiv F(1)^2, \quad \text{Mod. } \lambda.$$

Die complexe Zahl

$$F_1(\alpha) = F(\alpha).E_1(\alpha)^{M_1}.E_2(\alpha)^{M_2} \ldots E_{\mu-1}(\alpha)^{M_{\mu-1}},$$

in welcher die Exponenten der Einheiten durch die Congruenz

$$\frac{(-1)^{n+1}(\gamma^{2n}-1)B_n}{4n} M_n \equiv \frac{d_0^{2n} lF(e^v)}{dv^{2n}}, \quad \text{Mod. } \lambda$$

bestimmt sind, ist also eine solche, welche der vorgelegten Aufgabe genügt.

Ich füge noch die Bemerkung hinzu, dafs in der logarithmischen Entwicklung einer beliebigen complexen Zahl

$$-l\left(\frac{F(\alpha)}{F(1)}\right) \equiv \frac{d_0 lF(e^v)}{dv} X_1(\alpha) + \frac{d_0^2 lF(e^v)}{dv^2} X_2(\alpha) + \cdots + \frac{d_0^{\lambda-2} lF(e^v)}{dv^{\lambda-2}} X_{\lambda-2}(\alpha), \quad \text{Mod. } \lambda$$

die Glieder von ungeraden Stellenzeigern von den complexen Einheiten in der complexen Zahl $F(\alpha)$ ganz unabhängig sind, d. h., dafs sie ungeändert bleiben, wenn man $F(\alpha)$ in $F(\alpha).\varepsilon(\alpha)$ verwandelt, wo $\varepsilon(\alpha)$ eine beliebige Einheit bezeichnet. Nur das erste Glied ist von der einfachen Einheit α^k abhängig, mit welcher $F(\alpha)$ multiplicirt sein kann; es ist congruent Null, wenn $F(\alpha)$ der Bedingung genügt, dafs $F(\alpha)-F(1)$ durch $(1-\alpha)^2$ theilbar ist:

§. 5.

Lösung einer Aufgabe, betreffend die allgemeinen Reciprocitätsgesetze.

In den Monatsberichten der Königlichen Akademie der Wissenschaften zu Berlin vom Mai 1851 habe ich zuerst das allgemeine Reciprocitätsgesetz für complexe Primzahlen veröffentlicht, welches ich bereits einige Jahre früher gefunden und in einem Schreiben vom 20ten Januar 1848 Herrn *Lejeune-Dirichlet* und durch diesen *Jacobi* mitgetheilt hatte. Dieses Gesetz, welches ich durch gewisse, so zu sagen metaphysische Schlüsse fand und nachher durch ziemlich umfangreich berechnete Tabellen verificirte, stützt sich auf die am Schlusse des vorhergehenden Paragraphen gelösete Aufgabe: eine *complexe* Zahl durch Multiplication mit passenden Einheiten auf eine solche Form zu bringen, dafs sie, mit ihrer reciproken multiplicirt, ein Product giebt, welches, für den Mod. λ, einer *nichtcomplexen* ganzen Zahl congruent ist. Wenn eine complexe Zahl durch Multiplication mit Einheiten in dieser Art zubereitet ist, * und sie aufserdem auch der Bedingung genügt, dafs sie für den Mod. $(1-\alpha)^\lambda$ einer nichtcomplexen ganzen Zahl congruent ist, so nenne ich die complexe Zahl eine *primäre,* oder eine complexe Zahl in der *primären Form.* Die primäre Form der complexen Zahl $\varphi(\alpha)$ wird also durch folgende zwei Congruenzen definirt:

$$\varphi(\alpha)\varphi(\alpha^{-1}) \equiv \varphi(1)^2, \text{ Mod. } \lambda,$$
$$\varphi(\alpha) \equiv \varphi(1), \text{ Mod. } (1-\alpha)^2,$$

und es kann jede gegebene (nicht durch $1-\alpha$ theilbare) complexe Zahl durch Multiplication mit passenden Einheiten auf diese primäre Form gebracht werden, wenn λ nicht eine von den Ausnahmezahlen ist, welche in einer der ersten $\frac{1}{2}(\lambda-3)$ *Bernoulli*schen Zahlen als Factoren enthalten sind. Das von mir gefundene allgemeine Reciprocitätsgesetz, auf diese Definition der primären Form der complexen Zahlen sich stützend, lautet nun einfach folgendermaafsen:

Wenn die beiden complexen Primzahlen $f_1(\alpha)$ und $\varphi_1(\alpha)$ in der primären Form genommen werden, oder, falls sie ideal sind, wenn die zu *wirklichen* complexen Zahlen werdenden Potenzen derselben in der primären Form genommen werden: so ist

$$\left(\frac{f_1(\alpha)}{\varphi_1(\alpha)}\right) = \left(\frac{\varphi_1(\alpha)}{f_1(\alpha)}\right).$$

Sowohl wegen der am Anfange dieser Abhandlung gegebenen Definition des Symbols $\left(\frac{f_1(\alpha)}{\varphi_1(\alpha)}\right)$, als auch wegen der Forderung, dafs $f_1(\alpha)$ und $\varphi_1(\alpha)$ complexe Zahlen von der primären Form sein sollen, erstreckt sich dieses Gesetz nicht auf die Ausnahmefälle, wo λ ein Factor einer der ersten $\frac{1}{2}(\lambda-3)$ *Bernoulli*schen Zahlen ist.

Einen strengen und allgemeinen Beweis des Gesetzes habe ich bisher noch nicht gefunden, da die neuen Hülfsmittel, welche ich zu diesem Zwecke aufgesucht habe und welche ich ein anderesmal zu veröffentlichen gedenke, nur für den Beweis einiger besondern, bisher noch nirgend bewiesenen Fälle ausgereicht haben, (z. B. für den Fall, dafs $f_1(\alpha)$ und $\varphi_1(\alpha)$ conjugirte complexe Zahlen sind, oder $\varphi_1(\alpha)=f_1(\alpha^k)$ ist). Wenn gleich daher die allgemeine Gültigkeit dieses einfachen Reciprocitätsgesetzes noch problematisch ist, bis ein strenger Beweis davon gefunden sein wird, so ist doch so viel klar, dafs die von mir definirte *primäre* Form der complexen Zahlen, für welche wenigstens in vielen Fällen die Reciprocitätsgesetze sich am einfachsten darstellen, für dieselben eine besondere Bedeutung hat. Es wird deshalb nicht uninteressant sein, eine Vergleichung des Reciprocitätsgesetzes für die *primären complexen* Primzahlen mit dem Reciprocitätsgesetze für diejenigen *complexen* Primzahlen aufzustellen, welche den Bedingungen primärer Zahlen nicht unterworfen sind. Um sogleich vollkommen präcis auszusprechen, um was es sich hier handeln soll, fasse ich es in die folgende

Aufgabe. Wenn das Reciprocitätsgesetz für zwei complexe Primzahlen, für die primäre Form derselben als gegeben angenommen wird: daraus das Reciprocitätsgesetz für die complexen Primzahlen abzuleiten, welche der Bedingung, primär zu sein, nicht unterliegen.

Es seien $f(\alpha)$ und $\varphi(\alpha)$ zwei beliebige complexe Primzahlen, von welchen ich der Einfachheit wegen nur das Eine voraussetze, dafs sie durch Multiplication mit einer passenden Potenz der einfachen Einheit α so zubebereitet sind, dafs sie den Bedingungen

$$f(\alpha) \equiv f(1) \quad \text{und} \quad \varphi(\alpha) \equiv \varphi(1), \quad \text{Mod.}(1-\alpha)^2$$

genügen. Es seien ferner $f_1(\alpha)$ und $\varphi_1(\alpha)$ dieselben complexen Primzahlen in der *primären* Form, also gereinigt von den in den Zahlen $f(\alpha)$ und $\varphi(\alpha)$

beliebig vorkommenden complexen Einheiten, so hat man, wie im vorigen Paragraphen gezeigt worden:

$$f(\alpha)\,E_1(\alpha)^{M_1}.\,E_2(\alpha)^{M_2}\ldots E_{\mu-1}(\alpha)^{M_{\mu-1}} = f_1(\alpha),$$

wo die Exponenten M_1, M_2, ... $M_{\mu-1}$ durch die Congruenz

$$\frac{(-1)^{n+1}(\gamma^{2n}-1)B_n}{4n}M_n \equiv \frac{d_0^{2n}lf(e^v)}{dv^{2n}},\ \text{Mod.}\ \lambda$$

bestimmt sind. Eben so hat man auch

$$\varphi(\alpha)\,E_1(\alpha)^{N_1}.\,E_2(\alpha)^{N_2}\ldots E_{\mu-1}(\alpha)^{N_{\mu-1}} = \varphi_1(\alpha),$$

wo die Exponenten N_1, N_2, ... $N_{\mu-1}$ durch die Congruenz

$$\frac{(-1)^{n+1}(\gamma^{2n}-1)B_n}{4n}N_n \equiv \frac{d_0^{2n}l\varphi(e^v)}{dv^{2n}},\ \text{Mod.}\ \lambda$$

bestimmt werden. Es soll nun das Zeichen des Index Ind. auf den Mod. $f(\alpha)$, oder, was Dasselbe ist $f_1(\alpha)$, und das Zeichen ind. auf den Mod. $\varphi(\alpha)$ oder $\varphi_1(\alpha)$ sich beziehen, so dafs, wenn $F(\alpha)$ eine beliebige complexe Zahl bezeichnet:

$$F(\alpha)^{\frac{Nf(\alpha)-1}{\lambda}} \equiv \alpha^{\text{Ind.}F(\alpha)},\ \text{Mod.}\, f(\alpha),$$

$$F(\alpha)^{\frac{N\varphi(\alpha)-1}{\lambda}} \equiv \alpha^{\text{ind.}F(\alpha)},\ \text{Mod.}\ \varphi(\alpha)\ \text{ist.}$$

Nimmt man nun auf beiden Seiten der Gleichung, welche $f_1(\alpha)$ durch $f(\alpha)$ ausdrückt, die Indices in Beziehung auf den Modul $\varphi(\alpha)$, so erhält man

$$\text{ind.}f(\alpha)+M_1\,\text{ind.}\,E_1(\alpha)+M_2\,\text{ind.}\,E_2(\alpha)+\cdots+M_{\mu-1}\text{ind.}\,E_{\mu-1}(\alpha)\equiv\text{ind.}f_1(\alpha)$$

für den Mod. λ. Eben so, wenn man auf beiden Seiten der Gleichung, welche $\varphi_1(\alpha)$ durch $\varphi(\alpha)$ ausdrückt, die Indices in Beziehung auf den Mod. $f(\alpha)$ nimmt:

$$\text{Ind.}\varphi(\alpha)+N_1\,\text{Ind.}\,E_1(\alpha)+N_2\,\text{Ind.}\,E_2(\alpha)+\cdots+N_{\mu-1}\text{Ind.}\,E_{\mu-1}(\alpha)\equiv\text{Ind.}\varphi_1(\alpha).$$

Ich wende jetzt die in dem ersten und dritten Paragraphen der gegenwärtigen Abhandlung gefundenen Ausdrücke der Indices der Einheiten $E_n(\alpha)$ an; und zwar diejenigen, welche nur die complexe Primzahl selbst enthalten, auf die der Index sich bezieht, welche auch in allen Fällen, es mag der Modul eine zum Exponenten Eins oder zu einem andern Exponenten gehörende complexe Primzahl sein, dieselben sind, nämlich:

$$\left.\begin{aligned}\text{Ind.}\,E_n(\alpha) &\equiv \frac{(-1)^n(\gamma^{2n}-1)B_n}{4n}\cdot\frac{d_0^{\lambda-2n}lf(e^v)}{dv^{\lambda-2n}},\\ \text{ind.}\,E_n(\alpha) &\equiv \frac{(-1)^n(\gamma^{2n}-1)B_n}{4n}\cdot\frac{d_0^{\lambda-2n}l\varphi(e^v)}{dv^{\lambda-2n}},\end{aligned}\right\}\ldots\ \text{Mod.}\ \lambda.$$

Diese Ausdrücke, verbunden mit den Congruenzen, welche M_n und N_n be-

19*

stimmen, geben

$$\left.\begin{aligned} M_n \operatorname{ind.} E_n(\alpha) &\equiv -\frac{d_0^{2n} lf(e^v)}{dv^{2n}} \cdot \frac{d_0^{\lambda-2n} l\varphi(e^v)}{dv^{\lambda-2n}}, \\ N_n \operatorname{Ind.} E_n(\alpha) &\equiv -\frac{d_0^{2n} l\varphi(e^v)}{dv^{2n}} \cdot \frac{d_0^{\lambda-2n} lf(e^v)}{dv^{\lambda-2n}}, \end{aligned}\right\} \ldots \text{Mod. } \lambda.$$

Vermöge dieser Ausdrücke verwandeln sich die obigen Congruenzen in folgende:

$$\left.\begin{aligned} \operatorname{ind.} f(\alpha) &\equiv \operatorname{ind.} f_1(\alpha) + \sum_{1}^{\mu-1}{}_n \frac{d_0^{2n} lf(e^v)}{dv^{2n}} \cdot \frac{d_0^{\lambda-2n} l\varphi(e^v)}{dv^{\lambda-2n}}, \\ \operatorname{Ind.} \varphi(\alpha) &\equiv \operatorname{Ind.} \varphi_1(\alpha) + \sum_{1}^{\mu-1}{}_n \frac{d_0^{2n} l\varphi(e^v)}{dv^{2n}} \cdot \frac{d_0^{\lambda-2n} lf(e^v)}{dv^{\lambda-2n}}, \end{aligned}\right\} \ldots \text{Mod. } \lambda.$$

Nimmt man die Glieder der in der zweiten dieser Congruenzen vorkommenden Summe in umgekehrter Ordnung, so läfst sich dieselbe auch folgendermaafsen darstellen:

$$\operatorname{Ind.} \varphi(\alpha) \equiv \operatorname{Ind.} \varphi_1(\alpha) + \sum_{1}^{\mu-1}{}_n \frac{d_0^{2n+1} lf(e^v)}{dv^{2n+1}} \cdot \frac{d_0^{\lambda-2n-1} l\varphi(e^v)}{dv^{\lambda-2n-1}}.$$

Subtrahirt man dieselbe von der ersten, so erhält man

$$\operatorname{ind.} f(\alpha) - \operatorname{Ind.} \varphi(\alpha) \equiv \operatorname{ind.} f_1(\alpha) - \operatorname{Ind.} \varphi_1(\alpha) + \Sigma,$$

wo der einfache Buchstabe Σ als abgekürztes Zeichen für die Reihe

$$\Sigma = \sum_{2}^{\lambda-2}{}_h (-1)^h \frac{d_0^h lf(e^v)}{dv^h} \cdot \frac{d_0^{\lambda-h} l\varphi(e^v)}{dv^{\lambda-1}}$$

gesetzt ist.

Diese Congruenz enthält die vollständige Lösung der gestellten Aufgabe. Wendet man die dem *Legendre*schen Zeichen für die quadratischen Reste analogen Zeichen für die höheren Potenzreste an, welche mit den Zeichen ind. und Ind. so zusammenhangen, dafs

$$\left(\frac{F(\alpha)}{\varphi(\alpha)}\right) = \alpha^{\operatorname{ind.} F(\alpha)}, \quad \left(\frac{F(\alpha)}{f(\alpha)}\right) = \alpha^{\operatorname{Ind.} F(\alpha)},$$

so nimmt das gefundene Resultat folgende Form an:

$$\frac{\left(\frac{f(\alpha)}{\varphi(\alpha)}\right)}{\left(\frac{\varphi(\alpha)}{f(\alpha)}\right)} = \frac{\left(\frac{f_1(\alpha)}{\varphi_1(\alpha)}\right)}{\left(\frac{\varphi_1(\alpha)}{f_1(\alpha)}\right)} \cdot \alpha^{\Sigma}.$$

Der Ausdruck des Σ, welcher entwickelt folgendermaafsen geschrieben wird:

$$\Sigma = \frac{d_0^2 lf(e^v)}{dv^2} \cdot \frac{d_0^{\lambda-2} l\varphi(e^v)}{dv^{\lambda-2}} - \frac{d_0^3 lf(e^v)}{dv^3} \cdot \frac{d_0^{\lambda-3} l\varphi(e^v)}{dv^{\lambda-3}} + \ldots$$

$$\ldots - \frac{d_0^{\lambda-2} lf(e^v)}{dv^{\lambda-2}} \cdot \frac{d_0^2 l\varphi(e^v)}{dv^2},$$

enthält, da die complexen Zahlen $f(\alpha)$ und $\varphi(\alpha)$ als gegeben betrachtet werden, nur gegebene Gröfsen, und wenn nun auch das Reciprocitätsgesetz für die primären complexen Zahlen $f_1(\alpha)$ und $\varphi_1(\alpha)$ als gegeben angenommen wird, d. h., wenn das Verhältnifs $\left(\frac{f_1(\alpha)}{\varphi_1(\alpha)}\right):\left(\frac{\varphi_1(\alpha)}{f_1(\alpha)}\right)$ bekannt ist: so ist vermöge dieser Gleichung auch das Reciprocitäts-Verhältnifs $\left(\frac{f(\alpha)}{\varphi(\alpha)}\right):\left(\frac{\varphi(\alpha)}{f(\alpha)}\right)$ der complexen Primzahlen $f(\alpha)$ und $\varphi(\alpha)$ gegeben.

Angenommen, dafs das von mir gefundene, oben aufgestellte Reciprocitätsgesetz, nach welchem

$$\left(\frac{f_1(\alpha)}{\varphi_1(\alpha)}\right) = \left(\frac{\varphi_1(\alpha)}{f_1(\alpha)}\right)$$

sein soll, richtig sei, so hat man für die nichtprimären complexen Primzahlen $f(\alpha)$ und $\varphi(\alpha)$, welche nur der einen Bedingung genügen müssen, dafs $f(\alpha)-f(1)\equiv 0$ und $\varphi(\alpha)-\varphi(1)\equiv 0$ ist, nach dem Mod. $(1-\alpha)^2$, das Reciprocitätsgesetz:

$$\left(\frac{f(\alpha)}{\varphi(\alpha)}\right) = \left(\frac{\varphi(\alpha)}{f(\alpha)}\right).\alpha^{\Sigma}.$$

Mit dem Reciprocitätsgesetze für die *primären complexen* Primzahlen ist also zugleich auch das Reciprocitätsgesetz für beliebige *nichtprimäre complexe* Primzahlen gegeben.

In der schon oben erwähnten Abhandlung (im 39ten Bande S. 351 dieses Journals) hat Herr *Eisenstein* eine, zwar auf einer unbewiesenen Voraussetzung beruhende, jedoch sehr sinnreiche Methode entwickelt, um diejenige Potenz von α zu finden, welche dem Verhältnisse $\left(\frac{A}{B}\right):\left(\frac{B}{A}\right)$ gleich ist, wenn A und B complexe Zahlen sind. Der Ausdruck dieser mit dem Namen *Umkehrungsfactor* bezeichneten Potenz von α, welchen er daselbst giebt, würde, wie es mir scheint, gehörig entwickelt und vereinfacht, sich auf dieselbe Form bringen lassen, wie der in der hier gegebenen Reciprocitätsgleichung enthaltene Ausdruck des Umkehrungsfactors α^{Σ}, in welchem Σ durch die Differentialquotienten der Logarithmen der complexen Zahlen $f(\alpha)$ und $\varphi(\alpha)$, für $\alpha = e^{v}$, gegeben ist. Wenn wirklich, wie ich vermuthe, dieses Resultat der erwähnten Abhandlung mit dem hier aufgestellten übereinstimmt, so hätte Herr *Eisenstein* aus denselben auch das von mir gefundene einfache Reciprocitätsgesetz für die primären complexen Primzahlen finden können; denn dieses ist eben so eine Folge von jenem, als jenes eine Folge von diesem ist. Dies ist ihm aber nicht gelungen, obgleich er wirklich (S. 361) den Versuch macht, nach seiner Methode ein Reciprocitätsgesetz von der einfachsten Form $\left(\frac{A}{B}\right)=\left(\frac{B}{A}\right)$ durch passende Bestimmung der Einheiten in complexen Zahlen A und B zu erlangen; denn die Bedingungen, welche er dafür aufstellt, lassen sich, wie leicht zu zeigen, im Allgemeinen nicht erfüllen.

Breslau, den 30ten November 1851.

B e r i c h t i g u n g e n.

S. 103 Z. 17 v. u. bis S. 104 Z. 1 v. u. und S. 119 Z. 4 v. u. bis S. 120 Z. 6 v. u. l. ind. st. Ind.

S. 123 Z. 9 v. o. l. D_i st. D^i

* # Über die Irregularität von Determinanten

Monatsberichte der Königlichen Preußischen Akademie der Wissenschaften zu Berlin aus dem Jahre 1853, 194–200

Hr. **Lejeune Dirichlet** theilte folgenden Auszug aus einem von Hrn. **Kummer** in Breslau, Correspondenten der Akademie an ihn gerichteten Briefe mit.

Ich habe neulich im Verlaufe meiner zahlentheoretischen Untersuchungen den Schlüssel zu der sehr mysteriösen Irregularität der Determinanten gefunden, von welcher Gaufs für die quadratischen Formen in den *disqu. arith.* pag. 529 etc. einige Andeutungen giebt, und über welche er sich so ausdrückt: *Hoc argumentum, quod ad arithmeticae sublimioris mysteria maxime recondita pertinere, disquisitionibusque difficillimis locum relinquere videtur, paucis tantum observationibus hic illustrare possumus.* Da ich glaube, dafs dieses Problem auch für Dich von besonderem Interesse sein wird, so will ich Dir darüber eine kurze Mittheilung machen, deren Aufnahme in den Monatsbericht, wenn die Akademie dieselbe dazu für geeignet halten sollte, mir sehr erwünscht sein würde. Ich lege meiner Untersuchung nicht die quadratischen Formen zu Grunde, sondern die Normformen, oder nach meiner Anschauungsweise der Sache die verschiedenen Klassen idealer complexer Zahlen, welche aus λ^{ten} Wurzeln der Einheit gebildet sind. Die Primzahl λ hat hier die Rolle der Determinante, und sie ist eine regelmäfsige Determinante, wenn es eine ideale Zahl giebt, durch deren verschiedene Potenzen alle Klassen idealer Zahlen repräsentirt werden, eine unregelmäfsige aber, wenn es eine solche ideale Zahl nicht giebt. Ich bezeichne eine ideale Zahl $f(\alpha)$, deren d^{te} Potenz $f(\alpha)^d$, aber keine niedere Potenz, eine wirkliche complexe Zahl ist, als eine zum Exponenten d gehörende. Den gröfsten aller Exponenten, zu welchem ideale Zahlen gehören, bezeichne ich mit h, die Anzahl aller Klassen mit H, und nenne nach Gaufs den Quotienten $\frac{H}{h}$, welcher stets eine ganze Zahl ist, den Exponenten der Irregularität.

Es lassen sich nun zunächst aus dem Begriffe der Äquivalenz selbst, auf elementare Weise, folgende Sätze ohne Schwierigkeit beweisen:

1) Wenn es eine ideale Zahl giebt, die zum Exponenten d gehört, und eine andere, die zum Exponenten d' gehört, und wenn t die kleinste durch d und d' zugleich theilbare Zahl ist, so giebt es stets ideale Zahlen, welche zum Exponenten t gehören.

Hieraus folgten unmittelbar folgende Sätze:

2) Der gröfste aller Exponenten h, zu welchem ideale Zahlen gehören, ist ein Vielfaches eines jeden Exponenten, zu welchem ideale Zahlen gehören.
3) Der Exponent der Irregularität $\frac{H}{h}$ enthält keine anderen Primfactoren in sich, als solche, welche auch in h enthalten sind.
4) Nur wenn die Klassenanzahl H irgend welche Primfactoren mehrmals enthält, kann die Determinante λ eine irreguläre sein, und nur die in H mehrfach enthaltenen Primfactoren können Primfactoren des Exponenten der Irregularität sein.

Diese Sätze, welche auch Gaufs für die quadratischen Formen gegeben hat, gelten ganz allgemein für alle Systeme nichtäquivalenter idealer Zahlen, welche man nur bilden kann, auch wenn die complexen Zahlen nicht aus den Wurzeln der Gleichung $\alpha^\lambda = 1$, sondern aus den Wurzeln irgend einer anderen algebraischen Gleichung gebildet werden. Um nun aber für die aus λ^{ten} Wurzeln der Einheit gebildeten complexen Zahlen diesen Gegenstand tiefer zu ergründen, mache ich von dem gefundenen Ausdrucke der Klassenanzahl für dieselben Gebrauch, nämlich

$$H = \frac{P}{(2\lambda)^{\mu-1}} \cdot \frac{D}{\Delta},$$

wo $\mu = \frac{\lambda-1}{2}$, $P = \phi(\beta)\phi(\beta^3)\phi(\beta^5) \ldots . \phi(\beta^{\lambda-2})$,

$$\phi(\beta) = 1 + \gamma_1 \beta + \gamma_2 \beta^2 + \gamma_3 \beta^3 + \ldots . + \gamma_{\lambda-2} \beta^{\lambda-2};$$

wo β eine primitive Wurzel der Gleichung $\beta^{\lambda-1} = 1$, γ eine primitive Wurzel der Congruenz $\gamma^{\lambda-1} \equiv 1$, Mod. λ, und γ_n der kleinste positive Rest ist, welchen γ^n giebt, für den Modul λ, ferner D die Determinante des Systems der Logarithmen der Kreistheilungs-Einheiten, Δ die entsprechende Determinante für das System der Fundamental-Einheiten. Die beiden Factoren, aus welchen die Klassenzahl H besteht, sind, wie ich schon früher bemerkt habe, für sich ganze Zahlen, und haben auch einzeln jeder seine besondere Bedeutung als Klassenzahlen. Es ist nämlich der zweite Factor $\frac{D}{\Delta}$ für sich gleich der Anzahl der Klassen, der aus den zweigliedrigen Perioden $\alpha + \alpha^{-1}$, $\alpha^2 - \alpha^{-2}$ etc. gebildeten idealen Zahlen, oder was dasselbe ist, derjenigen, deren reciproke (durch Verwandlung des α in α^{-1} entstehende) stets in derselben Klasse enthalten sind. Es ist ferner der erste Factor $\frac{P}{(2\lambda)^{\mu-1}}$ für sich gleich der Anzahl derjenigen Klassen der idealen Zahlen, welche, mit ihren reciproken idealen Zahlen multiplicirt, wirkliche complexe Zahlen geben. Auf diese Klassenzahl, welche der erste Factor $\frac{P}{(2\lambda)^{\mu-1}}$ repräsentirt, will ich hier meine Untersuchung beschränken.

Sei p eine Primzahl von der Form $\nu\lambda + 1$, x eine primitive Wurzel der Gleichung $x^p = 1$, $f(\alpha)$ ein idealer Primfactor des p, g eine primitive Wurzel des p und

$$F(\alpha, x) = x + \alpha x^g + \alpha^2 x^{g^2} + \ldots . + \alpha^{p-2} x^{g^{p-2}},$$

so ist, wie ich früher bewiesen habe,

$$F(\alpha^{-1}, x)^\lambda = \pm f(\alpha) . f(\alpha^\gamma)^{\gamma-1} . f(\alpha^{\gamma^2})^{\gamma-2} \ldots . . f(\alpha^{\gamma^{\lambda-2}})^{\gamma_1};$$

also wenn die Logarithmen genommen werden, und α in α^{γ^n} verwandelt wird, so hat man, bei Anwendung des Summenzeichens Σ,

$$\lambda l F(\alpha^{-\gamma^\varkappa}, x) = \sum_0^{\lambda-2}{}_h\, \gamma_{-h}\, l f(\alpha^{\gamma^{\varkappa+h}}).$$

Multiplicirt man diese Gleichung mit $\beta^{(2m+1)\varkappa}$ und nimmt die Summe für $\varkappa = 0, 1, 2, \ldots \lambda-2$, so hat man

$$\lambda \sum_0^{\lambda-2}{}_\varkappa\, \beta^{(2m+1)\varkappa}\, l F(\alpha^{-\gamma^\varkappa}, x) = \sum_0^{\lambda-2}{}_\varkappa \sum_0^{\lambda-2}{}_h\, \gamma_{-h}\, \beta^{(2m+1)\varkappa}\, l f(\alpha^{\gamma^{\varkappa+h}}).$$

Nimmt man auf der rechten Seite $\varkappa-h$ statt $\varkappa$, so zerfällt diese Doppelsumme in das Produkt zweier einfacher Summen, und man erhält

$$\lambda \sum_0^{\lambda-2}{}_\varkappa\, \beta^{(2m+1)\varkappa}\, l F(\alpha^{-\gamma^\varkappa}, x) = \sum_0^{\lambda-2}{}_h\, \gamma_{-h}\, \beta^{-(2m+1)h} \,.\, \sum_0^{\lambda-2}{}_\varkappa\, \beta^{(2m+1)\varkappa}\, l f(\alpha^{\gamma^\varkappa}).$$

Es ist nun aber $\sum_0^{\lambda-2}{}_h\, \gamma_{-h}\, \beta^{-(2m+1)h}$ genau dieselbe Gröſse, welche in dem Ausdrucke des $\frac{P}{(2\lambda)^{\mu-1}}$ vorkommt, und oben durch $\varphi(\beta^{2m+1})$ bezeichnet worden ist, wird daher dieselbe Bezeichnung hier eingeführt, und durch $\varphi(\beta^{2m+1})$ dividirt, so wird

$$\frac{\lambda \sum_0^{\lambda-2}{}_\varkappa\, \beta^{(2m+1)\varkappa}\, l F(\alpha^{-\gamma^\varkappa}, x)}{\varphi(\beta^{2m+1})} = \sum_0^{\lambda-2}{}_\varkappa\, \beta^{(2m+1)\varkappa}\, l f(\alpha^{\gamma^\varkappa}).$$

Nimmt man endlich auf beiden Seiten die Summe in Beziehung auf $m = 0, 1, 2, \ldots \mu-1$, so erhält man

$$\sum_0^{\mu-1}{}_m \frac{\lambda}{\varphi(\beta^{2m+1})} \,.\, \sum_0^{\lambda+2}{}_\varkappa\, \beta^{(2m+1)\varkappa}\, l F(\alpha^{-\gamma^\varkappa}, x) = \mu\, l\left(\frac{f(\alpha)}{f(\alpha^{-1})}\right).$$

Ich mache jetzt Gebrauch von der kleinsten-nichtcomplexen ganzen Zahl, in welcher die complexen Zahlen $\varphi(\beta)$, $\varphi(\beta^3)$, $\varphi(\beta^5) \ldots \varphi(\beta^{\lambda-2})$ alle ohne Rest aufgehen; dieselbe ist, wie leicht zu erkennen, stets ein Vielfaches von 2 und auch von λ, und soll darum durch $2\lambda Q$ bezeichnet werden. Vermöge der Eigenschaft der Zahl $2\lambda Q$, daſs sie durch jede complexe Zahl von der Form $\varphi(\beta^{2m+1})$ ohne Rest theilbar ist, lassen sich alle Brüche von der Form $\frac{\lambda}{\varphi(\beta^{2m+1})}$ in Brüche verwandeln, deren Nenner gleich $2Q$ ist, und deren Zähler ganze complexe (β enthaltende) Zahlen sind. Denkt man sich, nach dem allen auf der linken Seite der obigen Gleichung vorkommenden Brüchen der gemeinschaftliche Nenner $2Q$ gegeben ist, die Summation in Beziehung auf m ausgeführt, so verschwindet β nothwendig aus dieser Gleichung, und man erhält eine Gleichung von der Form

$$\frac{\sum_{0}^{\lambda-2}{}_{\varkappa} A_{\varkappa}\, lF(\alpha^{-\gamma^{\varkappa}}, x)}{2Q} = \mu l\left(\frac{f(\alpha)}{f(\alpha^{-1})}\right).$$

in welcher die Coëfficienten $A_{\varkappa}$ ganze Zahlen sind. Multiplicirt man nun mit $2Q$, und geht von den Logarithmen zu den Zahlen zurück, so erhält man

$$\Pi_{\varkappa} F(\alpha^{-\gamma^{\varkappa}})^{A_{\varkappa}} = \left(\frac{f(\alpha)}{f(\alpha^{-1})}\right)^{(\lambda-1)Q}.$$

Aus dem Producte auf der linken Seite dieser Gleichung verschwindet die Wurzel x nothwendig von selbst, dasselbe ist eine wirkliche complexe Zahl, welche nur α enthält. Andererseits ist $\frac{f(\alpha)}{f(\alpha^{-1})}$ eine ideale complexe Zahl, welche mit ihrer reciproken multiplicirt wirklich wird, und sie ist auch der Repräsentant aller derartigen idealen complexen Zahlen, obgleich sie sich nur auf ideale Primfactoren von der Form $p = \nu\lambda + 1$ bezieht, denn nach einem von mir in Crelle's Journal Bd. 35. pag. 357 bewiesenen Satze bewirken die idealen Primfactoren der anderen lineären Formen angehörenden Primzahlen keine besonderen Klassen idealer Zahlen. Also:

Die $(\lambda-1)Q^{\text{te}}$ Potenz jeder idealen Zahl der Art, welche hier in Rede steht, ist eine wirkliche complexe Zahl.

Der Factor $\lambda-1$ in diesem Exponenten fällt von selbst überall da weg, wo die Klassenzahl $\frac{P}{(2\lambda)^{\mu-1}}$ zu $\lambda-1$ relative Primzahl ist, in jedem Falle aber kann man statt desselben nur den gröſsten gemeinschaftlichen Factor der Klassenzahl mit $\lambda-1$ nehmen. Nennt man diesen c, so ist, wie man hieraus erkennt, jede Determinante λ eine irreguläre, für welche cQ kleiner ist als die Klassenzahl. Die Anzahl aller Klassen idealer Zahlen ist, abgesehen von dem Divisor $(2\lambda)^{\mu-1}$ gleich dem Producte aller complexen Zahlen $\phi(\beta), \phi(\beta^3) \ldots . \phi(\beta^{\lambda-2})$;

derjenige Exponent aber, welcher hinreicht, um alle idealen Zahlen zu wirklichen zu machen, enthält aufser dem c nur diejenige ganze Zahl, in welcher dieselben complexen Zahlen alle ohne Rest aufgehen, und weil im allgemeinen die kleinste Zahl, in welcher eine Anzahl gegebener Zahlen ohne Rest theilbar sind, kleiner ist, als das Produkt aller dieser Zahlen, so sieht man hieraus, dafs die Irregularität der Determinanten, für die hier in Rede stehenden complexen Zahlen, die Regel bilden mufs, und dafs dagegen die regulären Determinanten nur die Ausnahmen bilden. Wenn dies auch für eine gewisse Anzahl kleiner Determinanten, welche als Beispiele berechnet werden können, sich nicht zu bestätigen scheint, so folgt doch aus dem, was ich hier bewiesen habe, unwiderleglich, dafs die Irregularität für gröfsere Determinanten immer häufiger werden mufs, eine Bemerkung, welche auch Gaufs für die Determinanten quadratischer Formen von negativer Determinante an seinen sehr weit fortgesetzten Tafeln berechneter Klassenanzahlen auf dem Wege der Induction gemacht hat.

Für die Primzahlen λ innerhalb des ersten Hundert habe ich nicht ohne grofse Mühe die Klassenzahlen berechnet und in Liouvilles Journal veröffentlicht. Unter diesen sind nur die Determinanten 29, 31, 41 und 71, welche irregulär sein können, da ihre Klassenanzahlen mehrfache Factoren enthalten. Diese habe ich nach der hier gegebenen Methode geprüft und habe nach derselben streng bewiesen, dafs die Determinante 29, deren Klassenzahl gleich 8 ist, eine irreguläre Determinante ist, mit dem Exponenten der Irregularität 4, so dafs schon das Quadrat jeder aus 29ten Wurzeln der Einheit gebildeten idealen complexen Zahl zu einer wirklichen wird. Die Determinante 31 dagegen, deren Klassenanzahl gleich 9 ist, ist eine reguläre, oder es giebt immer für dieselbe ideale Zahlen, deren neunte Potenz, aber keine niedere wirklich ist. Die Determinante 41, deren Klassenzahl gleich 121, ist eine irreguläre, und der Exponent der Irregularität ist gleich 11. Endlich die Determinante 71, für welche die Klassenzahl gleich 7.7.79241 ist, (nicht 7.7.29.3851, wie wegen eines Versehens in Liouvilles Journal sich findet) ist ohnerachtet des in dieser zweimal enthaltenen Factors 7 eine reguläre.

Note sur une expression analogue à la résolvante de Lagrange pour l'équation $z^p = 1$

Atti dell' Accademia Pontificia de Nuovi Lincei VI, 237-241 (1852-1853)

Soit proposée l'expression

$$(\beta, z) = z + \beta z^g + \beta^2 z^{g^2} + \beta^3 z^{g^3} + \ldots + \beta^{p^{a-1}(p-1)-1} z^{g^{p^{a-1}(p-1)-1}},$$

où z désigne une racine primitive de l'équation $z^{p^a} = 1$, p un nombre premier impair, g une racine primitive de la congruence $g^{p^{a-1}(p-1)} \equiv 1$, mod. p^a, et β une racine quelconque de l'équation $\beta^{p^{a-1}(p-1)} = 1$. La racine β peut être composée d'une racine quelconque ω de l'équation $\omega^{p-1} = 1$, et d'une *pième* puissance de la racine z; on pourra donc poser: $\beta = \omega z^{rp}$, r désignant un nombre entier quelconque. Ainsi l'expression proposée peut être mise sous cette forme

$$(\beta, z) = (\omega z^{rp}, z) = \Sigma\, \omega^h z^{rph + g^h},$$

où la somme indiquée par le signe Σ, s'étend sur les valeurs de $h = 0$, 1, 2, . . . , $p^{a-1}(p-1) - 1$. Le cas $a = 1$, dans lequel l'expression proposée donne la résolvante même de Lagrange, étant exclu de cette discussion, posons $h + p^{a-2}(p-1)i$ au lieu de h, et changeons le simple somme en une somme double; nous aurons

$$(\omega z^{rp}, z) = \Sigma\Sigma \omega^h z^{rph + rp^{a-1}(p-1)i + g^{h+p^{a-2}(p-1)i}},$$

pour $h = 0, 1, 2, \ldots\ldots, p^{a-2}(p-1) - 1$, $i = 0, 1, 2, \ldots\ldots, p-1$. Pour effectuer la sommation par rapport à i, on déterminera le nombre e' par l'équation

$$g^{p-1} = 1 + e'p,$$

d'où, en élevant à la puissance de l'exposant $p^{a-2}i$, et rejetant les multiples de p^a, on aura:

$$g^{p^{a-2}(p-1)i} \equiv 1 + e'p^{a-1}i, \ mod. \ p^a,$$

ce qui, substitué dans la double somme, donne

(*) Comunicata dal Segretario, nella sessione del 3 giugno 1855.

$$(\omega z^{rp}, z) = \Sigma\Sigma\, \omega^h z^{rph + g^h + (e'g^h - r)p^{a-1} i}.$$

Maintenant on voit, que la somme par rapport à i, est toujours égale à zéro, excepté le cas, que $e'g^h - r \equiv 0$, mod. p, dans lequel cette somme devient égale à p. On en conclut d'abord:

Si r est un multiple de p, l'expression $(\omega z^{rp}, z)$ est égale à zero; car alors la condition $e'g^h \equiv r$, mod. p, ne peut être satisfaite pour aucune valeur de h puisque, g étant racine primitive pour le module p^a, e' n'est pas divisible par p.

Soit donc r un nombre premier à p, soit de plus ρ un nombre satisfaisant à la congruence $e'g^\rho \equiv r$, mod. p, toutes les valeurs de h, pour lesquelles la somme, prise par rapport à i, n'est pas égale à zéro, mais égale à p, seront comprises dans la formule $\rho + (p-1)k$, et de là cette double somme se réduit à

$$(\omega z^{rp}, z) = p\omega^\rho \Sigma\, z^{rp\rho + rp(p-1)k + g^{\rho + (p-1)k}},$$

pour $k = 0, 1, 2, \ldots, p^{a-2} - 1$.

Si on fait $\omega = 1$, on aura comme cas spécial

$$(z^{rp}, z) = p\Sigma\, z^{rp\rho + rp(p-1)k + g^{\rho + (p-1)k}};$$

et, en substituant cette somme dans l'expression générale,

$$(\omega z^{rp}, z) = \omega^\rho (z^{rp}, z),$$

ce qui montre, *que la racine ω n'entre dans l'expression $(\omega z^{rp}, z)$, que comme simple facteur ω^ρ, et qu'elle n'y entre nullement si $\rho \equiv 0$, mod. $p-1$, c'est à dire si $r \equiv e'$, mod. p.*

L'expression trouvée de $(\omega z^{rp}, z)$ est encore susceptible de simplifications ultérieurs, qu'on obtient en mettant $k + p^{a-3} i$ au lieu de k, ce qui donne, en supposant $a > 3$.

$$(\omega z^{rp}, z) = p\omega^\rho \Sigma\Sigma z^{rp\rho + rp(p-1)k + rp^{a-2}(p-1)i + g^{\rho + (p-1)k + p^{a-3}(p-1)i}},$$

pour $k = 0, 1, 2, \ldots, p^{a-3} - 1$, $i = 0, 1, 2, \ldots, p-1$. Ayant déjà posé $g^{p-1} = 1 + e'p$, on a, en élevant à la puissance de l'exposant $p^{a-3} i$, et rejetant les multiples de p^a,

$$g^{p^{a-3}(p-1)i} \equiv 1 + e'p^{a-2}i - \frac{e'^2p^{a-1}i}{2} \text{ mod. } p^a,$$

ce qui, substitué dans la somme précédente, donne

$$(\omega z'^p, z) = p\omega^\rho \Sigma\Sigma\, z^{rp\rho+rp(p-1)k+rp^{a-2}(p-1)i+g^{\rho+(p-1)k}\left(1+e'p^{a-2}i-\frac{e'^2p^{a-1}i}{2}\right)}.$$

Ici la somme par rapport aux valeurs de $i = 0, 1, 2, \ldots, p-1$ est aussi égale à zéro, à l'exception des cas où l'on a

$$r(p-1) + g^{\rho+(p-1)k}\left(e' - \frac{e'^2p}{2}\right) \equiv 0 \text{ , mod. } p^2,$$

dans lesquels cette somme est égale à p. Soit donc ρ' une valeur de $\rho + (p-1)k$, satisfaisante à la condition

$$r(p-1) + g^{\rho'}\left(e' - \frac{e'^2p}{2}\right) \equiv 0, \text{ mod. } p^2,$$

toutes les autres valeurs convenables de $\rho + (p-1)k$, seront comprises dans la forme $\rho' + p(p-1)h$, donc l'expression donnée se réduit à

$$(\omega z^{rp}, z) = p^2\omega^\rho \Sigma z^{rp\rho' + rp^2(p-1)h + g^{\rho' \pm p(p-1)h}},$$

pour $h = 0, 1, 2, \ldots, p^{a-4} - 1$.

Si a est plus grand que 5, on changera de nouveau cette simple somme en une somme double, en prenant $h + p^{a-5}i$ au lieu de h, où les sommes sont relatives aux valeurs de $h = 0, 1, 2, \ldots, p^{a-5}-1$, $i = 0, 1, 2, \ldots, p-1$; alors effectuant la sommation par rapport à i, on trouvera

$$(\omega z^{rp}, z) = p^3\omega^\rho \Sigma\, z^{rp\rho + rp^3(p-1)k + g^{\rho'' + p^2(p-1)k}},$$

pour $k = 0, 1, 2, \ldots, p^{a-6}-1$, où le nombre ρ'' est déterminé par la congruence

$$r(p-1) + g^{\rho''}\left(e' - \frac{e'^2p}{2} + \frac{e'^3p^2}{3}\right) \equiv 0, \text{ mod. } p^3.$$

La même réduction réitérée indéfiniment, donne le resultat plus général

$$(\omega z^{rp}, z) = p^n\omega^\rho \Sigma z^{rp\rho + p^n(p-1)k + g^{\rho + p^{n-1}(p-1)^k}},$$

où le nombre n est supposé $\leqq \frac{a}{2}$, le nombre ρ doit satisfaire à la con-

31

dition

$$r(p-1)+g^{\rho}\left(e'-\frac{e'^2p}{2}+\frac{e'^3p^2}{3}-\dots\frac{(-1)^{n-1}e'^np^{n-1}}{n}\right.,\quad \text{mod.}\, p^n,$$

et la somme s'étend aux valeurs de $k=0, 1, 2, \dots, p^{a-2n}-1$.

Il faut maintenant distinguer les deux cas où a est pair, ou impair. Soit donc 1° a un nombre pair, on prendra $n=\frac{a}{2}$, alors, la somme par rapport à k, n'ayant qu'un seul terme, provenant de la valeur $k=0$, on aura

$$(\omega z^{rp}, z)=p^{\frac{a}{2}}\omega^{\rho}z^{rp\rho+g^{\rho}},$$

où le nombre ρ est déterminé par la congruence

$$r(p-1)+g^{\rho}\left(e'-\frac{e'^2p}{2}+\frac{e'^3p^2}{3}-\dots-\frac{(-1)^{\frac{a}{2}}e'^{\frac{a}{2}}p^{\frac{a}{2}-1}}{\frac{a}{2}}\right),\quad \text{mod.}\, p^{\frac{a}{2}}.$$

Soit 2° a un nombre impair, on fera $n=\frac{a-1}{2}$, et on aura

$$(\omega z^{rp}, z)=p^{\frac{a-1}{2}}\omega^{\rho}\Sigma z^{rp\rho+rp^{\frac{a-1}{2}}(p-1)k+g^{\rho}+p^{\frac{a-3}{2}}(p-1)k},$$

où la sommation est relative à $k=0, 1, 2, \dots, p-1$. Quoique le nombre ρ ne doive satisfaire qu'à la condition donnée plus haut, pour $n=\frac{a-1}{2}$, il convient ici de le borner un peu de plus, de manière qu'il satisfasse à la congruence

$$r(p-1)+g^{\rho}\left(e'-\frac{e'^2p}{2}+\frac{e'^3p^2}{3}-\dots\frac{(-1)^{\frac{a-1}{2}}e'^{\frac{a+1}{2}}p^{\frac{a-1}{2}}}{\frac{a+1}{2}}\right),\quad \text{mod.}\, p^{\frac{a+1}{2}}.$$

Alors si l' on développe la puissance $g^{p^{\frac{a-3}{2}}(p-1)k}$, et qu'on rejette les multiples de p^a, on aura

$$g^{p^{\frac{a+3}{2}}(p-1)k}\equiv 1+\frac{e'p^{a-1}k^2}{2}+p^{\frac{a-1}{2}}k\left(e'-\frac{e'^2p}{2}+\frac{e'^3p^2}{3}-\dots\frac{(-1)^{\frac{a-1}{2}}e'^{\frac{a+1}{2}}p^{\frac{a-1}{2}}}{\frac{a+1}{2}}\right),$$

pour le module p^a, multipliant par g^ρ, et faisant usage de la condition posée pour g^ρ, on en déduit

$$g^{\rho+p^{\frac{a-3}{2}}(p-1)k} \equiv g^\rho + \frac{rp^{a-1}k^2}{2} + rp^{\frac{a-1}{2}}(p-1)k\,, \quad \text{mod.}\, p^a,$$

ce qui, substitué dans l'expression trouvée de $(\omega z^{rp}, z)$, donne

$$(\omega z^{rp}, z) = p^{\frac{a-1}{2}} \omega^\rho\, z^{rp\rho+g^\rho} \Sigma z^{\frac{rp^{a-1}k^2}{2}}.$$

La somme contenue dans cette formule est généralement connue; pour la représenter sous une forme plus simple, et plus facile à reconnaître, on n'a qu'à poser $z^{\frac{rp^{a-1}}{2}} = x$, où x est une racine imaginaire de l'equation $x^p = 1$, d'où elle devient

$$\Sigma z^{\frac{rp^{a-1}k^2}{2}} = 1 + x^1 + x^4 + x^9 + \ldots . + x^{(p-1)^2} = \pm (-1)^{\frac{p-1}{4}} . \sqrt{p}.$$

La substitution de cette valeur au lieu de la somme, donne

$$(\omega z^{rp}, z) = \pm (-1)^{\frac{p-1}{4}} . p^{\frac{a}{2}} \omega^\rho z^{rp^\rho+g^\rho}.$$

Ainsi *l'expression proposée* $(\omega z^{rp}, z)$, *se réduit toujours à la* aième *puissance, de la racine quarrée du nombre premier* p, *multipliée par une certaine puissance, de la racine* ω, *et une puissance de la racine* z, *auxquels, dans le ca où* a *est impair, et* p *un nombre de la forme* 4m + 3, *il faut encore ajouter le facteur* $\pm \sqrt{-1}$.

Observons encore que les résultats trouvés, donnent immédiatement la formule digne de remarque

$$(\omega z^{rp}, z)\ (\omega^{-1} z^{-rp}, z^{-1}) = p^a.$$

Über eine besondere Art, aus complexen Einheiten gebildeter Ausdrücke

Journal für die reine und angewandte Mathematik 50, 212–232 (1855)

I.

In einer Abhandlung im (44ten Bande S. 106) dieses Journals, über die Ergänzungssätze des allgemeinen Reciprocitätsgesetzes für Potenzreste, habe ich eine Verallgemeinerung der Theorie der ***Kreistheilung*** gegeben, welche sich an die bekannten, von ***Jacobi*** durch $\psi(\alpha)$ bezeichneten complexen Zahlen der Kreistheilung anschliefst, und welche zur Lösung des in der genannten Abhandlung behandelten Problems grade ausreicht. Eine wesentlich andere Verallgemeinerung der Theorie der Kreistheilung, welche ich vor längerer Zeit zu dem Zwecke ersann, sie als Mittel zu einem Beweise der allgemeinen Reciprocitätsgesetze zu benutzen, schliefst sich an die bekannte ***Lagrangesche Resolvente der Kreistheilung*** an; und wenn gleich sie den Erwartungen in Betreff der Reciprocitätsgesetze nicht vollständig entsprach, ist sie doch an sich bemerkenswerth, und verbreitet über manche schwierigen Puncte der allgemeinen Theorie der complexen Zahlen ein unerwartetes Licht; weshalb ich dieselbe in dem Folgenden kurz entwickeln will.

Die der ***Lagrangeschen Resolvente der Kreistheilung*** analogen, aus complexen Einheiten gebildeten Ausdrücke, welche hier behandelt werden sollen, sind in folgender Form enthalten:

$$1+\varepsilon(x)+\varepsilon(x)\varepsilon(x^g)+\varepsilon(x)\varepsilon(x^g)\varepsilon(x^{g^2})+\cdots+\varepsilon(x)\,\varepsilon(x^g)\ldots\varepsilon\left(x^{g^{p-3}}\right),$$

in welcher p eine ***Primzahl,*** x eine ***imaginäre*** Wurzel der Gleichung $x^p=1$, g eine ***primitive*** Wurzel der Congruenz $g^{p-1}\equiv 1$, Mod. p, und $\varepsilon(x)$ eine ***complexe Einheit*** bezeichnet, deren ***Norm,*** in Beziehung auf die verschiedenen Werthe von x genommen, gleich ***Eins*** sei. Damit die Untersuchung sogleich die nöthige Allgemeinheit bekomme, soll hier nicht vorausgesetzt werden, dafs die in der complexen Einheit $\varepsilon(x)$ enthaltenen Coëfficienten ***ganze*** Zahlen sind; sie sollen vielmehr vorläufig ganz unbestimmt gelassen werden, können also auch selbst wieder irrational sein, jedoch immer nur so, dafs

$$N\varepsilon(x) = \varepsilon(x)\varepsilon(x^g)\varepsilon(x^{g^2})\ldots\varepsilon\left(x^{g^{p-2}}\right) = 1$$

und $\varepsilon(x)$ eine rationale Function in Beziehung auf x allein ist.

Da der oben aufgestellte Ausdruck durch die zu Grunde gelegte Einheit $\varepsilon(x)$ vollständig bestimmt ist, soll er als Function dieser Einheit angesehen und durch $P\varepsilon(x)$ bezeichnet werden, so dafs

$$P\varepsilon(x) = 1 + \varepsilon(x) + \varepsilon(x)\varepsilon(x^g) + \cdots + \varepsilon(x)\varepsilon(x^g)\ldots\varepsilon\left(x^{g^{p-3}}\right)$$

ist. Die $p-1$ Glieder, aus welchen dieser Ausdruck besteht, bilden eine vollständige Periode; denn wenn man sich dieselben nach dem herrschenden Gesetze weiter fortgesetzt vorstellt, so wird vermöge der Bedingung

$$N\varepsilon(x) = 1$$

das p^{te} Glied dem ersten gleich, das $p+1^{\text{te}}$ Glied dem zweiten u. s. f. Hieraus folgt, dafs wenn man in diesem Ausdrucke x in x^g verwandelt und sodann mit $\varepsilon(x)$ multiplicirt, derselbe ganz ungeändert bleibt, da auf diese Weise nur das erste Glied in das zweite, das zweite in das dritte und so weiter, und das letzte in das erste übergeht. Man hat daher, als erste Grund-Eigenschaft dieser Ausdrücke:

$$(1.)\quad \varepsilon(x)P\varepsilon(x^g) = P\varepsilon(x);$$

was, verallgemeinert, sogleich die folgende giebt:

$$(2.)\quad \varepsilon(x)\varepsilon(x^g)\ldots\varepsilon\left(x^{g^{h-1}}\right)P\varepsilon\left(x^{g^h}\right) = P\varepsilon(x).$$

Bei der Verwandlung des x in x^{g^h} ändert sich also dieser Ausdruck nur in so weit, dafs eine bestimmte complexe Einheit als Factor hinzutritt. Die Ausdrücke

$$P\varepsilon(x),\ P\varepsilon(x^g),\ P\varepsilon(x^{g^2}),\ \ldots\ P\varepsilon\left(x^{g^{p-2}}\right)$$

sind, wenn man von den dieselben begleitenden Einheiten absieht, alle als gleich zu erachten.

Setzt man in der Gleichung (2.) nach einander $h = 1, 2, 3 \ldots p-1$, und multiplicirt die so entstehenden Gleichungen mit einander, so ergiebt sich:

$$(3.)\quad \varepsilon(x)^{p-1}.\varepsilon(x^g)^{p-2}.\varepsilon(x^{g^2})^{p-3}\ldots\varepsilon\left(x^{g^{p-2}}\right)^1.NP\varepsilon(x) = (P\varepsilon(x))^{p-1},$$

woraus sogleich

$$(4.)\quad NP\varepsilon(x) = \varepsilon(x^g)^1\varepsilon(x^{g^2})^2\ldots\varepsilon\left(x^{g^{p-2}}\right)^{p-2}(P\varepsilon(x))^{p-1}$$

folgt.

Es sei jetzt $e(x)$ eine zweite, ganz beliebige complexe Einheit, so hat man für dieselbe ebenfalls:

$$e(x)Pe(x^g) = Pe(x),$$

und wenn diese Gleichung mit der entsprechenden (1.) multiplicirt wird:

$$\varepsilon(x)e(x)P\varepsilon(x^g)Pe(x^g) = P\varepsilon(x)Pe(x).$$

Da ferner das Product der beiden Einheiten $\varepsilon(x)e(x)$ selbst wieder eine Einheit ist, so hat man für diese gleichfalls:

$$\varepsilon(x)e(x)P(\varepsilon(x^g)e(x^g)) = P(\varepsilon(x)e(x)),$$

und wenn die vorige Gleichung durch diese dividirt wird:

$$\frac{P\varepsilon(x^g)Pe(x^g)}{P(\varepsilon(x^g)e(x^g))} = \frac{P\varepsilon(x)Pe(x)}{P(\varepsilon(x)e(x))}.$$

Auf dieselbe Weise wird aus der Gleichung (2.) bewiesen, dafs auch allgemein

$$\frac{P\varepsilon(x^{g^h})Pe(x^{g^h})}{P(\varepsilon(x^{g^h})e(x^{g^h}))} = \frac{P\varepsilon(x)Pe(x)}{P(\varepsilon(x)e(x))} \text{ ist.}$$

Der auf beiden Seiten dieser Gleichung vorkommende Ausdruck bleibt, wie hieraus zu sehen, vollständig ungeändert, wenn statt x irgend eine andere imaginäre Wurzel der Gleichung $x^p = 1$ gesetzt wird; woraus unmittelbar folgt, dafs derselbe die Wurzel x in Wahrheit gar nicht enthält, sondern nur die in den Coëfficienten der Einheiten $\varepsilon(x)$ und $e(x)$ vorkommenden Gröfsen. Man hat daher

$$\frac{P\varepsilon(x)Pe(x)}{P(\varepsilon(x)e(x))} = A,$$

oder

$$(5.) \quad P\varepsilon(x)Pe(x) = AP(\varepsilon(x)e(x));$$

wo A eine Gröfse ist, welche die Wurzel x *nicht* enthält. Dies ist die *zweite* allgemeine Grund-Eigenschaft dieser Ausdrücke.

Besonders zu bemerken ist noch das specielle Resultat, welches sich findet, wenn man $e(x) = \frac{1}{\varepsilon(x)}$ annimmt, nämlich:

$$(6.) \quad P\varepsilon(x).P\left(\frac{1}{\varepsilon(x)}\right) = B,$$

wo B eine von x unabhängige Gröfse ist.

Es ist zu bemerken, dafs für gewisse, zu Grunde gelegte Einheiten $\varepsilon(x)$ und $e(x)$ die Ausdrücke $P\varepsilon(x)$ und $Pe(x)$ gleich Null werden können, und auch wirklich gleich Null werden; in welchen Fällen die aufgestellten Grund-Eigenschaften derselben zwar nicht unrichtig, aber nichtssagend werden. Diese ungünstigen Fälle lassen sich aber nunmehr durch geringe, an den Einheiten $\varepsilon(x)$ oder $e(x)$ anzubringende Änderungen leicht vermeiden.

An den gefundenen Grund-Eigenschaften der Ausdrücke $P\varepsilon(x)$ erkennt man sogleich ihre Analogie mit dem *Lagrange*schen Ausdrucke der Kreistheilung

$$F(\alpha, x) = x + \alpha x^g + \alpha^2 x^{g^2} + \cdots + \alpha^{p-2} x^{g^{p-2}},$$

in welchem α irgend eine Wurzel der Gleichung $\alpha^{p-1} = 1$ bezeichnet. Die bekannten Eigenschaften desselben, nämlich

$$\alpha^h F(\alpha, x^{g^h}) = F(\alpha, x), \qquad \frac{F(\alpha, x) F(\alpha^r, x)}{F(\alpha^{r+1}, x)} = \Psi_r(\alpha),$$
$$F(\alpha, x) F(\alpha^{-1}, x) = \pm p$$

sind offenbar den durch die Gleichungen (2, 5 und 6.) ausgedrückten allgemeinen Eigenschaften der Ausdrücke $P\varepsilon(x)$ vollständig analog. Der Grund der Übereinstimmung in den Fundamental-Eigenschaften liegt einfach darin, dafs die *Lagrange*sche Resolvente der Kreistheilungsgleichung $F(\alpha, x)$ in der That nur ein specieller Fall des allgemeineren Ausdrucks $Pe(x)$ ist. Setzt man nämlich die Einheit $\varepsilon(x) = \alpha x^{g-1}$, deren Norm, in Beziehung auf x genommen, gleich α^{p-1}, also gleich *Eins* ist, so hat man:

$$x P(\alpha x^{g-1}) = x + \alpha x^g + \alpha^2 x^{g^2} + \cdots + \alpha^{p-2} x^{g^{p-2}} = F(\alpha, x).$$

II.

Für den Zweck des gegenwärtigen Aufsatzes sollen nun den zu Grunde gelegten Einheiten etwas speciellere Bestimmungen gegeben werden; indem festgesetzt wird, dafs die Coëfficienten derselben *complexe*, aus den Wurzeln der Gleichung $\alpha^\lambda = 1$ gebildete *ganze* Zahlen sein sollen, wo λ eine *Primzahl* ist, und zwar ein Factor von $p-1$, so dafs die Primzahl p von der Form $p = \nu\lambda + 1$ ist. Ich bezeichne demgemäfs jetzt die zu Grunde zu legenden complexen Einheiten durch $\varepsilon(\alpha, x)$ und $e(\alpha, x)$ und wiederhole, um Mifsverständnissen vorzubeugen, noch einmal, dafs dieselben *ganze rationale* Functionen der beiden Wurzeln α und x mit ganzzahligen Coëfficienten sein sollen, und dafs die in Beziehung auf x allein genommenen Normen dieser Einheiten gleich *Eins* sein müssen.

28 *

Der Ausdruck $P\varepsilon(\alpha, x)$, welcher als eine, die beiden verschiedenen Wurzeln der Einheit α und x enthaltende complexe ganze Zahl mit ganzzahligen Coëfficienten betrachtet werden kann, soll nun in seine idealen oder wirklichen Primfactoren zerlegt werden. Für diese allgemeinere Art der complexen Zahlen, und überhaupt für die aus den n^{ten} Wurzeln der Einheit (wo n eine zusammengesetzte Zahl ist) gebildeten complexen Zahlen existiren, eben so wie für die, für welche $n = \lambda$, gleich einer Primzahl ist (deren Theorie ich vollständig ausgearbeitet und veröffentlicht habe), bestimmte ideale Primfactoren, und für diese auch eben so die Hauptsätze: dafs jede gegebene complexe Zahl nur eine endliche Anzahl unveränderlich bestimmter idealer Primfactoren enthält: dafs zwei complexe Zahlen, welche genau dieselben idealen Primfactoren enthalten, sich nur durch eine Einheit unterscheiden, welche als Factor hinzutreten kann und: dafs eine bestimmte Potenz jeder idealen Zahl einer wirklichen complexen Zahl gleich ist. Aufser diesen allgemeinen Sätzen ist für die gegenwärtige Untersuchung nur noch die nähere Kenntnifs der idealen Primfactoren der Zahl p selbst, für die aus p^{ten} und λ^{ten} Wurzeln der Einheit zugleich gebildeten complexen Zahlen nöthig.

* Wenn p, wie vorausgesetzt, eine Primzahl von der Form $p = r\lambda + 1$ ist, so ist bekanntlich für die complexen Zahlen, welche die Wurzel x allein enthalten, $1 - x$ der einzige Primfactor von p. Ferner giebt es für die complexen Zahlen, welche nur α allein, aber nicht x enthalten, $\lambda - 1$ verschiedene ideale Primfactoren von p, welche durch

$$f(\alpha),\quad f(\alpha^{\gamma}),\quad f(\alpha^{\gamma^2}),\; \ldots\; f(\alpha^{\gamma^{\lambda-2}})$$

bezeichnet werden sollen; wo γ eine ***primitive Wurzel*** der Congruenz $\gamma^{\lambda-1} \equiv 1, \text{Mod.}\,\lambda$ ist. Wird nun ein idealer Primfactor von p für die complexen Zahlen, welche α und x zugleich enthalten, durch $f(\alpha, x)$ bezeichnet, so hat man:

$$f(\alpha, x) f(\alpha, x^{g}) f(\alpha, x^{g^2}) \ldots f(\alpha, x^{g^{p-2}}) = E(\alpha) f(\alpha),$$

$$f(\alpha, x) f(\alpha^{\gamma}, x) f(\alpha^{\gamma^2}, x) \ldots f(\alpha^{\gamma^{\lambda-2}}, x) = E(x)(1-x).$$

Die nur durch die verschiedenen Werthe des x sich unterscheidenden Primfactoren $f(\alpha, x)$, $f(\alpha, x^{g})$, $f(\alpha, x^{g^2})$, u. s. w. sind alle als gleich zu erachten; eben so wie die Factoren $1-x$, $1-x^{g}$, $1-x^{g^2}$, u. s. w., abgesehen von den Einheiten, einander gleich sind. Es sind daher in p nur $\lambda - 1$ wirklich von einander verschiedene ideale Primfactoren von der Form $f(\alpha^{\gamma^i}, x^{g^h})$

vorhanden, nämlich:

$$f(\alpha, x), \quad f(\alpha^\gamma, x), \quad f(\alpha^{\gamma^2}, x), \quad \ldots\ldots \quad f(\alpha^{\gamma^{\lambda-2}}, x).$$

Diesemnach ist auch

$$f(\alpha, x)^{p-1} = E(\alpha, x) f(\alpha),$$

wo $E(\alpha, x)$ eine Einheit bezeichnet.

Um zu finden, ob eine gegebene complexe Zahl $\varphi(\alpha, x)$ einen bestimmten der $\lambda - 1$ verschiedenen idealen Primfactoren von p, z. B. den idealen Primfactor $f(\alpha^{\gamma^h}, x)$ enthält, oder nicht, hat man nur zu untersuchen, ob dieselbe complexe Zahl, wenn in ihr ***Eins*** statt x gesetzt wird, den Primfactor $f(\alpha^{\gamma^h})$ enthält. Wenn nämlich $\varphi(\alpha, 1)$ den Primfactor $f(\alpha^{\gamma^h})$ enthält, so enthält auch $\varphi(\alpha, x)$ den idealen Primfactor $f(\alpha^{\gamma^h}, x)$, und wenn $\varphi(\alpha, 1)$ den Primfactor $f(\alpha^{\gamma^h})$ nicht enthält, so enthält auch $\varphi(\alpha, x)$ nicht den Primfactor $f(\alpha^{\gamma^h}, x)$.

Diese allgemeinen und besondern Sätze über die idealen Primfactoren der die beiden Wurzeln der Einheit α und x zugleich enthaltenden complexen Zahlen lassen sich nach denselben Principien beweisen, welche ich für die, nur die Wurzel α allein enthaltenden complexen Zahlen im (35ten Bande S. 327 sqq.) dieses Journals vollständig entwickelt habe. Aus diesem Grunde glaube ich die Ausführung der Beweise der hier angeführten Hülfssätze jetzt ersparen zu dürfen, und werde die Sätze sogleich auf die Zerlegung der complexen Zahl $P\varepsilon(\alpha, x)$ anwenden.

Zufolge der ersten Grund-Eigenschaft der obigen Ausdrücke hat man:

$$\varepsilon(\alpha, x) P\varepsilon(\alpha, x^g) = P\varepsilon(\alpha, x).$$

Nimmt man auf beiden Seiten die *Norm* in Beziehung auf α allein, d. h. das Product für alle Werthe α, α^γ, α^{γ^2}, ... $\alpha^{\gamma^{\lambda-2}}$, so hat man, wenn diese partielle Norm durch N_α bezeichnet wird:

$$N_\alpha \varepsilon(\alpha, x) . N_\alpha P\varepsilon(\alpha, x^g) = N_\alpha P\varepsilon(\alpha, x),$$

und wenn

$$N_\alpha \varepsilon(\alpha, x) = E(x), \qquad N_\alpha P\varepsilon(\alpha, x) = F(x)$$

gesetzt wird, so ist

$$E(x) F(x^g) = F(x).$$

Einer solchen Gleichung für complexe Zahlen, welche nur x allein enthalten, und keine andere Irrationalität, kann aber nach der bekannten Theorie dieser complexen Zahlen keine andere Zahl $F(x)$ genügen, als eine solche von der Form

$$F(x) = C.E(x)(1-x)^r,$$

in welcher C eine nicht complexe ganze Zahl, $E(x)$ eine complexe Einheit und r eine ganze Zahl ist. Es ergiebt sich also hiernach:

$$N_\alpha P\varepsilon(\alpha, x) = C.E(x)(1-x)^r.$$

Hieraus folgt, dafs die complexe Zahl $P\varepsilon(\alpha, x)$ selbst, nur Factoren von einer der folgenden drei Arten enthalten kann: nämlich *erstens* solche, deren Normen, in Beziehung auf α allein genommen, ganze, nicht complexe Zahlen sind, welche also kein x enthalten und nur von der Form $\varphi(\alpha)$ sein können; *zweitens* solche, deren Normen, in Beziehung auf α genommen, Einheiten sind, welche also selbst nur Einheiten von der Form $E(\alpha, x)$ sein können; endlich *drittens* solche, deren in Beziehung auf α genommene Normen gleich $1-x$ oder gleich einer Potenz von $1-x$ sind, also nur die idealen Primfactoren von $1-x$, oder, was Dasselbe ist, von p, nämlich:

$$f(\alpha, x),\quad f(\alpha^\gamma, x),\quad f(\alpha^{\gamma^2}, x),\quad \dots\quad f(\alpha^{\gamma^{\lambda-2}}, x).$$

Dem Ausdrucke $P\varepsilon(\alpha, x)$ kann also immer folgende Form gegeben werden:

$$P\varepsilon(\alpha, x) = \varphi(\alpha)E(\alpha, x)f(\alpha, x)^m.f(\alpha^\gamma, x)^{m_1}.f(\alpha^{\gamma^2}, x)^{m_2}\ \dots\ f(\alpha^{\gamma^{\lambda-2}}, x)^{m_{\lambda-2}}.$$

Für die Bestimmung der Exponenten $m, m_1, m_2, \dots m_{\lambda-2}$, welche jetzt allgemein ausgeführt werden soll, ist wesentlich zu bemerken, dafs alle Vielfachen von $p-1$, welche etwa in denselben enthalten sein sollten, gänzlich vernachlässigt werden können; weil nämlich, nach der Eigenschaft der idealen complexen Primfactoren von p,

$$f(\alpha^{\gamma^h}, x)^{p-1} = f(\alpha^\gamma)E(\alpha^{\gamma^h}, x)$$

die $(p-1)^{\text{te}}$ Potenz eines solchen Primfactors einer nur α allein enthaltenden complexen Zahl, multiplicirt mit einer Einheit, gleich ist; so dafs man diese immer mit dem Factor $\varphi(\alpha)$ und der Einheit $E(\alpha, x)$ verbunden annehmen kann. Mit Rücksicht hierauf werden die Exponenten $m, m_1, \dots m_{\lambda-2}$ am leichtesten dadurch bestimmt, dafs man den Ausdruck

$$P\left(\frac{1-x^{g^r}}{1-x}\varepsilon(\alpha, x)\right)$$

in Betracht zieht, und untersucht, unter welcher Bedingung derselbe einen der idealen Primfactoren von p, z. B. $f(\alpha^{\gamma^h}, x)$ nicht enthält. Die hierzu nothwendige und hinreichende Bedingung ist, nach Dem was oben bemerkt worden, die, dafs dieser Ausdruck, für $x=1$, nämlich

$$1+g^r\varepsilon(\alpha,1)+g^{2r}\varepsilon(\alpha,1)^2+\dots+g^{(p-2)r}\varepsilon(\alpha,1)^{p-2}$$

den Factor $f(\alpha^{\gamma^h})$ nicht enthalte. Jede, nur die eine Wurzel der Einheit α enthaltende complexe Zahl ist aber bekanntlich nach dem Modul $f(\alpha^{\gamma^h})$, welcher ein idealer Primfactor von p ist, einer nichtcomplexen ganzen Zahl congruent, welche sich wiederum als Potenz der primitiven Wurzel g darstellen läfst, wenn sie nicht congruent Null ist. Setzt man daher

$$\varepsilon(\alpha,1)\equiv g^s,\quad \text{Mod.}\, f(\alpha^{\gamma^h}),$$

so geht die obige Bedingung in folgende über:

$$1+g^{r+s}+g^{2r+2s}+\dots+g^{(p-2)(r+s)}\ \text{nicht}\equiv 0,$$

für den Modul $f(\alpha^{\gamma^h})$, also auch für den Modul p; aus welcher dann unmittelbar folgt, dafs

$$r+s\equiv 0,\quad \text{Mod.}\, p-1,$$

sein mufs. Nun hat man aber vermöge der Grund-Eigenschaft:

$$A\cdot P\Big(\frac{1-x^{g^r}}{1-x}\varepsilon(\alpha,x)\Big)=P\Big(\frac{1-x^{g^r}}{1-x}\Big).P\varepsilon(\alpha,x);$$

wo A von der Wurzel der Einheit x unabhängig ist. Ferner ist aus der Definition der mit $P\varepsilon(x)$ bezeichneten Ausdrücke selbst, leicht zu zeigen, dafs

$$P\Big(\frac{1-x^{g^r}}{1-x}\Big)=\frac{C}{(1-x)(1-x^g)\dots(1-x^{g^{r-1}})}=\frac{CE(x)}{(1-x)^r}$$

ist, wo C eine ganze Zahl und $E(x)$ eine Einheit ist. Der ideale Primfactor $f(\alpha^{\gamma^h},x)$ von p ist also in $P\Big(\frac{1-x^{g^r}}{1-x}\Big)$ genau $(p-1-r)$ mal enthalten, und da derselbe nach der oben angesetzten Form in $P\varepsilon(\alpha,x)$ genau m_h mal enthalten ist, so folgt, dafs

$$P\Big(\frac{1-x^{g^r}}{1-x}\varepsilon(\alpha,x)\Big)$$

diesen idealen Primfactor $(p-1-r+m_h)$mal enthält; wobei $p-1$, oder jedes beliebige Vielfache von $p-1$, hinzugefügt, oder auch weggelassen werden kann. Die Bedingung, dafs dieser Ausdruck den idealen Primfactor $f(\alpha^{\gamma^h},x)$ nicht enthalte, ist also:

$$m_h-r\equiv 0,\quad \text{Mod.}\, p-1,$$

und weil oben, für dieselbe Bedingung,

$$r+s\equiv 0,\ \text{Mod.}\, p-1$$

gefunden wurde, so ergiebt sich:

$$s\equiv -m_h,\ \text{Mod.}\, p-1,$$

also auch:

$$\varepsilon(\alpha,1)\equiv g^{-m_h},\ \text{Mod.}\, f(\alpha^{\gamma^h});$$

welche Congruenz die gesuchte Bestimmung der Exponenten $m, m_1, \dots m_{\lambda-2}$ enthält.

Um dieselbe in ihrer einfachsten Form darzustellen, verwandle ich α in $\alpha^{\gamma^{-h}}$. Dies giebt

$$\varepsilon(\alpha^{\gamma^{-h}},1)\equiv g^{-m_h},\ \text{Mod.}\, f(\alpha).$$

Ferner bezeichne ich allgemein durch Ind. $\varphi(\alpha)$ den Index der complexen Zahl $\varphi(\alpha)$, für den Modul $f(\alpha)$ und für die primitive Wurzel g; dann ist

$$(7.)\qquad m_h\equiv -\text{Ind.}\,\varepsilon(\alpha^{\gamma^{-h}},1),\quad \text{Mod.}\, p-1,$$

die gesuchte Bestimmung der Exponenten in dem Ausdrucke

$$(8.)\ P\varepsilon(\alpha,x)=\varphi(\alpha)E(\alpha,x).f(\alpha,x)^m.f(\alpha^\gamma,x)^{m_1}.f(\alpha^{\gamma^2},x)^{m_2}\dots f(\alpha^{\gamma^{\lambda-2}},x)^{m_{\lambda-2}}.$$

In dieser allgemeinen Formel ist unter andern auch die Zerlegung der *Lagrange*schen Resolvente der Kreistheilung, welche oben mit $F(\alpha,x)$ bezeichnet wurde, in ihre idealen Primfactoren enthalten, und es werden für dieselbe, wenn man die Gleichung $F(\alpha,x)F(\alpha^{-1},x)=p$ zu Hülfe nimmt, die Exponenten $m, m_1, \dots m_{\lambda-2}$ nicht nur durch Congruenzen für den Modul $p-1$, sondern vollständig bestimmt. Man erhält nämlich, wenn man $\varepsilon(\alpha,x)=\alpha^{-1}x^{g-1}$ annimmt:

$$F(\alpha^{-1},x)=E(\alpha,x)f(\alpha,x)^\nu.f(\alpha^\gamma,x)^{\nu\gamma_{-1}}f(\alpha^{\gamma^2},x)^{\nu\gamma_{-2}}\dots f(\alpha^{\gamma^{\lambda-2}},x)^{\nu\gamma_{-(\lambda-2)}},$$

wo $E(\alpha,x)$ eine Einheit und γ_{-h} die kleinste positive Zahl bezeichnet, welche der Potenz γ^{-h} congruent ist, für den Modul λ, $\nu=\frac{p-1}{\lambda}$; und dieser Ausdruck stimmt vollständig mit demjenigen überein, welchen ich früher für die λ^{te} Potenz dieser *Lagrange*schen Resolvente gegeben habe.

Für die Anwendung, welche hier von der Zerlegung des Ausdrucks $P\varepsilon(\alpha,x)$ in seine idealen Primfactoren gemacht werden soll, nehme man noch eine besondere Einheit für $\varepsilon(\alpha,x)$ an, welche nicht sowohl die Wurzeln x, x^g, u. s. w. selbst, sondern nur die λ Perioden derselben, von je ν Gliedern, enthält. Wird eine beliebige von diesen Perioden durch η_r bezeichnet, so dafs

$$\eta_r = x^{g^r} + x^{g^{\lambda+r}} + x^{g^{2\lambda+r}} + \dots + x^{g^{(\nu-1)\lambda+r}}$$

ist, so ist

$$(1-\alpha x)(1-\alpha x^{g^\lambda})(1-\alpha x^{g^{2\lambda}}) \dots (1-\alpha x^{g^{(\nu-1)\lambda}}) = e(\alpha, \eta)$$

eine Einheit, welche nur die Perioden $\eta, \eta_1, \dots \eta_{\lambda-1}$ enthält, und deren in Beziehung auf diese Perioden allein genommene Norm gleich ***Eins*** ist. Aus dieser setze man nun die folgende zusammen:

$$E_n(\alpha, \eta) = e(\alpha, \eta) e(\alpha^\gamma, \eta)^{\gamma^{-2n}} e(\alpha^{\gamma^2}, \eta)^{\gamma^{-4n}} \dots e(\alpha^{\gamma^{\lambda-2}}, \eta)^{\gamma^{-2(\lambda-2)n}},$$

und bilde den Ausdruck

$$PE_n(\alpha, \eta) = \varphi(\alpha) E(\alpha, x) f(\alpha, x)^m f(\alpha^\gamma, x)^{m_1} \dots f(\alpha^{\gamma^{\lambda-2}}, x)^{m_{\lambda-2}}.$$

Um allgemein den Exponenten m_h zu bestimmen, mufs man in der Einheit $E_n(\alpha^{\gamma^{-h}}, \eta)$ dem x den Werth ***Eins*** geben, wodurch dieselbe in

$$(1-\alpha^{\gamma^{-h}})^\nu (1-\alpha^{\gamma^{1-h}})^{\nu\gamma^{-2n}} (1-\alpha^{\gamma^{2-h}})^{\nu\gamma^{-4n}} \dots (1-\alpha^{\gamma^{\lambda-2-h}})^{\nu\gamma^{-2(\lambda-2)n}}$$

übergeht. Hieraus folgt sogleich, nach der Congruenz (7.):

$$m_h \equiv -\nu \sum_0^{\lambda-2} {}_i \gamma^{-2ni} \operatorname{Ind.}(1-\alpha^{\gamma^{i-h}}), \quad \text{Mod. } p-1,$$

und wenn in dieser Summe i in $i+h$ verwandelt wird:

$$m_h \equiv -\nu\gamma^{-2nh} \sum_0^{\lambda-2} {}_i \gamma^{-2ni} \operatorname{Ind.}(1-\alpha^{\gamma^i}), \quad \text{Mod. } p-1.$$

Wird nun der Kürze wegen

$$\sum_0^{\lambda-2} {}_i \gamma^{-2ni} \operatorname{Ind.}(1-\alpha^{\gamma^i}) \equiv -s, \quad \text{Mod. } \lambda,$$

gesetzt, so ist

$$m_h \equiv \nu s \gamma^{-2nh},$$

und diesemnach:

$$PE_n(\alpha, \eta) = \varphi(\alpha) E(\alpha, x) \{f(\alpha, x) f(\alpha^\gamma, x)^{\gamma^{-2n}} \dots f(\alpha^{\gamma^{\lambda-2}}, x)^{\gamma^{-2(\lambda-2)n}}\}^{\nu s}.$$

Die Seite rechts in dieser Gleichung läfst sich ebenfalls leicht als Function der λ Perioden $\eta, \eta_1, \dots \eta_{\lambda-1}$ darstellen. Setzt man nämlich

$$f(\alpha, x) f(\alpha, x^{g^\lambda}) f(\alpha, x^{g^{2\lambda}}) \dots f(\alpha, x^{g^{(\nu-1)\lambda}}) = f(\alpha, \eta),$$

so ist, weil, abgesehen von den Einheiten, $f(\alpha, x)$, $f(\alpha, x^{g^\lambda})$ u. s. w. einander gleich sind, eben so auch

$$f(\alpha, \eta)^\lambda = f(\alpha) \quad \text{und} \quad f(\alpha, \eta) = f(\alpha, x)^\nu,$$

und wenn man von diesen Ausdrücken Gebrauch macht, so erhält man

$$(9.)\qquad PE_n(\alpha,\eta) = \varphi(\alpha)E(\alpha,\eta)\{f(\alpha,\eta)f(\alpha^\gamma,\eta)^{\gamma^{-2n}} \;\ldots.\; f(\alpha^{\gamma^{\lambda-2}},\eta)^{\gamma^{-2(\lambda-2)n}}\}^s.$$

Nimmt man auf beiden Seiten die *Norm* in Beziehung auf die λ Perioden, und erwägt, dafs $Nf(\alpha,\eta)=f(\alpha)$ ist, so hat man:

$$(10.)\qquad NPE_n(\alpha,\eta) = \varphi(\alpha)^\lambda E(\alpha)\{f(\alpha)f(\alpha^\gamma)^{\gamma^{-2n}} \;\ldots.\; f(\alpha^{\gamma^{\lambda-2}})^{\gamma^{-2(\lambda-2)n}}\}^s.$$

Damit dieser Ausdruck für die in dem Folgenden davon zu machende Anwendung noch besser zubereitet werde, wende man eine, nur die Perioden $\eta, \eta_1, \ldots. \eta_{\lambda-1}$ enthaltende, von α ganz unabhängige Einheit an, nämlich:

$$\frac{(1-x^g)(1-x^{g^{\lambda+1}})(1-x^{g^{2\lambda+1}})\ldots.(1-x^{g^{(\nu-1)\lambda+1}})}{(1-x)(1-x^{g^\lambda})(1-x^{g^{2\lambda}})\ldots.(1-x^{g^{(\nu-1)\lambda+1}})} = e(\eta),$$

Für diese Einheit findet sich nach den oben gegebenen allgemeinen Regeln sehr leicht:

$$P(e(\eta)^s) = C.E(\eta)\{f(\alpha,\eta)f(\alpha^\gamma,\eta)f(\alpha^{\gamma^2},\eta) \;\ldots.\; f(\alpha^{\gamma^{\lambda-2}},\eta)\}^{-s},$$

und wenn man diesen Ausdruck mit dem obigen (9.) multiplicirt, so erhält man, unter Anwendung der Grund-Eigenschaft (5.), folgendes Resultat:

$$(11.)\qquad P(e(\eta)^s E_n(\alpha,\eta))$$
$$= \varphi(\alpha)E(\alpha,\eta)\{f(\alpha^\gamma,\eta)^{\gamma^{-2n}-1}.f(\alpha^{\gamma^2},\eta)^{\gamma^{-4n}-1} \;\ldots.\; f(\alpha^{\gamma^{\lambda-2}},\eta)^{\gamma^{-2(\lambda-2)n}-1}\}^s.$$

Nimmt man endlich noch die Norm in Beziehung auf die Perioden $\eta, \eta_1, \ldots. \eta_{\lambda-1}$, so ergiebt sich auch

$$(12.)\qquad NP(e(\eta)^s E_n(\alpha,\eta))$$
$$= \varphi(\alpha)^\lambda E(\alpha)\{f(\alpha^\gamma)^{\gamma^{-2n}-1} f(\alpha^{\gamma^2})^{\gamma^{-4n}-1} \;\ldots.\; f(\alpha^{\gamma^{\lambda-2}})^{\gamma^{-2(\lambda-2)n}-1}\}^s,$$

wo

$$s \equiv -\sum_0^{\lambda-2}{}_i\, \gamma^{-2ni}\,\mathrm{Ind}(1-\alpha^{\gamma^i}),\quad \text{Mod. } \lambda \text{ ist.}$$

Die Gröfse s ist aus meiner Abhandlung über die Ergänzungssätze zu den allgemeinen Reciprocitätsgesetzen (Band 44. dieses Journals) näher bekannt. Sie läfst sich, wie daselbst (S. 99) gezeigt, in folgende Form schreiben:

$$(13.)\qquad s \equiv \frac{-2\,\mathrm{Ind}\,E_n(\alpha)}{\gamma^{2n}-1},$$

wo $E_n(\alpha)$ die zusammengesetzte Kreistheilungs-Einheit bezeichnet, deren Index in der genannten Abhandlung gefunden wird, nämlich die Einheit

$$E_n(\alpha) = e(\alpha)e(\alpha^\gamma)^{\gamma^{-2n}} e(\alpha^{\gamma^2})^{\gamma^{-4n}} \;\ldots.\; e(\alpha^{\gamma^{\mu-1}})^{\gamma^{-2(\mu-1)n}},$$

für

$$\mu = \tfrac{1}{2}(\lambda - 1) \quad \text{und} \quad e(\alpha) = \sqrt{\frac{(1-\alpha^{\gamma})(1-\alpha^{-\gamma})}{(1-\alpha)(1-\alpha^{-1})}}.$$

III.

Es sollen jetzt die gefundenen Resultate auf denjenigen Fall der allgemeinen Reciprocitätsgesetze angewendet werden, wo zwei conjugirte complexe Primzahlen mit einander zu vergleichen sind.

Zu diesem Zwecke müssen zunächst die in der Formel (12.) vorkommenden idealen oder wirklichen complexen Primfactoren von p, nämlich $f(\alpha)$, $f(\alpha^{\gamma})$, $f(\alpha^{\gamma^2})$, u. s. w., in Betreff der Einheiten mit welchen sie multiplicirt sein können, näher bestimmt werden. Wenn sie ideal sind, so werden sie zunächst zu einer Potenz des Exponenten H erhoben; was sie zu wirklichen complexen Zahlen macht. Diese wirklichen complexen Zahlen werden sodann durch Multiplication mit passenden Einheiten in die Form gebracht, welche ich als die *primäre* bezeichne, in welcher nämlich die complexe Zahl den beiden Bedingungen genügen mufs: *erstens:* dafs sie für den Modul $(1-\alpha)^2$ einer nicht complexen ganzen Zahl congruent ist; *zweitens:* dafs sie, mit ihrer reciproken multiplicirt, ein Product giebt, welches, für den Modul λ, einer nicht complexen ganzen Zahl congruent ist. Auf diese primäre Form läfst sich auch, wie ich früher bewiesen habe, jede aus λ^{ten} Wurzeln der Einheit gebildete complexe Zahl bringen, aufser wenn λ eine von denjenigen Ausnahmezahlen ist, die in einer der ersten $\frac{1}{2}(\lambda-3)$ *Bernoulli*schen Zahlen als Factoren des Zählers vorkommen (M. s. dieses Journal Bd. 44. S. 138 u. S. 141). Diese Ausnahmezahlen, für welche auch der niedrigste Exponent H derjenigen Potenz von $f(\alpha)$, welche *wirklich* wird, durch λ theilbar sein kann, sollen aus diesem Grunde von der gegenwärtigen Untersuchung ausgeschlossen bleiben.

Unter der Voraussetzung, dafs die Primfactoren $f(\alpha)$, $f(\alpha^{\gamma})$, u. s. w. von p in der primären Form genommen werden, ist nun, wie leicht zu sehen, die H^{te} Potenz des Products

$$f(\alpha^{\gamma})^{\gamma^{-2n}-1} . f(\alpha^{\gamma^2})^{\gamma^{-4n}-1} \;\ldots\; f(\alpha^{\gamma^{\lambda-2}})^{\gamma^{-2(\lambda-2)n}-1}$$

einer nicht complexen ganzen Zahl c für den Modul λ congruent. Wenn man daher beide Seiten der Gleichung (12.) zur H^{ten} Potenz erhebt, und daraus eine Congruenz für den Modul λ bildet, so erhält man:

$$[NP(e(\eta)^s \mathrm{E}_n(\alpha, \eta))]^H \equiv cE(\alpha)^H, \quad \text{Mod. } \lambda.$$

29 *

Nach den oben in (§. I.) bewiesenen Grund-Eigenschaften der Ausdrücke $P\varepsilon(x)$ wird nun die Norm eines solchen Ausdrucks folgendermafsen durch eine Potenz desselben ausgedrückt:

* $$NP(e(\eta)^s \mathrm{E}_n(\alpha,\eta)) = \sum_0^{\lambda-1}{}_i\, e(\eta_i)^{si}\mathrm{E}_n(\alpha,\eta_i)^i.(Pe(\eta)^s\mathrm{E}_n(\alpha,\eta))^\lambda.$$

Beachtet man, dafs in Beziehung auf den Modul λ jede λ^{te} Potenz einer complexen Zahl, welche α enthält, einer von α unabhängigen Zahl congruent ist, so findet sich

* $$NP(e(\eta)^s\mathrm{E}_n(\alpha,\eta)) \equiv C\sum_0^{\lambda-1}{}_i\,\mathrm{E}_n(\alpha,\eta_i)^i, \quad \text{Mod. } \lambda,$$

wo C eine complexe Zahl ist, welche zwar die Perioden η, η_1, etc., nicht aber die Wurzel der Einheit α enthält. Aus dieser Congruenz, verglichen mit der vorhergehenden, folgt

* $$cE(\alpha)^H \equiv C^H\sum_0^{\lambda-1}{}_i\,\mathrm{E}_n(\alpha,\eta_i)^{iH}, \quad \text{Mod. } \lambda.$$

Hieraus soll nun die bisher unbestimmte Einheit $E(\alpha)$ näher bestimmt werden. Zu diesem Zwecke nehme ich auf beiden Seiten der Congruenz die *Logarithmen;* und zwar in dem Sinne, in welchem ich die Theorie der logarithmischen Ausdrücke für complexe Zahlen in der mehrmals erwähnten Abhandlung (dieses Journals Bd. 44. S. 130 etc.) entwickelt habe. Dies giebt, wenn man noch durch $\boldsymbol{H}$, welches nach der Annahme nicht durch λ theilbar ist, dividirt:

$$l\Big(\frac{E(\alpha)}{E(1)}\Big) \equiv \sum_0^{\lambda-1}{}_i\, i\, l\Big(\frac{E_n(\alpha,\eta_i)}{E_n(1,\eta_i)}\Big), \quad \text{Mod. } \lambda,$$

und da

$$E_n(\alpha,\eta) = e(\alpha,\eta)e(\alpha^\gamma,\eta)^{\gamma^{-2n}} \dots\dots e(\alpha^{\gamma^{\lambda-2}},\eta)^{\gamma^{-2(\lambda-2)n}},$$
$$e(\alpha,\eta) = (1-\alpha x)(1-\alpha x^{g^\lambda}) \dots\dots (1-\alpha x^{g^{(\nu-1)\lambda}})$$

ist, so verwandelt sich dieser Ausdruck leicht in den folgenden:

$$l\Big(\frac{E(\alpha)}{E(1)}\Big) \equiv \sum_0^{p-2}{}_i\sum_0^{\lambda-2}{}_h\, i\gamma^{-2nh}\, l\left(\frac{1-\alpha^{\gamma^h}x^{g^i}}{1-x^{g^i}}\right), \quad \text{Mod. } \lambda.$$

Ich mache jetzt von der allgemeinen Formel meiner Abhandlung (Bd. 44. S. 134) Gebrauch, nämlich von der Formel

$$\sum_0^{\lambda-2}\gamma^{-kh}\, l\Big(\frac{\varphi(\alpha^{\gamma^h})}{\varphi(1)}\Big) \equiv \frac{d_0^k l\varphi(e^v)}{dv^k}\boldsymbol{X}_k(\alpha), \quad \text{Mod. } \lambda,$$

in welcher

$$X_k(\alpha) = \alpha + \gamma^{-k}\alpha^{\gamma} + \gamma^{-2k}\alpha^{\gamma^2} + \cdots + \gamma^{-(\lambda-2)k}\alpha^{\gamma^{\lambda-2}}$$

ist, und vermöge deren, wenn $\varphi(\alpha)$ hier gleich $1-\alpha x^{g^i}$ angenommen wird, der gefundene Ausdruck sich in folgenden verwandelt:

$$l\Big(\frac{E(\alpha)}{E(1)}\Big) \equiv \sum_0^{p-2}{}_i \frac{i\,d_0^{2n}\,l(1-e^v x^{g^i})}{dv^{2n}} X_{2n}(\alpha), \quad \text{Mod. } \lambda,$$

oder, wenn der Kürze wegen

$$(14.) \quad \sum_0^{p-2}{}_i \frac{i\,d_0^{2n}\,l(1-e^v x^{g^i})}{dv^{2n}} \equiv D_n, \quad \text{Mod. } \lambda,$$

gesetzt wird, in

$$l\Big(\frac{E(\alpha)}{E(1)}\Big) \equiv D_n X_{2n}(\alpha), \quad \text{Mod. } \lambda.$$

Für die zusammengesetzte Kreistheilungs-Einheit $E_n(\alpha)$, dieselbe, welche oben in (§. II.) vollständig definirt wurde, hat man aber nach (S. 139.) der erwähnten Abhandlung:

$$l\Big(\frac{E_n(\alpha)}{E_n(1)}\Big) \equiv \frac{(-1)^{n+1}(\gamma^{2n}-1)B_n}{4n} X_{2n}(\alpha), \quad \text{Mod. } \lambda,$$

wo B_n die n^{te} *Bernoulli*sche Zahl bezeichnet. Also wenn M_n durch die Congruenz

$$(15.) \quad \frac{(-1)^{n+1}(\gamma^{2n}-1)B_n}{4n}.M_n \equiv D_n, \quad \text{Mod. } \lambda,$$

bestimmt wird, so ist

$$l\Big(\frac{E(\alpha)}{E(1)}\Big) \equiv M_n.\,l\Big(\frac{E_n(\alpha)}{E_n(1)}\Big);$$

woraus nach der bekannten Theorie dieser logarithmischen Ausdrücke und der Einheiten, ohne Schwierigkeit

$$(16.) \quad E(\alpha) = \mathrm{E}_n(\alpha)^{M_n},$$

folgt, und wo die gesuchte Bestimmung der Einheit $E(\alpha)$ in der Gleichung (12.) liegt.

Das besondere Reciprocitätsgesetz, welches hier entwickelt werden soll, liegt nur in der Congruenz (15.), welche M_n und D_n verbindet. Diese Gröfsen lassen sich nämlich beide durch die Indices der complexen Primzahlen $f(\alpha^{\gamma})$, $f(\alpha^{\gamma^2})$, ... $f(\alpha^{\gamma^{\lambda-2}})$, in Beziehung auf den Modul $f(\alpha)$ genommen, ausdrücken, so dafs diese Congruenz für die verschiedenen Werthe 1, 2, 3, ... $\frac{1}{2}(\lambda-3)$ von n, ebensoviele Reciprocitätsgleichungen giebt, aus welchen das einfache Gesetz, wie ich es längst durch Induction gefunden hatte, sich leicht entwickeln läfst. Ich beginne mit der Entwickelung der Gröfse

$$D_n \equiv \overset{p-2}{\underset{0}{\Sigma_i}} \frac{i d_0^{2n} l(1-e^v x^{g^i})}{dv^{2n}}, \quad \text{Mod. } \lambda.$$

Zunächst hat man

$$\frac{dl(1-e^v x)}{dv} = \frac{-e^v x}{1-e^v x};$$

was auf folgende Form gebracht werden kann:

$$\frac{-e^v x}{1-e^v x} = \frac{x+(1+e^v)x^2+(1+e^v+e^{2v})x^3+\cdots+(1+e^v+\cdots+e^{(p-1)v})x^p}{1+e^v+e^{2v}+\cdots+e^{(p-1)v}},$$

deren Richtigkeit sogleich durch Multiplication mit dem Nenner $1-e^v x$ ersichtlich wird. Bildet man nun den $(2n-1)^{\text{ten}}$ Differentialquotienten dieses Bruchs, und setzt darin $v=0$, so werden alle Differentialquotienten des Nenners $1+e^v+\cdots+e^{(p-1)v}$, vom ersten bis zum $(2n-1)^{\text{ten}}$, congruent *Null*, für den Modul λ; dieser Nenner selbst aber wird congruent *Eins*. Die Differentialquotienten dieses Bruchs sind daher für $v=0$ nur den Differentialquotienten seines Zählers für $v=0$ congruent. Man erhält also:

$$\begin{aligned}\frac{d_0^{2n} l(1-e^v x)}{dv^{2n}} \equiv\; & 1^{2n-1}x^2+(1^{2n-1}+2^{2n-1})x^3+\cdots \\ & +(1^{2n-1}+2^{2n-1}+\cdots+(p-1)^{2n-1})x^p,\end{aligned}$$

oder, wenn der Kürze wegen

$$1^{2n-1}+2^{2n-1}+3^{2n-1}+\cdots+(k-1)^{2n-1} = S_{2n-1}(k)$$

gesetzt wird, und wenn man beachtet, dafs $S_{2n-1}(p)\equiv 0$ für den Modul λ ist, so hat man

$$\frac{d_0^{2n} l(1-e^v x)}{dv^{2n}} \equiv \overset{p-1}{\underset{2}{\Sigma_k}} S_{2n-1}(k)x^k, \quad \text{Mod. } \lambda,$$

und demgemäfs:

$$D_n \equiv \overset{p-2}{\underset{0}{\Sigma_i}}\overset{p-1}{\underset{2}{\Sigma_k}} i\, S_{2n-1}(k)x^{kg^i}, \quad \text{Mod. } \lambda.$$

Setzt man in dem Exponenten von x, $g^{\text{Ind.}k}$ statt k und verwandelt i in $i-\text{Ind.}k$, wobei man dem veränderten i wieder genau dieselben Werthe $i=0,1,2,\ldots p-2$ lassen kann, so wird

$$D_n \equiv \overset{p-2}{\underset{0}{\Sigma_i}}\overset{p-1}{\underset{2}{\Sigma_k}} (i-\text{Ind.}k)S_{2n-1}(k)x^{g^i},$$

und da $\overset{p-1}{\underset{2}{\Sigma_k}} S_{2n-1}(k)\equiv 0$, Mod. λ, $\overset{p-2}{\underset{0}{\Sigma_i}} x^{g^i} = -1$ ist, so fällt x aus dieser Congruenz gänzlich weg, und es wird

$$D_n \equiv \sum_2^{p-1}{}_k S_{2n-1}(k)\,\text{Ind.}\,k, \quad \text{Mod.}\ \lambda.$$

Dieser Ausdruck wird zu dem vorliegenden Zwecke weiter auf folgende Weise eingerichtet. Zunächst wird die einfache Summe, indem man $k+h\lambda$ statt k setzt, in folgende Doppelsumme verwandelt:

$$D_n \equiv \sum_1^{\lambda-1}{}_k \sum_0^{\nu-1}{}_h S_{2n-1}(k)\,\text{Ind.}(k+h\lambda).$$

Sodann ist

$$\text{Ind.}(k+h\lambda) \equiv \text{Ind.}(k\nu+h\lambda\nu)-\text{Ind.}\,\nu,$$

und da $\lambda\nu=p-1$ ist, und da die Vielfachen von p innerhalb der Indices weggelassen werden können,

$$\text{Ind.}(k+h\lambda) \equiv \text{Ind.}(k\nu-h)-\text{Ind.}\,\nu,$$

also

$$\sum_0^{\nu-1}{}_h \text{Ind.}(k+h\lambda) \equiv \text{Ind.}\left(\frac{\Pi k\nu}{\Pi(k-1)\nu}\right)-\nu\,\text{Ind.}\,\nu,$$

wenn $\Pi(r)=1.2.3\ldots.r$ oder

$$\sum_0^{\nu-1}{}_h \text{Ind.}(k+h\lambda) \equiv \text{Ind.}\left(\frac{\Pi k\nu}{\Pi(k-1)\nu\,\Pi\nu}\right)-\nu\,\text{Ind.}\,\nu+\text{Ind.}\,\Pi\nu \quad \text{ist.}$$

Erwägt man nun, dafs $\sum_1^{\lambda-1}{}_k S_{2n-1}(k)\equiv 0$ ist, so folgt hieraus:

$$D_n \equiv \sum_1^{\lambda-1}{}_k S_{2n-1}(k)\,\text{Ind.}\left(\frac{\Pi k\nu}{\Pi(k-1)\nu\,\Pi\nu}\right), \quad \text{Mod.}\ \lambda.$$

Der in diesem Ausdrucke vorkommende Binomialcoëfficient kann durch eine der von *Jacobi* mit $\Psi_k(\alpha)$ bezeichneten complexen Kreistheilungszahlen ersetzt werden. Aus der Theorie der Kreistheilung (M. s. eine Notiz von *Jacobi* über die Kreistheilung und ihre Anwendung auf Zahlentheorie in den Monatsberichten der Berliner Akademie vom Jahre 1837, abgedruckt in diesem Journal Bd. 30. S. 166 etc.) hat man nämlich:

$$F(\alpha,x) = x+\alpha x^g+\alpha^2 x^{g^2}+\cdots+\alpha^{p-2}x^{g^{p-2}},$$

$$\frac{F(\alpha^{-1},x)\,F(\alpha^{-k},x)}{F(\alpha^{-(k+1)},x)} = \Psi_k(\alpha^{-1}),$$

und wenn g^ν statt α gesetzt wird:

$$\Psi_{k-1}(g^{-\nu}) \equiv -\frac{\Pi k\nu}{\Pi(k-1)\nu\,\Pi\nu}, \quad \text{Mod.}\ p.$$

Wird nun $f(\alpha)$ als derjenige ideale Primfactor von p betrachtet, welcher in $g^{\nu}-\alpha$ enthalten ist, so dafs

$$g^{\nu} \equiv \alpha, \quad \text{Mod.}\, f(\alpha)$$

ist, so ist

$$\Psi_{k-1}(\alpha^{-1}) \equiv -\frac{\Pi k\nu}{\Pi(k-1)\nu\,\Pi\nu}, \quad \text{Mod.}\, f(\alpha).$$

Also, wenn das Zeichen Ind. nicht mehr auf den Modul p, sondern nur auf den Modul $f(\alpha)$, Factor von p, bezogen wird, kann D_n folgendermafsen dargestellt werden:

$$D_n \equiv \sum_{2}^{\lambda-1}{}_k\, S_{2n-1}(k)\,\text{Ind.}\,\Psi_{k-1}(\alpha^{-1}), \quad \text{Mod.}\ \lambda.$$

Es ist noch die complexe Zahl $\Psi_{k-1}(\alpha^{-1})$ in ihre Primfactoren, welche zugleich Primfactoren von p sind, zu zerlegen. Aus der von mir angegebenen Zerlegung der λ^{ten} Potenz des *Lagrange*schen Ausdrucks $F(\alpha, x)$ der Kreistheilung, welche, wenn allgemein $\left|\frac{b}{a}\right|$ die kleinste positive Zahl bezeichnet, die der Congruenz $ax \equiv b$, Mod. λ, genügt, folgendermafsen dargestellt werden kann:

$$F(\alpha^{-1}, x)^{\lambda} = \pm\alpha^{5} f(\alpha)^{\left|\frac{1}{1}\right|}.f(\alpha^{2})^{\left|\frac{1}{2}\right|}.f(\alpha^{3})^{\left|\frac{1}{3}\right|} \ \ldots\ldots\ f(\alpha^{\lambda-2})^{\left|\frac{1}{\lambda-1}\right|},$$

erhält man sehr leicht:

$$\Psi_{k-1}(\alpha^{-1}) = \pm\prod_{1}^{\lambda-1}{}_h\, f(\alpha^{h})^{\left(\frac{\left|\frac{1}{h}\right|+\left|\frac{k-1}{h}\right|-\left|\frac{k}{h}\right|}{\lambda}\right)}$$

und demnach

$$D_n \equiv \sum_{2}^{\lambda-1}{}_k \sum_{1}^{\lambda-1}{}_h\, S_{2n-1}(k)\left(\frac{\left|\frac{1}{h}\right|+\left|\frac{k-1}{h}\right|-\left|\frac{k}{h}\right|}{\lambda}\right)\text{Ind.}\,f(\alpha^{h}), \quad \text{Mod.}\ \lambda,$$

wo, wenn $f(\alpha^{\gamma^{h}})$ ideal und H der kleinste Exponent ist, für welchen $f(\alpha^{\gamma^{h}})^{H}$ *wirklich* wird, die Bedeutung des Ind. $f(\alpha^{\gamma^{h}})$ als gleich $\frac{1}{H}$ Ind. $f(\alpha^{\gamma^{h}})^{H}$ anzusehen ist.

Man setze nun statt $S_{2n-1}(k)$ die in Beziehung auf den Modul λ congruente Reihe $S_{(2n-1)\lambda}(k)$ und multiplicire mit λ, so erhält man

$$\lambda D_n \equiv \sum_{2}^{\lambda-1}{}_k \sum_{1}^{\lambda-1}{}_h\, S_{(2n-1)\lambda}(k)\left(\left|\frac{1}{h}\right|+\left|\frac{k-1}{h}\right|-\left|\frac{k}{h}\right|\right)\text{Ind.}\,f(\alpha^{h}), \quad \text{Mod.}\ \lambda^{2}.$$

Nun ist aus den bekannten Formeln für die Potenzsummen der natürlichen Zahlen leicht zu zeigen, dafs

$$\sum_{2}^{\lambda-1}{}_k\, S_{(2n-1)\lambda}(k) \equiv -\sum_{1}^{\lambda-1}{}_k\, k^{(2n-1)\lambda+1} \quad \text{und}$$

$$\sum_{2}^{\lambda-1}{}_k S_{(2n-1)\lambda}(k)\left(\left|\frac{k-1}{h}\right|-\left|\frac{k}{h}\right|\right) \equiv \sum_{1}^{\lambda-1}{}_k k^{(2n-1)\lambda}\left|\frac{k}{h}\right| \equiv h^{(2n-1)\lambda}.\sum_{1}^{\lambda-1}{}_k k^{(2n-1)\lambda+1},$$

$$\sum_{1}^{\lambda-1}{}_k k^{(2n-1)\lambda+1} \equiv \frac{(-1)^{n+1}B_n\lambda}{2n}$$

für den Modul λ^2 ist. Setzt man diese Werthe hinein, so wird sich der gemeinschaftliche Factor λ hinwegheben, und man erhält so wieder eine Congruenz für den einfachen Modul λ. Setzt man alsdann noch für $h^{(2n-1)\lambda}$ das einfachere h^{2n-1} zurück, so ergiebt sich endlich:

$$(17.)\qquad D_n \equiv \frac{(-1)^{n+1}B_n}{2n}\sum_{2}^{\lambda-1}{}_h \frac{h^{2n}-1}{h}\operatorname{Ind.}f(\alpha^h),\quad \text{Mod. } \lambda.$$

Nachdem so der schwierigste Theil der gegenwärtigen Untersuchung vollendet ist, wende ich mich zur Bestimmung der Gröfse M_n, die aus der oben gefundenen Gleichung (12.)

$$NP(e(\eta)^s \mathrm{E}_n(\alpha,\eta))$$

$$= \varphi(\alpha)^\lambda E(\alpha)\left\{f(\alpha^\gamma)^{\gamma^{-2n}-1} f(\alpha^\gamma)^{\gamma^{-4n}-1} \ldots . f(\alpha^{\gamma^{\lambda-2}})^{\gamma^{-2(\lambda-2)n}-1}\right\}^s$$

entwickelt werden mufs, in welcher

$$E(\alpha) = \mathrm{E}_n(\alpha)^{M_n} \quad\text{und}\quad s \equiv \frac{-2\operatorname{Ind.}\mathrm{E}_n(\alpha)}{\gamma^{2n}-1},\quad \text{Mod. } \lambda \text{ ist.}$$

Man erhebe beide Seiten dieser Gleichung zur Potenz p, und mache daraus eine Congruenz für den Modul p. Für diesen Modul wird, wie bekannt, die p^{te} Potenz jeder complexen Zahl, welche α und x, oder auch die Perioden $\eta, \eta_1, \ldots . \eta_{\lambda-1}$ enthält, einer complexen Zahl congruent, welche nur α enthält; also wird die $p$te Potenz von $P(e(\eta)^s \mathrm{E}_n(\alpha,\eta))$ einer complexen Zahl $F(\alpha)$ congruent und mithin wird

$$N(Pe(\eta)^s \mathrm{E}_n(\alpha,\eta))^p \equiv F(\alpha)^\lambda,\quad \text{Mod. } p.$$

Die p^{te} Potenz einer nur α allein enthaltenden complexen Zahl aber ist dieser complexen Zahl selbst congruent, wenn nämlich, wie hier, $p=\nu\lambda+1$ ist. Die obige Gleichung giebt daher folgende Congruenz:

$$F(\alpha)^\lambda \equiv \varphi(\alpha)^\lambda \mathrm{E}_n(\alpha)^{M_n}\left\{f(\alpha^\gamma)^{\gamma^{-2n}-1}.f(\alpha^{\gamma^2})^{\gamma^{-4n}-1} \ldots . f(\alpha^{\gamma^{\lambda-2}})^{\gamma^{-2(\lambda-2)n}-1}\right\}^s$$

für den Modul p, und darum auch für den Modul $f(\alpha)$, welcher ein Primfactor von p ist. Nimmt man nun auf beiden Seiten dieser Congruenz die Indices in Beziehung auf den Modul $f(\alpha)$, und setzt für s seinen oben gefundenen Werth, so erhält man:

$$0 \equiv M_n \operatorname{Ind.E}_n(\alpha) - \frac{2 \operatorname{Ind.E}_n(\alpha)}{\gamma^{2n}-1} \cdot \overset{\lambda-2}{\underset{1}{\Sigma_i}} (\gamma^{-2ni}-1) \operatorname{Ind.} f(\alpha^{\gamma^i}), \quad \text{Mod. } \lambda.$$

Unter der Voraussetzung, dafs $\operatorname{Ind.E}_n(\alpha)$ nicht congruent Null ist, für den Modul λ, kann man $\operatorname{Ind.E}_n(\alpha)$ aufheben, und man findet dann folgenden Ausdruck von M_n:

$$(18.) \qquad M_n \equiv \frac{2}{\gamma^{2n}-1} \overset{\lambda-2}{\underset{1}{\Sigma_i}} (\gamma^{-2ni}-1) \operatorname{Ind.} f(\alpha^{\gamma^i}), \quad \text{Mod. } \lambda.$$

Die gefundenen Ausdrücke von D_n und M_n sind nun in die Congruenz

$$\frac{(-1)^{n+1}(\gamma^{2n}-1) B_n}{4n} \cdot M_n \equiv D_n, \quad \text{Mod. } \lambda,$$

zu setzen, wodurch dieselbe, nach Aufhebung der gemeinschaftlichen Factoren, folgende Gestalt annimmt:

$$* \qquad \overset{\lambda-2}{\underset{1}{\Sigma_i}} (\gamma^{-2ni}-1) \operatorname{Ind.} f(\alpha^{\gamma^i}) \equiv \overset{\lambda-1}{\underset{2}{\Sigma_h}} \frac{h^{2n}-1}{1} \operatorname{Ind.} f(\alpha^h).$$

Setzt man $h \equiv \gamma^{-i}$, so wird

$$\overset{\lambda-2}{\underset{1}{\Sigma_i}} (\gamma^{-2ni}-1)(\operatorname{Ind.} f(\alpha^{\gamma^i}) - \gamma^i \operatorname{Ind.} f(\alpha^{\gamma^{-i}})) \equiv 0.$$

Multiplicirt man mit γ^{2nk}, wo k jede beliebige der Zahlen $1, 2, 3, \ldots \lambda-1$ bedeuten kann, jedoch mit Ausschlufs von $k = \frac{1}{2}(\lambda-1)$, und nimmt die Summe in Beziehung auf $n = 0, 1, 2, \ldots \frac{1}{2}(\lambda-3)$, so wird dieselbe stets congruent Null; mit alleiniger Ausnahme des Falles $i = k$, für welchen sich

$$(19.) \qquad \operatorname{Ind.} f(\alpha^{\gamma^k}) \equiv \gamma^k \operatorname{Ind.} f(\alpha^{\gamma^{-k}}), \quad \text{Mod. } \lambda$$

ergiebt.

Diese Congruenz stellt das gesuchte einfache Reciprocitätsgesetz unter je zwei complexen Primfactoren einer und derselben Primzahl p dar. Dasselbe kann in den Zeichen, welche für λ^{te} Potenzreste den *Legendre*schen für quadratische nachgebildet sind, folgendermafsen dargestellt werden:

$$(20.) \qquad \left(\frac{f(\alpha^{\gamma^k})}{f(\alpha)}\right) = \left(\frac{f(\alpha)}{f(\alpha^{\gamma^k})}\right).$$

Nach der Definition des *Legendre*schen Zeichens hat man nämlich:

$$\left(\frac{f(\alpha^{\gamma^k})}{f(\alpha)}\right) = \alpha^{\operatorname{Ind.} f(\alpha^{\gamma^k})},$$

also auch

$$\left(\frac{f(\alpha^{\gamma^{-k}})}{f(\alpha)}\right) = \alpha^{\operatorname{Ind.} f(\alpha^{\gamma^{-k}})},$$

und wenn α in α^{γ^k} verwandelt wird:

$$\left(\frac{f(\alpha)}{f(\alpha^{\gamma^k})}\right) = \alpha^{\gamma^k \operatorname{Ind}. f(\alpha^{\gamma^{-k}})};$$

woraus die Übereinstimmung beider Darstellungen dieses Reciprocitätsgesetzes erhellet.

Da der Fall $k = \frac{1}{2}(\lambda - 1)$ oben ausgeschlossen werden mufste, so sagt dieses Reciprocitätsgesetz über den Index des einen complexen Primfactors $f(\alpha^{-1})$ nichts aus. Dieser Mangel läfst sich durch diejenige Reciprocitätsgleichung, welche die von mir gefundene Formel der Kreistheilung giebt, leicht ergänzen. Aus dem Ausdrucke

$$F(\alpha^{-1}, x)^\lambda = \pm \alpha^s f(\alpha)^1 f(\alpha^\gamma)^{\gamma_{-1}} f(\alpha^{\gamma^2})^{\gamma_{-2}} \ldots\ldots f(\alpha^{\gamma^{\lambda-2}})^{\gamma_{-(\lambda-2)}},$$

in welchem für primäre Werthe von $f(\alpha)$ die Einheit α^s gleich Eins wird, erhält man nämlich, wenn man auf beiden Seiten durch $p = f(\alpha).f(\alpha^\gamma)\ldots.f(\alpha^{\gamma^{\lambda-2}})$ dividirt und die Indices in Beziehung auf den Modul $f(\alpha)$ nimmt:

$$0 \equiv \sum_{1}^{\lambda-2}{}_h (\gamma^{-h} - 1) \operatorname{Ind}. f(\alpha^{\gamma^h}), \quad \text{Mod. } \lambda.$$

Fasset man in dieser Summe je zwei vom Anfange und Ende gleich weit entfernte Glieder zusammen, so wird, vermöge des oben bewiesenen Reciprocitätsgesetzes, die Summe derselben congruent Null, und es bleibt nur das mittelste Glied, nämlich $-2 \operatorname{Ind}. f(\alpha^{-1})$ übrig. Man erhält also

$$(21.) \quad \operatorname{Ind}. f(\alpha^{-1}) \equiv 0, \quad \text{Mod. } \lambda,$$

oder, nach der *Legendre*schen Bezeichnung:

$$(22.) \quad \left(\frac{f(\alpha^{-1})}{f(\alpha)}\right) = 1.$$

Nach ähnlichen Methoden, jedoch nicht ohne Anwendung einiger andern Hülfsmittel, welche ziemlich bedeutende Vorarbeiten erfordern, habe ich auch den Beweis des allgemeinen Reciprocitätsgesetzes für zwei verschiedene nicht congruente complexe Primzahlen gesucht. Dieser allgemeinere Beweis aber leidet, eben so wie der hier gegebene, an dem Mangel, dafs er, auch für Potenzreste von einem bestimmten Exponenten λ, ***nicht alle*** complexen Primzahlen umfasset, weil die obige Voraussetzung, dafs $\operatorname{Ind}. E_n(\alpha)$ für den Modul λ nicht congruent Null sein darf, und zwar für keinen der Werthe $n = 1, 2, 3, \ldots\ldots \frac{1}{2}(\lambda - 3)$, nicht für alle Primzahlen $f(\alpha)$ erfüllt wird. Für $\lambda = 5$, also für die *fünften* Potenzreste, giebt es folgende acht Primzahlen

30*

unter Tausend, für deren complexe Primfactoren Ind. $E_n(\alpha) \equiv 0$, Mod. 5 ist nämlich: $p = 211$, 281, 421, 461, 521, 691, 881 und 991, auf welche also auch der vorstehende Beweis des Reciprocitätsgesetzes unter conjugirten Primzahlen sich nicht erstreckt, obgleich, wie die wirkliche Ausrechnung mit Hülfe der Tafeln des Canon arithmeticus von *Jacobi* leicht ergiebt, dasselbe auch für diese vollständig richtig ist. Sollte es mir nicht gelingen, diese Unvollständigkeit der auf die Ausdrücke $P\varepsilon(\alpha, x)$ gegründeten Beweise der Reciprocitätsgesetze zu entfernen, so würde ich genöthigt sein, den Weg, welchen ich seit mehreren Jahren ausdauernd verfolgt habe, gänzlich zu verlassen und einen andern zu suchen.

Breslau, den 31. August 1854.

Über die den Gaussischen Perioden der Kreistheilung entsprechenden Congruenzwurzeln

Journal für die reine und angewandte Mathematik 53, 142–148 (1857)

Die Theorie der idealen Primfactoren derjenigen complexen Zahlen, welche als einzige Irrationalitäten die Wurzeln der Gleichung $\alpha = 1$ enthalten, wo λ eine ungerade Primzahl ist, beruht hauptsächlich darauf, dafs allen Perioden, welche aus den Wurzeln der Gleichung

$$\alpha^{\lambda-1} + \alpha^{\lambda-2} + \alpha^{\lambda-3} + \cdots + \alpha^2 + \alpha + 1 = 0$$

gebildet werden können, gewisse ganze Zahlen entsprechen, welche aus den die Perioden bestimmenden Gleichungen hervorgehen, wenn man dieselben als Congruenzen für bestimmte Moduln betrachtet. Ist e ein Theiler von $\lambda - 1$, $e.f = \lambda - 1$, und γ eine primitive Wurzel für den Modul λ, so giebt der Ausdruck:

$$\eta_k = \alpha^{\gamma^k} + \alpha^{\gamma^{k+e}} + \alpha^{\gamma^{k+2e}} + \cdots + \alpha^{\gamma^{k+(f-1)e}}$$

für $k = 0, 1, 2, \ldots e-1$, die e Perioden von je f Gliedern. Diese sind die Wurzeln einer Gleichung des e^{ten} Grades mit ganzzahligen Coefficienten

$$(1.)\quad y^e + A_1 y^{e-1} + A_2 y^{e-2} + \cdots + A_e = 0$$

und aufserdem besteht für dieselben ein System von e^2 Gleichungen von der Form

$$(2.)\quad \eta_r \eta_{r+k} = \overset{k}{\mu} f + \overset{k}{m} \eta_r + \overset{k}{m_1} \eta_{r+1} + \overset{k}{m_2} \eta_{r+2} + \cdots + \overset{k}{m}_{e-1} \eta_{r-1}$$

für $r = 0, 1, 2, \ldots e-1$ und $k = 0, 1, 2, \ldots e-1$, auf welchen die Rechnung mit den Perioden beruht. Wenn nun q eine Primzahl ist, deren f^{te} Potenz congruent Eins ist, für den Modul λ, für welche also

$$q^f \equiv 1, \quad \text{mod. } \lambda,$$

so hat die Congruenz:

$$(3.)\quad y^e + A_1 y^{e-1} + A_2 y^{e-2} + \cdots + A_e \equiv 0, \quad \text{mod. } q,$$

stets die e realen Congruenzwurzeln

$$y = u,\ u_1,\ u_2,\ \ldots\ u_{e-1},$$

welche jedoch auch zum Theil einander gleich sein können. Dieses habe ich zuerst in der Abhandlung „Über die Divisoren gewisser Formen u. s. w." * Bd. 30, pag. 107 dieses Journals streng bewiesen. In der Abhandlung „Über die Zerlegung der aus Wurzeln der Einheit gebildeten complexen Zahlen in ihre Primfactoren" Bd. 35, pag. 327 habe ich sodann gezeigt, dafs dieselben Congruenzwurzeln $u, u_1, u_2, \dots u_{e-1}$ auch dem Systeme der e^2 Congruenzen

$$(4.)\qquad u_r u_{r+k} \equiv \overset{k}{\mu} f + \overset{k}{m} u_r + \overset{k}{m_1} u_{r+1} + \overset{k}{m_2} u_{r+2} + \cdots \overset{k}{m_{e-1}} u_{r-1}, \quad \text{mod. } q,$$

für $x = 0, 1, 2, \dots e-1$, $k = 0, 1, 2, \dots e-1$, genügen, wenn sie in gehöriger, cyklisch vollständig bestimmter Ordnung genommen werden, und dafs ganz allgemein jede rationale Gleichung unter den Perioden, als Congruenz für den Modul q befriedigt wird, wenn anstatt der Perioden die entsprechenden Congruenzwurzeln eingesetzt werden. Dieser wichtige Punct scheint mir noch einer näheren Aufklärung und vollständigeren Begründung zu bedürfen, welche ich hier mit Hülfe einer sehr einfachen, auch für allgemeinere complexe Zahlen anwendbaren Methode, ausführen will.

Ich gehe zu diesem Zwecke wieder von der auch in meinen früheren Untersuchungen benutzten Congruenz aus:

$$x(x-1)(x-2)(x-3)\dots(x-q+1) \equiv x^q - x, \quad \text{mod. } q,$$

welche identisch für jeden Werth des x erfüllt wird, in der Art, dafs wenn das Product auf der einen Seite nach Potenzen von x entwickelt wird, die Coefficienten der gleichen Potenzen auf beiden Seiten nach dem Modul q congruent sind. Setzt man in dieser Congruenz $x = \eta_k$ und bemerkt, dafs für jede Primzahl q, welche der Congruenz $q^f \equiv 1$, mod. λ, genügt, die q^{te} Potenz einer jeden Periode von f Gliedern derselben einfachen Periode congruent ist, nach dem Modul q, dafs also

$$\eta_k^q \equiv \eta_k, \quad \text{mod. } q,$$

welche, so wie auch die folgenden Congruenzen in denen Perioden vorkommen, in dem Sinne aufzufassen ist, dafs wenn man beide Seiten in die lineäre Form $a\eta + a_1\eta_1 + \cdots + a_{e-1}\eta_{e-1}$ bringt, die Coefficienten der gleichen Perioden auf beiden Seiten einzeln congruent sind nach dem Modul q: so erhält man hieraus die Congruenz:

$$(5.)\qquad \eta_k(\eta_k - 1)(\eta_k - 2)\dots(\eta_k - q + 1) \equiv 0, \quad \text{mod. } q.$$

Wird nun der Kürze wegen das Product

$$(y-\eta)(y-\eta_1)(y-\eta_2)\dots(y-\eta_{e-1})$$

oder, was dasselbe ist, die linke Seite der Gleichung (1.) durch $\varphi(y)$ bezeichnet, so ergiebt diese Congruenz, wenn nach einander $k=0,1,2,\dots e-1$ gesetzt und das Product aller so erhaltenen Congruenzen gebildet wird:

$$\varphi(0).\varphi(1).\varphi(2)\dots.\varphi(q-1)\equiv 0,\ \text{mod.}\ q^e.$$

Hieraus folgt, dafs die Congruenz des e^{ten} Grades $\varphi(y)\equiv 0$, mod. q, stets wirkliche Congruenzwurzeln hat. Es sollen nun nur die *von einander verschiedenen* Wurzeln dieser Congruenz in Betracht gezogen werden, deren Anzahl gleich h sei, und welche durch a, a_1, a_2, ... a_{h-1} bezeichnet werden sollen. Wenn nun diese h Zahlen alle diejenigen erschöpfen, welche der Congruenz $\varphi(y)\equiv 0$, mod. q, genügen, so folgt, dafs es unter den Zahlen $0, 1, 2, 3, \dots q-1$ aufserdem $q-h$ Zahlen giebt, welche dieser Congruenz nicht genügen, welche durch b, b_1, b_2, ... b_{q-h-1} bezeichnet werden sollen. Die Zahlen a, a_1, ... a_{h-1} mit den Zahlen b, b_1, ... b_{q-h-1} zusammen, erschöpfen so vollständig alle Zahlen $0, 1, 2, \dots q-1$, deshalb kann die Congruenz (5.) auch folgendermafsen dargestellt werden:

$$(a-\eta_k)(a_1-\eta_k)\dots(a_{h-1}-\eta_k)(b-\eta_k)(b_1-\eta_k)\dots(b_{q-h-1}-\eta_k)\equiv 0,\ \text{mod.}\ q.$$

Multiplicirt man diese Congruenz mit dem Producte aller zu $b-\eta_k$ conjugirten Factoren, ebenso mit dem Producte aller zu $b_1-\eta_k$ conjugirten Factoren u. s. w., so geht dieselbe über in:

$$(a-\eta_k)(a_1-\eta_k)\dots(a_{h-1}-\eta_k).\varphi(b).\varphi(b_1)\dots\varphi(b_{q-h-1})\equiv 0,\ \text{mod.}\ q.$$

Die Factoren $\varphi(b)$, $\varphi(b_1)$, ..., welche nur ganze Zahlen sind, deren keine durch q theilbar ist, weil die Zahlen b, b_1, ... b_{q-h-1} der Congruenz $\varphi(y)\equiv 0$, mod. q, nicht genügen, können nun durch Division weggehoben werden, wodurch diese Congruenz folgende einfachere Gestalt erhält:

$$(6.)\qquad (a-\eta_k)(a_1-\eta_k)(a_2-\eta_k)\dots.(a_{h-1}-\eta_k)\equiv 0,\ \text{mod.}\ q.$$

Aus einer gewissen Anzahl der Factoren dieses Products und der mit denselben conjugirten, nämlich aus den Factoren:

$$\begin{array}{lllll} a-\eta, & a_1-\eta, & a_2-\eta, & \dots & a_{h-1}-\eta, \\ a-\eta_1, & a_1-\eta_1, & a_2-\eta_1, & \dots & a_{h-1}-\eta_1, \\ \dots & \dots & \dots & \dots & \dots \\ a-\eta_{e-1}, & a_1-\eta_{e-1}, & a_2-\eta_{e-1}, & \dots & a_{h-1}-\eta_{e-1}, \end{array}$$

deren Anzahl gleich $h.e$ ist, bilde ich nun ein Product $\Psi(\eta)$ von der Beschaffenheit, dafs es möglichst viele dieser $h.e$ Factoren enthalte, keinen mehr als einmal, ohne jedoch durch q theilbar zu sein, dafs es aber durch Hinzufügung eines jeden dieser Factoren, der noch nicht darin enthalten ist, durch

q theilbar werde. Um ein solches Product zu bilden nimmt man irgend einen dieser Faktoren und multiplicirt ihn mit einem zweiten, ist dieses Produkt nicht durch q theilbar, so multiplicirt man mit einem dritten und so fort; erhält man durch Hinzufügung eines neuen Faktors ein durch q theilbares Produkt, so ist dieser Faktor zu verwerfen und ein anderer der noch vorhandenen Faktoren zu nehmen, welcher dem Produkte hinzuzufügen oder zu verwerfen ist, jenachdem er dasselbe durch q nicht theilbar oder durch q theilbar macht. Sind auf diese Weise alle $h.e$ Faktoren an die Reihe gekommen und entweder dem Produkte hinzugefügt oder verworfen, so ist das gesuchte Product $\Psi(\eta)$, welches die angegebenen Bedingungen erfüllt, gefunden.

In diesem Produkte $\Psi(\eta)$ können nun von den h zusammengehörenden Faktoren, welche dieselbe Periode η_k enthalten, nämlich

$$a-\eta_k, \quad a_1-\eta_k, \quad a_2-\eta_k, \quad \ldots \quad a_{h-1}-\eta_k$$

nicht alle zugleich enthalten sein, weil sonst vermöge der Congruenz (6.) $\Psi(\eta)$ durch q theilbar sein würde; es mufs also für jede der Perioden η, η_1, η_2, ... η_{e-1}, oder was dasselbe ist, für jeden Werth des k wenigstens einen dieser Faktoren geben, welcher nicht in $\Psi(\eta)$ enthalten ist, welcher daher mit $\Psi(\eta)$ multiplicirt ein durch q theilbares Produkt giebt. Wird die in diesem Faktor vorkommende Congruenzwurzel a durch u_k bezeichnet, so dafs dieser Faktor gleich $u_k-\eta_k$ ist, so hat man die Congruenz:

$$(7.)\quad \begin{cases} \Psi(\eta)(u_k-\eta_k) \equiv 0, \quad \text{mod. } q, \\ \text{oder} \\ \Psi(\eta)\eta_k \equiv \Psi(\eta)u_k, \quad \text{mod. } q, \end{cases}$$

welche Congruenz für jeden beliebigen Werth des $k=0, 1, 2, \ldots e-1$ giltig ist, und somit ein System von e Congruenzen repräsentirt. In dieser Congruenz gehört auch zu jeder bestimmten Periode η_k nur eine einzige Congruenzwurzel u_k; denn wenn derselben auch eine von u_k verschiedene Zahl v_k genügte, so würde daraus folgen

$$\Psi(\eta)(u_k-v_k) \equiv 0, \quad \text{mod. } q,$$

welches unmöglich ist.

Bezeichnet nun $F(\eta)$ irgend eine ganze und ganzzahlige Function der Perioden, welche also nur aus Gliedern von der Form $c.\eta_r\eta_s\eta_t\ldots$ besteht, wo die Indices der Perioden auch einander gleich sein können, so hat man

19*

durch mehrmals wiederholte Anwendung der Congruenz (7.)

$$\Psi(\eta).\eta_r.\eta_s.\eta_t \cdots \equiv \Psi(\eta).u_r.u_s.u_t \ldots \text{ mod. } q$$

und demnach, wenn $F(u)$ diejenige ganze Zahl bedeutet, welche man aus $F(\eta)$ erhält, wenn man in dieser anstatt der Perioden $\eta, \eta_1, \ldots \eta_{e-1}$ die entsprechenden Congruenzwurzeln $u, u_1, \ldots u_{e-1}$ setzt,

$$(8.) \quad \Psi(\eta)F(\eta) \equiv \Psi(\eta)F(u), \quad \text{mod. } q.$$

Hat man nun irgend eine rationale Gleichung unter den Perioden, welche aufser diesen nur ganze Zahlen enthält, so kann man dieselbe immer auf die Form $F(\eta) = 0$ bringen, indem man die etwa darin enthaltenen Brüche durch Multiplication entfernt und alle Glieder auf eine Seite bringt. Aus $F(\eta) = 0$ folgt aber vermöge der Congruenz (8.) die Congruenz $F(u) \equiv 0$, mod. q, und somit hat man den vollständigen Beweis des Satzes, *dafs jede rationale Gleichung unter den Perioden, welche aufser diesen nur ganze Zahlen enthält, als Congruenz für den Modul q befriedigt wird, wenn anstatt der Perioden $\eta, \eta_1, \ldots \eta_{e-1}$ die ihnen entsprechenden Congruenzwurzeln $u, u_1, \ldots u_{e-1}$ gesetzt werden.* Aus dem Systeme der e^2 Fundamentalgleichungen für die Rechnung mit den Perioden, nämlich

$$\eta_r\eta_{r+k} = \overset{k}{\mu}f + \overset{k}{m}\eta_r + \overset{k}{m_1}\eta_{r+1} + \cdots + \overset{k}{m_{e-1}}\eta_{r-1},$$

folgt somit auch das System der e^2 Congruenzen unter den Zahlen $u, u_1, \ldots u_{e-1}$:

$$u_r u_{r+k} \equiv \overset{k}{\mu}f + \overset{k}{m}u_r + \overset{k}{m_1}u_{r+1} + \cdots + \overset{k}{m_{e-1}}u_{r-1}, \quad \text{mod. } q.$$

Die zu $\Psi(\eta)$ conjugirten complexen Zahlen sind, wie man aus dem angegebenen Bildungsgesetze der Zahl $\Psi(\eta)$ sogleich erkennt, ebenfalls solche complexe Zahlen, welche denselben Bedingungen genügen. Verwandelt man daher in der Congruenz (7.) die in den Perioden zu Grunde gelegte Wurzel α in α^{γ^r}, wodurch die Indices aller Perioden um r vermehrt werden, so hat man

$$(9.) \quad \Psi(\eta_r)(u_k - \eta_{k+r}) \equiv 0, \quad \text{mod. } q,$$

oder wenn k in $k-r$ verwandelt wird:

$$(10.) \quad \Psi(\eta_r)(u_{k-r} - \eta_k) \equiv 0, \quad \text{mod. } q,$$

woraus folgt, dafs die den Perioden $\eta, \eta_1, \ldots \eta_{e-1}$ entsprechenden Congruenzwurzeln $u, u_1, \ldots u_{e-1}$, mit Beibehaltung ihrer cyklischen Ordnung, denselben auf e verschiedene Weisen zugeordnet werden können, nämlich so, dafs für jeden Werth des k, der Periode η_k die Congruenzwurzel u_{k-r} entspricht, wo r einen jeden der Werthe $r = 0, 1, 2, \ldots e-1$ haben kann.

Die nach der oben angegebenen Regel gebildeten complexen Zahlen $\Psi(\eta)$ können mit dem besten Erfolge als Grundlage für die Theorie der aus den Wurzeln der Gleichung $\alpha^\lambda = 1$ gebildeten complexen Zahlen, namentlich für die Zerlegung derselben in ihre idealen Primfaktoren benutzt werden. Aus diesem Grunde will ich hier noch die wesentlichsten Eigenschaften derselben kurz entwickeln, indem ich erstens zeige, *dafs die conjugirten complexen Zahlen $\Psi(\eta)$, $\Psi(\eta_1)$, ... $\Psi(\eta_{e-1})$ alle wirklich von einander verschieden sind,* und zweitens, *dafs diese e conjugirten complexen Zahlen alle diejenigen erschöpfen, welche nach dem gegebenen Bildungsgesetze überhaupt gebildet werden können.*

Gesetzt es wären zwei dieser conjugirten Zahlen einander gleich, z. B. $\Psi(\eta_s) = \Psi(\eta_t)$, so würde hieraus, wenn die Indices der Perioden um s vermindert und $t-s = r$ gesetzt wird, folgen, dafs auch

$$\Psi(\eta) = \Psi(\eta_r)$$

sein müfste. Nun ist aber nach den Congruenzen (7.) und (9.)

$$\Psi(\eta)(u_k - \eta_k) \equiv 0 \quad \text{und} \quad \Psi(\eta_r)(u_k - \eta_{k+r}) \equiv 0, \quad \text{mod. } q,$$

also müfste auch sein:

$$\Psi(\eta)(\eta_k - \eta_{k+r}) \equiv 0, \quad \text{mod. } q.$$

Multiplicirt man nun mit $\alpha^{-\gamma^k}$ und nimmt die Summe für $k = 0, 1, 2, \ldots \lambda-2$, so hat man zunächst

$$\alpha^{-\gamma^k}\eta_k = 1 + \alpha^{\gamma^k(\gamma^e - 1)} + \alpha^{\gamma^k(\gamma^{2e}-1)} + \cdots + \alpha^{\gamma^k(\gamma^{(f-1)e}-1)}$$

$$\alpha^{-\gamma^k}\eta_{k+r} = \alpha^{\gamma^k(\gamma^r - 1)} + \alpha^{\gamma^k(\gamma^{r+e}-1)} + \alpha^{\gamma^k(\gamma^{r+2e}-1)} + \cdots + \alpha^{\gamma^k(\gamma^{r+(f-1)e}-1)}$$

also

$$\sum_0^{\lambda-2}{}_k\, \alpha^{-\gamma^k}\eta_k = \lambda - f \quad \text{und} \quad \sum_0^{\lambda-2}{}_k\, \alpha^{-\gamma^k}\eta_{k+r} = -f,$$

woraus folgen würde

$$\Psi(\eta)\lambda \equiv 0, \quad \text{mod. } q,$$

da dieses aber unmöglich ist, so folgt, dafs die conjugirten complexen Zahlen $\Psi(\eta)$, $\Psi(\eta_1)$, ... $\Psi(\eta_{e-1})$ wirklich alle von einander verschieden sind.

Betrachtet man nun irgend welche zwei dieser conjugirten Zahlen, so sieht man, dafs dieselben, weil sie wirklich verschieden sind, nicht genau aus denselben Factoren von der Form $a-\eta$ zusammengesetzt sein können, dafs vielmehr die eine irgend welche dieser Factoren enthalten mufs, welche die andere nicht enthält, woraus unmittelbar folgt, dafs die Producte je zweier alle durch q theilbar sein müssen.

Die Summe dieser e conjugirten Zahlen, nämlich

$$\Psi(\eta)+\Psi(\eta_1)+\cdots\Psi(\eta_{e-1}) = M$$

ist als symmetrische Function aller Perioden eine nicht complexe ganze Zahl. Dieselbe ist auch nicht durch q theilbar, denn weil die Produkte je zweier verschiedener dieser conjugirten Zahlen alle durch q theilbar sind, so hat man

$$\Psi(\eta).M \equiv \Psi(\eta)^2, \text{ mod. } q.$$

Wäre nun M durch q theilbar, so müfste auch $\Psi(\eta)^2$ und folglich auch $\Psi(\eta)$ selbst durch q theilbar sein, welches nicht der Fall ist.

Nimmt man nun an, es liefse sich nach der oben angegebenen Regel aus den Factoren von der Form $a-\eta$ eine von $\Psi(\eta)$ und deren conjugirten verschiedene Zahl $\Phi(\eta)$ bilden, so würde für jedes r die Congruenz Statt haben

$$\Phi(\eta)\Psi(\eta_r) \equiv 0, \text{ mod. } q,$$

denn $\Phi(\eta)$ müfste nothwendig irgend welche Factoren von der Form $a-\eta$ enthalten, welche $\Psi(\eta_r)$ nicht enthält und eben so $\Psi(\eta_r)$ irgend welche nicht in $\Phi(\eta)$ enthaltene. Wenn nun nach einander $r=0, 1, 2, \ldots e-1$ genommen wird, so würde die Summation dieser e Congruenzen ergeben

$$\Phi(\eta)M \equiv 0, \text{ mod. } q,$$

und weil M eine nicht complexe, durch q nicht theilbare ganze Zahl ist, so müfste sein

$$\Phi(\eta) \equiv 0, \text{ mod. } q,$$

welches unmöglich ist. Die e conjugirten complexen Zahlen

$$\Psi(\eta), \ \Psi(\eta_1), \ \ldots \ \Psi(\eta_{e-1})$$

sind also die einzigen, welche den für diese Art von complexen Zahlen festgesetzten Bestimmungen genügen.

Berlin, den 5. Juni 1856.

Anzeige einer Schrift des Herrn Reuschle in Stuttgart

Journal für die reine und angewandte Mathematik 53, 379 (1857)

Den Mathematikern, welche sich mit zahlentheoretischen Untersuchungen beschäftigen wird es vielleicht erwünscht sein von einer Schrift des Herrn Professor Dr. *Reuschle* in Stuttgart Nachricht zu erhalten, welche derselbe als Programm des dasigen Gymnasiums zu Michaelis 1856 hat erscheinen lassen und von welcher besondere Abdrücke durch die Antiquariats-Buchhandlung von *Liesching* et Comp. in Stuttgart zu beziehen sind. Diese Schrift enthält sehr werthvolle, mit Fleifs und Sachkenntnifs berechnete Tafeln zum Gebrauch für Zahlentheorie, nämlich: 1) Die Zerlegung der Zahlen 10^n-1 in ihre Primfaktoren für eine gewisse Anzahl von Werthen des n, bis $n=242$, für welche die Faktoren als wirkliche Primfaktoren constatirt sind. 2) Eine Anzahl ähnlicher Zerlegungen der in der Form a^n-1 enthaltenen Zahlen in ihre Primfactoren, für eine Reihe von Werthen der Zahlen a und n, für welche die in denselben vorkommenden Primfaktoren als solche haben constatirt werden können. 3) Eine Tafel der Zerlegung der Primzahlen $p=6n+1$ in die Formen $p=A^2+3B^2$ und $4p=C^2+27M^2$, welche für die Theorie der cubischen Reste, so wie für die Kreistheilung wichtige Dienste leistet. Diese Tafel erstreckt sich für beide Formen vollständig bis $p=5743$, sodann für die Form $p=A^2+3B^2$ allein weiter bis 13669, und endlich für diejenigen Werthe des p, für welche die Zahl 10 cubischer Rest ist, noch bis $p=49999$. 4) Eine Tafel der Zerlegung der Primzahlen $p=4n+1$ in die Form a^2+b^2, zugleich mit der Zerlegung der Primzahlen $p=8n+1$ in die Form c^2+2d^2 bis $p=12377$. An diese schliefst sich noch die Zerlegung derjenigen Primzahlen $p=4n+1$, für welche die Zahl 10 quadratischer Rest ist, in die Form a^2+b^2, bis $p=24917$, zugleich mit der Zerlegung derjenigen Primzahlen, für welche die Zahl 10 biquadratischer Rest ist, in die Form $p=c^2+2d^2$, bis $p=24889$. Der Nutzen welchen diese Zerlegungen, namentlich für die Theorie der biquadratischen Reste und auch für die Kreistheilung haben, wird den Kennern dieser Theorieen einleuchten. 5) Folgt noch eine Reihe von Tafeln, in welchen die kleinsten Exponenten berechnet sind, zu denen gegebene Zahlen gehören nach bestimmten Moduln, oder die Bestimmung der kleinsten Zahl e, welche der Congruenz $a^e\equiv 1$, mod. p, genügt. Der erste Theil dieser Tafel umfafst die sechs Werthe des $a=2, 3, 5, 6, 7, 10$, und als Moduln alle Primzahlen bis $p=997$. Von da an bis $p=4999$ erstreckt sich diese Tafel nur auf die beiden Grundzahlen $a=2$ und $a=10$, und endlich von da an bis zu $p=14983$ nur auf die eine Grundzahl 10. Diese Tafeln, welche unter andern auch die Frage nach der Anzahl der Ziffern der Periode eines aus einem gewöhnlichen Bruche entstandenen Decimalbruchs beantworten, geben aufserdem zu manchen interessanten, so zu sagen statistischen Bemerkungen Veranlassung, welche in der Einleitung zu den Tabellen angestellt sind.

Aus dem angegebenen Inhalte wird man ersehen, dafs die ganze Arbeit im Interesse der Wissenschaft und zum wirklichen Nutzen derselben unternommen und ausgeführt ist, wofür auch noch der Antheil Bürgschaft leistet, welchen der verewigte *Jacobi* an der Entstehung derselben gehabt und der Werth, welchen dieser grofse Mathematiker den berechneten Tafeln beigelegt hat. Herr Professor *Reuschle* ist nämlich wegen dieser Arbeit mit *Jacobi* in schriftlichen und mündlichen Verkehr getreten und die aus zwei Briefen von *Jacobi* und zwei Briefen von Prof. *Reuschle* bestehende Correspondenz über diesen Gegenstand, in welcher mehrere Fragen über Potenzreste erörtert werden, bildet einen interessanten Bestandtheil der Einleitung, welche Herr Professor *Reuschle* seinen Tafeln vorangeschickt hat.

Kummer.

Theorie der idealen Primfaktoren der complexen Zahlen, welche aus den Wurzeln der Gleichung $\omega^n = 1$ gebildet sind, wenn n eine zusammengesetzte Zahl ist

Mathematische Abhandlungen der Königlichen Akademie der Wissenschaften zu Berlin aus dem Jahre 1856, 1–47

[Gelesen in der Akademie der Wissenschaften am 18. December 1856.]

In mehreren früheren Abhandlungen habe ich die Theorie der aus λ^{ten} Wurzeln der Einheit gebildeten complexen Zahlen für den Fall, dafs λ eine Primzahl ist, vollständig behandelt und deren Anwendung auf die allgemeinen Reciprocitätsgesetze, auf den letzten Fermat'schen Satz und auf die Theorie der Kreistheilung gezeigt. Seitdem hat die Grundidee dieser Theorie, nämlich die Zerlegung der complexen Zahlen in ihre wahren idealen Primfaktoren auch in der Algebra ihre Anwendung gefunden, da z. B. das von Herrn Kronecker gefundene, der Akademie unter dem 14. April d. J. mitgetheilte Resultat: dafs die Wurzeln jeder Abel'schen Gleichung mit ganzzahligen Coefficienten nur rationale Funktionen von Wurzeln der Einheit sind, von demselben nur mit Hülfe der idealen Primfaktoren gefunden und bewiesen worden ist. Damit nun die Anwendungen dieser Theorie nicht nur auf die Fälle beschränkt bleiben, wo es sich um Einheitswurzeln handelt, deren Exponent eine Primzahl ist, habe ich dieselbe in dem Folgenden auf die Einheitswurzeln mit beliebig zusammengesetzten Wurzelexponenten ausgedehnt. Die hauptsächlich in den Unterscheidungen vieler einzelnen Fälle bestehenden Weitläufigkeiten, welche die Betrachtung zusammengesetzter Zahlen in der Zahlentheorie gewöhnlich mit sich führt, habe ich durch zweckmäfsige Wahl der Methoden überall vermeiden können, und da in dieser allgemeinen Untersuchung die Beweise der Sätze der Natur der Sache nach nur auf die in allen Fällen bleibenden und daher wesentlichen Eigenschaften gegründet werden konnten, so ist es mir gelungen, dieselbe sogar einfacher darzustellen als die früher behandelte speciellere Theorie.

§. 1.

Die sämmtlichen **primitiven** Wurzeln der Gleichung

$$\omega^n = 1,$$

in welcher n eine beliebig zusammengesetzte Zahl bedeutet, deren Anzahl bekanntlich gleich der Anzahl aller Zahlen ist, welche kleiner als n und relative Primzahlen zu n sind, also nach der von Gaufs eingeführten Bezeichnung gleich $\phi(n)$, sind die Wurzeln einer bekannten Gleichung des Grades $\phi(n)$ mit ganzzahligen Coefficienten, in welcher der Coefficient der höchsten Potenz gleich Eins ist. Hieraus folgt, dafs jede ganze complexe Zahl, als ganze und ganzzahlige rationale Funktion einer primitiven Wurzel ω, in die Form gesetzt werden kann

$$F(\omega) = a + a_1\omega + a_2\omega^2 + \ldots + a_{\phi(n)-1}\,\omega^{\phi(n)-1};$$

alle höheren Potenzen des ω können nämlich mit Hülfe der Gleichung des Grades $\phi(n)$ eliminirt werden. Aus der Irreduktibilität dieser Gleichung, deren Wurzeln die sämmtlichen **primitiven** Wurzeln der Gleichung $\omega^n = 1$ sind, welche zuerst von Kronecker in Liouville's Journal, Jahrgang 1854 pag. 177 sqq. bewiesen worden ist, folgt ferner unmittelbar, dafs eine bestimmte complexe Zahl $F(\omega)$ sich nur auf eine einzige Weise in die obige Form setzen läfst. Eben so folgt daraus weiter, dafs wenn $F(\omega)$ durch eine nichtcomplexe ganze Zahl c theilbar sein soll, alle $\phi(n)$ Coefficienten derselben, $a, a_1, a_2,$ etc. einzeln durch c theilbar sein müssen. Die Congruenz $F(\omega) \equiv 0$, mod. c, enthält daher $\phi(n)$ Congruenzen unter nichtcomplexen, ganzen Zahlen in sich und ist mit diesen vollkommen gleichbedeutend.

Setzt man in eine complexe Zahl $F(\omega)$ für ω nacheinander alle primitiven Wurzeln der Gleichung $\omega^n = 1$, so erhält man $\phi(n)$ zusammengehörige complexe Zahlen, welche **conjugirte** heifsen sollen. Diese conjugirten complexen Zahlen sind also alle von der Form $F(\omega^k)$, wo k relative Primzahl zu n ist; wenn aber k mit n einen gemeinschaftlichen Faktor hat, ω^k also nicht primitive Wurzel ist, so soll $F(\omega^k)$ nicht als conjugirte Zahl zu $F(\omega)$ angesehen werden, in diesem Falle ist auch die obige allgemeine Form für $F(\omega)$ oder $F(\omega^k)$ keine bestimmte mehr.

Das Produkt aller $\phi(n)$ mit $F(\omega)$ conjugirten complexen Zahlen, welches als symmetrische Funktion aller primitiven Wurzeln der Gleichung

$\omega^n = 1$ nothwendig eine nichtcomplexe ganze Zahl ist, soll die Norm von $F(\omega)$ heifsen und durch $NF(\omega)$ bezeichnet werden.

Da jede complexe Zahl $F(\omega)$ ein Faktor ihrer Norm, also Faktor einer nichtcomplexen ganzen Zahl ist, so folgt, dafs auch alle complexen Faktoren des $F(\omega)$ als Faktoren nichtcomplexer ganzer Zahlen angesehen werden können. Aus diesem Grunde werden wir auch die zu untersuchenden idealen Primfaktoren der complexen Zahlen überall als Primfaktoren nichtcomplexer ganzer Zahlen, und weil diese selbst nur aus nichtcomplexen Primzahlen zusammengesetzt sind, als ideale Primfaktoren der nichtcomplexen Primzahlen betrachten. Diese gewöhnlichen Primzahlen sind nun entweder solche, welche nicht in n enthalten sind, oder sie sind Faktoren des n. Demgemäfs werden auch zwei verschiedene Arten idealer Primfaktoren gesondert zu betrachten sein, nämlich erstens die idealen Primfaktoren der nicht in n enthaltenen gewöhnlichen Primzahlen und zweitens die idealen Primfaktoren der in n enthaltenen gewöhnlichen Primzahlen. Die idealen Primfaktoren der zweiten Art, welche stets nur einer endlichen beschränkten Anzahl gewöhnlicher Primzahlen angehören, werden hierbei gegen die der ersten Art nur als Ausnahmen anzusehen sein, deren besondere Eigenthümlichkeiten nach der allgemeinen Untersuchung der idealen Primfaktoren der ersten Art und mit Hülfe derselben behandelt werden sollen.

§. 2.

Es sei q eine Primzahl, welche nicht in n enthalten ist, und t der Exponent, zu welchem q gehört, für den Modul n, so dafs

$$q^t \equiv 1, \text{ mod. } n,$$

dafs aber kein kleinerer Exponent als t existire, für welchen diese Congruenz Statt habe. Es sind dann die Potenzen

$$1,\ q,\ q^2,\ q^3 \ldots\ldots q^{t-1}$$

in Beziehung auf den Modul n alle verschieden von einander. Ferner ist nach bekannten Principien

$$F(\omega)^q \equiv F(\omega^q), \text{ mod. } q,$$

und allgemein für jeden ganzzahligen Werth von h

$$F(\omega)^{q^h} \equiv F(\omega^{q^h}), \text{ mod. } q,$$

also wenn h gleich einem Vielfachen von t, $h = kt$ genommen wird, weil

A 2

$q^{kt} \equiv 1$, mod. n, ist

$$F(\omega)^{q^{kt}} \equiv F(\omega), \text{ mod. } q.$$

Hieraus folgt: Wenn $F(\omega)$ nicht durch q theilbar ist, so ist auch keine Potenz von $F(\omega)$ durch q theilbar. Wäre nämlich $F(\omega)^m$ durch q theilbar, so müfste jede höhere Potenz ebenfalls durch q theilbar sein, darum, wenn k so grofs genommen wird, dafs $q^{kt} > m$ ist, so müfste $F(\omega)^{q^{kt}}$ durch q theilbar sein, also vermöge der obigen Congruenz auch $F(\omega)$ theilbar durch q.

Nimmt man nacheinander $h = 0, 1, 2, \ldots . t-1$ und multiplicirt alle diese Congruenzen mit einander, so erhält man auch

$$F(\omega)^{1+q+q^2+\ldots+q^{t-1}} \equiv F(\omega)F(\omega^q)F(\omega^{q^2})\ldots F(\omega^{q^{t-1}}), \text{ mod } q.$$

Der aus den Wurzeln der Gleichung $\omega^n = 1$ gebildete Ausdruck

$$\varpi = \omega + \omega^q + \omega^{q^2} + \ldots . + \omega^{q^{t-1}}$$

oder allgemeiner

$$\varpi_k = \omega^k + \omega^{kq} + \omega^{kq^2} + \ldots . + \omega^{kq^{t-1}}$$

in welchem ω eine primitive Wurzel dieser Gleichung bezeichnet und der Index k alle möglichen ganzzahligen Werthe haben kann, welcher für den allgemeineren Fall, wo n eine zusammengesetzte Zahl ist, genau dieselbe Rolle spielt, als die Gaufsischen Perioden für den besonderen Fall, wo in der Gleichung $\omega^n = 1$ n eine Primzahl ist, soll nun ebenfalls mit dem Namen Periode bezeichnet werden.

Aus der Definition dieser Perioden ergiebt sich zunächst unmittelbar:

$$\varpi_{k+hn} = \varpi_k,$$

so dafs man nur die Perioden

$$\varpi_1, \varpi_2, \varpi_3, \ldots . \varpi_n,$$

zu betrachten hat. Diese zerfallen nun ihrer Natur nach in eben so viele Gruppen, als n Divisoren hat. Ist nämlich d irgend ein Divisor von n, und bezeichnen $s, s_1, s_2, \ldots .$ alle Zahlen kleiner als n, deren gröfster gemeinschaftlicher Faktor mit n gleich d ist, so bilden die Perioden

$$\varpi_s, \varpi_{s_1}, \varpi_{s_2}, \varpi_{s_3}, \ldots ..$$

eine besondere Gruppe, welche als die zum Divisor d gehörende bezeichnet werden soll. Die Perioden einer solchen Gruppe, deren Anzahl, wenn die

Gruppe zum Divisor d gehört, gleich $\phi(\frac{n}{d})$ ist, sind nicht alle verschieden von einander, denn man hat offenbar

$$\varpi_s = \varpi_{sq} = \varpi_{sq^2} = \ldots.$$

Ist nun sq^τ der erste Index dieser Reihe, welcher congruent s ist, für den Modul n, so ist, weil s mit n den gröfsten gemeinschaftlichen Faktor d hat, $q^\tau \equiv 1$, mod. $\frac{n}{d}$, und τ der kleinste Exponent, welcher dieser Congruenz genügt; also wenn q zum Exponenten τ gehört, für den Modul $\frac{n}{d}$, so sind je τ Perioden der zum Divisor d gehörenden Gruppe einander gleich, die Anzahl der verschiedenen daher nicht gröfser als $\frac{\phi(\frac{n}{d})}{\tau}$. Es läfst sich auch beweisen, dafs diese wirklich immer von einander verschieden sind, aufser wenn sie alle gleich Null sind. Dieser Fall, dafs alle Perioden einer Gruppe gleich Null werden, tritt allemal dann, aber auch nur dann ein, wenn n eine nicht aus lauter verschiedenen Primfaktoren zusammengesetzte, sondern irgend welche quadratische Faktoren enthaltende Zahl ist. Wir haben aber für unseren gegenwärtigen Zweck nicht nöthig zu beweisen, dafs die oben angegebene reducirte Anzahl der Perioden einer Gruppe, wenn diese nicht alle gleich Null sind, wirklich aus lauter verschiedenen besteht, noch auch zu untersuchen, in welchen Gruppen alle Perioden gleich Null sind; denn die Methode, welche wir gebrauchen werden, so wie auch die durch dieselbe zu erlangenden Resultate setzen nicht voraus, dafs die Perioden wirklich verschieden sind und gelten eben so, wenn gewisse Perioden gleich Null werden.

Die sämmtlichen Perioden einer Gruppe sind immer die Wurzeln einer Gleichung des so vielten Grades, als die Gruppe Perioden enthält, mit ganzzahligen Coefficienten, in welcher der Coefficient der höchsten Potenz der Unbekannten gleich Eins ist.

Bildet man nämlich das Produkt

$$(z - \varpi_s)(z - \varpi_{s_1})(z - \varpi_{s_2}) \ldots\ldots = \mathrm{P}(z)$$

für alle Perioden der zum Divisor d gehörenden Gruppe, wobei man sich auch nur auf diejenigen $\frac{\phi(\frac{n}{d})}{\tau}$ Perioden beschränken kann, die sich schon beim Anblick wenigstens als formal verschieden zeigen, so sind die symmetrischen Funktionen der Perioden ϖ_s, ϖ_{s_1}, ϖ_{s_2} .... zugleich symmetrische Funktionen aller primitiven Wurzeln der Gleichung $\omega^n = 1$, also nur ganze Zahlen. In den Fällen, wo alle Perioden der Gruppe gleich Null sind, wird $P(z)$ zu einer bloſsen Potenz von z.

Das Produkt zweier Perioden, sei es daſs sie einer und derselben, oder auch verschiedenen Gruppen angehören, läſst sich immer als lineäre Funktion der Perioden darstellen. Multiplicirt man nämlich die beiden Perioden

$$\varpi_h = \omega^h + \omega^{hq} + \omega^{hq^2} + \ldots . + \omega^{hq^{\prime}-1}$$

$$\varpi_k = \omega^k + \omega^{kq} + \omega^{kq^2} + \ldots . + \omega^{kq^{\prime}-1}$$

mit einander in der Art, daſs jedes Glied des Multiplicators zuerst mit dem in derselben Stelle stehenden des Multiplicandus, sodann mit dem darauf folgenden Gliede, sodann mit dem um zwei Stellen folgenden u. s. w. verbunden wird, so erhält man

$$\varpi_k \varpi_h = \varpi_{k+h} + \varpi_{k+hq} + \varpi_{k+hq^2} + \ldots . + \varpi_{k+hq^{\prime}-1}$$

Diese Gleichung, in welcher k und h alle Zahlen 1, 2, 3, ... n bezeichnen können, repräsentirt ein System von n^2 Gleichungen, auf welchem die Rechnung mit diesen Perioden und mit den dieselben enthaltenden complexen Zahlen beruht. Vermittelst dieses Systems von Gleichungen kann jede beliebige ganze rationale Funktion der Perioden als eine lineäre Funktion derselben dargestellt werden, auch können mit Hülfe desselben die Gleichungen gebildet werden, welche alle Perioden einer Gruppe zu Wurzeln haben.

Eine complexe Zahl, welche nur die Perioden ϖ_1, ϖ_2, ϖ_3 ... ϖ_n enthält, nicht aber auſserdem noch Wurzeln der Gleichung $\omega^n = 1$, welche also immer als lineäre Funktion dieser Perioden dargestellt werden kann, bezeichne ich im allgemeinen kurz durch $F(\varpi_1)$ anstatt $F(\varpi_1, \varpi_2, \ldots \varpi_n)$, so daſs

$$F(\varpi_1) = a + a_1\varpi_1 + a_2\varpi_2 \ldots . + a_n\varpi_n,$$

wo die erste Periode ϖ_1 als Repräsentant aller Perioden anzusehen ist. Die conjugirte complexe Zahl, welche aus $F(\varpi_1)$ entsteht, wenn die primitive Wurzel ω in ω^r verwandelt wird, soll demgemäſs durch $F(\varpi_r)$ bezeichnet werden, so daſs allgemein

$$F(\varpi_r) = a + a_1\varpi_r + a_2\varpi_{2r} + \ldots\ldots + a_n\varpi_{nr}$$

ist. Die Anzahl der **verschiedenen** zu $F(\varpi_1)$ conjugirten complexen Zahlen ist hier nicht gleich $\phi(n)$, sondern nur gleich $\frac{\phi(n)}{t}$, weil von den $\phi(n)$ complexen Zahlen, welche man aus $F(\varpi_1)$ erhält, wenn man für ω nach einander alle primitiven Wurzeln der Gleichung $\omega^n = 1$ setzt, offenbar je t einander gleich sind. Ich setze nun

$$\frac{\phi(n)}{t} = \nu,$$

welche abgekürzte Bezeichnung auch in dem Folgenden überall angewendet werden soll; ferner bezeichne ich durch

$$r_1, r_2, r_3, \ldots\ldots r_\nu$$

diejenigen ν ganzen Zahlen, welche kleiner als n sind und relative Primzahlen zu n, und von denen der Quotient zweier niemals einer Potenz von q congruent ist, für den Modul n: so sind:

$$\omega^{r_1}, \omega^{r_2}, \omega^{r_3}, \ldots\ldots \omega^{r_\nu}$$

alle primitiven Wurzeln, welche in der complexen Zahl $F(\varpi)$ anstatt ω substituirt, alle verschiedenen, mit $F(\varpi_1)$ conjugirten complexen Zahlen geben, nämlich:

$$F(\varpi_{r_1}), F(\varpi_{r_2}), F(\varpi_{r_3}) \ldots\ldots F(\varpi_{r_\nu}).$$

§. 3.

Es soll nun gezeigt werden, wie den in dem vorigen Paragraphen aufgestellten Perioden immer gewisse ganzzahlige Congruenzwurzeln für den Modul q entsprechen, in der Art, dafs jede rationale Gleichung unter den Perioden, als Congruenz für den Modul q aufgefafst, befriedigt wird, wenn anstatt der Perioden diese entsprechenden Congruenzwurzeln gesetzt werden. Zu diesem Zwecke wird zunächst bewiesen, **dafs die Gleichungen höherer Grade, deren Wurzeln die verschiedenen Perioden einer und derselben Gruppe sind, wenn sie als Congruenzen für den Modul q aufgefafst werden, stets reale Congruenzwurzeln haben.**

Macht man nämlich Gebrauch von der Congruenz

$$y(y-1)(y-2)\ldots\ldots(y-q+1) \equiv y^q - y, \text{ mod. } q,$$

welche identisch für jeden beliebigen Werth des y Statt hat, in der Art, dafs wenn das Produkt linker Hand nach Potenzen von y entwickelt wird, die Coefficienten der gleichen Potenzen von y auf beiden Seiten einzeln congruent sind, und setzt in derselben $y = \varpi_i$, wo i ein ganz beliebiger Index ist, also ϖ_i eine beliebige Periode bezeichnet, so hat man bekanntlich

$$\varpi_i^{\,q} \equiv \varpi_i, \text{ mod. } q,$$

und daher

$$\varpi_i(\varpi_i - 1)(\varpi_i - 2) \ldots\ldots (\varpi_i - q + 1) \equiv 0, \text{ mod. } q.$$

Setzt man nun für i nach einander alle Indices einer und derselben Gruppe der Perioden, welche s, s_1, s_2 .... sein mögen, so erhält man, wenn wie im vorigen Paragraphen

$$(z - \varpi_s)(z - \varpi_{s_1})(z - \varpi_{s_2}) \ldots\ldots = \mathrm{P}(z)$$

gesetzt wird, durch Multiplication aller dieser Congruenzen:

$$\mathrm{P}(0)\,\mathrm{P}(1)\,\mathrm{P}(2) \ldots\ldots \mathrm{P}(q-1) \equiv 0, \text{ mod. } q^{\mu},$$

wo μ die Anzahl der Perioden dieser Gruppe bezeichnet. Es müssen also von den ganzen Zahlen $\mathrm{P}(0)$, $\mathrm{P}(1)$, $\mathrm{P}(2)$.... immer einige, mindestens eine, durch q theilbar sein, d. h. die Congruenz

$$\mathrm{P}(z) \equiv 0, \text{ mod. } q,$$

hat immer mindestens eine reale Congruenzwurzel.

Es seien jetzt a^i, a^i_1, a^i_2, ... a^i_{h-1} alle von einander verschiedenen Wurzeln der Congruenz $\mathrm{P}(z) \equiv 0$, mod. q, und ϖ_i eine Wurzel der Gleichung $\mathrm{P}(z) = 0$, dagegen seien b^i, b^i_1, b^i_2 ... b^i_{q-h-1} alle Nichtwurzeln derselben Congruenz, so dafs die Zahlen a^i, a^i_1 ... a^i_{h-1}, b^i, b^i_1 .... b^i_{q-h-1} mit den Zahlen 0, 1, 2, ... $q - 1$, wenn auch in anderer Ordnung, vollständig übereinstimmen: so kann die Congruenz

$$\varpi_i(\varpi_i - 1)(\varpi_i - 2) \ldots\ldots (\varpi_i - q + 1) \equiv 0, \text{ mod } q$$

folgendermaafsen dargestellt werden:

$$(a^i - \varpi_i)(a^i_1 - \varpi_i) \ldots (a^i_{h-1} - \varpi_i) \cdot (b^i - \varpi_i)(b^i_1 - \varpi_i) \ldots (b^i_{q-h-1} - \varpi_i) \equiv 0, \text{mod. } q,$$

Da nun $b^i - \varpi_i$, $b^i_1 - \varpi_i$ u. s. w. Faktoren von $\mathrm{P}(b^i)$, $\mathrm{P}(b^i_1)$ u. s. w. sind, so folgt hieraus

$$(a^i - \varpi_i)(a^i_1 - \varpi_i) \ldots (a^i_{h-1} - \varpi_i).\ \mathrm{P}(b^i)\,\mathrm{P}(b^i_1) \ldots \mathrm{P}(b^i_{q-h-1}) \equiv 0, \text{ mod. } q,$$

und weil b^i, b^i_1 .... b^i_{q-h-1} Nichtwurzeln der Congruenz $\mathrm{P}(z) \equiv 0$, mod. q, sind, so sind $\mathrm{P}(b^i)$, $\mathrm{P}(b^i_1)$ .... $\mathrm{P}(b^i_{q-h-1})$ durch q nicht theilbare ganze Zahlen, durch welche diese Congruenz dividirt werden kann. Man hat demnach

$$(a^i - \varpi_i)(a^i_1 - \varpi_i) \ldots (a^i_{h-1} - \varpi_i) \equiv 0, \text{ mod. } q.$$

Ich bilde jetzt aus einer gewissen Anzahl von allen verschiedenen unter den $h.n$ Faktoren, welche entstehen wenn man in

$$a^i - \varpi_i, \; a^i_1 - \varpi_i, \; \ldots \; (a^i_{h-1} - \varpi_i)$$

dem i die Werthe 1, 2, 3, ... n beilegt, ein Produkt, so beschaffen, dafs es möglichst viele verschiedene dieser Faktoren enthalte, ohne jedoch durch q theilbar zu sein; dafs es also durch q theilbar wird, sobald irgend einer dieser Faktoren, der nicht schon darin enthalten ist, hinzugefügt wird. Bei der Bildung dieses Produkts hat man von den angegebenen $h.n$ Faktoren zunächst alle diejenigen zu verwerfen, welche etwa für sich selbst congruent Null sein sollten nach dem Modul q, welches offenbar immer der Fall ist, wenn die Periode ϖ_i der Null gleich ist; ferner hat man von allen denen unter diesen Faktoren, welche wegen der Gleichheit der Perioden ϖ_i und der Congruenzwurzeln a^i für verschiedene Werthe des i einander gleich sind, nur je einen beizubehalten; von den übrig bleibenden Faktoren nimmt man einen beliebigen, multiplicirt diesen mit einem zweiten und wenn das Produkt nicht durch q theilbar ist, so fügt man einen dritten hinzu u. s. w. Wird durch Hinzufügung eines neuen Faktors das Produkt durch q theilbar, so wird derselbe verworfen und ein anderer der noch vorhandenen genommen. Sind auf diese Weise alle verschiedene Faktoren an die Reihe gekommen, also entweder dem Produkte zugefügt oder verworfen, so ist das Produkt gefunden, welches die verlangte Eigenschaft hat, dafs es nicht durch q theilbar ist, dafs es aber durch Multiplikation mit einem jeden noch nicht darin enthaltenen der obigen Faktoren ein durch q theilbares Produkt giebt. Sollte der besondere Fall eintreten, dafs alle Faktoren, aus welchen dieses Produkt zu bilden ist, für sich durch q theilbar wären (welcher Fall wirklich eintritt, wenn n eine Primzahl ist und $t = n - 1$), so würde dieses Produkt, welches keinen dieser Faktoren enthalten dürfte, einfach gleich Eins zu nehmen sein. Das nach den angegebenen Regeln gebildete Produkt, welches in der ganzen gegenwärtigen Theorie die wichtigste Rolle spielt, bezeichne ich durch $\Psi(\varpi_1)$.

Ist nun ϖ_i eine beliebige Periode irgend einer Gruppe, und $P(z) = 0$ die Gleichung, deren Wurzeln alle Perioden dieser Gruppe sind, welcher ϖ_i angehört, und sind a^i, a^i_1 ... a^i_{h-1} alle verschiedenen Wurzeln der Congruenz $P(z) \equiv 0$, mod. q, so können die Faktoren

$$a^i - \varpi_i, \; a^i_1 - \varpi_i, \; \ldots \; a^i_{h-1} - \varpi_i$$

nicht alle zugleich in dem Produkte $\Psi(\varpi_1)$ enthalten sein, weil sonst vermöge der Congruenz

$$(a^i - \varpi_i)(a^i_1 - \varpi_i) \ldots (a^i_{h-1} - \varpi_i) \equiv 0, \text{ mod. } q,$$

$\Psi(\varpi_1)$ durch q theilbar sein würde. Es mufs also wenigstens einen dieser Faktoren geben, der in $\Psi(\varpi_1)$ nicht enthalten ist, welchen ich durch $u_i - \varpi_i$ bezeichne, wo u_i eine der Zahlen a^i, $a^i_1 \ldots a^i_{h-1}$, d. h. eine der verschiedenen Wurzeln der Congruenz $P(z) \equiv 0$, mod. q, vorstellt. Durch Multiplikation mit diesem Faktor $u_i - \varpi_i$ mufs nun $\Psi(\varpi_1)$ durch q theilbar werden; man hat also

$$\Psi(\varpi_1)(u_i - \varpi_i) \equiv 0, \text{ mod. } q,$$

oder

$$\Psi(\varpi_1)\varpi_i \equiv \Psi(\varpi_1) \cdot u_i, \text{ mod. } q.$$

welche Congruenz für jeden beliebigen Werth des Index $i = 1, 2, 3 \ldots n$ Statt hat, und darum wesentlich ein System von n Congruenzen repräsentirt. Aus diesem Systeme von Congruenzen, welches die hauptsächlichste Grundlage der folgenden Untersuchungen bildet, folgt nun zunächst der wichtige Satz:

Jede rationale Gleichung unter den Perioden ϖ_i, welche aufser diesen nur ganze Zahlen enthält, wird, wenn anstatt der Perioden ϖ_i die entsprechenden Congruenzwurzeln u_i gesetzt werden, als Congruenz für den Modul q befriedigt.

Denkt man sich nämlich die gegebene Gleichung unter den Perioden von den etwa darin vorkommenden Brüchen befreit und alle Glieder auf eine Seite gebracht, so dafs sie die Form $F(\varpi_1) = 0$ annimmt, und multiplicirt mit $\Psi(\varpi_1)$, so kann man anstatt $\Psi(\varpi_1)\,\varpi_i$ überall die für den Modul q congruenten Ausdrücke $\Psi(\varpi_1)\,u_i$ setzen, und erhält so

$$\Psi(\varpi_1)\,F(\varpi_1) \equiv \Psi(\varpi_1)\,F(u_1), \text{ mod. } q,$$

wo $F(u_1)$ den ganzzahligen Ausdruck bezeichnet, welchen man aus $F(\varpi_1)$ durch die Substitution der ganzen Zahlen u_i anstatt der entsprechenden Congruenzwurzeln ϖ_i erhält. Da nun $F(\varpi_1) = 0$ ist und $\Psi(\varpi_1)$ nicht $\equiv 0$, mod, q, so mufs nothwendig der ganzzahlige Ausdruck $F(u_1) \equiv 0$, mod. q, sein, was zu beweisen war.

Nach diesem Satze erhält man zum Beispiel aus dem Systeme der n^2 Fundamentalgleichungen für die Perioden, nämlich

$$\varpi_k \varpi_h = \varpi_{k+h} + \varpi_{k+hq} + \varpi_{k+hq^2} + \ldots + \varpi_{k+hq^{f-1}}$$

für $k = 1, 2, 3, \ldots n$, $h = 1, 2, 3, \ldots n$, augenblicklich das entsprechende System von n^2 Congruenzen, welchen die Zahlen $u_1, u_2, u_3, \ldots u_n$ genügen müssen:

$$u_k u_h \equiv u_{k+h} + u_{k+hq} + u_{k+hq^2} + \ldots + u_{k+hq^{t-1}}, \text{ mod. } q,$$

für $k = 1, 2, 3, \ldots n$, $h = 1, 2, 3, \ldots n$, welches System mit Erfolg zur direkten Berechnung der Congruenzwurzeln $u_1, u_2, u_3, \ldots u_n$ benutzt werden kann.

Die Congruenzwurzeln u_i können immer auf mehrere verschiedene Weisen den Perioden ϖ_i entsprechend zugeordnet werden. Verwandelt man nämlich in dem Systeme der Congruenzen:

$$\Psi(\varpi_1)\,\varpi_i \equiv \Psi(\varpi_1)\,u_i \text{ mod. } q,$$

die den Perioden zu Grunde gelegte primitive Wurzel ω der Gleichung $\omega^n = 1$ in eine andere, z. B. in ω^r, wo r relative Primzahl zu n ist, so geht dasselbe über in

$$\Psi(\varpi_r)\,\varpi_{r\,i} \equiv \Psi(\varpi_r)\,u_i \text{ mod. } q.$$

Es giebt nun $\phi(n)$ verschiedene primitive Wurzeln ω^r, oder was dasselbe ist, $\phi(n)$ verschiedene Werthe des r, für welche ω^r eine primitive Wurzel ist, von diesen geben aber, wie oben (§ 2) gezeigt worden, je t nur genau dieselben Perioden. Die Anzahl der verschiedenen Congruenzen dieser Art ist also nicht gröfser als $\nu = \frac{\phi(n)}{t}$, und diesen entsprechen eben so viele, nämlich ν verschiedene Arten und Weisen, wie die Congruenzwurzeln u_i den Perioden $\varpi_{r\,i}$ zugeordnet und anstatt derselben substituirt werden können, wenn Gleichungen unter den Perioden in Congruenzen unter den Zahlen u_i verwandelt werden sollen. Die Substitution, nach welcher die Perioden $\varpi_r, \varpi_{2r}, \varpi_{3r}, \ldots \varpi_{nr}$ beziehungsweise durch die Zahlen $u_1, u_2, u_3, \ldots u_n$ ersetzt werden, bezeichne ich einfach als die Substitution $\varpi_{r\,i} = u_i$, indem i jede der Zahlen $1, 2, 3, \ldots n$ repräsentirt. Genau dieselbe Substitution kann demgemäfs auch als Substitution $\varpi_i = u_{\varrho i}$ bezeichnet werden, wenn ϱ durch die Congruenz $r\varrho \equiv 1$ mod. n, bestimmt ist; die Identität dieser Substitution mit der anderen fällt sogleich in die Augen, wenn man in dieser ir statt i setzt. Bringt man in einer complexen Zahl $F(\varpi_1)$ die Substitution $\varpi_{r\,i} = u_i$ an, oder was dasselbe ist, die Substitution $\varpi_i = u_{\varrho i}$, so soll die daraus entstehende nicht complexe ganze Zahl künftig durch $F(u_\varrho)$ bezeichnet werden; dieselbe Substitution in $F(\varpi_s)$ angebracht wird demnach $F(u_{\varrho s})$ ergeben.

B 2

§. 4.

Es sollen nun die für den vorliegenden Zweck wesentlichen Eigenschaften der complexen Zahl $\Psi(\varpi_1)$ aufgezeigt werden, und zwar zunächst die folgende:

Die zu $\Psi(\varpi_1)$ conjugirten complexen Zahlen, deren Anzahl gleich ν ist, sind alle wirklich verschieden von einander.

Wären irgend zwei dieser conjugirten Zahlen einander gleich, z. B.

$$\Psi(\varpi_k) = \Psi(\varpi_h)$$

wo k und h irgend zwei verschiedene der oben (§. 2) genau bestimmten Zahlen $r_1, r_2, r_3, \ldots r_\nu$ bedeuten, so würde man durch eine passende Veränderung der primitiven Wurzel ω hieraus eine Gleichung

$$\Psi(\varpi_1) = \Psi(\varpi_r)$$

erhalten, in welcher r relative Primzahl zu n und keiner Potenz von q congruent wäre, nach dem Modul n. Da nun aber nach dem Hauptresultate des vorigen Paragraphen

$$\left.\begin{aligned} \Psi(\varpi_1)\,\varpi_i &\equiv \Psi(\varpi_1)\,u_i \\ \Psi(\varpi_r)\,\varpi_{r\,i} &\equiv \Psi(\varpi_r)\,u_i \end{aligned}\right\} \text{ mod. } q,$$

so würde auch

$$\Psi(\varpi_1)\,\varpi_i \equiv \Psi(\varpi_1)\,\varpi_{r\,i}, \quad \text{mod. } q,$$

sein, für alle Werthe des $i = 1, 2, 3, \ldots n$. Multiplicirt man diese Congruenz auf beiden Seiten mit ω^{-i} und nimmt die Summe für $i = 1, 2, 3, \ldots n$, so erhält man hieraus

$$\Psi(\varpi_1)\sum_1^n{}_i\,\omega^{-i}\,\varpi_i \equiv \Psi(\varpi_1)\sum_1^n{}_i\,\omega^{-i}\,\varpi_{r\,i}, \text{ mod. } q.$$

Setzt man nun anstatt der Perioden ihre Ausdrücke durch die Wurzel ω, so hat man

$$\begin{aligned} \omega^{-i}\,\varpi_i &= 1 + \omega^{i(q-1)} + \omega^{i(q^2-1)} + \ldots\ldots\ldots + \omega^{i(q^{t-1}-1)} \\ \omega^{-i}\,\varpi_{r\,i} &= \omega^{i(r-1)} + \omega^{i(rq-1)} + \omega^{i(rq^2-1)} + \ldots + \omega^{i(rq^{t-1}-1)} \end{aligned}$$

Nimmt man jetzt die Summen in Beziehung auf $i = 1, 2, 3, \ldots n$ und bemerkt, daſs $\sum_1^n{}_i\,\omega^{mi}$ immer gleich Null ist, mit Ausnahme des einen Falles, wo m durch n theilbar, daſs aber $q-1, q^2-1, \ldots q^{t-1}-1$ und $r-1, rq-1, rq^2-1, \ldots rq^{t-1}-1$ nicht durch n theilbar sind, so hat man

$$\sum_1^n {}_i\, \omega^{-i}\, \varpi_i \;=\; n,$$

$$\sum_1^n {}_i\, \omega^{-i}\, \varpi_{r\,i} = 0.$$

die obige Congruenz giebt daher

$$\Psi(\varpi_1) \cdot n \equiv 0, \text{ mod. } q,$$

welches unmöglich ist, weil weder die ganze Zahl n noch die complexe Zahl $\Psi(\varpi_1)$ durch q theilbar ist. Die Annahme, dafs zwei der ν mit $\Psi(\varpi_1)$ conjugirten complexen Zahlen einander gleich sein könnten, ist also eine falsche.

Das Produkt je zweier verschiedener zu $\Psi(\varpi_1)$ conjugirter complexer Zahlen ist stets durch q theilbar.

Dieser Satz folgt unmittelbar aus der Bildungsweise dieser complexen Zahlen, welche, wie wir oben gezeigt haben, aus Faktoren von der Form $a_e^i - \varpi_i$ so zusammengesetzt sind, dafs keiner derselben mehrmals darin enthalten ist und dafs durch das Hinzutreten eines jeden nicht schon in $\Psi(\varpi_1)$ enthaltenen dieser Faktoren das Produkt durch q theilbar wird. Sind nun zwei solche complexe Zahlen verschieden von einander, so enthalten sie nicht genau dieselben Faktoren von dieser Form, sondern eine mufs gewisse enthalten, welche in der andern nicht vorkommen, darum mufs ihr Produkt nothwendig durch q theilbar sein.

Nach dem Bildungsgesetze, welches wir oben für die durch Ψ bezeichneten Produkte aufgestellt haben, scheint es, als ob einer und derselben Primzahl q mehrere wesentlich verschiedene solche Produkte angehören könnten; dafs diefs aber in der That nicht der Fall ist, wird folgendermaafsen gegeigt.

Sei $\Psi(\varpi_1)$ irgend eines der nach dem im vorigen Paragraphen angegebenen Bildungsgesetze aus den Faktoren von der Form $a_e^i - \varpi_i$ zu bildenden Produkte, so werden auch die zu $\Psi(\varpi_1)$ conjugirten complexen Zahlen offenbar solche Produkte sein, welche sich ebenfalls aus diesen Faktoren bilden lassen und den gegebenen Bedingungen genügen. Es sei nun die Summe dieser mit $\Psi(\varpi_1)$ conjugirten complexen Zahlen gleich M, also

$$M = \Psi(\varpi_{r_1}) + \Psi(\varpi_{r_2}) + \Psi(\varpi_{r_3}) + \dots + \Psi(\varpi_{r_\nu})$$

so ist M, als symmetrische Funktion dieser Perioden, oder auch als symmetrische Funktion aller primitiven Wurzeln der Gleichung $\omega^n = 1$, eine nicht-complexe ganze Zahl. Die Zahl M ist auch nicht durch q theilbar, denn multiplicirt man dieselbe mit $\Psi(\varpi_{r_1})$ und beachtet, daſs das Produkt je zweier verschiedener conjugirter Ψ durch q theilbar ist, so hat man

$$\Psi(\varpi_{r_1})\, M \equiv \Psi(\varpi_{r_1})^2, \text{ mod. } q,$$

und weil $\Psi(\varpi_{r_1})$ nicht durch q theilbar ist, so ist auch das Quadrat dieser complexen Zahl nicht durch q theilbar, also auch M nicht durch q theilbar. Wäre nun $\Psi'(\varpi_1)$ irgend eine von $\Psi(\varpi_1)$ und von allen conjugirten derselben verschiedene complexe Zahl, welche ebenfalls nach dem angegebenen Bildungsgesetze aus den Faktoren von der Form $a_e^i - \varpi_i$ zusammengesetzt wäre, so würde $\Psi'(\varpi_1)\,\Psi(\varpi_r)$ nothwendig durch q theilbar sein für alle Werthe des $r = r_1, r_2, \ldots\ldots r_\nu$. Die Verschiedenheit des $\Psi'(\varpi_1)$ von $\Psi(\varpi_r)$ kann nämlich nicht darin liegen, daſs eines dieser beiden Produkte irgend einen der Faktoren von der Form $a_e^i - \varpi_i$ mehrmals enthält als das andere, weil nach der Voraussetzung keiner dieser Faktoren mehr als einmal in einem solchen Produkte vorkommen kann, diese Verschiedenheit müſste daher nur daher kommen, daſs eine der beiden complexen Zahlen $\Psi'(\varpi_1)$ und $\Psi(\varpi_r)$ irgend einen der Faktoren von der Form $a_e^i - \varpi_i$ enthält, welcher in der andern nicht vorkommt, woraus folgen würde, daſs das Produkt derselben durch q theilbar sein müſste. Da also alle einzelnen Theile, aus welchen M besteht, nach Multiplication mit $\Psi'(\varpi_1)$ durch q theilbar wären, so würde auch $\Psi'(\varpi_1)M$ durch q theilbar sein, welches unmöglich ist, weil weder die nichtcomplexe Zahl M, noch die complexe Zahl $\Psi'(\varpi_1)$ durch q theilbar ist. Das gefundene Resultat läſst sich nun folgendermaaſsen aussprechen:

Die Anzahl aller verschiedenen Produkte, welche nach dem festgesetzten Bildungsgesetze aus den Faktoren von der Form $a_e^i - \varpi_i$ gebildet werden können ist gleich $\nu = \frac{\phi(n)}{f}$ und dieselben sind conjugirte complexe Zahlen. Die complexe Zahl $\Psi(\varpi_1)$ ist also, abgesehen von der zu Grunde liegenden primitiven Wurzel ω, eine vollständig bestimmte für jede Primzahl q.

Eine andere wichtige Eigenschaft der complexen Zahl $\Psi(\varpi_1)$ erhält man aus der Congruenz

$$\Psi(\varpi_1)\,\Psi(\varpi_r) \equiv \Psi(\varpi_1)\,\Psi(u_r), \text{ mod. } q,$$

welche aus dem Systeme der Congruenzen

$$\Psi(\varpi_1)\,\varpi_i \equiv \Psi(\varpi_1)\,u_i, \text{ mod. } q,$$

unmittelbar folgt, da nämlich $\Psi(\varpi_1)\ \Psi(\varpi_r) \equiv 0$, mod. q, sobald r weder der Eins noch einer Potenz von q congruent ist, für den Modul n, so giebt obige Congruenz

$$\Psi(\varpi_1)\ \Psi(u_r) \equiv 0, \text{ mod. } q,$$

woraus folgt:

$$\Psi(u_r) \equiv 0, \text{ mod. } q,$$

für den Fall $r = 1$ aber, wo $\Psi(\varpi_1) \cdot \Psi(\varpi_1)$ nicht durch q theilbar ist, folgt eben so:

$$\Psi(u_1) \text{ nicht} \equiv 0, \text{ mod. } q.$$

Diese Resultate geben folgenden Satz:

Die complexe Zahl $\Psi(\varpi_1)$ wird durch die Substitution $\varpi_{r\,i} = u_i$, wenn r für den Modul n weder der Eins noch einer Potenz von q congruent ist, immer eine durch q theilbare, durch die Substitution $\varpi_i = u_i$ aber eine durch q nicht theilbare ganze Zahl.

Aus diesem Satze folgt fast unmittelbar auch der folgende:

Wenn für eine nur die Perioden enthaltende complexe Zahl $F(\varpi_1)$ die Congruenz

$$\Psi(\varpi_r)\ F(\varpi_1) \equiv 0, \text{ mod. } q,$$

Statt hat, so ist

$$F(u_\varrho) \equiv 0, \text{ mod. } q,$$

wo ϱ durch die Congruenz $r\varrho \equiv 1$, mod. n, bestimmt ist, und auch umgekehrt: wenn $F(u_\varrho) \equiv 0$, mod. q, so ist $\Psi(\varpi_r)\ F(\varpi_1) \equiv 0$ mod. q.

Durch Anwendung der Substitution $\varpi_{r\,i} = u_i$, oder was dasselbe ist, $\varpi_i = u_{\varrho\,i}$, hat man nämlich aus der Congruenz, welche die Voraussetzung des Satzes bildet, sogleich $\Psi(u_1)\,F(u_\varrho) \equiv 0$, mod. q, und weil $\Psi(u_1)$ nicht durch q theilbar ist, mufs $F(u_\varrho) \equiv 0$, mod. q, sein. Multiplicirt man aber $F(u_\varrho)$ mit $\Psi(\varpi_r)$ und ersetzt umgekehrt die Congruenzwurzeln durch die Perioden, nach der Substitution $\varpi_i = u_{\varrho\,i}$, so erhält man aus $F(u_\varrho) \equiv 0$, mod. q, sogleich $\Psi(\varpi_r)\ F(\varpi_1) \equiv 0$ mod. q.

Die Congruenzwurzeln $u_1, u_2, u_3 \ldots\ldots u_n$, welche den Perioden ϖ_1, $\varpi_2, \varpi_3 \ldots\ldots \varpi_n$ auf ν verschiedene Weisen entsprechen, können, wenn denselben passende Vielfache von q zugefügt werden, immer so gewählt werden, dafs sie, in einer rationalen Gleichung unter den Perioden substituirt, eine Congruenz geben, welche nicht nur für den einfachen Modul q, sondern sogar für eine beliebig hohe Potenz von q richtig ist. Setzt man nämlich die Congruenz

$$\Psi(\varpi_1)\,(u_i - \varpi_i) \equiv 0, \text{ mod. } q,$$

in die Form einer Gleichung

$$\Psi(\varpi_1)\,(u_i - \varpi_i) = q \,.\, Q(\varpi_1)$$

und addirt auf beiden Seiten $\Psi(\varpi_1) \,.\, q \,.\, h$, wo h eine ganze Zahl bezeichnet, und multiplicirt sodann nochmals mit $\Psi(\varpi_1)$, so erhält man

$$\Psi(\varpi_1)^2\,(u_i + hq - \varpi_i) = q\,\Psi(\varpi_1)\,(Q(\varpi_1) + h\Psi(\varpi_1))$$

und weil

$$\left.\begin{array}{l}\Psi(\varpi_1)\,Q(\varpi_1) \equiv \Psi(\varpi_1)\,Q(u_1)\\ \Psi(\varpi_1)\,\Psi(\varpi_1) \equiv \Psi(\varpi_1)\,\Psi(u_1)\end{array}\right\} \text{mod. } q,$$

so hat man

$$\Psi(\varpi_1)^2\,(u_i + hq - \varpi_i) \equiv q\Psi(\varpi_1)\,(Q(u_1) + h\Psi(u_1)) \text{ mod. } q^2.$$

Da nun $\Psi(u_1)$ nicht durch q theilbar ist, so giebt es stets einen Werth des h für welchen die Congruenz

$$Q(u_1) + h\Psi(u_1) \equiv 0, \text{ mod. } q,$$

erfüllt ist, für welchen also

$$\Psi(\varpi_1)^2\,(u_i + h\,q - \varpi_i) \equiv 0, \text{ mod. } q^2,$$

ist, also wenn $u_i + hq$ durch $\overset{2}{u}_i$ bezeichnet wird:

$$\Psi(\varpi_1)^2\,(\overset{2}{u}_i - \varpi_i) \equiv 0, \text{ mod. } q^2.$$

Ganz auf dieselbe Weise kann man hieraus weiter die Congruenz

$$\Psi(\varpi_1)^3\,(\overset{3}{u}_i - \varpi_i) \equiv 0, \text{ mod. } q^3,$$

ableiten, und so fortfahrend findet man allgemein für jeden ganzzahligen Werth des m

$$\Psi(\varpi_1)^m\,(\overset{m}{u}_i - \varpi_i) \equiv 0, \text{ mod } q^m,$$

welche auch so dargestellt werden kann:

$$\Psi(\varpi_1)^m\,\varpi_i \equiv \Psi(\varpi_1)^m\,\overset{m}{u}_i, \text{ mod. } q^m,$$

oder wenn die primitive Wurzel ω in ω^r verändert wird

$$\Psi(\varpi_r)^m\,\varpi_{r,i} \equiv \Psi(\varpi_r)^m \,.\, \overset{m}{u}_i, \text{ mod } q^m.$$

Aus dieser Congruenz, welche, weil sie für alle Werthe des $i = 1, 2, 3, \dots n$ gültig ist, ein System von n Congruenzen repräsentirt, folgt nun ganz auf dieselbe Weise, wie oben (§. 3) für den Fall $m = 1$ gezeigt worden, der allgemeine Satz:

Jede rationale Gleichung unter den Perioden ϖ_i, welche aufser diesen nur ganze Zahlen enthält, wird, wenn anstatt der Perioden $\varpi_r, \varpi_{2r}, \varpi_{3r} \ldots\ldots \varpi_{nr}$ beziehungsweise die ganzen Zahlen $\overset{m}{u}_1, \overset{m}{u}_2, \overset{m}{u}_3 \ldots\ldots \overset{m}{u}_n$ gesetzt werden, also durch die Substitution $\varpi_{ri} = \overset{m}{u}_i$, als Congruenz für den Modul q^m befriedigt.

Ferner hat man für die Zahlen $\overset{m}{u}_i$ und den Modul q^m genau ebenso den Satz:

Die Bedingung $\Psi(\varpi_r)^m F(\varpi^i) \equiv 0$, mod. q^m, ist gleichbedeutend mit $F(\overset{m}{u}_\varrho) \equiv 0$, mod. q^m, wo $r\varrho \equiv 1$, mod. n.

In Betreff derjenigen Perioden ϖ_i, in welchen i mit n einen gemeinschaftlichen Faktor hat, welche also eigentlich nicht die n^{ten}, sondern niedere Wurzeln der Einheit enthalten, ist hier noch zu bemerken, dafs es in einzelnen Fällen nöthig ist dieselben besonders zu betrachten, wenn gleich sie, wie in dem Vorhergehenden gezeigt worden ist, genau denselben allgemeinen Regeln unterworfen sind, als die nur die primitiven Wurzeln der Gleichung $\omega^n = 1$ enthaltenden. Wenn in dem Ausdrucke der Periode

$$\varpi_i = \omega^i + \omega^{iq} + \omega^{iq^2} + \ldots\ldots + \omega^{iq^{t-1}}$$

i mit n den gröfsten gemeinschaftlichen Faktor ν hat und $i = \nu i'$, $n = \nu n'$ gesetzt wird, wo i' und n' relative Primzahlen sind und wenn q zum Exponenten τ gehört in Beziehung auf den Modul n', so ist τ nothwendig ein Divisor von t. Setzt man daher $t = d\tau$, so werden nur die ersten τ Glieder dieser Periode verschieden von einander und dieselben wiederholen sich $d$mal, es wird also

$$\varpi_i = d\,(\omega^i + \omega^{iq} + \omega^{iq^2} + \ldots + \omega^{iq^{\tau-1}})$$

Diesen ganzzahligen Faktor d kann man nun in den niederen Perioden, sowohl bei der Ermittelung der den Perioden entsprechenden Congruenzwurzeln, als bei der Bildung der mit Ψ bezeichneten complexen Zahlen überall beibehalten, oder auch weglassen, wodurch in den entwickelten Methoden und Sätzen nichts geändert wird. In dem besonderen Falle aber, wo d durch die Primzahl q theilbar ist, ist es unbedingt nöthig, dafs dieser ganzzahlige Faktor weggelassen und die Periode nicht als aus t Gliedern von denen je d einander gleich sind, sondern nur als aus den τ verschiedenen Gliedern bestehend angenommen wird. In diesem Falle, welcher übrigens nur für gewisse besondere und zwar kleine Werthe des q eintreten kann, würde nämlich die Beibehaltung des ganzzahligen Faktors d eine Unbestimmtheit herbeiführen, welche wie leicht zu sehen ist daher kommen würde, dafs die Congruenzen nach dem Modul q für die Perioden und die entsprechenden Congruenzwurzeln den Faktor q selbst enthalten und indem sie so nur $0 \equiv 0$, mod. q, ergeben, nichtssagend sein würden. Man kann aber, um einer jeden besonderen Betrachtung dieser speciellen Werthe des q überhoben zu sein, allgemein die Periode ϖ_i auch definiren als die Summe

$$\omega^{i} + \omega^{iq} + \omega^{iq^2} + \omega^{iq^3} + \cdots\cdots$$

fortgesetzt bis zu demjenigen Gliede ausschliefslich, welches dem ersten Gliede gleich wird. Da der Unterschied dieser Definition von der oben gegebenen lediglich darin besteht, dafs der die niederen Perioden behaftende ganzzahlige Faktor überall ausgeschlossen wird, so ist klar, dafs wenn man dieselbe an die Stelle der im §. 2. gegebenen Definition setzt, die Methoden so wie die entwickelten Resultate überall dieselben bleiben.

§. 5.

Die leitende Grundidee bei der Einführung der idealen Primfaktoren der aus den Wurzeln der Gleichung $\omega^n = 1$ gebildeten complexen Zahlen ist dieselbe, welche ich in meinen früheren Abhandlungen über die speciellere Art complexer Zahlen umständlich auseinandergesetzt habe. Sie liegt im wesentlichen darin, dafs, weil die einfachsten wirklichen Faktoren der complexen Zahlen nicht immer die wahre Natur von Primfaktoren haben, nach welcher jede gegebene Zahl eine endliche Anzahl derselben in unveränderlich bestimmter Weise enthalten mufs, die Primfaktoren durch gewisse Congruenzbedingungen ersetzt werden, welche alle wesentlichen Eigenschaften wahrhafter Primfaktoren darstellen, und welche in allen den Fällen, wo wirkliche Primfaktoren vorhanden sind, mit diesen vollständig übereinstimmen, so dafs diese Congruenzbedingungen als die permanenten Eigenschaften der complexen Zahlen zu betrachten sind, die wirkliche Darstellbarkeit derselben aber, in Form von selbständigen ganzen complexen Faktoren, nur als eine accidentelle Eigenschaft aufzufassen ist. Einer weiteren Ausführung dieser Grundidee glaube ich mich hier enthalten zu dürfen und stelle sogleich die Definition der idealen Primfaktoren der aus den Wurzeln der Gleichung $\omega^n = 1$ gebildeten complexen Zahlen auf.

Wenn eine complexe Zahl $F(\omega)$ die Eigenschaft hat, dafs

$$\Psi(\varpi_r)\, F(\omega) \equiv 0, \text{ mod. } q,$$

so soll von dieser Zahl $F(\omega)$ ausgesagt werden: sie enthält einen idealen Primfaktor des q, und zwar denjenigen, welcher zur Substitution $\varpi_{r,i} = u_i$ gehört. Hat die complexe Zahl $F(\omega)$ ausserdem die Eigenschaft, dafs

$$\Psi(\varpi_r)^m F(\omega) \equiv 0, \text{ mod. } q^m,$$

aber

$$\Psi(\varpi_r)^{m+1} F(\omega) \text{ nicht} \equiv 0, \text{ mod. } q^{m+1},$$

so soll von ihr ausgesagt werden, dafs sie den idealen Primfaktor des q, welcher zur Substitution $\varpi_{r,i} = u_i$ gehört, genau m mal enthält.

Da die Anzahl der verschiedenen complexen Zahlen $\Psi(\varpi_r)$, für die verschiedenen Werthe des r, und darum auch die Anzahl der verschiedenen Substitutionen $\varpi_{r,i} = u_i$, wie oben gezeigt worden, gleich ν ist, so folgt:

Die Anzahl der verschiedenen idealen Primfaktoren der nichtcomplexen Primzahl q, welche zum Exponenten t gehört für den Modul n, ist gleich $\frac{\phi(n)}{t} = \nu$.

Von den hier definirten idealen Primfaktoren des q ist nun nachzuweisen, dafs sie alle wesentlichen Eigenschaften wahrhafter Primfaktoren besitzen, zu welchem Zwecke zunächst folgender Satz bewiesen werden soll:

Wenn zwei oder mehrere complexe Zahlen einen idealen Primfaktor des q nicht enthalten, so enthält das entwickelte Produkt derselben diesen idealen Primfaktor ebenfalls nicht.

Es seien $f(\omega)$ und $f_1(\omega)$ zwei complexe Zahlen, deren keine den zur Substitution $\varpi_{r,i} = u_i$ gehörenden idealen Primfaktor des q enthalte, so ist nach der Definition

$$\left.\begin{array}{l}\Psi(\varpi_r)\, f(\omega) \text{ nicht} \equiv 0, \\ \Psi(\varpi_r)\, f_1(\omega) \text{ nicht} \equiv 0,\end{array}\right. \text{ mod. } q.$$

Weil nun diese beiden complexen Ausdrücke nicht durch q theilbar sind, so ist auch keine Potenz derselben durch q theilbar; erhebt man dieselben daher zur Potenz des Exponenten $1 + q + q^2 + \ldots + q^{t-1}$ und bemerkt, dafs allgemein

$$F(\omega)^{1+q+q^2+\ldots+q^{t-1}} \equiv F(\omega)\, F(\omega^q)\, F(\omega^{q^2}) \ldots F(\omega^{q^{t-1}}), \text{ mod. } q.$$

C 2

ist, so hat man, wenn der Kürze wegen $1 + q + q^2 \ldots + q^{t-1} = Q$ gesetzt wird:

$$\Psi(\varpi_r)^Q . f(\omega) f(\omega^q) f(\omega^{q^2}) \ldots f(\omega^{q^{t-1}}) \text{ nicht} \equiv 0.$$

mod. q.

$$\Psi(\varpi_r)^Q . f_1(\omega) f_1(\omega^q) f_1(\omega^{q^2}) \ldots f_1(\omega^{q^{t-1}}) \text{ nicht} \equiv 0.$$

Setzt man nun:

$$f(\omega) f(\omega^q) f(\omega^{q^2}) \ldots f(\omega^{q^{t-1}}) = F(\varpi_1)$$

$$f_1(\omega) f_1(\omega^q) f_1(\omega^{q^2}) \ldots f_1(\omega^{q^{t-1}}) = F_1(\varpi_1),$$

welche Bezeichnungen als nur die Perioden enthaltende complexe Zahlen ihnen wirklich zukommen, weil sie als symmetrische Funktionen aller in einer Periode vorkommenden Wurzeln ω, ω^q, ... $\omega^{q^{t-1}}$ nur die Perioden, nicht aber ausser diesen die Wurzel ω enthalten können, wo die Perioden in dem Sinne aufzufassen sind, welcher denselben am Schlusse des vorigen Paragraphen gegeben worden ist, nämlich befreit von den sie etwa behaftenden ganzzahligen Faktoren: so hat man

$$\Psi(\varpi_r)^Q F(\varpi_1) \text{ nicht} \equiv 0,$$

mod. q.

$$\Psi(\varpi_r)^Q F_1(\varpi_1) \text{ nicht} \equiv 0,$$

Hieraus folgt weiter, dass auch die durch die Substitution $\varpi_{r,i} = u_i$ aus $F(\varpi_1)$ und $F_1(\varpi_1)$ entstehenden nichtcomplexen ganzen Zahlen $F(u_\varrho)$ und $F_1(u_\varrho)$ nicht durch q theilbar sind, also auch nicht das Produkt derselben $F(u_\varrho) . F_1(u_\varrho)$. Multiplicirt man dieses Produkt noch mit $\Psi(\varpi_r)$ und substituirt rückwärts die Perioden für die Congruenzwurzeln, nach der Substitution $\varpi_{r,i} = u_i$, so hat man

$$\Psi(\varpi_r)\, F(\varpi_1)\, F_1(\varpi_1) \text{ nicht} \equiv 0, \text{ mod. } q.$$

und daher auch

$$\Psi(\varpi_r) . f(\omega) . f_1(\omega) \text{ nicht} \equiv 0, \text{ mod. } q.$$

weil dieser Ausdruck nur ein Theiler des vorigen ist. Das Produkt der beiden complexen Zahlen $f(\omega)$ und $f_1(\omega)$ enthält also den zur Substitution $\varpi_{r,i} = u_i$ gehörenden idealen Primfaktor des q nicht. Da nun der Satz für ein Produkt von zwei Faktoren bewiesen ist, so folgt unmittelbar, dass er auch für ein Produkt beliebig vieler Faktoren ebenso gelten muss.

Dieser Satz dient nun zum Beweise des folgenden allgemeineren:

Das entwickelte Produkt zweier oder mehrerer complexer

Zahlen enthält genau dieselben idealen Primfaktoren des q, und jeden genau eben so oft, als die Faktoren zusammengenommen.

Es enthalte $f(\omega)$ den zur Substitution $\varpi_{r,i} = u_i$ gehörenden idealen Primfaktor des q genau m mal, $f_1(\omega)$ enthalte denselben genau m' mal, so ist nach der Definition

$$\Psi(\varpi_r)^m f(\omega) \equiv 0, \text{ mod. } q^m$$

$$\Psi(\varpi_r)^{m'} f_1(\omega) \equiv 0, \text{ mod. } q^{m'};$$

setzt man demnach

$$\Psi(\varpi_r)^m f(\omega) = q^m P(\omega),$$

$$\Psi(\varpi_r)^{m'} f_1(\omega) = q^{m'} P_1(\omega),$$

so ist

$$\Psi(\varpi_r)^{m+m'} f(\omega) f_1(\omega) = q^{m+m'} P(\omega) P_1(\omega);$$

weil ferner $f(\omega)$ den zur Substitution $\varpi_{r,i} = u_i$ gehörenden idealen Primfaktor des q nicht mehr als m mal, $f_1(\omega)$ denselben nicht mehr als m' mal enthält, so folgt, dafs $P(\omega)$ und $P_1(\omega)$ diesen idealen Primfaktor nicht enthalten, also vermöge des vorhergehenden Satzes, dafs das Produkt $P(\omega)$ $P_1(\omega)$ denselben auch nicht enthält, also $\Psi(\varpi_r)$ $P(\omega)$ $P_1(\omega)$ nicht $\equiv 0$ ist mod. q. Darum ist

$$\Psi(\varpi_r)^{m+m'} f(\omega) f_1(\omega) \equiv 0, \text{ mod. } q^{m+m'},$$

aber

$$\Psi(\varpi_r)^{m+m'+1} f(\omega) f_1(\omega) \text{ nicht} \equiv 0, \text{ mod. } q^{m+m'+1},$$

d. h. das Produkt $f(\omega)$ $f_1(\omega)$ enthält diesen idealen Primfaktor genau $m+m'$ mal. Was nun für diesen einen idealen Primfaktor bewiesen ist, gilt offenbar ebenso für alle idealen Primfaktoren der nicht in der Zahl n enthaltenen Primzahlen, ebenso versteht sich von selbst, dafs der Satz auch für ein Produkt beliebig vieler Faktoren giltig bleibt.

Wenn eine complexe Zahl alle verschiedenen idealen Primfaktoren des q enthält, und zwar jeden mindestens m mal, so ist sie durch q^m theilbar. Enthält sie jeden der verschiedenen idealen Primfaktoren des q genau m mal, so erschöpft q^m alle in ihr enthaltenen idealen Primfaktoren des q.

Die Bedingung, dafs $F(\omega)$ alle idealen Primfaktoren des q enthält, und zwar jeden mindestens m mal, wird durch folgende Gleichungen ausgedrückt:

$$\Psi(\varpi_{r_1})^m F(\omega) = q^m P_1(\omega)$$
$$\Psi(\varpi_{r_2})^m F(\omega) = q^m P_2(\omega)$$
$$\vdots$$
$$\Psi(\varpi_{r_\nu})^m F(\omega) = q^m P_\nu(\omega).$$

Durch Addition aller dieser Gleichungen hat man, wenn der Kürze wegen

$$\Psi(\varpi_{r_1})^m + \Psi(\varpi_{r_2})^m + \ldots + \Psi(\varpi_{r_\nu})^m = M$$

gesetzt wird

$$MF(\omega) = q^m (P_1(\omega) + P_2(\omega) + \ldots + P_\nu(\omega))$$

Nun ist aber M, als symmetrische Funktion aller primitiven Wurzeln der Gleichung $\omega^n = 1$, eine nichtcomplexe ganze Zahl, von der eben so, wie im vorigen §. für den besonderen Fall $m = 1$ geschehen ist, bewiesen werden kann, dafs sie nicht durch q theilbar ist. Es mufs daher $F(\omega)$ durch q^m theilbar sein. Setzt man also $F(\omega) = q^m f(\omega)$, so ist:

$$\Psi(\varpi_{r_1})^m f(\omega) = P_1(\omega)$$
$$\Psi(\varpi_{r_2})^m f(\omega) = P_2(\omega)$$
$$\vdots$$
$$\Psi(\varpi_{r_\nu})^m f(\omega) = P_\nu(\omega)$$

Wenn nun, wie in dem zweiten Theile des zu beweisenden Satzes vorausgesetzt wird, $F(\omega)$ jeden der idealen Primfaktoren des q genau m mal enthält, so ist keine der Zahlen $P_1(\omega)$, $P_2(\omega) \ldots P_\nu(\omega)$ durch q theilbar, also $f(\omega)$ enthält keinen idealen Primfaktor des q. Somit ist auch dieser zweite Theil des Satzes bewiesen.

Wenn eine complexe Zahl genau m ideale Primfaktoren des q enthält, dieselben mögen verschieden oder auch zum Theil oder alle einander gleich sein, so enthält die Norm derselben den Faktor q genau mt mal.

Es möge $F(\omega)$ genau m ideale Primfaktoren des q enthalten, so enthält die Norm $NF(\omega)$, welche aus den $\phi(n)$ conjugirten complexen Zahlen zusammengesetzt ist, deren genau $m\phi(n)$, und es ist leicht zu übersehen, dafs jeder der ν verschiedenen idealen Primfaktoren des q in der Norm gleich viel mal enthalten sein mufs, also jeder derselben $\frac{m\phi(n)}{\nu}$ mal, oder weil

$\nu = \frac{\phi(n)}{t}$, jeder mt mal. Die Norm ist daher, vermöge des vorhergehenden Satzes, theilbar durch $q^{m'}$, aber durch keine höhere Potenz von q.

§. 6.

Es ist nun auch die besondere Art der idealen Primfaktoren, welche den in der zusammengesetzten Zahl n enthaltenen Primzahlen angehören, in ähnlicher Weise zu behandeln. Sei p eine Primzahl, Faktor von n, und zwar a mal in n enthalten, so daſs $n = p^a n'$ und n' nicht weiter durch p theilbar ist, sei ferner $\omega = z\omega'$, wo z eine primitive Wurzel der Gleichung $z^{p^a} = 1$, ω' eine primitive Wurzel der Gleichung $\omega'^{n'} = 1$ bezeichnet. Die Primzahl p gehöre zum Exponenten Θ, für den Modul n', so daſs $p^\Theta \equiv 1$, mod. n', daſs aber keine niedere Potenz als die des Exponenten Θ dieser Congruenz genüge. Sei auch

$$\varpi'_r = \omega'^r + \omega'^{rp} \omega'^{rp^2} + \ldots + \omega'^{rp^{\Theta-1}}$$

ebenso bezeichne $\Psi(\varpi'_1)$ dieselbe complexe Zahl für die Wurzel ω' und die Primzahl p, als oben $\Psi(\varpi_1)$ für die Wurzel ω und die Primzahl q, alsdann entsprechen hier ebenfalls den Perioden ϖ'_i bestimmte Congruenzwurzeln u'_i, in der Art, daſs die Congruenz

$$\Psi(\varpi'_1)(\varpi'_i - u'_i) \equiv 0, \text{ mod. } p.$$

für alle Werthe des $i = 1, 2, 3, \ldots n'$ Statt hat.

Nimmt man in einer complexen Zahl $F(\omega)$ oder $F(z\omega')$ für z alle verschiedenen primitiven Wurzeln der Gleichung $z^{p^a} = 1$ und bildet das Produkt dieser verschiedenen complexen Zahlen, so soll dasselbe die in Beziehung auf die Wurzel z genommene partielle Norm der complexen Zahl $F(z\omega')$ genannt und durch $N_z F(z\omega')$ bezeichnet werden. Diese partielle Norm, als symmetrische Funktion aller primitiven Wurzeln der Gleichung $z^{p^a} = 1$, enthält die Wurzel z selbst nicht weiter in sich, und ist daher eine nur die Wurzel ω' enthaltende complexe Zahl. In gleicher Weise wird die in Beziehung auf die Wurzel ω' genommene partielle Norm, welche durch $N_{\omega'} F(z\omega')$ bezeichnet werden soll, das Produkt aller derjenigen complexen Zahlen vorstellen, welche man erhält, indem man für ω' alle verschiedenen primitiven Wurzeln der Gleichung $\omega'^{n'} = 1$ setzt, und wird eine nur die Wurzel z enthaltende complexe Zahl sein. Die vollständige Norm von

$F(z\omega')$ erhält man offenbar, indem man von der in Beziehung auf die Wurzel z genommenen Norm noch in Beziehung auf die Wurzel ω' die Norm nimmt, oder auch umgekehrt, indem man zuerst die partielle Norm in Beziehung auf die Wurzel ω' und von dieser die Norm in Beziehung auf z nimmt, welches folgendermaafsen dargestellt werden kann:

$$NF(z\omega') = N'_\omega N_z F(z\omega') = N_z N'_\omega F(z\omega').$$

Die Primzahl p läfst sich immer in wirkliche complexe Faktoren von der Form $f(z)$ zerlegen, denn man hat bekanntermaafsen:

$$p = \Pi_h (1 - z^h),$$

wo das Produkt Π_h über alle ganzzahligen Werthe des h zu erstrecken ist, welche kleiner als p^a und relative Primzahlen zu p sind, so dafs dieses Produkt aus $p^{a-1}(p-1)$ Faktoren besteht.

Weil ferner

$$1 - z^h = (1 - z)(1 + z + z^2 + \ldots . + z^{h-1})$$

und $1 + z + z^2 + \ldots + z^{h-1}$, wenn h nicht durch p theilbar ist, nur eine complexe Einheit darstellt, so hat man ebenfalls

$$p = (1 - z)^{p^{a-1}(p-1)} E(z),$$

wo $E(z)$ eine complexe Einheit bezeichnet.

Eine complexe Zahl $F(\omega')$, welche nur die Wurzel ω', nicht aber die Wurzel z enthält, kann nicht durch $1 - z$ theilbar sein, ohne dafs sie auch durch p theilbar ist.

Ist nämlich $F(\omega')$ durch $1 - z$ theilbar, so ist nothwendig die $p^{a-1}(p-1)$ te Potenz derselben durch $(1 - z)^{p^{a-1}(p-1)}$, also auch durch p theilbar. Da aber oben §. 2. bewiesen worden, dafs eine Potenz einer complexen Zahl der Wurzel ω der Gleichung $\omega^n = 1$ durch eine nicht in n enthaltene Primzahl q nicht theilbar sein kann, ohne dafs diese complexe Zahl selbst durch q theilbar ist, so folgt für den gegenwärtigen Fall, dafs eine Potenz der complexen Zahl $F(\omega')$ der Wurzel ω' der Gleichung $\omega'^{n'} = 1$ durch die nicht in n' enthaltene Primzahl p nicht theilbar sein kann, ohne dafs $F(\omega')$ selbst durch p theilbar ist. Aus $F(\omega') \equiv 0$, mod. $(1 - z)$, folgt daher nothwendigerweise $F(\omega') \equiv 0$, mod. p.

Wenn die complexe Zahl $F(z\omega')$ durch $1 - z$ theilbar ist, so ist offenbar die in Beziehung auf z genommene partielle Norm derselben durch p theilbar, weil $N_z(1 - z) = p$ ist. Die Umkehrung dieses Satzes, nämlich

Wenn die in Beziehung auf z genommene partielle Norm einer complexen Zahl $F(z\omega')$ durch p theilbar ist, so ist $F(z\omega')$ selbst durch $1-z$ theilbar,

wird folgendermaafsen bewiesen. Jede complexe Zahl $F(z\omega')$ genügt, weil $z = 1-(1-z)$ ist, der Congruenz

$$F(z\omega') \equiv F(\omega'), \text{ mod. } (1-z),$$

und hieraus, wenn auf beiden Seiten die Norm in Beziehung auf z genommen wird, folgt:

$$N_z\, F(z\omega') \equiv F(\omega')^{p^{\alpha-1}(p-1)}, \text{ mod. } (1-z),$$

also, weil nach der Voraussetzung des Satzes $N_z\, F(z\omega')$ durch p, und darum auch durch $1-z$ theilbar ist:

$$F(\omega')^{p^{\alpha-1}(p-1)} \equiv 0, \text{ mod. } (1-z),$$

woraus nach der Ausführung im vorigen Satze folgt, dafs $F(\omega') \equiv 0$, mod. p, also auch mod. $(1-z)$, und darum endlich $F(z\omega') \equiv 0$, mod. $(1-z)$.

Da die Primzahl p, abgesehen von einer Einheit, einer Potenz von $1-z$ gleich ist, so werden sämmtliche ideale Primfaktoren des p zugleich als ideale Primfaktoren des $1-z$ angesehen werden können. Dieselben sollen nun folgendermaafsen definirt werden:

Wenn p eine in n enthaltene Primzahl ist und a, n', z, ω', ϖ'_r, u'_i, $\Psi(\varpi'_1)$ die angegebenen Bedeutungen haben, so soll von einer complexen Zahl $F(z\omega')$, welche die Eigenschaft hat, dafs

$$\Psi(\varpi'_r)\, F(z\omega') \equiv 0, \text{ mod. } (1-z),$$

ausgesagt werden: sie enthält einen idealen Primfaktor des p, und zwar denjenigen, welcher zur Substitution $\varpi'_{r,i} = u'_i$ gehört. Hat die complexe Zahl $F(z\omega')$ aufserdem die Eigenschaft, dafs

$$\Psi(\varpi'_r)^k F(z\omega') \equiv 0, \text{ mod. } (1-z)^\mu,$$

aber

$$\Psi(\varpi'_r)^k F(z\omega') \text{ nicht} \equiv 0, \text{ mod. } (1-z)^{\mu+1},$$

wo

$$kp^{\alpha-1}(p-1) > \mu,$$

so soll von dieser complexen Zahl ausgesagt werden, dafs sie den zur Substitution $\varpi'_{r,i} = u'_i$ gehörenden idealen Primfaktor des p genau μ mal enthält.

Die Anzahl aller verschiedenen idealen Primfaktoren des p, für complexe Zahlen der Wurzel $\omega = z\acute{\omega}$ ist hiernach gleich der Anzahl aller verschiedenen Substitutionen von der Form $\varpi'_{r,i} = u_i$, oder was dasselbe ist gleich der Anzahl aller verschiedenen complexen Zahlen $\Psi(\varpi')$; sie ist also nach dem was oben (§. 4.) für die Anzahl der verschiedenen complexen Zahlen $\Psi(\varpi_r)$ gezeigt worden ist, hier, wo p zum Exponenten Θ gehört für den Modul $\acute{n}$, gleich $\frac{\phi(\acute{n})}{\Theta}$.

Die Bedingung, dafs die complexe Zahl $F(z\acute{\omega})$ den zur Substitution $\varpi'_{r,i} = u'$ gehörenden idealen Primfaktor des p genau μ mal enthalte, kann offenbar auch folgendermaafsen in Form einer Gleichung dargestellt werden:

$$\Psi(\varpi'_r)^k F(z\acute{\omega}) = (1-z)^{\mu}\chi(z\acute{\omega}),$$

wo $\chi(z\acute{\omega})$ nicht weiter durch $1-z$ theilbar ist und $kp^{a-1}(p-1) > \mu$, Nimmt man nun auf beiden Seiten die Norm in Beziehung auf z, so erhält man

$$\Psi(\varpi'_r)^{kp^{a-1}(p-1)} N_z F(z\acute{\omega}) = p^{\mu} N_z \chi(z\acute{\omega}),$$

und weil $\chi(z\acute{\omega})$ nicht durch $1-z$ theilbar ist, so ist auch $N_z \chi(z\acute{\omega})$ nicht durch p theilbar. Hieraus folgt der Satz:

Die complexe Zahl $F(z\acute{\omega})$ enthält den zur Substitution $\varpi'_{r,i} = u'_i$ gehörenden idealen Primfaktor des p, für complexe Zahlen der Wurzel $\omega = z\omega'$ genau eben so oft, als $N_z F(z\acute{\omega})$ den zur Substitution $\varpi'_{r,i} = \acute{u}_i$ gehörenden idealen Primfaktor des p für complexe Zahlen der Wurzel ω' enthält.

Weil $N_z F(z^h\omega') = N_z F(z\acute{\omega})$, wenn h nicht durch p theilbar ist, so folgt hieraus unmittelbar der Satz:

Die in der complexen Zahl $F(z\acute{\omega})$ enthaltenen idealen Primfaktoren des p bleiben für alle verschiedenen primitiven Wurzeln der Gleichung $z^{p^a} = 1$, welche man für z nehmen mag, stets dieselben.

Ferner gilt auch für die idealen Primfaktoren des p der Satz:

Das Produkt zweier oder mehrerer complexer Zahlen der Wurzel $\omega = z\omega'$ enthält genau dieselben idealen Primfaktoren des p als alle Faktoren zusammengenommen;
hat man nämlich

$$F(z\acute{\omega}) = f(z\acute{\omega}) \,.\, f_1(z\acute{\omega}) \,.\, f_2(z\acute{\omega}) \ldots$$

so ist auch

$$N_z F(z\omega') = N_z f(z\omega') \cdot N_z f_1(z\omega') \cdot N_z f_2(z\omega') \ldots.$$

Für die in dieser Gleichung enthaltenen Normen in Beziehung auf z, welche nur ω' enthaltende complexe Zahlen sind, gilt aber in Betreff der idealen Primfaktoren der in dem Wurzelexponenten n' nicht enthaltenen Primzahl p der in dem vorigen Paragraphen bewiesene Satz: daſs die idealen Primfaktoren des Produktes dieselben sind, als die der einzelnen Faktoren zusammengenommen, woraus vermöge des Satzes, welcher zeigt wie die idealen Primfaktoren des p, welche in $F(z\omega')$ enthalten sind, mit den in $N_z F(z\omega')$ enthaltenen übereinstimmen, die Richtigkeit des aufgestellten Satzes erhellt.

Wenn eine complexe Zahl $F(z\omega')$ alle verschiedenen idealen Primfaktoren des p enthält, und zwar jeden mindestens μ mal, so ist sie durch $(1-z)^\mu$ theilbar.

Nach der Voraussetzung dieses Satzes findet die Congruenz

$$\Psi(\varpi'_r)^k F(z\omega') \equiv 0, \text{ mod. } (1-z)^\mu,$$

statt, für alle Werthe des r, welche verschiedene conjugirte complexe Zahlen $\Psi(\varpi'_r)$ ergeben. Bezeichnet man nun die Summe der $k$ten Potenzen dieser verschiedenen conjugirten complexen Zahlen durch M', so hat man durch Addition dieser Congruenzen

$$M' F(z\omega') \equiv 0, \text{ mod. } (1-z)^\mu$$

und weil M' eine nichtcomplexe ganze Zahl ist, welche nicht durch p und folglich auch nicht durch $1-z$ theilbar ist, so ist:

$$F(z\omega') \equiv 0, \text{ mod. } (1-z)^\mu,$$

wie im Satze behauptet worden. Ich bemerke noch, daſs in dem besonderem Falle, wo μ ein Vielfaches von $p^{e-1}(p-1)$ also $\mu = cp^{e-1}(p-1)$, $F(\omega')$ durch p^c theilbar ist.

Wenn eine complexe Zahl $F(\omega)$ genau μ ideale Primfaktoren des p enthält, welche verschieden oder auch zum Theil oder alle einander gleich sein können, so enthält die vollständige Norm derselben den Faktor p genau $\mu\Theta$ mal.

Es ist nämlich

$$NF(\omega) = N_{\omega'} N_z F(z\omega')$$

und wenn $F(z\omega')$ μ ideale Primfaktoren des p enthält für complexe Zahlen der Wurzel $\omega = z\omega'$, so enthält $N_z F(z\omega')$ genau eben so viele ideale Primfaktoren des p für complexe Zahlen der Wurzel ω'; darum enthält nach dem ent-

D 2

sprechenden Satze des vorhergehenden Paragraphen $N_{\omega}N_{z}F(z\omega')$ den Faktor p genau $\mu\Theta$ mal.

§. 7.

Nachdem nun in den vorhergehenden beiden Paragraphen die Eigenschaften der beiden verschiedenen Arten idealer Primfaktoren, nämlich der allgemeineren, welche den in dem Wurzelexponenten n nicht enthaltenen mit q bezeichneten Primzahlen, und der besonderen, welche den in n enthaltenen, mit p bezeichneten Primzahlen angehören, für sich behandelt worden sind, sollen jetzt die gefundenen Resultate zusammengefafst werden.

Wenn die Norm einer complexen Zahl $F(\omega)$ durch irgend eine der Primzahlen q oder p, d. h. durch eine in dem Wurzelexponenten n nicht enthaltene oder durch eine darin enthaltene, theilbar ist, so enthält $F(\omega)$ nothwendig irgend einen idealen Primfaktor dieser Primzahl; denn die entwickelte ganzzahlige Norm, wenn sie ein Vielfaches von q oder p ist, enthält alle idealen Primfaktoren des q oder p, also die conjugirten complexen Zahlen zu $F(\omega)$, welche die Norm bilden, müssen ebenfalls irgend welche ideale Primfaktoren des q oder p enthalten, nach dem für beide Arten der idealen Primfaktoren geltenden Satze: dafs das entwickelte Produkt genau dieselben idealen Primfaktoren enthält als die Faktoren zusammengenommen. Andererseits ist in den vorhergehenden beiden Paragraphen bewiesen worden, dafs wenn eine complexe Zahl m ideale Primfaktoren des q enthält ihre Norm den Faktor q^{mt} enthalten mufs und ebenso, wenn sie μ ideale Primfaktoren des p enthält, dafs ihre Norm den Faktor $p^{\mu\Theta}$ enthalten mufs. Fafst man diese Resultate zusammen und bezeichnet irgend welche andere Primzahlen derselben Art als q oder p durch q', q'' .... p', p'' .... und die Exponenten zu welchen erstere für den Modul n, letztere für den Modul n' gehören, beziehungsweise durch t', t'' ... Θ', Θ'' ... so hat man den Satz:

Die Norm einer jeden complexen Zahl $F(\omega)$ ist eine Zahl von der Form

$$NF(\omega) = p^{\mu\Theta} \,.\, p'^{\mu'\Theta'} \ldots\ldots q^{mt} \,.\, q'^{m't'} \,.\, q''^{m''t''} \ldots\ldots$$

Da die Summe der Zahlen $\mu + \mu' + \ldots + m + m' + m'' + \ldots$, welche für jede bestimmte Zahl $F(\omega)$ nur eine endliche ist, der Anzahl aller in $F(\omega)$ enthaltenen idealen Primfaktoren gleich ist, und da durch die gegebenen Definitionen selbst unzweideutig bestimmt ist: ob und wie viel mal eine gegebene complexe Zahl einen bestimmten idealen Primfaktor enthält, so hat man den Hauptsatz:

Jede gegebene complexe Zahl enthält nur eine endliche Anzahl unveränderlich bestimmter idealer Primfaktoren.

Wenn umgekehrt nicht die complexe Zahl selbst gegeben ist, sondern nur alle idealen Primfaktoren, welche sie enthält, so ist sie dadurch noch nicht vollständig bestimmt, weil die complexen Einheiten, mit welchen sie multiplicirt sein kann, dabei unbestimmt bleiben. Es seien $F(\omega)$ und $F_1(\omega)$ zwei complexe Zahlen, welche beide genau dieselben idealen Primfaktoren enthalten sollen, so ist für dieselben

$$NF(\omega) = NF_1(\omega).$$

Setzt man nun

$$\frac{NF(\omega)}{F(\omega)} = \Phi(\omega),$$

so ist

$$\frac{F_1(\omega)}{F(\omega)} = \frac{F_1(\omega)\,\Phi(\omega)}{NF(\omega)}.$$

Wenn nun $F(\omega)$ so wie $F_1(\omega)$ μ ideale Primfaktoren des p, μ' ideale Primfaktoren des p', m ideale Primfaktoren des q, m' des q', u. s. w. enthält, so ist

$$NF(\omega) = NF_1(\omega) = p^{\mu\Theta} \,.\, p'^{\mu'\Theta'} \ldots q^{mt} \,.\, q'^{m't'}\, q''^{m''t''} \ldots.$$

also

$$\frac{F_1(\omega)}{F(\omega)} = \frac{F_1(\omega)\,\Phi(\omega)}{p^{\mu\Theta} \,.\, p'^{\mu'\Theta'} \,.\,.\, q^{mt} \,.\, q'^{\mu't'} \ldots.}$$

Da nun $F_1(\omega)$ genau dieselben idealen Primfaktoren enthält als $F(\omega)$, so enthält auch $F_1(\omega)\,\Phi(\omega)$ genau dieselben als $F(\omega)\,\Phi(\omega) = NF(\omega)$. Dieses Produkt enthält daher alle idealen Primfaktoren des p, jeden $\mu p^{\alpha-1}(p-1)$ mal und ist darum durch $p^{\mu\Theta}$ theilbar; es enthält alle idealen Primfaktoren des p', jeden $\mu' p'^{\alpha'-1}(p'-1)$ mal und ist darum durch $p'^{\mu'\Theta'}$ theilbar; es enthält alle idealen Primfaktoren des q, jeden mt mal, und ist darum durch q^{mt} theilbar; es enthält alle idealen Primfaktoren des q', jeden $m't'$ mal, und ist darum durch $q'^{m't'}$ theilbar u. s. w. Das Produkt $F_1(\omega)\,\Phi(\omega)$ ist darum durch $NF(\omega)$ theilbar, folglich

$$\frac{F_1(\omega)}{F(\omega)} = E(\omega)$$

wo $E(\omega)$ eine ganze complexe Zahl ist, oder

$$F_1(\omega) = E(\omega)\,F(\omega)$$

Nimmt man endlich auf beiden Seiten die Norm, so erhält man vermöge $NF_1(\omega) = NF(\omega)$

$$NE(\omega) = 1.$$

$E(\omega)$ ist also nur eine complexe Einheit. Diefs giebt den Satz:

Zwei complexe Zahlen, welche genau dieselben idealen Primfaktoren enthalten, unterscheiden sich nur durch complexe Einheiten, mit welchen sie multiplicirt sein können.

Wenn die complexe Zahl $F_1(\omega)$ alle idealen Primfaktoren des $F(\omega)$ enthält, aber aufser diesen vielleicht noch andere, so gelten alle Schlüsse für den Beweis des vorhergehenden Satzes bis dahin, dafs $\frac{F_1(\omega)}{F(\omega)}$ einer ganzen complexen Zahl gleich sein mufs; da aber in diesem Falle $NF_1(\omega)$ nicht gleich $NF(\omega)$ ist, so ist dieser Quotient nicht eine complexe Einheit, sondern er enthält genau alle idealen Primfaktoren, welche $F_1(\omega)$ mehr enthält als $F(\omega)$. Man hat also den Satz:

Eine complexe Zahl ist durch eine andere theilbar, wenn alle idealen Primfaktoren des Divisors auch im Dividendus enthalten sind. Der Quotient enthält alsdann genau den Überschuss der idealen Primfaktoren des Dividendus über die des Divisors.

Wenn es eine wirkliche complexe Zahl $f(\omega)$ giebt, von der Art, dafs sie nur einen einzigen idealen Primfaktor und zwar nur einmal enthält, wenn ferner $F(\omega)$ irgend eine complexe Zahl ist, welche denselben idealen Primfaktor enthält: so ist nach dem soeben bewiesenen Satze $F(\omega)$ theilbar durch $f(\omega)$, und wenn der Quotient durch $Q(\omega)$ bezeichnet wird, so ist

$$F(\omega) = f(\omega) \,.\, Q(\omega)\,.$$

$f(\omega)$ ist aber in diesem Falle nicht mehr ein idealer, sondern ein wirklicher complexer Primfaktor. Hieraus folgt:

Die durch die gegebenen Definitionen bestimmten idealen Primfaktoren der nichtcomplexen Primzahlen stimmen überall wo es wirkliche complexe Primzahlen giebt mit diesen vollkommen überein.

Die idealen Primfaktoren, für welche, wie in dem Vorhergehenden gezeigt worden ist, die Rechnungsregeln genau dieselben sind als für gewöhnliche Primzahlen, da das Produkt mehrerer Faktoren genau dieselben enthält als die Faktoren zusammengenommen, und da jede gegebene complexe Zahl eine endliche unveränderlich bestimmte Anzahl solcher idealer Primfaktoren enthält, welche auch in allen Fällen, wo es wirkliche complexe Primfaktoren giebt, mit diesen vollkommen übereinstimmen, besitzen somit alle wesentlichen Eigenschaften wahrhafter Primfaktoren.

§. 8.

Für die Klassifikation der idealen complexen Zahlen, welche in dem Folgenden gegeben werden soll, ist es von Wichtigkeit die Congruenzbedingungen, durch welche die idealen Primfaktoren bestimmt werden, etwas genauer zu untersuchen. Eine Congruenz unter complexen Zahlen der Wurzel ω, deren Modul eine nichtcomplexe ganze Zahl ist, ist wie oben (§. 1.) gezeigt worden mit $\phi(n)$ Congruenzen unter nichtcomplexen ganzen Zahlen für denselben Modul gleichbedeutend, welche in besonderen Fällen sich auch auf eine geringere Anzahl reduciren können.

Einen solchen besonderen Fall gewährt die Congruenz

$$\Psi(\varpi_r)^m F(\omega) \equiv 0, \text{ mod. } q^m,$$

welche ausdrückt, daſs die complexe Zahl $F(\omega)$ den zur Substitution $\varpi_{r,i} = u_i$ gehörenden idealen Primfaktor des q m mal enthält; die Anzahl der nothwendigen Congruenzbedingungen unter den Coefficienten von $F(\omega)$, welche mit dieser einen für complexe Zahlen gleichbedeutend sind, reducirt sich nämlich immer auf t. Um dieſs zu zeigen, betrachte ich die Gleichung, des Grades t, deren Wurzeln die in der Periode ϖ_1 enthaltenen Wurzeln ω, ω^q, ω^{q^2}, ... $\omega^{q^{t-1}}$ sind, welche ich folgendermaaſsen darstelle,

$$\omega^t + P_1 \omega^{t-1} + P_2 \omega^{t-2} + \ldots\ldots + P_t = 0.$$

Die Coefficienten dieser Gleichung, P_1, P_2, ... P_t, als symmetrische Funktionen aller in einer Periode enthaltenen Wurzeln können offenbaar nur Funktionen der Perioden ϖ_1, ϖ_2 .... ϖ_n sein. Ist nun $F(\omega)$ irgend eine complexe Zahl, so hat man

$$F(\omega) = a + a_1\,\omega + a_2\,\omega^2 + \ldots\ldots + a_{\phi(n)-1}\,\omega^{\phi(n)-1}$$

und man kann nun vermittelst der Gleichung des Grades t alle Potenzen von ω, welche höher sind als die $(t-1)$ te eliminiren und dadurch jede complexe Zahl $F(\omega)$ in die Form setzen:

$$F(\omega) = \Phi(\varpi_1) + \Phi_1(\varpi_1)\,\omega + \Phi_2(\varpi_1)\,\omega^2 + \ldots\ldots + \Phi_{t-1}(\varpi_1)\,\omega^{t-1},$$

in welcher $\Phi(\varpi_1)$, $\Phi_1(\varpi_1) \ldots \Phi_{t-1}(\varpi_1)$ nur die Perioden enthalten und in Beziehung auf die Coefficienten a, a_1, $a_2 \ldots$ der complexen Zahl $F(\omega)$ lineäre Funktionen sind. Hieraus folgt zunächst, daſs wenn

$$\Psi(\varpi_r)^m \, \Phi_k(\varpi_1) \equiv 0, \text{ mod. } q^m,$$

für alle Werthe $k = 0, 1, 2, \ldots t-1$, auch

$$\Psi(\varpi_r)^m \, F(\omega) \equiv 0, \text{ mod. } q^m$$

ist. Um nun zu beweisen, daſs auch umgekehrt aus dieser einen Congruenzbedingung nothwendig jene t Congruenzen folgen, verwandle ich in dem Ausdrucke des $F(\omega)$ nach einander ω in ω^q, $\omega^{q^2} \ldots \omega^{q^{t-1}}$, wobei die Perioden ϖ_1, $\varpi_2 \ldots \varpi_n$ alle ungeändert bleiben. Es entsteht so folgendes System von t Gleichungen:

$$\begin{aligned}
F(\omega) &= \Phi(\varpi_1) + \omega\,\Phi_1(\varpi_1) + \omega^2\,\Phi_2(\varpi_1) + \ldots + \omega^{t-1}\,\Phi_{t-1}(\varpi_1)\\
F(\omega^q) &= \Phi(\varpi_1) + \omega^q\,\Phi_1(\varpi_1) + \omega^{2q}\,\Phi_2(\varpi_1) + \ldots + \omega^{(t-1)q}\,\Phi_{t-1}(\varpi_1)\\
&\vdots\\
F(\omega^{q^{t-1}}) &= \Phi(\varpi_1) + \omega^{q^{t-1}}\,\Phi_1(\varpi_1) + \omega^{2q^{t-1}}\,\Phi_2(\varpi_1) + \ldots +\\
&\qquad \omega^{(t-1)q^{t-1}}\,\Phi_{t-1}(\varpi_1).
\end{aligned}$$

Betrachtet man $\Phi(\varpi_1)$, $\Phi_1(\varpi_1) \ldots \Phi_{t-1}(\varpi_1)$ als die Unbekannten dieses Systems und löst dasselbe auf, so erhält man allgemein:

$$D(\omega)\,\Phi_k(\varpi_1) = \overset{k}{A} F(\omega) + \overset{k}{A}_1 F(\omega^q) + \overset{k}{A}_2 F(\omega^{q^2}) + \ldots + \overset{k}{A}_{t-1} F(\omega^{q^{t-1}}),$$

wo die Determinante $D(\omega)$ nach bekannten Regeln der Algebra gleich dem Produkte aller Faktoren von der Form $\omega^{q^r} - \omega^{q^s}$ ist, für $r = 0, 1, 2, \ldots t-1$, $s = 0, 1, 2, \ldots t-1$, mit Ausschluſs der Werthe $r = s$, und wo $\overset{k}{A}$, $\overset{k}{A}_1 \ldots \overset{k}{A}_{t-1}$ complexe ganze Zahlen der Wurzel ω sind. Wenn nun

$$\Psi(\varpi_r)^m \, F(\omega) \equiv 0, \text{ mod. } q^m,$$

so ist auch allgemein

$$\Psi(\varpi_r)^m \, F(\omega^{q^h}) \equiv 0, \text{ mod. } q^m,$$

für alle Werthe $h = 0, 1, 2, \ldots t-1$, und darum vermöge der gefundenen Gleichung:

$$D(\omega) \,.\, \Psi(\varpi_r)^m \,.\, \Phi_k(\varpi_1) \equiv 0, \text{ mod. } q^m,$$

für alle Werthe des $k = 0, 1, 2, \ldots t-1$. Multiplicirt man diese Congruenz noch mit allen zu $D(\omega)$ conjugirten complexen Zahlen so hat man

$$ND(\omega) \,.\, \Psi(\varpi_r)^m \;\Phi_k(\varpi_1) \equiv 0, \text{ mod. } q^m.$$

Die Norm $ND(\omega)$, welche eine nichtcomplexe ganze Zahl ist, enthält, wie leicht zu zeigen ist, niemals den Faktor q; da nämlich $D(\omega)$ nur aus Faktoren von der Form $\omega^{q^r} - \omega^{q^s}$ zusammengesetzt ist, so kann $ND(\omega)$ keine anderen Primfaktoren enthalten als solche, welche in n vorkommen, also niemals den Primfaktor q. Es kann daher $ND(\omega)$ aus dieser Congruenz hinweggehoben werden, welche alsdann zeigt, dafs wenn

$$\Psi(\varpi_r)^m \; F(\omega) \equiv 0, \text{ mod. } q^m,$$

nothwendig auch

$$\Psi(\varpi_r)^m \;\Phi_k(\varpi_1) \equiv 0, \text{ mod. } q^m,$$

ist, für jeden der Werthe $k = 0, 1, 2, \ldots t-1$. Da ferner, wie oben (§. 4.) gezeigt worden, bei Anwendung der Substitution $\varpi_{r,i} = \overset{m}{u}_i$ diese Congruenzen gleichbedeutend sind mit

$$\Phi_k(\overset{m}{u}_\varrho) \equiv 0, \text{ mod. } q^m,$$

für $k = 0, 1, 2, \ldots t-1$, wo $r\varrho \equiv 1$, mod. n, so hat man folgenden Satz:

Die Bedingung, dafs die complexe Zahl $F(\omega)$ den zur Substitution $\varpi_{r,i} = u_i$ gehörenden idealen Primfaktor des q m mal enthalte, oder was dasselbe ist die Congruenz

$$\Psi(\varpi_r)^m \; F(\omega) \equiv 0, \text{ mod. } q^m,$$

ist gleichbedeutend mit den t Congruenzen

$$\Phi(\overset{m}{u}_\varrho) \equiv 0,\; \Phi_1(\overset{m}{u}_\varrho) \equiv 0, \;\ldots\; \Phi_{t-1}(\overset{m}{u}_\varrho) \equiv 0, \text{ mod. } q^m,$$

welche man erhält wenn man $F(\omega)$ in die Form

$$F(\omega) = \Phi(\varpi_1) + \omega\,\Phi_1(\varpi_1) + \omega^2\,\Phi_2(\varpi_1) + \ldots + \omega^{t-1}\,\Phi_{t-1}(\varpi_1)$$

setzt, und in diesen Coefficienten der einzelnen Potenzen von ω die Perioden durch die Congruenzwurzeln $\overset{m}{u}_i$ ersetzt, welche ihnen für den Modul q^m entsprechen.

Es sind also t Congruenzen, welche in Beziehung auf die Coefficienten einer complexen Zahl $F(\omega)$ nur lineär sind, für den Modul q^m, nothwendig und hinreichend, damit $F(\omega)$ einen idealen Primfaktor des q m mal enthalte.

In ähnlicher Weise sollen nun auch für die besondere Art der idealen Primfaktoren, welche den in n enthaltenen Primzahlen angehören die Congruenzbedingungen entwickelt werden.

Ordnet man die complexe Zahl $F(z\omega')$ nach Potenzen von z allein, so kann man sie in folgende Form setzen:

$$F(z\omega') = A(\omega') + A_1(\omega')\, z + A_2(\omega')\, z^2 + \ldots + A_{\phi-1}(\omega')\, z^{\phi-1},$$

wo der Kürze wegen $\phi(p^a) = p^{a-1}(p-1)$ einfach durch ϕ bezeichnet ist, alle höheren Potenzen von z können nämlich vermittelst der Gleichung

$$z^{p^{a-1}(p-1)} + z^{p^{a-1}(p-2)} + z^{p^{a-1}(p-3)} + \ldots + 1 = 0$$

eliminirt werden. Setzt man ferner $1-(1-z)$ statt z und ordnet nach Potenzen von $1-z$, so erhält man

$$F(z\omega') = B(\omega') + B_1(\omega')(1-z) + B_2(\omega')(1-z)^2 + \ldots + B_{\phi-1}(\omega')(1-z)^{\phi-1}$$

Die Bedingung, dafs $F(z\omega')$ den zur Substitution $\varpi'_{r,i} = u'_i$ gehörenden idealen Primfaktor des p μ mal enthalte, nämlich

$$\Psi(\varpi'_r)^k\, F(z\omega') \equiv 0, \text{ mod. } (1-z)^\mu,$$

wo $k\phi > \mu$, giebt daher, wenn der Kürze wegen

$$\Psi(\varpi'_r)^k\, B_h(\omega') = C_h$$

gesetzt wird:

$$C + C_1(1-z) + C_2(1-z)^2 + \ldots + C_{\phi-1}(1-z)^{\phi-1}, \text{ mod. } (1-z)^\mu.$$

Hieraus folgt zunächst, dafs C durch $1-z$ theilbar sein mufs, welches, weil C nur die Wurzel ω', nicht aber z enthält, nicht anders geschehen kann, als dafs C durch p theilbar ist, also den Faktor $1-z$ ϕ mal enthält. Hebt man nun aus dieser Congruenz und dem Modul den Faktor $1-z$ einmal hinweg, so sieht man ferner, dafs auch C_1 durch $1-z$ und also durch p theilbar sein mufs u. s. w. Ist nun $\mu \leqq \phi$, so müssen, wie auf diese Weise gezeigt wird, die ersten μ Coefficienten $C, C_1, C_2, \ldots C_{\mu-1}$ durch p theilbar sein, und wenn dieselben durch p theilbar sind so ist auch dieser Congruenz genügt. Ist aber $\mu > \phi$, so müssen zunächst alle Coefficienten $C, C_1, \ldots C_{\phi-1}$ durch p theilbar sein, man kann daher diesen gemeinschaftlichen Faktor p, oder was dasselbe ist $(1-z)^\phi$, aus der Congruenz und dem Modul hinwegheben und erhält so eine Gleichung derselben Form, für den Modul $(1-z)^{\mu-\phi}$, auf welche man dieselben Schlüsse von Neuem anwenden kann. Ist in dieser $\mu \leqq 2\phi$ so müssen die ersten $\mu-\phi$ Coefficienten nochmals durch p theilbar sein, also $C, C_1, \ldots C_{\mu-\phi-1}$ theilbar durch p^2, und $C_{\mu-\phi}, C_{\mu-\phi+1}, \ldots C_{\phi-1}$ theilbar durch p. Fährt man, wenn $\mu > 2\phi$ ist, in derselben Weise fort zu schliefsen, so sieht man leicht, dafs allgemein wenn μ in den Grenzen $c\phi$ und $(c+1)\phi$

liegt, also $\mu = c\phi + \mu_1$ ist, wo $\mu_1 < \phi$, die ersten μ_1 Coefficienten C, C', $\ldots C_{\mu_1 - 1}$, durch p^{c+1}, die übrigen aber C_{μ_1}, $C_{\mu_1 + 1} \ldots . C_{\phi - 1}$ durch p^c theilbar sein müssen, damit dieser Congruenz genügt werde. Dafs aber umgekehrt, wenn diese Bedingungen erfüllt sind dieser Congruenz wirklich genügt wird, ist von selbst klar, da p den Faktor $1 - z$ ϕ mal enthält. Betrachtet man noch, dafs die Bedingung $C_k \equiv 0$, mod. p^k nichts anderes ausdrückt, als dafs $B_h(\omega')$ den zur Substitution $\varpi'_{r,i} = u'_i$ gehörenden idealen Primfaktor des p, für complexe Zahlen der Wurzel ω', k mal enthält, so hat man folgendes Resultat.

Die Bedingung, dafs die complexe Zahl $F(z\omega')$ den zur Substitution $\varpi'_{r,i} = u'_i$ gehörenden idealen Primfaktor des p, für complexe Zahlen der Wurzel $\omega = z\omega'$, μ mal enthalte, wo $\mu = c\phi + \mu_1, \mu_1 < \phi$, $\phi = p^{\sigma - 1}(p - 1)$, ist gleichbedeutend damit, dafs von den complexen Zahlen

$$B(\omega'),\ B_1(\omega'),\ B_2(\omega') \ldots\ldots B_{\phi - 1}(\omega'),$$

welche man erhält indem man $F(z\omega')$ in die Form

$$F(z\omega') = B(\omega') + B_1(\omega')(1 - z) + B_2(\omega')(1 - z)^2 + \ldots + B_{\phi - 1}(\omega')(1 - z)^{\phi - 1}$$

setzt, die ersten μ_1 den zur Substitution $\varpi'_{r,i} = u'_i$ gehörenden idealen Primfaktor des p für complexe Zahlen der Wurzel ω', $c + 1$ mal, die übrigen $\phi - \mu_1$ aber denselben idealen Primfaktor c mal enthalten.

Die Bedingung, dafs $B_h(\omega')$ einen bestimmten idealen Primfaktor des p $c + 1$ mal oder c mal enthalte kann aber, wie oben gezeigt worden, da die Primzahl p in dem Wurzelexponenten n' der Wurzel ω' nicht enthalten ist, und p zum Exponenten Θ gehört, mod. n', durch Θ Congruenzen für den Modul p^{c+1} oder p^c ausgedrückt werden, welche in Beziehung auf die ganzzahligen Coefficienten der complexen Zahl $B_h(\omega')$ nur lineär sind. Da nun $B_h(\omega')$ die Coefficienten der ursprünglich gegebenen complexen Zahl $F(z\omega')$ nur in lineärer Weise enthält, so schliefst man:

Die Bedingung, dafs die complexe Zahl $F(z\omega')$ einen bestimmten idealen Primfaktor des p μ mal enthält, wird, wenn $\mu = c\phi + \mu_1$ ist und $\mu_1 < \phi$, $\phi = p^{\sigma - 1}(p - 1)$, durch $\mu_1 \Theta$ lineäre Congruenzen für den Modul p^{c+1}, und durch $(\phi - \mu_1)\Theta$ lineäre Congruenzen für den Modul p^c, unter den Coefficienten der complexen Zahl $F(z\omega')$, ausgedrückt.

E 2

§. 9.

Es sollen nun die idealen Primfaktoren und die aus denselben zusammengesetzten oder zusammenzusetzenden idealen complexen Zahlen auf dieselbe Weise bezeichnet werden als die wirklichen, so dafs $f(\omega)$ oder $F(\omega)$ in dem Folgenden complexe Zahlen bezeichnen sollen, welche gewisse gegebene ideale Primfaktoren enthalten, abgesehen davon, ob solche complexe Zahlen wirklich existiren oder nicht. Die Aufgabe, eine wirkliche complexe Zahl zu finden, welche gewisse gegebene ideale Primfaktoren enthält, aber nur diese allein, ist nämlich im Allgemeinen nicht lösbar, wenn man aber zuläfst, dafs dieselbe aufser diesen gegebenen auch andere ideale Primfaktoren enthalten darf, so hat sie stets unendlich viele Lösungen. Diefs kann auch so ausgesprochen werden: es giebt für jede ideale complexe Zahl $F(\omega)$ eine unendliche Anzahl verschiedener idealer Multiplikatoren $F_1(\omega)$ von der Art, dafs das Produkt $F(\omega)\,F_1(\omega)$ eine wirkliche complexe Zahl wird. Wählt man diese idealen Multiplikatoren so aus, dafs die Normen derselben, welche stets wirkliche nichtcomplexe Zahlen sind, möglichst klein werden, so findet man, dafs eine endliche bestimmte Anzahl idealer Multiplikatoren hinreichend ist um alle idealen complexen Zahlen zu wirklichen zu machen.

Um diesen Hauptpunkt der Theorie der hier behandelten complexen Zahlen zu beweisen, stelle ich die ideale complexe Zahl $F(\omega)$ als Produkt der in ihr enthaltenen idealen Primfaktoren dar, nämlich

$$F(\omega) = \phi(\omega)^{\mu} \,.\, \phi'(\omega)^{\mu'} \ldots f(\omega)^{m} \,.\, f'(\omega)^{m'} \,.\, f''(\omega)^{m''} \ldots\ldots$$

wo $\phi(\omega)$ ein bestimmter idealer Primfaktor der in n enthaltenen Primzahl p ist, welche für den Modul n' zum Exponenten Θ gehört, $\phi'(\omega)$ ein idealer Primfaktor, der ebenfalls in n enthaltenen Primzahl p', welche zum Exponenten Θ' gehört u. s. w., ferner $f(\omega)$ ein idealer Primfaktor, der nicht in n enthaltenen, zum Exponenten t gehörenden Primzahl q, $f'(\omega)$ Primfaktor von q', welche zum Exponenten t' gehört, u. s. w. Ferner sei

$$\Phi(\omega) = x + x_1\,\omega + x_2\,\omega^2 + \ldots\ldots + x_{\phi(n)-1}\,\omega^{\phi(n)-1}$$

eine wirkliche complexe Zahl, welche alle idealen Primfaktoren des $F(\omega)$ und somit diese ideale complexe Zahl selbst als Faktor enthält, so dafs

$$\Phi(\omega) = F(\omega)\,F_1(\omega),$$

wo $F_1(\omega)$ ebenfalls eine ideale complexe Zahl ist. Die Coefficienten der zu suchenden wirklichen complexen Zahl $\Phi(\omega)$, sind hier als die zu bestimmenden Unbekannten durch x, x_1, x_2 . . . . . bezeichnet worden. Die Bedingung, dafs $\Phi(\omega)$ den idealen Primfaktor $\phi(\omega)$ des p μ mal enthalte, wenn $\mu = cp^{a-1}(p-1) + \mu_1$, $\mu_1 < p^{a-1}(p-1)$ wird nun, wie im vorhergehenden §. gezeigt worden, durch $\mu_1 \Theta$ Congruenzen mod. p^{c+1} und durch $(p^{a-1}(p-1) - \mu_1)\,\Theta$ Congruenzen, mod. p^c, ausgedrückt, welche in Beziehung auf die Coefficienten x, x_1, x_2 . . . . der complexen Zahl $\Phi(\omega)$ nur lineär sind. Von derselben Art sind die Bedingungen dafür, dafs der ideale Primfaktor $\phi'(\omega)$ μ' mal in $\Phi(\omega)$ enthalten sei, u. s. w. Ferner die Bedingung dafür, dafs $\Phi(\omega)$ den idealen Primfaktor $f(\omega)$ der nicht in n enthaltenen Primzahl q m mal enthält wird, wie ebenfalls im vorhergehenden §. gezeigt worden, durch t Congruenzen für den Modul q^m ausgedrückt, ebenso die Bedingung dafs $\Phi(\omega)$ den idealen Primfaktor $f'(\omega)$ des q' m' mal enthält, durch t' Congruenzen für den Modul $q'^{m'}$ u. s. w. Also $\Phi(\omega)$ wird den ganzen idealen Faktor $F(\omega)$ enthalten, wenn gewisse

$\mu_1\,\Theta$ Congruenzen mod. p^{c+1} und $(p^{a-1}(p-1)-\mu_1)\,\Theta$ Congruenzen mod. p^c
$\mu'_1\,\Theta'$ - mod. $p'^{c'+1}$ - $(p'^{a'-1}(p'-1)-\mu'_1)\,\Theta'$ - - $p'^{c'}$
u. s. w. und

t Congruenzen mod. q^m
t' - - $q'^{m'}$
t'' - - $q''^{m''}$

u. s. w. erfüllt sind, welche Congruenzen in Beziehung auf die Coefficienten x, x_1, x_2, ... der complexen Zahl $\Phi(\omega)$ alle nur lineäre sind. Denkt man sich alle diese Congruenzen auf die Form $X \equiv 0$ gebracht, wo X eine lineäre Funktion von x, x_1, x_2, . . . bezeichnet, so werden, wenn man diesen Coefficienten beliebige ganzzahlige Werthe giebt, diese X im Allgemeinen nicht congruent Null werden, in Beziehung auf die ihnen zugehörenden Moduln, sondern irgend welche Reste geben, welche kleiner als diese Moduln genommen werden können. Die gröfstmögliche Anzahl verschiedener Reste, welche eine solche Congruenz geben kann, ist daher nur gleich dem Modul selbst. Das System aller der Congruenzen, welche die Bedingung ausdrücken, dafs $\Phi(\omega)$ den idealen Faktor $F(\omega)$ enthält, wird demnach höchstens so viele verschiedene Restensysteme geben können, als das Produkt sämmtlicher Moduln Einheiten enthält, also wie aus der Anzahl und den Muduln dieser Congruenzen leicht zu erkennen, nur

$$p^{\mu\Theta} \cdot p'^{\mu'\Theta'} \ldots q^{mt} \cdot q'^{m't'} q''^{m''t''} \ldots$$

welche Anzahl der Norm der idealen complexen Zahl $F(\omega)$ gleich ist. Es giebt also nicht mehr verschiedene Restensysteme der Gröfsen X als $NF(\omega)$. Giebt man nun jedem der $\phi(n)$ Coefficienten $x, x_1, x_2, \ldots$ alle Werthe $0, 1, 2, \ldots k-1$, so hat man $k^{\phi(n)}$ verschiedene Werthcombinationen dieser Coefficienten und jeder derselben gehört ein bestimmtes Restensystem der Gröfsen X zu. Nimmt man nun k so grofs, dafs

$$k^{\phi(n)} > NF(\omega),$$

so hat man mehr Werthsysteme dieser Coefficienten als Restensysteme der Gröfsen X, es können also diese Restensysteme, welche jenen Werthsystemen zugehören, nicht alle verschieden von einander sein, sondern für irgend welche zwei verschiedene Werthsysteme der Coefficienten $x, x_1, x_2, \ldots$ müssen alle Gröfsen X für ihre Moduln dieselben Reste geben. Seien nun

$$x = a, x_1 = a_1, x_1 = a_2, \ldots.$$

und

$$x = b, x_1 = b_1, x_2 = b_2, \ldots.$$

zwei Werthsysteme der Coefficienten, welche genau dasselbe Restensystem für die Gröfsen X geben, so ist, weil diese Gröfsen X nur lineäre Funktionen von $x, x_1, x_2, \ldots$ sind, das Werthsystem

$$x = a - b, x_1 = a_1 - b_1, x_2 = a_2 - b_2 \ldots.$$

offenbar ein solches, für welches alle Reste der X gleich Null, also alle obigen Congruenzen erfüllt werden. Da die Gröfsen $a, a_1, a_2 \ldots$ sowohl als $b, b_1, b_2, \ldots$ alle kleiner als k und nicht negativ sind, so folgt, dafs die Differenzen $a - b, a_1 - b_1, a_2 - b_2, \ldots.$ abgesehen von ihren Vorzeichen alle kleiner als k sein müssen. Also wenn k so grofs angenommen wird, dafs $k^{\phi(n)} > NF(\omega)$, so wird der Bedingung, dafs eine wirkliche complexe Zahl $\Phi(\omega)$ den idealen Faktor $F(\omega)$ enthält, stets durch solche Werthe der Coefficienten von $\Phi(\omega)$ genügt, welche abgesehen von den Vorzeichen alle kleiner als k sind.

Multiplicirt man nun die complexe Zahl

$$\Phi(\omega) = x + x_1\omega + x_2\omega^2 + \ldots. + x_{\phi(n)-1}\,\omega^{\phi(n)-1}$$

mit ihrer reciproken $\Phi(\omega^{-1})$ und nimmt die Summe dieser Produkte in Beziehung auf alle Wurzeln der Gleichung $\omega^n = 1$, (nicht in Beziehung auf die primitiven allein,) so erhält man ohne Schwierigkeit

$$\sum_{1}^{n}{}_h\, \Phi(\omega^h)\, \Phi(\omega^{-h}) = n\,(x^2 + x_1^2 + x_2^2 + \ldots + x_{\phi(n)-1}^2).$$

Da nun alle Glieder von der Form $\Phi(\omega^h)\,\Phi(\omega^{-h})$ nur positiv sein können, so ist klar, dafs wenn man die Summe nur auf alle primitiven Wurzeln ω erstreckt, diese kleiner sein mufs als diejenige, welche sich auf alle Wurzeln der Gleichung $\omega^n = 1$ erstreckt, man hat daher

$$\Sigma\,\Phi(\omega)\,\Phi(\omega^{-1}) < n\,(x^2 + x_1^2 + x_2^2 + \ldots + x_{\phi(n)-1}^2),$$

wo die Summe Σ nur auf alle primitiven Wurzeln ω zu beziehen ist. Wendet man nun den bekannten Satz an, dafs das Produkt von m positiven Gröfsen immer kleiner ist als die $m$te Potenz des arithmetischen Mittels derselben, und bemerkt, dafs das Produkt der $\phi(n)$ Gröfsen von der Form $\Phi(\omega)\,\Phi(\omega^{-1})$ gleich dem Quadrate der Norm $N\Phi(\omega)$ ist, so hat man

$$(N\Phi(\omega))^2 < n^{\phi(n)}\left(\frac{x^2 + x_1^2 + x_2^2 + \ldots + x_{\phi(n)-1}^2}{\phi(n)}\right).$$

Nimmt man nun für alle Coefficienten $x, x_1, x_2, \ldots$ die gröfsten Werthe, welche sie überhaupt haben können, nämlich $k - 1$, und zieht auf beiden Seiten die Quadratwurzel aus, so hat man

$$N\Phi(\omega) < n^{\frac{1}{2}\phi(n)} \,.\, (k-1)^{\phi(n)}.$$

Die Zahl k ist nach der Voraussetzung so grofs zu wählen, dafs

$$k^{\phi(n)} > NF(\omega),$$

wodurch ihre untere Gränze bestimmt wird; bestimmt man aufserdem die obere Gränze dadurch, dafs

$$(k-1)^{\phi(n)} < NF(\omega)$$

sein soll, so ist

$$N\Phi(\omega) < n^{\frac{1}{2}\phi(n)} \,.\, NF(\omega)$$

und weil $\Phi(\omega) = F(\omega)\,F_1(\omega)$, also $N\Phi(\omega) = NF(\omega)\,NF_1(\omega)$, so ist endlich

$$NF_1(\omega) < n^{\frac{1}{2}\phi(n)};$$

d. h. die idealen Multiplikatoren, welche mit einer gegebenen idealen complexen Zahl zusammengesetzt, wirkliche complexe Zahlen als Produkte geben, können immer so gewählt werden, dafs die Normen derselben kleiner sind als $n^{\frac{1}{2}\phi(n)}$. Dieser Bedingung aber kann offenbaar nur eine endliche Anzahl idealer Multiplikatoren genügen; man hat daher den Satz:

Alle idealen complexen Zahlen, deren Anzahl unendlich ist, können durch Zusammensetzung mit einer endlichen

Anzahl idealer Multiplikatoren zu wirklichen complexen Zahlen gemacht werden.

§. 10.

Auf die Multiplikatoren, welche in endlicher Anzahl ausreichen, um alle idealen complexen Zahlen zu wirklichen zu machen, wird nun die Eintheilung der idealen Zahlen in verschiedene Klassen gegründet, nämlich:

Alle diejenigen idealen Zahlen, welche mit einem und demselben idealen Multiplikator zusammengesezt wirkliche complexe Zahlen als Produkte ergeben, sollen *äquivalente* ideale Zahlen genannt und einer und derselben *Klasse* der ideale Zahlen zugerechnet werden.

Die wirklichen complexen Zahlen werden in diese Klassifikation mit einbegriffen, in welcher sie eine Klasse für sich ausmachen, welche die Hauptklasse genannt wird.

Jeder Multiplikator, welcher eine ideale Zahl zu einer wirklichen macht, macht auch alle dieser äquivalenten idealen Zahlen zu wirklichen. Die Klassifikation der idealen Zahlen ist von der zufälligen Wahl der Multiplikatoren unabhängig.
Sind nämlich $F(\omega)$ und $G(\omega)$ zwei äquivalente ideale Zahlen, so mufs es einen Multiplikator $M(\omega)$ geben, welcher beide zu wirklichen macht. Es sei nun $M_1(\omega)$ ein anderer Multiplikator, welcher $F(\omega)$ zu einer wirklichen complexen Zahl macht, so sind die Produkte $F(\omega)M(\omega)$, $G(\omega)M(\omega)$ und $F(\omega)M(\omega)$ wirkliche complexe Zahlen. Multiplicirt man nun das zweite und dritte dieser Produkte mit einander und dividirt durch das erste, so erhält man als Quotienten $G(\omega)M_1(\omega)$ welches also ebenfalls eine wirkliche complexe Zahl ist. Jeder Multiplikator, welcher eine von zwei äquivalenten idealen Zahlen zu einer wirklichen macht, macht also ebenfalls auch die andere zu einer wirklichen.

Zwei ideale Zahlen, welche einer und derselben dritten äquivalent sind, sind auch unter sich äquivalent.
Denn ein bestimmter idealer Multiplikator, welcher die erste und dritte dieser idealen Zahlen zu wirklichen macht, mufs auch die zweite zu einer wirklichen machen, da auch die zweite der dritten äquivalent ist.

Äquivalente ideale Zahlen mit äquivalenten zusammengesetzt, geben äquivalente Produkte.

Sind nämlich $F(\omega)$ und $G(\omega)$ zwei äquivalente ideale Zahlen, $M(\omega)$ ihr gemeinschaftlicher Multiplikator, und auch $F_1(\omega)$ und $G_1(\omega)$ äquivalent und $M_1(\omega)$ ihr gemeinschaftlicher Multiplikator, so ist offenbaar $M(\omega)\,M_1(\omega)$ ein Multiplikator, welcher sowohl $F(\omega)\,F_1(\omega)$ als auch $G(\omega)\,G_1(\omega)$ zu wirklichen complexen Zahlen macht, weshalb diese Produkte äquivalent sind.

Fasst man diese Resultate mit dem zusammen, dafs eine endliche Anzahl von Multiplikatoren hinreicht, um alle idealen Zahlen zu wirklichen zu machen, so schliefst man:

Die Klassen, in welche alle idealen Zahlen sich vertheilen, sind der Zahl nach endlich und vollkommen bestimmt, so dafs jede ideale Zahl nur einer bestimmten Klasse angehört.

Ist $F(\omega)$ irgend eine ideale Zahl, so können die idealen Zahlen der Reihe

$$F(\omega),\ F(\omega)^2,\ F(\omega)^3,\ F(\omega)^4 \,.\,.\,.\,.\,.\,.\,.$$

welche man in's Unendliche fortsetzen kann, nicht alle verschiedenen Klassen angehören, weil die Anzahl der Klassen nur eine endliche ist; es müssen also irgend welche verschiedene Glieder dieser Reihe äquivalent sein, d. h. $F(\omega)^r$ äquivalent $F(\omega)^s$, für gewisse verschiedene Werthe des r und s. Ist nun $M(\omega)$ ein Multiplikator dieser beiden äquivalenten Zahlen, so sind $F(\omega)^r\,M(\omega)$ und $F(\omega)^s\,M(\omega)$ beide wirklich, und darum ist auch der Quotient derselben, nämlich $F(\omega)^{r-s}$ eine wirkliche complexe Zahl. Man hat also den wichtigen Satz:

Jede ideale Zahl wird durch Erhebung zu einer bestimmten Potenz zu einer wirklichen.

Hieraus folgt ferner:

Jede ideale Zahl läfst sich als eine Wurzel aus einer wirklichen darstellen.

Wenn h der niedrigste Exponent ist, für welchen die Potenz $F(\omega)^h$ zu einer wirklichen complexen Zahl wird, so gehören die h Zahlen

$$1,\ F(\omega),\ F(\omega)^2,\ F(\omega)^3,\ .\,.\,.\,.\,.\ F(\omega)^{h-1}$$

wie leicht zu übersehen ist alle verschiedenen Klassen an. Diese h Klassen können nun entweder alle vorhandenen Klassen idealer Zahlen erschöpfen, oder auch nicht. Ist das letztere der Fall, so sei $G(\omega)$ eine ideale Zahl, welche keiner dieser h Klassen äquivalent ist, alsdann sind wie leicht zu zeigen auch folgende h ideale Zahlen

$$G(\omega),\ G(\omega)\,F(\omega),\ G(\omega)\,F(\omega)^2,\ \ldots.\ G(\omega)\,F(\omega)^{h-1}$$

weder unter sich noch den vorigen äquivalent. Wenn diese $2h$ Klassen nun noch nicht alle vorhandenen erschöpfen und $H(\omega)$ eine ideale Zahl ist, welche keiner derselben äquivalent ist, so wird ebenso leicht gezeigt, dafs folgende h ideale Zahlen

$$H(\omega),\ H(\omega)\,F(\omega),\ H(\omega)\,F(\omega)^2\ \ldots..\ H(\omega)F(\omega)^{h-1}$$

weder unter sich noch mit irgend einer der vorhergehenden äquivalent sein können. Fährt man so fort bis alle Klassen der idealen Zahlen erschöpft sind, so erkennt man, dafs dieselben in Gruppen von je h Klassen sich ordnen lassen und man hat den Satz:

Der Exponent der niedrigsten Potenz einer jeder idealen Zahl, welche zu einer wirklichen wird, ist immer ein genauer Theil der Anzahl aller Klassen der idealen Zahlen.

§. 11.

Die in dem Vorhergehenden gegebene Theorie der Zerlegung der complexen Zahlen in ihre idealen Primfaktoren will ich nun noch auf die complexen Zahlen der Kreistheilung anwenden, weil sie allgemeinere Resultate liefert als meine früheren Arbeiten in diesem Gebiete, welche sich auf den Fall beschränkten, dafs der Wurzelexponent der zu Grunde gelegten Wurzel der Einheit eine Primzahl war.

Ist x eine imaginäre Wurzel der Gleichung $x^p = 1$, p eine Primzahl, g eine primitive Wurzel derselben, ω eine primitive Wurzel der Gleichung $\omega^n = 1$, n irgend ein Divisor von $p-1$ und setzt man

$$(\omega,\ x) = x + \omega x^g + \omega^2 x^{g^2} + \ldots. + \omega^{p-2} x^{g^{p-2}}$$

so hat man bekanntlich

$$(\omega,\ x)\,(\omega^{-1},\ x) = \omega^{\frac{p-1}{2}}\,p$$

$$\frac{(\omega,\ x)\,(\omega^{\lambda},\ x)}{(\omega^{\lambda+1},\ x)} = \psi_{\lambda}(\omega)$$

wo $\psi_{\lambda}(\omega)$ eine nur die Wurzel ω, nicht aber x enthaltende ganze complexe Zahl ist, welche folgendermaafsen dargestellt werden kann:

$$\psi_s(\omega) = \Sigma\, \omega^{h + s \,\mathrm{Ind.}\,(g^h + 1)}$$

wo das Summenzeichen auf alle Werthe des $h = 0, 1, 2, \ldots p - 2$ zu beziehen ist, mit Ausschluss des Werthes $h = \frac{p-1}{2}$, und das Zeichen Ind. den Index andeutet für die primitive Wurzel g und den Modul p. Ferner ist:

$$\psi_s(\omega)\, \psi_s(\omega^{-1}) = p$$

und

$$(\omega, x)^n = \omega^{\frac{p-1}{2}} p\, \psi_1(\omega)\, \psi_2(\omega)\, \psi_3(\omega) \ldots \psi_{n-2}(\omega).$$

Ich untersuche nun zunächst die idealen Primfaktoren der complexen Zahl $\psi_s(\omega)$. Diese können, weil $\psi_s(\omega)$ ein Faktor von p ist, nur ideale Primfaktoren des p sein, welches eine nicht in dem Wurzelexponenten n der Wurzel ω enthaltene Primzahl ist. Diese Primzahl p gehört, weil n ein Divisor von $p - 1$ ist, zum Exponenten Eins, für den Modul n, so daſs in dem vorliegenden Falle $t = 1$ ist. Die Perioden ϖ_i werden also nur eingliedrig und sind den Wurzeln ω^i selbst gleich, die ihnen entsprechenden Congruenzwurzeln u_i sind nur die Wurzeln der Congruenz $u^n \equiv 1$, mod. p. Wird der Kürze wegen gesetzt $\frac{p-1}{n} = m$, so ist $u \equiv g^m$ eine primitive Wurzel dieser Congruenz, und wenn man festsetzt, daſs diese der Periode ϖ_1 oder was dasselbe ist der Wurzel ω entspreche, so ist klar, daſs allgemein die Congruenzwurzel g^{mi} die der Wurzel ω^i entsprechende sein wird. Es sei nun $f(\omega)$ ein idealer Primfaktor des p, und zwar der zur Substitution $\omega^i = g^{mi}$, gehörende, so wird der zur Substitution $\omega^{ri} = g^{mi}$ gehörende, wo r eine relative Primzahl zu n sein soll, durch $f(\omega^r)$ zu bezeichnen sein. Bedeutet ϱ diejenige Zahl, welche der Congruenz $r\varrho \equiv 1$, mod. n, genügt, so ist die Substitution $\omega^{ri} = g^{mi}$ gleichbedeutend mit $\omega^i = g^{m\varrho i}$, und weil $\omega^i - g^{m\varrho i}$ offenbar den idealen Primfaktor $f(\omega^r)$ enthält, so hat man die Congruenz

$$\omega^i \equiv g^{m\varrho i}, \text{ mod. } f(\omega^r),$$

für alle Werthe $i = 1, 2, 3, \ldots n$. Substituirt man nun diese Congruenzwurzeln anstatt der Wurzeln ω^i in dem oben gegebenen Ausdrucke des $\psi_s(\omega)$ so hat man:

$$\psi_s(\omega) \equiv \Sigma_h\, g^{m\varrho(h + s\,\mathrm{Ind.}\,(g^h + 1))}, \text{ mod. } f(\omega^r),$$

und weil

$$g^{\mathrm{Ind.}\,(g^h + 1)} \equiv g^h + 1$$

für den Modul p, folglich auch für den Modul $f(\omega^r)$, so ist

F 2

$$\psi_{,}(\omega) \equiv \Sigma_h\, g^{m\varrho h}\, (g^h + 1)^{m\varrho s}, \text{ mod. } f(\omega^r),$$

für $h = 0, 1, 2, \ldots p-2$, wo man nicht weiter nöthig hat den Werth $h = \frac{p-1}{2}$ auszuschliefsen, da das demselben zugehörende Glied dieser Summe congruent Null ist. Anstatt des Potenzexponenten $m\varrho s$ kann nun sein kleinster positiver Rest für den Modul $p-1$ gesetzt werden, welcher wenn σ den kleinsten positiven Rest von ϱs, für den Modul n, bezeichnet, gleich $m\sigma$ ist, so dafs man hat

$$\psi_{,}(\omega) \equiv \sum_0^{p-2}{}_h\, g^{m\varrho h}\, (g^h + 1)^{m\sigma}, \text{ mod. } f(\omega^r).$$

Wird nun die Potenz von $g^h + 1$ nach dem binomischen Lehrsatze entwickelt, so ist

$$(g^h + 1)^{m\sigma} = \sum_0^{m\sigma}{}_k \frac{\Pi(m\sigma)}{\Pi(k)\,\Pi(m\sigma - k)}\, g^{hk},$$

wo $\Pi(k)$ das Produkt $1.\,2.\,3.\ldots.\,k$ bezeichnet. Man hat daher

$$\psi_{,}(\omega) \equiv \sum_0^{m\sigma}{}_k \sum_0^{p-2}{}_h \frac{\Pi(m\sigma)}{\Pi(k)\,\Pi(m\sigma - k)}\, g^{(m\varrho + k)h}, \text{ mod. } f(\omega^r).$$

Summirt man nun in Beziehung auf alle Werthe des h, so wird diese Summe stets congruent Null, für den Modul p, also auch für den Modul $f(\omega^r)$, mit Ausnahme des einzigen Falles, wo $m\varrho + k$ ein Vielfaches von $p-1$ ist. Da aber der höchste Werth des k, nämlich $m\sigma$, kleiner als $p-1$ ist, und auch $m\varrho < p-1$, so kann dieser Fall nur Statt haben, wenn $m\varrho + k = p-1$ ist. Diese Bedingung kann, da $k \overline{\gtreqless} m\sigma$ und $p-1 = mn$ ist, auch als $\varrho + \sigma \overline{\gtreqless} n$ dargestellt werden. Ist diese Bedindung $\varrho + \sigma \overline{\gtreqless} n$ nicht erfüllt, so hat man

$$\psi_{,}(\omega) \equiv 0, \text{ mod. } f(\omega^r)$$

ist sie aber erfüllt, so wird für den Werth $k = p - 1 - m\varrho$

$$\sum_0^{p-2}{}_h\, g^{(m\varrho + k)h} \equiv -1, \text{ mod. } p,$$

also

$$\psi_{,}(\omega) \equiv \frac{-\Pi(m\sigma)}{\Pi m(n-\varrho)\,\Pi m(\varrho + \sigma - n)}, \text{ mod. } f(\omega^r),$$

welcher Binomialcoefficient nicht durch p, also auch nicht durch $f(\omega^r)$ theilbar ist. In dem Falle, dafs $\varrho + \sigma \geqq n$, ist also

$$\psi_{,}(\omega) \text{ nicht} \equiv 0, \text{ mod. } f(\omega^r).$$

Ich bezeichne nun allgemein durch $\left|\frac{b}{a}\right|$ den kleinsten positiven Werth des z, welcher der Congruenz $az \equiv b$, mod. n, genügt; wodurch die Zahlen ϱ und

σ als $\varrho = \left|\frac{1}{r}\right|$, $\sigma = \left|\frac{s}{r}\right|$ dargestellt werden, und die Bedingung $\varrho + \sigma < n$ als $\left|\frac{1}{r}\right| + \left|\frac{s}{r}\right| < n$, und bemerke, dafs vermöge der Gleichung $\psi_s(\omega)\,\psi_s(\omega^{-1}) = p$, $\psi_s(\omega)$ keinen idealen Primfaktor des p mehrfach enthalten kann. Demnach kann das gefundene Resultat folgendermaafsen ausgesprochen werden:

Die complexe Zahl $\psi_s(\omega)$ enthält alle diejenigen idealen Primfaktoren des p von der Form $f(\omega^r)$, für welche r, als relative Primzahl zu n und $< n$, der Bedingung

$$\left|\frac{1}{r}\right| + \left|\frac{s}{r}\right| < n$$

genügt, und zwar jeden derselben nur einmal, ausser diesen aber keinen anderen.

Demgemäfs kann man die Zerlegung des $\psi_s(\omega)$ in seine Primfaktoren auch so darstellen:

$$\psi_s(\omega) = E(\omega)\,\Pi f(\omega^r),$$

wo das durch Π angedeutete Produkt auf alle Werthe des r zu erstrecken ist, welche der Bedingung

$$\left|\frac{1}{r}\right| + \left|\frac{s}{r}\right| < n$$

genügen, und wo $E(\omega)$ eine complexe Einheit darstellt.

Aus der gefundenen Zerlegung der complexen Zahl $\psi_s(\omega)$ wird nun auch die Zerlegung von $(\omega, x)^n$ leicht zusammengesetzt, vermittelst der Formel

$$(\omega, x)^n = \omega^{\frac{p-1}{2}}\, p \,.\, \psi_1(\omega)\,\psi_2(\omega)\,\psi_3(\omega) \ldots\ldots \psi_{n-2}(\omega)$$

Die Anzahl, wie viel mal der ideale Primfaktor $f(\omega^r)$ in diesem Produkte enthalten ist, ist nach dem Obigen gleich der Anzahl der Werthe des s aus der Reihe $s = 1, 2, 3, \ldots\ldots n-2$, für welche $\left|\frac{1}{r}\right| + \left|\frac{s}{r}\right| < n$ ist, vermehrt um eine Einheit, welche daher kommt, dafs der Faktor p dieses Ausdrucks den idealen Primfaktor $f(\omega^r)$ einmal enthält. Es ist nun $\left|\frac{1}{r}\right| + \left|\frac{n-1}{r}\right| = n$, woraus folgt, dafs alle Werthe des $\left|\frac{s}{r}\right|$, welche kleiner sind als $\left|\frac{n-1}{r}\right|$ der Bedingung $\left|\frac{1}{r}\right| + \left|\frac{s}{r}\right| < n$ genügen müssen, aufser diesen aber keine. Die Anzahl dieser Werthe des $\left|\frac{s}{r}\right|$ und folglich auch die des s ist aber gleich $\left|\frac{n-1}{r}\right| - 1$, und wenn diese um eine Einheit vermehrt wird, so hat man $\left|\frac{n-1}{r}\right|$

gleich der Anzahl wie viel mal der ideale Primfaktor $f(\omega^r)$ in $(\omega, x)^n$ enthalten ist. Das gefundene Resultat kann nun in folgender Form dargestellt werden:

$$(\omega, x)^n = E(\omega)\, \Pi f(\omega^r)^{\left|\frac{n-1}{r}\right|}$$

wo das Produkt auf alle ganzzahligen Werthe des r zu erstrecken ist, welche kleiner als n und relative Primzahlen zu n sind und $E(\omega)$ eine complexe Einheit bedeutet.

Verwandelt man ω in ω^{-1} und setzt $n-r$ statt r, so kann man dieselbe Formel auch so darstellen:

$$(\omega^{-1}, x)^n = E(\omega^{-1})\, \Pi f(\omega^r)^{\left|\frac{1}{r}\right|}.$$

Es kann hier auch die Aufgabe gestellt werden den Lagrangeschen Ausdruck (ω, x) selbst, und nicht nur dessen $n$te Potenz, welche von x unabhängig ist, in die idealen Primfaktoren zu zerlegen. Dieser Ausdruck ist nämlich eine complexe Zahl der Wurzel $x\omega$ der Gleichung $(x\omega)^{pn} = 1$ und gehört somit den in der gegenwärtigen Theorie behandelten complexen Zahlen an. Derselbe kann nur ideale Primfaktoren des p enthalten, weil $(\omega, x)\,(\omega^{-1}, x) = \omega^{\frac{p-1}{2}}\, p$ ist, p ist hier aber eine in dem Wurzelexponenten pn enthaltene Primzahl, so dafs es sich also in diesem nur um die idealen Primfaktoren der besonderen Art handelt, deren Eigenschaften im §. 6. entwickelt worden sind. Die in Beziehung auf x allein genommene Norm von (ω, x) ist, wie aus der Eigenschaft dieses Ausdruks

$$(\omega, x^{g^h}) = \omega^{-h}\, (\omega, x)$$

folgt, gleich der $(p-1)$ten Potenz desselben, d. h.

$$N_x\, (\omega, x) = (\omega, x)^{p-1};$$

die oben gefundene Zerlegung des $(\omega, x)^n$ in seine idealen Primfaktoren giebt daher folgende Zerlegung der Norm

$$N_x\, (\omega, x) = E(\omega)^m\, \Pi f(\omega^r)^{m\left|\frac{n-1}{r}\right|},$$

wo $m = \left|\frac{p-1}{n}\right|$. Wendet man nun auf die complexe Zahl (ω, x) den Satz an, dafs eine solche complexe Zahl die idealen Primfaktoren des p für complexe Zahlen der Wurzel $x\omega$ genau eben so oft enthält, als ihre in Beziehung auf x allein genommene Norm die entsprechenden, derselben Substitution angehörenden idealen Primfaktoren des p für complexe Zahlen der Wurzel

ω enthält, und bezeichnet durch $f(x\omega^r)$ den idealen Primfaktor des p, welcher der Substitution $\omega^{ri} = g^{mi}$ angehört, welcher also dem idealen Primfaktor $f(\omega^r)$, für complexe Zahlen der Wurzel ω entspricht, so erkennt man aus der Zerlegung von $N_x(\omega, x)$ unmittelbar, daſs (ω, x) den idealen Primfaktor $f(x\omega^r)$ genau $m\left|\frac{n-1}{p}\right|$ mal enthalten muſs, und man hat folgende Zerlegung

$$(\omega, x) = E(x\omega)\,\Pi\, f(x\omega^r)^{m\left|\frac{n-1}{r}\right|}$$

wo das Produktzeichen ebenfalls auf alle Werthe des r zu beziehen ist, welche kleiner als n sind und relative Primzahlen zu n, und $E(x\omega)$ eine complexe Einheit bezeichnet.

Einige Sätze über die aus den Wurzeln der Gleichung $\alpha^\lambda = 1$ gebildeten complexen Zahlen, für den Fall, daß die Klassenanzahl durch λ theilbar ist, nebst Anwendung derselben auf einen weiteren Beweis des letzten Fermat'schen Lehrsatzes

Monatsberichte der Königlichen Preußischen Akademie der Wissenschaften zu Berlin aus dem Jahre 1857, 275–282

4. Mai. Sitzung der physikalisch-mathematischen Klasse.

Hr. Kummer las: Einige Sätze über die aus den Wurzeln der Gleichung $\alpha^\lambda = 1$ gebildeten complexen Zahlen, für den Fall, dafs die Klassenanzahl durch λ theilbar ist, nebst Anwendung derselben auf einen weiteren Beweis des letzten Fermatschen Lehrsatzes.

Auszug.

Der Beweis des Fermatschen Satzes, dafs die Gleichung

$$x^\lambda + y^\lambda = z^\lambda$$

in ganzen Zahlen unlösbar ist, welchen ich zuerst der Königlichen Akademie unter dem 11. April 1847 mitgetheilt und später in Liouville's Journal Bd. 16, pag. 377 sq. in meinem Mémoire sur la théorie des nombres complexes vollständig ausgeführt habe, erstreckt sich nicht auf diejenigen Werthe der Primzahl λ, für welche die Anzahl der nichtäquivalenten Klassen aller idealen complexen Zahlen durch λ theilbar ist, weil für diese besonderen Werthe des λ diejenigen allgemeinen Eigenschaften der complexen Zahlen nicht Statt haben, auf

[1857.] 21

welche er gegründet ist. Ich habe nun versucht durch Erforschung der besonderen Eigenschaften, welche die complexen Zahlen besitzen, wenn die Klassenanzahl durch λ theilbar ist, die Richtigkeit des Fermatschen Satzes auch für diese Werthe des Potenzexponenten λ zu ergründen; da diese Untersuchung jedoch in ihrer ganzen Allgemeinheit grofse Schwierigkeiten darbietet, welche ich bisher noch nicht vollständig habe überwinden können, so beschränke ich mich hier auf den einfachsten Fall dieser Art und beweise den Fermatschen Satz für eine neue Reihe von Werthen des λ, welche durch drei bestimmte Voraussetzungen vollständig charakterisirt wird. Dieser Reihe gehören namentlich auch die drei Zahlen $\lambda = 37$, $\lambda = 59$ und $\lambda = 67$ an, die einzigen innerhalb des ersten Hundert, für welche die Richtigkeit des Fermatschen Satzes bisher noch zweifelhaft war.

Im Allgemeinen können die Zahlen λ, für welche der neue Beweis des Fermatschen Satzes gegeben werden soll, als solche bezeichnet werden, für welche die Anzahl der nichtäquivalenten Klassen idealer Zahlen den Faktor λ einmal enthält, für welche also, wenn wie in meinem früheren Beweise der erste und der zweite Faktor der Klassenanzahl gesondert betrachtet werden, nach dem daselbst bewiesenen Satze: dafs der zweite Faktor der Klassenanzahl nur dann durch λ theilbar sein kann, wenn der erste Faktor durch λ theilbar ist, dieser erste Faktor durch λ theilbar ist und nur einmal, der zweite Faktor aber durch λ nicht theilbar ist. Es ist alsdann nur eine der ersten $\frac{\lambda - 3}{2}$ Bernoullischen Zahlen durch λ theilbar, welche durch B_ν bezeichnet werden soll. Ich beweise nun, dafs die Frage: ob auch der zweite Faktor der Klassenanzahl durch λ theilbar ist, durch die folgende Einheit

$$E_\nu(\alpha) = e(\alpha)\, e(\alpha^\gamma)^{\gamma^{-2\nu}}\, e(\alpha^{\gamma^2})^{\gamma^{-4\nu}} \dots e(\alpha^{\gamma^{\mu-1}})^{\gamma^{-2(\mu-1)\nu}}$$

in welcher $e(\alpha)$ die bekannte Kreistheilungseinheit, γ eine primitive Wurzel von λ, $\mu = \frac{\lambda - 1}{2}$ und ν diejenige Zahl ist, für welche die Bernoullische Zahl $B_\nu \equiv 0$, mod. λ ist, vollständig erledigt wird, in der Art, dafs wenn $E_\nu(\alpha)$ eine λ^{te} Potenz einer Einheit ist, der zweite Faktor der Klassenanzahl durch λ theilbar sein mufs, aber wenn $E_\nu(\alpha)$ nicht eine λ^{te} Potenz ist, durch λ nicht theilbar sein kann. Wenn nun $E_\nu(\alpha)$ für irgend einen Modul einer λ^{ten} Potenz nicht congruent ist, so kann es auch selbst nicht eine λ^{te} Potenz sein; also kann alsdann der zweite Faktor der Klassenanzahl auch nicht durch λ theilbar sein. Die drei Voraussetzungen über die Zahlen λ, für welche die folgenden Sätze über die complexen Zahlen und der auf dieselben zu gründende Beweis des Fermatschen Satzes gelten sollen, stelle ich nun in folgender Form auf:

Voraussetzung I. Der erste der beiden Faktoren, aus welchen die Klassenanzahl besteht, soll den Faktor λ einmal und nur einmal enthalten.

Voraussetzung II. Es soll irgend einen Modul geben, für welchen die Einheit $E_\nu(\alpha)$ einer λ^{ten} Potenz nicht congruent ist.

Voraussetzung III. Die $\nu\lambda^{\text{te}}$ Bernoullische Zahl soll nicht congruent Null sein für den Modul λ^3.

Für diejenigen complexen Zahlen, welche diesen Voraussetzungen entsprechen, werden nun folgende Sätze bewiesen:

Lehrsatz 1. Jede Einheit, welche einer nichtcomplexen ganzen Zahl congruent ist nach dem Modul λ^2, ist eine λ^{te} Potenz einer anderen Einheit.

Lehrsatz 2. Wenn $F(\alpha)$ eine nur die zweigliedrigen Perioden enthaltende complexe Zahl ist, also $F(\alpha) = F(\alpha^{-1})$ und wenn

$$\frac{d_0^{2\nu}\, lF(e^v)}{dv^{2\nu}} \equiv 0, \text{ mod. } \lambda,$$

so läfst sich die complexe Zahl $F(\alpha)$ durch Multiplication mit einer passenden Einheit in die Form bringen, dafs sie einer nichtcomplexen Zahl congruent wird nach dem Modul λ.

Lehrsatz 3. Jede beliebige Einheit $E(\alpha)$ hat die Eigenschaft dafs

$$\frac{d_0^{2\nu}\, lE(e^v)}{dv^{2\nu}} \equiv 0, \text{ mod. } \lambda.$$

Lehrsatz 4. Wenn die λ^{te} Potenz einer nur die zweigliedrigen Perioden enthaltenden complexen

21*

Zahl eine wirkliche complexe Zahl ist, so ist diese complexe Zahl selbst eine wirkliche.

Lehrsatz 5. Eine jede complexe Zahl $f(\alpha)$, deren λ^{te} Potenz $f(\alpha)^\lambda = F(\alpha)$ eine wirkliche complexe Zahl ist, ist selbst eine wirkliche complexe Zahl wenn

$$\frac{d_0^{\lambda-2\nu}\, lF(e^v)}{dv^{\lambda-2\nu}} \equiv 0, \text{ mod. } \lambda,$$

und sie ist eine ideale complexe Zahl, wenn dieser $(\lambda-2\nu)^{te}$ Differenzialquotient nicht congruent Null ist nach dem Modul λ.

Ich bemerke, dafs die Lehrsätze 2, 3, 4 und 5 auch unabhängig von der Voraussetzung III gültig sind, der Lehrsatz 2 sogar unabhängig von den beiden Voraussetzungen II und III. Der Beweis des ersten Lehrsatzes wird durch Anwendung der Methode der logarithmischen Entwickelungen der complexen Zahlen in Beziehung auf den Modul λ, oder eine Potenz von λ geleistet, welche ich im §. 4. meiner Abhandlung über die Ergänzungssätze zu den allgemeinen Reciprocitätsgesetzten in Crelle's Journal Bd. 44 pag. 130 sq. vollständig behandelt habe. Aus derselben Quelle fliefsen auch die Lehrsätze 2 und 3. Der Lehrsatz 4 ist, wie leicht zu sehen, in dem allgemeineren Lehrsatz 5 eigentlich mit enthalten, so dafs er als Zusatz aus diesem gefolgert werden kann. Der Beweis dieses letzten Satzes aber liegt ziemlich tief, weil zu demselben aufser der genannten Methode der Entwickelung der complexen Zahlen noch die formale Darstellung eines vollständigen Systems aller nichtäquivalenten idealen Zahlen nöthig ist, so wie auch die Ergänzungssätze zu den allgemeinen Reciprocitätsgesetzen, welche sich auf die Einheiten beziehen. Diese letzteren, welche ich in der erwähnten Abhandlung gegeben habe, müssen insofern sie dort nur für diejenigen λ^{ten} Potenzen giltig entwickelt worden sind, für welche die Klassenanzahl der zugehörigen complexen Zahlen durch λ nicht theilbar ist, so modificirt werden, dafs sie für den vorliegenden Fall gelten.

Der Beweis des Fermatschen Satzes für diejenigen Potenzexponenten λ, welche den drei aufgestellten Voraussetzungen entsprechen, beginnt nun mit dem Falle, dafs in der Gleichung

$$x^\lambda + y^\lambda = z^\lambda$$

keine der drei Zahlen x, y, z durch λ theilbar ist. Ich zeige, daſs wenn überhaupt in diesem Falle die Gleichung durch ganze Zahlen lösbar sein soll, nothwendig die $\frac{\lambda-3}{2}$te und auch die $\frac{\lambda-5}{2}$te Bernoullische Zahl congruent Null sein muſs nach dem Modul λ, daſs also dieser Fall hier, wo vermöge der Voraussetzung I nur eine der ersten $\frac{\lambda-3}{2}$ Bernoullischen Zahlen congruent Null ist, nicht Statt haben kann.

Um nun die Unmöglichkeit dieser Fermatschen Gleichung auch für den Fall zu beweisen, wo eine der drei Zahlen x, y, z durch λ theilbar ist, lege ich die allgemeinere Gleichung

$$U^\lambda + V^\lambda = E(\alpha)\,(2-\alpha-\alpha^{-1})^{m\lambda}\,W^\lambda \qquad (1.)$$

zu Grunde, in welcher U, V und W complexe Zahlen sein können, jedoch nur solche, welche aus den zweigliedrigen Perioden $\alpha+\alpha^{-1}$, $\alpha^\gamma+\alpha^{-\gamma}$, ... gebildet sind, welche also unverändert bleiben, wenn α in α^{-1} verwandelt wird, in welcher ferner $E(\alpha)$ eine beliebige complexe Einheit bezeichnet und $(2-\alpha-\alpha^{-1})^{m\lambda}$, worin m gröſser als Eins sein soll, die Stelle einer Potenz von λ vertritt. Die Unmöglichkeit dieser Gleichung zieht die der Gleichung

$$x^\lambda + y^\lambda = \lambda^{k\lambda} z^\lambda$$

nach sich, weil diese nur ein specieller Fall von jener ist.

Aus der gegebenen Gleichung (1.) folgen nach bekannten Principien die beiden Gleichungen

$$U + \alpha^r V = \varepsilon_r(\alpha)\,(1-\alpha^r)\,\Theta_r(\alpha)^\lambda, \qquad (2.)$$

$$U + V = \varepsilon(\alpha)\,(2-\alpha-\alpha^{-1})^{m\lambda-\mu}\,T(\alpha)^\lambda, \qquad (3.)$$

und es wird hier durch Anwendung der Lehrsätze 4 und 5 bewiesen, daſs $\Theta_r(\alpha)$ und $T(\alpha)$, deren λ^{te} Potenzen wirkliche complexe Zahlen sind, auch selbst wirkliche complexe Zahlen sein müssen. Verwandelt man nun in der Gleichung (2.) α in α^{-1} und eliminirt aus dieser und den beiden unveränderten Gleichungen (2.) und (3.) die complexen Zahlen U und V, so erhält man nach einigen leichten Reductionen:

$$\Theta_r(\alpha)^\lambda - \Theta_r(\alpha^{-1})^\lambda = E_r(\alpha)(1-\alpha)^{(2m-1)\lambda}\,T(\alpha)^\lambda. \qquad (4.)$$

Aus dieser erhält man wieder auf bekannte Weise die Gleichungen:

$$\Theta_r(\alpha) - \alpha\,\Theta_r(\alpha^{-1}) = \mathfrak{E}_1(\alpha)(1-\alpha)P(\alpha)^\lambda \qquad (5.)$$

$$\Theta_r(\alpha) - \Theta_r(\alpha^{-1}) = \mathfrak{E}(\alpha)(1-\alpha)^{2(m-1)\lambda+1}\,Q(\alpha)^\lambda \qquad (6.)$$

und hieraus, weil $m>1$ ist und darum die $(2(m-1)\lambda+1)^{\text{te}}$ Potenz von $1-\alpha$ den Faktor λ sogar mehrmals enthält, folgt leicht die Congruenz:

$$\Theta_r(\alpha)\Theta_r(\alpha^{-1}) \equiv \mathfrak{E}_1(\alpha)\mathfrak{E}_1(\alpha^{-1})\big(P(\alpha)P(\alpha^{-1})\big)^\lambda, \text{ mod. } \lambda.$$

Der Lehrsatz 4 zeigt nun unmittelbar, daſs $P(\alpha)\,P(\alpha^{-1})$ eine wirkliche complexe Zahl ist und diesem zufolge ergiebt der Lehrsatz 3, daſs

$$\frac{d_0^{2\nu}\, l\big(\Theta_r(e^v)\Theta_r(e^{-v})\big)}{d^{2\nu}_v} \equiv 0, \text{ mod. } \lambda$$

ist, woraus weiter vermöge des Lehrsatzes 2 folgt, daſs $\Theta_r(\alpha)\Theta_r(\alpha^{-1})$ durch Multiplication mit einer passenden Einheit so verändert werden kann, daſs es einer nichtcomplexen Zahl congruent wird für den Modul λ.

Nachdem diese für die folgenden Schlüsse wichtige Eigenschaft der complexen Zahl $\Theta_r(\alpha)\Theta_r(\alpha^{-1})$ festgestellt ist, kehre ich zu den Gleichungen (2.) und (3.) zurück, aus welchen, wenn man noch diejenige Gleichung hinzunimmt, welche durch die Verwandlung der beliebigen Zahl r in eine andere Zahl s, so wie diejenigen Gleichungen, welche durch Verwandlung des α in α^{-1} erhalten werden, ohne Schwierigkeit folgende Gleichung erhalten wird:

$$\varepsilon_r(\alpha)\varepsilon_r(\alpha^{-1})\big(\Theta_r(\alpha)\Theta_r(\alpha^{-1})\big)^\lambda - \varepsilon_s(\alpha)\varepsilon_s(\alpha^{-1})\big(\Theta_s(\alpha)\Theta_s(\alpha^{-1})\big)^\lambda =$$
$$\frac{\varepsilon(\alpha)^2(2-\alpha-\alpha^{-1})^{(2m-1)\lambda+1}\,(\alpha^r+\alpha^{-r}-\alpha^s-\alpha^s)\,T(\alpha)^{2\lambda}}{(2-\alpha^r-\alpha^{-r})(2-\alpha^s-\alpha^{-s})} \qquad (7.)$$

Es sei nun $A_r(\alpha)$ eine Einheit, welche bewirkt, daſs $A_r(\alpha)\,\Theta_r(\alpha)\,\Theta_r(\alpha^{-1})$ einer nichtcomplexen Zahl congruent ist nach dem Modul λ und ebenso sei $A_s(\alpha)$ eine Einheit, welche dasselbe für die complexe Zahl $\Theta_s(\alpha)\,\Theta_s(\alpha^{-1})$ bewirkt und es sei

$$A_r(\alpha)\,\Theta_r(\alpha)\,\Theta_r(\alpha^{-1}) = T_r(\alpha),$$
$$A_s(\alpha)\,\Theta_s(\alpha)\,\Theta_s(\alpha^{-1}) = T_s(\alpha),$$

so wird durch Einführung dieser complexen Zahlen $T_r(\alpha)$ und $T_s(\alpha)$ die Gleichung (7.) leicht in folgende Form verwandelt:

$$T_r(\alpha)^\lambda + \mathfrak{E}_s(\alpha)\,T_s(\alpha)^\lambda = E_1(\alpha)\,(2-\alpha-\alpha^{-1})^{(2m-1)\lambda}\,T(\alpha)^{2\lambda}. \quad (8.)$$

Weil nun $T_r(\alpha)$ und $T_s(\alpha)$ nichtcomplexen ganzen Zahlen congruent sind für den Modul λ, so schliefst man leicht, dafs die λ^{ten} Potenzen derselben $T_r(\alpha)^\lambda$ und $T_s(\alpha)^\lambda$ nichtcomplexen Zahlen congruent sind für den Modul λ^2 und da aufserdem die rechte Seite der Gleichung (8.) durch λ^2 theilbar ist, dafs die Einheit $\mathfrak{E}_s(\alpha)$ einer nichtcomplexen Zahl congruent ist für den Modul λ^2. Diese Einheit ist also nach dem Lehrsatze 1 eine λ^{te} Potenz einer Einheit. Man kann daher setzen

$$\mathfrak{E}_r(\alpha)\,T'_s(\alpha)^\lambda = V_1^\lambda$$

und wenn man aufserdem setzt

$$T_r(\alpha) = U_1,\quad T(\alpha)^2 = W_1$$

so hat man endlich

$$U_1^\lambda + V_1^\lambda = E_1(\alpha)\,(2-\alpha-\alpha^{-1})^{(2m-1)\lambda}\,W_1^\lambda \quad (9.)$$

eine Gleichung genau von derselben Form als die vorgegebene Gleichung (1). Es ist klar, dafs durch wiederholte Anwendung derselben Methode aus dieser Gleichung eine dritte Gleichung derselben Form erhalten werden mufs, aus dieser sodann eine vierte, fünfte und so fort ins Unendliche. Dafs eine unendliche Reihe solcher Gleichungen eine unendliche Reihe absoluter ganzer Zahlen ergeben müfste, in welcher von einer endlichen Zahl anfangend jede folgende nothwendig kleiner wäre als die vorhergehende, wird hier durch die Betrachtung der Anzahl aller verschiedener idealer Primfaktoren gezeigt, welche die complexen Zahlen W, W_1, W_2, ... enthalten, indem bewiesen wird, dafs jede folgende complexe Zahl dieser Reihe nothwendig weniger verschiedene ideale Primfaktoren enthalten mufs, als die vorhergehende.

Nachdem so die Richtigkeit des letzten Fermatschen Lehrsatzes für alle diejenigen Potenzen bewiesen ist, deren Exponenten den obigen drei Voraussetzungen entsprechen, untersuche ich in dieser Beziehung die drei Zahlen $\lambda = 37$, $\lambda = 59$

und $\lambda = 67$, die einzigen innerhalb des ersten Hundert für welche die Klassenanzahl durch λ theilbar ist und zeige durch genaue specielle Berechnung, deren numerische Resultate ich angebe, dafs diese drei Zahlen den obigen drei Voraussetzungen vollständig genügen.

Einige Sätze über die aus den Wurzeln der Gleichung $\alpha^\lambda = 1$ gebildeten complexen Zahlen, für den Fall, daß die Klassenanzahl durch λ theilbar ist, nebst Anwendung derselben auf einen weiteren Beweis des letzten Fermat'schen Lehrsatzes

Mathematische Abhandlungen der Königlichen Akademie der Wissenschaften zu Berlin aus dem Jahre 1857, 41–74

[Gelesen in der Akademie der Wissenschaften am 4. Mai 1857].

In der Theorie der aus λ^{ten} Wurzeln der Einheit gebildeten complexen Zahlen, wo λ Primzahl ist, sind, wenn man auf die etwas tiefer liegenden Untersuchungen eingeht, die beiden Fälle wesentlich zu unterscheiden, erstens wo die Anzahl der nicht äquivalenten Klassen der idealen Zahlen durch λ nicht theilbar ist, und zweitens wo diese Klassenzahl durch λ theilbar ist, welche beiden Fälle, wie ich früher gezeigt habe sich auch so unterscheiden lassen: erstens wenn keine der ersten $\frac{\lambda-3}{2}$ Bernoullischen Zahlen durch λ theilbar ist, und zweitens wenn unter diesen ersten $\frac{\lambda-3}{2}$ Bernoullischen Zahlen durch λ theilbare vorkommen. Der erste Fall ist der einfachere und leichter zu behandelnde, weil für denselben gewisse einfache, wichtige Sätze bestehen, welche allemal dann Ausnahmen erleiden oder ganz verloren gehen, wenn λ eine Primzahl ist, welche dem zweiten Falle angehört. Aus diesem Grunde erstreckt sich auch mein Beweis des Fermatschen Satzes: dafs die Summe zweier λ^{ten} Potenzen nicht einer λ^{ten} Potenz gleich sein kann, nur auf diejenigen Primzahlen λ, welche dem ersten Falle angehören, wo keine der ersten $\frac{\lambda-3}{2}$ Bernoullischen Zahlen durch λ theilbar ist. Durch eine genauere Erforschung der besonderen Eigenschaften, welche die complexen Zahlen besitzen, wenn λ dem zweiten Falle angehört, habe ich seitdem gesucht die Mittel zu erhalten, um die Richtigkeit dieses Fermatschen Satzes auch für diejenigen Fälle zu ergründen, auf welche der genannte Beweis sich nicht erstreckt, und wenn gleich ich auf diesem Wege einen vollkommenen, alle Fälle erschöpfenden Beweis noch nicht gefunden habe, so ist es mir doch gelungen den Fermatschen Satz auch für eine ganze Reihe solcher Potenzexponenten λ zu beweisen, welche diesem zweiten Falle angehören, in welcher Reihe namentlich auch die drei Zahlen $\lambda = 37$, $\lambda = 59$ und $\lambda = 67$ enthalten sind, die einzigen innerhalb des ersten Hundert, für welche die Richtigkeit dieses Satzes bisher noch zweifelhaft war. Ich werde nun zunächst die zu diesem Behufe nöthigen neuen Sätze aus der Theorie der complexen Zahlen entwickeln und dieselben sodann auf den Fermatschen Lehrsatz anwenden.

§. 1.

In dem Ausdrucke der Anzahl aller nichtäquivalenten Klassen der aus den Wurzeln der Gleichung $\alpha^\lambda = 1$ gebildeten complexen Zahlen, wie ich denselben in Crelle's Journal Bd. 40 pag. 110 und 117, und in Liouville's Journal Bd. 16, pag. 471 gegeben habe, nämlich

$$H = \frac{P}{(2\lambda)^{\mu-1}} \cdot \frac{D}{\Delta},$$

sind die beiden durch den Punkt geschiedenen Faktoren, welche ich als den ersten und den zweiten Faktor der Klassenzahl bezeichne, für sich ganze Zahlen und haben beide sehr verschiedene Eigenschaften, weshalb es hier, so wie in den meisten die Klassenanzahl betreffenden Untersuchungen nöthig ist diese beiden Faktoren derselben gesondert zu betrachten. Die hier folgende Untersuchung soll nun hauptsächlich nur diejenige Gattung der complexen Zahlen betreffen, für welche die Klassenanzahl durch λ theilbar ist, und auch von diesen nur die einfachste Art, welche durch zwei über dieselben zu machende Voraussetzungen charakterisirt wird, die ich jetzt angeben und näher erörtern will.

Erstens soll in dem Folgenden überall angenommen werden, dafs der erste der beiden Faktoren der Klassenanzahl den Faktor λ einmal und auch nur einmal enthält.

Aus dieser Annahme folgt zunächst, dafs eine der ersten Bernoullischen Zahlen durch λ theilbar sein mufs und auch nur eine; denn der erste Faktor der Klassenanzahl mufs, wie aus meiner Untersuchung der Theilbarkeit der Klassenanzahl durch λ, (Liouville's Journal Bd. 16, pag. 473 sq.) unmittelbar folgt, den Faktor λ mindestens so viel mal enthalten, als wie viele der ersten $\frac{\lambda-3}{2}$ Bernoullischen Zahlen durch λ theilbar sind. Es soll daher die ν^{te} Bernoullische Zahl, welche ich durch B_ν bezeichne, als die eine durch λ theilbare angenommen werden.

In der erwähnten Untersuchung habe ich ferner gezeigt, daſs wenn der zweite Faktor der Klassenanzahl durch λ theilbar sein soll, nothwendig eine Einheit $\varepsilon(\alpha)$ existiren muſs von der Art, daſs

$$\varepsilon(\alpha)^{h\lambda} = e(\alpha)^{r_1} \cdot e(\alpha^{\gamma})^{r_2}\, e(\alpha^{\gamma^2})^{r_3} \ldots\ldots e(\alpha^{\gamma^{\mu-2}})^{r_{\mu-1}},$$

wo $e(\alpha)$ die bekannte Kreistheilungseinheit ist, $\mu = \frac{\lambda-1}{2}$ und γ eine primitive Wurzel der Primzahl λ (m. s. Liouv. Jour. Bd. 16, pag. 480) in welcher Gleichung die Zahlen $r_1, r_2, \ldots r_{\mu-1}$ nicht alle durch λ theilbar sein dürfen, aber dem Systeme der Congruenzen

$$r_1 + \gamma^{2n} r_2 + \gamma^{4n} r_3 + \ldots\ldots + \gamma^{2(\mu-2)n} r_{\mu-1} \equiv 0, \text{ mod. } \lambda,$$

für alle diejenigen Werthe des n aus der Reihe der Zahlen $1, 2, 3, \ldots. \mu-1$ genügen müssen, für welche die n^{te} Bernoullische Zahl B_n durch λ nicht theilbar ist. Im gegenwärtigen Falle also, wo nur die eine Bernoullische Zahl B_ν durch λ theilbar ist, muſs diese Congruenz Statt haben für alle Werthe $n = 1, 2, 3, \ldots \mu-1$ mit Ausschluſs des Werthes $n = \nu$, so daſs man zur Bestimmung der $\mu - 1$ Zahlen $r_1, r_2, \ldots r_{\mu-1}$ nur $\mu - 2$ Congruenzen hat. Setzt man nun

$$r_1 + \gamma^{2\nu} r_2 + \gamma^{4\nu} r_3 + \ldots\ldots + \gamma^{2(\mu-2)\nu} r_{\mu-1} \equiv \mu m, \text{ mod. } \lambda,$$

nimmt diese Congruenz zu jenen $\mu - 2$ Congruenzen hinzu und löst dieses System von $\mu = 1$ Congruenzen und eben so vielen Unbekannten auf, welches sehr einfach dadurch geleistet wird, daſs man dieselben für $n = 1, 2, 3, \ldots \mu - 1$ der Reihe nach mit $\gamma^{-2k}, \gamma^{-4k}, \gamma^{-6k} \ldots \gamma^{-2(\mu-1)k}$ multiplicirt und addirt, so findet man

$$r_k \equiv m\gamma^{-2k\nu} - m\gamma^{2\nu}, \text{ mod. } \lambda,$$

oder wenn durch Hinzufügung eines Vielfachen von λ aus dieser Congruenz eine Gleichung gemacht wird:

$$r_k = m\gamma^{-2k\nu} - m\gamma^{2\nu} + \lambda s_k.$$

Setzt man die durch diese Gleichung bestimmten Werthe der Exponenten in den obigen Ausdruck des $\varepsilon(\alpha)^{h\lambda}$ ein, so erhält man vermittelst der Gleichung $Ne(\alpha) = 1$.

F 2

$$\varepsilon(\alpha)^{h\lambda} = \left(e(\alpha)\, e(\alpha^{\gamma})^{\gamma^{-2\nu}} e(\alpha^{\gamma^2})^{\gamma^{-4\nu}} \ldots\ldots e(\alpha^{\gamma^{\mu-1}})^{\gamma^{-2(\mu-1)\nu}}\right)^m U(\alpha)^{\lambda},$$

wo der Kürze wegen

$$e(\alpha)^{s_1} \cdot e(\alpha^{\gamma})^{s_2} \ldots\ldots e(\alpha^{\gamma^{\mu-2}})^{s_{\mu-1}} = U(\alpha)$$

gesetzt ist. Die in den Klammern stehende Einheit ist eine von denjenigen, welche ich in meiner Abhandlung über die Ergänzungssätze zu den allgemeinen Reciprocitätsgesetzen, Crelles Journal Bd. 44, vielfach angewendet habe, ich bezeichne dieselbe daher mit dem dort gewählten Zeichen, indem ich allgemein für jeden Werth das n setze

$$\mathrm{E}_n(\alpha) = e(\alpha)\, e(\alpha^{\gamma})^{\gamma^{-2n}} e(\alpha^{\gamma^2})^{\gamma^{-4n}} \ldots\ldots e(\alpha^{\gamma^{\mu-1}})^{\gamma^{-2(\mu-1)n}}$$

In der Gleichung

$$\varepsilon(\alpha)^{h\lambda} = \mathrm{E}_{\nu}(\alpha)^m\, U(\alpha)^{\lambda}$$

darf m nicht durch λ theilbar sein, weil sonst alle Exponenten r_1, r_2, ... $r_{\mu-1}$ durch λ theilbar sein würden, es läfst sich deshalb immer eine Zahl c bestimmen von der Art dafs $mc \equiv 1$, mod. λ, ist, oder $mc = 1 + d\lambda$. Erhebt man also zur c^{ten} Potenz, nimmt $mc = 1 + d\lambda$, dividirt durch $\mathrm{E}_{\nu}(\alpha)^{d\lambda}\, U(\alpha)^{\lambda}$ und setzt der Einfachheit wegen

$$\frac{\varepsilon(\alpha)^h}{\mathrm{E}_{\nu}(\alpha)^d\, U(\alpha)} = \mathfrak{E}(\alpha)$$

so hat man endlich

$$\mathrm{E}_{\nu}(\alpha) = \mathfrak{E}(\alpha)^{\lambda}.$$

Dieses Resultat wird in Form eines Satzes folgendermafsen ausgesprochen:

Wenn die Bernoullische Zahl $B_{\nu} \equiv 0$, mod. λ, ist, so kann der zweite Faktor $\frac{\Delta}{D}$ der Klassenanzahl nur dann durch λ theilbar sein, wenn die Einheit $\mathrm{E}_{\nu}(\alpha)$ eine λ^{te} Potenz einer Einheit ist.

Dieser Satz gilt auch umgekehrt, nämlich wenn $\mathrm{E}_{\nu}(\alpha)$ eine λ^{te} Potenz einer Einheit ist, so ist der zweite Faktor der Klassenanzahl nothwendig durch λ theilbar, wie sich ohne Schwierigkeit zeigen läfst, er wird aber in dem Folgenden nur in so weit Anwendung finden, als er hier bewiesen ist.

Zu der ersten oben aufgestellten allgemeinen Voraussetzung über die complexen Zahlen, welche hier behandelt werden sollen, will ich nun noch eine zweite hinzufügen, welche im wesentlichen darauf hinausläuft, dafs $\mathrm{E}_{\nu}(\alpha)$

nicht eine λ^{te} Potenz und mithin der zweite Faktor der Klassenanzahl nicht durch λ theilbar sein soll, welche ich aber in folgender anderer Form gebe:

Es soll zweitens in dem Folgenden überall angenommen werden, dafs es irgend eine complexe ideale Zahl giebt, in Beziehung auf welche als Modul die Einheit $E_\nu(\alpha)$ nicht λ^{ter} Potenzrest, d. h. einer λ^{ten} Potenz nicht congruent ist.

Es ist klar, dafs diese Voraussetzung die mit in sich begreift, dafs $E_\nu(\alpha)$ einer λ^{ten} Potenz nicht gleich sei und also auch, dafs der zweite Faktor der Klassenanzahl durch λ nicht theilbar sei.

§. 2.

Weil nach der ersten Voraussetzung die ν^{te} Bernoullische Zahl durch λ theilbar ist, so findet einer der Hauptsätze, auf welchen mein früherer Beweis des Fermatschen Satzes beruht, nämlich dafs jede Einheit, welche für den Modul λ einer nichtcomplexen Zahl congruent ist, eine λ^{te} Potenz einer Einheit sein mufs, hier nicht mehr Statt. Um nun einen entsprechenden Satz an die Stelle desselben zu setzen untersuche ich für den gegenwärtigen Fall die Einheiten welche in Beziehung auf den Modul λ^2 nichtcomplexen ganzen Zahlen congruent sind. Zu diesem Zwecke bediene ich mich der logarithmischen Entwickelungen der complexen Zahlen in Beziehung auf den Modul λ, oder eine Potenz von λ, deren Theorie ich in der schon oben erwähnten Abhandlung (Crelles Journal Bd. 44, §. 4.) vollständig entwickelt habe.

Wenn $E(\alpha)$ irgend eine Einheit ist, so läfst sich eine bestimmte Potenz derselben durch das unabhängige System der conjugirten Kreistheilungseinheiten ausdrücken, und man hat

$$E(\alpha)^t = \pm \alpha^s e(\alpha)^m e(\alpha^\gamma)^{m_1} e(\alpha^{\gamma^2})^{m_2} \ldots\ldots e(\alpha^{\gamma^{\mu-1}})^{m^{\mu-1}}$$

wo s, m, m_1, ... $m_{\mu-1}$ und t ganze Zahlen sind, und letztere als der kleinste Exponent der Potenz zu welcher eine Einheit erhoben werden mufs, um durch ganze Potenzen der Kreistheilungseinheiten darstellbar zu sein, nothwendig ein Divisor des zweiten Faktors der Klassenanzahl ist, mithin in der gegenwärtigen Untersuchung nicht theilbar durch λ. Wenn nun $E(\alpha) \equiv c$, mod. λ^2, sein soll, so mufs, wie leicht zu erkennen ist, wenn α in α^{-1} verwandelt wird, $\alpha^s = 1$ sein, also

$$E(\alpha)^{l} \equiv \pm\, e(\alpha)^{m}\, e(\alpha^{\gamma})^{m_1}\, e(\alpha^{\gamma^2})^{m_2} \ldots\ldots e(\alpha^{\gamma^{\mu-1}})^{m_{\mu-1}}$$

Nimmt man nun die Logarithmen in Beziehung auf λ^2, so hat man, weil $E(\alpha) \equiv c$, mod. λ^2 sein soll:

$$l\left(\frac{E(\alpha)}{E(1)}\right) \equiv 0, \quad \text{mod. } \lambda^2;$$

also

$$m\,l\left(\frac{e(\alpha)}{e(1)}\right) + m_1\, l\left(\frac{e(\alpha^{\gamma})}{e(1)}\right) + \ldots\ldots + m_{\mu-1}\, l\left(\frac{e(\alpha^{\gamma^{\mu-1}})}{e(1)}\right) \equiv 0, \quad \text{mod. } \lambda^2.$$

Setzt man nun in der allgemeinen Formel für die logarithmischen Entwickelungen, welche ich pag. 134 der erwähnten Abhandlung in Crelle's Journal Bd. 44 gegeben habe, $\varphi(\alpha) = e(\alpha^{\gamma^h})$, $n = 1$ und bemerkt, daſs wegen der Eigenschaft der Kreistheilungseinheiten, nach welcher $e(\alpha) = e(\alpha^{-1})$ ist, sämtliche Differenzialquotienten von $le(e^{v})$ mit ungradem Index, wenn in denselben $v = 0$ gesetzt wird gleich Null werden, so hat man bei Anwendung des Summenzeichens

$$(\lambda - 1)\, l\left(\frac{e(\alpha^{\gamma^h})}{e(1)}\right) \equiv -\, l\left(e(1)^{\lambda-1}\right) + \sum_{1}^{\mu-1}{}_{n}\, \gamma^{2hn\lambda}\, \frac{d_0^{\,2n\lambda}\, l\,e(e^{v})}{dv^{2n\lambda}}\, X_{2n}(\alpha), \quad \text{mod. } \lambda^2.$$

Multiplicirt man nun mit m_h, und nimmt die Summe für $h = 0, 1, 2, \ldots \mu - 1$, so hat man, wenn der Kürze wegen

$$\sum_{0}^{\mu-1}{}_{h}\, m_h\, \gamma^{2hn\lambda} = m + m_1 \gamma^{2n\lambda} + m_2 \gamma^{4n\lambda} + \ldots + m_{\mu-1} \gamma^{2(\mu-1)n\lambda} = M_n$$

gesetzt wird:

$$0 \equiv -\, M_0\, l\left(e(1)^{\lambda-1}\right) + \sum_{1}^{\mu-1}{}_{n}\, M_n\, \frac{d_0^{\,2n\lambda}\, l e(e^{v})}{dv^{2n\lambda}}\, X_{2n}(\alpha), \quad \text{mod. } \lambda^2.$$

Es müssen nun die Coefficienten aller mit $X_2(\alpha)$, $X_4(\alpha)$ .... $X_{2\mu-2}(\alpha)$ multiplicirten Glieder einzeln congruent Null sein, nach dem Modul λ^2, und man hat allgemein für jeden Werth des $n = 1, 2, 3 \ldots. (\mu - 1)$:

$$M_n\, \frac{d_0^{\,2n\lambda}\, l e(e^{v})}{dv^{2n\lambda}} \equiv 0, \quad \text{mod. } \lambda^2.$$

Aus der Entwickelung von $le(e^{v})$ nach Potenzen der Variabeln v, welche ich Crelles Journal Bd. 44, pag. 139 gegeben habe, nämlich

$$le(e^{v}) = l(\gamma) + \frac{(\gamma^2 - 1) B_1}{1.\,2.\,2.}\, v^2 - \frac{(\gamma^4 - 1) B_2}{1.\,2.\,3.\,4.\,4.}\, v^4 + \ldots..$$

folgt nun

$$\frac{d_0^{2n\lambda}\, l\, e(e^v)}{dv^{2n\lambda}} = \frac{(-1)^{n+1}(\gamma^{2n\lambda}-1)\, B_{n\lambda}}{2n\lambda},$$

$$\frac{d_0^{2n}\, l\, e(e^v)}{dv^{2n}} = \frac{(-1)^{n+1}(\gamma^{2n}-1)\, B_n}{2n},$$

und weil ganz allgemein, für jede wirkliche nicht durch $1-\alpha$ theilbare complexe Zahl $\phi(\alpha)$ die Congruenz

$$\frac{d_0^{k\lambda}\, l\, \phi(e^v)}{dv^{k\lambda}} \equiv \frac{d_0^{k}\, l\, \phi(e^v)}{dv^{k}}, \quad \text{mod. } \lambda,$$

Statt hat, welche Congruenz aus der Vergleichung der logarithmischen Entwickelung von $l\left(\frac{\phi(\alpha)}{\phi(1)}\right)$ nach dem Modul λ mit der nach dem Modul λ^2 unmittelbar folgt, so erkennt man, dafs der Differenzialquotient

$$\frac{d_0^{2n\lambda}\, l\, e(e^v)}{dv^{2n\lambda}}$$

nur dann durch λ theilbar sein kann und auch wirklich durch λ theilbar ist, wenn $B_n \equiv 0$, mod. λ, ist, also hier nur für den einen Werth $n=\nu$. Dieser Faktor kann also aus der obigen Congruenz für alle Werthe des n, mit Ausschlufs des einen Werthes $n=\nu$ weggelassen werden, man hat daher

$$M_n \equiv 0, \text{ mod. } \lambda^2,$$

für die $\mu-2$ Werthe des $n=1, 2, 3, \ldots \mu-1$, mit Ausschlufs von $n=\nu$. Für diesen besonderen Werth $n=\nu$ hat man, weil $B_\nu \equiv 0$, mod. λ ist

$$\frac{d_0^{2\nu\lambda}\, l\, e(e^v)}{dv^{2\nu\lambda}} \equiv 0, \text{ mod. } \lambda.$$

Es soll nun angenommen werden, dafs dieser $2\nu\lambda^{\text{te}}$ Differenzialquotient den Faktor λ nicht mehr als einmal enthält, oder was dasselbe ist, dafs

$$\frac{B_{\nu\lambda}}{\nu\lambda} \text{ nicht} \equiv 0 \text{ mod. } \lambda^2, \text{ also } B_{\nu\lambda} \text{ nicht} \equiv 0, \text{ mod. } \lambda^3$$

ist, alsdann ist nothwendig $M_\nu \equiv 0$, mod. λ, man kann daher setzen

$$M_\nu \equiv \mu a \lambda, \text{ mod. } \lambda^2.$$

Setzt man aufserdem noch

$$M \equiv \mu b, \text{ mod. } \lambda^2;$$

so erhält man, indem man M_n mit $\gamma^{-2nk\lambda}$ multiplicirt und die Summe nimmt für $n=0, 1, 2, \ldots \mu-1$

$$\mu m_k \equiv \mu b + \mu a \lambda \gamma^{-2\nu k\lambda}, \text{ mod. } \lambda^2$$

oder wenn der gemeinschaftliche Faktor μ weggehoben und die Congruenz als Gleichung geschrieben wird

$$m_k = b + a\lambda\gamma^{-2\nu k\lambda} + s_k\lambda^2.$$

Setzt man endlich diese gefundenen Werthe der Exponenten m_k in dem Ausdrucke des $E(\alpha)^t$ durch die Kreistheilungseinheiten ein, und bemerkt dafs vermöge der Gleichung $Ne(\alpha)=1$, die Zahl b gänzlich verschwindet, so hat man

$$E(\alpha)^t = \left(e(\alpha)\, e(\alpha^\gamma)^{\gamma^{-2\nu\lambda}} e(\alpha^{\gamma^2})^{\gamma^{-4\nu\lambda}} \ldots\ldots\, e(\alpha^{\gamma^{\mu-1}})^{\gamma^{-2(\mu-1)\nu\lambda}}\right)^{a\lambda} \varepsilon(\alpha)^{\lambda^2},$$

wo $\varepsilon(\alpha)$ eine aus den Kreistheilungseinheiten zusammengesetzte Einheit ist. Es ist also $E(\alpha)^t$ eine λ^{te} Potenz einer Einheit, oder

$$E(\alpha)^t = E_1(\alpha)^\lambda.$$

Der Exponent t ist, wie oben gezeigt worden, nicht theilbar durch λ, darum kann man die beiden Zahlen c und d so bestimmen, dafs $tc = 1 + d\lambda$ ist; erhebt man also $E(\alpha)^t$ zur Potenz c, nimmt $tc = 1 + d\lambda$, und dividirt durch $E(\alpha)^{d\lambda}$, so hat man

$$E(\alpha) = \left(\frac{E_1(\alpha)^c}{E(\alpha)^d}\right)^\lambda,$$

also $E(\alpha)$ selbst ist gleich der λ^{ten} Potenz einer Einheit. Das gefundene Resultat wird nun folgendermafsen als Lehrsatz ausgesprochen:

Wenn $B_\nu \equiv 0$, mod. λ, aber $B_{\nu\lambda}$ nicht $\equiv 0$, mod. λ^3, so ist eine jede Einheit, welche einer nichtcomplexen Zahl congruent ist nach dem Modul λ^2, eine λ^{te} Potenz einer Einheit.

§. 3.

Wenn $F(\alpha)$ eine nicht durch $1-\alpha$ theilbare wirkliche complexe Zahl ist, welche nur die zweigliedrigen Perioden $\alpha + \alpha^{-1}$, $\alpha^\gamma + \alpha^{-\gamma}$, etc. enthält, nicht aber die einzelnen Wurzeln α, α^γ, α^{γ^2} etc., welche also der Bedingung $F(\alpha) = F(\alpha^{-1})$ genügt, so hat man für den Logarithmus derselben, in Beziehung auf den Modul λ genommen, folgende Entwickelung:

$$-l\left(\frac{F(\alpha)}{F(1)}\right) \equiv \frac{d_0{}^2\, lF(e^v)}{dv^2}\,X_2(\alpha) + \frac{d_0{}^4\, lF(e^v)}{dv^4}\,X_4(\alpha) + \ldots + \frac{d_0{}^{\lambda-3} lF(e^v)}{dv^{\lambda-3}}\,X_{\lambda-3}(\alpha)$$

nach dem Modul λ, denn alle Glieder, welche Differenzialquotienten mit ungraden Indices enthalten würden verschwinden wegen der Eigenschaft der complexen Zahl $F(\alpha)$, dafs $F(\alpha) = F(\alpha^{-1})$ ist. Wendet man nun die zusammengesetzten Kreistheilungseinheiten $E_n(\alpha)$ an, für welche man, wie ich in Crelle's Journal Bd. 44, pag. 139 gezeigt habe, folgende logarithmische Entwickelungen hat:

$$l\left(\frac{E_n(\alpha)}{E_n(1)}\right) \equiv \frac{(-1)^{n+1}(\gamma^{2n}-1)B_n}{4n} X_{2n}(\alpha), \quad \text{mod. } \lambda,$$

und bestimmt die Zahlen N_n für $n = 1, 2, 3, \ldots \mu - 1$ mit Ausschlufs des Werthes $n = \nu$ durch die Congruenz

$$\frac{(-1)^{n+1}(\gamma^{2n}-1)B_n}{4n} \cdot N_n \equiv \frac{d_0^{\,2n} lF(e^v)}{dv^{2n}}, \quad \text{mod. } \lambda,$$

so hat man

$$N_n\, l\left(\frac{E_n(\alpha)}{E_n(1)}\right) \equiv \frac{d_0^{\,2n} lF(e^v)}{dv^{2n}} \cdot X_{2n}(\alpha), \quad \text{mod. } \lambda.$$

und da für den besonderen Werth $n = \nu$

$$l\left(\frac{E_\nu(\alpha)}{E_\nu(1)}\right) \equiv 0, \quad \text{mod } \lambda,$$

ist, und darum auch

$$N_\nu\, l\left(\frac{E_\nu(\alpha)}{E_\nu(1)}\right) \equiv 0, \quad \text{mod. } \lambda,$$

für jeden beliebigen ganzzahligen Werth des N_ν, so verwandelt sich die Entwickelung des Logarithmus von $F(\alpha)$ in folgende:

$$-l\left(\frac{F(\alpha)}{F(1)}\right) \equiv N_1\, l\left(\frac{E_1(\alpha)}{E_1(1)}\right) + N_2\, l\left(\frac{E_2(\alpha)}{E_2(1)}\right) + \ldots + N_{\mu-1}\, l\left(\frac{E_{\mu-1}(\alpha)}{E_{\mu-1}(1)}\right) +$$

$$+ \frac{d_0^{\,2\nu} lF(e^v)}{dv^{2\nu}} X_{2\nu}(\alpha), \quad \text{mod. } \lambda,$$

und wenn der Kürze wegen gesetzt wird

$$E_1(\alpha)^{N_1} \cdot E_2(\alpha)^{N_2} \ldots E_{\mu-1}(\alpha)^{N_{\mu-1}} = E(\alpha)$$

so hat man

$$l\left(\frac{E(\alpha)\,F(\alpha)}{E(1)\,F(1)}\right) \equiv -\frac{d_0^{\,2\nu} lF(e^v)}{dv^{2\nu}} X_{2\nu}(\alpha), \quad \text{mod. } \lambda.$$

Also wenn die complexe Zahl $F(\alpha)$ die Eigenschaft hat, dafs dieser $2\nu^{\text{te}}$ Differenzialquotient ihres Logarithmus congruent Null ist, so hat man

$$l\left(\frac{E(\alpha)\,F(\alpha)}{E(1)\,F(1)}\right) \equiv 0 \quad \text{mod. } \lambda.$$

und folglich auch

$$E(\alpha)\,F(\alpha) \equiv E(1)\,F(1)\,, \quad \text{mod. } \lambda,$$

unter dieser Bedingung also läſst sich $F(\alpha)$ durch Multiplication mit einer Einheit so verändern, daſs es einer nichtcomplexen Zahl congruent wird nach dem Modul λ. Man hat also für die durch die beiden allgemeinen Voraussetzungen charakterisirten complexen Zahlen folgenden Lehrsatz:

Wenn $F(\alpha)$ eine nur die zweigliedrigen Perioden enthaltende complexe Zahl ist, also $F(\alpha) = F(\alpha^{-1})$, und wenn

$$\frac{d_0{}^{2\nu} l\, F(e^{v})}{dv^{2\nu}} \equiv 0\,, \quad \text{mod. } \lambda,$$

so läſst sich $F(\alpha)$ durch Multiplication mit einer passenden Einheit in die Form bringen, daſs es einer nichtcomplexen Zahl congruent ist für den Modul λ.

Mit diesem Satze hängt eine merkwürdige Eigenschaft aller Einheiten zusammen, welche in dem hier zu gebenden Beweise des Fermatschen Satzes ebenfalls ihre Anwendung finden wird, nämlich daſs wenn $E(\alpha)$ eine beliebige Einheit ist der $2\nu^{\text{te}}$ Differentialquotient des Logarithmus von $E(e^{v})$, für $v = 0$, der Null congruent ist, nach dem Modul λ, für alle diejenigen Werthe des λ, welche den beiden Voraussetzungen des §. 1 entsprechen. Drückt man nämlich, wie im §. 2, die t^{te} Potenz der Einheit $E(\alpha)$ durch die Kreistheilungs-Einheiten aus, so hat man

$$E(\alpha)^{t} = \pm\, \alpha^{s}\, e(\alpha)^{m_1}\, e(\alpha^{\gamma})^{m_2}\, e(\alpha^{\gamma^2})^{m_3} \,\ldots\ldots\, e(\alpha^{\gamma^{\mu-1}})^{m_{\mu-1}}\,,$$

wo t durch λ nicht theilbar ist, und hieraus folgt die angegebene Eigenschaft der Einheit $E(\alpha)$ unmittelbar vermöge der Congruenz

$$\frac{d_0{}^{2\nu} l\, e(e^{v})}{dv^{2\nu}} \equiv \frac{(-1)^{\nu+1}(\gamma^{2\nu}-1)\,B_\nu}{2\nu} \equiv 0, \quad \text{mod. } \lambda.$$

§. 4.

Für den vorliegenden Zweck ist es noch erforderlich ein Kriterium aufzusuchen, durch welches leicht und unzweifelhaft entschieden werden könne, ob eine complexe Zahl, deren λ^{te} Potenz wirklich ist, selbst eine wirkliche ist, oder eine ideale. Dieses Kriterium liegt etwas tiefer als die obigen Lehrsätze und erfordert zu seiner vollständigen Begründung einige Sätze, welche aus den bisherigen Arbeiten über die Theorie der complexen Zahlen nicht unmittelbar zu entnehmen sind. Zunächst ist es hierzu nöthig auf die Darstellung eines vollständigen Systems aller nicht äquivalenten Klassen der idealen Zahlen einzugehen. Ich werde mich dabei einer schon früher von mir gebrauchten abgekürzten Ausdrucksweise bedienen, welche in folgender Erklärung gegeben wird: Eine ideale Zahl $f(\alpha)$, welche zur h^{ten} Potenz erhoben werden mufs um wirklich zu werden, so dafs $f(\alpha)^h$, aber keine niedere Potenz von $f(\alpha)$ eine wirkliche complexe Zahl ist, soll als eine zum Exponenten h gehörende ideale Zahl bezeichnet werden.

Es sei nun $\phi(\alpha)$ eine zum Exponenten h gehörende ideale Zahl, so gehören die in der Form

$$\phi(\alpha)^m, \text{ für } m = 0, 1, 2 \ldots\ldots h-1,$$

enthaltenen h complexen Zahlen alle verchiedenen Klassen an; denn wäre $\phi(\alpha)^r$ äquivalent $\phi(\alpha)^s$, wo r und s zwei verschiedene Zahlen aus der Reihe 0, 1, 2, ... $h-1$ bezeichnen, so müfste $\phi(\alpha)^{r-s}$ wirklich sein, welches weil $r-s<h$ ist, der Voraussetzung wiederspricht. Wenn nun diese h idealen Zahlen nicht sämtliche Klassen idealer Zahlen erschöpfen, so sei $\phi_1(\alpha)$ eine ideale Zahl, welche keiner der in der Form $\phi(\alpha)^m$ enthaltenen äquivalent ist, und es sei h_1 der kleinste Exponent, für welchen $\phi_1(\alpha)^{h_1}$ einer in dieser Form enthaltenen Zahl äquivalent wird, so sind alle in der Form

$$\phi(\alpha)^m\ \phi_1(\alpha)^{m_1} \text{ für } \begin{cases} m\ = 0, 1, 2, \ldots\ldots h-1 \\ m_1 = 0, 1, 2, \ldots\ldots h_1-1 \end{cases}$$

enthaltene ideale Zahlen unter einander nicht äquivalent; denn wäre $\phi(\alpha)^r$ $\phi_1(\alpha)^{r_1}$ äquivalent $\phi(\alpha)^s\ \phi_1(\alpha)^{s_1}$, so müfste auch $\phi(\alpha)^{r-s}$ äquivalent $\phi_1(\alpha)^{s_1-r_1}$ und weil, wenn s_1 als die gröfsere der beiden Zahlen s_1 und r_1 angenommen wird, $s_1-r_1<h_1$ ist, so würde eine niedere Potenz von $\phi_1(\alpha)$ als die h_1^{te} einer Potenz von $\phi(\alpha)$ äquivalent sein, gegen die Voraussetzung. Wenn nun die in dieser Form enthaltenen idealen Zahlen noch nicht alle Klassen erschöpfen, so sei $\phi_2(\alpha)$ eine ideale Zahl, welche keiner derselben äquivalent ist, sei auch die h_2^{te} Potenz $\phi_2(\alpha)^{h_2}$ die niedrigste Potenz von $\phi_2(\alpha)$, welche einer der in dieser Form enthaltenen äquivalent wird, so wird ebenso gezeigt, dafs alle in der Form

$$\phi(\alpha)^m\ \phi_1(\alpha)^{m_1}\ \phi_2(\alpha)^{m_2}, \text{ für } \begin{cases} m\ = 0, 1, 2, \ldots\ldots h\ -1 \\ m_1 = 0, 1, 2, \ldots\ldots h_1-1 \\ m_2 = 0, 1, 2, \ldots\ldots h_2-1 \end{cases}$$

enthaltenen idealen Zahlen, deren Anzahl gleich $h h_1 h_2$ ist, nur verschiede-

G 2

nen Klassen angehören. Fährt man auf diese Weise fort, bis alle Klassen idealer Zahlen vollständig erschöpft sind, so hat man das vollständige System aller nicht äquivalenten Klassen der idealen Zahlen in der Form

$$\phi(\alpha)^m \ \phi_1(\alpha)^{m_1} \ \phi_2(\alpha)^{m_2} \ \phi_3(\alpha)^{m_3} \ldots\ldots\ldots$$

für $m = 0, 1, 2, \ldots h-1$, $m_1 = 0, 1, 2, \ldots\ldots h_1 - 1$, $m_2 = 0, 1, 2, \ldots\ldots h_2 - 1$ u. s. w. und wenn H die Klassenanzahl bezeichnet, so ist

$$H = h\, h_1\, h_2\, h_3 \ldots\ldots$$

Die Zahlen h_1, h_2, h_3 ... sind nothwendig genaue Theile der Exponenten zu welchen die idealen Grundzahlen $\phi_1(\alpha)$, $\phi_2(\alpha)$, $\phi_3(\alpha)$ .... gehören, oder wenn diese Exponenten beziehungsweise durch k_1, k_2, k_3 .... bezeichnet werden, so ist h_1 ein Divisor von k_1, h_2 Divisor von k_2 u. s. w. Um dies allgemein zu beweisen sei $\phi_r(\alpha)$ die ideale Grundzahl welche zur Potenz h_r erhoben werden mufs um einer der in der Form $\phi(\alpha)^m \ \phi_1(\alpha)^{m_1} \ldots\ldots \phi_{r-1}(\alpha)^{m_{r-1}}$ enthaltenen idealen Zahlen äquivalent zu werden, und k_r der Exponent zu welchem die ideale Zahl $\phi_r(\alpha)$ gehört. Weil $\phi_r(\alpha)^{h_r}$ einer der in der angegebenen Form enthaltenen idealen Zahlen äquivalent ist, so ist klar dafs dasselbe mit $\phi_r(\alpha)^{ih_r}$ der Fall sein mufs, für jeden beliebigen Werth der Zahl i und weil $\phi_r(\alpha)^{k_r}$ eine wirkliche complexe Zahl ist, und folglich auch $\phi_r(\alpha)^{jk_r}$ wirklich, dafs auch $\phi_r(\alpha)^{ih_r - jk_r}$ einer der in der angegebenen Form enthaltenen idealen Zahlen äquivalent sein mufs, für alle beliebigen ganzzahligen Werthe von i und j. Wenn nun d_r der gröfste gemeinschaftliche Faktor von h_r und k_r ist, so kann man die Zahlen i und j so bestimmen dafs $ih_r - jk_r = d_r$ ist, es mufs also auch $\phi_r(\alpha)^{d_r}$ einer der in der angegebenen Form enthaltenen idealen Zahlen äquivalent sein. Es ist aber die Potenz $\phi_r(\alpha)^{h_r}$ die niedrigste für welche diefs Statt hat, also darf d_r nicht kleiner als h_r sein, und da d_r der gröfste gemeinschaftliche Theiler von h_r und k_r ist, so mufs $d_r = h_r$ sein, und k_r ein Vielfaches von h_r, was zu beweisen war.

Nach den Voraussetzungen des §. 1 enthält die Klassenanzahl H den Faktor λ, es mufs also eine der Zahlen h, h_1, h_2 ... durch λ theilbar sein. Hieraus folgt nun, dafs auch eine der idealen Grundzahlen $\phi(\alpha)$, $\phi_1(\alpha)$, $\phi_2(\alpha)$ ... zu einem durch λ theilbaren Exponenten gehören mufs, denn dieser Exponent ist, wie gezeigt worden, ein Vielfaches des betreffenden h. Es giebt also ideale Zahlen, welche zu einem durch λ theilbaren Exponenten gehören, also auch solche, welche zum Exponenten λ selbst gehören. Man kann nun die erste der Grundzahlen nämlich $\phi(\alpha)$ als eine zum Exponenten λ gehörende wählen, so dafs $h=\lambda$ ist. Gesetzt nun es gäbe noch irgend eine zum Exponenten λ gehörende ideale Zahl, welche keiner der in der Form $\phi(\alpha)^m$ für $m = 0, 1, 2, \ldots \lambda-1$ enthaltenen idealen Zahlen äquivalent wäre, so könnte man diese als die zweite Grundzahl $\phi_1(\alpha)$ wählen, und es müfste alsdann $h_1 = \lambda$ sein. Da aber $H = h h_1 h_2 h_3 \ldots$ ist und der Voraussetzung gemäfs den Faktor λ nur einmal enthält, so kann nicht $h_1 = \lambda$ sein, wenn $h=\lambda$ ist, und man hat das Resultat, dafs wenn $\phi(\alpha)$ eine der zum Exponenten λ gehörenden idealen Zahlen ist, die λ idealen Zahlen

$$1,\quad \phi(\alpha),\ \phi(\alpha)^2,\ \phi(\alpha)^3 \ldots\ldots \phi(\alpha)^{\lambda-1}$$

alle Klassen derjenigen idealen Zahlen erschöpfen, deren λ^{te} Potenzen wirklich werden.

§. 4.

Ich gehe nun von dem Ausdrucke des Index der Einheit $E_n(\alpha)$ aus, welchen ich in der Abhandlung über die Ergänzungssätze zu den allgemeinen Reciprocitätsgesetzen, Crelle's Journal Bd. 44 pag. 103 gegeben habe:

$$\text{Ind. } E_n(\alpha) \equiv \frac{\gamma^{2n}-1}{2(1+r^{\lambda-2n}-(r+1)^{\lambda-2n})} \cdot \frac{d_0^{\;\lambda-2n} l\,\Psi_r(e^v)}{dv^{\lambda-2n}}, \quad \text{mod. } \lambda, \qquad (A)$$

in welchem $f(\alpha)$ die ideale complexe Primzahl ist, auf die das Zeichen Ind. sich bezieht und $\Psi_r(\alpha)$ die aus der Kreistheilung bekannte complexe Zahl, welche $\frac{\lambda-1}{2}$ der zu $f(\alpha)$ conjugirten Primfaktoren enthält, als deren Produkt sie folgendermaafsen dargestellt wird

$$\Psi_r(\alpha) = \pm\, \alpha^k\, \Pi f(\alpha^{\gamma^h}),$$

wo das Produkt für alle diejenigen Werthe des h aus der Reihe der Zahlen $0, 1, 2, \ldots \lambda-2$ zu nehmen ist, für welche

$$\gamma_{\mu-h} + \gamma_{\mu-h+\text{ind } r} > \lambda$$

ist, wo $\gamma_\varkappa$ die kleinste positive Zahl bezeichnet, welche congruent $\gamma^\varkappa$ ist nach dem Modul λ, μ wie hier überall das abgekürzte Zeichen für $\frac{\lambda-1}{2}$, γ eine primitive Wurzel der Primzahl λ, und der Index ind. auf den Modul λ zu beziehen ist. Aus dieser Formel habe ich an dem angeführten Orte den Ausdruck

$$\text{Ind. } E_n(\alpha) \equiv \frac{(-1)^n(\gamma^{2n}-1)B_n}{4nh}\, \frac{d_0^{\;\lambda-2n} l\left(f(e^v)^h\right)}{dv^{\lambda-2n}}, \quad \text{mod. } \lambda$$

hergeleitet, unter der Voraussetzung, dafs der Exponent h der Potenz, zu welcher $f(\alpha)$ erhoben werden mufs, um wirklich zu werden, nicht durch λ theilbar ist. Für den vorliegenden Zweck handelt es sich nun aber grade um solche ideale Zahlen, für welche der Exponent zu dem sie gehören durch λ theilbar ist, darum mufs hier der entsprechende Ausdruck des Ind. $\mathrm{E}_n(\alpha)$ nach einer andern Methode aus dem obigen abgeleitet werden.

Zunächst soll anstatt des $(\lambda-2n)^{\text{ten}}$ Differenzialquotienten in der Formel (A) der $(\lambda-2n)\lambda^{\text{te}}$ Differentialquotient genommen werden welcher jenem congruent ist nach dem Modul λ. Ferner wenn die Klassenanzahl $H = H_1\lambda$ gesetzt wird, wo H_1 nicht weiter durch λ theilbar ist, so ist die $H_1\lambda^{\text{te}}$ Potenz jeder idealen Zahl eine wirkliche complexe Zahl; setzt man also

$$f(\alpha)^{H_1\lambda} = F(\alpha)$$

so ist $F(\alpha)$ wirklich und man hat:

$$\Psi_r(\alpha)^{H_1\lambda} = \pm\,\Pi F(\alpha^{\gamma^h})$$

also auch

$$H_1\lambda\;\frac{d_0{}^{(\lambda-2n)\lambda}\,l\,\Psi_r(e^v)}{dv^{(\lambda-2n)\lambda}} \equiv \Sigma\;\frac{d_0{}^{(\lambda-2n)\lambda}\,l\,F(e^{v\gamma^h})}{dv^{(\lambda-2n)\lambda}},\quad \text{mod }\lambda^2,$$

wo die Summe Σ in Beziehung auf dieselben oben angegebenen Werthe des h zu nehmen ist, als das Produkt Π; denn es ist allgemein, für alle nicht durch $\lambda-1$ theilbaren Werthe des m

$$\frac{d_0{}^{m\lambda^r}\,l\,\phi(e^v)}{dv^{m\lambda^r}} \equiv \frac{d_0{}^{m\lambda^r}\,l\,\phi_1(e^v)}{dv^{m\lambda^r}},\quad \text{mod. }\lambda^{r+1},$$

wenn $\phi(\alpha) = \phi_1(\alpha)$, oder auch nur $\phi(\alpha) \equiv \phi_1(\alpha)$, mod. λ^{r+1}, ist. Multiplicirt man nun die Congruenz (A) mit $H_1\lambda$ und verwandelt den in derselben enthaltenen Differenzialquotienten des Logarithmen von $\Psi_r(e^v)$ auf die hier angegebene Weise, indem man zugleich

$$\frac{d_0{}^{(\lambda-2n)\lambda}\,l\,F(e^{v\gamma^h})}{dv^{(\lambda-2n)\lambda}} \text{ durch } \gamma^{(\lambda-2n)\lambda h}\,\frac{d_0{}^{(\lambda-2n)\lambda}\,l\,F(e^v)}{dv^{(\lambda-2n)\lambda}}$$

ersetzt, welches demselben vollkommen gleich ist, so erhält man die Congruenz

$$H_1\lambda \text{ Ind } \mathrm{E}_n(\alpha) \equiv \frac{\gamma^{2n}-1}{2\left(1+r^{\lambda-2n}-(r+1)^{\lambda-2n}\right)}\cdot\frac{d_0{}^{(\lambda-2n)\lambda}\,l\,F(e^v)}{dv^{(\lambda-2n)\lambda}}\cdot\Sigma\gamma^{(\lambda-2n)\lambda h}$$

für den Modul λ^2. Es ist nun die auf der rechten Seite dieser Congruenz stehende Summe zu finden, welche sich auf alle diejenigen Werthe des h aus der Reihe $0, 1, 2, \ldots \lambda - 2$ erstreckt, für welche $\gamma^{\mu-h} + \gamma^{\mu-h+\operatorname{ind} r} > \lambda$ ist. Man erkennt sehr leicht, dafs der Ausdruck

$$\frac{1}{\lambda}\left(\gamma_{\mu-h} + \gamma_{\mu-h+\operatorname{ind} r} - \gamma_{\mu-h+\operatorname{ind}(r+1)}\right)$$

nur die beiden Werthe 1 und 0 hat und zwar den Werth 1, wenn jene Ungleichheitsbedingung erfüllt ist, aber den Werth 0, wenn dieselbe nicht erfüllt ist, (cfr. Crelle's Journal Bd. 44, pag. 104.) darum hat man

$$\Sigma \gamma^{(\lambda-2n)\lambda h} = \sum_0^{\lambda-2}{}_h \frac{1}{\lambda}\left(\gamma_{\mu-h} + \gamma_{\mu-h+\operatorname{ind} r} - \gamma_{\mu-h+\operatorname{ind}(r+1)}\right) \gamma^{(\lambda-2n)\lambda h}$$

Um diese Summe, welche nun so verwandelt ist, dafs sie sich auf alle Werthe $h = 0, 1, 2, \ldots\ldots \lambda - 2$ erstreckt, leichter bestimmen zu können, verwandle ich $\gamma^{(\lambda-2n)\lambda h}$ in $\gamma^{(\lambda-2n)\lambda^2 h}$, wodurch in Beziehung auf den Modul λ^2 nichts geändert wird, ich multiplicire sodann mit λ und habe so die folgende Congruenz:

$$\lambda \Sigma \gamma^{(\lambda-2n)\lambda h} \equiv \sum_0^{\lambda-2}{}_h \gamma_{\mu-h}\, \gamma^{(\lambda-2n)\lambda^2 h} + \sum_0^{\lambda-2}{}_h \gamma_{\mu-h+\operatorname{ind} r}\, \gamma^{(\lambda-2n)\lambda^2 h}$$
$$- \sum_0^{\lambda-2}{}_h \gamma_{\mu-h+\operatorname{ind}(r+1)}\, \gamma^{(\lambda-2n)\lambda^2 h}, \quad \text{mod. } \lambda^3.$$

Es ist nun hinreichend von den drei Summen auf der rechten Seite dieser Congruenz nur die zweite zu betrachten, weil die erste aus derselben entsteht, wenn $r = 1$ genommen wird, und die dritte, wenn r in $r + 1$ verwandelt wird. Ich setze in dieser zweiten Summe

$$\gamma_{\mu-h+\operatorname{ind} r} = k$$

so erhält, während h alle Werthe $h = 0, 1, 2, \ldots \lambda - 2$ durchläuft, k alle Werthe $k = 1, 2, 3, \ldots\ldots \lambda - 1$; ferner hat man

$$\gamma^{\mu-h+\operatorname{ind} r} \equiv k \text{ also } \gamma^h \equiv -\frac{r}{k}, \quad \text{mod. } \lambda$$

woraus nach bekannten Regeln geschlossen wird.

$$\gamma^{(\lambda-2n)\lambda^2 h} \equiv -\frac{r^{(\lambda-2n)\lambda^2}}{k^{(\lambda-2n)\lambda^2}}, \quad \text{mod } \lambda^3$$

und weil $k^{(\lambda-1)\lambda^2} \equiv 1$, mod. λ^3,

$$\gamma^{(\lambda-2n)\lambda^2 h} \equiv - r^{(\lambda-2n)\lambda^2}\, k^{(2n-1)\lambda^2}, \quad \text{mod. } \lambda^3.$$

Man hat also:

$$\sum_0^{\lambda-1}{}_h \gamma_{\mu-h+\operatorname{ind} r}\, \gamma^{(\lambda-2n)\lambda^2 h} \equiv - r^{(\lambda-2n)\lambda^2} \sum_1^{\lambda-1}{}_k k^{(2n-1)\lambda^2+1}, \quad \text{mod. } \lambda^3,$$

und demnach, wenn man einerseits $r = 1$ setzt und addirt andererseits r in $r+1$ verwandelt und subtrahirt, so hat man

$$\lambda \Sigma \gamma^{(\lambda-2n)\lambda h} \equiv -\left(1 + r^{(\lambda-2n)\lambda^2} - (r+1)^{(\lambda-2n)\lambda^2}\right) \overset{\lambda-1}{\underset{1}{\Sigma_k}} k^{(2n-1)\lambda^2+1}, \text{ mod. } \lambda^3.$$

Aus den bekannten Summenausdrücken für die Reihe der Potenzen der natürlichen Zahlen, welche die Bernoullischen Zahlen in den Conflicienten enthalten, hat man nun

$$\overset{\lambda-1}{\underset{1}{\Sigma_k}} k^{(2n-1)\lambda^2+1} \equiv (-1)^{n+1} \frac{B_{(2n-1)\lambda^2+1}}{2} \cdot \lambda, \quad \text{mod. } \lambda^3$$

Setzt man den gefundenen Werth ein und dividirt durch λ, so hat man die Congruenz

$$\Sigma \gamma^{(\lambda-2n)\lambda} \equiv \left(1 + r^{(\lambda-2n)\lambda} - (r+1)^{(\lambda-2n)\lambda}\right) \frac{(-1)^n B_{(2n-1)\lambda^2+1}}{2}, \quad \text{mod. } \lambda^2.$$

Setzt man noch der Kürze wegen

$$1 + r^{(\lambda-2n)\lambda^2} - (r+1)^{(\lambda-2n)\lambda^2} = R', \quad 1 + r^{\lambda-2n} - (r+1)^{\lambda-2n} = R,$$

so verwandelt sich der obige Ausdruck des $H_1 \lambda$ Ind. $E_n(\alpha)$ in folgenden:

$$H_1\, \lambda \text{ Ind } E_n(\alpha) \equiv \frac{(-1)^n (\gamma^{2n} - 1) R'}{2R} \cdot \frac{B_{(2n-1)\lambda^2+1}}{2} \cdot \frac{d_0{}^{(\lambda-2n)\lambda} lF(e^v)}{dv^{(\lambda-2n)\lambda}}, \quad \text{mod. } \lambda^2.$$

Die in diesem Ausdrucke vorkommende Bernoullische Zahl läſst sich noch auf eine andere mit kleinerem Stellenzeiger zurückführen, vermittelst der Congruenz

$$\frac{B_m}{m} \equiv \frac{(-1)^{s\mu} B_{m+s\lambda\mu}}{m + s\lambda\mu}, \quad \text{mod. } \lambda^2,$$

welche aus den beiden in meiner Abhandlung über eine allgemeine Eigenschaft der Entwickelungs-Conflicienten einer bestimmten Gattung analytischer Funktionen (Crelles Journal Bd. 41, pag. 321) entwickelten Congruenzen

$$\frac{B_m}{m} \equiv (-1)^{\mu} \frac{B_{m+\mu}}{m+\mu} \quad \text{mod. } \lambda$$

$$\frac{B_m}{m} - (-1)^{\mu} \frac{2 B_{m+\mu}}{m+\mu} + \frac{B_{m+2\mu}}{m+2\mu} \equiv 0 \quad \text{mod. } \lambda^2$$

leicht folgt, und für alle Werthe des m, welche gröſser als 1 und durch μ nicht theilbar, und für alle positiven Werthe des s giltig ist. Setzt man nämlich $m = \frac{(2n-1)\lambda+1}{2} = n\lambda - \mu$ und $s = 2n-1$, so hat man

$$\frac{B_{(2n-1)\lambda^2+1}}{2} \equiv \frac{(-1)^{\mu} B_{n\lambda-\mu}}{2(n\lambda-\mu)}, \quad \text{mod. } \lambda^2.$$

für den besonderen Werth $n = \nu$, wo $B_\nu \equiv 0$ mod. λ und darum auch $B_{n\,\lambda-\mu} \equiv 0$, mod. λ, hat man einfacher

$$\frac{B_{(2\nu-1)\lambda^2+1}}{2} \equiv (-1)^\mu B_{\nu\lambda-\mu}, \quad \text{mod. } \lambda^2.$$

Giebt man nun in dem Ausdrucke des $H_1\lambda$ Ind. $E_n(\alpha)$ dem n den besonderen Werth $n = \nu$, so kann man, weil die Bernoullische Zahl auf der anderen Seite der Congruenz durch λ theilbar ist, diesen gemeinschaftlichen Faktor hinwegheben und erhält so

$$H_1 \text{ Ind } E_\nu(\alpha) \equiv \frac{(-1)^\mu(\gamma^{2\nu}-1)R'}{2R} \cdot \frac{B_{\nu\lambda-\mu}}{\lambda} \frac{d_0^{(\lambda-2\nu)\lambda} lF(e^v)}{dv^{(\lambda-2\nu)\lambda}}, \quad \text{mod. } \lambda,$$

bemerkt man noch dafs $R' \equiv R$, mod. λ, und

$$\frac{d_0^{(\lambda-2\nu)\lambda} lF(e^v)}{dv^{(\lambda-2n)\lambda}} \equiv \frac{d_0^{\lambda-2\nu} lF(e^v)}{dv^{\lambda-2\nu}}, \quad \text{mod } \lambda,$$

so hat man endlich

$$\text{Ind } E_\nu(\alpha) \equiv \frac{(-1)^{n+\mu}(\gamma^{2\nu}-1)B_{\nu\lambda-\mu}}{2H_1\lambda} \cdot \frac{d_0^{\lambda-2\nu} lF(e^v)}{dv^{\lambda-2\nu}}, \quad \text{mod., } \lambda$$

wo $F(\alpha) = f(\alpha)^{H_1\lambda}$ ist und $f(\alpha)$ die complexe Primzahl, auf welche der Index zu beziehen ist.

§. 5.

Die im vorigen Paragraphen entwickelte Formel für Ind $E_\nu(\alpha)$ soll nun in so fern verallgemeinert werden, dafs die ideale complexe Zahl $f(\alpha)$, auf welche das Zeichen Ind. sich bezieht und von deren $H_1\lambda^{\text{ter}}$ Potenz $F(\alpha)$ der Differenzialquotient des Logarithmus zu nehmen ist, nicht mehr eine complexe Primzahl sein mufs, sondern dafs auch eine zusammengesetzte complexe Zahl an die Stelle derselben treten kann. Die Verallgemeinerung der Bedeutung des Zeichens Ind. welche zu diesem Zwecke angenommen werden soll, stimmt mit der von Jacobi eingeführten Verallgemeinerung des Legendreschen Zeichens für die quadratischen Reste im Wesentlichen überein. Bei Anwendung des dem Legendreschen $(\frac{p}{q})$ entsprechenden Zeichens für den gegenwärtigen Fall, wo es sich um λ^{te} Potenzreste handelt, hat man nämlich, wenn $f(\alpha)$ complexe Primzahl ist und $\Phi(\alpha)$ eine nicht durch $f(\alpha)$ theilbare wirkliche complexe Zahl

$$\Phi(\alpha)^{\frac{1}{\lambda}(Nf(\alpha)-1)} \equiv \alpha^i, \quad \text{mod. } f(\alpha)$$

$$\alpha^i = \left(\frac{\Phi(\alpha)}{f(\alpha)}\right) \text{ und } i \equiv \text{Ind } \Phi(\alpha), \text{ mod. } \lambda.$$

Es sei nun $\phi(\alpha)$ eine aus den Primfaktoren $f(\alpha)$, $f_1(\alpha)$, $f_2(\alpha)$ ...., unter denen auch gleiche vorkommen können, zusammengesetzte complexe Zahl, also $\phi(\alpha) = f(\alpha), f_1(\alpha), f_2(\alpha)$ ...., so soll das Legendresche Zeichen für den zusammengesetzten Modul $\phi(\alpha)$ definirt werden durch die Gleichung

$$\left(\frac{\Phi(\alpha)}{\phi(\alpha)}\right) = \left(\frac{\Phi(\alpha)}{f(\alpha)}\right) \cdot \left(\frac{\Phi(\alpha)}{f_1(\alpha)}\right) \cdot \left(\frac{\Phi(\alpha)}{f_2(\alpha)}\right) \ldots\ldots$$

der Index Ind $\Phi(\alpha)$ in Beziehung auf die zusammengesetzte Zahl $\phi(\alpha)$ ist demnach einfach als die Summe der in Beziehung auf alle Primfaktoren des $\phi(\alpha)$ genommenen Indices aufzufassen.

Verwandelt man nun in der Formel für Ind $E_\nu(\alpha)$ nach einander $f(\alpha)$ in $f_1(\alpha)$, $f_2(\alpha)$, u. s. w., wo $f(\alpha)$, $f_1(\alpha)$, $f_2(\alpha)$ ... die Primfaktoren von $\phi(\alpha)$ sind, und addirt die so erhaltenen Congruenzen, so erhält man auf der einen Seite die Summe der Indices von $E_\nu(\alpha)$, bezogen auf alle einzelnen Primfaktoren von $\phi(\alpha)$, also den Index von $E_\nu(\alpha)$, bezogen auf die zusammenmengesetzte complexe Zahl $\phi(\alpha)$. Auf der anderen Seite erhält man anstatt des Logarithmus $lF(e^v)$, in welchem $F(e^v) = f(e^v)^{H_1\lambda}$ ist, die Summe der Logarithmen von $f(e^v)^{H_1\lambda}$, $f_1(e^v)^{H_1\lambda}$, $f_2(e^v)^{H_1\lambda}$ ..... also den Logarithmus des Produkts derselben, d. i. den Logarithmus von $\phi(e^v)^{H_1\lambda}$. Hieraus folgt, dafs die Formel

$$\text{Ind } E_\nu(\alpha) \equiv \frac{(-1)^{\nu+\mu}(\gamma^{2\nu}-1)B_{\nu\lambda-\mu}}{2H_1\lambda} \cdot \frac{d_0{}^{\lambda-2\nu} lF(e^v)}{dv^{\lambda-2\nu}}, \text{ mod. } \lambda,$$

unverändert giltig bleibt, auch wenn $F(\alpha) = \phi(\alpha)^{H_1\lambda}$ ist und $\phi(\alpha)$ eine zusammengesetzte complexe Zahl, in Beziehung auf welche der Index zu nehmen ist.

Als erste Folgerung, welche aus dieser verallgemeinerten Formel zu ziehen ist, bemerke ich, dafs wenn die complexe Zahl $\phi(\alpha)$ eine wirkliche ist, oder auch nur eine solche ideale Zahl, welche zu einem nicht durch λ theilbaren Exponenten gehört, der betreffende Index der Einheit $E_\nu(\alpha)$ stets congruent Null ist nach dem Modul λ, oder was dasselbe ist, dafs für alle derartigen complexen Zahlen $\phi(\alpha)$

$$\left(\frac{E_\nu(\alpha)}{\phi(\alpha)}\right) = 1$$

ist. In der That, wenn $\phi(\alpha)^h$ wirklich ist und h ein Divisor von H_1, also nicht durch λ theilbar, und man setzt $H_1 = hh_1$, so ist

$$\phi(\alpha)^{H_1\lambda} = (\phi(\alpha)^h)^{h_1\lambda}$$

also auch

$$lF(e^v) = h_1\lambda\, l(\phi(e^v)^h)$$

und darum

$$\frac{d_0{}^{\lambda-2v} lF(e^v)}{dv^{\lambda-2v}} \equiv 0 \text{ und Ind } E_v(\alpha) \equiv 0, \quad \text{mod. } \lambda.$$

Ferner folgt hieraus, dafs wenn $\phi(\alpha)$ und $\phi_1(\alpha)$ zwei äquivalente ideale Zahlen sind, die Indices der Einheit $E_v(\alpha)$ in Beziehung auf die eine und die andere genommen einander gleich sind, oder was dasselbe ist, dafs wenn $\phi(\alpha)$ äquivalent $\phi_1(\alpha)$, auch

$$\left(\frac{E_v(\alpha)}{\phi(\alpha)}\right) = \left(\frac{E_v(\alpha)}{\phi_1(\alpha)}\right)$$

Wenn nämlich $\phi(\alpha)$ und $\phi_1(\alpha)$ äquivalent sind, so giebt es einen idealen Multiplicator, $M(\alpha)$, welcher beide zu wirklichen complexen Zahlen macht, so dafs $M(\alpha)\,\phi(\alpha)$ und $M(\alpha)\,\phi_1(\alpha)$ wirkliche complexe Zahlen sind. Für diese hat man daher

$$\left(\frac{E_v(\alpha)}{M(\alpha)\,\phi(\alpha)}\right) = 1 \quad \text{und} \quad \left(\frac{E_v(\alpha)}{M(\alpha)\,\phi_1(\alpha)}\right) = 1,$$

also nach der Definition dieses Zeichens für zusammengesetzte Moduln auch

$$\left(\frac{E_v(\alpha)}{M(\alpha)}\right)\left(\frac{E_v(\alpha)}{\phi(\alpha)}\right) = 1, \quad \left(\frac{E_v(\alpha)}{M(\alpha)}\right)\left(\frac{E_v(\alpha)}{\phi_1(\alpha)}\right) = 1,$$

woraus unmittelbar

$$\left(\frac{E_v(\alpha)}{\phi(\alpha)}\right) = \left(\frac{E_v(\alpha)}{\phi_1(\alpha)}\right)$$

folgt, was zu beweisen war.

§. 6.

Nach der zweiten allgemeinen Voraussetzung des §. 1. giebt es irgend einen complexen Modul in Beziehung auf welchen $E_v(\alpha)$ nicht λ^{ter} Potenzrest also auch Ind $E_v(\alpha)$ nicht $\equiv 0$ ist. Bezeichnet man einen solchen Modul mit $\chi(\alpha)$, so hat man

$$\left(\frac{E_v(\alpha)}{\chi(\alpha)}\right) = \alpha^i, \text{ wo } i \text{ nicht} \equiv 0, \quad \text{mod. } \lambda.$$

Nach dem ersten der §. 5 bewiesenen Sätze mufs ferner diese ideale complexe Zahl $\chi(\alpha)$ zu einem durch λ theilbaren Exponenten gehören, nimmt man denselben gleich $h\lambda$, so ist h nicht durch λ theilbar, weil $h\lambda$ nothwen-

H 2

dig ein Divisor der Klassenanzahl H ist, welche nach den Voraussetzungen des §. 1 den Faktor λ nur einmal enthält. Weil die ideale Zahl $\chi(\alpha)$ zum Exponenten $h\lambda$ gehört, so gehört $\chi(\alpha)^h$ nothwendig zum Exponenten λ, setzt man daher $\phi(\alpha) = \chi(\alpha)^h$, so hat man

$$\left(\frac{E_\nu(\alpha)}{\chi(\alpha)^h}\right) = \left(\frac{E_\nu(\alpha)}{\phi(\alpha)}\right) = \alpha^{hi}$$

und

$$\left(\frac{E_\nu(\alpha)}{\phi(\alpha)^m}\right) = \alpha^{him}, \text{ Ind } E_\nu(\alpha) \equiv him, \quad \text{mod. } \lambda,$$

wo der Index in Beziehung auf den Modul $\phi(\alpha)^m$ zu nehmen ist, und m alle beliebigen ganzzahligen Werthe haben kann. Wenn nun $f(\alpha)$ irgend eine complexe Zahl ist, deren λ^{te} Potenz wirklich ist, so ist sie nach dem im §. 3 bewiesenen Satze einer der λ idealen Zahlen

$$1, \quad \phi(\alpha), \quad \phi(\alpha)^2, \quad \phi(\alpha)^3, \quad \ldots\ldots \quad \phi(\alpha)^{\lambda-1}$$

äquivalent. Es sei also $f(\alpha)$ äquivalent $\phi(\alpha)^m$, so ist nach dem ersten Satze des §. 5.

$$\left(\frac{E_\nu(\alpha)}{f(\alpha)}\right) = \left(\frac{E_\nu(\alpha)}{\phi(\alpha)^m}\right)$$

also auch

$$\left(\frac{E_\nu(\alpha)}{f(\alpha)}\right) = \alpha^{him} \text{ oder Ind } E_\nu(\alpha) \equiv him, \quad \text{mod. } \lambda,$$

wo der Index in Beziehung auf den Modul $f(\alpha)$ zu nehmen ist. Da hi nicht durch λ theilbar ist, so ist dieser Index von $E_\nu(\alpha)$ nothwendig congruent Null, oder nicht congruent Null, je nachdem m congruent Null oder nicht congruent Null ist, welche Bedingung, weil $f(\alpha)$ äquivalent $\phi(\alpha)^m$, und $\phi(\alpha)$ zum Exponenten λ gehört, auch so ausgesprochen werden kann: je nachdem $f(\alpha)$ eine wirkliche oder eine ideale complexe Zahl ist. Wendet man nun den in §. 5 gegebenen Ausdruck des Index von $E_\nu(\alpha)$ an, welcher für den Fall, wo $f(\alpha)^\lambda = F(\alpha)$ eine wirkliche complexe Zahl ist folgendermaafsen dargestellt werden kann:

$$\text{Ind } E_\nu(\alpha) \equiv \frac{(-1)^{\nu+\mu}(\gamma^{2\nu}-1)\, B_{\nu\lambda-\mu}}{2\lambda} \cdot \frac{d_0^{\;\lambda-2\nu} l F(e^v)}{dv^{\lambda-2\nu}}, \quad \text{mod. } \lambda$$

und bemerkt, dafs der erste Faktor auf der rechten Seite, welcher von dem Modul $f(\alpha)$ ganz unabhängig ist, nicht congruent Null sein kann, mod. λ, weil sonst Ind $E_\nu(\alpha) \equiv 0$, mod. λ, sein würde, für alle beliebigen Moduln, gegen die zweite Voraussetzung des §. 1: so sieht man, dafs Ind $E_\nu(\alpha)$ congruent Null oder nicht congruent Null ist, mod. λ, je nachdem der Differenzialquotient $\frac{d_0{}^{\lambda-2\nu}\, lF(e^{v})}{dv^{\lambda-2\nu}}$ congruent Null oder nicht congruent Null ist, nach dem Modul λ. Vergleicht man endlich diese nothwendige und hinreichende Bedingung dafür, daſs Ind $E_\nu(\alpha) \equiv 0$, mod. λ ist, mit der anderen, so hat man das gesuchte Kriterium, vermittelst dessen man entscheiden kann, ob eine complexe Zahl deren λ^{te} Potenz wirklich ist, selbst eine wirkliche ist, oder eine ideale. Dieses Kriterium wird folgendermaaſsen ausgesprochen:

Wenn die λ^{te} Potenz einer complexen Zahl: $f(\alpha)^\lambda = F(\alpha)$ eine wirkliche complexe Zahl ist, und ν diejenige Zahl für welche $B_\nu \equiv 0$, mod. λ, so ist $f(\alpha)$ selbst eine wirkliche complexe Zahl, wenn

$$\frac{d_0{}^{\lambda-2\nu}\, lF(e^{v})}{dv^{\lambda-2\nu}} \equiv 0, \quad \text{mod. } \lambda,$$

und es ist $f(\alpha)$ eine ideale complexe Zahl, wenn

$$\frac{d_0{}^{\lambda-2\nu}\, lF(e^{v})}{dv^{\lambda-2\nu}} \text{ nicht} \equiv 0, \quad \text{mod. } \lambda.$$

In dem besonderen Falle, wo $f(\alpha)$ und somit auch $F(\alpha)$ eine nur die zweigliedrigen Perioden $\alpha + \alpha^{-1}$, $\alpha^\gamma + \alpha^{-\gamma}$, ... enthaltende complexe Zahl ist, wo also $F(\alpha) = F(\alpha^{-1})$, ist jeder ungrade Differenzialquotient von $lF(e^{v})$, für $v = 0$, nothwendig congruent Null, also auch der $(\lambda - 2\nu)^{\text{te}}$ Differenzialquotient. Man hat daher folgenden Zusatz:

Wenn die λ^{te} Potenz einer nur die zweigliedrigen Perioden $\alpha + \alpha^{-1}$, $\alpha^\gamma + \alpha^{-\gamma}$, ... enthaltenden complexen Zahl eine wirkliche complexe Zahl ist, so ist diese complexe Zahl selbst nur eine wirkliche.

§. 7.

Nachdem nun in dem Vorhergehenden alle die Sätze entwickelt sind, welche bei dem zu gebenden weiteren Beweise des Fermatschen Satzes Anwendung finden werden, gehe ich zu diesem Beweise selbst über und zeige zunächst, daſs die Gleichung

$$x^\lambda + y^\lambda + z^\lambda = 0$$

wenn x, y, z nichtcomplexe ganze Zahlen sind, welche von jedem gemeinschaftlichen Faktor befreit sein sollen, auch in dem gegenwärtigen Falle, wo eine der ersten $\frac{\lambda-3}{2}$ Bernoullischen Zahlen congruent Null für den Modul λ ist, nicht bestehen kann, wenn nicht eine der drei Zahlen x, y, z durch λ

theilbar ist. Es soll also zunächst angenommen werden, dafs keine der drei Zahlen x, y, z durch λ theilbar ist.

Zerlegt man $x^\lambda + y^\lambda$ in seine λ Faktoren ersten Grades, so hat man

$$(x+y)(x+\alpha y)(x+\alpha^2 y) \ldots\ldots (x+\alpha^{\lambda-1} y) = -z^\lambda$$

Damit dieses Produkt von λ complexen Faktoren, welche wie leicht zu sehen unter sich keine gemeinschaftlichen Faktoren haben können, einer λ^{ten} Potenz gleich sei, müssen nothwendig alle einzelnen Faktoren für sich λ^{te} Potenzen complexer Zahlen sein, welche mit complexen Einheiten multiplicirt sein können. Man hat also:

$$x + \alpha y = \varepsilon(\alpha) f(\alpha)^\lambda,$$

wo $\varepsilon(\alpha)$ eine Einheit ist und $f(\alpha)$ eine complexe Zahl, welche auch ideal sein kann, deren λ^{te} Potenz aber nothwendig wirklich ist. Verwandelt man α in α^{γ^h}, so hat man auch

$$x + \alpha^{\gamma^h} y = \varepsilon(\alpha^{\gamma^h}) f(\alpha^{\gamma^h})^\lambda.$$

Ich gebe nun dem h alle diejenigen unter den Werthen $0, 1, 2, \ldots \lambda-2$, für welche

$$\gamma_{\mu-h} + \gamma_{\mu-h+\text{ind}\, r} > \lambda$$

ist, und bilde das Produkt, entsprechend dem im §. 4 behandelten Produkte, so ist

$$\Pi\,(x+\alpha^{\gamma^h} y) = \Pi\,\varepsilon(\alpha^{\gamma^h}) \left(\Pi f(\alpha^{\gamma^h})\right)^\lambda.$$

Das Produkt $\Pi f(\alpha^{\gamma^h})$, für alle die angegebenen Werthe des h, hat die ausgezeichnete Eigenschaft, dafs es stets eine wirkliche complexe Zahl ist, auch wenn die demselben zu Grunde liegende complexe Zahl $f(\alpha)$ irgend eine ideale ist, m. s. Crelle's Journal Bd. 35, pag. 364, und das Produkt der Einheiten $\Pi\,\varepsilon(\alpha^{\gamma^h})$, für dieselben Werthe des h giebt stets eine der simplen Einheiten $\pm\alpha^k$, weil es, wegen der allgemeinen Eigenschaft aller Einheiten, nach welcher $\varepsilon(\alpha) = \alpha^r \varepsilon(\alpha^{-1})$ ist, und wegen des Umstandes, dafs von den beiden Zahlen h und $\lambda - h$ immer eine und nur eine der Bedingung $\gamma_{\mu-h} + \gamma_{\mu-h+\text{ind}\, r} > \lambda$ genügt, alle die conjugirten Faktoren

$$\varepsilon(\alpha),\ \varepsilon(\alpha^\gamma),\ \varepsilon(\alpha^{\gamma^2}) \ldots\ldots \varepsilon(\alpha^{\gamma^{\mu-1}})$$

und zwar jeden nur einmal enthält, welche Faktoren sich zu ± 1 zusammensetzen. Bezeichnet man daher die wirkliche complexe Zahl $\Pi f(\alpha^{\gamma^h})$ durch $\phi(\alpha)$, so hat man

$$\Pi(x + a^{\gamma^h} y) = \pm\, a^k\, \phi(a)^\lambda$$

Setzt man nun e^v statt a und nimmt den $(\lambda - 2n)^{\text{ten}}$ Differenzialquotienten des Logarithmus für den Werth $v = 0$ so hat man

$$\Sigma \frac{d_0{}^{\lambda-2n} l(x + e^{v\gamma^h} y)}{dv^{\lambda-2n}} \equiv 0, \quad \text{mod.}\ \lambda,$$

und folglich

$$\frac{d_0{}^{\lambda-2n} l(x + e^v y)}{dv^{\lambda-2n}}\ \Sigma\, \gamma^{(\lambda-2n)h} \equiv 0, \quad \text{mod.}\ \lambda,$$

wo die Summe Σ ebenfalls für alle diejenigen Werthe des h aus der Reihe der Zahlen $0, 1, 2, \ldots\ldots \lambda - 2$ zu nehmen ist, welche der Bedingung $\gamma_{\mu-h} + \gamma_{\mu-h+\text{ind.}r} > \lambda$ genügen, und wo n einen jeden der Werthe $n = 1, 2, 3, \ldots\ldots \mu - 1$ haben kann. Die Summe $\Sigma\gamma^{(\lambda-2n)h}$ erhält man nun unmittelbar aus der oben §. 4. gefundenen Summe:

$$\Sigma\, \gamma^{(\lambda-2n)\lambda h} \equiv (-1)^n \left(1 + r^{(\lambda-2n)\lambda} - (r+1)^{(\lambda-2n)\lambda}\right) \frac{B_{(2n-1)\lambda^2+1}}{2} \quad \text{mod.}\ \lambda^2$$

welche für den einfachen Modul λ in

$$\Sigma\gamma^{(\lambda-2n)h} \equiv (-1)^n \left(1 + r^{\lambda-2n} - (r+1)^{\lambda-2n}\right) \frac{B_n}{2n}, \quad \text{mod.}\ \lambda$$

übergeht. Denselben Werth dieser Summe habe ich auch in der Abhandlung über die Ergänzungssätze zu den allgemeinen Reciprocitätsgesetzen in Crelles Journal Bd. 44, pag. 106 direct hergeleitet. Weil man die beliebige Zahl r immer so wählen kann, dafs $1 + r^{\lambda-2n} - (r+1)^{\lambda-2n}$ nicht congruent Null ist für den Modul λ, so kann diese Summe nur in dem einen Falle congruent Null sein, wenn die Bernoullische Zahl B_n congruent Null ist, also nur für den einen Werth $n = \nu$, für alle anderen Werthe des n hat man nothwendig

$$\frac{d_0{}^{\lambda-2n} l(x + e^v y)}{dv^{\lambda-2n}} \equiv 0, \quad \text{mod.}\ \lambda.$$

Allen in dieser Form für $n = 1, 2, 3, \ldots\ldots \mu - 1$, mit Ausschlufs von $n = \nu$ enthaltenen $\mu - 2$ Congruenzen müssen also die Zahlen x und y genügen, wenn $x^\lambda + y^\lambda + z^\lambda = 0$ sein soll, ohne dafs eine der Zahlen x, y, z durch λ theilbar ist, auch ist klar, dafs sie ebenfalls den entsprechenden Congruenzen genügen müssen, welche man durch Vertauschung von x, y und z aus diesen erhält.

Für den gegenwärtigen Zweck reicht es hin nur die beiden äufsersten Werthe des n, nämlich $n = \mu - 1 = \frac{\lambda-3}{2}$ und $n = \mu - 2 = \frac{\lambda-5}{2}$ in Betracht zu ziehen, welchen die Congruenzen

$$\frac{d_0{}^3\, l(x+e^v y)}{dv^3} \equiv 0, \quad \text{mod. } \lambda$$

$$\frac{d^5\, l(x+e^v y)}{dv^5} \equiv 0, \quad \text{mod. } \lambda$$

entsprechen, deren erste nur in dem einen Falle, wo $\nu = \mu - 1$, die zweite wenn $\nu = \mu - 2$ ist, nicht nothwendig Statt haben mufs. Führt man die angedeuteten Differenziationen aus und setzt $v = 0$, so erhält man

$$\frac{d_0{}^3\, l(x+e^v y)}{dv^3} = \frac{xy\,(x-y)}{(x+y)^3}$$

$$\frac{d_0{}^5\, l(x+e^v y)}{dv^3} = \frac{xy\,(x-y)\,(x^2-10xy+y^2)}{(x+y)^5}$$

Wenn nun ν nicht gleich $\mu-1$ ist, so mufs nothwendig $xy\,(x-y) \equiv 0$ sein, und weil x und y nicht durch λ theilbar sind, $x \equiv y$ und durch Vertauschung des y und z erhält man hieraus auch $x \equiv z$; da aber aus der Gleichung $x^\lambda + y^\lambda + z^\lambda = 0$ unmittelbar die Congruenz $x + y + z \equiv 0$, mod. λ, folgt, so hat man $3x \equiv 0$, und auch $3y \equiv 0$, $3z \equiv 0$, welches mit Ausnahme des hier nicht in Betracht kommenden Falles $\lambda = 3$ unmöglich ist. Es mufs also nothwendig $\nu = \mu - 1$ sein d. h. es mufs die $\frac{\lambda-3}{2}$te Bernoullische Zahl diejenige sein, welche congruent Null ist, mod. λ, wenn die Gleichung $x^\lambda + y^\lambda + z^\lambda =$ bestehen soll, ohne dafs eine der Zahlen x, y, z durch λ theilbar ist. Diese Bedingung ist im wesentlichen dieselbe, welche Cauchy in dem Compte rendu vom J. 1847, 2tes Semester p. 181 zuerst gefunden und so ausgedrückt hat: dafs die Summe der Reihe $1 + 2^{\lambda-4} + 3^{\lambda-4} + \ldots \left(\frac{\lambda-1}{2}\right)^{\lambda-4}$ nothwendig durch λ theilbar sein mufs.

Wenn nun wirklich $\nu = \mu - 1$ ist, welcher Fall möglicherweise eintreten kann, so mufs man den 5ten Differenzialquotienten zu Hülfe nehmen, welcher alsdann nothwendig congruent Null sein mufs, weil nicht zugleich $\nu = \mu - 1$ und $\nu = \mu - 2$ sein kann. Dieser giebt

$$(x-y)\,(x^2 - 10xy + y^2) \equiv 0$$

und wenn y mit z vertauscht wird, auch

$$(x-z)\,(x^2 - 10xz + z^2) \equiv 0.$$

Es kann nun erstens nicht $x \equiv y$ sein, denn vermöge der Congruenz $x + y + z \equiv 0$ würde hieraus $z \equiv -2x$ folgen und durch Einsetzung dieses Werthes von z in die zweite Congruenz würde man $3 . 25 .\ x^3 \equiv 0$ erhalten, welches unmöglich ist, wenn man die beiden hier nicht in Betracht kommenden Fälle $\lambda = 3$ und $\lambda = 5$ ausschliefst.

Genau aus demselben Grunde kann auch nicht $x \equiv z$ sein. Es bleibt also nur der eine Fall übrig wo

$$x^2 - 10\,xy + y^2 \equiv 0 \quad \text{und} \quad x^2 - 10\,xz + z^2 \equiv 0.$$

Die zweite dieser Congruenzen giebt, wenn z vermittelst der Congruenz $x + y + z \equiv 0$ eliminirt wird

$$12\,x^2 + 12\,xy + y^2 \equiv 0$$

und wenn von dieser die erste Congruenz subtrahirt wird, so hat man $11x + 22xy \equiv 0$ also $11x\,(x + 2y) \equiv 0$. Sieht man von dem hier ebenfalls nicht in Betracht kommenden Falle $\lambda = 11$ ab, so mufs $x \equiv -2y$ sein und wenn dieser Werth des x in $x^2 - 10xy + y^2 \equiv 0$ eingesetzt wird, so erhält man $3.\ 25.\ y^2 \equiv 0$, welches ebenfalls unmöglich ist. Die nothwendigen Bedingungen, damit $x^\lambda + y^\lambda + z^\lambda = 0$ sei, ohne dafs eine der Zahlen x, y, z durch λ theilbar ist, können also, wenn eine einzige der ersten $\frac{\lambda-3}{2}$ Bernoullischen Zahlen congruent Null für den Modul λ ist, niemals erfüllt werden. Es bleibt daher nur noch der Fall zu erörtern, dafs eine der drei Zahlen x, y, z durch λ theilbar ist.

§. 8.

Anstatt der Gleichung

$$x^\lambda + y^\lambda + z^\lambda = 0,$$

in welcher nun eine der Zahlen x, y, z durch λ theilbar, also $z = \lambda^k z_1$ anzunehmen ist, lege ich der Untersuchung eine allgemeinere Gleichung für complexe Zahlen zu Grunde, nämlich folgende:

$$U^\lambda + V^\lambda = E(\alpha)\,(2 - \alpha - \alpha^{-1})^{m\lambda}\,W^\lambda,$$

in welcher die drei Zahlen U, V, W wirkliche complexe Zahlen sein sollen, und zwar solche, welche nicht die einfachen Wurzeln der Gleichung $\alpha^\lambda = 1$, sondern nur die zweigliedrigen Perioden derselben $\alpha + \alpha^{-1}$, $\alpha^2 + \alpha^{-2}$, u. s. w. enthalten, welche also unverändert bleiben, wenn α in α^{-1} verwandelt wird; in welcher ferner $E(\alpha)$ irgend eine Einheit bezeichnet und $2 - \alpha - \alpha^{-1} = (1 - \alpha)\,(1 - \alpha^{-1})$, einer von $\mu = \frac{\lambda-1}{2}$ gleichen Faktoren des λ, an die Stelle von λ getreten ist. Es ist klar, dafs die Gleichung $x^\lambda + y^\lambda + \lambda^{k\lambda} z_1^\lambda = 0$ als specieller Fall in dieser enthalten ist, nämlich wenn $U = x$, $V = y$, $W = -z_1$ und $m = k\mu$ genommen, und die Einheit $E(\alpha)$ so gewählt wird, daſs

$$E(\alpha)\,(2-\alpha-\alpha^{-1})^{k\mu\lambda} = \lambda^{k\lambda}$$

ist, welches immer geschehen kann, weil $(2-\alpha-\alpha^{-1})^{\mu}$, abgesehen von einer Einheit, gleich λ ist. Auſserdem ist von den drei complexen Zahlen U, V und W anzunehmen, daſs keine derselben durch $1-\alpha$, den Primfaktor des λ, theilbar ist, und von der Zahl m, daſs sie gröſser als Eins ist. Der Beweis der Unmöglichkeit dieser allgemeineren Gleichung wird daher nothwendig auch für jene speciellere gelten.

Zerlegt man nun $U^{\lambda}+V^{\lambda}$ in seine Faktoren ersten Grades, so hat man
$(U+V)\,(U+\alpha V)\,(U+\alpha^2 V)\ldots.(U+\alpha^{\lambda-1}V) = E(\alpha)\,(2-\alpha-\alpha^{-1})^{m\lambda}\,W^{\lambda}$.
Die λ Faktoren von der Form $U+\alpha^r V$ haben hier alle den gemeinschaftlichen Faktor $1-\alpha$; denn da das Produkt derselben diesen Faktor enthält, so muſs wenigstens eine der complexen Zahlen $U+\alpha^r V$ denselben enthalten, und aus der identischen Gleichung

$$U+\alpha^r V = U+\alpha^s V + (\alpha^r-\alpha^s)V$$

ersieht man, daſs wenn $U+\alpha^r V$ durch $1-\alpha$ theilbar ist, auch jede andere Zahl $U+\alpha^s V$ durch $1-\alpha$ theilbar sein muſs. Der Faktor $1-\alpha$ ist auch der gröſste gemeinschaftliche Faktor aller dieser complexen Zahlen; denn aus den beiden identischen Gleichungen

$$(U+\alpha^r V)-(U+\alpha^s V) = (\alpha^r-\alpha^s)V$$

$$\text{und } \alpha^r(U+\alpha^s V)-\alpha^s(U+\alpha^r V) = (\alpha^r-\alpha^s)U$$

folgt unmittelbar, daſs der gröſste gemeinschaftliche Faktor von $U+\alpha^r V$ und $U+\alpha^s V$ zugleich auch gemeinschaftlicher Faktor von $(\alpha^r-\alpha^s)V$ und
* $(\alpha^r-\alpha^s)U$ sein muſs, und weil U und V keinen gemeinschaftlichen Faktor haben, daſs $\alpha^r-\alpha^s$, oder was dasselbe ist, $1-\alpha$ der gröſste gemeinschaftliche Faktor ist. Ferner folgt leicht, daſs nur eine der complexen Zahlen $U+\alpha^r V$ den Faktor $1-\alpha$ mehrmals enthalten kann, und zwar nur die erste, nämlich $U+V$, denn setzt man

$$U+\alpha^r V \equiv 0, \quad \text{mod. } (1-\alpha)^2$$

so hat man, wenn α in α^{-1} verwandelt wird, wobei U und V unverändert bleiben und auch der Modul $(1-\alpha^{-1})^2$ derselbe ist als $(1-\alpha)^2$,

$$U+\alpha^{-r}V \equiv 0, \quad \text{mod. } (1-\alpha)^2$$

welche beide Congruenzen zugleich nur für den einen Werth $r=0$ bestehen können. Da nun das Produkt aller λ Faktoren von der Form $U+\alpha^r V$ den Faktor $2-\alpha-\alpha^{-1}$ genau $m\lambda$ mal, also den Faktor $1-\alpha$ genau $2m\lambda$ mal enthält, und aufser dem ersten $U+V$ alle übrigen $\lambda-1$ Faktoren den Faktor $1-\alpha$, jeder nur einmal, enthalten, so folgt, dafs $U+V$ den Faktor $1-\alpha$ genau $(2m\lambda-\lambda+1)$ mal enthalten mufs, oder was dasselbe ist, den Faktor $2-\alpha-\alpha^{-1}$ genau $(m\lambda-\mu)$ mal, wo $\mu=\frac{\lambda-1}{2}$. Weil nun, abgesehen von den Faktoren $1-\alpha$ oder $2-\alpha-\alpha^{-1}$, die λ Faktoren $U+V$, $U+\alpha V$, $U+\alpha^2 V$, ... relative Primzahlen sind und das Produkt derselben, wenn man auch von der Einheit $E(\alpha)$ absieht, eine λ^{te} Potenz ist, so müssen dieselben einzeln λ^{te} Potenzen sein, multiplicirt mit irgend welchen Einheiten, und man hat, indem man anstatt des in $U+\alpha^r V$ enthaltenen Faktors $1-\alpha$ den nur durch eine Einheit von ihm verschiedenen Faktor $1-\alpha^r$ nimmt,

$$U+\alpha^r V=\varepsilon_r(\alpha)\,(1-\alpha^r)\,\Theta_r(\alpha)^\lambda \tag{A}$$

für alle Werthe des $r=1, 2, 3, \ldots \lambda-1$, und

$$U+V=\varepsilon(\alpha)\,(2-\alpha-\alpha^{-1})^{m\lambda-\mu}\,\mathrm{T}(\alpha)^\lambda \tag{B}$$

wo $\varepsilon_r(\alpha)$ und $\varepsilon(\alpha)$ Einheiten sind, $\Theta_r(\alpha)$ eine die einzelnen Wurzeln der Gleichung $\alpha^\lambda=1$ enthaltende complexe Zahl, $\mathrm{T}(\alpha)$ aber, weil es unverändert bleibt, wenn α in α^{-1} verwandelt wird, eine nur die zweigliedrigen Perioden dieser Einheitswurzeln enthaltende complexe Zahl ist.

Eliminirt man U aus den beiden Gleichungen (A) und (B) so erhält man folgenden Ausdruck des V.

$$V=-\varepsilon_r(\alpha)\,\Theta_r(\alpha)^\lambda+\frac{\varepsilon(\alpha)\,(2-\alpha-\alpha^{-1})^{m\lambda-\mu}\,\mathrm{T}(\alpha)^\lambda}{1-\alpha^r}$$

Verwandelt man in dieser Gleichung α in e^v, wo v eine Variable ist, und nimmt den $(\lambda-2\nu)^{\text{ten}}$ Differenzialquotienten des Logarithmen für den Werth $v=0$, so erhält man

$$\frac{d_0{}^{\lambda-2\nu}lV(e^v)}{dv^{\lambda-2\nu}}\equiv\frac{d_0{}^{\lambda-2\nu}l\,\varepsilon_r(e^v)}{dv^{\lambda-2\nu}}+\frac{d_0{}^{\lambda-2\nu}l(\Theta_r(e^v)^\lambda)}{dv^{\lambda-2\nu}},\quad \text{mod. } \lambda.$$

Der $(\lambda-2\nu)^{\text{te}}$ Differenzialquotient des $lV(e^v)$, für $v=0$, ist aber da $\lambda-2\nu$ ungrade ist und $V(\alpha)=V(\alpha^{-1})$ nothwendig congruent Null für den Modul λ; ebenfalls mufs dieser Differenzialquotient von $l\varepsilon_r(e^v)$ congruent Null sein, wegen der allgemeinen Eigenschaft der Einheiten, nach welcher $\varepsilon^r(\alpha)=\alpha^k\varepsilon_r(\alpha^{-1})$ ist. Diese Congruenz ergiebt also

$$\frac{d_0{}^{\lambda-2\nu}l(\Theta_r(e^v)^\lambda)}{dv^{\lambda-2\nu}}\equiv 0,\quad \text{mod. } \lambda,$$

und hieraus folgt nach dem im §. 6 bewiesenen Kriterium, dafs die complexe Zahl $\Theta(\alpha)$, deren λ^{te} Potenz wirklich ist, selbst eine w i r k l i c h e complexe

I 2

Zahl ist. Dafs auch die complexe Zahl $\mathrm{T}(\alpha)$, welche nur die zweigliedrigen Perioden enthält eine **wirkliche** complexe Zahl ist, ergiebt sich unmittelbar aus dem Zusatze am Schlusse des §. 6.

Verwandelt man nun in dem gefundenen Ausdrucke des V, α in α^{-1} und bemerkt, dafs dabei die Einheit $\varepsilon(\alpha)$ ungeändert bleiben mufs, welches aus der Gleichung (B) unmittelbar zu erkennen ist, so erhält man aus der Vergleichung dieses veränderten Ausdrucks des V mit dem unveränderten

$$\varepsilon_r(\alpha)\,\Theta_r(\alpha)^\lambda - \varepsilon_r(\alpha^{-1})\,\Theta_r(\alpha^{-1})^\lambda = \frac{\varepsilon(\alpha)\,(2-\alpha-\alpha^{-1})^{m\lambda-\mu}(1+\alpha^r)\,\mathrm{T}(\alpha)^\lambda}{1-\alpha^r}.$$

Weil $\Theta_r(\alpha)$ eine **wirkliche** complexe Zahl ist, so ist die λ^{te} Potenz derselben einer nichtcomplexen ganzen Zahl congruent, nach dem Modul λ, diese sei c, so ist $\Theta_r(\alpha)^\lambda \equiv c$, und ebenso $\Theta_r(\alpha^{-1})^\lambda \equiv c$, macht man also aus dieser Gleichung eine Congruenz für den Modul λ, so erhält man, wenn der gemeinschaftliche nicht durch λ theilbare Faktor c weggehoben wird:

$$\varepsilon_r(\alpha) - \varepsilon_r(\alpha^{-1}) \equiv 0, \quad \text{mod. } \lambda,$$

woraus vermöge der allgemeinen Eigenschaft aller Einheiten, nach welcher $\varepsilon_r(\alpha) = \alpha^k\,\varepsilon_r(\alpha^{-1})$ ist, unmittelbar folgt

$$\varepsilon_r(\alpha) = \varepsilon_r(\alpha^{-1}).$$

Dividirt man nun die obige Gleichung durch $\varepsilon_r(\alpha)$, und nimmt $2-\alpha-\alpha^{-1} = \alpha^{-1}(1-\alpha)^2$ so erhält man

$$\Theta_r(\alpha)^\lambda - \Theta_r(\alpha^{-1})^\lambda = \varepsilon'(\alpha)\,(1-\alpha)^{(2m-1)\lambda}\,\mathrm{T}(\alpha)^\lambda \qquad (C)$$

wo $\varepsilon'(\alpha)$ eine Einheit ist. Da $\Theta_r(\alpha)$ in dieser Gleichung, so wie auch in den vorhergehenden nur zur λ^{ten} Potenz erhoben vorkommt, so ist diese complexe Zahl noch in so fern unbestimmt, als sie mit irgend einer λ^{ten} Wurzel der Einheit, α^k, behaftet angenommen werden kann. Diese Wurzel α^k soll nun so gewählt werden, dafs $\Theta_r(\alpha)$ einer nichtcomplexen ganzen Zahl congruent sei für den Modul $(1-\alpha)^2$ (m. vgl. Liouvilles Journal Bd. 16, pag. 489). Zerlegt man nun die linke Seite dieser Gleichung (C) in ihre lineären Faktoren von der Form $\Theta_r(\alpha) - \alpha^i\,\Theta_r(\alpha^{-1})$, so wird eben so wie oben gezeigt, dafs je zwei dieser λ Faktoren den gröfsten gemeinschaftlichen Faktor $1-\alpha$ haben, dafs also $\lambda-1$ derselben den Faktor $1-\alpha$ jeder **einmal** enthalten und nur ein einziger ihn mehrmals enthalten kann. Man erkennt leicht, dafs vermöge der Festsetzung dafs $\Theta_r(\alpha)$ für den Modul $(1-\alpha)^2$ einer nichtcomplexen Zahl congruent sein soll, dieser eine Faktor $\Theta_r(\alpha) - \Theta_r(\alpha^{-1})$ ist und dafs vermöge der Gleichung (C) dieser den Faktor $1-\alpha$ genau $(2m-1)\lambda - \lambda + 1 = (2m-2)\lambda + 1$ mal enthält. Weil nun endlich die λ linearen Faktoren in welche die linke Seite der Gleichung (C) zerfällt, abgesehen von den Faktoren $1-\alpha$ relative Primzahlen sind und das Produkt derselben eine λ^{te} Potenz sein soll, so müssen dieselben einzeln λ^{ten} Potenzen, mit Einheiten multiplicirt, gleich sein und man hat

$$\Theta_r(\alpha) - \alpha^t\,\Theta_r(\alpha^{-1}) = \varepsilon_t''(\alpha)\,(1-\alpha^t)\,P_t(\alpha)^\lambda$$

für $t = 1, 2, 3, \ldots\ldots \lambda - 1$ und

$$\Theta_r(\alpha) - \Theta_r(\alpha^{-1}) = \varepsilon''(\alpha)\,(1-\alpha)^{(2m-2)\lambda+1}\,Q(\alpha)^\lambda.$$

Eliminirt man $\Theta_r(\alpha^{-1})$ aus diesen beiden Gleichungen und schreibt das Resultat als Congruenz nach dem Modul $(1-\alpha)^{(2m-2)\lambda}$, so hat man

$$\Theta_r(\alpha) \equiv \varepsilon_t''(\alpha)\,P_t(\alpha)^\lambda, \quad \text{mod. } (1-\alpha)^{(2m-2)\lambda}$$

und wenn α in α^{-1} verwandelt wird:

$$\Theta_r(\alpha^{-1}) \equiv \varepsilon_t''(\alpha^{-1})\,P_t(\alpha^{-1})^\lambda, \quad \text{mod. } (1-\alpha)^{(2m-2)\lambda}$$

also durch Multiplication dieser beiden Congruenzen:

$$\Theta_r(\alpha)\,\Theta_r'(\alpha^{-1}) \equiv \varepsilon_t''(\alpha)\,\varepsilon_t''(\alpha^{-1})\,\Big(P_t(\alpha)\,P_t(\alpha^{-1})\Big)^\lambda, \quad \text{mod. } (1-\alpha)^{(2m-2)\lambda}$$

welche Congruenz, weil $m > 1$ ist und darum $(1-\alpha)^{(2m-2)\lambda}$ ein Vielfaches von λ, auch in Beziehung auf den einfacheren Modul λ genommen werden kann. Die complexe Zahl $P_t(\alpha)\,P_t(\alpha^{-1})$, deren λ^{te} Potenz wirklich ist, ist nach dem Zusatze am Schlusse des §. 6 selbst eine w i r k l i c h e. Ersetzt man nun die Wurzel α durch die Variable e^v und nimmt den $2\nu^{\text{ten}}$ Differenzialquotienten des Logarithmus für den Werth $v = 0$, so ist zunächst, weil $P_t(\alpha)\,P_t(\alpha^{-1})$ wirklich ist:

$$\frac{d_0^{\,2\nu} l\big(P_t(e^v)\,P_t(e^{-v})\big)^\lambda}{dv^{2\nu}} \equiv \lambda\,\frac{d_0^{\,2\nu} l\big(P_t(e^v)\,P_t(e^{-v})\big)}{dv^{2\nu}} \equiv 0, \quad \text{mod. } \lambda,$$

und daher

$$\frac{d_0^{\,2\nu} l\big(\Theta_r(e^v)\,\Theta_r(e^{-v})\big)}{dv^{2\nu}} \equiv \frac{d_0^{\,2\nu} l\big(\varepsilon_t''(e^v)\varepsilon_t''(e^{-v})\big)}{dv^{2\nu}}, \quad \text{mod. } \lambda$$

und weil nach dem zweiten der im §. 3 bewiesenen beiden Sätze der $2\nu^{\text{te}}$ Differenzialquotient des Logarithmus einer jeden Einheit, in welcher α in e^v verwandelt worden, für $v = 0$ der Null congruent ist nach dem Modul λ, so hat man

$$\frac{d_0^{\,2\nu} l\big(\Theta_r(e^v)\,\Theta_r e^{-v})\big)}{dv^{2\nu}} \equiv 0, \quad \text{mod. } \lambda$$

und hieraus folgt endlich vermöge des ersten der im §. 3 bewiesenen Sätze, dafs die complexe Zahl $\Theta_r(\alpha)\,\Theta_r(\alpha^{-1})$ durch Multiplikation mit einer passenden Einheit einer nichtcomplexen ganzen Zahl congruent gemacht werden kann nach dem Modul λ.

Nachdem nun dieser für den folgenden Beweis wichtige Punkt erledigt ist, wende ich mich zu den beiden Gleichungen (A) und (B) zurück. Wird in der Gleichung (A) α in α^{-1} verwandelt und die so erhaltene Gleichung mit der unveränderten multiplicirt, so erhält man:

$$U^2+(\alpha^r+\alpha^{-r})\,UV+V^2=\varepsilon_r(\alpha)\,\varepsilon_r(\alpha^{-1})\,(2-\alpha^r-\alpha^{-r})\left(\Theta_r(\alpha)\Theta_r(\alpha^{-1})\right)^\lambda$$

verwandelt man r in s, so hat man ebenfalls:

$$U^2+(\alpha^s+\alpha^{-s})\,UV+V^2=\varepsilon_s(\alpha)\,\varepsilon_s(\alpha^{-1})\,(2-\alpha^s-\alpha^{-s})\left(\Theta_s(\alpha)\Theta_s(\alpha^{-1})\right)^\lambda$$

und wenn die Gleichung (B) zum Quadrate erhoben wird:

$$U^2+2\,UV+V^2=\varepsilon(\alpha)^2\,(2-\alpha-\alpha^{-1})^{2m\lambda-2\mu}\,\mathrm{T}(\alpha)^{2\lambda}.$$

Eliminirt man nun aus diesen drei Gleichungen die beiden Gröfsen U^2+U^2 und UV, so erhält man nach einigen leichten Reductionen

$$\varepsilon_r(\alpha)\,\varepsilon_r(\alpha^{-1})\left(\Theta_r(\alpha)\,\Theta_r(\alpha^{-1})\right)^\lambda-\varepsilon_s(\alpha)\,\varepsilon_s(\alpha^{-1})\left(\Theta_s(\alpha)\Theta_s(\alpha^{-1})\right)^\lambda=\frac{\varepsilon(\alpha)^2\,(2-\alpha-\alpha^{-1})^{2m\lambda-2\mu}(\alpha^r+\alpha^{-r}-\alpha^s-\alpha^{-s})\,\mathrm{T}(\alpha)^{2\lambda}}{(2-\alpha^r-\alpha^{-r})\,(2-\alpha^s-\alpha^{-s})}.$$

Da nun, wie gezeigt worden ist, $\Theta_r(\alpha)\,\Theta_r(\alpha^{-1})$ und folglich auch $\Theta_s(\alpha)\,\Theta_s(\alpha^{-1})$ durch Multiplikation mit passenden Einheiten nichtcomplexen Zahlen congruent gemacht werden können, nach dem Modul λ, so seien $A_r(\alpha)$ und $A_s(\alpha)$ diese passenden Einheiten. Man setze ferner der Kürze wegen

$$A_r(\alpha)\,\Theta_r(\alpha)\,\Theta_r(\alpha^{-1})=\mathrm{T}_r(\alpha),\quad A_s(\alpha)\,\Theta_s(\alpha)\,\Theta_s(\alpha^{-1})=\mathrm{T}_s(\alpha)$$

und
$$\frac{\varepsilon_r(\alpha)\,\varepsilon_r(\alpha^{-1})}{A_r(\alpha)^\lambda}=\mathfrak{e}_r(\alpha),\quad \frac{\varepsilon_s(\alpha)\,\varepsilon_s(\alpha^{-1})}{A_s(\alpha)^\lambda}=\mathfrak{e}_s(\alpha)$$

so wird die gefundene Gleichung folgendermaafsen dargestellt

$$\mathfrak{e}_r(\alpha)\,\mathrm{T}_r(\alpha)^\lambda-\mathfrak{e}_s(\alpha)\,\mathrm{T}_s(\alpha)^\lambda=\frac{\varepsilon(\alpha)^2(2-\alpha-\alpha^{-1})^{2m\lambda-2\mu}(\alpha^r+\alpha^{-r}-\alpha^s-\alpha^{-s})\,\mathrm{T}(\alpha)^{2\lambda}}{(2-\alpha^r-\alpha^{-r})\,(2-\alpha^s-\alpha^{-s})}$$

Auf der rechten Seite dieser Gleichung enthält das, was mit $\mathrm{T}(\alpha)^{2\lambda}$ multiplicirt ist den Faktor $2-\alpha-\alpha^{-1}$ genau $(2m\lambda-2\mu-1)$ mal, oder weil $2\mu-1=\lambda$ ist, genau $(2m-1)\lambda$ mal, denn $\alpha^r+\alpha^{-r}-\alpha^s-\alpha^{-s}$, so wie auch $2-\alpha^r-\alpha^{-r}$ und $2-\alpha^s-\alpha^{-s}$ enthalten diesen Faktor $2-\alpha-\alpha^{-1}$ jede einmal; aufserdem enthält dieser Ausdruck nur noch Einheiten. Dividirt man also durch die Einheit $\mathfrak{e}_r(\alpha)$ so kann man dieselbe Gleichung auch so darstellen:

$$\mathrm{T}_r(\alpha)^\lambda - \frac{\mathfrak{e}_s(\alpha)}{\mathfrak{e}_r(\alpha)}\mathrm{T}_s(\alpha)^\lambda = \mathrm{E}_1(\alpha)(2-\alpha-\alpha^{-1})^{(2m-1)\lambda}\mathrm{T}(\alpha)^{2\lambda}$$

Macht man aus dieser Gleichung eine Congruenz nach dem Modul λ^2, so erhält man

$$\mathrm{T}_r(\alpha)^\lambda \equiv \frac{\mathfrak{e}_s(\alpha)}{\mathfrak{e}_r(\alpha)}\mathrm{T}_s(\alpha)^\lambda, \quad \text{mod. } \lambda^2.$$

Nun ist, wie oben gezeigt worden, $\mathrm{T}_r(\alpha)$ einer nichtcomplexen ganzen Zahl congruent für den Modul λ, woraus folgt, dafs $\mathrm{T}_r(\alpha)^\lambda$ einer nichtcomplexen Zahl congruent sein mufs, für den Modul λ^2; diese nichtcomplexe Zahl sei a und die entsprechende, welcher $\mathrm{T}_s(\alpha)^\lambda$ congruent sein mufs nach dem Modul λ^2 sei b, so hat man

$$a \equiv \frac{\mathfrak{e}_s(\alpha)}{\mathfrak{e}_r(\alpha)}b, \quad \text{mod. } \lambda^2$$

also die Einheit $\frac{\mathfrak{e}_s(\alpha)}{\mathfrak{e}_r(\alpha)}$ ist einer nichtcomplexen Zahl congruent nach dem Modul λ^2. Wenn nun zu den beiden Voraussetzungen, welche im §. 1 gemacht worden sind, noch als dritte Voraussetzung angenommen wird, d a f s d i e $\nu\lambda^{\text{te}}$ B e r n o u l l i s c h e Z a h l n i c h t d u r c h λ^3 t h e i l b a r s e i n s o l l, so folgt, vermöge des im §. 2 bewiesenen Satzes, dafs diese Einheit nothwendig eine λ^{te} Potenz einer andern Einheit sein mufs, also

$$\frac{\mathfrak{e}_s(\alpha)}{\mathfrak{e}_r(\alpha)} = \mathfrak{E}(\alpha)^\lambda.$$

Setzt man nun

$$\mathrm{T}_r(\alpha) = U_1, \quad -\mathfrak{E}(\alpha)\mathrm{T}_s(\alpha) = V_1, \quad \mathrm{T}(\alpha)^2 = W_1$$

so hat man endlich

$$U_1^\lambda + V_1^\lambda = E_1(\alpha)(2-\alpha-\alpha^{-1})^{(2m-1)\lambda}W_1^\lambda$$

eine Gleichung genau von derselben Form als die vorgegebene. Die vorgegebene Gleichung zieht nun nicht nur diese eine, sondern mit ihr zugleich eine unendliche Reihe von Gleichungen derselben Form nach sich, weil auf die neu entstandene Gleichung immer wieder dieselbe Methode angewendet werden kann. Dafs diese unendliche Reihe von Gleichungen auf einen Widerspruch führt, läfst sich hier minder leicht aus der Vergleichung der Gröfse der Normen der complexen Zahlen U, V, W mit den Normen von U_1, V_1, W_1 nachweisen, als durch die Vergleichung der Anzahl aller verschiedenen idealen Primfaktoren, welche die complexe Zahl W enthält mit der Anzahl aller verschiedenen in W_1 enthaltenen Primfaktoren, und zwar einfach durch den Nachweis, dafs W_1 nothwendig weniger verschiedene ideale Primfaktoren enthalten mufs als W. Dies ergiebt sich sehr leicht aus den beiden Gleichungen (A) und (B), aus welchen man zunächst

$$W = \mathrm{T}(\alpha)\,\Theta_1(\alpha)\,\Theta_2(\alpha)\,\Theta_3(\alpha)\,.....\,\Theta_{\lambda-1}(\alpha)$$

hat. Weil nun die complexen Zahlen $\mathrm{T}(\alpha)$, $\Theta_1(\alpha)$, $\Theta_2(\alpha)$ .... je zwei relative Primzahlen sind, so folgt dafs $\mathrm{T}(\alpha)$ nur in dem einen Falle, alle verschiedenen idealen Primfaktoren des W enthalten kann, wenn die $\lambda - 1$ complexen Zahlen $\Theta_1(\alpha)$, $\Theta_2(\alpha)$ .... $\Theta_{\lambda-1}(\alpha)$ alle nur complexe Einheiten sind, oder was dasselbe ist, wenn

$$\frac{U + \alpha^r V}{1 - \alpha^r}$$

eine complexe Einheit ist, für alle Werthe des $r = 1, 2, 3, \ldots \lambda - 1$. Wegen der allgemeinen Eigenschaft aller complexen Einheiten, nach welcher $E(\alpha) = \alpha^k\, E(\alpha^{-1})$ ist, müfste also sein

$$\frac{U + \alpha^r V}{1 - \alpha^r} = \frac{\alpha^k (U + \alpha^{-r} V)}{1 - \alpha^{-r}}$$

oder wenn man diese Gleichung vereinfacht:

$$U(1 + \alpha^{k+r}) + V(\alpha^k + \alpha^r) = 0,$$

und weil nach der Gleichung (B), $U + V \equiv 0$, mod. λ ist, so müfste

$$1 + \alpha^{k+r} - (\alpha^k + \alpha^r) \equiv 0, \quad \text{mod. } \lambda$$

sein, oder

$$(1 - \alpha^k)(1 - \alpha^r) \equiv 0, \quad \text{mod. } \lambda,$$

welches unmöglich ist, wenn man von dem einen Falle $\lambda = 3$ absieht, denn $k = 0$ kann nicht Statt haben, weil sonst $U + V = 0$ sein müfste. Also $\mathrm{T}(\alpha)$ enthält weniger verschiedene ideale Primfaktoren in sich als W, und weil $W_1 = \mathrm{T}(\alpha)^2$ ist, so enthält W_1 nur alle verschiedene Primfaktoren des $\mathrm{T}(\alpha)$, also weniger als W, was zu beweisen war. In der unendlichen Reihe von Gleichungen, welche die vorgegebene Gleichung nach sich zieht, müfste also in der Reihe der complexen Zahlen W, W_1, W_2, W_3 und so fort in's unendliche jede folgende eine kleinere Anzahl idealer Primfaktoren enthalten als die vorhergehende, es müfste also, weil die in W enthaltene Anzahl idealer Primfaktoren eine endliche ist, eine von dieser Zahl anfangende unendliche Reihe absoluter ganzer Zahlen geben, in welcher jede folgende kleiner als die vorhergehende wäre, welches unmöglich ist.

Hiermit ist also die Richtigkeit des Fermatschen Satzes bewiesen für alle diejenigen λ^{ten} Potenzen, in welchen λ folgenden drei Bedingungen genügt:

1. dafs der erste Faktor der Klassenanzahl der aus den λ^{ten} Wurzeln der Einheit gebildeten idealen complexen Zahlen den Faktor λ einmal enthält und mithin eine der ersten $\frac{\lambda-3}{2}$ Bernoullischen Zahlen (die ν^{te}) congruent Null ist für den Modul λ.
2. dafs es irgend einen complexen idealen Modul gebe, für welchen die bestimmte Einheit $E_\nu(\alpha)$ einer λ^{ten} Potenz nicht congruent sei.
3. dafs die $\nu\lambda^{\text{te}}$ Bernoullische Zahl nicht durch λ^3 theilbar sei.

Es ist nicht schwer, aber namentlich für gröfsere Werthe des λ etwas langwierig, diejenigen Zahlen λ, für die irgend welche der ersten $\frac{\lambda-3}{2}$ Bernoullischen Zahlen durch λ theilbar sind, zu prüfen, ob sie diesen drei Bedingungen genügen. Diese Prüfung habe ich für die drei Zahlen $\lambda = 37$, $\lambda = 59$ und $\lambda = 67$ wirklich ausgeführt, und habe gefunden, dafs für dieselben diese drei Bedingungen wirklich erfüllt sind, wodurch die Richtigkeit des Fermatschen Satzes nun für alle Potenzen, deren Exponenten innerhalb des ersten Hunderts liegen, bewiesen ist. Dafs diese drei Zahlen die erste Bedingung erfüllen, im ersten Faktor der Klassenanzahl einmal enthalten zu sein, geht unmittelbar aus den von mir früher berechneten Zahlenwerthen dieses ersten Faktors der Klassenanzahl für alle Primzahlen λ innerhalb des ersten Hunderts hervor, welche in Liouville's Journal Bd. 16, pag. 473 gegeben sind, es handelte sich also nur noch um die Untersuchung der zweiten und der dritten Bedingung. Für $\lambda = 37$, wo die sechzehnte Bernoullische Zahl durch 37 theilbar ist, also $\nu = 16$, ist der ideale Primfaktor der Zahl 149 ein solcher, für welchen als Modul Ind $E_{16}(\alpha)$ nicht congruent Null ist, und zwar ist Ind $E_{16}(\alpha) \equiv 24$, mod. 37, in Beziehung auf denjenigen idealen Primfaktor der Zahl 149, welcher zur Substitution $\alpha = 17$ gehört, wenn die in $E_{16}(\alpha)$ enthaltene primitive Wurzel von 37, $\gamma = 2$ gewählt wird. Ferner ist die 16. 37te Bernoullische Zahl nicht congruent Null, sondern congruent 35. 37. 37 nach dem Modul $(37)^3$. Für $\lambda = 59$, wo $\nu = 22$, ist in Beziehung auf denjenigen idealen Primfaktor der Primzahl 709, welcher zur Substitution $\alpha = 385$ gehört, und für $\gamma = 2$, Ind $E_{22}(\alpha) \equiv 50$; mod. 59, und die 22. 59te Bernoullische Zahl congruent 41. 59. 59 nach dem Modul $(59)^3$. Endlich für $\lambda = 67$, wo $\nu = 29$, ist für denjenigen idealen Primfaktor der Primzahl 269, welcher zur Substitution $\alpha = 47$ gehört und für $\gamma = 2$, Ind $E_{29}(\alpha) \equiv 4$ mod. 67, und die 29. 67te Bernoullische Zahl ist congruent 49. 67. 67 nach dem Modul $(67)^3$. Die Berechnung der Indices der Einheit $E_\nu(\alpha)$ habe ich mit Hülfe des *Canon arithmeticus* ausgeführt, die Untersuchung der sehr hohen Bernoullischen Zahlen aber durch besondere Kunstgriffe, welche hier näher zu entwickeln zu weitläufig sein würde. Alle diese Rechnungen sind entweder so eingerichtet worden, dafs sie die Controlle ihrer Richtigkeit in sich selbst hatten, oder sie sind auf zwei verschiedene Weisen ausgeführt worden, so dafs die Richtigkeit derselben verbürgt werden kann.

Über die allgemeinen Reciprocitätsgesetze der Potenzreste

Monatsberichte der Königlichen Preußischen Akademie der Wissenschaften zu Berlin aus dem Jahre 1858, 158–171

18. Februar. Gesammtsitzung der Akademie.

Hr. Kummer las den ersten Theil einer Abhandlung „Über die allgemeinen Reciprocitätsgesetze der Potenzreste."

In dieser Abhandlung giebt er den ersten strengen Beweis der Reciprocitätsgesetze in der Ausdehnung wie er sie, als durch Induktion gefunden, der Akademie bereits im Mai 1850 mitgetheilt hat, nämlich für die Reste und Nichtreste λter Potenzen, wo λ Primzahl und nur der einzigen Bedingung unterworfen ist, dafs sie nicht in einer der ersten $\frac{1}{2}(\lambda-3)$ Bernoullischen Zahlen als Faktor des Zählers enthalten sei. Dieser Beweis beruht der Hauptsache nach auf den Principien des zweiten Gaufsischen Beweises des quadratischen Reciprocitätsgesetzes in den *Disquisitiones arithmeticae, sectio quinta*, und kann als eine Verallgemeinerung desselben angesehen werden. So wie nämlich aus der Theorie der quadratischen Formen und zwar aus der Vergleichung der wirklich vorhandenen *Genera* derselben mit denen, welche möglicherweise Statt haben könnten bei Gaufs das *theorema fundamentale* gefolgert wird, so werden hier zum Beweise des unter den Resten und Nichtresten der λten Potenzen Statt findenden Reciprocitätsgesetzes bestimmte Formen des λten Grades mit λ Unbestimmten angewendet, oder vielmehr die solchen Formen entsprechenden wirklichen und idealen complexen Zahlen.

Die Theorie der für diesen Zweck zu benutzenden complexen Zahlen hat die Theorie der aus den Wurzeln der Gleichung $\alpha^\lambda = 1$ gebildeten zur Grundlage. Als höhere Irrationalitäten enthalten diese complexen Zahlen aufserdem die Wurzeln einer Gleichung des λten Grades, deren Coefficienten complexe Zahlen jener niederen Theorie sind, und zwar einer solchen Gleichung, welche sich auf die reine Gleichung des λten Grades

$$w^\lambda = D(\alpha) \tag{1}$$

zurückführen läfst. Die Zahl $D(\alpha)$, welche als eine wirkliche der Bedingung $D(\alpha) \equiv 1$, mod. $(1-\alpha)$, genügende complexe Zahl der niederen Theorie angenommen wird, ist die Determinante der höheren Theorie. Aus einer Wurzel w der Gleichung (1) wird nun folgender rationale und ganze Ausdruck gebildet

$$z = (1-\alpha)(1 + w + w^2 + \cdots\cdots + w^{\lambda-1}) \tag{2}$$

oder

$$z = \frac{(1-\alpha)\bigl(1-D(\alpha)\bigr)}{1-w},$$

welcher selbst eine Wurzel folgender Gleichung des λten Grades ist:

$$z^\lambda - \lambda\varrho z^{\lambda-1} + \frac{\lambda(\lambda-1)}{1\cdot 2}\varrho^2\bigl(1-D(\alpha)\bigr)z^{\lambda-2} - \ldots - \varrho^\lambda\bigl(1-D(\alpha)\bigr)^{\lambda-1} = 0 \tag{3}$$

wo der Kürze wegen $1-\alpha$ durch ϱ bezeichnet ist. Die λ verschiedenen Wurzeln dieser Gleichung sind in folgender Form enthalten:

$$z_\varkappa = \frac{\varrho\bigl(1-D(\alpha)\bigr)}{1-w\alpha^\varkappa} \tag{4}$$

für $\varkappa = 0, 1, 2 \ldots (\lambda-1)$. Es wird nun bewiesen, dafs jede ganze rationale Funktion der Wurzeln $z, z_1, z_2, \ldots . z_{\lambda-1}$ als lineare Funktion dieser Wurzeln sich darstellen läfst und demgemäfs wird den aus der Gleichung (3) hervorgehenden complexen Zahlen, welche kurz als complexe Zahlen in z bezeichnet werden, folgende Form gegeben:

5 $F(z) = C + Az + A_1 z_1 + A_2 z_2 + \ldots + A_{\lambda-1} z_{\lambda-1}$
wo $C, A, A_1 \ldots A_{\lambda-1}$ als einzige Irrationalität die Einheitswurzel α enthalten, also complexe Zahlen in α sind.

Jede ganze complexe Zahl $F(z)$ läſst sich wie leicht zu sehen auch in folgende Form bringen:

$$F(w) = B + B_1 w + B_2 w^2 + \ldots. + B_{\lambda-1} w^{\lambda-1}$$

wo $B, B_1, \ldots B_{\lambda-1}$ ganze complexe Zahlen in α sind. Nennt man nun einen Ausdruck $F(w)$ von dieser Form eine complexe Zahl in w, so ist jede ganze complexe Zahl in z zugleich eine ganze complexe Zahl in w. Man kann auch umgekehrt jede ganze complexe Zahl in w als complexe Zahl in z darstellen, aber im allgemeinen nicht als ganze, sondern nur als gebrochene, weil die Ausdrücke von $w, w^2 \ldots. w^{\lambda-1}$ durch $z, z_1, \ldots z_{\lambda-1}$ die Zahl λ in den Nennern enthalten. Damit eine ganze complexe Zahl in w auch eine ganze complexe Zahl in z sei, muſs unter den Coefficienten derselben ein System von $\lambda - 1$ Bedingungs-Congruenzen für die Moduln $\varrho, \varrho^2, \varrho^3 \ldots. \varrho^{\lambda-1}$, Statt haben. Die ganzen complexen Zahlen in w sind daher die allgemeineren, die ganzen complexen Zahlen in z die specielleren; alle Eigenschaften jener müssen auch diesen zukommen, aber diese haben auſserdem gewisse besondere Eigenschaften für sich, und zwar sind es grade diese besonderen Eigenschaften der complexen Zahlen in z, welche für die Ergründung der allgemeinen Reciprocitätsgesetze von der gröſsten Wichtigkeit sind.

Zu denjenigen Eigenschaften, welche die complexen Zahlen in z mit denen in w gemein haben, gehören die idealen Primfaktoren, welche für beide Theorien dieselben sind, bei deren Untersuchung daher die Form der complexen Zahlen in w als die einfachere zu Grunde gelegt wird. Bei der Frage wie die Primzahlen in α in der Theorie der complexen Zahlen in w weiter zu zerlegen sind, damit man die wahren Primfaktoren in dieser höheren Theorie erhalte, sind zunächst diejenigen bestimmten Primfaktoren von der Form $f(\alpha)$, welche in $\lambda D(\alpha)$ enthalten sind auszusondern und alsdann die übrigen in zwei besondere Arten zu unterscheiden, je nachdem für sie $D(\alpha)$ ein Nichtrest oder ein Rest einer λten Potenz ist. Von den complexen Primzahlen der ersten Art, welche durch $\psi(\alpha)$ bezeichnet werden, für welche also nach dem Legendre'schen Zeichen, wenn dasselbe auf λte Potenzreste angewendet wird, $\left(\frac{D(\alpha)}{\psi(\alpha)}\right)$ nicht $= 1$ ist, wird gezeigt, dafs sie auch in der Theorie der complexen Zahlen in w überall als Primzahlen anzusehen sind. Von den mit $\phi(\alpha)$ zu bezeichnenden complexen Primzahlen der zweiten Art, für welche $\left(\frac{D(\alpha)}{\phi(\alpha)}\right) = 1$ ist, wird gezeigt, wie sie in der Theorie der complexen Zahlen in w in λ ideale Primfaktoren zerlegt werden und wie jeder dieser Primfaktoren zu einer der λ Wurzeln der Congruenz

$$\xi^{\lambda} \equiv D(\alpha), \text{ mod. } \phi(\alpha), \qquad 6$$

gehört; auch werden die in einer gegebenen wirklichen complexen Zahl $F(w)$ oder $F(z)$ enthaltenen idealen Primfaktoren durch Congruenzbedingungen überall vollständig bestimmt.

Eine Definition der idealen Primfaktoren der in $\lambda D(\alpha)$ enthaltenen complexen Primzahlen von der Form $f(\alpha)$ wird nicht gegeben und zwar nicht allein darum weil sie für den vorliegenden Zweck entbehrlich ist, sondern hauptsächlich auch darum, weil diese complexen Primzahlen von der Form $f(\alpha)$ in der höheren Theorie der complexen Zahlen in w oder z im allgemeinen in wahre ideale Primfaktoren sich gar nicht zerlegen lassen, welche im vollen Sinne diesen Namen verdienen.

In Betreff der Äquivalenz der idealen Zahlen macht sich zuerst die Besonderheit der complexen Zahlen in z geltend. Wenn nämlich unter idealen complexen Zahlen in dieser höheren Theorie nur solche verstanden werden, welche aus den vollständig definirten idealen Primfaktoren zusammengesetzt sind, deren in Beziehung auf w oder z genommene Normen also keine Faktoren des $\lambda D(\alpha)$ enthalten, und wenn zwei ideale complexe Zahlen der höheren Theorie als äquivalent definirt werden, wenn sie mit einer und derselben dritten zusammengesetzt wirkliche complexe Zahlen als Produkte ergeben: so ist diese Bedingung eine andere wenn diese Produkte der idealen Zahlen wirkliche complexe Zahlen in z sein müssen, als wenn nur verlangt wird, dafs sie wirkliche complexe Zahlen in w sein sollen. Alle äquivalenten idealen Zahlen in z sind nothwendig auch äquivalent in w, aber nicht umgekehrt. In der ersteren Theorie sind mehr verschiedene Klassen idealer Zahlen vorhanden, als in der letzteren, denn alle diejenigen wirklichen Zahlen in w, welche in z sich nicht als ganze, sondern nur als gebrochene, mit dem Nenner λ oder $\varrho^{\varkappa}$ darstellen lassen, sind in der Theorie der Zahlen in z nur ideal. Die beiden Theorieen der complexen Zahlen in z und in w stehen überhaupt genau in demselben Verhältnisse zu einander, wie in der Gaufsischen Theorie der quadratischen Formen der *ordo primitivus* zu einem *ordo derivatus.*

Wenn nun alle äquivalenten idealen Zahlen in z, (welche nur aus den vollständig definirten idealen Primfaktoren bestehen) einer und derselben Klasse, nichtäquivalente aber verschiedenen Klassen zugetheilt werden, so erhält man für eine jede gegebene Determinante $D(\alpha)$ eine endliche bestimmte Anzahl verschiedener Klassen. Die weitere Eintheilung dieser Klassen in die *Genera* beruht alsdann auf den Charakteren, welche den verschiedenen Klassen der idealen Zahlen, sowohl in Beziehung auf den Modul λ, als auch in Beziehung auf alle verschiedenen Primzahlen, welche in der Determinante $D(\alpha)$ enthalten sind, zukommen. Wenn $\phi(z)$ eine ideale complexe Zahl in z bezeichnet und die Norm derselben, nämlich das Produkt der λ conjugirten zu $\phi(z)$, als complexe Zahl in α, durch $\phi(\alpha)$ bezeichnet wird, so haben alle äquivalenten idealen Zahlen $\phi(z)$, also alle einer und derselben Klasse angehörenden, folgende gemeinsame Charaktere in Beziehung auf den Modul λ:

$$\left.\begin{aligned} \frac{d_0^3 l\ \phi(e^v)}{dv^3} &\equiv c_3 \\ \frac{d_0^5 l\ \phi(e^v)}{dv^5} &\equiv c_5 \\ \frac{d_0^{\lambda-2} l\ \phi(e^v)}{dv^{\lambda-2}} &\equiv c_{\lambda-2} \\ \frac{1 - N\phi(\alpha)}{\lambda} &\equiv c_{\lambda-1} \end{aligned}\right\} \quad \text{mod. } \lambda \qquad 7$$

und folgende Charaktere in Beziehung auf alle verschiedenen Primfaktoren der Determinante $D(\alpha)$, welche durch $f(\alpha)$, $f_1(\alpha)$, ... bezeichnet werden sollen:

$$\left(\frac{\phi(\alpha)}{f(\alpha)}\right) = \alpha^{\varkappa}$$

$$\left(\frac{\phi(\alpha)}{f_1(\alpha)}\right) = \alpha^{\varkappa_1}$$

etc.

nämlich für alle idealen Zahlen derselben Klasse haben c_3, c_5, ... $c_{\lambda-2}$, $c_{\lambda-1}$ und $\varkappa$, $\varkappa_1$, ... dieselben Werthe nach dem Modul λ. Die Charaktere c_3, c_5, ... $c_{\lambda-2}$, $c_{\lambda-1}$ sind von den Einheiten, mit welchen $\phi(\alpha)$, als Norm der idealen complexen Zahl $\phi(z)$, beliebig behaftet sein kann, vollkommen unabhängig. Diese sind darum an sich feste Charaktere. Die Charaktere $\varkappa$, $\varkappa_1$, ... aber ändern im allgemeinen ihre Werthe, wenn $\phi(\alpha)$ mit anderen Einheiten behaftet gewählt wird; damit diese zu festen Charakteren werden, mufs über $\phi(\alpha)$ noch eine Festsetzung getroffen werden, und zwar die, dafs die Norm der idealen Zahlen $\phi(z)$ stets in der primären Form genommen werden soll, welche durch folgende zwei Bedingungen bestimmt ist: erstens, dafs $\phi(\alpha)$ einer nicht complexen ganzen Zahl congruent sei, nach dem Modul $(1-\alpha)^2 = \varrho^2$, zweitens dafs $\phi(\alpha)\,\phi(\alpha^{-1})$ einer nichtcomplexen ganzen Zahl congruent sei, nach dem Modul λ. Zu bemerken ist noch, dafs in den Fällen, wo die Norm $\phi(\alpha)$ eine in der Theorie der complexen Zahlen in α ideale Zahl ist, deren $h$te Potenz wirklich ist, für $\phi(\alpha)$ der Ausdruck $\sqrt[h]{\phi(\alpha)^h}$ zu nehmen ist; dieser giebt alle Charaktere unzweideutig, sobald h nicht durch λ theilbar ist, welches nach der Voraussetzung, dafs λ in keiner der ersten $\frac{\lambda-3}{2}$ Bernoullischen Zahlen enthalten ist, nicht Statt hat. Die so dargestellte ideale Zahl $\phi(\alpha)$ heifst primär, wenn $\phi(\alpha)^h$ primär ist.

Die Anzahl der Einzelcharaktere ist gleich $\frac{\lambda-1}{2}+r$, wenn r die Anzahl der verschiedenen in $D(\alpha)$ enthaltenen complexen Primzahlen bezeichnet. Die Anzahl der Gesammt-Charaktere ist demnach gleich der $\left(\frac{\lambda-1}{2}+r\right)$ten Potenz von λ, und dieses ist zugleich die Anzahl der angebbaren *Genera,* wenn alle Klassen, welche dieselben Charaktere haben, zu einem *Genus* gerechnet werden. Wegen des Umstandes aber, dafs gewisse dieser angebbaren *Genera* gar keine Klassen idealer Zahlen enthalten, sind von diesen die wirklich vorhandenen wohl zu unterscheiden. Die Ermittelung der Anzahl der wirklich vorhandenen *Genera,* welche für den zu gebenden Beweis des allgemeinen Reciprocitätsgesetzes nothwendig ist, bildet den schwie-

11*

rigsten Theil der gegenwärtigen Untersuchung. Sie wird ähnlich wie bei Gauſs mit Hülfe einer besonderen Art von Klassen der idealen Zahlen geleistet, welche den *Classes ancipites* analog sind und darum mit dem Namen der Ambigen bezeichnet werden.

Eine ambige complexe Zahl in z wird eine solche genannt, welche ihren conjugirten äquivalent ist, eine ambige Klasse eine solche, welche nur ambige complexe Zahlen enthält. Die Untersuchung dieser Ambigen stützt sich nun hauptsächlich auf folgenden Satz: Jede ideale Ambige ist in einer wirklichen complexen Zahl $f(z)$ so enthalten, daſs $f(z)$ alle idealen Primfaktoren enthält, aus welchen die Ambige besteht und auſser diesen keinen der definirten idealen Primfaktoren weiter, deren Norm aber auſser der Norm der Ambigen noch Faktoren von $\lambda D(\alpha)$ enthalten kann. Die idealen Ambigen werden nun als in diesem Sinne in den wirklichen complexen Zahlen enthalten betrachtet und es wird zunächst gezeigt, daſs wenn eine wirkliche complexe Zahl $f(z)$ eine Ambige enthält, und man stellt $f(z)$ als complexe Zahl in w dar, wobei eine Potenz von ϱ als gemeinschaftlicher Faktor aller Glieder heraustreten kann, so daſs

9 $$f(z) = \varrho^m f(w)$$

ist, die Norm von $f(w)$ den Faktor ϱ nicht weiter enthalten kann. Die complexe Zahl $f(w)$ ist nun eine solche, welche die in $f(z)$ enthaltene Ambige ebenfalls enthält und welche von den fremdartigen complexen Faktoren ϱ, der Zahl λ, vollständig gereinigt ist.

Um die Zahl $f(w)$, welche die Ambige enthält, weiter auch von den fremdartigen Faktoren des $D(\alpha)$ zu befreien, muſs man diese Determinante in ihre Primfaktoren zerlegen. Sei

10 $$D(\alpha) = f(\alpha)^h \cdot f_1(\alpha)^{h_1} \cdot f_2(\alpha)^{h_2} \ldots\ldots,$$

wo $f(\alpha)$, $f_1(\alpha)$ u. s. w. die in $D(\alpha)$ enthaltenen verschiedenen Primfaktoren bezeichnen, ferner

11 $$u^\lambda = f(\alpha)^h, \quad u_1^\lambda = f_1(\alpha)^{h_1}, \text{ etc.}$$

also $$w = u \cdot u_1 \cdot u_2 \ldots\ldots$$

Es wird nun bewiesen, daſs wenn $f(w)$ eine Ambige enthält, diese complexe Zahl in w sich in folgender Form darstellen lassen muſs:

12 $$f(w) = u^n\, u_1^{n_1}\, u_2^{n_2} \ldots f(u, u_1\, u_2 \ldots),$$

wo $f(u, u_1, u_2 \ldots)$ eine solche complexe Zahl in $u, u_1, u_2 \ldots$ ist, welche, wenn man von Faktoren α^n absieht, die hinzutreten, bei allen Veränderungen dieser λten Wurzeln nur λ verschiedene Werthe erhält und deren Norm, nämlich das Produkt dieser λ verschiedenen conjugirten Werthe, keinen Faktor der Determinante $D(\alpha)$ weiter enthält. Die complexe Zahl $f(u, u_1, u_2 \ldots)$ ist nun die in $f(z)$ und in $f(w)$ enthaltene ideale Ambige selbst, und zwar als wirkliche complexe Zahl in der Theorie der complexen Zahlen in $u, u_1, u_2 \ldots.$ dargestellt, welche als eine höhere Stufe der complexen Zahlen in w anzusehen ist. In dem besonderen Falle, wo $D(\alpha)$ nur eine Primzahl enthält, hat man $u = w$ und $u_1, u_2 \ldots$ kommen gar nicht vor, in diesem Falle also läfst sich jede ideale Ambige in z als wirkliche complexe Zahl in w darstellen.

Die Auffindung aller idealen Ambigen in z hat nun, da sie alle als wirkliche complexe Zahlen in $u, u_1, u_2 \ldots$ dargestellt werden können, keine besonderen Schwierigkeiten; die Untersuchung ihrer Äquivalenz aber, und die Aufstellung des vollständigen Systems der nichtäquivalenten Klassen der Ambigen, oder auch nur ihrer Anzahl, ist nicht so leicht durchzuführen. Hierzu gehört nämlich nothwendig die Kenntnifs der Hauptsätze über die Einheiten dieser complexen Theorie, welche aus den von Hrn. Dirichlet der Akademie im März des Jahres 1846 mitgetheilten Resultaten seiner Untersuchungen über die in der allgemeinen Theorie der zerlegbaren Formen vorkommenden Einheiten geschöpft wird. Mit Hülfe dieser Dirichletschen Sätze wird die genaue Anzahl der ambigen Klassen gleich $\lambda^{\frac{\lambda-1}{2}+r-1}$ gefunden, also genau gleich dem λten Theile der Gesammtcharaktere oder was dasselbe ist, der angebbaren *Genera.*

Nachdem dieser Hauptpunkt festgestellt ist, wird gezeigt, dafs wenn die Anzahl der ambigen Klassen in der Theorie der complexen Zahlen in z durch A und die Anzahl aller Klassen durch H bezeichnet wird, das *Genus principale,* nämlich dasjenige, dessen Charaktere alle congruent Null sind, nach dem Modul λ, mindestens $\frac{H}{A}$ Klassen enthalten mufs. Da ferner leicht zu zeigen ist, dafs alle wirklich vorhandenen *Genera* nothwendig gleich viele Klassen enthalten, also jedes mindestens $\frac{H}{A}$ Klassen, so folgt, dafs die Anzahl der wirklich vorhandenen *Genera* höchstens gleich A sein kann, also höchstens gleich dem λten Theile aller angebbaren *Genera* oder Gesammt-Charaktere.

Das allgemeine Reciprocitätsgesetz kann aus diesem Satze nicht auf so einfache Weise gefolgert werden, als das *theorema fundamentale* für die quadratischen Reste bei Gaufs aus dem entsprechenden Satze für die quadratischen Formen abgeleitet wird, namentlich deshalb, weil hier nicht nur die Reste von den Nichtresten, sondern auch $\lambda - 1$ verschiedene Arten von Nichtresten unter einander zu unterscheiden sind. Wegen dieses Umstandes reicht auch der gefundene Satz, dafs die Anzahl der wirklichen *Genera* nicht gröfser ist, als der λte Theil der möglichen, hier nicht aus, sondern es ist nöthig, dafs die Anzahl der wirklichen *Genera* genau gefunden werde, wozu andere Principien erforderlich sind, als welche Gaufs zur Ergründung dieser wichtigen Frage für die quadratischen Formen angewendet hat. Es ist hier wieder die Beziehung der complexen Zahlen in z zu den beiden ihr übergeordneten Stufen der complexen Zahlen in w und derer in $u, u_1, u_2 \ldots$, welche in Verbindung mit dem bereits gefundenen Satze, die genaue Anzahl der wirklich vorhandenen *Genera* ergiebt.

Da der besondere Fall, wo die Determinante $D(\alpha)$ nur eine Potenz einer einzigen Primzahl ist, multiplicirt mit einer beliebigen Einheit, also

13 $$D(\alpha) = E(\alpha) f(\alpha)^h,$$

wo der nicht durch λ theilbare Exponent h so zu wählen ist, dafs wenn $f(\alpha)$ ideal sein sollte, $f(\alpha)^h$ und somit die Determinante $D(\alpha)$, wirklich wird, zu dem Beweise des allgemeinen Reciprocitätsgesetzes fast vollständig ausreicht, so wird dieser zunächst behandelt. Es besteht in diesem Falle, in welchem nur die beiden Stufen der complexen Zahlen in z und derer in w Statt haben, aufser den $\frac{\lambda - 1}{2}$ Charakteren, welche mit $c_3, c_5, \ldots . c_{\lambda-2}, c_{\lambda-1}$ bezeichnet worden sind, nur ein einziger Charakter von der Form

$$\left(\frac{\phi(\alpha)}{f(\alpha)}\right) = \alpha^n,$$

die Anzahl der wirklich vorhandenen *Genera* ist also hier höchstens gleich $\lambda^{\frac{\lambda-1}{2}}$, und sie wird genau gleich $\lambda^{\frac{\lambda-1}{2}}$ sein, wenn für jedes beliebige System von Werthen der Charaktere $c_3, c_5, \ldots c_{\lambda-2}, c_{\lambda-1}$ eine ideale complexe Zahl gefunden werden kann welcher dieselben zukommen. Es wird nun untersucht, in wie weit die wirklichen complexen Zahlen in w, welche ideale Zahlen in z sind, ausreichen um diese *Genera* vollständig auszufüllen; weil aber die Aufgabe eine in w wirkliche complexe Zahl $\phi(w)$ zu finden, deren Norm, $N\phi(w) = \phi(\alpha)$, den ersten $\frac{\lambda-1}{2}$ Bedingungen, welche als Charaktere von $\phi(w)$ bezeichnet sind, nämlich:

$$\begin{aligned} \frac{d^3 l\,\phi(e^v)}{dv^3} &\equiv c_3 \\ \frac{d^5 l\,\phi(e^v)}{dv^5} &\equiv c_5 \\ \frac{d_0^{\lambda-2} l\,\phi(e^v)}{dv^{\lambda-2}} &\equiv c_{\lambda-2} \\ \frac{1 - N\,\phi(\alpha)}{\lambda} &\equiv c_{\lambda-1} \end{aligned} \qquad 14$$

nach dem Modul λ für alle beliebig gegebenen Werthe der Zahlen $c_3, c_5, \ldots c_{\lambda-1}$ genügt, im allgemeinen unendlich viele Lösungen hat, so wird diese Aufgabe noch weiter so beschränkt, dafs auch die Differenzialquotienten der Logarithmen von $\phi(\alpha)$

$$\begin{aligned} \frac{d_0 l\,\phi(e^v)}{dv} &\equiv c_1 \\ \frac{d_0^2 l\,\phi(e^v)}{dv^2} &\equiv c_2 \\ \frac{d_0^4 l\,\phi(e^v)}{dv^4} &\equiv c_4 \\ \frac{d_0^{\lambda-3} l\,\phi(e^v)}{dv^{\lambda-3}} &\equiv c_{\lambda-3} \end{aligned} \qquad 15$$

gegebene Werthe nach dem Modul λ haben sollen. Als die nothwendige und zugleich auch hinreichende Bedingung der Lösbarkeit dieser Aufgabe ergiebt sich, unter der Voraussetzung dafs die Determinante nicht die Eigenschaft hat, für den Modul $(1-\alpha)^2$ einer nicht complexen ganzen Zahl congruent zu sein, für die Zahlen $c_3, c_5 \ldots c_{\lambda-1}$ und $c_1, c_2, c_4 \ldots . c_{\lambda-3}$ folgende lineäre Congruenz:

16 $$c_1 d_{\lambda-1} - c_2 d_{\lambda-2} + c_3 d_{\lambda-3} - c_4 d_{\lambda-4} + \ldots - c_{\lambda-1} d_1 \equiv 0, \text{ mod. } \lambda,$$

in welcher die Coefficienten $d_1, d_2, d_3 \ldots d_{\lambda-1}$ folgende durch die Determinante $D(\alpha)$ allein bestimmte, den Gröfsen $c_1, c_2 \ldots c_{\lambda-1}$ vollständig entsprechende Werthe haben

17 $$d_n \equiv \frac{d_0^n l\, D(e^v)}{dv^n} \text{ mod. } \lambda, \quad \text{für } n = 1, 2, 3, \ldots . \lambda - 2, \text{ und} \quad d_{\lambda-1} \equiv \frac{1 - ND(\alpha)}{\lambda}, \text{ mod. } \lambda.$$

Die gemachte Voraussetzung, dafs $D(\alpha)$ nach dem Modul $(1-\alpha)^2$ einer nichtcomplexen Zahl nicht congruent sein soll, ist mit der identisch, dafs d_1 nicht congruent Null sei für den Modul λ.

Wenn nun die Charaktere $c_3, c_5, \ldots c_{\lambda-2}, c_{\lambda-1}$ allein gegeben sind, so kann durch passende Wahl der Zahlen $c_1, c_2, c_4, \ldots c_{\lambda-3}$ der Bedingungscongruenz (16) immer genügt werden, mit Ausnahme des einen Falles, wo die Determinante $D(\alpha)$ so beschaffen ist, dafs die $\frac{\lambda-1}{2}$ Zahlen $d_3, d_5, \ldots d_{\lambda-2}, d_{\lambda-1}$ alle congruent Null sind, in welchem die Charaktere $c_3, c_5, \ldots c_{\lambda-1}$ der Bedingungs-Congruenz

$$c_3 d_{\lambda-3} + c_5 d_{\lambda-5} + \ldots . - c_{\lambda-1} d_1 \equiv 0, \text{ mod. } \lambda,$$

genügen müssen, also nicht alle möglichen Werthe nach dem Modul λ annehmen können. Die Bedingung, dafs die Zahlen $d_3, d_5, \ldots d_{\lambda-1}$ alle congruent Null sind, stimmt, wie aus den in der Abhandlung in Crelle's Journal Bd. 44: Über die Ergänzungssätze zu den allgemeinen Reciprocitätsgesetzen bewiesenen Sätzen hervorgeht, vollkommen mit der überein, dafs $D(\alpha)$ Potenz einer solchen Primzahl $f(a)$ ist, in Beziehung auf welche alle Einheiten ohne Ausnahme λte Potenzreste sind. Das gefundene Resultat läfst sich also folgendermafsen aussprechen: Für alle beliebigen Werthe der Charaktere $c_3, c_5, \ldots . c_{\lambda-1}$ existiren wirkliche *Genera*, wenn die Determinante $D(\alpha)$ nicht die Eigenschaft hat, dafs für den in ihr enthaltenen Primfaktor alle Einheiten λte Potenzreste sind. Da die Anzahl dieser *Genera* gleich $\lambda^{\frac{\lambda-1}{2}}$ ist und, wie gezeigt worden, mehr wirkliche *Genera* nicht Statt haben, so folgt, dafs der besondere Charakter $\varkappa$ in der Gleichung

$$\left(\frac{\varphi(\alpha)}{f(\alpha)}\right) = \alpha^{\varkappa}$$

durch die Charaktere $c_3, c_5, \ldots. c_{\lambda-1}$ vollkommen bestimmt ist.

Es sei nun $\varphi(\alpha)$ irgend eine complexe Primzahl in α, in Beziehung auf welche die Determinante $D(\alpha)$ ein λter Potenzrest ist, welche also als Norm einer idealen Zahl $\varphi(z)$ angesehen werden kann, wenn sie, wie oben für die Normen der idealen Zahlen festgesetzt worden ist, primär genommen wird. Es seien auch die besonderen Werthe der Determinante $D(\alpha)$ ausgeschlossen, welche eine Ausnahme in dem Satze über die Anzahl der wirklich vorhandenen *Genera*, oder vielmehr nur in dem hier gegebenen Beweise desselben begründen würden. Die ideale Zahl $\varphi(z)$, deren Norm $\varphi(\alpha)$ ist, habe nun die beliebig bestimmten Charaktere $c_3, c_5, \ldots c_{\lambda-2}, c_{\lambda-1}$, so giebt es, wie gezeigt worden, eine in der Theorie der complexen Zahlen in w wirkliche Zahl $F(w)$, welche als ideale Zahl in der Theorie der complexen Zahlen in z aufgefafst, genau dieselben Charaktere hat als $\varphi(z)$, welche also auch den letzten Charakter $\varkappa$ mit ihr gemein haben mufs. Die Norm von $F(w)$, wie sie als Produkt der λ conjugirten in w wirklichen complexen Zahlen hervorgeht, ist aber im allgemeinen nicht primär, und mufs insofern $F(w)$, in seiner Eigenschaft als ideale Zahl in z, eine primäre Norm haben mufs, mit einer passenden Einheit multiplicirt werden. Man hat nun

$$\left(\frac{\varphi(\alpha)}{f(\alpha)}\right) = \left(\frac{\varepsilon(\alpha)\, NF(w)}{f(\alpha)}\right)$$

und da, wie leicht zu zeigen, die Norm einer jeden wirklichen complexen Zahl in w ein λter Potenzrest in Beziehung auf jeden Primfaktor der Determinante ist, so erhält man durch Anwendung der in der Abhandlung über die Ergänzungssätze u. s. w. bewiesenen Sätze:

$$\left(\frac{\phi(\alpha)}{f(\alpha)}\right) = \alpha^{\varkappa}$$

wo

$$h\varkappa \equiv c_1 d_{\lambda-1} - c_2 d_{\lambda-2} - c_4 d_{\lambda-4} - \ldots - c_{\lambda-3} d_3 .$$

Ferner hat man wegen der Bedingung, dafs $\phi(z)$ einer der λ idealen Primfaktoren des $\phi(\alpha)$ ist,

$$\left(\frac{D(\alpha)}{\phi(\alpha)}\right) = 1$$

und wenn nach derselben Methode dieses Legendre'sche Zeichen für das nicht primäre $D(\alpha)$ durch dasjenige, welches auf den als primär anzunehmenden Primfaktor $f(\alpha)$ sich bezieht, ausgedrückt wird, so hat man

$$\left(\frac{f(\alpha)}{\phi(\alpha)}\right) = \alpha^{l} ,$$

wo

$$hl \equiv d_1 c_{\lambda-1} - d_2 c_{\lambda-2} - d_4 c_{\lambda-4} - \ldots - d_{\lambda-3} c_3$$

also vermöge der Congruenz (16)

$$h\varkappa \equiv hl$$

und darum

$$\left(\frac{\phi(\alpha)}{f(\alpha)}\right) = \left(\frac{f(\alpha)}{\phi(\alpha)}\right).$$

Diese Gleichung giebt das Reciprocitätsgesetz für zwei primäre Primzahlen $f(\alpha)$ und $\phi(\alpha)$ und zwar unter der Bedingung, dafs $f(\alpha)$ nicht eine solche complexe Primzahl ist, in Beziehung auf welche alle complexen Einheiten λte Potenzreste sind und ferner, dafs $\left(\frac{D(\alpha)}{\phi(\alpha)}\right) = 1$ sei, welche letztere Bedingung durch passende Wahl der in der Determinante $D(\alpha)$ enthaltenen beliebigen Einheit $E(\alpha)$ für jede complexe Primzahl $\phi(\alpha)$ befriedigt werden kann, für welche nicht alle Einheiten λte Potenzreste sind.

Der gegebene Beweis des Reciprocitätsgesetzes gilt daher vollständig für je zwei primäre complexe Primzahlen, deren keine die Eigenschaft hat, dafs für sie alle Einheiten λte Potenzreste sind. Wenn aber eine der zu vergleichenden Primzahlen die genannte besondere Eigenschaft hat, oder auch beide, so zeigt er nur, dafs wenn die eine Zahl Rest der anderen ist, auch die andere Zahl Rest der ersten sein mufs und demzufolge auch, dafs wenn die eine Zahl Nichtrest der anderen ist, auch die andere Nichtrest der ersten sein mufs.

Es giebt in der That solche complexe Primzahlen $f(\alpha)$, für welche alle complexen Einheiten λte Potenzreste sind, dieselben sind jedoch namentlich für die complexen Primfaktoren der Primzahlen $p = 2m\lambda + 1$ ziemlich selten, da z. B. für $\lambda = 5$ unter den complexen Primfaktoren der nichtcomplexen Primzahlen von der Form $10m + 1$ im ersten Tausend gar keine vorkommen.

Um das Wenige zu ergänzen, was dem gegebenen Beweise noch an Vollständigkeit fehlt, mufs man Determinanten, welche zwei verschiedene Primfaktoren $f(\alpha)$ und $f_1(\alpha)$ enthalten, in Anwendung bringen. Da nämlich in diesem Beweise zwar die Reciprocitäts-Beziehung unter den Resten sich unmittelbar ergiebt, aber die unter den verschiedenen Arten der Nichtreste nur durch die der Determinante zugefügte beliebige Einheit hervorgebracht wird: so mufs in dem Falle wo alle Einheiten λte Potenzreste sind und darum keinen Unterschied im Reciprocitäts-Charakter bewirken, die Stelle dieser Einheiten durch andere passende der Determinante hinzuzufügende Zahlen ersetzt werden. Für die zwei verschiedene Primfaktoren enthaltenden Determinanten sind alsdann, wie oben gezeigt worden, complexe Zahlen in u und u_1 zu benutzen. Wendet man auf diese die in dem Vorhergehenden für den specielleren Fall entwickelte Methode mit den nöthigen Modifikationen an, indem man $f(\alpha)$ als eine solche complexe Primzahl nimmt, für welche alle Einheiten λte Potenzreste sind, $f_1(\alpha)$ aber passenden speciellen Bestimmungen unterwirft, so erhält man das Reciprocitätsgesetz zwischen den $\lambda - 1$ verschiedenen Arten der Nichtreste für die complexen Primzahlen $f(\alpha)$ und $\phi(\alpha)$, sowohl in dem Falle, dafs $f(\alpha)$ allein eine solche Primzahl ist, für welche alle Einheiten Reste sind, als auch für den Fall, dafs $f(\alpha)$ und $\phi(\alpha)$ beide diese besondere Eigenschaft haben.

Über die Ergänzungssätze zu den allgemeinen Reciprocitätsgesetzen

Journal für die reine und angewandte Mathematik 56, 270–279 (1859)

In der Abhandlung Bd. 44, pag. 93 dieses Journals, welche denselben Titel führt, habe ich gezeigt, dafs die auf irgend eine ideale oder wirkliche Primzahl $f(\alpha)$, in der Theorie der aus λ^{ten} Wurzeln der Einheit gebildeten complexen Zahlen, sich beziehenden Indices der Zahlen $1-\alpha$, λ und der complexen Einheiten durch die Zahlen der Kreistheilung in einfacher Weise bestimmt werden. Da aber diese Bestimmung nur auf die idealen Primfactoren der nichtcomplexen Primzahlen p, von der Form $p=m\lambda+1$, Anwendung findet, so habe ich durch eine Verallgemeinerung der Theorie der Kreistheilung die entsprechenden Resultate auch für alle diejenigen idealen Primzahlen $f(\alpha)$ gewonnen, welche Primfactoren nichtcomplexer Primzahlen von anderen linearen Formen sind. Ferner habe ich die Indices eines bestimmten unabhängigen Systems von Einheiten, durch welches alle Einheiten sich ausdrücken lassen, auch so dargestellt, dafs sie nicht mehr durch die Zahlen der Kreistheilung, sondern durch die complexe Primzahl $f(\alpha)$ selbst bestimmt werden, auf welche der Index sich bezieht. Seitdem habe ich bei Gelegenheit eines Beweises der allgemeinen Reciprocitätsgesetze für λ^{te} Potenzreste unter je zwei complexen Primzahlen, den ich im Februar 1858 in der Königlichen Akademie der Wissenschaften vorgetragen habe, auch für die Ergänzungssätze zu denselben einige neue, beachtenswerthe Ausdrücke gefunden, welche ich hier kurz entwickeln will.

1.

Setzt man in der bekannten *Lagrange*schen Resolvente der Kreistheilung

$$F(\alpha, x) = x+\alpha x^{g}+\alpha^{2} x^{g^{2}}+\cdots+\alpha^{p-2} x^{g^{p-2}},$$

in welcher α eine primitive Wurzel der Gleichung $\alpha^{\lambda}=1$, x eine primitive Wurzel der Gleichung $x^{p}=1$, λ eine ungrade Primzahl, p eine Primzahl der Form $p=m\lambda+1$ und g eine primitive Wurzel von p ist, statt der

Wurzel α die variable Exponentialgröfse e^v und bildet den λ^{ten} Differentialquotienten des Logarithmus von $F(e^v, x)$, so erhält man, nach Weglassung der mit dem Factor λ behafteten Glieder, die Congruenz:

$$(1.)\qquad \frac{d^\lambda lF(e^v,x)}{dv^\lambda} \equiv \frac{d^\lambda F(e^v,x)}{F(e^v,x)dv^\lambda} - \left(\frac{dF(e^v,x)}{F(e^v,x)dv}\right)^\lambda, \quad (\text{mod.}\ \lambda).$$

Man hat nämlich nach einer bekannten Formel der Differentialrechnung, wenn U eine Function von u, und u eine Function von v ist:

$$(2.)\qquad \frac{d^\lambda U}{dv^\lambda} = C_1\frac{dU}{du} + C_2\frac{d^2U}{du^2} + C_3\frac{d^3U}{du^3} + \cdots + C_\lambda\frac{d^\lambda U}{du^\lambda},$$

wo $C_1, C_2, C_3, \ldots C_\lambda$ aus den Differentialquotienten des u nach v zusammengesetzt sind, und namentlich

$$C_1 = \frac{d^\lambda u}{dv^\lambda}, \quad C_\lambda = \left(\frac{du}{dv}\right)^\lambda$$

ist, und wo, wenn λ Primzahl ist, die Gröfsen $C_2, C_3, \ldots C_{\lambda-1}$ alle den Zahlenfactor λ enthalten. Aus dieser Gleichung folgt die obige Congruenz unmittelbar, wenn $u = F(e^v, x)$, $U = lF(e^v, x) = l(u)$ genommen wird, und wenn man beachtet, dafs $1.2.3\ldots(\lambda-1) \equiv -1$, (mod. λ) ist.

Setzt man nun $v = 0$, so wird:

$$F(1,x) = x + x^g + x^{g^2} + \cdots\cdots\cdots + x^{g^{p-2}} = -1,$$

$$\frac{d_0F(e^v,x)}{dv} = x^g + 2x^{g^2} + 3x^{g^3} + \cdots + (p-2)\,x^{g^{p-2}} = \sum_0^{p-2}{}_h\, h\,x^{g^h},$$

$$\frac{d_0^\lambda F(e^v,x)}{dv^\lambda} = x^g + 2^\lambda x^{g^2} + 3^\lambda x^{g^3} + \cdots + (p-2)^\lambda x^{g^{p-2}} = \sum_0^{p-2}{}_h\, h^\lambda x^{g^h},$$

also vermöge der Congruenz $h^\lambda \equiv h$, (mod. λ),

$$\frac{d_0^\lambda F(e^v,x)}{dv^\lambda} \equiv \frac{d_0F(e^v,x)}{dv} \equiv \sum_0^{p-2}{}_h\, hx^{g^h}, \quad (\text{mod.}\ \lambda);$$

die Congruenz (1.) giebt daher:

$$(3.)\qquad \frac{d_0^\lambda lF(e^v,x)}{dv^\lambda} \equiv -\sum_0^{p-2}{}_h\, hx^{g^h} + \left(\sum_0^{p-2}{}_h\, hx^{g^h}\right)^\lambda, \quad (\text{mod.}\ \lambda).$$

Die λ^{te} Potenz eines Polynoms ist aber, wenn λ Primzahl ist, der Summe der λ^{ten} Potenzen der einzelnen Glieder desselben congruent; man hat daher

$$\left(\sum_0^{p-2}{}_h\, hx^{g^h}\right)^\lambda \equiv \sum_0^{p-2}{}_h\, h^\lambda x^{\lambda g^h}, \quad (\text{mod.}\ \lambda),$$

oder wenn λ, als Potenz der primitiven Wurzel dargestellt,

$$(4.)\qquad \lambda \equiv g^i, \quad (\text{mod.}\ p)$$

35 *

giebt, und wenn h statt h^λ gesetzt wird,

$$\left(\sum_0^{p-2}{}_h\, hx^{g^h}\right)^\lambda \equiv \sum_0^{p-2}{}_h\, hx^{g^{h+i}}, \quad (\text{mod. } \lambda).$$

Verwandelt man h in $h-i$, so wird

$$\sum_0^{p-2}{}_h\, hx^{g^{h+i}} = \sum_0^{p-2}{}_h\, hx^{g^h} - i\sum_0^{p-2}{}_h\, x^{g^h} = \sum_0^{p-2}{}_h\, hx^{g^h} + i,$$

die Congruenz (3.) giebt demnach

$$\frac{d_0^\lambda\, l F(e^v, x)}{dv^\lambda} \equiv i, \quad (\text{mod. } \lambda).$$

Es sei nun $f(\alpha)$ einer der $\lambda-1$ conjugirten idealen Primfactoren des p, und zwar derjenige, welcher zur Congruenzwurzel $\alpha = g^{\frac{p-1}{\lambda}}$ gehört, es bezeichne ferner Ind. den Index, welcher sich auf die Primzahl $f(\alpha)$ bezieht, so dafs

$$\left(\frac{\lambda}{f(\alpha)}\right) = \alpha^{\text{Ind.}\lambda} \equiv \lambda^{\frac{Nf(\alpha)-1}{\lambda}}, \quad (\text{mod. } f(\alpha))$$

ist, so hat man, weil $Nf(\alpha) = p$ und $\alpha \equiv g^{\frac{p-1}{\lambda}}$, (mod. $f(\alpha)$) ist,

$$\lambda^{\frac{p-1}{\lambda}} \equiv g^{\frac{p-1}{\lambda}\text{Ind.}\lambda}$$

für den Modul $f(\alpha)$, und darum auch für den Modul p. Andererseits giebt die Congruenz (4.), durch welche die Zahl i bestimmt ist,

$$\lambda^{\frac{p-1}{\lambda}} \equiv g^{\frac{p-1}{\lambda}i}, \quad (\text{mod. } p),$$

und aus der Vergleichung dieser beiden Congruenzen folgt:

$$i \equiv \text{Ind.}\lambda, \quad (\text{mod. } \lambda).$$

Die Congruenz (5.) giebt daher folgenden merkwürdigen Ausdruck des Index von λ in Beziehung auf $f(\alpha)$:

$$(6.) \qquad \text{Ind.}\lambda \equiv \frac{d_0^\lambda\, l F(e^v, x)}{dv^\lambda}, \quad (\text{mod. } \lambda).$$

2.

Aus dem gefundenen, dem Gebiete der Kreistheilung angehörenden Ausdrucke des Index von λ erhält man einen entsprechenden durch die ideale Primzahl $f(\alpha)$, auf welche der Index sich bezieht, unmittelbar bestimmten Ausdruck, vermittelst der Zerlegung der λ^{ten} Potenz von $F(\alpha, x)$ in ihre idealen Primfactoren, nach der Formel

$$(7.)\qquad F(\alpha, x)^\lambda = \pm\alpha^s f(\alpha^{-1})^{m_1} f(\alpha^{-2})^{m_2} f(\alpha^{-3})^{m_3} \dots f(\alpha^{-(\lambda-1)})^{m_{\lambda-1}},$$

in welcher die Exponenten m_h als die kleinsten positiven Wurzeln der Congruenzen von der Form

$$hm_h \equiv 1, \quad (\text{mod.}\,\lambda)$$

bestimmt sind und $f(\alpha)$ derjenige ideale Primfactor des p ist, welcher der Congruenzwurzel $\alpha \equiv g^{\frac{p-1}{\lambda}}$ angehört. Um die idealen Factoren dieses Products zu wirklichen zu machen, wird auf beiden Seiten zur H^{ten} Potenz erhoben, wo H der Exponent derjenigen Potenz sein soll, für welche $f(\alpha)^H = \varphi(\alpha)$ eine wirkliche complexe Zahl ist. Es soll auch angenommen werden, dafs H nicht durch λ theilbar ist, welche Bedingung bekanntlich immer erfüllt ist, wenn λ nicht ein Factor des Zählers einer der ersten $\frac{\lambda-3}{2}$ *Bernoulli*schen Zahlen ist. Verwandelt man alsdann die Wurzel α in die variable Exponentialgröfse e^v, so erhält man aus der Gleichung (7.) in bekannter Weise eine für jeden beliebigen Werth der Variabeln e^v geltende Gleichung von der Form

$$(8.)\qquad F(e^v, x)^{H\lambda} = \pm e^{Hsv}\prod_1^{\lambda-1}{}_h\,\varphi(e^{-hv})^{m_h} + V.W,$$

in welcher

$$V = 1 + e^v + e^{2v} + \dots + e^{(\lambda-1)v}$$

und W eine ganze rationale Function von e^v und von der Wurzel x ist. Nimmt man auf beiden Seiten die Logarithmen, so erhält man

$$H\lambda l F(e^v, x) = l(\pm 1) + Hsv + \sum_1^{\lambda-1}{}_h\, m_h l\varphi(e^{-hv}) + l(1 + VW_1),$$

wo

$$W_1 = \frac{W}{\pm e^{Hsv}\prod_h \varphi(e^{-hv})^{m_h}},$$

und wenn der λ^{te} Differentialquotient nach v genommen wird:

$$(9.)\qquad H\lambda\frac{d^\lambda l F(e^v, x)}{dv^\lambda} = \sum_1^{\lambda-1}{}_h\, m_h \frac{d^\lambda l\varphi(e^{-hv})}{dv^\lambda} + \frac{d^\lambda l(1 + VW_1)}{dv^\lambda}.$$

Nimmt man nun $v = 0$ und macht aus dieser Gleichung eine Congruenz nach dem Modul λ^2, so hat man zunächst

$$(10.)\qquad \frac{d_0^\lambda l(1 + VW_1)}{dv^\lambda} \equiv 0, \quad (\text{mod.}\,\lambda^2),$$

für jeden beliebigen Werth des W_1, welcher ganz oder gebrochen sein kann, wenn nur, wie es hier der Fall ist, der Nenner desselben für $v=0$ nicht den Factor λ enthält. Entwickelt man nämlich diesen Differentialquotienten des Logarithmus nach der allgemeinen Formel (2.), so erhält man

$$\frac{d^\lambda l(1+VW_1)}{dv^\lambda} = \frac{\frac{d^\lambda(VW_1)}{dv^\lambda}}{1+VW_1} - \frac{\lambda\frac{d^{\lambda-1}(VW_1)}{dv^{\lambda-1}}\frac{d(VW_1)}{dv}}{(1+VW_1)^2}+\cdots,$$

wo alle folgenden Glieder Producte von mindestens zwei der ersten $\lambda-2$ Differentialquotienten von VW_1 im Zähler enthalten. Diese $\lambda-2$ ersten Differentialquotienten sind aber für den Werth $v=0$ alle durch λ theilbar, weil V mit seinen ersten $\lambda-2$ Differentialquotienten für $v=0$ durch λ theilbar ist. Hieraus folgt, dafs für $v=0$ jedes der auf das erste folgenden Glieder dieser Entwicklung den Factor λ mindestens zweimal enthält, dafs also

$$\frac{d_0^\lambda l(1+VW_1)}{dv^\lambda} \equiv \frac{d_0^\lambda(VW_1)}{dv^\lambda}\cdot\frac{1}{1+\lambda W_1}, \quad (\text{mod. } \lambda^2).$$

Ferner hat man nach einer bekannten Formel für die Differentiation eines Products zweier Factoren:

$$\frac{d^\lambda(VW_1)}{dv^\lambda} = \frac{d^\lambda V}{dv^\lambda}W_1+\lambda\frac{d^{\lambda-1}V}{dv^{\lambda-1}}\frac{dW_1}{dv}+\frac{\lambda(\lambda-1)}{1.2}\frac{d^{\lambda-2}V}{dv^{\lambda-2}}\frac{d^2W_1}{dv^2}+\cdots+V\frac{d^\lambda W_1}{dv^\lambda},$$

und weil die Differentialquotienten des V bis zum $\lambda-2^{\text{ten}}$ einschliefslich für $v=0$ alle den Factor λ enthalten, und die Binomialcoefficienten der λ^{ten} Potenz denselben ebenfalls enthalten, so hat man

$$\frac{d_0^\lambda(VW_1)}{dv^\lambda} \equiv \frac{d_0^\lambda V}{dv^\lambda}W_1+\lambda\frac{d_0^{\lambda-1}V}{dv^{\lambda-1}}\frac{d_0W_1}{dv}+\lambda\frac{d_0^\lambda W_1}{dv^\lambda}, \quad (\text{mod. } \lambda^2).$$

Es ist aber

$$\frac{d_0^\lambda V}{dv^\lambda} = 1^\lambda+2^\lambda+3^\lambda+\cdots+(\lambda-1)^\lambda \equiv 0, \quad (\text{mod. } \lambda^2),$$

$$\frac{d_0^{\lambda-1}V}{dv^{\lambda-1}} = 1^{\lambda-1}+2^{\lambda-1}+3^{\lambda-1}+\cdots+(\lambda-1)^{\lambda-1} \equiv -1, \quad (\text{mod. } \lambda),$$

$$\frac{d_0^\lambda W_1}{dv^\lambda} \equiv \frac{d_0W_1}{dv}, \quad (\text{mod. } \lambda),$$

folglich

$$\frac{d_0^\lambda(VW_1)}{dv^\lambda} \equiv 0, \quad (\text{mod. } \lambda^2),$$

und darum auch

$$\frac{d_0^\lambda l(1+VW_1)}{dv^\lambda} \equiv 0, \quad (\text{mod. } \lambda^2),$$

wie behauptet worden.

Nimmt man nun die ideale Zahl $f(\alpha)$, als H^{te} Wurzel aus der *wirklichen* Zahl $\varphi(\alpha) = f(\alpha)^H$ dargestellt, in der Form $f(\alpha) = \sqrt[H]{\varphi(\alpha)}$, so kann man einfach

$$\frac{1}{H}\frac{d_0^\lambda l\varphi(e^{-hv})}{dv^\lambda} = \frac{d_0^\lambda lf(e^{-hv})}{dv^\lambda}$$

bezeichnen, und weil

$$\frac{d_0^\lambda lf(e^{-hv})}{dv^\lambda} = -h^\lambda\frac{d_0^\lambda lf(e^v)}{dv^\lambda}$$

ist, so giebt die Gleichung (9.) folgende Congruenz:

$$\frac{\lambda d_0^\lambda lF(e^v, x)}{dv^\lambda} \equiv -\frac{d_0^\lambda lf(e^v)}{dv^\lambda}\sum_1^{\lambda-1}{}_h h^\lambda m_h, \quad (\text{mod. } \lambda^2),$$

also vermöge der Formel (6.)

$$\lambda\,\text{Ind.}\,\lambda \equiv -\frac{d_0^\lambda lf(e^v)}{dv^\lambda}\sum_1^{\lambda-1}{}_h h^\lambda m_h, \quad (\text{mod. } \lambda^2).$$

Weil $hm_h \equiv 1$, (mod. λ), so hat man

$$\sum_1^{\lambda-1}{}_h h^\lambda m_h \equiv \sum_1^{\lambda-1}{}_h h^{\lambda-1} \equiv -1, \quad (\text{mod. } \lambda);$$

hieraus schliefst man zunächst, dafs $\frac{d_0^\lambda lf(e^v)}{dv^\lambda}$ durch λ theilbar sein mufs, welches auch anderweitig leicht zu beweisen ist; dividirt man also durch λ, so hat man folgenden neuen Ausdruck des Index von λ:

$$(11.)\quad \text{Ind.}\,\lambda \equiv \frac{1}{\lambda}\frac{d_0^\lambda lf(e^v)}{dv^\lambda}, \quad (\text{mod. } \lambda).$$

Obgleich die Methode der Herleitung dieses Ausdrucks voraussetzt, dafs $f(\alpha)$ ein complexer Primfactor einer nichtcomplexen Primzahl p der linearen Form $m\lambda+1$ ist, so gilt derselbe dennoch ebenfalls für alle anderen idealen Primzahlen, weil für alle Primzahlen $f(\alpha)$, welche complexe Primfactoren nichtcomplexer Primzahlen anderer linearer Formen sind, stets zugleich

$$(12.)\quad \text{Ind.}\,\lambda \equiv 0 \quad \text{und} \quad \frac{1}{\lambda}\frac{d_0^\lambda lf(e^v)}{dv^\lambda} \equiv 0, \quad (\text{mod. } \lambda),$$

statt hat. Es folgt diefs daraus, dafs die conjugirten complexen Primfactoren einer jeden nichtcomplexen Primzahl q, welche nicht von der Form $m\lambda+1$ ist, zum Theil einander gleich sind, dafs also für dieselben eine Gleichung von der Form

$$f(\alpha) = f(\alpha^r)$$

statt hat, in welcher r nicht congruent Eins nach dem Modul λ ist. Vermöge dieser Gleichung hat man

$$\frac{1}{\lambda}\,\frac{d_0^\lambda l f(e^v)}{dv^\lambda} \equiv \frac{1}{\lambda}\,\frac{d_0^\lambda l f(e^{rv})}{dv^\lambda} \equiv r^\lambda.\frac{1}{\lambda}\,\frac{d_0^\lambda l f(e^v)}{dv^\lambda}, \quad (\text{mod. } \lambda),$$

also, weil $r^\lambda \equiv r$, nicht $\equiv 1$ ist,

$$\frac{1}{\lambda}\,\frac{d_0^\lambda l f(e^v)}{dv^\lambda} \equiv 0, \quad (\text{mod. } \lambda).$$

Ferner hat man, weil $f(\alpha) = f(\alpha^r)$ ist,

$$\left(\frac{\lambda}{f(\alpha)}\right) = \alpha^{\text{Ind.}\lambda}, \qquad \left(\frac{\lambda}{f(\alpha^r)}\right) = \alpha^{\text{Ind.}\lambda},$$

und wenn man in der ersten dieser beiden Gleichungen α^r statt α setzt

$$\left(\frac{\lambda}{f(\alpha^r)}\right) = \alpha^{r\,\text{Ind.}\lambda},$$

und demnach

$$\alpha^{r\,\text{Ind.}\lambda} = \alpha^{\text{Ind.}\lambda}, \qquad r\,\text{Ind.}\lambda \equiv \text{Ind.}\lambda, \qquad (\text{mod. } \lambda),$$

woraus, weil r nicht congruent Eins ist, $\text{Ind.}\lambda \equiv 0$ folgt.

3.

Aus dem gefundenen Ausdrucke (11.) des Index von λ, dessen Allgemeingültigkeit für alle complexen Primzahlen $f(\alpha)$ hierdurch bewiesen ist, soll nun mit Hülfe der in der angeführten Abhandlung entwickelten Indices der Einheiten ein neuer für alle Primzahlen $f(\alpha)$ geltender Ausdruck des Index von $1-\alpha^x$ hergeleitet werden; in ähnlicher Weise, wie dies a. a. O. pag. 129 für den speciellen Fall ausgeführt ist, wo $\text{Ind.}\lambda \equiv 0$, (mod. λ) ist.

Aus der Formel

$$\lambda = (1-\alpha)(1-\alpha^\gamma)(1-\alpha^{\gamma^2})\dots(1-\alpha^{\gamma^{\lambda-2}}),$$

in welcher γ eine primitive Wurzel von λ bezeichnet, erhält man

$$\lambda = (1-\alpha)^{\lambda-1}\alpha^{-\mu}e(\alpha)^{\lambda-1}e(\alpha^\gamma)^{\lambda-2}e(\alpha^{\gamma^2})^{\lambda-3}\dots e(\alpha^{\gamma^{\lambda-2}})^1,$$

wo $\mu = \frac{\lambda-1}{2}$, und wo $e(\alpha)$ die Kreistheilungseinheit

$$e(\alpha) = \sqrt{\frac{(1-\alpha^\gamma)(1-\alpha^{-\gamma})}{(1-\alpha)(1-\alpha^{-1})}}$$

ist. Verwandelt man α in α^x und nimmt auf beiden Seiten die Indices nach dem Modul $f(\alpha)$, so erhält man

$$(13.)\quad \text{Ind.}\lambda \equiv -\text{Ind.}(1-\alpha^{\varkappa}) - \mu\varkappa\,\text{Ind.}\alpha - \sum_{0}^{\lambda-1}{}_h\, h\,\text{Ind.}e(\alpha^{\varkappa\gamma^{h-1}}),\ (\text{mod.}\lambda).$$

Drückt man nun die einfachen Kreistheilungseinheiten $e(\alpha)$, $e(\alpha^{\gamma})$, ... durch die zusammengesetzten Kreistheilungseinheiten $E_1(\alpha)$, $E_2(\alpha)$, ... $E_{\mu-1}(\alpha)$ aus, welche durch die Gleichung

$$E_n(\alpha) = e(\alpha)e(\alpha^{\gamma})^{\gamma^{-2n}} e(\alpha^{\gamma^2})^{\gamma^{-4n}} \dots e(\alpha^{\gamma^{\mu-1}})^{\gamma^{-2(\mu-1)n}}$$

bestimmt sind, aus welcher

$$\text{Ind.}E_n(\alpha) \equiv \sum_{0}^{\mu-1}{}_h\, \gamma^{-2nh}\text{Ind.}e(\alpha^{\gamma^h}),\quad (\text{mod.}\lambda)$$

und durch Umkehrung

$$\text{Ind.}e(\alpha^{\gamma^h}) \equiv -2\sum_{1}^{\mu-1}{}_n\, \gamma^{2nh}\text{Ind.}E_n(\alpha),\quad (\text{mod.}\lambda)$$

folgt, so erhält man aus der Congruenz (13.)

$$(14.)\quad \text{Ind.}\lambda \equiv -\text{Ind.}(1-\alpha^{\varkappa}) - \mu\varkappa\,\text{Ind.}\alpha + 2\sum_{1}^{\lambda-1}{}_h \sum_{1}^{\mu-1}{}_n\, h\gamma^{2n(h-1)}\text{Ind.}E_n(\alpha^{\varkappa}).$$

Führt man die Summation in Beziehung auf h aus und bringt $\text{Ind.}(1-\alpha^{\varkappa})$ allein auf eine Seite der Congruenz, so hat man

$$(15.)\quad \text{Ind.}(1-\alpha^{\varkappa}) \equiv -\text{Ind.}\lambda - \mu\varkappa\,\text{Ind.}\alpha - 2\sum_{1}^{\mu-1}{}_n \frac{\text{Ind.}E_n(\alpha^{\varkappa})}{\gamma^{2n}-1},\quad (\text{mod.}\lambda).$$

Macht man nun von dem a. a. O. pag. 128 gegebenen Index der Einheit $E_n(\alpha)$ Gebrauch:

$$(16.)\quad \text{Ind.}E_n(\alpha^{\varkappa}) \equiv (-1)^n(\gamma^{2n}-1)\frac{B_n\varkappa^{\lambda-2n}}{4n}\,\frac{d_0^{\lambda-2n}lf(e^{v})}{dv^{\lambda-2n}},\quad (\text{mod.}\lambda),$$

in welchem B_n die n^{te} *Bernoulli*sche Zahl ist, und setzt der Kürze halber

$$\frac{d_0^{\varkappa}lf(e^{v})}{dv^{\varkappa}} = D_{\varkappa},$$

so erhält man, indem man für Ind. λ den gefundenen Werth bei (11.) und für Ind.(α) seinen Werth $\frac{Nf(\alpha)-1}{\lambda}$ setzt, folgenden Ausdruck des Index von $1-\alpha^{\varkappa}$:

$$(17.)\quad \text{Ind.}(1-\alpha^{\varkappa}) \equiv$$

$$-\frac{D_{\lambda}}{\lambda} + \tfrac{1}{2}\frac{Nf(\alpha)-1}{\lambda}\varkappa + B_1D_{\lambda-2}\frac{\varkappa^2}{2} - B_2D_{\lambda-4}\frac{\varkappa^4}{4} + \dots + (-1)^{\mu}B_{\mu-1}D_3\frac{\varkappa^{\lambda-3}}{\lambda-3},$$

nach dem Modul λ.

4.

Aus den Indices der besonderen Einheiten $E_1(\alpha)$, $E_2(\alpha)$, ... $E_{\mu-1}(\alpha)$ läfst sich ein allgemeiner Ausdruck für den Index einer jeden beliebigen Einheit $E(\alpha)$ auf folgende Weise ableiten. Da das System dieser $\mu-1$ Einheiten ein vollständiges unabhängiges System ist, so läfst sich jede gegebene Einheit $E(\alpha)$ durch dasselbe ausdrücken in der Form

$$(18.)\qquad E(\alpha) = \alpha^m E_1(\alpha)^{m_1} E_2(\alpha)^{m_2} \ldots E_{\mu-1}(\alpha)^{m_{\mu-1}},$$

aus welcher man

$$(19.)\quad \text{Ind.}\,E(\alpha) = m\,\text{Ind.}\,\alpha + m_1\,\text{Ind.}\,E_1(\alpha) + m_2\,\text{Ind.}\,E_2(\alpha) + \cdots + m_{\mu-1}\,\text{Ind.}\,E_{\mu-1}(\alpha)$$

erhält, wo m_1, m_2, ... $m_{\mu-1}$ auch rationale Brüche sein können, aber nicht solche, welche in ihren Nennern λ enthalten, wenn nämlich, wie vorausgesetzt worden ist, λ nicht in einer der ersten $\frac{\lambda-3}{2}$ *Bernoulli*schen Zahlen als Factor des Zählers enthalten ist. Verwandelt man nun in der Form (18.) die Wurzel α in die Variable e^v und differentiirt logarithmisch, so werden alle ungraden Differentialquotienten des Logarithmus von $E_n(e^v)$ congruent Null, nach dem Modul λ, wegen der Eigenschaft dieser Einheit, nach welcher $E_n(\alpha) = E_n(\alpha^{-1})$ ist. Ferner hat man

$$\frac{d_0^{2\varkappa} l E_n(e^v)}{dv^{2\varkappa}} \equiv \sum_0^{\mu-1}{}_h \gamma^{-2nh} \frac{d_0^{2\varkappa} l e(e^{v\gamma^h})}{dv^{2\varkappa}} \equiv \frac{d_0^{2\varkappa} l e(e^v)}{dv^{2\varkappa}} \sum_0^{\mu-1}{}_h \gamma^{(2\varkappa-2n)h}, \text{ (mod. } \lambda),$$

und weil diese Summe stets congruent Null ist, mit Ausnahme des Falles $\varkappa = n$, in welchem sie congruent μ wird, nach dem Modul λ, so hat man

$$(20.)\quad \begin{cases} \dfrac{d_0^{2\varkappa} l E_n(e^v)}{dv^{2\varkappa}} \equiv 0, \text{ (mod. } \lambda), \text{ wenn } \varkappa \text{ nicht gleich } n, \\[2ex] \dfrac{d_0^{2n} l E_n(e^v)}{dv^{2n}} \equiv \mu \dfrac{d_0^{2n} l e(e^v)}{dv^{2n}}, \text{ (mod. } \lambda), \end{cases}$$

und weil

$$\frac{d_0^{2n} l e(e^v)}{dv^{2n}} \equiv (-1)^{n+1}(\gamma^{2n}-1)\frac{B_n}{2n}, \text{ (mod. } \lambda),$$

wie in der angeführten Abhandlung pag. 139 gezeigt worden ist, so ist

$$(21.)\qquad \frac{d_0^{2n} l E_n(e^v)}{dv^{2n}} \equiv (-1)^n(\gamma^{2n}-1)\frac{B_n}{4n}, \text{ (mod. } \lambda).$$

Man erhält demnach aus der Gleichung (18.) folgenden Werth des $2n^{\text{ten}}$ Differentialquotienten von $l E(e^v)$:

$$\frac{d_0^{2n} l E(e^v)}{dv^{2n}} \equiv m_n \frac{d_0^{2n} l E_n(e^v)}{dv^{2n}} \equiv m_n(-1)^n(\gamma^{2n}-1)\frac{B_n}{4n},$$

und wenn mit $\frac{d_0^{\lambda-2n} l f(e^v)}{dv^{\lambda-2n}}$ multiplicirt wird:

$$\frac{d_0^{2n} l E(e^v)}{dv^{2n}} \frac{d_0^{\lambda-2n} l f(e^v)}{dv^{\lambda-2n}} \equiv m_n(-1)^n(\gamma^{2n}-1)\frac{B_n}{4n}\frac{d_0^{\lambda-2n} l f(e^v)}{dv^{\lambda-2n}},$$

also vermöge der Formel (16.)

$$\frac{d_0^{2n} l E(e^v)}{dv^{2n}} \frac{d_0^{\lambda-2n} l f(e^v)}{dv^{\lambda-2n}} \equiv m_n \operatorname{Ind.} \mathrm{E}_n(\alpha).$$

Der Ausdruck des Index der beliebigen complexen Einheit $E(\alpha)$ wird daher, weil überdies $\frac{d_0 l E(e^v)}{dv} \equiv m$ ist:

$$(22.)\quad \operatorname{Ind.} E(\alpha) \equiv \frac{d_0 l E(e^v)}{dv} \frac{N f(\alpha)-1}{\lambda} + \sum_1^{\mu-1}{}_n \frac{d_0^{2n} l E(e^v)}{dv^{2n}} \frac{d_0^{\lambda-2n} l f(e^v)}{dv^{\lambda-2n}}, \pmod{\lambda}.$$

Berlin, im December 1858.

---

36 *

Über die allgemeinen Reciprocitätsgesetze unter den Resten und Nichtresten der Potenzen, deren Grad eine Primzahl ist

Mathematische Abhandlungen der Königlichen Akademie der Wissenschaften zu Berlin aus dem Jahre 1859, 19–159

[Gelesen in der Akademie der Wissenschaften am 18. Februar 1858 und am 5. Mai 1859.]

Die Reciprocitätsgesetze, welche unter den Resten und Nichtresten der Potenzen Statt haben, bilden gewissermaafsen den Schlufsstein der Lehre von den Potenzresten und eröffnen zugleich den Weg für weitere und tiefer liegende arithmetische Untersuchungen. Sie sind in diesen beiden Beziehungen für die Zahlentheorie von grofser Wichtigkeit, aber eine noch höhere Bedeutung haben sie in der geschichtlichen Entwickelung dieser mathematischen Disciplin dadurch erlangt, dafs die Beweise derselben, so weit sie überhaupt gefunden sind, fast durchgängig aus neuen, bis dahin noch unerforschten Gebieten haben geschöpft werden müssen, welche so der Wissenschaft aufgeschlossen worden sind. Wegen dieser Schwierigkeit der Beweise ist man in der Erkenntnifs der Reciprocitätsgesetze bisher nicht viel über die quadratischen, kubischen und biquadratischen hinausgekommen, obgleich mehrere der ausgezeichnetsten Mathematiker der neueren Zeit sie zum Gegenstande ihrer Forschungen gemacht haben.

Euler hat das Verdienst, das Reciprocitätsgesetz für quadratische Reste zuerst bemerkt zu haben, m. s. dessen *Commentationes arithmeticae collectae* Bd. 1, pg. 486, aber man kennt keinen Versuch, den er gemacht hätte dasselbe zu beweisen. Hierauf hat Legendre, von Euler unabhängig, dieses Gesetz ebenfalls gefunden, und weil er dessen grofse Wichtigkeit erkannte, einen sehr sinnreichen Beweis desselben aufgestellt, welcher nur in so fern unvollständig ist, als er voraussetzt, dafs zu einer jeden Primzahl von der Form $4n+1$ eine Primzahl der Form $4n+3$ gefunden werden

C 2

kann, in Beziehung auf welche jene quadratischer Nichtrest ist, welches Postulat leicht von dem allgemeineren Satze abhängig gemacht wird, dafs jede arithmetische Reihe, in welcher nicht alle Glieder einen gemeinschaftlichen Faktor haben, nothwendig Primzahlen enthalten mufs. Um diesen Mangel des Legendreschen Beweises zu heben, hat später Hr. Dirichlet diese Eigenschaft der arithmetischen Reihen streng bewiesen, und zwar durch die neuen, überaus fruchtbaren Methoden, durch deren Hülfe er auch die Klassenanzahl der quadratischen Formen gefunden hat. Diese berühmten Arbeiten des Hrn. Dirichlet können daher als solche betrachtet werden, welche der Beschäftigung mit den Reciprocitätsgesetzen ihre Entstehung verdanken.

Den ersten vollständigen und strengen Beweis des quadratischen Reciprocitätsgesetzes hat Gaufs in seinen *Disquisitiones arithmeticae* pg. 124, sq. gegeben, indem er gezeigt hat, dafs, wenn dieses Gesetz für alle Primzahlen bis zu einer bestimmten Gränze hin richtig ist, dasselbe auch richtig bleibt, wenn diese Gränze so weit erweitert wird, dafs sie eine Primzahl mehr umfafst, woraus nach dem bekannten Schlusse der Induktionsbeweise die Allgemeingültigkeit desselben folgt. Dieser Beweis, welcher vor allen übrigen sich dadurch auszeichnet, dafs er keine, dem Gebiete der Congruenzen zweiten Grades fremden Hülfsmittel braucht, ist später von Hrn. Dirichlet in einfacherer Weise dargestellt worden in Crelle's Journal, Bd. 47, pg. 139.

Der zweite Beweis des *theorema fundamentale,* wie Gaufs dieses Reciprocitätsgesetz der quadratischen Reste bezeichnet, findet sich ebenfalls in den *Disquisitiones arithmeticae,* pg. 414, sq., wo er aus der Theorie der quadratischen Formen abgeleitet wird. Die Grundgedanken desselben werde ich weiter unten genauer zu entwickeln Gelegenheit nehmen, da es diejenigen sind, welche ich für den ersten Beweis des allgemeinen Reciprocitätsgesetzes in Anwendung gebracht habe, den ich in der gegenwärtigen Abhandlung geben will.

Aufser den beiden, in den *Disquisitiones arithmeticae* enthaltenen Beweisen hat Gaufs später noch vier verschiedene Beweise desselben Satzes in den Commentarien der Göttinger Akademie gegeben. Zwei von diesen, nämlich der als dritter und der als fünfter von Gaufs bezeichnete, sind beinahe eben so elementar, als der erste Beweis, da sie nur in so fern das Gebiet der Congruenzen zweiten Grades verlassen, als ein Satz über die reinen

Congruenzen höherer Grade hinzugezogen wird. Beide Beweise stützen sich auch auf einen und denselben, nicht schwer zu beweisenden Satz, welcher in der Abzählung der kleinsten positiven Reste einer arithmetischen Reihe, die gröſser als die Hälfte des Moduls sind, ein Kriterium dafür giebt, ob eine gegebene Zahl quadratischer Rest oder Nichtrest dieses Moduls ist. Der Unterschied derselben liegt hauptsächlich nur in der Art der Abzählung dieser Reste, welche in dem einen Beweise für sich selbst betrachtet, in dem anderen aber durch die gröſsten Ganzen ausgedrückt werden, welche in gewissen gebrochenen Zahlen enthalten sind. Als eine Modification dieser Gauſsischen Beweise ist auch derjenige anzusehen, welchen Eisenstein in Crelle's Journal, Bd. 28, pg. 246, gegeben hat. Dieser Beweis unterscheidet sich nämlich von dem dritten Gauſsischen nur darin, daſs geometrische Anschauungen zu Hülfe genommen, und die in den Brüchen enthaltenen gröſsten Ganzen durch die Anzahl der in bestimmten Gränzen liegenden Gitterpunkte eines Netzes dargestellt werden.

Die anderen beiden Gauſsischen Beweise haben ihre Quelle in der Theorie der Kreistheilung. Diese zeigt, wie die Quadratwurzel einer jeden gegebenen Zahl durch Wurzeln der Einheit in ganzer rationaler Form dargestellt werden kann, wobei nur der eine Punkt unentschieden bleibt: ob diese Darstellung den Werth der Quadratwurzel mit dem positiven oder mit dem negativen Vorzeichen giebt. Die Bestimmung dieses Vorzeichens ist der hauptsächlichste Gegenstand der Gauſsischen Abhandlung, welche den vierten Beweis des quadratischen Reciprocitätsgesetzes enthält, und den Titel *Summatio quarundam serierum singularium* führt. Dieselbe, sowohl durch die Einfachheit des für Primzahlen und für zusammengesetzte Zahlen gleichmäſsig geltenden Resultats, als auch durch die Schwierigkeit des Beweises interessante Vorzeichenbestimmung, nebst dem darauf gegründeten Beweise des Reciprocitätsgesetzes, ist später von Hrn. Dirichlet, in einer vor der Akademie im Jahre 1835 vorgetragenen Abhandlung, nach einer anderen Methode, mit Anwendung bestimmter Integrale, ausgeführt worden.

Der sechste Gauſsische Beweis, welcher auch auf der Kreistheilung beruht, uud zwar auf demselben Ausdrucke der Quadratwurzel aus p durch $p$te Wurzeln der Einheit, unterscheidet sich wesentlich dadurch von dem anderen, daſs er die Bestimmung des Vorzeichens der Quadratwurzel nicht erfordert. Der eigentliche Kern dieses Beweises wird bei Gauſs dadurch etwas verhüllt, dafs anstatt der $p$ten Wurzel der Einheit eine unbestimmte Variable x angewendet wird, was zur Folge hat, dafs Congruenzen unter ganzen rationalen Funktionen von x nach dem Modul $1+x+x^2+\ldots+x^{p-1}$ angewendet werden müssen, anstatt deren man nur einfache Gleichungen erhält, wenn dem x der specielle Werth einer primitiven $p$ten Wurzel der Einheit gegeben wird. Diese Vereinfachung des sechsten Gaufsischen Beweises hat zuerst Jacobi ausgeführt und im Jahre 1827 an Legendre brieflich mitgetheilt, welcher sie im Jahre 1830 in die dritte Ausgabe seiner théorie des nombres aufgenommen hat. Mit dieser Jacobischen Darstellung des Gaufsischen Beweises stimmt auch ein von Cauchy in Férussac bulletin im Jahre 1829 gegebener Beweis des Reciprocitätsgesetzes im Wesentlichen überein. Eisenstein, welchem Jacobi's und Cauchy's Arbeiten über diesen Gegenstand unbekannt geblieben waren, hat denselben Beweis im Jahre 1844 in Crelle's Journal, Bd. 28, pag. 41, reproducirt.

Die aufserdem noch zu erwähnenden Beweise des quadratischen Reciprocitätsgesetzes hängen alle mit der Theorie der Kreistheilung zusammen. Am nächsten steht den von Jacobi und Cauchy gegebenen Beweisen derjenige, welchen Hr. Liouville im 12ten Bande seines Journal's, pg. 95, aufgestellt hat, dessen Unterschied von jenen hauptsächlich nur darin liegt, dafs Hr. Liouville nicht den Ausdruck der Quadratwurzel aus p durch eine Summe von $p$ten Wurzeln der Einheit, sondern den Ausdruck als Produkt von Differenzen dieser Einheitswurzeln anwendet, welcher, wenn man von den fertigen Resultaten der Kreistheilung keinen Gebrauch machen will, viel leichter unmittelbar herzustellen ist, und darum einen Vorzug vor jenem hat. Der Beweis von Hrn. Lebesgue, in Liouville's Journal, Bd. 3, pag. 134, beruht auf der vollständigen Potenzerhebung des Ausdrucks der Quadratwurzel aus p durch die Einheitswurzeln, wodurch die $\frac{1}{2}(q-1)$te Potenz von p gewonnen wird, ohne dafs die Vielfachen von q weggelassen werden. Die Coefficienten dieser Entwickelung, welche nur drei verschiedene Zahlen sind, werden in bekannter Weise durch die Anzahl der Auflösungen gewisser Congruenzen definirt, und aus dem so bestimmten Ausdrucke der $\frac{1}{2}(q-1)$ten Potenz von p wird das Reciprocitätsgesetz für die beiden Primzahlen p und q ohne Schwierigkeit erschlossen. An diesen Beweis von Hrn. Lebesgue schliefst sich der von Eisenstein, in Crelle's Journal, Bd. 27, pag. 322, gegebene Beweis sehr genau an, obgleich er scheinbar von ganz anderen Prinzipien ausgeht. Eisenstein gebraucht in demselben gewisse Zahlenausdrücke, welche auf combinatorischem Wege aus den bekannten Legendreschen Zeichen, deren Werthe nur $+1$ und -1 sind, je nachdem eine Zahl Rest oder Nichtrest der anderen ist, zusammengesetzt werden, und er stellt durch diese die $\frac{1}{2}(q-1)$te Potenz von p so dar, dafs durch Weglassung der Vielfachen von q das Reciprocitätsgesetz gewonnen wird. Betrachtet man aber diese Eisensteinschen Zahlenausdrücke näher, so bemerkt man leicht ihren Ursprung aus der Kreistheilung, welchen Eisenstein selbst verschwiegen hat, und man erkennt, dafs sie nichts anderes sind, als die Coefficienten der Entwickelung einer Potenz des Ausdrucks von $\sqrt{p}$ durch die $p$ten Wurzeln der Einheit, und dafs sie mit den von Hrn. Lebesgue durch die Anzahl der Auflösungen gewisser Congruenzen definirten Zahlen wesentlich übereinstimmen. Dieses hat auch Hr. Lebesgue, in Liouville's Journal, Bd. 12, pag. 457, nachgewiesen.

Für einen der schönsten Beweise dieses von den ausgezeichnetsten Mathematikern viel bewiesenen Theorems wird aber derjenige mit Recht gehalten, welchen Eisenstein in Crelle's Journal, Bd. 29, pag. 177, gegeben hat. In diesem wird das Legendresche Zeichen $\left(\frac{p}{q}\right)$ durch Kreisfunktionen so ausgedrückt, dafs bei der Vertauschung von p und q dieser Ausdruck, bis auf eine leicht zu bestimmende Änderung im Vorzeichen, ungeändert bleibt. Dieser Beweis hängt in so fern ebenfalls mit der Theorie der Kreistheilung zusammen, als der Ausdruck des $\left(\frac{p}{q}\right)$ nur die Sinus der Vielfachen von $\frac{2\pi}{p}$ enthält, welche mit den $p$ten Wurzeln der Einheit ganz auf derselben Stufe stehen, auch ist dieser Ausdruck mit dem von Hrn. Liouville in seinem Beweise angewendeten Produktausdrucke der Quadratwurzel aus p nahe verwandt. Wenn dieser Eisensteinsche Beweis schon wegen seiner vorzüglichen Eleganz beachtenswerth ist, so wird der Werth desselben noch dadurch erhöht, dafs er, wie Eisenstein selbst gezeigt hat, ohne besondere Schwierigkeit auch auf die biquadratischen und die kubischen Reciprocitätsgesetze angewendet werden kann, wenn anstatt der Kreisfunktionen elliptische Funktionen mit bestimmten Moduln angewendet werden.

Was nun in dem Gebiete der Reciprocitätsgesetze für die Reste und Nichtreste höherer Potenzen bisher geleistet worden ist, beschränkt sich zwar hauptsächlich nur auf die vollständigen Beweise dieser Gesetze für die biquadratischen und die kubischen Reste und Nichtreste und darüber hinaus nur auf gewisse sehr specielle Fälle; es ist aber grade dieses für die ganze weitere Entwickelung der Zahlentheorie von den bedeutendsten Folgen gewesen, weil dadurch das Gebiet dieser mathematischen Disciplin unendlichfach erweitert worden ist, nach Gaufs eigenen Worten: *ut campus Arithmeticae sublimioris infinities quasi promoveatur.* Der eben so einfache als vielumfassende Gedanke dieser Erweiterung der Zahlentheorie, nämlich die Einführung complexer ganzer Zahlen, welche unter denselben Gesichtspunkten betrachtet werden können, als die gewöhnlichen ganzen Zahlen, ist zuerst in der im Jahre 1828 erschienenen, aber der Göttinger Akademie schon im Jahre 1825 übergebenen Abhandlung von Gaufs, über die biquadratischen Reste, niedergelegt, und in einer zweiten Abhandlung über denselben Gegenstand vom Jahre 1832 weiter ausgeführt. Nach Jacobi's Meinung ist dieser Gedanke nicht aus dem Gebiete der Arithmetik allein erwachsen, sondern unter Mitwirkung der Theorie der elliptischen Funktionen, namentlich der lemniskatischen, für die eine complexe Multiplikation mit Zahlen von der Form $a+b\sqrt{-1}$ und die entsprechende Division Statt hat, welche Gaufs für sich schon über ein Vierteljahrhundert eher gekannt hat, als sie durch die Arbeiten von Abel und Jacobi ein Allgemeingut der Wissenschaft geworden ist. Zwar ist in der Gaufsischen Darstellung des Gedankens der complexen Zahlen keine Spur dieses von Jacobi vermutheten Ursprungs zu finden; da aber Gaufs in seinen Ahandlungen mehr darauf ausging, die mathematischen Wahrheiten kunstgerecht aufzubauen, als sie genetisch zu entwickeln, und da er nach seinem eigenen Ausdrucke das Gerüst abtrug, wenn der Bau vollendet war, so läfst sich schwer entscheiden, ob wirklich die Lemniskatenfunktionen mit zu den Balken des Gerüstes gehört haben, mit dessen Hülfe er dieses unvergängliche Werk errichtet hat. In der Theorie der biquadratischen Reste erscheint die Einführung der complexen Zahlen von der Form $a+b\sqrt{-1}$ dadurch motivirt, dafs die biquadratischen Reciprocitätsgesetze für gewöhnliche Primzahlen sehr complicirt sind, namentlich für die Primzahlen von der Form $4n+1$, welche sich als Summen zweier Quadratzahlen darstellen lassen, dafs diese Reciprocitätsgesetze aber die einfachste Form annehmen, wenn man die gewöhnlichen Primzahlen von der Form $p=a^2+b^2$ in die imaginären Faktoren $a+b\sqrt{-1}$ und $a-b\sqrt{-1}$ zerlegt, und unter diesen als den für die vorliegende Frage wahren Primzahlen die Reciprocitätsgesetze aufstellt. Dieses neue Princip ist es, welches die genannten beiden Gaufsischen Abhandlungen zu Dokumenten einer bedeutenden Epoche in der geschichtlichen Entwickelung der Zahlentheorie erhebt, welche auch auf die verwandten mathematischen Disciplinen einen grofsen Einflufs ausgeübt hat; die in den Abhandlungen enthaltenen neuen Sätze über biquadratische Reste treten dagegen in den Hintergrund. Gaufs hat nämlich in denselben nur diejenigen Sätze vollständig bewiesen, welche als die Ergänzungssätze zu dem biquadratischen Reciprocitätsgesetze bezeichnet werden müssen, da sie grade nur die biquadratischen Charaktere derjenigen Zahlen geben, auf welche der allgemeine Ausdruck dieses Gesetzes sich nicht erstreckt. Das vollständige Reciprocitätsgesetz für die Reste der vierten Potenzen hat er nur aufgestellt, und sein Beweis desselben, welcher in der dritten Abhandlung folgen sollte, ist niemals erschienen.

Als die erste der beiden genannten Gaufsischen Abhandlungen noch nicht erschienen war, sondern nur eine vorläufige Ankündigung derselben in den Göttinger gelehrten Anzeigen, aus welcher nicht mehr zu erfahren war, als dafs die Lösung der Frage: ob eine Zahl biquadratischer Rest einer gegebenen Primzahl ist, oder nicht, von den Zahlenwerthen der Unbestimmten gewisser quadratischer Formen abhängig sei, in welche der Modul gesetzt werden kann, veröffentlichte Hr. Dirichlet eine Abhandlung über die biquadratischen Reste, in Crelle's Journal, Bd. 3, pag. 35. In dieser ist er, ohne von dem noch nicht bekannten neuen Gaufsischen Prinzipe der complexen Zahlen Gebrauch machen zu können, durch blofse Anwendung der Theorie der quadratischen Formen und Reste schon sehr tief in die Theorie der biquadratischen Reste eingedrungen, ohne jedoch das Reciprocitätsgesetz derselben finden zu können.

In demselben Jahre 1827 fand Jacobi in der Theorie der Kreistheilung, die er bedeutend vereinfacht und weiter ausgebildet hatte, eine reiche Quelle für die Reciprocitätsgesetze der Potenzreste, aus welcher er nicht nur den oben bereits erwähnten Beweis des quadratischen Reciprocitätsgesetzes ableiten konnte, sondern auch die in Crelle's Journal, Bd. 2, pg. 66, von ihm aufgestellten Sätze über kubische Reste, aus welcher er auch einige Zeit später, durch Anwendung des neuen Gaufsischen Prinzip's, die vollständigen Reciprocitätsgesetze für die kubischen und die biquadratischen

Reste in ihrer einfachsten Gestalt hergeleitet und bewiesen hat. Die vollständige Entwickelung dieser Sätze hat Jacobi aber nur in seinen Vorlesungen über Zahlentheorie, in denen die Lehre von der Kreistheilung den eigentlichen Kern bildete, seinen Zuhörern in Königsberg mitgetheilt, durch deren Hefte sie sodann weiter verbreitet worden sind. Aufserdem hat Jacobi der hiesigen Akademie zwei Mittheilungen darüber in den Jahren 1837 und 1839 gemacht, welche in den Monatsberichten veröffentlicht sind. In der letzteren dieser Mittheilungen spricht er auch die Erwartung aus, dafs er aus derselben Quelle eben so die Reciprocitätsgesetze für die fünften und achten Potenzen werde ableiten können, eine Erwartung, welche nicht erfüllt werden konnte, wie bald näher gezeigt werden soll.

Aufser Gaufs, Jacobi und Dirichlet hat nur noch Eisenstein in diesem Gebiete der kubischen und biquadratischen Reste selbständig und mit Erfolg gearbeitet. Seine ersten Beweise, des kubischen Reciprocitätsgesetzes, in Crelle's Journal, Bd. 27, pag. 289, und des biquadratischen, ebendaselbst Bd. 28, pag. 53, sind zwar nur ganz dieselben, welche Jacobi mehr als zehn Jahre früher gefunden hatte, auch ist der Beweis des biquadratischen Reciprocitätsgesetzes in Crelle's Journal, Bd. 28, p. 233, welcher dem oben besprochenen ebendaselbst, Bd. 27, pag. 322, entspricht, und ebenso wie dieser seine Abstammung aus der Kreistheilung verbirgt, zwar scharfsinnig, wie alle Arbeiten Eisenstein's, aber doch mehr nur scheinbar als wirklich originell; aber Eisenstein ist bei diesen nicht stehen geblieben, sondern hat bald darauf neue Beweise dieser Gesetze gegeben, welche mit zu seinen vorzüglichsten Leistungen zu rechnen sind und mit Recht die Bewunderung der ersten Mathematiker erregt haben. Es sind diefs die schon oben beiläufig erwähnten Beweise, welche die quadratischen, kubischen und biquadratischen Reciprocitätsgesetze in ähnlicher Weise umfassen, in der Art, dafs zu dem Beweise des kubischen die elliptischen Funktionen mit dem Modul $k = \sin \frac{\pi}{12}$, und für die biquadratischen Reste die elliptischen Funktionen mit dem Modul $k = \sin \frac{\pi}{4} = \sqrt{\frac{1}{2}}$, die Lemniskatenfunktionen, angewendet werden, also in allen diesen Fällen periodische Funktionen, welche für die besonderen den aliquoten Theilen ihrer Perioden entsprechenden Werthe ihrer Variabeln zu Wurzeln algebraischer Gleichungen mit ganzzahligen Coefficienten werden. Es war sehr natürlich, dafs Eisenstein in dieser Anwendung der periodischen Funktionen die wahre Quelle auch für die Reciprocitätsgesetze höherer Potenzreste gefunden zu haben glaubte, so daſs er schon eine gröſsere Arbeit über dieselben ankündigte; es ist aber weder ihm selbst noch anderen bisher gelungen, mit Hülfe dieser Principien irgend welche höhere Reciprocitätsgesetze zu beweisen, oder auch nur aufzufinden. Seine eigenen Bemühungen in dieser Beziehung sind schon an den achten Potenzen gescheitert, für deren Reciprocitätsbeziehung er nur specielle Resultate hat gewinnen können. Ebenso hat Eisenstein in einer späteren Arbeit über die allgemeineu Reciprocitätsgesetze für die Reste der Potenzen, deren Exponent eine beliebige Primzahl ist, welche im Jahre 1850 durch Jacobi der Akademie mitgetheilt, und in den Monatsberichten derselben veröffentlicht ist, nur einen sehr beschränkten Fall ergründen können, nämlich denjenigen, in welchem eine der beiden zn vergleichenden Primzahlen eine nichtcomplexe ist. Er hat dazu auch nicht die Prinzipien gebraucht, auf welche er früher seine Hoffnung gesetzt hatte, sondern nur die Mittel der Kreistheilung angewendet, und zwar die von mir gefundenen Ausdrücke, welche die complexen Zahlen der Kreistheilung, in ihre wirklichen oder idealen Primfaktoren zerlegt, darstellen. Endlich ist noch eine Arbeit von Eisenstein zu erwähnen, in Crelle's Journal, Bd. 39, pg. 351, in welcher er darauf ausgeht, die allgemeinen Reciprocitätsgesetze durch Induktion zu finden. Der Weg, den er dabei einschlägt, hat aber nicht zum Ziele geführt und überhaupt kein Resultat ergeben.

Der wahre Grund, warum alle die hier genannten sehr verschiedenen, scharfsinnigen und für die quadratischen, kubischen und biquadratischen Reste auch durchaus sachgemäſsen Methoden auf die Erforschung der höheren Reciprocitätsgesetze entweder gar keine, oder doch nur eine sehr beschränkte Anwendung gestattet haben, liegt in einem eigenthümlichen Umstande, welcher für die, diesen Gesetzen zu Grunde zu legenden complexen Zahlen eintritt, sobald man über die vierten Potenzen hinausgeht, nämlich in der unendlichen Anzahl der Einheiten. Die complexen Primzahlen haben in Beziehung darauf, ob sie Reste oder Nichtreste sind, ganz andere Charaktere, je nachdem man die Einheiten, mit welchen sie behaftet sein können, anders und anders wählt; die einfachsten Reciprocitätsgesetze lassen sich darum erst dann aufstellen, wenn man diese Einheiten den richtigen Bestimmungen unterworfen hat, d. h. wenn man die complexen Primzahlen

D 2

um die es sich handelt, in derjenigen Form gewählt hat, welche für die vorliegende Frage die angemessenste ist. Da ferner die Reciprocitätsgesetze für die λten Potenzreste, wo λ als Primzahl angenommen wird, zwischen je zwei aus λten Wurzeln der Einheit gebildeten complexen Primzahlen aufgestellt werden müssen, so gehört zur Erforschung dieser Gesetze nothwendig eine vollständige Theorie dieser complexen Zahlen, namentlich ihrer Primfaktoren und Einheiten. Es gehört dazu auch wesentlich die von mir in die Zahlentheorie eingeführte Erweiterung des Gaufsischen Prinzips durch die idealen Zahlen, ohne welche die complexen Zahlen in den meisten Fällen gar keine wahren Primfaktoren haben würden, nämlich keine solchen, welche in einer gegebenen complexen Zahl als die unveränderlichen Elemente derselben enthalten sein müfsten. Es gehört selbst die Kenntnifs der Klassenanzahl dieser idealen Zahlen dazu, und die Unterscheidung derjenigen Exponenten λ, für welche diese Klassenanzahl durch λ nicht theilbar ist, von denjenigen, welchen eine durch λ theilbare Klassenanzahl zukommt, so wie auch die Kenntnifs der besonderen Eigenschaften, welche die Einheiten besitzen, je nachdem der Wurzelexponent λ der einen oder der anderen Art angehört. Erst nachdem ich diese ganze Theorie der aus λten Wurzeln der Einheit gebildeten complexen Zahlen, und zwar mit der schon in meiner ersten Schrift über diesen Gegenstand ausgesprochenen Absicht, sie für die höheren Reciprocitätsgesetze benutzen zu können, in hinreichender Vollständigkeit erarbeitet hatte, ging ich an die Erforschung dieser Gesetze selbst, und es gelang mir im Jahre 1847 dieselben für die Reste der λten Potenzen, wenn λ eine Primzahl ist, für welche die Klassenanzahl der idealen Zahlen durch λ nicht theilbar ist, in ihrer einfachsten Form aufzufinden. Nachdem ich dieselben durch berechnete Tafeln in ziemlich grofser Ausdehnung verificirt hatte, theilte ich sie im Januar 1848 an Hrn. Dirichlet und Jacobi und später, im Mai 1850, auch der Königl. Akademie mit, m. s. die Monatsberichte. Es blieb nun noch übrig, die gefundenen Gesetze zu beweisen. Das Mittel, zu welchem ich in dieser Absicht zuerst griff, war die Theorie der Kreistheilung, welche bereits die einfachsten Beweise aller bisher ergründeten Reciprocitätsgesetze geliefert hatte, und von welcher ich um so mehr erwarten konnte, da sie durch meine Arbeiten über complexe Zahlen wesentlich gefördert worden war. Ich fand auch in der That durch dieses Mittel die einfachen Beweise der

Ergänzungssätze zu dem allgemeinen Reciprocitätsgesetze, nämlich die Charaktere oder die Indices der Einheiten und der Zahl $1-\alpha$, des Primfaktors von λ, welche Beweise ich gleichzeitig mit dem, nur durch Induktion erhärteten, allgemeinen Reciprocitätsgesetze der Königl. Akademie mitgetheilt habe. Ich erkannte aber bald, dafs die Kreistheilung allein die vollständigen Reciprocitätsgesetze für λte Potenzreste zwischen je zwei complexen Primzahlen nicht geben könne, wenn λ gröfser als drei ist, oder was dasselbe ist, wenn aufser den einfachen Einheiten ± 1, $\pm\alpha$, $\ldots \pm\alpha^{\lambda-1}$ unendlich viele Einheiten existiren. Der Grund ist der, dafs in allen complexen Zahlen der Kreistheilung die conjugirten, wirklichen oder idealen Primfaktoren nur in solchen Verbindungen vorkommen, dafs, wenn man einen derselben mit einer Einheit $E(\alpha)$, für welche $E(\alpha) = E(\alpha^{-1})$ ist, und die ihm conjugirten Primfaktoren mit den entsprechenden conjugirten Einheiten multiplicirt, diese complexen Zahlen der Kreistheilung ganz ungeändert bleiben. Wegen dieses Umstandes kann die Kreistheilung keine Reciprocitätsgesetze geben, bei denen die verschiedene Wahl solcher Einheiten in den complexen Primzahlen einen Unterschied des Potenzcharakters bewirkt. Dieses der Theorie der Kreistheilung unübersteigliche Hindernifs für die Erkenntnifs der allgemeinen Reciprocitätsgesetze ist auch durch andere Methoden in Jacobi's und Eisenstein's Arbeiten selbst nicht in irgend einem besonderen Falle besiegt worden. Den ersten Schritt über diese Gränze hinaus habe ich in einem Beweise des Reciprocitätsgesetzes für λte Potenzreste, welches unter je zwei conjugirten complexen Primzahlen Statt hat, in Crelle's Journal, Bd. 50, pag. 212, gemacht, und zwar mit Hülfe gewisser, aus Einheiten gebildeter Ausdrücke, welche die Lagrangesche Resolvente der Kreistheilung als speciellen Fall in sich enthalten, und darum als eine Verallgemeinerung der Kreistheilung angesehen werden können. Aber auch dieses neue, für die Theorie der complexen Zahlen überhaupt sehr nützliche Instrument, welches in der gegenwärtigen Untersuchung ebenfalls vielfache Anwendung finden wird, hat mir die vollständigen Beweise der Reciprocitätsgesetze nicht gegeben, und ich habe mich endlich genöthigt gesehen, den bis dahin eingeschlagenen Weg der Verallgemeinerung der Kreistheilung aufzugeben und andere Mittel und Wege aufzusuchen. Ich wendete meine Aufmerksamkeit auf die Methode des zweiten Gaufsischen Beweises des *theorema fundamentale,* welcher auf der Theorie der quadratischen Formen beruht. Dieser Beweis, obgleich seine Methode bis dahin auf die quadratischen Reste beschränkt geblieben war, schien mir in seinen Prinzipien denjenigen Charakter der Allgemeinheit zu haben, welcher hoffen liefs, dafs dieselben mit Erfolg auch auf die Untersuchung der Reste höherer Potenzen möchten angewendet werden können, und diese meine Erwartung ist in der That erfüllt worden.

Der Hauptnerv dieses zweiten Gaufsischen Beweises liegt in der Eintheilung der Klassen der quadratischen Formen einer gegebenen Determinante in *Genera,* welche durch die Charaktere der Klassen bestimmt sind, und namentlich darin, dafs die Anzahl der wirklich vorhandenen *Genera* nur höchstens halb so grofs ist, als die Anzahl der angebbaren, d. h. derjenigen, welche vermöge der vorhandenen Charaktere der Klassen möglicherweise Statt haben könnten. Nachdem dieser Punkt bei Gaufs durch die Untersuchung der *Classes ancipites* festgestellt ist, wird der Beweis des Reciprocitätsgesetzes in der Art geführt, dafs gezeigt wird: wenn dasselbe nicht Statt hätte, so müfste die Anzahl der wirklich vorhandenen *Genera* gröfser sein, als die Hälfte der blofs angebbaren. Um nun nach diesen Prinzipien die Reciprocitätsgesetze der Reste und Nichtreste der λten Potenzen zu ergründen, hat man anstatt der Formen des zweiten Grades mit zwei Unbestimmten hier Formen des λten Grades mit λ Unbestimmten zu Grunde zu legen, und zwar Formen, deren Coefficienten nicht gewöhnliche ganze Zahlen, sondern aus λten Wurzeln der Einheit gebildete complexe ganze Zahlen sind. Die Theorie dieser Formen des λten Grades mufs auch bis zu dem Punkte ergründet werden, der in der Theorie der quadratischen Formen der Stelle entspricht, an welcher der Gaufsische Beweis seinen Platz gefunden hat, und sogar noch bedeutend weiter, weil der Umstand, dafs in der Theorie der λten Potenzreste nicht nur Reste von Nichtresten, sondern auch die Nichtreste von $\lambda - 1$ verschiedenen Arten zu unterscheiden sind, nöthig macht, dafs wenigstens in gewissen Hauptfällen die Anzahl der wirklich vorhandenen *Genera* genau ermittelt, und nicht blofs eine Gränze gefunden werde, welche diese Anzahl niemals überschreiten kann. Diese schwer zu bewältigende Arbeit hat offenbar der Anwendung der Principien dieses Gaufsischen Beweises auf die Untersuchung der höheren Potenzreste bisher entgegengestanden. Ich selbst würde auch nicht gewagt haben, diese Arbeit zu unternehmen, wenn ich nicht die Überzeugung gehabt hätte, dafs gewifse, für den vorliegenden Zweck passend zu wählende specielle Formen des λten

Grades mit complexen Coefficienten ausreichen möchten, und wenn ich nicht in meinem Principe der idealen Faktoren der complexen Zahlen ein Mittel gehabt hätte, die Betrachtung der zerlegbaren Formen des λten Grades durch die bei weitem einfachere Betrachtung der complexen Zahlen, welche aus den Wurzeln einer Gleichung des Grades λ gebildet sind, und der idealen Faktoren derselben, zu ersetzen.

Ich gebrauche zu dem vorliegenden Zwecke zwei über einander liegende Theorieen complexer Zahlen, deren niedere, nur die Wurzeln der Gleichung $\alpha^{\lambda}=1$ enthaltende, aus meinen früheren Arbeiten als bekannt gelten kann, und deren höhere aufser dieser λten Wurzel der Einheit noch die Wurzel einer Gleichung des λten Grades enthält. Diese höhere Theorie der complexen Zahlen wird alsdann weiter in drei verschiedene Stufen getheilt, welche zu einander in derselben Beziehung stehen, wie die *Ordines derivati* zu dem *Ordo primitivus,* welches Verhältnifs in der Theorie der complexen Zahlen die eigenthümliche Bedeutung hat, dafs gewisse complexe Zahlen, welche in der niederen Stufe als wirkliche ganze Zahlen nicht darstellbar sind, sondern nur als wirkliche gebrochene, und welche darum als ideale gelten müssen, innerhalb der höheren Stufe als wirkliche und ganze complexe Zahlen dargestellt werden können. Der Gedanke, welcher dieser Anwendung verschiedener einander übergeordneter Theorieen der complexen Zahlen zu Grunde liegt, nämlich dafs dasjenige, was in der Theorie der gewöhnlichen Zahlen schwer oder vielleicht gar nicht zu finden ist, in einer richtig gewählten complexen Theorie gesucht werden mufs, und dafs ferner dasjenige, was auch diese versagt, weiter in einer passenden höheren Theorie zu suchen ist und so fort, bis das vorgesteckte Ziel vollständig erreicht ist, darf übrigens nur als eine einfache Consequenz des ursprünglichen Gaufsischen Gedankens der Einführung complexer ganzer Zahlen überhaupt angesehen werden. Man hat auch bereits Beispiele des Aufsteigens von einer complexen Theorie zu einer höheren, denn wenn z. B. bei dem Jacobischen Beweise des kubischen Reciprocitätsgesetzes $\alpha^3=1$ und $x^p=1$ ist, und p Primzahl der Form $6n+1$, so ist die Anwendung der Lagrangeschen Resolvente der Kreistheilung, welche die beiden Wurzeln α und x zugleich enthält, nichts anderes, als das Aufsteigen von complexen Zahlen, welche α allein enthalten, zu complexen Zahlen der höheren, die Wurzeln α und x enthaltenden Theorie.

Nachdem ich nun den gegenwärtigen Standpunkt der Wissenschaft in der Theorie der Potenzreste, so wie auch die leitenden Gedanken für den in derselben zu machenden weiteren Fortschritt angegeben habe, werde ich in der gegenwärtigen Abhandlung die Theorie derjenigen complexen Zahlen, auf welche der hier zu gebende Beweis der allgemeinen Reciprocitätsgesetze sich gründet, in soweit entwickeln, als es für den vorliegenden Zweck nöthig ist, und sodann den Beweis der Reciprocitätsgesetze selbst folgen lassen, durch welchen dieselben genau in derjenigen Ausdehnung, in welcher ich sie im Mai 1850 der Königlichen Akademie ohne Beweise mitgetheilt habe, vollständig nnd streng begründet werden.

§. 1.

Definition und allgemeine Eigenschaften der complexen Zahlen, welche der gegenwärtigen Untersuchung zu Grunde gelegt werden.

Die in der folgenden Untersuchung in Anwendung kommenden complexen Zahlen sollen aufser den Wurzeln der Gleichung des $\lambda-1$ten Grades

(1.) $$\alpha^{\lambda-1}+\alpha^{\lambda-2}+\alpha^{\lambda-3}+\ldots+\alpha+1=0$$

auch die Wurzeln der Gleichung des λten Grades

(2.) $$w^{\lambda}=D(\alpha)$$

enthalten, in welcher $D(\alpha)$ eine nur die Wurzel α enthaltende ganze complexe Zahl ist. Wenn diese Zahl $D(\alpha)$, welche als Determinante der, die Wurzeln α und w enthaltenden, complexen Zahlen bezeichnet werden soll, nicht eine vollständige λte Potenz ist, so ist diese Gleichung (2.) eine irreductible, in dem Sinne, dafs sie nicht in Faktoren zerlegt werden kann, deren Coefficienten ganze nur die Wurzel α enthaltende complexe Zahlen sind.

Jede ganze rationale Funktion der Wurzeln α und w mit ganzzahligen Coefficienten soll als eine aus diesen Wurzeln gebildete ganze complexe Zahl angesehen, und kurz als complexe Zahl in w bezeichnet werden. Weil vermöge der Gleichung (2.) die Potenzen von w, welche höher sind, als die $\lambda-1$te, durch niedere ersetzt werden können, so folgt, dafs jede complexe Zahl in w in die Form

(3.) $$F(w)=A+A_1w+A_2w^2+\ldots+A_{\lambda-1}w^{\lambda-1}$$

gesetzt werden kann, in welcher die Coefficienten $A, A_1, A_2, \ldots A_{\lambda-1}$ nur die Wurzel α enthaltende complexe ganze Zahlen sind, welche zum Unterschiede von den aufserdem auch w enthaltenden kurz als complexe Zahlen in α bezeichnet werden. Aus der Irreductibilität der Gleichung (2.) folgt auch, dafs eine jede gegebene complexe Zahl in w nur auf eine einzige Weise in diese Form gesetzt werden kann.

Die zu einer complexen Zahl $F(w)$ conjugirten Zahlen sind diejenigen, welche man erhält, indem man der Wurzel w ihre λ verschiedenen Werthe $w, w\alpha, w\alpha^2, \ldots w\alpha^{\lambda-1}$ giebt. Das Produkt dieser λ conjugirten Zahlen

$$F(w)\, F(w\alpha)\, F(w\alpha^2) \ldots F(w\alpha^{\lambda-1}) = NF(w)$$

wird die Norm einer derselben genannt, und ist eine complexe Zahl in α.

Es soll ferner eine bestimmte Art dieser aus den Wurzeln der Gleichung (2.) gebildeten complexen Zahlen besonders betrachtet werden, in welcher diese Wurzeln nicht einzeln, sondern nur in folgenden bestimmten Verbindungen vorkommen:

$$\begin{aligned}
z_0 &= \varrho\,(1 + w + w^2 + \ldots + w^{\lambda-1}) \\
z_1 &= \varrho\,(1 + \alpha w + \alpha^2 w^2 + \ldots + \alpha^{\lambda-1} w^{\lambda-1}) \\
z_2 &= \varrho\,(1 + \alpha^2 w + \alpha^4 w^2 + \ldots + \alpha^{2\lambda-2} w^{\lambda-1}) \qquad (4.)\\
&\vdots \\
z_{\lambda-1} &= \varrho\,(1 + \alpha^{\lambda-1} w + \alpha^{2\lambda-2} w^2 + \ldots + \alpha^{(\lambda-1)^2} w^{\lambda-1}),
\end{aligned}$$

wo ϱ als abgekürztes Zeichen für die sehr häufig vorkommende Zahl $1-\alpha$ gesetzt ist, welche Bedeutung dieser Buchstabe auch in dem Folgenden überall behalten soll.

Der allgemeine Ausdruck

$$z_k = \varrho\,(1 + \alpha^k w + \alpha^{2k} w^2 + \ldots + \alpha^{(\lambda-1)k} w^{\lambda-1})$$

kann auch in folgende Form gesetzt werden:

* $$z_k = \frac{\rho\,(1 - D(\alpha))}{1 - \alpha_k w}$$

und giebt so

$$\alpha^k w = 1 - \frac{\rho\,(1 - D(\alpha))}{z_k} \quad .$$

Multiplicirt man mit z_k und erhebt beide Seiten dieser Gleichung zur λten Potenz, so erhält man folgende Gleichung des λten Grades:

$$D(\alpha)\, z_k^\lambda = \big(z_k - \varrho\,(1 - D(\alpha))\big)^\lambda,$$

welcher, weil k in den Coefficienten nicht vorkommt, alle λ Werthe $z_0, z_1,$

$z_2, \ldots z_{\lambda-1}$ als Wurzeln genügen müssen. Schreibt man also z statt z_k und entwickelt nach Potenzen von z, so erhält man

$$(5.)\qquad z^{\lambda}-\lambda\varrho z^{\lambda-1}+\frac{\lambda(\lambda-1)}{1.2}\varrho^2(1-D(\alpha))z^{\lambda-2}-\ldots-\varrho^{\lambda}(1-D(\alpha))^{\lambda-1}=0$$

welche Gleichung als die, der besonderen Art von complexen Zahlen in w zu Grunde liegende Gleichung anzusehen ist, in der Art, dafs eine jede rationale und ganze Funktion der Wurzeln derselben, welche ganze complexe Zahlen in α zu Coefficienten hat, als eine complexe Zahl dieser besonderen Art angesehen werden soll. Zur Unterscheidung von den allgemeineren complexen Zahlen in w sollen die aus den Wurzeln der Gleichung (5.) gebildeten, als complexe Zahlen in z bezeichnet werden.

Nimmt man zwei verschiedene Wurzeln der Gleichung (5.):

$$z_k=\frac{\varrho(1-D(\alpha))}{1-\alpha^k w},\quad z_h=\frac{\varrho(1-D(\alpha))}{1-\alpha^h w},$$

und eliminirt die Gröfse w aus diesen Ausdrücken, so erhält man

$$(6.)\qquad z_k z_h=\frac{\varrho(1-D(\alpha))}{\alpha^h-\alpha^k}(\alpha^h z_h-\alpha^k z_k).$$

Diese Formel zeigt, wie das Produkt zweier beliebigen, aber verschiedenen Wurzeln der Gleichung (5.) als lineäre Funktion derselben Wurzeln ausgedrückt wird, und zwar mit Coefficienten welche ganze complexe Zahlen in α sind, weil der Nenner $\alpha^h-\alpha^k$ gegen den Faktor $\varrho=1-\alpha$ des Zählers hinweggehoben werden kann. Es läfst sich aber auch das Quadrat einer jeden Wurzel z_k als lineäre Funktion aller Wurzeln darstellen; denn man hat aus der Gleichung (5.) die Summe aller Wurzeln

$$(7.)\qquad z_0+z_1+z_2+\ldots.+z_{\lambda-1}=\lambda\varrho,$$

also wenn man mit z_k multiplicirt und die Produkte zweier verschiedenen Wurzeln lineär ausdrückt, so erhält man z_k^2 als lineäre Funkion aller Wurzeln und zwar ebenfalls mit ganzen complexen Coefficienten. Da also das Produkt je zweier Wurzeln der Gleichung (5.), sie mögen verschieden oder auch dieselben sein, als ganze lineäre Funktion aller Wurzeln mit ganzen Coefficienten sich darstellen läfst, so folgt unmittelbar, dafs dasselbe auch für ein jedes Produkt beliebig vieler Wurzeln der Fall ist, und darum auch für jede ganze rationale Funktion der Wurzeln. Man hat daher folgenden Satz:

(I.) Jede ganze rationale Funktion der Wurzeln $z_0, z_1, \ldots. z_{\lambda-1}$ läfst sich als lineäre Funktion dieser Wurzeln darstellen, und wenn die Coefficienten dieser ganzen rationalen Funktion ganze complexe Zahlen in α sind, so sind auch die Coefficienten ihres Ausdrucks in der linearen Form nur ganze complexe Zahlen in α.

Die complexen Zahlen in z lassen sich also stets in folgender Form darstellen:

(8.) $$F(z) = C + Bz + B_1 z_1 + B_2 z_2 + \dots + B_{\lambda-1} z_{\lambda-1},$$

in welcher die Coefficienten C, B, B_1, ... $B_{\lambda-1}$ ganze complexe Zahlen in α sind. Man kann auch diese aus $\lambda + 1$ Gliedern bestehende Form mit Hülfe der Gleichung (7.) so vereinfachen, dafs sie ein Glied weniger enthält. Das erste Glied C läfst sich auf diese Weise im Allgemeinen nicht entfernen, ohne dafs die Coefficienten dieser Form Brüche mit dem Nenner $\lambda\varrho$ werden, jedes andere Glied aber kann weggeschafft werden, ohne dafs dieser Übelstand eintritt. Schafft man das letzte Glied weg, so erhält man die Form

(9.) $$F(z) = C + Bz + B_1 z_1 + B_2 z_2 + \dots + B_{\lambda-2} z_{\lambda-2}.$$

Diese Form ist eine solche, in welche eine gegebene complexe Zahl $F(z)$ sich nur auf *eine* Weise setzen läfst. Wenn nämlich die Zahl $F(z)$ zwei verschiedene Darstellungen derselben Form hätte, so würde durch Subtraktion derselben eine Gleichung von der Form

(10.) $$0 = c + bz + b_1 z_1 + b_2 z_2 + \dots + b_{\lambda-2} z_{\lambda-2}$$

entstehen, deren Coefficienten c, b, b_1, ... $b_{\lambda-2}$ nicht alle zugleich gleich Null wären. Drückt man nun vermittelst der Ausdrücke (4.) die Wurzeln z, z_1, ... $z_{\lambda-2}$ alle durch w aus, so erhält man eine Gleichung von folgender Form:

(11.) $$o = c + \varrho m + \varrho m_1 w + \varrho m_2 w^2 + \dots + \varrho m_{\lambda-1} w^{\lambda-1},$$

in welcher die Gröfsen m, m_1, m_2, ... $m_{\lambda-1}$ folgende Werthe haben:

$$\begin{aligned}
m &= b + b_1 + b_2 + \dots + b_{\lambda-2}\\
m_1 &= b + \alpha b_1 + \alpha^2 b_2 + \dots + \alpha^{\lambda-2} b_{\lambda-2}\\
m_2 &= b + \alpha^2 b_1 + \alpha^4 b_2 + \dots + \alpha^{2\lambda-4} b_{\lambda-2}\\
&\vdots\\
m_{\lambda-1} &= b + \alpha^{\lambda-1} b_1 + \alpha^{2\lambda-2} b_2 + \dots + \alpha^{(\lambda-1)(\lambda-2)} b_{\lambda-2}
\end{aligned}$$

Wegen der Irreductibilität der Gleichung $w^\lambda = D(\alpha)$ kann aber die Gleichung (11.) nicht anders bestehen, als wenn die Coefficienten der einzelnen Potenzen von w alle gleich Null sind, man hat daher

E 2

$$m_1 = 0, \; m_2 = 0, \; \dots \; m_{\lambda-1} = 0 \text{ und } c + \varrho m = 0.$$

Die ersten $\lambda - 1$ Gleichungen geben nothwendig

$$b = 0, \; b_1 = 0, \; b_2 = 0, \; \dots \; b_{\lambda-2} = 0,$$

weil die Determinante dieses Systems lineärer Gleichungen nicht gleich Null ist, hieraus folgt sodann, dafs auch $m = 0$ sein mufs, und darum auch $c = 0$. Die Gleichung (10.) kann also nicht bestehen, ohne dafs alle ihre Coefficienten einzeln gleich Null sind, woraus folgt, dafs die complexe Zahl $F(z)$ nur auf eine einzige Weise in die Form (9.) gesetzt werden kann.

Als die conjugirten complexen Zahlen zu $F(z)$ sollen diejenigen betrachtet werden, welche man aus dieser erhält, indem man die Indices aller Wurzeln $z, z_1, z_2, \dots z_{\lambda-2}$ um eine und dieselbe Zahl vermehrt, wobei, wenn diese Indices gröfser als $\lambda - 1$ werden, statt derselben nur ihre kleinsten Reste nach dem Modul λ zu nehmen sind. Die λ conjugirten Zahlen, welche man auf diese Weise erhält, sollen kurz durch $F(z)$, $F(z_1)$, $F(z_2)$, ... $F(z_{\lambda-1})$ bezeichnet werden, so dafs allgemein

$$F(z_k) = C + Bz_k + B_1 z_{k+1} + B_2 z_{k+2} + \dots + B_{\lambda-1} z_{k+\lambda-1}$$

eine jede conjugirte Zahl zu $F(z)$ darstellt. Das Produkt aller conjugirten Zahlen

$$F(z) \, F(z_1) \, F(z_2) \, \dots\dots \, F(z_{\lambda-1}) = NF(z),$$

welches die Norm einer derselben ausmacht, ist als symmetrische Funktion aller Wurzeln der Gleichung (2.) nur eine complexe Zahl in α.

§. 2.

Gegenseitiges Verhältnifs der complexen Zahlen in z und in w.

Alle ganzen complexen Zahlen in z sind zugleich auch ganze complexe Zahlen in w, denn die Wurzeln $z_0, z_1, z_2 \dots z_{\lambda-1}$ selbst sind ganze rationale Funktionen von w mit ganzen Coefficienten. Es läfst sich auch umgekehrt jede ganze rationale Funktion von w als lineäre Funktion der Wurzeln $z_0, z_1, z_2 \dots z_{\lambda-1}$ darstellen, jedoch im Allgemeinen nur so, dafs in den Coefficienten dieser lineären Funktion Brüche vorkommen. Durch Umkehrung des Systems der Gleichungen, welche $z_0, z_1 \dots z_{\lambda-1}$ als Funktionen von w geben, erhält man nämlich

$$
(1.)\quad \begin{aligned}
\lambda\varrho &= z_0 + z_1 + z_2 + \dots + z_{\lambda-1}\\
\lambda\varrho w &= z_0 + \alpha^{-1} z_1 + \alpha^{-2} z_2 + \dots + \alpha^{-\lambda+1} z_{\lambda-1}\\
\lambda\varrho w^2 &= z_0 + \alpha^{-2} z_1 + \alpha^{-4} z_2 + \dots + \alpha^{-2\lambda+2} z_{\lambda-1}\\
&\vdots\\
\lambda\varrho w^{\lambda-1} &= z_0 + \alpha^{-(\lambda-1)} z_1 + \alpha^{-(2\lambda-2)} z_2 + \dots + \alpha^{-(\lambda-1)(\lambda-1)} z_{\lambda-1}
\end{aligned}
$$

Multiplicirt man nun den allgemeinen Ausdruck einer ganzen complexen Zahl in w

$$F(w) = A + A_1 w + A_2 w^2 + \dots + A_{\lambda-1} w^{\lambda-1}$$

mit $\lambda\varrho$, und drückt die Gröſsen $\lambda\varrho$, $\lambda\varrho w$, $\lambda\varrho w^2$ .... $\lambda\varrho w^{\lambda-1}$ nach diesen Formeln durch die Wurzeln z_0, z_1 .... $z_{\lambda-1}$ aus, so erhält man einen Ausdruck des $\lambda\varrho F(w)$ als lineäre Funktion dieser Wurzeln mit ganzen Coefficienten, welcher durch Anwendung der Summenzeichen sich folgendermaaſsen darstellen läſst:

$$\lambda\varrho F(w) = \sum_{0}^{\lambda-1}{}_k \sum_{0}^{\lambda-1}{}_h A_k \alpha^{-kh} z_h.$$

Wenn nun mittelst der ersten der Gleichungen (1.) $z_{\lambda-1}$ weggeschafft und durch ϱ dividirt wird, so wird:

$$(2.)\qquad \lambda F(w) = \sum_{0}^{\lambda-1}{}_k \sum_{0}^{\lambda-2}{}_h \frac{A_k(\alpha^{-kh} - \alpha^k)}{1-\alpha} z_h + \lambda \sum_{0}^{\lambda-1}{}_k A_k.$$

Da der Nenner $1-\alpha$ gegen den Faktor $\alpha^{-kh} - \alpha^k$ des Zählers sich wegheben lässt, so sind alle Coefficienten dieses lineären Ausdrucks ganz, und man hat demnach den Satz:

(I.) Das λfache einer jeden ganzen complexen Zahl in w läſst sich stets als ganze complexe Zahl in z darstellen.

Für die speciellere complexe Zahl $(1-w)^n$, wo $n<\lambda$, hat man

$$A=1,\ A_1 = -\frac{n}{1}\ A_2 = +\frac{n(n-1)}{2}, \dots\dots$$

und folglich

$$\lambda(1-w)^n = \sum_{0}^{\lambda-2}{}_h \frac{(1-\alpha^{-h})^n - (1-\alpha)^n}{1-\alpha} z_h.$$

Weil die Zahl λ den Faktor $\varrho = 1-\alpha$ genau $\lambda-1$ mal enthält, und auſserdem nur Einheiten, so ist $\varrho^{\lambda-1} = \lambda E(\alpha)$, wo $E(\alpha)$ eine Einheit bezeichnet; multiplicirt man daher mit $E(\alpha)$, und dividirt durch ϱ^{n-1}, so erhält man:

$$\varrho^{\lambda-n}(1-w)^n = E(\alpha)\sum_0^{\lambda-2}{}_h\left(\left(\frac{1-\alpha^{-h}}{1-\alpha}\right)^n - 1\right)z_h\,.$$

Hieraus folgt, dafs $\varrho^{\lambda-n}(1-w)^n$ als ganze complexe Zahl in z sich darstellen läfst, wenn n eine ganze positive Zahl und kleiner als λ ist.

Die nothwendigen und hinreichenden Bedingungen dafür, dafs die allgemeine complexe Zahl $F(w)$ selbst, und nicht blofs das λfache derselben, als ganze complexe Zahl in z mit ganzen Coefficienten sich darstellen läfst, liegen vermöge der Gleichung (2.) darin, dafs die Congruenz

$$\sum_0^{\lambda-1}{}_k A_k\,(\alpha^{-kh} - \alpha^k) \equiv 0, \text{ mod. } \lambda\varrho,$$

für alle Werthe des $h = 0, 1, 2, \ldots \lambda-2$ Statt habe, welche, wenn h in $h-1$ verwandelt und durch α^k dividirt wird, auch so dargestellt werden kann:

$$(3.)\qquad \sum_1^{\lambda-1}{}_k A_k\,(\alpha^{-hk} - 1) \equiv 0, \text{ mod. } \lambda\varrho,$$

für $h = 1, 2, 3 \ldots \lambda-1$. Anstatt des Moduls $\lambda\varrho$ kann man, weil λ, abgesehen von einer Einheit, der $\lambda-1$ten Potenz von ϱ gleich ist, auch den Modul ϱ^λ wählen. Entwickelt man nun

$$\alpha^{-kh} = (1-(1-\alpha^{-h}))^k$$

nach dem binomischen Satze und setzt der Kürze wegen

$$\sum_1^{\lambda-1}{}_k \frac{k(k-1)\ldots\ldots(k-n+1)}{1.\,2\ldots\ldots n} A_k = G_k\,,$$

so geht die Congruenz (3.) in folgende über:

$$(4.)\qquad \begin{aligned} -G_1(1-\alpha^{-h}) + G_2(1-\alpha^{-h})^2 - G_3(1-\alpha^{-h})^3 + \ldots \\ + G_{\lambda-1}(1-\alpha^{-h})^{\lambda-1} \equiv 0, \text{ mod. } \varrho^\lambda. \end{aligned}$$

Hebt man aus dieser Congruenz und ihrem Modul den gemeinschaftlichen Faktor $1-\alpha^{-h}$ hinweg, und giebt sodann dem h alle Werthe $h = 1, 2, 3, \ldots\ldots \lambda-1$, welchen man auch den Werth $h = 0$ hinzufügen kann, weil für diesen die Congruenz identisch erfüllt ist, so erhält man durch Addition:

$$G_1 \equiv 0, \quad \text{mod. } \varrho^{\lambda-1}.$$

Läfst man nun das erste Glied aus der Congruenz (4.) hinweg, da dasselbe congruent Null ist nach dem Modul ϱ^λ, hebt sodann den gemeinschaftlichen

Faktor $(1 - \alpha^{-h})^2$ aus dieser Congruenz und dem Modul heraus, und bildet die Summe für alle Werthe $h = 0, 1, 2, \ldots \lambda - 1$, so erhält man

$$G_2 \equiv 0, \quad \text{mod. } \varrho^{\lambda-2}.$$

In derselben Weise weiter schliefsend erhält man allgemein

$$G_k \equiv 0, \quad \text{mod. } \varrho^{\lambda-k},$$

für $k = 1, 2, 3, \ldots \lambda - 1$. Diese $\lambda - 1$ Congruenzen müssen also nothwendig erfüllt sein, damit $F(w)$ als ganze complexe Zahl in z sich darstellen lasse. Dafs die Erfüllung dieser Congruenzen auch zugleich die hinreichende Bedingung hierfür ist, wird leicht gezeigt, wenn man $F(w)$ nach Potenzen von $1 - w$ entwickelt, wodurch man

$$F(w) = G + G_1(1 - w) + G_2(1 - w)^2 + \ldots + G_{\lambda-1}(1 - w)^{\lambda-1}$$

erhält. Wenn nämlich $G_k \equiv 0$, mod. $\varrho^{\lambda-k}$ ist, so ist nach dem oben bewiesenen Satze: dafs $\varrho^{\lambda-k}(1 - w)^k$ als ganze complexe Zahl in z sich darstellen läfst, nothwendig jeder einzelne Theil dieser Entwickelung von $F(w)$, und darum auch $F(w)$ selbst, als ganze complexe Zahl in z darstellbar. Das gefundene Resultat giebt folgenden Lehrsatz:

(II.) Die nothwendigen und hinreichenden Bedingungen dafür, dafs eine gegebene ganze complexe Zahl in w:

$$F(w) = A + A_1 w + A_2 w^2 + \ldots + A_{\lambda-1} w^{\lambda-1}$$

als ganze complexe Zahl in z sich darstellen läfst, sind in folgenden $\lambda - 1$ Congruenzen enthalten:

$$A_1 + 2A_2 + 3A_3 + 4A_4 + \ldots + (\lambda - 1)A_{\lambda-1} \equiv 0, \text{ mod. } \varrho^{\lambda-1},$$

$$A_2 + 3A_3 + 6A_4 + \ldots + \frac{(\lambda-1)(\lambda-2)}{1.\,2.} A_{\lambda-1} \equiv 0, \text{ mod. } \varrho^{\lambda-2},$$

$$A_3 + 4A_4 + \ldots + \frac{(\lambda-1)(\lambda-2)(\lambda-3)}{1.\,2.\,3} A_{\lambda-1} \equiv 0, \text{ mod. } \varrho^{\lambda-3},$$

$$\vdots$$

$$A_{\lambda-1} \equiv 0. \text{ mod. } \varrho.$$

§. 3.

Die den Gleichungswurzeln der complexen Zahlen in w entsprechenden Congruenzwurzeln.

Es sind nun zunächst die Bedingungen zu untersuchen, unter welchen eine gegebene complexe Primzahl in α, welche in dieser niederen Theorie wirklich oder auch ideal sein kann, ein Divisor der Norm einer complexen Zahl der höheren Theorie in w ist. Diese Untersuchung wird zugleich auch für die Normen der complexen Zahlen in z ausreichen, weil jede ganze complexe Zahl in z als eine ganze complexe Zahl in w dargestellt werden kann.

Sei $\phi(\alpha)$ irgend eine wirkliche oder ideale complexe Primzahl in α, jedoch nicht eine von denen, welche in der Determinante $D(\alpha)$ enthalten sind und auch nicht die besondere Primzahl $\varrho = 1 - \alpha$. Dieselbe sei ein Primfaktor der nicht complexen Primzahl q, und es sei t der Exponent, zu welchem q gehört, nach dem Modul λ, so dafs $q^t \equiv 1$, mod. λ ist, und t der kleinste Exponent welcher dieser Bedingung entspricht, alsdann hat man, wie aus der Theorie der complexen Zahlen in α bekannt ist:

$$N\phi(\alpha) = \phi(\alpha)\,\phi(\alpha^2)\,\phi(\alpha^3)\,\ldots.\,\phi(\alpha^{\lambda-1}) = q^t.$$

Ferner ist für jede nicht durch $\phi(\alpha)$ theilbare wirkliche complexe Zahl $D(\alpha)$

$$D(\alpha)^{\frac{1}{\lambda}(N\phi(\alpha)-1)} = D(\alpha)^{\frac{1}{\lambda}(q^t-1)} \equiv \alpha^i, \quad \text{mod. } \phi(\alpha),$$

wo i eine der Zahlen $0, 1, 2, \ldots. \lambda - 1$ ist, und es ist $D(\alpha)$ ein λter Potenzrest von $\phi(\alpha)$, wenn $i = 0$ ist, ein Nichtrest, wenn i nicht gleich Null ist, und zwar ein Nichtrest der $i$ten Klasse. Es sei endlich

$$F(w) = A + A_1 w + A_2 w^2 + \ldots. + A_{\lambda-1} w^{\lambda-1}$$

eine beliebige complexe Zahl in w.

Um nun zu untersuchen, ob $\phi(\alpha)$ ein Divisor von $NF(w)$ ist, wird $F(w)$ zur $q$ten Potenz erhoben, und zwar in der Art, dafs die den Faktor q enthaltenden Glieder dieser $q$ten Potenz weggelassen werden, wodurch man eine Congruenz nach dem Modul q erhält (¹). Weil in einem Polynom, welches zur $q$ten Potenz erhoben wird, wenn q Primzahl ist, aufser den $q$ten Potenzen der einzelnen Theile alle übrigen Glieder den Faktor q enthalten, so wird

$$(1.)\quad F(w)^q \equiv A^q + A_1^q w^q + A_2^q w^{2q} + \ldots. + A_{\lambda-1}^q w^{(\lambda-1)q}, \text{ mod. } q.$$

(¹) Die Congruenz zweier complexen Zahlen in w in Beziehung auf einen Modul, welcher eine nichtcomplexe Zahl, oder auch eine complexe Zahl in α ist, hat die Bedeutung, dafs wenn beide Seiten der Congruenz in die oben aufgestellte Normalform gesetzt werden, in welche eine gegebene complexe Zahl in w sich nur auf eine einzige Weise setzen läfst, die Coefficienten aller λ Glieder auf der einen Seite den entsprechenden auf der anderen Seite einzeln congruent sein müssen. Dasselbe gilt auch für die Congruenzen unter complexen Zahlen in z.

Erhebt man in derselben Weise t mal hinter einander zur $q$ten Potenz, so erhält man

(2.) $F(w)^{q^t} \equiv A^{q^t} + A_1^{q^t} w^{q^t} + A_2^{q^t} w^{2q^t} + \ldots + A_{\lambda-1}^{q^t} w^{(\lambda-1)q^t}$, mod. q.

Weil die Coefficienten $A, A_1, A_2 \ldots A_{\lambda-1}$ complexe Zahlen in α sind, und $q^t \equiv 1$, mod. λ, so hat man, wie aus der Theorie dieser complexen Zahlen bekannt ist:

$$A_h^{q^t} \equiv A_h\,,\ \text{mod. } q,$$

ferner ist

$$w^{q^t} = ww^{q^t-1} = w\,D(\alpha)^{\frac{1}{\lambda}(q^t-1)},$$

und weil

$$D(\alpha)^{\frac{1}{\lambda}(q^t-1)} \equiv \alpha^i\,,\ \text{mod. } \phi(\alpha),$$

so ist

$$w^{q^t} \equiv w\alpha^i,\ \text{mod. } \phi(\alpha).$$

Nimmt man nun in der Congruenz (2.) anstatt des Moduls q den Modul $\phi(\alpha)$, welcher ein Theiler von q ist, und setzt die gefundenen Ausdrücke für $A_h^{q^t}$ und w^{q^t} ein, so hat man

$$F(w)^{q^t} \equiv A + A_1 w\alpha^i + A_2 w^2\alpha^{2i} + \ldots + A_{\lambda-1} w^{\lambda-1}\alpha^{(\lambda-1)i},\ \text{mod. } \phi(\alpha),$$

welche Congruenz auch so dargestellt werden kann:

(3.) $$F(w)^{q^t} \equiv F(w\alpha^i),\ \text{mod. } \phi(\alpha).$$

Erhebt man beide Seiten dieser Congruenz zu wiederholten Malen zur Potenz q^t, so erhält man daraus die verallgemeinerte

(4.) $$F(w)^{q^{ht}} \equiv F(w\alpha^{hi}),\ \text{mod. } \phi(\alpha),$$

und wenn h gleich einem Vielfachen von λ genommen wird:

$$F(w)^{q^{k\lambda t}} \equiv F(w),\ \text{mod. } \phi(\alpha),$$

aus welcher Congruenz unmittelbar folgender Satz hervorgeht:

(I.) Wenn irgend eine Potenz einer complexen Zahl $F(w)$ durch eine nicht in $\varrho D(\alpha)$ enthaltene, wirkliche oder ideale

Primzahl $\phi(\alpha)$ theilbar ist, so ist auch diese Zahl $F(w)$ selbst durch $\phi(\alpha)$ theilbar.

Es sollen nun in der Congruenz (4.) die beiden Fälle: erstens wo i nicht congruent Null ist, und zweitens wo i congruent Null ist, nach dem Modul λ, besonders betrachtet werden.

Wenn erstens i nicht $\equiv 0$, mod. λ, also die Determinante $D(\alpha)$ ein Nichtrest von $\phi(\alpha)$ ist, so gebe man dem h in der Congruenz (4.) nach einander die Werthe 0, 1, 2, ... $\lambda - 1$, und multiplicire die so erhaltenen Congruenzen, so hat man:

$$F(w)^{1+q^t+q^{2t}+\dots+q^{(\lambda-1)t}} \equiv F(w)F(w\alpha^i)F(w\alpha^{2i}) \quad (5.)$$
$$\dots F(w\alpha^{(\lambda-1)i}), \text{ mod. } \phi(\alpha).$$

Setzt man der Kürze wegen

$$1 + q^t + q^{2t} + \dots + q^{(\lambda-1)t} = Q,$$

und bezeichnet das Produkt der λ conjugirten complexen Zahlen in w, als Norm, durch $NF(w)$, so ist

$$F(w)^Q \equiv NF(w), \text{ mod. } \phi(\alpha).$$

Hieraus folgt nun, dafs die Norm von $F(w)$ niemals durch $\phi(\alpha)$ theilbar sein kann, ohne dafs eine Potenz von $F(w)$ durch $\phi(\alpha)$ theilbar ist, und weil gezeigt worden ist, dafs eine Potenz von $F(w)$ nicht durch $\phi(\alpha)$ theilbar sein kann, ohne dafs $F(w)$ selbst durch $\phi(\alpha)$ theilbar ist, so hat man folgenden Lehrsatz:

(II.) Wenn $\phi(\alpha)$ eine complexe Primzahl in α ist, in Beziehung auf welche die Determinante $D(\alpha)$ ein Nichtrest ist, so enthält die Norm einer complexen Zahl $F(w)$ nur dann den Faktor $\phi(\alpha)$, wenn $F(w)$ selbst durch $\phi(\alpha)$ theilbar ist.

Hieraus folgt ferner, dafs, wenn die Norm $NF(w)$ einen Primfaktor $\phi(\alpha)$ enthält, für welchen die Determinante $D(\alpha)$ Nichtrest ist, diese Norm den Primfaktor $\phi(\alpha)$ nothwendig λ mal, oder $k\lambda$ mal enthalten mufs.

Es sei nun zweitens $i \equiv 0$, mod. λ, also $\phi(\alpha)$ eine solche Primzahl, in Beziehung auf welche die Determinante $D(\alpha)$ ein λter Potenzrest ist, so hat man:

(6.) $$D(\alpha) \equiv \xi^\lambda, \quad \text{mod. } \phi(\alpha),$$

wo ξ eine wirkliche complexe Zahl in α ist. Die λ Wurzeln dieser Congruenz des λten Grades sind, wenn eine derselben durch ξ bezeichnet wird:

$$\xi, \ \xi\alpha, \ \xi\alpha^2, \ \ldots. \ \xi\alpha^{\lambda-1}.$$

Diese Congruenzwurzeln entsprechen vollständig den Wurzeln

$$w, \ w\alpha, \ w\alpha^2, \ \ldots. \ w\alpha^{\lambda-1}$$

der Gleichung

$$D(\alpha) = w^\lambda,$$

und können denselben auf λ verschiedene Weisen zugeordnet werden, in der Art, dafs, wenn der bestimmten Gleichungswurzel w die Congruenzwurzel $\xi\alpha^k$ als die entsprechende zugeordnet wird, allgemein der Gleichungswurzel $w\alpha^h$ die Congruenzwurzel $\xi\alpha^{k+h}$ entspricht.

Hat man irgend eine ganze und rationale Gleichung unter ganzen complexen Zahlen in w, welche, wenn alle Glieder auf eine Seite gebracht werden, immer die Form

$$\Phi(w) = 0$$

annimmt, wo $\Phi(w)$ irgendwie aus Summen, Differenzen, Produkten oder Potenzen complexer Zahlen in w zusammengesetzt sein kann, so mufs $\Phi(w)$, wenn w als eine unbestimmte Gröfse aufgefafst wird, nothwendig den Faktor $w^\lambda - D(\alpha)$ enthalten, weil die Gleichung $w^\lambda - D(\alpha) = 0$ eine irreductible ist. Setzt man nun für w irgend eine der Congruenzwurzeln ξ, $\xi\alpha$, ... $\xi\alpha^{\lambda-1}$, so wird der Faktor $w^\lambda - D(\alpha)$, und mit ihm $\Phi(w)$ selbst, congruent Null nach dem Modul $\phi(\alpha)$. Man hat also folgenden Satz:

(III.) Aus einer jeden rationalen ganzen Gleichung unter complexen Zahlen in w erhält man eine richtige Congruenz nach einem Modul $\phi(\alpha)$, für welchen die Determinante λter Potenzrest ist, wenn man die Wurzel w durch irgend eine Wurzel der Congruenz

$$\xi^\lambda \equiv D(\alpha), \quad \text{mod. } \phi(\alpha),$$

ersetzt.

Wendet man diesen Satz auf die Norm einer complexen Zahl (Fw) an, so hat man aus der Gleichung

F 2

$$NF(w) = F(w)\,F(wa)\,F(wa^2)\,\ldots\ldots\,F(wa^{\lambda-1})$$

die Congruenz:

(7.) $\quad NF(w) \equiv F(\xi)\,F(\xi a)\,F(\xi a^2)\,\ldots\ldots\,F(\xi a^{\lambda-1}), \text{ mod. } \phi(a),$

und durch diese folgenden Satz:

(IV.) Wenn die Norm einer complexen Zahl $F(w)$ durch die complexe Primzahl $\phi(a)$ theilbar ist, für welche $D(a)$ ein λter Potenzrest ist, so ist nothwendig eine der nur a enthaltenden complexen Zahlen $F(\xi a^h)$, welche man erhält, indem man für w eine der entsprechenden Congruenzwurzeln setzt, durch $\phi(a)$ theilbar;

und umgekehrt:

Wenn eine complexe Zahl $F(w)$ die Eigenschaft hat, dafs, wenn man für w eine der entsprechenden Congruenzwurzeln setzt, die daraus entstehende complexe Zahl in a durch $\phi(a)$ theilbar ist, so ist $NF(w)$ theilbar durch $\phi(a)$.

Setzt man in dem Ausdrucke einer beliebigen complexen Zahl in w

$$F(w) = A + A_1 w + A_2 w^2 + \ldots. + A_{\lambda-1} w^{\lambda-1},$$

für w nach einander die λ Werthe $w, wa, wa^2, \ldots wa^{\lambda-1}$, multiplicirt die so erhaltenen Gleichungen der Reihe nach mit $1, a, a^{-h}, a^{-2h}, \ldots a^{-(\lambda-1)h}$ und addirt, so erhält man:

$$F(w) + a^{-h} F(wa) + a^{-2h} F(wa^2) + \ldots + a^{-(\lambda-1)h} F(wa^{\lambda-1}) = \lambda A_h w^h,$$

und wenn man anstatt w in dieser Gleichung die Congruenzwurzel ξ setzt, so hat man die Congruenz:

$$F(\xi) + a^{-h} F(\xi a) + a^{-2h} F(\xi a^2) + \ldots + a^{-(\lambda-1)h} F(\xi a^{-1})$$
$$\equiv \lambda A_h \xi^h, \quad \text{mod. } \phi(a).$$

Wenn nun die complexen Zahlen $F(\xi)$, $F(\xi a)$, $F(\xi a^2)$ .... alle durch $\phi(a)$ theilbar sind, so mufs nothwendig $A_h \equiv 0$ sein, nach dem Modul $\phi(a)$, für alle Werthe $h = 0, 1, 2, \ldots \lambda - 1$, also alle Coefficienten von $F(w)$ müssen durch $\phi(a)$ theilbar sein. Man hat daher folgenden Satz:

(V.) Wenn die complexe Zahl $F(w)$ die Eigenschaft hat, dafs alle die complexen Zahlen in a, welche man aus ihr erhält, indem man für w die λ entsprechenden Congruenzwurzeln setzt, den Faktor $\phi(a)$ enthalten, so enthält $F(w)$ selbst den Faktor $\phi(a)$.

§. 4.

Die idealen Primfaktoren der complexen Zahlen in w und in z.

Bei der Untersuchung der idealen Primfaktoren der complexen Zahlen in w hat man von den complexen Primzahlen in α auszugehen, welche innerhalb dieser niederen Theorie selbst ideal oder wirklich sein können, und man hat diese mit Hülfe der Wurzeln der Gleichung $w^\lambda = D(\alpha)$ weiter in diejenigen Faktoren zu zerlegen, welche innerhalb dieser höheren Theorie der complexen Zahlen in w als die wahren Primfaktoren anzusehen sind; denn die complexen Zahlen in w stehen zu denen in α genau in demselben Verhältnifs, wie diese selbst zu den gewöhnlichen ganzen Zahlen stehen: was in der niederen Theorie nothwendig als Primzahl angesehen werden mufs, wird in der höheren Theorie im Allgemeinen weiter zerlegbar, sei es in wirklicher oder in idealer Weise. Wenn also $\phi(\alpha)$ eine wirkliche oder ideale complexe Primzahl in der Theorie der complexen Zahlen in α ist, so handelt es sich darum von dem Standpunkte der Theorie der complexen Zahlen in w aus, diese Zahl $\phi(\alpha)$ weiter in diejenigen idealen oder wirklichen Faktoren zu zerlegen, welche innerhalb dieser höheren Theorie als die Primfaktoren anzusehen sind. Wenn nur diejenigen Primzahlen $\phi(\alpha)$ in Betracht gezogen werden, welche in $\varrho D(\alpha)$ nicht enthalten sind, so sind dieselben wieder in der Rücksicht besonders zu betrachten: ob für sie die Determinante $D(\alpha)$ Nichtrest oder Rest einer λten Potenz ist.

Ich stelle nun zunächst für diese beiden verschiedenen Arten der Primzahlen $\phi(\alpha)$ die Definitionen der idealen Primfaktoren innerhalb der Theorie der complexen Zahlen in w fest, und werde alsdann zeigen, dafs die so definirten idealen Primfaktoren die wesentliche und erschöpfende Eigenschaft wahrer Primfaktoren besitzen, nämlich die, dafs sie in einer jeden gegebenen complexen Zahl stets in unveränderlicher Weise enthalten sind, und dafs sie in allen Fällen, wo wirkliche Primfaktoren existiren, mit diesen vollständig übereinstimmen.

Definition: Wenn $\phi(\alpha)$ eine ideale oder wirkliche complexe Primzahl innerhalb der Theorie der complexen Zahlen in α ist, für welche $D(\alpha)$ ein Nichtrest einer λten Potenz ist, so soll $\phi(\alpha)$ auch in der Theorie der complexen Zahlen in w als Primzahl angesehen werden.

Für die Zerlegung der complexen Primzahlen der anderen Art, für welche $D(\alpha)$ ein λter Potenzrest ist, benutze ich folgende complexe Zahl:

$$\Psi(w) = (w - \xi\alpha)(w - \xi\alpha^2) \ldots\ldots (w - \xi\alpha^{\lambda-1}), \tag{1.}$$

in welcher ξ ebenso wie oben eine Wurzel der Congruenz $\xi^\lambda \equiv D(\alpha)$, mod. $\phi(\alpha)$, bezeichnet. Diese complexe Zahl kann auch so dargestellt werden:

$$\Psi(w) = w^{\lambda-1} + \xi w^{\lambda-2} + \xi^2 w^{\lambda-3} + \ldots\ldots \xi^{\lambda-1}, \tag{2.}$$

oder in Form eines Bruches:

$$\Psi(w) = \frac{w^\lambda - \xi^\lambda}{w - \xi}.$$

Mit Hülfe dieser complexen Zahl $\Psi(w)$ gebe ich nun folgende Definition der idealen Primfaktoren des $\phi(\alpha)$ in der Theorie der complexen Zahlen in w:

Definition: Wenn $\phi(\alpha)$ eine ideale oder wirkliche complexe Primzahl innerhalb der Theorie der complexen Zahlen in α ist, für welche $D(\alpha)$ ein λter Potenzrest ist, und man hat

$$\Psi(w\alpha^{-k})\, F(w) \equiv 0, \text{ mod. } \phi(\alpha),$$

so soll von $F(w)$ ausgesagt werden: es enthält einen idealen Primfaktor des $\phi(\alpha)$ und zwar denjenigen, welcher zur Congruenzwurzel $\xi\alpha^k$ gehört. Wenn ferner

$$\Psi(w\alpha^{-k})^m \cdot F(w) \equiv 0, \quad \text{mod. } \phi(\alpha)^m,$$

aber

$$\Psi(w\alpha^{-k})^{m+1} \cdot F(w) \text{ nicht} \equiv 0, \quad \text{mod. } \phi(\alpha)^{m+1},$$

so soll von der complexen Zahl $F(w)$ ausgesagt werden: sie enthält den zur Congruenzwurzel $\xi\alpha^k$ gehörenden idealen Primfaktor des $\phi(\alpha)$ genau m mal.

Nach dieser Definition giebt es in der Theorie der complexen Zahlen in w so viele verschiedene ideale Primfaktoren einer complexen Primzahl $\phi(\alpha)$ der niederen Theorie, für welche die Determinante $D(\alpha)$ ein λter Potenzrest ist, als die Congruenz

$$\xi^\lambda \equiv D(\alpha), \quad \text{mod. } \phi(\alpha)$$

verschiedene Wurzeln hat, also λ, welche als conjugirte ideale Primfaktoren des $\phi(\alpha)$ bezeichnet werden sollen, und deren Produkt, die Norm des idealen Primfaktors, überall der Primzahl $\phi(\alpha)$ selbst gleich genommen werden soll.

Ich beweise nun zunächst, dafs die so definirten idealen Primfaktoren der folgenden ersten Bedingung wahrer Primzahlen genügen:

(I.) Wenn zwei oder mehrere complexe Zahlen einen bestimmten idealen Primfaktor nicht enthalten, so enthält das Produkt derselben diesen Primfaktor ebenfalls nicht.

Es seien $F(w)$ und $G(w)$ zwei complexe Zahlen und $H(w)$ das Produkt derselben, also $F(w) \cdot G(w) = H(w)$. Es sei ferner zunächst $\phi(\alpha)$ eine complexe Primzahl der ersten Art, für welche $D(\alpha)$ Nichtrest ist, so ist nach der Definition $\phi(\alpha)$ selbst auch in der Theorie der complexen Zahlen in w eine Primzahl. Wenn nun $F(w)$ und $G(w)$ den Faktor $\phi(\alpha)$ nicht enthalten, so enthalten nach dem zweiten der im §. 3. bewiesenen Sätze $NF(w)$ und $NG(w)$ denselben ebenfalls nicht, und weil diese Normen nur complexe Zahlen in α sind, so enthält das Produkt derselben $NF(w) \cdot NG(w) = NH(w)$ diesen Primfaktor $\phi(\alpha)$ auch nicht, woraus nach demselben erwähnten Satze des §. 3. folgt, dafs auch $H(w)$ denselben nicht enthält. Es sei nun zweitens $\phi(\alpha)$ eine Primzahl der zweiten Art, für welche $D(\alpha)$ ein λter Potenzrest ist, und $F(w)$ so wie $G(w)$ enthalten den zur Congruenzwurzel $\xi\alpha^k$ gehörenden idealen Primfaktor des $\phi(\alpha)$ nicht, so hat man nach der Definition

$$\begin{array}{l} \Psi(w\alpha^{-k})\, F(w) \text{ nicht} \equiv 0, \\ \Psi(w\alpha^{-k})\, G(w) \text{ nicht} \equiv 0, \end{array} \text{mod. } \phi(\alpha).$$

Wenn nun eine complexe Zahl in w durch $\phi(\alpha)$ nicht theilbar ist, so mufs dieselbe nach dem letzten Satze des §. 3., wenn man anstatt w die Congruenzwurzeln ξ, $\xi\alpha$, $\xi\alpha^2$, ... $\xi\alpha^{\lambda-1}$ setzt, wenigstens für einen dieser Werthe durch $\phi(\alpha)$ nicht theilbar sein; da aber $\Psi(w)$, wie der Produktausdruck dieser complexen Zahl zeigt, congruent Null wird, sobald für w eine der Congruenzwurzeln $\xi\alpha$, $\xi\alpha^2$, ... $\xi\alpha^{\lambda-1}$ gesetzt wird, und nur für die eine Substitution der Congruenzwurzel ξ nicht congruent Null ist, nach dem Modul $\phi(\alpha)$, so folgt, dafs die beiden complexen Zahlen

$$\Psi(w\alpha^{-k})\, F(w) \quad \text{und} \quad \Psi(w\alpha^{-k})\, G(w)$$

für die Substitution der Congruenzwurzel $\xi\alpha^k$ anstatt w, nicht congruent Null werden, nach dem Modul $\phi(\alpha)$. Es ist also

$$\begin{matrix} \Psi(\xi)\, F(\xi\alpha^k) \text{ nicht} \equiv 0, \\ \Psi(\xi)\, G(\xi\alpha^k) \text{ nicht} \equiv 0, \end{matrix} \quad \text{mod. } \phi(\alpha),$$

also auch

$$F(\xi\alpha^k) \text{ nicht} \equiv 0, \quad G(\xi\alpha^k) \text{ nicht} \equiv 0, \text{ mod. } \phi(\alpha),$$

und weil diese nur complexe Zahlen in α sind, so ist auch

$$F(\xi\alpha^k)\, G(\xi\alpha^k) \text{ nicht} \equiv 0, \text{ mod. } \phi(\alpha).$$

Enthielte nun aber $F(w)\; G(w) = H(w)$ den idealen Primfaktor des $\phi(\alpha)$ welcher zur Congruenzwurzel $\xi\alpha^k$ gehört, so müſste nach der Definition sein:

$$\Psi(w\alpha^{-k})\, F(w)\; G(w) \equiv 0, \text{ mod. } \phi(\alpha),$$

also, wenn für w die Congruenzwurzel $\xi\alpha^k$ gesetzt wird, so müſste auch

$$\Psi(\xi)\, F(\xi\alpha^k)\, G(\xi\alpha^k) \equiv 0, \text{ mod. } \phi(\alpha),$$

sein, und weil $\Psi(\xi)$ nicht congruent Null ist, so müſste

$$F(\xi\alpha^k)\, G(\xi\alpha^k) \equiv 0, \text{ mod. } \phi(\alpha)$$

sein, welches nicht der Fall ist. Das Produkt $F(w)\; G(w) = H(w)$ enthält also keinen idealen Primfaktor, welcher nicht schon in einem der beiden Faktoren enthalten ist, und dieser für das Produkt zweier Faktoren bewiesene Satz wird ohne Schwierigkeit auch auf jedes Produkt einer beliebigen Anzahl von Faktoren ausgedehnt.

Auf diesen soeben bewiesenen specielleren Satz stützt sich nun der Beweis des folgenden allgemeineren Satzes:

(II.) D a s e n t w i c k e l t e P r o d u k t z w e i e r o d e r m e h r e r e r c o m p l e x e r Z a h l e n i n w e n t h ä l t g e n a u d i e s e l b e n i d e a l e n P r i m f a k t o r e n , u n d z w a r j e d e n d e r s e l b e n g e n a u s o o f t , a l s d i e F a k t o r e n d i e s e s P r o d u k t s z u s a m m e n g e n o m m e n.

Es sei wieder $F(w)\; G(w) = H(w)$, und es sei erstens $\phi(\alpha)$ eine complexe Primzahl der ersten Art, für welche $D(\alpha)$ Nichtrest einer λten Potenz ist, welche also auch in der Theorie der complexen Zahlen in w als Primzahl definirt ist. Diese Primzahl $\phi(\alpha)$ sei in $F(w)$ genau m mal und in $G(w)$ genau n mal enthalten, so daſs man setzen kann:

$$F(w) = \phi(\alpha)^m\, F_1(w), \quad G(w) = \phi(\alpha)^n\, G_1(w),$$

so ist

$$H(w) = \phi(\alpha)^{m+n}\, F_1(w)\, G_1(w),$$

also $H(w)$ enthält den Primfaktor $\phi(\alpha)$ $m+n$ mal. Dafs es denselben auch nicht mehr als $m+n$ mal enthält, folgt aber daraus, dafs weder $F_1(w)$ noch $G_1(w)$ denselben enthält, und folglich nach dem vorigen Satze auch das Produkt $F_1(w)\,G_1(w)$ ihn nicht enthalten kann.

Es sei nun zweitens $\phi(\alpha)$ eine complexe Primzahl der zweiten Art, für welche $D(\alpha)$ ein λter Potenzrest ist, und es enthalte $F(w)$ den zur Congruenzwurzel $\xi\alpha^k$ gehörenden idealen Primfaktor des $F(\alpha)$ genau m mal, $G(w)$ enthalte denselben genau n mal, so hat man:

$$\Psi(w\alpha^{-k})^m F(w) \equiv 0 \text{ mod. } \phi(\alpha)^m ,$$

$$\Psi(w\alpha^{-k})^n G(w) \equiv 0 \text{ mod. } \phi(\alpha)^n ,$$

also kann man setzen:

$$\Psi(w\alpha^{-k})^m F(w) = \phi(\alpha)^m P(w),$$

$$\Psi(w\alpha^{-k})^n G(w) = \phi(\alpha)^n Q(w),$$

woraus folgt:

$$\Psi(w\alpha^{-k})^{m+n} F(w)\,G(w) = \phi(\alpha)^{m+n} P(w)\,Q(w),$$

Diese Gleichung zeigt, dafs das Produkt $F(w)\,G(w) = H(w)$ den zur Congruenzwurzel $\xi\alpha^k$ gehörenden idealen Primfaktor des $\phi(\alpha)$ $m+n$ mal enthält. Dafs es denselben auch nicht mehr als $m+n$ mal enthält, wird folgendermaafsen bewiesen. Nach der Voraussetzung, dafs $F(w)$ den in Rede stehenden idealen Primfaktor nicht mehr als m mal, und $G(w)$ denselben nicht mehr als n mal enthält, hat man:

$$\Psi(w\alpha^{-k})^{m+1} F(w) \text{ nicht} \equiv 0, \text{ mod. } \phi(\alpha)^{m+1} ,$$

$$\Psi(w\alpha^{-k})^{n+1} G(w) \text{ nicht} \equiv 0, \text{ mod. } \phi(\alpha)^{n+1} ,$$

also

$$\Psi(w\alpha^{-k})\,\phi(\alpha)^m P(w) \text{ nicht} \equiv 0, \text{ mod. } \phi(\alpha)^{m+1} ,$$

$$\Psi(w\alpha^{-k})\,\phi(\alpha)^n Q(w) \text{ nicht} \equiv 0, \text{ mod. } \phi(\alpha)^{n+1} ,$$

und folglich auch:

$$\Psi(w\alpha^{-k})\,P(w) \text{ nicht} \equiv 0, \text{ mod. } \phi(\alpha),$$

$$\Psi(w\alpha^{-k})\,Q(w) \text{ nicht} \equiv 0, \text{ mod. } \phi(\alpha),$$

woraus folgt, dafs weder $P(w)$ noch $Q(w)$ diesen idealen Primfaktor enthält, und mithin das Produkt derselben ihn ebenfalls nicht enthalten kann, oder was dasselbe ist, dafs das Produkt

$$\Psi(w\alpha^{-k})\,P(w)\,Q(w)$$

den Faktor $\phi(\alpha)$ nicht enthalten kann. Aus der Gleichung

$$\Psi(w\alpha^{-k})^{m+n}\,F(w)\,G(w) = \phi(\alpha)^{m+n}\,P(w)\,Q(w)$$

folgt aber:

$$\Psi(w\alpha^{-k})^{m+n+1}\,F(w)\,G(w) = \phi(\alpha)^{m+n}\,\Psi(w\alpha^{-k})\,P(w)\,Q(w),$$

und hieraus:

$$\Psi(w\alpha^{-k})^{m+n+1}\,F(w)\,G(w) \text{ nicht} \equiv 0, \text{ mod. } \phi(\alpha)^{m+n+1};$$

also das Produkt $F(w)\,G(w)$ enthält den zur Congruenzwurzel $\xi\alpha^k$ gehörenden idealen Primfaktor des $\phi(\alpha)$ nicht mehr als $m+n$ mal. Eine wiederholte Anwendung dieses für zwei Faktoren bewiesenen Satzes zeigt, dafs derselbe ebenso für ein Produkt beliebig vieler Faktoren gültig ist.

§. 5.

Verhältnifs der idealen complexen Zahlen zu den wirklichen.

Der in dem vorhergehenden Paragraphen bewiesene Hauptsatz zeigt, dafs die idealen Primfaktoren, wie sie oben definirt sind, die erste Grundeigenschaft wahrer Primfaktoren haben, nämlich in einer und derselben Zahl in unveränderlicher Weise enthalten zu sein, und unabhängig davon, ob diese Zahl in entwickelter Form, oder in Form eines Produkts gegeben ist. Es sind nun weiter diejenigen Sätze zu entwickeln, welche die Übereinstimmung dieser idealen Primfaktoren mit den wirklichen, wo solche existiren, nachweisen, und welche zeigen, welchen Gebrauch man von den idealen Primfaktoren, vorzüglich in der Multiplikation und Division der complexen Zahlen in w machen kann.

(I.) Wenn eine complexe Zahl $F(w)$ alle λ idealen Primfaktoren einer Primzahl $\phi(\alpha)$ enthält, für welche die Determinante $D(\alpha)$ ein λter Potenzrest ist, und zwar jeden derselben mindestens m mal, so ist $F(w)$ durch $\phi(\alpha)^m$ theilbar.

Nach der Voraussetzung dieses Satzes hat man nämlich:

$$\Psi(w\alpha^{-k})^m F(w) \equiv 0, \quad \text{mod. } \phi(\alpha)^m,$$

für alle Werthe des $k = 0, 1, 2, \dots \lambda - 1$. Giebt man nun dem k nach einander alle diese λ Werthe und summirt, so erhält man:

$$\left(\Psi(w)^m + \Psi(w\alpha^{-1})^m + \dots + \Psi(w\alpha^{-(\lambda-1)})^m\right) F(w) \equiv 0, \text{ mod. } \phi(\alpha)^m.$$

Die Summe innerhalb der Klammern, als symmetrische Funktion aller Wurzeln der Gleichung $w^\lambda = D(\alpha)$, enthält die Wurzel w selbst nicht, und ist nur eine complexe Zahl in α, welche durch $\Phi(\alpha)$ bezeichnet werden mag. Nimmt man nun $F(w)$ in der Form

$$F(w) = A + A_1 w + A_2 w^2 + \dots + A_{\lambda-1} w^{\lambda-1},$$

so hat man:

$$\Phi(\alpha)\,(A + A_1 w + A_2 w^2 + \dots + A_{\lambda-1} w^{\lambda-1}) \equiv 0, \text{ mod. } \phi(\alpha)^m.$$

Vermöge der Irreductibilität der Gleichung $w^\lambda = D(\alpha)$ kann aber diese Congruenz nicht bestehen, ohne daſs folgende λ einzelnen Congruenzen Statt haben:

$$\Phi(\alpha)\, A \equiv 0, \ \Phi(\alpha)\, A_1 \equiv 0, \ \dots, \ \Phi(\alpha)\, A_{\lambda-1} \equiv 0, \text{ mod. } \phi(\alpha)^m.$$

$\Phi(\alpha)$ aber enthält den Faktor $\phi(\alpha)$ nicht; denn wenn man in der Gleichung

$$\Phi(\alpha) = \Psi(w)^m + \Psi(w\alpha^{-1})^m + \dots + \Psi(w\alpha^{-(\lambda-1)})^m$$

die Wurzel w durch die Congruenzwurzel ξ ersetzt, und bemerkt, daſs die in der Gleichung (2.) §. 4. gegebene Form des $\Psi(w)$

$$\Psi(\xi\alpha^{-h}) \equiv \xi^{\lambda-1}\,(\alpha^h + \alpha^{2h} + \alpha^{3h} + \dots + \alpha^{(\lambda-1)h} + 1), \text{ mod. } \phi(\alpha),$$

ergiebt, also allgemein:

$$\Psi(\xi\alpha^{-h}) \equiv 0, \text{ mod. } \phi(\alpha), \text{ wenn } h \text{ nicht} = 0,$$

aber für $h = 0$:

$$\Psi(\xi) \equiv \lambda\xi^{\lambda-1}, \text{ mod. } \phi(\alpha),$$

so hat man:

$$\Phi(\alpha) \equiv \lambda^m \xi^{(\lambda-1)m}, \text{ mod. } \phi(\alpha);$$

also $\Phi(\alpha)$ nicht durch $\phi(\alpha)$ theilbar. Hieraus folgt, daſs die Coefficienten

$A, A_1, A_2, \ldots A_{\lambda-1}$ alle einzeln den Faktor $\phi(\alpha)^m$ enthalten müssen, dafs also $F(w)$ durch $\phi(\alpha)^m$ theilbar sein mufs, w. z. b. w.

(II.) Wenn eine complexe Zahl $F(w)$ genau m ideale Primfaktoren einer Primzahl $\phi(\alpha)$ enthält, für welche die Determinante $D(\alpha)$ ein λter Potenzrest ist, dieselben mögen verschieden sein, oder auch nicht, so enthält die Norm von $F(w)$ den Faktor $\phi(\alpha)$ genau m mal.

Die Norm $NF(w)$, als das Produkt der λ conjugirten complexen Zahlen, mufs nach der Voraussetzung des Satzes jeden der λ verschiedenen idealen Primfaktoren des $\phi(w)$ genau m mal enthalten; also mufs sie, vermöge des vorigen Satzes, den Faktor $\phi(\alpha)^m$ enthalten. Dieselbe kann auch nicht eine höhere Potenz von $\phi(\alpha)$ enthalten, weil sie sonst einen jeden der λ idealen Primfaktoren des $\phi(\alpha)$ mehr als m mal enthalten müfste.

Dieser Lehrsatz, verbunden mit dem Lehrsatz II. §. 3., zeigt nicht nur, dafs die Anzahl der in einer gegebenen complexen Zahl $F(w)$ enthaltenen idealen Primfaktoren stets eine endliche bestimmte ist, sondern er gewährt auch ein leichtes Mittel um zu erkennen, wie viele ideale Primfaktoren dieselbe enthält, und von welchen Primzahlen in der niederen Theorie der complexen Zahlen in α sie herrühren. Bildet man nämlich die Norm $NF(w)$, zerlegt dieselbe als complexe Zahl in α in ihre, dieser niederen Theorie angehörenden Primfaktoren, und unterscheidet dabei diejenigen Primfaktoren, für welche die Determinante Nichtrest einer λten Potenz ist, von denen, wo sie Rest ist: so mufs erstens die Anzahl, wie viel mal eine solche Primzahl der ersten Art in $NF(w)$ vorkommt, ein Vielfaches von λ sein, und wenn dieselbe gleich $k\lambda$ ist, so ist dieser ideale Primfaktor, welcher in der höheren Theorie ebenfalls Primfaktor ist, genau k mal in $F(w)$ enthalten; wenn zweitens irgend eine Primzahl der zweiten Art genau m mal in der Norm vorkommt, so mufs $F(w)$ selbst nothwendig m ideale Primfaktoren derselben in der höheren Theorie der complexen Zahlen in w enthalten. Die Norm $NF(w)$ kann aufserdem noch Primfaktoren der Determinante $D(\alpha)$, oder auch den Primfaktor $1-\alpha$ enthalten, von deren zugehörenden idealen Primfaktoren in der Theorie der complexen Zahlen in w aber erst weiter unten die Rede sein wird.

(III.) Wenn $f(w)$ eine wirkliche complexe Zahl in w ist, deren Norm keinen gemeinschaftlichen Faktor mit $\varrho D(\alpha)$ hat, und $F(w)$ eine andere wirkliche complexe Zahl in w, welche alle idealen Primfaktoren des $f(w)$, jeden mindestens eben so oft, als $f(w)$ selbst enthält, so ist $F(w)$ durch $f(w)$ theilbar.

Man kann den Quotienten der beiden Zahlen $F(w)$ und $f(w)$ in folgende Form setzen:

$$\frac{F(w)}{f(w)} = \frac{F(w) \, . \, f(w\alpha) \, f(w\alpha^2) \, ... \, f(w\alpha^{\lambda-1})}{Nf(w)} \, .$$

Wenn nun erstens $f(w)$ irgend einen Primfaktor $\varphi(\alpha)$ n mal enthält, für welchen $D(\alpha)$ Nichtrest ist, welcher also in der niederen und in der höheren Theorie zugleich Primfaktor ist, so kommt derselbe in dem Nenner $Nf(w)$ nothwendig $n\lambda$ mal vor, und eben so viel mal mindestens kommt er auch in dem Zähler vor, weil jede der complexen Zahlen $f(w\alpha)$, $f(w\alpha^2)$, ... $f(w\alpha^{\lambda-1})$ denselben genau n mal, und $F(w)$ denselben nach der Voraussetzung des Satzes mindestens n mal enthält. Jeder solcher Faktor hebt sich also aus dem Nenner des Bruches vollständig hinweg. Wenn nun zweitens $f(w)$ irgend welche idealen Primfaktoren einer Primzahl der niederen Theorie $\varphi(\alpha)$ enthält, für welche $D(\alpha)$ Rest einer λten Potenz ist, und die Anzahl dieser idealen Primfaktoren ist m, so enthält der Nenner $Nf(w)$ den Faktor $\varphi(\alpha)$ genau m mal, der Zähler $F(w)$ $f(w\alpha)$ $f(w\alpha^2)$ .... $f(w\alpha^{\lambda-1})$ enthält aber alle idealen Primfaktoren des $\varphi(\alpha)$, jeden mindestens m mal, weil $F(w)$ alle idealen Primfaktoren des $f(w)$ enthält, also vermöge des Satzes (II.) enthält der Zähler den Faktor $\varphi(\alpha)$ nothwendig m mal, also auch ein jeder solcher Faktor mufs sich vollständig aus dem Nenner hinwegheben. Da endlich in $Nf(w)$ nach der Voraussetzung des Satzes keine anderen, als die hier untersuchten Faktoren vorkommen, indem die in $\varrho D(\alpha)$ enthaltenen ausgeschlossen sind, so folgt, dafs der ganze Nenner $Nf(w)$ gegen den Zähler sich hinwegheben mufs, und dafs der Quotient $\frac{F(w)}{f(w)}$ eine ganze complexe Zahhl ist, w. z. b. w.

(IV.) Wenn zwei wirkliche complexe Zahlen in w genau dieselben idealen Primfaktoren enthalten, und wenn ihre Normen keinen gemeinschaftlichen Faktor mit $\varrho D(\alpha)$ haben, so unterscheiden sich diese Zahlen nur durch Einheiten, welche als Faktoren hinzutreten können.

Die beiden wirklichen complexen Zahlen, von denen der Satz handelt, seien $F(w)$ und $f(w)$, so ist vermöge des vorhergehenden Satzes:

$$\frac{F(w)}{f(w)} = E(w), \quad F(w) = E(w)\, f(w),$$

wo $E(w)$ eine ganze complexe Zahl ist, und zwar eine wirkliche. Nimmt man nun auf beiden Seiten die Normen, so hat man:

$$NF(w) = NE(w) \cdot Nf(w).$$

Die beiden Normen $NF(w)$ und $Nf(w)$, welche complexe Zahlen in α sind, enthalten nun genau dieselben Primfaktoren, auch in dieser niederen Theorie. Dieselben können sich daher nur durch eine Einheit $E(\alpha)$ unterscheiden, und man hat:

$$NF(w) = E(\alpha)\, Nf(w),$$

und diese Gleichung, mit der vorhergehenden verbunden, ergiebt:

$$NE(w) = E(\alpha).$$

Die ganze complexe Zahl $E(w)$ ist also eine solche, deren Norm eine Einheit in α ist, sie ist daher eine Einheit in der Theorie der complexen Zahlen in w, weil überhaupt als Einheit in dieser Theorie eine jede ganze complexe Zahl, deren Norm eine Einheit der niederen Theorie der complexen Zahlen in α ist, definirt wird.

Aus den hier angenommenen Resultaten ergiebt sich nun leicht der Satz, welcher noch erfordert wird, um die gegebenen Definitionen der idealen Primfaktoren zu rechtfertigen, nämlich:

(V.) In jedem Falle, wo in der Theorie der complexen Zahlen in w ein Primfaktor einer Primzahl $\phi(\alpha)$ der niederen Theorie, für welche $D(\alpha)$ ein λter Potenzrest ist, als ein wirklicher existirt, und darum als eine complexe Zahl in w definirt werden kann, deren Norm, abgesehen von einer Einheit, gleich $\phi(\alpha)$ ist, stimmt die oben gegebene allgemeine Definition vollständig mit dieser beschränkteren zusammen.

Es sei $f(w)$ ein wirklicher complexer Primfaktor der Primzahl $\phi(\alpha)$ der niederen Theorie, also $Nf(w) = E(\alpha)\,\phi(\alpha)$, wo $E(\alpha)$ eine Einheit ist, so mufs für eine bestimmte Wurzel der Congruenz $\xi^\lambda \equiv D(\alpha)$, mod. $\phi(\alpha)$, welche statt w gesetzt wird, z. B. für die Wurzel $\xi\alpha^k$, wie oben §. 3. bewiesen worden, $f(\xi\alpha^k) \equiv 0$, mod. $\phi(\alpha)$, sein, und es ist alsdann $f(w)$ als der zur Congruenzwurzel $\xi\alpha^k$ gehörende Primfaktor des $\phi(\alpha)$ zu bezeichnen.

Wenn nun eine wirkliche complexe Zahl $F(w)$ alle idealen Primfaktoren des $f(w)$, im vorliegenden Falle also nur den einen zur Congruenzwurzel $\xi \alpha^k$ gehörenden Primfaktor des $\phi(\alpha)$ enthält, so ist, wie oben gezeigt worden, $F(w)$ durch $f(w)$ theilbar, also:

$$\frac{F(w)}{f(w)} = G(w) \text{ oder } F(w) = f(w)\, G(w),$$

wo $G(w)$ eine ganze und wirkliche complexe Zahl ist. Der in $F(w)$ enthaltene ideale Primfaktor tritt also in diesem Falle als der wirkliche heraus.

Es würde jetzt eigentlich noch übrig sein, auch von den idealen Primfaktoren derjenigen complexen Primzahlen in α zu handeln, welche in der Determinante $D(\alpha)$ enthalten sind, so wie über die Zerlegung von $1-\alpha=\varrho$ in der höheren Theorie der complexen Zahlen in w. Es treten aber bei der Zerlegung dieser Primzahlen in einfachere Faktoren der höheren Theorie ganz eigenthümliche Umstände ein, welche bewirken, dafs es im Allgemeinen unmöglich ist, wahre ideale Primfaktoren der in $\varrho D(\alpha)$ enthaltenen complexen Primzahlen anzugeben, welche im vollen Sinne diesen Namen verdienen, so wie die hier behandelten; namentlich treten diese störenden Umstände immer dann ein, wenn ein Primfaktor mehr als einmal in der Determinante enthalten ist. Für den vorliegenden Zweck ist es aber nicht nöthig, die Zerlegung der in $\varrho D(\alpha)$ enthaltenen Faktoren der niederen Theorie in die idealen Faktoren, welche sie innerhalb der höheren Theorie der complexen Zahlen in w haben möchten, näher zu untersuchen. Der Ausdruck „idealer Primfaktor" oder „ideale Primzahl" dieser höheren Theorie, wird darum überall nur von den im §. 4. definirten idealen Primfaktoren gebraucht werden, deren Normen ausschliefslich nur die in $\varrho D(\alpha)$ nicht enthaltenen Primzahlen der niederen Theorie sind. Ebenso soll auch der Ausdruck „ideale complexe Zahl" in der höheren Theorie nur von einer Zusammensetzung der definirten idealen Primfaktoren gebraucht werden, d. h. von dem gleichzeitigen Bestehen beliebig vieler, der die idealen Primfaktoren charakterisirenden Congruenzbedingungen, oder von dem Bestehen einer derjenigen Congruenzbedingungen, welche ausdrücken, dafs eine complexe Zahl einen idealen Primfaktor mehrmals enthält, niemals aber soll von idealen Zahlen die Rede sein, deren Normen irgend welche Faktoren mit $\varrho D(\alpha)$ gemein haben möchten. Dadurch soll jedoch die Anwendung wirklicher complexer Zahlen in w oder z nicht ausgeschlossen werden, deren

Normen den Faktor ϱ oder Primfaktoren der Determinante $D(\alpha)$ enthalten, von denen vielmehr in dem Folgenden mit grofsem Nutzen Gebrauch gemacht werden wird.

§. 6.

Eintheilung der idealen complexen Zahlen in die Klassen und Bestimmung der Klassenanzahl.

Die idealen Primfaktoren der complexen Zahlen in w, welche in den vorhergehenden beiden Paragraphen durch Congruenzbedingungen definirt und in ihren wesentlichen Grundeigenschaften betrachtet worden sind, bleiben vollkommen dieselben, wenn man die speciellere Theorie der complexen Zahlen in z zu Grunde legt, denn jede ganze complexe Zahl in z ist zugleich auch eine ganze complexe Zahl in w. Die Definition der Aequivalenz, nach welcher zwei ideale Zahlen äquivalent heifsen, wenn sie mit einer und derselben dritten idealen Zahl zusammengesetzt, wirkliche complexe Zahlen ergeben, so wie die allgemeinen Sätze über Zusammensetzung, Aequivalenz und Klassifikation der idealen Zahlen, sind sogar nicht nur für die beiden Arten der complexen Zahlen in z und in w, sondern für die idealen Zahlen aller complexen Theorieen, welche überhaupt existiren, vollständig dieselben. Es sind diefs die Sätze, welche ich im §. 5. meines *Memoire sur la théorie des nombres complexes* etc. in Liouville's Journal, Bd. 16, pg. 439 etc. von den aus λten Wurzeln der Einheit gebildeten, und später in einer vor der Königlichen Akademie vorgetragenen, in den Abhandlungen vom Jahre 1856 gedruckten Abhandlung, auch von den, aus beliebigen Wurzeln der Einheiten, deren Wurzelexponenten nicht Primzahlen sind, gebildeten complexen Zahlen vollständig bewiesen habe, nämlich folgende:

(I.) Es giebt stets eine endliche bestimmte Anzahl idealer Multiplikatoren, welche hinreichen, um alle idealen Zahlen zu wirklichen zu machen, wenn sie mit denselben zusammengesetzt werden, oder die Anzahl aller nichtäquivalenten Klassen der idealen Zahlen ist eine endliche bestimmte.

(II.) Die Eintheilung der nichtäquivalenten idealen Zahlen in die verschiedenen Klassen ist von der zufälligen Wahl der idealen Multiplikatoren ganz unabhängig.

(III.) Wenn zwei ideale Zahlen einer dritten äquivalent sind, so sind sie unter einander äquivalent.

(IV.) Aequivalente ideale Zahlen mit äquivalenten zusammengesetzt, geben äquivalente Produkte.

(V.) Jede ideale Zahl wird durch Erhebung zu einer bestimmten Potenz zu einer wirklichen, und kann daher als Wurzel aus einer wirklichen complexen Zahl dargestellt werden.

(VI.) Der Exponent der niedrigsten Potenz einer idealen Zahl, welche zu einer wirklichen wird, ist ein genauer Theil der Anzahl aller verschiedenen Klassen.

Die Beweise dieser Sätze für die Theorie der complexen Zahlen in w oder z will ich hier unterdrücken, da dieselben gröſstentheils fast wörtlich mit den an den angeführten Orten, für die aus Einheitswurzeln gebildeten, complexen Zahlen, gegebenen übereinstimmen würden. Nur in dem Beweise des ersten Satzes, über die endliche Anzahl der Klassen, müssen gewisse Modifikationen eintreten, welche jedoch eben so wenig neue principielle Schwierigkeiten darbieten. Ich kann in Betreff dieser, so wie überhaupt der allgemeinen Sätze, welche allen Theorieen complexer Zahlen gemein sind auch auf eine Arbeit von Hrn. Kronecker verweisen, welche nächstens erscheinen wird, in welcher die Theorie der allgemeinsten complexen Zahlen, in ihrer Verbindung mit der Theorie der zerlegbaren Formen aller Grade, vollständig und in groſsartiger Einfachheit entwickelt ist.

Man kann auch in ähnlicher Weise wie ich dieſs im §. VIII der angeführten Abhandlung in Liouville's Journal, Bd. 16, pag. 454 etc. für die aus Wurzeln der Einheit gebildeten, complexen Zahlen vollständig ausgeführt habe, nach den Dirichletschen Methoden einen Ausdruck für die Klassenanzahl der complexen Zahlen in w entwickeln, und ebenso den entsprechenden für die complexen Zahlen in z. Da dieser Ausdruck für eine der folgenden Untersuchungen von Wichtigkeit ist, so will ich denselben hier in der Kürze entwickeln, indem ich mich begnüge, die Hauptmomente der Methode anzugeben, die Ausführung im Einzelnen aber, in so weit sie keinerlei Schwierigkeit hat, übergehe.

Es wird folgende Reihe zu Grunde gelegt:

$$(1.)\qquad R = (s-1)\,\Sigma\,\frac{1}{(NNF(w))^s},$$

in welcher $F(w)$ alle verschiedenen idealen Zahlen in w repräsentirt, d. h. alle, welchen verschiedene Zerlegungen in ihre Primfaktoren zukommen, wo ferner $NNF(w)$ die Norm von $F(w)$, zuerst in Beziehung auf w, und sodann in Beziehung auf α genommen bedeutet, und s eine Zahl > 1 ist. Die in Beziehung auf w genommene Norm von $F(w)$, als complexe Zahl in α in ihre Primfaktoren dieser niederen Theorie zerlegt, hat stets folgende Form:

$$NF(w) = \phi(\alpha)^{m}\,\phi_1(\alpha)^{m_1}\ldots\,\psi(\alpha)^{n\lambda}\,\psi_1(\alpha)^{n_1\lambda}\ldots$$

wo $\phi(\alpha)$, $\phi_1(\alpha)$ ... Primzahlen sind, für welche $D(\alpha)$ ein λter Potenzrest ist, dagegen $\psi(\alpha)$, $\psi_1(\alpha)$ solche, deren Nichtrest $D(\alpha)$ ist. Es giebt nun genau

$$\frac{\lambda(\lambda+1)\ldots(\lambda+m-1)}{1.2.\ldots m}\cdot\frac{\lambda(\lambda+1)\ldots(\lambda+m_1-1)}{1.2.\ldots m_1}\ldots$$

verschiedene ideale Zahlen $F(w)$, welche diese selbe Norm haben, weil jede der Zahlen $\phi(\alpha)$ λ verschiedene ideale Primfaktoren hat, jede der Zahlen $\psi(\alpha)$ aber auch in der höheren Theorie selber Primzahl ist, also $\phi(\alpha)^m$ auf so viele Weisen entstehen kann, als man die λ idealen Primfaktoren des $\phi(\alpha)$ mit Wiederholungen, aber ohne Versetzungen zu je m verbinden kann, $\psi(\alpha)^n$ aber nur auf eine Weise entstehen kann. Demnach ist:

$$(2.)\qquad R = (s-1)\,\Sigma\,\frac{\dfrac{\lambda(\lambda+1)\ldots(\lambda+m-1)}{1.2.\ldots m}\cdot\dfrac{\lambda(\lambda+1)\ldots(\lambda+m_1-1)}{1.2.\ldots m_1}\ldots}{N\phi(\alpha)^{ms}\,N\phi_1(\alpha)^{m_1 s}\ldots N\psi(\alpha)^{n\lambda s}\,N\psi_1(\alpha)^{n_1\lambda s}\ldots},$$

wo die Summe auf alle ganzzahligen Werthe der Gröfsen m, m_1 ... n, n_1 ... von Null bis Unendlich sich bezieht. Führt man diese einzelnen Summationen nach dem binomischen Lehrsatze aus, so erhält man:

$$(3.)\qquad R = (s-1)\left(1-\frac{1}{N\phi(\alpha)^s}\right)^{-\lambda}\cdot\left(1-\frac{1}{N\phi_1(\alpha)^s}\right)^{-\lambda}\ldots\left(1-\frac{1}{N\psi(\alpha)^{\lambda s}}\right)^{-1}\ldots$$

Die beiden verschiedenen Arten der Faktoren dieses Produkts werden mit Hülfe des Legendreschen Zeichens $\left(\frac{D(\alpha)}{\phi(\alpha)}\right)$ leicht unter eine und dieselbe Form gebracht, denn das Produkt

$$\Pi_k{}_0^{\lambda-1}\left(1-\frac{\left(\frac{D(\alpha)}{\phi(\alpha)}\right)^k}{N\phi(\alpha)^s}\right)^{-1}$$

giebt, wenn $D(\alpha)$ Rest von $\phi(\alpha)$ ist, einen Faktor der ersten Art, und wenn $D(\alpha)$ Nichtrest von $\phi(\alpha)$ ist, einen Faktor der zweiten Art. Man erhält so, wenn der Kürze wegen das auf alle idealen Primzahlen in α, mit Ausschlufs der in $\varrho D(\alpha)$ enthaltenen, bezogene unendliche Produkt:

$$(4.)\qquad \Pi\left(1-\frac{\left(\frac{D(\alpha)}{\phi(\alpha)}\right)^k}{N\phi(\alpha)^s}\right)^{-1} = L_k$$

gesetzt wird:

$$(5.)\qquad R = (s-1)\, L_0\, L_1\, L_2 \,\ldots\, L_{\lambda-1}.$$

Durch Entwickelung der (-1)ten Potenz der zweitheiligen Gröfse und Ausführung der angedeuteten Multiplikation erhält man aus dem Produktausdrucke des L_k nach bekannter Methode folgenden Summenausdruck:

$$(6.)\qquad L_k = \Sigma\, \frac{\left(\frac{D(\alpha)}{F(\alpha)}\right)^k}{NF(\alpha)^s},$$

in welchem $F(\alpha)$ alle verschiedenen idealen Zahlen in der Theorie der complexen Zahlen in α bezeichnet, welche keinen gemeinschaftlichen Faktor mit $\varrho D(\alpha)$ haben, $\left(\frac{D(\alpha)}{F(\alpha)}\right)$ das in bekannter Weise für zusammengesetzte Moduln verallgemeinerte Legendresche Zeichen ist, und das Summenzeichen auf alle verschiedenen, idealen Zahlen $F(\alpha)$ zu beziehen ist. Der Werth der Reihe $(s-1)\, L_0$, für $s=1$, jedoch mit Zulassung der idealen Zahlen $F(\alpha)$, welche mit $\varrho D(\alpha)$ gemeinschaftliche Faktoren haben, ist in meiner Abhandlung in Liouville's Journal, Bd. 16, pag. 460 gegeben, und ist, um in den Zahlen $F(\alpha)$ die mit $\varrho D(\alpha)$ gemeinschaftlichen Faktoren auszuschliefsen, wenn $D(\alpha)$ die verschiedenen Primfaktoren $f(\alpha)$, $f_1(\alpha)$, ... enthält, nur mit

$$(7.)\qquad \left(1-\frac{1}{\lambda}\right)\left(1-\frac{1}{Nf(\alpha)}\right)\left(1-\frac{1}{Nf_1(\alpha)}\right)\ldots = C$$

zu multipliciren. Man hat daher:

$$(8.)\qquad (s-1)\, L_0 = \frac{2^{\mu-1}\,\pi^{\mu}\, C P D}{\lambda^{2\mu-\frac{1}{2}}},\ \text{für } s=1,$$

wo $\mu = \frac{\lambda-1}{2}$, und P und D die am angeführten Orte angegebenen Bedeutungen haben, und wenn dieser gefundene Werth der Einfachheit wegen mit K bezeichnet wird:

$$(9.)\qquad R = K\, L_1\, L_2\, L_3 \,\ldots\, L_{\lambda-1},\ \text{für } s=1.$$

H 2

Es wird nun der Werth der Reihe R für den Gränzwerth $s = 1$ nach einer anderen Methode gefunden, indem diese Reihe in so viel besondere Reihen zerlegt wird, als es verschiedene Klassen der idealen Zahlen $F(w)$ giebt. Sei H die Anzahl dieser Klassen, und $F_0(w)$, $F_1(w)$, ... $F_{H-1}(w)$ bezeichnen alle idealen Zahlen beziehungsweise der ersten, zweiten u. s. w. Klasse, so ist:

$$(10.)\qquad R = \Sigma\, \frac{s-1}{NNF_0(w)^s} + \Sigma\, \frac{s-1}{NNF_1(w)^s} + \ldots + \Sigma\, \frac{s-1}{NNF_{H-1}(w)^s}.$$

Es wird hier zunächst auf dieselbe Weise, wie in der genannten Abhandlung pag. 469 gezeigt, daſs für $s = 1$ alle diese H verschiedenen Summen denselben Werth erhalten, so daſs es hinreicht die erste derselben, in welcher $F_0(w)$ die erste Klasse, die der wirklichen complexen Zahlen in w repräsentirt, zu finden. Man kann nun jede wirkliche complexe Zahl $F_0(w)$, welche λ complexe Zahlen in α als Coefficienten hat, deren jeder wieder $\lambda - 1$ nichtcomplexe Coefficienten enthält, die als unbestimmte Zahlen mit $x_{h,k}$ bezeichnet werden sollen, so darstellen:

$$(11.)\qquad F_0(w) = \sum_{0}^{\lambda-2}{}_h \sum_{0}^{\lambda-1}{}_k\; x_{h,k}\, \alpha^h\, w^k.$$

Für alle möglichen Werthe der ganzzahligen Coefficienten $x_{h,k}$ erhält man nun aber nicht bloſs verschiedene wirkliche complexe Zahlen $F_0(w)$, sondern auch alle diejenigen, welche sich nur durch Einheiten unterscheiden. Um in dieser Form eine jede complexe Zahl nur einmal zu haben, muſs man mit Hülfe der Fundamental-Einheiten die Coefficienten den nöthigen Beschränkungen unterwerfen. Nach Hrn. Dirichlet's Untersuchungen über die allgemeinen Einheiten, giebt es aber für die complexen Zahlen in w $\frac{\lambda(\lambda-1)}{2} - 1$ Fundamental-Einheiten, über welche weiter unten vollständig gehandelt werden wird, und wenn man dieselben durch $\varepsilon_1(w)$, $\varepsilon_2(w)$, etc. bezeichnet, so sind alle Einheiten in der Form

$$(12.)\qquad \alpha_1\, \varepsilon_1(w)^{m_1}\, \varepsilon_2(w)^{m_2} \ldots\ldots \varepsilon_{\nu-1}(w)^{m_{\nu-1}}$$

enthalten, in welcher α_1 eine der $2\lambda^2$ Wurzeln der Gleichung $\alpha_1^{2\lambda^2} = 1$, $\nu = \frac{\lambda(\lambda-1)}{2}$ ist, und m_1, m_2, ... $m_{\nu-1}$ alle möglichen positiven und negativen ganzen Zahlen sein können. Bezeichnet man nun mit $MF(w)$ und $M\varepsilon_k(w)$ die analytischen Moduln der imaginären Gröſsen $F(w)$ und $\varepsilon_k(w)$, so erhält man in ähnlicher Weise, wie in der erwähnten Abhandlung, das Resultat: Wenn die Coefficienten $x_{h,k}$ in der complexen Zahl $F_0(w)$ so beschränkt werden; dafs in dem Systeme von Gleichungen, welches man aus der Gleichung

$$(13.)\quad \log. MF_0(w) = y_1 \log. M\varepsilon_1(w) + y_2 \log. M\varepsilon_2(w) + \ldots + y_{\nu-1} \log. M\varepsilon_{\nu-1}(w)$$

erhält, indem man zu den darin enthaltenen complexen Zahlen ihre conjugirten nimmt, die Gröfsen $y_1, y_2, \ldots y_{\nu-1}$ alle in den Gränzen 0 und 1 liegen müssen: so enthält die Form $F_0(w)$ alle verschiedenen wirklichen complexen Zahlen, jede genau $2\lambda^2$ mal. Unter den conjugirten werden aber hier alle diejenigen verstanden, welche man erhält, indem man dem w seine λ Werthe $w, w\alpha, \ldots w\alpha^{\lambda-1}$ giebt, und alsdann auch dem α (auch insofern es in $w = \sqrt[\lambda]{D(\alpha)}$ enthalten ist) die Werthe $\alpha, \alpha^2 \ldots \alpha^{\lambda-1}$ giebt. Dabei wird die eine Hälfte der so entstandenen $\lambda(\lambda-1)$ Gleichungen der andern Hälfte gleich, und ist deshalb zu verwerfen, von den übrig bleibenden ν Gleichungen ist alsdann noch eine beliebige, als mit den übrigen identisch, wegzulassen, so dafs genau $\nu - 1$ Gleichungen bleiben, mit ebenso vielen Gröfsen $y_1, y_2, \ldots y_{\nu-1}$.

Mit Hülfe der Dirichletschen Sätze wird nun der Gränzwerth, welchen die Reihe R für $s = 1$ annimmt, vollständig durch ein $\lambda(\lambda-1)$faches Integral bestimmt, nämlich:

$$(14.)\qquad R = \frac{C \cdot H}{2\lambda^2} \int dx_{0,0}\, dx_{0,1} \ldots dx_{\lambda-2,\lambda-1},$$

in welchen die Gröfsen $x_{0,0}$ etc. als continuirliche Variable auftreten, und die Integrationen auf alle Werthe derselben von $-\infty$ bis $+\infty$ sich erstrecken, für welche die Variabeln $y_1, y_2 \ldots y_{\nu-1}$ des obigen Systems in die Gränzen 0 und 1 zu liegen kommen, und für welche auch die Norm $NNF_0(w)$ in denselben Gränzen 0 und 1 liegt. Der Faktor C, welcher derselbe ist, als in der Gleichung (7.) rührt daher, dafs auch in $F_0(w)$ die Werthe ausgeschlossen sind, für welche $NF_0(w)$ durch ϱ, oder durch einen Primfaktor der Determinante $D(\alpha)$ theilbar sein würde.

Dieses $\lambda(\lambda-1)$fache Integral wird nun zunächst so transformirt, dafs anstatt der Variabeln $x_{0,0}$ etc. die $\lambda(\lambda-1)$ zu $F_0(w)$ conjugirten complexen Zahlen, welche durch die zu der Gleichung (11.) conjugirten Gleichungen mit den alten Variabeln verbunden sind, als neue Variable eingeführt werden. Die Funktional-Determinante dieser linearen Substitution erhält den einfachen Werth:

$$(15.) \qquad \delta = (-1)^{\frac{\mu}{2}} \lambda^{\frac{2\lambda^2-3\lambda}{2}} (ND(\alpha))^{\mu},$$

welcher nicht schwer zu finden ist.

Hierauf werden diese $\lambda(\lambda-1)$ Variabeln $F_0(w)$ mit allen conjugirten durch die Variabeln u_k und v_k ersetzt, welche aus jenen erhalten werden, indem man die Logarithmen derselben in die Form $u+\sqrt{-1}\,v$ setzt, wo u und v reale Gröfsen sind. Diese neue Substitution giebt:

$$(16.) \quad R = \frac{2^{\nu}\, C \cdot H}{2\lambda^2 \delta} \int e^{2u_1}\, e^{2u_2} \dots e^{2u_\nu}\, du_1\, du_2 \dots du_\nu\, dv_1\, dv_2 \dots dv_\nu.$$

Da die Integrationen in Beziehung auf die Variabeln $v_1, v_2, \dots v_\nu$ alle in den Gränzen $-\frac{\pi}{2}$ bis $+\frac{3\pi}{2}$ auszuführen sind, so hat man hieraus das νfache Integral:

$$(17.) \qquad R = \frac{2^{2\nu}\, \pi^{\nu}\, C \cdot H}{2\lambda^2 \delta} \int e^{2u_1}\, e^{2u_2} \dots e^{2u_\nu}\, du_1\, du_2 \dots du_\nu.$$

Wird die Integration in Beziehung auf u_ν in den, durch die Bedingung, dafs $NNF(w)$ positiv und kleiner als Eins sein mufs bestimmten Gränzen ausgeführt, so erhält man:

$$(18.) \qquad R = \frac{2^{2\nu}\, \pi^{\nu}\, C \cdot H}{4\lambda^2 \delta} \int du_1\, du_2 \dots du_{\nu-1}.$$

Endlich werden nun anstatt dieser $\nu-1$ Variabeln die Variabeln $y_1, y_2 \dots y_{\nu-1}$ aus dem Systeme der Gleichungen, welche aus (13.) entstehen, eingeführt, und weil diese nur in den Gränzen 0 und 1 zu integriren sind, so erhält man:

$$(19.) \qquad R = \frac{2^{2\nu}\, \pi^{\nu}\, CH\Delta}{4\lambda^2 \delta}, \quad \text{für } s = 1,$$

wo Δ die Determinante aus den Logarithmen der analytischen Moduln der Fundamentaleinheiten in w und ihrer conjugirten bezeichnet.

Aus der Vergleichung dieses Resultats mit (9.) hat man nun den Ausdruck der Klassenzahl H, für die hier betrachtete Theorie der complexen Zahlen in w:

$$(20.) \qquad H = \frac{4\lambda^2 \delta K}{2^{2\nu}\, \pi^{\nu}\, \Delta C} \cdot L_1\, L_2\, L_3 \dots L_{\lambda-1},$$

oder wenn für K und δ ihre Werthe bei (8.) und (15.) gesetzt werden:

$$(21.)\qquad H = \frac{\lambda^{2\nu-3\mu+2}\,(ND(\alpha))^{\mu}\,P\cdot D}{2^{2\nu-\mu-1}\,\pi^{\nu-\mu}\,\Delta}\,L_1\,L_2\,L_3\,\ldots\,L_{\lambda-1}.$$

Die Summation der unendlichen Reihen L_1, L_2 ... würde diesem Ausdrucke erst seine Vollendung geben; dieselbe scheint aber äufserst schwierig zu sein, und bietet wenigstens den gewöhnlichen Mitteln Trotz. Für die Anwendung, die in dem Folgenden von dem gefundenen Resultate gemacht werden soll, kommt es aber nur darauf an, dafs das Produkt L_1 L_2 ... $L_{\lambda-1}$, einen endlichen und von Null verschiedenen Werth hat, und dieses geht unmittelbar daraus hervor, dafs sowohl die Klassenanzahl H, als auch die Gröfsen Δ, P und D endliche, von Null verschiedene Werthe haben.

Die Klassenanzahl der complexen idealen Zahlen in z ist nur ein Vielfaches der gefundenen Klassenanzahl der idealen Zahlen in w, und zwar wird sie aus dieser durch Multiplikation mit einer Potenz von λ erhalten, deren nähere Bestimmung ich hier übergehe, weil sie für das Folgende nicht nöthig ist.

§. 7.

Eintheilung der verschiedenen Klassen der idealen Zahlen in z in ihre Gattungen.

Es soll nun die Theorie der complexen Zahlen in z in's Besondere, und zwar zunächst die Eintheilung der verschiedenen Klassen der idealen Zahlen dieser Theorie in ihre Gattungen (*Genera*) behandelt werden. Hierbei soll, wie überhaupt in allen folgenden Paragraphen dieser Abhandlung, angenommen werden, dafs die Primzahl λ nicht eine von denen ist, welche ich in meinen Untersuchungen über die complexen Zahlen in α als Ausnahmszahlen bezeichnet habe, also nicht eine solche, welche als Faktor des Zählers einer der ersten $\frac{\lambda-3}{2}$ Bernoullischen Zahlen vorkommt. Nach Ausschliessung dieser besonderen Werthe der Primzahl λ gelten für die complexen Zahlen und Einheiten dieser niederen Theorie folgende Sätze, welche ich in dem *Mémoire* in Liouville's Journal, Bd. 16, §. 9, und in der Abhandlung über die Ergänzungssätze zu den allgemeinen Reciprocitätsgesetzen in Crelle's Journal, Bd. 44, pag. 138 etc. bewiesen habe:

Die Klassenanzahl der idealen Zahlen in α ist nicht durch λ theilbar.

Der kleinste Exponent der Potenz, zu welcher eine ideale Zahl $f(\alpha)$ erhoben werden mufs, um zu einer wirklichen zu werden, ist nicht durch λ theilbar.

Jede Einheit $E(\alpha)$, welche einer nichtcomplexen ganzen Zahl congruent ist, nach dem Modul λ, ist eine λte Potenz einer anderen Einheit $\varepsilon(\alpha)$.

Jede nicht durch $1-\alpha$ theilbare, wirkliche complexe Zahl $F(\alpha)$ läfst sich durch Multiplikation mit einer passenden Einheit in die primäre Form bringen, in welcher sie die Bedingungen erfüllt: dafs erstens das Produkt $F(\alpha)\ F(\alpha^{-1})$ einer nichtcomplexen Zahl congruent ist, nach dem Modul λ, und dafs zweitens $F(\alpha)$ selbst einer nichtcomplexen Zahl congruent ist, nach dem Modul $(1-\alpha)^2$.

Es sei nun

$$F(z) = C + Bz + B_1 z_1 + \dots + B_{\lambda-1} z_{\lambda-1}$$

eine wirkliche complexe Zahl in z. Die Norm derselben, als Produkt aller conjugirten, ist eine symmetrische Funktion aller Wurzeln $z, z_1, \dots z_{\lambda-1}$ der Gleichung (5.), §. 1. In dieser Norm ist C^λ das einzige, kein z enthaltende Glied, aufser welchem noch symmetrische Funktionen der ersten, zweiten u. s. w. bis λten Dimension vorkommen. Alle diese symmetrischen Funktionen sind aber durch die Gleichungs-Coefficienten rational und ganz darstellbar, und müssen nothwendig alle den Faktor $\lambda\varrho$ enthalten, weil alle Gleichungscoefficienten denselben enthalten. Läfst man nun die durch $\lambda\varrho$ theilbaren Glieder der Norm weg, so hat man die Congruenz:

(1.) $$NF(z) \equiv C^\lambda, \text{ mod. } \lambda\varrho,$$

wo C eine ganze complexe Zahl in α ist. Bezeichnet man dieselbe durch $C(\alpha)$ und nimmt $NF(z) = F(\alpha)$, so hat man:

(2.) $$F(\alpha) \equiv C(\alpha)^\lambda, \text{ mod. } \lambda\varrho.$$

Weil die λte Potenz einer complexen Zahl in α einer nichtcomplexen Zahl congruent ist, nach dem Modul λ, so erkennt man hieraus, dafs die Norm jeder wirklichen complexen Zahl in z einer nichtcomplexen ganzen Zahl congruent ist, nach dem Modul λ.

Giebt man in der Congruenz (2.) dem α nach einander alle seine $\lambda-1$ Werthe, (wobei der Modul $\lambda\varrho$ wesentlich derselbe bleibt), und multiplicirt diese λ Congruenzen, so hat man:

$$NF(\alpha) \equiv (NC(\alpha))^\lambda, \text{ mod. } \lambda\varrho.$$

Weil nun $NC(\alpha)$, als Norm einer nicht durch ϱ theilbaren complexen Zahl in α, von der Form $1+m\lambda$ ist, die λte Potenz davon also von der Form $1+m\lambda^2$, weil ferner zwei nichtcomplexe ganze Zahlen, welche nach dem Modul $\lambda\varrho$ congruent sind, nothwendig auch nach dem Modul λ^2 congruent sein müssen, so hat man:

(3.) $$NF(\alpha) \equiv 1, \text{ mod. } \lambda^2.$$

Setzt man jetzt die wirkliche complexe Zahl $F(z)$ in die Form einer complexen Zahl in w:

$$F(z) = A + A_1 w + A_2 w^2 + \ldots + A_{\lambda-1} w^{\lambda-1},$$

und nimmt die Norm derselben, so ist diese Norm eine ganze rationale Funktion von w^λ, d. i. von $D(\alpha)$, und es ist A^λ das einzige Glied der Norm, welches w^λ nicht enthält. Läfst man nun alle Glieder weg, welche w^λ enthalten, so hat man die Congruenz:

(4.) $$NF(z) = F(\alpha) \equiv A^\lambda, \text{ mod. } D(\alpha).$$

Wenn nun die Determinante $D(\alpha)$ die von einander verschiedenen Primfaktoren $f(\alpha)$, $f_1(\alpha)$, $f_2(\alpha)$ ... enthält, welche ideal sein können, während $D(\alpha)$ als wirkliche complexe Zahl in α vorausgesetzt worden ist, so erkennt man aus dieser Congruenz, dafs in Beziehung auf alle diese Primfaktoren der Determinante die Norm der wirklichen complexen Zahl einer λten Potenz congruent ist, oder dafs man hat:

(5.) $$\left(\frac{F(\alpha)}{f(\alpha)}\right) = 1, \left(\frac{F(\alpha)}{f_1(\alpha)}\right) = 1, \left(\frac{F(\alpha)}{f_2(\alpha)}\right) = 1 \ldots\ldots$$

Aus den hier entwickelten Eigenschaften der Normen der wirklichen complexen Zahlen $F(z)$ ergeben sich nun die bestimmten Charaktere, welche die Normen der idealen Zahlen dieser Theorie besitzen. Die Norm einer idealen Zahl $F(z)$, als complexe Zahl in α, welche in dieser niederen Theorie auch selbst noch ideal sein kann, ist in Betreff der Einheiten in α, mit denen sie behaftet genommen werden kann, vollständig unbestimmt. Um diese Unbestimmtheit zu heben setze ich fest, dafs die Norm einer jeden idealen Zahl in z, als complexe Zahl in α, in der primären Form genommen werden soll, wie diese oben definirt ist, welche Bestimmung sich dadurch rechtfertigt, dafs sie für die Normen der wirklichen complexen Zahlen in z von selbst erfüllt ist, da jede solche Norm einer nicht complexen ganzen Zahl congruent ist, nach dem Modul λ, wie oben gezeigt worden. Wenn $F(\alpha)$, als Norm der idealen Zahl $F(z)$, selbst noch ideal ist, so wird die niedrigste Potenz derselben, welche wirklich wird, in der primären Form genommen.

Wenn nun $F(z)$ eine ideale Zahl in z bezeichnet und $F(\alpha)$ die Norm derselben, wobei der Fall, dafs $F(z)$ auch wirklich sein kann, nicht ausgeschlossen wird, wenn ferner $f(\alpha)$, $f_1(\alpha)$ .... die verschiedenen, in der Determinante $D(\alpha)$ enthaltenen Primfaktoren sind und wenn die Determinante den Faktor $\varrho = 1 - \alpha$ nicht enthält, welche Bestimmung auch in dem Folgenden überall beibehalten werden soll: so sollen die Zahlenwerthe folgender $\frac{\lambda-3}{2}$ Differenzialquotienten des Logarithmus von $F(e^v)$:

$$(6.)\qquad \begin{array}{ll} \dfrac{d_0^3\, l\, F(e^v)}{dv^3} \equiv C_3, & \\ \dfrac{d_0^5\, l\, F(e^v)}{dv^5} \equiv C_5, & \text{mod. } \lambda, \\ \vdots \qquad\qquad \vdots & \\ \dfrac{d_0^{\lambda-2}\, l\, F(e^v)}{dv^{\lambda-2}} \equiv C_{\lambda-2}, & \end{array}$$

ferner die Zahl

$$(7.)\qquad \frac{1 - NF(\alpha)}{\lambda} \equiv C_{\lambda-1}, \text{ mod. } \lambda.$$

und endlich auch die, durch die Legendreschen Zeichen:

$$(8.)\qquad \begin{array}{l} \left(\dfrac{F(\alpha)}{f(\alpha)}\right) = \alpha^{K} \\ \left(\dfrac{F(\alpha)}{f_1(\alpha)}\right) = \alpha^{K_1} \end{array}$$

u. s. w. bestimmten Zahlen K, K_1 .... als Charaktere der idealen Zahl $F(z)$, oder auch als Charaktere der Norm derselben $F(\alpha)$ bezeichnet werden. Aus dieser Erklärung der Charaktere ergiebt sich zunächst fast unmittelbar der Satz:

(I.) Die Charaktere des Produkts zweier oder mehrerer idealen Zahlen werden gefunden, wenn man die Charaktere der einzelnen Faktoren zu einander addirt.

Für die, als Differenzialquotienten der Logarithmen ausgedrückten Charaktere folgt die Richtigkeit dieses Satzes aus der analogen Eigenschaft der Logarithmen, und für die durch die Legendreschen Zeichen definirten aus der Eigenschaft dieser Zeichen, nach welcher

$$\left(\frac{F(\alpha)\cdot\phi(\alpha)}{f(\alpha)}\right) = \left(\frac{F(\alpha)}{f(\alpha)}\right)\left(\frac{\phi(\alpha)}{f(\alpha)}\right)$$

ist. Um dieselbe auch für den mit $C_{\lambda-1}$ bezeichneten Charakter nachzuweisen, kann man von der Congruenz

$$(1-NF(\alpha))\,(1-N\phi(\alpha)) \equiv 0, \text{ mod. } \lambda^2,$$

ausgehen, welche nothwendig Statt hat, weil sowohl $1-NF(\alpha)$ als $1-N\phi(\alpha)$ durch λ theilbar sind. Entwickelt man das Produkt dieser beiden Faktoren, bringt $NF(\alpha)\ N\phi(\alpha)$ auf die andere Seite, addirt auf beiden Seiten Eins und dividirt durch λ, so hat man:

(9.) $$\frac{1-NF(\alpha)}{\lambda} + \frac{1-N\phi(\alpha)}{\lambda} \equiv \frac{1-N(F(\alpha)\phi(\alpha))}{\lambda}, \text{ mod. } \lambda,$$

und der auf diese Weise für zwei Faktoren geführte Beweis wird durch bloſse Wiederholung auf beliebig viele Faktoren ausgedehnt, und giebt, wenn die Faktoren einander gleich angenommen werden:

(10.) $$\frac{k(1-NF(\alpha))}{\lambda} \equiv \frac{1-NF(\alpha)^k}{\lambda}, \text{ mod. } \lambda.$$

Die Benennung Charaktere kommt diesen Zahlen darum zu, weil sie nicht nur einzelnen idealen Zahlen angehören, sondern für alle idealen Zahlen einer Klasse dieselben bleiben. Um dieſs zu beweisen, bemerke ich zunächst, daſs für die Klasse der wirklichen complexen Zahlen in z alle diese Charaktere den Werth Null haben. In der That sind erstens die $\frac{\lambda-3}{2}$, als Differenzialquotienten von $lF(e^v)$ definirten, alle congruent Null, weil $F(\alpha)$ einer λten Potenz congruent ist, nach dem Modul λ; zweitens ist der Charakter $C_{\lambda-1}$ congruent Null, weil $1-NF(\alpha)$ durch λ^2 theilbar ist, und drittens sind auch $K, K_1, \ldots$ congruent Null, weil $F(\alpha)$ einer λten Potenz congruent ist, nach dem Modul $D(\alpha)$, und somit auch nach den Moduln $f(\alpha), f_1(\alpha) \ldots$. Es seien nun $F_r(z)$ und $F_s(z)$ zwei äquivalente ideale Zahlen, und $\Phi(z)$ ein Multiplikator, welcher beide zu wirklichen macht, also sowohl $\Phi(z)\ F_r(z)$ als auch $\Phi(z)\ F_s(z)$ wirklich. Es sei auch $NF_r(z) = F_r(\alpha)$, $NF_s(z) = F_s(\alpha)$ und $N\Phi(z) = \Phi(\alpha)$, so hat man:

$$\frac{d_0^{2n+1}\, l(\Phi(e^v)\cdot F_r(e^v))}{dv^{2n+1}} \equiv 0, \text{ mod. } \lambda,$$

$$\frac{d_0^{2n+1}\, l(\Phi(e^v)\cdot F_s(e^v))}{dv^{2n+1}} \equiv 0, \text{ mod. } \lambda,$$

also auch

I 2

(11.) $$\frac{d^{2n+1}\, l\, F_r(e^v)}{dv^{2n+1}} \equiv \frac{d^{2n+1}\, l\, F_s(e^v)}{dv^{2n+1}}, \text{ mod. } \lambda.$$

Ferner hat man für die Normen der wirklichen Zahlen $\Phi(z)\; F_r(z)$ und $\Phi(z)\; F_s(z)$:

$$N(\Phi(\alpha)\; F_r(\alpha)) \equiv 1, \text{ mod. } \lambda^2,$$
$$N(\Phi(\alpha)\; F_s(\alpha)) \equiv 1, \text{ mod. } \lambda^2,$$

woraus folgt:

$$NF_r(\alpha) \equiv NF_s(\alpha), \text{ mod. } \lambda^2,$$

und

(12.) $$\frac{NF_r(\alpha)-1}{\lambda} \equiv \frac{NF_s(\alpha)-1}{\lambda}, \text{ mod. } \lambda.$$

Endlich hat man auch für die Normen dieser wirklichen Zahlen:

$$\left(\frac{\Phi(\alpha)\; F_r(\alpha)}{f(\alpha)}\right) = 1,\; \left(\frac{\Phi(\alpha)\; F_s(\alpha)}{f(\alpha)}\right) = 1,$$

also

$$\left(\frac{\Phi(\alpha)}{f(\alpha)}\right)\cdot\left(\frac{F_r(\alpha)}{f(\alpha)}\right) = 1,\; \left(\frac{\Phi(\alpha)}{f(\alpha)}\right)\cdot\left(\frac{F_s(\alpha)}{f(\alpha)}\right) = 1,$$

demnach

(13.) $$\left(\frac{F_r(\alpha)}{f(\alpha)}\right) = \left(\frac{F_s(\alpha)}{f(\alpha)}\right),$$

und ebenso für alle verschiedenen Primfaktoren $f_1(\alpha)$, $f_2(\alpha)$ ... der Determinante. Man hat also den Satz:

(II.) Alle äquivalenten, einer und derselben Klasse angehörenden, idealen Zahlen haben gleiche Charaktere.

Die aufgestellten Charaktere sind somit nicht blofs Charaktere der einzelnen idealen Zahlen, sondern Charaktere der Klassen. Aus diesem Grunde wird auf sie die Eintheilung der Klassen in die Gattungen (*Genera*) gegründet, indem alle nicht äquivalenten Klassen, welche vollständig dieselben Charaktere haben, einer und derselben Gattung, diejenigen aber, für welche nicht alle Charaktere dieselben sind, verschiedenen Gattungen zugetheilt werden.

Wenn die Anzahl der verschiedenen, in der Determinante $D(\alpha)$ enthaltenen Primfaktoren gleich r ist, so giebt es für jede ideale Zahl in z $\frac{\lambda-1}{2}+r$ besondere Charaktere, deren jeder einen der λ Werthe 0, 1, 2, ...

$\lambda - 1$ haben kann. Die Anzahl aller möglichen Combinationen dieser Werthe der einzelnen Charaktere ist gleich der $(\frac{\lambda-1}{2}+r)$ten Potenz von λ, welches also die Anzahl aller Gesammtcharaktere ist die überhaupt möglicherweise Statt haben können, oder die Anzahl aller angebbaren Gattungen. Weil aber der Fall eintreten kann, und wie in dem Folgenden gezeigt werden wird auch wirklich eintritt, dafs gewisse dieser angebbaren Gattungen gar keine Klassen idealer Zahlen enthalten, so sind von diesen blofs angebbaren, die wirklich vorhandenen Gattungen wohl zu unterscheiden. Diejenige Gattung, deren Charaktere alle gleich Null sind, welcher, wie oben gezeigt worden ist die Klasse der wirklichen complexen Zahlen angehört, welche also immer eine wirklich vorhandene ist, soll die Hauptgattung (*Genus principale*) genannt werden.

Über die Vertheilung der einzelnen Klassen in die Gattungen wird nun zunächst folgender Satz bewiesen:

(III.) Alle wirklich vorhandenen Gattungen enthalten gleich viele Klassen idealer Zahlen.

Setzt man nämlich alle Klassen einer gegebenen Gattung mit einer bestimmten Klasse zusammen, so gehören alle dadurch entstehenden verschiedenen Klassen wieder einer und derselben Gattung an, weil sie, wie der Satz (I.) zeigt, alle dieselben Charaktere haben. Durch passende Wahl dieser einen Klasse kann man aber aus einer jeden gegebenen Gattung, welche n Klassen enthält eben so viele verschiedene Klassen einer jeden anderen gegebenen Gattung erzeugen. Wählt man nun für die eine Gattung, aus welcher alle übrigen erzeugt werden, eine solche, welche nicht weniger Klassen enthält als irgend eine andere, so folgt, dafs alle anderen Klassen nicht nur nicht mehr, sondern auch nicht weniger Klassen enthalten können als diese, wodurch die Richtigkeit des Satzes bewiesen ist.

Aufser der Eintheilung in die Gattungen ist noch eine andere Eintheilung der nichtäquivalenten Klassen beachtenswerth, welche von den verschiedenen Klassen der idealen Zahlen in α herrührt. Wenn h die Anzahl der nichtäquivalenten Klassen dieser niederen Theorie ist, und die idealen Zahlen $\phi(\alpha)$, $\phi_1(\alpha)$, $\phi_2(\alpha)$, ... $\phi_{h-1}(\alpha)$ repräsentiren diese verschiedenen Klassen, wenn ferner $F(z)$ eine ideale Zahl in z ist, so stellen

$$\phi(\alpha)F(z),\ \phi_1(\alpha)F(z),\ \ldots.\ \phi_{h-1}(\alpha)F(z)$$

* h nichtäquivalente Klassen dieser höheren Theorie dar, welche in gewissem Sinne als zusammengehörig zu betrachten sind, und eine Gruppe bilden. Ist $F_1(z)$ eine nicht in dieser Gruppe enthaltene ideale Zahl, so hat man aus ihr eine zweite Gruppe von Klassen:

$$\phi(\alpha)F_1(z),\ \ \phi_1(\alpha)F_1(z),\ \ldots.\ \ \phi_{h-1}(\alpha)F_1(z)\ ,$$

welche weder unter sich, noch mit den Klassen der ersten Gruppe äquivalent sind. In dieser Weise fortfahrend kann man alle Klassen der idealen Zahlen in z in solche Gruppen von je h Klassen zusammenfassen, woraus beiläufig folgt, dafs die Klassenanzahl der idealen Zahlen in z durch die Klassenanzahl der idealen Zahlen in α theilbar ist. Diese Eintheilung in die Gruppen wird bei einigen der zu erörternden Hauptfragen ihre Anwendung finden, in welchen der Unterschied der, einer und derselben Gruppe angehörenden Klassen nur als ein unwesentlicher anzusehen sein wird. Aus diesem Grunde sollen nur diejenigen Klassen der idealen Zahlen in z, welche verschiedenen Gruppen angehören, wesentlich verschiedene Klassen benannt werden. Nimmt man aus jeder Klasse eine einzige ideale Zahl, welche als Repräsentant der ganzen Klasse angesehen wird, so kann man dieselbe immer so wählen, dafs sie keinen wirklichen Faktor enthält, namentlich auch keinen wirklichen Faktor, welcher nur eine complexe Zahl in α ist, weil durch das Weglassen eines wirklichen Faktors an der Klasse, welcher eine ideale Zahl angehört, nichts geändert wird. Nimmt man ferner aus jeder der Gruppen von h Klassen eine der idealen Zahlen, welche diese Klassen repräsentiren, als Repräsentant der Gruppe, so kann man dieselbe immer so wählen, dafs sie niemals alle λ conjugirten idealen Primfaktoren irgend einer idealen Primzahl $\phi(\alpha)$ der niederen Theorie, und somit $\phi(\alpha)$ selbst als Faktor enthält; denn wenn man den idealen Faktor $\phi(\alpha)$ aus der idealen Zahl wegläfst, so erhält man eine ideale Zahl derselben Gruppe. Schliefst man, wie es bei gewissen Untersuchungen nützlich ist, diejenigen idealen Zahlen in z vollständig aus, welche ideale Faktoren $\phi(\alpha)$ der niederen Theorie enthalten, so gehören alle nichtäquivalenten Klassen nothwendig auch verschiedenen Gruppen an, und geben darum nur alle diejenigen Klassen, welche wir als wesentlich verschiedene bezeichnet haben. Die Anzahl dieser ist gleich dem $h$ten Theile der Anzahl aller nichtäquivalenten Klassen.

Die Eintheilung in die Gruppen ordnet sich der Eintheilung in die Gattungen vollständig unter, da alle Klassen einer und derselben Gruppe nothwendig auch einer und derselben Gattung angehören. Die Charaktere der idealen Zahlen $\phi(\alpha)$, $\phi_1(\alpha)$, ... $\phi_{h-1}(\alpha)$, welche hier als ideale Zahlen in z auftreten, sind nämlich alle gleich Null, weil die Normen dieser idealen Zahlen in Beziehung auf z nur die λten Potenzen derselben sind, ihre Charaktere also alle gleich Null, so dafs durch das Hinzutreten derselben an den Charakteren einer idealen Zahl $F(z)$ nichts geändert wird.

Conjugirte ideale Zahlen gehören im Allgemeinen verschiedenen Klassen und verschiedenen Gruppen, aber stets nur einer und derselben Gattung an, weil die Charaktere nur von der Norm abhängen, welche für alle conjugirten dieselbe ist. Wenn aber von conjugirten Zahlen zwei derselben Klasse angehören, also äquivalent sind, so sind alle λ conjugirten äquivalent; denn wenn $\phi(z)$ eine ideale Zahl ist, welche einer ihrer conjugirten $\phi(z_k)$ äquivalent ist, so giebt es einen Multiplikator $\psi(z)$, für welchen $\psi(z)\,\phi(z)$ und $\psi(z)\,\phi(z_k)$ beide zugleich wirklich sind; verwandelt man nun z in z_k, so sind auch $\psi(z_k)\,\phi(z_k)$ und $\psi(z_k)\,\phi(z_{2k})$ beide zugleich wirklich, woraus $\phi(z_k)$ äquivalent mit $\phi(z_{2k})$ geschlossen wird, und auf diese Weise weiter fortschliefsend sieht man, dafs alle conjugirten äquivalent sind.

Eine ideale Zahl, welche die Eigenschaft hat, dafs sie ihren conjugirten äquivalent ist, soll eine ambige ideale Zahl genannt werden, und die Klasse, welcher sie angehört, eine ambige Klasse, ähnlich wie bei Gaufs in der Theorie der quadratischen Formen, diejenige Klasse, welche eine ihrer entgegengesetzten, also conjugirten äquivalente Form enthält, als *Classis anceps* bezeichnet wird. In Betreff der Gruppen von je h Gliedern ist zu bemerken, dafs wenn eine Klasse einer Gruppe eine ambige ist, nothwendig alle h Klassen derselben ambige sein müssen.

Die Anzahl der ambigen Klassen, und namentlich der wesentlich verschiedenen, steht mit der Anzahl der wirklich vorhandenen Gattungen in einem sehr innigen Zusammenhange, welcher in dem Folgenden genauer erörtert werden wird. Für jetzt soll in dieser Beziehung nur der eine Hauptsatz bewiesen werden:

(IV.) Die Anzahl aller wirklich vorhandenen Gattungen ist nicht gröfser, als die Anzahl aller wesentlich verschiedenen, nicht äquivalenten ambigen Klassen.

Wenn $f(z)$ eine beliebige ideale Zahl ist, $f(z_1)$ ihre erste conjugirte, so giebt es immer eine ideale Zahl $F(z)$ von der Art, dafs $F(z)f(z_1)$ mit $f(z)$ äquivalent ist, welches ich so ausdrücke:

(14.) $$f(z) \text{ aeqv. } F(z)f(z_1).$$

Untersucht man nun, unter welcher Bedingung zwei verschiedene Klassen idealer Zahlen, welche durch $f(z)$ und $g(z)$ repräsentirt werden, in dieser Äquivalenz eine und dieselbe Klasse $F(z)$ ergeben, indem man diese Äquivalenz mit der folgenden:

$$g(z) \text{ aeqv. } F(z)g(z_1)$$

verbindet, so kann man die Klasse $g(z)$ durch $f(z)\,\phi(z)$ ersetzen, wo $\phi(z)$ stets so bestimmt werden kann, dafs $g(z)$ äquivalent $f(z)\,\phi(z)$ ist, woraus sodann geschlossen wird, dafs auch die conjugirte Zahl $g(z_1)$ der conjugirten $f(z_1)\,\phi(z_1)$ äquivalent ist. Man hat daher:

$$f(z)\,\phi(z) \text{ aeqv. } F(z)\,f(z_1)\,\phi(z_1)$$

und hieraus nach der Äquivalenz (14.):

$$\phi(z) \text{ aeqv. } \phi(z_1),$$

woraus folgt, dafs $\phi(z)$ eine Ambige sein mufs, und umgekehrt, wenn $\phi(z)$ eine Ambige ist, dafs $f(z)$ und $f(z)\,\phi(z)$ in der Äquivalenz (14.) dieselbe Klasse $F(z)$ ergeben. Also alle diejenigen verschiedenen Klassen $f(z)$, welche aus einer derselben entstehen, indem man dieselbe mit allen ambigen Klassen zusammensetzt, ergeben für $F(z)$ eine und dieselbe Klasse, diejenigen aber, welche nicht auf diese Weise aus einer einzigen erzeugt werden können, ergeben verschiedene Klassen für $F(z)$. Wenn man nun bei dieser Frage nur wesentlich verschiedene Klassen in Betracht zieht, und die Anzahl derselben mit $\mathfrak{H}$, die Anzahl der wesentlich verschiedenen Klassen aber mit $\mathfrak{A}$ bezeichnet, so hat man die Anzahl aller wesentlich verschiedenen Klassen $F(z)$, welche der Äquivalenz (14.) genügen, wenn für $f(z)$ alle verschiedenen Klassen genommen werden, gleich $\frac{\mathfrak{H}}{\mathfrak{A}}$.

Es gehört nun aber jede Klasse $F(z)$, welche der Äquivalenz (14.) genügen kann, nothwendig der Hauptgattung an, deren Charaktere alle gleich Null sind; denn die Charaktere des Produkts $F(z)f(z_1)$ findet man, indem man die Charaktere von $F(z)$ zu den entsprechenden von $f(z_1)$, die denen von $f(z)$ gleich sind, addirt, und weil die Charaktere von $F(z)f(z_1)$ denen von $f(z)$, wegen der Äquivalenz beider idealen Zahlen, gleich sind, so folgt, dafs die Charaktere von $F(z)$ alle gleich Null sein müssen.

Die Hauptgattung enthält also nothwendig diese $\frac{\mathfrak{H}}{\mathfrak{A}}$ wesentlich verschiedenen Klassen, wobei der Fall nicht ausgeschlossen ist, dafs sie aufserdem auch noch andere enthalten könnte. Da aber alle wirklich vorhandenen Gattungen gleich viele Klassen, und darum auch gleich viele der wesentlich verschiedenen Klassen enthalten, so folgt, dafs jede der vorhandenen Gattungen mindestens $\frac{\mathfrak{H}}{\mathfrak{A}}$ wesentlich verschiedene Klassen enthält, wodurch der aufgestellte Satz bewiesen ist.

§. 8.

Die idealen ambigen Zahlen, insofern sie in gewissen wirklichen complexen Zahlen in z enthalten sind.

Sei $\phi(z)$ eine ideale Ambige, und zwar eine solche, welche nicht alle λ conjugirten idealen Primfaktoren einer Primzahl $\phi(\alpha)$ der niederen Theorie und keinen complexen Primfaktor in α welcher in der Theorie der complexen Zahlen in z Primfaktor ist, also überhaupt keinen idealen oder wirklichen Faktor dieser niederen Theorie enthält. Sei ferner $\psi(z)$ ein idealer Multiplikator, welcher mit $\phi(z)$ und der conjugirten $\phi(z_1)$ zusammengesetzt, wirkliche complexe Zahlen ergiebt, so dafs

$$G(z) = \psi(z)\,\phi(z) \text{ und } G_1(z) = \psi(z)\,\phi(z_1)$$

* wirkliche complexe Zahlen sind. Da $G(z)$ und $G(z_1)$, lediglich durch die idealen Faktoren bestimmt sind, welche sie enthalten, so können sie beliebig mit Einheiten in z behaftet angenommen werden, diese können aber stets so gewählt werden, dafs die Normen der Zahlen $G(z)$ und $G_1(z)$ einander gleich werden. Die Normen $NG(z)$ und $NG_1(z)$ sind nämlich erstens genau aus denselben idealen Faktoren in z zusammengesetzt, und können sich daher nur durch eine Einheit in z unterscheiden, sie sind zweitens complexe Zahlen der niederen Theorie in α, darum kann diese Einheit, durch welche sie sich unterscheiden, nur eine Einheit $E(\alpha)$ sein, und man hat

$$NG(z) = E(\alpha)\, NG_1(z).$$

Die Normen aller wirklichen complexen Zahlen in z sind aber, nach dem Modul λ, nichtcomplexen ganzen Zahlen congruent, es mufs also auch die Einheit $E(\alpha)$ einer nichtcomplexen Zahl congruent sein, nach dem Modul λ, also in Folge des im §. 7. citirten Satzes, mufs sie eine λte Potenz einer Einheit sein, also:

$$E(\alpha) = \varepsilon(\alpha)^\lambda.$$

Nimmt man $\varepsilon(\alpha)\, G(z)$ anstatt $G(z)$, wozu man berechtigt ist, weil die Wahl der $G(z)$ und $G_1(z)$ behaftenden Einheiten völlig frei ist, so hat man:

(1.) $$NG(z) = NG_1(z).$$

Wenn nun $G(z)$ und $G_1(z)$ so gewählt sind, dafs sie dieser Bedingung genügen, so setze ich

(2.) $$\frac{G(z)}{G_1(z)} = E(z).$$

Aus dieser gebrochenen complexen Zahl $E(z)$, deren Norm gleich Eins ist, bilde ich einen der Ausdrücke, deren Theorie ich in Crelle's Journal, Bd. 50, pag. 212 behandelt habe, nämlich

(3.) $$P(E(z)) = 1 + E(z) + E(z)E(z_1) + E(z)E(z_1)E(z_2) + \cdots \\ \cdots + E(z)E(z_1)\ldots E(z_{\lambda-2}).$$

Dieser Ausdruck, welcher selbst eine wirkliche gebrochene complexe Zahl in z ist, kann, wenn die Wurzeln z, aus den Nennern der Brüche entfernt werden, in folgende Form gesetzt werden:

(4.) $$PE(z) = \frac{Af(z)}{B},$$

wo A und B wirkliche complexe Zahlen in α sind, und $f(z)$ eine wirkliche ganze complexe Zahl in z. Vermöge der ersten allgemeinen Grundeigenschaft des mit $PE(z)$ bezeichneten Ausdrucks, nämlich

$$E(z)\, P(E(z_1)) = P(E(z)),$$

hat man nun, wenn man die Form (4.) einsetzt, und $\frac{A}{B}$ als gemeinschaftlichen Faktor wegläfst:

$$E(z)\, f(z_1) = f(z),$$

oder, wenn für $E(z)$ sein Werth bei (2.) zurückgesetzt, und mit $G_1(z)$ multiplicirt wird:

(5.) $$G(z)f(z_1) = G_1(z)f(z).$$

Es ist hier zu bemerken, dafs der weggehobene gemeinschaftliche Faktor $\frac{A}{B}$ unter Umständen gleich Null sein kann, und dafs dadurch diese Gleichungen illusorisch werden können, nämlich wenn $PE(z)$ gleich Null ist. In diesem Falle kann man aber anstatt der Einheit $E(z)$ die Einheit $\alpha^k E(z)$ nehmen und die Zahl k so bestimmen, dafs $P(\alpha^k E(z))$ nicht gleich Null ist. Dafs unter den λ Werthen dieses Ausdrucks für $k = 0, 1, 2, \ldots \lambda - 1$, wenigstens einer nicht gleich Null ist, folgt unmittelbar daraus, dafs die Summe derselben gleich λ ist. Dieselbe Bemerkung ist ebenso auf alle Anwendungen zu beziehen, welche in dem Folgenden von diesen aus Einheiten zusammengesetzten Ausdrücken gemacht werden mögen.

Die ideale Ambige $\phi(z)$ hat die Eigenschaft, dafs ihre $h\lambda$te Potenz wirklich ist, wenn h die Klassenanzahl der idealen complexen Zahlen in α bezeichnet; denn weil die Ambige allen ihren conjugirten äquivalent ist, so hat man:

$$N\phi(z) \text{ aeqv. } \phi(z)^\lambda$$

* und wenn zur $h$ten Potenz erhoben wird, so wird $(NF(z))^h$ wirklich, weil $N\phi(z)$ nur eine complexe ideale Zahl in α ist, deren $h$te Potenz nothwendig wirklich ist; es ist also auch $\phi(z)^{h\lambda}$ wirklich. Setzt man nun

$$\phi(z)^{h\lambda} = \Phi(z), \quad \psi(z)^{h\lambda} = \Psi(z),$$

wo $\Phi(z)$ und $\Psi(z)$ wirklich sind, so hat man:

$$\begin{aligned} G(z)^{h\lambda} &= \Psi(z)\,\Phi(z)\,\varepsilon(z) \\ G_1(z)^{h\lambda} &= \Psi(z)\,\Phi(z_1)\,\varepsilon_1(z) \end{aligned}$$

und darum giebt die Gleichung (5.) zur $h\lambda$ten Potenz erhoben, wenn der gemeinschaftliche wirkliche Faktor $\Psi(z)$ hinweggehoben, und der Quotient der beiden Einheiten $\varepsilon(z)$ und $\varepsilon_1(z)$ durch $e(z)$ bezeichnet wird:

(6.) $$e(z)\,\Phi(z)f(z_1)^{h\lambda} = \Phi(z_1)f(z)^{h\lambda}.$$

Die Einheit $e(z)$ ist hier eine ganze Einheit, und zwar eine solche, deren Norm gleich Eins ist. Man hat nun:

K 2

(7.) $$\frac{f(z)^{h\lambda}}{\Phi(z)} = \frac{f(z)^{h\lambda}\,\Phi(z_1)\,\Phi(z_2)\,\ldots.\,\Phi(z_{\lambda-1})}{N\Phi(z)} = \frac{F(z)}{N\Phi(z)},$$

wenn dieser Zähler der Einfachheit wegen durch $F(z)$ bezeichnet wird, und demnach aus der Gleichung (6.):

(8.) $$e(z)\,F(z_1) = F(z).$$

Die complexe Zahl $F(z)$, welche dieser Gleichung (8.) genügt, hat die Eigenschaft, dafs sie zu jedem idealen Primfaktor, welchen sie enthält, auch alle seine conjugirten enthalten mufs. Ist nämlich $p(z)$ ein in $F(z)$ enthaltener idealer Primfaktor, so mufs vermöge der Gleichung (8.) derselbe auch in $F(z_1)$ enthalten sein, und wenn z in $z_{\lambda-1}$, z_1 in z, u. s. w. verwandelt wird, wodurch $F(z_1)$ in $F(z)$ übergeht, so folgt, dafs $F(z)$ auch den Primfaktor $p(z_{\lambda-1})$ enthalten mufs. Hieraus folgt weiter, vermöge der Gleichung (8.), dafs auch $F(z_1)$ den Primfaktor $p(z_{\lambda-1})$ enthalten mufs, und wenn wieder z in $z_{\lambda-1}$ verwandelt wird, dafs $F(z)$ den Faktor $p(z_{\lambda-2})$ enthalten mufs. So fortschliefsend findet man, dafs $F(z)$ alle λ conjugirten idealen Primfaktoren, und folglich die complexe Primzahl $p(\alpha)$ der niederen Theorie enthalten mufs, zu welcher sich dieselben zusammensetzen. Da dasselbe für alle definirten idealen Primfaktoren gilt, so folgt, dafs $F(z)$ sich in zwei Faktoren zerlegen läfst, deren einer nur eine complexe Zahl in α ist, der andere aber eine complexe Zahl in z, welche die Eigenschaft hat, keinen der definirten idealen Primfaktoren zu enthalten. Der erste dieser Faktoren könnte auch eine ideale Zahl in α sein, um ihn also gewifs zu einem wirklichen zu machen, erhebe ich $F(z)$ zur $h$ten Potenz, weil hierdurch, wenn h die Klassenanzahl der idealen Zahlen in α ist, der erste Faktor wirklich wird, so mufs der zweite auch wirklich werden, und es wird:

(9.) $$F(z)^h = C\Delta(z),$$

wo C eine wirkliche complexe Zahl in α ist, und $\Delta(z)$ eine wirkliche complexe Zahl in z, welche keinen der definirten idealen Primfaktoren enthält, deren Norm also lediglich aus den Faktoren der Determinante $D(\alpha)$ und einer Potenz von ϱ bestehen kann.

Ich erhebe nun die Gleichung (7.) zur $h$ten Potenz, und setze den gefundenen Werth des $F(z)^h$ ein, so wird:

(10.) $$f(z)^{h^2\lambda} = \frac{C\Delta(z)\,\Phi(z)^h}{(N\Phi(z))^h}.$$

Es sei nun $p(z)$ ein idealer Primfaktor von $\phi(z)$, also auch von $\Phi(z)$, so enthält $N\Phi(z)$ alle seine conjugirten, welche zusammen den Primfaktor $p(\alpha)$ der niederen Theorie bilden. $\Phi(z)^h$ im Zähler, enthält (nach der Voraussetzung, dafs $\phi(z)$ nicht einen idealen Faktor in α enthalten soll,) nicht alle conjugirten idealen Primfaktoren zu $p(z)$, $\Delta(z)$ enthält keinen derselben, also mufs C einen oder einige derselben enthalten, weil alle diese Faktoren des Nenners gegen die des Zählers sich hinwegheben müssen. C aber, als complexe Zahl in α, kann nicht ideale Primfaktoren in z enthalten, ohne dafs sie alle conjugirten zugleich, und folglich $p(\alpha)$ enthält. Der Faktor $p(\alpha)$ des Nenners $(N\Phi(z))^h$ hebt sich also vollständig gegen den Faktor C des Zählers hinweg, und weil dasselbe für alle Faktoren des Nenners gilt, so folgt, dafs C durch $(N\Phi(z))^h$ theilbar ist. Bezeichnet man diesen Quotienten mit K, so hat man:

$$(11.)\qquad f(z)^{h^2\lambda} = K\Delta(z)\,\Phi(z)^h$$

und wenn durch $\phi(z)^{h^2\lambda} = \Phi(z)^h$ dividirt wird:

$$(12.)\qquad \left(\frac{f(z)}{\phi(z)}\right)^{h^2\lambda} = K\Delta(z).$$

Hieraus folgt, dafs $f(z)$ alle idealen Primfaktoren des $\phi(z)$ enthalten mufs, und ferner, dafs zu jedem idealen Primfaktor in z, welchen es ausserdem enthalten könnte, alle conjugirten in $f(z)$ enthalten sein müssen, welche sich zu einer complexen Zahl in α zusammensetzen. Verbindet man diese complexe Zahl in α mit der Ambigen $\phi(z)$, wodurch dieselbe in eine andere, derselben Gruppe angehörende, also nicht wesentlich verschiedene Ambige übergeht, so läfst sich das gefundene Resultat so aussprechen:

(I.) Jede ideale Ambige $\phi(z)$, wenn sie von den Faktoren, welche sich zu idealen oder wirklichen complexen Zahlen in α zusammensetzen, befreit angenommen wird, ist in einer wirklichen complexen Zahl $f(z)$ so enthalten, dafs diese wirkliche Zahl $f(z)$ die ideale Ambige $\phi(z)$, aber ausserdem keinen der definirten idealen Primfaktoren weiter enthält.

Wenn alle in einer wirklichen complexen Zahl $f(z)$ enthaltenen, idealen Primfaktoren zusammen eine Ambige $\phi(z)$ ausmachen, so soll von der Zahl $f(z)$ ausgesagt werden: sie enthält die Ambige $\phi(z)$.

Wenn nun $f(z)$ die Ambige $\phi(z)$ enthält, so ist

$$(13.)\qquad Nf(z) = \Delta(\alpha)\cdot N\phi(z),$$

und es enthält $\Delta(\alpha)$ keine anderen Primfaktoren in α, als die in der Determinante $D(\alpha)$ enthaltenen, und ausserdem eine Potenz von ϱ.

Es sollen nun die nothwendigen und hinreichenden Bedingungen dafür gefunden werden, dafs eine wirkliche complexe Zahl in z eine ideale Ambige enthalte. Es sei also $\phi(z)$ eine ideale Ambige, welche in der wirklichen Zahl $f(z)$ enthalten ist, so ist der Complex aller in $f(z)^{h\lambda-1}f(z_1)$ enthaltenen idealen Faktoren gleich $\phi(z)^{h\lambda-1}\phi(z_1)$, welches, wegen der Bedingung $\phi(z)$ aeqv. $\phi(z_1)$, mit $\phi(z)^{h\lambda}$, also mit einer wirklichen complexen Zahl äquivalent, und darum selbst wirklich ist. Setzt man nun der Kürze wegen:

$$(14.)\qquad \phi(z)^{h\lambda-1}\phi(z_1) = \Phi(z),$$

so hat man:

$$(15.)\qquad f(z)^{h\lambda-1}f(z_1) = \Delta(z)\,\Phi(z),$$

wo $\Delta(z)$ und $\Phi(z)$ zwei wirkliche complexe Zahlen sind, deren erste keinen idealen Primfaktor in z enthält, und darum in ihrer Norm nur die in $\varrho D(\alpha)$ enthaltenen Primfaktoren haben kann, von denen dagegen die zweite nur aus idealen Primfaktoren in z zusammengesetzt, also ihre Norm zu $\varrho D(\alpha)$ relative Primzahl ist. Umgekehrt, wenn der Complex aller in $f(z)^{h\lambda-1}f(z_1)$ enthaltenen idealen Primfaktoren eine wirkliche complexe Zahl in z ist, und $\phi(z)$ stellt den Complex aller in $f(z)$ enthaltenen idealen Primfaktoren dar,
* so ist $\phi(z)^{h\lambda-1}\phi(z_1)$ wirklich, und hieraus folgt, dafs $\phi(z)$ äquivalent $\phi(z_1)$, also eine Ambige ist. Man hat daher folgenden Satz:

* (II.) Wenn die wirkliche complexe Zahl $\phi(z)$ eine Ambige enthält, so läfst sich $f(z)^{h\lambda-1}f(z_1)$ in zwei wirkliche Faktoren zerlegen, welche so beschaffen sind, dafs die Norm des einen nur die Primfaktoren von $\varrho D(\alpha)$, die Norm des andern dagegen keinen dieser Primfaktoren enthält, und umgekehrt: wenn diese complexe Zahl eine solche Zerlegung in zwei Faktoren gestattet, so enthält $f(z)$ eine Ambige.

Man kann die Bedingung dafür, dafs $f(z)$ eine Ambige enthält, auch noch auf eine etwas einfachere und zweckmäfsigere Weise ausdrücken. Da die $h\lambda$te Potenz einer jeden Ambigen wirklich wird, so folgt, daſs wenn $f(z)$ eine Ambige $\phi(z)$ enthält, $f(z)^{h\lambda}$ sich in zwei wirkliche Faktoren zerlegen läſst, deren einer keinen der idealen Primfaktoren, der andere nur ideale Primfaktoren enthält, also:

$$(16.)\qquad f(z)^{h\lambda} = d(z)\cdot\Psi(z)$$

wo $\Psi(z) = \phi(z)^{h\lambda}$ ist, und $Nd(z)$ keine anderen Primfaktoren, als ϱ und die Primfaktoren der Determinante enthält. Diese Gleichung mit (15.) verbunden giebt:

$$(17.)\qquad \Psi(z)f(z_1) = \frac{\Delta(z)}{d(z)}\,\Phi(z)f(z).$$

* Nimmt man auf beiden Seiten die Norm, und dividirt durch $Nf(\alpha)$, so hat man:

$$(18.)\qquad N\Psi(z) = N\left(\frac{\Delta(z)}{d(z)}\right) N\Phi(z).$$

Die Normen von $\Psi(z)$ und $\Phi(z)$ sind aber vollständig aus denselben idealen Faktoren zusammengesetzt, können sich also nur durch eine Einheit unterscheiden, welche eine Einheit in α sein muſs; dieselbe muſs auch einer nicbtcomplexen Zahl congruent sein, nach dem Modul λ, weil die Normen der wirklichen Zahlen $\Psi(z)$ und $\Phi(z)$ diese Eigenschaft haben, und hieraus wird geschlossen, daſs diese Einheit nur eine λte Potenz sein kann. Man hat daher

$$(19.)\qquad N\left(\frac{\Delta(z)}{d(z)}\right) = e(\alpha)^{\lambda},$$

und folglich:

$$(20.)\qquad N\left(\frac{\Delta(z)}{e(\alpha)\,d(z)}\right) = 1.$$

Bezeichnet man diese gebrochene complexe Zahl, deren Norm gleich Eins ist, mit $e(z)$, und bildet den Ausdruck:

$$(21.)\qquad Pe(z) = 1 + e(z) + e(z)\,e(z_1) + \ldots + e(z)\,e(z_1)\ldots e(z_{\lambda-2}),$$

welcher selbst eine gebrochene Zahl in z ist, und darum in die Form

$$(22.)\qquad Pe(z) = \frac{A\cdot\delta(z)}{B}$$

gesetzt werden kann, so hat man vermöge der Grundeigenschaft dieses Ausdrucks, nach welcher

$$e(z)\, Pe(z_1) = Pe(z)$$

ist, die Gleichung:

$$(23.) \qquad \Delta(z)\, \delta(z_1) = e(\alpha)\, d(z)\, \delta(z).$$

Aus dieser Gleichung wird nun leicht gefolgert, dafs wenn $\delta(z)$ irgend einen idealen Primfaktor in z enthält, es zugleich auch alle seine conjugirten enthalten mufs, welche sich zu einem Primfaktor in α zusammensetzen. Wenn nämlich $p(z)$ ein idealer Primfaktor des $\delta(z)$ ist, so mufs $\delta(z_1)$ denselben ebenfalls enthalten, weil $\Delta(z)$ überhaupt keinen idealen Primfaktor in z enthält. Weil nun $\delta(z_1)$ den Primfaktor $p(z)$ enthält, so folgt, dafs $\delta(z)$ auch den Primfaktor $p(z_{-1})$ enthalten mufs, denselben mufs darum auch wieder $\delta(z_1)$ enthalten, und folglich $\delta(z)$ auch den Primfaktor $p(z_{-2})$ u. s. w. Setzt man nun den Werth des $\Delta(z)$ aus (23.) in (17.) ein, so hat man:

$$(24.) \qquad \Psi(z) f(z_1)\, \delta(z_1) = e(\alpha)\, \Phi(z) f(z)\, \delta(z).$$

Man kann nun $\delta(z)$, welches keine anderen idealen Faktoren in z enthält, als welche sich zu Faktoren der niederen Theorie in α zusammensetzen, mit $f(z)$ verbinden, ohne dafs dadurch die in $f(z)$ enthaltene Ambige wesentlich geändert wird. Schreibt man also einfach $f(z)$ statt $f(z)\,\delta(z)$, so hat man:

$$\Psi(z) f(z_1) = e(\alpha)\, \Phi(z) f(z),$$

und wenn mit $\Psi(z_1)\, \Psi(z_2) \ldots\ldots \Psi(z_{\lambda-1})$ multiplicirt wird:

$$N\Psi(z) f(z_1) = e(\alpha)\, \Phi(z)\, \Psi(z_1)\, \Psi(z_2) \ldots \Psi(z_{\lambda-1}) f(z),$$

welche Gleichung in der einfacheren Form

$$(25.) \qquad L(\alpha) f(z_1) = M(z) f(z)$$

dargestellt werden kann, wo $M(z)$ eine wirkliche complexe Zahl in z ist, deren Norm ebenfalls keinen gemeinschaftlichen Faktor mit $\varrho D(\alpha)$ hat, und $L(\alpha)$ eine complexe Zahl in α, deren λte Potenz die Norm von $M(z)$ ist.

Dieser einfachen Gleichung (25.) also müssen alle wirklichen complexen Zahlen $f(z)$ genügen, welche Ambigen enthalten, und umgekehrt, jede wirkliche Zahl $f(z)$, welche einer solchen Gleichung genügt, enthält
* eine Ambige, wie man sogleich sieht, wenn man mit $f(z^{h\,\lambda-1})$ multiplicirt, wodurch man auf die Bedingung des Satzes (II.) zurückkommt. Diejenigen Ambigen, welche sich nur durch ideale oder wirkliche Faktoren der niede-

ren Theorie unterscheiden, sind dabei nur als eine und dieselbe gerechnet. Man hat also folgenden Satz:

Wenn eine wirkliche complexe Zahl $f(z)$ einer Gleichung

$$L(\alpha)f(z_1) = M(z)f(z)$$

genügt, in welcher $L(\alpha)$ eine wirkliche complexe Zahl in α ist, die mit $\varrho D(\alpha)$ keinen gemeinschaftlichen Faktor hat, $M(z)$ eine wirkliche complexe Zahl in z, deren Norm keinen gemeinschaftlichen Faktor mit $\varrho D(\alpha)$ hat, so enthält $f(z)$ eine Ambige. Umgekehrt: wenn $f(z)$ eine Ambige enthält, so genügt es stets einer solchen Gleichung.

§. 9.

Darstellung der ambigen idealen Zahlen in z, als wirkliche complexe Zahlen in $u, u_1, u_2 \ldots$.

Die wirkliche complexe Zahl $f(z)$, welche eine ideale Ambige enthalten soll, soll nun als eine complexe Zahl in w dargestellt werden. In dieser Form einer ganzen rationalen Funktion von w, des Grades $\lambda - 1$, tritt, wenn die Norm von $f(z)$ durch ϱ theilbar ist, eine Potenz von ϱ als gemeinschaftlicher Faktor aller Glieder heraus, auch wenn $f(z)$ in der Form einer lineären Funktion der Wurzeln $z, z_1, \ldots z_{\lambda-1}$ den Faktor ϱ nicht enthält. Man hat also:

$$f(z) = \varrho^{\nu} f(w),$$

und demgemäfs auch

$$f(z_1) = \varrho^{\nu} f(w\alpha),$$

und wenn diese Ausdrücke in die im Satz (III.) des vorigen Paragraphen gegebene Gleichung eingesetzt, und der Faktor ϱ^{ν} gehoben wird:

$$(1.) \qquad L(\alpha)f(w\alpha) = M(z)f(w).$$

Diese Gleichung kann nun anstatt der obigen als diejenige benutzt werden, welche alle idealen Ambigen gewährt, nämlich als in denjenigen wirklichen complexen Zahlen $f(w)$ enthalten, die dieser Gleichung genügen; denn die complexe Zahl $f(w)$ enthält genau dieselben idealen Primfaktoren, als $f(z)$, von welcher sie sich nur durch einen Faktor ϱ^{ν} unterscheidet.

In den folgenden Untersuchungen ist es nun vortheilhaft die Determinante $D(\alpha)$, welche der Theorie der complexen Zahlen in w und in z zu Grunde gelegt ist, einer neuen Beschränkung zu unterwerfen, welche darin besteht, dafs $(1-D(\alpha))$ durch ϱ theilbar sein soll, aber nicht durch ϱ^2. Diese Annahme über die Determinante soll hier, so wie in dem Folgenden überall gemacht werden.

Es sei nun

$$f(w) = A + A_1 w + A_2 w^2 + \dots + A_{\lambda-1} w^{\lambda-1},$$

sei ferner $M(z)$, in die Form einer complexen Zahl in w gesetzt:

$$M(z) = B + B_1 w + B_2 w^2 + \dots + B_{\lambda-1} w^{\lambda-1}.$$

Setzt man nun

$$M(z) f(w) = C + C_1 w + C_2 w^2 + \dots + C_{\lambda-1} w^{\lambda-1},$$

so hat man durch Ausführung der Multiplikation für die λ Coefficienten C, C_1, C_2 ... $C_{\lambda-1}$ ebensoviele Gleichungen, welche durch die eine

$$\begin{aligned} C_k = {} & A_k B + A_{k-1} B_1 + \dots + A B_k + \\ & w^\lambda (A_{\lambda-1} B_{k+1} + \dots + A_{k+1} B_{\lambda-1}), \end{aligned} \tag{2.}$$

für $k = 0, 1, 2, \dots \lambda-1$, repräsentirt werden. Die Gleichung (1.) giebt daher unter den Coefficienten von $M(z)$ und $f(w)$ ein System von λ Gleichungen, welches durch die folgende repräsentirt wird:

$$\begin{aligned} & A_k (B - \alpha^k L(\alpha)) + A_{k-1} B_1 + \dots + A B_k + \\ & w^\lambda (A_{\lambda-1} B_{k+1} + \dots + A_{k+1} B_{\lambda-1}) = 0, \end{aligned} \tag{3.}$$

für $k = 0, 1, 2, \dots \lambda-1$.

Ich mache nun aus diesem Systeme von λ Gleichungen ein System von Congruenzen, nach dem Modul ϱ^2. Die Coefficienten $B, B_1, B_2, \dots B_{\lambda-1}$, als Coefficienten einer complexen Zahl in w, welche zugleich eine * ganze complexe Zahl in z ist, müssen dem Systeme der Congruenzen (7.) §. 2. genügen, aus welchem unmittelbar folgt, erstens, dafs alle, mit Ausschlufs des ersten, durch ϱ theilbar sein müssen, und zweitens, dafs alle, mit Ausschlufs des ersten, unter einander congruent sein müssen, nach dem Modul ϱ^2. Die Gleichung (3.) giebt daher zunächst für den Modul ϱ die Congruenz:

$$A_k (B - \alpha^k L(\alpha)) \equiv 0, \quad \text{mod. } \varrho, \tag{4.}$$

für $k=0, 1, 2, \ldots \lambda-1$, und weil A_k nicht für alle Werthe des k durch ϱ theilbar sein soll, so folgt hieraus, dafs

(5.) $$B \equiv L(\alpha), \text{ mod. } \varrho,$$

sein mufs. Ferner giebt die Gleichung (3.) nach dem Modul ϱ^2, weil

$$B_1 \equiv B_2 \equiv B_3 \ldots\ldots \equiv B_{\lambda-1}, \text{ mod. } \varrho^2$$

ist, die Congruenz:

$$A_k\,(B-\alpha^k L(\alpha)) +$$
$$B_1\,(A_{k-1}+A_{k-2}+\ldots+A+A_{\lambda-1}+\ldots+A_{k+1}), \text{ mod. } \varrho^2,$$

welche, wenn zum zweiten Theile $B_1 A_k$ addirt, und dasselbe vom ersten Theile subtrahirt wird, auch so dargestellt werden kann:

$$A_k\,(B-\alpha^k L(\alpha)-B_1)+B_1\,(A+A_1+\ldots+A_{\lambda-1}) \equiv 0, \text{ mod. } \varrho^2.$$

Setzt man nun, da $B-L(\alpha)$ und B_1 beide durch ϱ theilbar sind:

$$B-L(\alpha) \equiv c\varrho, \quad B_1 \equiv b\varrho, \text{ mod. } \varrho^2,$$

und beachtet, dafs $w^\lambda - 1$ durch ϱ theilbar, und

$$\alpha^k \equiv 1-k\varrho, \quad \text{mod. } \varrho^2,$$

ist, so hat man, wenn der gemeinschaftliche Faktor ϱ hinweggehoben wird:

(6.) $$A_k\,(kB+c-b)+b\,(A+A_1+\ldots+A_{\lambda-1}) \equiv 0, \text{ mod. } \varrho.$$

Ich setze nun erstens den Fall, es sei

$$A+A_1+A_2+\ldots+A_{\lambda-1} \equiv 0, \text{ mod. } \varrho,$$

so wäre auch:

(7.) $$A_k\,(kB+c-b) \equiv 0, \text{ mod. } \varrho,$$

für $k=0, 1, 2, \ldots \lambda-1$; weil nun B nicht durch ϱ theilbar ist, so kann die Congruenz $kB+c-b \equiv 0$, mod. ϱ, nur für e i n e n Werth des k Statt haben, für alle übrigen Werthe des k müsste also A_k durch ϱ theilbar sein, und es müsste, weil die Summe aller Coefficienten $A, A_1, \ldots A_{\lambda-1}$ durch ϱ theilbar angenommen worden ist, sogar auch dieser eine Coefficient durch ϱ theilbar sein, also alle Coefficienten des $f(w)$ müssten einzeln durch ϱ theilbar sein, welches nicht Statt hat, weil bei der Verwandlung der complexen Zahl $f(z)$ in $f(w)$ aus letzterer ϱ^ν, als höchste Potenz von ϱ, welche sie enthalten kann, herausgehoben worden ist. Es folgt hieraus, dafs in einer Zahl $f(w)$, wel-

L 2

che der Gleichung (1.) genügt, die Summe ihrer Coefficienten nicht durch ϱ theilbar sein kann.

Die Congruenz

$$A + A_1 + A_2 + \ldots + A_{\lambda-1} \equiv 0, \text{ mod. } \lambda,$$

enthält aber genau die nothwendige und hinreichende Bedingung dafür, dafs die Norm von $f(w)$ durch ϱ theilbar sei. Entwickelt man nämlich diese Norm, als Produkt der λ conjugirten Faktoren, und läfst alle Vielfachen von λ weg, so erhält man:

$$(8.) \qquad Nf(w) \equiv A^\lambda + A_1^\lambda w^\lambda + A_2^\lambda w^{2\lambda} + \ldots + A_{\lambda-1}^\lambda w^{(\lambda-1)\lambda}, \text{ mod. } \lambda;$$

da nun $w^\lambda \equiv 1$, mod. ϱ, und da die λte Potenz einer jeden complexen Zahl in α dieser einfachen complexen Zahl congruent ist, nach dem Modul ϱ, so erhält man:

$$(9.) \qquad Nf(w) \equiv A + A_1 + A_2 + \ldots + A_{\lambda-1}, \text{ mod. } \varrho.$$

Da man also anstatt der Bedingung, dafs die Summe der Coefficienten von $f(w)$ nicht durch ϱ theilbar sei, die setzen kann, dafs die Norm von $f(w)$ nicht durch ϱ theilbar sei, so hat man folgenden Satz:

(I.) Eine wirkliche complexe Zahl $f(w)$, deren Norm durch ϱ theilbar ist, ohne dafs $f(w)$ selbst durch ϱ theilbar ist, kann niemals eine Ambige enthalten.

Nimmt man nun zweitens in der Congruenz (6.)

$$A + A_1 + A_2 + \ldots + A_{\lambda-1} \text{ nicht} \equiv 0, \text{ mod. } \varrho,$$

und bemerkt, dafs für einen bestimmten der λ Werthe des $k = 0, 1, 2, \ldots \lambda - 1$

$$(10.) \qquad kB + c - b \equiv 0, \text{ mod. } \varrho,$$

sein mufs, so hat man aus der Congruenz (6.) nothwendig $b \equiv 0$, mod. ϱ, und hieraus folgt weiter, dafs A_k, für alle Werthe des $k \equiv 0, 1, 2, \ldots \lambda - 1$, mit Ausschlufs des einen Werthes des k, welcher $kB + c - b \equiv 0$, mod. ϱ, giebt, durch ϱ theilbar sein mufs. Man hat daher folgenden Satz:

(II.) Wenn die wirkliche complexe Zahl $f(w)$ eine Ambige enthält, so müssen alle Coefficienten derselben, mit Ausschlufs eines einzigen, durch ϱ theilbar sein.

Es sollen nun die in der Determinante $D(\alpha)$ enthaltenen verschiedenen Primfaktoren in α in Betracht gezogen werden, welche mit $f(\alpha)$, $f_1(\alpha)$,

$f_2(\alpha)$ ... bezeichnet werden sollen, so dafs die Determinante, in ihre Primfaktoren zerlegt, folgenden Ausdruck hat:

$$(11.)\qquad D(\alpha) = E(\alpha) f(\alpha)^m f_1(\alpha)^{m_1} f_2(\alpha)^{m_2} \dots,$$

in welchem $E(\alpha)$ eine Einheit bezeichnet. Es soll auch angenommen werden, dafs in diesem Ausdrucke die Faktoren $f(\alpha)^m$, $f_1(\alpha)^{m_2}$ u. s. w. nur wirkliche complexe Zahlen in α sind, durch welche Annahme die Primfaktoren $f(\alpha)$, $f_1(\alpha)$, ... keiner einschränkenden Bedingung unterworfen werden, sondern lediglich die Exponenten m, m_1, ..., da derselben z. B. immer dadurch genügt werden kann, dafs für alle diese Exponenten beliebige Vielfache der Klassenanzahl h genommen werden.

* Macht man nun aus dem Systeme der λ Gleichungen bei (6.) ein System von Congruenzen, nach dem Modul $f(\alpha)^m$, indem bemerkt, dafs w^λ nach diesem Modul congruent Null ist, so hat man:

$$(12.)\qquad A_k\,(B - \alpha^k L(\alpha)) + A_{k-1}B_1 + \dots + AB_k \equiv 0, \text{ mod. } f(\alpha)^m,$$

für $k = 0, 1, 2, \dots \lambda - 1$.

Wenn nun angenommen wird, dafs von den Coefficienten des $f(w)$ die ersten n durch $f(\alpha)$ theilbar sind, der $(n+1)$te aber nicht theilbar, welches den Fall $n = 0$, wo der erste Coefficient A durch $f(\alpha)$ nicht theilbar ist, nicht ausschliefsen soll, so giebt die Congruenz (12.), für $k = n$, weil $A, A_1, \dots A_{n-1}$ congruent Null sind:

$$(13.)\qquad A_n\,(B - \alpha^n L(\alpha)) \equiv 0, \text{ mod } f(\alpha),$$

und weil A_n nicht durch $f(\alpha)$ theilbar ist:

$$B - \alpha^n L(\alpha) \equiv 0, \quad \text{mod. } f(\alpha).$$

Es kann aber $B - \alpha^k L(\alpha)$ nur für diesen einen Werth $k = n$ durch $f(\alpha)$ theilbar sein, denn hätte man zugleich

$$B - \alpha^n L(\alpha) \equiv 0, \quad \text{und } B - \alpha^\nu L(\alpha) \equiv 0,$$

so würde daraus folgen:

$$(\alpha^n - \alpha^\nu) L(\alpha) \equiv 0, \quad \text{mod. } f(\alpha),$$

und hieraus:

$$L(\alpha) \equiv 0, \quad \text{mod. } f(\alpha),$$

welches unmöglich ist, weil $L(\alpha)$ keinen Faktor der Determinante enthält. Die Congruenz (12.) giebt nun für $k = 0, 1, 2, \dots n-1$,

*

$$
(13.)\quad \begin{aligned}
&A(B - F(\alpha)) \equiv 0, \\
&A_1(B - \alpha F(\alpha)) + AB_1 \equiv 0, \\
&A_2(B - \alpha^2 F(\alpha)) + A_1 B_1 + AB_2 \equiv 0, \qquad \text{mod.} f(\alpha)^m. \\
&\vdots \\
&A_{n-1}(B - \alpha^{n-1} F(\alpha)) + A_{n-2} B_1 + \dots + AB_{n-1} \equiv 0.
\end{aligned}
$$

Da $B - F(\alpha)$ nicht durch $f(\alpha)$ theilbar ist, so folgt aus der ersten dieser Congruenzen, dafs A den Faktor $f(\alpha)^m$ enthalten mufs. Hieraus, und weil auch $B - \alpha F(\alpha)$ nicht durch $f(\alpha)$ theilbar ist, ergiebt die zweite Congruenz, dafs A_1 durch $f(\alpha)^m$ theilbar sein mufs, und so fortschliefsend erhält man:

$$A \equiv 0, \; A_1 \equiv 0, \; \dots . \; A_{n-1} \equiv 0, \quad \text{mod.} f(\alpha)^m.$$

Also wenn die ersten n Coefficienten durch $f(\alpha)$ theilbar sind, so müssen sie auch durch $f(\alpha)^m$ theilbar sein, und dieses für den einen Faktor der Determinante bewiesene Resultat gilt nothwendig eben so für alle anderen.

Ich setze nun:

$$
(14.)\quad \begin{aligned}
u^\lambda = e(\alpha) f(\alpha)^m, \; u_1{}^\lambda = e_1(\alpha) f_1(\alpha)^{m_1}, \; u_2{}^\lambda = e_2(\alpha) f_2(\alpha)^{m_2} \dots \\
e(\alpha) e_1(\alpha) e_2(\alpha) \dots = E(\alpha),
\end{aligned}
$$

so ist:

$$D(\alpha) = u^\lambda \cdot u_1^\lambda \cdot u_2^\lambda \dots .$$

also:

$$w = u\, u_1\, u_2 \dots .$$

Wenn nun die ersten n Coefficienten des $f(w)$ durch $f(\alpha)$, also auch durch $f(\alpha)^m$, oder was dasselbe ist, durch u^λ theilbar sind, ferner die ersten n_1 Coefficienten theilbar durch $f_1(\alpha)$, also auch durch $f_1(\alpha)^{m_1}$, oder u_1^λ, u. s. w. und man setzt $u\, u_1\, u_2 \dots$ statt w, so hebt sich aus $f(w)$ der Faktor u^n, ebenso der Faktor $u_1{}^{n_1}$ u. s. w. heraus, und man hat:

$$(15.)\qquad f(w) = u^n\, u_1^{n_1}\, u_2^{n_2} \dots f(u, u_1, u_2 \dots),$$

wo $f(u, u_1, u_2 \dots)$ eine aus den Irrationalitäten $u, u_1, u_2 \dots$ gebildete complexe Zahl von folgender Form ist:

$$(16.)\qquad f(u, u_1, u_2 \ldots) = \Sigma_k A_k\, u^{|k-n|}\, u_1^{|k-n_1|}\, u_2^{|k-n_2|} \ldots.$$

wenn allgemein $|c|$ den kleinsten, nicht negativen Rest der Zahl c, nach dem Modul λ, bezeichnet. Die Bedingung, daſs in dem Ausdrucke des $f(w)$ der Coefficient A_n der erste nicht durch $f(\alpha)$ theilbare, der Coeffiicient A_{n_1} der erste nicht durch $f_1(\alpha)$ theilbare, u. s. w. sein soll, ergiebt für diesen Ausdruck der complexen Zahl $f(u, u_1, u_2 \ldots)$, daſs auch A_n nicht durch $f(\alpha)$, A_{n_1} nicht durch $f_1(\alpha)$, A_{n_2} nicht durch $f_2(\alpha)$ u. s. w. theilbar sein darf.

Die zu $f(u, u_1, u_2, \ldots)$ conjugirten complexen Zahlen erhält man, wenn man nur einer einzigen der Wurzelgröſsen $u, u_1, u_2, \ldots$ ihre λ Werthe giebt. Die Norm von $f(u, u_1, u_2 \ldots)$, als Produkt dieser λ conjugirten, ist alsdann eine, von den Irrationalitäten $u, u_1, u_2 \ldots$ vollständig freie, complexe Zahl in α. Vermöge der Bedingungen, daſs A_n nicht durch $f(\alpha)$, A_{n_1} nicht durch $f_1(\alpha)$ u. s. w. theilbar ist, kann diese Norm von $f(u, u_1, u_2 \ldots)$ keinen der Primfaktoren $f(\alpha)$, $f_1(\alpha)$, $f_2(\alpha) \ldots$ enthalten. Um dieſs zu beweisen, bemerke ich, daſs in dem Ausdrucke (16.) das $n$te Glied, welches $u^{|n-n|}$, also nur u^0 enthält, das einzige Glied ist, welches u nicht enthält, und daſs demgemäſs in der Norm von $f(u, u_1, u_2 \ldots)$, in welcher u selbst nicht mehr vorkommt, sondern nur noch u^λ, die λte Potenz dieses $n$ten Gliedes das einzige Glied sein muſs, welches u^λ nicht enthält, daſs also:

$$Nf(u, u_1, u_2 \ldots) \equiv A_n^\lambda\, u_1^{|n-n_1|\lambda}\, u_2^{|n-n_2|\lambda} \ldots \quad \text{mod. } u^\lambda,$$

und weil $A_n, u_1^\lambda, u_2^\lambda, \ldots$ den Faktor $f(\alpha)$ nicht enthalten, daſs diese Norm den Faktor $f(\alpha)$ nicht enthält. In derselben Weise wird gezeigt, daſs sie auch keinen der übrigen Faktoren der Determinante enthalten kann.

Da nun zuerst im §. 8. gezeigt worden ist, daſs jede ideale Ambige in einer wirklichen Zahl $f(z)$ als Complex aller idealen Primfaktoren enthalten ist, so daſs die Norm dieser Zahl $f(z)$ auſser der Norm der in ihr enthaltenen Ambigen nur noch die Primfaktoren der Determinante und eine Potenz von ϱ enthalten kann; da ferner in dem gegenwärtigen Paragraphen gezeigt worden ist, daſs, wenn diese, die Ambige enthaltende Zahl $f(z)$ als complexe Zahl in w dargestellt, und von einer Potenz von ϱ, welche in dieser Form als gemeinschaftlicher Faktor aller ihrer Coefficienten heraustreten kann, befreit wird, aus derselben eine complexe Zahl $f(w)$ entsteht, deren

Norm nicht mehr ϱ enthält; da endlich gezeigt worden ist, dafs diese die Ambige enthaltende Zahl $f(w)$ durch Einführung der Wurzeln u, u_1, u_2 ..., und nachdem die Potenzen von u, u_1, u_2, ..., welche dabei als gemeinschaftliche Faktoren aller Glieder heraustreten, entfernt werden, eine wirkliche complexe Zahl $f(u, u_1, u_2, \ldots)$ ergiebt, deren Norm keinen Faktor der Determinante weiter enthält: so folgt, dafs $f(u, u_1, u_2 \ldots)$ die in $f(z)$ enthaltene Ambige nicht nur ebenfalls enthält, sondern dafs sie als diese Ambige selbst angesehen werden mufs, welche somit als ideale Zahl in z, in der Theorie der complexen Zahlen in u, u_1, u_2 ... als wirkliche complexe Zahl dargestellt werden kann. Also:

(III.) Jede ideale Ambige in z läfst sich als eine wirkliche complexe Zahl von der Form $f(u, u_1, u_2 \ldots)$ darstellen, welche so beschaffen ist, dafs sie durch Multiplikation mit $u^n u_1^{n_1} u_2^{n_2}$ ..., wenn die Exponenten n, n_1, n_2 ... passend bestimmt werden, in eine complexe Zahl in w übergeht.

§. 10.

Untersuchung aller wirklichen complexen Zahlen in $u, u_1, u_2, \ldots$, welche ideale ambige Zahlen in z darstellen.

Die in den vorhergehenden Paragraphen bewiesenen Sätze gewähren die Mittel, alle idealen Ambigen in der Theorie der complexen Zahlen in z zu finden, und zwar in der Form von wirklichen complexen Zahlen, welche aus den Wurzeln u, u_1, u_2 ... gebildet sind. Als den einfachsten Weg zu diesem Ziele zu gelangen, wähle ich den, zunächst alle wirklichen Zahlen $f(w)$ zu finden, welche einer Gleichung von der Form (1.) §. 9:

$$(1.)\qquad L(\alpha)f(w\alpha) = M(z)f(w)$$

genügen, in welcher Gleichung vorläufig in $L(\alpha)$ und in $NM(z)$ die Faktoren der Determinante $f(\alpha)$, $f_1(\alpha)$, ... zugelassen werden sollen, der Faktor ϱ aber ausgeschlossen sein soll. Diese Aufgabe läfst sich folgendermaafsen aussprechen:

Alle wirklichen complexen Zahlen $f(w)$ zu finden, welche der Bedingung genügen, dafs $\frac{f(w\alpha)}{f(w)}$ als eine gebrochene, wirkliche complexe Zahl in z sich darstellen lasse, in der Art, daſs die Norm des Nenners nicht durch ϱ theilbar sei.

Zunächst ist klar, daſs wenn $f(w)$ dieser Bedingung genügt, auch $w^k f(w)$ derselben genügen muſs, für jeden Werth des k. Da nun oben §. 9. im Satze (II.) bewiesen worden ist, daſs in einer jeden Zahl $f(w)$, welche den Bedingungen der vorliegenden Aufgabe genügt, alle Coefficienten, mit Ausschluſs eines einzigen, durch ϱ theilbar sein müssen, so kann man durch Multiplikation mit einer passenden Potenz von w das Glied, welches diesen Coefficienten hat, zum ersten Gliede machen, d. h. zu dem Gliede, welches die irrationale Wurzel w nicht enthält, wodurch $f(w)$ die Form

$$f(w) = C + \varrho\psi(w) \tag{2.}$$

erhält, wo C eine durch ϱ nicht theilbare complexe Zahl in α ist, welche auch als nichtcomplexe ganze Zahl angenommen werden kann. Es reicht also hin, nur die in dieser Form enthaltenen, der Aufgabe genügenden Zahlen $f(w)$ zu finden.

Ferner folgt unmittelbar, wenn $f(w)$ der Aufgabe genügt, daſs auch $f(w)\,F(z)$ derselben genügen muſs, wenn $F(z)$ eine wirkliche complexe Zahl in z ist, deren Norm nicht durch ϱ theilbar ist.

Endlich ergiebt sich auch sehr leicht, daſs wenn $f(w)$ der Aufgabe genügt, ebenso alle complexen Zahlen in w, welche congruent $f(w)$ sind, nach dem Modul λ, der Aufgabe genügen müssen; denn es ist:

$$f(w) + \lambda g(w) = f(w)\left(1 + \frac{\lambda g(w)}{f(w)}\right),$$

und wenn man den Bruch $\frac{g(w)}{f(w)}$ in die Form bringt, daſs sein Nenner eine complexe Zahl in α wird, also:

$$\frac{g(w)}{f(w)} = \frac{G(w)}{F(\alpha)},$$

und bemerkt, daſs nach dem Satze (I.) §. 2 $\lambda G(w)$ eine ganze complexe Zahl in z ist, so hat man:

$$f(w) + \lambda g(w) = \frac{f(w)\,\Phi(z)}{F(\alpha)}, \tag{3.}$$

wo $N\Phi(z)$ nicht durch ϱ theilbar ist, also:

$$\frac{f(w\alpha)+\lambda g(w\alpha)}{f(w)+\lambda g(w)} = \frac{f(w\alpha)\,\Phi(z_1)}{f(w)\,\Phi(z)}, \tag{4.}$$

woraus die Richtigkeit der aufgestellten Behauptung erhellt.

Da hiernach nur alle nach dem Modul λ incongruenten Zahlen $f(w)$ von der Form $C+\varrho\psi(w)$ zu suchen sind, welche den Bedingungen der Aufgabe genügen, so wird es zweckmäſsig sein, bei dieser Untersuchung die Logarithmen der complexen Zahlen, nach dem Modul λ, anstatt dieser complexen Zahlen selbst anzuwenden, in ähnlicher Weise, wie ich dieselben schon früher für die Theorie der complexen Zahlen in α mit Erfolg angewendet habe.

Zunächst ist zu bemerken, daſs eine jede Zahl von der Form $C+\varrho\psi(w)$ stets auf verschiedene Weisen in diese Form gesetzt werden kann, da man dem $\psi(w)$ beliebig eine complexe Zahl in α hinzufügen kann, wenn man dafür das ϱfache derselben von C hinwegnimmt. Um diese Willkürlichkeit auszuschlieſsen, setze ich fest: es soll $\psi(w)$ in dieser Form stets so gewählt werden, daſs es für $w=1$ gleich Null wird, welches immer geleistet werden kann, indem man von $\psi(w)$ die Summe aller seiner Coefficienten abzieht, und das ϱfache dieser Summe dem C zulegt.

Ich entwickele nun den Logarithmus

$$l\left(\frac{f(w)}{C}\right) = l\left(1+\frac{\varrho\psi(w)}{C}\right) \tag{5.}$$

so nach Potenzen von ϱ, daſs in dieser Entwickelung alle diejenigen Glieder, welche Vielfache von λ werden, wegfallen. Das $k$te Glied der Entwickelung dieses Logarithmus ist:

$$\frac{-(-1)^k\varrho^k\,\psi(w)^k}{k\,C^k}.$$

Da nun $\varrho^{\lambda-1}$ ein Vielfaches von λ ist, so folgt, daſs das $\lambda-1$te Glied wegfällt, so wie auch alle folgenden, insofern sie nicht λ auch im Nenner enthalten, also insofern nicht k durch λ theilbar ist. Wenn aber k ein Vielfaches von λ ist, so nehme man $k=m\lambda^n$, wo m nicht weiter durch λ theilbar sein soll; man hat alsdann $\varrho^{m\lambda^n}=\varrho^{m\lambda^{n-1}(\lambda-1)}\,\varrho^{m\lambda^{n-1}}$, also theilbar durch $\lambda^{m\lambda^{n-1}}$, und demnach muſs $\frac{1}{k}\varrho^k$ für einen solchen Werth des k den Faktor λ mindestens $m\lambda^{n-1}-n$ mal enthalten. Diese Anzahl ist aber stets gröſser als Eins, auſser in dem einen Falle, wo zugleich $n=1$ und $m=1$, also $k=\lambda$

ist. Von allen Gliedern dieser Entwickelung des Logarithmus bleiben also nur die ersten $\lambda - 2$, und aufserdem das λte Glied, welches den Faktor $\frac{1}{\lambda}\varrho^\lambda$ hat, der bekanntlich gleich ϱ, multiplicirt mit einer Einheit ist. Man hat daher:

$$(6.)\quad l\left(\frac{f(w)}{C}\right) \equiv \frac{\varrho\psi(w)}{1\cdot C} - \frac{\varrho^2\psi(w)^2}{2\cdot C^2} + \dots + \frac{\varrho^{\lambda-2}\psi(w)^{\lambda-2}}{(\lambda-2)\,C^{\lambda-2}} + K, \text{ mod. } \lambda,$$

wo der Kürze wegen

$$K = \frac{\varrho E(\alpha)\,\psi(w)^\lambda}{C^\lambda}$$

gesetzt ist. Denkt man sich nun diesen ganzen Ausdruck nach den Potenzen von w entwickelt, und als ganze rationale Funktion von w des $\lambda - 1$ten Grades dargestellt; denkt man sich ferner die Brüche, deren Nenner die Potenzen von C sind, durch die ganzen Zahlen ersetzt, welchen sie congruent sind, nach dem Modul λ; setzt man alsdann überall $1 - \varrho$ für α, und ordnet nach Potenzen von ϱ; setzt man endlich $1 - v$ statt w und ordnet das, was in die einzelnen Potenzen des ϱ multiplicirt ist, nach Potenzen von v: so erhält man eine Entwickelung von folgender Form:

$$(7.)\quad l\left(\frac{f(w)}{C}\right) \equiv \varrho\psi_1(v) + \varrho^2\psi_2(v) + \dots + \varrho^{\lambda-2}\psi_{\lambda-2}(v), \text{ mod. } \lambda,$$

in welcher $\psi_1(v)$, $\psi_2(v)$ u. s. w. ganze rationale Funktionen von v, vom Grade $\lambda - 1$ sind, mit nichtcomplexen ganzen Zahlen als Coefficienten. Von der ersten derselben $\psi_1(v)$ insbesondere ist noch zu bemerken, dafs für $v = 0$ auch $\psi_1(v) = 0$ werden mufs, vermöge der Festsetzung, dafs in $f(w) = C + \varrho\psi(w)$, für $w = 1$, $\psi(w) = 0$ sein soll.

Wenn nun die logarithmische Entwickelung einer complexen Zahl der Form $C + \varrho\psi(w)$ nach den angegebenen Regeln gebildet ist, so ist sie nach dem Modul λ eine vollständig bestimmte, d. h. eine gegebene complexe Zahl hat nur e i n e Entwickelung ihres Logarithmus, nach dem Modul λ. Wenn nun aber umgekehrt die logarithmische Entwickelung gegeben ist, so ist die Frage: in wie weit dadurch die complexe Zahl selbst bestimmt ist. Um diefs zu untersuchen, gehe ich von dem ganz speciellen Falle aus, wo die logarithmische Entwickelung congruent Null ist, also

$$(8.)\quad l\left(\frac{f(w)}{C}\right) \equiv l\left(\frac{C + \varrho\psi w}{C}\right) \equiv 0, \quad \text{mod. } \lambda.$$

M 2

In diesem Falle hat man aus der Congruenz (6.), wenn man dieselbe zunächst nur für den Modul ϱ^2 betrachtet:

$$0 \equiv \varrho\psi(w), \quad \text{mod. } \varrho^2,$$

also $\psi(w)$ durch ϱ theilbar. Setzt man nun $\psi(w) = \varrho\chi(w)$, und betrachtet die Congruenz (6.) nach dem Modul ϱ^3, so ergiebt sie, dafs $\chi(w)$ weiter durch ϱ theilbar sein mufs, also $\psi(w)$ theilbar durch ϱ^2, so fortschliefsend erhält man zuletzt: $\psi(w)$ theilbar durch $\varrho^{\lambda-2}$, also $\varrho\psi(w)$ ein Vielfaches von λ, d. h. die logarithmische Entwickelung von $f(w)$ ist nur dann congruent Null, wenn $f(w)$ einer complexen Zahl in α congruent ist, nach dem Modul λ. Wenn nun die logarithmischen Entwickelungen zweier Zahlen $f(w)$ und $f'(w)$ congruent sind, also der Unterschied derselben congruent Null, so hat man

$$l\left(\frac{f(w)}{C}\right) - l\left(\frac{f'(w)}{C'}\right) \equiv 0, \quad \text{mod. } \lambda,$$

also

$$l\left(\frac{C'f(w)}{Cf'(w)}\right) \equiv 0, \quad \text{mod. } \lambda,$$

woraus folgt, dafs $\frac{C'f(w)}{Cf'(w)}$ einer complexen Zahl in α congruent sein mufs, nach dem Modul λ, oder was dasselbe ist: $Af(w) \equiv f'(w)$, mod. λ, wo A eine complexe Zahl in α ist.

Nachdem diese allgemeinen Eigenschaften der logarithmischen Entwickelungen der complexen Zahlen von der Form $C + \varrho\psi(w)$ festgestellt sind, wende ich dieselben zum Zwecke der Lösung der vorliegenden Aufgabe an. Ich verwandle in der Congruenz (7.) w in $w\alpha$, wodurch $v = 1 - w$ in $1 - w\alpha = v + \varrho(1 - v)$ übergeht, entwickele die rationalen Funktionen von $v + \varrho(1 - v)$ nach dem Taylorschen Satze nach Potenzen von ϱ, und ziehe die unveränderte Congruenz (7.) von diesen ab, so ist:

$$(9.)\quad l\left(\frac{f(w\alpha)}{f(w)}\right) \equiv \varrho^2(1-v)\psi_1'(v) + \frac{\varrho^3(1-v)^2\psi_1''(v)}{1.\,2.} + \frac{\varrho^4(1-v)^3\psi_1'''(v)}{1.\,2.\,3.} + \dots$$
$$+ \varrho^3(1-v)\psi_2'(v) + \frac{\varrho^4(1-v)^2\psi_2''(v)}{1.\,2.} + \dots$$
$$+ \varrho^4(1-v)\psi_3'(v) + \dots$$

Ich entwickele nun den Logarithmus der complexen Zahl $\frac{M(z)}{L(\alpha)}$, welche nach den Bedingungen der Aufgabe gleich $\frac{f(w\alpha)}{f(w)}$ werden soll. Setzt man $M(z)$ zunächst in die Form einer complexen Zahl in w, so hat man, wie im §. 2 gezeigt worden:

$$(10.)\quad M(z) = G + G_1(1-w) + G_2(1-w)^2 + \ldots + G_{\lambda-1}(1-w)^{\lambda-1},$$

wo $G_{\lambda-1}$ durch ϱ, $G_{\lambda-2}$ durch ϱ^2, $G_{\lambda-3}$ durch ϱ^3 etc. theilbar ist. Dividirt man durch $L(\alpha)$, ersetzt die Brüche, mit dem Nenner $L(\alpha)$ durch die ganzen complexen Zahlen, denen sie nach dem Modul λ congruent sind, nimmt ferner $1-w=v$ und $1-\alpha=\varrho$ und ordnet nach Potenzen von ϱ, so hat man:

$$(11.)\quad \begin{aligned}\frac{M(z)}{L(\alpha)} \equiv 1 + \mathfrak{b}_1 v^{\lambda-1}\varrho + (\mathfrak{a}_2 + \mathfrak{b}_2 v^{\lambda-1}\mathfrak{c}_2 v^{\lambda-2}) +\\ + (\mathfrak{a}_3 + \mathfrak{b}_3 v^{\lambda-1} + \mathfrak{c}_3 v^{\lambda-2} + \mathfrak{d}_3 v^{\lambda-3})\varrho^3 + \ldots\end{aligned}$$

nach dem Modul λ, wo $\mathfrak{b}_1$, $\mathfrak{a}_2$, $\mathfrak{b}_2$, $\mathfrak{a}_3$ u. s. w. nichtcomplexe ganze Zahlen sind. Hieraus erhält man nach der obigen Methode folgende Form der Entwickelung des Logarithmus:

$$(12.)\quad \begin{aligned}l\left(\frac{M(z)}{L(\alpha)}\right) \equiv \mathfrak{B}_1 v^{\lambda-1}\varrho + (\mathfrak{A}_2 + \mathfrak{B}_2 v^{\lambda-1} + \mathfrak{C}_2 v^{\lambda-2})\varrho^2 +\\ + (\mathfrak{A}_3 + \mathfrak{B}_3 v^{\lambda-1} + \mathfrak{C}_3 v^{\lambda-2} + \mathfrak{D}_3 v^{\lambda-3})\varrho^3 + \ldots\end{aligned}$$

nach dem Modul λ, wo $\mathfrak{B}_1$, $\mathfrak{A}_2$, $\mathfrak{B}_2$ u. s. w. ebenfalls nichtcomplexe ganze Zahlen sind. Weil nun

$$\frac{f(w\alpha)}{f(w)} = \frac{M(z)}{L(\alpha)}$$

sein soll, so mufs der Logarithmus der einen dieser complexen Zahlen dem Logarithmus der anderen congruent sein. Die Vergleichung der einzelnen Glieder beider Entwickelungen ergiebt folgende Congruenzen:

$$(13.)\quad \begin{aligned}&(1-v)\psi_1'(v) \equiv \mathfrak{A}_2 + \mathfrak{B}_2 v^{\lambda-1} + \mathfrak{C}_2 v^{\lambda-2}\\ &(1-v)\psi_2'(v) + \frac{(1-v)^2\psi'(v)}{1.\,2.} \equiv \mathfrak{A}_3 + \mathfrak{B}_3 v^{\lambda-1} + \mathfrak{C}_3 v^{\lambda-2} + \mathfrak{D}_3 v^{\lambda-3}\\ &(1-v)\psi_3'(v) + \frac{(1-v)^2\psi_1''(v)}{1.\,2.} + \frac{(1-v)^3\psi_1'''(v)}{1.\,2.\,3}\\ &\qquad\equiv \mathfrak{A}_4 + \mathfrak{B}_4 v^{\lambda-1} + \mathfrak{C}_4 v^{\lambda-2} + \mathfrak{D}_4 v^{\lambda-3} + \mathfrak{E}_4 v^{\lambda-4}\end{aligned}$$

u. s. w., nach dem Modul λ. Setzt man nun

$$\Psi_1(v) = \alpha_1 v + \alpha_2 v^2 + \ldots + \alpha_{\lambda-1} v^{\lambda-1},$$

so hat man

$$(14.)\qquad (1-v)\psi_1'(v) = a_1 + (2a_2 - a_1)v + (3a_3 - 2a_2)v^2 + \dots - (\lambda-1)a_{\lambda-1}v^{\lambda-1},$$

und weil vermöge der ersten der Congruenzen (13.) die Glieder dieses Ausdrucks, welche v, v^2, ... $v^{\lambda-3}$ enthalten, congruent Null sein müssen, so hat man:

$$2a_2 - a_1 \equiv 0,\ 3a_3 - 2a_2 \equiv 0,\ \dots (\lambda-2)a_{\lambda-2} - (\lambda-3)a_{\lambda-3} \equiv 0,$$

woraus unmittelbar folgt:

$$a_1 \equiv \frac{c_1}{1},\quad a_2 \equiv \frac{c_1}{2},\quad a_3 \equiv \frac{c_1}{3} \dots a_{\lambda-2} \equiv \frac{c_1}{\lambda-2},$$

so daſs man für $\psi_1(v)$ folgenden Ausdruck erhält:

$$(15.)\qquad \psi_1(v) \equiv c_1\left(\frac{v}{1} + \frac{v^2}{2} + \frac{v^3}{3} + \dots + \frac{v^{\lambda-1}}{\lambda-1}\right) + B_1 v^{\lambda-1},$$

wo c_1 und B_1 beliebige ganze Zahlen sind. Mit Hülfe dieses gefundenen Ausdrucks des $\psi_1(v)$ findet man aus der zweiten der Congruenzen (13.) ohne Schwierigkeit folgenden Ausdruck des $\psi_2(v)$:

$$(16.)\qquad \psi_2(v) \equiv c_2\left(\frac{v}{1} + \frac{v^2}{2} + \dots + \frac{v^{\lambda-1}}{\lambda-1}\right) + A_2 + B_2 v^{\lambda-1} + C_2 v^{\lambda-2},$$

wo c_2, A_2, B_2, C_2 beliebige ganze Zahlen sind. Ebenso findet man weiter aus der dritten der Congruenzen (13.):

$$(17.)\qquad \psi_3(v) \equiv c_3\left(\frac{v}{1} + \frac{v^2}{2} + \dots + \frac{v^{\lambda-1}}{\lambda-1}\right) + A_3 + B_3 v^{\lambda-1} + C_3 v^{\lambda-2} + D_3 v^{\lambda-3}.$$

Allgemein hat $\psi_k(v)$ nach den Congruenzen (13.) einen Ausdruck, welcher sich von den hier für $k = 1, 2, 3$ gegebenen nur dadurch unterscheidet, daſs die Glieder, welche dem ersten Theile hinzuzufügen sind, bis zu dem Gliede mit $v^{\lambda-k}$ einschlieſslich gehen. Setzt man nun der Kürze wegen:

$$W \equiv \frac{v}{1} + \frac{v^2}{2} + \frac{v^3}{3} + \dots + \frac{v^{\lambda-1}}{\lambda-1},$$

so hat man vermöge der gefundenen Ausdrücke der $\psi_1(v)$, $\psi_2(v)$ u. s. w.

$$(18.)\qquad l\left(\frac{f(\omega)}{C}\right) \equiv (c_1\varrho + c_2\varrho^2 + \dots + c_{\lambda-2}\varrho^{\lambda-2})W + B_1 v^{\lambda-1}\varrho + (A_2 + B_2 v^{\lambda-1} + C_2 v^{\lambda-2})\varrho^2 + (A_3 + B_3 v^{\lambda-1} + C_3 v^{\lambda-2} + D_3 v^{\lambda-3})\varrho^3 + \dots$$

Vergleicht man den Ausdruck auf der rechten Seite dieser Congruenz, welcher auf das erste Glied folgt, mit der Congruenz (12.), so erkennt man, daſs derselbe den Logarithmus einer complexen Zahl in z darstellt. Man hat daher:

$$(19.)\quad l\left(\frac{f(w)}{C}\right) \equiv (c_1\varrho + c_2\varrho^2 + \dots + c_{\lambda-2}\varrho^{\lambda-2})W + l\left(\frac{F(z)}{F(\alpha)}\right), \text{ mod. } \lambda,$$

als nothwendige Bedingung dafür, daſs die Zahl $f(w)$ eine Gleichung von der Form

$$L(\alpha)f(w\alpha) = M(z)f(w)$$

genüge, in welcher $NM(z)$ und $L(\alpha)$ nicht durch ϱ theilbar sind. Daſs diese Bedingung auch eine hinreichende ist, folgt daraus, daſs wenn w in $w\alpha$ verwandelt wird, also v in $v + \varrho\,(1 - v)$, und von der so veränderten Congruenz (19.) die unveränderte abgezogen wird, der Logarithmus von $\frac{f(w\alpha)}{f(w)}$ in der That congruent dem Logarithmus einer complexen Zahl in z gefunden wird. Von dem mit W bezeichneten Ausdrucke bemerke ich noch, daſs derselbe, wenn $v = 1 - w$ gesetzt wird, und man nach Potenzen von w ordnet, in einen Ausdruck derselben Form übergeht, oder daſs

$$(20.)\quad W \equiv \frac{w}{1} + \frac{w^2}{2} + \frac{w^3}{3} + \dots + \frac{w^{\lambda-1}}{\lambda-1}, \text{ mod. } \lambda.$$

Vergleicht man den Ausdruck (19.) des Logarithmus einer der Gleichung (1.) genügenden Zahl $f(w)$ mit dem Ausdrucke (12.) des Logarithmus einer complexen Zahl in z, so erkennt man sogleich, daſs, wenn $f(w)$ eine complexe Zahl in z sein soll, nothwendig die $\lambda - 2$ Zahlen c_1, c_2, ... $c_{\lambda-2}$ alle congruent Null sein müssen, mod. λ; denn die Glieder $c_1\varrho v$, $c_2\varrho^2 v$, ... $c_{\lambda-2}\varrho^{\lambda-2}v$, welche im Logarithmus von $f(w)$ enthalten sind, kommen in dem Logarithmus einer complexen Zahl in z nicht vor. Umgekehrt, wenn c_1, c_2 ... $c_{\lambda-2}$ alle congruent Null sind, mod. λ, so ist der Logarithmus von $f(w)$ derselbe, als der Logarithmus einer complexen Zahl in z, und darum $f(w) \equiv A\,F(z)$, mod. λ, oder $f(w) = A\,F(z) + \lambda G(w)$, also weil $\lambda G(w)$ eine complexe Zahl in z ist, ist $f(w)$ nothwendig eine complexe Zahl in z.

Wenn nun zwei Zahlen $f(w)$ und $f'(w)$, welche beide den Bedingungen der Aufgabe genügen, in ihren Logarithmen dieselben Werthe der Zahlen c_1, c_2 ... $c_{\lambda-2}$ haben, nach dem Modul λ, so giebt die Differenz ihrer Logarithmen, also der Logarithmus ihres Quotienten, den Logarithmus einer complexen Zahl in z, und man hat:

$$(21.)\qquad l\left(\frac{C'f'(w)}{Cf(w)}\right) \equiv l\left(\frac{F(z)}{M}\right), \text{ mod. } \lambda.$$

Hieraus folgt, wie oben gezeigt worden, dafs die eine complexe Zahl der anderen, multiplicirt mit einer complexen Zahl in α, congruent sein mufs, also:

$$(22.)\qquad \frac{f'(w)}{f(w)} \equiv \frac{AF(z)}{M}, \text{ mod. } \lambda.$$

Macht man nun aus dieser Congruenz eine Gleichung, indem man das λfache einer complexen Zahl in w hinzufügt, welches ebenfalls eine complexe Zahl in z ist, so erhält man:

$$(23.)\qquad f'(w) = \frac{f(w)\,F'(z)}{M'},$$

wo $F'(z)$ eine ganze complexe Zahl in z, M' eine ganze complexe Zahl in α ist, welche den Faktor ϱ nicht enthält. Die eine dieser beiden Zahlen $f(w)$ und $f'(w)$ entsteht also aus der anderen durch Multiplikation mit einer gebrochenen complexen Zahl in z, deren Nenner kein ϱ enthält. Umgekehrt, wenn $f(w)$ und $f'(w)$ in dieser durch die Gleichung (23.) ausgedrückten Beziehung zu einander stehen, so hat man

$$(24.)\qquad l\left(\frac{f'(w)}{C'}\right) - l\left(\frac{f(w)}{C}\right) \equiv \frac{lF'(z)}{M'}, \text{ mod. } \lambda,$$

woraus folgt, dafs die Logarithmen von $f(w)$ und von $f'(w)$ dieselben Werthe der Zahlen $c_1, c_2, \ldots c_{\lambda-1}$ haben müssen, nach dem Modul λ.

Es kann nun eine jede der $\lambda-2$ Zahlen $c_1, c_2, \ldots c_{\lambda-2}$ die λ verschiedenen Werthe $0, 1, 2, \ldots \lambda-1$ erhalten, die Anzahl aller verschiedenen Werthverbindungen dieser Zahlen, und nur diese, geben aber solche der Aufgabe genügende Zahlen $f(w)$, welche sich nicht durch Multiplikation mit complexen Zahlen in z eine aus der andern erzeugen lassen. Dieses Resultat giebt folgenden Satz:

Es giebt genau $\lambda^{\lambda-2}$ ursprüngliche complexe Zahlen $f(w)$, welche der Bedingung genügen, dafs $\frac{f(w\alpha)}{f(w)}$ einer gebrochenen complexen Zahl in z gleich sei, deren Nenner ϱ nicht enthält, welche in der Art von einander unabhängig sind, dafs keine aus einer anderen durch Multiplikation mit einer gebrochenen complexen Zahl in z, deren Nenner ϱ nicht enthält, erzeugt werden kann.

§. 11.

Anzahl der wesentlich verschiedenen Ambigen.

Aus den Zahlen $f(w)$, welche der im vorigen Paragraphen gestellten und gelösten Aufgabe genügen, sollen nun die Ambigen selbst, als wirkliche complexe Zahlen in $u, u_1, u_2 \ldots$ hergeleitet werden.

Es sei

$$(1.) \qquad f(w) = C + \varrho C_1 w + \varrho C_2 w^2 + \ldots + \varrho C_{\lambda-1} w^{\lambda-1}$$

irgend eine der $\lambda^{\lambda-2}$ complexen Zahlen, welche als ursprüngliche Lösungen der Aufgabe bezeichnet worden sind, so ist $f(w) + \lambda g(w)$ ebenfalls eine der Aufgabe genügende Zahl, aber eine solche, welche aus der ursprünglichen $f(w)$ durch Multiplikation mit einer wirklichen complexen Zahl in z entsteht, und man hat:

$$(2.) \; f(w) + \lambda g(w) = C + \lambda B + (\varrho C_1 + \lambda B_1) w + (\varrho C_2 + \lambda B_2) w^2 + \ldots.$$

In dieser complexen Zahl kann und soll nun über die Zahlen $B, B_1, B_2, \ldots B_{\lambda-1}$ so verfügt werden, dafs die n ersten Glieder durch u^λ, d. h. durch $f(\alpha)^m$ theilbar werden, das $n+1$te Glied aber nicht durch $f(\alpha)$ theilbar, ferner dafs die ersten n_1 Glieder durch u_1^λ, d. i. durch $f_1(\alpha)^{m_1}$ theilbar werden, das n_1+1te Glied aber nicht durch $f_1(\alpha)$ theilbar; ferner dafs die n_2 ersten Glieder durch u_2^λ, d. i. $f_2(\alpha)^{m_2}$ theilbar werden, das n_2+1te Glied aber nicht durch $f_2(\alpha)$ theilbar u. s. f. Setzt man alsdann für w seinen Werth $w = u\, u_1\, u_2 \ldots$, so kann man die Faktoren u^n, $u_1^{n_1}$, $u_2^{n_2} \ldots$ heraus heben, und erhält so:

$$(3.) \qquad f(w) + \lambda g(w) = u^n\, u_1^{n_1}\, u_2^{n_2} \ldots f(u, u_1, u_2, \ldots).$$

Es ist nun, wie im §. 9. gezeigt worden, $f(u, u_1, u_2, \ldots)$ eine complexe Zahl in $u, u_1, u_2 \ldots$, deren Norm keinen Faktor der Determinante $D(\alpha)$ enthält, und auch nicht durch ϱ theilbar ist, welche also eine Ambige selbst darstellt, nämlich die in $f(w) + \lambda g(w)$ enthaltene Ambige, und welche einer Gleichung von der Form

$$(4.) \qquad L(\alpha) f(u\alpha, u_1, u_2 \ldots) = M(z) f(u, u_1, u_2 \ldots)$$

genügt, in der $L(\alpha)$ und $NM(z)$ keinen Faktor der Determinante und kein ϱ enthalten.

Wenn nun die Anzahl der in der Determinante $D(\alpha)$ enthaltenen verschiedenen Primfaktoren $f(\alpha)$, $f_1(\alpha)$ u. s. w. gleich r ist, so hat man r Zahlen $n, n_1, n_2, \dots n_{r-1}$, denen man einzeln alle Werthe 0, 1, 2, ... $\lambda-1$ geben kann, welche also λ^r verschiedene Werthverbindungen zulassen. Man erhält also aus jeder der $\lambda^{\lambda-2}$ ursprünglichen Zahlen $f(w)$ genau λ^r complexe Zahlen $f(u, u_1, u_2 \dots)$, welche eben so viele Ambigen darstellen. Die Anzahl aller Ambigen, welche auf diese Weise erhalten werden, ist also gleich $\lambda^{\lambda-2+r}$, welche in so fern ebenfalls als ursprünglich angesehen werden können, als keine derselben aus einer andern durch Multiplikation mit einer wirklichen complexen Zahl in z erzeugt werden kann. Alle anderen Ambigen aber können aus diesen $\lambda^{\lambda-2+r}$ durch Multiplikation mit wirklichen complexen Zahlen in z erzeugt werden.

Aus den gefundenen Ambigen sollen nun die nichtäquivalenten ambigen Klassen ermittelt werden. Aus der Definition der Äquivalenz, nach welcher zwei ideale complexe Zahlen in z äquivalent sind, wenn sie durch Zusammensetzung mit einer und derselben dritten idealen Zahl zu wirklichen complexen Zahlen in z werden, folgt zunächst, dafs alle Ambigen, welche aus einer einzigen $f(u, u_1, u_2, \dots)$ entstehen, indem diese mit wirklichen complexen Zahlen in z zusammengesetzt wird, nothwendig äquivalent sind; denn wenn $F(u, u_1, u_2 \dots)$ ein in der Theorie der complexen Zahlen in z idealer Multiplikator ist, welcher mit $f(u, u_1, u_2 \dots)$ zusammengesetzt, eine wirkliche complexe Zahl in z ergiebt, so ergiebt derselbe auch in seiner Zusammensetzung mit $f(u, u_1, u_2 \dots)\, G(z)$ eine wirkliche complexe Zahl in z, wenn $G(z)$ wirklich ist. Die nichtäquivalenten Klassen der Ambigen sind aus diesem Grunde nur unter den $\lambda^{\lambda-2+r}$ aufzusuchen, welche sich nicht durch Multiplikation mit wirklichen complexen Zahlen in z aus einander erzeugen lassen.

Wenn nun $f(u, u_1, u_2 \dots)$, als wirkliche complexe Zahl in der Theorie der aus den Wurzeln u, u_1, u_2 gebildeten complexen Zahlen, eine ideale Zahl in der Theorie der complexen Zahlen in z darstellt, und $E(u, u_1, u_2 \dots)$ bezeichnet eine Einheit in $u, u_1, u_2 \dots$, d. h. eine wirkliche complexe Zahl dieser Theorie, deren Norm eine Einheit in α ist, so mufs nothwendig $E(u, u_1, u_2 \dots)\, f(u, u_1, u_2 \dots)$ vollständig dieselbe ideale Zahl in z darstellen, als $f(u, u_1, u_2 \dots)$, weil eine ideale Zahl in Beziehung auf Einheiten, mit denen sie behaftet sein kann, vollständig unbestimmt ist. Zugleich ist klar, dafs $E(u, u_1, u_2 \dots) f(u, u_1, u_2 \dots)$ alle Darstellungen einer und derselben idealen Zahl als wirkliche complexe Zahl in $u, u_1, u_2 \dots$ erschöpft, weil die durch ihre idealen Primfaktoren definirten, wirklichen complexen Zahlen sich lediglich durch Einheiten unterscheiden können. Wenn es sich aber, wie in dem vorliegenden Falle nur um solche Zahlen $f(u, u_1, u_2 \dots)$ handelt, die einer Gleichung von der Form (4.) genügen, so mufs die Einheit $E(u, u_1, u_2 \dots)$, mit welcher $f(u, u_1, u_2 \dots)$ behaftet genommen werden kann, selbst einer Gleichung von derselben Form genügen. Diese Bedingung läfst sich so ausdrücken: der Quotient zweier conjugirten Einheiten $E(u\alpha, u_1, u_2 \dots)$ und $E(u, u_1, u_2 \dots)$ mufs einer wirklichen complexen Zahl in z gleich sein, der Quotient zweier Einheiten aber ist selbst wieder eine Einheit, es mufs also

$$\text{(5.)} \qquad \frac{E(u\alpha, u_1, u_2 \dots)}{E(u, u_1, u_2 \dots)} = E(z)$$

sein, wo $E(z)$ eine ganze Einheit in z ist. Eine Einheit $E(u, u_1, u_2 \dots)$, welche dieser Bedingung genügt, soll in dem Folgenden eine ambige Einheit genannt werden.

Die Untersuchung, ob unter den $\lambda^{\lambda-2+r}$ Ambigen der Form $f(u, u_1, u_2 \dots)$ noch äquivalente vorhanden sind, und die Zurückführung derselben auf das System der nichtäquivalenten wird nun in folgender Weise geleistet. Es seien $f(u, u_1, u_2 \dots)$ und $f'(u, u_1, u_2 \dots)$ zwei beliebige dieser $\lambda^{\lambda-2+r}$ Ambigen, so ist

$$\text{(6.)} \qquad \frac{Nf(u, u_1, u_2 \dots)}{f(u, u_1, u_2 \dots)} = F(u, u_1, u_2 \dots)$$

eine ganze complexe Zahl in $u, u_1, u_2 \dots$, welche mit $f(u, u_1, u_2 \dots)$ multiplicirt ein in der Theorie der complexen Zahlen in z wirkliches Produkt giebt. Wenn nun $f(u, u_1, u_2 \dots)$ mit $f'(u, u_1, u_2 \dots)$ äquivalent sein soll, so mufs dieselbe Zahl $F(u, u_1, u_2 \dots)$ auch mit der andern zusammengesetzt eine wirkliche complexe Zahl in z ergeben. Weil aber die Zahl $f'(u, u_1, u_2 \dots)$, in so fern sie eine ideale Zahl in z darstellt, und in so fern sie einer Gleichung von der Form (4.) genügen soll, mit einer beliebigen ambigen Einheit $E(u, u_1, u_2 \dots)$ multiplicirt genommen werden kann, so hat man

$$\text{(7.)} \qquad F(u, u_1, u_2 \dots)\, E(u, u_1, u_2 \dots)\, f'(u, u_1, u_2 \dots) = G(z)$$

gleich einer ganzen complexen Zahl in z. Multiplicirt man diese Gleichung

N 2

mit $f(u, u_1, u_2 \ldots)$ und dividirt durch $G(z)$, so kann man diese Bedingung auch so aussprechen:

(I.) Alle Ambigen von der Form $f(u, u_1, u_2 \ldots)$, welche aus einer derselben entstehen, indem diese mit einer ambigen Einheit und mit einer wirklichen complexen Zahl in z multiplicirt wird, (welche letztere gebrochen sein kann, aber so, dafs sie in der Norm ihres Nenners weder ϱ noch die Faktoren von $D(\alpha)$ enthält), sind mit dieser Ambigen äquivalent, und umgekehrt: alle Ambigen, welche mit einer derselben äquivalent sind, lassen sich auf die angegebene Weise aus dieser einen ableiten.

Es seien jetzt $E(u, u_1, u_2 \ldots)$ und $E_1(u, u_1, u_2 \ldots)$ zwei verschiedene ambige Einheiten, von der Art, dafs die eine aus der anderen nicht durch Multiplikation mit einer Einheit $E(z)$ entstehen kann; es sei ferner $f(u, u_1, u_2 \ldots)$ eine Ambige, so sind

$$E(u, u_1, u_2 \ldots) f(u, u_1, u_2 \ldots) \text{ und } E_1(u, u_1, u_2 \ldots) f(u, u_1, u_2 \ldots)$$

zwei Ambigen, welche durch Multiplikation mit einer complexen Zahl in z nicht aufeinander zurückgeführt werden können, welche also dem Systeme der oben als ursprünglich bezeichneten $\lambda^{\lambda-2+r}$ Ambigen angehören, und dabei äquivalent sind. Es folgt hieraus, dafs eine jede Ambige in diesem Systeme der ursprünglichen Ambigen genau so viele äquivalente hat, als es ambige Einheiten giebt, die durch Multiplikation mit Einheiten in z auf einander nicht zurückgeführt werden können. Also wenn die Anzahl der in diesem Sinne von einander unabhängigen ambigen Einheiten mit E bezeichnet wird, so sind genau je E dieser $\lambda^{\lambda-2+r}$ Ambigen äquivalent, und die Anzahl aller nicht äquivalenten Ambigen ist darum gleich dem Eten Theile von $\lambda^{\lambda-2+r}$. Es folgt hieraus, dafs E eine Potenz von λ sein mufs, und dafs $E = \lambda^e$ gesetzt werden kann. Das gefundene Resultat giebt folgenden Satz:

(II.) Wenn die Anzahl der ambigen Einheiten, welche durch Multiplikation mit Einheiten in z sich nicht auf einander zurückführen lassen, gleich λ^e ist, und die Anzahl der in der Determinante $D(\alpha)$ enthaltenen verschiedenen Primfaktoren gleich r, so ist die Anzahl aller nicht äquivalenten Klassen der Ambigen gleich $\lambda^{\lambda-2+r-e}$.

§. 12.

Die complexen Einheiten in w und in z.

Zur vollständigen Bestimmung der Anzahl aller nichtäquivalenten, ambigen Klassen in der Theorie der complexen Zahlen in z, ist jetzt nur noch erforderlich, die Anzahl $E = \lambda^{\epsilon}$ der von einander unabhängigen ambigen Einheiten zu finden. Hierbei mache ich von den Resultaten Gebrauch, welche Hr. Dirichlet in seinen Untersuchungen über die, in der Theorie der allgemeinen, zerlegbaren Formen vorkommenden Einheiten gefunden, und im März 1846 der Königlichen Akademie mitgetheilt hat, m. s. den Monatsbericht.

Wenn eine irreduktible Gleichung des 2νten Grades, mit nichtcomplexen, ganzzahligen Coefficienten zu Grunde gelegt wird, deren Wurzeln alle imaginär sind, so giebt es nach Dirichlet in der Theorie der complexen Zahlen, welche rationale Funktionen einer Wurzel der gegebenen Gleichung, mit ganzzahligen Coefficienten sind, genau $\nu - 1$ complexe Einheiten, welche ein vollständiges, unabhängiges System von Einheiten dieser Theorie bilden. Ein System von Einheiten wird ein unabhängiges genannt, wenn ein Produkt von Potenzen dieser Einheiten nicht gleich Eins sein kann, ohne dafs die Exponenten der Potenzen alle einzeln gleich Null sind, und ein solches System von unabhängigen Einheiten ist zugleich ein vollständiges, wenn demselben keine Einheit weiter hinzugefügt werden kann, ohne dafs die Eigenschaft der Unabhängkeit des Systems dadurch verloren geht. Ein unabhängiges und vollständiges System hat die Eigenschaft, dafs alle complexen Einheiten dieser Theorie als Produkte von Potenzen der unabhängigen Einheiten sich darstellen lassen, wenn noch gewisse einfache Einheiten, welche nur Einheitswurzeln sind, als Faktoren hinzugenommen werden. Die Exponenten derjenigen Potenzen der unabhängigen Einheiten, durch deren Produkte alle Einheiten dargestellt werden können, haben niemals andere als rationale Zahlenwerthe. Ein unabhängiges und vollständiges System, durch welches alle Einheiten sich so darstellen lassen, dafs die Exponenten der Potenzen nur ganze Zahlen sind, heifst ein System von Fundamentaleinheiten.

Um nun diese allgemeinen Resultate auf die Untersuchung der Einheiten in w und in z anzuwenden, welche aufser den irrationalen Gröfsen w oder z auch noch die Wurzel α enthalten, mufs man eine irreduktible Gleichung mit nichtcomplexen, ganzzahligen Coefficienten aufstellen, von der Art, dafs durch eine Wurzel derselben, so wohl α, als auch w, und demgemäfs auch z, sich rational darstellen lasse. Ich wähle hierzu diejenige Gleichung, deren Wurzel $y = \alpha + w$ ist. Diese giebt zunächst

$$(1.)\qquad (y-\alpha)^\lambda - D(\alpha) = 0,$$

und demgemäfs, wenn für α seine $\lambda - 1$ Werthe $\alpha, \alpha^2, \ldots \alpha^{\lambda-1}$ gesetzt werden, und das Produkt gebildet wird:

$$(2.)\quad ((y-\alpha)^\lambda - D(\alpha))\,((y-\alpha^2)^\lambda - D(\alpha^2)) \ldots ((y-\alpha^{\lambda-1})^\lambda - D(\alpha^{\lambda-1})) = 0,$$

welche Gleichung vom Grade $\lambda(\lambda-1)$, vollständig entwickelt, nichtcomplexe ganze Zahlen zu Coefficienten hat. Dafs diese Gleichung irreduktibel ist, folgt daraus, dafs die $\lambda - 1$ Faktoren, aus denen sie zusammengesetzt ist, einzeln in dem Sinne irreduktibel sind, dafs sie nicht in Faktoren niederer Grade, deren Coefficienten rational in α wären, sich zerlegen lassen, wenn nämlich die Determinante $D(\alpha)$, wie überall vorausgesetzt wird, nicht eine vollständige λte Potenz ist. Ein jeder Faktor der Gleichung (2.), welcher nichtcomplexe rationale Zahlen als Coefficienten hätte, müsste darum nur ein Produkt einer Anzahl jener $\lambda - 1$ Faktoren des λten Grades sein. Entwickelt man aber ein solches Produkt nach Potenzen von y, so erkennt man schon aus der Betrachtung des Coefficienten des zweiten Gliedes, dafs dasselbe nicht anders rationale Coefficienten haben kann, als wenn es alle diese $\lambda - 1$ Faktoren der Gleichung (2.) umfasst.

Um nun α als rationale Funktion von y zu bestimmen, hat man nur die gemeinsame Wurzel α der beiden Gleichungen

$$(y-\alpha)^\lambda - D(\alpha) = 0 \text{ und } 1 + \alpha + \alpha^2 + \ldots + \alpha^{\lambda-1} = 0$$

zu suchen. Diese Gleichungen haben nämlich nur die eine gemeinschaftliche Wurzel α; denn hätten sie aufserdem auch eine andere Wurzel α^k mit einander gemein, so müsste die Wurzel $y = \alpha + w$ der Wurzel $y = \alpha^k + \sqrt[\lambda]{D(\alpha^k)}$ gleich sein, und die Gleichung (2.) wäre nicht irreduktibel. Da also α eine rationale Funktion von y ist, so kann man demselben die Form

$$\alpha = \frac{f(y)}{\delta}$$

geben, wo δ ein von der Determinante $D(\alpha)$ abhängiger, nichtcomplexer ganzzahliger Nenner ist, nnd $f(y)$ eine ganze, rationale Funktion von y des Grades $\lambda(\lambda-1)-1$, mit ganzzahligen Coefficienten. Es ist alsdann $w = y - \alpha$ eine rationale Funktion derselben Art, mit demselben Nenner δ, und da demnach auch die Potenzen $\alpha, \alpha^2, \alpha^3, \ldots \alpha^{\lambda-1}$, so wie die Potenzen $w, w^2, w^3, \ldots w^{\lambda-1}$ sich als rationale Funktionen von y ausdrücken lassen, deren Nenner nur Potenzen von δ sind, so folgt, daſs jede ganze und ganzzahlige Funktion von α und w, d. i. jede complexe Zahl in w sich als gebrochene rationale Funktion von y darstellen läſst, deren Nenner eine bestimmte, nur von der Determinante $D(\alpha)$ abhängige, ganze nichtcomplexe Zahl Δ ist, daſs also

$$\Delta F(w, \alpha) = F(y) \tag{3.}$$

ist, wo $F(y)$, als ganze und ganzzahlige rationale Funktion von y, auch complexe Zahl in y genannt werden kann. Das Δfache einer jeden ganzen complexen Zahl in w läſst sich also als ganze complexe Zahl in y darstellen.

Eine ganze complexe Zahl in w enthält im Ganzen $\lambda(\lambda-1)$ Coefficienten, welche nichtcomplexe ganze Zahlen sind, wenn man alle ihre Glieder, welche verschiedene Potenzen von α und w enthalten, besonders betrachtet. In Beziehung auf einen gegebenen Modul Δ kann jeder dieser $\lambda(\lambda-1)$ Coefficienten die Δ verschiedenen Reste $0, 1, 2, \ldots \Delta-1$ geben, es giebt daher nach dem Modul Δ nicht mehr als $\Delta^{\lambda(\lambda-1)}$ incongruente complexe Zahlen in w. Erhebt man nun eine complexe Einheit in w: $E(w)$ zu Potenzen, so müssen in der Reihe

$$1, \quad E(w), \; E(w)^2, \; E(w)^3, \; E(w)^4 \ldots\ldots$$

wenn dieselbe so weit fortgesetzt wird, daſs sie mehr als $\Delta^{\lambda(\lambda-1)}$ Glieder enthält, nothwendig nach dem Modul Δ congruente Glieder vorkommen, man hat also

$$E(w)^m \equiv E(w)^n, \quad \text{mod. } \Delta,$$

wo m und n verschiedene Zahlen sind und nicht gröſser als $\Delta^{\lambda(\lambda-1)}$. Es hat nun $E(w)$, als Einheit, keinen gemeinschaftlichen Faktor mit dem Modul Δ, man kann also, wenn $n < m$ ist, diese Congruenz durch $E(w)^n$ dividiren, und erhält so

$$E(w)^{m-n} \equiv 1, \text{ mod. } \Delta,$$

oder als Gleichung geschrieben:

$$(4.) \qquad E(w)^{m-n} = 1 + \Delta F(w),$$

und weil $\Delta F(w)$, wie oben gezeigt worden, eine ganze complexe Zahl in y ist, so erkennt man hieraus, dafs eine bestimmte Potenz einer jeden complexen Einheit in w sich als ganze complexe Einheit in y darstellen läfst.

Aus der Definition eines vollständigen Systems von Einheiten geht hervor, dafs dasselbe diese beiden Eigenschaften behält, wenn man die einzelnen Einheiten, aus welchen es besteht, zu irgend welchen ganzen Potenzen erhebt; wenn man daher die Einheiten eines vollständigen, unabhängigen Systems in w zu denjenigen Potenzen erhebt, welche diese zu ganzen complexen Einheiten in y machen, so mufs man ein unabhängiges und vollständiges System von Einheiten in y erhalten, und weil die Anzahl der Einheiten aus welchen dieses besteht, nach dem Dirichletschen Satze gleich $\frac{\lambda(\lambda-1)}{2} - 1$ ist, so mufs ein vollständiges, unabhängiges System von Einheiten in w nothwendig genau eben so viele Einheiten enthalten. Weil ferner die λte Potenz einer jeden complexen Zahl in w eine complexe Zahl in z, also auch die λte Potenz einer jeden Einheit in w eine Einheit in z ist, so folgt, dafs auch ein vollständiges unabhängiges System von Einheiten in z genau $\frac{\lambda(\lambda-1)}{2} - 1$ Einheiten enthalten mufs.

Die complexen Einheiten der niederen Theorie in α sind unter den Einheiten der höheren Theorie in w oder in z als besondere mit enthalten, die unabhängigen Einheiten dieser niederen Theorie können also auch mit als unabhängige Einheiten der höheren Theorie benutzt werden. Setzt man nun, wie im §. 6. der Kürze halben $\mu = \frac{\lambda-1}{2}$, $\nu = \frac{\lambda(\lambda-1)}{2}$, so hat man $\mu - 1$ unabhängige Einheiten in α, und da die Anzahl aller unabhängigen Einheiten in w oder in z gleich $\nu - 1$ ist, so hat man zu diesen $\mu - 1$ Einheiten in α noch $\nu - \mu$ Einheiten in w oder in z hinzuzunehmen, damit das System vollständig werde. Diese $\nu - \mu$ Einheiten der höheren Theorie kann man immer so wählen, dafs ihre Normen, welche nicht complexe Einheiten in α sind, gleich Eins selbst werden. Wenn nämlich eine Einheit $E(w)$ diese Bedingung nicht erfüllt, sondern die Norm $\varepsilon(\alpha)$ hat, so nehme man anstatt derselben ihre λte Potenz, dividirt durch $\varepsilon(\alpha)$, wobei weder die Vollständigkeit noch die Unabhängigkeit des Systems beeinträchtigt wird, die an die Stelle von $E(w)$ tretende Einheit aber die verlangte Eigenschaft hat; wenn es sich um eine Einheit $E(z)$ handelt, so ist deren Norm schon von selbst eine λte Potenz, gleich $\varepsilon(\alpha)^\lambda$, man hat sie daher nur durch $\varepsilon(\alpha)$ zu dividiren. Also:

(I.) Es giebt genau $\nu - \mu = \frac{(\lambda-1)(\lambda-1)}{2}$ unabhängige complexe Einheiten in w oder in z, deren Normen gleich Eins sind.

Um die nun folgenden Sätze leichter beweisen und anschaulicher darstellen zu können, mache ich von einer eigenthümlichen Art der Bezeichnung Gebrauch, welche Hr. Kronecker in seiner Dissertation *de unitatibus complexis*. Berol. 1845. zuerst eingeführt hat, und welche in allen Fällen, wo man es mit Produkten von Potenzen conjugirter Einheiten zu thun hat, höchst vortheilhaft ist. Ich bezeichne, wie dieſs Hr. Kronecker in einer ähnlichen Untersuchung gethan hat, ein Produkt von der Form

$$E(w)^{a}\, E(w\alpha)^{a_1}\, E(w\alpha^2)^{a_2} \ldots\ldots E(w\alpha^{\lambda-1})^{a_{\lambda-1}},$$

gleichsam als Potenz mit einem complexen Exponenten durch

$$E(w)^{a + a_1\alpha + a_2\alpha^2 + \ldots\ldots + a_{\lambda-1}\alpha^{\lambda-1}}$$

oder noch kürzer, wenn der Exponent als complexe Zahl in α mit $a(\alpha)$ bezeichnet wird:

$$(5.) \qquad E(w)^{a}\, E(w\alpha)^{a_1}\, E(w\alpha^2)^{a_2} \ldots\ldots E(w\alpha^{\lambda-1})^{a_{\lambda-1}} = E(w)^{a(\alpha)}.$$

Ebenso, wenn es sich um complexe Einheiten in z handelt, hat man

$$(6.) \qquad E(z)^{a}\, E(z_1)^{a_1}\, E(z_2)^{a_2} \ldots\ldots E(z_{\lambda-1})^{a_{\lambda-1}} = E(z)^{a(\alpha)}.$$

Wenn den Potenzen der Einheiten mit complexen Exponenten dieser bestimmte Sinn gegeben wird, so kann man den complexen Exponenten $a(\alpha)$ mit Hülfe der Gleichung $1 + \alpha + \alpha^2 + \ldots + \alpha^{\lambda-1} = 0$ beliebig verändern, ohne daſs dieses Zeichen seinen Werth ändert, denn es ist,

$$(7.) \qquad E(w)^{1+\alpha+\alpha^2+\ldots+\alpha^{\lambda-1}} = E(w)\,E(w\alpha)\,E(w\alpha^2)\ldots E(w\alpha^{\lambda-1}) = 1.$$

Ferner findet für diese Potenzen mit complexen Exponenten genau dieselbe Regel der Multiplikation Statt, als für gewöhnliche Potenzen, denn man hat:

$$(8.) \qquad E(w)^{a(\alpha)}\, E(w)^{b(\alpha)} = E(w)^{a(\alpha)+b(\alpha)},$$

weil die eine, so wie die andere Seite nichts anderes darstellt, als das Produkt

$$E(w)^{a+b}\, E(w\alpha)^{a_1+b_1}\, E(w\alpha^2)^{a_2+b_2} \ldots E(w\alpha^{\lambda-1})^{a_{\lambda-1}+b_{\lambda-1}}.$$

Ebenso findet für diese Potenzen mit complexen Exponenten auch dieselbe Regel der Potenzerhebung einer Potenz Statt, wie für gewöhnliche Potenzen, denn man hat die Regel:

$$(9.)\qquad \left(E(w)^{a(\alpha)}\right)^{b(\alpha)} = E(w)^{a(\alpha)b(\alpha)},$$

von deren Richtigkeit man sich sogleich überzeugen kann, wenn man auf beiden Seiten dieser Gleichung die Potenzen der Einheit mit complexen Exponenten als Produkte von Potenzen der conjugirten Einheiten darstellt.

Mit Hülfe solcher Potenzen von Einheiten mit complexen Exponenten will ich nun zeigen, wie ein vollständiges System der $\nu-\mu$ unabhängigen Einheiten in w oder in z, deren Normen gleich Eins sind, durch μ verschiedene Einheiten dieser Art, mit ihren conjugirten dargestellt werden kann.

Wenn $E(w)$ irgend eine Einheit ist, deren Norm gleich Eins ist, so zeige ich zunächst, daſs die $\lambda-1$ conjugirten Einheiten

$$E(w),\ E(w\alpha),\ E(w\alpha^2),\ \ldots E(w\alpha^{\lambda-2})$$

von einander unabhängig sind, daſs also die Gleichung

$$(10.)\qquad E(w)^{m}\, E(w\alpha)^{m_1}\, E(w\alpha^2)^{m_2} \ldots E(w\alpha^{\lambda-2})^{m_{\lambda-2}} = 1,$$

in welcher die Exponenten $m,\ m_1,\ m_2,\ \ldots m_{\lambda-2}$ rationale, oder was hier dasselbe ist, ganze Zahlen sind, nicht bestehen kann, ohne daſs sie alle den Werth Null erhalten. Setzt man

$$m + m_1\alpha + m_2\alpha^2 + \ldots + m_{\lambda-2}\alpha^{\lambda-2} = m(\alpha),$$

und erhebt beide Seiten der Gleichung

$$E(w)^{m(\alpha)} = 1$$

zur Potenz $M(\alpha)$, wo $M(\alpha)$ durch die Gleichung $M(\alpha)\,m(\alpha) = Nm(\alpha)$ bestimmt ist, so erhält man

$$(11.)\qquad E(w)^{Nm(\alpha)} = 1.$$

Der Exponent Norm von $m(\alpha)$, ist eine nichtcomplexe Zahl, diese Potenz ist also eine Potenz im gewöhnlichen Sinne, welche, wenn $E(w)$ nicht eine einfache Wurzel der Einheit ist, nicht gleich Eins sein kann, ohne dafs $Nm(\alpha) = 0$ ist, also $m(\alpha) = 0$ und folglich $m = 0$, $m_1 = 0$, ... $m_{\lambda-2} = 0$, wegen der Irreduktibilität der Gleichung $1 + \alpha + \alpha^2 + \ldots + \alpha^{\lambda-1} = 0$. Die $\lambda - 1$ conjugirten Einheiten

$$E(w),\ E(w\alpha),\ E(w\alpha^2)\ \ldots.,\ E(w\alpha^{\lambda-2})$$

bilden also in der That ein unabhängiges, aber unvollständiges System. Es sei nun $E_1(w)$ eine andere Einheit, deren Norm gleich Eins ist, welche mit dieser zusammen ein unabhängiges System bilde, so behaupte ich, dafs die conjugirten $\lambda - 1$ Einheiten

$$E_1(w),\ E_1(w\alpha),\ E_1(w\alpha^2)\ \ldots.\ E_1(w\alpha^{\lambda-2})$$

mit den obigen zusammen ein unabhängiges System von $2(\lambda - 1)$ Einheiten bilden, dafs also die Gleichung

$$(12.)\qquad E(w)^{m(\alpha)}\, E_1(w)^{m_1(\alpha)} = 1$$

nicht bestehen kann, ohne dafs die complexen Exponenten $m(\alpha)$ und $m_1(\alpha)$ beide gleich Null sind. Erhebt man nämlich diese Gleichung zur Potenz $M_1(\alpha)$, wo $M_1(\alpha)$ durch die Gleichung $M_1(\alpha)\, m_1(\alpha) = Nm_1(\alpha)$ bestimmt ist, so hat man

$$(13.)\qquad E(w)^{M_1(\alpha)m(\alpha)}\, E_1(w)^{Nm_1(\alpha)} = 1,$$

welches, da $Nm_1(\alpha)$ eine nichtcomplexe Zahl ist, und darum $E_1(w)^{Nm_1(\alpha)}$ eine Potenz im gewöhnlichen Sinne des Wortes, der Voraussetzung widerspricht, nach welcher $E_1(w)$ eine Einheit ist, die mit den $\lambda - 1$ zu $E(w)$ conjugirten ein unabhängiges System bildet. Wenn nun $E_2(w)$ eine andere Einheit mit der Norm Eins ist, welche mit den bereits aufgestellten $2(\lambda - 1)$ Einheiten zusammen ein unabhängiges System bildet, so wird ganz auf dieselbe Weise bewiesen, dafs auch die conjugirten $\lambda - 1$ Einheiten

$$E_2(w),\ E_2(w\alpha),\ E_2(w\alpha^2),\ \ldots.\ E_2(w\alpha^{\lambda-2})$$

mit jenen $2(\lambda - 1)$ Einheiten zusammen ein unabhängiges System von $3(\lambda-1)$ Einheiten bilden. Auf diese Weise kann man nun fortfahren, bis man ein vollständiges unabhängiges System von Einheiten, deren Normen gleich

O 2

Eins sind, erlangt hat, welches, wie oben gezeigt worden ist, aus $\nu-\mu$ Einheiten besteht. Da $\nu-\mu=\frac{\lambda(\lambda-1)}{2}-\frac{\lambda-1}{2}=\mu(\lambda-1)$ ist, so folgt, dafs die Anzahl der Einheiten, welche mit ihren conjugirten zusammen ein vollständiges System bilden, gleich μ ist. Was hier von den Einheiten in w bewiesen ist, gilt nothwendig auch von den Einheiten der specielleren Theorie in z, man hat also folgenden Satz:

(II.) Es giebt genau μ Einheiten in w oder in z, welche, wenn man von einer jeden derselben $\lambda-1$ conjugirte nimmt, ein vollständiges unabhängiges System von $\mu(\lambda-1)$ Einheiten bilden, deren Normen gleich Eins sind.

Wenn nun die μ Einheiten:

$$E(w),\ E_1(w),\ E_2(w),\ \dots\ E_{\mu-1}(w)$$

mit ihren conjugirten ein vollständiges und unabhängiges System aller derjenigen Einheiten bilden, deren Normen gleich Eins sind, so kann aus diesem Systeme stets ein anderes hergestellt werden, dessen μ Einheiten

$$\varepsilon(w),\ \varepsilon_1(w),\ \varepsilon_2(w),\ \dots\ \varepsilon_{\mu-1}(w)$$

so beschaffen sind, dafs keine in der Form

$$\varepsilon(w)^{M}\ \varepsilon_1(w)^{M_1}\ \varepsilon_2(w)^{M_2}\ \dots\ \varepsilon_{\mu-1}(w)^{M_{\mu-1}}$$

enthaltene Einheit eine ϱte Potenz einer Einheit sein kann, ohne dafs die ganzen complexen Exponenten M, M_1, M_2 ... $M_{\mu-1}$ alle einzeln durch ϱ theilbar sind. Ein so beschaffenes System kann ein beziehungsweise fundamentales genannt werden, nämlich fundamental in Beziehung auf den Exponenten ϱ. Dasselbe wird aus dem gegebenen Systeme der Einheiten $E(w)$, $E_1(w)$ ... $E_{\mu-1}(w)$ auf folgende Weise hergeleitet.

Durch das gegebene System der Einheiten möge sich eine ϱ^k te Potenz einer Einheit so darstellen lassen, dafs die complexen Exponenten nicht alle durch ϱ theilbar sind, aber eine ϱ^{k+1}te Potenz einer Einheit soll nicht mehr in dieser Weise durch dieses System darstellbar sein, so hat man

$$E(w)^{A}\ E_1(w)^{A_1}\ E_2(w)^{A_2}\ \dots\ E_{\mu-1}(w)^{A_{\mu-1}}=\varepsilon(w)^{\varrho^k} \tag{14.}$$

für bestimmte Werthe der complexen Exponenten A, A_1, ... $A_{\mu-1}$, und weil wenigstens einer derselben durch ϱ nicht theilbar ist, so kann man den ersten A als einen solchen wählen. Durch die $\mu-1$ Einheiten $E_1(w)$,

$E_2(w)$, ... $E_{\mu-1}(w)$, mit Ausschlufs von $E(w)$, lasse sich eine ϱ^{k_1} te, aber nicht eine ϱ^{k_1+1} te Potenz einer Einheit so darstellen, dafs nicht alle complexen Exponenten dieser Einheiten durch ϱ theilbar sind, so hat man

$$(15.)\qquad E_1(w)^{B_1}\, E_2(w)^{B_2} \ldots E_{\mu-1}(w)^{B_{\mu-1}} = \varepsilon_1(w)^{\varrho^{k_1}},$$

und man kann hier ebenfalls den ersten Exponenten B_1 als einen der durch ϱ nicht theilbaren nehmen. Ebenso sei durch diese Einheiten, wenn weiter die Einheit $E_1(w)$ ausgeschlossen wird, noch eine ϱ^{k_2} te Potenz, aber nicht eine ϱ^{k_2+1} te Potenz einer Einheit darstellbar, so hat man

$$(16.)\qquad E_2(w)^{C_2}\, E_3(w)^{C_3} \ldots E_{\mu-1}(w)^{C_{\mu-1}} = \varepsilon_2(w)^{\varrho^{k_2}}.$$

wo wieder C_2 als einer der nicht durch ϱ theilbaren Exponenten genommen werden kann. So fortfahrend erhält man die μ Einheiten $\varepsilon(w)$, $\varepsilon_1(w)$, ... $\varepsilon_{\mu-1}(w)$, von denen nun bewiesen werden soll, dafs sie ein System bilden, welches die verlangte Eigenschaft besitzt.

Gesetzt es wäre für irgend welche bestimmte ganze complexe Exponenten M, M_1, ... $M_{\mu-1}$, die nicht alle durch ϱ theilbar sind,

$$(17.)\qquad \varepsilon(w)^{M}\, \varepsilon_1(w)^{M_1} \ldots \varepsilon_{\mu-1}(w)^{M_{\mu-1}} = e(w)^{\varrho},$$

und M_r der erste, nicht durch ϱ theilbare Exponent, so hätte man auch

$$(18.)\qquad \varepsilon_r(w)^{M_r}\, \varepsilon_{r+1}(w)^{M_{r+1}} \ldots \varepsilon_{\mu-1}(w)^{M_{\mu-1}} = e_1(w)^{\varrho}.$$

Wenn nun zur ϱ^{k_r} ten Potenz erhoben wird, so kann man, weil von den Zahlen k, k_1, k_2 ... keine folgende gröfser sein kann, als die vorhergehende, die in diesem Ausdrucke vorkommenden ϱ^{k_r} ten Potenzen der Einheiten $\varepsilon_r(w)$, $\varepsilon_{r+1}(w)$, ... $\varepsilon_{\mu-1}(w)$ alle durch die Einheiten $E_r(w)$, $E_{r+1}(w)$ .... $E_{\mu-1}(w)$ ausdrücken, und erhält so einen Ausdruck von folgender Form:

$$(19.)\qquad E_r(w)^{K_r}\, E_{r+1}(w)^{K_{r+1}} \ldots E_{\mu-1}(w)^{K_{\mu-1}} = e_1(w)^{\varrho^{k_r+1}},$$

in welchem der complexe Exponent K_r der ersten Einheit gleich dem Produkte des Exponenten M_r und des Exponenten H_r in dem Ausdrucke

$$(20.)\qquad E_r(w)^{H_r}\, E_{r+1}(w)^{H_{r+1}} \ldots E_{\mu-1}(w)^{H_{\mu-1}} = \varepsilon_r(w)^{\varrho^{k_r}}$$

ist, und weil M_r und H_r nach der Voraussetzung nicht durch ϱ theilbar sind, so ist $K_r = M_r H_r$ ebenfalls nicht durch ϱ theilbar. Es ist aber nach der Voraussetzung unmöglich, eine ϱ^{k_r+1}te Potenz einer Einheit durch die Form (19.) auszudrücken, ohne dafs alle complexen Exponenten K_r, K_{r+1}, ... $K_{\mu-1}$ durch ϱ theilbar sind, also ist es auch unmöglich, durch die Einheiten $\varepsilon(w)$, $\varepsilon_1(w)$ ... $\varepsilon_{\mu-1}(w)$ eine ϱte Potenz irgend einer Einheit in dieser Weise auszudrücken. Genau dieselben Schlüsse gelten auch, wenn man anstatt der Einheiten in w die specielleren Einheiten in z zu Grunde legt, man hat also folgenden Satz:

(III.) Es giebt für Einheiten in w (oder in z), deren Normen gleich Eins sind, ein vollständiges und unabhängiges System von μ Einheiten mit ihren conjugirten, welches so beschaffen ist, dafs ein Produkt von Potenzen dieser μ Einheiten mit ganzen complexen Exponenten nicht eine ϱte Potenz einer Einheit in w (oder in z) darstellen kann, ohne dafs die complexen Exponenten der Potenzen aus denen dieses Produkt besteht, alle einzeln durch ϱ theilbar sind.

Weil das System der μ Einheiten $\varepsilon(w)$, $\varepsilon_1(w)$, ... $\varepsilon_{\mu-1}(w)$ mit ihren conjugirten ein vollständiges unabhängiges System für alle Einheiten ist, deren Normen gleich Eins sind, so kann man durch dasselbe alle Einheiten, deren Normen gleich Eins sind, darstellen, wenn man in den Coefficienten der complexen Exponenten M, M_1, ... $M_{\mu-1}$ der Form

$$(21.)\qquad \varepsilon(w)^{M}\ \varepsilon_1(w)^{M_1}\ \ldots\ \varepsilon_{\mu-1}(w)^{M_{\mu-1}}$$

rationale Brüche zuläfst, oder was dasselbe ist, wenn man diese Exponenten als gebrochene complexe Zahlen in α passend bestimmt. Diese gebrochenen Exponenten können nun, vermöge der in dem Satze (III.) ausgesprochenen Eigenthümlichkeit des vorliegenden Systems der unabhängigen Einheiten, in ihren Nennern niemals den Faktor ϱ enthalten, wenn gemeinschaftliche Faktoren der Zähler und Nenner nicht Statt haben, aus welchem Grunde ein solches System oben als ein in Beziehung auf ϱ fundamentales bezeichnet worden ist. Gesetzt es wäre eine ganze complexe Einheit $e(w)$ durch die Form (21.) so darstellbar, dafs irgend welche der gebrochenen complexen Exponenten M, M_1, ... $M_{\mu-1}$ in ihren Nennern ϱ enthielten, aber nicht in ihren Zählern, und es wäre ϱ^h die höchste Potenz von ϱ, welche in einem dieser Nenner vorkommt, so würde der kleinste gemeinschaftliche Nenner, den man allen diesen gebrochenen Exponenten geben könnte, von der Form $\varrho^h N$ sein, wo N eine nicht weiter durch ϱ theilbare complexe Zahl in α wäre. Wenn man also eine durch die Form (21.) ausgedrückte Einheit zur $\varrho^h N$ten Potenz erhöbe, so würde man die Potenz, welche eine ϱte Potenz einer Einheit ist, durch dieselbe Form (21.) aber mit ganzen complexen Exponenten, welche alle nicht durch ϱ theilbar sind, ausgedrückt erhalten. Da dieses nicht möglich ist, so folgt, dafs die gebrochenen Exponenten M, M_1, ... $M_{\mu-1}$ in ihren Nennern niemals den Faktor ϱ enthalten können.

Wegen dieser Eigenschaft der Exponenten M, M_1, ... $M_{\mu-1}$ kann man von denselben die kleinsten nicht negativen, ganzen, nichtcomplexen Zahlen absondern, denen sie congruent sind nach dem Modul ϱ, und erhält so:

$$(22.)\quad M = a + \varrho M',\ M_1 = a_1 + \varrho M'_1,\ \ldots\ M_{\mu-1} = a_{\mu-1} + \varrho M'_{\mu-1},$$

wo die Zahlen a, a_1, ... $a_{\mu-1}$ nur die Werthe 0, 1, 2, ... $\lambda - 1$ haben können. Die Form (21.) giebt demnach folgende Form:

$$(23.)\quad \varepsilon(w)^{a}\,\varepsilon_1(w)^{a_1}\ldots\varepsilon_{\mu-1}(w)^{a_{\mu-1}}\Big(\varepsilon(w)^{M'}\,\varepsilon_1(w)^{M'_1}\ldots\varepsilon_{\mu-1}(w)^{M'_{\mu-1}}\Big)^{\varrho},$$

welche ebenfalls alle Einheiten darstellt, deren Normen gleich Eins sind. Also alle Einheiten, deren Normen gleich Eins sind, entstehen aus den in der Form

$$(24.)\quad \varepsilon(w)^{a}\,\varepsilon_1(w)^{a_1}\ldots\ldots\varepsilon_{\mu-1}(w)^{a_{\mu-1}}$$

enthaltenen, in welcher a, a_1, $a_{\mu-1}$ nur alle Werthe 0, 1, 2, ... $\lambda - 1$ haben, wenn diese mit ϱten Potenzen von Einheiten multiplicirt werden. Von zwei verschiedenen, in dieser Form (24.) enthaltenen Einheiten kann aber niemals eine aus der andern durch Multiplikation mit einer ϱten Potenz einer Einheit erzeugt werden; denn hätte man

$$\varepsilon(w)^{a}\,\varepsilon_1(w)^{a_1}\ldots\varepsilon_{\mu-1}(w)^{a_{\mu-1}} = \varepsilon(w)^{b}\,\varepsilon_1(w)^{b_1}\ldots\varepsilon_{\mu-1}(w)^{b_{\mu-1}}\,E(w)^{\varrho},$$

so wäre auch

$$(25.)\quad \varepsilon(w)^{a-b}\,\varepsilon_1(w)^{a_1-b_1}\ldots\ldots\varepsilon_{\mu-1}(w)^{a_{\mu-1}-b_{\mu-1}} = E(w)^{\varrho},$$

und weil durch diese Form keine ϱte Potenz einer Einheit ausgedrückt werden kann, ohne dafs alle Exponenten einzeln durch ϱ theilbar sind, so müssen $a-b$, a_1-b_1, ... $a_{\mu-1}-b_{\mu-1}$ alle durch ϱ, und weil es nichtcomplexe ganze Zahlen sind, alle durch λ theilbar sein, welches nur dann möglich ist, wenn $a=b$, $a_1=b_1$ ... $a_{\mu-1}=b_{\mu-1}$, also nur dann, wenn die beiden in der Form (24.) enthaltenen Einheiten dieselben sind. Da die Anzahl aller verschiedenen in dieser Form enthaltenen Einheiten gleich λ^μ ist, indem jeder der μ Exponenten a, a_1, ... $a_{\mu-1}$ alle λ Werthe 0, 1, 2, ... $\lambda-1$ annehmen kann, so hat man folgenden Satz:

(IV.) Es giebt genau λ^μ complexe Einheiten in w (oder in z), deren Normen gleich Eins sind, von denen keine aus der andern durch Multiplikation mit einer ϱten Potenz einer Einheit abgeleitet werden kann, welche alle als Produkte von Potenzen von μ, in Beziehung auf ϱ, fundamentalen Einheiten so dargestellt werden können, dafs den Potenzexponenten alle Werthe 0, 1, 2, ... $\lambda-1$ gegeben werden.

§. 13.

Die ambigen Einheiten und die nichtäquivalenten Ambigen. Schlufs auf die Anzahl der wirklich vorhandenen Gattungen.

Aus den Einheiten in z, deren Normen gleich Eins sind, werden nun die ambigen Einheiten in folgender Weise hergeleitet. Wenn $E(z)$ eine solche Einheit ist, so bilde man den Ausdruck:

$$PE(z) = 1 + E(z) + E(z)\,E(z_1) \;....\; E(z)\,E(z_1) \;...\; E(z_{\lambda-2}),$$

welcher als ganze complexe Zahl in z einfach durch $F(z)$ bezeichnet werden soll. Vermöge der Grundeigenschaft des Ausdrucks $PE(z)$, nach welcher

$$E(z)\,PE(z_1) = PE(z)$$

ist, hat man

$$E(z)\,F(z_1) = F(z). \tag{1.}$$

Die Zahl $F(z)$, welche einer solchen Gleichung genügt, hat aber, wie im §. 8. gezeigt worden ist, die Eigenschaft, dafs ihre $h$te Potenz, wenn h die Klassenanzahl der idealen Zahlen in α bezeichnet, sich in zwei Faktoren

$$F(z)^h = C \cdot \Delta(z) \tag{2.}$$

zerlegen läfst, deren einer, C, eine wirkliche, ganze complexe Zahl in α ist, der andere, $\Delta(z)$, eine wirkliche complexe Zahl in z, deren Norm keinen anderen Primfaktor enthält, als die, welche in $\varrho D(\alpha)$ vorkommen. Diese Zahl $\Delta(z)$ genügt vermöge der Gleichungen (1.) und (2.) der Gleichung:

$$(3.) \qquad E(z)^h\,\Delta(z_1) = \Delta(z).$$

Es ist nun oben im §. 9. gezeigt worden, dafs eine jede Zahl $f(z)$, welche einer Gleichung von der Form

$$L(\alpha)f(z_1) = M(z)f(z)$$

genügt, wenn sie zuerst als complexe Zahl in ω dargestellt, und sodann $\omega = u\,u_1\,u_2\ldots$ gesetzt wird, folgende Form annimmt:

$$(4.) \qquad f(z) = \varrho^{\nu}\,u^{n}\,u_1^{n_1}\,u_2^{n_2}\ldots f(u, u_1, u_2\ldots),$$

in welcher die Norm von $f(u, u_1, u_2\ldots)$ keinen Faktor mit $\varrho D(\alpha)$ gemein hat. Wendet man dieses Resultat auf die Zahl $\Delta(z)$ an, welche einer Gleichung derselben Form genügt, nämlich für $M(z) = E(z)^{-h}$ und $L(\alpha) = 1$, und beachtet, dafs die Norm von $\Delta(z)$ keinen anderen Primfaktor enthält, als ϱ und die Primfaktoren der Determinante $D(\alpha)$, so hat man

$$(5.) \qquad \Delta(z) = \varrho^{\nu}\,u^{n}\,u_1^{n_1}\,u_2^{n_2}\ldots E(u, u_1, u_2\ldots)$$

wo $E(u, u_1, u_2\ldots)$ eine Einheit ist, weil ihre Norm weder die in $\varrho D(\alpha)$ enthaltenen, noch auch die in $\varrho D(\alpha)$ nicht enthaltenen Primfaktoren haben darf. Verwandelt man ferner u in $u\alpha$, wodurch z in z_1 übergeht, so hat man

$$(6.) \qquad \Delta(z_1) = \varrho^{\nu}\,\alpha^{n}\,u^{n}\,u_1^{n_1}\,u_2^{n_2}\ldots E(u\alpha, u_1, u_2\ldots),$$

also vermöge der Gleichung (3.)

$$(7.) \qquad \frac{E(u\alpha, u_1, u_2\ldots)}{E(u, u_1, u_2\ldots)} = \alpha^{-n}\,E(z)^{-h}$$

Die Einheit $E(u, u_1, u_2\ldots)$ ist also eine ambige Einheit. Auf dieselbe Weise läfst sich aus jeder Einheit in z, deren Norm gleich Eins ist, eine ambige Einheit erzeugen.

Es ist nun weiter zu untersuchen, unter welchen Bedingungen zwei verschiedene Einheiten $E(z)$ und $E_1(z)$ zwei in der Art verschiedene ambige Einheiten erzeugen, dafs die eine aus der andern durch Multiplikation mit einer Einheit in z nicht hergeleitet werden kann. Aus den beiden Einheiten $E(z)$ und $E_1(z)$ erhalte man die Ambigen $E(u, u_1, u_2 \ldots)$ und $E_1(u, u_1, u_2 \ldots)$, so ist

$$(8.)\qquad \frac{E(u\alpha, u_1, u_2 \ldots)}{E(u, u_1, u_2 \ldots)} = \alpha^{-n} E(z)^{-h},$$

$$(9.)\qquad \frac{E_1(u\alpha, u_1, u_2 \ldots)}{E_1(u, u_1, u_2 \ldots)} = \alpha^{-n_1} E_1(z)^{-h}.$$

Wenn nun die beiden ambigen Einheiten so beschaffen sind, dafs

$$(10.)\qquad E_1(u, u_1, u_2 \ldots) = E(u, u_1, u_2 \ldots)\, e(z),$$

so erhält man aus den beiden vorhergehenden Gleichungen

$$(11.)\qquad E_1(z)^h = \alpha^{n-n_1} E(z)^h e(z)^\varrho.$$

Da h nicht durch λ theilbar ist, so kann man zwei ganze Zahlen b und c so bestimmen, dafs sie der Gleichung $bh = 1 + c\lambda$ genügen, erhebt man daher die Gleichung (11.) zur $b$ten Potenz und nimmt $bh = 1 + c\lambda$, so erhält man

$$(12.)\qquad E_1(z) = E(z)\alpha^{(n-n_1)b} E_1(z)^{-c\lambda} E(z)^{c\lambda} e(z)^{b\varrho},$$

und weil eine λte Potenz einer Einheit zugleich als eine ϱte Potenz angesehen werden kann, so folgt, dafs abgesehen von einer Potenz von α, welche zu jeder Einheit $E(z)$ beliebig hinzugenommen werden kann, die Einheit $E_1(z)$ aus der Einheit $E(z)$ entsteht, indem diese mit einer ϱten Potenz einer Einheit multiplicirt wird. Umgekehrt, wenn

$$(13.)\qquad E_1(z) = \alpha^k e(z)^\varrho E(z)$$

ist, so erhält man aus den beiden Gleichungen (8.) und (9.):

$$(14.)\qquad \frac{E_1(u\alpha, u_1, u_2 \ldots)}{E_1(u, u_1, u_2 \ldots)} = \alpha^{n-n_1-hk} \frac{E(u\alpha, u_1, u_2 \ldots)\, e(z_1)^h}{E(u, u_1, u_2 \ldots)\, e(z)^h},$$

setzt man also

$$(15.)\qquad \frac{E_1(u, u_1, u_2 \ldots)}{E(u, u_1, u_2 \ldots)\, e(z)^h} = \varepsilon(u, u_1, u_2 \ldots),$$

so hat man

$$(16.)\qquad \varepsilon(u\alpha, u_1, u_2 \ldots) = \alpha^{-n+n_1+hk} \varepsilon(u, u_1, u_2 \ldots).$$

Muliplicirt man nun mit einem passend gewählten Produkte von der Form $u^m u_1^{m_1} u_2^{m_2} \ldots$, um die complexe Einheit $\varepsilon(u, u_1, u_2 \ldots)$ in eine complexe Zahl in w zu verwandeln und setzt

(17.) $$u^m u_1^{m_1} u_2^{m_2} \ldots \varepsilon(u, u_1, u_2 \ldots) = f(w),$$

so giebt die Gleichung (16.) eine Gleichung von der Form:

(18.) $$f(w\alpha) = \alpha' f(w),$$

aus welcher folgt, dafs $f(w)$ nur eine Potenz von w sein kann, multiplicirt mit einer complexen Zahl in α. Man hat daher

$$u^m u_1^{m_1}, u_2^{m_2} \ldots \varepsilon(u, u_1, u_2 \ldots) = Cw^s$$

und wenn für w der Werth $u\,u_1\,u_2 \ldots$ gesetzt wird:

(19.) $$\varepsilon(u, u_1, u_2 \ldots) = Cu^{s-m} u_1^{s-m_1} u_2^{s-m_2} \ldots,$$

da aber $\varepsilon(u, u_1, u_2 \ldots)$ eine Einheit ist, und darum keinen Faktor u, u_1, $u_2 \ldots$ enthalten kann, so folgt, dafs $m = s$, $m_1 = s$, $m_2 = s \ldots$ sein mufs, und C eine complexe Einheit in α, also

(20.) $$\varepsilon(u, u_1, u_2 \ldots) = E(\alpha),$$

und folglich

(21.) $$E_1(u, u_1, u_2 \ldots) = E(u, u_1, u_2 \ldots)\, E(\alpha)\, e(z)^h$$

Die eine dieser ambigen Einheiten entsteht also aus der andern durch Multiplikation mit einer Einheit in z. Das gefundene Resultat giebt den Satz:

(I.) Alle ambigen Einheiten, welche durch Multiplikation mit Einheiten in z sich nicht auf einander zurückführen lassen, und nur diese, entstehen aus allen denjenigen Einheiten in z, mit der Norm gleich Eins, welche sich durch Multplikation mit ϱten Potenzen von Einheiten in z nicht auf einander zurückführen lassen.

Nach dem Satze (IV.) §. 12. ist die Anzahl derjenigen Einheiten in z mit der Norm Eins, welche durch Multiplikation mit ϱten Potenzen von Einheiten in z sich nicht auf einander zurückführen lassen, gleich λ^μ; daher ergiebt der gefundene Satz zugleich den folgenden:

P 2

(II.) Die Anzahl aller ambigen Einheiten, welche durch Multiplikation mit Einheiten in z sich nicht auf einander zurückführen lassen, ist gleich λ^{μ}.

Im §. 11. ist diese Anzahl der ambigen Einheiten mit $E = \lambda^{\varepsilon}$ bezeichnet worden, man hat also $\varepsilon = \mu = \frac{\lambda-1}{2}$, $\lambda - 2 + r - \varepsilon = \mu + r - 1$, und demnach ergiebt der Satz (II.) §. 11. den folgenden:

(III.) Wenn die Anzahl der in der Determinante $D(\alpha)$ enthaltenen verschiedenen Primfaktoren gleich r ist, so ist die Anzahl aller wesentlich verschiedenen, nicht äquivalenten Klassen der Ambigen gleich $\lambda^{\mu+r-1}$.

In Betreff der wirklich vorhandenen Gattungen der verschiedenen Klassen der idealen Zahlen in z giebt daher der Satz (V.) §. 7:

(IV.) Die Anzahl der wirklich vorhandenen Gattungen der idealen Klassen in der Theorie der complexen Zahlen in z ist nicht gröfser als $\lambda^{\mu+r-1}$.

Da die Anzahl der verschiedenen Charaktere der idealen Zahlen in z gleich $\mu + r$ ist, und mithin die Anzahl aller Gesamtcharaktere, oder die Anzahl aller angebbaren Gattungen gleich $\lambda^{\mu+r}$ ist, so kann man diesen Satz auch so aussprechen:

(V.) Die Anzahl der wirklich vorhandenen Gattungen ist nicht gröfser als der λte Theil aller blofs angebbaren Gattungen oder Gesamtcharaktere.

§. 14.

Congruenzbedingung für die Darstellbarkeit einer complexen Zahl in α als Norm einer wirklichen complexen Zahl in w.

Für die Anwendung dieser Theorie der complexen Zahlen auf den Beweis der allgemeinen Reciprocitätsgesetze reicht das gefundene Resultat über die Anzahl der Gattungen: dafs die Anzahl der wirklich vorhandenen nicht gröfser ist, als der λte Theil der angebbaren Gesamtcharaktere, nicht aus, es ist zu diesem Zwecke erforderlich auf die Frage: ob der λte Theil der angebbaren Gesamtcharaktere auch wirklich vorhandene Gattungen giebt, näher einzugehen und dieselbe wenigstens in gewissen besonderen Fällen vollständig zu lösen. Ein Mittel hierzu gewähren die complexen Zahlen in w und die complexen Zahlen in $u, u_1, u_2 \ldots$ Da nämlich die wirklichen complexen Zahlen in diesen Theorieen sich als ideale Zahlen in der Theorie der complexen Zahlen in z darstellen, so kann man untersuchen, in wie weit es gelingt, durch diese die $\lambda^{\mu+r-1}$ Gattungen, welche nach dem gefundenen Satze Statt haben können, auszufüllen, d. h. in wie weit man für jeden von den $\lambda^{\mu+r-1}$ Gesamtcharakteren eine entsprechende wirkliche complexe Zahl in w oder in $u, u_1, u_2 \ldots$ finden kann. Die Grundlage dieser Untersuchung bildet eine Congruenz unter den Charakteren $C_3, C_5, \ldots C_{\lambda-1}$ einer jeden idealen Zahl in z, welche als wirkliche complexe Zahl in w dargestellt werden kann, welche darum jetzt hergeleitet werden soll.

Zunächst ist es hierzu nöthig den Charakter

$$C_{\lambda-1} = \frac{NF(\alpha)-1}{\lambda}$$

in ähnlicher Weise wie die anderen ebenfalls durch Differenzialquotienten eines Logarithmus auszudrücken, welches vermittelst einer allgemeineren Formel geschehen kann, die folgendermaafsen gefunden wird: Man nehme eine ganze rationale Funktion der Variabeln x, $\varphi(x)$, vom $\lambda-1$ten Grade, deren Coefficienten beliebige Constanten sein können, und bilde das Produkt von λ Faktoren

$$(1.) \qquad \prod_{0}^{\lambda-1}{}_r\, \varphi(x+k(\alpha^r-1)) = \Phi(x,k).$$

Dasselbe ist, als Funktion von k betrachtet, eine ganze rationale Funktion vom Grade $\lambda(\lambda-1)$, welche die Eigenschaft hat, dafs alle Glieder, welche k^λ und die höheren Potenzen von k enthalten, Vielfache von λ^2 sind, wovon man sich leicht überzeugt, wenn man bemerkt, dafs mit jedem einfachen k ein Faktor α^r-1, also ein Faktor ϱ verbunden vorkommt, also mit k^n nothwendig der Faktor ϱ^n, statt dessen, wenn $n > \lambda-1$ ist, auch $\lambda\varrho^{n-\lambda+1}$ genommen werden kann. Jedes Glied, welches k^n enthält, ist also, wenn $n > \lambda-1$ ist, durch $\lambda\varrho$ theilbar, und weil das entwickelte Produkt die Wurzel α nicht enthält, so kann das mit k^n behaftete Glied derselben nicht durch $\lambda\varrho$ theilbar sein, ohne durch λ^2 theilbar zu sein. Bei der Untersuchung des Ausdrucks $\Phi(x,k)$ in Beziehung auf den Modul λ^2 kann man also alle Glieder, welche k^λ und höhere Potenzen des k enthalten, als durch λ^2 theilbar vernachlässigen. Ich entwickele nun den Logarithmus

$$l\phi(x+k(\alpha^r-1)) = l\phi(x) + \frac{(\alpha^r-1)k}{1}\,\frac{dl\phi(x)}{dx} + \frac{(\alpha^r-1)^2k^2}{1.\,2.}\,\frac{d^2l\phi(x)}{dx^2} + \\ + \ldots. + k^\lambda A,$$

wo $k^\lambda A$ den Rest der Reihe vorstellt, vom λten Gliede an. Giebt man dem r die Werthe 0, 1, 2 ... $\lambda-1$ und summirt, indem man bemerkt, dafs

$$(\alpha-1)^n + (\alpha^2-1)^n + (\alpha^3-1)^n + \ldots + (\alpha^{\lambda-1}-1)^n = (-1)^n\lambda$$

ist, für alle Werthe des n, welche kleiner als λ sind, so erhält man:

$$(2.)\qquad l\Phi(x,k) = \lambda\Big\{l\Phi(x) - \frac{k}{1}\,\frac{dl\phi(x)}{dx} + \frac{k^2}{1.\,2.}\,\frac{d^2l\phi(x)}{dx^2} - \ldots + \\ + \frac{k^{\lambda-1}}{1.\,2.\,\ldots(\lambda-1)}\,\frac{d^{\lambda-1}l\phi(x)}{dx^{\lambda-1}}\Big\} + k^\lambda B$$

Entwickelt man nun den Logarithmus von $\phi(x+k(e^v-1))$ in eine nach Potenzen von k geordnete Reihe, und nimmt von dieser den $\lambda-1$ten Differenzialquotienten nach v, für den besonderen Werth $v=1$, so erhält man:

$$\frac{d_0^{\lambda-1}l\phi(x+k(e^v-1))}{dv^{\lambda-1}} = \frac{k}{1}\,\frac{dl\phi(x)}{dx} + \frac{(2^{\lambda-1}-2.\,1^{\lambda-1})k^2}{1.\,2.}\,\frac{d^2l\phi(x)}{dx^2} + \\ \frac{(3^{\lambda-1}-3.\,2^{\lambda-1}+3.\,1^{\lambda-1})k^3}{1.\,2.\,3.}\,\frac{d^3l\phi(x)}{dx^3} + \ldots$$

welche Reihe eine endliche ist, und nur $\lambda-1$ Glieder hat. Die Zahlencoefficienten, welche die $\lambda-1$ten Differenzialquotienten von $(e^v-1)^n$ sind, für $v=0$, sind abwechselnd congruent $+1$ und -1, nach dem Modul λ, man hat daher:

$$(3.)\qquad \frac{d_0^{\lambda-1}l\phi(x+k(e^v-1))}{dv^{\lambda-1}} = \frac{k}{1}\,\frac{dl\phi(x)}{dx} - \frac{k^2}{1.\,2.}\,\frac{d^2l\phi(x)}{dx^2} + \\ + \ldots - \frac{k^{\lambda-1}}{1.\,2.\,\ldots(\lambda-1)}\,\frac{d^{\lambda-1}l\phi(x)}{dx^{\lambda-1}} + \lambda P,$$

wo P eine ganze rationale Funktion von k, des $\lambda-1$ten Grades ist, deren Zahlencoefficienten, insofern sie Brüche sind, in den Nennern kein λ enthalten. Die Vergleichung der beiden Ausdrücke (2.) und (3.) giebt:

$$l\Phi(x,k) = \lambda l\phi(x) - \lambda\,\frac{d_0^{\lambda-1}l\phi(x+k(e^v-1))}{dv^{\lambda-1}} + \lambda^2P + k^\lambda B.$$

Nimmt man nun die Exponentialgröfsen und entwickelt, indem man die Glieder wegläfst, welche λ^2 und die welche k^λ oder höhere Potenzen von k enthalten, so hat man

(4.) $$\Phi(x,k) \equiv \phi(x)^\lambda \left(1 - \lambda \frac{d_0^{\lambda-1} l\phi(x+k(e^v-1))}{dv^{\lambda-1}}\right), \text{ mod. } \lambda^2,$$

und demnach für $k = x$:

(5.) $$\phi(x)\,\phi(x\alpha)\ldots\phi(x\alpha^{\lambda-1}) \equiv \phi(x)^\lambda \left(1 - \frac{d_0^{\lambda-1} l\phi(xe^v)}{dv^{\lambda-1}}\right), \text{ mod. } \lambda^2.$$

Wenn die Coefficienten von $\phi(x)$ ganze Zahlen sind, deren Summe nicht durch λ theilbar ist, so hat man hieraus, wenn man $x = 1$ setzt und durch $\phi(1)$ dividirt:

(6.) $$N\phi(\alpha) \equiv \phi(1)^{\lambda-1}\left(1 - \lambda \frac{d_0^{\lambda-1} l\phi(e^v)}{dv^{\lambda-1}}\right), \text{ mod. } \lambda^2.$$

für jede complexe Zahl in α, welche den Faktor ϱ nicht enthält. Hieraus folgt weiter, weil $\phi(1)^{\lambda-1} \equiv 1$, mod. λ, ist:

(7.) $$\frac{1 - N\phi(\alpha)}{\lambda} \equiv \frac{d_0^{\lambda-1} l\phi(e^v)}{dv^{\lambda-1}} - \frac{\phi(1)^{\lambda-1} - 1}{\lambda}, \text{ mod. } \lambda,$$

welches den gesuchten Ausdruck des oben mit $C_{\lambda-1}$ bezeichneten Charakters durch den Differenzialquotienten des Logarithmus giebt.

Es sei nun $F(w)$ eine wirkliche complexe Zahl in w, und $F(\alpha)$ die Norm derselben, welche nicht durch ϱ theilbar sein soll, also wenn

$$F(w) = A + A_1 w + A_2 w^2 + \ldots + A_{\lambda-1} w^{\lambda-1}$$

gesetzt wird,

$$A + A_1 + A_2 + \ldots + A_{\lambda-1} \text{ nicht} \equiv 0, \text{ mod. } \varrho.$$

Es seien ferner $a, a_1, a_2 \ldots a_{\lambda-1}$ die nichtcomplexen ganzen Zahlen, welchen die complexen Coefficienten $A, A_1, A_2, \ldots A_{\lambda-1}$ congruent sind, nach dem Modul ϱ, und

$$\phi(w) = a + a_1 w + a_2 w^2 + \ldots a_{\lambda-1} w^{\lambda-1},$$

so hat man:

$$F(w) = \phi(w) + \varrho\psi(w),$$

wo $\psi(w)$ irgend eine complexe Zahl in w ist, aus welcher Gleichung ohne Schwierigkeit die Congruenz

(8.) $$NF(w) \equiv N\phi(w), \text{ mod. } \lambda\varrho,$$

gefolgert wird. Entwickelt man nun die Norm von $\phi(w)$, und zwar so, dafs in dieser Entwickelung vorläufig w als eine beliebige unbestimmte Gröfse betrachtet wird, so erhält man, von dem Fermatschen Satze $a^\lambda \equiv a$, mod. λ, Gebrauch machend, folgende Form:

$$* \quad (9.) \qquad N\phi(w) = a + a_1 w^\lambda + a^2 w^{2\lambda} + \dots a_{\lambda-1} w^{(\lambda-1)\lambda} + \lambda R(w^\lambda),$$

wo alle durch λ theilbaren Glieder als $\lambda R(w^\lambda)$ zusammengefafst sind. Bezeichnet man nun mit $g(\alpha)$ die ganze complexe Zahl:

$$g(\alpha) = a + a_1 \alpha + a_2 \alpha^2 + \dots + a^{\lambda-1}\alpha^{\lambda-1},$$

und nimmt $w = 1$, so giebt diese Gleichung:

$$(10.) \qquad g(1)\, Ng(\alpha) \doteq g(1) + \lambda R(1).$$

Giebt man aber dem w in der Gleichung (9.) seinen Werth als Wurzel der Gleichung $w^\lambda = D(\alpha)$, und macht aus derselben eine Congruenz nach dem Modul $\lambda\varrho$, so erhält man, weil $w^\lambda \equiv 1$, mod. ϱ ist, indem man für $N\phi(w)$ den nach dem Modul $\lambda\varrho$ congruenten Werth $NF(w) = F(\alpha)$ setzt:

$$(11.) \qquad \begin{aligned} F(\alpha) \equiv a + a_1 D(\alpha) + a_2 D(\alpha)^2 + \dots + a_{\lambda-1} D(\alpha)^{\lambda-1} + \\ + \lambda R(1), \text{ mod. } \lambda\varrho. \end{aligned}$$

Die Determinante hat nach der Voraussetzung die Eigenschaft, $D(\alpha) \equiv 1$, mod. ϱ, also für $\alpha = 1$, $D(1) \equiv 1$, mod. λ, man kann daher $D(1) = 1$ nehmen. Hierdurch wird vermöge der Congruenz (11.)

$$F(1) \equiv g(1) + \lambda R(1), \text{ mod. } \lambda^2,$$

und wenn zur $\lambda - 1$ten Potenz erhoben wird:

$$F(1)^{\lambda-1} \equiv g(1)^{\lambda-1} - \lambda g(1)^{\lambda-2} R(1), \text{ mod. } \lambda^2,$$

also

$$\frac{F(1)^{\lambda-1} - 1}{\lambda} \equiv \frac{g(1)^{\lambda-1} - 1}{\lambda} - g(1)^{\lambda-2} R(1), \text{ mod. } \lambda,$$

oder wenn für $R(1)$ sein Werth aus der Gleichung (10.) gesetzt wird; weil $g(1)^{\lambda-1} \equiv 1$, mod. λ ist:

$$\frac{F(1)^{\lambda-1} - 1}{\lambda} \equiv \frac{g(1)^{\lambda-1} - 1}{\lambda} + \frac{1 - Ng(\alpha)}{\lambda}, \text{ mod. } \lambda.$$

Vermöge der Congruenz (7.) erhält man hieraus

$$\frac{F(1)^{\lambda-1} - 1}{\lambda} \equiv \frac{d_0^{\lambda-1} lg(e^v)}{dv^{\lambda-1}}, \text{ mod. } \lambda,$$

und weil nach derselben Congruenz (7.)

$$C_{\lambda-1} \equiv \frac{1-NF(\alpha)}{\lambda} \equiv \frac{d_0^{\lambda-1} lF(e^v)}{dv^{\lambda-1}} - \frac{F(1)^{\lambda-1}-1}{\lambda}, \text{ mod. } \lambda,$$

so hat man endlich

$$(12.)\qquad C^1_{\lambda-1} \equiv \frac{d_0^{\lambda-1} lF(e^v)}{dv^{\lambda-1}} - \frac{d_0^{\lambda-1} lg(e^v)}{dv^{\lambda-1}}.$$

Ich bezeichne nun auch diejenigen Differenzialquotienten des Logarithmus von $F(e^v)$, welche nicht als Charaktere auftreten, ebenfalls mit C und dem entsprechenden Index, so dafs

$$C_n \equiv \frac{d_0^n lF(e^v)}{dv^n}, \text{ mod. } \lambda,$$

ist, für alle Werthe des $n = 1, 2, 3, \ldots \lambda - 2$. Ferner bezeichne ich die Differenzialquotienten des Logarithmus der Determinante $D(\alpha)$ mit dem Buchstaben D und dem zugehörigen Index, so dafs

$$D_n \equiv \frac{d_0^n lD(e^v)}{dv^n}, \text{ mod. } \lambda,$$

ist, für alle Werthe des $n = 1, 2, 3, \ldots \lambda - 2$ und

$$D_{\lambda-1} \equiv \frac{1-ND(\alpha)}{\lambda}, \text{ mod. } \lambda,$$

welcher Ausdruck vermöge der Congruenz (7.), da $D(1) = 1$ ist,

$$D_{\lambda-1} \equiv \frac{d_0^{\lambda-1} l D(e^v)}{dv^{\lambda-1}}, \text{ mod. } \lambda,$$

giebt.

Ich mache jetzt von den bekannten Formeln der Differenzialrechnung Gebrauch, welche zeigen, wie die Differenzialquotienten der Funktion einer Funktion gefunden werden. Sei y eine Funktion von x, und x eine Funktion von v, so hat man:

$$\frac{dy}{dv} = \frac{dx}{dv}\frac{dy}{dx}$$

$$(13.)\qquad \frac{d^2y}{dv^2} = \frac{d^2x}{dv^2}\frac{dy}{dx} + \left(\frac{dx}{dv}\right)^2 \frac{d^2y}{dx^2}$$

$$\frac{d^3y}{dx^3} = \frac{d^3x}{dv^3}\frac{dy}{dx} + 3\frac{dx}{dv}\frac{d^2x}{dv^2}\frac{d^2y}{dx^2} + \left(\frac{dx}{dv}\right)^3 \frac{d^3y}{dx^3}$$

und allgemein

$$\frac{d^ny}{dx^n} = V_1 \frac{dy}{dx} + V_2 \frac{d^2y}{dx^2} + V_3 \frac{d^3y}{dx^3} + \ldots + V_n \frac{d^ny}{dx^n}.$$

Die Gröfsen V_1, V_2, ... V_n können nach bekannten Regeln auf combinatorischem Wege für jeden Werth des n gefunden werden. Aus diesen Regeln ergiebt sich auch, was hier von Wichtigkeit ist, dafs wenn $n=\lambda$ eine Primzahl ist, die Ausdrücke V_2, V_3, ... $V_{\lambda-1}$ in allen ihren Gliedern den numerischen Faktor λ haben. Aufserdem ist zu bemerken, dafs für $n=\lambda$, $V_1 = \frac{d^\lambda x}{dv^\lambda}$ und $V_\lambda = \left(\frac{dx}{dv}\right)^\lambda$ ist.

Ich setze nun $y = l\,F(e^v)$ und $x = l\,D(e^v)$. Es ist so y in der That eine Funktion von x, wenn man den Ausdruck aus Congruenz (11.):

$$F(\alpha) \equiv a + a_1 D(\alpha) + a_2 D(\alpha)^2 + \ldots + a_{\lambda-1} D(\alpha)^{\lambda-1}, \text{ mod. } \lambda,$$

anwendet. Für diese Werthe des y und x geben die Gleichungen (13.), wenn nach geschehener Differenziation $v=0$ gesetzt wird, und wenn der Kürze wegen noch

$$\frac{d_0^n\, l g(e^v)}{dv^n} = g_n,$$

für $n = 1, 2, 3, \ldots \lambda - 1$, gesetzt wird:

*
$$\begin{aligned}
C_1 &\equiv D_1 g_1\\
C_2 &\equiv D_2 g_1 + D_1^2 g_2\\
C_3 &\equiv D_3 g_1 + 3D_1 D_2 g_2 + D_1^3 g_3\\
&\;\vdots\\
C_{\lambda-1} + g_{\lambda-1} &\equiv D_{\lambda-1} g_1 + V_2 g_2 + V_3 g_3 + \ldots + D_1^{\lambda-1} g_{\lambda-1}.
\end{aligned}$$

Aus der schon oben festgesetzten Bedingung, dafs $D(\alpha) - 1$ den Faktor ϱ einmal, aber nicht öfter enthalten soll, folgt, dafs $D(e^v) = 1 + (1-e^v)\psi(e^v)$ ist, wo $\psi(e^v)$, für $v=0$, nicht congruent Null ist, man hat daher $D_1 = \frac{d_0 l D(e^v)}{dv}$ offenbar nicht congruent Null, nach dem Modul λ, und folglich $D_1^{\lambda-1} \equiv 1$, mod. λ. Wegen dieser Congruenz fällt die Gröfse $g_{\lambda-1}$ aus der letzten Congruenz von selbst weg und man hat $\lambda - 1$ Congruenzen, aus welchen die $\lambda - 2$ Gröfsen g_1, g_2, g_3 ... $g_{\lambda-2}$ zu eliminiren sind, um die gesuchte Congruenz unter den Gröfsen C_1, C_2, C_3 ... $C_{\lambda-1}$ zu erhalten.

Diese Elimination wird am leichtesten auf folgende Weise ausgeführt. In der allgemeinen Formel (13.) setze man:

$$y = \int l F(e^v)\frac{dx}{dv}\cdot dv, \quad x = l D(e^v),$$

so hat man für $n = \lambda$ und $v = 0$:

$$(15.)\quad \frac{d_0^{\lambda-1}\left(lF(e^v)\frac{dlD(e^v)}{dv}\right)}{dv^{\lambda-1}} = V_1 lF(1) + V_2 g_2 + V_3 g_3 + \ldots + \\ + V_\lambda g_{\lambda-1},$$

entwickelt man nun denselben Differenzialquotienten nach der bekannten Formel für die Differenziation eines Produkts von zwei Faktoren, so erhält man für $v = 0$, indem man von den oben festgesetzten Bezeichnungen der Differenzialquotienten von $lF(e^v)$ und $lD(e^v)$ Gebrauch macht:

$$(16.)\quad \frac{d_0^{\lambda-1}\left(lF(e^v)\frac{dlD(e^v)}{dv}\right)}{dv^{\lambda-1}} = \frac{d_0^\lambda lD(e^v)}{dv^\lambda}\, lF(1) + \frac{\lambda-1}{1} D_{\lambda-1} C_1 + \\ + \frac{(\lambda-1)(\lambda-1)}{1.\,2} D_{\lambda-2} C_2 + \ldots + D_1 (C_{\lambda-1} + g_{\lambda-1}).$$

Vergleicht man nun diese beiden Ausdrücke mit einander, indem man beachtet, dafs

$$V_1 = \frac{d_0^\lambda\, lD(e^v)}{dv^\lambda},\quad V_\lambda = D_1^\lambda,$$

und dafs $V_2, V_3, \ldots V_{\lambda-1}$ alle den Faktor λ enthalten, so erhält man, wenn man die Vielfachen von λ wegläfst, folgende Congruenz:

$$(17.)\quad D_{\lambda-1} C_1 - D_{\lambda-2} C_2 + D_{\lambda-3} C_3 - \ldots - D_1 C_{\lambda-1} \equiv 0, \text{ mod. } \lambda,$$

welches die gesuchte Congruenz ist.

Es soll nun gezeigt werden, dafs diese Congruenz auch die einzige Bedingung enthält, welche diese Gröfsen $C_1, C_2, \ldots C_{\lambda-1}$ erfüllen müssen, damit eine wirkliche Zahl $F(w)$ existire, deren Norm $F(\alpha)$ irgend welche gegebene Werthe der Gröfsen $C_1, C_2, \ldots C_{\lambda-1}$ habe.

Wenn in den ersten $\lambda - 2$ Congruenzen bei (14.) die $\lambda - 2$ Gröfsen $C_1, C_2, \ldots C_{\lambda-2}$ als gegebene betrachtet werden, so lassen sich durch dieselben die $\lambda - 2$ Gröfsen $g_1, g_2, \ldots g_{\lambda-2}$ vollständig bestimmen, weil D_1 nicht congruent Null ist, durch diese Gröfsen $g_1, g_2, \ldots g_{\lambda-2}$ werden nun weiter die Zahlen $a, a_1, a_2, \ldots a_{\lambda-1}$ in folgender Weise bestimmt: Man hat, wenn u irgend eine Funktion von v ist, unter den Differenzialquotienten von u und denen des Logarithmus von u folgende Gleichungen:

Q 2

$$\begin{aligned}
&\frac{du}{dv} = u\frac{dlu}{dv}\\
(18.)\qquad &\frac{d^2u}{dv^2} = \frac{du}{dv}\frac{dlu}{dv} + u\frac{d^2lu}{dv^2}\\
&\frac{d^3u}{dv^3} = \frac{d^2u}{dv^2}\frac{dlu}{dv} + 2\frac{du}{dv}\frac{d^2lu}{dv^2} + u\frac{d^3lu}{dv^3}
\end{aligned}$$

u. s. w.

Nimmt man nun $u = g(e^v)$ und setzt nach geschehener Differenziation $v = 0$, so hat man, indem man von der Bezeichnung der Differenzialquotienten des Logarithmus von $g(e^v)$ durch $g_1, g_2, \dots g_{\lambda-1}$ Gebrauch macht, und außerdem der Kürze wegen

$$\frac{d_0^n g(e^v)}{dv^n} = b_n$$

nimmt, folgende Gleichungen:

$$\begin{aligned}
&b_1 = b_0 g_1\\
&b_2 = b_1 g_1 + b_0 g_2\\
(19.)\quad &b_3 = b_2 g_1 + 2b_1 g_2 + b_0 g_3\\
&\vdots \qquad \vdots\\
&b_{\lambda-2} = b_{\lambda-3} g_1 + \frac{\lambda-3}{1} b_{\lambda-4} g_2 + \frac{(\lambda-3)(\lambda-4)}{1.2} b_{\lambda-5} g_3 + \dots\\
&\qquad + b_0 g_{\lambda-2}.
\end{aligned}$$

Vermittelst dieser Gleichungen, welche man auch bloß als Congruenzen nach dem Modul λ aufzufassen braucht, kann man nun, wenn $g_1, g_2, \dots g_{\lambda-2}$ gegeben sind, die Größen $b_1, b_2, \dots b_{\lambda-2}$ vollständig bestimmen, wobei der Werth des b_0 unbestimmt bleibt. Endlich hat man aus dem Ausdrucke des $g(\alpha)$:

$$g(e^v) = a + a_1 e^v + a_2 e^{2v} + \dots + a_{\lambda-1} e^{(\lambda-1)v}$$

die Congruenzen:

$$\begin{aligned}
&b_0 - a \equiv a_1 + a_2 + a_3 + \dots + a_{\lambda-1}\\
&b_1 \equiv 1a_1 + 2a_2 + 3a_3 + \dots + (\lambda-1)a_{\lambda-1}\\
(20.)\quad &b_2 \equiv 1^2 a_1 + 2^2 a_2 + 3^2 a_3 + \dots + (\lambda-1)^2 a_{\lambda-1}\\
&\vdots \qquad \vdots\\
&b_{\lambda-2} \equiv 1^{\lambda-2} a_1 + 2^{\lambda-2} a_2 + 3^{\lambda-2} a_3 + \dots + (\lambda-1)^{\lambda-2} a_{\lambda-1},
\end{aligned}$$

aus denen, weil die Determinante dieses Systems nicht durch λ theilbar ist, die Zahlen $a_1, a_2, \dots a_{\lambda-1}$ vermittelst der gegebenen $b_1, b_2, \dots b_{\lambda-2}$ bestimmt werden, wobei a zugleich mit b_0 unbestimmt bleibt. Für alle gegebenen Werthe von C_1, C_2, C_3 ... $C_{\lambda-2}$ giebt es darum stets zugehörige Werthe der Zahlen a, a_1, a_2 ... $a_{\lambda-1}$, und demnach auch zugehörige Werthe der A, A_1, A_2 ... $A_{\lambda-1}$, welche nur dadurch bestimmt sind, dafs sie den Zahlen a, a_1, a_2 ... $a_{\lambda-1}$ nach dem Modul ϱ congruent sein müssen. Es giebt also auch wirkliche complexe Zahlen

$$F(w) = A + A_1 w + A_2 w^2 + \dots + A_{\lambda-1} w^{\lambda-1},$$

welchen die Gröfsen C_1, C_2, ... $C_{\lambda-2}$ angehören, für alle Werthe, die man denselben beilegen mag, und wenn die letzte dieser Gröfsen $C_{\lambda-1}$ aus den übrigen durch die Congruenz (17.) bestimmt ist, so gehört auch diese der complexen Zahl $F(w)$ an. Das Resultat dieser ganzen Untersuchung ist in dem folgenden Satze enthalten:

(I.) Wenn eine complexe Zahl $F(\alpha)$ die Norm einer wirklichen complexen Zahl in w, der Determinante $D(\alpha)$ ist, so mufs sie der Congruenz

$$D_{\lambda-1}C_1 - D_{\lambda-2}C_2 + D_{\lambda-3}C_3 - \dots - D_1 C_{\lambda-1} \equiv 0, \quad \text{mod. } \lambda,$$

genügen, in welcher

$$C_n = \frac{d_0^n\, l\, F(e^v)}{dv^n}, \quad D_n = \frac{d_0^n\, l\, D(e^v)}{dv^n},$$

für $n = 1, 2, 3, \lambda - 2$ und

$$C_{\lambda-1} = \frac{1 - NF(\alpha)}{\lambda}, \quad D_{\lambda-1} = \frac{1 - ND(\alpha)}{\lambda},$$

und wenn die Zahlenwerthe der Gröfsen C_1, C_2, C_3, ... $C_{\lambda-1}$ irgend wie gegeben sind, mit der alleinigen Bedingung, dafs sie dieser Congruenz genügen, so kann man stets eine wirkliche complexe Zahl $F(w)$ der Determinante $D(\alpha)$ angeben, deren Norm $F(\alpha)$ diese gegebenen Zahlenwerthe der Gröfsen C_1, C_2, C_3, ... $C_{\lambda-1}$ hat.

Die Congruenz (17.), welche für die Theorie der complexen Zahlen in w von grofser Bedeutung ist, und namentlich dadurch sich auszeichnet, dafs die Norm $F(\alpha)$ und die Determinante $D(\alpha)$ in denselben vollkommen symmetrisch vorkommen, kann auch in eine andere höchst einfache Form gebracht werden, welche ich mit Hülfe eines in meiner Abhandlung über die Ergänzungssätze zu den allgemeinen Reciprocitätsgesetzen,

Crelle's Journal, Bd. 44. bewiesenen Satzes hier entwickeln will. Ich habe daselbst, pag. 144, folgende Formel bewiesen:

$$\text{Ind. } \phi(\alpha) \equiv \text{Ind. } \phi_1(\alpha) + \sum_0^{\mu-1} {}_n \frac{d_0^{2n}\, l\phi(e^v)}{dv^{2n}} \cdot \frac{d_0^{\lambda-2n}\, lf(e^v)}{dv^{\lambda-2n}}, \text{ mod. } \lambda,$$

in welcher $\phi(\alpha)$ eine beliebige complexe Zahl ist, die nur der einen Bedingung unterworfen ist, dafs sie einer nichtcomplexen, von Null verschiedenen Zahl congruent ist, nach dem Modul ϱ^2, und $\phi_1(\alpha)$ dieselbe Zahl darstellt als $\phi(\alpha)$, aber primär genommen, wo ferner das Zeichen des Index: Ind. auf die Primzahl $f(\alpha)$ sich beziehend, durch die Gleichung

$$\left(\frac{\phi(\alpha)}{f(\alpha)}\right) = \alpha^{\text{Ind. } \phi(\alpha)}$$

definirt ist. Es ist nun für den gegenwärtigen Zweck nöthig, die complexe Zahl $\phi(\alpha)$ auch noch von der einen ihr auferlegten Beschränkung zu befreien, dafs sie nach dem Modul ϱ^2 einer nichtcomplexen Zahl congruent sein soll, welches geschieht, indem sie mit einer beliebigen Potenz von α multiplicirt wird. Man hat

$$\text{Ind. } (\alpha^k\, \phi(\alpha)) \equiv \text{Ind. } \phi(\alpha) + k \text{ Ind. } (\alpha),$$

ferner ist

$$\frac{d_0\, l(e^{vk}\, \phi(e^v))}{dv} \equiv k, \quad \frac{d_0^{2n}\, l(e^{vk}\, \phi(e^v))}{dv^{2n}} \equiv \frac{d_0^{2n}\, l\phi(e^v)}{dv^{2n}},$$

und weil nach der Definition des Legendreschen Zeichens oder des Index

$$\text{Ind. } (\alpha) \equiv \frac{Nf(\alpha) - 1}{\lambda}$$

ist, so hat man, wenn anstatt $\alpha^k\, \phi(\alpha)$ einfach $\phi(\alpha)$ geschrieben wird:

$$\text{(21.)} \quad \text{Ind. } \phi(\alpha) \equiv \text{Ind. } \phi_1(\alpha) - \frac{d_0\, l\phi(e^v)}{dv} \left(\frac{1 - Nf(\alpha)}{\lambda}\right) + \\ + \sum_0^{\mu-1} {}_n \frac{d_0^{2n}\, l\phi(e^v)}{dv^{2n}} \frac{d_0^{\lambda-2n}\, lf(e^v)}{dv^{\lambda-2n}},$$

in welcher Formel $\phi(\alpha)$ eine jede beliebige (nicht durch ϱ theilbare) complexe Zahl ist, und $\phi_1(\alpha)$ dieselbe in der primären Form.

Verallgemeinert man nun das Legendresche Zeichen in der Art, dafs es auch auf zusammengesetzte Moduln anwendbar ist, indem man definirt, dafs wenn $\psi(\alpha) = E(\alpha)\, f(\alpha)^m\, f_1(\alpha)^{m_1}\, f_2(\alpha)^{m_2} \ldots$ ist, wo $E(\alpha)$ eine Einheit bezeichnet,

$$\left(\frac{\phi(\alpha)}{\psi(\alpha)}\right) = \left(\frac{\phi(\alpha)}{f(\alpha)}\right)^{m} \left(\frac{\phi(\alpha)}{f_1(\alpha)}\right)^{m_1} \left(\frac{\phi(\alpha)}{f_2(\alpha)}\right)^{m_2} \cdots$$

sein soll, und verallgemeinert man demgemäſs auch das Zeichen des Index Ind., so daſs es auch für die zusammengesetzte Zahl $\psi(\alpha)$ durch die Gleichung

$$\left(\frac{\phi(\alpha)}{\psi(\alpha)}\right) = \alpha^{\text{Ind. } \phi(\alpha)}$$

definirt ist, und bemerkt, daſs für den angegebenen Werth des $\psi(\alpha)$

$$\frac{d_0^n l\psi(e^v)}{dv^n} \equiv m \frac{d_0^n lf(e^v)}{dv^n} + m_1 \frac{d_0^n lf_1(e^v)}{dv^n} + \cdots\cdots$$

$$\frac{1-N\psi(\alpha)}{\lambda} \equiv m \frac{1-Nf(\alpha)}{\lambda} + m_1 \frac{1-Nf_1(\alpha)}{\lambda} + \cdots\cdots$$

so erhält man die allgemeinere, für alle beliebigen, nicht durch ϱ theilbaren Zahlen $\phi(\alpha)$ und $\psi(\alpha)$ geltende Congruenz:

(22.) $$\text{Ind. } \phi(\alpha) \equiv \text{Ind. } \phi_1(\alpha) - \frac{d_0 l\phi(e^v)}{dv} \cdot \frac{1-N\psi(\alpha)}{\lambda} + \\ + \sum_{1}^{\mu-1}{}_n \frac{d_0^{2n} l\phi(e^v)}{dv^{2n}} \cdot \frac{d_0^{\lambda-2n} l\psi(e^v)}{dv^{\lambda-2n}},$$

welche auch so dargestellt werden kann:

(23.) $$\left(\frac{\phi(\alpha)}{\psi(\alpha)}\right) = \left(\frac{\phi_1(\alpha)}{\psi(\alpha)}\right) \alpha^{\Sigma},$$

wo

$$\Sigma \equiv - \frac{d_0 l\phi(e^v)}{dv} \cdot \frac{1-N\psi(\alpha)}{\lambda} + \\ + \sum_{1}^{\mu-1}{}_n \frac{d_0^{2n} l\phi(e^v)}{dv^{2n}} \cdot \frac{d_0^{\lambda-2n} l\psi(e^v)}{dv^{\lambda-2n}}, \quad \text{mod. } \lambda.$$

Diese Formel giebt auch einen neuen Ausdruck für den Index einer beliebigen Einheit $E(\alpha)$. Nimmt man nämlich in derselben $\phi_1(\alpha) = 1$, so ist $\phi(\alpha)$ eine ganz beliebige Einheit $E(\alpha)$, und man hat:

(24.) $$\text{Ind. } E(\alpha) \equiv - \frac{d_0 l E(e^v)}{dv} \cdot \frac{1-N\psi(\alpha)}{\lambda} + \\ + \sum_{1}^{\mu-1}{}_n \frac{d_0^{2n} l E(e^v)}{dv^{2n}} \cdot \frac{d_0^{\lambda-2n} l\psi(e^v)}{dv^{\lambda-2n}},$$

oder was dasselbe ist:

$$(25.)\quad \left(\frac{E(\alpha)}{\psi(\alpha)}\right) = \alpha^{\frac{-d_0 l E(e^v)}{dv} \cdot \frac{1 - N\psi(\alpha)}{\lambda} + \sum_{1}^{\mu-1}{}_n \frac{d_0^{2n} l E(e^v)}{dv^{2n}} \cdot \frac{d_0^{\lambda-2n} l \psi(e^v)}{dv^{\lambda-2n}}}$$

Um nun die gefundene Formel (22.) auf die Umformung der Congruenz (17.) anzuwenden, nehme ich zunächst $\phi(\alpha) = F(\alpha)$, $\psi(\alpha) = D(\alpha)$, und setze $F(\alpha) = E(\alpha)\,F_1(\alpha)$, wo $F_1(\alpha)$ primär sein soll, und $E(\alpha)$ eine Einheit ist. Für diese Zahlen $F(\alpha)$ und $D(\alpha)$ giebt die Congruenz (22.) unter Anwendung der eingeführten Zeichen für die Differenzialquotienten der Logarithmen der betreffenden complexen Zahlen:

$$(26.)\quad \text{ind. } E(\alpha) \equiv -C_1 D_{\lambda-1} + C_2 D_{\lambda-2} + C_4 D_{\lambda-4} + \dots + C_{\lambda-3} D_3,$$

wo das Zeichen des Index: ind. auf den Modul $D(\alpha)$ sich bezieht. Setzt man zweitens in der Formel (22.) $\phi(\alpha) = D(\alpha)$, $\psi(\alpha) = F(\alpha)$ und $D(\alpha) = \varepsilon(\alpha) D_1(\alpha)$, wo $D_1(\alpha)$ primär sein soll, und $\varepsilon(\alpha)$ diejenige Einheit, welche aus $D(\alpha)$ herausgehoben werden mufs, damit es primär werde, so hat man:

$$(27.)\quad \text{Ind. } \varepsilon(\alpha) \equiv -D_1 C_{\lambda-1} + D_2 C_{\lambda-2} + D_4 C_{\lambda-4} + \dots + D_{\lambda-3} C_3,$$

wo der Index: Ind. auf den Modul $F(\alpha)$ sich bezieht. Vergleicht man endlich die beiden Ausdrücke des ind. $E(\alpha)$ und des Ind. $\varepsilon(\alpha)$ mit der Congruenz (17.), so hat man die gesuchte Umformung derselben, nämlich:

$$(28.)\quad \text{ind. } E(\alpha) \equiv \text{Ind. } \varepsilon(\alpha), \quad \text{mod. } \lambda,$$

welche in den Legendreschen Zeichen so ausgedrückt wird:

$$(29.)\quad \left(\frac{E(\alpha)}{D(\alpha)}\right) = \left(\frac{\varepsilon(\alpha)}{F(\alpha)}\right).$$

Dieses Resultat kann nun vollständig so ausgesprochen werden:

(II.) Die Congruenz, welcher eine jede complexe Zahl $F(\alpha)$ genügen mufs, wenn sie als Norm einer wirklichen complexen Zahl in w der Determinante $D(\alpha)$ darstellbar ist, nämlich:

$$D_{\lambda-1} C_1 - D_{\lambda-2} C_2 + D_{\lambda-3} C_3 - \dots - D_1 C_{\lambda-1} \equiv 0, \quad \text{mod. } \lambda,$$

ist gleichbedeutend mit der Gleichung:

$$\left(\frac{E(\alpha)}{D(\alpha)}\right) = \left(\frac{\varepsilon(\alpha)}{F(\alpha)}\right),$$

in welcher $E(\alpha)$ und $\varepsilon(\alpha)$ diejenigen Einheiten bezeichnen, welche durch die Gleichungen

$$F(\alpha) = E(\alpha)\,F_1(\alpha), \quad D(\alpha) = \varepsilon(\alpha)\,D_1(\alpha)$$

bestimmt werden, in denen $F_1(\alpha)$ und $D_1(\alpha)$ primär sind.

§. 15.

Sätze über die genaue Anzahl der wirklich vorhandenen Gattungen der idealen Zahlen in z.

In den jetzt folgenden Untersuchungen wird eine besondere Art von complexen Primzahlen in α auftreten, welche in einigen allgemeinen Sätzen Ausnahmen begründen, oder doch eine besondere Behandlung erfordern, nämlich die complexen Primzahlen, welche die besondere Eigenschaft haben, daſs alle Einheiten für sie λte Potenzreste sind. Wenn $\psi(\alpha)$ eine complexe Primzahl dieser besonderen Art ist, so zeigt der bei (24.) §. 14. gegebene Ausdruck des Index einer beliebigen Einheit, daſs für dieselbe die Ausdrücke

$$\frac{d_0^3 l\psi(e^v)}{dv^3}, \quad \frac{d_0^5 l\psi(e^v)}{dv^5}, \quad \dots \quad \frac{d_0^{\lambda-2} l\psi(e^v)}{dv^{\lambda-2}}, \quad \frac{1-N\psi(\alpha)}{\lambda},$$

alle einzeln congruent Null sein müssen, nach dem Modul λ, und umgekehrt, wenn diese Ausdrücke alle congruent Null sind, daſs jede beliebige Einheit $E(\alpha)$ ein λter Potenzrest von $\psi(\alpha)$ ist. In Rücksicht auf dieses Verhalten gegen die Einheiten unterscheide ich darum zwei verschiedene Arten von complexen Primzahlen, und nenne diejenigen complexen Primzahlen in α, welche die besondere Eigenschaft haben, daſs für sie alle Einheiten in α λte Potenzreste sind: complexe Primzahlen der zweiten Art, diejenigen aber, welche diese besondere Eigenschaft nicht haben, complexe Primzahlen der ersten Art.

Ich bemerke, daſs in der Theorie der quadratischen Reste, wo es sich nur um gewöhnliche ganze Zahlen handelt, den hier als Primzahlen der ersten Art bezeichneten die Primzahlen von der Form $4n+3$ entsprechen, dagegen den als Primzahlen der zweiten Art bezeichneten die Primzahlen von der Form $4n+1$. Es giebt nämlich hier nur die beiden Einheiten $+1$ und -1, und diese sind für die Primzahlen der Form $4n+1$ beide quadratische Reste, für die Primzahlen der Form $4n+3$ aber ist -1 ein Nichtrest.

Ich wende nun den im vorigen Paragraphen gefundenen Satz (I.) auf die Bestimmung der genauen Anzahl der Gattungen der idealen Zahlen in z

an, und zwar zunächst für den Fall, daſs die Determinante $D(\alpha)$ nur einen Primfaktor enthält. Wenn die als Charaktere bezeichneten Gröſsen

$$C_3, C_5, C_7, \ldots C_{\lambda-2}, C_{\lambda-1},$$

deren jede in Beziehung auf den Modul λ die Werthe $0, 1, 2, \ldots \lambda-1$ haben kann, irgend wie gegeben sind, so kann man, da noch die Gröſsen $C_1, C_2, C_4, \ldots C_{\lambda-3}$ verfügbar bleiben, diese im allgemeinen so wählen, daſs der Congruenz

$$D_{\lambda-1}C_1 - D_{\lambda-2}C_2 + D_{\lambda-3}C_3 - \ldots - D_1 C_{\lambda-1} \equiv 0, \text{ mod. } \lambda,$$

genügt wird. Nur in dem einen besonderen Falle wird dieſs nicht möglich sein, wenn von den Gröſsen $C_1, C_2, C_4, \ldots C_{\lambda-3}$ keine in dieser Congruenz vorkommt, also nur dann, wenn die Determinante $D(\alpha)$ so beschaffen ist, daſs für dieselbe die Gröſsen

$$D_3, D_5, D_7, \ldots D_{\lambda-2}, D_{\lambda-1}$$

alle einzeln congruent Null sind. In diesem Falle können die genannten Charaktere einer idealen Zahl in z, welche sich als wirkliche complexe Zahl in w darstellen läſst, nicht mehr alle beliebigen Werthe haben, sondern nur diejenigen Werthe, welche der Congruenz

$$(1.)\qquad D_{\lambda-3}C_3 + D_{\lambda-5}C_5 + \ldots + D_2 C_{\lambda-2} - D_1 C_{\lambda-1} \equiv 0, \text{ mod. } \lambda,$$

genügen. Für alle anderen Determinanten $D(\alpha)$ aber, welche nicht diese besondere Eigenschaft haben, kann man, wie auch die Charaktere $C_3, C_5, \ldots C_{\lambda-2}, C_{\lambda-1}$ gegeben sein mögen, stets wirkliche complexe Zahlen in w finden, welche, als ideale Zahlen in z betrachtet, diese gegebenen Charaktere haben. Da die Anzahl dieser Einzelcharaktere gleich $\mu = \frac{\lambda-1}{2}$ ist, und jeder die λ Werthe $0, 1, 2 \ldots \lambda-1$ haben kann, so folgt, daſs die Anzahl aller Werthverbindungen derselben gleich λ^μ ist, daſs also λ^μ Gattungen wirklich existiren, da man zu jeder derselben ideale Zahlen angeben kann, welche sie enthält. Da ferner nach dem Satze (IV.) §. 13, in dem vorliegenden Falle, wo die Determinante nicht mehr als einen Primfaktor enthält, nicht mehr als λ^μ Gattungen existiren, so folgt, daſs der mit K bezeichnete Charakter durch die Charaktere $C_3, C_5, \ldots C_{\lambda-1}$ vollständig bestimmt ist, so daſs alle idealen Zahlen in z, welche dieselben Werthe dieser Charaktere haben, auch denselben Werth des Charakters K haben müssen. Der Ausnahmefall, in welchem $D_3, D_5, \ldots D_{\lambda-1}$ alle congruent Null sind, mod. λ, tritt nur dann, und immer dann ein, wenn die eine in der Determinante $D(\alpha)$ enthaltene Primzahl eine complexe Primzahl der zweiten Art ist, denn diese Bedingungen stimmen mit den oben für die Primzahlen der zweiten Art angegebenen vollständig überein. Man hat also folgende zwei Sätze:

(I.) Wenn die Determinante nur einen Primfaktor enthält, und zwar eine Primzahl der ersten Art, so ist die Anzahl der wirklich vorhandenen Gattungen der idealen Zahlen in z genau gleich λ^μ, also genau gleich dem λten Theile aller angebbaren Gesamtcharaktere, und der Charakter K ist durch die Charaktere C_3, C_5, ... $C_{\lambda-1}$ vollständig bestimmt.

(II.) Unter derselben Voraussetzung enthält jede wirklich vorhandene Gattung solche ideale Zahlen in z, welche sich als wirkliche complexe Zahlen in w darstellen lassen.

Es soll nun weiter auch in dem Falle, wo die Determinante zwei verschiedene Primfaktoren enthält, untersucht werden, in wie weit es gelingt, die $\lambda^{\mu+1}$ Gattungen, welche nach dem Satze (IV.), §. 13. noch Statt haben können, durch solche ideale Zahlen in z, welche sich als wirkliche complexe Zahlen in u, u_1 darstellen lassen, vollständig auszufüllen, und dadurch nachzuweisen, daſs die Anzahl der wirklich vorhandenen Gattungen in diesem Falle genau gleich $\lambda^{\mu+1}$ ist. Es sei nach der im §. 9. eingeführten Bezeichnung

$$u^\lambda = e(\alpha)f(\alpha)^m, \qquad u_1^\lambda = e_1(\alpha)f_1(\alpha)^{m_1}$$

$$\text{(2.)} \qquad D(\alpha) = u^\lambda u_1^\lambda, \qquad w = u u_1$$

$$F(u, u_1) = \sum_0^{\lambda-1}{}_k\, A_k\, u^{|k-n|}\, u_1^{|k-n_1|}$$

wo $|k-n|$ und $|k-n_1|$ die kleinsten nicht negativen Reste von $k-n$ und $k-n_1$, nach dem Modul λ, bezeichnen. Es sei ferner:

$$u^n u_1^{n_1} F(u, u_1) = F'(w)$$

$$\text{(3.)} \qquad NF'(w) = F'(\alpha), \qquad NF(u, u_1) = F(\alpha),$$

$$e(\alpha)f(\alpha)^m = d(\alpha), \qquad e_1(\alpha)f_1(\alpha)^{m_1} = \delta(\alpha),$$

R 2

so ist

$$(4.)\qquad F'(\alpha) = d(\alpha)^{n}\,\delta(\alpha)^{n_1}\,F(\alpha).$$

Weil nun $F'(w)$ eine wirkliche complexe Zahl in w ist, deren Norm den Faktor ϱ nicht enthält, so findet für dieselbe die Congruenz des Satzes (I.) §. 14. Statt:

$$(5.)\quad D_{\lambda-1}C'_1 - D_{\lambda-2}C'_2 + D_{\lambda-3}C'_3 - \dots - D_1 C'_{\lambda-1} \equiv 0, \text{ mod. } \lambda,$$

in welcher

$$C'_k = \frac{d^k_0\, l F'(e^v)}{dv^k},\quad C'_{\lambda-1} = \frac{1-NF'(\alpha)}{\lambda}$$

gesetzt ist. Bezeichnet man dem entsprechend die Differenzialquotienten der Logarithmen für die complexen Zahlen $d(\alpha)$ und $\delta(\alpha)$ durch

$$d_k = \frac{d^k_0\, l d(e^v)}{dv^k},\quad d_{\lambda-1} = \frac{1-Nd(\alpha)}{\lambda},$$

$$\delta_k = \frac{d^k_0\, l\delta(e^v)}{dv^k},\quad \delta_{\lambda-1} = \frac{1-N\delta(\alpha)}{\lambda},$$

so hat man vermöge der Gleichung (4.):

$$(6.)\qquad C'_k \equiv n d_k + n_1\delta_k + C_k,\quad \text{mod. } \lambda,$$

und vermöge der Gleichung $D(\alpha) = d(\alpha)\,\delta(\alpha)$:

$$(7.)\qquad D_k \equiv d_k + \delta_k,\quad \text{mod. } \lambda,$$

für alle Werthe $k = 1, 2, 3, \dots \lambda-1$. Setzt man diese Ausdrücke in die Congruenz (5.) ein, indem man der Einfachheit wegen die Summenzeichen gebraucht, so hat man

$$(8.)\qquad \sum_{1}^{\lambda-1}{}_k\, (-1)^k\,(d_{\lambda-k} + \delta_{\lambda-k})\,(C_k + nd_k + n_1\delta_k) \equiv 0, \text{ mod. } \lambda,$$

und wenn die Multiplikation unter dem Summenzeichen ausgeführt wird:

$$\Sigma\,(-1)^k\,(d_{\lambda-k} + \delta_{\lambda-k})\,C_k + n\Sigma(-1)^k\,d_{\lambda-k}d_k + n\Sigma(-1)^k\,\delta_{\lambda-k}d_k$$
$$+ n_1\Sigma(-1)^k\,d_{\lambda-k}\delta_k + n_1\Sigma(-1)^k\,\delta_{\lambda-k}\delta_k \equiv 0,\quad \text{mod. } \lambda.$$

Die zweite und fünfte dieser Summen verschwinden von selbst, weil in ihnen die vom Anfange und vom Ende gleich weit entfernten Glieder gleich sind und entgegengesetzte Vorzeichen haben, die vierte Summe aber ist gleich der dritten mit entgegengesetzten Vorzeichen, wie man sogleich sieht, wenn man in derselben $k-\lambda$ statt k setzt. Man hat daher folgende Congruenz, welcher alle complexen Zahlen $F(\alpha)$ genügen müssen, die sich als Normen von wirklichen complexen Zahlen in u, u_1 darstellen lassen:

$$(9.)\quad \Sigma(-1)^k\,(d_{\lambda-k}+\delta_{\lambda-k})\,C_k+(n-n_1)\,\Sigma(-1)^k\,\delta_{\lambda-k}\,d_k\equiv 0,\ \text{mod. }\lambda.$$

Um diese Congruenz in einer übersichtlicheren Form darzustellen, führe ich folgende abgekürzte Bezeichnungen der auch in dem Folgenden mehrfach wiederkehrenden Ausdrücke ein:

$$(10.)\quad \begin{aligned} mS &\equiv -d_{\lambda-1}C_1+d_{\lambda-2}C_2+d_{\lambda-4}C_4+\ldots+d_3C_{\lambda-3},\\ m_1S_1 &\equiv -\delta_{\lambda-1}C_1+\delta_{\lambda-2}C_2+\delta_{\lambda-4}C_4+\ldots+\delta_3C_{\lambda-3},\\ ms &\equiv -d_{\lambda-1}\delta_1+d_{\lambda-2}\delta_2+d_{\lambda-4}\,{}_4\delta+\ldots+d_3\delta_{\lambda-3},\quad \text{mod.}\lambda.\\ m_1s_1 &\equiv -\delta_{\lambda-1}d_1+\delta_{\lambda-2}d_2+\delta_{\lambda-4}d_4+\ldots+\delta_3d_{\lambda-3},\\ T &\equiv -(d_1+\delta_1)\,C_{\lambda-1}+(d_2+\delta_2)\,C_{\lambda-2}+(d_4+\delta_4)\,C_{\lambda-4}+\\ &\quad +\ldots+(d_{\lambda-3}+\delta_{\lambda-3})\,C_3. \end{aligned}$$

Vermittelst dieser Zeichen stellt sich die gefundene Congruenz (9.) in folgender Weise dar:

$$(11.)\quad T-mS-m_1S_1+(n-n_1)\,(ms-m_1s_1)\equiv 0,\quad \text{mod. }\lambda.$$

Es ist nun auch umgekehrt zu zeigen, wenn die Werthe der Gröfsen C_1, C_2, C_3, C_4, ... $C_{\lambda-1}$ und der Zahlen n und n_1 irgend wie gegeben sind, mit der einzigen Bedingung, dafs sie dieser Congruenz (11.) genügen, dafs man stets eine wirkliche complexe Zahl in u, u_1, $F(u, u_1)$ finden kann, welcher diese Werthe angehören. Man kann zunächst, wie oben gezeigt worden ist, immer wirkliche complexe Zahlen $F'(w)$ finden, denen die Werthe C'_1, C'_2, C'_3, ... $C'_{\lambda-1}$ angehören, wenn dieselben irgend wie so gegeben sind, dafs sie der Congruenz (5.) genügen. Ist nun $F'(w)$ eine solche passende Zahl, so ist allgemein $F'(w)+\varrho\psi(w)$ eine complexe Zahl, welche dieselben Werthe der Gröfsen C'_1, C'_2, C'_3, ... $C'_{\lambda-1}$ giebt, wo $\psi(w)$ vollständig beliebig ist; denn es ist die Norm von $F'(w)+\varrho\psi(w)$ der Norm von $F'(w)$ congruent, nach dem Modul $\lambda\varrho$. Die beliebige Zahl $\psi(w)$ kann nun immer so bestimmt werden, dafs in der complexen Zahl $F'(w)+\varrho\psi(w)$ die ersten n Coefficienten durch $u^\lambda=d(\alpha)$ theilbar sind, der $n+1$te aber nicht theilbar durch $f(\alpha)$, und dafs die ersten n_1 Coefficienten durch $u_1^\lambda=\delta(\alpha)$ theilbar sind, der n_1+1te aber nicht theilbar durch $f_1(\alpha)$. Hierdurch wird aber

$$F'(w)+\varrho\psi(w)=u^{n}\,u_1^{n_1}\,F(u, u_1),$$

und die Congruenz (5.) geht in die Congruenz (11.) über, welche letztere darum die nothwendige und hinreichende Bedingung giebt, dafs wirkliche complexe Zahlen $F(u, u_1)$ existiren, denen gegebene Werthe von C_1, C_2, C_3, ... $C_{\lambda-1}$, n und n_1 angehören.

Ich ziehe nun auch einen der beiden mit K und K_1 bezeichneten Charaktere mit in Betracht, welche in dem vorliegenden Falle Statt haben, wo die Determinante die beiden verschiedenen Primfaktoren $f(\alpha)$ und $f_1(\alpha)$ enthält. Der Charakter K ist definirt durch die Gleichung

$$(12.)\qquad \left(\frac{F_1(\alpha)}{f(\alpha)}\right) = \alpha^K,$$

in welcher $F_1(\alpha)$ die complexe Zahl $F(\alpha) = NF(u, u_1)$ in ihrer primären Form bezeichnet. Da nämlich die Charaktere für die Normen der idealen Zahlen in z Statt haben, von welchen festgesetzt worden ist, dafs sie stets in der primären Form genommen werden sollen, so ist in dieser Bestimmung des Charakters K die Norm der idealen Zahl in z nicht in der, gewöhnlich nicht primären Form zu nehmen, in welcher sie als Norm der wirklichen Zahl $F(u, u_1)$ erhalten wird, sondern auf die primäre Form zu bringen.

Aus der bei (2.) gegebenen Form der complexen Zahl $F(u, u_1)$, in welcher das $n+1$te Glied: $A_n\, u_1^{|n-n_1|}$ das einzige ist, welches u nicht enthält, folgt nun, dafs in der Norm $NF(u, u_1)$ das Glied $A_n^\lambda\, u^{|n-n_1|\lambda}$ das einzige sein mufs, welches u^λ nicht enthält, man hat daher die Congruenz:

$$F(\alpha) \equiv A_n^\lambda\, \delta(\alpha)^{|n-n_1|}, \quad \text{mod. } d(\alpha),$$

welche folgende Gleichung giebt:

$$(13.)\qquad \left(\frac{F(\alpha)}{d(\alpha)}\right) = \left(\frac{\delta(\alpha)}{d(\alpha)}\right)^{n-n_1}.$$

Man kann nun, weil $d(\alpha) = e(\alpha) f(\alpha)^m$ ist, die Gleichung (12.) auch in folgende Form setzen:

$$(14.)\qquad \left(\frac{F_1(\alpha)}{d(\alpha)}\right) = \alpha^{mK},$$

und erhält so, indem man die in diesen beiden Gleichungen enthaltenen Legendreschen Zeichen für das nicht primäre $F(\alpha)$ und für das zugehörige primäre $F_1(\alpha)$ nach der Formel (23.) §. 14. aufeinander zurückführt:

$$\left(\frac{\delta(\alpha)}{d(\alpha)}\right)^{n-n_1} = \alpha^{mK - d_{\lambda-1}C_1 + d_{\lambda-2}C_2 + d_{\lambda-4}C_4 + \,..\, + d_3 C_{\lambda-3}}$$

und wenn man von dem bei (10.) angegebenen abgekürzten Zeichen Gebrauch macht:

$$(15.)\qquad \left(\frac{\delta(\alpha)}{d(\alpha)}\right)^{n-n_1} = \alpha^{mK+mS}.$$

Ich nehme nun die in $\delta(\alpha)$ enthaltene Primzahl $f_1(\alpha)$ primär, also auch $f_1(\alpha)^{m_1}$ primär, und drücke wieder nach der Formel (23.) §. 14. das Legendresche Zeichen der Gleichung (15.) für das nichtprimäre $\delta(\alpha)$ durch das entsprechende Zeichen für das zugehörige primäre $f_1(\alpha)^{m_1}$ aus, so ist:

$$\left(\frac{\delta(\alpha)}{d(\alpha)}\right) = \left(\frac{f_1(\alpha)}{d(\alpha)}\right)^{m_1} \alpha^{-d_{\lambda-1}\delta_1 + d_{\lambda-2}\delta_2 + d_{\lambda-4}\delta_4 + \dots + d_3\delta_{\lambda-3}}$$

und wenn von dem bei (10.) angegebenen abgekürzten Zeichen Gebrauch gemacht wird:

$$(16.)\qquad \left(\frac{\delta(\alpha)}{d(\alpha)}\right) = \left(\frac{f_1(\alpha)}{d(\alpha)}\right)^{m_1} \alpha^{ms}.$$

Weil nun $d(\alpha) = e(\alpha)f(\alpha)^m$ ist, so ist

$$\left(\frac{f_1(\alpha)}{d(\alpha)}\right) = \left(\frac{f_1(\alpha)}{f(\alpha)}\right)^m,$$

und wenn

$$\left(\frac{f_1(\alpha)}{f(\alpha)}\right) = \alpha^i$$

gesetzt wird, so erhält man aus den Gleichungen (15.) und (16.) die Congruenz

$$(17.)\qquad K + S \equiv (n-n_1)(m_1 i + s), \text{ mod. } \lambda.$$

Vermittelst der beiden Congruenzen (11.) und (17.) kann nun die Frage: ob es möglich ist für jede der $\lambda^{\mu+1}$ Werthverbindungen, welche die Charaktere C_3, C_5, ... $C_{\lambda-1}$ und K haben können, eine ideale Zahl in z zu finden, welche sich als wirkliche complexe Zahl in u, u_1 darstellen läſst, vollständig gelöst werden; denn wenn für alle möglichen gegebenen Werthe der Charaktere C_3, C_5, ... $C_{\lambda-1}$ und K die verfügbar bleibenden Werthe der Gröſsen C_1, C_2, C_4, ... $C_{\lambda-3}$ und der Zahl $n-n_1$ sich so bestimmen lassen, daſs diesen beiden Congruenzen genügt wird, und auch nur unter dieser Bedingung, giebt es für alle Werthverbindungen dieser Charaktere auch wirkliche complexe Zahlen in u, u_1. Bei der groſsen Anzahl der verfügbar bleibenden Gröſsen, nämlich $\mu+1$, mittelst deren man nur zwei Congruenzen zu genügen hat, ist klar, dafs die Aufgabe im Allgemeinen immer lösbar sein wird, und dafs wieder nur gewisse besondere Werthe der Determinante $D(\alpha)$ Ausnahmen begründen können. Für den Zweck der gegenwärtigen Abhandlung ist es nicht erforderlich, diese Ausnahmen einzeln zu erörtern, es reicht vielmehr hin, nur den einen Fall vollständig zu untersuchen, wo von den beiden in der Determinante enthaltenen Primzahlen $f(\alpha)$ und $f_1(\alpha)$ in Beziehung auf die Einheiten die eine der ersten Art, die andere der zweiten Art angehört, uud wo die Primzahl der ersten Art $f_1(\alpha)$ Nichtrest der Primzahl der zweiten Art $f(\alpha)$ ist.

In diesem besonderen Falle hat man, weil in Beziehung auf $f(\alpha)$ und darum auch in Beziehung auf $d(\alpha)$ alle Einheiten λte Potenzreste sind:

$$d_3 \equiv 0, \quad d_5 \equiv 0, \quad \ldots d_{\lambda-2} \equiv 0, \quad d_{\lambda-1} \equiv 0, \quad \text{mod. } \lambda,$$

und darum

$$S \equiv 0 \quad \text{und} \quad s \equiv 0, \quad \text{mod. } \lambda,$$

die beiden Congruenzen (11.) und (17.) nehmen daher folgende einfachere Formen an:

$$\begin{aligned} &(18.) \qquad T - m_1 S_1 - (n-n_1) m_1 s_1 \equiv 0, \\ &(19.) \qquad K - (n-n_1) m_1 i \equiv 0, \end{aligned} \quad \text{mod. } \lambda.$$

Weil nach der Voraussetzung $f_1(\alpha)$ Nichtrest von $f(\alpha)$ ist, so ist i nicht $\equiv 0$, mod. λ, und da auch m_1 nicht durch λ theilbar ist, so folgt, dafs der Congruenz (19.) durch passende Bestimmung der Zahl $n - n_1$ immer genügt werden kann. In der Congruenz (18.) kommen nun die noch verfügbaren Gröfsen $C_1, C_2, C_4, \ldots C_{\lambda-3}$ alle nur lineär vor, sie wird also durch dieselben stets erfüllt werden können, wenn diese nicht alle aus der Congruenz dadurch verschwinden, dafs die Gröfsen, mit denen sie multiplicirt sind, alle einzeln congruent Null sind. Diese Gröfsen, mit denen sie behaftet vorkommen, sind folgende: $d_3 + \delta_3$, $d_5 + \delta_5$, ... $d_{\lambda-2} + \delta_{\lambda-2}$, $d_{\lambda-1} + \delta_{\lambda-1}$ oder weil $d_3, d_5, \ldots\ldots d_{\lambda-2}, d_{\lambda-1}$ alle congruent Null sind, $\delta_3, \delta_5, \ldots\ldots \delta_{\lambda-2}, \delta_{\lambda-1}$, und da $f_1(\alpha)$ nach der Voraussetzung eine Primzahl der ersten Art ist, so sind dieselben nicht alle congruent Null, es kann also auch der Congruenz (18.) immer genügt werden. Man kann also zu jeder der $\lambda^{\mu+1}$ Werthverbindungen, welche die Charaktere $C_3, C_5, \ldots C_{\lambda-1}$ und K haben können, zugehörige ideale Zahlen in z angeben, und zwar solche, welche als wirkliche complexe Zahlen in u, u_1 darstellbar sind. Es existiren also $\lambda^{\mu+1}$ Gattungen der idealen Zahlen in z als wirklich vorhandene, und da nach dem Satze (IV.), §. 13. in dem gegenwärtigen Falle, wo die Determinante zwei verschiedene Primzahlen enthält, nicht mehr als $\lambda^{\mu+1}$ wirklich vorhanden sein können, so folgt, dafs diefs die genaue Anzahl derselben ist. Man hat demnach folgende zwei Sätze:

(III.) Wenn die Determinante zwei verschiedene Primfaktoren enthält, und zwar einen der ersten Art $f_1(\alpha)$ und einen der zweiten Art $f(\alpha)$, und wenn $f_1(\alpha)$ in der primären Form ein Nichtrest von $f(\alpha)$ ist, so ist die Anzahl der wirklich vorhandenen Gattungen der idealen Zahlen in z genau gleich $\lambda^{\mu+1}$, also genau gleich dem λten Theile aller angebbaren Gesamtcharaktere, und der Charakter K_1 ist durch die Charaktere C_3, C_5, ... $C_{\lambda-1}$ und K vollständig bestimmt.

(IV.) Unter denselben Voraussetzungen enthält jede wirklich vorhandene Gattung solche ideale Zahlen in z, welche sich als wirkliche complexe Zahlen in u, u_1 darstellen lassen.

§. 16.

Allgemeine Bestimmung der genauen Anzahl der wirklich vorhandenen Gattungen für die idealen Zahlen in z.

Die vollständige Erledignng der Frage nach der wahren Anzahl der wirklich vorhandenen Gattungen der idealen Zahlen in z, namentlich auch für diejenigen Determinanten, welche in den im vorigen Paragraphen gegebenen Sätzen Ausnahmen begründen, kann mit den daselbst angewandten Mitteln, nämlich mit Hülfe derjenigen idealen Zahlen in z, welche sich als wirkliche complexe Zahlen in w oder u, u_1, u_2 ... darstellen lassen, nicht geleistet werden; weil für gewisse Determinanten in der That solche Gattungen existiren, denen keine, als wirkliche Zahl in w oder in u, u_1, u_2 ... darstellbare, ideale Zahl in z angehört. Die wahre Anzahl der wirklich vorhandenen Gattungen ist auch in diesen Fällen immer genau gleich dem λten Theile aller angebbaren Gesamtcharaktere, nach der Methode aber, durch welche ich die Richtigkeit dieses Satzes begründet habe, sind zu dem

Beweise desselben die Reciprocitätsgesetze selbst erforderlich. Dessenungeachtet will ich diese Untersuchung hier so weit durchführen, dafs sie durch den nachher zu gebenden Beweis der Reciprocitätsgesetze vollständig abgeschlossen wird, und zwar hauptsächlich aus dem Grunde, weil der Hauptsatz, auf welchem diese Methode beruht, auch in dem Beweise der Reciprocitätsgesetze seine Anwendung finden wird. Dieser jetzt zu beweisende Satz ist folgender:

(I.) Wenn $F(\alpha)$, $F_1(\alpha)$, $F_2(\alpha)$ ... $F_{n-1}(\alpha)$ wirkliche complexe Zahlen sind, welche die eine Bedingung erfüllen, dafs das Produkt

$$F(\alpha)^m \, F_1(\alpha)^{m_1} \, F_2(\alpha)^{m_2} \ldots. \, F_{n-1}(\alpha)^{m_{n-1}}$$

für ganzzahlige Werthe der Exponenten m, m_1, m_2 .... m_{n-1} nicht anders eine λte Potenz werden kann, als wenn diese Exponenten alle congruent Null sind, nach dem Modul λ, so giebt es stets unendlich viele verschiedene Primzahlen $\phi(\alpha)$, in Beziehung auf welche die Indices der complexen Zahlen $F(\alpha)$, $F_1(\alpha)$, ... $F_{n-1}(\alpha)$ beliebig gegebenen Zahlen proportional sind, nach dem Modul λ.

Um diesen Satz zu beweisen, wende ich die Methode von Hrn. Dirichlet an, durch welche derselbe den Satz bewiesen hat, dafs in jeder arithmetischen Reihe, in welcher nicht alle Glieder einen gemeinschaftlichen Faktor haben, unendlich viele Primzahlen enthalten sind. Ich setze

$$(1.) \qquad F(\alpha)^b \, F_1(\alpha)^{b_1} \, F_2(\alpha)^{b_2} \ldots F_{n-1}(\alpha)^{b_{n-1}} = D(\alpha)$$

und bilde das unendliche Produkt:

$$(2.) \qquad \Pi \left(1 - \frac{\left(\frac{D(\alpha)}{\phi(\alpha)}\right)^k}{(N\phi(\alpha))^s}\right)^{-1} = L_k ,$$

in welchem das Produktzeichen Π auf alle unendlich vielen verschiedenen Primzahlen $\phi(\alpha)$ sich bezieht, mit Ausschlufs der in $\varrho D(\alpha)$ enthaltenen, welches Produkt bereits im §. 6., bei der Untersuchung der Klassenanzahl der idealen complexen Zahlen in w, seine Anwendung gefunden hat.

Der Logarithmus von L_k entwickelt giebt:

(3.) $$\log. L_k = \Sigma \frac{\left(\frac{D(\alpha)}{\phi(\alpha)}\right)^k}{(N\phi(\alpha))^s} + \frac{1}{2}\Sigma \frac{\left(\frac{D(\alpha)}{\phi(\alpha)}\right)^{2k}}{(N\phi(\alpha))^{2s}} + \frac{1}{3}\Sigma \frac{\left(\frac{D(\alpha)}{\phi(\alpha)}\right)^{3k}}{(N\phi(\alpha))^{3s}} + \ldots.$$

wo die Summenzeichen auf alle verschiedenen, nicht in $D(\alpha)$ enthaltenen Primzahlen $\phi(\alpha)$ zu beziehen sind. Alle diese Summen, mit Ausschlufs der ersten, haben die Eigenschaft, dafs sie für $s = 1$ nur endliche bestimmte Werthe haben, selbst dann, wenn $D(\alpha)$ eine λte Potenz ist, wo $\frac{D(\alpha)}{\phi(\alpha)}$ stets den Werth Eins hat, denn in den Nennern der Brüche, aus welchen sie bestehen, kommen nur die Quadrate oder Cuben oder höhere Potenzen der nichtcomplexen Primzahlen vor, und zwar jede derselben nur λ mal, weil es nicht mehr als λ conjugirte Primzahlen $\phi(\alpha)$ giebt, welche dieselbe Norm haben. Wenn daher mit $G(s)$ eine jede beliebige eindeutige Funktion von s bezeichnet wird, welche in den Grenzen $s = 1$ bis $s = \infty$ continuirlich ist, und auch für $s = 1$ einen endlichen bestimmten Werth behält, so kann die Gleichung (3.) einfacher so dargestellt werden:

(4.) $$\log. L_k = \Sigma \frac{\left(\frac{D(\alpha)}{\phi(\alpha)}\right)^k}{(N\phi(\alpha))^s} + G(s).$$

Wenn nun der Kürze wegen folgender, aus den in $D(\alpha)$ enthaltenen Zahlen $b, b_1, b_2, \ldots b_{n-1}$ und aus anderen n Zahlen $c, c_1, c_2, \ldots c_{n-1}$ gebildete Ausdruck:

(5.) $$bc + b_1 c_1 + b_2 c_2 + \ldots b_{n-1} c_{n-1}$$

mit C bezeichnet wird, und man multiplicirt die Gleichung (4.) mit α^{-C}, und nimmt sodann die Summe für alle Werthe des $b = 0, 1, 2, \ldots \lambda - 1$, $b_1 = 0, 1, 2, \ldots \lambda - 1$ u. s. w. so erkennt man zunächst, weil

$$\alpha^{-C}\left(\frac{D(\alpha)}{\phi(\alpha)}\right)^k = \alpha^{-bc}\left(\frac{F(\alpha)}{\phi(\alpha)}\right)^{bk} \cdot \alpha^{-b_1 c_1}\left(\frac{F_1(\alpha)}{\phi(\alpha)}\right)^{b_1 k} \ldots. \ldots. \alpha^{-b_{n-1} c_{n-1}}\left(\frac{F_{n-1}(\alpha)}{\phi(\alpha)}\right)^{b_{n-1} k},$$

dafs die, in Beziehung auf die angegebenen Werthe der Zahlen $b, b_1, b_2,$

S 2

... b_{n-1} genommene Summe dieses Ausdrucks immer gleich Null wird, wenn nicht gleichzeitig

$$(6.)\qquad \left(\frac{F(\alpha)}{\phi(\alpha)}\right)^k = \alpha^c\,,\quad \left(\frac{F_1(\alpha)}{\phi(\alpha)}\right)^k = \alpha^{c_1} \;\ldots\ldots\; \left(\frac{F_{n-1}(\alpha)}{\phi(\alpha)}\right)^k = \alpha^{c_{n-1}}\,,$$

dafs aber, wenn diese Gleichungen zugleich Statt haben, diese Summe gleich λ^n ist. Es folgt diefs unmittelbar daraus, dafs die Legendreschen Zeichen nur Potenzen von α sind, und dafs $1 + \alpha^r + \alpha^{2r} + \ldots + \alpha^{(\lambda-1)r}$ gleich Null ist, wenn r nicht durch λ theilbar ist, aber gleich λ, wenn r durch λ theilbar ist. Man hat daher:

$$(7.)\qquad S\,\alpha^{-C} \log.\, L_k = \lambda^n\, \Sigma\, \frac{1}{(N\phi(\alpha))^s} + G(s),$$

wo das Summenzeichen S auf alle Werthe der in C und L_k enthaltenen Zahlen $b = 0, 1, 2, \ldots \lambda - 1$, $b_1 = 0, 1, 2, \ldots \lambda - 1$ u. s. w. sich bezieht, das Summenzeichen Σ aber auf alle diejenigen Primzahlen $\phi(\alpha)$, welche den bei (6.) gegebenen Bedingungen genügen, und wo $G(s)$ eine Funktion von s von derselben Beschaffenheit ist, als die oben mit demselben Zeichen belegte. Endlich gebe ich noch der unbestimmten Zahl k die Werthe 1, 2, 3 ... $\lambda - 1$ und summire, so wird:

$$(8.)\qquad S\,\alpha^{-C} \log.\,(L_1, L_2, L_3 \ldots L_{\lambda-1}) = \lambda^n\, \Sigma\, \frac{1}{(N\phi(\alpha))^s} + G(s),$$

wo das Summenzeichen Σ auf alle diejenigen complexen Primzahlen $\psi(\alpha)$ sich erstreckt, welche den Bedingungen (6.), für irgend welche Werthe des $k = 1, 2, 3, \ldots \lambda - 1$ genügen.

Das Produkt $L_1\, L_2\, L_3 \ldots L_{\lambda-1}$ ist, wie im §. 6. gezeigt worden ist, für den Werth $s = 1$ ein Faktor der Klassenanzahl aller idealen Zahlen in w der Determinante $D(\alpha)$, welche den bei (1.) angegebenen Werth hat. Dieses Produkt ist daher für $s = 1$ immer endlich, sobald $D(\alpha)$ die einzige Bedingung erfüllt, welche bei der Entwickelung dieser Klassenanzahl gemacht worden ist, nämlich, dafs die Determinante nicht eine vollständige λte Potenz sei, in welchem Falle von complexen Zahlen in w gar nicht die Rede sein kann. Nach der Voraussetzung des Satzes ist aber $D(\alpha)$ niemals eine λte Potenz, wenn nicht alle Exponenten $b, b_1, b_2, \ldots b_{n-1}$ einzeln congruent Null sind, nach dem Modul λ; darum ist das eine Glied der Summe S, für welches $b = 0$, $b_1 = 0$, $b_2 = 0$, $b_{n-1} = 0$ ist, das einzige, welches für $s = 1$ nicht einen endlichen Werth behalten mufs. Wirft man nun alle die Glieder dieser Summe, von denen fest steht, dafs sie für $s = 1$ endliche Werthe behalten, auf die andere Seite der Gleichung und vereinigt sie dort mit den durch $G(s)$ bezeichneten Theilen, so hat man, weil das unendliche Produkt L_k für $D(\alpha) = 1$ dasselbe ist als L_0:

$$(9.)\qquad (\lambda - 1) \log. L_0 = \lambda^n \Sigma \frac{1}{(N\phi(\alpha))^s} + G(s).$$

Läfst man nun s bis zur Gränze $s = 1$ abnehmen, so wird L_0 unendlich grofs, wie im §. 6. gezeigt worden ist, $G(s)$ aber bleibt endlich; darum mufs die Summe Σ, welche sich auf die Primzahlen $\phi(\alpha)$ bezieht unendlich grofs werden, es mufs also unendlich viele Primzahlen $\phi(\alpha)$ geben, welche den bei (6.) angegebenen Bedingungen, für gewisse Werthe des $k = 1, 2, 3, \ldots \lambda - 1$ genügen. Setzt man diese Bedingungen, indem man anstatt der Legendreschen Zeichen die Indices anwendet, welche sich auf die Primzahl $\phi(\alpha)$ als Modul beziehen, in die Form

$$(10.)\qquad k \,\text{Ind.}\, F(\alpha) \equiv c,\; k \,\text{Ind.}\, F_1(\alpha) \equiv c_1,\; \ldots k \,\text{Ind.}\, F_{n-1}(\alpha) \equiv c_{n-1},\; \text{mod}\, \lambda.$$

so hat man die Indices dieser complexen Zahlen $F(\alpha)$, $F_1(\alpha)$ .... $F_{n-1}(\alpha)$ proportional den beliebig gegebenen Zahlen $c, c_1, \ldots c_{n-1}$, nach dem Modul λ, was zu beweisen war.

Um den gefundenen Satz auf die Frage wegen der wahren Anzahl der wirklich vorhandenen Gattungen der idealen Zahlen in z anzuwenden, nehme ich für die Zahlen $F(\alpha)$, $F_1(\alpha)$, $F_2(\alpha)$ ... $F_{n-1}(\alpha)$ folgende $\mu + r$:

$$(11.)\qquad E_1(\alpha), E_2(\alpha), \ldots E_{\mu-1}(\alpha), f(\alpha)^m, f_1(\alpha)^{m_1}, \ldots f_{r-1}(\alpha)^{m_{r-1}}$$

wo $E_n(\alpha)$ folgende zusammengesetzte Kreistheilungseinheit bezeichnet:

$$(12.)\qquad E_n(\alpha) = e(\alpha)\, e(\alpha^{\gamma})^{\gamma^{-2n}}\, e(\alpha^{\gamma^2})^{\gamma^{-4n}} \ldots e(\alpha^{\gamma^{\mu-1}})^{\gamma^{-2(\mu-1)n}},$$

von welcher ich in der Abhandlung über die Ergänzungssätze zu den allgemeinen Reciprocitätsgesetzen bewiesen habe, dafs ihr Index in Beziehung auf die Primzahl $\phi(\alpha)$ folgenden Werth hat:

$$(13.)\qquad \text{Ind.}\, E_n(\alpha) \equiv (-1)^n (\gamma^{2n} - 1) \frac{B_n}{4n} \frac{d_0^{\lambda-2n} l\phi(e^v)}{dv^{\lambda-2n}},\; \text{mod.}\, \lambda.$$

wenn B_n die $n$te Bernoullische Zahl ist, und γ die in $E_n(\alpha)$ enthaltene, primitive Wurzel der Primzahl λ; wo ferner $f(\alpha)$, $f_1(\alpha)$, ... $f_{r-1}(\alpha)$ verschiedene Primzahlen sind, welche alle primär genommen werden sollen, und die Exponenten $m, m_1, \ldots m_{r-1}$ so beschaffen, dafs diese Potenzen wirklich werden, wenn die complexen Zahlen selbst ideal sind. Diese $\mu + r$ wirklichen complexen Zahlen erfüllen die in dem Satze ausgesprochene Bedingung, dafs ein Produkt von Potenzen derselben nicht eine λte Potenz sein kann, ohne dafs die Exponenten der Potenzen alle einzeln durch λ theilbar sind. Wenn nämlich das Produkt:

$$(14.)\qquad \alpha^{b}\, E_1(\alpha)^{b_1}\, E_2(\alpha)^{b_2} \ldots E_{\mu-1}(\alpha)^{b_{\mu-1}}\, f(\alpha)^{b_\mu m}\, f_1(\alpha)^{b_{\mu+1} m_1} \ldots\ldots f_{r-1}(\alpha)^{b_{\mu+r-1} m_{r-1}}$$

eine λte Potenz sein soll, so müssen zunächst, weil $f(\alpha), f_1(\alpha), \ldots f_{r-1}(\alpha)$ verschiedene Primzahlen sind, $b_\mu, b_{\mu+1}, \ldots b_{\mu+r-1}$ alle Vielfache von λ sein, weil $m, m_1 \ldots m_{r-1}$ nicht durch λ theilbar sind; ferner mufs b ein Vielfaches von λ sein, weil die λte Potenz einer wirklichen complexen Zahl in α nothwendig einer nichtcomplexen Zahl congruent ist nach dem Modul λ und weil, wenn b nicht durch λ theilbar ist, dieses Produkt selbst nicht für den Modul ϱ^2 einer nichtcomplexen Zahl congruent sein kann. Es bleibt also nur noch zu zeigen, dafs auch das Produkt

$$E_1(\alpha)^{b_1}\, E_2(\alpha)^{b_2} \ldots E_{\mu-1}(\alpha)^{b_{\mu-1}}$$

nicht eine λte Potenz sein kann, ohne dafs $b_1, b_2, \ldots b_{\mu-1}$ alle durch λ theilbar sind. Dieses wird am leichtesten mit Hülfe des in der genannten Abhandlung pag. 134. gegebenen Ausdrucks des Logarithmus der Einheit $E_n(\alpha)$ gezeigt, nach welchem

$$(15.)\qquad l\left(\frac{E_n(\alpha)}{E(1)}\right) \equiv -(-1)^n\,(\gamma^{2n}-1)\,\frac{B_n}{4n}\,\chi_{2n}(\alpha),\ \text{mod. } \lambda,$$

$$\chi_{2n}(\alpha) = \alpha + \gamma^{-2n}\,\alpha^{\gamma} + \gamma^{-4n}\,\alpha^{\gamma^2} + \ldots + \gamma^{-2(\lambda-2)n}\,\alpha^{\gamma^{\lambda-2}}.$$

Vermöge dieses Ausdrucks wird der Logarithmus des obigen Produkts congruent:

$$(16.)\qquad -b_1\,\beta_1\,\chi_2(\alpha) - b_2\,\beta_2\,\chi_4(\alpha) - \ldots - b_{\mu-1}\,\beta_{\mu-1}\,\chi_{2\mu-2}(\alpha),$$

nach dem Modul λ, wo der Kürze wegen

$$(-1)^n (\gamma^{2n} - 1) \frac{B_n}{4n} = \beta_n$$

gesetzt ist. Dieser Ausdruck des Logarithmus des Produkts mufs congruent Null sein, wenn das Produkt eine λte Potenz ist, welches nicht anders geschehen kann, als dafs alle Glieder einzeln congruent Null sind, also $b_1 \beta_1 \equiv 0$, $b_2 \beta_2 \equiv 0$, $b_{\mu-1} \beta_{\mu-1} \equiv 0$, und weil keine der Zahlen β_1, β_2, ... $\beta_{\mu-1}$ durch λ theilbar ist, so müssen die Zahlen b_1, b_2, ... $b_{\mu-1}$ alle durch λ theilbar sein.

Da also die bei (11.) angenommenen complexen Zahlen den Bedingungen des Satzes (I.) entsprechen, so folgt, dafs es unendlich viele Primzahlen $\phi(\alpha)$ giebt, für welche

$$k \text{ Ind. } (\alpha) \equiv -C_{\lambda-1}, \; k \text{ Ind. } E_1(\alpha) \equiv \beta_1 C_{\lambda-2}, \; k \text{ Ind. } E_2(\alpha) \equiv \beta_2 C_{\lambda-4}$$

$$\text{(17.)} \qquad \dots k \text{ Ind. } E_{\mu-1}(\alpha) \equiv \beta_{\mu-1} C_3, \; k \text{ Ind. } f(\alpha)^m \equiv mK,$$

$$\dots k \text{ Ind. } f_{\mu-1}(\alpha)^{m_{r-1}} \equiv m_{r-1} K_{r-1}$$

ist, für alle beliebig gegebenen Werthe der Zahlen $C_{\lambda-1}$, $C_{\lambda-2}$, $C_{\lambda-4}$ ... C_3, K, K_1, ... K_{r-1}. Diese Congruenzen können vermöge des bei (13.) gegebenen Ausdrucks des Index der Einheit $E_n(\alpha)$, und vermittelst des Legendreschen Zeichens auch so dargestellt werden:

$$k \frac{1 - N\phi(\alpha)}{\lambda} \equiv C_{\lambda-1}, \qquad k \frac{d_0^{\lambda-2} l \phi(e^v)}{dv^{\lambda-2}} \equiv C_{\lambda-2},$$

$$\text{(18.)} \qquad k \frac{d_0^{\lambda-4} l \phi(e^v)}{dv^{\lambda-4}} \equiv C_{\lambda-4} \; \dots \qquad k \frac{d_0^3 l \phi(e^v)}{dv^3} \equiv C_3,$$

$$\left(\frac{f(\alpha)}{\phi(\alpha)}\right)^k \equiv \alpha^K \; \dots \; \left(\frac{f_{r-1}(\alpha)}{\phi(\alpha)}\right)^k \equiv \alpha^{K_{r-1}}.$$

Es ist nun die Bedingung einzuführen, dafs die Primzahl $\phi(\alpha)$ die Norm einer idealen Primzahl $\phi(z)$ der Determinante

$$D(\alpha) = \varepsilon(\alpha) f(\alpha)^m f_1(\alpha)^{m_1} \dots f_{r-1}(\alpha)^{m_{r-1}}$$

ist, welche darin besteht, dafs $\left(\frac{D(\alpha)}{\phi(\alpha)}\right) = 1$ sei. Setzt man der Kürze wegen:

$$\frac{d_0^n l \varepsilon(e^v)}{dv^n} = \varepsilon_n,$$

so hat man zunächst nach der §. 14., bei (25.) gegebenen Formel,

$$\left(\frac{\varepsilon(\alpha)}{\phi(\alpha)}\right) = \alpha^{-\varepsilon_1 C_{\lambda-1} + \varepsilon_2 C_{\lambda-2} + \varepsilon_4 C_{\lambda-4} + \dots + \varepsilon_{\lambda-3} C_3},$$

die Bedingung $\left(\frac{D(\alpha)}{\phi(\alpha)}\right) = 1$ giebt daher:

$$(19.) \quad \begin{aligned} -\varepsilon_1 C_{\lambda-1} + \varepsilon_2 C_{\lambda-2} + \varepsilon_4 C_{\lambda-4} + \dots + \varepsilon_{\lambda-3} C_3 + mK + m_1 K_1 \\ + \dots + m_{r-1} K_{r-1} \equiv 0, \quad \text{mod. } \lambda. \end{aligned}$$

Vermöge dieser Congruenz wird eine der $\mu + r$ Gröfsen $C_{\lambda-1}, C_{\lambda-2}, \dots C_3, K, K_1, \dots K_{r-1}$, als welche die letzte K_{r-1} genommen werden kann, durch die übrigen $\mu + r - 1$ bestimmt, deren jede einzelne alle Werthe $0, 1, 2, \dots \lambda - 1$ annehmen kann, so dafs im Ganzen $\lambda^{\mu+r-1}$ Werthverbindungen der Gröfsen $C_{\lambda-1}, C_{\lambda-2}, \dots C_3, K, K_1, \dots K_{r-1}$ bestehen, für welche $\phi(\alpha)$ die Norm einer idealen Zahl $\phi(z)$ ist.

Wenn nun, wie in dem Folgenden bewiesen werden wird, zwischen je zwei complexen primären Primzahlen $f(\alpha)$ und $\phi(\alpha)$ das Reciprocitätsgesetz

$$\left(\frac{f(\alpha)}{\phi(\alpha)}\right) = \left(\frac{\phi(\alpha)}{f(\alpha)}\right)$$

besteht, so kann man demgemäfs statt

$$\left(\frac{f(\alpha)}{\phi(\alpha)}\right), \left(\frac{f_1(\alpha)}{\phi(\alpha)}\right), \dots \left(\frac{f_{r-1}(\alpha)}{\phi(\alpha)}\right)$$

die Ausdrücke

$$\left(\frac{\phi(\alpha)}{f(\alpha)}\right), \left(\frac{\phi(\alpha)}{f_1(\alpha)}\right), \left(\frac{\phi(\alpha)}{f_{r-1}(\alpha)}\right)$$

nehmen und die Congruenzen folgendermaafsen darstellen:

$$(20.) \quad \begin{aligned} &\frac{1 - N\phi(\alpha)^k}{\lambda} \equiv C_{\lambda-1}, \quad \frac{d_0^{\lambda-2} l\phi(e^v)^k}{dv^{\lambda-2}} \equiv C_{\lambda-2}, \quad \frac{d_0^{\lambda-4} l\phi(e^v)^k}{dv^{\lambda-4}} \equiv C_{\lambda-4} \dots \\ &\frac{d_0^3 l\phi(e^v)^k}{dv^3} \equiv C_3, \quad \left(\frac{\phi(\alpha)^k}{f(\alpha)}\right) = \alpha^K, \dots \left(\frac{\phi(\alpha)^k}{f_{r-1}(\alpha)}\right) = \alpha^{K_{r-1}}. \end{aligned}$$

Die ideale Zahl $\phi(z)^k$, deren Norm gleich $\phi(\alpha)^k$, ist also eine solche, welche die Charaktere $C_{\lambda-1}, C_{\lambda-2}, C_{\lambda-4}, \dots C_3, K, K_1, \dots K_{r-1}$ hat, welche der einzigen Bedingung (19.) unterworfen sind, und wie auch die Werthe dieser Charaktere gewählt werden mögen, wenn sie nur der in der Congruenz (19.) gegebenen Bedingung genügen, so giebt es stets ideale Zahlen

$\phi(z)^k$, welchen diese Charaktere zukommen. Die Anzahl der angebbaren Gesamtcharaktere, welche gleich $\lambda^{\mu+r}$ ist, weil jeder der $\mu + r$ Charaktere die λ Werthe $0, 1, 2, \ldots \lambda - 1$ haben kann, wird durch die Congruenz (19.) genau auf den λten Theil eingeschränkt, diese $\lambda^{\mu+r-1}$ Gesamtchararaktere geben aber ebensoviele wirklich vorhandene Gattungen, weil jedem derselben ideale Zahlen in z angehören.

Hiermit ist, unter der Voraussetzung, daſs unter je zwei primären complexen Primzahlen $f(\alpha)$ und $\phi(\alpha)$ das Reciprocitätsgesetz $\left(\frac{f(\alpha)}{\phi(\alpha)}\right) = \left(\frac{\phi(\alpha)}{f(\alpha)}\right)$ gültig ist, der Satz bewiesen:

(II.) Die Anzahl der wirklich vorhandenen Gattungen der idealen Zahlen in z ist genau gleich dem λten Theile aller angebbaren Gesamtcharaktere.

§. 17.

Beweis der allgemeinen Reciprocitätsgesetze.

In dem Beweise der allgemeinen Reciprocitätsgesetze zwischen je zwei complexen Primzahlen in α, welcher sich auf die in dem Vorhergehenden entwickelte Theorie der complexen Zahlen in z, und namentlich auf die Eintheilung der idealen Zahlen dieser Theorie in die Gattungen stützt, sind diejenigen complexen Primzahlen in α, in Beziehung auf welche alle Einheiten λte Potenzreste sind, die als complexe Primzahlen der zweiten Art bezeichneten, von denen, welche diese besondere Eigenschaft nicht haben, den complexen Primzahlen der ersten Art, zu unterscheiden und namentlich folgende drei Fälle besonders zu behandeln: erstens der Fall, wo beide zu vergleichenden complexen Primzahlen der ersten Art angehören, zweitens, wo eine der ersten Art, die andere der zweiten Art angehört und drittens, wo beide der zweiten Art angehören.

Für den ersten dieser drei Fälle reicht es hin, nur solche complexe Zahlen in z anzuwenden, deren Determinante nicht mehr als einen idealen Primfaktor enthält. Es sei also:

$$D(\alpha) = e(\alpha) f(\alpha)^m,$$

$f(\alpha)$ eine complexe Primzahl der ersten Art, welche primär angenommen werden soll, m ein nicht durch λ theilbarer Exponent, welcher bewirkt, dafs $f(\alpha)^m$ wirklich ist, wenn $f(\alpha)$ selbst ideal sein sollte, und $e(\alpha)$ eine beliebige Einheit, welche jedoch durch die Bedingung, dafs $D(\alpha)-1$ durch ϱ, aber nicht durch ϱ^2 theilbar sein soll, einer gewissen Beschränkung unterworfen ist.

Unter diesen Voraussetzungen ist von den $\mu+1$ Charakteren C_3, C_5, ... $C_{\lambda-1}$ und K, welche überhaupt Statt haben, der letzte durch die übrigen vollständig bestimmt, in der Art, dafs alle idealen Zahlen in z, welche dieselben Werthe der Charaktere C_3, C_5, ... $C_{\lambda-1}$ haben, auch denselben Werth des Charakters K haben müssen. (Satz (I.) §. 16.). Wenn ferner $\varphi(z)$ irgend eine ideale Primzahl in z ist, so giebt es stets eine wirkliche complexe Zahl in w, $F(w)$, welche als ideale Zahl in z betrachtet, dieselben Werthe der Charaktere C_3, C_5, ... $C_{\lambda-1}$ hat, als $\varphi(z)$, (Satz (II.) §. 16.), und welche darum auch denselben Werth des letzten Charakters K haben mufs, als diese. Wird nun die Norm der idealen Primzahl $\varphi(z)$ mit $\varphi(\alpha)$ bezeichnet, wo $\varphi(\alpha)$ nach der allgemeinen Festsetzung über die Normen der idealen Zahlen in z primär ist; wird ferner die Norm der wirklichen complexen Zahl $F(w)$ mit $F(\alpha)$ bezeichnet, und dieselbe in der primären Form genommen durch $F_1(\alpha)$, so hat man:

$$(1.)\qquad \left(\frac{F_1(\alpha)}{f(\alpha)}\right)=\left(\frac{\varphi(\alpha)}{f(\alpha)}\right)=\alpha^K,$$

oder, was nach der §. 14. gegebenen Definition dieses Legendreschen Zeichens für zusammengesetzte Moduln dasselbe ist:

$$(2.)\qquad \left(\frac{F_1(\alpha)}{D(\alpha)}\right)=\alpha^{mK}.$$

Nimmt man nun

$$F(w)=A+A_1w+A_2w^2+\ldots.+A_{\lambda-1}w^{\lambda-1},$$

und entwickelt die Norm von $F(w)$, so ist A^λ das einzige Glied dieser Norm, welches w^λ nicht enthält, man hat also

$$NF(w)=F(\alpha)\equiv A^\lambda,\quad \text{mod. } D(\alpha),$$

oder was dasselbe ist:

$$(3.)\qquad \left(\frac{F(\alpha)}{D(\alpha)}\right)=1.$$

Drückt man nun nach Formel (23.), §. 14. das Legendresche Zeichen für das nicht primäre $F(\alpha)$ durch das entsprechende Zeichen für das primäre $F_1(\alpha)$ aus, so hat man

$$\left(\frac{F(\alpha)}{D(\alpha)}\right) = \left(\frac{F_1(\alpha)}{D(\alpha)}\right) \alpha^{-C_1 D_{\lambda-1} + C_2 D_{\lambda-2} + C_4 D_{\lambda-4} + \dots + C_{\lambda-3} D_3},$$

also vermöge der Gleichungen (2.) und (3.):

$$(4.)\quad mK - C_1 D_{\lambda-1} + C_2 D_{\lambda-2} + C_4 D_{\lambda-4} + \dots + C_{\lambda-3} D_3 \equiv 0, \text{ mod. } \lambda.$$

Ferner hat man nach derselben Formel (23.), §. 14, weil $f(\alpha)^m$ die complexe Zahl $D(\alpha)$ in ihrer primären Form darstellt:

$$\left(\frac{D(\alpha)}{\phi(\alpha)}\right) = \left(\frac{f(\alpha)}{\phi(\alpha)}\right)^m \alpha^{-D_1 C_{\lambda-1} + D_2 C_{\lambda-2} + C_4 D_{\lambda-4} + \dots + D_{\lambda-3} C_3}.$$

Setzt man nun

$$\left(\frac{f(\alpha)}{\phi(\alpha)}\right) = \alpha^{K'}$$

und beachtet, dafs $\phi(\alpha)$ als Norm einer idealen Zahl in z der Determinante $D(\alpha)$ der Bedingung

$$(5.)\qquad \left(\frac{D(\alpha)}{\phi(\alpha)}\right) = 1$$

genügen mufs, so hat man

$$(6.)\quad mK' - D_1 C_{\lambda-1} + D_2 C_{\lambda-2} + D_4 C_{\lambda-4} + \dots + D_{\lambda-3} C_3 \equiv 0, \text{ mod. } \lambda.$$

Verbindet man nun die beiden Congruenzen (4.) und (6.) mit der Congruenz

$$(7.)\quad D_{\lambda-1} C_1 - D_{\lambda-2} C_2 + D_{\lambda-3} C_3 - D_{\lambda-4} C_4 + \dots + D_1 C_{\lambda-1} \equiv 0, \text{ mod. } \lambda,$$

welche nach dem Satze (I.) §. 14. Statt haben mufs, weil $F(\alpha)$ die Norm der wirklichen complexen Zahl $F(w)$ ist, so erhält man:

$$mK \equiv mK', \text{ mod. } \lambda,$$

und weil m nicht durch λ theilbar ist:

$$K \equiv K', \text{ mod. } \lambda,$$

also

T 2

(8.) $$\left(\frac{\phi(\alpha)}{f(\alpha)}\right) = \left(\frac{f(\alpha)}{\phi(\alpha)}\right).$$

Diese Gleichung giebt das Reciprocitätsgesetz unter den beiden primären Primzahlen $f(\alpha)$ und $\phi(\alpha)$, deren erste eine complexe Primzahl der ersten Art ist, und die andere $\phi(\alpha)$ eine Primzahl, welche der einen in der Gleichung (5.) enthaltenen Bedingung unterworfen ist, dafs

(9.) $$\left(\frac{e(\alpha)}{\phi(\alpha)}\right)\left(\frac{f(\alpha)}{\phi(\alpha)}\right)^m = 1$$

sein mufs.

Wenn nun $f(\alpha)$ ebenfalls eine complexe Primzahl der ersten Art ist, also Einheiten $e(\alpha)$ existiren, für welche $\left(\frac{e(\alpha)}{\phi(\alpha)}\right)$ nicht gleich Eins ist, so kann man, welchen Werth auch $\left(\frac{f(\alpha)}{\phi(\alpha)}\right)$ habe, die Einheit $e(\alpha)$ in der Regel so wählen, dafs die Bedingung (9.) erfüllt wird, woraus folgt, dafs in der Reciprocitätsgleichung (8.) die Primzahl $\phi(\alpha)$ eine jede primäre Primzahl der ersten Art darstellt. In einem ganz besonderen Falle jedoch wird durch die Bedingung, welcher die Determinante $D(\alpha)$ unterworfen ist, dafs $D(\alpha)-1$ durch ϱ aber nicht durch ϱ^2 theilbar sein soll, eine Ausnahme begründet. Der primäre Faktor der Determinante: $f(\alpha)^m$ hat als solcher die Eigenschaft einer nichtcomplexen Zahl congruent zu sein, nach dem Modul ϱ^2; ferner, wenn die Einheit $e(\alpha)$ in die Form $\alpha^k\varepsilon(\alpha)$ gesetzt wird, wo $\varepsilon(\alpha)$ eine, nur die zweigliedrigen Perioden $\alpha+\alpha^{-1}$, $\alpha^2+\alpha^{-2}$, ... enthaltende Einheit ist, welche Form einer jeden Einheit in α gegeben werden kann, so hat auch $\varepsilon(\alpha)$ die Eigenschaft, einer nichtcomplexen Zahl congruent zu sein, nach dem Modul ϱ^2; da aber $D(\alpha)$ diese Eigenschaft nicht haben darf, so folgt, dafs α^k dieselbe Eigenschaft nicht haben darf. Es ist aber $\alpha^k \equiv 1-k\varrho$, mod. ϱ^2, woraus folgt, dafs k nicht durch ϱ theilbar, oder was dasselbe ist, k nicht gleich Null sein darf. Wenn nun die Primzahl $\phi(\alpha)$ die ganz besondere Eigenschaft hat, dafs für dieselbe alle aus den zweigliedrigen Perioden gebildeten Einheiten $\varepsilon(\alpha)$ λte Potenzreste sind, und wenn zugleich auch $f(\alpha)$ ein λter Potenzrest von $\phi(\alpha)$ ist, so ist die Bedingung (8.) nicht zu befriedigen, weil in $e(\alpha)=\alpha^k\varepsilon(\alpha)$ die Zahl k nicht gleich Null sein darf, es folgt aber auch, dafs diefs der einzige Ausnahmefall ist. Es ist indessen leicht auch für diese besonderen Primzahlen $\phi(\alpha)$ die Gültigkeit der Reciprocitätsgleichung (8.) zu erschliefsen. Weil nämlich dieser Ausnahmefall niemals eintritt, sobald

$f(\alpha)$ ein Nichtrest von $\phi(\alpha)$ ist, so zeigt die Gleichung (8.) zunächst, dafs, wenn eine der beiden Primzahlen der ersten Art $f(\alpha)$ und $\phi(\alpha)$ ein Nichtrest der anderen ist, auch diese andere Nichtrest der ersten sein mufs, und hieraus folgt sodann, dafs wenn die eine Rest der anderen ist, auch die andere Rest der ersten sein mufs; denn wäre die zweite ein Nichtrest der ersten, so müsste auch die erste ein Nichtrest der zweiten sein. Die Beschränkung der Determinante, dafs $D(\alpha)-1$ nicht durch ϱ^2 theilbar sein darf, begründet also keine Ausnahme in der Allgemeingültigkeit des Reciprocitätsgesetzes (8.) für je zwei beliebige complexe Primzahlen der ersten Art, und man hat das Resultat:

(I.) Wenn $f(\alpha)$ und $\phi(\alpha)$ zwei primäre complexe Primzahlen der ersten Art sind, so besteht unter ihnen das Reciprocitätsgesetz:

$$\left(\frac{\phi(\alpha)}{f(\alpha)}\right)=\left(\frac{f(\alpha)}{\phi(\alpha)}\right).$$

Wenn nun zweitens die Primzahl $\phi(\alpha)$ in der Gleichung (8.) eine complexe Primzahl der zweiten Art ist, für welche alle Einheiten λte Potenzreste sind, so folgt aus der Gleichung (9.), dafs $\phi(\alpha)$ die Bedingung $\left(\frac{f(\alpha)}{\phi(\alpha)}\right)=1$ erfüllen mufs, und dafs, wenn diese Bedingung erfüllt ist, die Reciprocitätsgleichung (8.) Statt hat. Man hat also folgenden Satz:

(II.) Wenn eine primäre complexe Primzahl der ersten Art, $f(\alpha)$, ein λter Potenzrest einer primären complexen Primzahl der zweiten Art, $\phi(\alpha)$ ist, so ist auch $\phi(\alpha)$ ein λter Potenzrest von $f(\alpha)$.

Dafs die Umkehrung dieses Satzes ebenfalls richtig ist, kann aus dem Vorhergehenden noch nicht erschlossen werden.

Um nun die Reciprocitätsgesetze auch für die Fälle, wo von den beiden zu vergleichenden Primzahlen die eine der ersten Art, die andere der zweiten Art angehört, und wo beide der zweiten Art angehören, vollständig zu entwickeln, wende ich complexe Zahlen in z an, deren Determinante zwei verschiedene Primfaktoren $f(\alpha)$ und $f_1(\alpha)$ enthält, für welche also

$$D(\alpha)=e(\alpha)f(\alpha)^{m}\cdot e_1(\alpha)f_1(\alpha)^{m_1}$$

ist. Es sollen auch hier $f(\alpha)$ und $f_1(\alpha)$ als primär angenommen, und m und

m_1 so gewählt werden, dafs $f(\alpha)^m$ und $f_1(\alpha)^{m_1}$ wirklich werden, wenn $f(\alpha)$ oder $f_1(\alpha)$ ideal sind; ferner soll, entsprechend den Voraussetzungen des Satzes (III), §. 15. festgesetzt werden, dafs $f(\alpha)$ eine Primzahl der zweiten Art, $f_1(\alpha)$ aber eine Primzahl der ersten Art, und dafs $f_1(\alpha)$ Nichtrest von $f(\alpha)$ sei.

Unter diesen Voraussetzungen findet nach dem Satze (IV.) §. 15. unter den Charakteren C_3, C_5, ... $C_{\lambda-1}$, K und K_1 einer jeden idealen Zahl $\phi(z)$ die Beziehung Statt, dafs der Charakter K_1 durch die übrigen Charaktere vollständig bestimmt ist, in der Art, dafs alle idealen Zahlen in z, welche dieselben Werthe der Charaktere C_3, C_5, ... $C_{\lambda-1}$ und K haben, auch denselben Werth des Charakters K_1 haben müssen. Es giebt auch zu jeder idealen Zahl $\phi(z)$ eine wirkliche complexe Zahl $F(u, u_1)$, welche als ideale Zahl in z betrachtet vollständig dieselben Werthe der Charaktere hat, als $\phi(z)$. Wenn nun wieder $\phi(z)$ als ideale Primzahl angenommen wird, und $\phi(\alpha)$ die Norm derselben, also primär ist; wenn ferner $F(\alpha)$ die Norm von $F(u, u_1)$ bezeichnet und $F_1(\alpha)$ die primäre Form der Zahl $F(\alpha)$, und wenn auch die übrigen im §. 15. festgesetzten Bezeichnungen beibehalten werden, so hat man:

$$\left(\frac{F_1(\alpha)}{f(\alpha)}\right) = \left(\frac{\phi(\alpha)}{f(\alpha)}\right) = \alpha^K, \tag{10.}$$

$$\left(\frac{F_1(\alpha)}{f_1(\alpha)}\right) = \left(\frac{\phi(\alpha)}{f_1(\alpha)}\right) = \alpha^{K_1}. \tag{11.}$$

Aus dem Ausdrucke der complexen Zahl $F(u, u_1)$ folgt ferner für die Norm derselben $F(\alpha)$, aufser der schon im §. 15. entwickelten Gleichung:

$$\left(\frac{F(\alpha)}{f(\alpha)}\right) = \left(\frac{\delta(\alpha)}{f(\alpha)}\right)^{n-n_1}, \tag{12.}$$

ebenso auch die Gleichung:

$$\left(\frac{F(\alpha)}{f_1(\alpha)}\right) = \left(\frac{d(\alpha)}{f_1(\alpha)}\right)^{n_1-n}. \tag{13.}$$

Wenn nun nach der schon mehrmals benutzten Formel (23.), §. 14. die in diesen beiden Gleichungen vorkommenden Legendreschen Zeichen für die nichtprimären Zahlen $F(\alpha)$, $\delta(\alpha)$ und $d(\alpha)$ durch die entsprechenden für die primären Zahlen $F_1(\alpha)$, $f_1(\alpha)^{m_1}$ und $f(\alpha)^m$ ausgedrückt werden, und

$$\left(\frac{f_1(\alpha)}{f(\alpha)}\right) = \alpha^{i}, \quad \left(\frac{f(\alpha)}{f_1(\alpha)}\right) = \alpha^{i'}$$

gesetzt wird, so erhält man die beiden Congruenzen:

$$\begin{aligned} &(14.) \quad K \equiv (n-n_1)\, m_1 i, \\ &(15.) \quad K_1 + S_1 \equiv (n_1 - n)\,(mi' + s_1), \end{aligned} \quad \text{mod. } \lambda,$$

deren erstere schon im §. 15. hergeleitet worden ist.

Aus der Bedingung, dafs $\phi(z)$ ein idealer Primfaktor von $\phi(\alpha)$ ist, hat man ferner

$$\left(\frac{D(\alpha)}{\phi(\alpha)}\right) = \left(\frac{e(\alpha)\, e_1(\alpha)}{\phi(\alpha)}\right) \cdot \left(\frac{f(\alpha)}{\phi(\alpha)}\right)^{m} \cdot \left(\frac{f_1(\alpha)}{\phi(\alpha)}\right)^{m_1} = 1,$$

und wenn man das Legendresche Zeichen für die nichtprimäre Zahl $D(\alpha)$ durch das entsprechende für die primäre Zahl $f(\alpha)^m f_1(\alpha)^{m_1}$ ausdrückt, und der Kürze wegen

$$\left(\frac{f(\alpha)}{\phi(\alpha)}\right) = \alpha^{K'}, \quad \left(\frac{f_1(\alpha)}{\phi(\alpha)}\right) = \alpha^{K'_1}$$

setzt, so erhält man aus der Bedingung, dafs $D(\alpha)$ Rest von $\phi(\alpha)$ ist, die Congruenz:

$$(16.) \quad T + mK' + m_1 K'_1 \equiv 0, \quad \text{mod. } \lambda.$$

Endlich, weil $F(\alpha)$ die Norm der wirklichen complexen Zahl $F(u, u_1)$ ist, hat man noch die §. 15, bei (18.) entwickelte Congruenz:

$$(17.) \quad T - m_1 S_1 - (n - n_1)\, m_1 s_1 \equiv 0, \quad \text{mod. } \lambda.$$

Aus den vier Congruenzen (14.), (15.), (16.) und (17.) erhält man nun durch Elimination der drei Gröfsen T, S_1 und $n - n_1$, durch welche s_1 von selbst mit weggeht, die Congruenz:

$$(18.) \quad m(iK' - i'K) + m_1 i(K'_1 - K_1) \equiv 0, \text{ mod. } \lambda,$$

aus welcher das Reciprocitätsgesetz für die beiden Fälle: erstens, wo eine der beiden Zahlen $f(\alpha)$ und $\phi(\alpha)$ der ersten Art, die andere der zweiten Art angehört, und zweitens, wo beide der zweiten Art angehören, entwickelt werden soll.

Ich nehme zuerst $\phi(\alpha)$ als eine Primzahl der ersten Art. Für eine solche Primzahl gilt, weil $f_1(\alpha)$ ebenfalls der ersten Art angehört, nach dem Satze (I.) das Reciprocitätsgesetz:

$$\left(\frac{\phi(\alpha)}{f_1(\alpha)}\right) = \left(\frac{f_1(\alpha)}{\phi(\alpha)}\right),$$

man hat also

$$K_1 \equiv K_1', \quad \text{mod. } \lambda,$$

Aus der Congruenz (18.) fällt daher das zweite Glied hinweg, und dieselbe giebt, wenn durch m dividirt wird, welches den Faktor λ nicht enthält:

(19.) $$iK' \equiv i'K, \quad \text{mod. } \lambda,$$

oder

(20.) $$\left(\frac{f(\alpha)}{\phi(\alpha)}\right)^i = \left(\frac{\phi(\alpha)}{f(\alpha)}\right)^{i'}.$$

Es ist nun auch hier zunächst zu ermitteln, in wie weit die Primzahl der ersten Art $\phi(\alpha)$ von den beiden Primzahlen $f(\alpha)$ und $f_1(\alpha)$ unabhängig ist. Dieselbe ist der einzigen Bedingung unterworfen, dafs $\left(\frac{D(\alpha)}{f(\alpha)}\right) = 1$ sein mufs, welche, da $\phi(\alpha)$ eine Primzahl der ersten Art ist, durch passende Wahl der in $D(\alpha)$ enthaltenen beliebigen Einheit $e(\alpha)\, e_1(\alpha)$ immer erfüllt werden kann, wenn nicht, ebenso wie in dem obigen ersten Falle, die Bedingung, dafs $D(\alpha) - 1$ nicht durch ϱ^2 theilbar sein darf, eine Ausnahme begründet. Dieses kann nur dann der Fall sein, wenn $\phi(\alpha)$ die ganz besondere Eigenschaft hat, dafs alle aus den zweigliedrigen Perioden gebildeten Einheiten λte Potenzreste von $\phi(\alpha)$ sind, die Einheit α aber Nichtrest ist, und wenn aufserdem

(21.) $$\left(\frac{f(\alpha)}{\phi(\alpha)}\right)^m \cdot \left(\frac{f_1(\alpha)}{\phi(\alpha)}\right)^{m_1} = 1$$

ist. Weil ferner die Zahlen m und m_1 nur in so weit bestimmt sind, dafs sie nicht durch λ theilbar sein dürfen und dafs $f(\alpha)^m$ und $f_1(\alpha)^{m_1}$ wirkliche complexe Zahlen sein sollen, so kann man anstatt m auch km setzen, wo k eine jede der Zahlen $1, 2, 3, \ldots \lambda - 1$ vorstellt. Hieraus folgt, dafs man die Zahlen m und m_1 immer so wählen kann, dafs die Gleichung (21.) nicht Statt hat, ausgenommen in dem Falle, dafs $\left(\frac{f(\alpha)}{\phi(\alpha)}\right)$ und $\left(\frac{f_1(\alpha)}{\phi(\alpha)}\right)$ beide einzeln den Werth Eins haben. Die Primzahl der ersten Art $\phi(\alpha)$ kann also in der Gleichung (20.) namentlich alle diejenigen Werthe ohne Ausnahme erhalten, für welche $\left(\frac{f(\alpha)}{\phi(\alpha)}\right)$ nicht gleich Eins ist, und es wird hinreichen, diese allein in Betracht zu ziehen.

Die erste Folgerung, welche ich aus der Gleichung (20.) ziehe, ist die, dafs, wenn $\left(\frac{f(\alpha)}{\phi(\alpha)}\right)$ nicht gleich Eins ist, i' nicht congruent Null sein kann; da nämlich nach der Voraussetzung i nicht congruent Null ist, so ist auch $\left(\frac{f(\alpha)}{\phi(\alpha)}\right)^i$ nicht gleich Eins und darum i' nicht congruent Null. Also wenn $f_1(\alpha)$ Nichtrest von $f(\alpha)$ ist, so ist auch $f(\alpha)$ Nichtrest von $f_1(\alpha)$. Die Gültigkeit dieses Schlusses hängt jedoch davon ab, dafs man immer eine Primzahl $\phi(\alpha)$ der ersten Art finden kann, für welche eine beliebig gegebene Primzahl $f(\alpha)$ der zweiten Art Nichtrest ist, welches Postulat demjenigen vollständig analog ist, dafs zu jeder gegebenen Primzahl der Form $4n+1$, eine Primzahl der Form $4n+3$ gefunden werden kann, in Beziehung auf welche jene quadratischer Nichtrest ist, welches L e g e n d r e in seinem Beweise des quadratischen Reciprocitätsgesetzes gemacht und unbewiesen gelassen hat. Aus dem im §. 16. bewiesenen Satze (I.) folgt aber fast unmittelbar, dafs es stets unendlich viele Primzahlen $\phi(\alpha)$ giebt, welche dieser Forderung genügen. Betrachtet man nämlich in diesem Satze nur zwei gegebene, wirkliche complexe Zahlen, und nimmt für die eine eine Einheit $E(\alpha)$, für die andere aber $f(\alpha)^m$, so zeigt derselbe, dafs es unendlich viele Primzahlen $\phi(\alpha)$ von der Art giebt, dafs

$$\left(\frac{E(\alpha)}{\phi(\alpha)}\right)^k = \alpha^c, \qquad \left(\frac{f(\alpha)}{\phi(\alpha)}\right)^{mk} = \alpha^{c_1}$$

ist, wo c und c_1 beliebig gegebene Zahlen sind, und k nicht durch λ theilbar. Wählt man also c und ebenso auch c_1 nicht durch λ theilbar, so ist $\left(\frac{E(\alpha)}{\phi(\alpha)}\right)$ nicht gleich Eins, also $\phi(\alpha)$ eine Primzahl der ersten Art, und auch $\left(\frac{f(\alpha)}{\phi(\alpha)}\right)$ nicht gleich Eins, wodurch die Existenz unendlich vieler, der Forderung entsprechender Primzahlen bewiesen ist. Die somit streng bewiesene Folgerung: wenn eine Primzahl der ersten Art Nichtrest einer Primzahl der zweiten Art ist, so ist auch diese Nichtrest von jener, bildet die Ergänzung des Satzes (II.) und giebt folgenden vollständigeren Satz:

(III.) Wenn von zwei primären complexen Primzahlen, deren eine der ersten, die andere der zweiten Art angehört, die eine λter Potenzrest der andern ist, so ist auch diese λter Potenzrest von jener.

Um nun für den gegenwärtigen Fall, wo die eine der beiden zu vergleichenden Zahlen der ersten Art angehört, die andere aber der zweiten Art, das Reciprocitätsgesetz auch für die $\lambda-1$ verschiedenen Klassen der Nichtreste in derselben einfachen Form zu erhalten wie im ersten Falle, zeige ich, dafs in der Gleichung (20.) nothwendig $i=i'$ sein mufs. Zu diesem Zwecke wende ich die eine Reciprocitätsgleichung an, welche die Kreistheilung gewährt, nämlich folgende:

$$(22.)\quad \left(\frac{\phi(\alpha)}{f(\alpha)}\right)\cdot\left(\frac{\phi(\alpha)}{f(\alpha^{\gamma})}\right)\cdots\left(\frac{\phi(\alpha)}{f(\alpha^{\gamma^{e-1}})}\right)=\left(\frac{f(\alpha)}{\phi(\alpha)}\right)\cdot\left(\frac{f(\alpha^{\gamma})}{\phi(\alpha)}\right)\cdots\left(\frac{f(\alpha^{\gamma^{e-1}})}{\phi(\alpha)}\right),$$

in welcher $\phi(\alpha)$ ein complexer Primfaktor der Primzahl p von der Form $n\lambda+1$ ist, $f(\alpha)$ ein complexer Primfaktor der Primzahl q, welche zum Exponenten f gehört, nach dem Modul λ, $ef=\lambda-1$ und γ eine primitive Wurzel von λ, so dafs $f(\alpha), f(\alpha^{\gamma}) \dots f(\alpha^{\gamma^{e-1}})$ die e verschiedenen idealen Primfaktoren des q sind. Man sehe die Abhandlung von Eisenstein in den Monatsberichten der Akademie vom Mai 1850.

Ich wähle nun die Primzahl $\phi(\alpha)$ in der Gleichung (22.) so, dafs die $e-1$ zu $f(\alpha)$ conjugirten Zahlen $f(\alpha^{\gamma}), f(\alpha^{\gamma^2}), \dots f(\alpha^{\gamma^{e-1}})$ λte Potenzreste für $\phi(\alpha)$ sind, die Zahl $f(\alpha)$ selbst aber ein Nichtrest von $\phi(\alpha)$. Dafs es stets Primzahlen $\phi(\alpha)$ giebt, welche diesen Bedingungen genügen, folgt unmittelbar aus dem Satze (I.) §. 16., wenn in demselben für $F(\alpha), F_1(\alpha), \dots$ die e wirklichen complexen Zahlen $f(\alpha)^h, f(\alpha^{\gamma})^h, \dots f(\alpha^{\gamma^{e-1}})^h$ und irgend eine Einheit $E(\alpha)$ genommen, und die Zahlen, welchen die Indices derselben proportional sein sollen, mit Ausschlufs des Index von $f(\alpha)$ und des Index der Einheit $E(\alpha)$ alle gleich Null gewählt werden. Wenn die Primzahl $\phi(\alpha)$ in dieser Weise bestimmt ist, so erfüllt sie die eine Bedingung: dafs die Norm derselben eine nichtcomplexe Primzahl p von der Form $n\lambda+1$ sei, von selbst; denn die nichtcomplexe Zahl $q=f(\alpha)f(\alpha^{\gamma}) \dots\dots f(\alpha^{\gamma^{e-1}})$ ist ein Nichtrest von $\phi(\alpha)$ und eine nichtcomplexe Zahl kann nur für solche complexe Primzahlen Nichtrest sein, welche zum Exponenten Eins gehören, das heifst deren Normen Primzahlen der Form $n\lambda+1$ sind.

Wenn nun die complexen Primzahlen $f(\alpha^{\gamma}), f(\alpha^{\gamma^2}) \dots f(\alpha^{\gamma^{e-1}})$ λte Potenzreste von $\phi(\alpha)$ sind, so ist nach dem Satze (III.) auch umgekehrt $\phi(\alpha)$ ein λter Potenzrest für jene, und die Gleichung (22.) giebt, wenn für die

Legendreschen Zeichen, deren Werth gleich Eins ist, dieser Werth gesetzt wird:

(23.) $$\left(\frac{\phi(\alpha)}{f(\alpha)}\right) = \left(\frac{f(\alpha)}{\phi(\alpha)}\right).$$

Für eine solche speciell bestimmte Primzahl $\phi(\alpha)$ ist also in der Gleichung (20.) $i \equiv i'$ und weil i und i' für alle complexen Primzahlen $\phi(\alpha)$ der ersten Art, für welche $f(\alpha)$ ein Nichtrest ist, dieselben Werthe haben müssen, so folgt, dafs für alle diese ebenfalls $i \equiv i'$ oder

$$\left(\frac{f(\alpha)}{\phi(\alpha)}\right)^i = \left(\frac{\phi(\alpha)}{f(\alpha)}\right)^i,$$

woraus unmittelbar folgt, dafs auch

(24.) $$\left(\frac{f(\alpha)}{\phi(\alpha)}\right) = \left(\frac{\phi(\alpha)}{f(\alpha)}\right)$$

ist, wodurch dieses einfache Reciprocitätsgesetz für die verschiedenen $\lambda - 1$ Klassen der Nichtreste bewiesen ist. Da dasselbe nach dem Satze (III.) für die Reste bereits fest steht, so hat man den allgemeineren Satz:

(IV.) Wenn von zwei primären complexen Primzahlen $\phi(\alpha)$ und $f(\alpha)$ die eine der ersten Art, die andere der zweiten Art angehört, so besteht unter denselben das Reciprocitätsgesetz:

$$\left(\frac{f(\alpha)}{\phi(\alpha)}\right) = \left(\frac{\phi(\alpha)}{f(\alpha)}\right).$$

Es bleibt nun noch übrig, das Reciprocitätsgesetz auch für den dritten Fall zu entwickeln, wo die zu vergleichenden primären Primzahlen $f(\alpha)$ und $\phi(\alpha)$ beide der zweiten Art angehören. Nimmt man zu diesem Zwecke in der Congruenz (18.) $\phi(\alpha)$ als eine primäre Primzahl der zweiten Art, so hat man, weil das Reciprocitätsgesetz für zwei primäre complexe Primzahlen, deren eine der ersten die andere der zweiten Art angehört, gültig ist:

(25.) $$\left(\frac{f_1(\alpha)}{f(\alpha)}\right) = \left(\frac{f(\alpha)}{f_1(\alpha)}\right), \quad \left(\frac{f_1(\alpha)}{\phi(\alpha)}\right) = \left(\frac{\phi(\alpha)}{f_1(\alpha)}\right),$$

also

(26.) $$i \equiv i', \qquad K_1 \equiv K_1' \quad \text{mod. } \lambda.$$

Die Congruenz (18.) giebt daher, weil i nicht $\equiv 0$ ist:

$$K' \equiv K, \quad \text{mod. } \lambda,$$

U 2

und mithin

$$(27.)\qquad \left(\frac{f(\alpha)}{\phi(\alpha)}\right) = \left(\frac{\phi(\alpha)}{f(\alpha)}\right),$$

welches Resultat für jede complexe Primzahl der zweiten Art $\phi(\alpha)$ gültig ist, die der Bedingung $\left(\frac{D(\alpha)}{\phi(\alpha)}\right) = 1$ genügt. Diese Bedingung giebt, weil in Beziehung auf $\phi(\alpha)$ als Primzahl der zweiten Art jede Einheit ein λter Potenzrest ist:

$$(28.)\qquad \left(\frac{f(\alpha)}{\phi(\alpha)}\right)^{m} \left(\frac{f_1(\alpha)}{\phi(\alpha)}\right)^{m_1} = 1,$$

welcher man durch passende Wahl der Zahlen m und m_1 immer genügen kann, wenn keines der beiden Zeichen $\left(\frac{f(\alpha)}{\phi(\alpha)}\right)$ und $\left(\frac{f_1(\alpha)}{\phi(\alpha)}\right)$ den Werth Eins hat. Wenn man also die Primzahl der ersten Art $f_1(\alpha)$ so wählt, dafs $\left(\frac{f_1(\alpha)}{\phi(\alpha)}\right)$ nicht gleich Eins ist, und dafs auch $\left(\frac{f_1(\alpha)}{f(\alpha)}\right)$ nicht gleich Eins ist, so gilt unter den beiden Primzahlen der zweiten Art $f(\alpha)$ und $\phi(\alpha)$ das Reciprocitätsgesetz (27.) für den Fall, dafs die eine Nichtrest der andern ist; wenn dasselbe aber für die Nichtreste besteht, so folgt von selbst, dafs es auch gelten mufs, wenn die eine Primzahl Rest der andern ist. Es bleibt also nur noch zu zeigen, dafs man, wie auch die beiden Primzahlen der zweiten Art $f(\alpha)$ und $\phi(\alpha)$ gegeben sein mögen, stets eine Primzahl der ersten Art finden kann, welche Nichtrest von $f(\alpha)$ und auch Nichtrest von $\phi(\alpha)$ ist. Wählt man in dem Satze (I.) §. 16. für $F(\alpha)$, $F_1(\alpha)$ ... die wirklichen complexen Zahlen $f(\alpha)^h$, $\phi(\alpha)^h$ und eine Einheit $E(\alpha)$, so zeigt derselbe unmittelbar, dafs es Primzahlen der ersten Art $f_1(\alpha)$ giebt, für welche $f(\alpha)$ und $\phi(\alpha)$ Nichtreste sind, und hieraus folgt nach dem Satze (IV.), dafs auch umgekehrt diese Zahlen $f_1(\alpha)$ sowohl für $f(\alpha)$, als auch für $\phi(\alpha)$ Nichtreste sind. Somit ist die Gültigkeit des einfachen Reciprocitätsgesetzes auch für je zwei Primzahlen der zweiten Art bewiesen, und man hat den Satz:

(V.) Wenn zwei primäre complexe Primzahlen $f(\alpha)$ und $\phi(\alpha)$ beide der zweiten Art angehören, so besteht unter denselben das Reciprocitätsgesetz:

$$\left(\frac{f(\alpha)}{\phi(\alpha)}\right) = \left(\frac{\phi(\alpha)}{f(\alpha)}\right).$$

Es besteht also in allen drei unterschiedenen Fällen: wenn beide Primzahlen der ersten Art angehören, wenn eine der ersten Art, die andere der zweiten Art angehört, und wenn beide der zweiten Art angehören, dasselbe einfache Reciprocitätsgesetz. Das Resultat dieser Untersuchung oder das allgemeine Reciprocitätsgesetz, in so weit es hier streng bewiesen worden ist, kann daher vollständig so ausgesprochen werden:

(VI.) Wenn λ eine ungrade Primzahl ist, welche in keiner der $\frac{\lambda-3}{2}$ ersten Bernoullischen Zahlen als Faktor des Zählers enthalten ist, so findet unter je zwei, aus λten Wurzeln der Einheit gebildeten, wirklichen oder idealen, primären complexen Primzahlen $f(\alpha)$ und $\phi(\alpha)$ das Reciprocitätsgesetz Statt:

$$\left(\frac{f(\alpha)}{\phi(\alpha)}\right) = \left(\frac{\phi(\alpha)}{f(\alpha)}\right);$$

wo da's dem Legendreschen analog gebildete Zeichen der Reste und Nichtreste der λten Potenzen durch folgende Congruenz bestimmt ist:

$$\left(\frac{\phi(\alpha)}{f(\alpha)}\right) \equiv \phi(\alpha)^{\frac{Nf(\alpha)-1}{\lambda}} \equiv \alpha^{K}, \quad \text{mod. } f(\alpha),$$

oder wenn $\phi(\alpha)$ ideal, aber $\phi(\alpha)^h$ wirklich ist, durch die Congruenz:

$$\left(\frac{\phi(\alpha)}{f(\alpha)}\right)^h \equiv (\phi(\alpha)^h)^{\frac{Nf(\alpha)-1}{\lambda}} \equiv \alpha^{hK}, \quad \text{mod. } f(\alpha),$$

und wo die Bedingung, dafs eine complexe Zahl $\phi(\alpha)$ primär ist, durch die beiden Congruenzen:

$$\phi(\alpha)\,\phi(\alpha^{-1}) \equiv \phi(1)^2, \text{ mod. } \lambda,$$
$$\phi(\alpha) \equiv \phi(1), \text{ mod. } \varrho^2,$$

oder wenn $\phi(\alpha)$ ideal, aber $\phi(\alpha)^h$ wirklich ist, durch die Congruenzen:

$$\phi(\alpha)^h\,\phi(\alpha^{-1})^h \equiv (\phi(1)^h)^2, \text{ mod. } \lambda,$$
$$\phi(\alpha)^h \equiv \phi(1)^h, \text{ mod. } \varrho^2,$$

bestimmt ist.

Schliefslich bemerke ich, dafs ich aufser dem hier gegebenen Beweise noch zwei andere Beweise des allgemeinen Reciprocitätsgesetzes gefunden habe, welche insofern einfacher sind, als sie bedeutend weniger Vorarbeiten erfordern. Dieselben stützen sich nämlich auf die Theorie der complexen Zahlen in w allein, so dafs die ganze Theorie der complexen Zahlen in z, die Eintheilung der idealen Zahlen dieser Theorie in die Klassen und Gattungen, die ganze Lehre von den ambigen Klassen und die schwierige Bestimmung der Anzahl der wirklich vorhandenen Gattungen erspart wird. Ich habe aber den hier gegebenen Beweis, auch nachdem ich die beiden kürzeren gefunden hatte, nicht unterdrücken wollen, weil er die allgemeine Anwendbarkeit der Principien des entsprechenden Gaufsischen Beweises für die quadratischen Reste ins Licht stellt, und weil er, wenn die nöthigen Sätze aus der Theorie der complexen Zahlen in z einmal entwickelt sind, eben so einfach ist, als diese beiden neuen Beweise, welche ich der Königlichen Akademie bei einer anderen Gelegenheit vorzutragen gedenke.

Inhaltsverzeichnifs.

Bemerkungen über die aus 29ten Einheitswurzeln gebildeten complexen Zahlen

Monatsberichte der Königlichen Preußischen Akademie der Wissenschaften zu Berlin aus dem Jahre 1860, 734–735

Hieran knüpfte Hr. Kummer folgende Bemerkungen:

Die aus 29ten Einheitswurzeln gebildeten complexen Zahlen haben darum ein besonderes Interesse, weil bei ihnen zuerst die Irregularität auftritt, dafs schon die Quadrate aller idealen Zahlen wirklich sind, während die Klassenanzahl nicht zwei, sondern acht ist, wie ich in einem Aufsatze in den Monatsberichten vom Juni 1853 nachgewiesen habe. Durch die von Hrn. Prof. Reuschle berechneten idealen Primfaktoren der Zahlen 59, 233, 349, 523, 929 bin ich nun in den Stand gesetzt worden zu untersuchen, wie die je achtundzwanzig conjugirten idealen Zahlen sich unter die vorhandenen acht Klassen vertheilen. Es geschieht diefs in der Art, dafs eine jede der sieben Klassen, welche die idealen Zahlen enthalten, vier von den achtundzwanzig conjugirten erhält und zwar stets diejenigen vier, deren Einheitswurzeln zusammen eine viergliedrige Periode bilden:

$$f(\alpha^{\gamma^k}),\; f(\alpha^{\gamma^{k+7}}),\; f(\alpha^{\gamma^{k+14}}),\; f(\alpha^{\gamma^{k+21}}),$$

wo γ eine primitive Wurzel von 29 ist und k eine der sieben Zahlen 0, 1, 2, 3, 4, 5, 6.

Die Methoden, nach denen Hr. Prof. Reuschle diese idealen Primfaktoren berechnet hat, welche er anderweitig zu veröffentlichen gedenkt, haben ihn namentlich für $p = 349$ und $p = 523$, nicht zu den einfachsten Ausdrücken geführt. Nach einer andern Art der Berechnung habe ich für diese idealen Primfaktoren folgende einfachere Ausdrücke gefunden:

$$p = 349, \quad u = 228, \quad (f(\alpha))^2 = 1 + \alpha + \alpha^2 + \alpha^3 + \alpha^4 + \alpha^8 + \alpha^{10} + \alpha^{22} + \alpha^{25},$$

$$p = 523, \quad u = 226, \quad (f(\alpha))^2 = 1 + \alpha^3 + \alpha^9 + \alpha^{16} + \alpha^{22} + \alpha^{24} - \alpha^8,$$

welche aus den von Hrn. Prof. Reuschle gegebenen auch durch Multiplikation mit passend gewählten Einheiten würden gewonnen werden können.

Zwei neue Beweise der allgemeinen Reciprocitätsgesetze unter den Resten und Nichtresten der Potenzen, deren Grad eine Primzahl ist

Journal für die reine und angewandte Mathematik 100, 10–50 (1887)

Einleitung.

Der erste Beweis der allgemeinen Reciprocitätsgesetze für Potenzreste, deren Grad eine Primzahl λ ist, welchen ich in den Abhandlungen der Akademie vom Jahre 1859 gegeben habe, stützt sich auf die Theorie gewisser Formen des λ^{ten} Grades mit λ Unbestimmten, in denen die Coefficienten sowohl, als die Unbestimmten aus λ^{ten} Einheitswurzeln gebildete complexe Zahlen sind. Diese Formen des λ^{ten} Grades, welche ich als Normen complexer Zahlen einer höheren Ordnung behandle, stehen mit den allgemeinen Reciprocitätsgesetzen für λ^{te} Potenzreste in einem eben so innigen Zusammenhange, als die quadratischen Formen mit den Reciprocitätsgesetzen für die quadratischen Reste, und man kann durch die Theorie derselben nicht nur auf dem Wege, welchen ich in der angeführten Abhandlung nach Analogie des zweiten *Gauss*ischen Beweises des *theorema fundamentale* eingeschlagen und durchgeführt habe, sondern auch auf anderen und zwar kürzeren Wegen zu den allgemeinen Reciprocitätsgesetzen gelangen. Die beiden neuen Beweise, welche ich aus dieser Quelle herleiten werde, haben ebenso wie jener zuerst gegebene ihre analogen, aus der Theorie der quadratischen Formen zu schöpfenden Beweise der quadratischen Reciprocitätsgesetze, deren kurze Entwickelung ich als Einleitung voranschicken will, um an diesen einfachsten Beispielen den Gedankengang der folgenden allgemeineren Untersuchung darzulegen.

Der erste der in dem Folgenden auszuführenden Beweise der allgemeinen Reciprocitätsgesetze beruht wesentlich nur auf der Theorie der Einheiten der, aus den Wurzeln der Gleichungen $\alpha^\lambda = 1$ und $\psi^\lambda = D(\alpha)$ gebildeten, complexen Zahlen, der analoge Beweis des quadratischen Reciprocitätsgesetzes also auf der *Pell*schen Gleichung $t^2 - Du^2 = 1$. *Legendre* hat bekanntlich den einen Fall, wo die zu vergleichenden Primzahlen beide von der Form $4n+3$ sind, durch diese *Pell*sche Gleichung bewiesen, während er für die anderen Fälle andere Hülfsmittel aus der Theorie der quadratischen Formen anwendet; es lässt sich aber dasselbe Princip consequent für alle Fälle durchführen, wenn man einige Hülfssätze über die Existenz von Primzahlen, die gewisse Bedingungen erfüllen müssen, anwendet, welche jetzt mit Hülfe der *Dirichlet*schen Methoden vollkommen streng bewiesen werden können.

Nimmt man in der *Pell*schen Gleichung

$$t^2 - Du^2 = 1$$

die Determinante D als eine Zahl von der Form $4n+1$, so ist nothwendig t ungrade und u grade, und diese Gleichung in die Form

$$(t+1)(t-1) = Du^2$$

gesetzt, ergiebt:

$$t+1 = 2m\varkappa^2, \quad t-1 = 2m'\lambda^2;$$

wo

$$mm' = D \quad \text{und} \quad 2\varkappa\lambda = u$$

ist. Man erhält hieraus

$$1 = m\varkappa^2 - m'\lambda^2.$$

Es sind nun so viele verschiedene Fälle zu betrachten, als es verschiedene Zerlegungen der Zahl D in zwei Factoren m und m' giebt. Nimmt man für t und u die kleinsten der *Pell*schen Gleichung genügenden positiven Zahlen, so findet von allen diesen Fällen stets nur ein einziger wirklich Statt, auch wird dadurch der Fall, wo $m = 1$, $m' = D$ ist, ausgeschlossen, weil $\varkappa$ und λ kleiner sind, als t und u. Für die eine Gleichung der Form $1 = m\varkappa^2 - m'\lambda^2$, welche wirklich Statt hat, muss m quadratischer Rest von m', also auch von allen Primfactoren des m' sein, und ebenso $-m'$ quadratischer Rest von m, also auch von allen Primfactoren des m.

Es sei nun erstens $D = pp'$, gleich dem Producte zweier verschiedenen Primzahlen der Form $4n+3$, so fällt ausser $1 = \varkappa^2 - pp'\lambda^2$ auch $1 = pp'\varkappa^2 - \lambda^2$ weg, und es bleiben nur folgende zwei Fälle übrig:

$$1 = p\varkappa^2 - p'\lambda^2, \quad \text{wenn} \quad \left(\frac{p}{p'}\right) = +1, \quad \left(\frac{p'}{p}\right) = -1,$$

$$1 = p'\varkappa^2 - p\lambda^2, \quad \text{wenn} \quad \left(\frac{p'}{p}\right) = +1, \quad \left(\frac{p}{p'}\right) = -1;$$

2*

man hat daher:

$$\text{Wenn } \left(\frac{p}{p'}\right) = +1, \text{ so ist } \left(\frac{p'}{p}\right) = -1.$$

$$\text{Wenn } \left(\frac{p}{p'}\right) = -1, \text{ so ist } \left(\frac{p'}{p}\right) = +1.$$

Es sei zweitens $D = pp'q$, und p und p' Primzahlen der Form $4n+3$, aber q eine Primzahl der Form $4n+1$, so sind acht Zerfällungen der Determinante D in zwei Factoren vorhanden, und demgemäss acht verschiedene Fälle besonders zu betrachten, deren erster $m = 1$, $m' = pp'q$ wegfällt, weil t und u die kleinsten der *Pell*schen Gleichung genügenden Zahlen sein sollen. Wählt man nun die Primzahl p' so, dass

$$\left(\frac{p'}{p}\right) = -1 \text{ und } \left(\frac{p'}{q}\right) = -1$$

ist, so ist nach dem bereits bewiesenen Falle des Reciprocitätsgesetzes $\left(\frac{p}{p'}\right) = +1$, und es bleiben alsdann, wenn man die diesen Bedingungen widersprechenden Fälle ausschliesst, nur folgende drei übrig:

$$1 = p\varkappa^2 - p'q\lambda^2, \text{ wenn } \left(\frac{p}{q}\right) = +1, \left(\frac{q}{p}\right) = +1,$$

$$1 = q\varkappa^2 - pp'\lambda^2, \text{ wenn } \left(\frac{q}{p}\right) = +1, \left(\frac{q}{p'}\right) = +1, \left(\frac{p}{q}\right) = -1,$$

$$1 = pp'\varkappa^2 - q\lambda^2, \text{ wenn } \left(\frac{q}{p}\right) = -1, \left(\frac{q}{p'}\right) = -1, \left(\frac{p}{q}\right) = -1.$$

Wenn $\left(\frac{p}{q}\right) = +1$ ist, so findet nur der erste dieser drei Fälle Statt, welcher zugleich $\left(\frac{q}{p}\right) = +1$ giebt; wenn ferner $\left(\frac{q}{p}\right) = -1$ ist, so findet nur der dritte Statt, welcher zugleich $\left(\frac{p}{q}\right) = -1$ giebt, also:

$$\text{Wenn } \left(\frac{p}{q}\right) = +1, \text{ so ist } \left(\frac{q}{p}\right) = +1.$$

$$\text{Wenn } \left(\frac{q}{p}\right) = -1, \text{ so ist } \left(\frac{p}{q}\right) = -1.$$

Bestimmt man die Primzahl p' in anderer Weise, und zwar so, dass

$$\left(\frac{p}{p'}\right) = +1 \text{ und } \left(\frac{q}{p'}\right) = -1 \text{ *)}$$

---

*) Nur an *dieser* Stelle wird ähnlich wie bei *Legendre* von der Voraussetzung Gebrauch gemacht, dass zu einer jeden Primzahl q von der Form $4n+1$ eine Primzahl p' von der Form $4n+3$ gefunden werden kann, für welche jene quadratischer Nichtrest ist. (Vgl. S. 14 oben.) *Kronecker.*

ist, woraus nach den bereits bewiesenen Fällen des Reciprocitätsgesetzes $\left(\frac{p'}{p}\right) = -1$ und $\left(\frac{p'}{q}\right) = -1$ folgt, so bleiben nur die beiden Fälle

$$1 = p\varkappa^2 - p'q\lambda^2, \quad \text{wenn} \quad \left(\frac{p}{q}\right) = +1, \quad \left(\frac{q}{p}\right) = +1,$$

$$1 = pp'\varkappa^2 - q\lambda^2, \quad \text{wenn} \quad \left(\frac{p}{q}\right) = -1, \quad \left(\frac{q}{p}\right) = -1$$

übrig, aus welchen folgt:

$$\text{Wenn} \quad \left(\frac{p}{q}\right) = -1, \quad \text{so ist} \quad \left(\frac{q}{p}\right) = -1.$$

$$\text{Wenn} \quad \left(\frac{q}{p}\right) = +1, \quad \text{so ist} \quad \left(\frac{p}{q}\right) = +1.$$

Es sei nun drittens $D = pp'qq'$, wo p und p' Primzahlen von der Form $4n+3$, q und q' Primzahlen der Form $4n+1$ sind. Weil hier D auf sechzehn verschiedene Weisen in zwei Factoren zerlegt werden kann, so sind sechzehn Fälle besonders zu betrachten, deren erster $m = 1$, $m' = pp'qq'$ jedoch aus dem oben angegebenen Grunde wegfällt. Wählt man nun die Primzahlen p und p' in der Art, dass

$$\left(\frac{p}{q}\right) = -1, \quad \left(\frac{p}{q'}\right) = +1, \quad \left(\frac{p'}{q}\right) = +1, \quad \left(\frac{p'}{q'}\right) = -1,$$

woraus nach dem bereits Bewiesenen

$$\left(\frac{q}{p}\right) = -1, \quad \left(\frac{q'}{p}\right) = +1, \quad \left(\frac{q}{p'}\right) = +1, \quad \left(\frac{q'}{p'}\right) = -1$$

folgt, so bleiben nach Ausschliessung der diesen Bedingungen widersprechenden Fälle nur folgende drei bestehen:

$$1 = pq\varkappa^2 - p'q'\lambda^2, \quad \text{wenn} \quad \left(\frac{p}{p'}\right) = +1, \quad \left(\frac{q}{q'}\right) = +1, \quad \left(\frac{q'}{q}\right) = +1,$$

$$1 = p'q\varkappa^2 - pq'\lambda^2, \quad \text{wenn} \quad \left(\frac{p'}{p}\right) = -1, \quad \left(\frac{q}{q'}\right) = -1, \quad \left(\frac{q'}{q}\right) = -1,$$

$$1 = p'q'\varkappa^2 - pq\lambda^2, \quad \text{wenn} \quad \left(\frac{p'}{p}\right) = +1, \quad \left(\frac{q'}{q}\right) = +1, \quad \left(\frac{q}{q'}\right) = +1.$$

Wenn nun $\left(\frac{q}{q'}\right) = -1$ ist, so findet nur der zweite dieser drei Fälle Statt, und man hat $\left(\frac{q'}{q}\right) = -1$; wenn aber $\left(\frac{q}{q'}\right) = +1$ ist, so kann nur der erste oder dritte dieser Fälle Statt haben; der eine so wie der andere ergiebt aber alsdann auch $\left(\frac{q'}{q}\right) = +1$, also:

$$\text{Wenn } \left(\frac{q}{q'}\right) = -1, \quad \text{so ist } \left(\frac{q'}{q}\right) = -1.$$

$$\text{Wenn } \left(\frac{q}{q'}\right) = +1, \quad \text{so ist } \left(\frac{q'}{q}\right) = +1.$$

Zur Vervollständigung dieses Beweises würde noch der Nachweis gehören, dass immer Primzahlen der Art existiren, wie sie hier als Hülfszahlen angewendet worden sind, ebenso wie zur Vollständigkeit des von *Legendre* gegebenen Beweises noch der Beweis des Satzes gehörte, dass zu einer jeden Primzahl von der Form $4n+1$ eine Primzahl der Form $4n+3$ gefunden werden kann, für welche jene quadratischer Rest ist; da ich aber hier nicht beabsichtige, einen neuen Beweis der quadratischen Reciprocitätsgesetze aufzustellen, sondern lediglich ein einfaches und genaues Schema des analogen Beweises der allgemeinen Reciprocitätsgesetze zu geben, so würde es überflüssig sein, auf den Beweis dieser Hülfssätze einzugehen, welcher durch die *Dirichlet*schen Methoden ausgeführt werden kann.

Der andere neue Beweis der allgemeinen Reciprocitätsgesetze entspricht keinem der bisher bekannten Beweise für das quadratische Reciprocitätsgesetz; das Princip, auf welchem derselbe beruht, ist aber ebenso auf diesen besonderen Fall anwendbar, und liefert einen neuen und zwar sehr einfachen Beweis dieses Fundamentaltheorems, welchen ich hier ebenfalls ausführen will, um in ihm die Umrisse der betreffenden allgemeineren Untersuchung einfach und klar zu zeigen.

Es seien p und p' Primzahlen der Form $4n+3$, q und q' Primzahlen der Form $4n+1$, und r eine Primzahl, welche sich durch eine quadratische Form der Determinante $-p$ darstellen lässt, welche also der Bedingung

$$\left(\frac{-p}{r}\right) = 1$$

genügt: so wird im allgemeinen r nicht durch die Hauptklasse darstellbar sein, sondern durch irgend eine andere Klasse der Formen dieser Determinante. Es ist aber alsdann nothwendig eine Potenz von r durch die Hauptform x^2+py^2 darstellbar, und der Exponent der niedrigsten, durch die Hauptform darstellbaren Potenz von r ist ein Theiler der Klassenanzahl der quadratischen Formen der Determinante $-p$, m. s. *Gauss disq. arithm. art.* 305. Diese Klassenanzahl ist aber nothwendig eine ungrade Zahl, weil für die Determinante $-p$ ausser der Hauptklasse keine *classis anceps* existirt und ausser dieser einen Hauptklasse alle anderen Klassen, wenn man zu jeder ihre entgegengesetzte hinzunimmt, paarweise vorkommen, m. s. *Gauss disq. arith. art.* 303. Es ist daher auch der Exponent der Potenz, zu welcher r erhoben werden muss, um durch die Hauptform darstellbar zu sein, nothwendig ungrade. Bezeichnet man denselben mit $2h+1$, so hat man

$$x^2+py^2 = r^{2h}.r,$$

und diese Gleichung als Congruenz nach dem Modul p betrachtet, zeigt, ass $\left(\frac{r}{p}\right)=1$ sein muss, also:

$$\text{Wenn } \left(\frac{-p}{r}\right)=+1, \quad \text{so ist } \left(\frac{r}{p}\right)=+1.$$

Nimmt man nun erstens $r=p'$ von der Form $4n+3$, und zweitens $r=q$ von der Form $4n+1$, so hat man:

$$\text{Wenn } \left(\frac{p}{p'}\right)=-1, \quad \text{so ist } \left(\frac{p'}{p}\right)=+1.$$

$$\text{Wenn } \left(\frac{p}{q}\right)=+1, \quad \text{so ist } \left(\frac{q}{p}\right)=+1.$$

Es sei zweitens r irgend eine Primzahl, welche sich durch quadratische Formen der Determinante $+q$ darstellen lässt, für welche also

$$\left(\frac{q}{r}\right) = +1$$

ist. Es ist alsdann nothwendig eine bestimmte Potenz von r, deren Exponent ein Divisor der Klassenanzahl der quadratischen Formen der Determinante $+q$ ist, durch die Hauptform x^2-qy^2 darstellbar, und weil für die Determinante $+q$ ebenfalls nur die eine Hauptklasse als *classis anceps* existirt, so ist die Klassenanzahl und folglich auch der Exponent der durch die Hauptform darstellbaren Potenz von r eine ungrade Zahl. Bezeichnet man denselben mit $2h+1$, so hat man

$$x^2-qy^2 = r^{2h}.r,$$

und diese Gleichung, als Congruenz nach dem Modul q betrachtet, giebt $\left(\frac{r}{q}\right)=1$, also:

$$\text{Wenn } \left(\frac{q}{r}\right)=+1, \quad \text{so ist } \left(\frac{r}{q}\right)=+1.$$

Nimmt man nun r einerseits als Primzahl p von der Form $4n+3$, andererseits als Primzahl q' von der Form $4n+1$, so hat man:

$$\text{Wenn} \quad \left(\frac{q}{p}\right) = +1, \quad \text{so ist} \quad \left(\frac{p}{q}\right) = +1.$$

$$\text{Wenn} \quad \left(\frac{q}{q'}\right) = +1, \quad \text{so ist} \quad \left(\frac{q'}{q}\right) = +1.$$

Die hier bewiesenen vier Fälle erschöpfen die quadratischen Reciprocitätsgesetze unter je zwei ungraden Primzahlen vollständig, weil die vier anderen Fälle, nämlich:

$$\text{Wenn} \quad \left(\frac{p}{p'}\right) = +1, \quad \text{so ist} \quad \left(\frac{p'}{p}\right) = -1.$$

$$\text{Wenn} \quad \left(\frac{p}{q}\right) = -1, \quad \text{so ist} \quad \left(\frac{q}{p}\right) = -1.$$

$$\text{Wenn} \quad \left(\frac{q}{p}\right) = -1, \quad \text{so ist} \quad \left(\frac{p}{q}\right) = -1.$$

$$\text{Wenn} \quad \left(\frac{q}{q'}\right) = -1, \quad \text{so ist} \quad \left(\frac{q'}{q}\right) = -1,$$

aus denselben unmittelbar folgen *).

Der Vortheil, welchen dieser sehr einfache Beweis vor dem obigen voraus hat, dass er keine Hülfssätze über die Existenz von Primzahlen, welche gewisse Bedingungen erfüllen, nöthig hat, geht in dem allgemeineren Falle verloren, wo es sich um λ^{te} Potenzreste handelt. Es scheint überhaupt, dass die allgemeinen Reciprocitätsgesetze aus der von mir zu Grunde gelegten Theorie complexer Zahlen ohne Anwendung solcher Hülfs-Prim-

---

*) Dass, wenn $\left(\frac{p}{p'}\right) = +1$ ist, nothwendig $\left(\frac{p'}{p}\right) = -1$ sein muss, ist hier aus Versehen als eine Folge der unmittelbar vorhergehenden Deduction bezeichnet worden; es ist dies aber gerade die erste und einfachste Schlussfolgerung, welche oben auf S. 12 aus der Existenz von Lösungen der *Pell*schen Gleichung $t^2 - pp'u^2 = 1$ gezogen worden ist. Die Benutzung der *Pell*schen Gleichung auch bei diesem zweiten *Kummer*schen Beweise des Reciprocitätsgesetzes, der im Uebrigen sich nur auf die Endlichkeit der Klassenanzahl der quadratischen Formen stützt, ist aber auch nothwendig; denn nur auf die *beiden Lagrange*schen Fundamentalsätze der Theorie der quadratischen Formen, nämlich auf die Endlichkeit der Klassenanzahl und auf die Auflösbarkeit der *Pell*schen Gleichung, kann ein Beweis des Reciprocitätsgesetzes gegründet werden, weil auch umgekehrt diese beiden *Lagrange*schen Fundamentalsätze sich als eine Folge des Reciprocitätsgesetzes darstellen lassen (Monatsbericht der hiesigen Akademie vom 12. Mai 1864 S. 294 und 295). In dem ersten auf die Lösung der *Pell*schen Gleichung basirten *Kummer*schen Beweise wird die Voraussetzung der Endlichkeit der Klassenanzahl bei jener Voraussetzung über die Existenz von Primzahlen p' gebraucht, für welche eine gegebene Primzahl q Nichtrest ist. Ich werde die hier angedeutete Beziehung des Reciprocitätsgesetzes und der Fundamentalsätze der Theorie der quadratischen Formen in einem besonderen Aufsatze näher darlegen.

Kronecker.

zahlen sich nicht möchten vollständig beweisen lassen; aber diese Theorie liefert zugleich auch den Beweis für die Existenz aller dieser Hülfs-Primzahlen, so dass der vollkommenen Strenge der Beweise durch deren Anwendung nichts vergeben wird.

Die beiden neuen Beweise der allgemeinen Reciprocitätsgesetze sind besonders auch darum einfacher als der erste, in den Abhandlungen der Akademie vom Jahre 1859 gegebene, weil sie die Betrachtung derjenigen besonderen Art complexer Zahlen, welche ich dort als complexe Zahlen in z bezeichnet habe, nicht erfordern, sondern lediglich auf der Theorie der, aus den Wurzeln der Gleichungen $\alpha^\lambda = 1$ und $w^\lambda = D(\alpha)$ gebildeten complexen Zahlen in w beruhen. Die Resultate der in jener Abhandlung entwickelten Theorie der complexen Zahlen in w werden daher in der gegenwärtigen Untersuchung überall ihre Anwendung finden. Da dieselben aber für den gegenwärtigen Zweck noch nicht ausreichen, so soll diese Theorie zunächst noch in so weit vervollständigt und weiter ausgeführt werden, dass alsdann die allgemeinen Reciprocitätsgesetze mit Leichtigkeit aus ihr gefolgert werden können.

§ 1.

Die complexen Zahlen in w, deren Normen keine anderen Primfactoren enthalten, als die der Determinante $D(\alpha)$ und $1-\alpha$ und Einheiten in α.

Die complexen Zahlen in w, welche keinen der im § 4 der ersten Abhandlung über die allgemeinen Reciprocitätsgesetze vom Jahre 1859 definirten idealen Primfactoren enthalten, deren Normen also lediglich aus den Primfactoren der Determinante $D(\alpha)$, dem Factor $1-\alpha = \varrho$ und Einheiten in α zusammengesetzt sind, stehen mit den Einheiten in w im genausten Zusammenhange. Dieselben werden aus dem, pag. 111 der genannten Abhandlung aufgestellten Systeme der in der Form

$$(1.)\qquad \varepsilon(w)^a \varepsilon_1(w)^{a_1} \ldots \varepsilon_{\mu-1}(w)^{a_{\mu-1}}$$

enthaltenen Einheiten abgeleitet, deren Normen gleich Eins sind, und welche in so fern von einander unabhängig sind, dass der Quotient je zweier derselben niemals eine ϱ^{te} Potenz einer Einheit sein kann.

Die in dieser Form (1.) enthaltenen Einheiten, deren Anzahl gleich λ^μ ist, weil jeder der μ Exponenten $a, a_1, \ldots a_{\mu-1}$ alle λ Werthe 0, 1, 2, ... $\lambda-1$ erhalten kann, haben noch eine gewisse Unbestimmtheit an sich, da sie mit den Potenzen der einfachen Einheit α beliebig multiplicirt genommen werden können. Wenn

$$E(w) = A+A_1w+A_2w^2+\cdots+A_{\lambda-1}w^{\lambda-1}$$

irgend eine dieser λ^μ Einheiten ist, so hat man

$$NE(w) = 1 \equiv A^\lambda \quad (\text{mod. } D(\alpha)),$$

also

$$(A-1)(A\alpha-1)(A\alpha^2-1)\ldots(A\alpha^{\lambda-1}-1) \equiv 0 \quad (\text{mod. } D(\alpha)).$$

Wenn nun $f(\alpha)$ irgend ein beliebig zu wählender Primfactor der Determinante $D(\alpha)$ ist, so muss nothwendig einer, und auch nur einer der Factoren dieses Products durch $f(\alpha)$ theilbar sein, man hat daher

$$A\alpha^k-1 \equiv 0 \quad (\text{mod. } f(\alpha)),$$

für einen bestimmten Werth des k. Nimmt man nun die Einheit $\alpha^kE(w)$ statt $E(w)$, so ist $A\alpha^k$ der erste Coefficient der Entwickelung, welcher demnach congruent Eins ist, nach dem Modul $f(\alpha)$. Man kann also jede Einheit, deren Norm gleich Eins ist, durch Multiplication mit einer passenden Einheit α^k so einrichten, dass in ihrer Darstellung als ganze rationale Function von w das erste, w nicht enthaltende Glied congruent Eins ist, für irgend einen beliebig zu wählenden Primfactor der Determinante als Modul.

Es sei nun $E(w)$ irgend eine in der Form (1.) enthaltene Einheit, so gewählt, dass in dem Ausdrucke

$$E(w) = A+A_1w+A_2w^2+\cdots+A_{\lambda-1}w^{\lambda-1}$$

das erste Glied A congruent Eins sei, nach dem als Primfactor in der Determinante $D(\alpha)$ enthaltenen Modul $f(\alpha)$, so hat der Ausdruck

$$PE(w) = 1+E(w)+E(w)E(w\alpha)+\cdots+E(w)E(w\alpha)\ldots E(w\alpha^{\lambda-2})$$

die Eigenschaft, dass

$$E(w)PE(w\alpha) = PE(w)$$

ist. Bezeichnet man diesen Ausdruck $PE(w)$, als ganze rationale Function von w oder complexe Zahl in w, mit $F(w)$, so hat man

$$(2.) \quad E(w)F(w\alpha) = F(w)$$

Diese Gleichung würde nichtssagend sein, wenn $PE(w)$ d. i. $F(w)$ gleich Null wäre; es ist aber leicht zu zeigen, dass dieser Fall niemals eintreten kann. Ordnet man nämlich den Ausdruck $PE(w)$ nach Potenzen von w, ohne von der Gleichung $w^\lambda = D(\alpha)$ Gebrauch zu machen, so erhält man

$$PE(w) = C+C_1w+C_2w^2+C_3w^3+\cdots,$$

wo

$$C = 1+A+A^2+A^3+\cdots+A^{\lambda-1},$$

und man hat

$$NPE(w) \equiv C^{\lambda} \quad (\text{mod. } D(\alpha)).$$

Nach dem Modul $f(\alpha)$, Primfactor von $D(\alpha)$, ist aber

$$A \equiv 1 \quad (\text{mod. } f(\alpha)),$$

und folglich

$$C \equiv \lambda \quad (\text{mod. } f(\alpha)),$$

also

$$NPE(w) \equiv \lambda^{\lambda} \quad (\text{mod. } f(\alpha)).$$

Da nun die Norm von $PE(w)$ nicht congruent Null ist, nach dem Modul $f(\alpha)$, so kann sie auch nicht gleich Null sein, und es kann darum auch $PE(w)$, d. h. $F(w)$ selbst nicht gleich Null sein.

Wenn nun die complexe Zahl $F(w)$ irgend einen der definirten idealen Primfactoren $\varphi(w)$ enthält, so zeigt die Gleichung (2.), dass auch $F(w\alpha)$ denselben enthalten muss, und wenn w in $w\alpha$, $w\alpha^2$, ... verwandelt wird, dass auch alle zu $F(w)$ conjugirten complexen Zahlen denselben enthalten müssen. Es muss darum $F(w)$ ausser dem idealen Primfactor $\varphi(w)$ auch alle zu demselben conjugirten enthalten, welche sich zu einem complexen Factor in α zusammensetzen. Weil in derselben Weise alle in $F(w)$ enthaltenen definirten idealen Primfactoren sich zu complexen Factoren in α zusammensetzen, so hat man

$$(3.) \qquad F(w) = \Psi(\alpha)F_1(w),$$

wo $F_1(w)$ keinen definirten idealen Primfactor weiter enthält. Wenn nun $\Psi(\alpha)$ eine *wirkliche* complexe Zahl in α ist, so muss auch $F_1(w)$ eine wirkliche complexe Zahl in w sein; wenn aber $\Psi(\alpha)$ *ideal* ist, und h bezeichnet die Klassenzahl der idealen Zahlen in α, so ist $\Psi(\alpha)^h$ wirklich, und darum auch $F_1(w)^h$ wirklich. Die Gleichungen (2.) und (3.) geben alsdann

$$E(w)^h F_1(w\alpha)^h = F_1(w)^h.$$

Bestimmt man nun die Zahlen i und k so, dass

$$hk = 1+i\lambda$$

ist, welches stets möglich ist, weil h nicht durch λ theilbar, und erhebt zur k^{ten} Potenz, so wird

$$E(w)^{1+i\lambda} F_1(w\alpha)^{hk} = F(w)^{hk}.$$

3*

Es ist nun $E(w)^{i\lambda}$, als λ^{te} Potenz einer Einheit, deren Norm gleich Eins ist, zugleich eine ϱ^{te} Potenz einer solchen Einheit, man kann daher setzen:

$$E(w)^{i\lambda} = E_1(w)^{\varrho} = \frac{E_1(w)}{E_1(w\alpha)},$$

man hat demnach

$$E(w)\frac{F_1(w\alpha)^{hk}}{E_1(w\alpha)} = \frac{F_1(w)^{hk}}{E_1(w)},$$

und wenn

$$\frac{F_1(w)^{hk}}{E_1(w)} = \varDelta(w)$$

gesetzt wird:

$$(4.)\qquad E(w)\varDelta(w\alpha) = \varDelta(w),\quad E(w) = \frac{\varDelta(w)}{\varDelta(w\alpha)},$$

wo $\varDelta(w)$ eine wirkliche und ganze complexe Zahl in w ist, welche keinen definirten Primfactor enthält, und welche von allen Factoren in α, den wirklichen, so wie den idealen befreit anzunehmen ist. Man hat daher folgenden Satz:

I. ***Jede Einheit $E(w)$, deren Norm gleich Eins ist, lässt sich als Quotient zweier conjugirten, wirklichen Zahlen $\varDelta(w)$ und $\varDelta(w\alpha)$ darstellen, deren Normen keine anderen Factoren enthalten, als die der Determinante $D(\alpha)$ und ausserdem $1-\alpha$ und Einheiten in α.***

Zu bemerken ist noch, dass, wenn $E(w)$ eine der λ^{μ} in der Form (1.) enthaltenen Einheiten ist, $\varDelta(w)$ niemals eine Einheit sein kann, deren Norm gleich Eins ist.

Um die complexen Zahlen $\varDelta(w)$ noch näher zu bestimmen, setze ich:

$$\begin{aligned}\varDelta(w) &= B + B_1 w + B_2 w^2 + \cdots + B_{\lambda-1} w^{\lambda-1},\\ E(w) &= A + A_1 w + A_2 w^2 + \cdots + A_{\lambda-1} w^{\lambda-1},\\ E(w)\varDelta(w\alpha) &= C + C_1 w + C_2 w^2 + \cdots + C_{\lambda-1} w^{\lambda-1}.\end{aligned}$$

Die Ausführung der Multiplication ergiebt alsdann:

$$\begin{aligned}C &= AB + w^{\lambda}(A_1 B_{\lambda-1}\alpha^{\lambda-1} + A_2 B_{\lambda-2}\alpha^{\lambda-2} + \cdots + A_{\lambda-1} B_1 \alpha),\\ C_1 &= AB_1\alpha + A_1 B + w^{\lambda}(A_2 B_{\lambda-1}\alpha^{\lambda-1} + \cdots + A_{\lambda-1} B_2 \alpha^2),\\ C_2 &= AB_2\alpha^2 + A_1 B_1 \alpha + A_2 B + w^{\lambda}(A_3 B_{\lambda-1}\alpha^{\lambda-1} + \cdots + A_{\lambda-1} B_3 \alpha^3),\end{aligned}$$

etc. etc.

Die Gleichung (4.) giebt demnach folgendes System von λ Congruenzen für den Modul $w^{\lambda} = D(\alpha)$:

$$(5.)\quad \begin{cases} (A\ \ -1)B \equiv 0, \\ (A\alpha\ -1)B_1 + A_1 B \equiv 0, \\ (A\alpha^2 - 1)B_2 + A_1 B_1 \alpha + A_2 B \equiv 0, \\ (A\alpha^3 - 1)B_3 + A_1 B_2 \alpha^2 + A_2 B_1 \alpha + A_3 B \equiv 0, \\ \text{etc. etc.} \end{cases} \qquad (\text{mod. } D(\alpha)).$$

Es seien nun eben so, wie im § 9, pag. 86 der früheren Abhandlung: $f(\alpha)$, $f_1(\alpha), \dots f_{r-1}(\alpha)$ die verschiedenen, in der Determinante $D(\alpha)$ enthaltenen Primfactoren, und

$$D(\alpha) = E(\alpha).f(\alpha)^m.f_1(\alpha)^{m_1}\dots f_{r-1}(\alpha)^{m_{r-1}}.$$

Ferner sei

$$\begin{aligned} u^\lambda &= e(\alpha)f(\alpha)^m = \delta(\alpha), \\ u_1^\lambda &= e_1(\alpha)f_1(\alpha)^{m_1} = \delta_1(\alpha), \\ &\vdots \\ u_{r-1}^\lambda &= e_{r-1}(\alpha)f_{r-1}(\alpha)^{m_{r-1}} = \delta_{r-1}(\alpha), \end{aligned}$$

$$E(\alpha) = e(\alpha)e_1(\alpha)\dots e_{r-1}(\alpha), \quad D(\alpha) = \delta(\alpha)\delta_1(\alpha)\dots\delta_{r-1}(\alpha), \quad w = uu_1\dots u_{r-1},$$

wo die Exponenten m, $m_1, \dots m_{r-1}$ nicht durch λ theilbar sind, und nur solche Werthe haben, dass die Factoren der Determinante $\delta(\alpha)$, $\delta_1(\alpha), \dots \delta_{r-1}(\alpha)$ alle einzeln wirklich sind, auch für ideale Primfactoren $f(\alpha)$, $f_1(\alpha), \dots f_{r-1}(\alpha)$.

Wenn die Congruenzen (5.), welche für den Modul $D(\alpha)$ gelten, in Beziehung auf den Factor $\delta_s(\alpha)$ oder den Primfactor $f_s(\alpha)$ aufgefasst werden, so muss, wie oben gezeigt worden, eine der Zahlen $A-1$, $A\alpha-1, \dots A\alpha^{\lambda-1}-1$, und nur eine durch $f_s(\alpha)$ theilbar sein. Diese eine sei $A\alpha^{n_s}-1$, so ergeben die Congruenzen (5.), dass die ersten n_s Coefficienten

$$B, \quad B_1, \quad \dots \quad B_{n_s-1}$$

alle durch $\delta_s(\alpha)$ theilbar sein müssen, dass aber der Coefficient B_{n_s} nicht durch $f_s(\alpha)$ theilbar sein kann, weil sonst alle Coefficienten von $\Delta(w)$ durch $f_s(\alpha)$ theilbar wären, und darum $\Delta(w)$ selbst diesen complexen Factor in α enthalten würde.

Da nun die n ersten Coefficienten des $\Delta(w)$ durch $\delta(\alpha) = u^\lambda$ theilbar sind, der $(n+1)^{\text{te}}$ aber nicht durch $f(\alpha)$ theilbar, ferner die n_1 ersten Coefficienten durch $\delta_1(\alpha) = u_1^\lambda$ theilbar, der $(n_1+1)^{\text{te}}$ aber nicht durch $f_1(\alpha)$ theilbar u. s. w., so kann man aus $\Delta(w)$ die Factoren u^n, $u_1^{n_1}$, u. s. w. herausheben, und erhält demnach:

$$(6.)\qquad \varDelta(w) = u^{n}.u_1^{n_1}\ldots u_{r-1}^{n_{r-1}} f(u, u_1, \ldots u_{r-1}),$$

wo $f(u, u_1, \ldots u_{r-1})$ eine aus den Irrationalitäten $u, u_1, \ldots u_{r-1}$ gebildete complexe Zahl von folgender Form ist:

$$(7.)\qquad f(u, u_1, \ldots u_{r-1}) = \Sigma_k a_k u^{|k-n|}.u_1^{|k-n_1|}\ldots u_{r-1}^{|k-n_{r-1}|},$$

für $k = 0, 1, 2, \ldots \lambda-1$, und $|k-n|, |k-n_1|, \ldots |k-n_{r-1}|$ die kleinsten nicht negativen Reste der Zahlen $k-n, k-n_1, \ldots k-n_{r-1}$, nach dem Modul λ bezeichnen, und wo die Norm von $f(u, u_1, \ldots u_{r-1})$ keinen der Primfactoren der Determinante $f(\alpha), f_1(\alpha), \ldots f_{r-1}(\alpha)$ weiter enthält, also nur noch aus einer Potenz von ϱ und einer Einheit $E(\alpha)$ bestehen kann; man vergleiche pag. 86 und 87 der früheren Abhandlung.

Nach der oben für die Einheit $E(w)$ festgesetzten Bestimmung ist $A-1$ durch $f(\alpha)$ theilbar und darum $n = 0$, man hat also

$$\varDelta(w) = u_1^{n_1} u_2^{n_2}\ldots u_{r-1}^{n_{r-1}} f(u, u_1, \ldots u_{r-1})$$

und

$$N\varDelta(w) = \delta_1(\alpha)^{n_1}\ldots\delta_{r-1}(\alpha)^{n_{r-1}}\varrho^k.\alpha^c\varepsilon_1(\alpha)^{c_1}\varepsilon_2(\alpha)^{c_2}\ldots\varepsilon_{\mu-1}(\alpha)^{c_{\mu-1}},$$

wenn $\varepsilon_1(\alpha), \varepsilon_2(\alpha), \ldots \varepsilon_{\mu-1}(\alpha)$ ein vollständiges System von Fundamentaleinheiten in α bezeichnen. Sucht man nun für jede der λ^{μ} verschiedenen, in der Form (1.) enthaltenen Einheiten die zugehörige complexe Zahl $\varDelta(w)$, so erhält man λ^{μ} solche Zahlen $\varDelta(w)$, welche alle wesentlich von einander verschieden sind, in der Art, dass in den Normen derselben die Exponenten $n, n_1, \ldots n_{r-1}, k, c, c_1, \ldots c_{\mu-1}$ nicht nur nicht gleich, sondern auch nicht alle congruent sein können, nach dem Modul λ, oder was dasselbe ist, dass der Quotient der Normen zweier solcher Zahlen nicht eine λ^{te} Potenz sein kann.

Es seien $E(w)$ und $E_1(w)$ zwei verschiedene Einheiten der Form (1.) und $\varDelta(w)$ und $\varDelta_1(w)$ die ihnen zugehörigen complexen Zahlen, so dass

$$E(w) = \frac{\varDelta(w)}{\varDelta(w\alpha)},\quad E_1(w) = \frac{\varDelta_1(w)}{\varDelta_1(w\alpha)},$$

sei ferner

$$\frac{E(w)}{E_1(w)} = E_2(w),$$

so ist $E_2(w)$ ebenfalls eine in der Form (1.) enthaltene Einheit. Dieselbe gebe

$$E_2(w) = \frac{\varDelta_2(w)}{\varDelta_2(w\alpha)},$$

so hat man

$$\frac{\Delta_1(w\alpha)\Delta_2(w\alpha)}{\Delta(w\alpha)} = \frac{\Delta_1(w)\Delta_2(w)}{\Delta(w)}.$$

Da dieser Ausdruck ungeändert bleibt, wenn w in $w\alpha$ verwandelt wird, so folgt wegen der Irreductibilität der Gleichung $w^\lambda - D(\alpha) = 0$, dass er eine symmetrische Function aller Wurzeln dieser Gleichung sein muss und darum eine complexe Zahl in α, welche jedoch auch eine gebrochene sein kann. Es ist demnach

$$\frac{\Delta_1(w)\Delta_2(w)}{\Delta(w)} = \Psi(\alpha),$$

und hieraus folgt

$$N\Delta_1(w)N\Delta_2(w) = \Psi(\alpha)^\lambda N\Delta(w).$$

Wäre nun der Quotient der Normen $N\Delta(w)$ und $N\Delta_1(w)$ einer λ^{ten} Potenz gleich, so müsste auch $N\Delta_2(w)$ eine λ^{te} Potenz sein, und zwar die λ^{te} Potenz einer ganzen complexen Zahl in α, und weil $N\Delta_2(w)$ dieselbe Form hat, wie $N\Delta(w)$, nämlich

$$N\Delta_2(w) = \delta_1(\alpha)^{n_1}\ldots\delta_{r-1}(\alpha)^{n_{r-1}}\varrho^k\alpha^c\varepsilon_1(\alpha)^{c_1}\ldots\varepsilon_{\mu-1}(\alpha)^{c_{\mu-1}},$$

so müssten die Exponenten $n_1, \ldots n_{r-1}, k, c, c_1, \ldots c_{\mu-1}$ alle einzeln durch λ theilbar sein. Da die Exponenten $n_1, n_2, \ldots n_{r-1}$ kleiner als λ sind, so müssten diese nothwendig gleich Null sein. Was den Exponenten k betrifft, so müsste dieser ebenfalls gleich Null sein. Um dieses letztere vollständig zu beweisen, setze ich die complexe Zahl $\Delta_2(w)$ in die Form:

$$\Delta_2(w) = G + G_1(1-w) + G_2(1-w)^2 + \cdots + G_{\lambda-1}(1-w)^{\lambda-1}.$$

Wenn nun G_i der erste nicht durch ϱ theilbare Coefficient dieser Form ist, so hat man:

$$\Delta_2(w) = (1-w)^i\Phi(w) + \varrho\Psi(w),$$

wo

$$\Phi(w) = G_i + G_{i+1}(1-w) + \cdots + G_{\lambda-1}(1-w)^{\lambda-i-1},$$

also $N\Phi(w)$ nicht durch ϱ theilbar. Aus dieser Form des $\Delta_2(w)$ folgt

$$N\Delta_2(w) \equiv N(1-w)^i N\Phi(w) \pmod{\varrho^\lambda},$$

also $N\Delta_2(w)$ enthält den Factor ϱ genau ebenso oft, als $N(1-w)^i$, also genau i-mal, und weil $i < \lambda$ ist, so folgt, dass $N\Delta_2(w)$ den Factor ϱ nur dann mehr als $(\lambda-1)$-mal enthalten kann, wenn $\Delta_2(w)$ selbst den Factor ϱ enthält, welcher Fall dadurch ausgeschlossen ist, dass $\Delta_2(w)$ überhaupt keinen Factor in α enthalten soll. Es folgt hieraus, dass k, welches durch λ theilbar und kleiner als λ sein muss, nur den Werth Null haben kann. Man hat demnach

$$N\Delta_2(w) = E(\alpha)^\lambda,$$

und wenn $\frac{\Delta_2(w)}{E(\alpha)}$, als ganze complexe Zahl, deren Norm gleich Eins ist, mit $\mathrm{E}(w)$ bezeichnet wird, so hat man

$$E_2(w) = \frac{\Delta_2(w)}{\Delta_2(w\alpha)} = \frac{\mathrm{E}(w)}{\mathrm{E}(w\alpha)} = \mathrm{E}(w)^\varrho.$$

Die Einheit $E_2(w)$ ist also nothwendig eine ϱ^{te} Potenz einer Einheit, woraus folgt, dass, wenn der Quotient der Normen $N\Delta(w)$ und $N\Delta_1(w)$ eine λ^{te} Potenz ist, der Quotient der entsprechenden Einheiten $E(w)$ und $E_1(w)$ eine ϱ^{te} Potenz einer Einheit ist, und folglich $E(w)$ und $E_1(w)$ nicht zwei verschiedene Einheiten der Form (1.).

Die λ^μ wesentlich verschiedenen complexen Zahlen $\Delta(w)$, welche man aus den λ^μ verschiedenen Einheiten der Form (1.) erhält, haben alle die Eigenschaft, dass sie den einen Primfactor $f(\alpha)$ der Determinante $D(\alpha)$ nicht enthalten. Will man diesen Primfactor ebenfalls in den Normen zulassen, so hat man nur jede der λ^μ Zahlen $\Delta(w)$ mit den Grössen 1, w, $w^2, \ldots w^{\lambda-1}$ zu multipliciren, wodurch man $\lambda^{\mu+1}$ Zahlen $\Delta(w)$ erhält, deren Normen von der Form

$$(8.)\qquad N\Delta(w) = \delta(\alpha)^n \delta_1(\alpha)^{n_1} \ldots \delta_{r-1}(\alpha)^{n_{r-1}} \varrho^k \alpha^c \varepsilon_1(\alpha)^{c_1} \ldots \varepsilon_{\mu-1}(\alpha)^{c_{\mu-1}},$$

nur verschiedene, d. i. nach dem Modul λ nicht congruente Exponenten $n, n_1, \ldots n_{r-1}, k, c, c_1, \ldots c_{\mu-1}$ haben, und welche die Eigenschaft besitzen, dass

$$\frac{\Delta(w)}{\Delta(w\alpha)} = E(w)$$

gleich einer Einheit ist, deren Norm gleich Eins ist. Da nur $\lambda^{\mu+1}$ wesentlich verschiedene complexe Zahlen $\Delta(w)$ existiren, welche aus den Einheiten der Form (1.) erzeugt werden können, und da die in $N\Delta(w)$ enthaltenen $\mu+1+r$ Exponenten $n, n_1, \ldots n_{r-1}, k, c, c_1, \ldots c_{\mu-1}$, $\lambda^{\mu+1+r}$ verschiedene, nach dem Modul λ nicht congruente Werthverbindungen zulassen, so folgt, dass diese nicht alle Statt haben können, sondern dass unter diesen Exponenten gewisse einschränkende Bedingungen Statt haben müssen, welche jetzt entwickelt werden sollen.

Wenn man die Zahl $\Delta(w)$, deren Norm den bei (8.) gegebenen Ausdruck hat, mit $w^{\lambda-n_s}$ multiplicirt, wo n_s der Exponent der in der Norm enthaltenen Potenz $\delta_s^{n_s}$ ist, so wird $w^{\lambda-n_s}\Delta(w)$ durch $\delta_s(\alpha)$ theilbar; wenn nun

$$\frac{w^{\lambda-n_s}\Delta(w)}{\delta_s(\alpha)} = \Delta'(w)$$

gesetzt wird, so kommt in der Norm von $\Delta'(w)$ der Factor $\delta_s(\alpha)$ nicht vor, und die Exponenten der Potenzen von $\delta(\alpha)$, $\delta_1(\alpha)$, ... $\delta_{r-1}(\alpha)$, mit Ausschluss von $\delta_s(\alpha)$ werden $\lambda+n-n_s$, $\lambda+n_1-n_s$, ... $\lambda+n_{r-1}-n_s$. Nimmt man nun in Beziehung auf den in $\delta_s(\alpha)$ enthaltenen Primfactor $f_s(\alpha)$ den Index, welcher durch Ind_s bezeichnet werden soll, so dass allgemein für jede complexe Zahl $\varphi(\alpha)$

$$\varphi(\alpha)^{\frac{Nf_s(\alpha)-1}{\lambda}} \equiv \left(\frac{\varphi(\alpha)}{f_s(\alpha)}\right) \equiv \alpha^{\mathrm{Ind}_s\varphi(\alpha)} \pmod{f_s(\alpha)}$$

ist, und bemerkt, dass die Norm einer jeden complexen Zahl in w in Beziehung auf jeden Primfactor der Determinante $D(\alpha)$ einer λ^{ten} Potenz congruent, also der Index derselben congruent Null ist (mod. λ), so erhält man:

$$(9.)\quad \begin{cases} 0 \equiv (n-n_s)\mathrm{Ind}_s\delta(\alpha)+(n_1-n_s)\mathrm{Ind}_s\delta_1(\alpha)+\cdots+(n_{r-1}-n_s)\mathrm{Ind}_s\delta_{r-1}(\alpha) \\ \qquad +k\,\mathrm{Ind}_s\varrho+c\,\mathrm{Ind}_s(\alpha)+c_1\mathrm{Ind}_s\varepsilon_1(\alpha)+\cdots+c_{\mu-1}\mathrm{Ind}_s\varepsilon_{\mu-1}(\alpha) \end{cases}$$

nach dem Modul λ, welche Congruenz, da s die Werthe 0, 1, 2, ... $r-1$ haben kann, ein System von r Congruenzen repräsentirt.

Wenn diese r Congruenzen so beschaffen sind, dass eine Anzahl r der Zahlen n, n_1, ... n_{r-1}, k, c, c_1, ... $c_{\mu-1}$ durch die $\mu+1$ übrigen vollständig bestimmt wird, d. h. wenn diese r Congruenzen nicht identisch und von einander unabhängig sind,. so bleiben von den $\mu+r+1$ Exponenten nur $\mu+1$ beliebig, die Anzahl aller verschiedenen Normen von der Form (8.) ist alsdann gleich $\lambda^{\mu+1}$, und da dieses genau die Anzahl aller wesentlich verschiedenen complexen Zahlen $\Delta(w)$ ist, welche aus den Einheiten der Form (1.) erzeugt werden können, so folgt, dass diese alle in der That als Normen von complexen Zahlen in w darstellbar sein müssen.

Man hat demnach folgenden Satz:

II. *Wenn die Determinante* $D(\alpha)=\delta(\alpha).\delta_1(\alpha)\ldots\delta_{r-1}(\alpha)$ r *verschiedene Primfactoren* $f(\alpha)$, $f_1(\alpha)$, ... $f_{r-1}(\alpha)$ *enthält, und*

$$\delta(\alpha)=e(\alpha)f(\alpha)^m,\quad \delta_1(\alpha)=e_1(\alpha)f_1(\alpha)^{m_1},\ \ldots,$$

wo die Exponenten m, m_1, ... m_{r-1} *nicht durch* λ *theilbar, und so beschaffen sind, dass* $\delta(\alpha)$, $\delta_1(\alpha)$, ... *wirklich sind, auch für ideale Primzahlen* $f(\alpha)$, $f_1(\alpha)$, ..., *so sind alle complexen Zahlen der Form*

$$\delta(\alpha)^n.\delta_1(\alpha)^{n_1}\ldots\delta_{r-1}(\alpha)^{n_{r-1}}\varrho^k\alpha^c\varepsilon_1(\alpha)^{c_1}\ldots\varepsilon_{\mu-1}(\alpha)^{c_{\mu-1}},$$

in welcher die Exponenten n, n_1, ... n_{r-1}, k, c, c_1, ... $c_{\mu-1}$ *dem Systeme der* r *Congruenzen*

$$0\equiv(n-n_s)\operatorname{Ind}_s\delta(\alpha)+(n_1-n_s)\operatorname{Ind}_s\delta_1(\alpha)+\cdots+(n_{r-1}-n_s)\operatorname{Ind}_s\delta_{r-1}(\alpha)$$
$$+k\operatorname{Ind}_s\varrho+c\operatorname{Ind}_s(\alpha)+c_1\operatorname{Ind}_s\varepsilon_1(\alpha)+\cdots+c_{\mu-1}\operatorname{Ind}_s\varepsilon_{\mu-1}(\alpha),$$

nach dem Modul λ, *für* $s=0, 1, 2, \ldots r-1$ *genügen, unter der Bedingung, dass diese* r *Congruenzen nicht identisch und von einander unabhängig sind, als Normen von wirklichen complexen Zahlen in* w *darstellbar, und diese lassen sich in die Form*

$$\Delta(w) = u^n u_1^{n_1}\ldots u_{r-1}^{n_{r-1}}f(u, u_1, \ldots u_{r-1})$$

setzen, in welcher $Nf(u, u_1, \ldots u_{r-1})$ *nur eine Potenz von* ϱ *und Einheiten in* α *enthält.*

§ 2.

Die idealen Ambigen der complexen Zahlen in w.

Eine ideale Zahl in w, welche ihren conjugirten äquivalent ist, soll eine ideale Ambige in dieser Theorie der complexen Zahlen in w genannt werden.

Wenn $\varphi(w)$ eine Ambige ist, also $\varphi(w)$ äquivalent $\varphi(w\alpha)$, und allgemein $\varphi(w)$ äquivalent $\varphi(w\alpha^k)$, und es ist $\psi(\alpha)$ irgend eine ideale Zahl in α, so ist $\psi(\alpha)\varphi(w)$ ebenfalls eine Ambige; wenn daher die Anzahl der nicht äquivalenten Klassen der idealen Zahlen in α gleich h ist, und

$$\psi(\alpha),\ \psi_1(\alpha),\ \psi_2(\alpha),\ \ldots\ \psi_{h-1}(\alpha)$$

sind die Repräsentanten der h verschiedenen Klassen, so sind

$$\psi(\alpha)\varphi(w),\ \psi_1(\alpha)\varphi(w),\ \ldots\ \psi_{h-1}(\alpha)\varphi(w)$$

h nicht äquivalente Ambige, welche alle zu einer und derselben Gruppe gerechnet werden sollen, weil man nur eine derselben zu kennen braucht, um alle übrigen dieser Gruppe zu haben.

Es sei nun $\varphi(w)$ eine ideale Ambige, welche von jedem idealen Factor in α befreit angenommen werden soll, so giebt es einen idealen

Multiplicator $\psi(w)$, welcher die beiden äquivalenten idealen Zahlen $\varphi(w)$ und $\varphi(w\alpha)$ zu wirklichen macht, und man hat

$$\psi(w)\varphi(w) = G(w),$$
$$\psi(w)\varphi(w\alpha) = G_1(w),$$

wo $G(w)$ und $G_1(w)$ wirkliche complexe Zahlen in w bezeichnen, welche in Betreff der Einheiten, die sie enthalten können, ganz unbestimmt sind, so dass ihnen Einheiten nach Belieben zugefügt werden können. Die Norm von $G(w)$ und die Norm von $G_1(w)$ enthalten genau dieselben idealen Factoren in w; dieselben können sich daher nur durch eine Einheit unterscheiden, welche eine Einheit in α sein muss, weil diese Normen complexe Zahlen in α sind. Man hat daher

$$NG_1(w) = E(\alpha)NG(w),$$

und weil die Normen von wirklichen complexen Zahlen in w in Beziehung auf jeden Factor der Determinante $D(\alpha)$ λ^{ten} Potenzen congruent sind, so erhält man hieraus $E(\alpha)$ einer λ^{ten} Potenz congruent, nach dem Modul $f_s(\alpha)$, und wenn man den Index in Beziehung auf den Primfactor $f_s(\alpha)$ anwendet:

$$\operatorname{Ind}_s E(\alpha) \equiv 0 \pmod{\lambda},$$

oder wenn $E(\alpha)$ durch die Fundamental-Einheiten ausgedrückt wird:

$$E(\alpha) = \alpha^c \varepsilon_1(\alpha)^{c_1} \dots \varepsilon_{\mu-1}(\alpha)^{c_{\mu-1}},$$

so hat man

$$(1.)\quad 0 \equiv c\operatorname{Ind}_s(\alpha) + c_1 \operatorname{Ind}_s \varepsilon_1(\alpha) + \dots + c_{\mu-1}\operatorname{Ind}_s \varepsilon_{\mu-1}(\alpha) \pmod{\lambda},$$

für $s = 0, 1, 2, \dots r-1$. Es soll nun angenommen werden, dass die Determinante $D(\alpha)$ so beschaffen ist, dass das System der r Congruenzen bei (9.) im § 1 ein unabhängiges ist, alsdann lassen sich, nach dem daselbst bewiesenen Satze, alle complexen Zahlen der Form

$$\delta(\alpha)^n . \delta_1(\alpha)^{n_1} \dots \delta_{r-1}(\alpha)^{n_{r-1}} \varrho^k \alpha^c \varepsilon_1(\alpha)^{c_1} \dots \varepsilon_{\mu-1}(\alpha)^{c_{\mu-1}},$$

welche diesen Congruenzen genügen, als Normen von wirklichen complexen Zahlen in w darstellen. Nimmt man nun $n = n_1 = \dots = n_{r-1} = 0$ und $k = 0$, so genügt die Einheit $E(\alpha)$, wie die Congruenzen bei (1.) zeigen, jenen Congruenzen bei (9.) § 1, darum muss die Einheit $E(\alpha)$ sich als Norm einer wirklichen complexen Zahl in w darstellen lassen, welche hier nur eine Einheit in w sein kann; man hat daher

$$E(\alpha) = NE(w).$$

4*

Verbindet man nun die Einheit $E(w)$ mit der wirklichen complexen Zahl $G(w)$, und schreibt einfach $G(w)$ statt $E(w)G(w)$, so hat man

$$(2.)\qquad NG(w) = NG_1(w),$$

also

$$N\left(\frac{G(w)}{G_1(w)}\right) = 1.$$

Aus diesem Quotienten, dessen Norm gleich Eins ist, wird nun der Ausdruck

$$P\left(\frac{G(w)}{G_1(w)}\right) = 1+\frac{G(w)}{G_1(w)}+\frac{G(w)G(w\alpha)}{G_1(w)G_1(w\alpha)}+\cdots+\frac{G(w)G(w\alpha)\ldots G(w\alpha^{\lambda-2})}{G_1(w)G_1(w\alpha)\ldots G_1(w\alpha^{\lambda-2})}$$

gebildet, welcher, wenn $NG_1(w) = B$ als gemeinschaftlicher Nenner aller Glieder genommen wird, in die Form

$$P\left(\frac{G(w)}{G_1(w)}\right) = \frac{AF(w)}{B}$$

gesetzt werden kann, wo A und B ganze und wirkliche complexe Zahlen in α sind, und $F(w)$ von jedem nur α enthaltenden Factor befreit genommen werden soll. Die identische Gleichung

$$\frac{G(w)}{G_1(w)}P\left(\frac{G(w\alpha)}{G_1(w\alpha)}\right) = P\left(\frac{G(w)}{G_1(w)}\right)$$

giebt alsdann

$$(3.)\qquad G(w)F(w\alpha) = G_1(w)F(w).$$

Erhebt man diese Gleichung zur $h\lambda^{\text{ten}}$ Potenz und setzt

$$\varphi(w)^{h\lambda} = \Phi(w),\quad \psi(w)^{h\lambda} = \Psi(w),$$

so sind $\Phi(w)$ und $\Psi(w)$ wirkliche complexe Zahlen, und man hat

$$G(w)^{h\lambda} = \Psi(w)\Phi(w)\varepsilon(w),$$
$$G_1(w)^{h\lambda} = \Psi(w)\Phi(w\alpha)\varepsilon_1(w),$$

wo $\varepsilon(w)$ und $\varepsilon_1(w)$ beliebige Einheiten sind. Die Gleichung (3.) giebt demnach, wenn der gemeinschaftliche und wirkliche Factor $\Psi(w)$ hinweggehoben, und

$$\frac{\varepsilon_1(w)}{\varepsilon(w)} = e(w)$$

gesetzt wird:

$$\Phi(w)F(w\alpha)^{h\lambda} = \Phi(w\alpha)F(w)^{h\lambda}e(w).$$

Nimmt man auf beiden Seiten die Normen, so erkennt man, dass die Einheit $e(w)$ eine solche ist, deren Norm gleich Eins ist; man hat daher

$$e(w) = \frac{\Delta(w)}{\Delta(w\alpha)},$$

wo $\Delta(w)$ keinen definirten idealen Factor enthält, und wenn dieser Ausdruck der Einheit eingesetzt wird, so ist

$$\frac{F(w\alpha)^{h\lambda}\Delta(w\alpha)}{\Phi(w\alpha)} = \frac{F(w)^{h\lambda}\Delta(w)}{\Phi(w)}.$$

Weil der Ausdruck auf der rechten Seite dieser Gleichung bei der Verwandlung von w in $w\alpha$ ungeändert bleibt, so folgt in derselben Weise, wie in dem analogen Falle im vorigen Paragraphen, dass derselbe nur eine complexe Zahl in α sein kann, und wenn diese mit C bezeichnet wird, so hat man

$$F(w)^{h\lambda}\Delta(w) = C\Phi(w).$$

Da die Ambige $\varphi(w)$ frei von jedem, auch idealen Factor in α angenommen worden ist, so enthält auch $\Phi(w) = \varphi(w)^{h\lambda}$ keinen complexen Factor in α; es muss darum C eine *ganze* complexe Zahl in α sein; denn enthielte es einen Nenner, so müsste derselbe sich gegen $\Phi(w)$ hinwegheben, welches unmöglich ist. Weil ferner $\Phi(w)$ nur definirte ideale Primfactoren enthält, aber $\Delta(w)$ keinen solchen, so muss $F(w)^{h\lambda}$ durch $\Phi(w)$ theilbar sein, und folglich $F(w)$ die ideale Ambige $\varphi(w)$ als Factor enthalten. Setzt man

$$\frac{F(w)^{h\lambda}}{\Phi(w)} = F_1(w),$$

so hat man

$$F_1(w)\Delta(w) = C,$$

woraus weiter folgt, dass, wenn $F_1(w)$ noch definirte ideale Primfactoren enthielte, diese sich zu complexen Zahlen in α zusammensetzen müssten. Es kann darum auch $F(w)$ ausser der Ambigen $\varphi(w)$ keine anderen definirten idealen Factoren enthalten, als solche, welche sich zu complexen Factoren in α zusammensetzen; fügt man diese nun der Ambigen $\varphi(w)$ hinzu, wodurch man eine andere Ambige derselben Gruppe erhält, so ist diese ideale Ambige in $F(w)$ enthalten, und ausser dieser kein anderer definirter idealer Factor. Man hat daher den Satz:

I. *Wenn die Determinante* $D(\alpha)$ *so beschaffen ist, dass die* r *Congruenzen* (9.) § 1 *nicht identisch und von einander unabhängig sind, so ist jede ideale Ambige* $\varphi(w)$, *bei passender Wahl der idealen Zahl in* α, *mit welcher sie behaftet sein kann, in einer wirklichen complexen Zahl* $F(w)$ *enthalten, welche ausser dieser Ambigen keine anderen definirten idealen Factoren enthält.*

Hat man eine wirkliche complexe Zahl $F(w)$ gefunden, welche die ideale Ambige $\varphi(w)$ und ausser dieser keine anderen definirten idealen Factoren enthält, so kann man aus derselben sogleich eine ganze Reihe anderer herleiten, welche dieselbe Eigenschaft haben, nämlich dadurch, dass man die Zahl $F(w)$ mit den $\lambda^{\mu+1}$ wirklichen complexen Zahlen $\varDelta(w)$ multiplicirt, welche keine definirten idealen Primfactoren enthalten. Wenn die r Congruenzen (9.) § 1 erfüllt werden können, ohne dass $k=0$ sein muss, so giebt es unter den $\lambda^{\mu+1}$ Zahlen $\varDelta(w)$ auch solche, deren Normen alle verschiedenen Potenzen von ϱ enthalten. Wenn ferner die Norm von $F(w)$ den Factor ϱ genau ν-mal enthält, und man multiplicirt $F(w)$ mit einer Zahl $\varDelta(w)$, deren Norm den Factor ϱ genau $(\lambda-\nu)$-mal enthält, so enthält die Norm von $\varDelta(w)F(w)$ den Factor ϱ genau λ-mal, woraus folgt, dass $\varDelta(w)F(w)$ durch ϱ theilbar sein muss, und wenn dieser Factor hinweggehoben wird, so hat man eine, die Ambige $\varphi(w)$ enthaltende Zahl, deren Norm den Factor ϱ nicht enthält.

Die Gleichung (3.)

$$G(w)F(w\alpha) = G_1(w)F(w),$$

welcher jede, eine Ambige enthaltende wirkliche Zahl $F(w)$ genügt, wird durch Multiplication mit

$$G_1(w\alpha)G_1(w\alpha^2)\dots G_1(w\alpha^{\lambda-1})$$

in folgende Form gebracht:

$$(4.)\qquad H(w)F(w\alpha) = KF(w),$$

wo $K=NG_1(w)$ eine complexe Zahl in α ist, welche keinen Factor der Determinante $D(\alpha)$ und auch keinen Factor ϱ enthält. Setzt man nun

$$\begin{aligned} H(w) &= A+A_1w+A_2w^2+\cdots+A_{\lambda-1}w^{\lambda-1}, \\ F(w) &= B+B_1w+B_2w^2+\cdots+B_{\lambda-1}w^{\lambda-1}, \end{aligned}$$

so erhält man aus der Gleichung (4.) in derselben Weise wie oben bei (5.) § 1 die Congruenzen

$$(5.)\quad \begin{cases} (A\alpha-K)B\equiv 0, \\ (A\alpha-K)B_1+A_1B\equiv 0, & \pmod{D(\alpha)} \\ (A\alpha^2-K)B_2+A_1B_1\alpha+A_2B\equiv 0, \\ (A\alpha^3-K)B_3+A_1B_2\alpha^2+A_2B_1\alpha+A_3B\equiv 0, \\ \qquad\text{etc.}\quad\text{etc.} \end{cases}$$

aus welchen eben so wie oben geschlossen wird, dass $F(w)$ die Form

$$F(w) = u^n u_1^{n_1} \dots u_{r-1}^{n_{r-1}} f(u, u_1, \dots u_{r-1})$$

hat, wo die Norm von $f(u, u_1, \dots u_{r-1})$ keinen Factor der Determinante $D(\alpha)$ enthält. Hat man nun $F(w)$ so gewählt, dass die Norm dieser Zahl den Factor ϱ nicht enthält, welches, wie gezeigt worden ist, immer geschehen kann, wenn das System der Congruenzen (9.) § 1 unabhängig ist, und nicht nothwendig $k = 0$ ergiebt, so enthält $f(u, u_1, \dots u_{r-1})$ nur alle in $F(w)$ enthaltenen definirten idealen Factoren, also genau die in $F(w)$ enthaltene Ambige, welche somit als wirkliche complexe Zahl in $u, u_1, \dots u_{r-1}$ dargestellt ist. Man hat also folgenden Satz:

II. *Wenn die r Congruenzen bei* (9.) § 1 *nicht identisch und unabhängig sind, und wenn dieselben erfüllt werden können, ohne dass $k = 0$ ist, so ist jede Ambige, wenn die ideale Zahl in α, mit welcher sie behaftet sein kann, passend gewählt wird, als wirkliche complexe Zahl in $u, u_1, \dots u_{r-1}$ darstellbar.*

§ 3.

Die Bedingungen, unter welchen die Exponenten derjenigen Potenzen, zu welchen die idealen Zahlen in w erhoben werden müssen, um wirkliche Zahlen in $u, u_1, \dots u_{r-1}$ zu werden, nicht durch λ theilbar sind.

Wenn es eine ideale Zahl in w giebt, welche zu einer Potenz, deren Exponent durch λ theilbar ist, erhoben werden muss, um als wirkliche complexe Zahl in $u, u_1, \dots u_{r-1}$ darstellbar zu sein, so muss es auch eine ideale Zahl in w geben, deren λ^{te} Potenz in dieser Weise als wirkliche Zahl darstellbar ist; denn wenn $f(w)$ eine ideale Zahl ist, deren $k\lambda^{\text{te}}$ Potenz als wirkliche Zahl in $u, u_1, \dots u_{r-1}$ darstellbar ist, aber keine niedere, so ist offenbar $f(w)^k$ eine ideale Zahl, deren λ^{te} Potenz diese Eigenschaft hat.

Es sei nun $\varphi(w)$ eine ideale Zahl, deren λ^{te} Potenz als wirkliche Zahl in $u, u_1, \dots u_{r-1}$ darstellbar ist, ohne dass sie selbst in dieser Weise dargestellt werden kann, so haben die conjugirten Zahlen $\varphi(w\alpha), \varphi(w\alpha^2), \dots$ offenbar dieselbe Eigenschaft. Bildet man nun folgende Reihe von idealen Zahlen:

$$(1.)\quad \begin{cases} \varphi_1(w) = \varphi(w)^{\lambda-1}\varphi(w\alpha), \\ \varphi_2(w) = \varphi_1(w)^{\lambda-1}\varphi_1(w\alpha), \\ \varphi_3(w) = \varphi_2(w)^{\lambda-1}\varphi_2(w\alpha), \\ \quad\text{etc.} \qquad \text{etc.} \end{cases}$$

so hat man allgemein für jeden Werth des n:

$$\varphi_n(w) = \varphi(w)^{(\lambda-1)^n}\varphi(w\alpha)^{\frac{n}{1}(\lambda-1)^{n-1}}.\varphi(w\alpha^2)^{\frac{n(n-1)}{1.2}(\lambda-1)^{n-2}}\dots\varphi(w\alpha^n),$$

und wenn $n=\lambda$ genommen wird, so hat man $\varphi_\lambda(w)$ gleich einer λ^{ten} Potenz einer complexen Zahl, welche nur aus den conjugirten idealen Zahlen $\varphi(w)$, $\varphi(w\alpha)$, $\varphi(w\alpha^2)$, ... zusammengesetzt ist. Es ist daher $\varphi_\lambda(w)$ als wirkliche Zahl in $u, u_1, \dots u_{r-1}$ darstellbar.

Da nun in der Reihe der complexen Zahlen

$$\varphi(w),\quad \varphi_1(w),\quad \varphi_2(w),\quad \dots\quad \varphi_\lambda(w)$$

die erste als wirkliche Zahl in $u, u_1, \dots u_{r-1}$ nicht darstellbar, die letzte aber als solche darstellbar ist, so muss eine bestimmte Zahl dieser Reihe die erste sein, welche diese Eigenschaft besitzt. Dieselbe sei $\varphi_{s+1}(w)$, so ist $\varphi_s(w)$ nicht als wirklich in $u, u_1, \dots u_{r-1}$ darstellbar, aber $\varphi_s(w)^{\lambda-1}\varphi_s(w\alpha)$ ist darstellbar. Weil ferner die λ^{te} Potenz einer jeden wirklichen complexen Zahl in $u, u_1, \dots u_{r-1}$ eine wirkliche complexe Zahl in w ist, so ist $\varphi_s(w)^\lambda$ eine wirkliche Zahl in w, und weil $\varphi_s(w)^{\lambda(\lambda-1)}$ eben so wie $\varphi_s(w)^{\lambda-1}\varphi_s(w\alpha)$ als wirklich in $u, u_1, \dots u_{r-1}$ darstellbar ist, so ist auch das Product derselben, nämlich $\varphi_s(w)^{\lambda^2-1}\varphi_s(w\alpha)$, eine wirkliche complexe Zahl in $u, u_1, \dots u_{r-1}$. Setzt man nun einfach $f(w)$ statt $\varphi_s(w)$, so hat man:

$$(2.)\quad \begin{cases} f(w)^{\lambda^2} = G(w), \\ f(w)^{\lambda^2-1}f(w\alpha) = G_1(u, u_1, \dots u_{r-1}), \end{cases}$$

wo $G(w)$ und $G_1(u, u_1, \dots u_{r-1})$ beide wirklich sind. Die Normen dieser beiden complexen Zahlen sind, wie die Gleichungen (1.) zeigen, genau aus denselben idealen Factoren zusammengesetzt; dieselben können sich also nur durch Einheiten unterscheiden, welche nothwendig Einheiten in α sein müssen, da diese Normen selbst complexe Zahlen in α sind. Man hat also

$$(3.)\quad E(\alpha)NG(w) = NG_1(u, u_1, \dots u_{r-1}).$$

Nimmt man nun

$$G(w) = A + A_1w + A_2w^2 + \dots + A_{\lambda-1}w^{\lambda-1},$$
$$G_1(u, u_1, \dots u_{r-1}) = \Sigma_k a_k u^{|k-n|}u_1^{|k-n_1|}\dots u_{r-1}^{|k-n_{r-1}|},$$

so hat man in Beziehung auf den Modul $\delta_s(\alpha)$ oder $f_s(\alpha)$

$$NG(w) \equiv A^\lambda,\quad NG_1(u, u_1, \dots u_{r-1}) \equiv a_{n_s}^\lambda.\delta(\alpha)^{|n_s-n|}:\delta_1(\alpha)^{|n_s-n_1|}\dots\delta_{r-1}(\alpha)^{|n_s-n_{r-1}|},$$

also

$$E(\alpha)A^\lambda \equiv a_{n_s}^\lambda.\delta(\alpha)^{|n_s-n|}.\delta_1(\alpha)^{|n_s-n_1|}\dots\delta_{r-1}(\alpha)^{|n_s-n_{r-1}|},$$

und wenn man die Indices nimmt in Beziehung auf $f_s(\alpha)$:

(4.) $\operatorname{Ind}_s E(\alpha) \equiv (n_s - n)\operatorname{Ind}_s \delta(\alpha) + (n_s - n_1)\operatorname{Ind}_s \delta_1(\alpha) + \cdots + (n_s - n_{r-1})\operatorname{Ind}_s \delta_{r-1}(\alpha)$,

nach dem Modul λ, für $s = 0, 1, 2, \ldots r-1$. Drückt man die Einheit $E(\alpha)$ durch die Fundamental-Einheiten aus:

$$E(\alpha) = \alpha^c \varepsilon_1(\alpha)^{c_1} \varepsilon_2(\alpha)^{c_2} \ldots \varepsilon_{\mu-1}(\alpha)^{c_{\mu-1}},$$

so stimmt dieses System von r Congruenzen mit dem Systeme bei (9.) § 1 für den besonderen Werth $k = 0$ vollständig überein; also wenn die Determinante $D(\alpha)$ so beschaffen ist, dass dieses System ein unabhängiges ist, so giebt es eine complexe Zahl $\Delta(w)$ von der Art, dass

$$\Delta(w) = u^n u_1^{n_1} \ldots u_{r-1}^{n_{r-1}} E(u, u_1, \ldots u_{r-1})$$

und

$$N\Delta(w) = \delta(\alpha)^n . \delta_1(\alpha)^{n_1} \ldots \delta_{r-1}(\alpha)^{n_{r-1}} E(\alpha).$$

Die Gleichung

$$NE(u, u_1, \ldots u_{r-1}) = E(\alpha),$$

mit (3.) verbunden, giebt

$$NG(w) = N\left(\frac{G_1(u, u_1, \ldots u_{r-1})}{E(u, u_1, \ldots u_{r-1})}\right);$$

ferner ist $u^n u_1^{n_1} \ldots u_{r-1}^{n_{r-1}} G_1(u, u_1, \ldots u_{r-1})$ eine wirkliche complexe Zahl in w, und diese, durch $\Delta(w)$ dividirt, zeigt, dass

$$\frac{G_1(u, u_1, \ldots u_{r-1})}{E(u, u_1, \ldots u_{r-1})}$$

ebenfalls eine wirkliche complexe Zahl in w ist, und zwar eine ganze, weil der Nenner dieses Bruches nur eine Einheit ist. Setzt man nun

$$\frac{G_1(u, u_1, \ldots u_{r-1})}{E(u, u_1, \ldots u_{r-1})} = G_1(w),$$

so hat man

$$NG(w) = NG_1(w),$$

und

$$f(w)^{\lambda^2-1} f(w\alpha) = E(u, u_1, \ldots u_{r-1}) G_1(w).$$

Weil $f(w)^{\lambda^2-1} f(w\alpha)$, als wirkliche complexe Zahl in $u, u_1, \ldots u_{r-1}$ nur durch die idealen Factoren bestimmt ist, welche sie enthält, so kann man derselben beliebige Einheiten in $u, u_1, \ldots u_{r-1}$ hinzufügen, oder von derselben wegnehmen; man kann daher die Einheit $E(u, u_1, \ldots u_{r-1})$ weglassen, und hat so in $G_1(w)$ eine wirkliche complexe Zahl in w, welche genau dieselben idealen Factoren hat, als $f(w)^{\lambda^2-1} f(w\alpha)$.

Die beiden Gleichungen

$$f(w)^{\lambda^s} = G(w),$$
$$f(w)^{\lambda^s-1}f(w\alpha) = G_1(w)$$

zeigen, dass $f(w)$ äquivalent $f(w\alpha)$ ist, da beide mit $f(w)^{\lambda^s-1}$ multiplicirt wirklich sind, dass also $f(w)$ eine Ambige ist. Nach dem Satze II. § 2 ist also, unter den daselbst angegebenen Bedingungen, $f(w)$ als wirkliche complexe Zahl in $u, u_1, \dots u_{r-1}$ darstellbar, woraus folgt, dass es überhaupt keine ideale Zahl $f(w)$ giebt, welche zur λ^{ten} Potenz erhoben als wirklich in $u, u_1, \dots u_{r-1}$ dargestellt werden kann, ohne selbst eine wirkliche Zahl in $u, u_1, \dots u_{r-1}$ zu sein. Also:

Wenn die r Congruenzen (9.) § 1 *nicht identisch und unabhängig sind, und wenn dieselben erfüllt werden können, ohne dass k nothwendig gleich Null ist, so sind die Exponenten der niedrigsten Potenzen, zu welchen die idealen Zahlen in w erhoben werden müssen, um zu wirklichen complexen Zahlen in* $u, u_1, \dots u_{r-1}$ *zu werden, niemals durch* λ *theilbar.*

§ 4.

Zweiter Beweis der allgemeinen Reciprocitätsgesetze.

Es sei $\Delta(w)$ eine von den $\lambda^{\mu+1}$ verschiedenen wirklichen complexen Zahlen, welche im § 1 vollständig behandelt worden sind, und zwar sei es eine solche, deren Norm den Factor ϱ nicht enthält, so ist

$$(1.)\qquad N\Delta(w) = \delta(\alpha)^n\delta_1(\alpha)^{n_1}\dots\delta_{r-1}(\alpha)^{n_{r-1}}\alpha^c\varepsilon_1(\alpha)^{c_1}\dots\varepsilon_{\mu-1}(\alpha)^{c_{\mu-1}}$$

und die Exponenten $n, n_1, \dots n_{r-1}, c, c_1, \dots c_{\mu-1}$ genügen den r Congruenzen:

$$(2.)\qquad \begin{cases} 0 \equiv (n-n_s)\operatorname{Ind}_s\delta(\alpha)+(n_1-n_s)\operatorname{Ind}_s\delta_1(\alpha)+\dots+(n_{r-1}-n_s)\operatorname{Ind}_s\delta_{r-1}(\alpha) \\ \qquad +c\operatorname{Ind}_s(\alpha)+c_1\operatorname{Ind}_s\varepsilon_1(\alpha)+\dots+c_{\mu-1}\operatorname{Ind}_s\varepsilon_{\mu-1}(\alpha), \end{cases}$$

mod. λ, für $s=0, 1, 2, \dots r-1$. Nach dem Satze II. § 1 sind dies auch die einzigen Bedingungen, denen diese Exponenten unterworfen sind, sobald dieses System von r Congruenzen ein unabhängiges ist.

Setzt man für $\delta(\alpha), \delta_1(\alpha), \dots$ ihre Werthe $e(\alpha)f(\alpha)^m, e_1(\alpha)f_1(\alpha)^{m_1}, \dots,$ in welchen die Primzahlen $f(\alpha), f_1(\alpha), \dots$ *in der primären Form* angenommen werden sollen, und setzt ausserdem der Kürze halber

$$\alpha^c\varepsilon_1(\alpha)^{c_1}\dots\varepsilon_{\mu-1}(\alpha)^{c_{\mu-1}} = E(\alpha),$$

so kann man die r Congruenzen (2.) auch so darstellen:

$$(3.)\qquad 0 \equiv \Sigma(n_t-n_s)m_t\,\mathrm{Ind}_s f_t(\alpha)+\Sigma(n_t-n_s)\,\mathrm{Ind}_s e_t(\alpha)+\mathrm{Ind}_s E(\alpha),$$

mod. λ, für $s=0,\ 1,\ 2,\ \dots\ r-1$, wo die Summenzeichen Σ sich auf die Werthe $t=0,\ 1,\ 2,\ \dots\ r-1$ beziehen.

Andererseits besteht für $\varDelta(w)$, als wirkliche complexe Zahl in w, die im § 14 der Abhandlung v. J. 1859 entwickelte allgemeine Gleichung (29.), welche hier, wo $f(\alpha)^{mn}.f_1(\alpha)^{m_1n_1}\dots f_{r-1}(\alpha)^{m_{r-1}n_{r-1}}$ die primäre Zahl $N\varDelta(w)$ ist, und $e(\alpha)^n e_1(\alpha)^{n_1}\dots e_{r-1}(\alpha)^{n_{r-1}}E(\alpha)$ die begleitende Einheit, und wo die Determinante $D(\alpha)$ in der primären Form gleich $f(\alpha)^m f_1(\alpha)^{m_1}\dots f_{r-1}(\alpha)^{m_{r-1}}$ ist, mit der begleitenden Einheit $e(\alpha)e_1(\alpha)\dots e_{r-1}(\alpha)$, durch das verallgemeinerte *Legendre*sche Zeichen folgendermassen dargestellt wird:

$$(4.)\qquad \left(\frac{e(\alpha)^n e_1(\alpha)^{n_1}\dots e_{r-1}(\alpha)^{n_{r-1}}E(\alpha)}{f(\alpha)^m f_1(\alpha)^{m_1}\dots f_{r-1}(\alpha)^{m_{r-1}}}\right)=\left(\frac{e(\alpha)e_1(\alpha)\dots e_{r-1}(\alpha)}{f(\alpha)^{mn}f_1(\alpha)^{m_1n_1}\dots f_{r-1}(\alpha)^{m_{r-1}n_{r-1}}}\right).$$

Wendet man anstatt des verallgemeinerten *Legendre*schen Zeichens die Indices an, so wird diese Gleichung folgendermassen als Congruenz dargestellt:

$$(5.)\qquad \Sigma\Sigma m_s n_t\,\mathrm{Ind}_s e_t(\alpha)+\Sigma m_s\,\mathrm{Ind}_s E(\alpha)\equiv\Sigma\Sigma m_s n_s\,\mathrm{Ind}_s e_t(\alpha),$$

mod. λ, wo die beiden Summenzeichen sich auf die Werthe $s=0,\ 1,\ 2,\ \dots\ r-1$ und $t=0,\ 1,\ 2,\ \dots\ r-1$ beziehen.

Verbindet man diese Congruenz mit der Congruenz (3.), indem man den Werth des $\mathrm{Ind}_s E(\alpha)$ aus jener in diese einsetzt, so erhält man:

$$\Sigma\Sigma m_s n_t\,\mathrm{Ind}_s e_t(\alpha)-\Sigma\Sigma(n_t-n_s)m_t m_s\,\mathrm{Ind}_s f_t(\alpha)$$
$$-\Sigma\Sigma(n_t-n_s)m_s\,\mathrm{Ind}_s e_t(\alpha)\equiv\Sigma\Sigma m_s n_s\,\mathrm{Ind}_s e_t(\alpha),$$

wo die beiden Summenzeichen sich ebenfalls auf alle Werthe $s=0,\ 1,\ 2,\ \dots\ r-1$ und $t=0,\ 1,\ 2,\ \dots\ r-1$ beziehen. Hebt man nun die Glieder hinweg, welche sich gegenseitig vernichten, so bleibt

$$\Sigma\Sigma(n_t-n_s)m_t m_s\,\mathrm{Ind}_s f_t(\alpha)\equiv 0 \quad (\text{mod. } \lambda),$$

für $t=0,\ 1,\ 2,\ \dots\ r-1$, $s=0,\ 1,\ 2,\ \dots\ r-1$, oder wenn die je zwei Glieder, welche durch Vertauschung von s und t in einander übergehen, in eins zusammengefasst werden, so erhält man schliesslich:

$$(6.)\qquad \Sigma\Sigma(n_t-n_s)m_t m_s\big(\mathrm{Ind}_s f_t(\alpha)-\mathrm{Ind}_t f_s(\alpha)\big)\equiv 0,$$

wo die beiden Summenzeichen nur auf die Werthe $t=1,\ 2,\ 3,\ \dots\ r-1$ und $s=0,\ 1,\ 2,\ \dots\ t-1$ zu beziehen sind.

5*

Um nun aus dieser allgemeinen Congruenz die einfachen Reciprocitätsgesetze zu entwickeln, ist es nöthig, nicht allein die complexen Primzahlen der ersten Art, für welche nicht alle Einheiten λ^{te} Potenzreste derselben sind, zu unterscheiden (m. vergl. pag. 129 der Abhandlung v. J. 1859), sondern auch noch gewisse Beziehungen, welche unter den Primzahlen der ersten Art Statt haben können, besonders zu betrachten.

Zwei Primzahlen der ersten Art $f(\alpha)$ und $f_1(\alpha)$, welche in der besonderen Beziehung zu einander stehen, dass für jede beliebige Einheit $E(\alpha)$ die Indices $\operatorname{Ind} E(\alpha)$ und $\operatorname{Ind}_1 E(\alpha)$ dasselbe Verhältniss nach dem Modul λ zu einander haben, sollen ***ähnliche Primzahlen*** genannt werden, wenn aber die Verhältnisse dieser beiden Indices nicht für alle Einheiten dieselben sind, so sollen sie als ***unähnliche Primzahlen*** bezeichnet werden. Da alle Einheiten durch die $\mu-1$ Fundamentaleinheiten $\varepsilon_1(\alpha)$, $\varepsilon_2(\alpha)$, ... $\varepsilon_{\mu-1}(\alpha)$ und die einfache Einheit α ausgedrückt werden können, so können die ähnlichen und unähnlichen Primzahlen auch so definirt werden: Wenn den μ Congruenzen

$$\begin{aligned} \operatorname{Ind}\alpha &\equiv z \operatorname{Ind}_1(\alpha), \\ \operatorname{Ind}\varepsilon_1(\alpha) &\equiv z \operatorname{Ind}_1 \varepsilon_1(\alpha), \qquad (\text{mod. } \lambda), \\ &\;\vdots \\ \operatorname{Ind}\varepsilon_{\mu-1}(\alpha) &\equiv z \operatorname{Ind}_1 \varepsilon_{\mu-1}(\alpha), \end{aligned}$$

durch denselben Werth des z genügt wird, so sind die beiden Primzahlen $f(\alpha)$ und $f_1(\alpha)$, auf welche die Indices Ind und Ind_1 sich beziehen, ähnliche Primzahlen, wenn nicht, unähnliche.

Es wird nun zunächst angenommen, dass die Determinante $D(\alpha)$ nur zwei verschiedene Primfactoren $f(\alpha)$ und $f_1(\alpha)$ enthält, welche überdies Primzahlen der ersten Art, und zwar unähnliche sein sollen, so ist

$$N\Delta(w) = \delta(\alpha)^n \delta_1(\alpha)^{n_1} \alpha^c \varepsilon_1(\alpha)^{c_1} \ldots \varepsilon_{\mu-1}(\alpha)^{c_{\mu-1}},$$

und die Exponenten n, n_1, c, c_1, ... $c_{\mu-1}$ genügen den beiden Congruenzen

$$(7.)\quad \begin{cases} 0 \equiv (n-n_1)\operatorname{Ind}\,\delta_1(\alpha) + c\operatorname{Ind}\,\alpha + c_1 \operatorname{Ind}\,\varepsilon_1(\alpha) + \cdots + c_{\mu-1}\operatorname{Ind}\,\varepsilon_{\mu-1}(\alpha), \\ 0 \equiv (n_1-n)\operatorname{Ind}_1\,\delta(\alpha) + c\operatorname{Ind}_1\alpha + c_1\operatorname{Ind}_1\varepsilon_1(\alpha) + \cdots + c_{\mu-1}\operatorname{Ind}_1\varepsilon_{\mu-1}(\alpha). \end{cases}$$

Diese beiden Congruenzen sind nicht identisch und von einander unabhängig; denn wenn eine derselben identisch erfüllt sein sollte, so müssten nothwendig die Indices aller Einheiten gleich Null sein, also die betreffende Primzahl $f(\alpha)$ oder $f_1(\alpha)$ eine Primzahl der zweiten Art, und wenn sie von einander abhängig sein sollten, so müsste die eine, mit einer bestimmten Zahl multiplicirt, der anderen gleich sein, und darum müssten $f(\alpha)$ und $f_1(\alpha)$ ähnliche Primzahlen sein. Aus der Unabhängigkeit dieser beiden Congruenzen folgt nach dem Satze II. § 1, dass für die Exponenten n, n_1, c, c_1, ... $c_{\mu-1}$ alle Werthe statthaft sind, welche diesen Congruenzen genügen, unter denen namentlich auch solche sind, für welche n nicht gleich n_1 ist. Die Congruenz (6.), welche in diesem Falle nur aus einem Gliede besteht, nämlich

$$(n_1-n)m_1 m(\operatorname{Ind} f_1(\alpha) - \operatorname{Ind}_1 f(\alpha)) \equiv 0,$$

giebt nun, weil n_1-n nicht congruent Null ist, und auch m_1 und m nicht congruent Null:

$$(8.) \qquad \operatorname{Ind} f_1(\alpha) \equiv \operatorname{Ind}_1 f(\alpha),$$

oder in den *Legendre*schen Zeichen:

$$\left(\frac{f_1(\alpha)}{f(\alpha)}\right) = \left(\frac{f(\alpha)}{f_1(\alpha)}\right).$$

Also: *unter je zwei unähnlichen Primzahlen der ersten Art* $f(\alpha)$ *und* $f_1(\alpha)$ *besteht das Reciprocitätsgesetz*

$$\left(\frac{f_1(\alpha)}{f(\alpha)}\right) = \left(\frac{f(\alpha)}{f_1(\alpha)}\right).$$

Um das Reciprocitätsgesetz auch zwischen zwei einander ähnlichen Primzahlen der ersten Art $f(\alpha)$ und $f_1(\alpha)$ zu beweisen, muss man einen dritten Primfactor der Determinante $f_2(\alpha)$ hinzunehmen. Es sei also

$$D(\alpha) = \delta(\alpha)\delta_1(\alpha)\delta_2(\alpha),$$
$$\delta(\alpha) = e(\alpha)f(\alpha)^m, \quad \delta_1(\alpha) = e_1(\alpha)f_1(\alpha)^{m_1}, \quad \delta_2(\alpha) = e_2(\alpha)f_2(\alpha)^{m_2}.$$

Es giebt alsdann λ^μ verschiedene Zahlen $\Delta(w)$, für welche

$$N\Delta(w) = \delta(\alpha)^n \delta_1(\alpha)^{n_1} \delta_2(\alpha)^{n_2} E(\alpha),$$

wo

$$E(\alpha) = \alpha^c \varepsilon_1(\alpha)^{c_1} \ldots \varepsilon_{\mu-1}(\alpha)^{c_{\mu-1}},$$

in denen die Exponenten n, n_1, n_2, c, c_1, ... $c_{\mu-1}$ folgenden drei Congruenzen genügen müssen:

$$(9.) \quad \begin{cases} (n-n_1)\operatorname{Ind} \delta_1(\alpha) + (n-n_2)\operatorname{Ind} \delta_2(\alpha) + \operatorname{Ind} E(\alpha) \equiv 0, \\ (n_1-n) \operatorname{Ind}_1 \delta(\alpha) + (n_1-n_2)\operatorname{Ind}_1 \delta_2(\alpha) + \operatorname{Ind}_1 E(\alpha) \equiv 0, \\ (n_2-n) \operatorname{Ind}_2 \delta(\alpha) + (n_2-n_1)\operatorname{Ind}_2 \delta_1(\alpha) + \operatorname{Ind}_2 E(\alpha) \equiv 0. \end{cases}$$

Weil $f(\alpha)$ und $f_1(\alpha)$ ähnliche Primzahlen sind, so giebt es eine bestimmte

Zahl z, welche der Congruenz

$$\operatorname{Ind} E(\alpha) \equiv z \operatorname{Ind}_1 E(\alpha) \pmod{\lambda}$$

genügt, für alle verschiedenen Einheiten $E(\alpha)$, also für alle möglichen Werthe der Exponenten $c, c_1, \ldots c_{\mu-1}$; multiplicirt man daher die zweite Congruenz mit z und subtrahirt sie von der ersten, so erhält man

$$(10.)\quad \begin{cases} (n-n_1)(\operatorname{Ind}\delta_1(\alpha)+z\operatorname{Ind}_1\delta(\alpha)\delta_2(\alpha)) \\ +(n-n_2)(\operatorname{Ind}\delta_2(\alpha)-z\operatorname{Ind}_1\delta_2(\alpha)) \equiv 0. \end{cases}$$

Es soll nun die Primzahl $f_2(\alpha)$ so gewählt werden, dass sie eine Primzahl der ersten Art sei, und mit $f(\alpha)$, also auch mit $f_1(\alpha)$ unähnlich, und dass

$$\operatorname{Ind}_2 f(\alpha) - z\operatorname{Ind}_2 f_1(\alpha) \text{ nicht } \equiv 0$$

sei. Dass es in der That Primzahlen $f_2(\alpha)$ giebt, welche diesen Bedingungen genügen, folgt unmittelbar aus dem Satze I. § 16 der Abhandlung vom Jahre 1859, nach welchem es unendlich viele Primzahlen giebt, für welche die Indices der wirklichen complexen Zahlen

$$f(\alpha)^m,\quad f_1(\alpha)^{m_1},\quad \alpha,\quad \varepsilon_1(\alpha),\quad \ldots\quad \varepsilon_{\mu-1}(\alpha)$$

beliebig gegebenen Zahlenwerthen proportional sind, nach dem Modul λ.

Setzt man an die Stelle der ersten der drei Congruenzen (9.) die Congruenz (10.), welche aus der Verbindung der ersten mit der zweiten entstanden ist, und beachtet, dass $f_1(\alpha)$ und $f_2(\alpha)$ unähnliche Primzahlen der ersten Art sind, für welche die Indices der Einheiten nicht alle gleich Null sind, auch nicht alle in demselben Verhältniss stehen, so erkennt man, dass diese drei Congruenzen von einander unabhängig sind, wenn nur die Congruenz (10.) nicht identisch ist. Dass dieses letztere aber nicht der Fall ist, folgt aus $\operatorname{Ind}_2 f(\alpha) - z\operatorname{Ind}_2 f_1(\alpha)$ nicht gleich Null, welche Bedingung, weil nach dem bereits bewiesenen Falle des Reciprocitätsgesetzes unter je zwei unähnlichen Primzahlen der ersten Art $\operatorname{Ind}_2 f(\alpha) \equiv \operatorname{Ind} f_2(\alpha)$ und $\operatorname{Ind}_2 f_1(\alpha) \equiv \operatorname{Ind}_1 f_2(\alpha)$ ist,

$$\operatorname{Ind} f_2(\alpha) - z\operatorname{Ind}_2 f_2(\alpha) \text{ nicht } \equiv 0,$$

und wenn mit m_2 multiplicirt, und $\operatorname{Ind} e(\alpha) - z\operatorname{Ind}_1 e(\alpha) \equiv 0$ hinzuaddirt wird,

$$\operatorname{Ind}\delta_2(\alpha) - z\operatorname{Ind}_1\delta_2(\alpha) \text{ nicht } \equiv 0$$

ergiebt. Die lineare Congruenz (10.), in welcher der Coefficient von $n-n_2$ nicht verschwindet, ist darum keine identische.

Wegen der Unabhängigkeit dieser drei Congruenzen können die Zahlen n, n_1, n_2, c, c_1, $\dots$ $c_{\mu-1}$ alle diejenigen Werthe erhalten, welche diesen Congruenzen genügen, und unter diesen sind immer solche, für welche $n-n_1$ nicht congruent Null ist.

Die Congruenz (6.) giebt in dem gegenwärtigen Falle, wo $r=3$ ist:

$$(11.)\quad \begin{cases} (n_1-n)\, m_1 m\, (\operatorname{Ind} f_1(\alpha)-\operatorname{Ind}_1 f(\alpha)) \\ +(n_2-n)\, m_2 m\, (\operatorname{Ind} f_2(\alpha)-\operatorname{Ind}_2 f(\alpha)) \\ +(n_2-n_1)\, m_2 m_1 (\operatorname{Ind}_1 f_2(\alpha)-\operatorname{Ind}_2 f_1(\alpha)) \equiv 0. \end{cases}$$

Weil nun vermöge des bereits bewiesenen Falles des Reciprocitätsgesetzes zwischen unähnlichen Primzahlen der ersten Art $\operatorname{Ind}_1 f_2(\alpha) \equiv \operatorname{Ind}_2 f_1(\alpha)$ und $\operatorname{Ind} f_2(\alpha) \equiv \operatorname{Ind}_2 f(\alpha)$ ist, so bleibt

$$(n_1-n) m_1 m (\operatorname{Ind} f_1(\alpha)-\operatorname{Ind}_1 f(\alpha)) \equiv 0,$$

und weil n und n_1 so gewählt werden können, dass n_1-n nicht congruent Null ist, auch m_1 und m_2 nicht congruent Null sind, so folgt

$$(12.)\quad \operatorname{Ind} f_1(\alpha) \equiv \operatorname{Ind}_1 f(\alpha),$$

oder nach den *Legendre*schen Zeichen:

$$\left(\frac{f_1(\alpha)}{f(\alpha)}\right) = \left(\frac{f(\alpha)}{f_1(\alpha)}\right).$$

Also auch *unter je zwei ähnlichen primären Primzahlen der ersten Art* $f(\alpha)$ *und* $f_1(\alpha)$ *besteht das Reciprocitätsgesetz:*

$$\left(\frac{f_1(\alpha)}{f(\alpha)}\right) = \left(\frac{f(\alpha)}{f_1(\alpha)}\right).$$

Es sei jetzt ebenfalls $r=3$ und $f(\alpha)$ eine beliebige Primzahl der zweiten Art, $f_1(\alpha)$ eine beliebige Primzahl der ersten Art, $f_2(\alpha)$ eine mit $f_1(\alpha)$ unähnliche Primzahl der ersten Art, für welche $\operatorname{Ind}_2 f(\alpha)$ nicht congruent Null ist.

Die erste der drei Congruenzen (9.) giebt hier, weil die in Beziehung auf $f(\alpha)$ genommenen Indices aller Einheiten congruent Null sind:

$$(13.)\quad (n-n_1) m_1 \operatorname{Ind} f_1(\alpha) + (n-n_2) m_2 \operatorname{Ind} f_2(\alpha) \equiv 0.$$

Die Congruenz (11.) giebt, weil $f_2(\alpha)$ und $f_1(\alpha)$ Primzahlen der ersten Art sind, für welche das Reciprocitätsgesetz bereits bewiesen ist,

$$(14.)\quad (n_1-n) m_1 (\operatorname{Ind} f_1(\alpha)-\operatorname{Ind}_1 f(\alpha)) + (n_2-n) m_2 (\operatorname{Ind} f_2(\alpha)-\operatorname{Ind}_2 f(\alpha)) \equiv 0,$$

und wenn die vorhergehende zu dieser addirt wird:

$$(15.)\quad (n-n_1) m_1 \operatorname{Ind}_1 f(\alpha) + (n-n_2) m_2 \operatorname{Ind}_2 f(\alpha) \equiv 0.$$

Es giebt nun unter den λ^μ verschiedenen complexen Zahlen $\varDelta(w)$, deren Normen den Factor ϱ nicht enthalten, nothwendig auch solche, für welche nicht zugleich $n\equiv n_1$ und $n\equiv n_2$ ist; denn diese beiden Congruenzen, verbunden mit den aus (9.) folgenden:

$$c\,\mathrm{Ind}_1(\alpha)+c_1\mathrm{Ind}_1\varepsilon_1(\alpha)+\cdots+c_{\mu-1}\mathrm{Ind}_1\varepsilon_{\mu-1}(\alpha)\equiv 0,$$
$$c\,\mathrm{Ind}_2(\alpha)+c_1\mathrm{Ind}_2\varepsilon_1(\alpha)+\cdots+c_{\mu-1}\mathrm{Ind}_2\varepsilon_{\mu-1}(\alpha)\equiv 0,$$

welche nicht identisch und von einander unabhängig sind, weil $f_1(\alpha)$ und $f_2(\alpha)$ unähnliche Primzahlen der ersten Art sind, würden, weil unter den $\mu+3$ Exponenten n, n_1, n_2, c, c_1, ... $c_{\mu-1}$ vier unabhängige Congruenzen (mod. λ) beständen, die Anzahl aller wesentlich verschiedenen Zahlen $\varDelta(w)$, deren Normen den Factor ϱ nicht enthalten, auf $\lambda^{\mu-1}$ einschränken.

Wenn daher die Zahl $\varDelta(w)$ so gewählt wird, dass nicht zugleich $n\equiv n_1$ und $n\equiv n_2$ ist, und wenn zunächst der Fall betrachtet wird, wo $\mathrm{Ind}_2 f(\alpha)\equiv 0$ ist, so zeigt die Congruenz (15.), dass in diesem Falle $n\equiv n_2$ sein muss, und weil alsdann n nicht congruent n_1 ist, so folgt aus der Congruenz (14.) $\mathrm{Ind} f_1(\alpha)\equiv 0$. Also: *wenn irgend eine Primzahl der zweiten Art Rest einer Primzahl der ersten Art ist, so ist auch diese Rest von jener.*

Es sei ferner $\mathrm{Ind}_1 f(\alpha)$ nicht congruent Null, so ist nach Congruenz (15.) auch $n-n_2$ nicht congruent Null, und ebenso $n-n_1$ nicht congruent Null. Die Congruenz (14.) zeigt alsdann, dass, wenn $\mathrm{Ind} f_2(\alpha)\equiv\mathrm{Ind}_2 f(\alpha)$ ist, auch $\mathrm{Ind} f_1(\alpha)\equiv\mathrm{Ind}_1 f(\alpha)$ sein muss, also dass, wenn das Reciprocitätsgesetz für $f_2(\alpha)$ und $f(\alpha)$ gilt, dasselbe auch für $f_1(\alpha)$ und $f(\alpha)$ gelten muss.

Ich unterwerfe nun die Primzahl $f_2(\alpha)$ ausser den bereits festgesetzten Bedingungen, dass sie eine mit $f_1(\alpha)$ unähnliche Primzahl der ersten Art sei, und dass $\mathrm{Ind}_2 f(\alpha)$ nicht congruent Null sei, noch der Bedingung, dass alle zu $f(\alpha)$ conjugirten Primzahlen Reste von $f_2(\alpha)$ sein sollen, $f(\alpha)$ selbst aber ein Nichtrest. Ist $f(\alpha)$ ein Primfactor von q, und q eine Primzahl, welche zum Exponenten f gehört, nach dem Modul λ, so dass $q^f\equiv 1$ (mod. λ), aber keine niedere Potenz von q der Eins congruent ist, so besteht bekanntlich $f(\alpha)$ nur aus den Perioden von je f Gliedern, welche aus den Wurzeln α, α^2, ... $\alpha^{\lambda-1}$ gebildet werden können, und wenn $\lambda-1=ef$ ist, so giebt es e conjugirte Zahlen $f(\alpha)$, $f(\alpha^\gamma)$, ... $f(\alpha^{\gamma^{e-1}})$, wo γ eine primitive Wurzel von λ bezeichnet. Die neu hinzukommenden Bedingungen, welche $f_2(\alpha)$ erfüllen soll, sind also die, dass $\mathrm{Ind}_2 f(\alpha^\gamma)$, $\mathrm{Ind}_2 f(\alpha^{\gamma^2})$, ... $\mathrm{Ind}_2 f(\alpha^{\gamma^{e-1}})$ congruent Null sein sollen, aber $\mathrm{Ind}_2 f(\alpha)$ nicht congruent Null. Aus dem allgemeinen Satze I. § 16 der früheren Abhandlung folgt unmittelbar, dass es unendlich viele Primzahlen $f_2(\alpha)$ giebt, welche diesen Bedingungen und den obigen zugleich genügen, weil die complexen Zahlen

$$f(\alpha),\ f(\alpha^\gamma),\ \dots\ f(\alpha^{\gamma^{e-1}}),\ \alpha,\ \varepsilon_1(\alpha),\ \dots\ \varepsilon_{\mu-1}(\alpha),$$

auf welche sich diese Bedingungen beziehen, die in dem Satze verlangte Eigenschaft haben, dass ein Product von Potenzen derselben nicht anders eine λ^{te} Potenz werden kann, als dass alle Potenzexponenten einzeln Vielfache von λ sind.

Aus den für $f_2(\alpha)$ festgesetzten Bedingungen folgt, dass diese complexe Zahl Primfactor einer Primzahl p von der Form $n\lambda+1$ sein muss; denn die nicht complexe Zahl $q = f(\alpha)f(\alpha^\gamma)\dots f(\alpha^{\gamma^{e-1}})$ ist ein Nichtrest von $f_2(\alpha)$, da der erste Factor $f(\alpha)$ ein Nichtrest ist, die übrigen Factoren aber Reste, und eine nicht complexe Zahl kann nur für solche complexe Primzahlen Nichtrest sein, deren Normen Primzahlen der Form $n\lambda+1$ sind.

Wendet man nun die eine Reciprocitätsgleichung an, welche die Kreistheilung gewährt, nämlich

$$(16.)\quad \left(\frac{f_2(\alpha)}{f(\alpha)}\right)\left(\frac{f_2(\alpha)}{f(\alpha^\gamma)}\right)\cdots\left(\frac{f_2(\alpha)}{f(\alpha^{\gamma^{e-1}})}\right) = \left(\frac{f(\alpha)}{f_2(\alpha)}\right)\left(\frac{f(\alpha^\gamma)}{f_2(\alpha)}\right)\cdots\left(\frac{f(\alpha^{\gamma^{e-1}})}{f_2(\alpha)}\right),$$

so hat man nach den für $f_2(\alpha)$ festgesetzten Bedingungen

$$\left(\frac{f(\alpha^\gamma)}{f_2(\alpha)}\right) = 1,\quad \left(\frac{f(\alpha^{\gamma^2})}{f_2(\alpha)}\right) = 1,\quad \dots\quad \left(\frac{f(\alpha^{\gamma^{e-1}})}{f_2(\alpha)}\right) = 1,$$

und weil oben bewiesen worden, dass, wenn eine Primzahl der zweiten Art Rest einer Primzahl der ersten Art ist, auch diese Rest von jener sein muss, so ist auch

$$\left(\frac{f_2(\alpha)}{f(\alpha^\gamma)}\right) = 1,\quad \left(\frac{f_2(\alpha)}{f(\alpha^{\gamma^2})}\right) = 1,\quad \dots\quad \left(\frac{f_2(\alpha)}{f(\alpha^{\gamma^{e-1}})}\right) = 1,$$

und folglich

$$\left(\frac{f_2(\alpha)}{f(\alpha)}\right) = \left(\frac{f(\alpha)}{f_2(\alpha)}\right),$$

oder

$$\operatorname{Ind} f_2(\alpha) \equiv \operatorname{Ind}_2 f(\alpha).$$

Die Congruenz (14.) giebt daher:

$$(17.)\quad (n_1-n)\,m_1\left(\operatorname{Ind} f_1(\alpha) - \operatorname{Ind}_1 f(\alpha)\right) \equiv 0,$$

und weil n_1-n und m_1 nicht congruent Null sind:

$$\operatorname{Ind} f_1(\alpha) \equiv \operatorname{Ind}_1 f(\alpha),$$

oder

$$\left(\frac{f_1(\alpha)}{f(\alpha)}\right) = \left(\frac{f(\alpha)}{f_1(\alpha)}\right),$$

wo $f(\alpha)$ und $f_1(\alpha)$ keinen anderen Einschränkungen unterworfen sind, als dass die eine eine Primzahl der zweiten Art, die andere der ersten Art ist, und dass $\mathrm{Ind}_1 f(\alpha)$ nicht congruent Null ist. Diese Gleichung zeigt, dass, wenn eine der beiden Primzahlen Nichtrest der andern ist, auch diese Nichtrest von jener sein muss, und hieraus folgt von selbst, dass, wenn die eine Rest der andern ist, auch diese Rest von jener sein muss, so dass diese Reciprocitätsgleichung auch gültig bleiben muss, wenn $\mathrm{Ind}_1 f(\alpha)$ congruent Null ist. Also: *unter je zwei primären Primzahlen $f(\alpha)$ und $f_1(\alpha)$, deren eine der ersten, die andere der zweiten Art angehört, besteht das Reciprocitätsgesetz*

$$\left(\frac{f_1(\alpha)}{f(\alpha)}\right) = \left(\frac{f(\alpha)}{f_1(\alpha)}\right).$$

Um nun endlich noch das Reciprocitätsgesetz unter je zwei Primzahlen der zweiten Art zu beweisen, muss man complexe Zahlen $\varDelta(w)$ anwenden, für welche die Determinante $D(\alpha)$ vier verschiedene Primfactoren enthält. Es sei also $D(\alpha) = \delta(\alpha)\delta_1(\alpha)\delta_2(\alpha)\delta_3(\alpha)$, $f(\alpha)$ und $f_1(\alpha)$ zwei beliebige Primzahlen der zweiten Art, und $f_2(\alpha)$ und $f_3(\alpha)$ zwei Primzahlen der ersten Art,

$$N\varDelta(w) = \delta(\alpha)^n \delta_1(\alpha)^{n_1} \delta_2(\alpha)^{n_2} \delta_3(\alpha)^{n_3} E(\alpha), \quad E(\alpha) = \alpha^c \varepsilon_1(\alpha)^{c_1} \dots \varepsilon_{\mu-1}(\alpha)^{c_{\mu-1}}.$$

Weil nach den bereits bewiesenen Fällen des Reciprocitätsgesetzes

$$\mathrm{Ind}_s f_t(\alpha) - \mathrm{Ind}_t f_s(\alpha) \equiv 0$$

ist, sobald $f_s(\alpha)$ und $f_t(\alpha)$ nicht beide Primzahlen der zweiten Art sind, so bleibt von der allgemeinen Congruenz (6.) hier nur das erste Glied stehen, und man hat:

$$(18.)\qquad (n_1 - n) m_1 m (\mathrm{Ind} f_1(\alpha) - \mathrm{Ind}_1 f(\alpha)) \equiv 0,$$

welche Congruenz das gesuchte Reciprocitätsgesetz giebt, wenn nicht $n_1 \equiv n$ ist. Es bleibt also nur noch zu beweisen, dass die complexe Zahl $\varDelta(w)$ stets so gewählt werden kann, dass die Exponenten n und n_1 nicht congruent sind.

Zu diesem Zwecke unterwerfe ich die beiden Primzahlen der ersten Art $f_2(\alpha)$ und $f_3(\alpha)$ noch folgenden näheren Bestimmungen: Die Primzahl $f_2(\alpha)$ soll so gewählt werden, dass $\mathrm{Ind}_2 f(\alpha) \equiv 0$ ist, aber $\mathrm{Ind}_2 f_1(\alpha)$ nicht congruent Null, und die Primzahl $f_3(\alpha)$ soll so gewählt werden, dass sie der Primzahl $f_2(\alpha)$ unähnlich ist, und dass $\mathrm{Ind}_3 f(\alpha)$ nicht congruent Null ist, aber $\mathrm{Ind}_3 f_1(\alpha) \equiv 0$. Die Existenz solcher Primzahlen $f_2(\alpha)$ und $f_3(\alpha)$ folgt wieder unmittelbar aus dem allgemeinen Satze I. § 16 der früheren Abhandlung. Weil unter je zwei Primzahlen, deren eine der ersten, die andere der zweiten Art angehört, das Reciprocitätsgesetz gültig ist, so folgt aus diesen für $f_2(\alpha)$ und $f_3(\alpha)$ festgesetzten Bestimmungen auch $\mathrm{Ind} f_2(\alpha) \equiv 0$, $\mathrm{Ind}_1 f_2(\alpha)$ nicht congruent Null, und $\mathrm{Ind} f_3(\alpha)$ nicht congruent Null, $\mathrm{Ind}_1 f_3(\alpha) \equiv 0$.

Die beiden ersten der vier Congruenzen, durch welche die Exponenten $n, n_1, n_2, n_3, c, c_1, \ldots c_{\mu-1}$ bestimmt werden, geben nun, wenn alle Glieder weggelassen werden, welche vermöge der über die Primzahlen $f(\alpha)$, $f_1(\alpha)$, $f_2(\alpha)$, $f_3(\alpha)$ festgesetzten Bestimmungen congruent Null sind:

$$(n_1-n)m_1 \mathrm{Ind}\, f_1(\alpha) + (n_3-n)m_3 \mathrm{Ind}\, f_3(\alpha) \equiv 0,$$
$$(n-n_1)m\, \mathrm{Ind}_1 f(\alpha) + (n_2-n_1)m_2 \mathrm{Ind}_1 f_2(\alpha) \equiv 0.$$

Wenn nun $n \equiv n_1$ wäre, so müsste nach diesen beiden Congruenzen auch $n \equiv n_3$ und $n_1 \equiv n_2$ sein, also auch $n \equiv n_2$. Die dritte und vierte der Congruenzen, durch welche die Exponenten $n, n_1, n_2, n_3, c, c_1, \ldots c_{\mu-1}$ bestimmt werden, geben in diesem Falle:

$$0 \equiv c\,\mathrm{Ind}_2(\alpha) + c_1 \mathrm{Ind}_2 \varepsilon_1(\alpha) + \cdots + c_{\mu-1} \mathrm{Ind}_2 \varepsilon_{\mu-1}(\alpha),$$
$$0 \equiv c\,\mathrm{Ind}_3(\alpha) + c_1 \mathrm{Ind}_3 \varepsilon_1(\alpha) + \cdots + c_{\mu-1} \mathrm{Ind}_3 \varepsilon_{\mu-1}(\alpha),$$

welches zwei nicht identische und von einander unabhängige Congruenzen unter den Exponenten $c, c_1, \ldots c_{\mu-1}$ sind, weil $f_2(\alpha)$ und $f_3(\alpha)$ Primzahlen der ersten Art und einander unähnlich sind. Diese beiden Congruenzen mit den drei Congruenzen $n \equiv n_1$, $n \equiv n_2$, $n \equiv n_3$ würden aber fünf unabhängige Congruenzen unter den $\mu+4$ Exponenten $n, n_1, n_2, n_3, c, c_1, \ldots c_{\mu-1}$ ausmachen, durch welche die Anzahl aller möglichen Werthverbindungen derselben auf $\lambda^{\mu-1}$ eingeschränkt werden würde; es würde also nur $\lambda^{\mu-1}$ wesentlich verschiedene complexe Zahlen $\Delta(w)$ geben, welches absurd ist, da die Anzahl derselben gleich λ^{μ} ist. Es giebt also nothwendig auch solche complexe Zahlen $\Delta(w)$, für welche $n-n_1$ nicht congruent Null ist. Aus der Congruenz (18.) folgt daher nothwendig:

$$\mathrm{Ind} f_1(\alpha) \equiv \mathrm{Ind}_1 f(\alpha),$$

oder in den *Legendre*schen Zeichen:

$$\left(\frac{f_1(\alpha)}{f(\alpha)}\right) = \left(\frac{f(\alpha)}{f_1(\alpha)}\right).$$

6*

Also auch *unter je zwei primären Primzahlen der zweiten Art* $f(\alpha)$ *und* $f_1(\alpha)$ *gilt dieses Reciprocitätsgesetz.*

Fasst man die bewiesenen besonderen Fälle zusammen, so hat man den vollständigen Beweis des allgemeinen Reciprocitätsgesetzes für je zwei beliebige primäre complexe Primzahlen $f(\alpha)$ und $f_1(\alpha)$:

$$\left(\frac{f_1(\alpha)}{f(\alpha)}\right) = \left(\frac{f(\alpha)}{f_1(\alpha)}\right).$$

§ 5.

Dritter Beweis der allgemeinen Reciprocitätsgesetze.

Wendet man den im § 3 bewiesenen allgemeinen Satz zunächst auf den besonderen Fall an, wo die Determinante $D(\alpha)$ der complexen Zahlen in w nur *einen* Primfactor enthält, also $D(\alpha) = e(\alpha)f(\alpha)^m$ ist, so zeigt derselbe, dass die Exponenten der niedrigsten Potenzen, zu welchen die idealen Zahlen in w erhoben werden müssen, um zu wirklichen complexen Zahlen in w zu werden, niemals durch λ theilbar sind, sobald die Congruenz

$$0 \equiv k\,\mathrm{Ind}\,\varrho + c\,\mathrm{Ind}(\alpha) + c_1\,\mathrm{Ind}\,\varepsilon_1(\alpha) + \cdots + c_{\mu-1}\,\mathrm{Ind}\,\varepsilon_{\mu-1}(\alpha)$$

nicht identisch erfüllt ist, und für k auch andere Werthe als $k = 0$ zulässt. Diese beiden Bedingungen sind offenbar erfüllt, wenn $f(\alpha)$ eine Primzahl der ersten Art ist.

Es sei also $f(\alpha)$ eine primäre complexe Primzahl der ersten Art, ferner sei $\varphi(\alpha)$ eine beliebige andere primäre complexe Primzahl der ersten Art, so kann man die in $D(\alpha) = e(\alpha)f(\alpha)^m$ enthaltene Einheit $e(\alpha)$ stets so bestimmen, dass

$$(1.)\quad \left(\frac{e(\alpha)f(\alpha)^m}{\varphi(\alpha)}\right) = 1$$

ist, welche Gleichung die nothwendige und hinreichende Bedingung dafür ist, dass $\varphi(\alpha)$ in λ ideale, complexe Primfactoren in w zerlegbar sei. Es sei demnach $\varphi(w)$ ein idealer Primfactor von $\varphi(\alpha)$, und H der Exponent derjenigen Potenz, zu welcher $\varphi(w)$ erhoben werden muss, damit $\varphi(w)^H$ eine wirkliche complexe Zahl in w sei, so ist H nicht durch λ theilbar. Es sei ferner

$$N\varphi(w)^H = \varphi(\alpha)^H E(\alpha),$$

so hat man vermöge der im § 14 der Abhandlung v. J. 1859 entwickelten allgemeinen Bedingung (29.), welcher alle wirklichen complexen Zahlen in w genügen müssen:

$$\left(\frac{E(\alpha)}{f(\alpha)^m}\right) = \left(\frac{e(\alpha)}{\varphi(\alpha)^H}\right),$$

oder was dasselbe ist:

$$(2.)\quad \left(\frac{E(\alpha)}{f(\alpha)}\right)^m = \left(\frac{e(\alpha)}{\varphi(\alpha)}\right)^H,$$

und weil die Norm der wirklichen complexen Zahl $\varphi(w)^H$ in Beziehung auf den Primfactor $f(\alpha)$ der Determinante einer λ^{ten} Potenz congruent ist, so ist

$$(3.)\quad \left(\frac{\varphi(\alpha)^H E(\alpha)}{f(\alpha)}\right) = 1.$$

Aus den drei Gleichungen (1.), (2.) und (3.) folgt nun durch Elimination von $\left(\frac{e(\alpha)}{f(\alpha)}\right)$ und $\left(\frac{E(\alpha)}{f(\alpha)}\right)$

$$\left(\frac{f(\alpha)}{\varphi(\alpha)}\right)^{mH} = \left(\frac{\varphi(\alpha)}{f(\alpha)}\right)^{mH},$$

und weil weder m noch H durch λ theilbar ist:

$$(4.)\quad \left(\frac{f(\alpha)}{\varphi(\alpha)}\right) = \left(\frac{\varphi(\alpha)}{f(\alpha)}\right),$$

wodurch das Reciprocitätsgesetz für je zwei primäre complexe Primzahlen $f(\alpha)$ *und* $\varphi(\alpha)$ *der ersten Art bewiesen ist.*

Es ist hierbei zu bemerken, dass in dem ganz besonderen Falle, wo $\varphi(\alpha)$ eine solche Primzahl der ersten Art ist, für welche alle aus den zweigliedrigen Perioden $\alpha+\alpha^{-1}$, $\alpha^2+\alpha^{-2}$ gebildeten Einheiten λ^{te} Potenzreste sind, und nur die Einheit α^k ein Nichtrest, und wo zugleich $\left(\frac{f(\alpha)}{\varphi(\alpha)}\right)=1$ ist, die Einheit $e(\alpha)$ sich nicht so bestimmen lässt, dass der Gleichung (1.), und zugleich auch der für die Determinante $D(\alpha)$ allgemein festgesetzten Eigenschaft, nach welcher $D(\alpha)-1$ durch ϱ, aber nicht durch ϱ^2 theilbar ist, genügt werde; dass dieser Umstand jedoch keine Lücke in dem gegebenen Beweise begründet, weil es hinreicht, das Reciprocitätsgesetz für alle Nichtreste bewiesen zu haben, da alsdann die Gültigkeit desselben für die Reste eine unmittelbare Folge ist. Man vergleiche den ersten Beweis in der Abhandlung v. J. 1859, p. 148.

Nimmt man $\varphi(\alpha)$ als eine complexe Primzahl der zweiten Art, so dass $\left(\frac{e(\alpha)}{\varphi(\alpha)}\right)=1$ ist, für jede beliebige Einheit $e(\alpha)$, so wird die Gleichung (1.) noch in dem Falle befriedigt, dass $\left(\frac{f(\alpha)}{\varphi(\alpha)}\right)=1$ ist. In diesem Falle findet also ebenfalls das Reciprocitätsgesetz Statt: *Wenn eine primäre Primzahl der ersten Art ein λ^{ter} Potenzrest einer primären Primzahl der zweiten Art ist, so ist auch diese ein λ^{ter} Potenzrest von jener.*

Zu dem Beweise der übrigen Fälle des allgemeinen Reciprocitätsgesetzes sind complexe Zahlen in w anzuwenden, deren Determinante zwei verschiedene Primfactoren enthält. Es sei also

$$D(\alpha) = \delta(\alpha)\delta_1(\alpha) = e(\alpha)f(\alpha)^m.e_1(\alpha)f_1(\alpha)^{m_1},$$

wo $f(\alpha)$ und $f_1(\alpha)$ primär, m und m_1 nicht durch λ theilbar, und so gewählt sein sollen, dass $f(\alpha)^m$ und $f_1(\alpha)^{m_1}$ wirklich sind. Ferner soll $f(\alpha)$ eine Primzahl der zweiten Art sein, $f_1(\alpha)$ eine Primzahl der ersten Art, und zwar eine solche, welche Nichtrest von $f(\alpha)$ ist, so dass $\operatorname{Ind} f_1(\alpha)$ nicht congruent Null ist.

Nach dem im § 3 bewiesenen allgemeinen Satze sind nun die kleinsten Exponenten der Potenzen, zu welchen die idealen Zahlen in w erhoben werden müssen, um als wirkliche complexe Zahlen in u, u_1 darstellbar zu werden, niemals durch λ theilbar, sobald die beiden Congruenzen

$$0 \equiv (n_1-n)\operatorname{Ind}\delta_1(\alpha)+k\operatorname{Ind}\varrho+c\operatorname{Ind}(\alpha)+c_1\operatorname{Ind}\varepsilon_1(\alpha)+\cdots+c_{\mu-1}\operatorname{Ind}\varepsilon_{\mu-1}(\alpha),$$
$$0 \equiv (n-n_1)\operatorname{Ind}_1\delta(\alpha)+k\operatorname{Ind}_1\varrho+c\operatorname{Ind}_1(\alpha)+c_1\operatorname{Ind}_1\varepsilon_1(\alpha)+\cdots+c_{\mu-1}\operatorname{Ind}_1\varepsilon_{\mu-1}(\alpha)$$

nicht identisch und von einander unabhängig sind, und für k auch andere Werthe zulassen als $k\equiv 0$. Die erste dieser Congruenzen giebt, weil $f(\alpha)$ eine Primzahl der zweiten Art ist, für welche die Indices aller Einheiten congruent Null sind:

$$0 \equiv (n_1-n)m_1\operatorname{Ind} f_1(\alpha)+k\operatorname{Ind}\varrho;$$

dieselbe ist nicht identisch erfüllt, weil $\operatorname{Ind} f_1(\alpha)$ nicht congruent Null ist. Ebenso ist auch die zweite nicht identisch erfüllt, weil $f_1(\alpha)$ eine Primzahl der ersten Art ist, für welche die Indices der Einheiten α, $\varepsilon_1(\alpha), \ldots\ \varepsilon_{\mu-1}(\alpha)$ nicht alle congruent Null sind, und aus demselben Grunde ist auch die zweite von der ersten unabhängig. Nach den über $f(\alpha)$ und $f_1(\alpha)$ festgesetzten Bestimmungen sind also die Exponenten der niedrigsten Potenzen, zu welchen die idealen Zahlen in w erhoben werden müssen, um als wirkliche complexe Zahlen in u, u_1 darstellbar zu sein, hier niemals durch λ theilbar.

Es sei nun $\varphi(\alpha)$ eine complexe Primzahl, welche als primär angenommen werden soll, und welche der Bedingung $\left(\frac{D(\alpha)}{\varphi(\alpha)}\right) = 1$ genügt, also

$$(5.)\quad \left(\frac{e(\alpha)f(\alpha)^m e_1(\alpha)f_1(\alpha)^{m_1}}{\varphi(\alpha)}\right) = 1,$$

so ist $\varphi(\alpha)$ in λ ideale Primfactoren in w zerlegbar, und wenn $\varphi(w)$ einer dieser Primfactoren des $\varphi(\alpha)$ ist, so ist $\varphi(w)^H$ als wirkliche complexe Zahl in u, u_1 darstellbar, und H nicht durch λ theilbar. Es sei demnach

$$\varphi(w)^H = F(u, u_1),$$

so ist

$$NF(u, u_1) = \varphi(\alpha)^H E(\alpha),$$

wo $E(\alpha)$ eine Einheit ist. Die allgemeine Form einer jeden wirklichen complexen Zahl in u, u_1:

$$F(u, u_1) = \Sigma a_k u^{|k-n|} u_1^{|k-n_1|}$$

giebt, wenn die Norm nach dem Modul $\delta(\alpha)$ oder $f(\alpha)$ und nach dem Modul $\delta_1(\alpha)$ oder $f_1(\alpha)$ betrachtet wird:

$$NF(u, u_1) \equiv a_n^\lambda \delta_1(\alpha)^{n-n_1} \quad (\text{mod. } \delta(\alpha)),$$
$$NF(u, u_1) \equiv a_{n_1}^\lambda \delta(\alpha)^{n_1-n} \quad (\text{mod. } \delta_1(\alpha)),$$

und diese beiden Congruenzen geben

$$(6.)\quad \left(\frac{\varphi(\alpha)^H E(\alpha)}{f(\alpha)}\right) = \left(\frac{e_1(\alpha)f_1(\alpha)^{m_1}}{f(\alpha)}\right)^{n-n_1},$$

$$(7.)\quad \left(\frac{\varphi(\alpha)^H E(\alpha)}{f_1(\alpha)}\right) = \left(\frac{e(\alpha)f(\alpha)^m}{f_1(\alpha)}\right)^{n_1-n}.$$

Endlich, weil $u^n u_1^{n_1} F(u, u_1)$ eine wirkliche complexe Zahl in w ist, deren Norm gleich $\delta(\alpha)^n \delta_1(\alpha)^{n_1} \varphi(\alpha)^H E(\alpha)$ den primären Theil $f(\alpha)^{mn} f_1(\alpha)^{m_1 n_1} \varphi(\alpha)^H$ und die Einheit $e(\alpha)^n e_1(\alpha)^{n_1} E(\alpha)$ enthält, so gilt für dieselbe die allgemeine Gleichung (29.) § 14 der Abhandlung v. J. 1859, welche im gegenwärtigen Falle

$$(8.)\quad \left(\frac{e(\alpha)^n e_1(\alpha)^{n_1} E(\alpha)}{f(\alpha)^m f_1(\alpha)^{m_1}}\right) = \left(\frac{e(\alpha)e_1(\alpha)}{f(\alpha)^{mn} f_1(\alpha)^{m_1 n_1} \varphi(\alpha)^H}\right)$$

giebt.

Wendet man anstatt der *Legendre*schen Zeichen die Zeichen der Indices an, und bezeichnet den in Beziehung auf die Primzahl $\varphi(\alpha)$ genommenen Index durch ind, die in Beziehung auf $f(\alpha)$ und $f_1(\alpha)$ genommenen wie oben durch Ind und Ind_1, so kann man die gefundenen vier Gleichungen (5.), (6.), (7.) und (8.) auch so darstellen:

$$(9.)\quad \mathrm{ind}\,e(\alpha)+\mathrm{ind}\,e_1(\alpha)+m\,\mathrm{ind}f(\alpha)+m_1\mathrm{ind}f_1(\alpha)\equiv 0,$$

$$(10.)\quad H\,\mathrm{Ind}\,\varphi(\alpha)+\mathrm{Ind}\,E(\alpha)-(n-n_1)\mathrm{Ind}\,e_1(\alpha)-(n-n_1)m_1\mathrm{Ind}f_1(\alpha)\equiv 0,$$

$$(11.)\quad H\,\mathrm{Ind}_1\varphi(\alpha)+\mathrm{Ind}_1E(\alpha)+(n-n_1)\mathrm{Ind}_1e(\alpha)+(n-n_1)m\,\mathrm{Ind}_1f(\alpha)\equiv 0,$$

$$(12.)\quad \begin{cases} m_1(n-n_1)\mathrm{Ind}_1e(\alpha)-m(n-n_1)\mathrm{Ind}\,e_1(\alpha)+m\,\mathrm{Ind}\,E(\alpha) \\ +m\,\mathrm{Ind}_1E(\alpha)-H\mathrm{ind}\,e(\alpha)-H\mathrm{ind}\,e_1(\alpha)\equiv 0. \end{cases}$$

Diese vier Congruenzen, der Reihe nach mit H, $-m$, $-m_1$, 1 multiplicirt und addirt, geben:

$$(13.)\quad \begin{cases} mH(\mathrm{ind}f(\alpha)-\mathrm{Ind}\,\varphi(\alpha))+m_1H(\mathrm{ind}f_1(\alpha)-\mathrm{Ind}_1\varphi(\alpha)) \\ +mm_1(n-n_1)(\mathrm{Ind}f_1(\alpha)-\mathrm{Ind}_1f(\alpha))\equiv 0. \end{cases}$$

Es sei nun zunächst $\varphi(\alpha)$ eine Primzahl der ersten Art, so gilt für die beiden Primzahlen $f_1(\alpha)$ und $\varphi(\alpha)$ das Reciprocitätsgesetz $\mathrm{ind}f_1(\alpha) \equiv \mathrm{Ind}_1\varphi(\alpha)$. Die Congruenz (13.) wird daher

$$(14.)\quad H(\mathrm{ind}f(\alpha)-\mathrm{Ind}\,\varphi(\alpha))+m_1(n-n_1)(\mathrm{Ind}f_1(\alpha)-\mathrm{Ind}_1f(\alpha))\equiv 0.$$

Ferner, weil $f(\alpha)$ eine Primzahl der zweiten Art ist, für welche die Indices aller Einheiten congruent Null sind, giebt die Congruenz (10.)

$$(15.)\quad H\,\mathrm{Ind}\,\varphi(\alpha)-m_1(n-n_1)\mathrm{Ind}f_1(\alpha)\equiv 0,$$

und wenn diese Congruenz zur vorhergehenden addirt wird:

$$(16.)\quad H\mathrm{ind}f(\alpha)-m_1(n-n_1)\mathrm{Ind}_1f(\alpha)\equiv 0.$$

Die Primzahl $\varphi(\alpha)$ wird nun ausser den bereits festgesetzten Bedingungen, dass sie eine Primzahl der ersten Art sei, und dass $\left(\frac{D(\alpha)}{\varphi(\alpha)}\right)=1$ sei, d. i. $\mathrm{ind}\,D(\alpha)\equiv 0$, noch der Bedingung unterworfen, dass alle zu $f(\alpha)$ conjugirten Zahlen Reste von $\varphi(\alpha)$ sein sollen, aber $f(\alpha)$ selbst ein Nichtrest, welchen Bedingungen vermöge des allgemeinen Satzes I § 16 der Abhandlung v. J. 1859 stets genügt werden kann. Die eine Reciprocitätsgleichung zwischen $\varphi(\alpha)$ und $f(\alpha)$, welche die Kreistheilung gewährt (16.) § 4, giebt alsdann in derselben Weise, wie dies in dem ersten Beweise pag. 154 der Abhandlung von 1859, so wie auch in dem betreffenden Passus des zweiten Beweises näher entwickelt worden ist:

$$\mathrm{ind}f(\alpha)\equiv\mathrm{Ind}\,\varphi(\alpha).$$

Die Congruenz (14.) wird demnach

$$(17.)\quad m_1(n-n_1)(\mathrm{Ind}f_1(\alpha)-\mathrm{Ind}_1f(\alpha))\equiv 0.$$

Es ist nun $n-n_1$ nicht congruent Null; denn wäre $n-n_1 \equiv 0$, so müsste, da H nicht congruent Null ist, vermöge Congruenz (16.), $\operatorname{ind} f(\alpha) \equiv 0$ sein, welches der Voraussetzung widerspricht, dass $f(\alpha)$ Nichtrest von $\varphi(\alpha)$ ist. Da überdies m_1 nicht durch λ theilbar ist, so folgt:

$$(18.)\qquad \operatorname{Ind} f_1(\alpha) \equiv \operatorname{Ind}_1 f(\alpha),$$

oder nach den *Legendre*schen Zeichen:

$$\left(\frac{f_1(\alpha)}{f(\alpha)}\right) = \left(\frac{f(\alpha)}{f_1(\alpha)}\right),$$

wo die Primzahl der zweiten Art $f(\alpha)$ ganz beliebig, die Primzahl der ersten Art $f_1(\alpha)$ aber nur der einen Bedingung unterworfen ist, dass $\operatorname{Ind} f_1(\alpha)$ nicht congruent Null ist. Dieselbe Congruenz (18.) ist aber, nach dem bereits bewiesenen besonderen Falle des Reciprocitätsgesetzes zwischen einer Primzahl der zweiten und einer der ersten Art, auch richtig, wenn $\operatorname{Ind} f_1(\alpha) \equiv 0$ ist, so dass auch diese Einschränkung wegfällt. Also: ***unter je zwei primären complexen Zahlen $f(\alpha)$ und $f_1(\alpha)$, deren eine der ersten Art, die andere der zweiten Art angehört, besteht das Reciprocitätsgesetz***

$$\left(\frac{f_1(\alpha)}{f(\alpha)}\right) = \left(\frac{f(\alpha)}{f_1(\alpha)}\right).$$

Um nun noch für je zwei primäre Primzahlen der zweiten Art das Reciprocitätsgesetz zu beweisen, nehme ich $\varphi(\alpha)$ als eine Primzahl der zweiten Art, während $f(\alpha)$ und $f_1(\alpha)$ die ihnen oben beigelegte Bedeutung behalten, nach welcher $f(\alpha)$ Primzahl der zweiten Art, $f_1(\alpha)$ Primzahl der ersten Art ist und $\operatorname{Ind} f_1(\alpha)$ nicht congruent Null.

Die Congruenz (13.) giebt nun, weil nach dem so eben bewiesenen Falle des Reciprocitätsgesetzes $\operatorname{Ind} f_1(\alpha) \equiv \operatorname{Ind}_1 f(\alpha)$ und $\operatorname{ind} f_1(\alpha) \equiv \operatorname{Ind}_1 \varphi(\alpha)$ ist:

$$mH(\operatorname{ind} f(\alpha) - \operatorname{Ind} \varphi(\alpha)) = 0,$$

und weil weder m noch H durch λ theilbar ist:

$$\operatorname{ind} f(\alpha) = \operatorname{Ind} \varphi(\alpha),$$

oder nach den *Legendre*schen Zeichen:

$$\left(\frac{f(\alpha)}{\varphi(\alpha)}\right) = \left(\frac{\varphi(\alpha)}{f(\alpha)}\right).$$

Die Zahl $\varphi(\alpha)$ ist hier der Bedingung $\left(\frac{D(\alpha)}{\varphi(\alpha)}\right) = 1$ unterworfen, welche, wenn für $D(\alpha)$ sein Werth $e(\alpha) f(\alpha)^m e_1(\alpha) f_1(\alpha)^{m_1}$ gesetzt und beachtet wird, dass für die Primzahl der zweiten Art $\varphi(\alpha)$ die Einheiten $e(\alpha)$ und $e_1(\alpha)$

λ^{te} Potenzreste sind,

$$\left(\frac{f(\alpha)}{\varphi(\alpha)}\right)^m \left(\frac{f_1(\alpha)}{\varphi(\alpha)}\right)^{m_1} = 1$$

giebt. Dieser Bedingung kann durch passende Wahl der Primzahl $f_1(\alpha)$ der ersten Art und ihres Exponenten m_1 immer genügt werden, für alle beliebigen Primzahlen $f(\alpha)$ und $\varphi(\alpha)$ der zweiten Art (man vergleiche pag. 166 des ersten Beweises). Also: *unter je zwei primären complexen Primzahlen der zweiten Art $f(\alpha)$ und $\varphi(\alpha)$ besteht das Reciprocitätsgesetz*

$$\left(\frac{f(\alpha)}{\varphi(\alpha)}\right) = \left(\frac{\varphi(\alpha)}{f(\alpha)}\right).$$

Die bewiesenen besonderen Fälle: erstens wo beide Primzahlen der ersten Art angehören, zweitens wo eine der ersten Art, die andere der zweiten Art angehört, und drittens wo beide der zweiten Art angehören, geben zusammengefasst das allgemeine Reciprocitätsgesetz für je zwei beliebige primäre complexe Primzahlen $f(\alpha)$ und $\varphi(\alpha)$:

$$\left(\frac{f(\alpha)}{\varphi(\alpha)}\right) = \left(\frac{\varphi(\alpha)}{f(\alpha)}\right).$$

Über die Klassenanzahl der aus *n*ten Einheitswurzeln gebildeten complexen Zahlen

Monatsberichte der Königlichen Preußischen Akademie der Wissenschaften zu Berlin aus dem Jahre 1861, 1051–1053

9. Decemb. Sitzung der physikalisch-mathematischen Klasse.

Hr. Kummer las über die Klassenanzahl der aus $n$ten Einheitswurzeln gebildeten complexen Zahlen.

Die Arbeiten des Hrn. Prof. Reuschle über die Zerlegung der Zahlen in ihre complexen Primfaktoren, welche derselbe gegenwärtig auch für den allgemeineren Fall ausführt, wo der Wurzelexponent der Einheitswurzeln eine zusammengesetzte Zahl ist, haben mich veranlafst, für diese complexen Zahlen die Klassenanzahl zu bestimmen. Der gefundene Ausdruck derselben, dessen Entwickelung ich der Königlichen Akademie nächstens mitzutheilen gedenke, läfst sich auf eine Form bringen, welche der, für den besonderen Fall, wo der Wurzelexponent eine Primzahl ist, früher von mir gegebenen ganz analog ist. Namentlich besteht dieser allgemeinere Ausdruck ebenso aus zwei verschiedenen Faktoren, welche als erster und zweiter Faktor der Klassenanzahl zu unterscheiden sind, deren ersterer nur aus Einheitswurzeln und ganzen Zahlen, der andere aber aus den Logarithmen der Kreistheilungseinheiten und der Fundamentaleinheiten zusammengesetzt ist. Wenn n eine beliebige ganze Zahl ist, und ω eine primitive $n$te Wurzel der Einheit, so ist der zweite Faktor für sich selbst die Klassenanzahl der, aus den Gröfsen $\omega+\omega^{-1}$, $\omega^2+\omega^{-2}$, $\omega^3+\omega^{-3}$, ... gebildeten com-

72*

plexen Zahlen, derselbe ist also nothwendig eine ganze Zahl. In dem früher von mir behandelten speciellen Falle, wo n eine Primzahl ist, so wie auch noch in den Fällen, wo n nur Potenz einer Primzahl ist, hat der erste Faktor der Klassenanzahl ebenfalls die Eigenschaft ganzzahlig zu sein; wenn aber n eine aus verschiedenen Primzahlen zusammengesetzte Zahl ist, so tritt der merkwürdige Umstand ein, dafs dieser erste Faktor nicht mehr ganzzahlig ist. Die Klassenanzahl der aus den Gröfsen $\omega + \omega^{-1}$, $\omega^2 + \omega^{-2}$, $\omega + \omega^{-3}$ ... gebildeten complexen Zahlen ist also nicht mehr ein genauer Theil der Klassenanzahl der aus den einfachen Einheitswurzeln ω, ω^2, ω^3 ... gebildeten complexen Zahlen. Die Nenner, welche in dem ersten Faktor vorkommen, sind aber keine anderen, als Potenzen der Zahl 2, deren Höhe hauptsächlich (jedoch nicht allein) von der Anzahl der in n enthaltenen verschiedenen Primzahlen abhängt.

Da es mir zunächst darauf ankam, zu bestimmen, welche Idealitäten bei der Zerlegung der Zahlen in ihre aus $n$ten Einheitswurzeln gebildeten complexen Primfaktoren vorkommen müssen, so habe ich die numerischen Werthe des ersten Faktors der Klassenanzahl, für alle Werthe bis $n = 100$, ausgerechnet; aufserdem hat auch der Hr. Dr. Fuchs hierselbst diese Rechnung vollständig durchgeführt, und aus der Vergleichung dieser von einander ganz unabhängig gewonnenen Resultate, ist die folgende, von Rechnungsfehlern freie Tafel festgestellt worden, durch welche die früher von mir für den Fall, wo n eine Primzahl ist, in Liouvilles Journal Bd. 16 pag. 473 gegebenen Zahlenwerthe des ersten Faktors vervollständigt werden.

$P'(2) = 1.$
$P'(3) = 1.$
$P'(4) = 1.$
$P'(5) = 1.$
$P'(6) = 1.$
$P'(7) = 1.$
$P'(8) = 1.$
$P'(9) = 1.$
$P'(10) = 1.$
$P'(11) = 1.$
$P'(12) = \frac{1}{2}.$
$P'(13) = 1.$
$P'(14) = 1.$
$P'(15) = \frac{1}{2}.$
$P'(16) = 1.$
$P'(17) = 1.$
$P'(18) = 1.$
$P'(19) = 1.$
$P'(20) = \frac{1}{2}.$
$P'(21) = \frac{1}{2}.$
$P'(22) = 1.$
$P'(23) = 3.$

$P'(24) = \frac{1}{2}$.
$P'(25) = 1$.
$P'(26) = 1$.
$P'(27) = 1$.
$P'(28) = \frac{1}{2}$.
$P'(29) = 2^3$.
$P'(30) = \frac{1}{2}$.
$P'(31) = 3^2$.
$P'(32) = 1$.
$P'(33) = \frac{1}{2}$.
$P'(34) = 1$.
$P'(35) = \frac{1}{2}$.
$P'(36) = \frac{1}{2}$.
$P'(37) = 37$.
$P'(38) = 1$.
$P'(39) = 1$.
$P'(40) = \frac{1}{2}$.
$P'(41) = 11^2$.
$P'(42) = \frac{1}{2}$.
$P'(43) = 211$.
$P'(44) = \frac{1}{2}$.
$P'(45) = \frac{1}{2}$.
$P'(46) = 3$.
$P'(47) = 5.\ 139$.
$P'(48) = \frac{1}{2}$.
$P'(49) = 43$.
$P'(50) = 1$.
$P'(51) = \frac{1}{2}.\ 5$.
$P'(52) = \frac{1}{2}.\ 3$.
$P'(53) = 4889$.
$P'(54) = 1$.
$P'(55) = 5$.
$P'(56) = 1$.
$P'(57) = \frac{1}{2}.\ 3^2$.
$P'(58) = 2^3$.
$P'(59) = 3.\ 59.\ 233$.
$P'(60) = \frac{1}{2}$.
$P'(61) = 41.\ 1861$.
$P'(62) = 3^2$.
$P'(63) = \frac{1}{2}.\ 7$.
$P'(64) = 17$.
$P'(65) = 2^5$.
$P'(66) = \frac{1}{2}$.
$P'(67) = 67.\ 12739$.
$P'(68) = 2^3$.
$P'(69) = \frac{1}{2}.\ 3.\ 23$.
$P'(70) = \frac{1}{2}$.
$P'(71) = 7.\ 7.\ 79241$.
$P'(72) = \frac{1}{2}.\ 3$.
$P'(73) = 89.\ 134353$.
$P'(74) = 37$.
$P'(75) = \frac{1}{2}.\ 11$.
$P'(76) = \frac{1}{2}.\ 19$.
$P'(77) = 2^7.\ 5$.
$P'(78) = 1$.
$P'(79) = 5.\ 53.\ 377911$.
$P'(80) = \frac{1}{2}.\ 5$.
$P'(81) = 2593$.
$P'(82) = 11^2$.
$P'(83) = 3.\ 279405653$.
$P'(84) = \frac{1}{2}$.
$P'(85) = \frac{1}{2}.\ 5.\ 17.\ 73$.
$P'(86) = 211$.
$P'(87) = 2^8.\ 3$.
$P'(88) = \frac{1}{2}.\ 5.\ 11$.
$P'(89) = 113.\ 118401449$.
$P'(90) = \frac{1}{2}$.
$P'(91) = 2^3.\ 7.\ 13.\ 37$.
$P'(92) = \frac{1}{2}.\ 3.\ 67^2$.
$P'(93) = \frac{1}{2}.\ 3^2.\ 5.\ 151$.
$P'(94) = 5.\ 139$.
$P'(95) = 2.\ 13.\ 19.\ 109$.
$P'(96) = \frac{1}{2}.\ 3^2$.
$P'(97) = 3457.\ 118982593$.
$P'(98) = 43$.
$P'(99) = \frac{1}{2}.\ 3.\ 31^2$.
$P'(100) = \frac{1}{2}.\ 5.\ 11$.

Über die Klassenanzahl der aus zusammengesetzten Einheitswurzeln gebildeten idealen complexen Zahlen

Monatsberichte der Königlichen Preußischen Akademie der Wissenschaften zu Berlin aus dem Jahre 1863, 21–28

8. Januar. Gesammtsitzung der Akademie.

Hr. **Kummer** las über **die Klassenanzahl der aus zusammengesetzten Einheitswurzeln gebildeten idealen complexen Zahlen.**

Die vorliegende Arbeit kann als eine Fortsetzung der in den Abhandlungen der Akademie vom Jahre 1856 niedergelegten Theorie der idealen Primfaktoren der complexen Zahlen, welche aus den Wurzeln der Gleichung $\omega^n = 1$ gebildet sind, angesehen werden. Die daselbst gefundenen Resultate über die Natur und die Eigenschaften der idealen Primfaktoren dieser Theorie sind vollkommen hinreichend um die Dirichletschen Methoden zur Bestimmung der Klassenanzahl derselben mit Erfolg anwenden zu können; auch sind die in der Summation gewisser unendlicher Reihen und in der Werthbestimmung eines vielfachen Integrales bestehenden analytischen Schwierigkeiten nicht gröfser, als in dem besonderen Falle, wo n eine einfache Primzahl ist, welchen Fall ich ausführlich in Liouville's Journal Bd. XVI behandelt habe. Es tritt aber hier eine besondere Schwierigkeit anderer Art ein, nämlich die Complication, welche die Betrachtung zusammengesetzter Zahlen gewöhnlich mit sich führt, und welche bei der Bestimmung dieser Klassenanzahl auch aus dem Endresultate nicht leicht zu entfernen ist. Wenn nämlich n eine zusammengesetzte Zahl ist, so enthält die Klassenanzahl der aus $n$ten Einheitswurzeln gebildeten idealen complexen Zahlen eigentlich so viele verschiedenartige Bestandtheile in sich, als die Zahl n verschiedene Divisoren hat, mit Ausschlufs des Divisors Eins. So z. B. wenn $n = pp_1$ das Produkt zweier verschiedenen ungraden Primzahlen p und p_1 ist, hat die Klassenanzahl drei verschiedene Bestandtheile, deren erster und zweiter für sich die Klassenanzahlen der aus $p$ten und der aus $p_1$ten Einheitswurzeln gebildeten complexen Zahlen sind, der dritte aber gewissermafsen als ein primitiver Faktor für die pp_1ten Einheitswurzeln anzusehen ist. Alle diese verschiedenartigen Bestandtheile lassen sich aber unter eine gemeinsame Form vereinigen, und zwar in der Art, dafs sie aus dieser allgemeinen Form mit Leichtigkeit als individuell verschiedene Bestandtheile dargestellt werden können.

In dem Falle, dafs n eine grade, und zwar durch 4 theilbare Zahl ist, ist der Ausdruck der Klassenanzahl etwas verschieden von dem für ungrade Werthe des n geltenden, weshalb es zweckmäfsig ist ihn für diese beiden Fälle besonders aufzustellen. Der Fall aber, wo n eine grade, nicht durch 4 theilbare Zahl ist, kann ganz unberücksichtigt gelassen werden, weil in diesem Falle die $n$ten Wurzeln der Einheit von den $\frac{n}{2}$ten Einheitswurzeln nur in den Vorzeichen verschieden sind, also die aus $n$ten Einheitswurzeln gebildeten complexen Zahlen keine anderen, als die aus $\frac{n}{2}$ten Einheitswurzeln gebildeten, und demnach die Klassenanzahlen beider dieselben.

Es sei nun erstens n eine ungrade Zahl

$$n = p^{\pi} \cdot p_1^{\pi_1} \cdot p_2^{\pi_2} \ldots$$

wo $p, p_1, p_2 \ldots$ verschiedene ungrade Primzahlen sind und $\pi, \pi_1, \pi_2 \ldots$ beliebige nicht negative ganze Zahlen. Seien $\mathrm{Ind}, \mathrm{Ind}_1, \mathrm{Ind}_2, \ldots$ die Indices für die Moduln $p^{\pi}, p_1^{\pi_1}, p_2^{\pi_2}, \ldots$ und für irgend welche primitive Wurzeln dieser Primzahlpotenzen; seien $\xi, \xi_1, \xi_2 \ldots$ primitive Wurzeln der Gleichungen

$$\xi^{p^{\pi-1}(p-1)} = 1, \quad \xi_1^{p_1^{\pi_1-1}(p_1-1)} = 1, \quad \xi_2^{p_2^{\pi_2-1}(p_2-1)} = 1, \ \ldots$$

ω eine primitive Wurzel der Gleichung $\omega^n = 1$ und $e(\omega^k)$ die sogenannte Kreistheilungseinheit

$$e(\omega^k) = \sqrt{(1-\omega^k)(1-\omega^{-k})}.$$

Es bezeichne ferner $K(c, c_1, c_2 \ldots)$ folgende Summe

$$K(c, c_1, c_2, \ldots) = \Sigma_k\, \xi^{c\,\mathrm{Ind}\,k} \xi_1^{c_1\,\mathrm{Ind}_1\,k} \xi_2^{c_2\,\mathrm{Ind}_2\,k} \ldots k,$$

in welcher das Summenzeichen sich auf alle diejenigen ganzzahligen Werthe des $k = 1, 2, \ldots n-1$ bezieht, für welche die in der Summe wirklich vorkommenden Indices nicht absurd werden, so dafs in dem Falle, wo keine der Zahlen $c, c_1, c_2 \ldots$ gleich Null ist, also alle Indices in der Formel wirklich vorkommen, k nur alle diejenigen Werthe erhalten mufs, welche kleiner als n sind und relative Primzahlen zu n, dafs aber, wenn z. B. c gleich Null ist, und darum der Index Ind, welcher sich auf den Modul p^{π} bezieht, in der Formel nicht wirklich vorkommt, für k auch die durch p theilbaren Werthe zugelassen sind; ebenso wenn $c_1 = 0$ ist, dafs die durch p_1 theilbaren Werthe des k zugelassen sind u. s. w. In ähnlicher Weise bezeichne $L(c, c_1, c_2 \ldots)$ die Summe:

$$L(c, c_1, c_2 \ldots) = \Sigma_k\, \xi^{c\,\mathrm{Ind}\,k} \xi_1^{c_1\,\mathrm{Ind}_1\,k} \xi_2^{c_2\,\mathrm{Ind}_2\,k} \ldots l e(\omega^k)$$

in welcher $l\,e(\omega^k)$ der natürliche Logarithmus der Kreistheilungseinheit ist, und wo das Summenzeichen auf alle diejenigen Werthe des k zwischen 0 und $\frac{n}{2}$ sich bezieht, für welche die in der Formel wirklich vorkommenden Indices nicht absurd werden.

Es sei ferner P das Produkt aller derjenigen $\frac{1}{2}\phi(n)$ Summen $K(c, c_1, c_2, \ldots)$, [$\phi(n)$ Anzahl der Zahlen, welche kleiner als n und relative Primzahlen zu n sind] welche man erhält, indem man den Zahlen $c, c_1, c_2 \ldots$ alle diejenigen der Werthe $c = 0, 1, \ldots \phi(p^{\pi}) - 1$, $c_1 = 0, 1, \ldots \phi(p_1^{\pi_1}) - 1$, $c_2 = 0, 1, \ldots \phi(p_2^{\pi_2}) - 1 \ldots$ giebt, welche der Bedingung genügen, dafs $c + c_1 + c_2 + \ldots$ eine ungrade Zahl ist. In ähnlicher Weise sei Θ das Produkt aller derjenigen $\frac{1}{2}\phi(n) - 1$ Summen $L(c, c_1, c_2, \ldots)$, welche man erhält indem man den Zahlen $c, c_1, c_2 \ldots$ von den angegebenen Werthen alle diejenigen giebt, welche der Bedingung genügen, dafs $c + c_1 + c_2 + \ldots$ eine grade Zahl ist, mit Ausschlufs der einen Werthverbindung $c = 0$, $c_1 = 0$, $c_2 = 0, \ldots$

Endlich sei Δ die Determinante der natürlichen Logarithmen der $\frac{1}{2}\phi(n)-1$ Fundamentaleinheiten und der zu denselben conjugirten, so wird die gesuchte Klassenanzahl H durch folgenden Ausdruck dargestellt:

$$H=\frac{P}{(2n)^{\frac{1}{2}\phi(n)-1}}\cdot\frac{\Theta}{\Delta}.$$

Es sei zweitens n eine grade, den Faktor 2 mindestens zweimal enthaltende Zahl, also

$$n=2^{\mu}.p^{\pi}.p_1^{\pi_1}\ldots$$

wo $\mu\geqq 2$ ist. Werden für diesen Fall die bereits festgesetzten Bezeichnungen der auf die Moduln p^{π}, $p_1^{\pi_1}$, ... bezüglichen Indices und der Wurzeln ξ, ξ_1, ... beibehalten, und wird aufserdem in Beziehung auf den Modul 2^{μ} ein Index ind durch die Congruenz

$$5^{\text{ind}\,k}\equiv\pm k,\quad \text{mod. } 2^{\mu}$$

definirt, und η als eine primitive Wurzel der Gleichung

$$\eta^{2^{\mu-2}}=1,$$

so sind hier für die im vorigen Falle mit K und L bezeichneten beiden Summen die folgenden zu nehmen:

$$K(a,b,c,c_1\ldots)=\Sigma_k(-1)^{\frac{a(k-1)}{2}}\eta^{b\,\text{ind}\,k}\xi^{c\,\text{Ind}\,k}\xi_1^{c_1\,\text{Ind}_1\,k}\ldots k$$

$$L(a,b,c,c_1\ldots)=\Sigma_k(-1)^{\frac{a(k-1)}{2}}\eta^{b\,\text{ind}\,k}\xi^{c\,\text{Ind}\,k}\xi_1^{c_1\,\text{Ind}_1\,k}\ldots l\,e(\omega^k)$$

in denen ebenso dem k alle diejenigen Werthe zwischen 0 und n in der ersteren aber nur zwischen 0 und $\frac{n}{2}$ in der letzteren zu geben sind, für welche die in den Summen wirklich vorkommenden Indices nicht absurd werden, und für welche $\frac{a(k-1)}{2}$ eine ganze Zahl ist, also im allgemeinen nur diejenigen Werthe des k, welche relative Primzahlen zu n sind, aber in den besondern Fällen, wo zugleich $a=0$ und $b=0$, auch die durch 2 theilbaren, wenn $c=0$ auch die durch p theilbaren u. s. w. hinzuzunehmen sind. Ferner ist für das oben mit P bezeichnete Produkt das Produkt aller derjenigen $\frac{1}{2}\phi(n)$ Summen $K(a, b, c, c_1, ..)$ zu nehmen, welche man erhält, indem man den Zahlen $a, b, c, c_1 ...$ alle diejenigen der Werthe $a = 0, 1, b = 0, 1, ... 2^{\mu-2} - 1$, $c = 0, 1, ... \phi(p^\pi) - 1$, $c_1 = 0, 1, ... \phi(p_1^{\pi_1}) - 1$ u. s. w. giebt, welche der Bedingung genügen, dafs $a + c + c_1 + ...$ eine ungrade Zahl ist. Ebenso ist für Θ das Produkt aller derjenigen $\frac{1}{2}\phi(n) - 1$ Summen $L(a, b, c, c_1, ...)$ zu nehmen, welche man erhält, indem man den Zahlen $a, b, c, c_1, ...$ von den angegebenen Werthen alle diejenigen giebt, für welche $a + c + c_1 + ...$ eine grade Zahl ist, mit Auschlufs der einen Werthverbindung $a = 0$, $b = 0$, $c = 0$, $c_1 = 0$, ... Die Klassenanzahl H, für grade, durch 4 theilbare Werthe des n, hat alsdann folgenden Ausdruck:

$$H = \frac{P}{2 \cdot (2n)^{\frac{1}{2}\phi(n) - 1}} \cdot \frac{\Theta}{\Delta}.$$

Diese Ausdrücke der Klassenanzahl haben ganz dieselbe Form, als der früher für den besonderen Fall, wo n eine Primzahl ist von mir gegebene; sie enthalten ebenso zwei getrennte Faktoren, den ersten, welcher als Norm einer complexen Zahl aufgefafst werden kann und den zweiten, welcher aus den Logarithmen der Kreistheilungseinheiten und der Fundamentaleinheiten gebildet ist. Für den ersten dieser beiden Faktoren habe ich die von Hrn. Dr. Fuchs und mir berechneten Zahlenwerthe, für alle Zahlen n im ersten Hundert, der Akademie bereits im December 1861 mitgetheilt[1]); auch habe ich daselbst bemerkt, dafs der zweite Faktor für sich selbst genommen die Klassenanzahl der aus den zweigliedrigen Perioden $\omega + \omega^{-1}$, $\omega^2 + \omega^{-2}$, ... gebildeten complexen Zahlen ist, und darum nothwendig eine ganze Zahl, während der erste Faktor im all-

[1]) Ich bemerke hierbei, dafs Hr. Prof. Reuschle, welcher diese Tafel der Werthe des ersten Faktors der Klassenanzahl selbständig auch über $n = 100$ hinaus ausgerechnet hat, mir ein kleines Versehen angezeigt hat, welches in derselben vorkommt, nämlich dafs $P'(92) = \frac{1}{2} . 3 . 67$ ist, nicht $\frac{1}{2} . 3 . 67^2$ wie jene Tafel irrthümlich angiebt, da der Faktor 67 nicht zweimal, sondern nur einmal in $P'(92)$ enthalten ist.

gemeinen eine gebrochene Zahl ist, welche jedoch im Nenner nur eine Potenz von 2 enthalten kann. Zur numerischen Berechnung eignen sich aber die hier gegebenen Ausdrücke weniger, als diejenigen, welche man aus denselben erhält, indem man die verschiedenartigen Bestandtheile von einander trennt, die sie in einer gemeinsamen Form enthalten, wodurch man zugleich den Vortheil erhält, dafs die überflüssigen gemeinsamen Faktoren des Zählers und Nenners entfernt werden.

Bezeichnet man den ersten Faktor der Klassenanzahl mit $P'(n)$, so sind $P'(2^\mu)$, $P'(p^\pi)$ und überhaupt die Zahlenwerthe des ersten Faktors in allen Fällen wo n eine einfache Primzahl oder eine Potenz einer solchen ist, stets ganze Zahlen. Wenn nun n nur zwei verschiedene ungrade Primzahlen enthält, also $n = p^\pi p_1{}^{\pi_1}$ ist, so läfst sich dieser erste Faktor in folgende Form setzen:

$$P'(p^\pi p_1{}^{\pi_1}) = \tfrac{1}{2} P'(p^\pi) P'(p_1{}^{\pi_1})\, Q(p^\pi p_1{}^{\pi_1})$$

in welcher

$$Q(p^\pi p_1{}^{\pi_1}) = \Pi_c\, \Pi_{c_1}\, \Sigma_h\, \Sigma_{h_1}\, \xi^{c\,\mathrm{Ind}\,h}\, \xi_1{}^{c_1\,\mathrm{Ind}_1 h_1}$$

ist, wo die beiden Produktzeichen sich auf alle diejenigen Werthe des $c = 1, 2, \ldots \phi(p^\pi) - 1$, $c_1 = 1, 2, \ldots \phi(p_1{}^{\pi_1}) - 1$ beziehen, für welche $c + c_1$ ungrade ist, die beiden Summenzeichen aber auf alle diejenigen Werthe des h und h_1, welche beziehungsweise kleiner als p^π und relative Primzahlen zu p^π, und kleiner als $p_1{}^{\pi_1}$ uud relative Primzahlen zu $p_1{}^{\pi_1}$ sind, und zugleich der Bedingung genügen, dafs

$$\frac{h}{p^\pi} + \frac{h_1}{p_1{}^{\pi_1}} < 1$$

ist.

Wenn $n = p^\pi p_1{}^{\pi_1} p_2{}^{\pi_2}$ drei verschiedene ungrade Primfaktoren enthält, so findet man in ähnlicher Weise:

$$P'(p^\pi p_1{}^{\pi_1} p_2{}^{\pi_2}) = \frac{1}{2^2} P'(p^\pi) P'(p_1{}^{\pi_1}) P'(p_2{}^{\pi_2})\, Q(p^\pi p_1{}^{\pi_1}) .$$
$$Q(p^\pi p_2{}^{\pi_2})\, Q(p_1{}^{\pi_1} p_2{}^{\pi_2})\, R(p^\pi p_1{}^{\pi_1} p_2{}^{\pi_2}),$$

wo $Q(p^\pi p_1^{\pi_1})$, $Q(p^\pi p_2^{\pi_2})$, $Q(p_1^{\pi_1} p_2^{\pi_2})$ die im vorigen Falle definirten Gröfsen sind, aber

$$R(p^\pi p_1^{\pi_1} p_2^{\pi_2})$$

$$= \Pi_c \, \Pi_{c_1} \, \Pi_{c_2} \, \Sigma_h \, \Sigma_{h_1} \, \Sigma_{h_2} \, \xi^{c\,\mathrm{Ind}\,h} \, \xi_1^{c_1 \mathrm{Ind}_1 h_1} \, \xi_2^{c_2 \mathrm{Ind}_2 h_2},$$

in welchem Ausdrucke die drei Produktenzeichen sich auf alle Werthe des $c = 1, 2, \ldots \phi(p^\pi) - 1$, $c_1 = 1, 2, \ldots \phi(p_1^{\pi_1}) - 1$, $c_2 = 1, 2, \ldots \phi(p_2^{\pi_2}) - 1$ beziehen, welche der Bedingung genügen, dafs $c + c_1 + c_2$ eine ungrade Zahl ist, die drei Summenzeichen aber auf alle diejenigen Werthe des h, h_1 und h_2, welche beziehungsweise kleiner als p^π und relative Primzahlen zu p^π, kleiner als $p_1^{\pi_1}$ und relative Primzahlen zu $p_1^{\pi_1}$ und kleiner als $p_2^{\pi_2}$ und relative Primzahlen zu $p_2^{\pi_2}$ sind, und welche der Bedingung genügen, dafs

$$\frac{h}{p^\pi} + \frac{h_1}{p_1^{\pi_1}} + \frac{h_2}{p_2^{\pi_2}} < 1$$

ist.

Die entsprechenden Ausdrücke des ersten Faktors der Klassenanzahl für den Fall, dafs n eine durch 4 theilbare Zahl ist, sind:

$$P'(2^\mu p^\pi) = \tfrac{1}{2} P'(2^\mu) \, P'(p^\pi) \, Q(2^\mu p^\pi),$$

$$Q(2^\mu p^\pi) = \Pi_a \, \Pi_b \, \Pi_c \, \Sigma_k \, \Sigma_h \, (-1)^{\frac{a(k-1)}{2}} \, \eta^{b\,\mathrm{Ind}\,k} \, \xi^{c\,\mathrm{Ind}\,h},$$

wo die drei Produktzeichen auf alle diejenigen Werthe von $a = 0, 1$, $b = 0, 1, \ldots 2^{\mu-2} - 1$, $c = 1, 2 \ldots \phi(p^\pi) - 1$ sich beziehen, für welche $a + c$ ungrade ist, und nicht a und b beide zugleich gleich Null sind, die Summenzeichen aber auf alle Werthe des $k = 1, 3, 5, \ldots 2^\mu - 1$ und $h = 1, 2, \ldots \phi(p^\pi) - 1$, mit Ausschlufs der durch p theilbaren, welche der Bedingung genügen, dafs

$$\frac{k}{2^\mu} + \frac{h}{p^\pi} < 1$$

ist. Ferner:

$$P'(2^{\mu}p^{\pi}p_1^{\pi_1}) = \frac{1}{2^2} P'(2^{\mu})\, P'(p^{\pi})\, P'(p_1^{\pi_1}) \,.\, Q(2^{\mu}p^{\pi})$$

$$Q(2^{\mu}p_1^{\pi_1})\, Q(p^{\pi}p_1^{\pi_1})\, R(2^{\mu}p^{\pi}p_1^{\pi_1})$$

$$R(2^{\mu}p^{\pi}p_1^{\pi_1})$$

$$= \Pi_a \Pi_b \Pi_c \Pi_{c_1} \Sigma_k \Sigma_h \Sigma_{h_1} (-1)^{\frac{a(k-1)}{2}} \eta^{b \operatorname{ind} k} \xi^{c \operatorname{Ind} h} \xi_1^{c_1 \operatorname{Ind}_1 h_1},$$

wo die vier Produktzeichen auf alle diejenigen Werthe des $a = 0, 1$, $b = 0, 1, \ldots 2^{\mu-2} - 1$, $c = 1, 2, \ldots \phi(p^{\pi}) - 1$, $c_1 = 1, 2, \ldots \phi(p_1^{\pi_1}) - 1$ sich beziehen, für welche $a + c + c_1$ ungrade ist und nicht zugleich $a = 0$ und $b = 0$, die Summenzeichen aber auf die Werthe des k, h und h_1 welche beziehungsweise kleiner als 2^{μ} und nicht durch 2 theilbar, kleiner als p^{π} und nicht durch p theilbar, kleiner als $p_1^{\pi_1}$ und nicht durch p_1 theilbar sind, und der Bedingung

$$\frac{k}{2^{\mu}} + \frac{h}{p^{\pi}} + \frac{h_1}{p_1^{\pi_1}} < 1$$

genügen.

Ebenso lassen sich auch die entsprechenden Verwandlungen des ersten Faktors der Klassenanzahl allgemein ausführen, wenn n nicht nur zwei oder drei, sondern eine beliebige Anzahl r verschiedener Primzahlen enthält; derselbe zerfällt alsdann in ein Produkt von $2^r - 1$ besonderen Faktoren, welche für sich ganzzahlig sind und die Potenz von 2, welche als Nenner bleibt, ist 2^{r-1}. Der erste Faktor der Klassenanzahl enthält also den Faktor Zwei als Nenner höchstens $r - 1$ mal.

Über die einfachste Darstellung der aus Einheitswurzeln gebildeten complexen Zahlen, welche durch Multiplication mit Einheiten bewirkt werden kann

Monatsberichte der Königlichen Preußischen Akademie der Wissenschaften zu Berlin aus dem Jahre 1870, 409–420

16. Juni. Gesammtsitzung der Akademie.

Hr. Kummer las über die einfachste Darstellung der aus Einheitswurzeln gebildeten complexen Zahlen, welche durch Multiplikation mit Einheiten bewirkt werden kann.

Unter den complexen Primfaktoren, welche Hr. Reuschle ausgerechnet und der Akademie übergeben hat, befindet sich ein idealer Primfaktor, dessen neunte Potenz wirklich ist und zwar ist dies ein idealer Primfaktor der Zahl 2, für die aus 31ten Einheitswurzeln gebildeten complexen Zahlen. Die neunte Potenz dieses idealen Primfaktors der Zahl 2 stellt sich, weil $2^5 \equiv 1 \bmod. 31$ ist, als wirklich complexe Zahl dar, welche nur die fünfgliedrigen Perioden der 31ten Wurzeln der Einheit enthält. Bezeichnet man die 31te Wurzel der Einheit mit α und nimmt die sechs fünfgliedrigen Perioden:

29*

$$\begin{aligned}
\eta &= \alpha + \alpha^{16} + \alpha^{8} + \alpha^{4} + \alpha^{2} \\
\eta_1 &= \alpha^{3} + \alpha^{17} + \alpha^{24} + \alpha^{12} + \alpha^{6} \\
\eta_2 &= \alpha^{9} + \alpha^{20} + \alpha^{10} + \alpha^{5} + \alpha^{18} \\
\eta_3 &= \alpha^{27} + \alpha^{29} + \alpha^{30} + \alpha^{15} + \alpha^{23} \\
\eta_4 &= \alpha^{19} + \alpha^{25} + \alpha^{28} + \alpha^{14} + \alpha^{7} \\
\eta_5 &= \alpha^{26} + \alpha^{13} + \alpha^{22} + \alpha^{11} + \alpha^{21}
\end{aligned}$$

so läſst sich die von Hrn. Reuschle gefundene neunte Potenz des idealen Primfaktors der 2 in der einfachsten Form darstellen als:

$$f(\eta)^9 = 3 + \eta_2 + \eta_3 + \eta_5\,, \qquad (1.)$$

welche complexe Zahl wirklich die Bedingung erfüllt, daſs ihre Norm gleich $2^9 = 512$ ist und daſs sie nur einen der sechs conjugirten idealen Primfaktoren neunmal enthält. Ich bemerke noch, daſs dieselbe neunte Potenz der idealen Zahl in gebrochener Form sich auch so darstellen läſst:

$$f(\eta)^9 = \frac{(1-\eta_4)^2}{1+\eta_4}\,. \qquad (2.)$$

Da dieser eine gefundene ideale Primfaktor zur Auffindung aller derjenigen aus 31ten Einheitswurzeln gebildeten idealen Primfaktoren, deren neunte Potenzen wirklich werden, den Weg eröffnet, so habe ich versucht mit Hülfe desselben auch einen von denjenigen idealen Primfaktoren auszurechnen, welche nicht aus Perioden, sondern aus den 31ten Einheitswurzeln selbst gebildet sind, welche also 30 conjugirte ideale Primfaktoren haben. Nach den aus dem Canon arithmeticus zu entnehmenden 30 Congruenzwurzeln, welche für $p = 311$ den Einheitswurzeln entsprechen, findet man sogleich, daſs die complexe Zahl

$$1 + \alpha^6 - \alpha^{16}$$

einen idealen Primfaktor der Zahl 311 enthält. Bildet man nun die Norm, so findet man

$$N(1 + \alpha^6 - \alpha^{16}) = 2^5 . 311\,, \qquad (3.)$$

woraus folgt, daſs diese complexe Zahl auſser dem einen idealen Primfaktor von 311 nur noch einen idealen Primfaktor von 2 enthält, und zwar, wie die für diesen vorhandenen Congruenzbedingungen zeigen, denselben, dessen neunte Potenz oben dargestellt ist. Bezeichnet man nun den idealen Primfaktor von 311 mit $\varphi(\alpha)$, so hat man

$$\varphi(\alpha)^9 = \frac{(1+\alpha^6-\alpha^{16})^9}{3+\eta_2+\eta_3+\eta_5}\,. \tag{4.}$$

Hiermit ist die neunte Potenz des gesuchten idealen Primfaktors als wirkliche complexe Zahl dargestellt, aber noch in gebrochener Form; um dieselbe als ganze complexe Zahl darzustellen, mufs man Zähler und Nenner mit der complexen Zahl $\psi(\eta)$ multipliciren, welche das Produkt der fünf zu $3+\eta_2+\eta_3+\eta_5$ conjugirten complexen Zahlen ist und daher die Eigenschaft hat, dafs

$$\psi(\eta)\,(3+\eta_2+\eta_3+\eta_5) = 2^9$$

ist und ausgerechnet folgenden Werth ergiebt:

$$\psi(\eta) = 101+51\eta-31\eta_1-6\eta_2-58\eta_3+35\eta_4\,. \tag{5.}$$

Hiernach erhält man

$$\varphi(\alpha)^9 = \frac{(1+\alpha^6-\alpha^{16})^9\,\psi(\eta)}{2^9}\,. \tag{6.}$$

Nach Ausführung der Potenzerhebung und Multiplikation im Zähler hebt sich der Nenner 2^9 von selbst hinweg und man erhält folgendes Resultat:

$$\begin{aligned}\varphi(\alpha)^9 = &-254+26\alpha+792\alpha^2+135\alpha^3-414\alpha^4-354\alpha^5\\ &-695\alpha^6+44\alpha^7+629\alpha^8+10\alpha^9-108\alpha^{10}-458\alpha^{11}\\ &-831\alpha^{12}+197\alpha^{13}+480\alpha^{14}+185\alpha^{15}+285\alpha^{16}\\ &-515\alpha^{17}-634\alpha^{18}+316\alpha^{19}+330\alpha^{20}+541\alpha^{21}\\ &+502\alpha^{22}-521\alpha^{23}-383\alpha^{24}+172\alpha^{25}+150\alpha^{26}\\ &+801\alpha^{27}+403\alpha^{28}-517\alpha^{29}-295\alpha^{30}\,.\end{aligned} \tag{7.}$$

Die Prüfung der Richtigkeit der numerischen Rechnung ergiebt sich zum Theil schon daraus, dafs der Nenner 2^9 sich wirklich hinweghebt, ich habe aber aufserdem auch in allen einzelnen Stadien dieser und auch der folgenden Rechnung die Congruenzen für den Modul 31 angewendet, welche alle Gleichungen erfüllen müssen, wenn $\alpha = 1$ gesetzt wird. Endlich habe ich das gefundene Resultat auch dadurch geprüft, dafs $\phi(\alpha)^9 \equiv 0$, mod. 311 sein mufs, wenn für die Einheitswurzeln die entsprechenden Congruenzwurzeln gesetzt werden. Die wirkliche Berechnung der Norm des gefundenen Ausdrucks von $\phi(\alpha)^9$ würde eine unverhältnifsmäfsig grofse Arbeit erfordern.

Da eine jede complexe Zahl, insofern sie nur durch die in ihr enthaltenen (idealen) Primfaktoren bestimmt ist, mit Einheiten ganz beliebig behaftet sein, und so in unendlich vielen verschiedenen Gestalten dargestellt werden kann, unter denen diejenigen, welche möglichst kleine Zahlen als Coëfficienten enthalten, offenbar den Vorzug verdienen, so habe ich durch Multiplication mit passend gewählten Einheiten die gefundene complexe Zahl zu vereinfachen gesucht und bin so bis zu folgender einfacheren Darstellung gelangt:

$$\begin{aligned} \phi(\alpha)^9 = & -5 - 2\alpha + 5\alpha^2 + 8\alpha^3 + 7\alpha^4 - 4\alpha^5 + 4\alpha^6 + \alpha^7 \\ & + 5\alpha^9 + 5\alpha^{10} - 6\alpha^{11} - 2\alpha^{12} + \alpha^{13} - 2\alpha^{14} - \alpha^{15} \\ & + 4\alpha^{16} - \alpha^{18} - 2\alpha^{19} + 2\alpha^{20} - 4\alpha^{21} - 10\alpha^{22} + 2\alpha^{23} \\ & - 2\alpha^{24} - 5\alpha^{25} + 3\alpha^{26} + 7\alpha^{27} - 2\alpha^{28} - 2\alpha^{29} - 2\alpha^{30}. \end{aligned} \qquad (8.)$$

Da auf dem bis dahin von mir eingeschlagenen Wege der nach einem bestimmten Principe angestellten Versuche eine weitere Vereinfachung sich nicht erreichen liefs, und da ich dessenungeachtet die Überzeugung hatte, dafs dies noch nicht die einfachste Form dieser complexen Zahl sei, so suchte ich eine Methode, durch welche man in den Stand gesetzt würde in directer Weise die einfachste Form einer jeden gegebenen complexen Zahl zu finden. Diese Methode will ich hier auseinandersetzen.

Wenn wir in dem Vorhergehenden diejenige complexe Zahl als die einfachere angesehen haben, deren Coëfficienten kleinere Zahlen sind, so ist diese Bestimmung insofern ungenau, als von zwei gegebenen Complexen von je n Zahlen sich nicht immer mit Bestimmtheit angeben läfst, welcher von ihnen die gröfseren oder die kleineren Zahlen enthält; es ist darum zunächst genau zu definiren, welche Form der complexen Zahl als die einfachere oder einfachste anzusehen ist. Diese Bestimmung ist an die wesentlicheren Eigenschaften der complexen Zahl anzuknüpfen.

Es sei λ eine Primzahl, $\alpha^\lambda = 1$, und $f(\alpha)$ eine aus λten Wurzeln der Einheit gebildete complexe Zahl, so ist das Produkt $f(\alpha)f(\alpha^{-1})$, sowie auch alle seine conjugirten, stets real und positiv. Setzt man nun der Kürze halber $\frac{\lambda-1}{2} = \mu$ und bezeichnet mit γ eine primitive Wurzel der Primzahl λ, so ist die Summe dieser μ conjugirten complexen Zahlen

$$M = f(\alpha)f(\alpha^{-1}) + f(\alpha^\gamma)f(\alpha^{-\gamma}) + \cdots + f(\alpha^{\gamma^{\mu-1}})f(\alpha^{-\gamma^{\mu-1}}) \quad (9.)$$

als symmetrische Funktion aller Wurzeln $\alpha, \alpha^2, \ldots \alpha^{\lambda-1}$ eine nichtcomplexe ganze Zahl. Diese Summe M nimmt andere und andere Werthe an, wenn $f(\alpha)$ mit anderen und anderen Einheiten multiplicirt wird, das Produkt dieser μ conjugirten complexen Zahlen, welches gleich der Norm $Nf(\alpha)$ ist, ist aber von den Einheiten, mit welchen $f(\alpha)$ multiplicirt werden kann, ganz unabhängig. Da das Produkt dieser μ stets positiven Gröſsen unverändert bleibt, so wird nach einem bekannnten Satze ihre Summe M den kleinsten Werth erhalten, wenn die einzelnen Theile derselben möglichst nahe einander gleich werden und umgekehrt, wenn M den möglichst kleinsten Werth erhält, werden die conjugirten complexen Zahlen, aus welchen diese Summe zusammengesetzt ist, möglichst nahe einander gleich werden. Da die möglichst nahe Gleichheit der Werthe dieser conjugirten complexen Zahlen, die wesentlichste Bedingung der Einfachheit der complexen Zahl $f(\alpha)$ ausmacht, so definire ich:

> Unter allen complexen Zahlen $f(\alpha)$, welche nur durch hinzugefügte Einheiten sich unterscheiden, soll diejenige als die einfachste betrachtet werden, für welche die Summe M der mit $f(\alpha)f(\alpha^{-1})$ conjugirten μ complexen Zahlen den kleinsten Werth erhält.

Nimmt man

$$f(\alpha) = a + a_1\alpha + a_2\alpha^2 + \cdots + a_{\lambda-1}\alpha^{\lambda-1},$$

so erhält man für die Summe M folgenden Ausdruck

$$2M = \lambda(a^2 + a_1^2 + a_2^2 + \cdots + a_{\lambda-1}^2)^2 - (a + a_1 + a_2 + \cdots + a_{\lambda-1})^2 \quad (10.)$$

m. vergl. meine Abhandlung in Lionvilles Journal Bd. XVI p 442, welcher auch so dargestellt werden kann:

$$\begin{aligned}2M = (a-a_1)^2 + (a-a_2)^2 + (a-a_3)^2 + \cdots + (a-a_{\lambda-1})^2 \\ +(a_1-a_2)^2+(a_1-a_3)^2+\cdots+(a_1-a_{\lambda-1})^2 \quad (11.) \\ +(a_2-a_3)^2+\cdots+(a_2-a_{\lambda-1})^2 \\ \vdots \\ +(a_{\lambda-2}-a_{\lambda-1})^2.\end{aligned}$$

Man hat daher mit der obigen Definition vollkommen übereinstimmend auch die folgende:

> Unter allen complexen Zahlen, welche nur durch hinzugefügte Einheiten sich unterscheiden, soll diejenige als die einfachste betrachtet werden, für welche die Summe der Quadrate der Unterschiede je zweier ihrer λ Coëfficienten den kleinsten Werth hat.

Die Aufgabe für eine gegebene complexe Zahl $f(\alpha)$ die einfachste Form zu finden, d. h. eine Einheit $E(\alpha)$ von der Art zu finden, daſs für $E(\alpha)f(\alpha)$ die Summe der Quadrate der Differenzen je zweier Coëfficienten den kleinsten Werth erhalte, wird nun durch folgende direkte Methode gelöst:

Es sei $e_1, e_2, e_3, \ldots e_{\mu-1}$ ein System von Fundamentaleinheiten, so daſs jede beliebige Einheit sich in der Form

$$\pm\, \alpha^k e_1^{x_1} e_2^{x_2} e_3^{x_3} \ldots e_{\mu-1}^{x_{\mu-1}}$$

darstellen läſst, so handelt es sich darum die Exponenten $x_1, x_2, \ldots x_{\mu-1}$ so zu bestimmen, daſs

$$e_1^{x_1} e_2^{x_2} \ldots e_{\mu-1}^{x_{\mu-1}} f(\alpha) = f'(\alpha) \qquad (12.)$$

die einfachste Form erhalte. Es wird nun, weil die Einheiten unverändert bleiben, wenn α in α^{-1} verwandelt wird

$$e_1^{2x_1} e_2^{2x_2} \ldots e_{\mu-1}^{2x_{\mu-1}} f(\alpha)f(\alpha^{-1}) = f'(\alpha)f'(\alpha^{-1})$$

und wenn die Logarithmen genommen werden:

$$\begin{aligned}x_1\, l(e_1^2) + x_2\, l(e_2^2) + \cdots + x_{\mu-1}\, l(e_{\mu-1}^2) \\ = l(f'(\alpha)f'(\alpha^{-1})) - l(f(\alpha)f(\alpha^{-1})),\end{aligned} \qquad (13.)$$

welche Gleichung, da statt der Wurzel α auch $\alpha^{\gamma}, \alpha^{\gamma^2}, \ldots \alpha^{\gamma^{\mu-1}}$ genommen werden kann, ein System von μ Gleichungen repräsentirt, von denen jedoch nur $\mu - 1$ unabhängig sind, da die Summe aller μ-Gleichungen identisch $0 = 0$ ergiebt.

Wenn man nun vorläufig darauf verzichtet, dafs die Gröfsen $x_1, x_2, \ldots x_{\mu-1}$ ganze Zahlen sein sollen, so kann man dieselben so bestimmen, dafs die numerischen Werthe der μ conjugirten complexen Zahlen

$$f'(\alpha)f'(\alpha^{-1}), \quad f'(\alpha^{\gamma})f'(\alpha^{-\gamma}), \quad \ldots f'(\alpha^{\gamma^{\mu-1}})f'(\alpha^{-\gamma^{\mu-1}}) \quad (14.)$$

nicht nur möglichst nahe, sondern sogar vollständig einander gleich werden, dafs also, da ihr Produkt gleich der Norm $Nf(\alpha)$ ist, jede derselben den Werth $\sqrt[\mu]{Nf(\alpha)}$ erhält. Man erhält so zur Bestimmung der $\mu - 1$ Gröfsen $x_1, x_2, \ldots x_{\mu-1}$ ein System von $\mu - 1$ unabhängigen lineären Gleichungen, welches durch

$$\begin{aligned} x_1 l(e_1^2) + x_2 l(e_2^2) + \cdots + x_{\mu-1} l(e_{\mu-1}^2) \\ = \frac{1}{\mu} lNf(\alpha) - l(f(\alpha)f(\alpha^{-1})) \end{aligned} \quad (15.)$$

repräsentirt wird, wo die Einheitswurzel α die $\mu - 1$ verschiedenen Werthe $\alpha, \alpha^{\gamma}, \alpha^{\gamma^2} \ldots \alpha^{\gamma^{\mu-1}}$ annimmt. Da nun die aus diesem Systeme von $\mu - 1$ unabhängigen lineären Gleichungen zu bestimmenden, nicht ganzzahligen Werthe der Gröfsen $x_1, x_2, \ldots x_{\mu-1}$ die vollständige Gleichheit der μ conjugirten complexen Zahlen (14) ergeben, so wird man die nahe Gleichheit derselben und somit einen sehr kleinen Werth ihrer Summe M erlangen, wenn man für die Exponenten $x_1, x_2, \ldots x_{\mu-1}$ diejenigen ganzen Zahlen nimmt, welche diesen gefundenen nicht ganzzahligen Werthen am nächsten liegen, namentlich diejenigen, welche sich nur um weniger als eine halbe Einheit von ihnen unterscheiden. Man kann jedoch nicht mit Sicherheit darauf rechnen, dafs man durch Multiplikation der complexen Zahl $f(\alpha)$ durch die nach dieser Methode bestimmte Einheit die absolut einfachste Darstellung derselben erhält, für welche M den absolut kleinsten Werth hat, sondern nur darauf, dafs man eine Darstellung der complexen Zahl erhält, welche der absolut einfachsten sehr nahe liegt.

Der mehr oder minder günstige Erfolg dieser Methode hängt nothwendig auch von der Wahl des Systems der Fundamentaleinheiten ab, durch welche die zu findende Einheit ausgedrückt wird.

Aus dem Systeme der Gleichungen (13) ersieht man unmittelbar, dafs diejenigen Fundamentaleinheiten die vortheilhaftesten sein werden, für welche kleine Änderungen der Gröfsen $x_1, x_2, \ldots x_{\mu-1}$ nur möglichst kleine Änderungen der Werthe von $\frac{1}{\mu} l N f(\alpha) - f(\alpha) f(\alpha^{-1})$ zur Folge haben und dies ist offenbar der Fall, wenn die Gröfsen

$$l(e_1^2)\,,\quad l(e_2^2)\,,\quad \ldots\; l(e_{\mu-1}^2)$$

und ihre conjugirten die möglichst kleinsten Werthe haben, d. h. dem Werthe 0 möglichst nahe kommen. Hieraus folgt, dafs die Quadrate der zu Grunde zu legenden Fundamentaleinheiten und der ihnen conjugirten, welche zum Theil gröfser und zum Theil kleiner als Eins sind, alle dem Werthe Eins möglichst nahe liegen müssen, dafs also für eine jede dieser Fundamentaleinheiten die oben mit M bezeichnete Zahl den möglichst kleinsten Werth erhalten mufs, dafs also diejenigen Fundamentaleinheiten zu wählen sind, welche in dem oben definirten Sinne selbst als die einfachsten anzusehen sind.

Da man in der Theorie der hier behandelten complexen Zahlen bis jetzt noch in keinem einzigen Falle ein fundamentaleres System unabhängiger Einheiten kennt, als das der conjugirten Kreistheilungseinheiten, so wird man für jetzt nothwendig nur ein solches zu Grunde zu legen haben; aber auch diese werden nach dem oben Bemerkten nicht alle gleich vortheilhaft sein, und man wird in jedem Falle denjenigen den Vorzug zu geben haben, für welche die Zahl M, also die Summe der Quadrate der Differenzen je zweier Coëfficienten den kleinsten Werth erhält. In dem Falle, wo ± 2 eine primitive Wurzel der Primzahl λ ist, hat man das unabhängige System der zu $\alpha + \alpha^{-1}$ conjugirten Einheiten zu wählen, für welches die Zahl M den Werth $\lambda - 2$ hat; wenn ± 3 die kleinste primitive Wurzel von λ ist, so hat man die zu $1 + \alpha + \alpha^{-1}$ conjugirten Einheiten zu wählen, für welche $M = \frac{3(\lambda - 3)}{2}$ ist u. s. w.

Das System lineärer Gleichungen, durch welche die Exponenten $x_1, x_2, \ldots x_{\mu-1}$ bestimmt werden, hat in dem Falle, wo ein System conjugirter Kreistheilungseinheiten zu Grunde gelegt wird, eine sehr einfache Auflösung. Nimmt man

$$e_h^2 = \frac{(1-\alpha^{\gamma^{h+1}})(1-\alpha^{-\gamma^{h+1}})}{(1-\alpha)(1-\alpha^{-1})}, \tag{16.}$$

wo $\pm\gamma$ eine primitive Wurzel der Primzahl λ ist, so bilden

$$e,\ e_1,\ e_2 \ldots e_{\mu-1}$$

ein System conjugirter Kreistheilungseinheiten, welches, da unter denselben nur die eine Gleichung

$$e \,.\, e_1 \,.\, e_2 \ldots e_{\mu-1} = 1$$

besteht, ein System von $\mu - 1$ unabhängigen Einheiten ist. Nimmt man nun

$$E = e^x e_1^{x_1} e_2^{x_2} \ldots e_{\mu-1}^{x_{\mu-1}} \tag{17.}$$

als die Einheit mit welcher $f(\alpha)$ zu multipliciren ist, damit es in der einfachsten Form dargestellt werde, so kann man ohne diese Einheit zu ändern die μ-Exponenten $x, x_1, \ldots x^{-1}$ alle um eine und dieselbe Gröfse vermehren oder vermindern, sodafs einer derselben, oder wenn man will die Summe aller unbestimmt bleibt und beliebig gewählt werden kann. Setzt man nun zur Vereinfachung

$$l(e_h^2) = \varepsilon_h\,,\quad \frac{1}{\mu} l N f(\alpha) - l\big(f(\alpha^{\gamma^h}) f(\alpha^{-\gamma^h})\big) - A_h\,,$$

so hat man folgendes System von Gleichungen:

$$\begin{aligned}
\varepsilon x + \varepsilon_1 x_1 + \varepsilon_2 x_2 + \cdots + \varepsilon_{\mu-1} x_{\mu-1} &= A \\
\varepsilon_1 x + \varepsilon_2 x_1 + \varepsilon_3 x_2 + \cdots + \varepsilon x_{\mu-1} &= A_1 \\
\varepsilon_2 x + \varepsilon_3 x_1 + \varepsilon_4 x_2 + \cdots + \varepsilon_1 x_{\mu-1} &= A_2 \\
\vdots \qquad\qquad\qquad\qquad &\quad \vdots \\
\varepsilon_{\mu-1} x + \varepsilon\ x_1 + \varepsilon_1 x_2 + \cdots + \varepsilon_{\mu-2} x_{\mu-1} &= A_{\mu-1}
\end{aligned} \tag{18.}$$

wo

$$\begin{aligned}
\varepsilon + \varepsilon_1 + \varepsilon_2 + \cdots + \varepsilon_{\mu-1} &= 0 \\
A + A_1 + A_2 + \cdots + A_{\mu-1} &= 0
\end{aligned} \tag{19.}$$

ist, sodafs nur $\mu - 1$ dieser μ-Gleichungen von einander unabhängig sind und eine derselben eine Folge der übrigen ist. Bezeichnet man nun mit β eine primitive Wurzel der Gleichung

$$\beta^{\mu} = 1,$$

als welche

$$\beta = \cos\frac{2\pi}{\mu} + i\sin\frac{2\pi}{\mu}$$

gewählt werden soll, so erhält man durch Multiplikation dieser lineären Gleichungen mit $1, \beta^h, \beta^{2h} \dots \beta^{(\mu-1)h}$ und Addition:

$$(\varepsilon + \varepsilon_1 \beta^h + \dots + \varepsilon_{\mu-1}\beta^{(\mu-1)h})(x + x_1\beta^{-h} + \dots + x_{\mu-1}\beta^{-(\mu-1)h})$$
$$= A + A_1\beta^h + \dots + A_{\mu-1}\beta^{(\mu-1)h}$$

also

$$x + \beta^{-h}x_1 + \dots + \beta^{-(\mu-1)h}x_{\mu-1}$$
$$= \frac{A + \beta^h A_1 + \dots + \beta^{(\mu-1)h}A_{\mu-1}}{\varepsilon + \beta^h\varepsilon_1 + \dots + \beta^{(\mu-1)h}\varepsilon_{\nu-1}}$$

und hieraus, wenn man mit β^{kh} multiplicirt und für $h = 1, 2, \dots \mu - 1$ die Summe nimmt:

$$\mu x_k - S = \sum_1^{\mu-1}{}_h \frac{\beta^{kh}(A + \beta^h A_1 + \dots + \beta^{(\mu-1)h}A_{\mu-1}}{\varepsilon + \beta^h\varepsilon_1 + \dots + \beta^{(\mu-1)h}\varepsilon_{\mu-1}} \qquad (20.)$$

wo $S = x + x_1 + \dots + x_{\mu-1}$ die Summe aller Exponenten bezeichnet, welche, wie oben gezeigt worden ist, beliebig gewählt werden kann. Hieraus folgt weiter, dafs die Werthe der Exponenten $x, x_1, \dots x_{\mu-1}$ in folgende Form gesetzt werden können

$$\begin{aligned}
\mu x &= S + CA + C_1A_1 + \dots + C_{\mu-1}A_{\mu-1} \\
\mu x_1 &= S_1 + C_1A + C_2A_1 + \dots + CA_{\mu-1} \\
&\vdots \\
\mu x_{\mu-1} &= S + C_{\mu-1}A + CA_1 + \dots + C_{\mu-2}A_{\mu-1}.
\end{aligned} \qquad (21.)$$

Die Coëfficienten $C, C_1, \dots C_{\mu-1}$ sind in realer Form durch folgenden Ausdruck gegeben

$$C_n = \sum_1^{\mu-1}{}_h \frac{\cos\frac{2nh\pi}{\mu}E_h + \sin\frac{2nh\pi}{\mu}E'_h}{E_h^2 + E_h'^2} \qquad (22.)$$

wo

$$\varepsilon + \beta^h\varepsilon_1 + \dots + \beta^{(\mu-1)h}\varepsilon_{\mu-1} = E_h + iE'_h.$$

Die Summen von $\mu-1$ Gliedern, durch welche die $C, C_1 \dots C_{\mu-1}$ zu berechnen sind, reduciren sich auf die Hälfte der Glieder, weil je zwei vom Anfange und Ende gleich abstehende Glieder einander gleich sind, welches daraus folgt, dafs

$$E_{\mu-h} = E_h \qquad \cos\left(\frac{2n(\mu-h)\pi}{\mu}\right) = \cos\frac{2nh\pi}{\mu},$$

$$E'_{\mu-h} = -E'_h \qquad \sin\frac{2n(\mu-h)\pi}{\mu} = -\sin\frac{2nh\pi}{\mu}.$$

Da die Gröfse S in dem Ausdrucke (22.) ganz beliebig gewählt werden kann, oder, was dasselbe ist, da man die Werthe der $x, x_1, \dots x_{\mu-1}$ alle gleichzeitig um eine und dieselbe beliebige Gröfse vermehren und vermindern kann, ohne das Resultat zu ändern, so folgt, dafs es nicht blofs ein einziges bestimmtes System von ganzzahligen Werthen dieser Exponenten giebt, welche sich von den gebrochenen Werthen um weniger als eine halbe Einheit unterscheiden, sondern dafs es im Allgemeinen μ solcher Werthsysteme giebt, welche man mit gleichem Rechte wählen könnte. Unter diesen hat man daher schliefslich noch dasjenige auszusuchen, welches den kleinsten Werth der Summe M ergiebt.

Nach dieser Methode habe ich nun für die oben gegebene complexe Zahl, welche die neunte Potenz eines idealen Primfaktors von $p=311$ für $\lambda=31$ giebt, die nöthigen Rechnungen vollständig ausgeführt und gefunden, dafs dieselbe folgende sehr einfache Form annimmt.

$$\begin{aligned}\phi(\alpha)^9 = 2 \; * \; &-2\alpha^2-2\alpha^3-\alpha^4+2\alpha^5+\alpha^6 \; * \; -\alpha^8 \\ &* \; * \; +3\alpha^{11} \; * \; * \; +\alpha^{14}-\alpha^{15}+2\alpha^{16} \; * \\ &+2\alpha^{18}+\alpha^{19}+\alpha^{20} \; * \; -\alpha^{22}-\alpha^{23} \; * \; -\alpha^{25} \\ &-\alpha^{26}+3\alpha^{27}+2\alpha^{28}+3\alpha^{29}+2\alpha^{30}.\end{aligned} \tag{23.}$$

Aus der bei (8.) gegebenen schon etwas vereinfachten Form erhält man diese einfache durch Multiplikation mit der Einheit

$$E = e^3 e_1^4 e_2^3 e_3^2 e_4^2 e_5^3 e_6^3 e_7^2 e_8^2 e_9^2 e_{11}^4 e_{12}^5 e_{13}^4 e_{14}^4, \tag{24.}$$

wo

$$e_h = 1+\alpha^{\gamma^h}+\alpha^{-\gamma^h}.$$

Aus der bei (7.) gegebenen, ursprünglich gefundenen Form geht dieselbe einfache Form hervor durch Multiplikation mit der Einheit

$$E = e^5 e_1^7 e_2^5 e_3^1 e_4^2 e_5^4 e_6^7 e_7^5 e_8^5 e_9^5 e_{11}^4 e_{12}^5 e_{13}^4 e_{14}^7 \,. \tag{25.}$$

Die Zahl M, durch welche der Grad der Einfachheit der complexen Zahl bestimmt wird, ist für die ursprüngliche Form (7.) $M = 96625316$, für die etwas vereinfachte Form (8.) $M = 8039$ und für die einfache Form (23.) $M = 987$. Wenn durch Multiplikation mit Einheiten die Theile der Summe M nicht blofs angenähert, sondern vollständig gleich gemacht werden könnten, so würde der Werth des M sich bis auf $\sqrt[\mu]{Nf(\alpha)}$ herabbringen lassen, also in dem vorliegenden Falle bis auf $\sqrt[15]{(311)^9} = 496,621\ldots$

Bei Ausführung der numerischen Rechnungen mit Hülfe der Logarithmen hat man nur denjenigen Grad der Genauigkeit einzuhalten, welcher dafür bürgt, dafs die gefundenen Werthe der Exponenten $x, x_1, \ldots x_{\mu-1}$ in den Ganzen, Zehnteln und Hunderteln genau erlangt werden, es werden also im Allgemeinen Logarithmentafeln von einer sehr geringen Stellenzahl ausreichen. Es tritt aber, wenn die ursprünglich gegebene complexe Zahl $f(\alpha)$ wenig einfach ist, allemal der Umstand ein, dafs von den zu $f(\alpha)f(\alpha^{-1})$ conjugirten complexen Zahlen, deren Werthe man berechnen mufs, eine oder einige aufserordentlich klein werden, wodurch ihre Berechnung und die Berechnung ihrer Logarithmen, welche man auf einige Stellen genau kennen mufs, aufserordentlich mühsam werden würde, wenn man ihre Ausdrücke als Summen von $\mu + 1$ Gliedern zu Grunde legen wollte. Bei der Durchführung der Rechnung, deren Resultat ich hier gegeben habe, liefs sich diese Unzuträglichkeit dadurch vermeiden, dafs bei der Berechnung dieser Zahlenwerthe nicht die entwickelte Form (7.), sondern die unentwickelte gebrochene Form (4.) zu Grunde gelegt wurde.

Schliefslich bemerke ich noch, dafs diese Methode der Reinigung der complexen Zahlen von den sie behaftenden Einheiten ohne Schwierigkeit auch auf die nicht aus den Einheitswurzeln selbst, sondern aus den Perioden gebildeten complexen Zahlen sich anwenden läfst.

Über die aus 31sten Wurzeln der Einheit gebildeten complexen Zahlen

Monatsberichte der Königlich Preußischen Akademie der Wissenschaften zu Berlin aus dem Jahre 1870, 755–766

10. October. Sitzung der physikalisch-mathematischen Klasse.

Hr. Kummer las

Über die aus 31ten Wurzeln der Einheit gebildeten complexen Zahlen.

Für die aus 31ten Einheitswurzeln gebildeten complexen Zahlen ist, wie ich früher nachgewiesen habe, der erste Faktor der Klassenzahl gleich 9. Da diese Zahl ein Quadrat ist, so bleibt es unentschieden, ob es ideale complexe Zahlen giebt, deren neunte Potenz, und keine niedere, wirklich wird, oder ob schon die dritten Potenzen aller hierhin gehörenden idealen Zahlen wirklich sind, also ob in Beziehung auf diese Klassenzahl 9 im Gaufsischen Sinne Regularität Statt hat, oder Irregularität. Die Analogie mit den aus 29ten Einheitswurzeln gebildeten complexen Zahlen, für welche der erste Faktor der Klassenzahl gleich 8 ist, für welche aber, wie ich aus der Theorie der Kreistheilung bewiesen habe, schon die zweiten Potenzen aller idealen Zahlen wirklich sind, giebt einen Anlafs zu der Vermuthung, dafs dies in dem hier zu betrachtenden Falle in ähnlicher Weise Statt haben möchte. Da nun Hr. Reuschle für die aus 31ten Einheitswurzeln gebildeten idealen Primfaktoren der Zahl 2 eine wirkliche Darstellung der

[1870] 52

neunten Potenz gefunden hat, während es ihm nicht gelungen ist, die dritte Potenz dieses idealen Primfaktors in wirklicher Form darzustellen, und da von der Zerlegung der einen Zahl 2 die Zerlegungen aller in dieselbe Kategorie gehörenden Primzahlen abhängig sind, so schien es mir namentlich auch für das von ihm herauszugebende Werk über die Zerlegung der Zahlen in ihre complexen Primfaktoren von Wichtigkeit vollständig zu ergründen, ob keine niedere als die neunte Potenz des idealen Primfaktors von 2 in dieser Theorie wirklich wird, oder was dasselbe ist: ob die dritte Potenz dieses idealen Primfaktors als wirkliche complexe Zahl sich darstellen läſst, oder nicht.

Bezeichnet man mit $\overset{5}{\eta}, \overset{5}{\eta}_1, \overset{5}{\eta}_2, \overset{5}{\eta}_3, \overset{5}{\eta}_4, \overset{5}{\eta}_5$ die sechs fünfgliedrigen, aus 31ten Einheitswurzeln gebildeten Perioden, geordnet nach der primitiven Wurzel 3, und die drei zehngliedrigen Perioden

$$\overset{5}{\eta} + \overset{5}{\eta}_3 = \overset{10}{\eta}, \quad \overset{5}{\eta}_1 + \overset{5}{\eta}_4 = \overset{10}{\eta}_1, \quad \overset{5}{\eta}_2 + \overset{5}{\eta}_5 = \overset{10}{\eta}_2$$

so hat man unter denselben die Gleichungen:

$$\begin{array}{llllllll}
\overset{5}{\eta}\overset{5}{\eta} = & \ast & + \overset{5}{\eta} & + 2\overset{5}{\eta}_1 & + 2\overset{5}{\eta}_2 & \ast & \ast & \ast \\
\overset{5}{\eta}\overset{5}{\eta}_1 = & \ast & + \overset{5}{\eta} & \ast & + \overset{5}{\eta}_2 & \ast & + 2\overset{5}{\eta}_4 & + \overset{5}{\eta}_5 \\
\overset{5}{\eta}\overset{5}{\eta}_2 = & \ast & \ast & + \overset{5}{\eta}_1 & + \overset{5}{\eta}_2 & \ast & + \overset{5}{\eta}_4 & + 2\overset{5}{\eta}_5 \\
\overset{5}{\eta}\overset{5}{\eta}_3 = & 5 & + \overset{5}{\eta} & + \overset{5}{\eta}_1 & \ast & + \overset{5}{\eta}_3 & + \overset{5}{\eta}_4 & \ast \\
\overset{5}{\eta}\overset{5}{\eta}_4 = & \ast & + \overset{5}{\eta} & \ast & + \overset{5}{\eta}_2 & + 2\overset{5}{\eta}_3 & \ast & + \overset{5}{\eta}_5 \\
\overset{5}{\eta}\overset{5}{\eta}_5 = & \ast & \ast & + \overset{5}{\eta}_1 & \ast & + 2\overset{5}{\eta}_3 & + \overset{5}{\eta}_4 & + \overset{5}{\eta}_5
\end{array}$$

und

$$\begin{array}{lllll}
\overset{10}{\eta}\overset{10}{\eta} = & 10 & + 3\overset{10}{\eta} & + 4\overset{10}{\eta}_1 & + 2\overset{10}{\eta}_2 \\
\overset{10}{\eta}\overset{10}{\eta}_1 = & \ast & + 4\overset{10}{\eta} & + 2\overset{10}{\eta}_1 & + 4\overset{10}{\eta}_2 \\
\overset{10}{\eta}\overset{10}{\eta}_2 = & \ast & + 2\overset{10}{\eta} & + 4\overset{10}{\eta}_1 & + 4\overset{10}{\eta}_2
\end{array}$$

Ferner hat man die nach dem Modul 2 den Perioden entsprechenden Congruenzwurzeln:

$$\overset{5}{\eta} \equiv 1\,, \quad \overset{5}{\eta}_1 \equiv 1\,, \quad \overset{5}{\eta}_2 \equiv 0\,, \quad \overset{5}{\eta}_3 \equiv 0\,, \quad \overset{5}{\eta}_4 \equiv 1\,, \quad \overset{5}{\eta}_5 \equiv 0$$

welche zu dem idealen Primfaktor $f(\overset{5}{\eta})$ der Zahl 2 gehören sollen.

Betrachtet man nun die complexe Zahl

$$1 + \overset{5}{\eta} + \overset{5}{\eta}_3 = 1 + \overset{10}{\eta}\,,$$

so findet man vermöge dieser Congruenzbedingungen, dafs sie die beiden idealen Primfaktoren von 2 $f(\overset{5}{\eta})$ und $f(\overset{5}{\eta}_3)$ enthält und weil

$$(1 + \overset{10}{\eta})\,(1 + \overset{10}{\eta}_1)\,(1 + \overset{10}{\eta}_2) = 2$$

ist, so folgt, dafs sie aufserdem keine anderen Primfaktoren enthält. Man hat daher die ideale Zerlegung

$$1 + \overset{10}{\eta} = f(\overset{5}{\eta})f(\overset{5}{\eta}_3).$$

Wenn nun die dritte Potenz des idealen Primfaktors der 2 sich als wirkliche complexe Zahl $F(\overset{5}{\eta})$ darstellen liefse, so würde auch die dritte Potenz von $1 + \overset{10}{\eta}$, multiplicirt mit einer passenden Einheit, sich als Produkt der beiden wirklichen complexen Zahlen $F(\overset{5}{\eta})$ und $F(\overset{5}{\eta}_3)$ darstellen lassen, man würde also haben

$$(1 + \overset{10}{\eta})^3 E(\overset{10}{\eta}) = F(\overset{5}{\eta})\,F(\overset{5}{\eta}_3),$$

wo $E(\overset{10}{\eta})$ irgend eine Einheit bezeichnet, welche nothwendig nur die zehngliedrigen Perioden enthält, weil alle Einheiten der aus den fünfgliedrigen Perioden gebildeten complexen Zahlen nur die zehngliedrigen Perioden enthalten können. Die drei conjugirten Kreistheilungseinheiten sind hier:

52*

$$e(\overset{10}{\eta}) = 7 + 4\overset{10}{\eta} + 2\overset{10}{\eta}_1$$

$$e(\overset{10}{\eta}_1) = 7 + 4\overset{10}{\eta}_1 + 2\overset{10}{\eta}_2$$

$$e(\overset{10}{\eta}_2) = 7 + 4\overset{10}{\eta}_2 + 2\overset{10}{\eta}$$

die numerischen Werthe der drei zehngliedrigen Perioden sind:

$$\overset{10}{\eta} = +3{,}08387\ ,\quad \overset{10}{\eta}_1 = -0{,}78680\ ,\quad \overset{10}{\eta}_2 = -3{,}29707$$

und demnach

$$e(\overset{10}{\eta}) = +17{,}76188\ ,\quad e(\overset{10}{\eta}_1) = -2{,}74134\ ,\quad e(\overset{10}{\eta}_2) = -0{,}02054\ ,$$

$$1 + \overset{10}{\eta} = +4{,}08387\ ,\quad 1 + \overset{10}{\eta}_1 = +0{,}21320\ ,\quad 1 + \overset{10}{\eta}_2 = -2{,}29707.$$

Es folgt hieraus, daſs

$$-(1 + \overset{10}{\eta})^3\, e(\overset{10}{\eta}_1)\ ,\quad -(1 + \overset{10}{\eta}_1)^3\, e(\overset{10}{\eta}_2)\ ,\quad -(1 + \overset{10}{\eta}_2)^3\, e(\overset{10}{\eta})$$

alle drei positive Werthe haben. Sondert man nun von der Einheit $E(\overset{10}{\eta})$ die Einheit $-e(\overset{10}{\eta}_1)$ ab, indem man setzt

$$E(\overset{10}{\eta}) = -e(\overset{10}{\eta}_1)\, E'(\overset{10}{\eta})\ ,$$

so daſs

$$-(1 + \overset{10}{\eta})^3\, e(\overset{10}{\eta}_1)\, E'(\overset{10}{\eta}) = F(\overset{5}{\eta})\, F(\overset{5}{\eta}_3)\ ,$$

so muſs, weil $F(\overset{5}{\eta})\, F(\overset{5}{\eta}_3)$ überhaupt nur positiv sein kann, diese Einheit $E'(\overset{10}{\eta})$ die Eigenschaft haben, daſs sie mit ihren beiden conjugirten zugleich nur positive Werthe hat. Eine solche Einheit $E'(\overset{10}{\eta})$ muſs aber nothwendig das Quadrat einer Einheit sein, wie leicht folgendermaſsen gezeigt wird. Man kann zwar nicht eine jede Einheit selbst, aber doch eine gewisse Potenz einer jeden Einheit als Produkt von Potenzen der Kreistheilungseinheiten ausücken. Man hat daher

$$E'(\overset{10}{\eta})^n = \pm\, e(\overset{10}{\eta})^\alpha \,.\, e(\overset{10}{\eta}_1)^\beta\ ,$$

wo n, α, β ganze Zahlen sind, welche nicht alle drei grade sein können, weil sonst schon die $\frac{n}{2}$ Potenz von $E'(\overset{10}{\eta})$ sich durch Kreistheilungseinheiten ausdrücken liefse. Aus den numerischen Werthen der $e(\overset{10}{\eta})$, $e(\overset{10}{\eta_1})$, $e(\overset{10}{\eta_2})$ ersieht man nun sogleich, dafs die drei Gröfsen

$$e(\overset{10}{\eta})^{\alpha}\, e(\overset{10}{\eta_1})^{\beta}\,,\quad e(\overset{10}{\eta_1})^{\alpha}\, e(\overset{10}{\eta_2})^{\beta}\,,\quad e(\overset{10}{\eta_2})^{\alpha}\, e(\overset{10}{\eta})^{\beta}$$

nicht alle drei ein und dasselbe Vorzeichen haben können, aufser wenn α und β beide grade sind und darum n ungrade, $n = 2\nu + 1$. Es mufs also $E'(\overset{10}{\eta})^{2\nu+1}$ das Quadrat einer Einheit sein und darum auch $E'(\eta)$ selbst das Quadrat einer Einheit $E'(\overset{10}{\eta}) = (\varepsilon(\overset{10}{\eta}))^2$. Setzt man nun $F(\overset{5}{\eta})\,\varepsilon(\overset{10}{\eta})$ statt $F(\overset{5}{\eta})$ und demgemäfs auch $F(\overset{5}{\eta_3})\,\varepsilon(\overset{10}{\eta})$ statt $F(\overset{5}{\eta_3})$, so erhält man die Gleichung

$$-(1+\overset{10}{\eta})^3\, e(\overset{10}{\eta_1}) = F(\overset{5}{\eta})\, F(\overset{5}{\eta_3})\,,$$

oder entwickelt:

$$-134\eta - 113\eta_1 - 155\eta_2 = F(\overset{5}{\eta})\, F(\overset{5}{\eta_3})\,.$$

Die nothwendige und hinreichende Bedingung dafür, dafs die dritte Potenz eines idealen Primfaktors der 2 eine wirklich complexe Zahl sei, liegt also darin, dafs es eine wirkliche complexe Zahl $F(\overset{5}{\eta})$ gebe, welche dieser Gleichung genügt.

Setzt man nun

$$F(\overset{5}{\eta}) = a\overset{5}{\eta} + a_1\overset{5}{\eta_1} + a_2\overset{5}{\eta_2} + a_3\overset{5}{\eta_3} + a_4\overset{5}{\eta_4} + a_5\overset{5}{\eta_5}$$

und entwickelt das Produkt in die Form

$$F(\overset{5}{\eta})\, F(\overset{5}{\eta_3}) = -A\overset{10}{\eta} - A_1\overset{10}{\eta_1} - A_2\overset{10}{\eta_2}\,,$$

so erhält man, weil $A = 134$, $A_1 = 113$, $A_2 = 155$ sein mufs, folgende drei Gleichungen:

$$\text{(A.)}\quad \begin{aligned} 134 &= 5Q - P^2 + a_1^2 + a_4^2 - (a - a_3)(a_1 - a_4) \\ &\qquad + 2(a a_5 + a_3 a_2) + a a_3\,, \\ 113 &= 5Q - P^2 + a_2^2 + a_5^2 - (a_1 - a_4)(a_2 - a_5) \\ &\qquad + 2(a_1 a + a_4 a_3) + a_1 a_4\,, \\ 155 &= 5Q - P^2 + a_3^2 + a^2 - (a_2 - a_5)(a_3 - a) \\ &\qquad + 2(a_2 a_1 + a_5 a_4) + a_2 a_5\,, \end{aligned}$$

wo der Abkürzung halber

$$P = a + a_1 + a_2 + a_3 + a_4 + a_5\,,$$
$$Q = a^2 + a_1^2 + a_2^2 + a_3^2 + a_4^2 + a_5^2\,,$$

gesetzt ist. Addirt man diese drei Gleichungen und multiplicirt mit 2, so erhält man

$$\text{(B.)}\qquad 804 = 31Q - 5P^2.$$

Setzt man aufserdem

$$\begin{aligned} R = (a-a_1)^2 + (a-a_2)^2 + (a-a_3)^2 + (a-a_4)^2 + (a-a_5)^2 \\ + (a_1-a_2)^2 + (a_1-a_3)^2 + (a_1-a_4)^2 + (a_1-a_5)^2 \\ + (a_2-a_3)^2 + (a_2-a_4)^2 + (a_2-a_5)^2 \\ + (a_3-a_4)^2 + (a_3-a_5)^2 \\ + (a_4-a_5)^2 \end{aligned}$$

so hat man die identische Gleichung

$$31Q - 5P^2 = Q + 5R\,,$$

also auch

$$\text{(C.)}\qquad 804 = Q + 5R\,.$$

Nachdem so die ganze Frage darauf reducirt ist: ob die drei Gleichungen (A.) mit 6 unbestimmten Gröfsen in ganzen Zahlen lösbar sind, oder nicht, untersuche ich zunächst die Congruenzbedingungen für den Modul 2 und sodann für den Modul 8, welche diese sechs Zahlen erfüllen müssen.

Da die Zahl $F(\eta)$ den idealen Primfaktor $f(\eta)$ der Zahl 2 enthalten soll und da sie keinen der übrigen fünf conjugirten enthalten darf, so hat man nach den oben angegebenen Congruenzwurzeln, welche den Perioden für den Modul 2 entsprechen:

$$\left.\begin{array}{l} a\ +a_1+a_4\equiv 0\,,\\ a_1+a_2+a_5\equiv 1\,,\\ a_2+a_3+a\ \equiv 1\,,\\ a_3+a_4+a_1\equiv 1\,,\\ a_4+a_5+a_2\equiv 1\,,\\ a_5+a\ +a_3\equiv 1\,,\end{array}\right\}\quad \text{mod. } 2.$$

woraus folgt, dafs die drei Zahlen a, a_2, a_5 grade sein müssen und die drei Zahlen a_1, a_3, a_4 ungrade, oder

$$a=2b\,,\ a_1=2b_1+1\,,\ a_2=2b_2\,,\ a_3=2b_3+1\,,$$
$$a_4=2b_4+1\,,\ a_5=2b_5\,.$$

Um weiter die nothwendigen Congruenzbedingungen nach dem Modul 8 zu entwickeln, setze ich diese gefundenen Werthe der a, a_1.. in die Gleichungen (A.) ein und erhalte so zunächst:

$$\begin{array}{l} 4(b\ -b_3)(b_1-b_4)+4b\,b_3+4b_5\equiv 2-2b+2b_1-2b_4\,,\\ 4(b_1-b_4)(b_2-b_5)+4b_1b_4+4b_3\equiv -2b_1+2b_4\,,\qquad \text{mod. } 8.\\ 4(b_2-b_5)(b_3-b)\ +4b_2b_5+4\ \ \equiv -2b_2+2b_5\,,\end{array}$$

aus welchen Congruenzen zunächst folgt, dafs

$$b_1\equiv b_4\,,\quad b_2\equiv b_5\,,\quad b\equiv 1\,,\quad \text{mod. } 2.$$

sein mufs, wodurch diese Congruenzen sich weiter vereinfachen:

$$\begin{array}{r} 4b_1+4b_5\equiv 2-2b\,,\\ 4b_3\equiv 2b_1+2b_4\,,\quad \text{mod. } 8.\\ 4\equiv 2b_2+2b_5\,,\end{array}$$

und wenn statt der Zahlen b wieder die Zahlen a eingeführt werden:

$$\begin{aligned} 2a_1 + 2a_5 &\equiv 4 - a\,, \\ 2a_3 &\equiv a_1 + a_4\,, \quad \text{mod. } 8. \\ 4 &\equiv a_2 + a_5\,, \end{aligned}$$

Macht man nun die Gleichung (B.) zu einer Congruenz nach dem Modul 31, so hat man

$$2 \equiv 5P^2\,, \quad \text{mod. } 31.$$

also

$$P^2 \equiv 19\,, \quad P \equiv \pm 9\,, \quad \text{mod. } 31.$$

Da man alle Vorzeichen der 6 Zahlen $a, a_1 \ldots$ gleichzeitig ändern kann, so reicht es hin P positiv zu nehmen; beachtet man aufserdem, dafs P ungrade ist, so erhält man für P folgende Reihe möglicher Werthe:

$$P = 9\,,\ 53\,,\ 71\,,\ 115\,,\ \ldots$$

Die aus (B.) zu berechnenden zugehörenden Werthe des Q sind

$$Q = 39\,,\ 479\,,\ 839\,,\ 2159\,,\ \ldots$$

und die aus (C.) zu berechnenden zugehörenden Werthe des R

$$R = 153\,,\ 65\,,\ -7\,,\ -272\,,\ \ldots$$

Da aber R als Summe von Quadraten nothwendig positiv ist, so bleiben nur die beiden Fälle übrig:

$$\begin{aligned} &1)\quad P = 9\,, \quad Q = 39\,, \quad R = 153\,, \\ &2)\quad P = 53\,, \quad Q = 479\,, \quad R = 65\,. \end{aligned}$$

Da in beiden Fällen $P \equiv 1$, mod. 4 ist, so hat man

$$a + a_1 + a_2 + a_3 + a_4 + a_5 \equiv 1 \quad \text{mod. } 4.$$

also

$$2b + 2b_1 + 2b_2 + 2b_3 + 2b_4 + 2b_5 \equiv 2 \quad \text{mod. } 4.$$

also nach den oben gegebenen Congruenzen für den Modul 8:

$$b_3 \equiv 0\,, \quad \text{mod. } 2 \quad , \quad a_3 \equiv 1\,, \quad \text{mod. } 4.$$

Diese Congruenzen ergeben deshalb folgende Resultate:

$$
(D.) \quad \begin{array}{llll} a \equiv 2\,, & \text{mod. } 4 \quad, & a_2 + a_5 \equiv 4\,, & \text{mod. } 8\,, \\ a_3 \equiv 1\,; & \text{mod. } 4 \quad, & a_1 + a_4 \equiv 2\,, & \text{mod. } 8\,. \end{array}
$$

Ich untersuche nun zunächst den ersten der beiden unterschiedenen Fälle, nämlich

$$
a + a_1 + a_2 + a_3 + a_4 + a_5 = 9\,,
$$

$$
a_2 + a_1^2 + a_2^2 + a_3^2 + a_4^2 + a_5^2 = 39\,.
$$

Zerlegt man die Zahl 39 auf alle möglichen Weisen in die Summe von 6 Quadraten, und wählt man die Vorzeichen so, daſs die Summe der sechs Wurzeln gleich 9 ist, so erhält man folgende acht verschiedene Fälle für die Werthe der Zahlen $a, a_1, a_., a_3, a_4, a_5$, welche diesen beiden Gleichungen genügen:

1) $+6\,,\ +1\,,\ +1\,,\ +1\,,\ \ 0\,,\ \ 0\,,$

2) $+5\,,\ +3\,,\ +2\,,\ \ 0\,,\ \ 0\,,\ -1\,,$

3) $+5\,,\ +2\,,\ +2\,,\ +2\,,\ -1\,,\ -1\,,$

4) $+5\,,\ +2\,,\ +2\,,\ +1\,,\ +1\,,\ -2\,,$

5) $+4\,,\ +4\,,\ +2\,,\ +1\,,\ -1\,,\ -1\,,$

6) $+4\,,\ +4\,,\ +1\,,\ +1\,,\ +1\,,\ -2\,,$

7) $+4\,,\ +3\,,\ +3\,,\ +1\,,\ \ 0\,,\ -2\,,$

8) $+3\,,\ +3\,,\ +2\,,\ +2\,,\ +2\,,\ -3\,.$

Die Fälle 1, 2, 5 und 6 sind aber mit den für die drei graden Zahlen a, a_2, a_5 bestehenden beiden Congruenzbedingungen (D.) unvereinbar. Ferner sind die Fälle 3, 7 und 8 mit den unter den drei ungraden Zahlen a_1, a_3, a_4 bestehenden beiden Congruenzbedingungen (D.) unvereinbar. Es bleibt also nur noch der Fall 4 übrig, welcher mit diesen vier Congruenzbedingungen bestehen kann, wenn

$$
a = -2\,,\ a_1 = +1\,,\ a_2 = +2\,,\ a_3 = +5\,,\ a_4 = +1\,,
$$

$$
a_5 = +2
$$

genommen wird. Um zu sehen, ob diese Werthe der Aufgabe wirklich genügen, muſs man zu den Gleichungen (A.) zurückgehen. Man erhält für diese Werthe:

$$5Q - P^2 + a_1^2 + a_4^2 - (a - a_3)(a_1 - a_4)$$
$$+ 2(aa_5 + a_3 a_2) + aa_3 = 118 ,$$

sie genügen also schon der ersten dieser drei Gleichungen nicht, nach welcher dieser Ausdruck den Werth 134 haben mufs. Es giebt also in dem ersten Hauptfalle wo $P = 9$ $Q = 39$ sein mufs überhaupt keine den drei Gleichungen (A.) genügenden Zahlen.

Es bleibt nun noch übrig auch den zweiten Hauptfall zu untersuchen, wo

$$P = 53 \quad , \quad Q = 479 \quad , \quad R = 65 .$$

Es sei m die kleinste der sechs Zahlen $a, a_1, a_2, a_3, a_4, a_5$; die übrigen fünf seien $m + c_1$, $m + c_2$, $m + c_3$, $m + c_4$, $m + c_5$, so sind c_1, c_2, c_3, c_4, c_5 positive Zahlen, bei welchen jedoch auch der Werth 0 nicht auszuschliefsen ist. Setzt man nun zur Abkürzung

$$c_1 + c_2 + c_3 + c_4 + c_5 = p ,$$
$$c_1^2 + c_2^2 + c_3^2 + c_4^2 + c_5^2 = q ,$$
$$(c_1 - c_2)^2 + (c_1 - c_3)^2 + (c_1 - c_4)^2 + (c_1 - c_5)^2 + (c_2 - c_3)^2$$
$$+ (c_2 - c_4)^2 + (c_2 - c_5)^2 + (c_3 - c_4)^2 + (c_3 - c_5)^2$$
$$+ (c_4 - c_5)^2 = r ,$$

so hat man

$$53 = 6m + p ,$$
$$479 = 6m^2 + 2mp + q ,$$
$$65 = q + r ,$$

und wenn man aus diesen drei Gleichungen p und q eliminirt:

$$414 = 106m - 6m^2 - r ,$$

und weil p, q, r positive Zahlen sind, so ist

$$53 > 6m ,$$
$$414 < 106m - m^2 ,$$

woraus folgt, dafs m nur die drei Werthe $m = 8$, $m = 7$ und $m = 6$ erhalten kann.

Nimmt man zuerst $m = 8$, so ist für diesen Werth

$$p = 5 \quad , \quad q = 15 \quad , \quad r = 50 .$$

Die einzige Art wie die Zahl $q = 15$ in fünf Quadrate zerlegt werden kann ist aber

$$15 = 3^2 + 2^2 + 1^2 + 1^2 + 0^2 ,$$

welche, weil keine der Zahlen c_1, c_2, c_3, c_4, c_5 negativ ist, nicht $p = 5$, sondern $p = 7$ giebt. Der Fall $m = 8$ giebt also keine Auflösung der Aufgabe.

Nimmt man zweitens $m = 7$, so hat man

$$p = 11 \quad , \quad q = 31 \quad , \quad r = 34 .$$

Die Zahl $q = 31$ läfst sich aber nur auf folgende drei Arten als Summe von 5 Quadratzahlen darstellen:

$$31 = 5^2 + 2^2 + 1^2 + 1^2 + 0^2 ,$$
$$31 = 4^2 + 3^2 + 2^2 + 1^2 + 1^2 ,$$
$$31 = 3^2 + 3^2 + 3^2 + 2^2 + 0^2 .$$

Die erste derselben ist zu verwerfen, weil sie nicht $p = 11$, sondern $p = 9$ ergiebt; die zweite und dritte sind mit dieser Bedingung im Einklange und ergeben für die fünf Zahlen c_1, c_2, c_3, c_4, c_5 die Werthe

$$4 , 3 , 2 , 1 , 1 ,$$
$$3 , 3 , 3 , 2 , 0 ,$$

und demgemäfs für die sechs Zahlen $a, a_1, a_2, a_3, a_4, a_5$ die Werthe

$$11 , 10 , 9 , 8 , 8 , 7 ,$$
$$10 , 10 , 10 , 9 , 7 , 7 .$$

Die ersteren sind aber mit den unter den drei graden Zahlen a, a_2, a_5 nothwendigen beiden Congruenzbedingungen (D.) und die anderen mit den unter den drei ungraden Zahlen a_1, a_3, a_4 nothwendig Statt habenden Congruenzbedingungen (D.) unvereinbar. Der Fall $m = 7$ giebt also ebenfalls keine Lösung der Aufgabe.

Nimmt man endlich $m = 6$, so hat man:

$$p = 17 \quad , \quad q = 59 \quad , \quad r = 6 .$$

Da r eine Summe von 10 Quadraten ist, so ist die Zahl 6 als Summe von 10 Quadraten darzustellen, welches auf folgende zwei verschiedene Arten möglich ist:

$$6 = 2^2 + 1^2 + 1^2 + 7.0^2,$$

$$6 = 1^2 + 1^2 + 1^2 + 1^2 + 1^2 + 1^2 + 4.0^2.$$

Es ist aber unmöglich, daſs von den 10 Differenzen je zweier der fünf Gröſsen c_1, c_2, c_3, c_4, c_5, aus deren Quadraten $r = 6$ besteht, genau 7 gleich Null sind; denn wenn selbst vier dieser Zahlen einander gleich wären, so würden nur 6 dieser Differenzen gleich Null sein. Es bleibt also nur die zweite Darstellung von $r = 6$ zu betrachten, für welche 3 von den fünf Zahlen c_1, c_2, c_3, c_4, c_5 einander gleich sein müssen und die übrigen beiden auch einander gleich und wo die Differenzen der ersten drei gleichen von den anderen zwei gleichen gleich Eins ist. Da die Summe p dieser Zahlen gleich 17 sein muſs, so genügen keine anderen Werthe der c_1, c_2, c_3, c_4, c_5 als

$$4, 4, 3, 3, 3.$$

Die zugehörenden Werthe der 6 Zahlen $a, a_1, a_2, a_3, a_4, a_5$ sind demnach:

$$10, 10, 9, 9, 9, 6,$$

welche sich in der That den einzelnen Zahlen $a, a_1, a_2, a_3, a_4, a_5$ so zuordnen lassen, daſs den vier Congruenzbedingungen (D.) genügt wird, und zwar nur auf folgende Weise:

$$a = 6, \quad a_1 = 9, \quad a_2 = 10, \quad a_3 = 9, \quad a_4 = 9, \quad a_5 = 10.$$

Aber auch diese Werthe genügen den drei Gleichungen (A.) nicht, denn man erhält für dieselben

$$5Q - P^2 + a_1^2 + a_4^2 - (a - a_3)(a_1 - a_4) + 2(aa_5 + a_3 a_2) + aa_3 = 102,$$

und nicht 134, welchen Werth dieser Ausdruck vermöge der ersten dieser Gleichungen haben muſs.

Es ist also in allen Fällen unmöglich, die sechs Zahlen $a, a_1, a_2, a_3, a_4, a_5$ so zu bestimmen, daſs sie den drei Gleichungen (A.) genügen und darum ist es unmöglich die dritte Potenz eines idealen Primfaktors der Zahl 2 als wirkliche complexe Zahl $F(\eta)$ darzustellen; es giebt also keine niedere Potenz des idealen Primfaktors der 2, als die neunte, welche wirklich ist.

Über eine Eigenschaft der Einheiten der aus den Wurzeln der Gleichung $\alpha^\lambda = 1$ gebildeten complexen Zahlen, und über den zweiten Factor der Klassenzahl

Monatsberichte der Königlich Preußischen Akademie der Wissenschaften zu Berlin aus dem Jahre 1870, 855–880

1. December. Gesammtsitzung der Akademie.

Hr. Kummer las:

Über eine Eigenschaft der Einheiten der aus den Wurzeln der Gleichung $\alpha^\lambda = 1$ gebildeten complexen Zahlen und über den zweiten Faktor der Klassenzahl.

Die Einheiten der complexen Zahlen sind, wenn man von den einfachen Einheitswurzeln absieht, welche als Faktoren hinzutreten können, stets reale Gröfsen. Betrachtet man eine beliebige solche Einheit mit allen ihren conjugirten zusammen, so erhält man eine Reihe von realen Gröfsen, welche im Allgemeinen zum Theil positiv, zum Theil negativ sein werden, welche aber in dem besonderen Falle, wo die gegebene Einheit ein Quadrat ist, nothwendig alle positiv sind. Hieran knüpft sich nun die Frage, ob auch umgekehrt alle diejenigen Einheiten, welche die Eigenschaft haben, dafs sie mit allen ihren conjugirten nur positive Werthe haben, vollständige Quadrate von Einheiten sein müssen, oder wenn dies nicht der Fall ist, welche weitere Bedingungen hierzu nöthig sind. Diese Frage ist es, welche ich hier für die aus λ ten Einheitswurzeln gebildeten complexen Zahlen erörtern will; sie hat auch darum ein besonderes Interesse, weil ihre Lösung eine neue Eigenschaft des schwer zugänglichen zweiten Faktors der Klassenzahl ergiebt, nämlich eine Bedingung seiner Theilbarkeit durch Zwei.

Es sei γ eine primitive Wurzel der Primzahl λ, ferner sei γ_k der kleinste positive Rest von γ^k, nach dem Modul λ, und $\alpha^\lambda = 1$, so wird das System der conjugirten Kreistheilungseinheiten dargestellt durch

$$e_k = \frac{\alpha^{\gamma_{k+1}} - \alpha^{-\gamma_{k+1}}}{\alpha^{\gamma_k} - \alpha^{-\gamma_k}},$$

für $k = 0, 1, 2, \ldots \mu - 1$, wo μ, wie auch in dem Folgenden, gleich $\frac{\lambda - 1}{2}$ ist. Weil

$$\alpha = \cos\frac{2\pi}{\lambda} + i\sin\frac{2\pi}{\lambda},$$

so hat man auch

$$e_k = \frac{\sin\left(\frac{2\gamma_{k+1}\pi}{\lambda}\right)}{\sin\left(\frac{2\gamma_k\pi}{\lambda}\right)},$$

woraus man ersieht, dafs e_k positiv ist, wenn γ_k und γ_{k+1} beide zugleich kleiner als $\frac{\lambda}{2}$, oder beide zugleich gröfser als $\frac{\lambda}{2}$ sind und dafs e_k negativ ist, wenn von den beiden Zahlen γ_k und γ_{k+1} die eine gröfser als $\frac{\lambda}{2}$, die andere aber kleiner als $\frac{\lambda}{2}$ ist. Da

$$e\, e_1\, e_2 \ldots\ldots e_{\mu-1} = -1$$

ist, so folgt, dafs die Anzahl der negativen unter den conjugirten Kreistheilungseinheiten eine ungrade ist, dafs diese also niemals alle positiv sind.

Es soll nun weiter untersucht werden, unter welchen Bedingungen eine aus den Kreistheilungseinheiten zusammengesetzte Einheit, welche sich als ein Produkt von Potenzen der conjugirten Kreistheilungseinheiten darstellt, die Eigenschaft haben kann, dafs sie mit allen ihren conjugirten nur positive Werthe hat. Es sei die zu betrachtende Einheit

$$E = e^x e_1^{x_1} e_2^{x_2} \ldots\ldots e_{\mu-1}^{x_{\mu-1}},$$

wo $x, x_1, \ldots x_{\mu-1}$ irgend welche ganze Zahlen sind, so handelt es sich darum diese Exponenten so zu bestimmen, dafs allgemein

$$E_k = e_k^{x}\, e_{k+1}^{x_1}\, e_{k+2}^{x_2} \ldots e_{k-1}^{x_{\mu-1}}$$

positiv sei, für jeden der Werthe $k = 0, 1, 2, \ldots \mu - 1$. Ich bestimme nun die Zahl c_k so, dafs sie für jeden Werth des k nur einen der beiden Werthe 0 oder 1 habe und zwar:

$c_k = 0$ für die Werthe des k, für welche e_k positiv,

$c_k = 1$ für die Werthe des k, für welche e_k negativ ist.

Die Bedingung, dafs E_k positiv sei, ist alsdann gleichbedeutend mit der, dafs

$$c_k x + c_{k+1} x_1 + c_{k+2} x_2 + \cdots + c_{k-1} x_{\mu-1}$$

eine grade Zahl ist und weil diese Bedingung für jeden der μ Werthe des k erfüllt sein soll, so hat man das System der Congruenzen:

$$\text{(C.)} \quad \left.\begin{array}{llllll} cx & + c_1 x_1 & + c_2 x_2 & + \cdots & + c_{\mu-1} x_{\mu-1} & \equiv 0, \\ c_1 x & + c_2 x_1 & + c_3 x_2 & + \cdots & + c x_{\mu-1} & \equiv 0, \\ c_2 x & + c_3 x_1 & + c_4 x_2 & + \cdots & + c_1 x_{\mu-1} & \equiv 0, \\ \vdots & & & & \vdots & \\ c_{\mu-1} x & + c x_1 & + c_1 x_2 & + \cdots & + c_{\mu-2} x_{\mu-1} & \equiv 0. \end{array}\right\} \text{mod. } 2.$$

Dieses System läfst sich wie bekannt durch die μten Wurzeln der Einheit auflösen; bezeichnet man mit w eine jede beliebige primitive oder nicht primitive Wurzel der Gleichung $w^\mu = 1$, multiplicirt diese Congruenzen der Reihe nach mit $1, w, w^2, \ldots w^{\mu-1}$ und addirt, so erhält man

$$\text{(D)} \, (c + c_1 w + c_2 w^2 + \cdots + c_{\mu-1} w^{\mu-1})(x + x_1 w^{-1} + x_2 w^{-2} + \cdots + x_{\mu-1} w) \equiv 0, \quad \text{mod. } 2.$$

Ich setze nun zur Abkürzung

59*

$$c + c_1 w + c_2 w^2 + \dots + c_{\mu-1} w^{\mu-1} = \psi(w),$$

so ist die Determinante dieses Systems von Congruenzen gleich der vollständigen, über alle primitiven und nicht primitiven Wurzeln der Gleichung $w^\mu = 1$ sich erstreckenden Norm von $\psi(w)$, welche ich durch $N\psi(w)$ bezeichne. Wenn nun diese Determinante $N\psi(w)$ nicht congruent Null ist, mod. 2, so müssen bekanntlich alle Werthe der Unbekannten $x, x_1, \dots x_{\mu-1}$ einzeln congruent Null sein, nach dem Modul 2, also die Einheit E mufs in diesem Falle ein vollständiges Quadrat sein.

Wenn nur E eine Einheit ist, welche sich nicht als ein Produkt von Potenzen der Kreistheilungseinheiten darstellen läfst, so läfst sich nach einem bekannten Satze doch stets eine bestimmte Potenz von E in dieser Weise ausdrücken und man hat allgemein für jede Einheit E eine Gleichung von der Form

$$E^n = e^x e_1^{x_1} e_2^{x_2} \dots e_{\mu-1}^{x_{\mu-1}},$$

wo $n, x, x_1, \dots x_{\mu-1}$ ganze Zahlen sind, deren eine man gleich Null nehmen kann, und welche nicht alle zugleich einen gemeinschaftlichen Faktor haben. Wenn nun $N\psi(w)$ nicht durch 2 theilbar ist, so kann E^n mit seinen conjugirten nicht stets positiv sein, ohne dafs $x, x_1, \dots x_{\mu-1}$ alle grade sind; alsdann mufs n, welches nicht mit allen diesen einen gemeinschaftlichen Faktor haben soll, ungrade sein und weil E^n ein Quadrat ist und n ungrade, so mufs E selbst ein Quadrat sein. Man hat demnach folgenden Satz:

(I.) „Für alle diejenigen Werthe der Primzahl λ, für welche „die vollständige, über alle der Gleichung $w^\mu = 1$ genügenden μ Werthe des w sich erstreckende Norm „$N\psi(w)$ nicht durch 2 theilbar ist, ist eine jede aus „λten Einheitswurzeln gebildete Einheit, welche mit „allen ihren conjugirten nur positive Werthe hat, nothwendig ein Quadrat einer Einheit.

Die Bedingung, dafs die vollständige Norm $N\psi(w)$ nicht durch Zwei theilbar sei, ist identisch mit der Bedingung, dafs der erste Faktor der Klassenzahl der aus λten Einheitswurzeln gebildeten complexen Zahlen nicht durch Zwei theilbar sei. Um dies zu zeigen, verwandle ich den Ausdruck der complexen Zahl $\psi(w)$ in folgender Weise:

Es sei r der Index von 2, für die primitive Wurzel γ, oder $\gamma^r \equiv 2$, mod. λ, so ist

$$2\gamma_k = \gamma_{k+r} \quad , \qquad \text{wenn } \gamma_k < \tfrac{\lambda}{2} ,$$
$$2\gamma_k = \gamma_{k+r} + \lambda , \qquad \text{wenn } \gamma_k > \tfrac{\lambda}{2} ,$$

also wenn γ_k und γ_{k+1} beide zugleich $< \frac{\lambda}{2}$ oder beide zugleich $> \frac{\lambda}{2}$ sind, d. i. wenn $c_k = 0$ ist, so hat man

$$2\gamma_k - 2\gamma_{k+1} = \gamma_{k+r} - \gamma_{k+1+r} ,$$

wenn aber von den beiden Zahlen γ_k und γ_{k+1} die eine gröfser, die andere kleiner als $\frac{\lambda}{2}$ ist, d. i. wenn $c_k = 1$ ist, so hat man

$$2\gamma_k - 2\gamma_{k+1} = \gamma_{k+r} - \gamma_{k+1+r} \pm \lambda ,$$

also in beiden Fällen hat man allgemein

$$2\gamma_k - 2\gamma_{k+1} = \gamma_{k+r} - \gamma_{k+1+r} \pm c_k\lambda$$

und demgemäfs

$$c_k \equiv \gamma_{k+r} - \gamma_{k+1+r} , \quad \text{mod. } 2.$$

Hieraus folgt

$$\psi(w) = \sum_0^{\mu-1}{}_k c_k w^k \equiv \sum_0^{\mu-1}{}_k (\gamma_{k+r} - \gamma_{k+1+r}) w^k , \quad \text{mod. } 2.$$

und weil für jeden Werth des k, $\gamma_{k+\mu} = \lambda - \gamma_k$ und demgemäfs $\gamma_{k+\mu} - \gamma_{k+1+\mu} \equiv \gamma_k - \gamma_{k+1}$, mod. 2, ist und $w^{k+\mu} = w^k$, so kann man diese Summe auch so darstellen:

$$\text{(E.)} \qquad \psi(w) \equiv w^{-r} \sum_0^{\mu-1}{}_k (\gamma_k - \gamma_{k+1}) w^k , \quad \text{mod. } 2.$$

Betrachtet man nun andererseits den ersten Faktor der Klassenzahl, welchen ich (Crelle's Journal Bd. 40 p. 110) so dargestellt habe:

$$P' = \frac{\varphi(\beta)\varphi(\beta^3)\varphi(\beta^5) \ldots\ldots \varphi(\beta^{\lambda-2})}{2^{\mu-1}\lambda^{\mu-1}} ,$$

wo β eine primitive Wurzel der Gleichung $\beta^{\lambda-1} = 1$ ist und

$$\phi(\beta) = 1 + \gamma_1\beta + \gamma_2\beta^2 + \cdots + \gamma_{\lambda-2}\beta^{\lambda-1}$$

und welcher, wenn β eine jede Wurzel der Gleichung $\beta^\mu = -1$ bezeichnet, auch so dargestellt werden kann:

$$P' = \frac{N\phi(\beta)}{2^{\mu-1}\lambda^{\mu-1}},$$

wo die Norm über alle μ Wurzeln der Gleichung $\beta^\mu = -1$ sich erstreckt; so hat man zunächst

$$(1-\beta^{-1})\phi(\beta) = \sum_0^{\lambda-2}(\gamma_k - \gamma_{k+1})\beta^k$$

und weil

$$(\gamma_{k+\mu} - \gamma_{k\,1+\mu})\beta^{k+\mu} = (\gamma_k - \gamma_{k+1})\beta^k,$$

so ist

$$(1-\beta^{-1})\phi(\beta) = 2\sum_{0}^{\mu-1}{}_k(\gamma_k - \gamma_{k+1})\beta^k,$$

also wenn durch 2 dividirt und mit β^{-r} multiplicirt wird:

$$\tfrac{1}{2}\beta^{-r}(1-\beta^{-1})\phi(\beta) = \psi(\beta),$$

wo

$$\psi(\beta) = \beta^{-r}\sum_0^{\mu-1}(\gamma_k - \gamma_{k+1})\beta^k.$$

Nimmt man nun die vollständige Norm in Beziehung auf alle Werthe, welche der Gleichung $\beta^\mu = -1$ genügen, so hat man $N(1-\beta^{-1}) = 2$ und demgemäfs

$$\lambda^{\mu-1}P' = N\psi(\beta),$$

oder als Congruenz nach dem Modul 2:

$$P' \equiv N\psi(\beta), \quad \text{mod. } 2.$$

Die complexe Zahl $\psi(\beta)$ hat nach dem Modul 2 ganz dieselben Coefficienten, als die obige complexe Zahl $\psi(w)$. Da ferner in den beiden Gleichungen μten Grades

$$\beta^{\mu} = -1 \quad \text{und} \quad w^{\mu} = +1$$

alle Coëfficienten der einen den Coëfficienten der andern congruent sind, nach dem Modul 2, so folgt, dafs auch eine jede symmetrische Funktion aller Wurzeln der Gleichung $\beta^{\mu} = -1$ derselben symmetrischen Funktion der Wurzeln der Gleichung $w^{\mu} = 1$ congruent sein mufs, nach dem Modul 2. Es ist daher

$$N\psi(w) \equiv N\psi(\beta) \equiv P', \quad \text{mod. } 2.$$

Die Bedingung, dafs $N\psi(w)$ nicht durch 2 theilbar sei, ist also identisch mit der, dafs der erste Faktor der Klassenzahl nicht durch 2 theilbar sei. Der obige Satz läfst sich daher auch so aussprechen:

(II.) „Für alle diejenigen Primzahlen λ, für welche der „erste Faktor der Klassenzahl nicht durch Zwei theil„bar ist, ist jede complexe Einheit, welche mit ihren „conjugirten nur positive Werthe hat, ein Quadrat einer „Einheit.

Hieraus ergiebt sich nun unmittelbar die Bedingung dafür, dafs der zweite Faktor der Klassenzahl nicht durch 2 theilbar sei. Wenn nämlich der zweite Faktor der Klassenzahl durch Zwei theilbar ist, so giebt es nothwendig eine Einheit E von der Art, dafs

$$E^2 = e^{x} e_1^{x_1} e_2^{x_2} \dots e_{\mu-1}^{x_{\mu-1}},$$

wo $x, x_1 \dots x_{\mu-1}$ ganze Zahlen sind, deren eine gleich Null genommen werden kann, und welche nicht alle den gemeinschaftlichen Faktor 2 haben. Wenn aber der erste Faktor der Klassenzahl nicht durch 2 theilbar ist, so giebt es keine solche Einheit

$$e^{x} e_1^{x_1} e_2^{x_2} \dots e_{\mu-1}^{x_{\mu-1}},$$

welche mit allen ihren conjugirten positive Werthe hat, ein solches Produkt von Potenzen von Kreistheilungseinheiten kann also nicht ein Quadrat, also nicht gleich E^2 sein. Hieraus folgt:

(III.) „Der zweite Faktor der Klassenzahl ist niemals durch „Zwei theilbar, wenn nicht zugleich auch der erste „Faktor der Klassenzahl durch Zwei theilbar ist.

Dieser Satz über die Theilbarkeit der Klassenzahl durch 2 ist vollkommen analog dem früher von mir bewiesenen Satze über die Theilbarkeit der Klassenzahl durch λ.

Wenn der erste Faktor der Klassenzahl durch 2 theilbar ist, also die Bedingung der Gültigkeit der oben aufgestellten Sätze (I.) und (II.) nicht erfüllt ist, so giebt es stets Einheiten von der Form

$$E = e^{x} e_1^{x_1} e_2^{x_2} \ldots . e_{\mu-1}^{x_{\mu-1}},$$

welche mit allen ihren conjugirten nur positive Werthe haben, ohne dafs die Exponenten $x, x_1 \ldots .$ alle durch 2 theilbar sind. Eine solche Einheit E ist nur in dem Falle ein vollständiges Quadrat, wo nicht nur der erste, sondern auch der zweite Faktor der Klassenzahl durch 2 theilbar ist, welches im Allgemeinen nicht der Fall ist, wie die folgenden ausgeführten Beispiele zeigen.

Unter den Primzahlen λ, welche im ersten Hundert liegen, giebt es nur eine, für welche der erste Faktor der Klassenzahl durch 2 theilbar ist, nämlich $\lambda = 29$. Unter den 14 conjugirten Kreistheilungseinheiten sind, wenn die primitive Wurzel $\gamma = 3$ zu Grunde gelegt wird, nur folgende fünf negativ:

$$e_2, e_4, e_8, e_9, e_{10},$$

es ist also

$$c_2 = c_4 = c_8 = c_9 = c_{10} = 1,$$

$$c = c_1 = c_3 = c_5 = c_6 = c_7 = c_{11} = c_{12} = c_3 = 0.$$

Setzt man diese Werthe der Gröfsen c in das System der Congruenzen (C.) ein, so erhält man durch Auflösung desselben alle Werthe der Exponenten x, die demselben genügen, dargestellt durch

$$\left.\begin{array}{l} x \equiv 1, x_1 \equiv 1, x_2 \equiv 1, x_3 \equiv 0, x_4 \equiv 1, x_5 \equiv 0, x_6 \equiv 0 \\ x_7 \equiv 1, x_8 \equiv 1, x_9 \equiv 1, x_{10} \equiv 0, x_{11} \equiv 1, x_{12} \equiv 0, x_{13} \equiv 0 \end{array}\right\} \text{mod. } 2,$$

und durch die cyklischen Vertauschungen derselben, deren es nur 7 verschiedene giebt. Es folgt hieraus, dafs für $\lambda = 29$ alle Einheiten, welche mit ihren conjugirten nur positive Werthe haben, durch die eine Einheit

$$E = e\, e_1\, e_2\, e_4\, e_7\, e_8\, e_9\, e_{11}$$

und durch ihre conjugirten gegeben sind, wenn man von den Quadraten von Kreistheilungseinheiten absieht, welche beliebig hinzutreten können, weil die Exponenten x nur nach dem Modul 2 bestimmt sind. Die Ausdrücke der Kreistheilungseinheiten sind hier:

$$\begin{aligned}
e &= 1 + \alpha^2 + \alpha^{-2} \; , \\
e_1 &= 1 + \alpha^6 + \alpha^{-6} \; , \\
e_2 &= 1 + \alpha^{11} + \alpha^{-11} \; , \\
e_3 &= 1 + \alpha^4 + \alpha^{-4} \; , \\
e_4 &= 1 + \alpha^{12} + \alpha^{-12} \; , \\
e_5 &= 1 + \alpha^7 + \alpha^{-7} \; , \\
e_6 &= 1 + \alpha^8 + \alpha^{-8} \; , \\
e_7 &= 1 + \alpha^5 + \alpha^{-5} \; , \\
e_8 &= 1 + \alpha^{14} + \alpha^{-14} \; , \\
e_9 &= 1 + \alpha^{13} + \alpha^{-13} \; , \\
e_{10} &= 1 + \alpha^{10} + \alpha^{-10} \; , \\
e_{11} &= 1 + \alpha + \alpha^{-1} \; , \\
e_{12} &= 1 + \alpha^3 + \alpha^{-3} \; , \\
e_{13} &= 1 + \alpha^9 + \alpha^{-9} \; .
\end{aligned}$$

Aus diesen folgt:

$$e\, e_1\, e_2 = -(\alpha + \alpha^{-1}) \quad , \quad e_7\, e_8\, e_9 = -(\alpha^{12} + \alpha^{-12}) \; ,$$

also

$$E = (\alpha + \alpha^{-1})(\alpha^{12} + \alpha^{-12})(1 + \alpha + \alpha^{-1})(1 + \alpha^{12} + \alpha^{-12}) \; .$$

Die Einheit E ist, wie man hieraus ersieht, nur von den viergliedrigen Perioden der Wurzeln der Gleichung $\alpha^{29} = 1$ abhängig, bezeichnet man diese, nach der primitiven Wurzel $\gamma = 3$ geordnet, durch $\eta, \eta_1, \eta_2, \eta_3, \eta_4, \eta_5, \eta_6$, so erhält man durch Ausführung der Multiplikation

$$E(\eta) = \eta_5(\eta + \eta_5 + 1)$$

und hieraus weiter

$$E(\eta) = 4 + 2\eta + \eta_1 + 2\eta_3 + \eta_4 + \eta_5 + \eta_6 \; .$$

Um nun zu untersuchen, ob diese Einheit $E(\eta)$ ein vollständiges Quadrat ist, oder ob nicht, reicht es hin von einer Congruenzbedingung nach dem Modul 4 Gebrauch zu machen, welche jede complexe Zahl $f(\alpha)$ erfüllen mufs, wenn sie ein vollständiges Quadrat sein soll, nämlich die Bedingung

$$f(\alpha)^2 - f(\alpha^2) \equiv 0\,, \quad \text{mod. } 4.$$

Wenn nämlich $f(\alpha) = \varphi(\alpha)^2$ ist, so ist $f(\alpha) \equiv \varphi(\alpha^2)$, mod. 2, also $f(\alpha)^2 \equiv \varphi(\alpha^2)^2$, mod. 4, also auch $f(\alpha)^2 \equiv f(\alpha^2)$, mod. 4. Damit $E(\eta)$ ein Quadrat sei, mufs also $E(\eta)^2 - E(\eta_3) \equiv 0$, mod. 4, sein. Die Ausführung der Rechnung ergiebt aber

$$E(\eta)^2 - E(\eta_3) \equiv 2\eta_1 + 2\eta_4\,, \quad \text{mod. } 4,$$

also nicht $\equiv 0$. Die Einheit $E(\eta)$ ist also nicht ein Quadrat; also für $\lambda = 29$ ist der zweite Faktor der Klassenzahl nicht durch 2 theilbar.

Um noch ein zweites Beispiel dieser Art zu erhalten, habe ich auch einige Primzahlen λ im zweiten Hundert untersucht und unter diesen $\lambda = 113$ als eine solche gefunden, deren erster Faktor der Klassenzahl durch 2 theilbar ist.

Für $\lambda = 113$ wird, wenn die primitive Wurzel $\gamma = 10$ genommen wird, e_k negativ und folglich $c_k = 1$ für folgende 29 Werthe des k:

$$k = 1,\ 3,\ 4,\ 9,\ 14,\ 16,\ 17,\ 18,\ 19,\ 20,\ 22,\ 24,\ 26,\ 27,\ 30,$$
$$31,\ 34,\ 35,\ 36,\ 37,\ 39,\ 42,\ 44,\ 45,\ 46,\ 49,\ 50,\ 51,\ 53.$$

Für die übrigen 27 Werthe des k ist $c_k = 0$. Die Auflösung der Congruenzen (C.) ergiebt nun folgende Werthe der Exponenten x:

$$\begin{aligned} x_k \equiv 1 \text{ für } k = {} & 0,\ 7,\ 14,\ 21,\ 28,\ 35,\ 42,\ 49 \\ & 1,\ 8,\ 15,\ 22,\ 29,\ 36,\ 43,\ 50 \\ & 2,\ 9,\ 16,\ 23,\ 30,\ 37,\ 44,\ 51 \\ & 5,\ 12,\ 19,\ 26,\ 33,\ 40,\ 47,\ 54 \end{aligned}$$

für die übrigen 24 Werthe des k ist $x_k \equiv 0$, mod. 2. Setzt man nun

$$H_k = e_k e_{k+7} e_{k+14} e_{k+21} e_{k+28} e_{k+35} e_{k+42} e_{k+49},$$

so ist, abgesehen von Quadraten der Kreistheilungseinheiten,

$$E = H H_1 H_2 H_5$$

mit ihren conjugirten die einzige, als Produkt von Potenzen der Kreistheilungseinheiten darstellbare Einheit, welche nur positive Werthe hat. Um diese Einheit E, welche wie man hieraus ersieht nur aus den 7 Perioden von je 16 der Wurzeln der Gleichung $\alpha^{113} = 1$ zusammengesetzt ist, die ich nach der primitiven Wurzel 10 geordnet mit $\eta, \eta_1, \eta_2, \eta_3, \eta_4, \eta_5, \eta_6$ bezeichne, bemerke ich zunächst, dafs

$$e\, e_1\, e_2 \ldots . e_{51} = \frac{\alpha^{+2} - \alpha^{-2}}{\alpha - \alpha^{-1}} = \alpha + \alpha^{-1}$$

ist. Wird nun der Abkürzung wegen

$$\alpha^{\gamma^k} + \alpha^{-\gamma^k} = e'_k$$

gesetzt, und

$$\varepsilon_k = e'_k e'_{k+7} e'_{k+14} e'_{k+21} e'_{k+28} e'_{k+35} e'_{k+42} e'_{k+49},$$

so zeigt die Ausführung der Multiplikation, dafs ε_k gleich dem Produkte zweier Perioden

$$\varepsilon_k = \eta_k \eta_{k+5}$$

ist, und dafs durch die Einheiten $H, H_1 \ldots$ ausgedrückt

$$\varepsilon_k = - H_k H_{k+1} H_{k+2}$$

ist. Hieraus folgt weiter

$$\varepsilon_1 \varepsilon_3 \varepsilon_4 \varepsilon_5 = H H_1 H_2 H_5 (H_3 H_4 H_5 H_6)^2,$$

also, wenn von dem quadratischen Faktor abgesehen wird,

$$\varepsilon_1 \varepsilon_3 \varepsilon_4 \varepsilon_5 = E$$

und durch die Perioden ausgedrückt, wird

$$\varepsilon_1 \varepsilon_3 \varepsilon_4 \varepsilon_5 = \eta_2 \eta_4 \eta_5 (\eta_1 \eta_3)^2,$$

also, wenn wieder von dem quadratischen Faktor abgesehen wird, stellt sich die Einheit, welche mit ihren conjugirten nur positive Werthe hat, dar als

$$E(\eta) = \eta_2\,\eta_4\,\eta_5\,\eta_6\,.$$

Die Ausführung der Multiplikation ergiebt:

$$E(\eta) = 12 - \eta + \eta_1 + 2\eta_2 + \eta_4 - 4\eta_6\,.$$

Wenn diese Etnheit ein Quadrat sein sollte, so müſste, wie im vorigen Beispiele gezeigt worden,

$$E(\eta)^2 - E(\eta_3) \equiv 0\,, \quad \text{mod. } 4$$

sein; man findet aber

$$E(\eta)^2 - E(\eta_3) \equiv 2\eta + 2\eta_1 + 2\eta_4 + 2\eta_5 + 2\eta_6\,, \quad \text{mod. } 4.$$

Es ist also $E(\eta)$ nicht ein Quadrat, und darum auch für $\lambda = 113$ der zweite Faktor der Klassenzahl nicht durch Zwei theilbar.

Nach dieser Methode läſst sich auch allgemein eine neue Bedingung dafür ableiten, daſs der zweite Faktor der Klassenzahl durch 2 theilbar sei. Wenn dieser Fall statt haben soll, so muſs es eine Einheit

$$E(\alpha) = e(\alpha)^{x}\,e(\alpha^{\gamma})^{x_1}\,\ldots\ldots\,e(\alpha^{\gamma^{\mu-1}})^{x_{\mu-1}}$$

geben, welche ein Quadrat einer Einheit ist, ohne daſs die Exponenten $x, x_1 \ldots x_{\mu-1}$ alle durch 2 theilbar sind, wenn einer derselben gleich Null genommen wird, wo $e(\alpha), e(\alpha^{\gamma}) \ldots\ldots e(\alpha^{\gamma^{\mu-1}})$ die oben mit $e, e_1 \ldots\ldots e_{\mu-1}$ bezeichneten Kreistheilungseinheiten sind, also

$$e(\alpha) = \frac{\alpha^{\gamma} - \alpha^{-\gamma}}{\alpha - \alpha^{-1}}\,.$$

Hieraus erhält man

$$e(\alpha)^2 - e(\alpha^2) = \frac{\alpha^{2\gamma} + \alpha^{-2\gamma} - 2}{\alpha^2 + \alpha^{-2} - 2} - \frac{\alpha^{2\gamma} - \alpha^{-2\gamma}}{\alpha^2 - \alpha^{-2}}$$

$$= \frac{2\left((\alpha^{2\gamma} - \alpha^{-2\gamma})(1 - \alpha^{-2}) - (\alpha^2 - \alpha^{-2})(1 - \alpha^{-2\gamma})\right)}{(\alpha^2 + \alpha^{-2} - 2)(\alpha^2 - \alpha^{-2})}$$

also wenn gesetzt wird

$$e(\alpha)^2 \equiv e(\alpha^2)\left(1+2f(\alpha)\right), \quad \text{mod. } 4,$$

so erhält man

$$f(\alpha) \equiv \frac{-1}{1-\alpha^2} + \frac{1}{1-\alpha^{2\gamma}}, \quad \text{mod. } 2.$$

Demnach wird

$$E(\alpha)^2 \equiv E(\alpha^2)\left(1+2\left(xf(\alpha)+x_1 f(\alpha^\gamma)+\cdots+x_{\mu-1}f(\alpha^{\gamma^{\mu-1}})\right)\right), \quad \text{mod. } 4.$$

Wenn nun $E(\alpha)$ ein Quadrat sein soll, so mufs, wie oben gezeigt worden,

$$E(\alpha)^2 \equiv E(\alpha^2). \quad \text{mod. } 4$$

sein, folglich auch

$$xf(\alpha)+x_1 f(\alpha^\gamma)+\cdots\cdot+x_{\mu-1}f(\alpha^{\gamma^{\mu-1}}) \equiv 0, \quad \text{mod. } 2,$$

welche Congruenz, weil sie ebenso für die μ mit α conjugirten Wurzeln $\alpha, \alpha^\gamma, \ldots \alpha^{\gamma^{\mu-1}}$ gelten mufs, ein System von μ Congruenzen repräsentirt. Setzt man für $f(\alpha)$ seinen Werth ein, und wendet das Summenzeichen an, so kann man dieses System von Congruenzen auch so darstellen:

$$\sum_0^{\mu-1}{}_k x_k\left(\frac{-1}{1-\alpha^{2\gamma^k}} + \frac{1}{1-\alpha^{2\gamma^{k+1}}}\right) \equiv 0, \quad \text{mod. } 2.$$

Multiplicirt man nun mit $\alpha^{-2\gamma^{h+1}}$ und summirt in Beziehung auf alle $\lambda-1$ verschiedenen Werthe der Wurzel α, so hat man

$$\sum_0^{\mu-1}{}_k x_k\left(\Sigma_\alpha \frac{-\alpha^{-2\gamma^{h+1}}}{1-\alpha^{2\gamma^k}} + \Sigma_\alpha \frac{\alpha^{-2\gamma^{h+1}}}{1-\alpha^{2\gamma^{k+1}}}\right) \equiv 0, \quad \text{mod. } 2.$$

In den beiden Summen, welche sich über alle Werthe der Wurzel α erstrecken, kann man statt dieser Wurzel eine beliebige andere zu Grunde legen; setzt man daher in der ersteren Summe α statt $\alpha^{2\gamma^k}$ und in der anderen α statt $\alpha^{2\gamma^{k+1}}$, so erhält man

$$\sum_{0}^{\mu-1}{}_k x_k \left(\Sigma_\alpha \frac{-\alpha^{-\gamma^{h-k+1}}}{1-\alpha} + \Sigma_\alpha \frac{\alpha^{-\gamma^{h-k}}}{1-\alpha} \right) \equiv 0, \quad \text{mod. } 2.$$

Aus der Gleichung

$$\frac{-\lambda}{1-\alpha} = \alpha + 2\alpha^2 + 3\alpha^3 + \cdots + (\lambda-1)\alpha^{\lambda-1},$$

welche auch so dargestellt werden kann:

$$\frac{-\lambda}{1-\alpha} = \alpha + \gamma_1 \alpha^{\gamma} + \gamma_2 \alpha^{\gamma^2} + \cdots + \gamma_{\lambda-2} \alpha^{\gamma^{\lambda-2}},$$

folgt aber, wenn mit $\alpha^{-\gamma^n}$ multiplicirt und in Beziehung auf alle $\lambda - 1$ verschiedenen Werthe der Wurzel α der Gleichung $\alpha^{\lambda-1} + \alpha^{\lambda-2} + \cdots + \alpha + 1 = 0$ summirt wird:

$$-\lambda \Sigma_\alpha \frac{\alpha^{-\gamma^n}}{1-\alpha} = -1 - \gamma_1 - \gamma_2 - \cdots - \gamma_{\lambda-2} + \lambda \gamma_n$$

und weil

$$1 + \gamma_1 + \gamma_2 + \cdots + \gamma_{\lambda-2} = \frac{\lambda(\lambda-1)}{2},$$

so folgt, wenn durch λ dividirt wird:

$$\Sigma_\alpha \frac{\alpha^{-\gamma^n}}{1-\alpha} = \frac{\lambda-1}{2} - \gamma_n.$$

Macht man von dieser Summation Gebrauch, so hat man:

$$\sum_{0}^{\mu-1}{}_k x_k (\gamma_{h-k+1} - \gamma_{h-k}) \equiv 0, \quad \text{mod. } 2.$$

Durch Multiplikation mit w^{-h}, wo w eine jede primitive oder nicht primitive Wurzel der Gleichung $w^\mu = 1$ bezeichnet und durch Summation in Beziehung auf die Werthe $h = k, k+1, \ldots k+\mu-1$ wird dieses System von μ Congruenzen in folgende Form gebracht:

$$\sum_{0}^{\mu-1}{}_k \sum_{k}^{k+\mu-1}{}_h (\gamma_{h-k+1} - \gamma_{h-k}) w^{-h+k} x_k w^{-k} \equiv 0, \quad \text{mod. } 2,$$

woraus endlich, wenn h in $h+k$ verwandelt wird, folgt

$$\sum_0^{\mu-1}{}_h (\gamma_{h+1} - \gamma_h)\, w^{-h} . \sum_0^{\mu-1}{}_k x^k w^{-k} \equiv 0\,, \quad \text{mod. } 2.$$

Nach der oben gegebenen Congruenz (E.) ist aber

$$\sum_0^{\mu-1}{}_h (\gamma_{h+1} - \gamma_h)\, w^{-h} \equiv -w^{-r} \psi(w^{-1})\,, \quad \text{mod. } 2,$$

also hat man

(F.) $$\psi(w^{-1}) . \sum_0^{\mu-1}{}_k x_k w^{-k} \equiv 0\,, \quad \text{mod. } 2,$$

als neue Bedingung für die Bestimmung der Exponenten $x, x_1 \ldots x_{\mu-1}$, während die nach der anderen Methode gefundene bei (D.) aufgestellte

$$\psi(w) . \sum_0^{\mu-1}{}_k x_k w^{-k} \equiv 0\,, \quad \text{mod. } 2,$$

war. Die schon oben hieraus abgeleitete nothwendige Bedingung dafür, dafs nicht alle Exponenten $x, x_1 \ldots x_{\mu-1}$ congruent Null sein müssen, nach dem Modul 2, dafs also der zweite Faktor der Klassenzahl durch 2 theilbar sein könne, nämlich dafs die vollständige Norm von $\psi(w)$ durch 2 theilbar sein mufs, läfst sich aber auch so aussprechen, dafs die complexe Zahl $\psi(w)$ einen complexen (idealen) Primfaktor von 2 enthalten mufs. Die Vergleichung der Congruenz (F.) mit der obigen (D.) ergiebt nun, dafs die complexe Zahl $\psi(w^{-1})$ d e n s e l b e n complexen Primfaktor von 2 enthalten mufs als $\psi(w)$. Also

(IV.) „Wenn der zweite Faktor der Klassenzahl durch Zwei „theilbar ist, so enthält die complexe Zahl $\psi(w^{-1})$ „nothwendig d e n s e l b e n complexen Primfaktor von „Zwei, welchen $\psi(w)$ enthält.

Es ist leicht zu erkennen, dafs die nothwendige Bedingung dieses Satzes für die Werthe $\lambda = 29$ und $\lambda = 113$ nicht erfüllt ist, dafs also der zweite Faktor der Klassenzahl für dieselben nicht durch Zwei theilbar ist, was oben durch specielle Ausrechnungnachgewiesen worden ist.

Die Methode, nach welcher diese nothwendige Bedingung der Theilbarkeit des zweiten Faktors der Klassenzahl durch 2 gefunden worden ist, läfst sich mit demselben Erfolge auch auf die Theilbarkeit dieses zweiten Faktors der Klassenzahl durch irgend eine ungrade von λ verschiedene Primzahl q anwenden. Es mufs hier die Einheit

$$E(\alpha) = e(\alpha)^{x} e(\alpha^{\gamma})^{x_1} \dots e(\alpha^{\gamma^{\mu-1}})^{x_{\mu-1}}$$

eine $q$te Potenz einer Einheit sein, ohne dafs die Exponenten $x, x_1 \dots x_{\mu-1}$ alle durch q theilbar sind, wenn einer derselben gleich Null genommen wird. Eine nothwendige Bedingung dafür, dafs $E(\alpha)$ eine $q$te Potenz sei, ist aber

$$E(\alpha)^q \equiv E(\alpha^q)\ , \quad \text{mod. } q^2;$$

denn setzt man $E(\alpha) = \psi(\alpha)^q$, so hat man bekanntlich

$$\psi(\alpha)^q \equiv \psi(\alpha^q)\ , \quad \text{mod. } q$$

oder

$$\psi(\alpha)^q = \psi(\alpha^q) + q\chi(\alpha)\ ,$$

und wenn man diese Gleichung auf beiden Seiten zur $q$ten Potenz erhebt und die Vielfachen von q^2 wegläfst, so hat man

$$\psi(\alpha)^{q^2} \equiv \psi(\alpha^q)^q\ , \quad \text{mod. } q^2,$$

und wenn für $\psi(\alpha)^q$ sein Werth $E(\alpha)$ und für $\psi(\alpha^q)^q$ ebenso $E(\alpha^q)$ zurückgesetzt wird, so erhält man die aufgestellte Congruenz.

Um dieselbe anzuwenden, ist zunächst die $q$te Potenz von

$$e(\alpha) = \frac{\alpha^{\gamma} - \alpha^{-\gamma}}{\alpha - \alpha^{-1}}$$

nach dem Modul q^2 zu entwickeln. Erhebt man $\alpha - \alpha^{-1}$ zur $q$ten Potenz, so erhält man

$$(\alpha - \alpha^{-1})^q \equiv \alpha^q - \alpha^{-q} + q\varphi(\alpha)\ , \quad \text{mod. } q^2\ ,$$

wo

$$\phi(\alpha) \equiv \alpha^{q-2} - \alpha^{-(q-2)} + \tfrac{1}{2}(\alpha^{q-4} - \alpha^{-(q-4)}) + \cdots + \frac{1}{\frac{q-1}{2}}(\alpha - \alpha^{-1}),$$
mod. q,

und demnach

$$e(\alpha)^q \equiv \frac{\alpha^{q\gamma} - \alpha^{-q\gamma} + q\,\phi(\alpha^\gamma)}{\alpha^q - \alpha^{-q} + q\,\phi(\alpha)}\,, \quad \text{mod. } q^2,$$

welches man auch in folgende Form setzen kann:

$$e(\alpha)^q \equiv e(\alpha^q)\big(1 + q f(\alpha)\big)\,, \quad \text{mod. } q^2,$$

wo

$$f(\alpha) \equiv \frac{\phi(\alpha)}{\alpha^q - \alpha^{-q}} - \frac{\phi(\alpha^\gamma)}{\alpha^{q\gamma} - \alpha^{-q\gamma}}\,, \quad \text{mod. } q.$$

Hieraus folgt ohne Schwierigkeit

$$E(\alpha)^q \equiv E(\alpha^q)\big(1 + q F(\alpha)\big)\,, \quad \text{mod. } q^2,$$

wo

$$F(\alpha) \equiv x f(\alpha) + x_1 f(\alpha^\gamma) + \cdots + x_{\mu-1} f(\alpha^{\gamma^{\mu-1}})\,, \quad \text{mod. } q,$$

und weil

$$E(\alpha)^q \equiv E(\alpha^q)\,, \quad \text{mod. } q^2$$

sein soll, so folgt hieraus

$$F(\alpha) \equiv 0\,, \quad \text{mod. } q.$$

Die complexe Zahl $F(\alpha)$ hat die Eigenschaft, dafs sie unverändert bleibt, wenn α in α^{-1} verwandelt wird, sie enthält daher nur die zweigliedrigen Perioden und kann in die Form gesetzt werden:

$$F(\alpha) = C(\alpha + \alpha^{-1}) + C_1(\alpha^\gamma + \alpha^{-\gamma}) + \cdots + C_{\mu-1}(\alpha^{\gamma^{\mu-1}} + \alpha^{-\gamma^{\mu-1}}).$$

Die Bedingung, dafs $F(\alpha)$ congruent Null sei, nach dem Modul q, erfordert also, dafs die Coefficienten $C, C_1 \ldots C_{\mu-1}$ alle einzeln congruent Null seien nach dem Modul q. Multiplicirt man $F(\alpha)$ mit $\alpha^{\gamma^h} + \alpha^{-\gamma^h} - 2$ und nimmt die Summe in Beziehung auf alle $\lambda - 1$ verschiedenen Wurzeln α, so erhält man

$$2\lambda . C_h = \Sigma_\alpha (\alpha^{\gamma^h} + \alpha^{-\gamma^h} - 2) F(\alpha) .$$

Multiplicirt man nun weiter mit w^h, wo w eine jede beliebige Wurzel der Gleichung $w^\mu = 1$ bezeichnet, mit alleiniger Ausnahme von $w = 1$, und nimmt die Summe für $h = 0, 1, 2, \ldots \mu - 1$, so erhält man

$$2\lambda \sum_{0}^{\mu-1}{}_h C_h w^h = \Sigma_\alpha \sum_{0}^{\mu-1}{}_h (\alpha^{\gamma^h} + \alpha^{-\gamma^h}) w^h F(\alpha) .$$

Es ist aber

$$\sum_{0}^{\mu-1}{}_h (\alpha^{\gamma^h} + \alpha^{-\gamma^h}) w^h = \sum_{0}^{\lambda-2}{}_h \alpha^{\gamma^h} w^h = (w, \alpha)$$

die bekannte Lagrangesche Resolvente der Kreistheilung, also

$$2\lambda \sum_{0}^{\mu-1}{}_h C_h w^h = \Sigma_\alpha (w, \alpha) F(\alpha) .$$

Setzt man nun den oben gegebenen Werth des $F(\alpha)$, und in demselben den Werth des $f(\alpha)$ ein und bemerkt, dafs nach einer bekannten Eigenschaft der Lagrangeschen Resolvente der Kreistheilung

$$(w, \alpha) = w^k (w, \alpha^{\gamma^k}) = w^{k+1} (w, \alpha^{\gamma^{k+1}}) ,$$

so erhält man:

$$2\lambda \sum_{0}^{\mu-1}{}_h C_h w^h =$$

$$\sum_{0}^{\mu-1}{}_k \Sigma_\alpha \frac{x_k w^k (w, \alpha^{\gamma^k}) \phi(\alpha^{\gamma^k})}{\alpha^{p\gamma^k} - \alpha^{-p\gamma^k}} - \sum_{0}^{\mu-1}{}_k \Sigma_\alpha \frac{x_k w^{k+1} (w, \alpha^{\gamma^{k+1}}) \phi(\alpha^{\gamma^{k+1}})}{\alpha^{p\gamma^{k+1}} - \alpha^{-p\gamma^{k+1}}} .$$

Setzt man in der ersten dieser beiden Summen α statt α^{γ^k} und in der zweiten α statt $\alpha^{\gamma^{k+1}}$, wodurch nichts geändert wird, weil alle Werthe des α genau dieselben sind, als alle Werthe des α^{γ^k} oder $\alpha^{\gamma^{k+1}}$, so wird:

$$2\lambda \sum_{0}^{\mu-1}{}_h C_h w^h = (1 - w) \sum_{0}^{\mu-1}{}_k x_k w^k . \Sigma_\alpha \frac{(w, \alpha) \phi(\alpha)}{a^q - \alpha^{-q}} .$$

Ich setze nun

$$2\lambda\Psi(w) = \Sigma_\alpha \frac{(w,\alpha)\,\phi(\alpha)}{\alpha^q - \alpha^{-q}},$$

so ist, wenn für $\phi(\alpha)$ der oben angegebene Werth eingesetzt wird

$$2\lambda\Psi(w) = \sum_0^{\mu-1}{}_h \sum_1^{\frac{q-1}{2}}{}_i \frac{w^h}{i}\, \Sigma_\alpha \frac{(\alpha^{\gamma^h} + \alpha^{-\gamma^h})\,(\alpha^{q-2i} - \alpha^{-(q-2i)})}{\alpha^q - \alpha^{-q}}.$$

Um den einfachsten Ausdruck des $\Psi(w)$ zu finden, betrachte ich die allgemeine Summe

$$S = \Sigma_\alpha \frac{(\alpha^c + \alpha^{-c})\,(\alpha^b - \alpha^{-b})}{\alpha^a - \alpha^{-a}},$$

ich setze in derselben α statt α^a und bezeichne mit $\left|\frac{b}{a}\right|$ die kleinste positive Wurzel der Congruenz

$$ax \equiv b, \quad \text{mod. } \lambda,$$

so wird

$$S = \Sigma_\alpha \frac{\left(\alpha^{\left|\frac{c}{a}\right|} + \alpha^{-\left|\frac{c}{a}\right|}\right)\left(\alpha^{\left|\frac{b}{a}\right|} - \alpha^{-\left|\frac{b}{a}\right|}\right)}{\alpha - \alpha^{-1}}$$

und weil

$$\frac{\alpha^{\left|\frac{b}{a}\right|} - \alpha^{-\left|\frac{b}{a}\right|}}{\alpha - \alpha^{-1}} = \alpha^{\left|\frac{b}{a}\right|-1} + \alpha^{\left|\frac{b}{a}\right|-3} + \cdots + \alpha^{-\left|\frac{b}{a}\right|+3} + \alpha^{-\left|\frac{b}{a}\right|+1},$$

so erhält man

$$S = \Sigma_\alpha \left\{\begin{array}{l} \alpha^{\left|\frac{b}{a}\right|+\left|\frac{c}{a}\right|-1} + \alpha^{\left|\frac{b}{a}\right|+\left|\frac{c}{a}\right|-3} + \cdots + \alpha^{-\left|\frac{b}{a}\right|+\left|\frac{c}{a}\right|+1} \\ + \alpha^{\left|\frac{b}{a}\right|-\left|\frac{c}{a}\right|-1} + \alpha^{\left|\frac{b}{a}\right|-\left|\frac{c}{a}\right|-3} + \cdots + \alpha^{-\left|\frac{b}{a}\right|-\left|\frac{c}{a}\right|+1} \end{array}\right\}$$

Die in Beziehung auf alle Werthe des α zu nehmenden Summen der einzelnen Theile, deren Anzahl gleich $2\left|\frac{b}{a}\right|$ ist, werden, wenn

60*

der Exponent der Potenz von α nicht durch λ theilbar ist, gleich -1, wenn aber dieser Exponent durch λ theilbar ist, so geben sie $\lambda-1$. Es wird aber einer der Exponenten in der ersten Zeile und zugleich der entsprechende gleiche, aber negative Exponent der zweiten Zeile nur in folgenden zwei Fällen durch λ theilbar: erstens wenn $\left|\frac{b}{a}\right|+\left|\frac{c}{a}\right|$ grade ist und gröſser als λ, zweitens wenn $\left|\frac{b}{a}\right|-\left|\frac{c}{a}\right|$ ungrade ist und gröſser als Null. Da der erste Fall auch $\left|\frac{b}{a}\right|-\left(\lambda-\left|\frac{c}{a}\right|\right)$ ungrade und gröſser als Null ausgesprochen werden kann, oder was dasselbe ist, $\left|\frac{b}{a}\right|-\left|\frac{-c}{a}\right|$ ungrade und gröſser als Null, da ferner die eine der beiden Zahlen $\left|\frac{b}{a}\right|-\left|\frac{c}{a}\right|$ und $\left|\frac{b}{a}\right|-\left|\frac{-c}{a}\right|$ stets grade, die andere aber ungrade ist, so erhält man folgenden Werth der Summe S:

$$S = 2\lambda\varepsilon - 2\left|\frac{b}{a}\right|,$$

wo $\varepsilon = 1$ ist, wenn die ungrade der beiden Zahlen $\left|\frac{b}{a}\right|-\left|\frac{c}{a}\right|$ und $\left|\frac{b}{a}\right|-\left|\frac{-c}{a}\right|$ positiv ist, und wo im entgegengesetzten Falle ε den Werth Null hat.

Um dieses Resultat auf den vorliegenden Ausdruck des $\Psi(w)$ anzuwenden, nehme ich $a=q$, $b=q-2i$, $c=\gamma^h$, so wird

$$\left|\frac{b}{a}\right| = \left|\frac{q-2i}{q}\right|, \quad \left|\frac{c}{a}\right| = \left|\frac{\gamma^h}{q}\right| = \gamma_{h-\varrho},$$

wenn $q \equiv \gamma^\varrho$, mod. λ ist. Es wird demnach

$$2\lambda.\Psi(w) = \sum_{0}^{\mu-1}{}_h \sum_{1}^{\frac{q-1}{2}}{}_i \frac{w^h}{i}\left(2\lambda\varepsilon - 2\left|\frac{q-2i}{q}\right|\right)$$

und weil $\left|\frac{q-2i}{q}\right|$ von h unabhängig und $\sum_{0}^{\mu-1}{}_h w^h = 0$ ist, so fällt dieser Theil weg und man hat, wenn der gemeinschaftliche Factor 2λ weggehoben wird:

$$\Psi(w) = \sum_0^{\mu-1}{}_h \sum_1^{\frac{q-1}{2}}{}_i \frac{\varepsilon\, w^h}{i}.$$

Die Gröfse ε hat nur die beiden Werthe 1 und 0 und zwar ist, wenn $\left|\frac{q-2i}{q}\right|$ einfach durch ν_i bezeichnet wird, $\varepsilon = 1$, wenn von den beiden Zahlen $\nu_i - \gamma_{h-\varrho}$ und $\nu_i - \gamma_{h-\varrho+\mu}$, deren eine nothwendig grade, die andere ungrade ist, die ungrade zugleich positiv ist, im entgegengesetzten Falle ist $\varepsilon = 0$. Setzt man $\gamma_{h-\varrho} = n$, so wird $h - \varrho \equiv \operatorname{ind} n$, mod. $\lambda - 1$, setzt man ferner $\gamma_{h-\varrho+\mu} = n'$, so wird $h - \varrho + \mu \equiv \operatorname{ind} n'$, mod. $\lambda - 1$, also $\operatorname{ind} n' \equiv \operatorname{ind} n$, mod. μ, also wenn man von der in Beziehung auf h zu nehmenden Summe nur diejenigen Glieder beibehält, für welche ε nicht gleich 0, sondern gleich 1 ist, so hat man:

$$(\text{G.})\quad \Psi(w) = w^{\varrho} \sum_1^{\frac{q-1}{2}}{}_i \frac{1}{i}\left(w^{\operatorname{ind}(\nu_i-1)} + w^{\operatorname{ind}(\nu_i-3)} + w^{\operatorname{ind}(\nu_i-5)} + \cdots\right),$$

welche Reihe, wenn ν_i grade ist, bis zu dem Gliede $w^{\operatorname{ind}(1)}$, und wenn ν_i ungrade ist, bis zum Gliede $w^{\operatorname{ind}(2)}$ fortzusetzen ist.

Die nothwendige Bedingung dafür, dafs die zusammengesetzte Kreistheilungseinheit $E(\alpha)$ eine $q$te Potenz einer Einheit sei, welche oben darauf zurückgeführt ist, dafs die Coëfficienten C_h alle congruent Null, mod. p, sein müssen, stellt sich demnach dar, als:

$$(\text{H.})\qquad \Psi(w)\,.\sum_0^{\mu-1}{}_k x_k w^k, \text{ mod. } q,$$

welche Congruenz für jeden Werth der Wurzel w der Gleichung

$$w^{\mu-1} + w^{\mu-2} + \cdots + w + 1 = 0$$

Statt haben mufs. Es folgt hieraus, dafs wenn $\Psi(w)$ keinen complexen (idealen) Primfaktor des q enthält, also die vollständige Norm $N\Psi(w)$ nicht durch q theilbar ist, nothwendig der andere Faktor $\sum_0^{\mu-1}{}_k x_k w^k$ alle complexen Primfaktoren des q enthalten und folglich durch q theilbar sein mufs, für jeden der $\mu - 1$ Werthe des w. Die aus den Wurzeln w der Gleichung $w^{\mu-1} + w^{\mu-2} + \cdots + w + 1 = 0$ gebildete complexe Zahl $\sum_{0}^{\mu-1}{}_k\, x_k\, w^k$, welche vermittelst dieser Gleichung auf den $\mu - 2$ten Grad erniedrigt wird, kann aber nicht für alle Wurzeln w congruent Null sein, nach dem Modul q, wenn nicht die $\mu - 1$ Coëfficienten $x - x_{\mu-1}, x_1 - x_{\mu-1}, \ldots x_{\mu-2} - x_{\mu-1}$ einzeln congruent Null sind, oder, was dasselbe ist, wenn nicht die μ Exponenten $x, x_1, \ldots x_{\mu-1}$ alle einer und derselben Zahl congruent sind, für welche man auch die Null nehmen kann, weil man einen beliebigen derselben gleich Null setzen kann. Also:

„Wenn die Einheit

$$E(\alpha) = e(\alpha)^{x} e(\alpha^{\gamma})^{x_1} \ldots . e(\alpha^{\gamma^{\mu-1}})^{x_{\mu-1}}$$

(V.) „eine $q$te Potenz einer anderen, fundamentaleren Einheit ist, so dafs die Exponenten $x, x_1 \ldots x_{\mu-1}$ der „Kreistheilungseinheiten nicht alle congruent Null sind, „nach dem Modul q, wenn einer derselben $= 0$ genommen wird, so mufs die complexe Zahl $\Psi(w)$ einen „complexen (idealen) Primfaktor von q enthalten und „demgemäfs die vollständige Norm von $\Psi(w)$ durch q „theilbar sein.

Hieraus folgt sodann unmittelbar der Satz:

(VI.) „Eine ungrade Primzahl q kann nicht Theiler des zweiten Faktors der Klassenzahl sein, wenn nicht die „complexe Zahl $\Psi(w)$ einen complexen Primfaktor von „q enthält, also die vollständige Norm von $\Psi(w)$ durch „q theilbar ist.

Wenn die für die Theilbarkeit des zweiten Faktors der Klassenzahl durch die Primzahl q nothwendige, aber nicht hinreichende Bedingung erfüllt ist, dafs $\Psi(w)$ einen idealen Primfaktor von q enthält, so kann der Fall eintreten, dafs dieser Primfaktor des q in $\Psi(w)$ nicht für die primitiven Wurzeln w der Gleichung $w^{\mu} = 1$ vorhanden ist, sondern für gewisse nicht primitive Wurzeln, welche der Gleichung niederen Grades $w'^{m} = 1$ angehören, wo m ein Factor von μ ist. Es sei $\mu = mm'$ und die Norm von $\Psi(w')$, für alle primitiven Wurzeln der Gleichung $w'^m = 1$ sei durch q theilbar, so zeigt die Congruenz

$$\Psi(w) \, . \sum_0^{\mu-1}{}_k \, x_k \, w^k \equiv 0 \,, \quad \text{mod. } q,$$

dafs für alle diejenigen Werthe des w, für welche $\Psi(w)$ keinen complexen Primfaktor des q enthält, $\sum\limits_0^{\mu-1}{}_k \, x_k \, w^k$ congruent Null sein mufs, nach dem Modul q. Es sind dies die Werthe des w, welche der Gleichung $w^{m\,m'} = 1$ genügen, ohne der Gleichung $w^m = 1$ zu genügen, also die Werthe des w, welche der Gleichung

$$\frac{1 - w^{m\,m'}}{1 - w^m} = 1 + w^m + w^{2m} + \cdots + w^{(m'-1)m} = 0$$

genügen. Hieraus schliefst man, dafs

$$\sum_0^{\mu-1}{}_k \, x_k \, w^k \equiv (1 + w^m + w^{2m} + \cdots + w^{(m'-1)m}) \, F(w)$$

sein mufs, wo $F(w)$ nur bis zum Grade $m - 1$ in w aufsteigt und hieraus folgert man weiter, dafs

$$x_k \equiv x_{k+m} \equiv x_{k+2m} \cdots \equiv x_{k+(m'-1)m} \,, \quad \text{mod. } q,$$

sein mufs. Man hat daher folgenden Satz:

(VII.) „Wenn die complexe Zahl $\Psi(w)$ nicht für die primitiven Wurzeln w der Gleichung $w^\mu = 1$, sondern für „die primitiven Wurzeln der Gleichung $w^m = 1$, wo „$\mu = m\,m'$, einen idealen Primfaktor von q enthält, so „kann die fundamentalere Einheit, deren $q$te Potenz „sich als Produkt von Potenzen der Kreistheilungseinheiten ausdrücken läfst, nur die m Perioden von je „$2m'$ Gliedern der Wurzeln der Gleichung $\alpha^\lambda = 1$ enthalten.

Ein einfaches Beispiel für den Fall, wo der zweite Faktor der Klassenzahl nicht gleich Eins ist, ist $\lambda = 229$. Für diesen Werth des λ haben schon die aus den zwei Perioden von je 114 Gliedern gebildeten complexen Zahlen drei verschiedene Klassen, welche durch die drei verschiedenen quadratischen Formen $x^2 + xy - 57y^2$, $3x^2 + xy - 19y^2$ und $3x^2 - xy - 19y^2$ repräsentirt werden. Der zweite Faktor der Klassenzahl mufs darum für $\lambda = 229$ durch drei theilbar sein. Bezeichnet man die beiden je 114 Glieder enthaltenden Perioden mit η und η_1, so findet man die aus der Entwickelung des Produktes $e\, e_2\, e_4 \ldots\, e_{112}$ zu bildende Kreistheilungseinheit gleich

$$1823 + 226\eta = (8 + \eta)^3.$$

Die vollständige Norm der complexen Zahl $\Psi(w)$ ist hier in der That durch 3 theilbar, da für den Werth $w = -1$ $\Psi(w) = 6$ wird. Die fundamentalere Einheit $8 + \eta$, deren dritte Potenz sich durch die Kreistheilungseinheit ausdrücken läfst, enthält auch, wie der letzte Satz es verlangt, nur die 2 Perioden von je 114 Gliedern.

Der Werth $\lambda = 257$ giebt ein zweites Beispiel derselben Art, wo der zweite Faktor der Klassenzahl durch 3 theilbar ist.

Ein Beispiel anderer Art giebt $\lambda = 163$. Dasselbe ist darum besonders bemerkenswerth, weil hier nicht wie im vorigen Beispiele die quadratische Form, sondern die cubische Form, in welche die Normform gesetzt werden kann, bewirkt, dafs der zweite Faktor der Klassenzahl nicht gleich Eins ist, sondern durch Zwei theilbar. Für $\lambda = 163$ und für die primitive Wurzel $\gamma = 70$ des Canon arithmeticus erhält man den Werth der bei (E.) gegebenen complexen Zahl $\Psi(w)$:

$$\psi(w) \equiv w^{-71} \left\{ \begin{array}{l} 1 + w^6 + w^8 + w^9 + w^{10} + w^{12} + w^{14} \\ + w^{16} + w^{18} + w^{19} + w^{21} + w^{23} + w^{24} + w^{25} \\ + w^{30} + w^{31} + w^{35} + w^{38} + w^{44} + w^{46} + w^{47} \bmod. 2 \\ + w^{51} + w^{52} + w^{55} + w^{58} + w^{59} + w^{61} + w^{62} \\ + w^{63} + w^{64} + w^{65} + w^{66} + w^{67} + w^{74} + w^{76} \end{array} \right.$$

wo w eine beliebige primitive oder nicht primitive Wurzel der Gleichung $w^{81} = 1$ ist. Nimmt man nun für w eine dritte Wurzel der Einheit, so dafs $w^3 = 1$ ist, so erhält man

$$\psi(w) \equiv 13 + 11w + 11w^2\,, \quad \text{mod. } 2,$$

also

$$\psi(w) \equiv 0\,, \quad \text{mod. } 2.$$

Es sind also hier die beiden nothwendigen Bedingungen der Theilbarkeit des zweiten Faktors der Klassenzahl durch 2 erfüllt, daſs $\Psi(w)$ einen complexen Faktor von 2 enthalte, und daſs $\Psi(w^{-1})$ eben denselben enthalte. Der zweite Faktor der Klassenzahl kann also für $\lambda = 163$ durch 2 theilbar sein. Daſs dies auch wirklich der Fall ist, wird nun aus der Betrachtung der Kreistheilungseinheiten nachgewiesen, welche in der That als Quadrate von fundamentaleren Einheiten sich darstellen. Diese Einheiten können hier, nach dem oben bewiesenen Satze (VII.) nur die drei Perioden von je 54 Gliedern enthalten. Werden dieselben, nach der primitiven Wurzel $\gamma = 70$ geordnet, mit η, η_1, η_2 bezeichnet, so hat man für die Rechnung mit denselben die Formeln

$$\begin{aligned} \eta^2 &= 54 + 20\eta + 16\eta_1 + 17\eta_2\,, \\ \eta\eta_1 &= \phantom{54 + {}} 16\eta + 17\eta_1 + 21\eta_2\,, \\ \eta\eta_2 &= \phantom{54 + {}} 17\eta + 21\eta_1 + 16\eta_2\,. \end{aligned}$$

Als das System der unabhängigen Kreistheilungseinheiten kann hier gewählt werden:

$$e_k = \alpha^{\gamma^k} + \alpha^{-\gamma^k}.$$

Bildet man nun das Produkt

$$E = -e.e_3.e_6\ldots.e_{78}\,,$$

so erhält man nach Ausführung der Multiplikation, die aus den 3 Perioden von je 54 Gliedern gebildete Kreistheilungseinheit

$$E = -63\eta - 62\eta_1 - 49\eta_2\,,$$

welche, wenn ihre conjugirten mit E_1 und E_2 bezeichnet werden,

$$E.E_1E_2 = +1$$

giebt. Die nothwendige und zugleich auch hinreichende Bedingung dafür, dafs für $\lambda = 163$ der zweite Faktor der Klassenzahl durch 2 theilbar sei, ist nun die, dafs die zusammengesetzte Kreistheilungseinheit

$$E^{x}.E_1^{x_1}E_2^{x_2}$$

gleich einem Quadrate einer Einheit sei, für irgend welche Werthe der x, x_1, x_2, welche nur gleich 0 oder 1 zu nehmen sind. Es ist aber hier schon E selbst ein vollständiges Quadrat, denn man hat

$$-63\eta - 62\eta_1 - 49\eta_2 = (5 + \eta_2)^2 ,$$

wie vermittelst der Formeln für die Multiplikation der Perioden leicht nachgewiesen wird.

Für $\lambda = 163$ ist also der zweite Faktor der Klassenzahl durch Zwei theilbar und man hat in diesem Falle die Einheiten $5 + \eta$, $5 + \eta_1$, $5 + \eta_2$, welche fundamentaler sind, als die Kreistheilungseinheiten.

Ein anderes Beispiel dieser Art, wo die kubische Form bewirkt, dafs der zweite Faktor der Klassenzahl durch Zwei theilbar ist, giebt $\lambda = 937$.

Über diejenigen Primzahlen λ, für welche die Klassenzahl der aus λten Einheitswurzeln gebildeten complexen Zahlen durch λ teilbar ist

Monatsberichte der Königlich Preußischen Akademie der Wissenschaften zu Berlin aus dem Jahre 1874, 239–248

19. März. Gesammtsitzung der Akademie.

Hr. **Kummer** las:

Über diejenigen Primzahlen λ, für welche die Klassenzahl der aus λten Einheitswurzeln gebildeten complexen Zahlen durch λ theilbar ist.

In der Theorie der aus λten Einheitswurzeln gebildeten complexen Zahlen sind diejenigen Primzahlen λ, für welche die Klassenzahl durch λ theilbar ist, in vielen wesentlichen Beziehungen von denen verschieden, für welche die Klassenzahl den Faktor λ nicht enthält. Auch in den Anwendungen dieser Theorie auf den Beweis des Fermat'schen Lehrsatzes für λte Potenzen und auf die allgemeinen Reciprocitätsgesetze unter den Resten und Nichtresten von λten Potenzen tritt diese Verschiedenheit so bedeutend auf, dass die von mir für diese beiden Probleme gegebenen Beweismethoden nur für den Fall gelten, wo λ eine Primzahl ist, welche in der Klassenzahl als Faktor nicht enthalten ist. Eine nähere Untersuchung der besonderen Eigenschaften derjenigen Primzahlen λ, für welche die Klassenzahl durch λ theilbar ist, hat mich schon oft und dauernd beschäftigt. Ich habe gezeigt, dass diese besonderen Primzahlen die Eigenschaft haben, dass eine der ersten $\frac{\lambda-3}{2}$ Bernoullischen Zahlen durch λ theilbar sein muss und dass, wenn die Klassenzahl der aus λten Einheitswurzeln gebildeten complexen Zahlen überhaupt durch λ theilbar ist, nothwendig der erste der beiden Faktoren, aus welchem die Klassenzahl besteht, durch λ theilbar sein muss. Ferner habe ich gezeigt, dass im ersten Hundert der Zahlen nur die drei Primzahlen $\lambda = 37$, $\lambda = 59$ und $\lambda = 67$ vorkommen, welche diese besondere Eigenschaft besitzen und dass für $\lambda = 37$ die 16te, für $\lambda = 59$ die 22te, für $\lambda = 67$ die 29te Bernoullische Zahl durch diese Primzahl λ theilbar ist.

Um nun das auf diese genannten drei Primzahlen beschränkte kleine Gebiet der über diese besondere Art von Primzahlen zu machenden Erfahrungen etwas zu erweitern, habe ich die vollständige Berechnung des ersten Faktors der Klassenzahl über die im ersten Hundert befindlichen Primzahlen hinaus fortgesetzt bis zur Primzahl 163. Die Resultate dieser mühsamen Rechnungen, welche in so fern auf Zuverlässigkeit Anspruch machen können, als keine derselben ohne mehrfache Controle ausgeführt ist, will ich hier mittheilen.

Der erste Faktor der Klassenzahl hat den Ausdruck

$$P' = \frac{P}{(2\lambda)^{\frac{\lambda-3}{2}}},$$

wo

$$P = \varphi(\beta)\varphi(\beta^3)\varphi(\beta^5)\,.....\,\varphi(\beta^{\lambda-2})$$

ist und

$$\varphi(\beta) = 1 + \gamma_1\beta + \gamma_2\beta^2 + \gamma_3\beta^3 + \cdots + \gamma_{\lambda-2}\beta^{\lambda-1},$$

β eine primitive Wurzel der Gleichung $\beta^{\lambda-1} = 1$ ist und $\gamma_1, \gamma_2, \gamma_3, \ldots$ die kleinsten positiven Reste der Potenzen einer primitiven Wurzel $\gamma, \gamma^2, \gamma^3, \ldots$ sind, nach dem Modul λ.

Das Produkt P zerfällt stets in eine Anzahl verschiedener rationaler Faktoren, welche Normen complexer Zahlen sind, genommen in Beziehung auf alle primitiven Wurzeln einer Gleichung von der Form

$$\beta^{\frac{\lambda-1}{m}} = 1.$$

Wenn nämlich $1, m, m', m''$ ..... alle ungeraden Divisoren von $\lambda - 1$ sind, so ist:

$$P = N\varphi(\beta) \,.\, N\varphi(\beta^m) \,.\, N\varphi(\beta^{m'}) \,.....$$

Das Produkt P besteht also aus eben so vielen rationalen und ganzzahligen Faktoren, als es ungerade Divisoren der Zahl $\lambda - 1$ giebt.

Die complexe Zahl $\varphi(\beta^m)$ enthält stets den Faktor λ, ausser für den Fall $m = 1$, wo ihr stets einer der idealen Primfaktoren des λ fehlt; deshalb enthält $N\varphi(\beta^m)$ den Faktor λ nothwendig so vielmal als die Gleichung $\beta^{\frac{\lambda-1}{m}} = 1$ primitive Wurzeln hat, also sovielmal, als es Zahlen giebt, welche kleiner als $\frac{\lambda - 1}{m}$ und zu $\frac{\lambda - 1}{m}$ relative Primzahlen sind. Nur für $m = 1$ enthält $\varphi(\beta)$ nicht den Faktor λ und $N\varphi(\beta)$ enthält den Faktor λ im Allgemeinen nur einmal weniger, als die Anzahl der relativen Primzahlen zu $\lambda - 1$, welche kleiner sind als $\lambda - 1$. Ebenso enthält auch $\varphi(\beta^m)$ stets den Faktor 2, mit alleiniger Ausnahme des Falles, wo m der grösste ungerade Divisor von $\lambda - 1$ ist. $N\varphi(\beta^m)$ enthält also den Faktor 2 nothwendig ebenso vielmal als die Anzahl der primitiven Wurzeln der Gleichung $\beta^{\frac{\lambda-1}{m}} = 1$ beträgt und nur in dem Falle, wo m der grösste ungerade Divisor von $\lambda - 1$ ist, ist die Anzahl dieser Faktoren 2 um eine Einheit niedriger. Man sieht hieraus, wie die in P nothwendig enthaltenen Faktoren 2 und λ, deren jeder darin $\frac{\lambda - 3}{2}$ mal enthalten sein muss, sich auf die einzelnen Normen vertheilen, aus welchen P zusammengesetzt ist. Ausser diesen in P nothwendig vorkommenden Faktoren 2 und λ können diese beiden Faktoren für besondere Werthe des λ noch öfter in P enthalten sein, und es sind grade die Fälle, wo λ noch ausserdem in P, also in P' enthalten sind, welche hier unsere besondere Aufmerksamkeit fesseln.

Da es für manche Untersuchungen von besonderer Wichtigkeit ist nicht nur den ersten Faktor der Klassenzahl selbst, sondern auch die Werthe der einzelnen Normen zu kennen, aus welchen P' sich zusammensetzt, so werde ich diese besonders mit angeben. Für die primitiven Wurzeln habe ich bei der Berechnung überall diejenigen gewählt, welche im Canon arithmeticus von Jacobi zu Grunde gelegt sind.

Für $\lambda = 101$

sind 1, 5, 25 die ungeraden Divisoren von 100 und es ist:

$$N\varphi(\beta) = 2^{40}.(101)^{39}.5^{2}.1135169401;$$
$$N\varphi(\beta^{5}) = 2^{8}.(101)^{8}.5^{2};$$
$$N\varphi(\beta^{25}) = 2.(101)^{2}.5;$$

also

$$P' = 5^{5}.1135169401.$$

Diese Klassenzahl ist durch $\lambda = 101$ theilbar, und man hat

$$P' = 5^{5}.101.11239301.$$

Hieraus folgt, dass die Primzahl 101 als Faktor des Zählers einer der ersten 49 Bernoullischen Zahlen vorkommen muss, und es ergiebt sich aus bekannten Sätzen der Theorie der complexen Zahlen, dass es die 34te Bernoullische Zahl sein muss, welche den Faktor 101 enthält.

Für $\lambda = 103$

sind 1, 3, 17, 51 die ungeraden Divisoren von 102 und es ist:

$$N\varphi(\beta) = 2^{32}.(103)^{31}.2816812173;$$
$$N\varphi(\beta^3) = 2^{16}.(103)^{16}.1021;$$
$$N\varphi(\beta^{17}) = 2^2.(103)^2.1;$$
$$N\varphi(\beta^{51}) = 103.5;$$

also

$$P' = 5.1021.2816812173.$$

Diese Klassenzahl ist durch $\lambda = 103$ theilbar und man hat

$$P' = 5.1021.103.27347691.$$

Es ist die 12te Bernoullische Zahl, welche durch 103 theilbar ist.

Für $\lambda = 107$

sind 1, 53 die beiden ungeraden Divisoren von 106 und es ist:

$$N\varphi(\beta) = 2^{52}.(107)^{51}.21144977847541;$$
$$N\varphi(\beta^{53}) = 107.3;$$

also

$$P' = 3.21144977847541.$$

Diese Klassenzahl ist durch $\lambda = 107$ nicht theilbar.

Für $\lambda = 109$

sind 1, 3, 9, 27 die ungeraden Divisoren von 108 und es ist:

$$N\varphi(\beta) = 2^{36}.(109)^{35}.9431866153;$$
$$N\varphi(\beta^3) = 2^{12}.(109)^{12}.1009;$$
$$N\varphi(\beta^9) = 2^4.(109)^4.1;$$
$$N\varphi(\beta^{27}) = 2.(109)^2.17;$$

also

$$P' = 17.1009.9431866153.$$

Diese Klassenzahl ist durch $\lambda = 109$ nicht theilbar.

Für $\lambda = 113$

sind 1, 7 die ungeraden Divisoren von 112 und es ist:

$$N\varphi(\beta) = 2^{48}.(113)^{47}.11853470598257.2^{3};$$
$$N\varphi(\beta^{7}) = 2^{7}.(113)^{8}.17;$$

also

$$P' = 17.2^{3}.11853470598257.$$

Diese Klassenzahl ist durch $\lambda = 113$ nicht theilbar.

Für $\lambda = 127$

sind 1, 3, 7, 9, 21,63 die ungeraden Divisoren von 126 und es ist:

$$N\varphi(\beta) = 2^{36}.(127)^{35}.553286917;$$
$$N\varphi(\beta^{3}) = 2^{12}.(127)^{12}.547;$$
$$N\varphi(\beta^{7}) = 2^{6}.(127)^{6}.3079;$$
$$N\varphi(\beta^{9}) = 2^{6}.(127)^{6}.43;$$
$$N\varphi(\beta^{21}) = 2^{2}.(127)^{2}.13;$$
$$N\varphi(\beta^{63}) = 127.5;$$

also

$$P' = 5.13.43.3079.547.553286917.$$

Diese Klassenzahl ist durch $\lambda = 127$ nicht theilbar.

Für $\lambda = 131$

sind 1, 5, 13, 65 die ungeraden Divisoren von 130 und es ist:

$$N\varphi(\beta) = 2^{48}.(131)^{47}.796544506758131;$$
$$N\varphi(\beta^{5}) = 2^{12}.(131)^{12}.3^{3}.53;$$
$$N\varphi(\beta^{13}) = 2^{4}.(131)^{4}.5;$$
$$N\varphi(\beta^{65}) = 131.5;$$

also

$$P' = 5.5.3^3.53.796544506758131.$$

Diese Klassenzahl ist durch 131 theilbar und man hat

$$P' = 5.5.3^3.53.131.6080492418001.$$

Es ist die 11te Bernoullische Zahl, welche durch 131 theilbar ist.

Für $\lambda = 137$

sind 1, 17 die ungeraden Divisoren von 136 und es ist:

$$N\varphi(\beta) = 2^{64}.(137)^{63}.17.2238413737630453177;$$
$$N\varphi(\beta^{14}) = 2^3.(137)^4.17;$$

also

$$P' = 17.17.2238413737630453177.$$

Diese Klassenzahl ist nicht durch $\lambda = 137$ theilbar.

Für $\lambda = 139$

sind 1, 3, 23, 69 die ungeraden Divisoren von 138 und es ist:

$$N\varphi(\beta) = 2^{44}.(139)^{43}.318474147982831;$$
$$N\varphi(\beta^3) = 2^{22}.(139)^{22}.623209;$$
$$N\varphi(\beta^{23}) = 2^2.(139)^2.3;$$
$$N\varphi(\beta^{69}) = 139.3;$$

also

$$P' = 3.3.623209.318474147982831.$$

Diese Klassenzahl ist durch $\lambda = 139$ nicht theilbar.

Für $\lambda = 149$

sind 1, 37 die ungeraden Divisoren von 148 und es ist:

$$N\varphi(\beta) = 2^{72}.(149)^{71}.7643198440968608001 3689;$$
$$N\varphi(\beta^{37}) = 2.(149)^{2}.3^{2};$$

also

$$P' = 3^{2}.76431984409686080013689.$$

Diese Klassenzahl ist durch $\lambda = 149$ theilbar und es ist

$$P' = 3^{2}.149.512966338320040805461.$$

Es ist die 65te Bernoullische Zahl, welche durch 149 theilbar ist.

Für $\lambda = 151$

sind 1, 3, 5, 15, 25, 75 die ungeraden Divisoren von 150 und es ist:

$$N\varphi(\beta) = 2^{40}.(151)^{39}.377809313842801;$$
$$N\varphi(\beta^{3}) = 2^{20}.(151)^{20}.25951;$$
$$N\varphi(\beta^{5}) = 2^{8}.(151)^{8}.11.11;$$
$$N\varphi(\beta^{15}) = 2^{4}.(151)^{4}.281;$$
$$N\varphi(\beta^{25}) = 2^{2}.(151)^{2}.1;$$
$$N\varphi(\beta^{75}) = 151.7;$$

also

$$P' = 7.281.11.11.25951.377809313842801.$$

Diese Klassenzahl ist durch $\lambda = 151$ nicht theilbar.

Für $\lambda = 157$

sind 1, 3, 13, 39 die ungeraden Divisoren von 156 und es ist:

$$N\varphi(\beta) = 2^{60}.(157)^{59}.13.21136212634488121\,;$$
$$N\varphi(\beta^3) = 2^{24}.(157)^{24}.3148601\,;$$
$$N\varphi(\beta^{13}) = 2^4.(157)^4.13\,;$$
$$N\varphi(\beta^{39}) = 2.(157)^2\,;$$

also

$$P' = 5.13.3148601.13.21136212634488121.$$

Diese Klassenzahl enthält den Faktor $\lambda = 157$ und zwar zweimal, sie giebt das erste Beispiel dieser Art. Es ist

$$P' = 5.13.3148601.13.157.157.857487631729.$$

Die beiden Bernoullischen Zahlen unter den ersten 77, welche den Faktor 137 enthalten, sind die 31te und die 55te Bernoullische Zahl.

Für $\lambda = 163$

sind 1, 3, 9, 27, 81 die ungeraden Divisoren von 162 und es ist:

$$N\varphi(\beta) = 2^{54}.(163)^{53}.1023624204620784393\,;$$
$$N\varphi(\beta^3) = 2^{18}.(163)^{18}.365473\,;$$
$$N\varphi(\beta^9) = 2^6.(163)^6.73\,;$$
$$N\varphi(\beta^{27}) = 2^2.(163)^2.1\,;$$
$$N\varphi(\beta^{81}) = 163.1\,;$$

also

$$P' = 73.365473.1023624204620784393.$$

Diese Klassenzahl ist durch $\lambda = 163$ nicht theilbar.

Aus den hier gegebenen Resultaten der Berechnung des ersten Faktors der Klassenzahl ergiebt sich, dass unter den ersten 13 Primzahlen im zweiten Hundert von $\lambda = 101$ bis $\lambda = 163$ fünf Primzahlen λ vorkommen, für welche die Klassenzahl durch λ theilbar ist, von denen eine diesen Faktor sogar zweimal enthält, während unter den 24 ungeraden Primzahlen innerhalb des ersten Hunderts sich nur drei Zahlen dieser Art befinden. Es scheint also, als ob die Häufigkeit des Vorkommens dieser besonderen Art von Primzahlen eine mit der Grösse der Primzahlen wachsende sei. Diese Häufigkeit könnte sogar vielleicht so stark wachsen, dass von einer bestimmten Gränze an alle Primzahlen nur dieser besonderen Art angehören möchten, oder was dasselbe ist, dass es vielleicht nur eine endliche bestimmte Anzahl von Primzahlen geben könnte, für welche die Klassenzahl nicht durch λ theilbar wäre. Es scheint dieses jedoch nicht der Fall zu sein, vielmehr kann man nach einfachen Principien der Wahrscheinlichkeitsrechnung, deren Anwendbarkeit auf die vorliegende Frage jedoch zweifelhaft bleibt, schliessen, oder vielmehr nur vermuthen, dass die Häufigkeit dieser besonderen Art von Primzahlen nur so weit wächst, bis sie schliesslich im Vergleich zu der Häufigkeit der Primzahlen, welche dieser besonderen Art nicht angehören, das Verhältniss von Eins zu Zwei asymptotisch erreicht.

Notes

p. 25 and p. 26, l. 2–3: these items have to be rectified according to footnote (1), *Introduction*, p. 4.

p. 26: Here Lampe's bibliography omits paper [40a]: Über die Irregularität von Determinanten, Berliner Monatsber. 1853, 194–200, gelesen 14. März 1853 (cf. the note to p. 539). Moreover, paper [41] is incorrectly listed under 1853; it is dated 3. VI. 1855.

p. 143: read $z^3 = 3pz + pt$.

p. 195, l. 12 from b.: the wrong argument offered here by Kummer will occur again and again (cf. [27] and [39]). The confusion seems to lie in the concept of "multiple root," which is meaningful in the ring of polynomials over the field of integers modulo q, but has no validity over the ring modulo q^e. All that Kummer is entitled to conclude here is that the congruence $\varphi(y) \equiv 0$ has at least one root in the prime field modulo q. As the sequel shows, Kummer has a dim feeling that there is something wrong with his argument when q divides $N(\eta - \eta_r)$; this is precisely the case when the congruence $\varphi(y) \equiv 0$ has multiple roots over the prime field. Eventually Kummer was to convince himself that such primes are in no way "exceptional" and behave like all others; but for a long time the argument he was to offer for this had also no validity (cf. the notes to pp. 213 and 575).

p. 195, l. 9 from b.: for q, read p.

p. 205: cf. the note to p. 195.

p. 213: here begins Kummer's attempt to dispose of the "exceptional primes" which he had encountered in [23]; but his argument, based on an illegitimate use of elimination theory, has no more validity than his reference to [23] concerning the roots of the congruence (8). Luckily for him, all the conclusions he derives from these arguments are quite correct; perhaps he depended less on the latter than on the experience he had acquired by treating numerical cases. Cf. the note to p. 575.

p. 320: from here on, until the end of this paper, Kummer tries to prove that the class-number of the field $k = Q(\alpha)$ is a multiple of the class-number for any subfield k'. In modern language, his "proof" amounts to this: if H, H' are the groups of ideal classes for k, k', then there is a natural homomorphism of H' onto some subgroup H'' of H, and "therefore" the order of H is a multiple of that of H'. We see the fallacy at once, and we see at the same time how inexperienced Kummer was still at this stage. In other contexts, of course, he had many examples of non-principal ideals in a given field which become principal in some bigger field. Curiously enough, he seems to repeat essentially the same mistake much later (see note to p. 750).

p. 323: for G, read P.

p. 324: in the left-hand side, for g, read G.

p. 327: for $2^{2\mu} - 1$, read $g^{2\mu} - 1$.

p. 328: for k, read K.

p. 337: here there is an unaccountable lapse. Kummer expresses himself as if u, v, w could always be divided by any common divisor that they might have; the whole point of ideal-theory is that this cannot be done unless the common divisor is a principal ideal. Because of this, the validity of Kummer's proof has to be restricted to the case when u, v, w have no common divisor. Without that assumption, the argument has to be somewhat modified; cf. e.g. Hilbert, *Zahlbericht*, Chapter XXXVI.

p. 353: lines 14 from b. and 13 from b., and, on line 11 from b., the words "für den Modul $p-1$, also auch" should be omitted and the word "genügen" inserted (cf. [40], where this slight lapse is rectified).

p. 395 and p. 396: still the same arguments as in [23] and [27]; cf. the notes to those papers and to [46], p. 575.

p. 399: for (mod. λ), read (mod. q).

p. 451: for $\mathrm{N}f(\alpha) > 1$, read $\mathrm{N}f(\alpha) < 1$.

p. 466: the last term should read $a_{\lambda-2}\alpha^{\lambda-2}$.

p. 484: here Kummer repeats the same mistake he has made in [36], p. 337 (cf. the note), even though, on p. 474, l. 14, he has expressly stated that u, v, w are assumed to be without a common factor.

p. 487 and p. 488: for $r-1$, $s-1$, read $\nu-1$.

p. 491: the first factor in the right-hand side should read $(\gamma^{2n}-1)$ and not (γ^{2n}).

p. 493: for ε^{v}, read e^{v}.

p. 504: for $q^{\frac{1}{2}t}$, read $\pm q^{\frac{1}{2}t}$; as the reader may discover, the determination of the sign is a non-trivial exercise.

p. 514: read "eine ganze rationale Function von e^{v} und e^{-v}".

p. 515, l. 10 from b.: the term Ind. $(g^{k}+1)$ should have the sign $-$, not $+$.

p. 515, last line: in the last factor; read h' for h.

p. 523: the last term should be $A_{\lambda-1}(1-\alpha)^{\lambda-1}$.

p. 525: the left-hand side should read $X_{k}(\alpha^{m})$.

p. 528: for $(\varphi\alpha)$, read $\varphi(\alpha)$.

p. 533: for $(1-\alpha)^{\lambda}$, read $(1-\alpha)^{2}$.

p. 539: The omission of this paper from Lampe's bibliography [L] is perhaps not accidental. Lampe may have noticed (or may even have learnt from Kummer himself) that the group-theoretic reasoning in that paper needed some mending, and may have found it best to omit it altogether. Actually that reasoning is only valid modulo the 2-primary component of the group of ideal classes. Nevertheless, it points to a very interesting phenomenon, which is best described in terms of Kummer's example $K=\mathrm{Q}(\alpha)$ with $\alpha^{41}=1$. The class-number of K is 11^{2}. What Kummer shows is that one can attach to each prime ideal $\mathfrak{p}$ a so-called "Jacobi sum" $J(\mathfrak{p})$ (a product of powers of the Gaussian sums belonging to $\mathfrak{p}$), taking its values in K, and such that, for each $\mathfrak{p}$, the principal ideal $J(\mathfrak{p}))$ is $\mathfrak{p}^{11n}\bar{\mathfrak{p}}^{-11n}$, where $\bar{\mathfrak{p}}$, is the transform of $\mathfrak{p}$ under the automorphism $\alpha\to\alpha^{-1}$ of K, and where n is an integer prime to 11. From this one concludes at once, as Kummer does, that all ideal-classes are of order 11, and therefore that the group of classes is non-cyclic (i.e. "irregular" in the sense of Gauss).

p. 556: read $p=\nu\lambda+1$.

p. 564, lines 4, 8, 12: for Σ, read Π.

p. 570: the denominator of the fraction in the right-hand side should read h, not 1.

p. 575: the references are to [23] and to [27]; but, as we have seen (see the note to p. 195) the words "streng bewiesen" are too optimistic, while Kummer himself indicates that he is dissatisfied with [27], since he says now that he will offer "eine vollständigere Begründung" (an obvious euphemism for the replacement of a defective argument by a correct one). As to the first point, Kummer is now only going to use the fact that the polynomial $\varphi(y)$ defined by (1) has *at least* one root in the prime field of integers modulo p; the argument in [23] did indeed establish this. The remainder of Kummer's treatment in the present paper is quite correct, even though modern readers may think it unnecessarily complicated; once one has the general ideal theory (in arbitrary algebraic numberfields) at one's disposal, most of it becomes superfluous. It is clear that the occasion for Kummer's going back to this topic was his elaboration of the ideal-theory for general cyclotomic fields (see [44]).

p. 664: here, too (cf. the note to p. 337), Kummer ought to have assumed that U, V, W have no common divisor.

p. 713: the denominator should read $1 - \alpha^k w$.

p. 750, l. 1: there seems to be no reason why the group H of ideal-classes of $Q(\alpha)$ (the "numbers in α") should be mapped injectively into the group H_z of ideal-classes of "numbers in z". Nevertheless, Kummer expresses himself as if this was so, and as if this were obvious (cf. the note to p. 320). Fortunately, this seems to be of no consequence in what follows; one should merely replace H, whenever needful, by its image H' in H_z, and h by the order of H'.

p. 753: for $G(z_1)$, read $G_1(z)$.

p. 755: for $(NF(z))^h$, read $(N\varphi(z))^h$.

p. 758, l. 11 from b.: Kummer's argument here is of questionable validity, and consequently, in (II), the second statement (lines 4–3 from b.) must remain in doubt. As it seems to be used nowhere except on p. 760, l. 2 from b., and as its use there can easily be replaced by a simple direct argument, this is of no consequence.

p. 758: for $\varphi(z)$, read $f(z)$.

p. 759: for $Nf(\alpha)$, read $Nf(z)$.

p. 760, l. 3 from b.: $f(z^{h\lambda-1})$ should read $f(z)^{h\lambda-1}$; anyway, lines 3–2 from b. should be omitted (see the note to p. 758, l. 11 from b.) and replaced by a direct argument.

p. 762: for (7), read (II); the reference is to the congruences on p. 719.

p. 765: for (6), read (3).

p. 766, lines 3, 4, 5, 6, 7: for $F(\alpha)$, read $L(\alpha)$.

p. 798: for $v = 1$, read $v = 0$.

p. 800: for a^2, read a_2.

p. 802, l. 15–18: these congruences should bear the number (14).

p. 921: the last factor should be $(e_{k+\mu-1})^{x_{\mu-1}}$.

Zeitfracht Medien GmbH
Ferdinand-Jühlke-Straße 7
99095 Erfurt, Deutschland
produktsicherheit@kolibri360.de